CONNECTIONS, SPRAYS AND FINSLER STRUCTURES

CONNECTIONS, SPRAYS AND FINSLER STRUCTURES

József Szilasi
Rezső L Lovas
Dávid Cs Kertész

University of Debrecen, Hungary

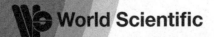

NEW JERSEY • LONDON • SINGAPORE • BEIJING • SHANGHAI • HONG KONG • TAIPEI • CHENNAI

Published by

World Scientific Publishing Co. Pte. Ltd.

5 Toh Tuck Link, Singapore 596224

USA office: 27 Warren Street, Suite 401-402, Hackensack, NJ 07601

UK office: 57 Shelton Street, Covent Garden, London WC2H 9HE

British Library Cataloguing-in-Publication Data
A catalogue record for this book is available from the British Library.

ISBN 978-981-4440-09-7

Printed in Singapore

IN MEMORIAM

Ludwig Berwald
1883–1942

András Rapcsák
1914–1993

Makoto Matsumoto
1920–2005

Preface

The Tao that can be told is not the eternal Tao.
The name that can be named is not the eternal name.
The nameless is the beginning of heaven and earth.
The named is the mother of ten thousand things.

Lao Tsu (Tao Te Ching [65], Ch. 1)

Mathematics is the music of science and real analysis is the Bach of mathematics. There are many foolish things I could say about the subject of this book, but the foregoing will give the reader an idea of where my heart lies.

Sterling K. Berberian [11]

1. The three concepts in the title lie at the heart of our book. The following diagram shows that the order in which they appear is also important:

	more special $\longrightarrow$	
manifolds with connections	$\supset$ manifolds with sprays	$\supset$ manifolds with Finsler structures
	$\longleftarrow$ *more general*	

We note already at this point that by a connection we always mean a so-called *Ehresmann connection*. As a first approach, specifying an Ehresmann connection means the fixing of a direct summand of the (canonical) vertical subbundle of the double tangent bundle. Thus, it is a geometric notion, and indeed a very simple one.

Our main goal was to give a comprehensive introduction to the theory of *Finsler manifolds*, i.e., manifolds endowed with a Finsler structure or Finsler function. The scheme above shows that the appropriate approach is the study of manifolds endowed with connections and sprays. Above all, we had in mind the needs of PhD students in Finsler geometry, and we tried to summarize the fundamentals of differential geometry together with a reasonable amount of the prerequisites from algebra and analysis in a single volume and in a coherent manner, and to expose the rudiments of Finsler geometry on these foundations. Experienced readers will notice that this approach makes it possible to derive several classical theorems from general principles in a simple and unified way.

The table of contents comprises all the topics included in the book. In the following we only point out a few important features of our method.

2. As we have already mentioned, the first aim of our book is to serve as a textbook. To illustrate this, we note the following features:

(1) In principle, the reading of the book requires only the knowledge of undergraduate linear algebra and analysis. However, readers with a background in classical differential geometry or elementary Riemannian geometry will assimilate the material and perceive the subtleties more easily.

(2) We define our technical terms unambiguously and use them in this spirit throughout. We state all our assertions in a clear and explicit manner, and we give detailed proofs for them. There are only a few exceptions to the latter, but in these few cases we nearly always give explicit references to the literature. In our proofs we carefully explain all steps, and we never leave non-trivial details to the reader (or at least we do not intend to do so).

(3) A course in 'Advanced Calculus' is an organic part of the book. Such a course could be formed from the relevant parts, arranged logically as follows: *Appendix A, Chapter 1, Appendix B, Appendix C.1, Subsections 4.2.1, 3.2.1, Section 9.1.* A non-standard feature of this course is the study of Finsler vector spaces, which is in close connection with convexity. Our teaching experience shows that the interest of upper undergraduate students in Finsler geometry may be aroused in this way.

We note that Appendix C.2, C.3 and Subsection 6.1.10 can easily be made independent of manifold theory, perhaps with some tutorial help. In this way we obtain an introduction to the classical theory of hypersurfaces of a finite-dimensional real vector space.

3. The conceptual framework for the study of sprays and Finsler structures is provided by the theory of *manifolds* and *vector bundles*. It must be clear from the foregoing that we do not explicitly rely on the reader's knowledge of manifold theory. However, our discussion of manifolds is rather concise, we usually restrict ourselves to the presentation of the most important concepts and facts, and we omit difficult proofs. We do so not only due to the limited size of the book, but also because there are plenty of excellent textbooks on manifold theory, e.g., Barden and Thomas [9], Jeffrey M. Lee [66], John M. Lee [69], Michor [76] and Tu [96], just to mention a few recent ones. There are also good course materials available on the Internet. Our main guides were such classical works as the first volume of the monograph of Greub, Halperin and Vanstone [52] and 'Semi-Riemannian Geometry' by O'Neill [79].

We only need a few special types of vector bundles in this book. First of all, of course,

$$\text{the } \textit{tangent bundle } \tau \colon TM \to M$$

of a manifold M, and also some vector bundles over the tangent manifold TM:

$$\tau_{TM} \colon TTM \to TM, \quad \tau_* \colon TTM \to TM,$$

$$\tau_{TM}^{\vee} \colon TTM \to TM - \text{the } \textit{vertical subbundle},$$

$$\pi \colon TM \times_M TM \to TM - \text{the } \textit{Finsler bundle}.$$

However, for the sake of transparency, and to avoid repetition, we introduce the basic concepts connected with bundles at the level of generality of fibre bundles and 'abstract vector bundles'. Even a part of the theory of Ehresmann connections might be discussed in such generality, but we think that this theory is easier to understand in the framework of the tangent bundle, and this approach also makes it possible to expound Ehresmann connections in their full complexity.

We introduce *covariant derivatives* at the general level of vector bundles. The reason is that we need them on different vector bundles, and thus the common definition ensures greater clarity.

The covariant derivatives on a manifold (more precisely, on its tangent bundle) are in a bijective correspondence with a special class of Ehresmann connections, the *linear Ehresmann connections*. Thus, these two objects show the two sides of the same coin: Ehresmann connections have a purely geometric character, whereas covariant derivatives act as differential operators. We discuss this fundamental relationship between covariant derivatives and linear Ehresmann connections in detail.

4. The reader will see that Ehresmann connections are indeed the cornerstones of the monumental edifice of differential geometry. Moreover, they also make it possible to give a visualizable geometric formulation and proof of several theorems in classical analysis and differential geometry.

The appearance of *sprays* on the scene makes the picture even more complex. Informally, a spray defines a special class of second-order differential equations over a manifold. The 'spray coefficients', which actually depend on the chart we are working with, can be interpreted as the equivalents of forces in Newtonian mechanics. Each homogeneous Ehresmann connection determines a spray in a natural manner. On the other hand, a homogeneous Ehresmann connection comes from a spray if, and only if, its torsion vanishes. Sprays provide a framework for a unified and systematic discussion of geodesics.

Now we are in a position to complete the explanation of the scheme sketched in paragraph 1. Each Finsler function gives rise to a spray in a canonical manner, which carries much information about the original structure. Thus an essential part of Finsler geometry may be discussed in the setting of spray geometry. The converse question of the 'Finsler metrizability' of sprays, i.e., when does a given spray come from a Finsler function as its canonical spray, leads to a jungle of difficult problems.

5.   Besides the carefully chosen conceptual background, we also need a substantial and flexible technical apparatus. The major part of this is developed in Chapters 3, 4 and 6. We apply traditional tensor calculus, tensor derivations (covariant derivatives in particular), the classical graded derivations of a Grassmann algebra (Lie derivatives, substitution operators and exterior derivative) and the graded derivations induced by vector-valued forms and described by Frölicher – Nijenhuis theory. The canonical objects and constructions on the tangent bundle and on the Finsler bundle (canonical involution, vertical endomorphism, vertical and complete lifts, etc.) may also be regarded as parts of our technical apparatus. Finally, the theorems about existence, uniqueness and smooth dependence of solutions of ordinary differential equations also belong to our indispensable tools.

6.   The heart of our book is made up of Chapters 5, 7, 8 and 9. We discuss Ehresmann connections and sprays in such a systematic and detailed manner that the relevant chapters, together with Chapter 4, can almost be considered as a monograph embedded in the textbook. The same cannot be said about the chapter discussing Finsler structures. This theory has grown so vast in the last two decades and it is still developing so intensely that we are able to present only some of the most important aspects of

Finsler manifolds.

To follow through the path

$$\text{connections} \rightarrow \text{sprays} \rightarrow \text{Finsler structures}$$

the reader has to climb a lot of mountains. That is what we imply by quoting the Korean proverb at the beginning of Section 3.4. We think, however, that the reaching of each peak will give the reader its own reward.

7. Hewitt and Stromberg wrote the following in the preface of their book 'Real and Abstract Analysis': 'exercises are to mathematicians what Czerny is to a pianist'. Although we completely agree with them, the reader will not find any explicit exercise in our book. The reason is that many of our assertions may be regarded as 'exercises', which are immediately followed by 'solutions', in the form of proofs. These assertions and their proofs usually have a technical character, which, of course, does not mean lack of quality. Most of them are easily recognizable, and we recommend ambitious readers to try to prove these assertions for themselves. If they find simpler or more elegant solutions than ours, then they are on the best way to profound understanding.

We try to apply the index-free or 'intrinsic' method in most of our proofs. Doing these proofs with the method of classical tensor calculus after choosing a chart may supply the reader with further useful exercises. (This way will often prove shorter.) Sometimes we only give a proof which uses coordinates. In these cases the reader may try to find an index-free argument.

We give very few numerical examples, since a large number of these would make the book too long, moreover, the most interesting and important examples are easily accessible. Here we refer to Z. Shen's monograph [88].

For the readers' convenience, errata and addenda will be available on the first author's homepage (http://www.math.unideb.hu/~szilasi) as soon as we find it necessary.

J. Szilasi
R. L. Lovas
D. Cs. Kertész

Acknowledgments

We would like to thank our colleague Ms Bernadett Aradi, who carefully read many chapters of the book, and who contributed to their final formulation with her useful comments.

We would like to thank the Publisher for publishing our book in a form of high quality, and also Ms Lai Fun Kwong, who helped us throughout the long process of writing this book.

The first two authors were supported by the OTKA project NK-81402 (Hungary).

The third author was supported by TÁMOP-4.2.2/B-10/1-2010-0024. That programme is subsidized by the European Union and by the European Social Fund.

Contents

Chapter 1

Modules, Algebras and Derivations

Throughout the book, we use the notation and terminology of naive set theory as they are summarized in Appendix A.1 and A.2. Concerning the basic concepts of algebra, we follow the conventions of Appendix A.3 and A.4. In this chapter we assume that R is a fixed nonzero commutative ring (see A.4.1). The elements of R will often be mentioned as scalars and will be denoted by Greek letters.

1.1 Modules and Vector Spaces

> Linear algebra is both one of the oldest and one of the newest branches of mathematics.
>
> Nicolas Bourbaki

> Except for boolean algebra there is no theory more universally employed in mathematics than linear algebra; and there is hardly any theory which is more elementary, in spite of the fact that generations of professors and textbook writers have obscured its simplicity by preposterous calculations with matrices.
>
> Jean Dieudonné

1.1.1 *Basic Definitions and Facts*

Definition 1.1.1. A *module* over R (or simply an R-*module*) is a commutative group V (written additively) together with a mapping

$$R \times V \to V, \ (\alpha, v) \mapsto \alpha v$$

called *scalar multiplication* which satisfies

1

(i) $\alpha(u + v) = \alpha u + \alpha v$ for all $\alpha \in \mathsf{R}$ and $u, v \in V$;

(ii) $(\alpha + \beta)u = \alpha u + \beta u$ for all $\alpha, \beta \in \mathsf{R}$ and $u \in V$;

(iii) $(\alpha\beta)u = \alpha(\beta u)$ for all $\alpha, \beta \in \mathsf{R}$ and $u \in V$;

(iv) $1u = u$ for all $u \in V$.

Then the fixed ring R is called the *ground ring*, and the elements of V are also mentioned as *vectors*. When the ground ring is a field F, we speak of a *vector space over* F (or an F-*vector space* for short). In particular, a vector space over $\mathbb{R}$ is called a *real vector space*, and a vector space over $\mathbb{C}$ is a *complex vector space*.

Remark 1.1.2. We assume the readers' familiarity with the rudiments of classical linear algebra, whose central objects are (finite-dimensional) real and complex vector spaces. There are strong analogies between vector spaces and modules at the level of concepts. However, the analogy is quite limited (and can be misleading) when we turn to theorems.

Modules are at least as important in differential geometry (and in many branches of mathematics) as vector spaces in analysis, and the reader will meet examples of them throughout this book. In the next paragraphs, for later use, we transfer some basic definitions and results known from linear algebra to the more general context of R-modules.

Definition 1.1.3. A *submodule* of an R-module V is a subset H of V satisfying the following conditions:

(i) H is a subgroup of the additive group V,

(ii) H is closed under scalar multiplication, i.e., $\lambda u \in H$ for all $\lambda \in \mathsf{R}$, $u \in H$.

Remark 1.1.4. (a) If H is a submodule of V, then H is an R-module in its own right, but with respect to the same addition and scalar multiplication as V. A submodule of a vector space is called a *vector subspace* (*linear subspace* or simply *subspace*).

(b) Every R-module V has at least two, not necessarily distinct, submodules, V itself and $\{0\}$; the latter is called the *trivial submodule*. Any submodule of V different from these two is mentioned as a *proper submodule*.

(c) Let $(H_i)_{i \in I}$ be a family of submodules of an R-module V. Then the intersection $\bigcap_{i \in I} H_i$ is again a submodule of V. If S is any subset of V, then

$$\mathrm{span}(S) := \bigcap \{H \in \mathcal{P}(V) \mid H \text{ is a submodule and } S \subset H\}$$

is the smallest submodule containing S. We say that span(S) is the submodule *spanned* or *generated* by S. Provided that S is nonempty, there is an equivalent definition of span(S) 'from below'. Namely,

$$\text{span}(S) = \left\{ \sum_{i=1}^{n} \alpha^i u_i \in V \; \middle| \; n \in \mathbb{N}^*; \; \alpha^1, \ldots, \alpha^n \in \mathsf{R}; \; u_1, \ldots, u_n \in S \right\}.$$

Definition 1.1.5. By the *sum* of a family $(H_i)_{i \in I}$ of submodules of an R-module V we mean the submodule spanned by the union of the family, and denote it by $\sum_{i \in I} H_i$. When $I = \{1, 2\}$, we simply write $H_1 + H_2$, and we use a similar notation for any finite set of indices. If $H_1 \cap H_2 = \{0\}$, we call $H_1 + H_2$ the *direct sum* of H_1 and H_2, and we denote this direct sum by $H_1 \oplus H_2$. More generally, the sum of the family $(H_i)_{i \in I}$ is direct, and written as $\bigoplus_{i \in I} H_i$, if, for each $j \in I$, we have

$$\left(\sum_{i \in I \setminus \{j\}} H_i \right) \cap H_j = \{0\}.$$

Remark 1.1.6. (a) A finite sum $H_1 + \cdots + H_k$ $(k \geq 2)$ of submodules of V is direct if, and only if, each $v \in H_1 + \cdots + H_k$ can be written *uniquely* in the form

$$v = u_1 + \cdots + u_k; \quad u_i \in H_i, \; i \in J_k.$$

A submodule H of V is said to be a *direct summand* of V if there exists a submodule K of V such that $V = H \oplus K$. Then K is called a *complementary submodule* of H in V.

(b) A *linear combination* of a finite family $(v_i)_{i=1}^{n}$ of elements in V is a vector of the form

$$\alpha^1 v_1 + \cdots + \alpha^n v_n = \sum_{i=1}^{n} \alpha^i v_i,$$

where $\alpha^1, \ldots, \alpha^n \in \mathsf{R}$. Using this terminology,

$$\text{span}(v_1, \ldots, v_n) := \text{span}\{v_1, \ldots, v_n\}$$

is just the set of all linear combinations of the family $(v_i)_{i=1}^{n}$.

(c) A family $(v_i)_{i=1}^{k}$ of elements in an R-module V is *linearly dependent* if there exist scalars $\alpha^1, \ldots, \alpha^k \in \mathsf{R}$, *not all zero*, such that

$$\alpha^1 v_1 + \cdots + \alpha^k v_k = 0.$$

Otherwise, $(v_i)_{i=1}^{k}$ is called *linearly independent*.

Definition 1.1.7. Let V be an R-module.

(a) We say that V is *finitely generated* if there exists a finite family $(a_i)_{i=1}^k$ of elements in V such that

$$\mathrm{span}(a_1, \ldots, a_k) = V.$$

Then $(a_i)_{i=1}^k$ is called a *generator family* (or a *system of generators*) of V. When the ground ring is a field, a finitely generated module is said to be a *finite-dimensional vector space*.

(b) A family $(b_i)_{i \in I}$ of elements in V is a *basis* of V if

for each $v \in V$ there exists a unique finite family $\left(\lambda^j\right)_{j \in J}$ of scalars in R such that $J \subset I$, $\lambda^j \neq 0$ for all $j \in J$ and $\sum_{j \in J} \lambda^j b_j = v$.

(c) A module is called *free* if it has a basis. If this basis is finite, the module is said to be a *finitely generated free module*. We adopt the convention that the trivial module $\{0\}$ has a (unique) finite basis, the empty family.

Example 1.1.8. *A finitely generated module need not be a finitely generated* free *module.* To illustrate this phenomenon, we present a simple but typical example.

Let V be a finite commutative group, written additively. Make V into a $\mathbb{Z}$-module as follows: for any $n \in \mathbb{Z}$ and $v \in V$ define

$$nv := \begin{cases} v + \cdots + v \ (n \text{ times}) & \text{if } n > 0 \\ 0 & \text{if } n = 0 \\ (-n)(-v) & \text{if } n < 0. \end{cases}$$

The $\mathbb{Z}$-module so obtained is clearly finitely generated, however, it does not admit any finite basis. Indeed, if there were a finite basis $(b_i)_{i=1}^n$ $(n \in \mathbb{N}^*)$ of V, then the mapping

$$(k_1, \ldots, k_n) \in \mathbb{Z}^n \mapsto k_1 b_1 + \cdots + k_n b_n \in V$$

would be bijective, which is impossible, since $\mathbb{Z}^n$ is an infinite set, while V is finite.

Lemma 1.1.9. *Let V be an R-module.*

(i) *If V is a free module, then any basis of V is a minimal generator family.*

(ii) *If V is a finitely generated free module, then any basis of V is finite.*

Proof. (i) Let $(b_i)_{i \in I}$ be a basis of V, and suppose that $(b_j)_{j \in J}$ is a proper subfamily of $(b_i)_{i \in I}$. Consider a vector b_k, where $k \in I \setminus J$. By the definition of a basis, b_k cannot be expressed as a linear combination of a finite subfamily of $(b_j)_{j \in J}$, therefore the subfamily $(b_j)_{j \in J}$ cannot be a generator family of V.

(ii) Now suppose that $(a_i)_{i=1}^k$ is a generator family, and $(b_i)_{i \in I}$ is a basis of V. Then there is a finite subfamily $(b_{i_j})_{j=1}^l$ of the basis such that

$$\{a_1, \ldots, a_k\} \subset \text{span}(b_{i_1}, \ldots, b_{i_l}).$$

From this we conclude that

$$V = \text{span}(a_1, \ldots, a_k) \subset \text{span}(\text{span}(b_{i_1}, \ldots, b_{i_l})) = \text{span}(b_{i_1}, \ldots, b_{i_l}).$$

Thus the subfamily $(b_{i_j})_{j=1}^l$ is a generator family of V, therefore, by part (i), $(b_{i_j})_{j=1}^l = (b_i)_{i \in I}$. $\square$

The following result is of fundamental importance.

Proposition 1.1.10. *Any two bases of a finitely generated free R-module have the same cardinality. In particular, every finite-dimensional vector space has a (finite) basis, and any two of its bases have the same number of elements.*

Here the second assertion is very well-known: it is one of the basic theorems of linear algebra. Accordingly, a proof for it is available in any good book on the subject. For a proof of the general case, first we refer to [50], pp. 369–370. Here we find a direct argument that uses determinants. There is, however, a more elegant, but trickier proof. The idea is the following. For the given R-module V we construct a vector space $\tilde{V}$ with the property that, for any basis of V, there is a basis of the same cardinality for $\tilde{V}$. Then we can appeal to the classical linear algebra result formulated in the second part of the proposition. For an elaboration of this plan we refer to [87], pp. 127–128.

Proposition 1.1.10 allows us to introduce the following notion.

Definition 1.1.11. Let V be a finitely generated free R-module. We say that V is *n-dimensional* or that n is the *dimension of V* if V has a basis consisting of n vectors. Then we write $\dim_R V = n$, or simply, $\dim V = n$ if the ground ring is understood.

Remark 1.1.12. (a) In module theory the term 'rank' is commonly used instead of dimension. However, we do not follow this practice in this book.

(b) It can be shown that *every vector space has a basis, and every linearly independent family in a vector space can be extended to a basis.* These results depend on the axiom of choice (see A.2.5). For a proof we refer to [51], pp. 12–13.

(c) If V is a finitely generated free R-module, and $(b_i)_{i=1}^n$ is a basis of V, then for each $v \in V$ there exists a unique n-tuple $(\lambda^1, \ldots, \lambda^n)$ of elements of R such that

$$v = \lambda^1 b_1 + \cdots + \lambda^n b_n.$$

The scalars $\lambda^1, \ldots, \lambda^n$ are called the *coordinates* or the *components of v with respect to the basis* $(b_i)_{i=1}^n$ of V.

(d) Observe that forming a linear combination above, we have tacitly adopted the convention that vectors are indexed with *sub*scripts and their coefficients are indexed with *super*scripts. Let us take the opportunity at this stage to introduce the *summation convention* first employed by Einstein:

> *Whenever a term contains a repeated index, one as a subscript and the other as a superscript, summation is implied over this index. The range of indices is either determined by context or has been declared previously.*

Example 1.1.13. Let n be a positive integer, and consider the Cartesian product R^n, i.e., the set of n-tuples of elements of our ground ring R. Define addition and scalar multiplication in R^n 'componentwise', i.e., by the rules

$$a + b = \left(\alpha^1, \ldots, \alpha^n\right) + \left(\beta^1, \ldots, \beta^n\right) := \left(\alpha^1 + \beta^1, \ldots, \alpha^n + \beta^n\right),$$

$$\lambda a = \lambda \left(\alpha^1, \ldots, \alpha^n\right) := \left(\lambda \alpha^1, \ldots, \lambda \alpha^n\right), \quad \lambda \in \mathsf{R}.$$

It is easy to check that these two operations make R^n into a module over R. In the following, whenever we refer to R^n as an R-module, it is the module structure defined above that is meant. Notice that this construction allows us to regard $\mathsf{R} = \mathsf{R}^1$ as a module over itself; then the 'scalars' are also 'vectors'. We extend the definition of the R-module R^n to the case $n = 0$ in the only possible way: $\mathsf{R}^0 := \{0\}$.

The R-module R^n is a finitely generated free module; the vectors

$$e_i := (\underbrace{0, \ldots, 0, 1, 0, \ldots, 0}_{i}), \quad i \in J_n$$

form a basis, called the *canonical basis* of R^n. If $v = \left(\nu^1, \ldots, \nu^n\right) \in \mathsf{R}^n$, the relation

$$v = \nu^1 e_1 + \cdots + \nu^n e_n$$

shows that the coordinates of v with respect to the canonical basis are just the scalars $\nu^1, \ldots, \nu^n$.

When we use matrix (or tensor) calculus, it is more convenient and consistent to write the elements of R^n as column vectors, i.e., to identify R^n with $M_{n \times 1}(\mathsf{R})$. (For the notation see also A.2.9.) So in this context,

$$v = \begin{pmatrix} \nu^1 \\ \vdots \\ \nu^n \end{pmatrix} = {}^t\!\left(\nu^1 \ \ldots \ \nu^n \right) \in \mathsf{R}^n \cong M_{n \times 1}(\mathsf{R}),$$

and the members of the canonical basis of R^n are

$$e_1 = \begin{pmatrix} 1 \\ 0 \\ 0 \\ \vdots \\ 0 \end{pmatrix}, \ e_2 = \begin{pmatrix} 0 \\ 1 \\ 0 \\ \vdots \\ 0 \end{pmatrix}, \ \ldots, e_n = \begin{pmatrix} 0 \\ 0 \\ \vdots \\ 0 \\ 1 \end{pmatrix}.$$

This formalism will be applied first in Lemma 1.1.20 below.

If we take

$$\mathsf{R} := \mathbb{R} = \textit{the field of real numbers},$$

we get the n-dimensional real vector space $\mathbb{R}^n$, called the *Euclidean n-space*.

Remark 1.1.14 (Vector Subspaces). To conclude this subsection, we collect here some basic facts about vector subspaces.

Let V be a vector space over a field.

(i) *Every subspace of V has at least one complementary subspace. That is, if K is a subspace of V, then there exists a subspace L of V such that $V = K \oplus L$.*

This is an easy consequence of the results formulated in Remark 1.1.12(b).

Now suppose that K and L are finite-dimensional subspaces of V.

(ii) *If $K \subset L$, then $\dim K \leq \dim L$ and $\dim K = \dim L$ implies $K = L$.*

(iii) *Both $K + L$ and $K \cap L$ are finite-dimensional and*

$$\dim(K + L) + \dim(K \cap L) = \dim K + \dim L \quad \text{(Grassmann's formula)}.$$

If $\dim V = n$, then

$$\dim(K \cap L) \geq \dim K + \dim L - n \quad \text{(Sylvester's inequality)}.$$

1.1.2   *Homomorphisms*

Definition 1.1.15. Let V and W be R-modules. A mapping $\varphi\colon V \to W$ is an R-*module homomorphism* or an R-*linear mapping* (simply a *homomorphism* or a *linear mapping*) if

(i) $\varphi(u + v) = \varphi(u) + \varphi(v)$ for all $u, v \in V$,
(ii) $\varphi(\lambda u) = \lambda\varphi(u)$ for all $\lambda \in \mathsf{R}$, $u \in V$.

An R-module homomorphism is an *isomorphism* (*of* R-*modules*) if it is bijective. The modules V and W are called *isomorphic*, denoted by $V \cong W$, if there exists an isomorphism of V onto W. An *automorphism* of V is an isomorphism of V onto itself.

Remark 1.1.16. (a) If $\varphi\colon V \to W$ is an R-module homomorphism, then the image of any submodule of V under φ is a submodule of W, and the inverse image under φ of any submodule of W is a submodule of V. In particular,

$$\mathrm{Ker}(\varphi) := \{v \in V \mid \varphi(v) = 0\} = \varphi^{-1}(0),$$

the *kernel* of φ, is a submodule of V, and $\mathrm{Im}(\varphi)$, the *image* of φ, is a submodule of W. A homomorphism $\varphi\colon V \to W$ is injective if, and only if, $\mathrm{Ker}(\varphi) = \{0\}$.

  (b) Suppose, in particular, that V and W are *vector spaces* over the same field. If V is finite-dimensional and $\varphi\colon V \to W$ is a linear mapping, then

$$\dim \mathrm{Im}(\varphi) + \dim \mathrm{Ker}(\varphi) = \dim V.$$

Here $\dim \mathrm{Im}(\varphi)$ is called the *rank* of φ and $\dim \mathrm{Ker}(\varphi)$ is called the *nullity* of φ. In these terms, the message of this important result can be compressed as *rank* + *nullity* = *dimension*.

  (c) If $\varphi\colon V \to W$ and $\psi\colon V \to W$ are R-module homomorphisms and $\lambda \in \mathsf{R}$, then the mappings

$$\varphi + \psi\colon V \to W, \ v \mapsto (\varphi + \psi)(v) := \varphi(v) + \psi(v)$$

and

$$\lambda\varphi\colon V \to W, \ v \mapsto (\lambda\varphi)(v) := \lambda\varphi(v)$$

are also homomorphisms. The set

$$\mathrm{Hom}_{\mathsf{R}}(V, W) \ \text{or} \ L_{\mathsf{R}}(V, W)$$

of all R-module homomorphisms of V into W, together with the above addition and scalar multiplication is an R-module. When there is no danger of confusion, we denote this module by $\mathrm{Hom}(V, W)$ or $L(V, W)$.

Let U also be an R-module. If $\psi \in \mathrm{Hom}_R(U, V)$ and $\varphi \in \mathrm{Hom}_R(V, W)$, then $\varphi \circ \psi \in \mathrm{Hom}_R(U, W)$. In particular,

$$\mathrm{End}_R(V) := \mathrm{Hom}_R(V, V)$$

with addition as above and multiplication defined as composition of mappings is a ring. The ring $\mathrm{End}_R(V)$ is called the *endomorphism ring* of V and is denoted simply by $\mathrm{End}(V)$ when the ground ring is clear from the context. Elements of $\mathrm{End}(V)$ are mentioned as (R-*linear*) *endomorphisms* or (R-)*linear transformations*. An endomorphism $\varphi \in \mathrm{End}_R(V)$ is called a *projection* (or *projector*) if $\varphi^2 := \varphi \circ \varphi = \varphi$.

The set of automorphisms of V forms a group under composition of mappings. We write $\mathrm{GL}(V)$ for this group and call it the *general linear group* of the module V. Notice that $\mathrm{GL}(V)$ is just the group of invertible elements of the ring $\mathrm{End}(V)$ (see A.4.1), i.e.,

$$\mathrm{GL}(V) = (\mathrm{End}(V))^{\times}. \tag{1.1.1}$$

Definition 1.1.17. Let U, V, W be R-modules. A sequence

$$\{0\} \to U \overset{\psi}{\to} V \overset{\varphi}{\to} W \to \{0\}$$

of R-module homomorphisms is called a *short exact sequence* if

$$\psi \text{ is injective, } \varphi \text{ is surjective, and } \mathrm{Im}(\psi) = \mathrm{Ker}(\varphi).$$

Here $\{0\} \to U$ is the unique homomorphism $0 \in \{0\} \mapsto 0 \in U$, and $W \to \{0\}$ is the unique homomorphism $w \in W \mapsto 0 \in \{0\}$. In this context, by an abuse of notation, we simply write 0 for the zero module $\{0\}$.

A short exact sequence

$$0 \to U \overset{\psi}{\to} V \overset{\varphi}{\to} W \to 0$$

of R-module homomorphisms is said to be *split exact* (or simply *split*) if there is an R-module homomorphism $\mu \colon W \to V$ which is a section of φ, i.e., $\varphi \circ \mu = 1_W$. Then μ is called a *splitting homomorphism* for the sequence.

Example 1.1.18. Given two R-modules U and V, make the Cartesian product $U \times V$ into an R-module by the componentwise addition

$$(u_1, v_1) + (u_2, v_2) := (u_1 + u_2, v_1 + v_2); \quad u_1, u_2 \in U; \; v_1, v_2 \in V$$

and scalar multiplication

$$\lambda(u,v) := (\lambda u, \lambda v); \quad \lambda \in \mathsf{R}, \ (u,v) \in U \times V.$$

The R-module so obtained is called the *direct product* or the *external direct sum* of U and V and is denoted by $U \oplus V$.

Now let $(U_i)_{i \in I}$ be a family of R-modules. The external direct sum of this family, denoted by $\bigoplus_{i \in I} U_i$, consists of all elements $(u_i)_{i \in I}$ of the Cartesian product $\prod_{i \in I} U_i$ such that only a finite number of the vectors u_i is different from zero. Addition and scalar multiplication in $\bigoplus_{i \in I} U_i$ is defined componentwise, as above.

Consider again two R-modules U and V. The projections

$$\pi_U : U \oplus V \to U, \ (u,v) \mapsto u,$$
$$\pi_V : U \oplus V \to V, \ (u,v) \mapsto v$$

and the inclusions

$$i_U : U \to U \oplus V, \ u \mapsto (u,0),$$
$$i_V : V \to U \oplus V, \ v \mapsto (0,v)$$

are homomorphisms which satisfy the identities

$$\begin{cases} \pi_U \circ i_U = 1_U, \ \pi_U \circ i_V = 0, \\ \pi_V \circ i_U = 0, \ \pi_V \circ i_V = 1_V, \\ i_U \circ \pi_U + i_V \circ \pi_V = 1_{U \oplus V}. \end{cases}$$

It follows that the sequence

$$0 \to U \xrightarrow{i_U} U \oplus V \xrightarrow{\pi_V} V \to 0$$

is a split exact sequence, for which the mapping i_V is a splitting homomorphism.

Now we show that every split exact sequence of R-modules is 'essentially' of this form.

Lemma 1.1.19. *Let U, V, W be R-modules. A short exact sequence*

$$0 \to U \xrightarrow{\psi} V \xrightarrow{\varphi} W \to 0 \tag{1.1.2}$$

is split if, and only if, $\psi(U)$ is a direct summand in V, i.e., there is a submodule H of V such that $V = \psi(U) \oplus H$. Then H is mapped isomorphically onto W by φ.

Proof. Suppose first that (1.1.2) is split, and let $\mu\colon W \to V$ be a splitting homomorphism for the sequence. Then $H := \mu(W)$ is a submodule of V; we show that $V = \psi(U) \oplus H$.

Every element v in V can be written in the form

$$v = (v - \mu \circ \varphi(v)) + \mu \circ \varphi(v).$$

Here the second term $\mu \circ \varphi(v)$ belongs to H. The first term is an element of $\psi(U)$, since

$$\varphi(v - \mu \circ \varphi(v)) = \varphi(v) - (\varphi \circ \mu \circ \varphi)(v) = \varphi(v) - (1_W \circ \varphi)(v) = \varphi(v) - \varphi(v) = 0$$

and $\operatorname{Ker} \varphi = \operatorname{Im} \psi$. Thus $V = \psi(U) + H$, where, obviously, H is mapped isomorphically onto W by φ. If $a \in \psi(U) \cap H$, then $a = \psi(u) = \mu(w)$ for a unique $u \in U$ and $w \in W$. Then

$$w = 1_W(w) = \varphi(\mu(w)) = \varphi(\psi(u)) = 0,$$

whence $a = \mu(w) = 0$. So $\psi(U) \cap H = \{0\}$, therefore $V = \psi(U) \oplus H$, as desired.

Conversely, suppose that $V = \psi(U) \oplus H$ for some submodule H of V. Then

$$\varphi \restriction H \colon H \to W$$

is an isomorphism of R-modules (since $\varphi \circ \psi = 0$). Let $\mu := (\varphi \restriction H)^{-1}$. Then μ is a splitting homomorphism for the sequence

$$0 \to U \xrightarrow{\psi} \psi(U) \oplus H \xrightarrow{\varphi} W \to 0.$$

$\square$

In the following we review some basic results concerning the homomorphisms of a finitely generated free module.

Lemma 1.1.20. *Let V be a finitely generated free R-module, and let W be an R-module. Suppose that $(b_i)_{i=1}^n$ is a basis of V and $(w_i)_{i=1}^n$ is a family of elements in W. Then there exists a unique homomorphism $\varphi\colon V \to W$ such that*

$$\varphi(b_i) = w_i, \quad i \in J_n. \tag{1.1.3}$$

This homomorphism is injective if, and only if, $(w_i)_{i=1}^n$ is linearly independent; surjective if, and only if, $(w_i)_{i=1}^n$ is a system of generators.

Proof.   Assume that $\varphi \in \mathrm{Hom}_{\mathsf{R}}(V, W)$ and (1.1.3) is satisfied. Then for each $v = \sum_{i=1}^{n} \nu^i b_i \in V$,

$$\varphi(v) = \sum_{i=1}^{n} \nu^i \varphi(b_i) = \sum_{i=1}^{n} \nu^i w_i,$$

therefore φ, if it exists, is unique. To prove the existence of φ, define φ just by the formula above. Then it is easily checked that φ is an R-linear mapping and satisfies (1.1.3).

Now suppose that φ is injective. If $\sum_{i=1}^{n} \nu^i w_i = 0$, then the vector $v := \sum_{i=1}^{n} \nu^i b_i$ belongs to $\mathrm{Ker}(\varphi) = \{0\}$. This implies that $\nu^i = 0$ for each $i \in J_n$, hence the family $(w_i)_{i=1}^{n}$ is linearly independent.

Conversely, if $(w_i)_{i=1}^{n}$ is linearly independent, then

$$v = \sum_{i=1}^{n} \nu^i b_i \in \mathrm{Ker}(\varphi)$$

implies that $\sum_{i=1}^{n} \nu^i w_i = 0$, whence $v = 0$. Thus $\mathrm{Ker}(\varphi) = \{0\}$, which is equivalent to the injectiveness of φ.

It is similarly easy to show that φ is surjective if, and only if, $(w_i)_{i=1}^{n}$ is a generator family of W.   □

Corollary 1.1.21. *A non-trivial R-module V is a finitely generated free module if, and only if, there exists a positive integer n such that V is isomorphic to the R-module R^n. More precisely, let $(v_i)_{i=1}^{n}$ be a family of elements in V, and let $(e_i)_{i=1}^{n}$ be the canonical basis of R^n. Then $(v_i)_{i=1}^{n}$ is a basis of V if, and only if, there exists an isomorphism $\kappa \colon V \to \mathsf{R}^n$ such that*

$$\kappa(v_i) = e_i \quad \text{for all } i \in J_n.$$ □

The isomorphism κ depends, of course, on the choice of a basis in V. If the chosen basis is $\mathcal{B} = (b_i)_{i=1}^{n}$, we use the more precise notation $\kappa_{\mathcal{B}}$. Then, writing the elements of R^n as column vectors, for each $v = \sum_{i=1}^{n} \nu^i b_i \in V$ we have

$$\kappa_{\mathcal{B}}(v) = \sum_{i=1}^{n} \nu^i \kappa_{\mathcal{B}}(b_i) = \sum_{i=1}^{n} \nu^i e_i = \begin{pmatrix} \nu^1 \\ \vdots \\ \nu^n \end{pmatrix}.$$

The $n \times 1$ matrix

$$[v]_{\mathcal{B}} := \kappa_{\mathcal{B}}(v) = \begin{pmatrix} \nu^1 \\ \vdots \\ \nu^n \end{pmatrix}$$

is called the *coordinate (column) vector* of v with respect to the basis $\mathcal{B}$.

Example 1.1.22. Consider the R-modules

$$\mathsf{R}^n \cong M_{n\times 1}(\mathsf{R}), \ \mathsf{R}^m \cong M_{m\times 1}(\mathsf{R}); \quad m, n \in \mathbb{N}^*.$$

Let $A = (\alpha_j^i) \in M_{m\times n}(\mathsf{R})$. We can associate with A the R-linear mapping

$$\varphi_A \colon \mathsf{R}^n \to \mathsf{R}^m, \ v = \begin{pmatrix} \nu^1 \\ \vdots \\ \nu^n \end{pmatrix} \mapsto \varphi_A(v) := Av = \left(\sum_{j=1}^{n} \alpha_j^i \nu^j \right) \in \mathsf{R}^m.$$

Observe that

$$\varphi_A(e_j) = Ae_j = \begin{pmatrix} \alpha_j^1 \\ \vdots \\ \alpha_j^m \end{pmatrix} = j\text{th column of } A.$$

Conversely, let $\varphi \in \operatorname{Hom}_\mathsf{R}(\mathsf{R}^n, \mathsf{R}^m)$. Form the matrix

$$[\varphi] := A \in M_{m\times n}(\mathsf{R})$$

whose columns are the images of the members of the canonical basis $(e_i)_{i=1}^n$ of R^n:

$$[\varphi] := A := \begin{pmatrix} \vdots & \vdots & & \vdots \\ \varphi(e_1) & \varphi(e_2) & \dots & \varphi(e_n) \\ \vdots & \vdots & & \vdots \end{pmatrix}.$$

Then

$$\varphi(e_j) = Ae_j = \varphi_A(e_j) \text{ for all } j \in J_n,$$

so Lemma 1.1.20 implies that $\varphi = \varphi_A$. We have, moreover,

$$[\varphi_A] = A \text{ for all } A \in M_{m\times n}(\mathsf{R}).$$

We call $[\varphi]$ *the matrix associated with the homomorphism* φ, or *the matrix of φ with respect to the canonical bases* of R^n and R^m.

Lemma 1.1.23. *The mapping*

$$\varphi \in \operatorname{Hom}_\mathsf{R}(\mathsf{R}^n, \mathsf{R}^m) \mapsto [\varphi] \in M_{m\times n}(\mathsf{R})$$

is an isomorphism of R-*modules whose inverse is the mapping*

$$A \in M_{m\times n}(\mathsf{R}) \mapsto \varphi_A \in \operatorname{Hom}_\mathsf{R}(\mathsf{R}^n, \mathsf{R}^m).$$

If $\psi \in \operatorname{Hom}_\mathsf{R}(\mathsf{R}^k, \mathsf{R}^n)$, $\varphi \in \operatorname{Hom}_\mathsf{R}(\mathsf{R}^n, \mathsf{R}^m)$, then

$$[\varphi \circ \psi] = [\varphi][\psi].$$

On the other hand, for any $A \in M_{m\times n}(\mathsf{R})$ and any $B \in M_{n\times k}(\mathsf{R})$ we have

$$\varphi_A \circ \varphi_B = \varphi_{AB}. \qquad \triangle$$

Now we turn to the general case.

Lemma and Definition 1.1.24. *Suppose that V and W are non-trivial finitely generated free modules over* R, *and let* $\varphi\colon V \to W$ *be a homomorphism. Choose a basis* $\mathcal{B} = (b_i)_{i=1}^n$ *for* V *and a basis* $\mathcal{C} = (c_j)_{j=1}^m$ *for* W. *Then the mapping*

$$\kappa_{\mathcal{C}} \circ \varphi \circ \kappa_{\mathcal{B}}^{-1} : \mathsf{R}^n \to \mathsf{R}^m$$

is R-*linear; the matrix associated to this linear mapping is called the matrix of* φ *with respect to the bases* $\mathcal{B}$ *and* $\mathcal{C}$ *and it is denoted by* $[\varphi]_{\mathcal{C}}^{\mathcal{B}}$:

$$[\varphi]_{\mathcal{C}}^{\mathcal{B}} := \left[\kappa_{\mathcal{C}} \circ \varphi \circ \kappa_{\mathcal{B}}^{-1} \right]. \qquad\qquad \triangle$$

Remark 1.1.25. Equivalently,

$$[\varphi]_{\mathcal{C}}^{\mathcal{B}} = \left(\alpha_j^i \right) \in M_{m \times n}(\mathsf{R}) \text{ if, and only if, } \varphi(b_j) = \sum_{i=1}^m \alpha_j^i c_i \text{ for all } j \in J_n.$$

The matrix $[\varphi]_{\mathcal{C}}^{\mathcal{B}} = \left(\alpha_j^i \right)$ 'represents' φ in the sense that for any vector $v = \sum_{j=1}^n \nu^j b_j \in V$ we have

$$[\varphi(v)]_{\mathcal{C}} = [\varphi]_{\mathcal{C}}^{\mathcal{B}} [v]_{\mathcal{B}} = \left(\sum_{j=1}^n \alpha_j^i \nu^j \right) \in \mathsf{R}^m. \qquad (1.1.4)$$

If, in particular, $\varphi \in \text{End}(V)$, then the matrix

$$[\varphi]_{\mathcal{B}} := [\varphi]_{\mathcal{B}}^{\mathcal{B}}$$

is said to be *the matrix of* φ *with respect to the basis* $\mathcal{B}$. When the chosen basis is clear from the context, we simply write $[\varphi]$ instead of $[\varphi]_{\mathcal{B}}$.

Lemma 1.1.26. *Let* U, V, W *be finitely generated, non-trivial* R-*modules, and let* $\mathcal{A}$, $\mathcal{B}$, $\mathcal{C}$ *be bases of* U, V, W, *respectively. Then:*

(i) *The mapping*

$$\varphi \in \text{Hom}_{\mathsf{R}}(V, W) \mapsto [\varphi]_{\mathcal{C}}^{\mathcal{B}} := \left[\kappa_{\mathcal{C}} \circ \varphi \circ \kappa_{\mathcal{B}}^{-1} \right] \in M_{m \times n}(\mathsf{R})$$

is an isomorphism of R-*modules.*

(ii) *If* $\psi \in \text{Hom}_{\mathsf{R}}(U, V)$, $\varphi \in \text{Hom}_{\mathsf{R}}(V, W)$, *then*

$$[\varphi \circ \psi]_{\mathcal{C}}^{\mathcal{A}} = [\varphi]_{\mathcal{C}}^{\mathcal{B}} [\psi]_{\mathcal{B}}^{\mathcal{A}}. \qquad (1.1.5)$$

This can be displayed by the following diagram:

$$
\begin{array}{ccccc}
U & \xrightarrow{\;\psi\;} & V & \xrightarrow{\;\varphi\;} & W \\
{\scriptstyle \kappa_{\mathcal{A}}}\big\downarrow & & {\scriptstyle \kappa_{\mathcal{B}}}\big\downarrow & & {\scriptstyle \kappa_{\mathcal{C}}}\big\downarrow \\
\mathsf{R}^k & \xrightarrow[{[\psi]_{\mathcal{B}}^{\mathcal{A}}}]{} & \mathsf{R}^n & \xrightarrow[{[\varphi]_{\mathcal{C}}^{\mathcal{B}}}]{} & \mathsf{R}^m
\end{array} \quad .
$$

(iii) *A homomorphism $\varphi \in \operatorname{Hom}_R(V, W)$ is an isomorphism if, and only if,
$\dim V = \dim W$ and the matrix $[\varphi]_C^B$ is invertible. In this case,*

$$[\varphi^{-1}]_B^C = ([\varphi]_C^B)^{-1}. \qquad \triangle$$

Definition 1.1.27. Let $V \neq \{0\}$ be a finitely generated free module. If
$\mathcal{B} = (b_i)_{i=1}^n$ and $\mathcal{B}' = (b_i')_{i=1}^n$ are bases of V, then the matrix

$$T = (\tau_j^i) := [1_V]_B^{B'} = \left[\kappa_B \circ \kappa_{B'}^{-1}\right]$$

is said to be the *transition matrix for the change of basis from $\mathcal{B}$ to $\mathcal{B}'$.*

Remark 1.1.28. Clearly, the transition matrix T is invertible, that is,
$T \in \operatorname{GL}_n(\mathsf{R})$ (see A.4.2(c)). The inverse of T is

$$T^{-1} = [1_V]_{B'}^B.$$

In terms of the entries of T,

$$b_j' = \sum_{i=1}^n \tau_j^i b_i, \quad j \in J_n. \tag{1.1.6}$$

We have the following *change of basis formulae*:

$[v]_{B'} = T^{-1} [v]_B$ – for the coordinate vector of $v \in V$,
$[\varphi]_{B'} = T^{-1} [\varphi]_B T$ – for the matrix representative of $\varphi \in \operatorname{End}(V)$.

Indeed,

$$[v]_{B'} = [1_V(v)]_{B'} \overset{(1.1.4)}{=} [1_V]_{B'}^B [v]_B = T^{-1} [v]_B;$$

$$[\varphi]_{B'} = [1_V \circ \varphi \circ 1_V]_{B'}^{B'} \overset{(1.1.5)}{=} [1_V]_{B'}^B [\varphi \circ 1_V]_B^{B'}$$

$$\overset{(1.1.5)}{=} [1_V]_{B'}^B [\varphi]_B^B [1_V]_B^{B'} = T^{-1} [\varphi]_B T.$$

Definition and Lemma 1.1.29. Let V be an R-module. The R-module
$\operatorname{Hom}_R(V, \mathsf{R})$ of *linear forms* on V is called the *dual module* of V, and is
denoted by V^*. The dual

$$V^{**} := (V^*)^* = \operatorname{Hom}_R(V^*, \mathsf{R})$$

of the dual of V^* is called the *second dual* or *bidual* of V. We have a
canonical R-linear mapping

$$v \in V \mapsto \tilde{v} \in V^{**}, \quad \tilde{v}(\alpha) := \alpha(v) \text{ for all } \alpha \in V^*. \tag{1.1.7}$$

If this mapping is an isomorphism of R-modules, then V is said to be
reflexive. In this case we may freely identify V with V^{**}. $\qquad \triangle$

Remark 1.1.30. (a) There is no consensus concerning the notation of the dual of an R-module: one also sees V', V^d, V^v, etc. instead of V^*. Our choice is in some conflict with the notation $\mathbb{R}^*$, $\mathbb{Q}^*$, ... , but the slight inconsistency is acceptable.

(b) Let V and W be R-modules. If $\varphi \in \mathrm{Hom}_\mathsf{R}(V, W)$, then for each $f \in W^*$ the composite $f \circ \varphi$ is a linear form on V, i.e., $f \circ \varphi \in V^*$:

$$
\begin{array}{ccc}
V & \xrightarrow{\;\varphi\;} & W \\
& {\scriptstyle f\circ\varphi}\searrow & \downarrow{\scriptstyle f} \\
& & \mathsf{R}
\end{array}.
$$

It may be checked immediately that the mapping

$$f \in W^* \mapsto f \circ \varphi \in V^* \tag{1.1.8}$$

is R-linear. This mapping is called the *transpose* of φ and is denoted by ${}^t\varphi$. If φ is an isomorphism of V onto W, then ${}^t\varphi$ is an isomorphism of W^* onto V^*.

Lemma 1.1.31. *Suppose that V is a finitely generated free module, and let $(b_i)_{i=1}^n$ be a basis of V. Then there exists a unique basis $\left(b^i\right)_{i=1}^n$ of V^* such that*

$$b^i(b_j) = \delta_j^i \text{ (Kronecker delta)}; \quad i, j \in J_n. \tag{1.1.9}$$

Proof. Lemma 1.1.20 assures that there exists a unique family $\left(b^i\right)_{i=1}^n$ of linear forms in V^* satisfying (1.1.9); our only task is to show that this family is a basis of V^*.

Observe first that every $v \in V$ can be written uniquely in the form

$$v = \sum_{i=1}^n b^i(v)b_i. \tag{1.1.10}$$

Indeed, if $v = \sum_{j=1}^n \nu^j b_j$, then $b^i(v) = \sum_{j=1}^n \nu^j b^i(b_j) \overset{(1.1.9)}{=} \nu^i$ for each $i \in J_n$. Now let $\ell \in V^*$. Then

$$\ell(v) = \ell\left(\sum_{i=1}^n b^i(v)b_i\right) = \left(\sum_{i=1}^n \ell(b_i)b^i\right)(v)$$

for every $v \in V$, therefore

$$\ell = \sum_{i=1}^n \ell(b_i)b^i. \tag{1.1.11}$$

This proves that $\left(b^i\right)_{i=1}^n$ is a generator family of V^*.
If

$$\lambda_1 b^1 + \lambda_2 b^2 + \cdots + \lambda_n b^n = 0 \in V^*$$

for some scalars $\lambda_1, \ldots, \lambda_n$ in R, then applying this linear form to $b_1, \ldots, b_n$, we find that $\lambda_1 = \cdots = \lambda_n = 0$. Thus the family $\left(b^i\right)_{i=1}^n$ is also linearly independent, and hence it is a basis of V^*. $\qquad\square$

Remark 1.1.32. (a) The basis $\left(b^i\right)_{i=1}^n$ of V^* associated to the basis $(b_i)_{i=1}^n$ of V by Lemma 1.1.31 is called the *dual basis* to $(b_i)_{i=1}^n$. Now it follows immediately that *every finitely generated free module is reflexive*. Indeed, let $(b_i)_{i=1}^n$ be a basis of V, and $\left(b^i\right)_{i=1}^n$ the dual basis to $(b_i)_{i=1}^n$. Consider the family $\left(\tilde{b}_i\right)_{i=1}^n$ of elements of V^{**}, where the linear forms $\tilde{b}_i$ are given by (1.1.7). Then

$$\tilde{b}_j\left(b^i\right) := b^i(b_j) = \delta_j^i; \quad i, j \in J_n;$$

therefore $\left(\tilde{b}_i\right)_{i=1}^n$ is the basis of V^{**} dual to $\left(b^i\right)_{i=1}^n$.

(b) Here is the appropriate moment to extend the conventions formulated in Remark 1.1.12(d):

> For a family of elements in the dual V^* of an R-module V, we use *superscripts* to enumerate them. The components of an element of V^* with respect to a basis are written with indices as *subscripts*.

Lemma 1.1.33. *Let* $\mathcal{B} = (b_i)_{i=1}^n$ *be a basis of an R-module* V, *and let* $\mathcal{B}^* = (b^i)_{i=1}^n$ *be the corresponding dual basis of* V^*. *If* $\bar{\mathcal{B}} = (\bar{b}_i)_{i=1}^n$ *is another basis of* V *and* $\bar{\mathcal{B}}^* = (\bar{b}^i)_{i=1}^n$ *is the dual basis of* $\bar{\mathcal{B}}$, *then*

$$\bar{b}^i = \sum_{j=1}^n \sigma_j^i b^j, \quad i \in J_n, \tag{1.1.12}$$

where $(\sigma_j^i) \in \mathrm{GL}_n(\mathsf{R})$ *is the* inverse *of the transition matrix from* $\mathcal{B}$ *to* $\bar{\mathcal{B}}$.

Proof. By Remark 1.1.28, (σ_j^i) is the transition matrix from $\bar{\mathcal{B}}$ to $\mathcal{B}$, so we have

$$b_j = \sum_{k=1}^n \sigma_j^k \bar{b}_k, \quad j \in J_n. \tag{$*$}$$

Applying (1.1.11),

$$\bar{b}^i = \sum_{j=1}^n \bar{b}^i(b_j) b^j \overset{(*)}{=} \sum_{j=1}^n \sum_{k=1}^n \bar{b}^i(\sigma_j^k \bar{b}_k) b^j = \sum_{j=1}^n \sum_{k=1}^n \sigma_j^k \delta_k^i b^j = \sum_{j=1}^n \sigma_j^i b^j,$$

for all $i \in J_n$. This is what was to be shown. $\qquad\square$

Lemma 1.1.34. *Let V, W be finitely generated free modules over R with bases $\mathcal{B}$, $\mathcal{C}$, respectively. Let $\mathcal{B}^*$ be the dual basis of $\mathcal{B}$, and $\mathcal{C}^*$ the dual basis of $\mathcal{C}$. If $\varphi \in \mathrm{Hom}_\mathsf{R}(V, W)$, then*

$$[{}^t\varphi]_{\mathcal{B}^*}^{\mathcal{C}^*} = {}^t\big([\varphi]_{\mathcal{C}}^{\mathcal{B}}\big).$$

The proof is an easy exercise.

Example 1.1.35. Let n be a positive integer, and consider the R-module R^n defined in Example 1.1.13. If $(e^i)_{i=1}^n$ is the dual of the canonical basis $(e_i)_{i=1}^n$ of R^n, then for every $v = (\nu^1, \ldots, \nu^n) \in \mathsf{R}^n$ we have

$$e^i(v) = \nu^i, \quad i \in J_n.$$

Thus e^i is the 'ith coordinate function' on R^n that assigns to an element of R^n its ith coordinate. If, in particular, the ground ring is the field $\mathbb{R}$ of real numbers, and hence the module is the Euclidean n-space $\mathbb{R}^n$, then we say that $(e^i)_{i=1}^n$ is the *canonical coordinate system* on $\mathbb{R}^n$, and the functions $e^1, \ldots, e^n$ are the *natural* (or *Euclidean*) *coordinate functions* of $\mathbb{R}^n$.

Let us return to the general case of an arbitrary ground ring R. As we have already mentioned, in the context of tensor calculus we regard R^n as the R-module of $n \times 1$ matrices. Then the dual $(\mathsf{R}^n)^*$ of R^n may be interpreted as the module $M_{1 \times n}(\mathsf{R})$ of $1 \times n$ matrices, whose elements act on R^n by matrix multiplication: if $a = (\alpha_1 \ \ldots \ \alpha_n) \in (\mathsf{R}^n)^*$, $v = {}^t(\nu^1 \ \ldots \ \nu^n) \in \mathsf{R}^n$, then

$$a(v) := av = (\alpha_1 \ \ldots \ \alpha_n) \begin{pmatrix} \nu^1 \\ \vdots \\ \nu^n \end{pmatrix} = \sum_{i=1}^n \alpha_i \nu^i.$$

1.1.3 Cosets and Affine Mappings

Definition 1.1.36. Let V be a real vector space, and let a be a fixed vector in V. The mapping

$$t_a : V \to V, \quad v \mapsto t_a(v) := a + v$$

is called the *translation of V by a*. The image of a subspace U of V, i.e., the subset

$$t_a(U) = \{a + u \in V \mid u \in U\} =: a + U$$

of V is said to be a *coset* (or *translated subspace*) of U in V.

Lemma 1.1.37. *The translations of a vector space V have the following properties:*

(i) $t_0 = 1_V$, *where* 0 *is the zero vector in* V;

(ii) $t_a \circ t_b = t_b \circ t_a = t_{a+b}$ *for all* $a, b \in V$;

(iii) t_a *is invertible, and its inverse is* t_{-a}.

Thus the translations of V form a commutative subgroup in $\mathrm{Bij}(V)$. *This group is isomorphic to the additive group of V.*

The proof is immediate.

Lemma and Definition 1.1.38. *Let V be a real vector space and U a subspace of V.*

(i) *The set of all cosets of U in V is a real vector space under the operations*

$$(a + U) + (b + U) := (a + b) + U, \quad \lambda(a + U) = \lambda a + U \ (\lambda \in \mathbb{R}).$$

This vector space is called the quotient vector space *of V modulo U and is denoted by V/U, (cf. A.3.3).*

(ii) *The* natural projection

$$\pi \colon V \to V/U, \ v \mapsto \pi(v) := v + U$$

is a surjective linear mapping, and $\mathrm{Ker}(\pi) = U$. *If V is finite-dimensional, then*

$$\dim V/U = \dim V - \dim U.$$

(iii) *(The first isomorphism theorem). If $\varphi \colon V \to W$ is a linear mapping, then the mapping*

$$\bar{\varphi} \colon V/\mathrm{Ker}(\varphi) \to W, \ v + \mathrm{Ker}(\varphi) \mapsto \varphi(v)$$

is an injective linear mapping and yields a natural isomorphism $V/\mathrm{Ker}(\varphi) \cong \mathrm{Im}(\varphi)$.

These are standard results in linear algebra.

Definition 1.1.39. Let V and W be real vector spaces. We say that a mapping $f \colon V \to W$ is *affine* if there exists a linear mapping $\varphi \in L(V, W)$ and a vector $b \in W$ such that

$$f = t_b \circ \varphi, \ \text{i.e.,} \ f(v) = \varphi(v) + b \ \text{for every} \ v \in V. \tag{1.1.13}$$

Then t_b and φ are called the *translation part* and the *linear part* of f, respectively. We also say that t_b is the translation and φ is the linear mapping *associated* to the affine mapping f.

Remark 1.1.40. (a) Let $f\colon V \to W$ be an affine mapping given by (1.1.13). Then

$$f(0) = \varphi(0) + b = b, \quad \varphi = t_{-b} \circ f = t_{-f(0)} \circ f,$$

therefore the decomposition $f = t_b \circ \varphi$ is uniquely determined by f.

(b) Since every translation is bijective, it follows that an affine mapping is injective (or surjective, or bijective) if, and only if, the same is true for its linear part. It can easily be checked that the bijective affine transformations of a vector space V form a group under composition of mappings. This group is called the *general affine group* of V and it is denoted by $\mathrm{GA}(V)$.

(c) Every affine mapping from the real line into a real vector space V is of the form

$$t \in \mathbb{R} \mapsto ta + b \in V,$$

where a, b are vectors in V. Indeed, if $\varphi\colon \mathbb{R} \to V$ is a *linear* mapping, then for every $t \in \mathbb{R}$ we have

$$\varphi(t) = \varphi(t \cdot 1) = t\varphi(1) = ta, \quad a := \varphi(1).$$

1.2 Tensors

Entering into tensor calculus has certainly its conceptual difficulties, apart of the fear of indices, which has to be overcome. But formally this calculus is of extreme simplicity, much simpler, for example, than elementary vector calculus. Two operations: multiplication and contraction, i.e. juxtaposition of components of tensors with distinct indices and identification of two indices, one up one down, with implicit summation. It has often been attempted to introduce an invariant notation for tensor calculus... But then one needs such a mass of symbols and such an apparatus of rules of calculation (if one does not want to go back to components after all) that the net effect is very much negative. One has to protest strongly against these orgies of formalism, with which one now starts to bother even engineers.

Hermann Weyl

The quotation is from Weyl's classic work *Raum, Zeit, Materie* (Springer, Berlin, 1918), translated by Wulf Rossmann. The careful reader will immediately notice that Weyl's attitude is in sharp contrast with Dieudonné's viewpoint quoted in the previous section. We share Dieudonné's opinion, and we give definite priority to the coordinate- and index-free notation. In fact, neither did Weyl persist in his opinion. Rossmann comments the citation above with the following words: 'Elsewhere one can find similar sentiments expressed in similar words with reversed casting of hero and villain.'

1.2.1 Tensors as Multilinear Mappings

We begin with a general remark. Let S be a nonempty set and W an R-module. Then the set $\text{Map}(S, W)$ can be made into an R-module: for $\varphi, \psi \in \text{Map}(S, W)$ and $\lambda \in \text{R}$ we define the mappings $\varphi + \psi$ and $\lambda\varphi$ in the usual manner, by the rules

$$(\varphi + \psi)(s) := \varphi(s) + \psi(s), \quad (\lambda\varphi)(s) := \lambda\varphi(s); \quad s \in S.$$

If V is also an R-module, then the module $\text{Map}(V, W)$ contains the module $\text{Hom}(V, W)$ introduced in Remark 1.1.16(c) as a submodule.

Now let $V_1, \ldots, V_m$ $(m \in \mathbb{N}^*)$ be R-modules. A mapping

$$A : V_1 \times \cdots \times V_m \to W$$

is called m-*multilinear* (or m-*linear*) if it is (R-)linear in each of its arguments, i.e., if for any fixed index $i \in J_m$ and any fixed elements $a_j \in V_j$, $j \in J_m \setminus \{i\}$, the mapping

$$v \in V_i \mapsto A(a_1, \ldots, a_{i-1}, v, a_{i+1}, \ldots, a_m) \in W$$

is R-linear. A *bilinear* (resp. *trilinear*) mapping means 2-linear (resp. 3-linear). When m is clear from the context we often speak simply about a multilinear mapping. A multilinear mapping with the ring R of scalars as range is called a *multilinear form* (or an m-*linear form*). The set $L(V_1, \ldots, V_m; W)$ of all m-linear mappings from $V_1 \times \cdots \times V_m$ into W is an R-module, namely a submodule of the module $\text{Map}(V_1 \times \cdots \times V_m, W)$.

In the most important special case each of the modules V_i is either V or V^*. When $V_1 = \cdots = V_m =: V$, we write

$$L^m(V, W) := L(\underbrace{V, \ldots, V}_{m \text{ copies}}; W).$$

If $k, l \in \mathbb{N}$, $(k, l) \neq (0, 0)$, then we say that

$$T_l^k(V, W) := L(\underbrace{V^*, \ldots, V^*}_{k \text{ copies}}, \underbrace{V, \ldots, V}_{l \text{ copies}}; W) \tag{1.2.1}$$

is the module of W-*valued tensors on V, contravariant of order k and covariant of order l*, or simply, of *type* (k, l). We make the convention

$$T_0^0(V, W) := W.$$

If R $= W$, we write

$$T_l^k(V) := T_l^k(V, \mathsf{R})$$

and drop the adjective 'R-valued'.

A covariant tensor $A \in T_l^0(V, W)$ is said to be

> *symmetric* if $\sigma A = A$ for all $\sigma \in S_l$,
>
> *alternating* if $\sigma A = \varepsilon(\sigma) A$ for all $\sigma \in S_l$

(for the action of S_l on $T_l^0(V, W)$ see A.3.7). Both the symmetric and the alternating tensors form a submodule in $T_l^0(V, W)$. For the latter (which will be applied more frequently) we use the notation $A_l(V, W)$. In particular, we write $A_l(V) := A_l(V, \mathsf{R})$. A tensor in $A_l(V)$ is called an *alternating l-tensor*, an *alternating l-form*, or simply an *l-form*. Analogously, we denote by $S_l(V, W)$ the submodule of symmetric tensors in $T_l^0(V, W)$, and $S_l(V)$ stands for $S_l(V, \mathsf{R})$.

Given two tensors $A \in T_{l_1}^{k_1}(V)$, $B \in T_{l_2}^{k_2}(V)$, we define their *tensor product* (briefly *product*) $A \otimes B \in T_{l_1+l_2}^{k_1+k_2}(V)$ by

$$A \otimes B \left(\alpha^1, \dots, \alpha^{k_1+k_2}, v_1, \dots, v_{l_1+l_2} \right)$$
$$:= A \left(\alpha^1, \dots, \alpha^{k_1}, v_1, \dots, v_{l_1} \right) B \left(\alpha^{k_1+1}, \dots, \alpha^{k_1+k_2}, v_{l_1+1}, \dots, v_{l_1+l_2} \right),$$
$$(1.2.2)$$

where $\alpha^i \in V^*$, $v_j \in V$; $(i, j) \in J_{k_1+k_2} \times J_{l_1+l_2}$.

It is immediately clear that the so defined multiplication

$$(A, B) \in T_{l_1}^{k_1}(V) \times T_{l_2}^{k_2}(V) \mapsto A \otimes B \in T_{l_1+l_2}^{k_1+k_2}(V)$$

is R-*bilinear* and *associative*: if $C \in T_{l_3}^{k_3}(V)$ is a further tensor, then

$$(A \otimes B) \otimes C = A \otimes (B \otimes C). \qquad (1.2.3)$$

1.2.2   Substitution Operators and Pull-back

Let V be an R-module and let V^* be the dual of V. Given a vector $v \in V$, for every $k \in \mathbb{N}$ and for every positive integer l we define the *substitution operator*

$$i_v \colon T_l^k(V) \to T_{l-1}^k(V), \ A \mapsto i_v A$$

as follows:

$$(i_v A)(\alpha^1, \dots, \alpha^k, u_1, \dots, u_{l-1}) := A(\alpha^1, \dots, \alpha^k, v, u_1, \dots, u_{l-1}). \quad (1.2.4)$$

We agree that $i_v A := 0$ if $A \in T_0^k(V)$.

Similarly, let $\beta \in V^*$, and suppose that $k, l \in \mathbb{N}$, $k \geq 1$. Then the substitution operator

$$i^\beta \colon T_l^k(V) \to T_l^{k-1}(V), \quad A \mapsto i^\beta A$$

is given by

$$(i^\beta A)(\alpha^1, \ldots, \alpha^{k-1}, u_1, \ldots, u_l) := A(\beta, \alpha^1, \ldots, \alpha^{k-1}, u_1, \ldots, u_l). \quad (1.2.5)$$

By convention, $i^\beta A := 0$ if $A \in T_l^0(V)$. Obviously, i_v and i^β are R-linear mappings, and so are

$$v \in V \mapsto i_v \in \operatorname{Hom}_{\mathsf{R}}\left(T_l^k(V), T_{l-1}^k(V)\right)$$

and

$$\beta \in V^* \mapsto i^\beta \in \operatorname{Hom}_{\mathsf{R}}\left(T_l^k(V), T_l^{k-1}(V)\right).$$

Let a homomorphism $\varphi \in \operatorname{Hom}_{\mathsf{R}}(U, V)$ be given. The *pull-back* of a covariant tensor $A \in T_l^0(V)$ ($l \in \mathbb{N}^*$) by φ is the tensor $\varphi^* A \in T_l^0(U)$ given by

$$\varphi^* A(u_1, \ldots, u_l) := A(\varphi(u_1), \ldots, \varphi(u_l)); \quad u_1 \ldots, u_l \in U. \quad (1.2.6)$$

The pull-back operation has the following properties:

$$\varphi^* \in \operatorname{Hom}_{\mathsf{R}}\left(T_l^0(V), T_l^0(U)\right),$$

$$(\psi \circ \varphi)^* = \varphi^* \circ \psi^* \text{ if } \psi \in \operatorname{Hom}_{\mathsf{R}}(V, W),$$

$$\varphi^*(A_1 \otimes A_2) = \varphi^*(A_1) \otimes \varphi^*(A_2) \text{ if } A_1 \in T_{l_1}^0(V), \ A_2 \in T_{l_2}^0(V).$$

1.2.3 Canonical Isomorphisms

Lemma 1.2.1. *Let a type $(0,2)$ tensor $B \in T_2^0(V)$ be given. Then the mapping*

$$j_B \colon V \to V^*, \quad v \mapsto j_B(v) := j_v B, \quad j_v B(u) := B(u, v) \quad (1.2.7)$$

is R-linear, and the mapping

$$j \colon T_2^0(V) \to \operatorname{Hom}_{\mathsf{R}}(V, V^*), \quad B \mapsto j_B \quad (1.2.8)$$

is a canonical isomorphism of R-modules.

Proof. The R-linearity of the mappings j_B and j is clear. We show that the R-linear mapping

$$\sim \colon \operatorname{Hom}_{\mathsf{R}}(V, V^*) \to T_2^0(V), \quad f \mapsto \tilde{f}, \quad \tilde{f}(u, v) := f(v)(u)$$

is the inverse of j.

Let $B \in T_2^0(V)$. Then for all $u, v \in V$,

$$\widetilde{j_B}(u, v) := j_B(v)(u) = j_v B(u) = B(u, v),$$

therefore $\widetilde{j_B} = B$, and hence $\sim \circ j = 1_{T_2^0(V)}$.

Now let $f \in \mathrm{Hom}_\mathsf{R}(V, V^*)$. Then

$$j_{\tilde{f}}(v)(u) := j_v \tilde{f}(u) = \tilde{f}(u, v) = f(v)(u)$$

for each $u, v \in V$, which implies that $j_{\tilde{f}} = f$, whence $j \circ \sim = 1_{\mathrm{Hom}_\mathsf{R}(V, V^*)}$. $\square$

For the rest of this section we assume
that V is a finitely generated free R-module.

Under the finiteness condition the module V is reflexive, so we have

$$T_0^1(V) := L(V^*, \mathsf{R}) = V^{**} \cong V.$$

Thus the elements of V may be regarded as contravariant tensors of first order on V; the identification of V and $T_0^1(V)$ is given by (1.1.7):

$$v \in V \mapsto \tilde{v} \in T_0^1(V), \ \tilde{v}(\alpha) := \alpha(v) \text{ for all } \alpha \in V^*.$$

As a rule, we shall not make a notational distinction between the elements of $T_0^1(V)$ and V.

There exists a further important canonical isomorphism

$$T_l^k(V, V) \cong T_l^{k+1}(V)$$

defined as follows:

$$\begin{cases} A \in T_l^k(V, V) \mapsto \widetilde{A} \in T_l^{k+1}(V), \\ \widetilde{A}\left(\beta, \alpha^1, \ldots, \alpha^k, v_1, \ldots, v_l\right) := \beta\left(A\left(\alpha^1, \ldots, \alpha^k, v_1, \ldots, v_l\right)\right). \end{cases} \quad (1.2.9)$$

In particular, $\mathrm{End}_\mathsf{R}(V) = T_1^0(V, V) \cong T_1^1(V)$.

1.2.4 Tensor Components

Let $v_i \in V$, $\alpha^j \in V^*$; $(i, j) \in J_k \times J_l$. Then

$$\tilde{v}_1 \otimes \cdots \otimes \tilde{v}_k \otimes \alpha^1 \otimes \cdots \otimes \alpha^l,$$

or simply, according to our above remark,

$$v_1 \otimes \cdots \otimes v_k \otimes \alpha^1 \otimes \cdots \otimes \alpha^l$$

is a type (k, l) tensor on V which acts by the rule

$$v_1 \otimes \cdots \otimes v_k \otimes \alpha^1 \otimes \cdots \otimes \alpha^l \left(\beta^1, \ldots, \beta^k, u_1, \ldots, u_l\right)$$
$$= \beta^1(v_1) \ldots \beta^k(v_k) \alpha^1(u_1) \ldots \alpha^l(u_l).$$

If, in particular, $\mathcal{B} = (b_i)_{i=1}^n$ is a basis of V, and $\mathcal{B}^* = \left(b^i\right)_{i=1}^n$ is the dual basis of V^*, then the family

$$b_{i_1} \otimes \cdots \otimes b_{i_k} \otimes b^{j_1} \otimes \cdots \otimes b^{j_l}; \quad i_1, \ldots, i_k, j_1 \ldots, j_l \in J_n$$

is a basis for $T_l^k(V)$. In terms of this basis, any tensor $A \in T_l^k(V)$ may be expressed as

$$A = \sum_{(i)(j)} A_{j_1 \ldots j_l}^{i_1 \ldots i_k} b_{i_1} \otimes \cdots \otimes b_{i_k} \otimes b^{j_1} \otimes \cdots \otimes b^{j_l}, \tag{1.2.10}$$

where

$$A_{j_1 \ldots j_l}^{i_1 \ldots i_k} = A\left(b^{i_1}, \ldots, b^{i_k}, b_{j_1}, \ldots, b_{j_l}\right). \tag{1.2.11}$$

We say that the scalars $A_{j_1 \ldots j_l}^{i_1 \ldots i_k}$ are the *components of A with respect to the basis* $(b_i)_{i=1}^n$.

Let $(\tau_j^i) \in \mathrm{GL}_n(\mathbb{R})$ be the transition matrix from $\mathcal{B}$ to a basis $\bar{\mathcal{B}} = (\bar{b}_i)_{i=1}^n$. If $\bar{\mathcal{B}}^* = (\bar{b}^i)_{i=1}^n$ is the dual basis of $\bar{\mathcal{B}}$, then by (1.1.12) we have

$$\bar{b}^i = \sum_{j=1}^n \sigma_j^i b^j; \quad i \in J_n, \ (\sigma_j^i) = (\tau_j^i)^{-1}.$$

If the components of A with respect to $\bar{\mathcal{B}}$ are $\bar{A}_{j_1 \ldots j_l}^{i_1 \ldots i_k}$, then we immediately obtain the *tensor transformation rule*

$$\bar{A}_{j_1 \ldots j_l}^{i_1 \ldots i_k} = A_{s_1 \ldots s_l}^{r_1 \ldots r_k} \sigma_{r_1}^{i_1} \ldots \sigma_{r_k}^{i_k} \tau_{j_1}^{s_1} \ldots \tau_{j_l}^{s_l}.$$

Observe that here we used the summation convention with respect to the indices $r_1, \ldots, r_k$ and $s_1, \ldots, s_l$.

Lemma 1.2.2. *Let $B \in T_2^0(V)$. Choose a basis $\mathcal{B} = (b_i)_{i=1}^n$ for V and let $\mathcal{B}^* = \left(b^i\right)_{i=1}^n$ be its dual. If*

$$[B]_{\mathcal{B}} := (B(b_i, b_j)) =: (B_{ij}) \in M_n(\mathsf{R}),$$

then

$$[j_B]_{\mathcal{B}^*}^{\mathcal{B}} = [B]_{\mathcal{B}}. \tag{1.2.12}$$

Proof. Suppose that $[j_B]_{\mathcal{B}^*}^{\mathcal{B}} = (A_{ij}) \in M_n(\mathsf{R})$. Then

$$j_B(b_l) = \sum_{i=1}^n A_{il} b^i, \quad l \in J_n.$$

Thus, for each $k, l \in J_n$,

$$j_B(b_l)(b_k) = \sum_{i=1}^n A_{il} b^i(b_k) = \sum_{i=1}^n A_{il} \delta_k^i = A_{kl}.$$

On the other hand,

$$j_B(b_l)(b_k) \overset{(1.2.7)}{=} j_{b_l} B(b_k) = B(b_k, b_l) = B_{kl},$$

whence our claim follows. $\qquad \square$

The moral is that the elements of the canonically isomorphic R-modules $T_2^0(V)$ and $\text{Hom}_\mathsf{R}(V, V^*)$ become perfectly identical in a coordinate description.

Example 1.2.3. (a) The evaluation mapping

$$\delta \colon V^* \times V \to \mathsf{R}, \quad (f, v) \mapsto \delta(f, v) := f(v)$$

is clearly R-bilinear, so it is a type $(1, 1)$ tensor on V, called also the *Kronecker delta tensor* (*Kronecker tensor* or the *unit tensor*). (Obviously, this definition does not require the finiteness condition prescribed above.) If $\mathcal{B} = (b_i)_{i=1}^n$ is a basis of V, then the components of δ with respect to $\mathcal{B}$ are

$$\delta\left(b^i, b_j\right) := b^i(b_j) = \delta_j^i,$$

which justifies the term.

Observe that in this example the tensor components do not depend on the choice of the basis $\mathcal{B}$.

(b) Let $\varphi \in \text{End}_\mathsf{R}(V)$, and consider the $(1, 1)$ tensor $\tilde{\varphi} \in T_1^1(V)$ given by

$$\tilde{\varphi}(\alpha, v) := \alpha(\varphi(v)); \quad \alpha \in V^*, \ v \in V$$

(see (1.2.9)). Choose a basis $\mathcal{B} = (b_i)_{i=1}^n$ for V, and let

$$[\varphi]_\mathcal{B} = \left(A_j^i\right) \in M_n(\mathsf{R}).$$

Then the components of $\tilde{\varphi}$ with respect to $\mathcal{B}$ are

$$\tilde{\varphi}_j^i := \tilde{\varphi}\left(b^i, b_j\right) = b^i(\varphi(b_j)) = b^i \left(\sum_{k=1}^n A_j^k b_k\right)$$

$$= \sum_{k=1}^n A_j^k b^i(b_k) = \sum_{k=1}^n A_j^k \delta_k^i = A_j^i \quad (i, j \in J_n),$$

therefore at the level of tensor components, resp. matrix representatives, $\tilde{\varphi}$ and φ are indistinguishable (cf. the remark after Lemma 1.2.2).

(c) Let $T \in T_l^0(V)$ ($l \in \mathbb{N}^*$), $\varphi \in \text{Hom}_\mathsf{R}(U, V)$. Choose a basis $\mathcal{A} = (a_j)_{j=1}^m$ for U and a basis $\mathcal{B} = (b_i)_{i=1}^n$ for V. Let $[\varphi]_\mathcal{B}^\mathcal{A} = \left(A_j^i\right) \in M_{n \times m}(\mathsf{R})$. Then, applying the summation convention,

$$\varphi^* T(a_{j_1}, \dots, a_{j_l}) = T(\varphi(a_{j_1}), \dots, \varphi(a_{j_l})) = T\left(A_{j_1}^{i_1} b_{i_1}, \dots, A_{j_l}^{i_l} b_{i_l}\right)$$

$$= T(b_{i_1}, \dots, b_{i_l}) A_{j_1}^{i_1} \dots A_{j_l}^{i_l},$$

therefore if the components of T with respect to $\mathcal{B}$ are $T_{i_1 \dots i_l}$, then the components of the pull-back $\varphi^* T \in T_l^0(U)$ with respect to $\mathcal{A}$ are

$$(\varphi^* T)_{j_1 \dots j_l} = T_{i_1 \dots i_l} A_{j_1}^{i_1} \dots A_{j_l}^{i_l}, \ \left(A_j^i\right) = [\varphi]_\mathcal{B}^\mathcal{A}. \tag{1.2.13}$$

1.2.5 Contraction and Trace

Definition and Lemma 1.2.4. Let $A \in T_l^k(V)$, where $k, l \in \mathbb{N}^*$. Choose a basis $(b_i)_{i=1}^n$ of V, and let $(b^i)_{i=1}^n$ be its dual.

(a) Given a pair $(r, s) \in J_k \times J_l$ of indices, define a mapping

$$C_s^r A \colon (V^*)^{k-1} \times V^{l-1} \to \mathbb{R}$$

by

$$C_s^r A \left(\alpha^1, \ldots, \alpha^{k-1}, v_1, \ldots, v_{l-1} \right)$$

$$:= \sum_{i=1}^n A\left(\alpha^1, \ldots, b^i, \ldots, \alpha^{k-1}, v_1, \ldots, b_i, \ldots, v_{l-1} \right), \quad (1.2.14)$$

where b^i occupies the rth contravariant slot and b_i occupies the sth covariant slot. Then $C_s^r A$ is well-defined (i.e., independent of the choice of basis) and multilinear. Thus $C_s^r A$ is a tensor of type $(k-1, l-1)$ called the *contraction of A over r, s*. If, in particular, $A \in T_1^1(V)$, then the scalar

$$\mathrm{tr}(A) := C_1^1 A \in \mathbb{R} \quad (1.2.15)$$

is said to be the *trace* of A.

(b) The mapping

$$C_s^r \colon T_l^k(V) \to T_{l-1}^{k-1}(V), \quad A \mapsto C_s^r A$$

is $\mathbb{R}$-linear, called the *contraction of the rth contravariant index with the sth covariant index*, or for short, the *(r, s)-contraction*. For all $v_1, \ldots, v_k \in V$ and $\alpha^1, \ldots, \alpha^l \in V^*$, we have

$$C_s^r \left(v_1 \otimes \cdots \otimes v_k \otimes \alpha^1 \otimes \cdots \otimes \alpha^l \right)$$

$$= \alpha^s(v_r) v_1 \otimes \cdots \otimes \breve{v}_r \otimes \cdots \otimes v_k \otimes \alpha^1 \otimes \cdots \otimes \breve{\alpha}^s \otimes \cdots \otimes \alpha^l, \quad (1.2.16)$$

where breve over a vector or a linear form means that it is omitted. The (r, s) contraction C_s^r is the only linear mapping from $T_l^k(V)$ into $T_{l-1}^{k-1}(V)$ with this property. In particular,

$$\mathrm{tr}(v \otimes \alpha) := C_1^1(v \otimes \alpha) = \alpha(v); \quad v \in V, \ \alpha \in V^*. \quad (1.2.17)$$

(c) If the components of A with respect to the basis $(b_i)_{i=1}^n$ are $A_{j_1 \ldots j_l}^{i_1 \ldots i_k}$, then the components of $C_s^r A$ are

$$(C_s^r A)_{j_1 \ldots \breve{j}_s \ldots j_l}^{i_1 \ldots \breve{i}_r \ldots i_k} = \sum_{m=1}^n A_{j_1 \ldots j_{s-1} m \, j_{s+1} \ldots j_l}^{i_1 \ldots i_{r-1} m \, i_{r+1} \ldots i_k}. \quad (1.2.18)$$

Here, as above, the breve means omission. $\triangle$

Example 1.2.5. (a) Consider the Kronecker delta tensor $\delta \in T_1^1(V)$. Then

$$\operatorname{tr}(\delta) = C_1^1 \delta := \sum_{i=1}^{n} \delta\left(b^i, b_i\right) = \sum_{i=1}^{n} b^i(b_i) = \sum_{i=1}^{n} \delta_i^i = n.$$

(b) By the *trace of an endomorphism* $\varphi \in \operatorname{End}_\mathsf{R}(V)$ we mean the trace of the type $(1,1)$ tensor $\widetilde{\varphi}$ which corresponds to φ under the isomorphism given by (1.2.9):

$$\operatorname{tr}(\varphi) := \operatorname{tr}(\widetilde{\varphi}) := C_1^1 \widetilde{\varphi}.$$

If $\mathcal{B} = (b_i)_{i=1}^n$ is a basis of V, and $[\varphi]_\mathcal{B} = A = \left(A_j^i\right) \in M_n(\mathsf{R})$, then

$$\operatorname{tr}(\varphi) := C_1^1 \widetilde{\varphi} := \sum_{i=1}^{n} \widetilde{\varphi}\left(b^i, b_i\right) \overset{(1.2.9)}{=} \sum_{i=1}^{n} b^i \varphi(b_i) = \sum_{i=1}^{n} b^i \left(\sum_{j=1}^{n} A_i^j b_j\right)$$

$$= \sum_{i=1}^{n} \sum_{j=1}^{n} A_i^j b^i(b_j) = \sum_{i=1}^{n} \sum_{j=1}^{n} A_i^j \delta_j^i = \sum_{i=1}^{n} A_i^i =: \operatorname{tr}(A).$$

So we have

$$\operatorname{tr}(\varphi) = \operatorname{tr}[\varphi], \tag{1.2.19}$$

where $[\varphi]$ is an arbitrary matrix representative of φ.

(c) More generally, the trace of a V-valued tensor $T \in T_l^0(V, V)$ $(l \geq 1)$ is the contraction with respect to the pair $(1,1)$ of the corresponding type $(1,l)$ tensor $\widetilde{T} \in T_l^1(V)$:

$$\operatorname{tr} T := C_1^1 \widetilde{T} \in T_{l-1}^0(V).$$

The *components* $T_{j_1 \dots j_l}^i$ of T with respect to $\mathcal{B}$ are defined by

$$T(b_{j_1}, \dots, b_{j_l}) = \sum_{i=1}^{n} T_{j_1 \dots j_l}^i b_i; \quad j_1, \dots, j_l \in J_n.$$

Then, as above,

$$\operatorname{tr} T(b_{j_2}, \dots, b_{j_l}) := C_1^1 \widetilde{T}(b_{j_2}, \dots, b_{j_l}) := \sum_{i=1}^{n} \widetilde{T}\left(b^i, b_i, b_{j_2}, \dots, b_{j_l}\right)$$

$$\overset{(1.2.9)}{=} \sum_{i=1}^{n} b^i(T(b_i, b_{j_2}, \dots, b_{j_l})) = \sum_{i=1}^{n} b^i \left(\sum_{k=1}^{n} T_{ij_2 \dots j_l}^k b_k\right)$$

$$= \sum_{i=1}^{n} \sum_{k=1}^{n} T_{ij_2 \dots j_l}^k \delta_k^i = \sum_{i=1}^{n} T_{ij_2 \dots j_l}^i.$$

Thus the components of $\operatorname{tr} T$ are

$$(\operatorname{tr} T)_{j_2 \ldots j_l} = \sum_{i=1}^{n} T^i_{i j_2 \ldots j_l}; \quad j_2, \ldots, j_l \in J_n. \tag{1.2.20}$$

Lemma 1.2.6. *Let* $T \in T^0_l(V, V)$, $l \geq 2$. *Then*

$$\operatorname{tr} T(v_1, \ldots, v_{l-1}) = \operatorname{tr}(v \mapsto T(v, v_1, \ldots, v_{l-1})) \tag{1.2.21}$$

for all $v_1, \ldots, v_{l-1} \in V$, *and we have the 'recursive formula'*

$$i_v \operatorname{tr} T = \operatorname{tr}(j_v T), \quad v \in V, \tag{1.2.22}$$

where $j_v T \in T^0_{l-1}(V, V)$ *is given by*

$$j_v T(v_1, \ldots, v_{l-1}) := T(v_1, v, \ldots, v_{l-1}). \tag{1.2.23}$$

Proof. Choose a basis $(b_i)_{i=1}^{n}$ of V. Define a mapping φ by

$$\varphi(v) := T(v, v_1, \ldots, v_{l-1}), \quad v \in V,$$

where $v_1, \ldots, v_{l-1}$ are arbitrary, but fixed elements of V. Then $\varphi \in \operatorname{End}_{\mathsf{R}}(V)$, and

$$\operatorname{tr}(v \mapsto T(v, v_1, \ldots, v_{l-1})) = \operatorname{tr}\varphi := C^1_1 \tilde{\varphi} := \sum_{i=1}^{n} \tilde{\varphi}\left(b^i, b_i\right)$$

$$= \sum_{i=1}^{n} b^i(\varphi(b_i)) = \sum_{i=1}^{n} b^i T(b_i, v_1, \ldots, v_{l-1}) = \sum_{i=1}^{n} \tilde{T}\left(b^i, b_i, v_1, \ldots, v_{l-1}\right)$$

$$=: \left(C^1_1 \tilde{T}\right)(v_1, \ldots, v_{l-1}) =: \operatorname{tr} T(v_1, \ldots, v_{l-1}).$$

This proves (1.2.21). Similarly,

$$\operatorname{tr}(j_v T)(v_1, \ldots, v_{l-1}) := C^1_1 \widetilde{j_v T}(v_1, \ldots, v_{l-1})$$

$$= \sum_{i=1}^{n} \widetilde{j_v T}\left(b^i, b_i, v_1, \ldots, v_{l-1}\right) = \sum_{i=1}^{n} b^i(j_v T(b_i, v_1, \ldots, v_{l-1}))$$

$$= \sum_{i=1}^{n} b^i(T(b_i, v, v_1, \ldots, v_{l-1})) = \sum_{i=1}^{n} \tilde{T}\left(b^i, b_i, v, v_1, \ldots, v_{l-1}\right)$$

$$=: \left(C^1_1 \tilde{T}\right)(v, v_1, \ldots, v_{l-1}) =: \operatorname{tr} T(v, v_1, \ldots, v_{l-1})$$

$$= i_v \operatorname{tr} T(v_1, \ldots, v_{l-1}),$$

which proves (1.2.22). $\qquad\square$

It is important to note that formula (1.2.22) enables us to calculate the trace of a tensor $T \in T^0_l(V; V) \cong T^1_l(V)$, where $l \geq 1$, step by step.

1.3   Algebras and Derivations

1.3.1   *Basic Definitions*

Let R be a commutative ring. An R-module A is said to be an *algebra over* R, briefly an R-*algebra*, if it is endowed with an R-*bilinear mapping* A × A → A called *multiplication* and denoted by $(a, b) \mapsto a \cdot b$, or simply by juxtaposition. We also say that $a \cdot b = ab$ is the *product* of the elements $a, b \in$ A. As an immediate consequence of the bilinearity of multiplication we have

$$0 \cdot a = a \cdot 0 = 0 \text{ for all } a \in \mathsf{A}.$$

An R-algebra A is *associative*, resp. *commutative*, if multiplication in A is associative, resp. commutative. We say that A is a *unital algebra* if there is an element e of A such that

$$ea = ae = a \text{ for all } a \in \mathsf{A}.$$

Then e is necessarily unique; it is called the *identity element* of A. If A is an associative unital algebra, then it is a ring at the same time with respect to the addition (given in A as in an R-module) and multiplication.

Let H be a submodule of an R-algebra A. If H is closed under multiplication, i.e., $h_1, h_2 \in$ H implies $h_1 h_2 \in$ H, then the restriction to H × H of the multiplication of A makes H into an R-algebra (with the R-module structure on H). The R-algebra so obtained is called a *subalgebra* of A.

We say that a subset S *generates* the algebra A or that S is a *set of generators* for A if every element of A can be expressed as an (R-)linear combination of products of elements of S. We speak of a *generator family* of A in the same sense. The algebra A is *finitely generated* if it has a finite set (or family) of generators.

Suppose B is also an R-algebra. An R-linear mapping $\varphi \colon$ A → B is called an R-*algebra homomorphism* (or simply a *homomorphism* (of algebras)) if φ preserves products, i.e.,

$$\varphi(ab) = \varphi(a)\varphi(b) \text{ for all } a, b \in \mathsf{A}.$$

Endomorphisms, isomorphisms and automorphisms are special cases of homomorphisms and are defined in the usual way.

Example 1.3.1. The module $\mathrm{End}_{\mathsf{R}}(V)$ of R-linear transformations of an R-module V is an associative, unital R-algebra if the multiplication is defined as composition of mappings (see Remark 1.1.16). Obviously, the R-module

$M_n(\mathsf{R})$ of $n \times n$ matrices over R together with the matrix multiplication is also an associative, unital R-algebra. If V is a finitely generated free module and $\mathcal{B} = (b_i)_{i=1}^n$ is a basis of V, then it follows from (1.1.5) that the mapping

$$\mathrm{End}_\mathsf{R}(V) \to M_n(\mathsf{R}), \ \varphi \mapsto [\varphi]_\mathcal{B}$$

is an algebra isomorphism.

1.3.2 Derivations

Let R be a commutative ring and A an R-algebra. A *derivation* of A is an R-linear endomorphism D of A satisfying the *product rule* (or *Leibniz rule*)

$$D(ab) = (Da)b + a(Db) \text{ for all } (a, b) \in \mathsf{A} \times \mathsf{A}. \tag{1.3.1}$$

If D_1, D_2, D are derivations of A, and we define the sum $D_1 + D_2$ and the scalar multiple λD ($\lambda \in \mathsf{R}$) in the usual way, then $D_1 + D_2$ and λD are derivations. This implies that the set $\mathrm{Der}(\mathsf{A})$ of derivations of A is a submodule of the module $\mathrm{End}_\mathsf{R}(\mathsf{A})$ of R-linear transformations of the R-module A. Hence, in its own right, $\mathrm{Der}(\mathsf{A})$ *is an R-module.*

Lemma 1.3.2. *If $D_1, D_2 \in \mathrm{Der}(\mathsf{A})$, then their* commutator

$$[D_1, D_2] := D_1 \circ D_2 - D_2 \circ D_1$$

is again a derivation of A.

Indeed, the R-linearity of $[D_1, D_2]$ is obvious, and an easy calculation shows that it also satisfies the product rule.

Lemma 1.3.3. *Let A be an R-algebra and $D \in \mathrm{Der}(\mathsf{A})$.*

(i) *If A has a nonzero identity element e, then $D(e) = 0$.*
(ii) *If A is associative, then for all $D \in \mathrm{Der}(\mathsf{A})$ and $a_1, \ldots, a_n \in \mathsf{A}$ we have*

$$D(a_1 a_2 \ldots a_n) = \sum_{i=1}^n a_1 \ldots a_{i-1}(Da_i)a_{i+1} \ldots a_n. \tag{1.3.2}$$

Proof. We have

$$D(e) = D(e \cdot e) \overset{(1.3.1)}{=} (De)e + e(De) = D(e) + D(e),$$

whence $D(e) = 0$. Using (1.3.1), relation (1.3.2) can be obtained by induction on n. $\qquad\square$

Corollary 1.3.4. *Let* A *be an associative algebra, and let* G *be a set of generators for* A. *If* $D_1, D_2 \in \mathrm{Der}(\mathsf{A})$ *and* $D_1 \upharpoonright G = D_2 \upharpoonright G$, *then* $D_1 = D_2$.

$\triangle$

Example 1.3.5. Consider an associative algebra A and let $a \in \mathsf{A}$. Then the mapping

$$\mathrm{ad}(a)\colon \mathsf{A} \to \mathsf{A}, \ u \mapsto \mathrm{ad}(a)(u) := [a, u] := au - ua$$

is a derivation of A, called an *inner derivation*.

1.3.3 *Lie Algebras*

Let $\mathfrak{g}$ be an R-algebra in which the product of two elements a, b is denoted by $[a, b]$. We say that the mapping

$$[\,,\,]\colon \mathfrak{g} \times \mathfrak{g} \to \mathfrak{g}, \ (a, b) \mapsto [a, b]$$

is a *Lie bracket* on $\mathfrak{g}$, and the algebra $(\mathfrak{g}, [\,,\,])$ (or simply $\mathfrak{g}$) is a *Lie algebra over* R if the following conditions are satisfied:

(L1) $[a, a] = 0$ for all $a \in \mathfrak{g}$;
(L2) $[a, [b, c]] + [b, [c, a]] + [c, [a, b]] = 0$ for all $a, b, c \in \mathfrak{g}$.

The identity (L2) is called the *Jacobi identity*. By a *real Lie algebra* we mean a Lie algebra over $\mathbb{R}$.

Condition (L1) implies that

$$[a, b] = -[b, a] \text{ for all } a, b \in \mathfrak{g}. \tag{1.3.3}$$

Indeed, $0 \overset{(L1)}{=} [a + b, a + b] = [a, a] + [b, a] + [a, b] + [b, b] \overset{(L1)}{=} [b, a] + [a, b]$, whence our claim.

Example 1.3.6. (a) If A is an associative R-algebra, then the commutator

$$(a, b) \in \mathsf{A} \times \mathsf{A} \mapsto [a, b] := ab - ba$$

makes A into a Lie algebra. We denote this Lie algebra by A_L. Here we mention two important special cases:

$\mathfrak{gl}(V) := (\mathrm{End}_{\mathsf{R}} V)_L$ – the Lie algebra obtained from the associative algebra $\mathrm{End}_{\mathsf{R}}(V)$ (Example 1.3.1);
$\mathfrak{gl}_n(\mathsf{R}) := (M_n(\mathsf{R}))_L$ – the Lie algebra arising from the associative algebra $M_n(\mathsf{R})$.

(b) If A is an R-algebra, then the R-module $\mathrm{Der}(\mathsf{A})$ endowed with the commutator defined in Lemma 1.3.2 is a Lie algebra.

Lemma 1.3.7. *Let* $(\mathfrak{g}, [\ ,\ ])$ *be a Lie algebra over* R. *For every* $a \in \mathfrak{g}$, *the mapping*

$$\mathrm{ad}(a) \colon \mathfrak{g} \to \mathfrak{g}, \ b \mapsto \mathrm{ad}(a)(b) := [a, b] \qquad (1.3.4)$$

is a derivation of $\mathfrak{g}$, *and the mapping*

$$\mathrm{ad} \colon \mathfrak{g} \to \mathrm{Der}(\mathfrak{g}) \subset \mathfrak{gl}(\mathfrak{g}), \ a \mapsto \mathrm{ad}(a) \qquad (1.3.5)$$

is a homomorphism of Lie algebras. For any $a \in \mathfrak{g}$ *and any* $D \in \mathrm{Der}(\mathfrak{g})$ *we have*

$$[D, \mathrm{ad}(a)] = \mathrm{ad}(Da). \qquad (1.3.6)$$

Proof. Since the Lie bracket $[\ ,\ ]$ is an R-bilinear mapping, $\mathrm{ad}(a)$ is R-linear. For any $b, c \in \mathfrak{g}$ we have

$$\mathrm{ad}(a)([b, c]) := [a, [b, c]] \overset{(L2)}{=} -[b, [c, a]] - [c, [a, b]]$$
$$\overset{(1.3.3)}{=} [[a, b], c] + [b, [a, c]] = [\mathrm{ad}(a)(b), c] + [b, \mathrm{ad}(a)(c)],$$

therefore $\mathrm{ad}(a)$ satisfies the product rule (1.3.1).

We obtain by a similar calculation that

$$\mathrm{ad}([a, b]) = [\mathrm{ad}(a), \mathrm{ad}(b)]; \quad a, b \in \mathfrak{g},$$

which means that the mapping ad is a homomorphism of the Lie algebra $\mathfrak{g}$ into the Lie algebra $\mathrm{Der}(\mathfrak{g})$.

Finally, for all $v \in \mathfrak{g}$,

$$[D, \mathrm{ad}(a)](v) = D([a, v]) - [a, Dv] = [Da, v] = \mathrm{ad}(Da)(v),$$

which proves (1.3.6). $\qquad\qquad\qquad\qquad\qquad\qquad\qquad\qquad\qquad\Box$

We say that a derivation of $\mathfrak{g}$ of the form $\mathrm{ad}(a)$ $(a \in \mathfrak{g})$ is an *inner derivation* (cf. Example 1.3.5). The homomorphism $\mathrm{ad} \colon \mathfrak{g} \to \mathfrak{gl}(\mathfrak{g})$ is called the *adjoint representation* of $\mathfrak{g}$.

1.3.4 *Graded Algebras and Graded Derivations*

An R-algebra A is said to be *graded* by the natural numbers (or a *graded algebra of type* $\mathbb{N}$) if it is the direct sum of a sequence $(\mathsf{A}_i)_{i \in \mathbb{N}}$ of its submodules (see Example 1.1.18), satisfying the condition

$$\mathsf{A}_i \mathsf{A}_j \subset \mathsf{A}_{i+j} \text{ for all } i, j \in \mathbb{N},$$

which means that

$$a \in \mathsf{A}_i \text{ and } b \in \mathsf{A}_j \text{ imply } ab \in \mathsf{A}_{i+j}.$$

We say that the elements of A_i are *homogeneous of degree i*. Then the zero element of A is homogeneous of all degrees, but every nonzero homogeneous element belongs to only one submodule of the sequence. Thus we may write

$$\deg(a) := i \text{ if } a \in A_i \setminus \{0\}.$$

We agree that $A_k := \{0\}$ if $k \in \mathbb{Z} \setminus \mathbb{N}$. With this convention, $A = \bigoplus_{n \in \mathbb{Z}} A_n$ is a graded algebra of type $\mathbb{Z}$. A graded algebra is called *graded commutative* if for any two nonzero homogeneous elements $a, b \in A$,

$$ab = (-1)^{\deg(a)\deg(b)} ba.$$

The algebra A is *graded anticommutative* if

$$ab = -(-1)^{\deg(a)\deg(b)} ba,$$

for all homogeneous elements $a, b \in A \setminus \{0\}$.

Suppose that $A = \bigoplus_{n \in \mathbb{Z}} A_n$ is a graded algebra over R in which the product of two elements a, b is denoted by $[a, b]$. We say that the mapping

$$[\ ,\ ] : A \times A \to A, \quad (a, b) \mapsto [a, b]$$

is a *graded Lie bracket* on A, and the algebra $(A, [\ ,\ ])$ (or simply A) is a *graded Lie algebra* over R if the following conditions are satisfied:

$$[a, b] = -(-1)^{\deg(a)\deg(b)} [b, a], \tag{1.3.7}$$

$$(-1)^{\deg(a)\deg(c)} [a, [b, c]] + (-1)^{\deg(b)\deg(a)} [b, [c, a]]$$
$$+ (-1)^{\deg(c)\deg(b)} [c, [a, b]] = 0. \tag{1.3.8}$$

In these conditions a, b, c are arbitrary (nonzero) homogeneous elements of A. The identity (1.3.8) is called the *graded Jacobi identity*.

Consider a graded R-algebra $A = \bigoplus_{n \in \mathbb{Z}} A_n$ and let $r \in \mathbb{Z}$. An endomorphism D of A is said to be of *degree r* if

$$D(A_n) \subset A_{n+r} \text{ for all } n \in \mathbb{N}.$$

By a *graded derivation of degree r* of A we mean a mapping $D \in \mathrm{End}_R(A)$ of degree r such that

$$D(ab) = (Da)b + (-1)^{r\deg(a)} a(Db) \tag{1.3.9}$$

for any homogeneous elements a, b of A.

Notice that a graded derivation of *even* degree r is just a derivation of degree r.

Lemma 1.3.8. *Let A be a graded R-algebra. If D_1 and D_2 are graded derivations of A of degree r and s, respectively, then their graded commutator*

$$[D_1, D_2] := D_1 \circ D_2 - (-1)^{rs} D_2 \circ D_1 \tag{1.3.10}$$

is a graded derivation of A *of degree* $r + s$. *The set of all graded derivations of* A *forms a graded Lie algebra with respect to the usual addition and scalar multiplication of* R-*linear mappings, and with respect to the graded commutator as multiplication.*

The proof is just a (little lengthy but straightforward) calculation.

1.3.5 The Exterior Algebra of an R-module

For the rest of this section we assume that $k!$ *is a unit, i.e., invertible in our ground ring* R *for all* $k \in \mathbb{N}$.

Let V and W be R-modules, and consider the R-module

$$T_k^0(V, W) \overset{(1.2.1)}{=} L^k(V, W)$$

of W-valued covariant tensors of order k on V.

Lemma 1.3.9. *If* $\alpha \in T_k^0(V, W)$ *and*

$$\operatorname{Sym}(\alpha) := \frac{1}{k!} \sum_{\sigma \in S_k} \sigma\alpha, \quad \operatorname{Alt}(\alpha) := \frac{1}{k!} \sum_{\sigma \in S_k} \varepsilon(\sigma)\sigma\alpha, \qquad (1.3.11)$$

then $\operatorname{Sym}(\alpha) \in S_k(V, W)$, $\operatorname{Alt}(\alpha) \in A_k(V, W)$, *i.e.,* $\operatorname{Sym}(\alpha)$ *is a symmetric,* $\operatorname{Alt}(\alpha)$ *is an alternating* W-*valued type* $(0, k)$ *tensor on* V. *In particular,*

$$\operatorname{Sym}(\alpha) = \alpha \ \textit{if } \alpha \textit{ is symmetric}, \quad \operatorname{Alt}(\alpha) = \alpha \ \textit{if } \alpha \textit{ is alternating}.$$

The mappings Sym *and* Alt *are (*R-*linear) projections of* $T_k^0(V, W)$ *onto* $S_k(V, W)$ *and* $A_k(V, W)$, *respectively.*

Proof. We check only that $\operatorname{Alt}(\alpha)$ is indeed an alternating tensor, the remainder of the lemma can also be proved easily. Let $\varrho \in S_k$. Since the mapping

$$\alpha \in T_k^0(V, W) \mapsto \varrho\alpha \in T_k^0(V, W)$$

defined by (A.3.2) is R-linear, we obtain

$$\varrho(\operatorname{Alt}(\alpha)) = \varrho\left(\frac{1}{k!} \sum_{\sigma \in S_k} \varepsilon(\sigma)\sigma\alpha\right) = \frac{1}{k!} \sum_{\sigma \in S_k} \varepsilon(\sigma)\varrho(\sigma\alpha)$$

$$\overset{(A.3.3)}{=} \frac{1}{k!} \sum_{\sigma \in S_k} \varepsilon(\sigma)(\varrho \circ \sigma)\alpha = \frac{1}{k!} \sum_{\sigma \in S_k} \varepsilon(\varrho \circ \sigma)\varepsilon(\varrho)(\varrho \circ \sigma)\alpha$$

$$= \varepsilon(\varrho)\frac{1}{k!} \sum_{\varrho \circ \sigma \in S_k} \varepsilon(\varrho \circ \sigma)(\varrho \circ \sigma)\alpha = \varepsilon(\varrho)\operatorname{Alt}(\alpha),$$

as was to be shown. $\qquad\qquad\square$

The mappings Sym and Alt are called *symmetrizer* and *alternator*, respectively. We also say that $\mathrm{Sym}(\alpha)$ is the *symmetrization* of α and $\mathrm{Alt}(\alpha)$ is the *skew-symmetrization* of α.

By the *wedge* (or *exterior*) *product* of two alternating tensors $\alpha \in A_k(V)$, $\beta \in A_l(V)$ we mean the alternating $(k+l)$-tensor

$$\alpha \wedge \beta := \frac{(k+l)!}{k!l!} \, \mathrm{Alt}(\alpha \otimes \beta) = \frac{1}{k!l!} \sum_{\sigma \in S_{k+l}} \varepsilon(\sigma)\sigma(\alpha \otimes \beta). \qquad (1.3.12)$$

Thus we obtain an R-bilinear mapping

$$\wedge \colon A_k(V) \times A_l(V) \to A_{k+l}(V), \ (\alpha, \beta) \mapsto \alpha \wedge \beta.$$

Now we form the *(external) direct sum*

$$A(V) := \bigoplus_{k \in \mathbb{N}} A_k(V) \quad (A_0(V) := \mathsf{R})$$

of the R-modules $A_k(V)$, $k \in \mathbb{N}$ as follows:

(i) The elements of $A(V)$ are the sequences

$$(\alpha_i)_{i \in \mathbb{N}} =: (\alpha_0, \alpha_1, \ldots, \alpha_n, \ldots) =: \sum_{i \in \mathbb{N}} \alpha_i,$$

where only finitely many α_i are different from zero, and $\alpha_i \in A_i(V)$ $(i \in \mathbb{N})$.

(ii) The addition and scalar multiplication are defined componentwise, i.e., by

$$\sum_{i \in \mathbb{N}} \alpha_i + \sum_{i \in \mathbb{N}} \beta_i := \sum_{i \in \mathbb{N}} (\alpha_i + \beta_i), \quad \lambda \sum_{i \in \mathbb{N}} \alpha_i := \sum_{i \in \mathbb{N}} \lambda \alpha_i, \qquad (1.3.13)$$

where $\lambda \in \mathsf{R}$.

The operations (1.3.13) make $A(V)$ into an R-module. The wedge product defined by (1.3.12) leads to an R-bilinear multiplication $\wedge \colon A(V) \times A(V) \to A(V)$ by putting

$$\begin{cases} \left(\sum_{i \in \mathbb{N}} \alpha_i \right) \wedge \sum_{i \in \mathbb{N}} \beta_i := \sum_{k \in \mathbb{N}} \left(\sum_{i+j=k} \alpha_i \wedge \beta_j \right), \\ \lambda \wedge \alpha := \lambda\alpha \text{ if } \lambda \in A_0(V) = \mathsf{R}, \quad \alpha \in A_j(V). \end{cases} \qquad (1.3.14)$$

Then $(A(V), \wedge)$ is an R-algebra, called the *exterior algebra* or the *Grassmann algebra* of the R-module V.

Lemma 1.3.10. *The exterior algebra of an R-module is a unital, associative and graded commutative graded algebra.*

Proof. Let V be an R-module, and consider its exterior algebra $A(V) := \bigoplus_{k \in \mathbb{N}} A_k(V)$.

(a) The identity element of $A(V)$ is the sequence $(1, 0, \ldots, 0, \ldots)$, which may be identified with the identity element $1 \in \mathsf{R} = A_0(V)$ of the ground ring.

(b) To prove the associativity, note first that

$$\text{Alt}(\sigma f) = \varepsilon(\sigma) \, \text{Alt}(f) \tag{1.3.15}$$

for all $f \in A_k(V, W)$, $\sigma \in S_k$. Indeed,

$$\text{Alt}(\sigma f) := \frac{1}{k!} \sum_{\varrho \in S_k} \varepsilon(\varrho) \varrho(\sigma f) = \frac{1}{k!} \sum_{\varrho \in S_k} \varepsilon(\sigma) \varepsilon(\varrho \circ \sigma)(\varrho \circ \sigma) f$$

$$= \frac{1}{k!} \varepsilon(\sigma) \sum_{\varrho \circ \sigma \in S_k} \varepsilon(\varrho \circ \sigma)(\varrho \circ \sigma) f = \varepsilon(\sigma) \, \text{Alt}(f).$$

Now let $\alpha \in A_k(V)$, $\beta \in A_l(V)$, $\gamma \in A_m(V)$. Identify the group S_{k+l} with the subgroup of S_{k+l+m} formed by the permutations which fix the elements $k+l+1, \ldots, k+l+m$. Then

$$(\alpha \wedge \beta) \wedge \gamma := \frac{(k+l+m)!}{(k+l)!m!} \, \text{Alt}((\alpha \wedge \beta) \otimes \gamma)$$

$$= \frac{(k+l+m)!}{(k+l)!m!} \, \text{Alt}\left(\frac{1}{k!l!} \sum_{\sigma \in S_{k+l}} \varepsilon(\sigma) \sigma((\alpha \otimes \beta) \otimes \gamma) \right)$$

$$\overset{(1.2.3)}{=} \frac{(k+l+m)!}{k!l!m!(k+l)!} \sum_{\sigma \in S_{k+l}} \varepsilon(\sigma) \, \text{Alt}(\sigma(\alpha \otimes \beta \otimes \gamma))$$

$$\overset{(1.3.15)}{=} \frac{(k+l+m)!}{k!l!m!(k+l)!} \sum_{\sigma \in S_{k+l}} \text{Alt}(\alpha \otimes \beta \otimes \gamma)$$

$$= \frac{(k+l+m)!}{k!l!m!} \, \text{Alt}(\alpha \otimes \beta \otimes \gamma).$$

We obtain in the same way that

$$\alpha \wedge (\beta \wedge \gamma) = \frac{(k+l+m)!}{k!l!m!} \, \text{Alt}(\alpha \otimes \beta \otimes \gamma),$$

therefore $(\alpha \wedge \beta) \wedge \gamma = \alpha \wedge (\beta \wedge \gamma)$. From this associativity property and from the definition (1.3.14) it follows immediately that the wedge product is associative in the algebra $A(V)$.

(c) We show that the algebra $(A(V), \wedge)$ is graded commutative. To do this, it is enough to check that

$$\alpha \wedge \beta = (-1)^{kl} \beta \wedge \alpha \text{ for all } \alpha \in A_k(V), \ \beta \in A_l(V).$$

Consider the permutation $\sigma \in S_{k+l}$ given by

$$1 \mapsto k+1, \ \ldots, \ l \mapsto k+l, \ l+1 \mapsto 1, \ \ldots, \ l+k \mapsto k.$$

Then $\varepsilon(\sigma) = (-1)^{kl}$, and for any vectors $v_1, \ldots, v_{k+l} \in V$ we have

$$\sigma(\beta \otimes \alpha)(v_1, \ldots, v_{k+l}) := \beta \otimes \alpha \left(v_{\sigma(1)}, \ldots, v_{\sigma(l)}, v_{\sigma(l+1)}, \ldots, v_{\sigma(l+k)} \right)$$

$$= \beta \otimes \alpha(v_{k+1}, \ldots, v_{k+l}, v_1, \ldots, v_k) = \beta(v_{k+1}, \ldots, v_{k+l})\alpha(v_1, \ldots, v_k)$$

$$= \alpha \otimes \beta(v_1, \ldots, v_{k+l}),$$

therefore $\alpha \otimes \beta = \sigma(\beta \otimes \alpha)$. Thus we obtain

$$\mathrm{Alt}(\alpha \otimes \beta) = \mathrm{Alt}(\sigma(\beta \otimes \alpha)) \overset{(1.3.15)}{=} \varepsilon(\sigma) \mathrm{Alt}(\beta \otimes \alpha) = (-1)^{kl} \mathrm{Alt}(\beta \otimes \alpha),$$

which implies the desired relation $\alpha \wedge \beta = (-1)^{kl} \beta \wedge \alpha$. $\square$

Corollary 1.3.11. *Let V be an R-module, and let $\ell^1, \ldots, \ell^k \in V^*$. Then*

$$\ell^1 \wedge \cdots \wedge \ell^k = k! \, \mathrm{Alt}\left(\ell^1 \otimes \cdots \otimes \ell^k\right). \tag{1.3.16}$$

Proof. We have seen in part (b) of the preceding proof that

$$\alpha \wedge \beta \wedge \gamma = \frac{(k+l+m)!}{k!l!m!} \, \mathrm{Alt}(\alpha \otimes \beta \otimes \gamma)$$

for $\alpha \in A_k(V)$, $\beta \in A_l(V)$, $\gamma \in A_m(V)$. Then, in particular,

$$\ell^1 \wedge \ell^2 \wedge \ell^3 = 3! \, \mathrm{Alt}\left(\ell^1 \otimes \ell^2 \otimes \ell^3\right),$$

and (1.3.16) follows by induction on k. $\square$

Remark 1.3.12. All that we said above can be repeated (mutatis mutandis) by replacing alternating tensors by symmetric tensors, and the alternating operator Alt by the symmetrizing operator Sym. So we define the *symmetric product* of two symmetric tensors $\alpha \in S_k(V)$ and $\beta \in S_l(V)$ by the formula

$$\alpha \odot \beta := \frac{(k+l)!}{k!l!} \, \mathrm{Sym}(\alpha \otimes \beta) = \frac{1}{k!l!} \sum_{\sigma \in S_{k+l}} \sigma(\alpha \otimes \beta). \tag{1.3.17}$$

Evidently, $\alpha \odot \beta \in S_{k+l}(V)$. The symmetric product is associative and commutative:

$$(\alpha \odot \beta) \odot \gamma = \alpha \odot (\beta \odot \gamma) \ (\gamma \in S_m(V)) \text{ and } \alpha \odot \beta = \beta \odot \alpha.$$

Now consider the direct sum

$$S(V) := \bigoplus_{k \in \mathbb{N}} S_k(V) \quad (S_0(V) := \mathsf{R}).$$

The symmetric product can be extended in a unique way to multiplication in the R-module $S(V)$, denoted by the same symbol $\odot$. Thus $S(V)$ becomes an associative and commutative R-algebra, called the *(covariant) symmetric algebra* over V.

1.3.6 Determinants

We recall an important characterization of the determinant function

$$\det\colon M_n(\mathsf{R}) \to \mathsf{R}, \ A = \left(\alpha_j^i\right) \mapsto \det(A) := \sum_{\sigma \in S_n} \varepsilon(\sigma)\alpha_1^{\sigma(1)} \ldots \alpha_n^{\sigma(n)} \quad (1.3.18)$$

introduced in A.4.2(c). Write the elements of R^n as columns. Then $M_n(\mathsf{R})$ can be naturally identified with $(\mathsf{R}^n)^n$ by the R-module isomorphism

$$\begin{cases} A = (\alpha_j^i) \in M_n(\mathsf{R}) \mapsto (A_1, \ldots, A_n) \in (\mathsf{R}^n)^n, \\ A_j := {}^t(\alpha_j^1 \ldots \alpha_j^n), \quad j \in J_n. \end{cases}$$

Keeping this in mind, we have the following interpretation:

Proposition 1.3.13. *The function* $\det\colon M_n(\mathsf{R}) \to \mathsf{R}$ *given by* (A.4.2) *is the unique function satisfying the following two axioms:*

(i) $\det$ *is an alternating n-linear form on* R^n, *i.e.,* $\det \in A_n(\mathsf{R}^n)$, *where the n-tuples are the n columns of the matrices in* $M_n(\mathsf{R})$;

(ii) $\det(1_n) = 1$.

For a proof, see, e.g., [50], 8.3.6–8.3.8.

Lemma 1.3.14. *Let V be an R-module and let* $\ell^1, \ldots, \ell^k \in V^*$. *Then*

$$\ell^1 \wedge \cdots \wedge \ell^k(v_1, \ldots, v_k) = \det\left(\ell^i(v_j)\right) \quad (v_1, \ldots, v_k \in V), \quad (1.3.19)$$

$$\ell^{\sigma(1)} \wedge \cdots \wedge \ell^{\sigma(k)} = \varepsilon(\sigma)\ell^1 \wedge \cdots \wedge \ell^k \quad (\sigma \in S_k), \quad (1.3.20)$$

$$\ell^1 \wedge \cdots \wedge \ell^k = \sum_{\sigma \in S_k} \varepsilon(\sigma)\ell^{\sigma(1)} \otimes \cdots \otimes \ell^{\sigma(k)}. \quad (1.3.21)$$

Proof. Applying (1.3.16) we obtain

$$\ell^1 \wedge \cdots \wedge \ell^k(v_1, \ldots, v_k) = k!\, \mathrm{Alt}\left(\ell^1 \otimes \cdots \otimes \ell^k\right)(v_1, \ldots, v_k)$$

$$= k!\frac{1}{k!} \sum_{\sigma \in S_k} \varepsilon(\sigma)\ell^1\left(v_{\sigma(1)}\right) \ldots \ell^k\left(v_{\sigma(k)}\right) = \det\left(\ell^i(v_k)\right),$$

which proves (1.3.19). From this it follows that

$$\ell^{\sigma(1)} \wedge \cdots \wedge \ell^{\sigma(k)}(v_1, \ldots, v_k) = \det\left(\ell^{\sigma(i)}(v_k)\right)$$

$$\overset{\text{Proposition 1.3.13}}{=} \varepsilon(\sigma)\det\left(\ell^i(v_k)\right) = \varepsilon(\sigma)\ell^1 \wedge \cdots \wedge \ell^k(v_1, \ldots, v_k),$$

whence (1.3.20). Finally, evaluating the left-hand side of (1.3.21) on $(v_1, \ldots, v_k) \in V^k$ and taking into account (1.3.19) we find that (1.3.21) is also true. $\square$

Proposition 1.3.15. *Let V be a non-trivial finitely generated free R-module. If $(b_i)_{i=1}^n$ is a basis of V and $(b^i)_{i=1}^n$ is its dual basis, then the family*

$$\left(b^{i_1} \wedge \cdots \wedge b^{i_k}\right)_{1 \leq i_1 < \cdots < i_k \leq n} \tag{1.3.22}$$

is a basis of $A_k(V)$ for all $k \in \mathbb{N}^*$. Therefore,

$$\dim A_k(V) = \begin{cases} \binom{n}{k} & \text{if } k \in J_n, \\ 0 & \text{if } k > n. \end{cases} \tag{1.3.23}$$

In particular, $\dim A_n(V) = 1$, and $(b^1 \wedge \cdots \wedge b^n)$ is a basis of $A_n(V)$ such that $b^1 \wedge \cdots \wedge b^n(b_1, \ldots, b_n) = 1$.

Proof. **Step 1.** Let $k \in \mathbb{N}^*$, and consider an arbitrary tensor $\alpha \in A_k(V)$. In view of (1.2.10) and (1.2.11), this tensor can be written as

$$\alpha = \sum_{(i)} \alpha_{i_1 \ldots i_k} b^{i_1} \otimes \cdots \otimes b^{i_k},$$

where $\alpha_{i_1 \ldots i_k} = \alpha(b_{i_1}, \ldots, b_{i_k})$. Applying the alternator to both sides of this relation, and taking into account that $\mathrm{Alt}(\alpha) = \alpha$ by Lemma 1.3.9, we find that

$$\alpha = \sum_{(i)} \alpha_{i_1 \ldots i_k} \mathrm{Alt}\left(b^{i_1} \otimes \cdots \otimes b^{i_k}\right) \overset{(1.3.16)}{=} \frac{1}{k!} \sum_{(i)} \alpha_{i_1 \ldots i_k} b^{i_1} \wedge \cdots \wedge b^{i_k}.$$

If $k > n$, then every sequence $(b^{i_1}, \ldots, b^{i_k})$ contains at least two equal terms, so by the alternating property (1.3.20) of the wedge product it follows that $b^{i_1} \wedge \cdots \wedge b^{i_k} = 0$. Thus $\alpha = 0$, and hence $A_k(V) = \{0\}$ if $k > n$. Then $\dim A_k(V) = 0$, and the family (1.3.22) is a basis (and the only basis) of $A_k(V)$, because it is the empty family.

Now suppose that $k \in J_n$. Then the above relation reduces to

$$\alpha = \sum_< \alpha_{i_1 \ldots i_k} b^{i_1} \wedge \cdots \wedge b^{i_k},$$

where the symbol $<$ means that the summation is over the sequences $(i_1, \ldots, i_k)$ such that $i_1 < \cdots < i_k$. This completes the proof that the family (1.3.22) is a *generator family* of $A_k(V)$ for all $k \in \mathbb{N}^*$.

Step 2. We show that the family (1.3.22) is linearly independent when $k \in J_n$. Suppose that

$$\sum_< \lambda_{i_1 \ldots i_k} b^{i_1} \wedge \cdots \wedge b^{i_k} = 0. \tag{$*$}$$

Then for each $(v_1, \ldots, v_k) \in V^k$ we have

$$0 = \left(\sum_< \lambda_{i_1 \ldots i_k} b^{i_1} \wedge \cdots \wedge b^{i_k} \right) (v_1, \ldots, v_k)$$

$$\overset{(1.3.19)}{=} \sum_< \lambda_{i_1 \ldots i_k} \det \begin{pmatrix} b^{i_1}(v_1) & \ldots & b^{i_1}(v_k) \\ \vdots & & \vdots \\ b^{i_k}(v_1) & \ldots & b^{i_k}(v_k) \end{pmatrix}. \qquad (**)$$

Now choose a fixed family $(m_1, \ldots, m_k)$ of indices such that

$$1 \le m_1 < \cdots < m_k \le n.$$

Let $v_i := b_{m_i}$, $i \in J_n$. Then $(**)$ yields

$$\sum_< \lambda_{i_1 \ldots i_k} \det \begin{pmatrix} \delta^{i_1}_{m_1} & \ldots & \delta^{i_1}_{m_k} \\ \vdots & & \vdots \\ \delta^{i_k}_{m_1} & \ldots & \delta^{i_k}_{m_k} \end{pmatrix} = 0.$$

Here the determinant is equal to 1 if $i_j = m_j$ for all $j \in J_k$, otherwise it is zero. Thus $(*)$ implies that

$$\lambda_{m_1 \ldots m_k} = 0, \quad 1 \le m_1 < \cdots < m_k \le n,$$

whence the family (1.3.22) is linearly independent. This completes the proof that the family (1.3.22) is a basis of $A_k(V)$ if $k \in J_n$. Then $\dim A_k(V) = \binom{n}{k}$. In particular, $\dim A_n(V) = 1$; in this case the basis (1.3.22) has the only element $b^1 \wedge \cdots \wedge b^n$, and

$$b^1 \wedge \cdots \wedge b^n(b_1, \ldots, b_n) = \det(b^i(b_j)) = \det(1_n) \overset{\text{Prop. 1.3.13(ii)}}{=} 1. \qquad \square$$

Remark 1.3.16. Any two bases of $A_n(V)$ in the preceding proposition differ only by a unit in R. So if $\beta_0 = b^1 \wedge \cdots \wedge b^n$ and (β) is a basis of $A_n(V)$, then $\beta = \lambda \beta_0$ for some $\lambda \in$ R, and λ must be a unit.

Lemma and Definition 1.3.17. *Let V be a free R-module, and suppose that V has a basis consisting of n elements $(n \in \mathbb{N}^*)$. Let (β_0) be a basis of $A_n(V)$. For every endomorphism $\varphi \in \mathrm{End}(V)$, there exists a unique scalar $\det(\varphi) \in$ R such that*

$$\varphi^* \beta_0 = \det(\varphi) \beta_0. \qquad (1.3.24)$$

In a less condensed form,

$$\beta_0(\varphi(v_1), \ldots, \varphi(v_n)) = \det(\varphi) \beta_0(v_1, \ldots, v_n),$$

for all $(v_1, \ldots, v_n) \in V^n$. The scalar $\det(\varphi)$ does not depend on the choice of β_0. It is called the determinant *of φ.*

Proof. It may be checked immediately that $\varphi^*\beta_0 \in A_n(V)$. Since (β_0) is a basis of $A_n(V)$, we obtain (1.3.24). Suppose that (β) is another basis of $A_n(V)$. Then, by Remark 1.3.16, $\beta = \lambda\beta_0$, where $\lambda \in \mathsf{R}$ is a unit. Thus

$$\varphi^*\beta = \varphi^*(\lambda\beta_0) = \lambda\varphi^*(\beta_0) \overset{(1.3.24)}{=} \det(\varphi)\lambda\beta_0 = \det(\varphi)\beta,$$

as was to be shown. □

Remark 1.3.18. We continue to assume that V is an n-dimensional R-module ($n \in \mathbb{N}^*$). Let (β_0) be any basis of $A_n(V)$.

(a) If $\varphi = \lambda 1_V$ ($\lambda \in \mathsf{R}$), then $\varphi^*\beta_0 = \lambda^n\beta_0$, therefore $\det(\varphi) = \lambda^n$. Thus, in particular, $\det(1_V) = 1$, and the determinant of the zero transformation in $\mathrm{End}(V)$ is zero.

(b) If $\varphi, \psi \in \mathrm{End}(V)$, then

$$(\varphi \circ \psi)^*\beta_0 = \psi^*(\varphi^*\beta_0) = \psi^*(\det(\varphi)\beta_0) = \det(\varphi)\det(\psi)\beta_0,$$

whence $\det(\varphi \circ \psi) = \det(\varphi)\det(\psi)$. This implies immediately that if φ is an automorphism, then $\det(\varphi) \in \mathsf{R}^\times$. The converse may also be easily shown.

These observations yield the following important result.

Corollary 1.3.19. *If V is a non-trivial finitely generated free R-module, then*

$$\mathrm{GL}(V) \overset{(1.1.1)}{=} (\mathrm{End}(V))^\times = \left\{\varphi \in \mathrm{End}(V) \mid \det(\varphi) \in \mathsf{R}^\times\right\}.$$

In particular, if R is a field, then

$$\mathrm{GL}(V) = \{\varphi \in \mathrm{End}(V) \mid \det(\varphi) \neq 0\}. \qquad \square$$

We note that an analogous characterization of the general linear group $\mathrm{GL}_n(\mathsf{R})$ is formulated in A.4.2(c).

Proposition 1.3.20. *Let V be an n-dimensional R-module ($n \in \mathbb{N}^*$). If $\varphi \in \mathrm{End}(V)$, and A is the matrix of φ with respect to an arbitrarily chosen basis of V, then $\det(\varphi) = \det(A)$.*

Proof. Let $\mathcal{A} = (a_i)_{i=1}^n$ be a basis of V, and let $(a^i)_{i=1}^n$ be its dual basis. Then $\alpha_0 := a^1 \wedge \cdots \wedge a^n$ is a basis of $A_n(V)$, and, as we have already seen, $\alpha_0(a_1, \ldots, a_n) = 1$.

Suppose that $A = (A_j^i) = [\varphi]_{\mathcal{A}} \in M_n(\mathsf{R})$. Then we obtain

$$\det(\varphi) = \det(\varphi)\alpha_0(a_1,\ldots,a_n) \overset{(1.3.24)}{=} (\varphi^*\alpha_0)(a_1,\ldots,a_n)$$

$$:= \alpha_0(\varphi(a_1),\ldots,\varphi(a_n)) = \alpha_0 \left(\sum_{i_1=1}^n A_1^{i_1} a_{i_1}, \ldots, \sum_{i_n=1}^n A_n^{i_n} a_{i_n} \right)$$

$$= \sum_{(i)} A_1^{i_1} \ldots A_n^{i_n} \alpha_0(a_{i_1},\ldots,a_{i_n})$$

$$= \left(\sum_{\sigma \in S_n} \varepsilon(\sigma) A_1^{\sigma(1)} \ldots A_n^{\sigma(n)} \right) \alpha_0(a_1,\ldots,a_n) = \det(A),$$

as was to be shown. $\qquad\qquad\qquad\qquad\qquad\qquad\qquad\qquad\qquad\qquad\Box$

1.3.7 Volume Forms and Orientation

Let V be an n-dimensional real vector space, $n \geq 1$. Then the nonzero elements of the one-dimensional vector space $A_n(V)$ are called *volume forms* on V. Define a relation $\sim$ on the set $A_n(V) \setminus \{0\}$ of volume forms as follows:

$\mu_1 \sim \mu_2$ *if there is a positive real number* λ *such that* $\mu_2 = \lambda\mu_1$.

It is easy to check that this relation is an equivalence relation and there are two equivalence classes. These classes are called *orientations* on V. If $\mu \in A_n(V) \setminus \{0\}$ then

$$[\mu] := \{\lambda\mu \in A_n(V) \setminus \{0\} \mid \lambda \in \mathbb{R},\ \lambda > 0\}$$

is the *orientation represented by* μ, and $[-\mu]$ is the *reversed* or *opposite* orientation.

An *oriented vector space* is a pair $(V, \mathcal{O})$, where V is a non-trivial finite-dimensional real vector space, and $\mathcal{O} = [\mu]$ $(\mu \in A_n(V) \setminus \{0\}, n = \dim V)$ is an orientation of V. A basis $(b_i)_{i=1}^n$ of an oriented vector space $(V, \mathcal{O})$ is called *positive* if $\mu(b_1,\ldots,b_n) > 0$ for one (and hence for every) volume form $\mu \in \mathcal{O}$, otherwise the basis is *negative*. If $(b_i)_{i=1}^n$ is a positive basis of $(V, \mathcal{O})$ and $\sigma \in S_n$, then the basis $\left(b_{\sigma(i)}\right)_{i=1}^n$ is positive if, and only if, the permutation σ is even.

The *canonical orientation* of the real vector space $\mathbb{R}^n$ $(n \geq 1)$ is the orientation represented by $e^1 \wedge \cdots \wedge e^n$, where $\left(e^i\right)_{i=1}^n$ is the dual of the canonical basis of $\mathbb{R}^n$.

Let $(V, [\mu_V])$ and $(W, [\mu_W])$ be oriented vector spaces of the same positive dimension. A linear isomorphism $\varphi\colon V \to W$ is called

$$\text{orientation preserving} \quad \text{if} \quad [\varphi^*\mu_W] = [\mu_V],$$
$$\text{orientation reversing} \quad \text{if} \quad [\varphi^*\mu_W] = [-\mu_V].$$

Lemma and Definition 1.3.17 implies that *a linear automorphism* $\varphi \in$ GL(V) *is orientation preserving if, and only if,* $\det(\varphi) > 0$.

The set of all orientation preserving linear automorphisms forms a group, which will be denoted by $\mathrm{GL}^+(V)$. This group is clearly independent of the orientation of V.

Lemma 1.3.21. *Let V be a non-trivial finite-dimensional real vector space. A two-form $\omega \in A_2(V)$ is non-degenerate if, and only if, V has even dimension, say* $\dim V = 2n$, *and*

$$\omega^n := \omega \wedge \cdots \wedge \omega \quad (n \text{ factors})$$

is a volume form on V.

For a proof we refer to [1], 3.1.3 Proposition.

1.4 Orthogonal Spaces

1.4.1 *Scalar Product and Non-degeneracy*

Let V be a non-trivial real vector space. If $g \in S_2(V)$, i.e., g is a symmetric bilinear form on V, then we also say that g is a *scalar product* on V. A *(real) orthogonal space* is a non-trivial real vector space together with a scalar product. (Here we follow the terminology of [83].) An orthogonal space whose scalar product is the zero bilinear form is said to be *isotropic*. If H is a non-trivial subspace of an orthogonal space (V, g), then H, equipped with the scalar product $g \restriction H \times H$, is also an orthogonal space.

By the *nullspace* of a scalar product $g \in S_2(V)$ we mean the kernel of the linear mapping

$$j_g \colon V \to V^*, \ v \mapsto j_g(v) := j_v g$$

introduced in Lemma 1.2.1. The nullspace of g will be denoted by $N(g)$. Thus

$$N(g) := \mathrm{Ker}(j_g) = \{v \in V \mid j_g(v) = 0 \in V^*\}$$
$$= \{v \in V \mid g(u, v) = 0 \text{ for all } u \in V\}. \tag{1.4.1}$$

The scalar product g, and also the orthogonal space (V, g), is called *non-degenerate* if $N_g = \{0\}$, otherwise we say that g (and (V, g)) is *degenerate*.

In the finite-dimensional case by the *rank* of g we mean the rank of the mapping $j_g \in \mathrm{Hom}(V, V^*)$, i.e.,

$$\mathrm{rank}(g) := \mathrm{rank}(j_g) := \dim(\mathrm{Im}(j_g)).$$

As an immediate consequence of the rank + nullity = dimension theorem for linear mappings, Lemma 1.1.26(iii) and Lemma 1.2.2, we obtain

Corollary 1.4.1. *For a finite-dimensional orthogonal space* (V, g) *the following conditions are equivalent:*

(i) *The space is non-degenerate.*

(ii) *The mapping* $j_g \colon V \to V^*$ *is a linear isomorphism.*

(iii) *The matrix of* g *with respect to one (and hence every) basis of* V *is invertible.* $\qquad\square$

A finite-dimensional non-degenerate orthogonal space will be called a *semi-Euclidean vector space*.

1.4.2   The Associated Quadratic Form

Let (V, g) be an orthogonal space. The function

$$Q \colon V \to \mathbb{R}, \quad v \mapsto Q(v) := g(v, v)$$

is called the *quadratic form associated with* g. The scalar product can be reconstructed from Q by the *polarization identity*

$$g(u, v) = \frac{1}{2}(Q(u + v) - Q(u) - Q(v)); \quad u, v \in V. \tag{1.4.2}$$

The norm of a vector $v \in V$ with respect to g is

$$\|v\|_g := \sqrt{|g(v, v)|} = \sqrt{|Q(v)|}.$$

We simply write $\|v\|$ if the scalar product is clear from the context. A *unit vector* is a vector of norm 1. The set of unit vectors is said to be the *unit sphere* in V.

The scalar product g, and also the orthogonal space (V, g) is called

positive definite if $Q(v) > 0$ for all $v \in V \setminus \{0\}$,
positive semidefinite if $Q(v) \geq 0$ for all $v \in V$.

Negative definiteness and negative semidefiniteness are defined analogously. A scalar product is called *indefinite* if it is neither positive semidefinite nor negative semidefinite. By a *Euclidean vector space* we mean a finite-dimensional, positive definite orthogonal space. The Euclidean n-space $\mathbb{R}^n$

introduced in Example 1.1.13 is a Euclidean vector space with the scalar product $\langle\,,\,\rangle$ defined by

$$\langle a, b \rangle := \sum_{i=1}^{n} \alpha^i \beta^i \quad \text{if } a = (\alpha^1, \ldots, \alpha^n),\ b = (\beta^1 \ldots \beta^n). \tag{1.4.3}$$

This scalar product is called the *canonical scalar product* (or the *dot product*) on $\mathbb{R}^n$.

Returning to the general case, a vector v of (V, g) is said to be *spacelike* if $Q(v) > 0$, *isotropic* if $Q(v) = 0$ and $v \neq 0$, and *timelike* if $Q(v) < 0$.

If (V, g) is a *positive semidefinite* orthogonal space, then the scalar product satisfies the *Schwarz inequality*

$$(g(u, v))^2 \leq Q(u)Q(v); \quad u, v \in V. \tag{1.4.4}$$

We obtain equality when the vectors u and v are linearly dependent, but equality may also hold for linearly independent vectors. If g is a *positive definite* scalar product, then the linear dependence of u and v is a necessary and sufficient condition for equality in (1.4.4).

A positive or negative definite orthogonal space is clearly non-degenerate. *If (V, g) is positive semidefinite and has an isotropic vector v, then g is degenerate.* Indeed, the Schwarz inequality implies that

$$(g(u, v))^2 \leq g(u, u)g(v, v) = 0, \quad u \in V;$$

whence $g(u, v) = 0$ for all $u \in V$.

Lemma and Definition 1.4.2. *Let (V, g) be a positive definite orthogonal space. Then V is a normed space (Definition B.4.2) with the norm*

$$v \in V \mapsto \|v\| := \|v\|_g := \sqrt{g(v, v)} \in \mathbb{R}. \tag{1.4.5}$$

In particular, every Euclidean vector space is a normed space with the norm given by (1.4.5), which is called then the Euclidean *norm.*

Proof. We clearly have (N1). For any $\lambda \in \mathbb{R}$ and any $v \in V$,

$$\|\lambda v\| = \sqrt{g(\lambda v, \lambda v)} = \sqrt{\lambda^2 g(v, v)} = |\lambda| \|v\|,$$

and so (N2) is satisfied. To verify (N3), let $u, v \in V$. Then

$$\|u + v\|^2 = g(u + v, u + v) = g(u, u) + 2g(u, v) + g(v, v)$$

$$\overset{(1.4.4)}{\leq} \|u\|^2 + 2\|u\|\|v\| + \|v\|^2 = (\|u\| + \|v\|)^2.$$

Taking square roots of both sides, we obtain (N3). $\qquad\square$

1.4.3 Orthonormal Bases

We continue to assume that (V, g) is an orthogonal space. Two vectors u and v in V are called *(mutually) orthogonal* (or *perpendicular*) if $g(u, v) = 0$. Then we write $u \perp_g v$, or simply $u \perp v$ if the scalar product is clear from the context. If S is a nonempty subset of V, then

$$S^{\perp_g} := \{v \in V \mid g(u, v) = 0 \text{ for all } u \in S\}$$

is a subspace of V, called the *(g-)orthogonal complement* of S. When there is no danger of confusion, we write $S^{\perp}$ instead of $S^{\perp_g}$. Also, we write $v \perp_g S$ (or $v \perp S$) if $v \in S^{\perp_g}$. Obviously, $\{0\}^{\perp} = V$. Observe that

$$V^{\perp} \overset{(1.4.1)}{=} N(g) = \mathrm{Ker}(j_g).$$

If (V, g) is finite-dimensional, a basis $(b_i)_{i=1}^{n}$ of V is said to be *(g-)orthogonal* if $g(b_i, b_j) = 0$ for $i \neq j$. We agree that in a one-dimensional orthogonal space every basis is orthogonal. This property holds automatically whenever (V, g) is isotropic. An orthogonal basis $(b_i)_{i=1}^{n}$ is called *orthonormal* if

$$g(b_i, b_i) \in \{1, 0, -1\} \text{ for all } i \in J_n,$$

i.e., if it consists of unit vectors or isotropic vectors.

Lemma 1.4.3. *Any finite-dimensional orthogonal space has an orthonormal basis.*

Proof. Let (V, g) be an n-dimensional orthogonal space. We may suppose that V is not isotropic and $n \geq 2$. (If $n = 1$ and $g(v, v) \neq 0$, then $b := \frac{1}{\|v\|} v$ forms an orthonormal basis for V.)

Step 1. *There exists an element v of V such that $g(v, v) \neq 0$.*

To see this, we argue by contradiction. Suppose that $g(v, v) = 0$ for every $v \in V$. Then, for all $u, v \in V$,

$$0 = g(u + v, u + v) = g(v, u) + g(u, v) = 2g(u, v),$$

whence $g = 0$. But this is impossible, since (V, g) is not isotropic.

Step 2. *If $g(v, v) \neq 0$, then $V = \mathrm{span}(v) \oplus (\mathrm{span}(v))^{\perp}$.*

We can prove this as follows. Choose an arbitrary vector u of V. Since $g(v, v)$ is a nonzero real number, $g(u, v) = \lambda g(v, v)$ for some $\lambda \in \mathbb{R}$. Then $g(u - \lambda v, v) = 0$, hence $u - \lambda v \in (\mathrm{span}(v))^{\perp}$, and the expression

$$u = \lambda v + (u - \lambda v)$$

shows that $V = \operatorname{span}(v) + (\operatorname{span}(v))^{\perp}$. Since $v \notin (\operatorname{span}(v))^{\perp}$ (because $g(v, v) \neq 0$), it follows that $\operatorname{span}(v) \cap (\operatorname{span}(v))^{\perp} = \{0\}$, so the sum $\operatorname{span}(v) + (\operatorname{span}(v))^{\perp}$ is direct.

Step 3. By Step 1, choose an element v_1 of V such that $g(v_1, v_1) \neq 0$. Then $(\operatorname{span}(v_1))^{\perp}$ is also an orthogonal space, whose dimension is $n - 1$. By induction, $(\operatorname{span}(v_1))^{\perp}$ has an orthogonal basis $(v_j)_{j=2}^{n}$. Then $(v_i)_{i=1}^{n}$ is an orthogonal basis of V, from which we obtain the orthonormal basis $(b_i)_{i=1}^{n}$, where

$$b_i := \begin{cases} v_i \text{ if } v_i \text{ is isotropic,} \\ \dfrac{1}{\|v_i\|} v_i \text{ if } v_i \text{ is not isotropic.} \end{cases} \qquad \square$$

In a Euclidean vector space (V, g) each member of a basis is a space-like vector, and we have a canonical way of getting an orthonormal basis, starting with an arbitrary basis $(v_i)_{i=1}^{n}$ of V. Let

$$b_1 := \frac{1}{\|v_1\|} v_1$$

and

$$w_2 := v_2 - g(v_2, b_1)b_1, \quad b_2 := \frac{1}{\|w_2\|} w_2.$$

Then $\|b_1\| = \|b_2\| = 1$ and $b_1 \perp b_2$. Proceeding inductively, we define

$$w_k := v_k - \sum_{i=1}^{k-1} g(v_k, b_i)b_i, \quad b_k := \frac{1}{\|w_k\|} w_k \quad (k > 2).$$

In this way we obtain an orthonormal basis $(b_i)_{i=1}^{n}$ of (V, g). The construction just described is known as the *Gram – Schmidt orthogonalization process*. It works equally well in infinite-dimensional positive definite orthogonal spaces.

We now state (without proof) a basic result concerning finite-dimensional orthogonal spaces:

Sylvester's law of inertia. *The number of spacelike, isotropic and time-like vectors in an orthonormal basis of a finite-dimensional orthogonal space does not depend on the choice of the basis.* $\triangle$

If $(b_i)_{i=1}^{n}$ is an orthonormal basis of (V, g) such that

$$\begin{aligned} g(b_i, b_i) &= 1 && \text{for } i \in \{1, \ldots, p\}, \\ g(b_{p+j}, b_{p+j}) &= -1 && \text{for } j \in \{1, \ldots, q\}, \\ g(b_{r+k}, b_{r+k}) &= 0 && \text{for } k \in \{1, \ldots, n-r\}, \ r := p + q, \end{aligned}$$

then p is called the *coindex*, q is called the *index* of g (or of (V, g)). Clearly, $r = p + q$ is the rank of g, and $n - r = \dim(N(g))$ is the *nullity* of g. The pair of numbers (p, q) is said to be the *signature* of the scalar product and of the orthogonal space.

1.4.4 Orthogonal Mappings and the Adjoint

Let (V, g) and $(\widetilde{V}, \tilde{g})$ be orthogonal spaces. A linear mapping $\varphi \colon V \to \widetilde{V}$ is said to be *orthogonal* if it preserves the scalar product, i.e.,

$$\tilde{g}(\varphi(u), \varphi(v)) = g(u, v) \text{ for all } u, v \in V; \tag{1.4.6}$$

briefly, $\varphi^* \tilde{g} = g$. A bijective orthogonal mapping is called an *orthogonal iso-morphism*. Two orthogonal spaces are *isomorphic* if there is an orthogonal isomorphism between them.

Lemma 1.4.4. *Let (V, g) and $(\widetilde{V}, \tilde{g})$ be as above. A linear mapping $\varphi \colon V \to \widetilde{V}$ is orthogonal if, and only if,*

$$ {}^t\varphi \circ j_{\tilde{g}} \circ \varphi = j_g, \tag{1.4.7}$$

that is, if, and only if, the diagram

$$
\begin{array}{ccc}
V & \xrightarrow{\;\varphi\;} & \widetilde{V} \\
{\scriptstyle j_g}\downarrow & & \downarrow{\scriptstyle j_{\tilde{g}}} \\
V^* & \xleftarrow[\;{}^t\varphi\;]{} & \widetilde{V}^*
\end{array}
$$

commutes.

Proof. Given two vectors u, v in V, we have on the one hand

$${}^t\varphi \circ j_{\tilde{g}} \circ \varphi(v)(u) = {}^t\varphi \circ j_{\varphi(v)}\tilde{g}(u) = j_{\varphi(v)}\tilde{g} \circ \varphi(u) = \tilde{g}(\varphi(u), \varphi(v)).$$

On the other hand,

$$j_g(v)(u) = j_v g(u) = g(u, v),$$

therefore the conditions (1.4.6) and (1.4.7) are equivalent. □

Corollary 1.4.5. *Any orthogonal mapping from a non-degenerate orthogonal space into an arbitrary orthogonal space is injective.*

Proof. Let (V, g) and $(\widetilde{V}, \tilde{g})$ be orthogonal spaces. Suppose that (V, g) is non-degenerate, and let $\varphi \colon V \to \widetilde{V}$ be an orthogonal mapping. Then $N(g) = \{0\}$, so, by (1.4.1), $j_g \overset{(1.4.7)}{=} ({}^t\varphi \circ j_{\tilde{g}}) \circ \varphi$ is injective, which implies that φ is injective. □

Corollary 1.4.6. *If (V, g) is a finite-dimensional, non-degenerate orthogonal space, then the set $O(V)$ of all orthogonal transformations of V is a group under the composition of mappings.*

Proof. Clearly, the identity transformation of V is orthogonal, and composition of orthogonal mappings remains orthogonal. By our hypotheses, the previous corollary implies that $O(V)$ consists of linear automorphisms of V. Finally, applying, e.g., (1.4.7), it follows readily that the inverse of an orthogonal transformation is also orthogonal. $\square$

The group $O(V)$ is said to be the *orthogonal group* of (V, g). The orientation preserving orthogonal transformations form a subgroup of $O(V)$, denoted by $SO(V)$. The elements of $SO(V)$ are called *special orthogonal transformations* or *rotations*.

Lemma 1.4.7. *Let (V, g) be a finite-dimensional, non-degenerate orthogonal space and φ an endomorphism of V. There exists a unique endomorphism φ^* of V such that*

$$g(\varphi(u), v) = g(u, \varphi^*(v)). \tag{1.4.8}$$

Proof. First we show that only one endomorphism of V has the desired property. Suppose that $\psi_1, \psi_2 \in \text{End}(V)$ are such that

$$g(u, \psi_1(v)) = g(u, \psi_2(v)); \quad u, v \in V.$$

This relation can also be written in the form

$$j_g(\psi_1(v))(u) = j_g(\psi_2(v))(u)$$

by (1.2.7). Thus

$$j_g(\psi_1(v)) = j_g(\psi_2(v))$$

for every $v \in V$. Since, in particular, j_g is injective by Corollary 1.4.1, it follows that $\psi_1 = \psi_2$, proving the uniqueness part of the lemma.

Now we show that there exists an endomorphism $\varphi^* \in \text{End}(V)$ satisfying (1.4.8). Since the linear mapping $j_g : V \to V^*$ is in fact bijective by Corollary 1.4.1, we can form the composite linear mapping

$$\varphi^* := j_g^{-1} \circ {}^t\varphi \circ j_g. \tag{1.4.9}$$

Then for all $u, v \in V$,

$$g(u, \varphi^*(v)) \overset{(1.2.7)}{=} j_g(\varphi^*(v))(u) \overset{(1.4.9)}{=} {}^t\varphi \circ j_g(v)(u)$$

$$\overset{(1.1.8)}{=} j_g(v)(\varphi(u)) \overset{(1.2.7)}{=} g(\varphi(u), v),$$

which shows that (1.4.9) defines the required linear mapping. $\square$

The unique linear endomorphism $\varphi^* \in \mathrm{End}(V)$ related to $\varphi \in \mathrm{End}(V)$ by (1.4.8) is called the *adjoint* of φ (with respect to the given non-degenerate scalar product g). The endomorphism φ is said to be *symmetric* (or *self-adjoint*) if $\varphi = \varphi^*$.

Corollary 1.4.8. *A linear automorphism of a finite-dimensional, non-degenerate orthogonal space is an orthogonal transformation if, and only if, its inverse is equal to its adjoint.*

Proof. Let (V, g) be an orthogonal space with the required properties, and let $\varphi \in \mathrm{GL}(V)$. Then the following are equivalent:

(i) φ is an orthogonal transformation of V.
(ii) ${}^t\varphi \circ j_g \circ \varphi = j_g$ (by Lemma 1.4.4).
(iii) $(j_g^{-1} \circ {}^t\varphi \circ j_g) \circ \varphi = 1_V$ (since j_g is bijective).
(iv) $\varphi^* \circ \varphi = 1_V$ (by (1.4.9)).
(v) The inverse of φ is its adjoint. $\qquad\qquad\square$

1.4.5 Modules with Scalar Product

Let V be an R-module. If $g \colon V \times V \to \mathsf{R}$ is a symmetric R-bilinear form, we also say (as in the case of real vector spaces in 1.4.1) that g is a *scalar product* on V. Given a scalar product g on V, we have a natural mapping j_g from V to its dual V^* defined just as in Lemma 1.2.1:

$$j_g \colon V \to V^*, \quad j_g(v)(u) := j_v g(u) := g(u, v); \quad u, v \in V. \qquad (1.4.10)$$

The *nullspace* of g is $N(g) := \mathrm{Ker}(j_g)$; g is *non-degenerate* if $N(g) = \{0\}$, i.e., if

$$g(u, v) = 0 \text{ for all } u \in V \text{ implies } v = 0.$$

Lemma 1.4.9. *Let V be an R-module and consider its dual module V^*. Suppose that there exists an isomorphism $f \colon V \to V^*$ of R-modules such that the bilinear form*

$$g \colon V \times V \to \mathsf{R}, \quad (u, v) \mapsto g(u, v) := f(u)(v) \qquad (1.4.11)$$

is symmetric. Then

(i) *g is a non-degenerate scalar product on V;*
(ii) *for any endomorphism $\varphi \in \mathrm{End}_\mathsf{R}(V)$ there exists a unique endomorphism $\widetilde{\varphi} \in \mathrm{End}_\mathsf{R}(V)$ such that φ and $\widetilde{\varphi}$ are related by*

$$g(\varphi(u), v) = g(u, \widetilde{\varphi}(v)); \quad u, v \in V. \qquad (1.4.12)$$

Proof. (i) Let v be a fixed vector of V, and suppose that $g(u,v) = 0$ for all $u \in V$. Then, by the symmetry of g,

$$0 = g(v,u) := f(v)(u) \text{ for all } u \in V,$$

therefore $f(v) = 0 \in V^*$. Since, in particular, f is injective, it follows that $v = 0$, thus proving that $N(g) = 0$.

(ii) First we show that only one endomorphism of V has the desired behaviour. Suppose that φ_1, φ_2 are such that

$$g(\varphi(u), v) = g(u, \varphi_1(v)) = g(u, \varphi_2(v))$$

for all $u, v \in V$. Then

$$g(u, \varphi_1(v) - \varphi_2(v)) = g(\varphi_1(v) - \varphi_2(v), u) = 0,$$

i.e.,

$$f(\varphi_1(v) - \varphi_2(v))(u) = 0, \quad u, v, \in V.$$

This implies that $f(\varphi_1(v) - \varphi_2(v)) = 0 \in V^*$ and hence $\varphi_1(v) - \varphi_2(v) = 0$ for all $v \in V$. Thus $\varphi_1 = \varphi_2$ as claimed.

Now we prove that there exists an endomorphism of V with the required property. On the analogy of (1.4.9), let

$$\widetilde{\varphi} := f^{-1} \circ {}^t\varphi \circ f.$$

Then $f \circ \widetilde{\varphi} = {}^t\varphi \circ f$, and for all $u, v \in V$ we have

$$g(u, \widetilde{\varphi}(v)) = g(\widetilde{\varphi}(v), u) \overset{(1.4.11)}{:=} f(\widetilde{\varphi}(v))(u) = {}^t\varphi(f(v))(u)$$

$$= f(v)(\varphi(u)) = g(v, \varphi(u)) = g(\varphi(u), v),$$

as desired. $\square$

Remark 1.4.10. (a) The endomorphism $\widetilde{\varphi} \in \mathrm{End}(V)$ defined by (1.4.12) may also be called the *adjoint* of φ with respect to the scalar product g given by (1.4.11).

(b) Concerning the assumptions of Lemma 1.4.9, we note that *an R-module is not necessarily isomorphic to its dual*. As an example, consider the commutative group $\mathbb{Z}/2\mathbb{Z}$ of integers modulo 2. The elements of $\mathbb{Z}/2\mathbb{Z}$ are the residue classes

$$\overline{0} := \{2k \in \mathbb{Z} \mid k \in \mathbb{Z}\} \text{ and } \overline{1} := \{2k + 1 \in \mathbb{Z} \mid k \in \mathbb{Z}\},$$

and the group operation is the addition

$$\overline{a} + \overline{b} := \overline{a + b}; \quad \overline{a}, \overline{b} \in \mathbb{Z}/2\mathbb{Z}.$$

According to Example 1.1.8, $\mathbb{Z}/2\mathbb{Z}$ is a $\mathbb{Z}$-module. If ℓ is a linear form on $\mathbb{Z}/2\mathbb{Z}$, then

$$0 = \ell(\overline{0}) = \ell(\overline{1} + \overline{1}) = \ell(\overline{1}) + \ell(\overline{1}) = 2\ell(\overline{1}),$$

whence $\ell(\overline{1}) = 0$. Thus the dual of $\mathbb{Z}/2\mathbb{Z}$ consists only of the zero linear form, therefore it is not isomorphic to $\mathbb{Z}/2\mathbb{Z}$.

To conclude this subsection, we formulate and prove a purely algebraic result taken from the monograph [53] (pp. 344–345), which will be surprisingly efficient later, in differential geometric context; see the proof of Proposition 6.1.43.

Proposition 1.4.11. *Keep the notation and hypotheses of Lemma 1.4.9, and suppose in addition that the mapping $\lambda \in \mathsf{R} \mapsto 2\lambda := \lambda + \lambda \in \mathsf{R}$ is bijective. Given an alternating R-bilinear mapping $\beta \colon V \times V \to V$, i.e., a tensor $\beta \in A_2(V, V)$, there exists a unique R-linear mapping*

$$\omega \colon V \to \mathrm{End}_{\mathsf{R}}(V), \ u \mapsto \omega(u)$$

such that

(i) *$\omega(u) \in \mathrm{End}_{\mathsf{R}}(V)$ is skew-symmetric with respect to the scalar product g on V, i.e.,*

$$g(\omega(u)(v), w) + g(v, \omega(u)(w)) = 0; \quad v, w \in V;$$

(ii) *$\beta(u, v) = \omega(u)(v) - \omega(v)(u); \ u, v \in V$.*

Proof. *Existence.* For any fixed $u \in V$, let β_u be the linear mapping

$$V \to V, \ v \mapsto \beta_u(v) := \beta(u, v),$$

and let $\widetilde{\beta}_u$ be the adjoint of β_u with respect to the scalar product g given by (1.4.12):

$$g(\beta_u(v), w) = g(v, \widetilde{\beta}_u(w)); \quad v, w \in V.$$

Define a mapping

$$\omega \colon V \to \mathrm{End}_{\mathsf{R}}(V), \ u \mapsto \omega(u)$$

by

$$\omega(u)(v) := \frac{1}{2}\big(\beta(u, v) - \widetilde{\beta}_u(v) - \widetilde{\beta}_v(u)\big), \quad v \in V.$$

Then

$$g(\omega(u)(v), v) = \frac{1}{2}\big(g(\beta(u, v), v) - g(\widetilde{\beta}_u(v), v) - g(\widetilde{\beta}_v(u), v)\big)$$

$$= \frac{1}{2}\big(g(\beta(u, v), v) - g(\beta(u, v), v) - g(u, \beta(v, v))\big) = 0,$$

taking into account the skew-symmetry of β. So it follows that for all $u, v, w \in V$,

$$0 = g\big(\omega(u)(v + w), v + w\big) = g(\omega(u)(v), v) + g(\omega(u)(w), v)$$

$$+ g(\omega(u)(v), w) + g(\omega(u)(w), w) = g(\omega(u)(v), w) + g(v, \omega(u)(w)),$$

which proves that ω satisfies (i). Condition (ii) is also satisfied, since for all $u, v \in V$,

$$2\omega(u)(v) := \beta(u,v) - \widetilde{\beta}_u(v) - \widetilde{\beta}_v(u),$$

$$2\omega(v)(u) := \beta(v,u) - \widetilde{\beta}_v(u) - \widetilde{\beta}_u(v),$$

and so, subtracting the second relation from the first and dividing by 2, we obtain

$$\beta(u,v) = \omega(u)(v) - \omega(v)(u).$$

Uniqueness. It is sufficient to verify that $\beta = 0$ implies $\omega = 0$. If $\beta = 0$, then (ii) reduces to

$$\omega(u)(v) = \omega(v)(u); \quad u, v \in V, \tag{$*$}$$

and we find

$$g\big(\omega(u)(v), w\big) \overset{\text{(i)}}{=} -g(v, \omega(u)(w)) \overset{(*)}{=} -g(v, \omega(w)(u))$$

$$\overset{\text{(i)}}{=} g(\omega(w)(v), u) \overset{(*)}{=} g(\omega(v)(w), u)$$

$$\overset{\text{(i)}}{=} -g(w, \omega(v)(u)) \overset{(*)}{=} -g(\omega(u)(v), w).$$

Thus, for all $u, v, w \in V$,

$$0 = g(\omega(u)(v), w) \overset{(1.4.11)}{=} f(\omega(u)(v))(w),$$

whence $f(\omega(u)(v)) = 0 \in V^*$, $\omega(u)(v) = 0$ for all $u, v \in V$, and, finally, $\omega = 0$. $\qquad\square$

Chapter 2

Manifolds and Bundles

Beginning with this chapter, we shall assume the basics of point set topology. The most important facts we need are collected in Appendix B. This and the forthcoming chapters also presuppose the readers' familiarity with the rudiments of multivariable calculus. These prerequisites are summarized in Appendix C.1.

2.1 Smooth Manifolds and Mappings

La notion générale de variété est assez difficile à définir avec précision. (The general notion of a manifold is quite difficult to define with precision.)

Élie Cartan [29], p. 56

Do not think that Cartan was without a clear concept of manifold. ... A correct definition of a manifold is not difficult today if one uses modern general topology and differential calculus.

Marcel Berger [12], p. 144

2.1.1 *Charts, Atlases, Manifolds*

Definition 2.1.1. (a) A *topological manifold of dimension n* (or simply a *topological n-manifold*) is a second countable Hausdorff topological space M that is locally homeomorphic to $\mathbb{R}^n$, i.e., every point of M has an open neighbourhood which is homeomorphic to an open subset of $\mathbb{R}^n$. In this case we write $\dim M := n$.

(b) A *chart* for a topological n-manifold M is a pair $(\mathcal{U}, u)$ where $\mathcal{U}$ is an open subset of M and u is a homeomorphism of $\mathcal{U}$ onto an open subset of $\mathbb{R}^n$. We call the set $\mathcal{U}$ the *domain* of the chart, or a *coordinate neighbourhood* (of any of its points). The mapping u is called a *coordinate map*, its *ith coordinate function* is $u^i = e^i \circ u$ ($(e^i)_{i=1}^n$ is the dual of the canonical basis of $\mathbb{R}^n$, see Example 1.1.35). We say that $(u^i)_{i=1}^n$ is a *local coordinate system* on M, and instead of $(\mathcal{U}, u)$ we also write $(\mathcal{U}, (u^i)_{i=1}^n)$.

(c) Let $(\mathcal{U}, u)$ be a chart for a topological n-manifold M. If its domain contains the point $p \in M$, we say that $(\mathcal{U}, u)$ is a *chart at p* (or *around p*). In the special case when $u(p) = 0 \in \mathbb{R}^n$, the chart is called *centred on p*.

(d) A chart for a topological manifold M is said to be *global* if its domain is M.

Definition 2.1.2. Let M be a topological n-manifold.

(a) Two charts $(\mathcal{U}, \varphi)$ and $(\mathcal{V}, \psi)$ of M are said to be *smoothly compatible* if either $\mathcal{U} \cap \mathcal{V} = \emptyset$ or the *transition homeomorphism*

$$\psi \circ \varphi^{-1} \colon \varphi(\mathcal{U} \cap \mathcal{V}) \to \psi(\mathcal{U} \cap \mathcal{V})$$

and its inverse

$$\varphi \circ \psi^{-1} \colon \psi(\mathcal{U} \cap \mathcal{V}) \to \varphi(\mathcal{U} \cap \mathcal{V})$$

are smooth.

(b) A *smooth atlas* for M is a set

$$\mathcal{A} = \{(\mathcal{U}_i, u_i) \mid i \in I\}$$

of charts of M, where I is an indexing set, such that

(A_1) the chart domains $\mathcal{U}_i$ form a covering of M,

(A_2) any two charts of $\mathcal{A}$ are smoothly compatible.

(c) A chart of M is called *compatible* with a smooth atlas $\mathcal{A}$ for M if it is compatible with all charts of the atlas $\mathcal{A}$.

(d) A smooth atlas $\mathcal{A}$ for M is *maximal* if every chart compatible with $\mathcal{A}$ is already in $\mathcal{A}$. A maximal atlas for a topological manifold is also referred to as a *smooth structure*.

Lemma 2.1.3. *Let M be a topological manifold.*

(a) *Every smooth atlas for M is contained in a unique maximal smooth atlas.*

(b) *Two smooth atlases for M determine the same smooth structure if, and only if, their union is a smooth atlas.*

For a proof see for example [9] or [69].

Definition 2.1.4. An n-dimensional *smooth manifold* is a topological n-manifold endowed with a smooth structure.

Remark 2.1.5. (a) Strictly speaking, an n-dimensional smooth manifold is a *pair* $(M, \mathcal{A})$, where M is a topological n-manifold and $\mathcal{A}$ is a smooth structure on M. However, by a typical abuse of notation, we usually write M for the smooth manifold, the presence of the smooth structure $\mathcal{A}$ being understood.

(b) Due to Lemma 2.1.3(a), we may specify a smooth structure on a topological manifold by *any* convenient small smooth atlas. In practice, we always try to minimize the number of charts that we use to determine a smooth structure.

(c) Henceforth *we shall use the term 'manifold' in the sense of a smooth manifold*. If M is a manifold, an *atlas of M* is any atlas contained in the smooth structure of M, and a *chart of M* is any chart belonging to an atlas of M.

2.1.2 Examples of Manifolds

Example 2.1.6 (Zero-dimensional Manifolds). Since $\mathbb{R}^0$ is by definition a single point, a zero-dimensional topological manifold M is just a countable set endowed with the discrete topology. If (p_n) is the sequence of points of M, the charts for M are the pairs $(\{p_i\}, \varphi_i)$, where φ_i is the unique mapping of $\{p_i\}$ onto $\mathbb{R}^0$. The set of all charts on M trivially satisfies the smooth compatibility condition, and forms a maximal smooth atlas for M. Thus every zero-dimensional topological manifold has a unique smooth structure.

This example seems to be pathological, but in fact it does occur in practice. However, in our geometric considerations we shall generally meet manifolds of positive dimension. For the only exception, see Remark 3.4.22.

Example 2.1.7 (Open Submanifolds). Let M be a manifold, $\mathcal{A}$ an atlas for M, and N an open subset of M. Let

$$\mathcal{A}_N := \{(N \cap \mathcal{U}_i, u_i \upharpoonright (N \cap \mathcal{U}_i)) \mid (\mathcal{U}_i, u_i) \in \mathcal{A}\}.$$

Then the members of $\mathcal{A}_N$ are charts for N, and N is covered by the domains of charts in $\mathcal{A}_N$, because every point of N is in the domain of some chart of $\mathcal{A}$. It is easy to verify that $\mathcal{A}_N$ also satisfies condition (A_2) in Definition 2.1.2. Thus $\mathcal{A}_N$ is a smooth atlas for N, and therefore defines a smooth structure on N, depending only on that of M. This structure on N is said to be *induced* by the smooth structure of M. Any open subset of a manifold endowed with the induced smooth structure is called an *open submanifold*.

Example 2.1.8 (Products of Manifolds). Let M and N be manifolds of dimension m and n, respectively. Consider the product set $M \times N$, endowed with the product topology (Example B.1.2(c)). Then $M \times N$ is also a second countable Hausdorff topological space. We check first that $M \times N$ is locally homeomorphic to $\mathbb{R}^{m+n}$. Given a point (a, b) of $M \times N$, we can choose a chart $(\mathcal{U}, \varphi)$ of M at a and a chart $(\mathcal{V}, \psi)$ of N at b. Then the product mapping

$$\varphi \times \psi \colon \mathcal{U} \times \mathcal{V} \to \mathbb{R}^{m+n}, (p, q) \mapsto (\varphi(p), \psi(q))$$

is a homeomorphism onto its image, which is an open subset in $\mathbb{R}^{m+n}$. Thus $M \times N$ is a topological $(m+n)$-manifold.

Now let $\mathcal{A} = \{(\mathcal{U}_\alpha, \varphi_\alpha) \mid \alpha \in A\}$ be an atlas of M, $\mathcal{B} = \{(\mathcal{V}_i, \psi_i) \mid i \in I\}$ an atlas of N, and let, by an abuse of notation,

$$\mathcal{A} \times \mathcal{B} := \{(\mathcal{U}_\alpha \times \mathcal{V}_i, \varphi_\alpha \times \psi_i) \mid \alpha \in \mathcal{A}, i \in I\}.$$

We claim that $\mathcal{A} \times \mathcal{B}$ is a smooth atlas for $M \times N$. Indeed, the covering condition is clearly satisfied. If $(\mathcal{U}_\alpha \times \mathcal{V}_i, \varphi_\alpha \times \psi_i)$ and $(\mathcal{U}_\beta \times \mathcal{V}_j, \varphi_\beta \times \psi_j)$ are two charts in $\mathcal{A} \times \mathcal{B}$ such that $(\mathcal{U}_\alpha \times \mathcal{V}_i) \cap (\mathcal{U}_\beta \times \mathcal{V}_j) \neq \emptyset$, then on $\varphi_\beta \times \psi_j((\mathcal{U}_\alpha \times \mathcal{V}_i) \cap (\mathcal{U}_\beta \times \mathcal{V}_j))$ we have

$$(\varphi_\alpha \times \psi_i) \circ (\varphi_\beta \times \psi_j)^{-1} = (\varphi_\alpha \circ \varphi_\beta^{-1}) \times (\psi_i \circ \psi_j^{-1}),$$

which is a smooth mapping. Thus the condition of smooth compatibility is also satisfied.

The smooth structure defined by $\mathcal{A} \times \mathcal{B}$ depends only on the smooth structures of M and N. The topological manifold $M \times N$ endowed with this structure is called the *product* of the manifolds M and N. This construction can obviously be extended to the product of any finite number of manifolds.

Example 2.1.9 (Finite-dimensional Vector Spaces). Let n be a positive integer, and V an n-dimensional real vector space. Endow V with the unique Hausdorff topology which is compatible with its vector space structure (see B.4). Let $\varphi \colon V \to \mathbb{R}^n$ be a bijective linear mapping. Then φ is a homeomorphism, so the atlas consisting of the single chart (V, φ) defines a smooth structure on V. This structure is independent of the choice of the linear isomorphism φ. To see this, let $\psi \colon V \to \mathbb{R}^n$ be another bijective linear mapping of V onto $\mathbb{R}^n$. Then $\psi \circ \varphi^{-1} \colon \mathbb{R}^n \to \mathbb{R}^n$ is a linear bijection of $\mathbb{R}^n$ onto itself, hence it is a diffeomorphism. Thus the charts (V, φ) and (V, ψ) are smoothly compatible, hence $\{(V, \varphi), (V, \psi)\}$ is also a smooth atlas for V. This implies by Lemma 2.1.3(b) that (V, φ) and (V, ψ) define the same smooth structure. We call this the *canonical smooth structure* on V.

In particular, the canonical smooth structure on $\mathbb{R}^n$ is determined by the single chart $(\mathbb{R}^n, 1_{\mathbb{R}^n})$.

Example 2.1.10 (The General Linear Group of a Vector Space).
Let V be an n-dimensional real vector space $(n \geq 1)$. The space $\text{End}(V)$ of linear transformations of V is an n^2-dimensional real vector space, so it has the canonical smooth structure described in the preceding example. In view of Corollary 1.3.19, the general linear group $\text{GL}(V)$ of V is

$$\text{GL}(V) = \{\varphi \in \text{End}(V) \mid \det(\varphi) \neq 0\} = \det^{-1}(\mathbb{R}^*).$$

Since the function $\det\colon \text{End}(V) \to \mathbb{R}$ is continuous, it follows that $\text{GL}(V)$ is an open subset of $\text{End}(V)$, and hence, by Example 2.1.7, a manifold.

If we choose a basis of V and assign to each element of $\text{End}(V)$ its matrix representative with respect to that basis, $\text{End}(V)$ may be identified, together with its canonical smooth structure, with the real vector space $M_n(\mathbb{R}) \cong \mathbb{R}^{n^2}$ of the $n \times n$ matrices with entries in $\mathbb{R}$, and $\text{GL}(V)$ with the group $\text{GL}_n(\mathbb{R})$ of the invertible matrices in $M_n(\mathbb{R})$. Thus the *(real) general linear group* $\text{GL}_n(\mathbb{R})$ is also a manifold, an open submanifold of $\mathbb{R}^{n^2}$. We have, in particular, a natural identification of $\text{GL}_n(\mathbb{R})$ with $\text{GL}(\mathbb{R}^n)$.

Example 2.1.11 (The n-dimensional Sphere). Consider the Euclidean $(n+1)$-space $\mathbb{R}^{n+1}$ $(n \in \mathbb{N}^*)$, and let $\|\ \|$ be the Euclidean norm on $\mathbb{R}^{n+1}$ (see (1.4.3) and (1.4.5)). The *unit sphere* in $\mathbb{R}^{n+1}$ is

$$\mathbb{S}^n := \{p \in \mathbb{R}^{n+1} \mid \|p\| = 1\}.$$

Then $\mathbb{S}^n$ is a second countable Hausdorff topological space with the subspace topology.

First we show that $\mathbb{S}^n$ is an n-dimensional topological manifold. Let us identify $\mathbb{R}^{n+1}$ with $\mathbb{R}^n \times \mathbb{R}$ by the canonical mapping

$$(\alpha^1, \ldots, \alpha^n, \alpha^{n+1}) \in \mathbb{R}^{n+1} \mapsto ((\alpha^1, \ldots, \alpha^n), \alpha^{n+1}) \in \mathbb{R}^n \times \mathbb{R},$$

and $\mathbb{R}^n$ with the subspace $\mathbb{R}^n \times \{0\}$ of $\mathbb{R}^{n+1}$ via the embedding

$$a \in \mathbb{R}^n \mapsto (a, 0) \in \mathbb{R}^{n+1}.$$

Then we can write the elements of $\mathbb{R}^{n+1}$ simply in the form

$$P = (p, t); \quad p \in \mathbb{R}^n,\ t \in \mathbb{R}.$$

We call the unit vector $N := (\mathbf{0}, 1)$ the *north pole* of the sphere, and $S := -N$ the *south pole* $(\mathbf{0} := (0, \ldots, 0) \in \mathbb{R}^n)$. Let

$$\mathcal{U}_N := \mathbb{S}^n \setminus \{N\}, \quad \mathcal{U}_S := \mathbb{S}^n \setminus \{S\}.$$

We define the *stereographic projection*

$$\varphi_N \colon \mathcal{U}_N \to \mathbb{R}^n, \ P \mapsto \varphi_N(P)$$

of $\mathbb{S}^n$ from the north pole by the rule

$$\varphi_N(P) := \overleftrightarrow{NP} \cap (\mathbb{R}^n \times \{0\}) \in \mathbb{R}^n,$$

where $\overleftrightarrow{NP}$ is the straight line containing N and P. Similarly, the stereographic projection of $\mathbb{S}^n$ from the south pole is the mapping

$$\varphi_S \colon \mathcal{U}_S \to \mathbb{R}^n, \ P \mapsto \varphi_S(P) := \overleftrightarrow{SP} \cap (\mathbb{R}^n \times \{0\}).$$

There is a unique scalar $\lambda \in \mathbb{R}$ such that the point $\varphi_N(P)$ of the line $\overleftrightarrow{NP}$ can be written in the form

$$\varphi_N(P) = N + \lambda(N - P) = (\mathbf{0},1) + \lambda\big((\mathbf{0},1) - (p,t)\big) = \big(-\lambda p, 1 + \lambda(1 - t)\big).$$

Since $\varphi_N(P) \in \mathbb{R}^n \times \{0\}$, we have $1 + \lambda(1 - t) = 0$, whence $\lambda = -1/(1 - t)$ and

$$\varphi_N(P) = \frac{1}{1 - t}p, \ P = (p, t).$$

In the same way,

$$\varphi_S(P) = \frac{1}{1 + t}p.$$

Both mappings are continuous and bijective; we obtain by an immediate calculation that their inverses are given by

$$\varphi_N^{-1}(q) = \left(\frac{2}{\|q\|^2 + 1}q, \frac{\|q\|^2 - 1}{\|q\|^2 + 1} \right),$$

$$\varphi_S^{-1}(q) = \left(\frac{2}{\|q\|^2 + 1}q, \frac{1 - \|q\|^2}{\|q\|^2 + 1} \right); \quad q \in \mathbb{R}^n.$$

We see that φ_N^{-1} and φ_S^{-1} are also continuous, therefore $(\mathcal{U}_N, \varphi_N)$ and $(\mathcal{U}_S, \varphi_S)$ are charts for $\mathbb{S}^n$. Since $\mathcal{U}_N = \mathbb{S}^n \setminus \{N\}$ and $\mathcal{U}_S = \mathbb{S}^n \setminus \{-N\}$ cover $\mathbb{S}^n$, we have defined an atlas for $\mathbb{S}^n$. Furthermore,

$$\varphi_N(\mathcal{U}_N \cap \mathcal{U}_S) = \varphi_S(\mathcal{U}_N \cap \mathcal{U}_S) = \mathbb{R}^n \setminus \{\mathbf{0}\},$$

and for each $v \in \mathbb{R}^n \setminus \{\mathbf{0}\}$ we have

$$\varphi_S \circ \varphi_N^{-1}(v) = \frac{1}{\|v\|^2}v = \varphi_N \circ \varphi_S^{-1}(v).$$

Therefore, the charts $(\mathcal{U}_N, \varphi_N)$ and $(\mathcal{U}_S, \varphi_S)$ are smoothly (in fact, analytically) compatible. (Observe that the transition homeomorphism $\varphi_S \circ \varphi_N^{-1}$, and hence its inverse $\varphi_N \circ \varphi_S^{-1}$, is just the *inversion* with respect to $\mathbb{S}^n$.) Thus $\{(\mathcal{U}_N, \varphi_N), (\mathcal{U}_S, \varphi_S)\}$ is a smooth atlas, and so defines a smooth structure on $\mathbb{S}^n$. We call this its *standard smooth structure*.

Example 2.1.12 (Tori). Let d be a positive integer. The d-*torus*

$$\mathbf{T}^d := \underbrace{\mathbb{S}^1 \times \cdots \times \mathbb{S}^1}_{d \text{ times}}$$

is a d-dimensional manifold. Indeed, by the preceding example, $\mathbb{S}^1$ is a 1-dimensional manifold and then Example 2.1.8, with the construction applied successively, implies our claim.

Example 2.1.13 (Submanifolds). Let M be an n-dimensional manifold with $n \geq 1$, and let $d \leq n$ be a positive integer. We say that a subset S of M is a d-*dimensional submanifold* of M if for each point $p \in S$ there exists a chart $(\mathcal{U}, \varphi)$ of M at p which has the following *submanifold property*:

$$\varphi(\mathcal{U} \cap S) = \varphi(\mathcal{U}) \cap (\mathbb{R}^d \times \{0\}), \tag{SM}$$

where $\mathbb{R}^d \times \{0\}$ is considered as a subspace of $\mathbb{R}^n$ via the canonical inclusion

$$\mathbb{R}^d \times \{0\} \to \mathbb{R}^d \times \mathbb{R}^{n-d} = \mathbb{R}^n.$$

The integer $n - d$ is called the *codimension* of S (in M). Submanifolds of codimension 1 are said to be *hypersurfaces* in M.

If S is a d-dimensional submanifold of M, then S is a second countable Hausdorff topological space with the subspace topology, because these properties of M are inherited by a subspace. Condition (SM) implies immediately that S is locally homeomorphic to $\mathbb{R}^d$. If $(\mathcal{U}, \varphi)$ and $(\mathcal{V}, \psi)$ are two charts satisfying (SM) and $\mathcal{U} \cap \mathcal{V} \cap S \neq \emptyset$, then

$$\psi \circ \varphi^{-1} \restriction (\mathcal{U} \cap \mathcal{V} \cap S) = (\psi \restriction \mathcal{V} \cap S) \circ (\varphi \restriction \mathcal{U} \cap S)^{-1} \restriction \varphi(\mathcal{U} \cap \mathcal{V} \cap S),$$

therefore $\psi \circ \varphi^{-1} \restriction (\mathcal{U} \cap \mathcal{V} \cap S)$ is a smooth mapping. Thus

$$\{(\mathcal{U} \cap S, \varphi \restriction \mathcal{U} \cap S) \mid (\mathcal{U}, \varphi) \text{ is a chart of } M \text{ with property (SM)}\}$$

is a smooth atlas for S. In this way, *any submanifold of a manifold has a natural smooth structure*. Notice that the open submanifolds (Example 2.1.7) are just the submanifolds of codimension zero.

2.1.3 *Mappings of Class C^r*

Definition 2.1.14. Let M and N be manifolds of dimension m and n, respectively, and let $r \in \mathbb{N}^* \cup \{\infty\}$.

(a) A continuous mapping $f \colon M \to N$ is said to be *of class C^r* (or r *times continuously differentiable* or a C^r *mapping*) if, for any point $p \in M$, there exist a chart $(\mathcal{U}, x)$ of M at p and a chart $(\mathcal{V}, y)$ of N at $f(p)$ such that the mapping

$$y \circ f \circ x^{-1} \colon x(\mathcal{U} \cap f^{-1}(\mathcal{V})) \subset \mathbb{R}^m \to y(\mathcal{V}) \subset \mathbb{R}^n$$

is of class C^r. Mappings of class C^∞ are called *smooth mappings*.

(b) Let S be a proper *open subset* of M. A continuous mapping $f \colon S \to N$ is said to be of class C^r if it is a C^r mapping of S, as an *open submanifold* of M, into N.

(c) Suppose S is an *arbitrary subset* of M, endowed with the subspace topology. A continuous mapping $f \colon S \to N$ is called of class C^r if for every point $s \in S$ there is an open subset $\mathcal{U}$ of M containing s and a C^r mapping $\tilde{f} \colon \mathcal{U} \to N$ such that $\tilde{f} \upharpoonright (S \cap \mathcal{U}) = f \upharpoonright (S \cap \mathcal{U})$.

Remark 2.1.15. Notation and hypotheses as in the definition above.

(a) The mapping $y \circ f \circ x^{-1}$ is called the *local representative* (or *coordinate expression*) of f with respect to the charts $(\mathcal{U}, x)$ and $(\mathcal{V}, y)$. The following statements are equivalent:

(i) $y \circ f \circ x^{-1}$ is defined at $q \in x(\mathcal{U})$,
(ii) $f(x^{-1}(q)) \in \mathcal{V}$,
(iii) $x^{-1}(q) \in f^{-1}(\mathcal{V})$,
(iv) $q \in x(\mathcal{U} \cap f^{-1}(\mathcal{V}))$;

thus the domain of definition of $y \circ f \circ x^{-1}$ is indeed the subset $x(\mathcal{U} \cap f^{-1}(\mathcal{V}))$ of $\mathbb{R}^m$. Since f is continuous, and $\mathcal{V}$ is open in N, it follows that $\mathcal{U} \cap f^{-1}(\mathcal{V})$ is open in M, therefore $x(\mathcal{U} \cap f^{-1}(\mathcal{V}))$ *is open in* $\mathbb{R}^m$ (because x is a homeomorphism). Hence we can speak about the differentiability of $y \circ f \circ x^{-1}$.

(b) *If f is of class C^r, then all of its representatives are of class C^r.*

To see this, let $(\tilde{\mathcal{U}}, \tilde{x})$ be a chart of M and $(\tilde{\mathcal{V}}, \tilde{y})$ a chart of N such that $f(\tilde{\mathcal{U}}) \cap \tilde{\mathcal{V}} \neq \emptyset$. It is sufficient to show that $\tilde{y} \circ f \circ \tilde{x}^{-1}$ is of class C^r in a neighbourhood of every point $\tilde{x}(p)$, where $p \in \tilde{\mathcal{U}} \cap f^{-1}(\tilde{\mathcal{V}})$. Let the charts $(\mathcal{U}, x)$ and $(\mathcal{V}, y)$ be as in Definition 2.1.14(a). By the continuity of f, we may suppose that $f(\mathcal{U}) \subset \mathcal{V}$ and $f(\tilde{\mathcal{U}}) \subset \tilde{\mathcal{V}}$; then $f(\mathcal{U} \cap \tilde{\mathcal{U}}) \subset \mathcal{V} \cap \tilde{\mathcal{V}}$. Since $\mathcal{U} \cap \tilde{\mathcal{U}}$ and $\mathcal{V} \cap \tilde{\mathcal{V}}$ are open neighbourhoods of p and $f(p)$, respectively, it follows that

$\tilde{x}(\mathcal{U} \cap \tilde{\mathcal{U}})$ is an open neighbourhood of $\tilde{x}(p)$,
$\tilde{y}(\mathcal{V} \cap \tilde{\mathcal{V}})$ is an open neighbourhood of $\tilde{y}(f(p))$.

On the neighbourhood $\tilde{x}(\mathcal{U} \cap \tilde{\mathcal{U}})$ we have

$$\tilde{y} \circ f \circ \tilde{x}^{-1} = (\tilde{y} \circ y^{-1}) \circ (y \circ f \circ x^{-1}) \circ (x \circ \tilde{x}^{-1}).$$

On the right-hand side of this relation

$x \circ \widetilde{x}^{-1}$ is smooth by the compatibility of $(\mathcal{U}, x)$ and $(\widetilde{\mathcal{U}}, \widetilde{x})$,

$y \circ f \circ x^{-1}$ is of class C^r by condition,

$\widetilde{y} \circ y^{-1}$ is smooth by the compatibility of $(\mathcal{V}, y)$ and $(\widetilde{\mathcal{V}}, \widetilde{y})$,

therefore $\widetilde{y} \circ f \circ \widetilde{x}^{-1}$ is of class C^r on $\widetilde{x}(\mathcal{U} \cap \widetilde{\mathcal{U}})$.

(c) We write $C^r(\mathcal{U}, N)$ for the set of all C^r mappings from an open subset $\mathcal{U}$ of M to N. In particular, $C^r(\mathcal{U}, \mathbb{R})$ (where $\mathbb{R}$ is endowed with the canonical smooth structure described in Example 2.1.9) is abbreviated to $C^r(\mathcal{U})$. Thus $C^r(M)$ stands for the set of C^r functions on M. Notice that, as a special case of our definition above,

> $f \in C^r(M)$ *if, and only if,* f *is continuous and for each point* $p \in M$ *there is a chart* $(\mathcal{U}, u)$ *at* p *such that* $f \circ u^{-1} \colon u(\mathcal{U}) \to \mathbb{R}$ *is a* C^r *function.*

If $f, g \in C^r(\mathcal{U})$, $\lambda \in \mathbb{R}$, and for any $p \in \mathcal{U}$,

$$(f + g)(p) := f(p) + g(p), \quad (\lambda f)(p) := \lambda f(p), \quad (fg)(p) := f(p)g(p),$$

then these 'pointwise operations' turn $C^\infty(\mathcal{U})$ into a commutative ring and also an algebra over $\mathbb{R}$.

Proposition 2.1.16. *Let* M *and* N *be manifolds,* $f \colon M \to N$ *a mapping and* $r \in \mathbb{N}^* \cup \{\infty\}$.

(i) *If* f *is of class* C^r *and* $\mathcal{U}$ *is an open subset of* M, *then* $f \restriction \mathcal{U} \colon \mathcal{U} \to N$ *is also a* C^r *mapping.*

(ii) *If the restriction of* f *to each element of an open covering of* M *is of class* C^r, *then* $f \in C^r(M, N)$. $\qquad\qquad \triangle$

Example 2.1.17. (a) If M is an n-dimensional manifold and $(\mathcal{U}, u)$ is a chart of M, then $u \colon \mathcal{U} \to \mathbb{R}^n$ is a smooth mapping, and the coordinate functions $u^i := e^i \circ u \colon \mathcal{U} \to \mathbb{R}$ (see Definition 2.1.1(b)) are smooth functions.

(b) *Compositions of* C^r *mappings are* C^r *mappings.* To see this, let M, N and S be manifolds, $f \in C^r(M, N)$, $g \in C^r(N, S)$. Given a point $p \in M$, let $q := f(p)$. Since $g \colon N \to S$ is a C^r mapping, there is a chart $(\mathcal{V}, y)$ of N at q and a chart $(\mathcal{W}, z)$ of S at $g(q)$, such that $g(\mathcal{V}) \subset \mathcal{W}$ and the mapping

$$z \circ g \circ y^{-1} \colon y(\mathcal{V}) \to z(\mathcal{W})$$

is of class C^r. Now, for any chart $(\mathcal{U}, x)$ of M at p, we have

$$g \circ f\big(\mathcal{U} \cap f^{-1}(\mathcal{V})\big) = g\big(f(\mathcal{U}) \cap \mathcal{V}\big) \subset g(\mathcal{V}) \subset \mathcal{W},$$

and, over $x\big(\mathcal{U} \cap f^{-1}(\mathcal{V})\big)$,

$$z \circ (g \circ f) \circ x^{-1} = (z \circ g \circ y^{-1}) \circ (y \circ f \circ x^{-1}).$$

Here the right-hand side is of class C^r by Proposition C.1.18, so with the charts $\left(\mathcal{U} \cap f^{-1}(\mathcal{V}), x \upharpoonright \mathcal{U} \cap f^{-1}(\mathcal{V})\right)$ and $(\mathcal{W}, z)$, the mapping $g \circ f \colon M \to S$ satisfies the condition of C^r differentiability.

(c) Let M and N be manifolds of dimension m and n, respectively. The canonical projections

$$\mathrm{pr}_1 \colon M \times N \to M \text{ and } \mathrm{pr}_2 \colon M \times N \to N$$

are smooth. We check this for the first projection pr_1. Let (p, q) be a point of $M \times N$, and take a chart $(\mathcal{U} \times \mathcal{V}, \varphi \times \psi)$ at (p, q) as in Example 2.1.8. Then $\mathrm{pr}_1(\mathcal{U} \times \mathcal{V}) = \mathcal{U}$, and for each point

$$(a, b) \in (\varphi \times \psi)(\mathcal{U} \times \mathcal{V}) \subset \mathbb{R}^{m+n} = \mathbb{R}^m \times \mathbb{R}^n$$

we have

$$\varphi \circ \mathrm{pr}_1 \circ (\varphi \times \psi)^{-1}(a, b) = \varphi \circ \mathrm{pr}_1 \left(\varphi^{-1}(a), \psi^{-1}(b)\right) = a.$$

Thus on a suitable open subset of $\mathbb{R}^m \times \mathbb{R}^n$ the mapping $\varphi \circ \mathrm{pr}_1 \circ (\varphi \times \psi)^{-1}$ acts as the canonical projection $\mathbb{R}^m \times \mathbb{R}^n \to \mathbb{R}^m$, which is a smooth mapping.

(d) The set theoretic constructions mentioned at the end of A.2.2 yield differentiable mappings if we start from differentiable mappings. Namely, let M, M_1, M_2, N_1, N_2 be manifolds. If

$$f_i \in C^r(M_i, N_i), \ g_i \in C^r(M, N_i) \quad (i \in \{1, 2\}),$$

then the mappings

$$f_1 \times f_2 \colon M_1 \times M_2 \to N_1 \times N_2, \ (p, q) \mapsto (f_1(p), f_2(q))$$

and

$$(g_1, g_2) \colon M \to N_1 \times N_2, \ p \mapsto (g_1(p), g_2(p))$$

are of class C^r.

Definition 2.1.18. Let $(\mathcal{U}, u) = (\mathcal{U}, (u^i)_{i=1}^n)$ be a chart of a manifold M with $p \in \mathcal{U}$. Suppose that f is a real-valued function of class C^r defined in an open neighbourhood of p. Then the *partial derivative* of f with respect to the ith coordinate function at p is

$$\frac{\partial f}{\partial u^i}(p) := \mathrm{D}_i(f \circ u^{-1})(u(p)),$$

where D_i denotes the operator of the standard ith partial derivative on $\mathbb{R}^n$ (see Definition C.1.13).

Lemma 2.1.19. *Let M be an n-dimensional manifold and $(\mathcal{U}, (u^i)_{i=1}^n)$ a chart of M at a point p. Write $C_p^r(M)$ for the set of all real-valued functions of class C^r defined in some open neighbourhood of p.*

(i) *The function $f \in C_p^r(M) \mapsto \frac{\partial f}{\partial u^i}(p) \in \mathbb{R}$ is $\mathbb{R}$-linear:*

$$\frac{\partial(\alpha f + \beta g)}{\partial u^i}(p) = \alpha \frac{\partial f}{\partial u^i}(p) + \beta \frac{\partial g}{\partial u^i}(p);$$

and satisfies the Leibniz rule:

$$\frac{\partial(fg)}{\partial u^i}(p) = f(p) \frac{\partial g}{\partial u^i}(p) + g(p) \frac{\partial f}{\partial u^i}(p)$$

($f, g \in C_p^r(M)$; $\alpha, \beta \in \mathbb{R}$).

(ii) *For every $(i,j) \in J_n \times J_n$,*

$$\frac{\partial u^i}{\partial u^j}(p) = \delta_j^i \ \text{(Kronecker delta)}.$$

(iii) *If $(\widetilde{\mathcal{U}}, (\widetilde{u}^i)_{i=1}^n)$ is another chart of M at p and $f \in C_p^r(M)$, then we have*

$$\frac{\partial f}{\partial \widetilde{u}^j}(p) = \sum_{i=1}^n \frac{\partial u^i}{\partial \widetilde{u}^j}(p) \frac{\partial f}{\partial u^i}(p), \quad j \in J_n.$$

(iv) *Let N be a further manifold, $\varphi \in C^r(M, N)$, and let $h \in C_{\varphi(p)}^r(N)$. Choose a chart $(\mathcal{V}, (x^j)_{j=1}^k)$ at $\varphi(p)$ for N. Then*

$$\frac{\partial(h \circ \varphi)}{\partial u^i}(p) = \sum_{j=1}^n \frac{\partial h}{\partial x^j}(\varphi(p)) \frac{\partial(x^j \circ \varphi)}{\partial u^i}(p), \quad i \in J_n \qquad (2.1.1)$$

(chain rule for partial derivatives). $\triangle$

Remark 2.1.20. If $(\mathcal{U}, (u^i)_{i=1}^n)$ is a chart of the manifold M and $f \in C^r(\mathcal{U})$, then the functions

$$\frac{\partial f}{\partial u^i} : \mathcal{U} \to \mathbb{R}, \ q \mapsto \frac{\partial f}{\partial u^i}(q), \quad i \in J_n$$

are of class C^{r-1}.

Definition 2.1.21. Let M and N be manifolds.

(a) A homeomorphism $f : M \to N$ is said to be a C^r *diffeomorphism* if both it and its inverse are of class C^r. A 'smooth diffeomorphism' is shortened to a *diffeomorphism*. Two manifolds are *diffeomorphic* if there exists a diffeomorphism between them.

(b) A C^r mapping $f : M \to N$ is called a *local diffeomorphism at a point* $p \in M$ if there exists an open neighbourhood $\mathcal{U}$ of p in M such that $f \restriction \mathcal{U}$ is a homeomorphism of $\mathcal{U}$ onto an open neighbourhood of $f(p)$ and the inverse homeomorphism is of class C^r. If this property holds at every point of M, we say that f is a *local diffeomorphism*.

Remark 2.1.22. (a) It is easy to show that a C^r mapping between manifolds is a C^r diffeomorphism if, and only if, it is a bijective local diffeomorphism.

(b) We denote by $\mathrm{Diff}^r(M, N)$ the set of C^r diffeomorphisms from M to N, and abbreviate $\mathrm{Diff}^r(M, M)$ to $\mathrm{Diff}^r(M)$. By Example 2.1.17(b), $\mathrm{Diff}^r(M)$ is a group with respect to composition. In the case $r = \infty$, we simply write $\mathrm{Diff}(M, N)$ and $\mathrm{Diff}(M)$. By the *diffeomorphism group of M* we mean the group $\mathrm{Diff}(M)$.

Theorem 2.1.23. *If M is a connected manifold, then the group* $\mathrm{Diff}(M)$ *acts transitively on M.*

For a proof the reader is referred to [13], pp. 65–66.

Definition 2.1.24. Let $r \in \mathbb{N}^* \cup \{\infty\}$.

(a) A *curve* of class C^r in a manifold M is a C^r mapping $\alpha\colon I \to M$, where $I \subset \mathbb{R}$ is an open (but nonempty) interval considered as an open submanifold of $\mathbb{R}$. (We allow I to be unbounded.)

(b) If $[a, b]$ is a closed interval in $\mathbb{R}$ such that $a < b$, then a *curve segment* $\alpha\colon [a, b] \to M$ of class C^r is a mapping that has a C^r extension to an open interval containing $[a, b]$ (cf. 2.1.14(c)).

(c) A mapping $\alpha\colon [a, b] \to M$ is a *piecewise C^r differentiable curve segment* provided that there is a partition

$$a = t_0 < t_1 < \cdots < t_k = b$$

of $[a, b]$ such that the restrictions

$$\alpha \restriction [t_{i-1}, t_i], \quad i \in J_k$$

are curve segments of class C^r.

(d) If $I \subset \mathbb{R}$ is a nonempty open interval, a mapping $\alpha\colon I \to M$ is said to be a *piecewise C^r differentiable curve* provided that for all $a < b$ in I the restriction $\alpha \restriction [a, b]$ is a piecewise C^r curve segment.

Remark 2.1.25. Suppose that I and $\widetilde{I}$ are nonempty open intervals in $\mathbb{R}$. Let $\alpha\colon I \to M$ be a curve of class C^r, and let $\theta\colon \widetilde{I} \to I$ be a C^r function. Then the composite mapping

$$\widetilde{\alpha} := \alpha \circ \theta\colon \widetilde{I} \to I \to M$$

is also a curve of class C^r called the *reparametrization* of α by θ. The reparametrization is *positive* (resp. *negative*) if $\theta'(t) > 0$ (resp. $\theta'(t) < 0$) for all $t \in \widetilde{I}$. It is called *affine* if $\theta'' = 0$. In this case there exist fixed scalars $a, b \in \mathbb{R}$ such that $\theta(t) = at + b$ for all $t \in \widetilde{I}$.

Proposition 2.1.26. *A manifold M is connected if, and only if, it is smoothly path connected, i.e., for any two points p, q in M there exists a smooth curve segment $\alpha \colon [a, b] \to M$ such that $\alpha(a) = p$, $\alpha(b) = q$.*

For a proof, see [52], p. 35.

Definition 2.1.27. A manifold M is called *contractible* if there is a point $a \in M$ and a smooth mapping $h \colon \mathbb{R} \times M \to M$ such that

$$h(0, p) = p \quad \text{and} \quad h(1, p) = a \quad \text{for all } p \in M.$$

2.1.4 Smooth Partitions of Unity

Definition 2.1.28. Let f be a real-valued or vector-valued function on a topological space S. The *support* of f, denoted by $\operatorname{supp}(f)$, is the closure of the subset of S on which f takes nonzero values:

$$\operatorname{supp}(f) := \operatorname{cl}\{p \in S \mid f(p) \neq 0\}.$$

We say that f has *compact support* if $\operatorname{supp}(f)$ is compact in S. If $\operatorname{supp}(f)$ is contained in some set $\mathcal{U}$, f is called *supported in $\mathcal{U}$*.

Definition 2.1.29. A *smooth partition of unity* on a manifold M is a family $(f_i)_{i \in I}$ of smooth functions on M satisfying the following conditions:

(i) For all $p \in M$ and $i \in I$ we have $f_i(p) \geq 0$.
(ii) The family $(\operatorname{supp}(f_i))_{i \in I}$ of supports is *locally finite*, that is, for each point $p \in M$ there exists a neighbourhood $\mathcal{U}$ of p such that $\mathcal{U} \cap \operatorname{supp}(f_i) = \emptyset$ except for finitely many indices $i \in I$.
(iii) For each point $p \in M$ we have $\sum_{i \in I} f_i(p) = 1$. (By (ii), this is a finite sum.)

A partition of unity $(f_i)_{i \in I}$ on M is said to be *subordinate to* an open covering $(\mathcal{U}_i)_{i \in I}$ of M if $\operatorname{supp}(f_i) \subset \mathcal{U}_i$ for all $i \in I$.

Theorem 2.1.30. *Given an open covering of a manifold M, there exists a smooth partition of unity on M subordinate to this covering.*

A carefully elaborated proof of this fundamental result (for partitions of unities of class C^r) may be found in [13], pp. 107–109.

Definition 2.1.31. Let M be a manifold, K a closed subset of M and $\mathcal{U}$ an open neighbourhood of K. A continuous function $f \colon M \to \mathbb{R}$ is called a *bump function for K supported in $\mathcal{U}$* if

(i) $0 \leq f(p) \leq 1$ for $p \in M$;

(ii) $f(q) = 1$ for $q \in K$;

(iii) $\mathrm{supp}(f) \subset \mathcal{U}$.

If K contains an open neighbourhood of a point $p_0 \in \mathcal{U}$, we say that f is a *bump function at p_0 supported in $\mathcal{U}$*.

Corollary 2.1.32 (Existence of Smooth Bump Functions). *Let M be a manifold. For any closed subset K of M and any open neighbourhood $\mathcal{U}$ of K there exists a smooth bump function for K supported in $\mathcal{U}$.*

Proof. In view of Theorem 2.1.30, there exists a smooth partition of unity (f, g) on M subordinate to the open covering $(\mathcal{U}, M \setminus K)$ of M. Then

$$\begin{cases} \mathrm{supp}(f) \subset \mathcal{U}, \ \mathrm{supp}(g) \subset M \setminus K; \\ 0 \leq f(p), g(p); \ f(p) + g(p) = 1 \text{ for } p \in M, \end{cases}$$

therefore $g(p) = 0$, and hence $f(p) = 1$, if $p \in K$. Thus the function f has the required properties. □

Corollary 2.1.33 (Principle of Extension). *Let K be a closed subset of a manifold M, and let $f \colon K \to \mathbb{R}^d$ be a smooth mapping. For any open neighbourhood $\mathcal{U}$ of K, there exists a smooth mapping $\widetilde{f} \colon M \to \mathbb{R}^d$ such that $\widetilde{f} \upharpoonright K = f$ and $\mathrm{supp}(\widetilde{f}) \subset \mathcal{U}$.*

For a proof, see [69], p. 56.

2.2 Fibre Bundles

2.2.1 *Fibre Bundles, Bundle Maps, Sections*

Definition 2.2.1. Let F, M and E be manifolds, and let $\pi \colon E \to M$ be a smooth mapping. The triple (E, π, M) is said to be a *fibre bundle* with *typical fibre* F if the following condition of *local triviality* is satisfied:

(LT) For each point $p \in M$ there exists an open neighbourhood $\mathcal{U}$ of p in M and a diffeomorphism $\psi \colon \mathcal{U} \times F \to \pi^{-1}(\mathcal{U})$ such that

$$\pi \circ \psi = \mathrm{pr}_1 \colon \mathcal{U} \times F \to \mathcal{U}, \ (q, v) \mapsto q,$$

that is, the diagram

$$\mathcal{U} \times F \xrightarrow{\ \psi\ } \pi^{-1}(\mathcal{U})$$
$$\mathrm{pr}_1 \searrow \ \ \swarrow \pi$$
$$\mathcal{U}$$

is commutative.

Then E, M and π are called the *total manifold*, the *base manifold* and the *bundle projection* (or simply the *projection*) of the fibre bundle, respectively. For each point $p \in M$, the set $E_p := \pi^{-1}(p) \subset E$ is the *fibre over p* (or *through p*). We say that the diffeomorphism ψ in (LT) is a *trivializing map* for π (or for E). It is also mentioned as a *trivialization* of $E_{\mathcal{U}} := \pi^{-1}(\mathcal{U})$. A family $(\mathcal{U}_i, \psi_i)_{i \in I}$ is called a *trivializing covering* for π (or for E by abuse of language) if $(\mathcal{U}_i)_{i \in I}$ is an open covering of M, and the mappings

$$\psi_i \colon \mathcal{U}_i \times F \to \pi^{-1}(\mathcal{U}_i), \quad i \in I$$

are trivializing maps for π.

Remark 2.2.2. Depending on what we wish to emphasize, we sometimes omit some ingredients from the notation of a fibre bundle and we write '$\pi \colon E \to M$ is a fibre bundle', 'π is an F-bundle over M', 'π is a fibre bundle'. By an abuse of language, we also say that 'E is an F-bundle', or 'E is a bundle'.

Definition 2.2.3. Let $\lambda_i = (E_i, \pi_i, M_i)$ be F_i-bundles over M_i, $i \in \{1, 2\}$.
 (a) A smooth mapping $\varphi \colon E_1 \to E_2$ is said to be a *bundle map* if it is *fibre preserving*, that is,

$$\pi_1(z_1) = \pi_1(z_2) \text{ implies } \pi_2(\varphi(z_1)) = \pi_2(\varphi(z_2)) \quad (z_1, z_2 \in E_1).$$

 (b) A bundle map is called a *bundle isomorphism* if it is a diffeomorphism. The bundles λ_1 and λ_2 are *isomorphic* if there exists a bundle isomorphism $\varphi \colon E_1 \to E_2$. In the case $\lambda_1 = \lambda_2$ a bundle isomorphism is called a *bundle automorphism*.

Lemma 2.2.4. *Let (E_1, π_1, M_1) and (E_2, π_2, M_2) be fibre bundles as above. A smooth mapping $\varphi \colon E_1 \to E_2$ is a bundle map if, and only if, there exists a smooth mapping $\varphi_B \colon M_1 \to M_2$ between the base manifolds such that*

$$\pi_2 \circ \varphi = \varphi_B \circ \pi_1, \tag{2.2.1}$$

that is, the diagram

$$E_1 \xrightarrow{\varphi} E_2$$
$$\pi_1 \downarrow \quad \varphi_B \quad \downarrow \pi_2$$
$$M_1 \xrightarrow{\varphi_B} M_2$$

commutes.

Proof. The sufficiency of the condition is obvious. To show the necessity, suppose that $\varphi \colon E_1 \to E_2$ is a fibre preserving smooth mapping. Define a mapping $\varphi_B \colon M_1 \to M_2$ by

$$\varphi_B(p) := \pi_2(\varphi(z)) \text{ if } z \in (E_1)_p.$$

Since φ is fibre preserving, $\varphi_B(p)$ does not depend on the choice of the element $z \in (E_1)_p$. So φ_B is well-defined, and it satisfies requirement (2.2.1) automatically. It remains only to check the smoothness of φ_B. Let $p \in M$, and choose a trivializing map $\psi \colon \mathcal{U} \times F_1 \to (\pi_1)^{-1}(\mathcal{U})$ for π_1 with $p \in \mathcal{U}$. Then, for any fixed $v \in F_1$,

$$\varphi_B(p) = \varphi_B(\pi_1(\psi(p,v))) \stackrel{(2.2.1)}{=} (\pi_2 \circ \varphi \circ \psi)(p,v),$$

which implies the smoothness of φ_B at the arbitrarily chosen point p in M. □

Definition 2.2.5. With the notation already introduced, the smooth mapping $\varphi_B \colon M_1 \to M_2$ satisfying (2.2.1) is called the mapping *induced* by the bundle map φ. Then φ is also mentioned as a bundle map *along* (or *over*) φ_B. If the bundles have the same base manifold M and $\varphi_B = 1_M$, then we say that φ is a *strong bundle map* or a *bundle map over* M. Two fibre bundles E_1, E_2 over the same base manifold M are said to be *equivalent* if there exists a strong bundle isomorphism $\varphi \colon E_1 \to E_2$, that is, a bundle map over M which is also a diffeomorphism.

Remark 2.2.6. (a) If $\varphi \colon E_1 \to E_2$ is a bundle isomorphism, then the induced mapping $\varphi_B \colon M_1 \to M_2$ is also a diffeomorphism.

(b) Let $\pi \colon E \to M$ be a fibre bundle, and denote the set of all bundle automorphisms $\varphi \colon E \to E$ by $\mathrm{Aut}(E)$. Under the composition of mappings, $\mathrm{Aut}(E)$ is a group, and the map

$$\gamma \colon \mathrm{Aut}(E) \to \mathrm{Diff}(M), \quad \varphi \mapsto \gamma(\varphi) := \varphi_B$$

is a group homomorphism. Its kernel consists of the self-equivalences of the bundle, thus, by (2.2.1),

$$\mathrm{Ker}(\gamma) = \{\varphi \in \mathrm{Aut}(E) \mid \pi \circ \varphi = \pi\}.$$

The notation $\mathrm{Gau}(E) := \mathrm{Ker}(\gamma)$ is frequently used, and $\mathrm{Gau}(E)$ is called the group of *gauge transformations* of the bundle. Obviously, the sequence

$$\{1_E\} \to \mathrm{Gau}(E) \to \mathrm{Aut}(E) \xrightarrow{\gamma} \mathrm{Diff}(M)$$

is an exact sequence of group homomorphisms.

Definition 2.2.7. Let $\pi\colon E \to M$ be a fibre bundle.

(a) If A is a (nonempty) subset of M, a mapping $s\colon A \to E$ is said to be a *section* of $\pi\colon E \to M$, briefly of π (or, by an abuse of language, of E) *over* A if $\pi \circ s = 1_A$, i.e., $s(a) =: s_a \in E_a$ for all $a \in A$. If $A = M$, s is called a *global section*, or simply a *section of* π (or of E).

(b) The set of all *smooth* sections of π over A is denoted by $\Gamma_A(\pi)$ or by $\Gamma_A(E)$, i.e.,

$$\Gamma_A(\pi) = \Gamma_A(E) := \{s \in C^\infty(A, E) \mid \pi \circ s = 1_A\}.$$

In particular, $\Gamma(\pi)$ (or $\Gamma(E)$) stands for the set of all (global, smooth) sections of π.

(c) If $\mathcal{U}$ is an open submanifold of M, elements of $\Gamma_{\mathcal{U}}(\pi)$ are also mentioned as *local sections* of π. If $p \in M$, then the set of all local sections of π whose domains contain p is denoted by $\Gamma_p(\pi)$ (or $\Gamma(p)$).

Example 2.2.8 (Trivial Bundles). If M and F are two manifolds, the triple $(M \times F, \mathrm{pr}_1, M)$ is an F-bundle over M, called a *trivial F-bundle*. Its fibres $\{p\} \times F$ are *canonically* diffeomorphic to the typical fibre F. Each diffeomorphism $\varphi \in \mathrm{Diff}(M)$ lifts to a bundle automorphism $\widetilde{\varphi} \in \mathrm{Aut}(M \times F)$ given by

$$\widetilde{\varphi}(p, v) := (\varphi(p), v), \quad (p, v) \in M \times F.$$

The mapping

$$\mathrm{Diff}(M) \to \mathrm{Aut}(M \times F), \quad \varphi \mapsto \widetilde{\varphi}$$

is a group homomorphism. The smooth sections of $M \times F$ are of the form

$$s\colon M \to M \times F, \quad p \mapsto s(p) := (p, \underline{s}(p)),$$

where $\underline{s} \in C^\infty(M, F)$. Thus we have a canonical bijection

$$\Gamma(M \times F) \to C^\infty(M, F), \quad s \mapsto \underline{s}.$$

Remark 2.2.9. More generally, an F-bundle $\lambda = (E, \pi, M)$ is said to be *trivializable* if it is equivalent to a trivial fibre bundle $(M \times F, \mathrm{pr}_1, M)$. Then a strong bundle isomorphism $E \to M \times F$ is also called a *trivialization* of λ.

Example 2.2.10 (Pull-backs). Let $\lambda = (E, \pi, M)$ be an F-bundle, N a manifold, and $f\colon N \to M$ a smooth mapping. Then the *fibre product*

$$N \times_M E = f^*E := \{(q, z) \in N \times E \mid f(q) = \pi(z)\}$$

is a closed submanifold of $N \times E$. If

$$\pi_1 := \mathrm{pr}_1 \restriction N \times_M E,$$

then $f^*\lambda := (f^*E, \pi_1, N)$ is an F-bundle over N, called the *pull-back* of λ (or of E) by f. It is also denoted by $f^*\pi$ or (see Remark 2.2.2) by f^*E. For every point $q \in N$, the fibre $(f^*E)_q$ is equal to $\{q\} \times E_{f(q)}$, so it is canonically diffeomorphic to the fibre $E_{f(q)}$ of λ. If

$$\widetilde{f} := \mathrm{pr}_2 \restriction N \times_M E, \tag{2.2.2}$$

then $\widetilde{f}$ is a bundle map along f, that is, the diagram

$$
\begin{array}{ccc}
f^*E = N \times_M E & \xrightarrow{\ \widetilde{f}\ } & E \\
{\scriptstyle \pi_1}\big\downarrow & & \big\downarrow {\scriptstyle \pi} \\
N & \xrightarrow[\ \ f\ \ ]{} & M
\end{array}
\tag{2.2.3}
$$

is commutative.

Definition 2.2.11. Let $\pi\colon E \to M$ be a fibre bundle and $f\colon N \to M$ a smooth mapping. A *section of π along f* is a smooth mapping $\sigma\colon N \to E$ such that $\pi \circ \sigma = f$, i.e., $\sigma(q) \in E_{f(q)}$ for all $q \in N$. This can be displayed by the diagram

$$
\begin{array}{ccc}
 & E & \\
{\scriptstyle \sigma}\nearrow & \big\downarrow {\scriptstyle \pi} & \\
N \xrightarrow[\ f\ ]{} & M &
\end{array}.
$$

Then the mapping $\sigma \in C^\infty(N, E)$ is also called a *lift of f to E*. The set of sections along f is denoted by $\Gamma_f(\pi)$ or $\Gamma_f(E)$.

Lemma 2.2.12. *Let a fibre bundle $\pi\colon E \to M$ and a smooth mapping $f\colon N \to M$ be given. There is a natural bijection between $\Gamma(f^*E)$ and $\Gamma_f(E)$ given by*

$$s \in \Gamma(f^*E) \mapsto \underline{s} := \widetilde{f} \circ s \in \Gamma_f(E), \tag{2.2.4}$$

where the mapping $\widetilde{f}\colon N \times_M E \to E$ is defined by (2.2.2).

Proof. Since $s \in C^\infty(N, N \times_M E)$ and $\tilde{f} \in C^\infty(N \times_M E, E)$, it follows that $\underline{s} = \tilde{f} \circ s \in C^\infty(N, E)$. We also have the property $\pi \circ \underline{s} = f$, because

$$\pi \circ \underline{s} = \pi \circ \tilde{f} \circ s \overset{(2.2.3)}{=} f \circ \pi_1 \circ s = f.$$

Thus $\underline{s} = \tilde{f} \circ s$ is indeed a section of π along f. It may be checked immediately that the mapping

$$\underline{s} \in \Gamma_f(E) \mapsto (1_N, \underline{s}) \in \Gamma(f^*E) \tag{2.2.5}$$

is the inverse of the mapping given by (2.2.4), whence our claim. $\quad\Box$

Remark 2.2.13. By the preceding lemma, any smooth section of the pullback bundle (f^*E, π_1, N) is of the form

$$s: q \in N \mapsto s(q) = (q, \underline{s}(q)) \in N \times_M E, \tag{2.2.6}$$

where $\underline{s} \in \Gamma_f(E)$, i.e., $\underline{s} \in C^\infty(N, E)$ such that $\pi \circ \underline{s} = f$. We also say that $\underline{s}$ is the *principal part* of the section s. In view of Lemma 2.2.12, a smooth section of f^*E may be identified canonically with its principal part. In particular, for each section $\sigma \in \Gamma(E)$, the mapping

$$\hat{\sigma}: N \to N \times_M E, \quad q \mapsto \hat{\sigma}(q) := (q, \sigma \circ f(q)) \tag{2.2.7}$$

is a smooth section of f^*E with principal part $\sigma \circ f \in \Gamma_f(E)$.

Lemma 2.2.14. *Let (E, π, M) be an F-bundle over M, and let $(\mathcal{U}_i, \psi_i)_{i \in I}$ be a trivializing covering for π. Then*

(i) *for each $i \in I$ and $p \in \mathcal{U}_i$, the mapping*

$$(\psi_i)_p: F \to E_p, \quad v \mapsto (\psi_i)_p(v) := \psi_i(p, v)$$

is a bijection;

(ii) *for each $p \in \mathcal{U}_{ij} := \mathcal{U}_i \cap \mathcal{U}_j$, the mapping*

$$(\psi_j)_p^{-1} \circ (\psi_i)_p: F \to F$$

is a diffeomorphism.

Proof. Assertion (i) is immediate from condition (LT) in Definition 2.2.1. To prove (ii), observe that the mappings $\psi_i \upharpoonright \mathcal{U}_{ij} \times F$ and $\psi_j \upharpoonright \mathcal{U}_{ij} \times F$ are diffeomorphisms from $\mathcal{U}_{ij} \times F$ onto $\pi^{-1}(\mathcal{U}_{ij})$ (also by (LT)), therefore

$$\psi_{ji} := \psi_j^{-1} \circ \psi_i \in \mathrm{Diff}(\mathcal{U}_{ij} \times F).$$

However, for every $(p, v) \in \mathcal{U}_{ij} \times F$,

$$\psi_{ji}(p, v) = (p, (\psi_j)_p^{-1} \circ (\psi_i)_p(v)),$$

which implies that $(\psi_j)_p^{-1} \circ (\psi_i)_p$ is a smooth mapping of F onto itself, and interchanging the role of i and j shows that $(\psi_j)_p^{-1} \circ (\psi_i)_p \in \mathrm{Diff}(F)$. $\quad\Box$

Proposition 2.2.15. *Let* M *and* F *be manifolds,* E *a set, and let* $\pi\colon E \to M$ *be a surjective mapping. Suppose that there is an open covering* $(\mathcal{U}_i)_{i \in I}$ *of* M *and a family* $(\psi_i)_{i \in I}$ *of bijective mappings*

$$\psi_i\colon \mathcal{U}_i \times F \to \pi^{-1}(\mathcal{U}_i)$$

satisfying the following conditions:

(i) $\pi \circ \psi_i(q, v) = q$ *for all* $i \in I$ *and* $(q, v) \in \mathcal{U}_i \times F$*;*
(ii) $\psi_{ji} := \psi_j^{-1} \circ \psi_i \in \mathrm{Diff}(\mathcal{U}_{ij} \times F)$ *for every* $(i, j) \in I \times I$ *such that* $\mathcal{U}_{ij} := \mathcal{U}_i \cap \mathcal{U}_j \neq \emptyset$.

Then there exists a unique topology and smooth structure on E *for which* (E, π, M) *is an* F*-bundle over* M *with trivializing covering* $(\mathcal{U}_i, \psi_i)_{i \in I}$.

For a proof, see [52], p. 39.

2.2.2 Vector Bundles

Definition 2.2.16. Let V be a finite-dimensional real vector space. A fibre bundle (E, π, M) with typical fibre V is said to be a (real) *vector bundle* if the following conditions are satisfied:

(i) for each $p \in M$, the fibre $E_p := \pi^{-1}(p)$ is a real vector space;
(ii) there is a trivializing covering $(\mathcal{U}_i, \psi_i)_{i \in I}$ for π such that the mappings

$$(\psi_i)_p\colon V \to E_p, \quad v \mapsto (\psi_i)_p(v) := \psi_i(p, v) \quad (i \in I, p \in \mathcal{U}_i)$$

are linear isomorphisms.

The dimension of V is called the *rank* of the vector bundle. A trivializing covering for π satisfying (ii) is said to be a *trivializing covering* for the *vector* bundle π. In the context of vector bundles this term (and the term 'trivializing map') will always be used in this sense.

Definition 2.2.17. Let (E_1, π_1, M_1) and (E_2, π_2, M_2) be vector bundles. A bundle map $\varphi\colon E_1 \to E_2$ is said to be a *vector bundle map* or *homomorphism of vector bundles* if it is *fibrewise linear*, i.e., for each point $p \in M_1$, the restriction

$$\varphi_p := \varphi \restriction (E_1)_p\colon (E_1)_p \to (E_2)_{\varphi_B(p)}$$

is linear (φ_B is the mapping induced by φ between the base manifolds, see Definition 2.2.5).

Remark 2.2.18. Having the concept of a vector bundle map, we also have the corresponding concepts of vector bundle isomorphism and automorphism, strong vector bundle map, equivalence of vector bundles and trivializability of vector bundles.

For simplicity,

> we agree that in the context of vector bundles by a 'bundle map' we mean a 'vector bundle map'.

Proposition 2.2.19. *Let* (E_1, π_1, M_1) *and* (E_2, π_2, M_2) *be vector bundles. Let* $\varphi \colon E_1 \to E_2$ *be a vector bundle map inducing the smooth mapping* $\varphi_B \colon M_1 \to M_2$ *between the base manifolds. Then* φ *is a vector bundle isomorphism if, and only if, the following conditions are satisfied:*

(i) $\varphi_B \colon M_1 \to M_2$ *is a diffeomorphism,*

(ii) *for every* $p \in M_1$, *the mapping* $\varphi_p \colon (E_1)_p \to (E_2)_{\varphi_B(p)}$ *is a linear isomorphism.*

Proof. If φ is a vector bundle isomorphism (i.e., according to 2.2.3(b), it is a diffeomorphism), then φ_B is also a diffeomorphism (see Remark 2.2.6(a)), and the restrictions $\varphi_p \colon (E_1)_p \to (E_2)_{\varphi_B(p)}$ $(p \in M_1)$ are obviously linear isomorphisms.

Conversely, suppose that φ satisfies (i) and (ii). Then φ is bijective; our only task is to show that the inverse mapping $\varphi^{-1} \colon E_2 \to E_1$ is also smooth. Since the question is local, using trivializations we may reduce the problem to the case

$$M_1 = M_2 =: M; \ E_1 = M \times V_1, \ E_2 = M \times V_2 \ (V_1 \text{ and } V_2 \text{ are}$$
finite-dimensional real vector spaces), $\varphi_B = 1_M$.

Then

$$\Phi \colon M \to L(V_1, V_2), \ p \mapsto \Phi(p) := \varphi_p$$

is a smooth mapping, and φ^{-1} is given by

$$(p, w) \in M \times V_2 \mapsto \varphi^{-1}(p, w) = (p, (\Phi(p))^{-1}(w)),$$

which is a smooth mapping. $\qquad\square$

Proposition 2.2.20 (Vector Bundle Construction). *Let M be a manifold and V a finite-dimensional real vector space. Suppose that*

(i) $(\mathcal{U}_i)_{i \in I}$ *is an open covering of M,*

(ii) $(E_p)_{p \in M}$ *is a family of pairwise disjoint real vector spaces,*

(iii) $(\psi_i)_p \colon V \to E_p$ is a linear isomorphism for each index $i \in I$ and point $p \in \mathcal{U}_i$

satisfying the following condition:

(S) The mappings
$$\gamma_{ij} \colon \mathcal{U}_i \cap \mathcal{U}_j \to \mathrm{GL}(V), \quad p \mapsto \gamma_{ij}(p) := (\psi_i)_p^{-1} \circ (\psi_j)_p$$
are smooth.

Let $E := \bigcup_{p \in M} E_p$ (disjoint union), and let $\pi \colon E \to M$ be the mapping defined by
$$\pi(z) := p \text{ if } z \in E_p.$$
Then there is a unique topology and smooth structure on E which make the triple (E, π, M) into a vector bundle with typical fibre V and trivializing covering $(\mathcal{U}_i, \psi_i)_{i \in I}$, where
$$\psi_i \colon \mathcal{U}_i \times V \to \pi^{-1}(\mathcal{U}_i), \quad (p, v) \mapsto \psi_i(p, v) := (\psi_i)_p(v), \quad i \in I.$$

This can easily be seen by applying the general construction principle formulated in Proposition 2.2.15.

Remark 2.2.21. If $\pi \colon E \to M$ is a vector bundle with typical fibre V, and $(\mathcal{U}_i, \psi_i)_{i \in I}$ is a trivializing covering for π, then the smooth mappings
$$\gamma_{ij} \colon \mathcal{U}_i \cap \mathcal{U}_j \to \mathrm{GL}(V), \quad p \mapsto \gamma_{ij}(p) := (\psi_i)_p^{-1} \circ (\psi_j)_p$$
(cf. condition (S)) are called the *transition maps* associated with the covering $(\mathcal{U}_i, \psi_i)_{i \in I}$. These mappings satisfy the following 'cocycle conditions':

(i) $\gamma_{ii}(p) = 1_V$ for $p \in \mathcal{U}_i$,
(ii) $\gamma_{ij}(p) = (\gamma_{ji}(p))^{-1}$ for $p \in \mathcal{U}_i \cap \mathcal{U}_j$,
(iii) $\gamma_{ij}(p) \circ \gamma_{jk}(p) = \gamma_{ik}(p)$ for $p \in \mathcal{U}_i \cap \mathcal{U}_j \cap \mathcal{U}_k$

$(i, j, k \in I)$. The transition maps, in fact, determine the vector bundle E. More precisely, we have the following version of the construction principle formulated in Proposition 2.2.20:

Proposition 2.2.22. Let a manifold M and a finite-dimensional real vector space V be given. Suppose that $(\mathcal{U}_i)_{i \in I}$ is an open covering of M, and let $(\gamma_{ij})_{(i,j) \in I \times I}$ be a family of smooth mappings $\gamma_{ij} \colon \mathcal{U}_i \cap \mathcal{U}_j \to \mathrm{GL}(V)$ satisfying the cocycle conditions (i) – (iii). Then there exists a V-vector bundle E over M and a trivializing covering $(\mathcal{U}_i, \psi_i)_{i \in I}$ for E such that the transition maps associated with the covering form the given family (γ_{ij}).

$\triangle$

Lemma 2.2.23. *If* (E, π, M) *is a vector bundle, then the set* $\Gamma(\pi) = \Gamma(E)$ *of smooth sections of* π *has a structure of module over the ring* $C^\infty(M)$ *with the pointwise operations*

$$(s_1 + s_2)(p) := s_1(p) + s_2(p), \quad (fs)(p) := f(p)s(p) \quad (p \in M),$$

where $s, s_1, s_2 \in \Gamma(\pi)$, $f \in C^\infty(M)$. *The zero element of this module is the zero section*

$$o \colon p \in M \mapsto o(p) := 0_p := \text{the zero vector of } E_p. \qquad \triangle$$

Lemma 2.2.24. *Let* (E, π, M) *be a vector bundle with typical fibre* V.

(i) *Choose a point* $p \in M$. *For each* $z \in E_p$ *there exists a smooth section* $s \in \Gamma(\pi)$ *such that* $s(p) = z$.

(ii) *Let* $\mathcal{U}$ *be an open subset of* M, $\sigma \in \Gamma_\mathcal{U}(\pi)$, *and* $p \in \mathcal{U}$. *There is a smooth section* $s \in \Gamma(\pi)$ *such that* s *coincides with* σ *in an open neighbourhood of* p.

Proof. (i) Let $\psi \colon \mathcal{U} \times V \to \pi^{-1}(\mathcal{U})$ be a trivializing map for the vector bundle (E, π, M) with $p \in \mathcal{U}$. Choose a bump function $f \in C^\infty(M)$ at p supported in $\mathcal{U}$. There is a unique vector $v \in V$ such that $\psi(p, v) = z$. Define a section s of π by

$$s(q) := \begin{cases} \psi(q, f(q)v) & \text{if } q \in \mathcal{U}, \\ 0 & \text{if } q \in M \setminus \mathcal{U}. \end{cases}$$

Then $s \in \Gamma(\pi)$ and $s(p) = z$.

(ii) The proof of this assertion is similar to the preceding one and is left to the reader. □

Definition 2.2.25. Let (E, π, M) be a vector bundle of rank k.

(a) A basis of a fibre E_p is called a *frame* at the point p.

(b) Let $\mathcal{U}$ be an open subset of M. A family $(s_i)_{i=1}^k$ of local sections $s_i \in \Gamma_\mathcal{U}(\pi)$ is said to be a *(local) frame field* of E over $\mathcal{U}$ if $(s_i(p))_{i=1}^k$ is a basis of E_p for all $p \in \mathcal{U}$. The frame field is called *global* if $\mathcal{U} = M$.

Lemma 2.2.26. *Let* (E, π, M) *be a vector bundle of rank* k *with typical fibre* V. *Choose a fixed basis* $(b_i)_{i=1}^k$ *for* V.

(i) *If* $\psi \colon \mathcal{U} \times V \to \pi^{-1}(\mathcal{U})$ *is a trivializing map for* π, *then the mappings*

$$s_i \colon \mathcal{U} \to E, \quad p \mapsto s_i(p) := \psi_p(b_i) = \psi(p, b_i), \quad i \in J_k$$

form a local frame field of E *over* $\mathcal{U}$.

(ii) *Conversely, let $(s_i)_{i=1}^k$ be a local frame field of E over an open subset $\mathcal{U}$ of M. Then the mapping*

$$\begin{cases} \psi : \mathcal{U} \times V \to \pi^{-1}(\mathcal{U}), \ (p,v) \mapsto \psi(p,v) := \sum_{i=1}^k v^i s_i(p) \\ \text{if } v = \sum_{i=1}^k v^i b_i \end{cases} \qquad (2.2.8)$$

is a trivialization of $E_{\mathcal{U}} := \pi^{-1}(\mathcal{U})$.

The proof is easy and left to the reader.

Example 2.2.27. Let M be a manifold, V a k-dimensional real vector space, and consider the trivial vector bundle $\mathrm{pr}_1 : M \times V \to M$. Then $C^\infty(M, V)$ is a $C^\infty(M)$-module under pointwise addition and multiplication by smooth real-valued functions on M, and the canonical bijective mapping

$$\begin{cases} \Gamma(M \times V) \to C^\infty(M,V), \ s \mapsto \underline{s} \\ s(p) = (p, \underline{s}(p)), \quad p \in M \end{cases}$$

given in Example 2.2.8 becomes an isomorphism of $C^\infty(M)$-modules. We say, as in Remark 2.2.13, that the mapping $\underline{s} \in C^\infty(M, V)$ is the *principal part* of the section $s \in \Gamma(M \times V)$. A section $s \in \Gamma(M \times V)$ is called *constant* if its principal part is a constant mapping, i.e., if there is a vector $v \in V$ such that

$$s(p) = (p, v) \text{ for all } p \in M.$$

In particular, if $(b_i)_{i=1}^k$ is a basis of V, then the constant sections

$$s_i : p \in M \mapsto s_i(p) := (p, b_i) \in M \times V, \quad i \in J_k$$

form a global frame field of $M \times V$. The family $(s_i)_{i=1}^k$ is a basis for $\Gamma(M \times V)$, thus it is a finitely generated free $C^\infty(M)$-module.

Conversely, if a vector bundle admits a global frame field, then it is trivializable by Lemma 2.2.26.

Lemma 2.2.28. *Let $\pi : E \to M$ be a vector bundle of rank k with typical fibre V. Suppose that $(s_i)_{i=1}^k$ is a local frame field of E over an open subset $\mathcal{U} \subset M$. Then for any section σ of π over $\mathcal{U}$ there exists a unique k-tuple $(\sigma^1, \ldots, \sigma^k)$ of functions $\sigma^i : \mathcal{U} \to \mathbb{R}$ such that*

$$\sigma = \sum_{i=1}^k \sigma^i s_i. \qquad (2.2.9)$$

The section σ is smooth if, and only if, the functions σ^i are smooth.

Proof. For any point $p \in \mathcal{U}$ there exist uniquely determined real numbers $\sigma^1(p), \ldots, \sigma^k(p)$ such that $\sigma(p) = \sum_{i=1}^k \sigma^i(p) s_i(p)$, since $(s_i(p))_{i=1}^k$ is a basis of E_p. Thus we obtain a unique k-tuple of functions

$$\sigma^i : \mathcal{U} \to \mathbb{R}, \quad p \mapsto \sigma^i(p); \quad i \in J_k$$

such that $\sigma = \sum_{i=1}^k \sigma^i s_i$. If the functions σ^i are smooth, then the section $\sigma : \mathcal{U} \to E$ is also smooth.

Suppose, conversely, that $\sigma : \mathcal{U} \to E$ is a smooth section of π over $\mathcal{U}$. Let $(b_i)_{i=1}^k$ be a basis of V, and let

$$\psi : \mathcal{U} \times V \to \pi^{-1}(\mathcal{U})$$

be the trivialization of $\pi^{-1}(\mathcal{U})$ given by (2.2.8). Then, at any point p of $\mathcal{U}$,

$$\psi \left(p, \sum_{i=1}^k \sigma^i(p) b_i \right) = \sum_{i=1}^k \sigma^i(p) s_i(p) = \sigma(p).$$

If $\mathrm{pr}_2 : \mathcal{U} \times V \to V$ is the canonical projection and $(b^i)_{i=1}^k$ is the dual of the basis $(b_i)_{i=1}^k$ of V, then we obtain that

$$\sigma^i = b^i \circ \mathrm{pr}_2 \circ \psi^{-1} \circ \sigma, \quad i \in J_k.$$

This proves the smoothness of the functions σ^i. $\qquad\square$

Remark 2.2.29. The functions σ^i in (2.2.9) are called the *components* of σ with respect to the local frame field $(s_i)_{i=1}^k$. With this terminology, Lemma 2.2.28 may be reformulated as follows:

If $(s_i)_{i=1}^k$ is a local frame field of a vector bundle E over an open subset $\mathcal{U}$ of the base manifold, then a section $\sigma : \mathcal{U} \to E$ is smooth if, and only if, its component functions with respect to $(s_i)_{i=1}^k$ are smooth.

Definition 2.2.30. Let π_1 and π_2 be vector bundles over the same base manifold M. A mapping $\Phi : \Gamma(\pi_1) \to \Gamma(\pi_2)$ is said to be *tensorial* if it is $C^\infty(M)$-linear, i.e.,

$$\Phi(s_1 + s_2) = \Phi(s_1) + \Phi(s_2), \quad \Phi(fs) = f\Phi(s)$$

for all $s, s_1, s_2 \in \Gamma(\pi_1)$, $f \in C^\infty(M)$.

Proposition 2.2.31 (Characterization of Tensorial Mappings). *With the notation already introduced, a mapping $\Phi : \Gamma(\pi_1) \to \Gamma(\pi_2)$ is tensorial if, and only if, there exists a strong bundle map $\varphi : E_1 \to E_2$ such that $\Phi(s) = \varphi \circ s$ for all $s \in \Gamma(\pi_1)$.*

Proof. If $\varphi \colon E_1 \to E_2$ is a strong bundle map, then

$$\Phi \colon s \in \Gamma(\pi_1) \mapsto \Phi(s) := \varphi \circ s \in \Gamma(\pi_2)$$

is clearly a tensorial mapping. The difficult part is the proof of the converse statement. We only sketch the main steps of a possible proof. Details in a well-elaborated form can be found in [69], pp. 117–118.

Step 1. *Tensorial mappings are local*: if $\Phi \colon \Gamma(\pi_1) \to \Gamma(\pi_2)$ is tensorial and $s \in \Gamma(\pi_1)$ vanishes in some open subset $\mathcal{U} \subset M$, then $\Phi(s) \restriction \mathcal{U} = 0$. (To see this, apply a bump function at a point $p \in \mathcal{U}$ supported in $\mathcal{U}$.)

Step 2. *Tensorial mappings are in fact pointwise*: if $s \in \Gamma(\pi_1)$ and $s(p) = 0$, then $\Phi(s)(p) = 0$. (Consider a local frame field in some open neighbourhood of p, and extend its members to global smooth sections by Lemma 2.2.24(ii).)

Step 3. Given a tensorial mapping $\Phi \colon \Gamma(\pi_1) \to \Gamma(\pi_2)$, define a mapping $\varphi \colon E_1 \to E_2$ as follows:

$$\varphi(z) := \Phi(s)(p), \text{ if } z \in (E_1)_p, \text{ and } s \in \Gamma(\pi_1) \text{ such that } s(p) = z.$$

Then $\Phi(s)(p) \in (E_2)_p$ is well-defined, the mapping $\varphi \restriction (E_1)_p \colon (E_1)_p \to (E_2)_p$ is linear, and we clearly have $\varphi \circ s(p) = \Phi(s)(p)$. It remains to show that φ is a smooth mapping, for which we refer to John M. Lee's text cited above. $\square$

Remark 2.2.32. (a) With the notation as in Definition 2.2.30, let $\varphi, \psi \colon E_1 \to E_2$ be strong bundle maps and let $f \in C^\infty(M)$. Define the mappings $\varphi + \psi$ and $f\varphi$ pointwise by

$$(\varphi + \psi)(z) := \varphi(z) + \psi(z), \ (f\varphi)(z) := f(\pi_1(z))\varphi(z); \quad z \in E_1.$$

Under these operations the set of strong bundle maps from E_1 to E_2 becomes a $C^\infty(M)$-module, which we denote by $\mathrm{Bun}_M(\pi_1, \pi_2)$. The tensorial mappings from $\Gamma(\pi_1)$ to $\Gamma(\pi_2)$ also form a $C^\infty(M)$-module under the addition and scalar multiplication given by

$$(\Phi_1 + \Phi_2)(s) := \Phi_1(s) + \Phi_2(s), \ (f\Phi)(s) := f\Phi(s); \quad s \in \Gamma(\pi_1).$$

This module is denoted by $\mathrm{Hom}_M(\Gamma(\pi_1), \Gamma(\pi_2))$. Proposition 2.2.31 implies that *the mapping*

$$\begin{cases} \#\colon \mathrm{Bun}_M(\pi_1, \pi_2) \to \mathrm{Hom}_M(\Gamma(\pi_1), \Gamma(\pi_2)), \ \varphi \mapsto \varphi_\#; \\ \varphi_\#(s) := \varphi \circ s \text{ for all } s \in \Gamma(\pi_1) \end{cases}$$

is an isomorphism of $C^\infty(M)$-modules.

In the following we shall frequently identify a tensorial mapping $\Phi \in \mathrm{Hom}_M(\Gamma(\pi_1), \Gamma(\pi_2))$ with the corresponding strong bundle map

$\varphi \in \operatorname{Bun}_M(\pi_1, \pi_2)$ and write $\varphi(s)$ or simply φs rather than $\varphi_\#(s) = \varphi \circ s$ ($s \in \Gamma(\pi_1)$).

(b) Now we reject the condition that the considered vector bundles have a common base manifold. Let, therefore, $\pi_1 \colon E_1 \to M_1$ and $\pi_2 \colon E_2 \to M_2$ be two vector bundles. Suppose that $\varphi \colon E_1 \to E_2$ is a bundle map which induces a diffeomorphism $\varphi_B \colon M_1 \to M_2$ between the base manifolds. Define a mapping $\varphi_\# \colon \Gamma(\pi_1) \to \Gamma(\pi_2)$ by

$$\varphi_\#(s) := \varphi \circ s \circ \varphi_B^{-1}, \quad s \in \Gamma(\pi_1). \tag{2.2.10}$$

The diagram of mappings is

$$
\begin{array}{ccc}
E_1 & \xrightarrow{\ \varphi\ } & E_2 \\
{\scriptstyle s}\uparrow & & \uparrow{\scriptstyle \varphi_\#(s)} \\
M_1 & \xleftarrow[\ \varphi_B^{-1}\ ]{} & M_2
\end{array}
\ .
$$

Then $\varphi_\#(s)$ is smooth, and

$$\pi_2 \circ \varphi_\#(s) = \pi_2 \circ \varphi \circ s \circ \varphi_B^{-1} \overset{(2.2.1)}{=} \varphi_B \circ \pi_1 \circ s \circ \varphi_B^{-1}$$
$$= \varphi_B \circ 1_{M_1} \circ \varphi_B^{-1} = 1_{M_2},$$

so we indeed have $\varphi_\#(s) \in \Gamma(\pi_2)$. For any sections $s, s_1, s_2 \in \Gamma(\pi_1)$ and any function $f \in C^\infty(M_1)$,

$$\varphi_\#(s_1 + s_2) = \varphi \circ (s_1 + s_2) \circ \varphi_B^{-1} = \varphi \circ s_1 \circ \varphi_B^{-1} + \varphi \circ s_2 \circ \varphi_B^{-1}$$
$$= \varphi_\#(s_1) + \varphi_\#(s_2);$$
$$\varphi_\#(fs) = \varphi \circ (fs) \circ \varphi_B^{-1} = \varphi \circ (f \circ \varphi_B^{-1})(s \circ \varphi_B^{-1})$$
$$\overset{(*)}{=} (f \circ \varphi_B^{-1})(\varphi \circ s \circ \varphi_B^{-1}) = (f \circ \varphi_B^{-1})\varphi_\#(s),$$

taking into account in step $(*)$ that φ is fibrewise linear. Since the mapping

$$(\varphi_B^{-1})^* \colon C^\infty(M_1) \to C^\infty(M_2), \ f \mapsto (\varphi_B^{-1})^*(f) := f \circ \varphi_B^{-1}$$

is an isomorphism of rings, it follows that *the mapping*

$$s \in \Gamma(\pi_1) \mapsto \varphi_\#(s) := \varphi \circ s \circ \varphi_B^{-1} \in \Gamma(\pi_2)$$

is a 'homomorphism of modules under the ring isomorphism $(\varphi_B^{-1})^*$'. We say that $\varphi_\#(s)$ is the *push-forward* of the section $s \in \Gamma(\pi_1)$ by the bundle map φ.

2.2.3   Examples and Constructions

Example 2.2.33 (Trivial Vector Bundles). (a) Let M be a manifold and V a k-dimensional real vector space. Consider the trivial V-bundle $(M \times V, \mathrm{pr}_1, M)$ (see Example 2.2.8). Transport the vector space structure of the typical fibre onto the fibres $\{p\} \times V = (\mathrm{pr}_1)^{-1}(p)$ by means of the canonical bijections

$$v \in V \mapsto (p, v) \in \{p\} \times V, \quad p \in M;$$

i.e., by the rules

$$(p, v_1) + (p, v_2) := (p, v_1 + v_2), \quad \lambda(p, v) := (p, \lambda v)$$

$(v, v_1, v_2 \in V; \ \lambda \in \mathbb{R})$. Then $\{(M, 1_{M \times V})\}$ is a trivializing covering for $(M \times V, \mathrm{pr}_1, M)$, which becomes a vector bundle of rank k over M in this way.

(b) Consider the trivial vector bundles $\pi_1 \colon M \times V_1 \to M$ and $\pi_2 \colon M \times V_2 \to M$ over M. For any strong bundle map $\varphi \colon M \times V_1 \to M \times V_2$ there exists a unique smooth mapping $\widetilde{\varphi} \colon M \to L(V_1, V_2)$ such that

$$\varphi(p, v) = (p, \widetilde{\varphi}(p)(v)) \text{ for all } (p, v) \in M \times V_1.$$

Thus we obtain a canonical isomorphism

$$\mathrm{Bun}_M(\pi_1, \pi_2) \to C^\infty(M, L(V_1, V_2)), \quad \varphi \mapsto \widetilde{\varphi}$$

of $C^\infty(M)$-modules.

(c) It may be shown that *over a contractible base manifold* (Definition 2.1.27) *every vector bundle is trivializable.* This is a difficult theorem. For a proof, we refer to [2], Supplement 3.4B or [53], Corollary II, section 7.18. (The latter proof requires more preparation.)

Example 2.2.34 (Restriction). Let $\pi \colon E \to M$ be a vector bundle and $\mathcal{U}$ an open submanifold of M. Then $(\pi^{-1}(\mathcal{U}), \pi \restriction \pi^{-1}(\mathcal{U}), \mathcal{U})$ is also a vector bundle over $\mathcal{U}$ with the same typical fibre as the original bundle. We call this the *restriction of π (or of E) over $\mathcal{U}$.*

Example 2.2.35 (Vector Subbundles). (a) Let $\pi \colon E \to M$ be a vector bundle of rank r. Suppose that $k \in J_r$, and E' is a subset of E satisfying the following conditions:

(i) For every point $p \in M$, $E'_p := E' \cap E_p$ is a vector subspace of E_p.
(ii) For every point $p \in M$, there exists an open neighbourhood $\mathcal{U}$ of p and a local frame field $(s_i)_{i=1}^r$ of E over $\mathcal{U}$ such that for each $q \in \mathcal{U}$, $(s_i(q))_{i=1}^k$ is a basis of E'_q.

Then E' is a closed submanifold of E, and the triple $(E', \pi \restriction E', M)$ is a vector bundle of rank k over M called a *vector subbundle* of (E, π, M).

To prove these assertions, we may replace E' by $E' \cap \pi^{-1}(\mathcal{U})$. Then, applying Lemma 2.2.26(ii) and the definition of a submanifold (Example 2.1.13), a simple argument leads to the desired conclusion.

(b) Let $\pi_1 \colon E_1 \to M$ and $\pi_2 \colon E_2 \to M$ be vector bundles over M. A strong bundle map $\varphi \colon E_1 \to E_2$ is said to have *constant rank k* if the restrictions

$$\varphi_p \colon (E_1)_p \to (E_2)_p, \quad p \in M$$

are of rank k. In this case

$$\mathrm{Ker}(\varphi) := \bigcup_{p \in M} \mathrm{Ker}(\varphi_p) \subset E_1 \text{ and } \mathrm{Im}(\varphi) := \bigcup_{p \in M} \mathrm{Im}(\varphi_p) \subset E_2$$

are subbundles of E_1 and E_2, respectively. More precisely,

$\big(\mathrm{Ker}(\varphi), \pi_1 \restriction \mathrm{Ker}(\varphi), M\big)$ is a vector subbundle of (E_1, π_1, M),

$\big(\mathrm{Im}(\varphi), \pi_2 \restriction \mathrm{Im}(\varphi), M\big)$ is a vector subbundle of (E_2, π_2, M).

These assertions follow from the *rank theorem for vector bundle maps*, see, e.g., [21], pp. 24–25 or [66], pp. 272–273.

Example 2.2.36 (Product Bundle). Let $\pi_i \colon E_i \to M_i$ be vector bundles with typical fibres V_i, $i \in \{1, 2\}$. Then

$$\pi_1 \times \pi_2 \colon E_1 \times E_2 \to M_1 \times M_2$$

is a vector bundle with typical fibre $V_1 \oplus V_2$ and fibres $(E_1)_{p_1} \oplus (E_2)_{p_2}$ at $(p_1, p_2) \in M_1 \times M_2$. (For the definition of the product mapping see Example 2.1.17.) Indeed, if $(\mathcal{U}_i, \psi_i^1)_{i \in I}$ and $(\mathcal{V}_j, \psi_j^2)_{j \in J}$ are trivializing coverings for π_1 and π_2, respectively, then

$$(\mathcal{U}_i \times \mathcal{V}_j, \psi_i^1 \times \psi_j^2)_{(i,j) \in I \times J}$$

is a trivializing covering for $\pi_1 \times \pi_2$. We call the vector bundle $\pi_1 \times \pi_2$ the *(Cartesian) product* of π_1 and π_2.

Example 2.2.37 (Pull-backs). Let $\pi \colon E \to M$ be a vector bundle with typical fibre V and let $f \colon N \to M$ be a smooth mapping. Consider the pull-back bundle

$$\pi_1 \colon f^* E = N \times_M E \to N$$

(Example 2.2.10). Each fibre

$$(f^* E)_q = \{q\} \times E_{f(q)}, \quad q \in N$$

of π_1 has a natural vector space structure which makes the bundle map $\widetilde{f} \colon N \times_M E \to E$ a vector bundle homomorphism, moreover, the restrictions

$$\widetilde{f}_q := \widetilde{f} \upharpoonright (f^*E)_q \colon (f^*E)_q \to E_{f(q)}$$

become linear isomorphisms. Then

$$(q, z_1) + (q, z_2) = (q, z_1 + z_2), \quad \lambda(q, z) = (q, \lambda z)$$

for all $z, z_1, z_2 \in E_{f(q)}$, $\lambda \in \mathbb{R}$. If $(\mathcal{U}_i, \psi_i)_{i \in I}$ is a trivializing covering for π, and we define the mappings

$$f^*\psi_i \colon f^{-1}(\mathcal{U}_i) \times V \to \pi_1^{-1}(f^{-1}(\mathcal{U}_i))$$

by

$$(f^*\psi_i)(q, v) := (q, \psi_i(f(q), v)); \quad q \in f^{-1}(\mathcal{U}_i), \quad v \in V,$$

then $(f^{-1}(\mathcal{U}_i), f^*\psi_i)_{i \in I}$ is a trivializing covering for π_1. Thus *the pull-back of a V-vector bundle (E, π, M) by a smooth mapping $f \colon N \to M$ has a natural vector bundle structure with typical fibre V.* The $C^\infty(N)$-module $\Gamma(f^*E)$ of smooth sections of f^*E is canonically isomorphic to the module of sections of E along f via the mapping given by (2.2.4).

Example 2.2.38 (Direct Sum). Let $\pi_1 \colon E_1 \to M$ and $\pi_2 \colon E_2 \to M$ be vector bundles with typical fibres V_1 and V_2, respectively. We create a new vector bundle over M whose typical fibre is $V_1 \oplus V_2$, and whose fibre over a point $p \in M$ is $(E_1)_p \oplus (E_2)_p$. We sketch two equivalent constructions.

(a) Let $(\mathcal{U}_i, \psi_i^1)_{i \in I}$ and $(\mathcal{U}_i, \psi_i^2)_{i \in I}$ be trivializing coverings for π_1 and π_2 respectively. (We may assume that they have the same family of open subsets $\mathcal{U}_i \subset M$: otherwise we take a common refinement.) Let (γ_{ij}^1) and (γ_{ij}^2) be the families of transition maps associated with the coverings. For every $i \in I$ and $p \in \mathcal{U}_i$ define the linear isomorphism

$$\begin{cases} (\psi_i)_p := (\psi_i^1)_p \times (\psi_i^2)_p \colon V_1 \oplus V_2 \to (E_1)_p \oplus (E_2)_p, \\ (v_1, v_2) \mapsto \big((\psi_i^1)_p(v_1), (\psi_i^2)_p(v_2)\big) \end{cases}$$

(see (A.2.2)). Then the mappings

$$\gamma_{ij} \colon p \in \mathcal{U}_i \cap \mathcal{U}_j \mapsto \gamma_{ij}(p) := (\psi_i)_p^{-1} \circ (\psi_j)_p \in \mathrm{GL}(V_1 \oplus V_2)$$

act by the rule

$$\begin{aligned} \gamma_{ij}(p)(v_1, v_2) &= \big((\psi_i^1)_p^{-1} \circ (\psi_j^1)_p(v_1), (\psi_i^2)_p^{-1} \circ (\psi_j^2)_p(v_2)\big) \\ &= \big(\gamma_{ij}^1(p)(v_1), \gamma_{ij}^2(p)(v_2)\big), \quad (v_1, v_2) \in V_1 \times V_2, \end{aligned}$$

and so they are smooth. Thus condition (S) of Proposition 2.2.20 is satisfied, therefore, if $E_1 \oplus E_2 := \bigcup_{p \in M}((E_1)_p \oplus (E_2)_p)$, and π is the natural projection (the 'foot mapping'), then

$$(E_1 \oplus E_2, \pi, M)$$

is a vector bundle over M with typical fibre $V_1 \oplus V_2$. This vector bundle is called the *direct sum* (or *Whitney sum*) of $\pi_1 \colon E_1 \to M$ and $\pi_2 \colon E_2 \to M$.

(b) In the second construction, we first build the Cartesian product bundle

$$\pi_1 \times \pi_2 \colon E_1 \times E_2 \to M \times M$$

(Example 2.2.36). Next we consider the diagonal mapping

$$\Delta \colon M \to M \times M, \ p \mapsto \Delta(p) := (p, p)$$

and form the pull-back bundle $\Delta^*(E_1 \times E_2)$ over M. Then

$$\begin{aligned}
\Delta^*(E_1 \times E_2) &= \{(p, (z_1, z_2)) \in M \times (E_1 \times E_2) \mid \Delta(p) = (\pi_1 \times \pi_2)(z_1, z_2)\} \\
&= \{(p, (z_1, z_2)) \in M \times (E_1 \times E_2) \mid (p, p) = (\pi_1(z_1), \pi_2(z_2))\} \\
&= \{(p, (z_1, z_2)) \in M \times (E_1 \times E_2) \mid \pi_1(z_1) = \pi_2(z_2) = p\},
\end{aligned}$$

therefore the mapping

$$\Delta^*(E_1 \times E_2) \to E_1 \oplus E_2, \ (p, (z_1, z_2)) \mapsto (z_1, z_2)$$

is a natural strong bundle map. Thus the direct sum of the vector bundles π_1 and π_2 may equivalently be defined as the pull-back of the product bundle $\pi_1 \times \pi_2$ by the diagonal mapping Δ:

$$\pi_1 \oplus \pi_2 = \Delta^*(\pi_1 \times \pi_2).$$

Notice that the direct sum of two trivial bundles is again a trivial bundle, but the direct sum of nontrivial bundles can also be trivial.

Remark 2.2.39. With the notation as in the previous example, consider the direct sum

$$\pi_1 \oplus \pi_2 \colon E_1 \oplus E_2 \to M$$

of the vector bundles π_1 and π_2. For each $p \in M$, we have the canonical inclusions

$$(i_1)_p \colon (E_1)_p \to (E_1)_p \oplus (E_2)_p, \ u \mapsto (u, 0),$$
$$(i_2)_p \colon (E_2)_p \to (E_1)_p \oplus (E_2)_p, \ v \mapsto (0, v)$$

and the natural projections

$$(\mathrm{pr}_1)_p \colon (E_1)_p \oplus (E_2)_p \to (E_1)_p, \ (u,v) \mapsto u,$$

$$(\mathrm{pr}_2)_p \colon (E_1)_p \oplus (E_2)_p \to (E_2)_p, \ (u,v) \mapsto v.$$

They induce the strong bundle maps

$$i_k \colon E_k \to E_1 \oplus E_2, \ \mathrm{pr}_k \colon E_1 \oplus E_2 \to E_k, \quad k \in \{1,2\}$$

such that

$$i_k \restriction (E_k)_p = (i_k)_p, \ \mathrm{pr}_k \restriction (E_1)_p \oplus (E_2)_p = (\mathrm{pr}_k)_p, \quad p \in M.$$

Example 2.2.40 (Bundles of Linear Mappings). Data and notations as in the preceding example. Now we construct a vector bundle over M with typical fibre $L(V_1, V_2)$ whose fibres are $L((E_1)_p, (E_2)_p)$ ($p \in M$). Given an index $i \in I$, for every $p \in \mathcal{U}_i$ define the mapping

$$(\psi_i)_p \colon L(V_1, V_2) \to L((E_1)_p, (E_2)_p)$$

by

$$(\psi_i)_p(A) := (\psi_i^2)_p \circ A \circ (\psi_i^1)_p^{-1}.$$

This definition is more vividly displayed by the diagram

$$
\begin{array}{ccc}
(E_1)_p & \xrightarrow{\ (\psi_i)_p(A)\ } & (E_2)_p \\
{\scriptstyle (\psi_i^1)_p^{-1}} \big\downarrow & & \big\uparrow {\scriptstyle (\psi_i^2)_p} \\
V_1 & \xrightarrow{\quad A \quad} & V_2
\end{array}\ .
$$

Then $(\psi_i)_p$ is a linear isomorphism, and its inverse is given by

$$(\psi_i)_p^{-1}(\varphi_p) = (\psi_i^2)_p^{-1} \circ \varphi_p \circ (\psi_i^1)_p, \quad \varphi_p \in L((E_1)_p, (E_2)_p).$$

It remains to check the smoothness of the mappings

$$\gamma_{ij} \colon \mathcal{U}_i \cap \mathcal{U}_j \to \mathrm{GL}(L(V_1, V_2)), \ p \mapsto (\psi_i)_p^{-1} \circ (\psi_j)_p.$$

This is, however, an easy consequence of the smoothness of the γ_{ij}^1's and γ_{ij}^2's, because for each $p \in \mathcal{U}_i \cap \mathcal{U}_j$ and $A \in L(V_1, V_2)$ we have

$$
\begin{aligned}
\gamma_{ij}(p)(A) :={}& (\psi_i)_p^{-1} \circ (\psi_j)_p(A) := (\psi_i)_p^{-1}((\psi_j^2)_p \circ A \circ (\psi_j^1)_p^{-1}) \\
={}& (\psi_i^2)_p^{-1} \circ (\psi_j^2)_p \circ A \circ (\psi_j^1)_p^{-1} \circ (\psi_i^1)_p \\
={}& \gamma_{ij}^2(p) \circ A \circ (\gamma_{ij}^1(p))^{-1}.
\end{aligned}
$$

Thus the construction principle formulated in Proposition 2.2.20 yields the vector bundle $(L(E_1, E_2), \pi, M)$, where $L(E_1, E_2) := \cup_{p \in M} L((E_1)_p, (E_2)_p)$ and π is the foot mapping. This is called the *bundle of linear mappings* $(E_1)_p \to (E_2)_p$, briefly a *Lin-bundle*, and is also denoted by $\mathrm{Lin}_M(\pi_1, \pi_2)$. If $\pi_1 = \pi_2 =: \pi$, we write $\mathrm{End}(\pi) := \mathrm{Lin}_M(\pi, \pi)$. Since for a bundle $\pi \colon E \to M$ we frequently write simply (and by abuse of notation) E, we also write $\mathrm{End}(E)$ instead of $\mathrm{End}(\pi)$.

Example 2.2.41 (Short Exact Sequences). Let E_0, E_1, E_2 be vector bundles over a manifold M. Denote by 0 the trivial vector bundle with zero-dimensional typical fibre over M. A sequence

$$0 \longrightarrow E_0 \xrightarrow{\varphi_1} E_1 \xrightarrow{\varphi_2} E_2 \longrightarrow 0 \qquad (2.2.11)$$

of strong bundle maps, where the first arrow is the mapping

$$(p,0) \in M \times \{0\} \mapsto 0_p \in E_0$$

and the last arrow is the zero mapping, is said to be a *short exact sequence* if it restricts to a short exact sequence on each fibre, i.e., if

$$0 \longrightarrow (E_0)_p \xrightarrow{(\varphi_1)_p} (E_1)_p \xrightarrow{(\varphi_2)_p} (E_2)_p \longrightarrow 0 \qquad (2.2.12)$$

is a short exact sequence of linear mappings in the sense of Definition 1.1.17 for all $p \in M$. The short exact sequence (2.2.11) is called *split exact* if there exists a strong bundle map $\psi \colon E_2 \to E_1$ such that $\varphi_2 \circ \psi = 1_{E_2}$. Then we say that ψ is a *splitting bundle map* for (2.2.11). In this case, for each $p \in M$,

$$\psi_p := \psi \restriction (E_2)_p \colon (E_2)_p \longrightarrow (E_1)_p$$

is a splitting linear mapping for the sequence (2.2.12) (cf. Definition 1.1.17). If ψ is a splitting map for (2.2.11), then

$$E_1 = \mathrm{Im}(\varphi_1) \oplus \mathrm{Im}(\psi);$$

cf. Lemma 1.1.19.

Example 2.2.42 (Dual Bundles). Let $\pi \colon E \to M$ be a vector bundle with typical fibre V, and consider the trivial vector bundle $\mathrm{pr}_1 \colon M \times \mathbb{R} \to M$ of rank 1. Then the vector bundle $\mathrm{Lin}_M(\pi, \mathrm{pr}_1)$ is called the *dual* of π. The fibre of $\mathrm{Lin}_M(\pi, \mathrm{pr}_1)$ over a point $p \in M$ is

$$L(E_p, \{p\} \times \mathbb{R}) \cong L(E_p, \mathbb{R}) = E_p^*,$$

so the total manifold of the bundle may be identified with the disjoint union

$$E^* := \bigcup_{p \in M} E_p^*$$

of the dual vector spaces of the fibres of π. The bundle projection is the foot mapping

$$\widehat{\pi} \colon E^* \to M, \ z^* \mapsto \widehat{\pi}(z^*) := p \text{ if } z^* \in E_p^*.$$

In the following for the dual vector bundle of π we shall use the more transparent notation $\widehat{\pi} \colon E^* \to M$ (and its variants, see Remark 2.2.2, 'by abuse of notation').

If $(s_i)_{i=1}^k$ is a local frame field of E over an open subset $\mathcal{U} \subset M$, then there exists a unique frame field $(s^i)_{i=1}^k$ of E^* over $\mathcal{U}$ such that

$$s^i(s_j) = \delta_j^i; \quad i, j \in J_k.$$

The frame field (s^i) is called the *dual coframe* of (s_i).

Lemma 2.2.43. *If $\pi_1 \colon E_1 \to M$ and $\pi_2 \colon E_2 \to M$ are vector bundles over M, then we have a canonical isomorphism*

$$\Gamma(E_1 \oplus E_2) \cong \Gamma(E_1) \oplus \Gamma(E_2)$$

of $C^\infty(M)$-modules, namely, with the notation of Remark 2.2.39, the mappings

$$s \in \Gamma(E_1 \oplus E_2) \mapsto (\mathrm{pr}_1 \circ s, \mathrm{pr}_2 \circ s) \in \Gamma(E_1) \oplus \Gamma(E_2),$$

$$(s_1, s_2) \in \Gamma(E_1) \oplus \Gamma(E_2) \mapsto i_1 \circ s_1 + i_2 \circ s_2 \in \Gamma(E_1 \oplus E_2)$$

are $C^\infty(M)$-homomorphisms, inverses to each other.

Proof. Immediate verification by the identities established in Example 1.1.18. □

Proposition 2.2.44. *Let $\pi_1 \colon E_1 \to M$ and $\pi_2 \colon E_2 \to M$ be vector bundles. There exists a canonical isomorphism*

$$s \in \Gamma\big(\mathrm{Lin}_M(\pi_1, \pi_2)\big) \mapsto \widetilde{s} \in \mathrm{Hom}_M\big(\Gamma(\pi_1), \Gamma(\pi_2)\big)$$

of $C^\infty(M)$-modules given by

$$\widetilde{s}(\sigma)(p) := s(p)(\sigma(p)) \text{ for all } \sigma \in \Gamma(\pi_1), \ p \in M.$$

In particular, $\Gamma(E^*) \cong (\Gamma(E))^*$.

Sketch of proof. One verifies trivially that the mapping $\widetilde{s}$ is indeed in $\mathrm{Hom}_M\big(\Gamma(\pi_1), \Gamma(\pi_2)\big)$. Equally trivially, one checks that the mapping

$$\varphi \colon s \in \Gamma\big(\mathrm{Lin}_M(\pi_1, \pi_2)\big) \mapsto \varphi(s) := \widetilde{s} \in \mathrm{Hom}_M\big(\Gamma(\pi_1), \Gamma(\pi_2)\big)$$

is also a homomorphism of $C^\infty(M)$-modules.

To see that φ is actually a module-isomorphism, choose an element $A \in \mathrm{Hom}_M\big(\Gamma(\pi_1), \Gamma(\pi_2)\big)$, and define $s \in \Gamma\big(\mathrm{Lin}_M(\pi_1, \pi_2)\big)$ by

$$s(p)(z) := A(\sigma)(p) \text{ for } z \in (E_1)_p, \quad p \in M,$$

where $\sigma \in \Gamma(\pi_1)$ such that $\sigma(p) = z$. The existence of σ is guaranteed by Lemma 2.2.24(i). We need, however, to show that s is well-defined, i.e., does not depend on the choice of the section σ. To prove this, it suffices to check that

$$\sigma(p) = 0 \text{ implies } A(\sigma)(p) = 0.$$

The method is standard: we choose a local frame field of E_1 over an open neighbourhood $\mathcal{U}$ of p, extend it by a bump function at p supported in $\mathcal{U}$ – and so forth; for details we refer to [66], p. 286. Finally, it is easy to see that the constructed mapping

$$\psi \colon \text{Hom}_M\big(\Gamma(\pi_1), \Gamma(\pi_2)\big) \to \Gamma\big(\text{Lin}_M(\pi_1, \pi_2)\big), \; A \mapsto \psi(A) := s$$

is a homomorphism of $C^\infty(M)$-modules and $\psi = \varphi^{-1}$. $\qquad\square$

Combining this result with Remark 2.2.32(a), we conclude

Corollary 2.2.45. *If $\pi_1 \colon E_1 \to M$ and $\pi_2 \colon E_2 \to M$ are vector bundles, then the $C^\infty(M)$-modules*

$$\text{Bun}_M(\pi_1, \pi_2), \; \Gamma\big(\text{Lin}_M(\pi_1, \pi_2)\big) \text{ and } \text{Hom}_M\big(\Gamma(\pi_1), \Gamma(\pi_2)\big)$$

are canonically isomorphic. $\qquad\square$

2.2.4 π-tensors and π-tensor fields

Lemma and Definition 2.2.46. *Let $\pi \colon E \to M$ be a vector bundle of rank r with typical fibre V.*

(i) *For natural numbers k, l not both zero, form the disjoint union*

$$T_l^k(E) := \bigcup_{p \in M} T_l^k(E_p)$$

of spaces of type (k, l) tensors (see 1.2.1) of the fibres of π, and let

$$\pi_l^k \colon T_l^k(E) \to M$$

be the natural projection (the foot mapping).
There exist a unique topology and smooth structure on $T_l^k(E)$ which make the triple

$$\big(T_l^k(E), \pi_l^k, M\big)$$

into a vector bundle of rank r^{k+l} with typical fibre $T_l^k(V)$. $T_l^k(E)$ is called the bundle of type (k, l) tensors for E, or briefly, the bundle of type (k, l) π-tensors.

(ii) *The $C^\infty(M)$-module $\Gamma(\pi_l^k) = \Gamma(T_l^k(E))$ of smooth sections of π_l^k (or, by a slight abuse of language, of $T_l^k(E)$) is said to be the module of type (k, l) π-tensor fields on M.*

Remark 2.2.47. (a) Any π-tensor field $A \in \Gamma(T_l^k(E))$ induces a $C^\infty(M)$-multilinear mapping

$$\widetilde{A}\colon \underbrace{(\Gamma(E))^* \times \cdots \times (\Gamma(E))^*}_{k} \times \underbrace{\Gamma(E) \times \cdots \times \Gamma(E)}_{l} \to C^\infty(M),$$

i.e., a tensor $\widetilde{A} \in T_l^k(\Gamma(E))$ such that

$$\widetilde{A}(\alpha^1, \ldots, \alpha^k, \sigma_1, \ldots, \sigma_l)(p) = A_p\big(\alpha^1(p), \ldots, \alpha^k(p), \sigma_1(p), \ldots, \sigma_l(p)\big)$$

for all $\alpha^1, \ldots, \alpha^k \in (\Gamma(E))^* \cong \Gamma(E^*)$; $\sigma_1, \ldots, \sigma_l \in \Gamma(E)$ and $p \in M$.

Conversely, any tensor $B \in T_l^k(\Gamma(E))$ is induced by a unique π-tensor field $A \in \Gamma(T_l^k(E))$ defined as follows:

for each $p \in M$; $a^1, \ldots, a^k \in E_p^*$; $z_1, \ldots, z_l \in E_p$

$$A_p(a^1, \ldots, a^k, z_1, \ldots, z_l) := B(\alpha^1, \ldots, \alpha^k, \sigma_1, \ldots, \sigma_l)(p),$$

where $\alpha^1, \ldots, \alpha^k \in (\Gamma(E))^*$; $\sigma_1, \ldots, \sigma_l \in \Gamma(E)$ such that $\alpha^i(p) = a^i$, $i \in J_k$; $\sigma_j(p) = z_j$, $j \in J_l$.

(Of course, one has to show that A is well-defined. For the idea of a proof we refer to [79], pp. 37–38.) So there is a canonical isomorphism

$$T_l^k(\Gamma(E)) \cong \Gamma(T_l^k(E))$$

of $C^\infty(M)$-modules, and we have the following multilinear version of Proposition 2.2.31:

Criterion for Tensoriality. *A mapping*

$$\underbrace{(\Gamma(E))^* \times \cdots \times (\Gamma(E))^*}_{k} \times \underbrace{\Gamma(E) \times \cdots \times \Gamma(E)}_{l} \to C^\infty(M)$$

is induced by a π-tensor field of type (k, l) if, and only if, it is $C^\infty(M)$-multilinear.

To sum up, it can be stated that by a π-tensor field of type (k, l) on M we can equivalently mean a tensor in the $C^\infty(M)$-module $T_l^k(\Gamma(\pi))$. In the following, by a slight abuse of language,

instead of 'a π-tensor field on M' we often simply say 'a π-tensor on M'.

(b) **Local characterization of π-tensor fields.** In view of Lemma 2.2.26, a local frame field $(s_i)_{i=1}^r$ of E over an open subset $\mathcal{U}$ of M may be identified with a trivializing map

$$\psi \colon \mathcal{U} \times \mathbb{R}^r \to \pi^{-1}(\mathcal{U}), \ (p, (\nu^1, \dots, \nu^r)) \mapsto \sum_{i=1}^r \nu^i s_i(p),$$

and vice versa. So a trivializing map may be written simply in the form

$$\psi = (s_1, \dots, s_r) \colon \mathcal{U} \times \mathbb{R}^r \to \pi^{-1}(\mathcal{U}),$$

where $(s_i)_{i=1}^r$ is a local frame field over $\mathcal{U}$, and we understand that the action of ψ is given by the above formula. Now, it is not difficult to prove the following result, which opens the door to the world of traditional tensor calculus.

A section A of π_l^k, i.e., a mapping

$$A \colon M \to T_l^k(E) \ \text{such that} \ \pi_l^k \circ A = 1_M$$

is a π-tensor field on M of type (k, l) if, and only if, for any open subset $\mathcal{U}$ of M and any frame field $(s_i)_{i=1}^r$ over $\mathcal{U}$, there exist smooth functions

$$A_{j_1 \dots j_l}^{i_1 \dots i_k} \in C^\infty(\mathcal{U}); \quad i_1, \dots, i_k, j_1, \dots, j_l \in J_r$$

such that

$$A \upharpoonright \mathcal{U} = \sum_{(i)(j)} A_{j_1 \dots j_l}^{i_1 \dots i_k} s_{i_1} \otimes \cdots \otimes s_{i_k} \otimes s^{j_1} \otimes \cdots \otimes s^{j_l}, \quad (2.2.13)$$

where $(s^j)_{j=1}^r$ is the dual coframe associated with the local frame field $(s_i)_{i=1}^r$ (see Example 2.2.42).

The functions $A_{j_1 \dots j_l}^{i_1 \dots i_k}$ are called the *components* of A with respect to the frame field $(s_i)_{i=1}^r$. (The tensor products applied in the local representation of $A \upharpoonright \mathcal{U}$ are defined in 1.2.1.)

2.2.5 *Vector Bundles with Additional Structures*

Example 2.2.48 (The Bundle of Alternating π-tensors). Consider a vector bundle $\pi \colon E \to M$ of rank $r \geq 1$. Choose an integer $k \in J_r$, and let

$$A_k(E) := \bigcup_{p \in M} A_k(E_p) \quad \text{(disjoint union)},$$

where $A_k(E_p)$ is the vector space of alternating k-linear functions on E_p. Then $A_k(E) \subset T_k^0(E)$, and for each $p \in M$, $A_k(E) \cap T_k^0(E_p) = A_k(E_p)$ is a vector subspace of $T_k^0(E_p)$. It may easily be shown that the triple

$$\left(A_k(E), \pi_k^0 \restriction A_k(E), M \right)$$

is a vector subbundle of $\left(T_k^0(E), \pi_k^0, M \right)$, called the *bundle of alternating π-tensors* of degree k. The rank of this bundle is $\binom{r}{k}$; in particular, $A_r(E)$ is of rank 1. The canonical module isomorphism established in Remark 2.2.47(a) implies that

$$\Gamma(A_k(E)) \cong A_k(\Gamma(E)).$$

The elements of these isomorphic $C^\infty(M)$-modules are said to be π-*forms of degree k on M*. We recall (see Remark 2.2.47(a) again and 1.2.1) that $A_k(\Gamma(E))$ consists of $C^\infty(M)$-multilinear mappings $\alpha \colon (\Gamma(E))^k \to C^\infty(M)$ such that $\sigma \alpha = \varepsilon(\sigma)\alpha$ for all $\sigma \in S_k$, that is,

$$\alpha(s_{\sigma(1)}, \ldots, s_{\sigma(k)}) = \varepsilon(\sigma)\alpha(s_1, \ldots, s_k); \quad s_1, \ldots, s_k \in \Gamma(\pi).$$

Definition 2.2.49. Let $\pi \colon E \to M$ be a vector bundle of rank $r \geq 1$. We say that π is *orientable* if there exists a π-form $\omega \in A_r(\Gamma(E))$ such that $\omega(p) \neq 0$ for all $p \in M$. Then ω is called a *volume form* (or a *determinant function*) on π.

Remark 2.2.50. We continue to assume that $\pi \colon E \to M$ is a vector bundle of rank $r \geq 1$.

(a) π *is orientable if, and only if, the rank 1 vector bundle $A_r(E)$ is trivializable*. Indeed, if π is orientable and ω is a volume form on π, then the mapping

$$\psi \colon M \times \mathbb{R} \to A_r(E), \ (p, t) \mapsto t\omega(p)$$

is a strong bundle isomorphism between the trivial vector bundle $M \times \mathbb{R}$ and $A_r(E)$, hence $A_r(E)$ is trivializable (see Remark 2.2.9). Conversely, if $A_r(E)$ is trivializable and the strong bundle isomorphism

$$\widetilde{\psi} \colon A_r(E) \to M \times \mathbb{R}$$

is a trivialization of $A_r(E)$, then the mapping

$$M \to A_r(E), \ p \mapsto \widetilde{\psi}^{-1}(p, 1)$$

is a volume form on π.

(b) Suppose that ω_1 and ω_2 are volume forms on π. Then there exists a unique, nowhere vanishing smooth function f on M such that $\omega_2 = f\omega_1$. Define a relation $\sim$ on the set of volume forms on π by

$\omega_1 \sim \omega_2$ *if there exists a positive real-valued function f on M such that $\omega_2 = f\omega_1$*

(cf. 1.3.7). This relation is clearly an equivalence relation. We denote the equivalence class of a volume form ω by $[\omega]$ and call an equivalence class $[\omega]$ an *orientation* on π. An *oriented vector bundle* is an orientable vector bundle π together with an orientation on π.

(c) Let $(\pi, [\omega])$ be an oriented vector bundle. Then each fibre E_p is an oriented vector space with the orientation $[\omega_p]$. A local frame field $(\sigma_i)_{i=1}^r$ of π over an open subset $\mathcal{U}$ of M is called *positively oriented* if $(\sigma_i(p))_{i=1}^r$ is a positive basis of E_p for each $p \in \mathcal{U}$.

(d) *If $\pi \colon E \to M$ is orientable and the base manifold M is connected, then there are exactly two orientations on π.* Indeed, let ω_1 and ω_2 be two volume forms on π. Then $\omega_2 = f\omega_1$, where $f \in C^\infty(M)$, and $f(p) \neq 0$ for all $p \in M$. Since M is connected, the intermediate-value theorem implies that

$$f(p) > 0 \text{ for all } p \in M \text{ or else } f(p) < 0 \text{ for all } p \in M.$$

Thus $\omega_1 \sim \omega_2$ or $\omega_1 \sim -\omega_2$, whence our claim.

Proposition 2.2.51. *A vector bundle $\pi \colon E \to M$ with typical fibre V is orientable if, and only if, there exists a trivializing covering $(\mathcal{U}_i, \psi_i)_{i \in I}$ for π such that the transition mappings take values in the group $\mathrm{GL}^+(V)$, i.e., for all $i, j \in I$ we have*

$$\gamma_{ij}(p) := (\psi_i)_p^{-1} \circ (\psi_j)_p \in \mathrm{GL}^+(V), \quad p \in \mathcal{U}_i \cap \mathcal{U}_j.$$

For a proof, see [52], pp. 64–65.

Definition 2.2.52. (a) Let $\pi \colon E \to M$ be a vector bundle of rank at least one. A type $(0, 2)$ π-tensor field g on M is said to be a *scalar product* on π if for each point $p \in M$, the bilinear form

$$g_p \colon E_p \times E_p \to \mathbb{R}$$

is symmetric.

(b) A vector bundle $\pi \colon E \to M$ with scalar product g is called

semi-Euclidean if $g_p \in T_2^0(E_p)$ is a non-degenerate scalar product,
Euclidean if $g_p \in T_2^0(E_p)$ is a positive definite scalar product

for all $p \in M$.

(c) Suppose that $\pi_1 \colon E_1 \to M_1$ and $\pi_2 \colon E_2 \to M_2$ are vector bundles with scalar products g_1 and g_2, respectively. A bijective bundle map $\varphi \colon E_1 \to E_2$ is called *isometric* (or an *isometry*) if the restrictions

$$\varphi_p := \varphi \upharpoonright (E_1)_p \colon (E_1)_p \to (E_2)_{\varphi_B(p)}, \quad p \in M_1$$

are orthogonal isomorphisms.

Remark 2.2.53. Equivalently, a scalar product on a vector bundle $\pi \colon E \to M$ is a mapping

$$g \colon p \in M \to g_p \in S_2(E_p)$$

such that the function

$$g(\sigma_1, \sigma_2) \colon M \to \mathbb{R}, \quad p \mapsto g(\sigma_1, \sigma_2)(p) := g_p(\sigma_1(p), \sigma_2(p))$$

is smooth for all $\sigma_1, \sigma_2 \in \Gamma(\pi)$.

In the context of vector bundles, Corollary 1.4.1 yields the following

Corollary 2.2.54. *Let* $\pi \colon E \to M$ *be a semi-Euclidean vector bundle with scalar product* g*, and let* $\widehat{\pi} \colon E^* \to M$ *be the dual of* π*. Then the mapping*

$$j_g \colon E \to E^*, \quad z \mapsto j_g(z) := j_{g_p}(z) = j_z(g_p) \text{ if } z \in E_p$$

is a strong bundle isomorphism. $\square$

Remark 2.2.55. We recall that

$$j_z(g_p)(w) = g_p(w, z) \text{ for all } w \in E_p$$

by (1.2.7). Since g_p is symmetric, we can also write

$$j_z(g_p)(w) = g_p(z, w).$$

In the sequel we shall use the less pedantic, but more expressive notation

$$\flat \colon E \to E^*, \quad z \mapsto z^\flat, \quad z^\flat(w) := g_p(z, w); \quad z, w \in E_p \qquad (2.2.14)$$

for the isomorphism j_g, and we shall denote its inverse by $\sharp$. Then

$$\sharp \colon E^* \to E, \quad \alpha \mapsto \alpha^\sharp; \quad g_p(\alpha^\sharp, v) = \alpha(v) \quad \text{if } \alpha \in E_p^*, v \in E_p. \qquad (2.2.15)$$

We read the expressions $z^\flat$ and $\alpha^\sharp$ as 'z flat' and 'α sharp', respectively, and we say that $\sharp$ and $\flat$ are the *musical isomorphisms* with respect to g. In classical notation the isomorphism $\flat$ lowers, its inverse $\sharp$ raises indices; see later (6.1.56) and (6.1.57).

Lemma 2.2.56. *Every vector bundle admits a positive definite scalar product.*

Proof. Consider first a trivial vector bundle

$$\mathrm{pr}_1 \colon M \times V \to M, \quad (p, v) \mapsto p,$$

where V is a finite-dimensional real vector space. We may suppose that $\dim V \geq 1$. Clearly, V can be equipped with a positive definite scalar product $\langle\ ,\ \rangle$: given a basis $(v_i)_{i=1}^k$ of V, let

$$\langle u, v \rangle := \sum_{i=1}^k \lambda^i \mu^i \text{ if } u = \sum_{i=1}^k \lambda^i v_i, \ v = \sum_{i=1}^k \mu^i v_i.$$

Then the mapping

$$g \colon p \in M \mapsto g_p \in T_2^0(\{p\} \times V) \cong T_2^0(V)$$

given by

$$g_p(u, v) := \langle u, v \rangle; \quad u, v \in V$$

is a positive definite scalar product on pr_1.

Now let $\pi \colon E \to M$ be an arbitrary vector bundle with typical fibre V, $\dim V \geq 1$. Choose a trivializing covering $(\mathcal{U}_i, \psi_i)_{i \in I}$ for π, and let $(f_i)_{i \in I}$ be a subordinate partition of unity. Then

$$\pi_i \colon \mathcal{U}_i \times V \to \mathcal{U}_i, \quad (p, v) \mapsto p, \quad i \in I$$

are trivial vector bundles, so there is a positive definite scalar product g_i on π_i for all $i \in I$. From these we obtain a positive definite scalar product $g := \sum_{i \in I} f_i g_i$ on π. $\qquad \square$

Chapter 3

Vector Fields, Tensors and Integration

3.1 Tangent Bundle and Vector Fields

The crucial step in generalizing calculus from $\mathbb{R}^n$ to an arbitrary manifold is the following elegant definition, which axiomatizes the *directional derivative* aspect of Euclidean tangent vectors.

Barrett O'Neill

I don't understand why many differential geometers (...) take the definition of a tangent vector as a differential operator.

Serge Lang

3.1.1 *Tangent Vectors and Tangent Space*

Definition 3.1.1. Let M be a manifold and p a point of M. A *tangent vector to M at p* is a *derivation of the real algebra $C^\infty(M)$ at p*, i.e., an $\mathbb{R}$-linear function $v\colon C^\infty(M) \to \mathbb{R}$ which satisfies the *Leibniz rule*

$$v(fg) = v(f)g(p) + f(p)v(g) \text{ for all } f,g \in C^\infty(M). \qquad (3.1.1)$$

The point p is called the *footpoint* of v. The set of all tangent vectors to M at p is said to be the *tangent space* of M at p, and it is denoted by T_pM.

Remark 3.1.2. At each point $p \in M$, *the tangent space T_pM is a real vector space* under the linear operations

$$(v + w)(f) := v(f) + w(f), \quad (\lambda v)(f) = \lambda v(f),$$

where $v, w \in T_pM$, $f \in C^\infty(M)$, $\lambda \in \mathbb{R}$.

Lemma 3.1.3 (Tangent Vectors are Local I). *Let M be a manifold and $v \in T_pM$. If $f, g \in C^\infty(M)$ are equal on an open neighbourhood of p, then $v(f) = v(g)$.*

Proof. Let $o \in C^\infty(M)$ be the zero function. Then
$$v(o) = v(o + o) = v(o) + v(o),$$
whence $v(o) = 0$.

Now suppose that $f \restriction \mathcal{U} = g \restriction \mathcal{U}$ on an open neighbourhood $\mathcal{U}$ of p. By linearity it suffices to show that $v(f - g) = 0$. Choose a bump function $\varrho \in C^\infty(M)$ at p, supported in $\mathcal{U}$ (Definition 2.1.31, Corollary 2.1.32). Then $\varrho(f - g) = 0$, hence, by our above observation and the Leibniz rule (3.1.1),
$$0 = v(\varrho(f - g)) = v(\varrho)(f(p) - g(p)) + \varrho(p)v(f - g) = v(f - g). \qquad \square$$

Corollary 3.1.4. *If a function $f \in C^\infty(M)$ is constant on an open neighbourhood of a point $p \in M$, then $v(f) = 0$ for all $v \in T_pM$.*

Proof. Let **1** denote the identity element of the ring $C^\infty(M)$ (cf. A.4.2(b)). Observe that $v(\mathbf{1}) = 0$, since
$$v(\mathbf{1}) = v(\mathbf{1} \cdot \mathbf{1}) \overset{(3.1.1)}{=} v(\mathbf{1})\mathbf{1}(p) + \mathbf{1}(p)v(\mathbf{1}) = 2v(\mathbf{1}).$$

Now suppose that f has a constant value λ on an open neighbourhood of p. Then, by the preceding lemma,
$$v(f) = v(\lambda \mathbf{1}) = \lambda v(\mathbf{1}) = 0. \qquad \square$$

Corollary 3.1.5 (Tangent Vectors are Local II). *Let M be a manifold and $\mathcal{U}$ an open submanifold of M. Then, for each $p \in \mathcal{U}$, there is a canonical isomorphism between the tangent spaces $T_p\mathcal{U}$ and T_pM given by*
$$v \in T_p\mathcal{U} \mapsto \tilde{v} \in T_pM, \ \tilde{v}(f) := v(f \restriction \mathcal{U}) \text{ for all } f \in C^\infty(M).$$

We leave the proof to the reader.

Remark 3.1.6. (a) In what follows, we identify the canonically isomorphic vector spaces $T_p\mathcal{U}$ and T_pM, by simply writing $T_p\mathcal{U} = T_pM$.

(b) Let $(\mathcal{U}, (u^i)_{i=1}^n)$ be a chart at a point p of a manifold M. Then the functions
$$\left(\frac{\partial}{\partial u^i}\right)_p : f \in C^\infty(M) \mapsto \left(\frac{\partial}{\partial u^i}\right)_p (f) := \frac{\partial f}{\partial u^i}(p) := D_i(f \circ u^{-1})(u(p)) \in \mathbb{R}$$
(see Definition 2.1.18) are tangent vectors to M at p, called the *coordinate vectors* of the local coordinate system $(u^i)_{i=1}^n$ at this point.

In the next result (and in Remark 3.2.4) we use the well-known concept of convexity. A formal definition will be given later, in subsection 9.1.1.

Lemma 3.1.7 (Hadamard). *Assume $\mathcal{U}$ is a convex open subset of $\mathbb{R}^n$, and let $p = (p^1, \ldots, p^n)$ be a fixed point in $\mathcal{U}$. For any smooth function f on $\mathcal{U}$ there exist smooth functions $f_i \in C^\infty(\mathcal{U})$ such that*

$$f_i(p) = D_i f(p) \quad \text{for all } i \in J_n$$

and

$$f = f(p)\mathbf{1} + \sum_{i=1}^n (e^i - p^i\mathbf{1})f_i.$$

In this formula $\mathbf{1}$ denotes the constant function on $\mathcal{U}$ with the value 1, and $(e^i)_{i=1}^n$ is the dual of the canonical basis of $\mathbb{R}^n$.

Proof. Let $q \in \mathcal{U}$ be an arbitrarily fixed point, and define the curve segment

$$\gamma_q \colon [0,1] \to \mathcal{U}, \quad t \mapsto \gamma_q(t) := p + t(q - p).$$

This curve segment is constant if $q = p$. Otherwise, γ_q is a parametrized straight line segment, so $\text{Im}(\gamma_q) \subset \mathcal{U}$ by the convexity of $\mathcal{U}$. Then

$$f(q) = f(p) + \int_0^1 (f \circ \gamma_q)'.$$

We calculate the derivative of $f \circ \gamma_q$. For every $t \in \,]0,1[$,

$$(f \circ \gamma_q)'(t) \overset{(C.1.4)}{=} f'(\gamma_q(t))\gamma_q'(t) = f'(\gamma_q(t))(q - p)$$

$$= f'(\gamma_q(t))\left(\sum_{i=1}^n e^i(q-p)e_i\right) = \sum_{i=1}^n e^i(q-p)f'(\gamma_q(t))(e_i)$$

$$= \sum_{i=1}^n e^i(q-p)D_i f(\gamma_q(t)),$$

therefore

$$(f \circ \gamma_q)' = \sum_{i=1}^n e^i(q-p)(D_i f \circ \gamma_q).$$

Thus

$$f(q) = f(p) + \sum_{i=1}^n (e^i - p^i\mathbf{1})(q)\int_0^1 (D_i f) \circ \gamma_q, \quad q \in \mathcal{U}.$$

Define the functions

$$f_i \colon \mathcal{U} \to \mathbb{R}, \quad q \mapsto f_i(q) := \int_0^1 (D_i f) \circ \gamma_q, \quad i \in J_n.$$

Then, obviously, $f_i(p) = D_i f(p)$ for all $i \in J_n$. Since f is of class C^∞, each of its partial derivatives are of class C^∞, and it follows from Lemma C.1.23 that the functions f_i are also of class C^∞. This concludes the proof. $\square$

Proposition 3.1.8 (The Basis Theorem). *If $(\mathcal{U}, (u^i)_{i=1}^n)$ is a chart of a manifold M at a point $p \in M$, then the family*

$$\left(\left(\frac{\partial}{\partial u^i} \right)_p \right)_{i=1}^n$$

of the coordinate vectors at p is a basis of $T_p M$. Every vector $v \in T_p M$ has a unique representation of the form

$$v = \sum_{i=1}^n v(u^i) \left(\frac{\partial}{\partial u^i} \right)_p \tag{3.1.2}$$

with respect to this basis.

Proof. Applying Lemma 2.1.19(ii), it may be checked immediately that the coordinate vectors are linearly independent.

We show that $\left(\left(\frac{\partial}{\partial u^i} \right)_p \right)_{i=1}^n$ is a generator family of $T_p M$. We may assume without loss of generality that $u(\mathcal{U}) = B_\varepsilon(0)$ for some positive $\varepsilon \in \mathbb{R}$, and that $u(p) = 0$. Then, for any $f \in C^\infty(M)$, $\widetilde{f} := f \circ u^{-1} \in C^\infty(B_\varepsilon(0))$ and by Lemma 3.1.7,

$$\widetilde{f} = \widetilde{f}(0)\mathbf{1} + \sum_{i=1}^n e^i \widetilde{f}_i,$$

where $\widetilde{f}_i \in C^\infty(B_\varepsilon(0))$, $\widetilde{f}_i(0) = D_i \widetilde{f}(0)$ $(i \in J_n)$. Thus

$$f \upharpoonright \mathcal{U} = \widetilde{f} \circ u = \widetilde{f}(0)(\mathbf{1} \circ u) + \sum_{i=1}^n (e^i \circ u)(\widetilde{f}_i \circ u)$$

$$= f(p)(\mathbf{1} \circ u) + \sum_{i=1}^n u^i(\widetilde{f}_i \circ u).$$

Hence, for any $v \in T_p M$, we have

$$v(f) \overset{\text{Cor. 3.1.5}}{=} v(f \upharpoonright \mathcal{U}) \overset{\text{Cor. 3.1.4}}{=} v \left(\sum_{i=1}^n u^i(\widetilde{f}_i \circ u) \right)$$

$$= \sum_{i=1}^{n} \left(v(u^i) \widetilde{f}_i(u(p)) + u^i(p) v(\widetilde{f}_i \circ u) \right)$$

$$\overset{u(p)\,=\,0}{=} \sum_{i=1}^{n} v(u^i) \widetilde{f}_i(0) = \sum_{i=1}^{n} v(u^i) D_i(f \circ u^{-1})(u(p))$$

$$= \sum_{i=1}^{n} v(u^i) \frac{\partial f}{\partial u^i}(p) = \left(\sum_{i=1}^{n} v(u^i) \left(\frac{\partial}{\partial u^i} \right)_p \right) (f).$$

Since we have chosen the function $f \in C^\infty(M)$ arbitrarily, it follows that $v = \sum_{i=1}^{n} v(u^i) \left(\frac{\partial}{\partial u^i} \right)_p$, and the proposition is proved. $\square$

Corollary 3.1.9 (Transformation Rule of Coordinate Vectors). *Let M be a manifold and $p \in M$. If $(u^i)_{i=1}^{n}$ and $(\widetilde{u}^i)_{i=1}^{n}$ are local coordinate systems of M at p, then*

$$\left(\frac{\partial}{\partial \widetilde{u}^j} \right)_p = \sum_{i=1}^{n} \frac{\partial u^i}{\partial \widetilde{u}^j}(p) \left(\frac{\partial}{\partial u^i} \right)_p, \quad j \in J_n; \qquad (3.1.3)$$

therefore the transition matrix from the basis $\left(\left(\frac{\partial}{\partial u^i} \right)_p \right)_{i=1}^{n}$ to the basis $\left(\left(\frac{\partial}{\partial \widetilde{u}^i} \right)_p \right)_{i=1}^{n}$ is

$$\left(\frac{\partial u^i}{\partial \widetilde{u}^j}(p) \right) = \left(D_j(e^i \circ u \circ \widetilde{u}^{-1})(\widetilde{u}(p)) \right) \in \mathrm{GL}_n(\mathbb{R}),$$

i.e., the Jacobi matrix of the transition mapping $u \circ \widetilde{u}^{-1}$ at $\widetilde{u}(p)$.

Proof. Apply (3.1.2) to the tangent vector $\left(\frac{\partial}{\partial \widetilde{u}^j} \right)_p$ or/and have a look at Lemma 2.1.19(iii). $\square$

3.1.2 The Derivative of a Differentiable Mapping

Lemma 3.1.10. *Let M and N be manifolds, and let $\varphi \colon M \to N$ be a C^k mapping. Choose a point $p \in M$. If $v \in T_pM$ then the function*

$$(\varphi_*)_p(v) \colon C^\infty(N) \to \mathbb{R}, \quad h \mapsto (\varphi_*)_p(v)(h) := v(h \circ \varphi) \qquad (3.1.4)$$

is a tangent vector at $\varphi(p)$, and the mapping $v \mapsto (\varphi_*)_p(v)$ is a linear mapping from T_pM into $T_{\varphi(p)}N$.

The proof is a straightforward verification, and is left to the reader.

Definition 3.1.11. Let $\varphi \colon M \to N$ be a mapping of class C^k. For each point $p \in M$, the linear mapping

$$(\varphi_*)_p \colon T_p M \to T_{\varphi(p)} N, \quad v \mapsto (\varphi_*)_p(v)$$

defined by (3.1.4) is said to be the *derivative of φ at p*.

Remark 3.1.12. With notation as in Lemma 3.1.10, let $(\mathcal{U}, (x^i)_{i=1}^m)$ be a chart of M at p and $(\mathcal{V}, (y^j)_{j=1}^n)$ a chart of N at $\varphi(p)$. Then

$$(\varphi_*)_p \left(\frac{\partial}{\partial x^j} \right)_p = \sum_{i=1}^n \frac{\partial(y^i \circ \varphi)}{\partial x^j}(p) \left(\frac{\partial}{\partial y^i} \right)_{\varphi(p)}, \quad j \in J_m. \tag{3.1.5}$$

Indeed, if $w_j := (\varphi_*)_p \left(\frac{\partial}{\partial x^j} \right)_p$, then by the basis theorem,

$$w_j = \sum_{i=1}^n w_j(y^i) \left(\frac{\partial}{\partial y^i} \right)_{\varphi(p)}.$$

Since

$$w_j(y^i) := (\varphi_*)_p \left(\frac{\partial}{\partial x^j} \right)_p (y^i) := \left(\frac{\partial}{\partial x^j} \right)_p (y^i \circ \varphi) = \frac{\partial(y^i \circ \varphi)}{\partial x^j}(p),$$

we obtain (3.1.5). Thus the matrix of $(\varphi_*)_p$ with respect to the bases

$$\left(\left(\frac{\partial}{\partial x^i} \right)_p \right)_{i=1}^m \quad \text{and} \quad \left(\left(\frac{\partial}{\partial y^j} \right)_{\varphi(p)} \right)_{j=1}^n$$

is

$$\left(\frac{\partial(y^i \circ \varphi)}{\partial x^j}(p) \right) \in M_{n \times m}(\mathbb{R}),$$

called the *Jacobian matrix* of $(\varphi_*)_p$ with respect to the local coordinate systems $(x^i)_{i=1}^m$ and $(y^j)_{j=1}^n$. Since

$$\frac{\partial(y^i \circ \varphi)}{\partial x^j}(p) = D_j(y^i \circ \varphi \circ x^{-1})(x(p)) = D_j(e^i \circ y \circ \varphi \circ x^{-1})(x(p))$$

$$= \left(J_{y \circ \varphi \circ x^{-1}}(x(p)) \right)^i_j,$$

it follows that *the Jacobian matrix of $(\varphi_*)_p$ with respect to (x^i) and (y^j) is just the Jacobian matrix of the coordinate representative $y \circ \varphi \circ x^{-1}$ of φ at $x(p)$*.

Lemma 3.1.13 (Chain Rule). *If $\varphi \colon M \to N$ and $\psi \colon N \to S$ are differentiable mappings, then*

$$((\psi \circ \varphi)_*)_p = (\psi_*)_{\varphi(p)} \circ (\varphi_*)_p, \quad p \in M. \tag{3.1.6}$$

Proof. For any tangent vector $v \in T_p M$ and any smooth function h on S,

$$((\psi \circ \varphi)_*)_p(v)(h) \overset{(3.1.4)}{=} v(h \circ \psi \circ \varphi) \overset{(3.1.4)}{=} (\varphi_*)_p(v)(h \circ \psi)$$

$$= (\psi_*)_{\varphi(p)} \circ (\varphi_*)_p(v)(h),$$

whence our claim. $\square$

Corollary 3.1.14. *Let M and N be manifolds. Then $((1_M)_*)_p = 1_{T_p M}$ for all $p \in M$. If $\varphi \in \mathrm{Diff}^k(M, N)$, then $(\varphi_*)_p$ is a linear isomorphism with inverse $((\varphi^{-1})_*)_{\varphi(p)}$.* $\square$

Remark 3.1.15. Consider the real line $\mathbb{R} = \mathbb{R}^1$ as a manifold (Example 2.1.9). The canonical coordinate system of $\mathbb{R}$ is $(r) := (1_\mathbb{R})$, the coordinate vector of this coordinate system at $t \in \mathbb{R}$ is $\left(\frac{d}{dr}\right)_t$. Then

$$\left(\frac{d}{dr}\right)_t (h) = h'(t) \text{ for all } h \in C^\infty(\mathbb{R}). \tag{3.1.7}$$

Definition 3.1.16. Let M be a manifold, $I \subset \mathbb{R}$ a nonempty open interval, and consider a curve $\alpha \colon I \to M$ of class C^k. The *tangent vector* (or *velocity vector*) of α at $t \in I$ is

$$\dot{\alpha}(t) := (\alpha_*)_t \left(\frac{d}{dr}\right)_t \in T_{\alpha(t)} M. \tag{3.1.8}$$

The curve α is called *regular* if $\dot{\alpha}(t) \neq 0$ for all $t \in I$.

Remark 3.1.17. With the hypotheses and notation of the preceding definition, we list some basic properties of the tangent vector of a curve.

(a) As an immediate consequence of (3.1.8) and (3.1.7), we have

$$\dot{\alpha}(t)(f) = (f \circ \alpha)'(t), \quad f \in C^\infty(M). \tag{3.1.9}$$

(b) If $\tilde{I} \subset \mathbb{R}$ is an open interval and $\theta \colon \tilde{I} \to I$ is a differentiable function, then the tangent vector of the *reparametrization* $\alpha \circ \theta \colon \tilde{I} \to M$ of α at $s \in \tilde{I}$ is

$$(\alpha \circ \theta)'(s) = \theta'(s)\dot{\alpha}(\theta(s)). \tag{3.1.10}$$

(c) If $\varphi \colon M \to N$ is a mapping of class C^k, then $\varphi \circ \alpha \colon I \to N$ is a curve of class C^k in N, whose tangent vector at $t \in I$ is

$$(\varphi \circ \alpha)'(t) = (\varphi_*)_{\alpha(t)}(\dot{\alpha}(t)). \tag{3.1.11}$$

Thus *the derivatives of a differentiable mapping send the tangent vectors of a curve to the tangent vectors of the composite curve.*

(d) Let $(\mathcal{U}, (u^i)_{i=1}^n)$ be a chart of M at $\alpha(t)$. Then

$$\dot{\alpha}(t) = \sum_{i=1}^n (u^i \circ \alpha)'(t) \left(\frac{\partial}{\partial u^i}\right)_{\alpha(t)}. \tag{3.1.12}$$

Lemma 3.1.18. *Every tangent vector to a manifold can be obtained as a tangent vector to a curve in the manifold.*

Proof. Let M be a manifold, p a point in M and $v \in T_pM$. Choose a chart $(\mathcal{U}, (u^i)_{i=1}^n)$ for M, centred on p. By the basis theorem

$$v = \sum_{i=1}^n \nu^i \left(\frac{\partial}{\partial u^i} \right)_p, \quad \nu^i = v(u^i).$$

Consider the curve

$$\alpha \colon\,]-1, 1[\, \to M, \ t \mapsto \alpha(t) := u^{-1}(t\nu^1, \ldots, t\nu^n).$$

Then

$$u^i \circ \alpha(t) = e^i \circ u \circ u^{-1}(t\nu^1, \ldots, t\nu^n) = t\nu^i,$$

hence

$$\dot{\alpha}(0) \overset{(3.1.12)}{=} \sum_{i=1}^n (u^i \circ \alpha)'(0) \left(\frac{\partial}{\partial u^i} \right)_{\alpha(0)} = \sum_{i=1}^n \nu^i \left(\frac{\partial}{\partial u^i} \right)_p = v.$$

$\square$

Lemma 3.1.19. *Let V be a finite-dimensional real vector space endowed with its canonical smooth structure. For every $p \in V$, V may be naturally identified with its tangent space T_pV by the mapping*

$$\begin{cases} \iota_p \colon V \to T_pV, \ v \mapsto \iota_p(v) := \dot{\alpha}(v), \\ \text{where } \alpha \colon \mathbb{R} \to V, \ t \mapsto \alpha(t) := p + tv. \end{cases}$$

If $y \colon V \to \mathbb{R}^n$ is a linear isomorphism, and hence $(V, y) = (V, (y^1, \ldots, y^n))$ is a global chart for V, then

$$\iota_p(v) = \sum_{i=1}^n y^i(v) \left(\frac{\partial}{\partial y^i} \right)_p. \tag{3.1.13}$$

Proof. For every $v \in V$,

$$\iota_p(v) = \dot{\alpha}(0) \overset{(3.1.12)}{=} \sum_{i=1}^n y^i(v) \left(\frac{\partial}{\partial y^i} \right)_{\alpha(0)} = \sum_{i=1}^n y^i(v) \left(\frac{\partial}{\partial y^i} \right)_p.$$

So ι_p is linear and injective, therefore it is a linear isomorphism. $\square$

3.1.3 Some Local Properties of Differentiable Mappings

Definition 3.1.20. Let M and N be manifolds. A mapping $\varphi\colon M \to N$ of class C^k is called an *immersion* at a point $p \in M$ if $(\varphi_*)_p$ is injective and a *submersion* if $(\varphi_*)_p$ is surjective. We say that φ is an *immersion* or a *submersion of M into N* if the relevant property holds at every point of M.

Corollary 3.1.21 (Inverse Mapping Theorem). *Let $\varphi\colon M \to N$ be a mapping of class C^k. The derivative of φ at a point $p \in M$ is a linear isomorphism if, and only if, φ is a local diffeomorphism of class C^k at the point p.*

Proof. The assertion is an immediate consequence of the classical inverse mapping theorem (Theorem C.1.24). □

Definition 3.1.22. A smooth mapping $\varphi\colon M \to N$ is said to be an *embedding* of M into N if it is an immersion, and the induced mapping

$$\bar{\varphi}\colon M \to \varphi(M), \quad p \mapsto \bar{\varphi}(p) := \varphi(p)$$

is a homeomorphism when $\varphi(M)$ is endowed with the subspace topology.

Proposition 3.1.23. (i) *If S is a submanifold of a manifold M (Example 2.1.13), then the canonical inclusion of S into M is an embedding.*

(ii) *Let M and N be manifolds, and let $\varphi\colon N \to M$ be a smooth mapping. If φ is an immersion at a point $p \in N$, then there exists an open neighbourhood $\mathcal{U}$ of p such that $\varphi \restriction \mathcal{U}$ is an embedding of $\mathcal{U}$ into M. Then $\varphi(\mathcal{U})$ is a submanifold of M and $\varphi \restriction \mathcal{U}$ is a diffeomorphism of $\mathcal{U}$ onto $\varphi(\mathcal{U})$.*

(iii) *If $\varphi\colon N \to M$ is an embedding, then $\varphi(N)$ is a submanifold of M, and φ is a diffeomorphism between N and $\varphi(N)$.*

(iv) *If N is a compact manifold and φ is an injective immersion from N into a manifold M, then φ is an embedding, and hence $\varphi(N)$ is a submanifold of M.*

For a proof we refer to [13], p. 88. An equally careful treatment of these delicate results can also be found, e.g., in [69] and [66].

Remark 3.1.24. Suppose that S is a submanifold of a manifold M, and let $j\colon S \to M$ be the canonical inclusion. Then for every $f \in C^\infty(M)$,

$$f \restriction S = f \circ j\colon S \to M \to \mathbb{R},$$

so $f \restriction S \in C^\infty(S)$. On the other hand, the derivatives

$$(j_*)_p\colon T_pS \to T_pM, \quad p \in S$$

are (canonical) injective linear mappings. Thus $(j_*)_p$ can be ignored, and we can consider the tangent space T_pS as a vector subspace of T_pM.

Definition 3.1.25. Let $\psi \colon M \to N$ be a smooth mapping. A point $q \in N$ is called a *regular value* of ψ if for each $p \in \psi^{-1}(q)$, the derivative $(\psi_*)_p \colon T_pM \to T_{\psi(p)}N$ is surjective. (Then, in particular, the points of $N \setminus \psi(M)$ are regular values.)

Proposition 3.1.26 (Regular Value Theorem). *Let M be an n-dimensional, N a k-dimensional manifold, and let $\psi \colon M \to N$ be a smooth mapping. If $q \in \psi(M)$ is a regular value of ψ, then $\psi^{-1}(q)$ is a submanifold of M of dimension $n - k$.*

For a proof, see [69], p. 182.

3.1.4 The Tangent Bundle of a Manifold

Lemma 3.1.27. *Let M be an n-dimensional manifold $(n \geq 1)$. Form the disjoint union $TM := \cup_{p \in M} T_pM$ of tangent spaces to M, and consider the 'footpoint mapping'*

$$\tau \colon TM \to M, \ v \mapsto \tau(v) := p \ if \ v \in T_pM.$$

The topology and smooth structure of M induces a unique topology and smooth structure on TM such that for each chart $(\mathcal{U}, u) = (\mathcal{U}, (u^i)_{i=1}^n)$ on M,

$$\begin{cases} (\tau^{-1}(\mathcal{U}), (x, y)) = \left(\tau^{-1}(\mathcal{U}), ((x^i)_{i=1}^n, (y^i)_{i=1}^n)\right), \\ x^i := u^i \circ \tau, \ y^i(v) := v(u^i), \ i \in J_n \end{cases} \tag{3.1.14}$$

is a chart of TM. Then the mapping τ is smooth.

Sketch of proof. (a) The mapping

$$(x, y) \colon \tau^{-1}(\mathcal{U}) \to u(\mathcal{U}) \times \mathbb{R}^n \subset \mathbb{R}^n \times \mathbb{R}^n, \ v \mapsto \left(u(\tau(v)), (y^1(v), \dots, y^n(v))\right)$$

is clearly bijective, and its inverse is given by

$$\left(u(p), (\alpha^1, \dots, \alpha^n)\right) \mapsto \sum_{i=1}^n \alpha^i \left(\frac{\partial}{\partial u^i}\right)_p.$$

Choose the topology on TM in such a way that for every chart $(\mathcal{U}, u)$ of an atlas of M the mapping (x, y) becomes a homeomorphism. Then this topology is a second countable Hausdorff topology.

(b) Let $(\mathcal{U}, u)$ and $(\widetilde{\mathcal{U}}, \widetilde{u})$ be two charts on M such that $\mathcal{U} \cap \widetilde{\mathcal{U}} \neq \emptyset$. Consider the induced charts

$$(x, y) \colon \tau^{-1}(\mathcal{U}) \to u(\mathcal{U}) \times \mathbb{R}^n \text{ and } (\widetilde{x}, \widetilde{y}) \colon \tau^{-1}(\widetilde{\mathcal{U}}) \to \widetilde{u}(\widetilde{\mathcal{U}}) \times \mathbb{R}^n$$

on TM given by (3.1.14). The transition homeomorphism $(x, y) \circ (\widetilde{x}, \widetilde{y})^{-1}$ maps the open subset

$$(\widetilde{x}, \widetilde{y})\big(\tau^{-1}(\mathcal{U}) \cap \tau^{-1}(\widetilde{\mathcal{U}})\big) = \widetilde{u}(\mathcal{U} \cap \widetilde{\mathcal{U}}) \times \mathbb{R}^n$$

onto the open subset $u(\mathcal{U} \cap \widetilde{\mathcal{U}}) \times \mathbb{R}^n$. Let $(\widetilde{e}^{\,i})_{i=1}^{2n}$ be the canonical coordinate system of $\mathbb{R}^{2n} = \mathbb{R}^n \times \mathbb{R}^n$. We obtain by a simple calculation that for each $i \in J_n$,

$$\widetilde{e}^{\,i} \circ (x, y) \circ (\widetilde{x}, \widetilde{y})^{-1} = \widetilde{e}^{\,i} \circ u \circ \widetilde{u}^{-1} \circ \mathrm{pr}_1,$$

$$\widetilde{e}^{\,n+i} \circ (x, y) \circ (\widetilde{x}, \widetilde{y})^{-1} = \sum_{j=1}^{n} (\widetilde{e}^{\,n+j} \circ \mathrm{pr}_2) \left(\frac{\partial u^i}{\partial \widetilde{u}^j} \circ \widetilde{u}^{-1} \circ \mathrm{pr}_1 \right),$$

where pr_1 and pr_2 are the first and the second projections of $\mathbb{R}^n \times \mathbb{R}^n$ onto $\mathbb{R}^n$. These relations prove the smoothness of $(x, y) \circ (\widetilde{x}, \widetilde{y})^{-1}$.

(c) The coordinate expression of $\tau \colon TM \to M$ with respect to the charts $(\tau^{-1}(\mathcal{U}), (x, y))$ and $(\mathcal{U}, u)$ is

$$u \circ \tau \circ (x, y)^{-1} = \mathrm{pr}_1 \restriction u(\mathcal{U}) \times \mathbb{R}^n,$$

therefore τ is a smooth mapping. $\qquad\qquad\square$

Remark 3.1.28. In what follows we assume that TM is endowed with the smooth structure described above. The manifold so obtained is called the *tangent manifold* of M. If $(\mathcal{U}, u)$ is a chart of M, then the chart $(\tau^{-1}(\mathcal{U}), (x, y))$ defined by (3.1.14) will be mentioned as the chart *induced by* $(\mathcal{U}, u)$ on TM.

Lemma 3.1.29. *Let M be an n-dimensional manifold ($n \geq 1$) and consider its tangent manifold TM. Then the triple (TM, τ, M) (or the footpoint mapping $\tau \colon TM \to M$) is a vector bundle with typical fibre $\mathbb{R}^n$ and with the tangent spaces $T_p M$ ($p \in M$) as fibres.*

Proof. It is sufficient to show that for every chart $(\mathcal{U}, u) = (\mathcal{U}, (u^i)_{i=1}^n)$ of M, the mapping

$$\psi \colon \mathcal{U} \times \mathbb{R}^n \to \tau^{-1}(\mathcal{U}), \quad (q, (\alpha^1, \ldots, \alpha^n)) \mapsto \sum_{i=1}^{n} \alpha^i \left(\frac{\partial}{\partial u^i} \right)_q$$

is a trivializing map for τ such that for any fixed $p \in \mathcal{U}$, the mapping

$$\psi_p \colon \mathbb{R}^n \to T_pM, \ a \mapsto \psi_p(a) := \psi(p, a)$$

is a linear isomorphism (see Definitions 2.2.1 and 2.2.16).

The latter condition obviously holds, and it is also clear that $\tau \circ \psi = \mathrm{pr}_1$. Let $(\tau^{-1}(\mathcal{U}), (x, y))$ be the chart induced by $(\mathcal{U}, u)$. Then

$$\psi = (x, y)^{-1} \circ (u \times 1_{\mathbb{R}^n}),$$

where $u \times 1_{\mathbb{R}^n}(q, a) \overset{(A.2.2)}{:=} (u(q), 1_{\mathbb{R}^n}(a)) = (u(q), a)$. Since both (x, y) and $u \times 1_{\mathbb{R}^n}$ are diffeomorphisms, so is ψ. $\qquad \square$

Remark 3.1.30. (a) The vector bundle (TM, τ, M) is called the *tangent bundle* of the manifold M. In this context we say that the footpoint mapping τ is the *tangent bundle projection*. Sometimes we use the more precise notation τ_M instead of τ.

(b) In the same way, we may construct the tangent bundle

$$(TTM, \tau_{TM}, TM) \text{ or } \tau_{TM} \colon TTM \to TM$$

of the tangent manifold TM. However, as we shall see in Proposition 4.1.10, TTM carries another, quite distinct, vector bundle structure with the same base manifold TM.

(c) Concerning the sections of the tangent bundle of a manifold, we use the terminology and partly the notation introduced in Definition 2.2.7. However, the sections of class C^r ($r \geq 0$) of $\tau \colon TM \to M$ will be called *vector fields of class C^r* on M. In particular, the smooth sections (resp. smooth local sections) of τ will be mentioned as *vector fields* (resp. *local vector fields*) on M. The $C^\infty(M)$-module of vector fields on M will be denoted by $\mathfrak{X}(M)$. Thus

$$\mathfrak{X}(M) := \Gamma(TM) := \big\{ X \in C^\infty(M, TM) \mid \tau \circ X = 1_M \big\}.$$

If $\mathcal{U}$ is an open subset of M, then

$$\mathfrak{X}(\mathcal{U}) := \Gamma_{\mathcal{U}}(TM) := \big\{ X \in C^\infty(\mathcal{U}, TM) \mid \tau \circ X = 1_{\mathcal{U}} \big\}.$$

(d) We now specialize to the case of the tangent bundle the general concept of a section along a smooth mapping introduced in Definition 2.2.11. Let M be a manifold and $\gamma \colon I \to M$ a smooth curve, where I is a nonempty open interval in $\mathbb{R}$. By a *vector field along γ* we mean a section of TM along γ, i.e., a smooth mapping $X \colon I \to TM$ such that $\tau \circ X = \gamma$. Then $\Gamma_\gamma(TM)$ is a $C^\infty(I)$-module, which we denote also by $\mathfrak{X}_\gamma(M)$. Thus

$$\mathfrak{X}_\gamma(M) := \big\{ X \in C^\infty(I, TM) \mid \tau \circ X = \gamma \big\}.$$

Obviously, if $Y \in \mathfrak{X}(M)$, then $Y \circ \gamma \in \mathfrak{X}_\gamma(M)$.

Example 3.1.31 (Coordinate Vector Fields). Let M be a manifold, and let $(\mathcal{U}, (u^i)_{i=1}^n)$ be a chart of M. Then the mappings

$$\frac{\partial}{\partial u^i} : \mathcal{U} \to TM, \quad p \mapsto \left(\frac{\partial}{\partial u^i}\right)(p) := \left(\frac{\partial}{\partial u^i}\right)_p, \quad i \in J_n$$

are local vector fields, called the *coordinate vector fields* of the given chart. Indeed, $\frac{\partial}{\partial u^i}$ is a local section of τ for all $i \in J_n$. To see its smoothness, consider the chart $(\tau^{-1}(\mathcal{U}), ((x^i)_{i=1}^n, (y^i)_{i=1}^n))$ induced by $(\mathcal{U}, (u^i)_{i=1}^n)$ on TM. Then for each $k \in J_n$,

$$x^k \circ \frac{\partial}{\partial u^i} \circ u^{-1} = e^k \upharpoonright u(\mathcal{U}), \quad y^k \circ \frac{\partial}{\partial u^i} \circ u^{-1} = \delta_i^k \mathbf{1} \circ u^{-1},$$

where $\mathbf{1}$ is the identity element of the ring $C^\infty(\mathcal{U})$. This implies the desired property.

Thus $\left(\frac{\partial}{\partial u^i}\right)_{i=1}^n$ is a frame field of TM over $\mathcal{U}$ (Definition 2.2.25). It follows from Lemma 2.2.28 that a (local) section $X : \mathcal{U} \to TM$ is a vector field on $\mathcal{U}$ (i.e., $X \in C^\infty(\mathcal{U}, TM)$) if, and only if, there exists a family $(X^i)_{i=1}^n$ of smooth functions on $\mathcal{U}$ such that $X = \sum_{i=1}^n X^i \frac{\partial}{\partial u^i}$.

Lemma 3.1.32. *Let M be a manifold, $\mathcal{V}$ an open subset of M, and consider a section $X : \mathcal{V} \to TM$. For each $f \in C^\infty(\mathcal{U})$, where $\mathcal{U} \subset \mathcal{V}$ is an open set, define a function Xf on $\mathcal{U}$ by*

$$(Xf)(p) := X(p)(f) = X_p(f), \quad p \in \mathcal{U}. \tag{3.1.15}$$

Then the following are equivalent:

(i) $X \in C^\infty(\mathcal{V}, TM)$, *therefore X is a vector field on M.*
(ii) *If $(\mathcal{U}, (u^i)_{i=1}^n)$ is a chart with $\mathcal{U} \subset \mathcal{V}$, then $Xu^i \in C^\infty(\mathcal{U})$ for all $i \in J_n$.*
(iii) *If $f \in C^\infty(\mathcal{V})$, then $Xf \in C^\infty(\mathcal{V})$.*

Proof. It can be checked immediately that

$$X \upharpoonright \mathcal{U} = \sum_{i=1}^n (Xu^i) \frac{\partial}{\partial u^i}. \tag{3.1.16}$$

Since $\left(\frac{\partial}{\partial u^i}\right)_{i=1}^n$ is a frame field of TM over $\mathcal{U}$, the preceding example (or, directly, Lemma 2.2.28) implies the equivalence of statements (i)–(iii). $\square$

Lemma 3.1.33. *Let $\varphi : M \to N$ be a smooth mapping, and define a mapping $\varphi_* : TM \to TN$ by*

$$\varphi_*(v) := (\varphi_*)_p(v) \text{ if } v \in T_p M.$$

Then φ_* is a bundle map from τ_M into τ_N inducing the mapping φ between the base manifolds.

Proof. Obviously, φ_* is fibre preserving and fibrewise linear. To show the smoothness of φ_*, choose a chart $(\mathcal{U}, (u^i)_{i=1}^n)$ on M, a chart $(\widetilde{\mathcal{U}}, (\widetilde{u}^j)_{j=1}^k)$ on N such that $\varphi(\mathcal{U}) \cap \widetilde{\mathcal{U}} \neq \emptyset$, and consider the chart $(\tau^{-1}(\mathcal{U}), ((x^i)_{i=1}^n, (y^i)_{i=1}^n))$ induced by $(\mathcal{U}, u)$ on TM. Applying (3.1.5), we obtain by a straightforward calculation that

$$\varphi_* \upharpoonright \tau^{-1}(\mathcal{U}) = \sum_{i=1}^k \sum_{j=1}^n y^j \left(\frac{\partial(\widetilde{u}^i \circ \varphi)}{\partial u^j} \circ \tau \right) \left(\frac{\partial}{\partial \widetilde{u}^i} \circ \varphi \circ \tau \right), \qquad (3.1.17)$$

hence φ_* is smooth. $\qquad\qquad\qquad\qquad\qquad\qquad\qquad\qquad\qquad\qquad\square$

Remark 3.1.34. The mapping $\varphi_* \colon TM \to TN$ is called the *derivative* of φ. If $\varphi \in \mathrm{Diff}(M)$, then $\varphi_* \in \mathrm{Aut}(TM)$, $\varphi_{**} := (\varphi_*)_* \in \mathrm{Aut}(TTM)$ (see the notation introduced in Remark 2.2.6). From (3.1.17) it is easy to deduce that φ_{**} can be described locally as follows:

$$\begin{cases} (\varphi_{**})_v \left(\dfrac{\partial}{\partial x^j} \right)_v = \left(\dfrac{\partial(\widetilde{u}^i \circ \varphi)}{\partial u^j} \circ \tau \right)(v) \left(\dfrac{\partial}{\partial \widetilde{x}^i} \right)_{\varphi_*(v)} \\ \qquad\qquad + y^k(v) \left(\dfrac{\partial^2(\widetilde{u}^i \circ \varphi)}{\partial u^j \partial u^k} \circ \tau \right)(v) \left(\dfrac{\partial}{\partial \widetilde{y}^i} \right)_{\varphi_*(v)} \\ (\varphi_{**})_v \left(\dfrac{\partial}{\partial y^j} \right)_v = \left(\dfrac{\partial(\widetilde{u}^i \circ \varphi)}{\partial u^j} \circ \tau \right)(v) \left(\dfrac{\partial}{\partial \widetilde{y}^i} \right)_{\varphi_*(v)} \end{cases}, \qquad (3.1.18)$$

where $v \in \tau^{-1}(\mathcal{U})$, $j \in J_n$.

Given a further smooth mapping $\psi \colon N \to S$, the chain rule, formulated in Lemma 3.1.13 pointwise, takes the elegant form

$$(\psi \circ \varphi)_* = \psi_* \circ \varphi_*. \qquad (3.1.19)$$

Thus the diagram

$$\begin{array}{ccc} M & \xrightarrow{\ \varphi\ } & N \\ {\scriptstyle \psi \circ \varphi} \searrow & & \swarrow {\scriptstyle \psi} \\ & S & \end{array}$$

induces the commutative diagram

$$\begin{array}{ccc} TM & \xrightarrow{\ \varphi_*\ } & TN \\ {\scriptstyle (\psi \circ \varphi)_*} \searrow & & \swarrow {\scriptstyle \psi_*} \\ & TS & \end{array} \ .$$

Example 3.1.35. Let X be a vector field on M. We give a coordinate expression for the derivative $X_* \colon TM \to TTM$. To make our formulae shorter, we shall use the summation convention introduced in Remark 1.1.12(d).

Choose a chart $(\mathcal{U}, (u^i)_{i=1}^n)$ on M and consider the induced chart $(\tau^{-1}(\mathcal{U}), ((x^i)_{i=1}^n, (y^i)_{i=1}^n))$ on TM. If $X \underset{(\mathcal{U})}{=} X^i \frac{\partial}{\partial u^i}$, then

$$x^i \circ X = u^i, \quad y^i \circ X = X^i \quad (i \in J_n). \tag{$*$}$$

Indeed, $x^i \circ X \overset{(3.1.14)}{=} u^i \circ \tau \circ X = u^i$, and for each $p \in \mathcal{U}$,

$$y^i \circ X(p) = y^i(X(p)) \overset{(3.1.14)}{=} X(p)(u^i) = (Xu^i)(p) \overset{(3.1.16)}{=} X^i(p).$$

Now, applying (3.1.5) and taking into account $(*)$, for any $v = v^i \left(\frac{\partial}{\partial u^i}\right)_p$ $(p \in \mathcal{U})$ we have

$$(X_*)_p(v) = (X_*)_p\left(v^i \left(\frac{\partial}{\partial u^i}\right)_p\right) = v^i (X_*)_p \left(\frac{\partial}{\partial u^i}\right)_p$$

$$= v^j \left(\frac{\partial (x^i \circ X)}{\partial u^j}(p) \left(\frac{\partial}{\partial x^i}\right)_{X(p)} + \frac{\partial (y^i \circ X)}{\partial u^j}(p) \left(\frac{\partial}{\partial y^i}\right)_{X(p)}\right)$$

$$= v^j \left(\left(\frac{\partial}{\partial x^j}\right)_{X(p)} + \frac{\partial X^i}{\partial u^j}(p) \left(\frac{\partial}{\partial y^i}\right)_{X(p)}\right).$$

Thus

$$X_* \upharpoonright \tau^{-1}(\mathcal{U}) = y^j \left(\frac{\partial}{\partial x^j} \circ X \circ \tau + \left(\frac{\partial X^i}{\partial u^j} \circ \tau\right)\left(\frac{\partial}{\partial y^i} \circ X \circ \tau\right)\right). \tag{3.1.20}$$

Lemma 3.1.36 (Tangent Vectors to a Product Manifold). *Let M and N be manifolds and consider the product manifold $M \times N$. Let pr_1 and pr_2 be the canonical projections of $M \times N$ onto M and N, respectively. For each $(a, b) \in M \times N$, the tangent space of $M \times N$ at (a, b) is canonically isomorphic to the direct sum $T_a M \oplus T_b N$ via the mapping*

$$\varphi_{a,b} \colon T_{(a,b)}(M \times N) \to T_a M \oplus T_b N$$

given by

$$\varphi_{a,b}(w) := ((\mathrm{pr}_1)_*(w), (\mathrm{pr}_2)_*(w)), \quad w \in T_{(a,b)}(M \times N). \tag{3.1.21}$$

Proof. Obviously, $\varphi_{(a,b)}$ is a linear mapping. We show that it has a (necessarily unique, linear) inverse. Introduce the mappings

$$_a j \colon N \to M \times N, \quad q \mapsto {}_a j(q) := (a, q),$$

$$j_b \colon M \to M \times N, \quad p \mapsto j_b(p) := (p, b),$$

$$_a c \colon N \to M, \quad q \mapsto {}_a c(q) := a,$$

$$c_b \colon M \to N, \quad p \mapsto c_b(p) := b.$$

(We say that $_a j$ and j_b are the *inclusion mappings opposite* $a \in M$ and $b \in N$, respectively.) Then we have

$$\mathrm{pr}_1 \circ {_a j} = {_a c}, \ \mathrm{pr}_1 \circ j_b = 1_M, \ \mathrm{pr}_2 \circ {_a j} = 1_N, \ \mathrm{pr}_2 \circ j_b = c_b,$$

and the chain rule (3.1.19) leads to the relations

$$\left. \begin{array}{ll} (\mathrm{pr}_1)_* \circ ({_a j})_* = 0, & (\mathrm{pr}_1)_* \circ (j_b)_* = 1_{TM}, \\ (\mathrm{pr}_2)_* \circ ({_a j})_* = 1_{TN}, & (\mathrm{pr}_2)_* \circ (j_b)_* = 0. \end{array} \right\} \tag{3.1.22}$$

Now define the mapping

$$\psi_{a,b} \colon T_a M \oplus T_b N \to T_{(a,b)}(M \times N)$$

by

$$\psi_{a,b}(u, v) := (j_b)_*(u) + ({_a j})_*(v), \quad (u, v) \in T_a M \oplus T_b N. \tag{3.1.23}$$

Then $\psi_{a,b}$ is a well-defined linear mapping. Composing $\varphi_{a,b}$ on the right with $\psi_{a,b}$, we find

$$\varphi_{a,b} \circ \psi_{a,b}(u, v) = \varphi_{(a,b)}((j_b)_*(u) + ({_a j})_*(v))$$
$$= ((\mathrm{pr}_1)_*((j_b)_*(u) + ({_a j})_*(v)), (\mathrm{pr}_2)_*((j_b)_*(u) + ({_a j})_*(v)))$$
$$\overset{(3.1.22)}{=} (u + 0_a, 0_b + v) = (u, v)$$

for all $(u, v) \in T_a M \oplus T_b N$. Thus

$$\varphi_{a,b} \circ \psi_{a,b} = 1_{T_a M \oplus T_b N}. \tag{$*$}$$

Since both $T_{(a,b)}(M \times N)$ and $T_a M \oplus T_b N$ are $(\dim M + \dim N)$-dimensional real vector spaces, $(*)$ implies that $\varphi_{a,b}$ and $\psi_{a,b}$ are inverse isomorphisms. $\square$

Remark 3.1.37. By the lemma, we can freely identify the vector spaces $T_{(a,b)}(M \times N)$ and $T_a M \oplus T_b N$ via $\varphi_{a,b}$ or $\psi_{a,b}$. Thus for any tangent vector

$$w = (w_1, w_2) := ((\mathrm{pr}_1)_*(w), (\mathrm{pr}_2)_*(w)) \in T_a M \oplus T_b N$$

and any function $f \in C^\infty(M \times N)$ we have

$$w(f) = w_1(f \circ j_b) + w_2(f \circ {_a j}). \tag{3.1.24}$$

Indeed,

$$w(f) = \psi_{a,b}(w_1, w_2)(f) \overset{(3.1.23)}{=} (j_b)_*(w_1)(f) + ({_a j})_*(w_2)(f)$$
$$\overset{(3.1.4)}{=} w_1(f \circ j_b) + w_2(f \circ {_a j}).$$

Example 3.1.38. Let M be a manifold. Consider the tangent bundle $\tau \colon TM \to M$ of M and the tangent bundle $\tau_{TM} \colon TTM \to TM$ of TM (the latter is also mentioned as the *double tangent bundle*). Choose a chart $(\mathcal{U}, (u^i)_{i=1}^n)$ on M and let $\left(\tau^{-1}(\mathcal{U}), ((x^i)_{i=1}^n, (y^i)_{i=1}^n)\right)$ be the induced chart on TM.

(a) We denote by μ_λ and by $\widetilde{\mu}_\lambda$ the multiplication by a real number λ on TM and on TTM, respectively. Then, for example,

$$\mu_\lambda \colon TM \to TM, \quad v \mapsto \mu_\lambda(v) := \lambda v.$$

If $\lambda \neq 0$, μ_λ is an automorphism of TM. We say that μ_λ is the *dilation* of TM by λ. In particular, the dilation

$$\varrho := \mu_{-1} \colon v \in TM \mapsto -v \in TM$$

is called the *reflection map* of TM.

The component functions of μ_λ with respect to the local coordinate system $((x^i), (y^i))$ are

$$x^i \circ \mu_\lambda = x^i, \quad y^i \circ \mu_\lambda = \lambda y^i; \quad i \in J_n.$$

Thus, applying (3.1.5), we find that

$$(\mu_\lambda)_* \left(\frac{\partial}{\partial x^i}\right)_v = \sum_{j=1}^n \left(\frac{\partial x^j}{\partial x^i}(v)\left(\frac{\partial}{\partial x^j}\right)_{\lambda v} + \lambda \frac{\partial y^j}{\partial x^i}(v)\left(\frac{\partial}{\partial y^j}\right)_{\lambda v}\right) = \left(\frac{\partial}{\partial x^i}\right)_{\lambda v},$$

$$(\mu_\lambda)_* \left(\frac{\partial}{\partial y^i}\right)_v = \sum_{j=1}^n \left(\frac{\partial x^j}{\partial y^i}(v)\left(\frac{\partial}{\partial x^j}\right)_{\lambda v} + \lambda \frac{\partial y^j}{\partial y^i}(v)\left(\frac{\partial}{\partial y^j}\right)_{\lambda v}\right) = \lambda \left(\frac{\partial}{\partial y^i}\right)_{\lambda v}$$

for all $v \in \tau^{-1}(\mathcal{U})$. So we have

$$(\mu_\lambda)_* \circ \frac{\partial}{\partial x^i} = \frac{\partial}{\partial x^i} \circ \mu_\lambda, \quad (\mu_\lambda)_* \circ \frac{\partial}{\partial y^i} = \widetilde{\mu}_\lambda \circ \frac{\partial}{\partial y^i} \circ \mu_\lambda; \quad i \in J_n. \quad (3.1.25)$$

These relations immediately imply that

$$(\mu_\alpha)_* \circ \widetilde{\mu}_\beta = \widetilde{\mu}_\beta \circ (\mu_\alpha)_*; \quad \alpha, \beta \in \mathbb{R}. \quad (3.1.26)$$

(b) Now we describe the action of τ_* on the coordinate vectors of the local coordinate system $((x^i), (y^i))$. For each $i \in J_n$ and $v \in \tau^{-1}(\mathcal{U})$,

$$\tau_* \left(\frac{\partial}{\partial x^i}\right)_v \overset{(3.1.5)}{=} \sum_{j=1}^n \frac{\partial(u^j \circ \tau)}{\partial x^i}(v)\left(\frac{\partial}{\partial u^j}\right)_{\tau(v)}$$

$$= \sum_{j=1}^n \frac{\partial x^j}{\partial x^i}(v)\left(\frac{\partial}{\partial u^j}\right)_{\tau(v)} = \left(\frac{\partial}{\partial u^i} \circ \tau\right)(v).$$

Similarly,

$$\tau_* \left(\frac{\partial}{\partial y^i} \right)_v = \sum_{j=1}^n \frac{\partial (u^j \circ \tau)}{\partial y^i}(v) \left(\frac{\partial}{\partial u^j} \right)_{\tau(v)} = \sum_{j=1}^n \frac{\partial x^j}{\partial y^i}(v) \left(\frac{\partial}{\partial u^j} \right)_{\tau(v)} = 0.$$

We thus obtain

$$\tau_* \circ \frac{\partial}{\partial x^i} = \frac{\partial}{\partial u^i} \circ \tau, \ \tau_* \circ \frac{\partial}{\partial y^i} = 0; \quad i \in J_n. \tag{3.1.27}$$

These imply that for any vector

$$w = \sum_{i=1}^n \left(w^i \left(\frac{\partial}{\partial x^i} \right)_v + w^{n+i} \left(\frac{\partial}{\partial y^i} \right)_v \right) \in T_v TM,$$

where $v \in T_p M$, $p \in \mathcal{U}$, we have

$$\tau_*(w) = \sum_{i=1}^n w^i \left(\frac{\partial}{\partial u^i} \right)_p \in T_{\tau(v)} M = T_p M. \tag{3.1.28}$$

Since

$$\tau(\tau_*(w)) = p \text{ and } \tau(\tau_{TM}(w)) = \tau(v) = p, \tag{3.1.29}$$

it follows that the diagram

$$
\begin{array}{ccc}
TTM & \xrightarrow{\tau_*} & TM \\
\tau_{TM} \downarrow & & \downarrow \tau \\
TM & \xrightarrow[\tau]{} & TM
\end{array}
$$

is commutative.

Example 3.1.39 (Velocity and Acceleration). Let M be a manifold, $I \subset \mathbb{R}$ a nonempty open interval, and consider a smooth curve $\gamma \colon I \to M$. By the *velocity vector field* (or simply *velocity*) of γ we mean the vector field

$$\dot{\gamma} = \gamma_* \circ \frac{d}{dr} \colon I \to TM \tag{3.1.30}$$

along γ. (Since $\dot{\gamma}(t) \in T_{\gamma(t)} M$ for all $t \in I$ by (3.1.8), we indeed have $\dot{\gamma} \in \Gamma_\gamma(TM) = \mathfrak{X}_\gamma(M)$.) We also say that $\dot{\gamma}$ is the *canonical lift* of γ into TM. The velocity vector field of $\dot{\gamma}$ is called the *acceleration vector field* (briefly the *acceleration*) of γ. We denote it by $\ddot{\gamma}$. Thus

$$\ddot{\gamma} \in \Gamma_{\dot{\gamma}}(TTM) = \mathfrak{X}_{\dot{\gamma}}(TM). \tag{3.1.31}$$

The velocity and the acceleration vector field of γ are related by

$$\tau_* \circ \ddot{\gamma} = \dot{\gamma}. \tag{3.1.32}$$

Indeed,

$$\tau_* \circ \ddot{\gamma} \overset{(3.1.30)}{=} \tau_* \circ \left(\gamma_* \circ \frac{d}{dr}\right)_* \circ \frac{d}{dr} \overset{(3.1.19)}{=} \left(\tau \circ \gamma_* \circ \frac{d}{dr}\right)_* \circ \frac{d}{dr}$$

$$\overset{(3.1.30)}{=} (\tau \circ \dot{\gamma})_* \circ \frac{d}{dr} = \gamma_* \circ \frac{d}{dr} =: \dot{\gamma}.$$

Using the definition of the velocity vector field of a smooth curve and the chain rule (3.1.19), we obtain by a similar calculation that

$$(\varphi \circ \gamma)^{\cdot} = \varphi_* \circ \dot{\gamma}, \quad (\varphi \circ \gamma)^{\cdot\cdot} = \varphi_{**} \circ \ddot{\gamma} \quad \text{for all } \varphi \in C^\infty(M, M). \quad (3.1.33)$$

Coordinate description. Let $(\mathcal{U}, (u^i)_{i=1}^n)$ be a chart on M such that $\tilde{I} := \gamma^{-1}(\mathcal{U}) \neq \emptyset$, and consider the induced chart $(\tau^{-1}(\mathcal{U}), (x^i), (y^i))$ on TM. Let $\gamma^i := u^i \circ \gamma$, $i \in J_n$. Then

$$\dot{\gamma} \restriction \tilde{I} = \sum_{i=1}^n (\gamma^i)' \left(\frac{\partial}{\partial u^i} \circ \gamma\right), \quad (3.1.34)$$

$$\ddot{\gamma} \restriction \tilde{I} = \sum_{i=1}^n \left((\gamma^i)' \left(\frac{\partial}{\partial x^i} \circ \dot{\gamma}\right) + (\gamma^i)'' \left(\frac{\partial}{\partial y^i} \circ \dot{\gamma}\right)\right). \quad (3.1.35)$$

Indeed, the first relation is an immediate consequence of (3.1.12). Then

$$\ddot{\gamma} \restriction \tilde{I} = \sum_{i=1}^n \left((x^i \circ \dot{\gamma})' \left(\frac{\partial}{\partial x^i} \circ \dot{\gamma}\right) + (y^i \circ \dot{\gamma})' \left(\frac{\partial}{\partial y^i} \circ \dot{\gamma}\right)\right).$$

Since

$$x^i \circ \dot{\gamma} = u^i \circ \tau \circ \dot{\gamma} = u^i \circ \gamma = \gamma^i$$

and, for all $t \in \tilde{I}$,

$$y^i \circ \dot{\gamma}(t) = y^i(\dot{\gamma}(t)) = \dot{\gamma}(t)(u^i) = \sum_{j=1}^n (\gamma^j)'(t) \frac{\partial u^i}{\partial u^j}(\gamma(t))$$

$$= \sum_{j=1}^n (\gamma^j)'(t) \delta_j^i = (\gamma^i)'(t),$$

relation (3.1.35) is also true.

For later reference, we fix the following formulae:

$$x^i \circ \dot{\gamma} = u^i \circ \gamma = \gamma^i, \quad y^i \circ \dot{\gamma} = (\gamma^i)'; \quad i \in J_n. \quad (3.1.36)$$

Lemma 3.1.40. *A smooth curve $c: I \to TM$ is the canonical lift (i.e., the velocity vector field) of a smooth curve $\gamma: I \to M$ if, and only if, $c = \tau_* \circ \dot{c}$.*

Proof. If $c = \dot{\gamma}$, then $\tau_* \circ \dot{c} = \tau_* \circ \ddot{\gamma} \overset{(3.1.32)}{=} \dot{\gamma} = c$, as we claimed. Conversely, suppose that $c = \tau_* \circ \dot{c}$. Then $\gamma := \tau \circ c: I \to M$ is a smooth curve, whose velocity vector field is

$$\dot{\gamma} = (\tau \circ c)^{\cdot} \overset{(3.1.11)}{=} \tau_* \circ \dot{c} \overset{\text{cond.}}{=} c.$$

So c is the canonical lift of γ. $\square$

3.1.5   The Lie Algebra of Vector Fields

Lemma 3.1.41 (Vector Fields as Derivations). *Let M be a manifold. Then the mapping*

$$\begin{cases} \mathfrak{X}(M) \to \mathrm{Der}(C^\infty(M)), \ X \mapsto \mathcal{L}_X \\ \mathcal{L}_X f := Xf, \ (Xf)(p) \overset{(3.1.15)}{=} X_p(f) \quad (p \in M) \end{cases} \tag{3.1.37}$$

is a canonical isomorphism of the real vector space of vector fields on M onto the real vector space of derivations of the real algebra $C^\infty(M)$.

Proof.   (a) The Leibnizian property (3.1.1) of tangent vectors implies immediately that $\mathcal{L}_X \in \mathrm{Der}(C^\infty(M))$ for all $X \in \mathfrak{X}(M)$. The mapping (3.1.37) is clearly $\mathbb{R}$-linear, and a straightforward verification shows that it is injective.

(b) We show that *every derivation of $C^\infty(M)$ comes from a vector field* according to (3.1.37). Let a derivation $D \in \mathrm{Der}(C^\infty(M))$ be given. Define a mapping $X \colon M \to TM, \ p \mapsto X_p$ by

$$X_p(f) := (Df)(p) \text{ for all } f \in C^\infty(M).$$

Then the derivation properties of D (see 1.3.2) imply that X_p is indeed a tangent vector to M at p, so X is a section of TM. It follows from the definition of X that $Xf = Df$ for all $f \in C^\infty(M)$. Since $Df \in C^\infty(M)$, this implies the smoothness of X by Lemma 3.1.32. Thus $X \in \mathfrak{X}(M)$ and $D = \mathcal{L}_X$. $\qquad\square$

Remark 3.1.42. In what follows, we shall freely identify $\mathfrak{X}(M)$ with $\mathrm{Der}(C^\infty(M))$, i.e., *we shall consider vector fields to be derivations of $C^\infty(M)$* if it is so convenient. We also say that $\mathcal{L}_X f = Xf$ is the *Lie derivative of f with respect to X*.

Corollary 3.1.43. *Let M be a manifold. Given two vector fields X, Y on M, there is a unique vector field $[X, Y]$ on M such that*

$$\mathcal{L}_{[X,Y]} = [\mathcal{L}_X, \mathcal{L}_Y] = \mathcal{L}_X \circ \mathcal{L}_Y - \mathcal{L}_Y \circ \mathcal{L}_X. \tag{3.1.38}$$

The mapping

$$\mathfrak{X}(M) \times \mathfrak{X}(M) \to \mathfrak{X}(M), \ (X, Y) \mapsto [X, Y]$$

is a Lie bracket (cf. 1.3.3) that makes $\mathfrak{X}(M)$ into a real Lie algebra. $\qquad\triangle$

Remark 3.1.44. (a) As a vector field on M, the Lie bracket $[X, Y]$ assigns to each point $p \in M$ the tangent vector $[X, Y]_p$ such that

$$[X, Y]_p(f) = X_p(Yf) - Y_p(Xf) \quad \text{for all } f \in C^\infty(M).$$

Later we shall also write $\mathcal{L}_X Y := [X, Y]$. Then we say that $\mathcal{L}_X Y$ is the *Lie derivative of Y with respect to X*.

(b) For any chart $(\mathcal{U}, u) = (\mathcal{U}, (u^i)_{i=1}^n)$ on a manifold M, the Lie brackets of coordinate vector fields associated to the chart vanish identically. This is just a reformulation of the equality of mixed partial derivatives in the context of manifolds.

Indeed, applying Definition 2.1.18, we obtain for any $i, j \in J_n$ and $f \in C^\infty(M)$

$$\left[\frac{\partial}{\partial u^i}, \frac{\partial}{\partial u^j}\right](f) = \frac{\partial}{\partial u^i}\left(D_j(f \circ u^{-1}) \circ u\right) - \frac{\partial}{\partial u^j}\left(D_i(f \circ u^{-1}) \circ u\right)$$

$$= D_i\big(D_j(f \circ u^{-1})\big) \circ u - D_j\big(D_i(f \circ u^{-1})\big) \circ u \overset{\text{Theorem C.1.19}}{=} 0.$$

Lemma 3.1.45. *Let M be a manifold. For any two vector fields X, Y on M and for any function $f \in C^\infty(M)$, we have*

$$[fX, Y] = f[X, Y] - (Yf)X, \quad [X, fY] = f[X, Y] + (Xf)Y. \tag{3.1.39}$$

$$\triangle$$

Remark 3.1.46. (a) If $(\mathcal{U}, u) = (\mathcal{U}, (u^i)_{i=1}^n)$ is a chart on M, $X = \sum_{i=1}^n X^i \frac{\partial}{\partial u^i}$ and $Y = \sum_{i=1}^n Y^i \frac{\partial}{\partial u^i}$ are vector fields on $\mathcal{U}$, then an easy calculation shows that

$$[X, Y] = \sum_{i,j=1}^n \left(X^j \frac{\partial Y^i}{\partial u^j} - Y^j \frac{\partial X^i}{\partial u^j}\right) \frac{\partial}{\partial u^i}. \tag{3.1.40}$$

(b) Perhaps this is the best place to mention that we shall sometimes act with a vector field that is not smooth on a function that is not smooth, in the sense that either of them is only of class C^r with some $r < \infty$. If $(\mathcal{U}, (u^i)_{i=1}^n)$ is a chart of M at $p \in M$, $v \in T_p M$ and $f \in C^\infty(M)$, then, by the basis theorem, we have

$$v(f) = \sum_{i=1}^n v(u^i) \left(\frac{\partial}{\partial u^i}\right)_p (f) = \sum_{i=1}^n v(u^i) D_i(f \circ u^{-1})(u(p)).$$

Note first that the formula on the right-hand side makes sense perfectly even if f is only of class C^r with $r \in \mathbb{N}^*$, thus we may consider it as the definition of $v(f)$ in this case, which is, in addition, clearly independent of the choice of the chart.

Now let $r, s \in \mathbb{N}$, $s \geq 1$, X a vector field on M of class C^r, and f a function on M of class C^s. Then a new function Xf on M may be defined by (3.1.15), which will be of class C^k, where $k := \min\{r, s - 1\}$. The proof is completely analogous to that of Lemma 3.1.32.

Similarly, if X is a vector field of class C^r and Y is a vector field of class C^s with $r, s \geq 1$, then we may define a new vector field $[X, Y]$ over the domain $\mathcal{U}$ of a chart by (3.1.40). Since the construction is local and independent of the choice of the chart, we obtain a global vector field on M in this way, which will be of class C^k with $k := \min\{r - 1, s - 1\}$.

Definition 3.1.47. Let M and N be manifolds. If $\varphi \colon M \to N$ is a smooth mapping, we say that two vector fields $X \in \mathfrak{X}(M)$ and $Y \in \mathfrak{X}(N)$ are *φ-related* if

$$\varphi_* \circ X = Y \circ \varphi, \tag{3.1.41}$$

i.e., the diagram

$$
\begin{array}{ccc}
TM & \xrightarrow{\varphi_*} & TN \\
X\uparrow & & \uparrow Y \\
M & \xrightarrow[\varphi]{} & N
\end{array}
$$

commutes. In this case we write $X \underset{\varphi}{\sim} Y$.

Remark 3.1.48. If $\varphi \colon M \to N$ is a surjective smooth mapping, then for every $X \in \mathfrak{X}(M)$ there exists at most one $Y \in \mathfrak{X}(N)$ such that $X \underset{\varphi}{\sim} Y$.

Indeed, if X is φ-related to $Y_1, Y_2 \in \mathfrak{X}(N)$, then we have $Y_1 \circ \varphi = Y_2 \circ \varphi$ by (3.1.41) and hence $Y_1 = Y_2$ by the surjectivity of φ.

Lemma 3.1.49 (Criterion for φ-relatedness). *Two vector fields X on M and Y on N are φ-related if, and only if,*

$$X(h \circ \varphi) = Yh \circ \varphi \quad \text{for all } h \in C^\infty(N). \tag{3.1.42}$$

The proof is easy; see, e.g., [79], p. 14.

Remark 3.1.50. A smooth mapping $\varphi \colon M \to N$ induces an $\mathbb{R}$-algebra homomorphism

$$\varphi^* \colon C^\infty(N) \to C^\infty(M), \quad h \mapsto \varphi^* h := h \circ \varphi.$$

With the help of this homomorphism, the preceding lemma may be formulated as follows:

$$X \underset{\varphi}{\sim} Y \text{ if, and only if, } X \circ \varphi^* = \varphi^* \circ Y,$$

i.e., the diagram

$$
\begin{array}{ccc}
C^\infty(M) & \xrightarrow{X} & C^\infty(M) \\
\varphi^*\uparrow & & \uparrow\varphi^* \\
C^\infty(N) & \xrightarrow[Y]{} & C^\infty(N)
\end{array}
$$

commutes.

Lemma 3.1.51 (Related Vector Field Lemma). *With the same notation as above, let X and Y be vector fields on M, X_1 and Y_1 vector fields on N. If $X \underset{\varphi}{\sim} X_1$, $Y \underset{\varphi}{\sim} Y_1$, then*

(i) $\lambda X + \mu Y \underset{\varphi}{\sim} \lambda X_1 + \mu Y_1$, $\lambda, \mu \in \mathbb{R}$;

(ii) $(\varphi^* h) X \underset{\varphi}{\sim} h X_1$, $h \in C^\infty(N)$;

(iii) $[X, Y] \underset{\varphi}{\sim} [X_1, Y_1]$.

Proof. We prove part (iii); parts (i) and (ii) are left as easy exercises.

Let $h \in C^\infty(N)$. Then, by Lemma 3.1.49,

$$Y(X_1 h \circ \varphi) = Y_1(X_1 h) \circ \varphi \quad \text{since} \quad Y \underset{\varphi}{\sim} Y_1,$$
$$Y(X_1 h \circ \varphi) = Y(X(h \circ \varphi)) \quad \text{since} \quad X \underset{\varphi}{\sim} X_1,$$

therefore

$$Y(X(h \circ \varphi)) = Y_1(X_1 h) \circ \varphi.$$

In the same way,

$$X(Y(h \circ \varphi)) = X_1(Y_1 h) \circ \varphi.$$

From the last two relations, by subtraction, we obtain that

$$[X, Y](h \circ \varphi) = ([X_1, Y_1]h) \circ \varphi, \quad h \in C^\infty(N).$$

Applying Lemma 3.1.49 again, we conclude $[X, Y] \underset{\varphi}{\sim} [X_1, Y_1]$, as was to be shown. $\qquad \square$

Remark 3.1.52. Let M and N be manifolds, and let $\varphi \colon M \to N$ be a smooth mapping.

(a) A vector field X on M is called *projectable* (by φ) if there exists a vector field Y on N such that $X \underset{\varphi}{\sim} Y$.

(b) If φ is a *diffeomorphism*, then the general push-forward construction described in Remark 2.2.32(b) assigns to each vector field X on M a vector field $\varphi_\# X$ on N. Now we show this by using a somewhat different reasoning. Since φ is a diffeomorphism, the mapping $\varphi^* \colon C^\infty(N) \to C^\infty(M)$ is an isomorphism of rings, so for every vector field X on M there exists a unique vector field $\varphi_\# X$ on N such that

$$(\varphi_\# X)(h) = (\varphi^*)^{-1}(X(\varphi^* h)) \text{ for all } h \in C^\infty(N).$$

Equivalently,

$$\varphi^* \circ \varphi_\# X = X \circ \varphi^*.$$

This shows by Remark 3.1.50 that $X \underset{\varphi}{\sim} \varphi_\# X$. Thus we have $\varphi_* \circ X = \varphi_\# X \circ \varphi$, whence

$$\varphi_\# X = \varphi_* \circ X \circ \varphi^{-1}. \qquad (3.1.43)$$

We say that the vector field $\varphi_\# X$ is the *push-forward* of X by φ. The mapping

$$\varphi_\# : \mathfrak{X}(M) \to \mathfrak{X}(N), \quad X \mapsto \varphi_\# X$$

so obtained is an *isomorphism of Lie algebras*. Indeed, the $\mathbb{R}$-linearity is clear. To check that $\varphi_\#$ preserves the Lie bracket, let X and Y be vector fields on M. Since $X \underset{\varphi}{\sim} \varphi_\# X$ and $Y \underset{\varphi}{\sim} \varphi_\# Y$, Lemma 3.1.51(iii) implies that $[X, Y] \underset{\varphi}{\sim} [\varphi_\# X, \varphi_\# Y]$, i.e., $\varphi_* \circ [X, Y] = [\varphi_\# X, \varphi_\# Y] \circ \varphi$, whence

$$\varphi_\#[X, Y] = [\varphi_\# X, \varphi_\# Y]. \qquad (3.1.44)$$

We recall that

$$\varphi_\#(fX) = (f \circ \varphi^{-1})\varphi_\# X, \quad f \in C^\infty(M) \qquad (3.1.45)$$

(see the next to last assertion in Remark 2.2.32(b)).

(c) A vector field X on M is said to be *invariant* under a diffeomorphism $\psi \in \mathrm{Diff}(M)$ if it is ψ-related to itself, i.e., $\psi_* \circ X = X \circ \psi$, or, equivalently $\psi_\# X = X$. By Remark 3.1.50,

$$\psi_\# X = X \text{ if, and only if, } X \circ \psi^* = \psi^* \circ X.$$

Example 3.1.53. (a) Consider the standard coordinate vector field $\frac{d}{dr}$ on the real line ($r := 1_\mathbb{R}$). We claim that $\frac{d}{dr}$ *is invariant under any translation*

$$\theta_s : \mathbb{R} \to \mathbb{R}, \quad t \mapsto \theta_s(t) := s + t \quad (s \in \mathbb{R})$$

of the real line, i.e.,

$$(\theta_s)_* \circ \frac{d}{dr} = \frac{d}{dr} \circ \theta_s \text{ for all } s \in \mathbb{R}. \qquad (3.1.46)$$

To see this, let $h \in C^\infty(\mathbb{R})$. Then for each $t \in \mathbb{R}$,

$$\left(\frac{d}{dr}(h \circ \theta_s) \right)(t) = \left(\frac{d}{dr} \right)_t (h \circ \theta_s) = (h \circ \theta_s)'(t) = h'(\theta_s(t))\theta_s'(t)$$

$$= h'(\theta_s(t)) = \left(\frac{d}{dr} \right)_{\theta_s(t)} h = \left(\frac{d}{dr} h \right) \circ \theta_s(t),$$

whence

$$\frac{d}{dr}(h \circ \theta_s) = \left(\frac{d}{dr} h \right) \circ \theta_s, \quad h \in C^\infty(\mathbb{R}).$$

In view of Lemma 3.1.49, this proves our assertion.

(b) Let M and N be manifolds, and consider their product manifold $M \times N$. As earlier, we denote by pr_1 and pr_2 the canonical projections of $M \times N$ onto M and N, respectively. Every vector field X on M determines a vector field $i_M X$ on $M \times N$ given by

$$i_M X(p,q) := (X(p), 0_q) \in T_p M \oplus T_q N \cong T_{(p,q)}(M \times N),$$

where $(p,q) \in M \times N$. Equivalently, we have

$$(i_M X)(p,q) := (j_q)_* X(p), \tag{3.1.47}$$

since

$$(j_q)_* X(p) = (j_q)_* X(p) + (_p j)_* (0_q) \overset{(3.1.23)}{=} (X(p), 0_q).$$

Then

$$i_M X \underset{\mathrm{pr}_1}{\sim} X, \quad i_M X \underset{\mathrm{pr}_2}{\sim} o, \tag{3.1.48}$$

where $o \in \mathfrak{X}(N)$ is the zero vector field. Indeed, for any $(p,q) \in M \times N$ we have

$$(\mathrm{pr}_1)_* (i_M X(p,q)) = (\mathrm{pr}_1)_* (X(p), 0_q)$$

$$\overset{(3.1.23)}{=} (\mathrm{pr}_1)_* ((j_q)_* (X(p)) + (_p j)_* (0_q)) \overset{(3.1.22)}{=} X(p) = X \circ \mathrm{pr}_1(p,q),$$

whence $(\mathrm{pr}_1)_* \circ i_M X = X \circ \mathrm{pr}_1$, as desired. The second relation can be verified in the same way. Similarly, if $Y \in \mathfrak{X}(N)$ and for each $(p,q) \in M \times N$

$$i_N Y(p,q) := (0_p, Y(q)) = (_p j)_* Y(q) \in T_p M \oplus T_q N \cong T_{(p,q)}(M \times N), \tag{3.1.49}$$

then $i_N Y \in \mathfrak{X}(M \times N)$ and

$$i_N Y \underset{\mathrm{pr}_1}{\sim} o, \quad i_N Y \underset{\mathrm{pr}_2}{\sim} Y. \tag{3.1.50}$$

We also have the following relations:

$$i_M X(f \circ \mathrm{pr}_1) = (Xf) \circ \mathrm{pr}_1, \quad i_N Y(f \circ \mathrm{pr}_1) = 0, \quad f \in C^\infty(M); \tag{3.1.51}$$

$$i_N Y(h \circ \mathrm{pr}_2) = (Yh) \circ \mathrm{pr}_2, \quad i_M X(h \circ \mathrm{pr}_2) = 0, \quad h \in C^\infty(N). \tag{3.1.52}$$

We verify the first relation of (3.1.51), the remainder is left to the reader. Let $(p,q) \in M \times N$. Then

$$i_M X(f \circ \mathrm{pr}_1)(p,q) = (i_M X)(p,q)(f \circ \mathrm{pr}_1) = (X(p), 0_q)(f \circ \mathrm{pr}_1)$$

$$\overset{(3.1.24)}{=} X(p)(f \circ \mathrm{pr}_1 \circ j_q) + 0_q(f \circ \mathrm{pr}_1 \circ _p j)$$

$$= X(p)(f) = (Xf)(p) = (Xf) \circ \mathrm{pr}_1(p,q),$$

whence $i_M X(f \circ \mathrm{pr}_1) = (Xf) \circ \mathrm{pr}_1$.

(c) Let S be a submanifold of a manifold M. A vector field X on M is called *tangent* to S if $X_p \in T_p S$ for all $p \in S$. (According to Remark 3.1.24, we regard $T_p S$ as a subspace of $T_p M$.) The following assertions can easily be verified:

(i) If $X \in \mathfrak{X}(M)$ is tangent to S, then $X \upharpoonright S \in \mathfrak{X}(S)$.

(ii) If $X, Y \in \mathfrak{X}(M)$ are tangent to S, then $[X, Y]$ is also tangent to S and

$$[X, Y] \upharpoonright S = [X \upharpoonright S, Y \upharpoonright S]. \qquad (3.1.53)$$

3.2   First-order Differential Equations

Ever since the 17th century, almost all natural phenomena that involve certain quantities varying continuously as functions of a parameter (usually time) have led to problems about differential equations, and these problems have been constant sources of stimulation for the mathematical theory.

Jean Dieudonné [45]

3.2.1   *Basic Existence and Uniqueness Theorems*

Definition 3.2.1. Let $I \subset \mathbb{R}$ be an open interval, $\mathcal{U} \subset \mathbb{R}^n$ an open set and $f \colon I \times \mathcal{U} \to \mathbb{R}^n$ a continuous mapping.

(a) We say that the relation

$$x'(t) = f(t, x(t)) \qquad (3.2.1)$$

is a *time-dependent differential equation*. A *solution of* (3.2.1) is a differentiable mapping α of an open subinterval J of I into $\mathcal{U}$ such that

$$\alpha'(t) = f(t, \alpha(t)) \text{ for all } t \in J.$$

If $J = I$, then the solution α is said to be *global*. If $t_0 \in J$ and $\alpha(t_0) = p \in \mathcal{U}$, then we say that α is a *solution satisfying the initial condition* $x(t_0) = p$, or is a solution of the *initial value problem*

$$x'(t) = f(t, x(t)), \quad x(t_0) = p. \qquad (3.2.2)$$

(b) If $\tilde{J}$ is an open interval containing J, then a solution $\tilde{\alpha} \colon \tilde{J} \to \mathcal{U}$ of the differential equation is said to be an *extension* of α if $\tilde{\alpha} \upharpoonright J = \alpha$. By a *maximal solution* we mean a solution which has no proper extension.

Remark 3.2.2. If $I = \mathbb{R}$, and f does not depend on its first variable, i.e., there is a mapping $\tilde{f} \colon \mathcal{U} \to \mathbb{R}^n$ such that $f(t, p) = \tilde{f}(p)$ for each $t \in \mathbb{R}$ and $p \in \mathcal{U}$, then we speak of a *time-independent* or *autonomous* differential equation. In this case, we can simply write f rather than $\tilde{f}$, and instead of the original differential equation we write the simpler form

$$x'(t) = f(x(t)) \qquad (3.2.3)$$

or even

$$x' = f \circ x. \tag{3.2.4}$$

Of course, all results of this subsection are also valid for this special case.

Theorem 3.2.3 (Local Existence and Uniqueness). *Let $I \subset \mathbb{R}$ be an open interval, $\mathcal{U} \subset \mathbb{R}^n$ an open set and $f : I \times \mathcal{U} \to \mathbb{R}^n$ a continuous mapping. Suppose that f is uniformly Lipschitz continuous in its second variable, i.e., there is a positive constant L such that*

$$\|f(t, p) - f(t, q)\| \le L\|p - q\|$$

for any $t \in I$ and $p, q \in \mathcal{U}$. Let $t_0 \in I$, $p \in \mathcal{U}$. Then there exists an open interval $J \subset I$ with $t_0 \in J$ such that the initial value problem (3.2.2) has a unique solution on J.

For a proof, we refer the reader to [60], Theorem 3.10, or [63], p. 367, or [64], pp. 542–543.

Remark 3.2.4. If f is of class C^r with $r \ge 1$, then uniform Lipschitz continuity holds, at least locally, and the assertion of the theorem remains valid. Indeed, let $J \subset I$ be an open bounded interval such that $\mathrm{cl}(J) \subset I$ and $t_0 \in J$, $\mathcal{V}$ an open bounded convex set such that $\mathrm{cl}(\mathcal{V}) \subset \mathcal{U}$ and $p \in \mathcal{V}$, and

$$L := \max\{\|(f_t)'(q)\| \mid t \in \mathrm{cl}(J), q \in \mathrm{cl}(\mathcal{V})\},$$

where the mapping f_t is given by the rule $q \in \mathcal{U} \mapsto f_t(q) := f(t, q) \in \mathbb{R}$ for any fixed $t \in I$. Then $(f_t)'(q) \in \mathrm{End}(\mathbb{R}^n)$, and we assume that $\mathrm{End}(\mathbb{R}^n)$ is endowed with the usual operator norm (Proposition and Definition B.4.8). Since the set $\mathrm{cl}(J) \times \mathrm{cl}(\mathcal{V})$ is compact, and the function $(t, q) \mapsto \|(f_t)'(q)\|$ is continuous, the maximum L exists. Now let $t \in J$ and $q_1, q_2 \in \mathcal{V}$. By the convexity of $\mathcal{V}$, the line segment

$$\overline{q_1 q_2} := \{(1 - t)q_1 + tq_2 \in \mathbb{R}^n \mid t \in [0, 1]\}$$

lies entirely in $\mathcal{V}$, so it follows from the mean value inequality (Lemma C.1.21) that

$$\|f(t, q_1) - f(t, q_2)\| \le \|q_1 - q_2\| \sup_{\theta \in [0,1]} \|(f_t)'(q_1 + \theta(q_2 - q_1))\| \le L\|q_1 - q_2\|.$$

Thus the conditions of Theorem 3.2.3 are satisfied on $J \times \mathcal{V}$.

Theorem 3.2.5 (Existence and Uniqueness of Maximal Solutions). *Under the conditions of Theorem 3.2.3 the initial value problem (3.2.2) has a unique maximal solution.*

Proof.   Let $\alpha_1 \colon J \to \mathcal{U}$ and $\alpha_2 \colon J \to \mathcal{U}$ be two solutions of (3.2.2). We show that $\alpha_1 = \alpha_2$. Let $A := \{ t \in J \mid \alpha_1(t) = \alpha_2(t) \}$. It is enough to prove that A is both open and closed in J, then the connectedness of J implies that $A = J$ because $A \neq \emptyset$ by hypothesis. The closedness of A follows from the following simple observation:

> If f_1 and f_2 are two continuous mappings of a topological space T into a Hausdorff space, then the set $\{ t \in T \mid f_1(t) = f_2(t) \}$ is closed.

To prove that A is open, let $s \in A$ and $q := \alpha_1(s) = \alpha_2(s)$. Then both α_1 and α_2 are solutions of the initial value problem

$$x'(t) = f(t, x(t)), \ x(s) = q,$$

thus, by Theorem 3.2.3, they coincide in an open neighbourhood of s, and therefore A is open.

Now we show that there exists a maximal solution of (3.2.2). Let $\mathcal{S}$ be the set of all solutions of (3.2.2), and let J_α denote the domain of a fixed solution $\alpha \in \mathcal{S}$. Then $J := \cup_{\alpha \in \mathcal{S}} J_\alpha$ is also an open interval containing t_0. Since, as we have just seen,

$$\alpha \restriction J_\alpha \cap J_\beta = \beta \restriction J_\alpha \cap J_\beta \text{ for all } \alpha, \beta \in \mathcal{S},$$

all the mappings in $\mathcal{S}$ define a single solution $\alpha \colon J \to \mathcal{U}$ of (3.2.2) such that $\alpha \restriction J_\beta = \beta$ for every $\beta \in \mathcal{S}$. Obviously, α is a maximal solution of (3.2.2).

Finally, if α and β are both maximal solutions of (3.2.2), then they coincide on the intersection of their domains, and, by maximality, they actually coincide. $\qquad\qquad\qquad\qquad\qquad\qquad\qquad\qquad\qquad\qquad\qquad\square$

Definition 3.2.6. Let $f \colon I \times \mathcal{U} \to \mathbb{R}^n$ be as in Theorem 3.2.3. For a fixed point $(t_0, p) \in I \times \mathcal{U}$, denote the domain of the maximal solution of the initial value problem (3.2.2) by $J_{(t_0, p)}$ and the maximal solution itself by $\alpha_{(t_0, p)} \colon J_{(t_0, p)} \to \mathcal{U}$. Let W be the subset of $I \times I \times \mathcal{U}$ given by

$$W := \left\{ (t, t_0, p) \in I \times I \times \mathcal{U} \mid t \in J_{(t_0, p)} \right\}.$$

By the *flow* of the differential equation $x'(t) = f(t, x(t))$ we mean the mapping

$$\varphi \colon W \to \mathcal{U}, \quad (t, t_0, p) \mapsto \varphi(t, t_0, p) := \alpha_{(t_0, p)}(t).$$

Theorem 3.2.7 (Smooth Dependence on the Initial Condition). *Let $f \colon I \times \mathcal{U} \to \mathbb{R}^n$ be as in Theorem 3.2.3, and let $r \in \mathbb{N}^* \cup \{\infty\}$. If f is of class C^r, then the domain W of the flow is open in $I \times I \times \mathcal{U}$, and the flow itself is of class C^r.*

This is just a reformulation of (10.8.2) in [42].

Remark 3.2.8. Let $\mathcal{U} \subset \mathbb{R}^n$ be an open set, $f: \mathcal{U} \to \mathbb{R}^n$ a continuous mapping, and consider the time-independent differential equation $x' = f \circ x$. Then the uniform Lipschitz continuity in Theorem 3.2.3 reduces to the following simple condition: there is a positive constant L such that

$$\|f(p) - f(q)\| \leq L\|p - q\|$$

for any $p, q \in \mathcal{U}$. Let $p \in \mathcal{U}$, $t_0 \in \mathbb{R}$, and consider the maximal solution $\alpha_{(0,p)}: J_{(0,p)} \to \mathcal{U}$. If θ_{-t_0} denotes the translation of the real line by t_0 as in the proof of Theorem 3.2.5, then

$$\left(\alpha_{(0,p)} \circ \theta_{-t_0}\right)' = \alpha'_{(0,p)} \circ \theta_{-t_0} = f \circ \alpha_{(0,p)} \circ \theta_{-t_0},$$

and $\alpha_{(0,p)} \circ \theta_{-t_0}(t_0) = \alpha_{(0,p)}(0) = p$, thus $\alpha_{(0,p)} \circ \theta_{-t_0}$ is a solution which satisfies the initial condition $x(t_0) = p$. It can be easily seen that $\alpha_{(0,p)} \circ \theta_{-t_0}$ is maximal as well, thus $\alpha_{(t_0,p)} = \alpha_{(0,p)} \circ \theta_{-t_0}$. So it follows that *all maximal solutions of a time-independent differential equation are determined by the maximal solutions with initial values at $t_0 = 0$.* Hence, in the case of a time-independent differential equation, we usually simply write $I_p := J_{(0,p)}$, $\alpha_p := \alpha_{(0,p)}$, and by the flow of the differential equation we mean the mapping

$$\varphi: W \to \mathcal{U}, \quad (t,p) \mapsto \varphi(t,p) := \alpha_p(t),$$

where

$$W := \{(t,p) \in \mathbb{R} \times \mathcal{U} \mid t \in I_p\}.$$

The mapping defined in this way, although it has only two variables rather than three, contains the same information as the flow in our original sense in Definition 3.2.6.

Theorem 3.2.9 (Global Existence and Uniqueness). *Let a continuous mapping $f: I \times \mathbb{R}^n \to \mathbb{R}^n$ be given. Suppose that f is globally uniformly Lipschitz continuous in its second variable, i.e., there is a positive constant L such that*

$$\|f(t,p) - f(t,q)\| \leq L\|p - q\|$$

for any $t \in I$ and $p, q \in \mathbb{R}^n$. Let $t_0 \in I$, $p \in \mathbb{R}^n$. Then the initial value problem (3.2.2) has a unique global solution, i.e., there is a unique solution $\alpha: I \to \mathbb{R}^n$ of the differential equation with the initial condition $\alpha(t_0) = p$.

Proof.   Let δ be a real number such that $0 < \delta < 1/L$. First we show that our initial value problem has a unique solution on the interval $[t_0 - \delta, t_0 + \delta]$. Of course, we may assume that $[t_0 - \delta, t_0 + \delta] \subset I$.

Consider the set $C([t_0 - \delta, t_0 + \delta], \mathbb{R}^n)$ of all continuous mappings of $[t_0 - \delta, t_0 + \delta]$ into $\mathbb{R}^n$. Define the norm of a mapping $\alpha \in C([t_0 - \delta, t_0 + \delta], \mathbb{R}^n)$ by

$$\|\alpha\|_\infty := \sup\{\|\alpha(t)\| \mid t \in [t_0 - \delta, t_0 + \delta]\}.$$

With this norm, as it is well-known, $C([t_0 - \delta, t_0 + \delta], \mathbb{R}^n)$ is a Banach space (see, e.g., [42], (7.1.3)). We define a mapping

$$\varphi \colon C([t_0 - \delta, t_0 + \delta], \mathbb{R}^n) \to C([t_0 - \delta, t_0 + \delta], \mathbb{R}^n)$$

by the rule

$$\varphi(\alpha)(t) := p + \int_{t_0}^t f(s, \alpha(s))ds \quad (t \in I; \alpha \in C([t_0 - \delta, t_0 + \delta], \mathbb{R}^n)).$$

It can be easily seen that α is a solution of $(*)$ if, and only if, it is a fixed point of φ. Thus, by the contraction principle (Lemma B.2.4), it is sufficient to show that φ is a contraction.

If $\alpha, \beta \in C([t_0 - \delta, t_0 + \delta], \mathbb{R}^n)$, then we have

$$\|\varphi(\alpha) - \varphi(\beta)\|_\infty = \sup_{t_0 - \delta \le t \le t_0 + \delta} \|\varphi(\alpha)(t) - \varphi(\beta)(t)\|$$

$$= \sup_{t_0 - \delta \le t \le t_0 + \delta} \left\| \int_{t_0}^t (f(s, \alpha(s)) - f(s, \beta(s)))ds \right\|$$

$$\le \sup_{t_0 - \delta \le t \le t_0 + \delta} \int_{t_0}^t \|f(s, \alpha(s)) - f(s, \beta(s))\|ds$$

$$\le \sup_{t_0 - \delta \le t \le t_0 + \delta} \int_{t_0}^t L\|\alpha(s) - \beta(s)\|ds \le L\delta\|\alpha - \beta\|_\infty,$$

thus, since $L\delta < 1$, φ is indeed a contraction, and (3.2.2) has a unique solution on $[t_0 - \delta, t_0 + \delta]$.

Taking into account that δ does not depend on t_0, the above argument can be repeated by replacing t_0 with both endpoints of the interval $[t_0 - \delta, t_0 + \delta]$. Proceeding in this way, we obtain that the initial value problem has a unique solution on the whole of I. $\square$

Remark 3.2.10. The previous theorem is due to A. Bielecki [18]. The proof presented here is different from Bielecki's original proof, and its idea was taken from the book [60].

3.2.2 Integral Curves

Throughout this subsection M will stand for an n-dimensional manifold, where $n \geq 1$. By a 'curve' we shall mean a 'smooth curve'.

Definition 3.2.11. Let $I \subset \mathbb{R}$ be an open interval containing the zero.

(a) A curve $\gamma\colon I \to M$ is called an *integral curve* of a vector field $X \in \mathfrak{X}(M)$ if

$$\dot\gamma = X \circ \gamma, \text{ i.e., } \dot\gamma(t) = X(\gamma(t)) \text{ for all } t \in I. \tag{3.2.5}$$

An integral curve γ of X *starts at* $p \in M$ if $\gamma(0) = p$.

(b) If $\widetilde I$ is an open interval containing I, then an integral curve $\widetilde\gamma\colon \widetilde I \to M$ is said to be an *extension* of $\gamma\colon I \to M$ if $\widetilde\gamma \restriction I = \gamma$. By a *maximal integral curve* of a vector field we mean an integral curve which does not admit a proper extension.

(c) We say that a vector field is *complete* if each of its maximal integral curves is defined on the entire real line.

Remark 3.2.12. (a) Suppose that $\gamma\colon I \to M$ is an integral curve of $X \in \mathfrak{X}(M)$. Given $a, b \in \mathbb{R}$, define a function $\theta\colon \widetilde I \to I$ by $\theta(t) := at + b$. Then $\widetilde\gamma := \gamma \circ \theta\colon \widetilde I \to M$ is an integral curve of aX. Indeed

$$\dot{\widetilde\gamma} = (\gamma \circ \theta)^{\cdot} \stackrel{(3.1.10)}{=} \theta'(\dot\gamma \circ \theta) = a(X \circ \gamma \circ \theta) = (aX) \circ \widetilde\gamma,$$

as we claimed.

(b) Choose a chart $(\mathcal{U}, u) = (\mathcal{U}, (u^i)_{i=1}^n)$ on M. A curve $\gamma\colon I \to M$ is an integral curve of a vector field $X \in \mathfrak{X}(M)$ on $\gamma^{-1}(\mathcal{U})$ if, and only if, the curve

$$(\gamma^1, \ldots, \gamma^n)\colon \gamma^{-1}(\mathcal{U}) \subset \mathbb{R} \to \mathbb{R}^n, \ \gamma^i := u^i \circ \gamma \quad (i \in J_n)$$

satisfies the autonomous differential equation (3.2.4) on $u(\mathcal{U}) \subset \mathbb{R}^n$, where

$$f = (f^1, \ldots, f^n); \ f^i := X^i \circ u^{-1}, \ X^i = Xu^i \quad (i \in J_n).$$

Indeed, for each $t \in \gamma^{-1}(\mathcal{U})$ we have

$$\dot\gamma(t) \stackrel{(3.1.12)}{=} \sum_{i=1}^n (\gamma^i)'(t) \left(\frac{\partial}{\partial u^i} \right)_{\gamma(t)}$$

and

$$X(\gamma(t)) = \left(\sum_{i=1}^{n} X^i \frac{\partial}{\partial u^i}\right)(\gamma(t)) = \sum_{i=1}^{n} X^i(\gamma(t))\left(\frac{\partial}{\partial u^i}\right)_{\gamma(t)}$$

$$= \sum_{i=1}^{n} (X^i \circ u^{-1})\big(u(\gamma(t))\big)\left(\frac{\partial}{\partial u^i}\right)_{\gamma(t)}$$

$$= \sum_{i=1}^{n} f^i \circ (\gamma^1, \dots, \gamma^n)(t)\left(\frac{\partial}{\partial u^i}\right)_{\gamma(t)},$$

therefore the following are equivalent:

$X \circ \gamma = \dot{\gamma}$ on $\gamma^{-1}(\mathcal{U})$,

$f^i \circ (\gamma^1, \dots, \gamma^n) = (\gamma^i)'$ for all $i \in J_n$,

$(\gamma^1, \dots, \gamma^n)$ is a solution of $(*)$.

Thus, by using charts, the results of the preceding subsection can be translated immediately into manifold terminology.

Theorem 3.2.13 (Existence and Uniqueness of Integral Curves).
Let X be a vector field on M. Then for each $p \in M$ there exists a unique maximal integral curve $\gamma_p \colon I_p \to M$ of X such that $\gamma_p(0) = p$.

The proof is completely analogous to that of Theorem 3.2.5, so we leave the details to the reader.

Remark 3.2.14. (a) The local theory of ordinary differential equations makes it possible to refine the existence-uniqueness result formulated in the preceding theorem in the following way:
Let X be a vector field on M. For each point $p \in M$ there exists an open neighbourhood $\mathcal{V}$ of p and an open interval J containing $0 \in \mathbb{R}$ such that

(i) *there is an integral curve $\alpha \colon J \to M$ of X starting at any point $q \in \mathcal{V}$;*
(ii) *if β is an integral curve of X starting at q, then β coincides with α in a neighbourhood of 0.*

For more details, see, e.g., [20], Proposition 8.2.2.
  (b) We shall sometimes need the integral curves of a vector field of class C^r with $r \in \mathbb{N}^*$ (cf. Remark 3.1.46(b)). Let X be a vector field on M of class C^r. A curve $\gamma \colon I \to M$ of class C^1 is an integral curve of X if it satisfies (3.2.5). It can be easily seen by induction that γ is automatically of class C^{r+1} in this case. In Remark 3.2.4 we noted that the uniform Lipschitz

continuity required in Theorems 3.2.3 and 3.2.5 holds for mappings of class C^r, thus their 'manifold counterpart', i.e., Theorem 3.2.13 also holds for vector fields of class C^r.

For more details, see again [20], Proposition 8.2.2.

Lemma 3.2.15. *Let X be a vector field on M. Given a point p of M, consider the maximal integral curve $\gamma_p \colon I_p \to M$ of X starting at p. Then, for any fixed $s \in I_p$, $\gamma_p \circ \theta_s$ is the maximal integral curve of X starting at $\gamma_p(s)$ with domain $\theta_{-s}(I_p)$. (Here, as above, θ_s and θ_{-s} are translations of $\mathbb{R}$ by s and $-s$, respectively.)*

Proof. It is clear that the domain of $\gamma_p \circ \theta_s$ is the open interval $\theta_{-s}(I_p)$, which also contains the zero. From Remark 3.2.12(a) it follows that $\gamma_p \circ \theta_s$ is an integral curve of X. Since $\gamma_p \circ \theta_s(0) = \gamma_p(s)$, the integral curve $\gamma_p \circ \theta_s$ starts at $q := \gamma_p(s)$.

Now we verify that

$$\gamma_p \circ \theta_s = \gamma_q, \quad q = \gamma_p(s).$$

By the maximality of the integral curve $\gamma_q \colon I_q \to M$ we obviously have:

$$\text{domain of } \gamma_p \circ \theta_s = \theta_{-s}(I_p) \subset I_q. \qquad (*)$$

Then, in particular, $-s \in I_q$. So, by Remark 3.2.12(a), $\gamma_q \circ \theta_{-s}$ is an integral curve of X starting at $\gamma_q(-s) = p$ with domain $\theta_s(I_q)$. Hence:

$$\text{domain of } \gamma_q \circ \theta_{-s} = \theta_s(I_q) \subset I_p. \qquad (**)$$

Relations $(*)$ and $(**)$ imply that $I_q = \theta_{-s}(I_p)$. Thus the domain of $\gamma_p \circ \theta_s$ is equal to the domain of the maximal integral curve γ_q, whence $\gamma_p \circ \theta_s = \gamma_q$. $\qquad \square$

Definition 3.2.16. A nonconstant curve $\gamma \colon \mathbb{R} \to M$ is said to be *periodic* if there is a positive number c such that $\gamma(t+c) = \gamma(t)$ for each $t \in \mathbb{R}$. The smallest such positive c is called the *period* of γ. If, in addition, there is a point $a \in \mathbb{R}$ such that $\gamma \restriction [a, a+c[$ is injective, then we say that γ is *simply periodic*.

Remark 3.2.17. The existence of the smallest positive c above follows by the following simple argument. If there is no such smallest c, then there exists a sequence $(c_n)_{n \in \mathbb{N}}$ of positive numbers with this property tending to zero, which easily implies that γ is constant on a dense subset of $\mathbb{R}$. Thus, γ is constant by continuity, which is a contradiction.

Corollary 3.2.18. *Let X be a vector field on M. Then every maximal integral curve of X is either injective, simply periodic, or constant.*

Proof.   Suppose that γ is not injective. Then there are points a, b in $\mathbb{R}$ with $a < b$ such that $\gamma(a) = \gamma(b)$. If we fix a, then there are two possibilities: either there is a smallest such $b > a$, or there is not.

(1) In the first case let $c := b - a$. By Lemma 3.2.15, $\gamma \circ \theta_c$ is the maximal integral curve of X with $(\gamma \circ \theta_c)(a) = \gamma(a + c) = \gamma(b) = \gamma(a)$, thus, by the uniqueness assertion of Theorem 3.2.13, $\gamma \circ \theta_c = \gamma$. By the choice of b, $\gamma \upharpoonright [a, b[$ is injective, and γ is simply periodic.

(2) In the second case there is a sequence $(b_n)_{n \in \mathbb{N}}$ converging to a such that $b_n > a$ and $\gamma(a) = \gamma(b_n)$ for each $n \in \mathbb{N}$. If f is a smooth function on M, then

$$\dot{\gamma}(a)f \overset{(3.1.9)}{=} (f \circ \gamma)'(a) = \lim_{t \to a} \frac{f(\gamma(t)) - f(\gamma(a))}{t - a}$$

$$= \lim_{n \to \infty} \frac{f(\gamma(b_n)) - f(\gamma(a))}{b_n - a} = 0,$$

thus $\dot{\gamma}(a) = 0$ and $X(\gamma(a)) = 0$. It follows that the constant curve defined on $\mathbb{R}$ with value a is also a maximal integral curve of X, and, again by the uniqueness assertion of Theorem 3.2.13, γ coincides with this constant curve. This concludes the proof. $\square$

Proposition 3.2.19. *A vector field is complete if, and only if, there exists an open interval I containing $0 \in \mathbb{R}$ such that each maximal integral curve of the vector field is defined on I.*

Proof.   The condition is clearly necessary. To show the sufficiency suppose that for a vector field $X \in \mathfrak{X}(M)$ there exists an open interval I satisfying the requirements of the proposition. There is then a positive number ε such that the domain of each maximal integral curve of X contains the closed interval $[-\varepsilon, \varepsilon]$. Now, arguing by contradiction, suppose that there exists a point $p \in M$ such that the domain I_p of the maximal integral curve γ_p of X starting at p differs from $\mathbb{R}$. If I_p has a least upper bound b, consider the point $q := \gamma_p(b - \varepsilon)$. Then, by Lemma 3.2.15, $I_q = \theta_{-b+\varepsilon}(I_p)$. So each element of I_q is less than ε, and, therefore, I_q does not contain $[-\varepsilon, \varepsilon]$. We arrived at a contradiction. The argument is similar when I_p has a greatest lower bound. $\square$

Corollary 3.2.20. *Every vector field on a compact manifold is complete.*

Proof. Suppose that M is a compact manifold and let $X \in \mathfrak{X}(M)$. By Remark 3.2.14(a), with each point $p \in M$ we can associate an open neighbourhood $\mathcal{V}(p)$ of p and an open interval $J(p)$ around 0 such that conditions (i),(ii) of the cited remark are satisfied. Since M is compact, there is a finite sequence $(p_i)_{i=1}^k$ of points of M such that $(\mathcal{V}(p_i))_{i=1}^k$ is a covering of M. If $I := \bigcap_{i=1}^k (J(p_i))$, then I is an open interval around 0 which is contained in the domain of each maximal integral curve of X. Thus, by Proposition 3.2.19, the vector field X is complete. $\square$

Definition 3.2.21. A smooth function f on M is called a *first integral* of a vector field X on M if for every integral curve γ of X, the function $f \circ \gamma$ is constant.

Lemma 3.2.22. *Let $X \in \mathfrak{X}(M)$. A smooth function f on M is a first integral of X if, and only if, $Xf = 0$.*

Proof. Let $\gamma \colon I \to M$ be an integral curve of X. Then for every $t \in I$,

$$Xf(\gamma(t)) = X_{\gamma(t)}f = (X \circ \gamma)(t)f = \dot{\gamma}(t)(f) \overset{(3.1.9)}{=} (f \circ \gamma)'(t).$$

Therefore, $f \circ \gamma$ is constant for every integral curve γ of X if, and only if, $(Xf) \circ \gamma = 0$. In view of Theorem 3.2.13 the latter condition is equivalent to the vanishing of Xf. $\square$

3.2.3 *Flows*

We continue to assume that M is an at least 1-dimensional manifold and 'curves' mean smooth curves.

Definition 3.2.23. (a) A subset W of the product manifold $\mathbb{R} \times M$ is called *radial* if for each $p \in M$,

$$W \cap (\mathbb{R} \times \{p\}) = I_p \times \{p\} \text{ or } W \cap (\mathbb{R} \times \{p\}) = \emptyset,$$

where I_p is an open interval containing 0.

(b) Let W be a radial open neighbourhood of $\{0\} \times M$ in $\mathbb{R} \times M$. A smooth mapping

$$\varphi \colon W \to M, \quad (t,p) \mapsto \varphi(t,p)$$

is said to be a *local flow on M* if the following conditions are satisfied:

(Fl$_1$) $\varphi(0,p) = p$ for all $p \in M$,

(Fl$_2$) $\varphi(t,\varphi(s,p)) = \varphi(t+s,p)$ whenever both sides are defined.

We say that a local flow is *global* if its domain is $\mathbb{R} \times M$. A local flow is called *maximal* if it admits no proper extension as a flow, i.e., there is no local flow $\tilde{\varphi}\colon \widetilde{W} \to M$ such that W is a proper subset of $\widetilde{W}$, and $\tilde{\varphi} \restriction W = \varphi$.

(c) If $\varphi\colon W \to M$ is a local flow and $p \in M$, then the curve

$$\alpha_p\colon I_p \to M, \ t \mapsto \alpha_p(t) := \varphi(t, p)$$

is called the *flow line* of p (or a flow line of φ). The mapping

$$X^\varphi\colon M \to TM, \ p \mapsto X^\varphi(p) := \dot{\alpha}_p(0) \tag{3.2.6}$$

is the *velocity field* or the *infinitesimal generator* of the local flow.

Proposition 3.2.24. *If $\varphi\colon W \to M$ is a local flow on M, then its velocity field X^φ is a vector field on M, and the flow lines of φ are integral curves of X^φ. In particular, if the local flow φ is maximal, then it is uniquely determined by the vector field X^φ.*

Proof. Let, for brevity, $X := X^\varphi$.

(1) For each $p \in M$,

$$X(p) := \dot{\alpha}_p(0) \in T_{\alpha_p(0)} M \overset{\mathrm{(Fl_1)}}{=} T_p M,$$

so X is a section of TM.

(2) We show that $X \in C^\infty(M, TM)$. Given any smooth function f on M, we have, on the one hand,

$$(Xf)(p) = X(p)(f) = \dot{\alpha}_p(0)(f) \overset{(3.1.9)}{=} (f \circ \alpha_p)'(0), \quad p \in M.$$

On the other hand, we can express $(f \circ \alpha_p)'(0)$ more directly in terms of the local flow φ. By Example 3.1.53(b), the standard coordinate vector field $\frac{d}{dr} \in \mathfrak{X}(\mathbb{R})$ induces the vector field $i_\mathbb{R}\frac{d}{dr}$ on $\mathbb{R} \times M$ given by

$$(t, q) \in \mathbb{R} \times M \mapsto i_\mathbb{R}\frac{d}{dr}(t, q) := \left(\left(\frac{d}{dr} \right)_t, 0_q \right) \in T_t\mathbb{R} \oplus T_q M \cong T_{(t,q)}(\mathbb{R} \times M).$$

Acting by $i_\mathbb{R}\frac{d}{dr}$ on the smooth function $f \circ \varphi\colon W \subset \mathbb{R} \times M \to M \to \mathbb{R}$, we find

$$\left(i_\mathbb{R}\frac{d}{dr} \right)(f \circ \varphi)(0, p) = \left(i_\mathbb{R}\frac{d}{dr} \right)(0, p)(f \circ \varphi) = \left(\left(\frac{d}{dr} \right)_0, 0_p \right)(f \circ \varphi)$$

$$\overset{(3.1.24)}{=} \left(\frac{d}{dr} \right)_0 (f \circ \varphi \circ j_p) + 0_p(f \circ \varphi \circ 0j)$$

$$= (f \circ \varphi \circ j_p)'(0) = (f \circ \alpha_p)'(0),$$

taking into account that

$$\varphi \circ j_p(t) = \varphi(t, p) = \alpha_p(t) \text{ for all } (t, p) \in W$$

(see the proof of Lemma 3.1.36). Since $(0, p) = {}_0 j(p)$ for all $p \in M$, we conclude that

$$X f = \left(i_{\mathbb{R}} \frac{d}{dr} \right) (f \circ \varphi) \circ {}_0 j,$$

whence $Xf \in C^\infty(M)$. By Lemma 3.1.32, this proves the smoothness of X.

(3) Let $p \in M$, and consider the flow line

$$\alpha_p \colon I_p \to M, \ t \mapsto \alpha_p(t) := \varphi(t, p)$$

of p. We show that $\dot\alpha_p = X \circ \alpha_p$. Since

$$\dot\alpha_p(0) =: X(p) \overset{(\mathrm{Fl}_1)}{=} X \circ \alpha_p(0),$$

the desired equality holds automatically at $0 \in I_p$. Now let $t \in I_p$ be arbitrary, and write $q := \alpha_p(t)$. Then for all $s \in I_q$,

$$\alpha_q(s) = \varphi(s, q) = \varphi(s, \alpha_p(t)) = \varphi(s, \varphi(t, p)) \overset{(\mathrm{Fl}_2)}{=} \varphi(s + t, p) = \alpha_p(s + t),$$

whence $\dot\alpha_q(0) = \dot\alpha_p(t)$. So we have

$$X \circ \alpha_p(t) = X(q) := \dot\alpha_q(0) = \dot\alpha_p(t),$$

as was to be shown.

(4) Owing to the property just proved, the uniqueness part of Theorem 3.2.13 guarantees that a maximal local flow φ is uniquely determined by its infinitesimal generator X^φ. $\qquad\square$

Theorem 3.2.25 (Integrability Theorem for Vector Fields). *Every vector field on a manifold is the velocity field of a unique maximal local flow, on a compact manifold even a global one. Namely, if $X \in \mathfrak{X}(M)$, and*

$$\mathcal{D}_X := \bigcup_{p \in M} (I_p \times \{p\}), \quad \varphi^X \colon \mathcal{D}_X \to M, \ (t, p) \mapsto \varphi^X(t, p) := \gamma_p(t),$$

where $\gamma_p \colon I_p \to M$ is the maximal integral curve of X starting at p, then φ^X is a maximal local flow on M, and it is the unique maximal local flow such that

$$X^{(\varphi^X)} = X.$$

For a proof, we refer to [59], Theorem 8.5.12.

Remark 3.2.26. We note that the theorem above holds for vector fields of class C^r, with the obvious modification that the flow whose velocity vector field is the given vector field is also of class C^r. For a proof of this slightly more general result, see [44], (18.2.5).

Remark 3.2.27. (a) If X is a vector field on M, then we shall call the unique maximal local flow φ^X whose existence is guaranteed by the previous theorem the *maximal local flow generated by* X or the *maximal local flow of* X or simply the *flow of* X.

(b) Suppose that X is a complete vector field on M. Then, by the above theorem, X generates the global flow

$$\varphi^X : \mathbb{R} \times M \to M, \ (t,p) \mapsto \varphi^X(t,p) := \gamma_p(t)$$

on M. For any fixed $t \in \mathbb{R}$, the mappings

$$\varphi_t^X : M \to M, \ p \mapsto \varphi_t^X(p) := \varphi^X(t,p)$$

have the following properties:

(i) $\varphi_t^X \circ \varphi_s^X = \varphi_{t+s}^X$ for all $s, t \in \mathbb{R}$;
(ii) $\varphi_0^X = 1_M$.

Thus the mapping

$$\varphi^X : \mathbb{R} \times M \to M, \ (t,p) \mapsto t \cdot p := \varphi_t^X(p)$$

is an *action* of the additive group $\mathbb{R}$ on M in the sense of A.3.5. This action is called a *smooth action* since the mapping φ^X is smooth. Continuing the analogy with A.3.5,

$$\varphi_t^X \in \text{Diff}(M) \text{ with } (\varphi_t^X)^{-1} = \varphi_{-t}^X \text{ for all } t \in \mathbb{R},$$

so we obtain a group homomorphism

$$\mathbb{R} \to \text{Diff}(M), \ t \mapsto \varphi_t^X.$$

In this way *the smooth actions of* $\mathbb{R}$ *on* M *are in a bijective correspondence with the complete vector fields on* M.

Since the diffeomorphisms φ_t^X ($t \in \mathbb{R}$) form a group, it is also reasonable (and quite standard) to call the family $(\varphi_t^X)_{t \in \mathbb{R}}$ the *one-parameter group generated by* X or the *one-parameter group of* X. If the generating vector field is clear from the context, we can write $(\varphi_t)_{t \in \mathbb{R}}$ or simply (φ_t).

(c) More generally, let X be any vector field on M, and consider its maximal local flow $\varphi^X \colon \mathcal{D}_X \to M$. It can be shown that the sets

$$M_t = \{p \in M \mid (t,p) \in \mathcal{D}_X\}$$

are open subsets of M. (Actually, this is a part of the proof of Theorem 3.2.25.) The mappings

$$\varphi_t^X \colon M_t \to M, \ p \mapsto \varphi_t^X(p) := \varphi^X(t,p)$$

are injective, $\mathrm{Im}(\varphi_t^X) = M_{-t}$, and φ_t^X is a diffeomorphism onto M_{-t} with inverse φ_{-t}^X.

On the analogy above, the family (φ_t^X) (or (φ_t) if there is no danger of confusion) is often called the *local one-parameter group of X*.

(d) A vector field $X \in \mathfrak{X}(M)$ is the velocity vector field of a maximal local flow $\varphi \colon W \subset \mathbb{R} \times M \to M$ if, and only if,

$$\varphi_* \circ i_{\mathbb{R}} \frac{d}{dr} = X \circ \varphi, \tag{3.2.7}$$

where $i_{\mathbb{R}} \frac{d}{dr} \in \mathfrak{X}(\mathbb{R} \times M)$ is the vector field defined in Example 3.1.53(b).

To see this, fix a point $p \in M$, and consider the inclusion

$$j_p \colon \mathbb{R} \to \mathbb{R} \times M, \ t \mapsto j_p(t) := (t,p)$$

opposite p. If $\alpha_p \colon I_p \to M$ is the flow line of p, then for each $t \in I_p$ we have

$$\dot{\alpha}_p(t) = (\varphi \circ j_p)^{\cdot}(t) = \varphi_* \circ (j_p)_* \left(\frac{d}{dr}\right)_t \overset{(3.1.47)}{=} \varphi_* \circ i_{\mathbb{R}} \frac{d}{dr}(t,p).$$

Thus, α_p is the maximal integral curve of X starting at p, i.e.,

$$\dot{\alpha}_p(t) = X \circ \alpha_p(t) = X \circ \varphi(t,p), \quad t \in I_p$$

if, and only if, (3.2.7) is satisfied.

Theorem 3.2.28 (Smooth Dependence Theorem). *Let $\mathcal{V}$ be an open subset of a finite-dimensional real vector space, and let $\psi \colon \mathcal{V} \to \mathfrak{X}(M)$ be a mapping, such that*

$$\widetilde{\psi} \colon \mathcal{V} \times M \to TM, \ (v,p) \mapsto \widetilde{\psi}(v,p) := \psi(v)(p)$$

is a smooth mapping ('the vector field $\psi(v)$ depends smoothly on v'). Then for each $(p_0, v_0) \in M \times \mathcal{V}$ there exist

an open neighbourhood $\mathcal{U}$ of p_0 in M,
an open interval I containing 0,
an open neighbourhood $\mathcal{W}$ of v_0 in $\mathcal{V}$,
a smooth mapping $\varphi \colon I \times \mathcal{U} \times \mathcal{W} \to M$

such that for each $(p, v) \in \mathcal{U} \times \mathcal{W}$ the curve

$$\varphi_p^v \colon I \to M, \ t \mapsto \varphi_p^v(t) := \varphi(t, p, v)$$

is an integral curve of the vector field $\psi(v)$ starting at p, i.e.,

$$\dot{\varphi}_p^v = \psi(v) \circ \varphi_p^v, \quad \varphi_p^v(0) = p.$$

Proof. We follow the argument of [59], Proposition 7.5.15. Form the product manifold $N := \mathcal{V} \times M$. Then

$$Y \colon (v, p) \in \mathcal{V} \times M \mapsto Y(v, p) := (0_v, \psi(v)(p)) \in T(\mathcal{V} \times M)$$

is a vector field on $\mathcal{V} \times M$, whose integral curves are of the form

$$t \in I_{(v,p)} \mapsto \gamma(t) = (v, \gamma_{(v,p)}(t)),$$

where $\gamma_{(v,p)} \colon I_{(v,p)} \to M$ is an integral curve of $\psi(v) \in \mathfrak{X}(M)$. Now the assertion follows from the smoothness of the local flow of $Y \in \mathfrak{X}(\mathcal{V} \times M)$ guaranteed by Theorem 3.2.25. $\qquad\square$

3.2.4 Commuting Flows

First, we point out that the Lie derivative of a function (see Remark 3.1.42) and the Lie bracket of two vector fields can be described in terms of flows.

Lemma 3.2.29. Let X be a vector field on M, and let $\varphi^X \colon \mathcal{D}_X \to M$ be its maximal local flow. If $(t, p) \in \mathcal{D}_X$, then for every smooth function $f \in C^\infty(M)$ we have

$$(\mathcal{L}_X f)(p) = X_p(f) = \lim_{t \to 0} \frac{1}{t}(f \circ \varphi_t^X(p) - f(p)), \qquad (3.2.8)$$

or, by a slight abuse of notation,

$$\mathcal{L}_X f = Xf = \lim_{t \to 0} \frac{1}{t}(f \circ \varphi_t^X - f). \qquad (3.2.9)$$

Proof. Indeed, the flow lines of φ^X are the maximal integral curves of X. In particular, $X_p = \dot{\gamma}_p(0)$, where $\gamma_p \colon I_p \to M$ is the maximal integral curve of X starting at p. Thus

$$X_p(f) = \dot{\gamma}_p(0)(f) \overset{(3.1.9)}{=} (f \circ \gamma_p)'(0) = \lim_{t \to 0} \frac{1}{t}(f \circ \gamma_p(t) - f \circ \gamma_p(0))$$

$$= \lim_{t \to 0} \frac{1}{t}(f \circ \varphi_t^X(p) - f(p)),$$

as we asserted. $\qquad\square$

Proposition 3.2.30. *Let X and Y be vector fields on M, and let $\varphi^X \colon \mathcal{D}_X \to M$ be the local flow of X. At each point $p \in M$, the limit*

$$\lim_{t \to 0} \frac{1}{t}\left((\varphi^X_{-t})_* \circ Y \circ \varphi^X_t(p) - Y(p)\right) = \lim_{t \to 0} \frac{1}{t}\left(((\varphi^X_{-t})_\# Y)(p) - Y(p)\right)$$

exists, and the mapping

$$M \to TM, \ p \mapsto \lim_{t \to 0} \frac{1}{t}\left(((\varphi^X_{-t})_\# Y)(p) - Y(p)\right)$$

is a vector field on M, namely the Lie bracket $[X,Y]$. So we have

$$[X,Y](p) = \lim_{t \to 0} \frac{1}{t}\left(((\varphi^X_{-t})_\# Y)(p) - Y(p)\right) \tag{3.2.10}$$

at every point p of M.

For a nice proof of (3.2.10) we refer to O'Neill's book [79], pp. 31–32; see also [66], 2.8.2 and [69], Ch. 18. We shall frequently abbreviate (3.2.10) as

$$[X,Y] = \lim_{t \to 0} \frac{1}{t}\left((\varphi^X_{-t})_\# Y - Y\right) = \lim_{t \to 0} \frac{1}{t}\left(Y - (\varphi^X_t)_\# Y\right). \tag{3.2.11}$$

Lemma 3.2.31. *Let $F \colon M \to N$ be a smooth mapping into a further manifold N. Two vector fields $X \in \mathfrak{X}(M)$ and $Y \in \mathfrak{X}(N)$ are F-related if, and only if,*

$$\varphi^Y_t \circ F = F \circ \varphi^X_t \quad (t \in \mathbb{R}) \tag{3.2.12}$$

whenever both sides are defined.

Proof. We recall that for any fixed point $p \in M$, the smooth curve

$$\varphi^X_p \colon I_p \to M, \ t \mapsto \varphi^X_p(t) := \varphi^X(t,p)$$

is the maximal integral curve of X starting at p. In terms of these integral curves, (3.2.12) takes the form

$$\varphi^Y_{F(p)}(t) = F \circ \varphi^X_p(t), \quad t \in I_p. \tag{3.2.13}$$

Now suppose that $X \underset{F}{\sim} Y$, i.e., $F_* \circ X = Y \circ F$. Consider the curve

$$\alpha := F \circ \varphi^X_p \colon I_p \to N$$

in N. The velocity of α at $t \in I_p$ is

$$\dot{\alpha}(t) = (F \circ \varphi^X_p)\dot{\,}(t) \overset{(3.1.11)}{=} (F_*)_{\varphi^X_p(t)}(\dot\varphi^X_p(t)) = (F_*)_{\varphi^X_p(t)}\left(X(\varphi^X_p(t))\right)$$

$$= F_* \circ X \circ \varphi^X_p(t) \overset{\text{cond.}}{=} Y \circ F \circ \varphi^X_p(t) = Y \circ \alpha(t),$$

therefore α is an integral curve of Y starting at $\alpha(0) = F(\varphi_p^X(0)) = F(p)$. Thus, by Theorem 3.2.13, the maximal integral curve $\varphi_{F(p)}^Y$ of Y must be defined at least on the interval I_p, and in I_p the equality $F \circ \varphi_p^X = \varphi_{F(p)}^Y$ must hold. This proves (3.2.13).

Conversely, assume (3.2.13). Then for all $p \in M$ we have

$$F_* \circ X(p) \stackrel{(3.2.6)}{=} F_*(\dot{\varphi}_p^X(0)) \stackrel{(3.1.11)}{=} (F \circ \varphi_p^X)\dot{\,}(0) \stackrel{(3.2.13)}{=} \dot{\varphi}_{F(p)}^Y(0) = Y \circ F(p),$$

whence $X \underset{F}{\sim} Y$. $\square$

Proposition 3.2.32. *Let X and Y be vector fields on M with flows $\varphi^X \colon \mathcal{D}_X \to M$ and $\varphi^Y \colon \mathcal{D}_Y \to M$, respectively. Then the following are equivalent:*

(i) *The Lie bracket $[X, Y]$ vanishes.*

(ii) *$(\varphi_{-t}^X)_\# Y = Y \restriction M_t$ whenever M_t is not empty – 'the vector field Y is invariant under the flow of X'.*

(iii) *$\varphi_t^X \circ \varphi_s^Y = \varphi_s^Y \circ \varphi_t^X$ whenever either side is defined – 'the flows of X and Y commute'.*

Proof. Obviously, (ii) implies (i) by (3.2.11). Now we prove the converse implication. Given a point $p \in M$, consider the curve

$$\beta \colon I_p \to T_p M, \; t \mapsto \beta(t) := (\varphi_{-t}^X)_* \circ Y \circ \varphi_t^X(p),$$

where I_p is the domain of the maximal integral curve of X starting at p. Observe that $\beta(0) = Y(p)$. We are going to show that $\beta' = 0$. Then β is constant, whence

$$(\varphi_{-t}^X)_* \circ Y \circ \varphi_t^X(p) = Y(p) \text{ for all } t \in I_p,$$

so we arrive at the desired conclusion. Let $t \in I_p$. Then we find

$$\beta'(t) = \lim_{s \to 0} \frac{1}{s}(\beta(t + s) - \beta(t))$$

$$= \lim_{s \to 0} \frac{1}{s}\left((\varphi_{-t-s}^X)_* \circ Y \circ \varphi_{t+s}^X(p) - (\varphi_{-t}^X)_* \circ Y \circ \varphi_t^X(p)\right)$$

$$\stackrel{(\mathrm{Fl}_2)}{=} \lim_{s \to 0} \frac{1}{s}\left((\varphi_{-t}^X)_* \circ (\varphi_{-s}^X)_* \circ Y\left(\varphi_s^X(\varphi_t^X(p))\right) - (\varphi_{-t}^X)_* \circ Y(\varphi_t^X(p))\right)$$

$$= (\varphi_{-t}^X)_* \lim_{s \to 0} \frac{1}{s}\left((\varphi_{-s}^X)_* Y\left(\varphi_s^X(\varphi_t^X(p))\right) - Y(\varphi_t^X(p))\right)$$

$$= (\varphi_{-t}^X)_*[X, Y](\varphi_t^X(p)) \stackrel{\text{cond.}}{=} 0,$$

as wanted.

To prove the equivalence of (ii) and (iii), note first that (ii) can be reformulated as follows:

$$(\varphi_t^X)_* \circ Y = Y \circ \varphi_t^X \quad \textit{whenever defined.}$$

Since $\varphi_t^X \in \mathrm{Diff}(M_t, M_{-t})$ (see Remark 3.2.27(c)), this equality is equivalent to the relation

$$Y \restriction M_t \underset{\varphi_t^X}{\sim} Y \restriction M_{-t}.$$

Now, applying Lemma 3.2.31 with the cast

$$X := Y \restriction M_t, \quad Y := Y \restriction M_{-t}, \quad F := \varphi_t^X,$$

it follows that the last relation is equivalent to

$$\varphi_s^Y \circ \varphi_t^X = \varphi_t^X \circ \varphi_s^Y \quad \textit{whenever defined.}$$

This concludes the proof. $\qquad\qquad\qquad\qquad\qquad\qquad\qquad\qquad\square$

Definition 3.2.33. Let m be a positive integer, and consider the real vector space $\mathbb{R}^m$ together with its canonical coordinate system $(e^i)_{i=1}^m$. Given an open subset $\mathcal{U}$ of $\mathbb{R}^m$, let $\varphi \colon \mathcal{U} \to M$ be a smooth mapping into a manifold M. Then the mapping

$$\dot{\varphi}_i := \varphi_* \circ \frac{\partial}{\partial e^i} \colon \mathcal{U} \to TM \quad (i \in J_m) \qquad (3.2.14)$$

is called the *ith partial derivative* of φ. The *second partial derivatives* of φ are

$$\ddot{\varphi}_{ij} = ((\dot{\varphi}_i)\dot{})_j \colon \mathcal{U} \to TTM; \quad i, j \in J_m.$$

Remark 3.2.34. With the notation of the definition, consider the parametrized straight line

$$\alpha \colon \mathbb{R} \to \mathbb{R}^m, \quad t \mapsto \alpha(t) := a + t e_i,$$

where $a \in \mathcal{U}$ is a fixed point, $(e_j)_{j \in J_m}$ is the canonical basis of $\mathbb{R}^m$, and $i \in J_m$. Then $\gamma := \varphi \circ \alpha$ is a smooth curve in M, and by (3.1.11) we obtain that

$$\dot{\gamma} = \dot{\varphi}_i \circ \alpha, \quad \ddot{\gamma} = \ddot{\varphi}_{ii} \circ \alpha. \qquad (3.2.15)$$

Observe that in the case $m = 1$, φ itself is a smooth curve in M, and $\dot{\varphi}_1 = \dot{\varphi}$ is its velocity, $\ddot{\varphi}_{11} = \ddot{\varphi}$ is its acceleration.

If $(\mathcal{U}, (u^i)_{i=1}^n)$ is a chart on M and $\varphi^i := u^i \circ \varphi$, then (3.1.5) yields immediately that

$$\dot{\varphi}_i = \sum_{j=1}^n \frac{\partial \varphi^j}{\partial e^i} \left(\frac{\partial}{\partial u^j} \circ \varphi \right), \quad i \in J_m. \qquad (3.2.16)$$

Proposition 3.2.35. *Let $\mathcal{U}$ be an open subset of a manifold M. Suppose that $(X_i)_{i=1}^k$ is a family of (pointwise) linearly independent vector fields on $\mathcal{U}$ satisfying*

$$[X_i, X_j] = 0 \quad \text{for all} \quad i, j \in J_k.$$

Then for every point p in $\mathcal{U}$ there is an open neighbourhood $\mathcal{V}$ of $0 \in \mathbb{R}^k$ together with an immersion $\varphi \colon \mathcal{V} \to \mathcal{U}$ such that $\varphi(0) = p$ and

$$\frac{\partial}{\partial e^i} \underset{\varphi}{\sim} X_i, \quad i \in J_k,$$

where $(e^i)_{i=1}^k$ is the canonical coordinate system on $\mathcal{V}$.

Proof.   Choose and fix a point p in $\mathcal{U}$. Let $\mathcal{V}$ be a suitable small open neighbourhood of $0 \in \mathbb{R}^k$, and define the mapping

$$\varphi \colon \mathcal{V} \to \mathcal{U}, \ q = (q^1, \dots, q^k) \mapsto \varphi(q) := \varphi_{q^1}^{X_1} \circ \cdots \circ \varphi_{q^k}^{X_k}(p).$$

Then φ is clearly smooth. Given a point $a = (a^1, \dots, a^k)$ in $\mathcal{V}$ and an index $i \in J_k$, consider the parametrized straight line

$$\alpha \colon \mathbb{R} \to \mathbb{R}^k, \ t \mapsto \alpha(t) := a + t e_i.$$

Then we have

$$\varphi \circ \alpha(t) = \varphi(a^1, \dots, a^i + t, \dots, a^k) := \varphi_{a^1}^{X_1} \circ \cdots \circ \varphi_{a^i + t}^{X_i} \circ \cdots \circ \varphi_{a^k}^{X_k}(p)$$

$$\overset{\text{(Fl}_2)}{=} \varphi_{a^1}^{X_1} \circ \cdots \circ \varphi_t^{X_i} \circ \varphi_{a^i}^{X_i} \circ \cdots \circ \varphi_{a^k}^{X_k}(p)$$

$$\overset{\text{Prop. } 3.2.32}{=} \varphi_t^{X_i}(\varphi(a)) = \varphi^{X_i}(t, \varphi(a)),$$

whenever $(t, \varphi(a)) \in \mathcal{D}_{X_i}$. Thus $\varphi \circ \alpha$ is the flow line of φ^{X_i} through $\varphi(a)$, therefore

$$(\varphi_*)_a \left(\frac{\partial}{\partial e^i} \right)_a \overset{(3.2.14)}{=} \dot{\varphi}_i(a) \overset{(3.2.15)}{=} (\varphi \circ \alpha)^{\cdot}(0) = X_i(\varphi(a)).$$

Since the tangent vectors $X_i(\varphi(a)) \in T_{\varphi(a)}M$, $i \in J_k$, are linearly independent, it follows that the mapping $(\varphi_*)_a$ is injective. Thus, by the arbitrariness of $a \in \mathcal{U}$, we conclude that φ is an immersion, and $\varphi_* \circ \frac{\partial}{\partial e^i} = X_i \circ \varphi$. $\square$

Corollary 3.2.36. *Let $\mathcal{V}$ be an open subset of a manifold M, and let $(X_i)_{i=1}^n$ be a local frame field of TM over $\mathcal{V}$ consisting of mutually commuting vector fields. Then at every point p of $\mathcal{V}$ there exists a chart $(\mathcal{U}, (u^i)_{i=1}^n)$ of M such that*

$$\mathcal{U} \subset \mathcal{V} \quad \text{and} \quad X_i \upharpoonright \mathcal{U} = \frac{\partial}{\partial u^i}, \quad i \in J_n.$$

Proof. According to the previous proposition, there exists an open neighbourhood $\mathcal{W}$ of $0 \in \mathbb{R}^n$ together with an immersion $\varphi \colon \mathcal{W} \to M$ such that $\varphi(0) = p$ and $\frac{\partial}{\partial e^i} \underset{\varphi}{\sim} X_i$ for all $i \in J_n$. Then $(\varphi_*)_0$ is a linear isomorphism of $T_0 \mathbb{R}^n$ onto $T_p M$. The inverse mapping theorem (Corollary 3.1.21) guarantees the existence of an open neighbourhood $\widetilde{\mathcal{W}}$ of $0 \in \mathbb{R}^n$ and an open neighbourhood $\mathcal{U}$ of p such that $\widetilde{\varphi} := \varphi \upharpoonright \widetilde{\mathcal{W}} \colon \widetilde{\mathcal{W}} \to \mathcal{U}$ is a diffeomorphism. Obviously, $\mathcal{U} \subset \mathcal{V}$. Let $u := \widetilde{\varphi}^{-1}$. Then $(\mathcal{U}, u)$ is a chart of M, and $(u_*)_q$ sends $X_i(q)$ to $\left(\frac{\partial}{\partial e^i}\right)_{u(q)}$ for all $q \in \mathcal{U}$, $i \in J_n$. Thus, for any $i \in J_n$ and $q \in \mathcal{U}$,

$$
du^j \circ X_i(q) = X_i(q)(e^j \circ u) = (u_*)_q(X_i(q))(e^j) = \left(\frac{\partial}{\partial e^i}\right)_{u(q)}(e^j) = \delta_i^j,
$$

whence $X_i \upharpoonright \mathcal{U} = \frac{\partial}{\partial u^i}$, $i \in J_n$. This concludes the proof. $\qquad\square$

3.3 Tensors and Differential Forms

3.3.1 The Cotangent Bundle of a Manifold

Definition 3.3.1. Let M be a manifold. The dual bundle

$$
\widehat{\tau} \colon T^*M \to M, \quad T^*M := \bigcup_{p \in M} T_p^*M := \bigcup_{p \in M} (T_p M)^*
$$

of the tangent bundle $\tau \colon TM \to M$ is said to be the *cotangent bundle* of M. By a *one-form* of class C^r ($r \geq 0$) on M we mean a section of class C^r of $\widehat{\tau} \colon T^*M \to M$. The smooth one-forms are simply mentioned as *one-forms*.

Remark 3.3.2. For the $C^\infty(M)$-module $\Gamma(\widehat{\tau}) = \Gamma(T^*M)$ we use the notation $\mathcal{A}_1(M)$. By Proposition 2.2.44,

$$
\mathcal{A}_1(M) := \Gamma(T^*M) \cong (\Gamma(TM))^* = (\mathfrak{X}(M))^* =: \mathfrak{X}^*(M).
$$

The natural isomorphism between $\mathcal{A}_1(M)$ and $\mathfrak{X}^*(M)$ is given by

$$
\omega \in \mathcal{A}_1(M) \mapsto \widetilde{\omega} \in \mathfrak{X}^*(M) = \mathrm{Hom}\left(\mathfrak{X}(M), C^\infty(M)\right),
$$
$$
\widetilde{\omega}(X)(p) := \omega_p(X_p); \quad X \in \mathfrak{X}(M), \ p \in M.
$$

As always in a similar situation, we make no notational distinction between the elements of $\mathcal{A}_1(M)$ and $\mathfrak{X}^*(M)$.

If $\mathcal{U}$ is an open subset of M, then

$$
\mathcal{A}_1(\mathcal{U}) := \Gamma_{\mathcal{U}}(T^*M) = \left\{\omega \in C^\infty(\mathcal{U}, T^*M) \,\middle|\, \widehat{\tau} \circ \omega = 1_{\mathcal{U}}\right\} \cong \mathfrak{X}^*(\mathcal{U}).
$$

Example 3.3.3. (a) Let f be a smooth function on M. The *differential of f at a point* $p \in M$ is the linear form $(df)_p \in T_p^*M$ defined by

$$(df)_p(v) := v(f), \quad v \in T_pM. \tag{3.3.1}$$

Then the mapping

$$df \colon M \to T^*M, \ p \mapsto (df)_p$$

is a section of $\hat{\tau} \colon T^*M \to M$; we show that df is a smooth section, and hence a one-form on M. Let X be a vector field on M, and define the function $df(X)$ by

$$df(X)(p) := (df)_p(X_p), \quad p \in M.$$

Since $(df)_p(X_p) := X_p(f) = (Xf)(p)$, we obtain

$$df(X) = Xf, \quad X \in \mathfrak{X}(M). \tag{3.3.2}$$

The mapping $X \in \mathfrak{X}(M) \mapsto Xf \in C^\infty(M)$, where $f \in C^\infty(M)$ is fixed, is clearly $C^\infty(M)$-linear. Thus, by the isomorphism

$$\mathrm{Hom}\left(\mathfrak{X}(M), C^\infty(M)\right) \cong \mathcal{A}_1(M),$$

it follows that $df \in \mathcal{A}_1(M)$.

Since the vector fields act as derivations (Lemma 3.1.41), relation (3.3.2) implies immediately that the mapping

$$d \colon C^\infty(M) \to \mathcal{A}_1(M), \ f \mapsto df$$

(i) is $\mathbb{R}$-linear,
(ii) satisfies the product rule $d(fg) = g\,df + f\,dg$ $(f, g \in C^\infty(M))$.

The one-form df is called the *differential* of f.

(b) Let $(\mathcal{U}, (u^i)_{i=1}^n)$ be a chart of M. Then the differentials $(du^i)_p$, $i \in J_n$, form a basis of T_p^*M, which is the dual of the basis $\left(\left(\frac{\partial}{\partial u^i}\right)_p\right)_{i=1}^n$ of T_pM. If follows that any linear form $v^* \in T_p^*M$ can uniquely be written in the form

$$v^* = \sum_{i=1}^n v^*\left(\frac{\partial}{\partial u^i}\right)_p (du^i)_p,$$

cf. (1.1.11). In particular, if $f \in C^\infty(M)$, then

$$(df)_p = \sum_{i=1}^n (df)_p\left(\frac{\partial}{\partial u^i}\right)_p (du^i)_p = \sum_{i=1}^n \frac{\partial f}{\partial u^i}(p)(du^i)_p,$$

hence

$$df \upharpoonright \mathcal{U} = \sum_{i=1}^{n} \frac{\partial f}{\partial u^i} du^i. \tag{3.3.3}$$

More generally, for any one-form $\omega \in \mathcal{A}_1(M)$ we have

$$\omega \upharpoonright \mathcal{U} = \sum_{i=1}^{n} \omega \left(\frac{\partial}{\partial u^i} \right) du^i,$$

therefore the family $(du^i)_{i=1}^{n}$ is a frame field of T^*M over $\mathcal{U}$.

Proposition 3.3.4. *The $C^\infty(M)$-module $\mathcal{A}_1(M)$ is generated by the differentials of smooth functions on M.*

For a proof of this delicate result we refer to [52], pp. 117–118.

Lemma 3.3.5. *Let M be an at least two-dimensional manifold and f a smooth function on M. A real number $\lambda \in \mathrm{Im}(f)$ is a regular value of f if, and only if, $(df)_p \neq 0$ for all $p \in S$. In this case $S := f^{-1}(\lambda)$ is a hypersurface of M.*

Proof. Let $(r) = (1_{\mathbb{R}})$ be the canonical coordinate system of $\mathbb{R}$ (Example 2.1.9). Let $p \in S$, $v \in T_pS$. We recall that T_pS can be regarded as a subspace of T_pM (see Remark 3.1.24). Then

$$(df)_p(v) = v(f) = v(r \circ f) \overset{(3.1.4)}{=} (f_*)_p(v)(r),$$

so by the basis theorem (Proposition 3.1.8) we obtain that

$$(f_*)_p(v) = (df)_p(v) \left(\frac{d}{dr} \right)_{f(p)}.$$

From this we see immediately that $(f_*)_p$ is surjective if, and only if, $(df)_p \neq 0$. The second assertion follows from Proposition 3.1.26. $\square$

Corollary 3.3.6. *Let M be a manifold with $\dim M \geq 2$, and let f be a smooth function on M. Suppose that $\lambda \in \mathrm{Im}(f)$ is a regular value of f. Consider the hypersurface $S := f^{-1}(\lambda)$, and let $j \colon S \to M$ be the canonical inclusion. Then for every $p \in S$,*

$$(j_*)_p(T_pS) = \mathrm{Ker}(df)_p,$$

or simply $T_pS = \mathrm{Ker}(df)_p$ (see again Remark 3.1.24).

Proof. Let $p \in S$, $v \in T_pS$. By Lemma 3.1.18, there is a smooth curve $\alpha \colon]-1, 1[\to S$ such that $\dot{\alpha}(0) = v$. Then the smooth function

$$f \circ j \circ \alpha \colon]-1, 1[\to S \to M \to \mathbb{R}$$

is constant, therefore

$$0 = (f \circ j \circ \alpha)'(0) \overset{(3.1.9)}{=} (j \circ \alpha)'(0)(f) \overset{(3.1.11)}{=} (j_*)_{\alpha(0)}(\dot{\alpha}(0))(f)$$

$$= (j_*)_p(v)(f) = (df)_p((j_*)_p(v)).$$

Thus $\mathrm{Im}(j_*)_p = (j_*)_p(T_pS) \subset \mathrm{Ker}(df)_p$. Here $(j_*)_p$ is injective, so we have

$$\dim((j_*)_p(T_pS)) = n - 1.$$

On the other hand, by the previous lemma, $(df)_p \neq 0$, so the linear function $(df)_p \colon T_pM \to \mathbb{R}$ is surjective. From this we conclude that $\mathrm{Ker}(df_p)$ has also dimension $n - 1$. Summing up, we found that the $(n-1)$-dimensional vector space $(j_*)_p(T_pS)$ is a subspace of the $(n-1)$-dimensional vector space $\mathrm{Ker}(df)_p$, which implies the desired equality. $\qquad\square$

3.3.2　Tensors on a Manifold

In Section 1.2 we introduced tensors in the abstract setting of R-modules. In subsection 2.2.4 we transposed the tensor concept to the context of vector bundles: taking a vector bundle $\pi \colon E \to M$ as our starting-point, we defined π-tensors and π-tensor fields. In the case of the tangent bundle $\tau \colon TM \to M$ we shall use a special terminology and notation for 'τ-tensors' and 'τ-tensor fields'. In this subsection, for the readers' convenience, we present a summary of the material of 1.2 and 2.2.4 in the context of the tangent bundle.

Definition 3.3.7. Let M be a manifold and consider its tangent bundle $\tau \colon TM \to M$. The vector bundle

$$\tau_l^k \colon T_l^k(TM) \to M$$

is said to be the bundle of (k, l)-tensors on M. The $C^\infty(M)$-module

$$\mathcal{T}_l^k(M) := \Gamma(\tau_l^k) = \Gamma(T_l^k(TM))$$

of the smooth sections of τ_l^k is called the module of type (k, l) tensor fields (or, by an abuse of language, of type (k, l) tensors) on M.

Remark 3.3.8. (a) By the tensoriality criterion formulated in Remark 2.2.47, we have a natural isomorphism

$$T_l^k(M) \cong T_l^k(\mathfrak{X}(M))$$

of $C^\infty(M)$-modules. Thus a type (k,l) tensor field on M may be freely considered as a $C^\infty(M)$-multilinear mapping

$$A\colon \underbrace{\mathfrak{X}^*(M) \times \cdots \times \mathfrak{X}^*(M)}_{k} \times \underbrace{\mathfrak{X}(M) \times \cdots \times \mathfrak{X}(M)}_{l} \to C^\infty(M).$$

Now we mention two frequently used interpretations. First,

$$T_0^1(M) \cong T_0^1(\mathfrak{X}(M)) := (\mathfrak{X}^*(M))^* \cong \mathfrak{X}(M), \tag{3.3.4}$$

the mapping

$$X \in \mathfrak{X}(M) \mapsto \widetilde{X} \in T_0^1(M); \ \widetilde{X}(\theta) := \theta(X) \ (\theta \in \mathfrak{X}^*(M))$$

being a canonical isomorphism of $C^\infty(M)$-modules (cf. 1.2.3). Second,

$$T_l^1(M) \cong T_l^1(\mathfrak{X}(M)) \cong T_l^0(\mathfrak{X}(M), \mathfrak{X}(M)), \tag{3.3.5}$$

where $l \geq 1$. To understand this well, we recall that $T_l^0(\mathfrak{X}(M), \mathfrak{X}(M))$ denotes the $C^\infty(M)$-module of the $C^\infty(M)$-multilinear mappings

$$B\colon \underbrace{\mathfrak{X}(M) \times \cdots \times \mathfrak{X}(M)}_{l} \to \mathfrak{X}(M)$$

(see 1.2.1). Keeping this in mind, the canonical isomorphism (3.3.5) is given by

$$B \in T_l^0(\mathfrak{X}(M), \mathfrak{X}(M)) \mapsto \widetilde{B} \in T_l^1(\mathfrak{X}(M)),$$

$$\widetilde{B}(\theta, X_1, \ldots, X_l) := \theta(B(X_1, \ldots, X_l)).$$

In particular,

$$T_1^1(M) \cong \mathrm{End}(\mathfrak{X}(M)). \tag{3.3.6}$$

The *Kronecker tensor* or *unit tensor field on* M is the tensor $\delta \in T_1^1(M)$ which corresponds to the identity element $1_{\mathfrak{X}(M)}$ under the isomorphism (3.3.6). Then (cf. Example 1.2.3(a))

$$\delta(\theta, X) = \theta(X) \text{ for all } \theta \in \mathfrak{X}^*(M), \ X \in \mathfrak{X}(M).$$

(b) Let $A \in T_{l_1}^{k_1}(M)$, $B \in T_{l_2}^{k_2}(M)$. We define the *tensor product* $A \otimes B \in T_{l_1+l_2}^{k_1+k_2}(M)$ by the rule (1.2.2). If, in particular, $k_2 = l_2 = 0$, then $B := f \in C^\infty(M)$, and

$$A \otimes f = f \otimes A := fA.$$

(c) Let $\varphi\colon M \to N$ be a smooth mapping. With the help of the derivative $\varphi_*\colon TM \to TN$ of φ, we may carry out the pull-back construction described in 1.2.2 fibrewise: if $B \in \mathcal{T}_l^0(N)$ ($l \geq 1$), then we define its *pull-back* $\varphi^* B \in \mathcal{T}_l^0(M)$ by

$$(\varphi^* B)_p(v_1, \ldots, v_l) := B_{\varphi(p)}\big(\varphi_*(v_1), \ldots, \varphi_*(v_l)\big) \qquad (3.3.7)$$

for all $p \in M$; $v_1, \ldots, v_l \in T_p M$. If $B := f \in \mathcal{T}_0^0(N) = C^\infty(N)$, then $\varphi^* f := f \circ \varphi \in C^\infty(M)$ as in Remark 3.1.50. The following properties of the pull-back are easy consequences of the definition.

(i) The mapping $\varphi^*\colon \mathcal{T}_l^0(N) \to \mathcal{T}_l^0(M)$ is $\mathbb{R}$-linear.

(ii) $(1_M)^*$ is the identity transformation of $\mathcal{T}_l^0(M)$.

(iii) If $\varphi \in \mathrm{Diff}(M, N)$, then $\varphi^*\colon \mathcal{T}_l^0(N) \to \mathcal{T}_l^0(M)$ is an $\mathbb{R}$-linear isomorphism.

(iv) If $\psi\colon N \to S$ is also a smooth mapping, then

$$(\psi \circ \varphi)^* = \varphi^* \circ \psi^*\colon \mathcal{T}_l^0(S) \to \mathcal{T}_l^0(M) \text{ for all } l \in \mathbb{N}. \qquad (3.3.8)$$

(d) Let $k, l \in \mathbb{N}^*$, $(r, s) \in J_k \times J_l$. There exists a unique $C^\infty(M)$-linear mapping

$$C_s^r\colon \mathcal{T}_l^k(M) \to \mathcal{T}_{l-1}^{k-1}(M), \ A \mapsto C_s^r A,$$

called the *contraction* of the rth contravariant and the sth covariant index, given by

$$(C_s^r A)_p := (C_s^r)_p A_p, \quad p \in M,$$

where $(C_s^r)_p\colon T_l^k(T_p M) \to T_{l-1}^{k-1}(T_p M)$ is the contraction operation introduced in 1.2.5. Then we have

$$C_s^r(X_1 \otimes \cdots \otimes X_k \otimes \theta^1 \otimes \cdots \otimes \theta^l)$$
$$= \theta^s(X_r) X_1 \otimes \cdots \otimes \check{X}_r \otimes \cdots \otimes X_k \otimes \theta^1 \otimes \cdots \otimes \check{\theta}^s \otimes \cdots \otimes \theta^l,$$

for all $X_1, \ldots, X_k \in \mathfrak{X}(M)$, $\theta^1, \ldots, \theta^l \in \mathfrak{X}^*(M)$.

(e) Let $(\mathcal{U}, (u^i)_{i=1}^n)$ be a chart on M. The *components* of a tensor field $A \in \mathcal{T}_l^k(M)$ with respect to this chart are the real-valued smooth functions

$$A_{j_1 \ldots j_l}^{i_1 \ldots i_k} := A\left(du^{i_1}, \ldots, du^{i_k}, \frac{\partial}{\partial u^{j_1}}, \ldots, \frac{\partial}{\partial u^{j_l}}\right)$$

on $\mathcal{U}$, where all indices run from 1 to n. Then, locally, A can be expressed as

$$A \underset{(\mathcal{U})}{=} \sum_{(i)(j)} A_{j_1 \ldots j_l}^{i_1 \ldots i_k} \frac{\partial}{\partial u^{i_1}} \otimes \cdots \otimes \frac{\partial}{\partial u^{i_k}} \otimes du^{j_1} \otimes \cdots \otimes du^{j_l}$$

$$= A_{j_1 \ldots j_l}^{i_1 \ldots i_k} \frac{\partial}{\partial u^{i_1}} \otimes \cdots \otimes \frac{\partial}{\partial u^{i_k}} \otimes du^{j_1} \otimes \cdots \otimes du^{j_l};$$

$$(3.3.9)$$

cf. (1.2.10) and Remark 2.2.47(b).

3.3.3 Tensor Derivations

Definition 3.3.9. A *tensor derivation* on a manifold M is a family

$$\mathcal{D}_l^k \colon \mathcal{T}_l^k(M) \to \mathcal{T}_l^k(M), \quad (k, l) \in \mathbb{N} \times \mathbb{N},$$

of $\mathbb{R}$-linear mappings, all denoted by the same symbol $\mathcal{D}$ for convenience, such that

(i) $\mathcal{D}$ obeys the *Leibniz rule*

$$\mathcal{D}(A \otimes B) = (\mathcal{D}A) \otimes B + A \otimes (\mathcal{D}B); \quad A \in \mathcal{T}_{l_1}^{k_1}(M), \ B \in \mathcal{T}_{l_2}^{k_2}(M);$$

(ii) $\mathcal{D}$ commutes with contractions, i.e.,

$$\mathcal{D}(C_s^r A) = C_s^r(\mathcal{D}A); \quad A \in \mathcal{T}_l^k(M), \quad (r, s) \in J_k \times J_l.$$

Remark 3.3.10. If $(k, l) = (0, 0)$, then $\mathcal{D} = \mathcal{D}_0^0$ is a derivation of $\mathcal{T}_0^0(M) = C^\infty(M)$ so, by Lemma 3.1.41 there is a unique vector field X on M such that

$$\mathcal{D}f = Xf \text{ for all } f \in C^\infty(M).$$

Lemma 3.3.11. Tensor derivations are local operators: *if $\mathcal{D}$ is a tensor derivation on a manifold M, $\mathcal{U}$ is an open subset of M, and A is a tensor field on M such that $A \upharpoonright \mathcal{U} = 0$, then $\mathcal{D}A \upharpoonright \mathcal{U} = 0$.*

Proof. Let p be any point in $\mathcal{U}$. Corollary 2.1.32 implies that there exists a smooth function f on M such that $f(p) = 0$ and $f(q) = 1$ if $q \in M \setminus \mathcal{U}$. Then $A = fA = f \otimes A$, and applying the Leibniz rule we obtain

$$\mathcal{D}A = \mathcal{D}(f \otimes A) = \mathcal{D}f \otimes A + f \otimes \mathcal{D}A = (\mathcal{D}f)A + f(\mathcal{D}A).$$

Thus

$$(\mathcal{D}A)_p = (\mathcal{D}f)(p)A_p + f(p)(\mathcal{D}A)_p = (\mathcal{D}f)(p) \cdot 0 + 0 \cdot (\mathcal{D}A)_p = 0.$$

Since the choice of $p \in \mathcal{U}$ was arbitrary, it follows that $\mathcal{D}A \upharpoonright \mathcal{U} = 0$. $\qquad\square$

Lemma 3.3.12. Tensor derivations are natural with respect to restrictions: *if $\mathcal{D}$ is a tensor derivation on a manifold M and $\mathcal{U}$ is an open subset of M, then there exists a unique tensor derivation $\mathcal{D}_\mathcal{U}$ on (the open submanifold) $\mathcal{U}$ such that*

$$\mathcal{D}_\mathcal{U}(A \upharpoonright \mathcal{U}) = (\mathcal{D}A) \upharpoonright \mathcal{U} \text{ for all tensor fields } A \text{ on } M. \tag{3.3.10}$$

Sketch of proof. Let B be a tensor field defined on $\mathcal{U}$. Choose a point $p \in \mathcal{U}$ and a bump function $f \in C^{\infty}(M)$ at p supported in $\mathcal{U}$. Let A be a tensor field on M given by

$$A := fB \text{ on } \mathcal{U}, \ A := 0 \text{ outside of } \mathcal{U}.$$

Define

$$(\mathcal{D}_{\mathcal{U}}B)(p) := (\mathcal{D}A)(p).$$

This definition is independent of the choice of the bump function f. To see this, let $h \in C^{\infty}(M)$ be any other bump function at p supported in $\mathcal{U}$, and consider the tensor field $\widetilde{A}$ on M given by

$$\widetilde{A} := hB \text{ on } \mathcal{U}, \ \widetilde{A} := 0 \text{ outside of } \mathcal{U}.$$

Then there is an open neighbourhood $\mathcal{V}$ of p such that $f(q) = h(q) = 1$ for all $q \in \mathcal{V}$ (see Definition 2.1.31). On this neighbourhood of p we have $(A - \widetilde{A}) \restriction \mathcal{V} = 0$. Hence, by Lemma 3.3.11, $\mathcal{D}(A - \widetilde{A}) \restriction \mathcal{V} = 0$, therefore $(\mathcal{D}A)(p) = (\mathcal{D}\widetilde{A})(p)$.

We leave to the reader to check the following statements:

(i) $\mathcal{D}_{\mathcal{U}}$ is a tensor derivation on the open submanifold $\mathcal{U}$.

(ii) $\mathcal{D}_{\mathcal{U}}$ satisfies the naturality condition (3.3.10).

(iii) $\mathcal{D}_{\mathcal{U}}$ is uniquely determined by the naturality condition (3.3.10). $\qquad\square$

Remark 3.3.13. Condition (3.3.10) can be displayed by the following commutative diagram:

$$
\begin{array}{ccc}
\mathcal{T}_l^k(M) & \xrightarrow{\restriction_{\mathcal{U}}} & \mathcal{T}_l^k(\mathcal{U}) \\
{\scriptstyle \mathcal{D}}\downarrow & & \downarrow{\scriptstyle \mathcal{D}_{\mathcal{U}}} \\
\mathcal{T}_l^k(M) & \xrightarrow[\restriction_{\mathcal{U}}]{} & \mathcal{T}_l^k(\mathcal{U})
\end{array} \ .
$$

We say that $\mathcal{D}_{\mathcal{U}}$ is the *restriction* of $\mathcal{D}$ to $\mathcal{U}$. However, for simplicity, we usually write $\mathcal{D}$ for $\mathcal{D}_{\mathcal{U}}$.

Proposition 3.3.14 (The Product Rule). *Let $\mathcal{D}$ be a tensor derivation on a manifold M. Then for every tensor field $A \in \mathcal{T}_l^k(M)$ we have*

$$
\begin{aligned}
(\mathcal{D}A)(\theta^1,\ldots,\theta^k,X_1,\ldots,X_l) = {} & \mathcal{D}\big(A(\theta^1,\ldots,\theta^k,X_1,\ldots,X_l)\big) \\
& - \sum_{i=1}^{k} A\big(\theta^1,\ldots,\mathcal{D}\theta^i,\ldots,\theta^k,X_1,\ldots,X_l\big) \\
& - \sum_{j=1}^{l} A\big(\theta^1,\ldots,\theta^k,X_1,\ldots,\mathcal{D}X_j,\ldots,X_l\big).
\end{aligned}
\tag{3.3.11}
$$

For a proof we refer to [79], p. 44.

Corollary 3.3.15. *A tensor derivation is uniquely determined by its action on the smooth functions and on the vector fields of the underlying manifold.*

Proof. Let $\mathcal{D}$ be a tensor derivation on a manifold M. If θ is a one-form, and X is a vector field on M, then

$$\mathcal{D}(\theta(X)) = \mathcal{D}\big(C_1^1(\theta \otimes X)\big) = C_1^1\big(\mathcal{D}(\theta \otimes X)\big)$$
$$= C_1^1\big((\mathcal{D}\theta) \otimes X + \theta \otimes (\mathcal{D}X)\big) = (\mathcal{D}\theta)(X) + \theta(\mathcal{D}X),$$

whence

$$(\mathcal{D}\theta)(X) = \mathcal{D}(\theta(X)) - \theta(\mathcal{D}X).$$

Thus the action of $\mathcal{D}$ on the one-forms on M is uniquely determined by its action on $C^\infty(M)$ and on $\mathfrak{X}(M)$. Since $\mathcal{D}$ must obey formula (3.3.11), this implies our assertion. $\qquad\qquad\qquad\qquad\qquad\qquad\qquad\qquad\qquad\quad\square$

Proposition 3.3.16 (Willmore's Theorem). *Let M be a manifold. Given a vector field X on M and an $\mathbb{R}$-linear mapping $\mathcal{D}_0\colon \mathfrak{X}(M) \to \mathfrak{X}(M)$ such that*

$$\mathcal{D}_0(fY) = (Xf)Y + f\mathcal{D}_0(Y) \ \text{for all } f \in C^\infty(M),\ Y \in \mathfrak{X}(M), \quad (3.3.12)$$

there exists a (necessarily unique) tensor derivation $\mathcal{D}$ on M such that

$$\mathcal{D}f = Xf \ \text{if } f \in C^\infty(M) \ \text{and } \mathcal{D}Y = \mathcal{D}_0Y \ \text{if } Y \in \mathfrak{X}(M).$$

For a proof, we refer again to [79], p. 45.

Example 3.3.17 (Lie Derivative on a Manifold). Let M be a manifold.

(a) Given a vector field X on M, there exists a unique tensor derivation d_X on M such that

$$d_Xf = \mathcal{L}_Xf \overset{(3.1.37)}{=} Xf \ \text{for all } f \in C^\infty(M),$$
$$d_XY = [X,Y] \ \text{for all } Y \in \mathfrak{X}(M).$$

Indeed, define the mapping $\mathcal{D}_0\colon \mathfrak{X}(M) \to \mathfrak{X}(M)$ by $\mathcal{D}_0(Y) := [X,Y]$. Then $\mathcal{D}_0$ satisfies condition (3.3.12), since for every $f \in C^\infty(M)$,

$$\mathcal{D}_0(fY) := [X, fY] \overset{(3.1.39)}{=} f[X,Y] + (Xf)Y = (Xf)Y + f\mathcal{D}_0(Y).$$

Thus Willmore's theorem guarantees the existence and Corollary 3.3.15 the uniqueness of the desired tensor derivation. In what follows we shall use the more conventional notation $\mathcal{L}_X := d_X$, and we say that $\mathcal{L}_X$ is the *Lie derivative* on M with respect to X.

(b) For all $X, Y \in \mathfrak{X}(M)$,

$$\mathcal{L}_{[X,Y]} = [\mathcal{L}_X, \mathcal{L}_Y] := \mathcal{L}_X \circ \mathcal{L}_Y - \mathcal{L}_Y \circ \mathcal{L}_X. \tag{3.3.13}$$

To show this, by Corollary 3.3.15 it is enough to check that the effect of both sides of (3.3.13) is the same on $C^\infty(M)$ and $\mathfrak{X}(M)$. The first requirement holds automatically by the definition of the Lie bracket of vector fields (Corollary 3.1.43). Since $\mathfrak{X}(M)$ is a (real) Lie algebra, applying the Jacobi identity ((L2) in 1.3.3) we find that for all $Z \in \mathfrak{X}(M)$,

$$[\mathcal{L}_X, \mathcal{L}_Y](Z) = [X, [Y, Z]] - [Y, [X, Z]] = [X, [Y, Z]] + [Y, [Z, X]]$$

$$= -[Z, [X, Y]] = [[X, Y], Z] = \mathcal{L}_{[X,Y]}(Z),$$

so $\mathcal{L}_{[X,Y]}$ and $[\mathcal{L}_X, \mathcal{L}_Y]$ agree also on vector fields.

(c) Let $A \in \mathcal{T}_l^0(M)$, $B \in \mathcal{T}_l^1(M) \cong T_l^0(\mathfrak{X}(M), \mathfrak{X}(M))$, where $l \geq 1$. Let X be a vector field on M. Then the product rule (3.3.11) yields the following:

$$(\mathcal{L}_X A)(X_1, \dots, X_l) = X\big(A(X_1, \dots, X_l)\big) - \sum_{i=1}^{l} A\big(X_1, \dots, [X, X_i], \dots, X_l\big),$$

$$\tag{3.3.14}$$

$$(\mathcal{L}_X B)(X_1, \dots, X_l) = [X, B(X_1, \dots, X_l)] - \sum_{i=1}^{l} B\big(X_1, \dots, [X, X_i], \dots, X_l\big)$$

$$\tag{3.3.15}$$

for any vector fields $X_1, \dots, X_l$ on M.

If, in particular, $B \in \mathrm{End}(\mathfrak{X}(M))$, then

$$[B, Y] := -\mathcal{L}_Y B \in \mathrm{End}(\mathfrak{X}(M))$$

is said to be the *Frölicher–Nijenhuis bracket* of the vector-valued type $(1,1)$-tensor field B and the vector field Y. From (3.3.15) we obtain

$$[B, Y]X = [BX, Y] - B[X, Y]; \quad X, Y \in \mathfrak{X}(M). \tag{3.3.16}$$

(d) Let A and B be type $(1,1)$ tensor fields on M considered as $C^\infty(M)$-linear endomorphisms of $\mathfrak{X}(M)$. Define a mapping

$$[A, B] \colon \mathfrak{X}(M) \times \mathfrak{X}(M) \to \mathfrak{X}(M), \ (X, Y) \mapsto [A, B](X, Y)$$

by

$$[A, B](X, Y) := [AX, BY] + [BX, AY] + (A \circ B + B \circ A)([X, Y])$$

$$- A[BX, Y] - A[X, BY] - B[AX, Y] - B[X, AY]. \tag{3.3.17}$$

It may be shown by a straightforward calculation that $[A, B]$ is $C^\infty(M)$-bilinear. Thus $[A, B] \in T_2^0\big(\mathfrak{X}(M), \mathfrak{X}(M)\big)$, i.e., $[A, B]$ is a (vector-valued) type $(1, 2)$ tensor field on M. It is easy to see that

$$[A, B](X, Y) = -[A, B](Y, X), \quad X, Y \in \mathfrak{X}X(M),$$

and $[A, B]$ is graded anticommutative, i.e., $[A, B] = [B, A]$ (cf. (1.3.7), and see later Remark 3.3.47(iii)). We say that $[A, B]$ is the *Fröhlicher–Nijenhuis bracket* of A and B. In particular, $N_A := \frac{1}{2}[A, A]$ is called the *Nijenhuis tensor* (or *Nijenhuis torsion*) of A. Then (3.3.17) gives

$$N_A(X, Y) = [AX, AY] + A^2[X, Y] - A[AX, Y] - A[X, AY]. \quad (3.3.18)$$

3.3.4 Differential Forms

First we repeat the construction of the bundle of alternating π-tensors (see Example 2.2.48) in the tangent bundle and fit our terminology to this situation.

Let M be an n-dimensional manifold and $k \in J_n$. At each point $p \in M$ consider the vector space $A_k(T_pM)$ of alternating k-linear functions $(T_pM)^k \to \mathbb{R}$. Form the disjoint union

$$A_k(TM) := \bigcup_{p \in M} A_k(T_pM),$$

and let $\widehat{\tau}_k \colon A_k(TM) \to M$ be the footpoint mapping given by $\widehat{\tau}_k(\Phi) := p$ if $\Phi \in A_k(T_pM)$. Then $(A_k(TM), \widehat{\tau}_k, M)$ is a vector bundle of rank $\binom{n}{k}$, called the *k-form bundle* over M. We agree that $A_0(TM) := M \times \mathbb{R}$. The *exterior algebra bundle* over M is the direct sum

$$A(TM) := \bigoplus_{k=0}^{n} A_k(TM)$$

of the k-form bundles over M.

Definition 3.3.18. Let M be a manifold. A *differential form of degree k* on M (or a *differential k-form*, simply a *k-form*, on M) is a smooth section of the k-form bundle $A_k(TM)$. We denote the $C^\infty(M)$-module of k-forms by $\mathcal{A}_k(M)$, thus

$$\mathcal{A}_k(M) := \Gamma(A_k(TM)).$$

If $\alpha \in \mathcal{A}_k(M)$, $\beta \in \mathcal{A}_l(M)$, then their *exterior product* (or *wedge product*) is the $(k + l)$-form $\alpha \wedge \beta$ given by

$$(\alpha \wedge \beta)_p := \alpha_p \wedge \beta_p, \quad p \in M.$$

Remark 3.3.19. Let M be an n-dimensional manifold.

(a) By (1.3.12), (A.3.2) and (1.2.2), the exterior product $\alpha \wedge \beta$ can explicitly be given as follows: if $p \in M$; $v_1, \ldots, v_{k+l} \in T_p M$, then

$$(\alpha \wedge \beta)_p(v_1, \ldots, v_{k+l}) := (\alpha_p \wedge \beta_p)(v_1, \ldots, v_{k+l})$$

$$= \frac{1}{k! l!} \left(\sum_{\sigma \in S_{k+l}} \varepsilon(\sigma) \sigma(\alpha_p \otimes \beta_p) \right) (v_1, \ldots, v_{k+l})$$

$$= \frac{1}{k! l!} \sum_{\sigma \in S_{k+l}} \varepsilon(\sigma) (\alpha_p \otimes \beta_p) \big(v_{\sigma(1)}, \ldots, v_{\sigma(k+l)} \big)$$

$$= \frac{1}{k! l!} \sum_{\sigma \in S_{k+l}} \varepsilon(\sigma) \alpha_p \big(v_{\sigma(1)}, \ldots, v_{\sigma(k)} \big) \beta_p \big(v_{\sigma(k+1)}, \ldots, v_{\sigma(k+l)} \big),$$

that is, we have

$$(\alpha \wedge \beta)_p(v_1, \ldots, v_{k+l})$$
$$= \frac{1}{k! l!} \sum_{\sigma \in S_{k+l}} \varepsilon(\sigma) \alpha_p \big(v_{\sigma(1)}, \ldots, v_{\sigma(k)} \big) \beta_p \big(v_{\sigma(k+1)}, \ldots, v_{\sigma(k+l)} \big). \quad (3.3.19)$$

The exterior product makes the direct sum $\mathcal{A}(M) := \bigoplus_{k=0}^n \mathcal{A}_k(M)$ into an associative and graded commutative graded algebra over the ring $C^\infty(M)$ (cf. Lemma 1.3.10), called the *Grassmann algebra* of the manifold M.

(b) Any differential k-form α on M determines an alternating $C^\infty(M)$-multilinear mapping $(\mathfrak{X}(M))^k \to C^\infty(M)$ denoted by the same symbol and given by

$$\alpha(X_1, \ldots, X_k)(p) := \alpha_p \big(X_1(p), \ldots, X_k(p) \big), \quad p \in M,$$

where $X_1, \ldots, X_k \in \mathfrak{X}(M)$. In this way we obtain canonical isomorphisms

$$\mathcal{A}_k(M) \cong A_k(\mathfrak{X}(M)) \ (k \in J_n) \text{ and } \mathcal{A}(M) \cong A(\mathfrak{X}(M))$$

of $C^\infty(M)$-modules, cf. Example 2.2.48. We shall identify these isomorphic modules without any comments.

Definition 3.3.20. Let $\pi \colon E \to M$ be a vector bundle of rank r $(r \in \mathbb{N}^*)$, and let $k \in J_n$ $(n = \dim M)$. Form the vector bundle $A_k(TM, E)$ over M whose fibre at a point $p \in M$ consists of the alternating k-linear mappings

$$T_p M \times \cdots \times T_p M \to E_p,$$

and so

$$A_k(TM, E) = \bigcup_{p \in M} A_k(T_p M, E_p).$$

By an *E-valued k-form* on M we mean a smooth section of the bundle $A_k(TM, E)$. If, in particular, π is the trivial bundle $M \times V \to M$, where V is an r-dimensional real vector space, we speak of V-*valued* (or *vector-valued*) *forms* as the smooth sections of the bundles

$$A_k(TM, V) \to M, \ A_k(TM, V) := \bigcup_{p \in M} (A_k(T_pM, V)), \quad k \in J_n.$$

Remark 3.3.21. (a) The E-valued k-forms on M form a $C^\infty(M)$-module, which we denote by $\mathcal{A}_k(TM, E)$. Thus

$$\mathcal{A}_k(TM, E) := \Gamma(A_k(TM, E)).$$

We agree that $\mathcal{A}_0(TM, E) := \Gamma(\pi)$.

Each element A of $\mathcal{A}_k(TM, E)$ determines an alternating $C^\infty(M)$-multilinear mapping $\bar{A} \colon (\mathfrak{X}(M))^k \to \Gamma(\pi)$ given by

$$\bar{A}(X_1, \ldots, X_k)(p) := A_p((X_1)_p, \ldots, (X_k)_p), \quad p \in M,$$

where $X_1, \ldots, X_k \in \mathfrak{X}(M)$. Thus we obtain a canonical isomorphism

$$\mathcal{A}_k(TM, E) \cong A_k(\mathfrak{X}(M), \Gamma(\pi)), \ A \mapsto \bar{A}$$

of $C^\infty(M)$-modules. Henceforth we shall identify these modules under this isomorphism, and we shall write A also for $\bar{A}$.

(b) Let

$$\mathcal{A}(TM, E) := \bigoplus_{k=0}^{n} \mathcal{A}_k(TM, E), \quad n = \dim M.$$

We define a left action

$$\mathcal{A}(M) \times \mathcal{A}(TM, E) \to \mathcal{A}(TM, E), \ (\alpha, A) \mapsto \alpha \wedge A$$

of the Grassmann algebra $\mathcal{A}(M)$ on $\mathcal{A}(TM, E)$ as follows: if $\alpha \in \mathcal{A}_k(M)$, $A \in \mathcal{A}_l(TM, E)$, then $\alpha \wedge A \in \mathcal{A}_{k+l}(TM, E)$ such that for all $p \in M$; $v_1, \ldots, v_{k+l} \in T_pM$,

$$(\alpha \wedge A)_p(v_1, \ldots, v_{k+l})$$
$$:= \frac{1}{k!l!} \sum_{\sigma \in S_{k+l}} \varepsilon(\sigma) \alpha_p(v_{\sigma(1)}, \ldots, v_{\sigma(k)}) A_p(v_{\sigma(k+1)}, \ldots, v_{\sigma(k+l)}). \quad (3.3.20)$$

Equivalently, applying the interpretations

$$\mathcal{A}(M) \cong A(\mathfrak{X}(M)), \ \mathcal{A}(TM, E) \cong A(\mathfrak{X}(M), \Gamma(\pi)) = \bigoplus_{k=0}^{n} A_k(\mathfrak{X}(M), \Gamma(\pi)),$$

we can write

$$(\alpha \wedge A)(X_1, \ldots, X_{k+l})$$
$$:= \frac{1}{k!l!} \sum_{\sigma \in S_{k+l}} \varepsilon(\sigma)\alpha(X_{\sigma(1)}, \ldots, X_{\sigma(k)})A(X_{\sigma(k+1)}, \ldots, X_{\sigma(k+l)}), \quad (3.3.21)$$

where $X_1, \ldots, X_{k+l} \in \mathfrak{X}(M)$. In this way $\mathcal{A}(TM, E)$ becomes a graded (left) module over the algebra $\mathcal{A}(M)$. (Formally, a module over an algebra and a graded module over an algebra can be defined in the same way as a module over a commutative ring and a graded R-algebra, see Definition 1.1.1 and subsection 1.3.4.)

(c) Consider the End-bundle $\mathrm{End}(E) := \bigcup_{p \in M} \mathrm{End}(E_p) \to M$ of the vector bundle $\pi\colon E \to M$ (see Example 2.2.40). Then

$$\mathcal{A}_k(TM, \mathrm{End}(E)) \cong A_k(\mathfrak{X}(M), \Gamma(\mathrm{End}(E)))$$
$$\overset{\text{Proposition 2.2.44}}{\cong} A_k(\mathfrak{X}(M), \mathrm{Hom}_M(\Gamma(\pi), \Gamma(\pi)))$$
$$= A_k(\mathfrak{X}(M), \mathrm{End}(\Gamma(\pi))).$$

We define the *wedge product* of an $\mathrm{End}(E)$-valued k-form Φ on M and an E-valued l-form A on M as the E-valued $(k + l)$-form

$$\Phi \wedge A \in \mathcal{A}_{k+l}(TM, E) \cong A_{k+l}(\mathfrak{X}(M), \Gamma(\pi))$$

given by

$$(\Phi \wedge A)(X_1, \ldots, X_{k+l})$$
$$:= \frac{1}{k!l!} \sum_{\sigma \in S_{k+l}} \varepsilon(\sigma)\Phi(X_{\sigma(1)}, \ldots, X_{\sigma(k)})(A(X_{\sigma(k+1)}, \ldots, X_{\sigma(k+l)})) \quad (3.3.22)$$

for all $X_1, \ldots, X_{k+l} \in \mathfrak{X}(M)$. Here

$$A(X_{\sigma(k+1)}, \ldots, X_{\sigma(k+l)}) \in \Gamma(\pi), \quad \Phi(X_{\sigma(1)}, \ldots, X_{\sigma(k)}) \in \mathrm{End}(\Gamma(\pi)),$$

so we indeed have $(\Phi \wedge A)(X_1, \ldots, X_{k+l}) \in \Gamma(\pi)$.

To illustrate the formalism, let

$$\Phi \in \mathcal{A}_2(TM, \mathrm{End}(\pi)), \quad \sigma \in \Gamma(\pi) = \mathcal{A}_0(TM, \pi).$$

Then $\Phi \wedge \sigma \in \mathcal{A}_2(TM, \pi)$, and for any vector fields X, Y on M we have

$$(\Phi \wedge \sigma)(X, Y) := \frac{1}{2}(\Phi(X, Y)(\sigma) - \Phi(Y, X)(\sigma)) = \Phi(X, Y)(\sigma). \quad (3.3.23)$$

3.3.5 The Classical Graded Derivations of $\mathcal{A}(M)$

Lemma 3.3.22. *Let M be a manifold and consider its Grassmann algebra $\mathcal{A}(M)$.*

(i) Graded derivations of $\mathcal{A}(M)$ are local operators: *if $\mathcal{D}\colon \mathcal{A}(M) \to \mathcal{A}(M)$ is a graded derivation, $\mathcal{U}$ is an open subset of M, and $\alpha \in \mathcal{A}(M)$ is a differential form such that $\alpha \restriction \mathcal{U} = 0$, then $\mathcal{D}\alpha \restriction \mathcal{U} = 0$.*

(ii) Graded derivations of $\mathcal{A}(M)$ are natural with respect to restrictions: *with the notation already introduced, there exists a unique graded derivation $\mathcal{D}_{\mathcal{U}}\colon \mathcal{A}(\mathcal{U}) \to \mathcal{A}(\mathcal{U})$ such that the following diagram is commutative:*

$$\begin{array}{ccc} \mathcal{A}(M) & \xrightarrow{\restriction_{\mathcal{U}}} & \mathcal{A}(\mathcal{U}) \\ {\scriptstyle \mathcal{D}}\big\downarrow & & \big\downarrow{\scriptstyle \mathcal{D}_{\mathcal{U}}} \\ \mathcal{A}(M) & \xrightarrow[\restriction_{\mathcal{U}}]{} & \mathcal{A}(\mathcal{U}) \end{array}.$$

The proofs of these assertions are essentially identical with the proofs of the analogous Lemma 3.3.11 and Lemma 3.3.12.

Lemma 3.3.23. *Every graded derivation of the Grassmann algebra of a manifold is uniquely determined by its action on the smooth functions of the manifold and on their differentials.*

Proof. Let M be a manifold and let $\mathcal{D}\colon \mathcal{A}(M) \to \mathcal{A}(M)$ be a graded derivation of degree r. Choose an arbitrary chart $(\mathcal{U}, (u^i)_{i=1}^n)$ on M, and, by part (ii) of the previous lemma, consider the induced derivation $\mathcal{D}_{\mathcal{U}}\colon \mathcal{A}(\mathcal{U}) \to \mathcal{A}(\mathcal{U})$. It is sufficient to check that $\mathcal{D}_{\mathcal{U}}$ is uniquely determined by its action on $C^\infty(\mathcal{U})$ and on $\{df \in \mathcal{A}_1(\mathcal{U}) \mid f \in C^\infty(\mathcal{U})\}$. Since $\mathcal{D}_{\mathcal{U}}$ is additive, it is enough to consider differential forms which can be represented as

$$\alpha = f\, du^{i_1} \wedge \cdots \wedge du^{i_k}, \quad f \in C^\infty(\mathcal{U}).$$

Applying induction on k we find

$$\mathcal{D}(du^{i_1} \wedge \cdots \wedge du^{i_k}) = \sum_{j=1}^{k} (-1)^{r(j-1)} du^{i_1} \wedge \cdots \wedge \mathcal{D}(du^{i_j}) \wedge \cdots \wedge du^{i_k},$$

so it follows that

$$\mathcal{D}\alpha = (\mathcal{D}f) \wedge du^{i_1} \wedge \cdots \wedge du^{i_k} + f\,\mathcal{D}(du^{i_1} \wedge \cdots \wedge du^{i_k})$$

$$= (\mathcal{D}f) \wedge du^{i_1} \wedge \cdots \wedge du^{i_k}$$

$$+ f\sum_{j=1}^{k}(-1)^{r(j-1)}du^{i_1} \wedge \cdots \wedge \mathcal{D}(du^{i_j}) \wedge \cdots \wedge du^{i_k}.$$

This concludes the proof. □

Definition 3.3.24. A graded derivation of the Grassmann algebra of a manifold is called *algebraic* if it acts as the zero operator on the smooth functions of the manifold.

Remark 3.3.25. If $\mathcal{D}: \mathcal{A}(M) \to \mathcal{A}(M)$ is an algebraic derivation, then it is $C^\infty(M)$-linear: for any differential form $\alpha \in \mathcal{A}(M)$ and smooth function f on M,

$$\mathcal{D}(f\alpha) = \mathcal{D}(f \wedge \alpha) = (\mathcal{D}f) \wedge \alpha + (-1)^{0 \cdot r}f \wedge \mathcal{D}\alpha = f \wedge \mathcal{D}\alpha = f(\mathcal{D}\alpha)$$

(r is the degree of $\mathcal{D}$). This justifies the attribute 'algebraic'.

Corollary 3.3.26. *An algebraic derivation is uniquely determined by its action on the differentials of smooth functions.* □

Remark 3.3.27. (a) In Example 3.3.17 we have already defined the Lie derivative of a covariant or of a vector-valued covariant tensor field with respect to a vector field. Thus, in particular, if M is a manifold and $X \in \mathfrak{X}(M)$, then the Lie derivative $\mathcal{L}_X$ acts on $\mathcal{A}(M)$. So for any k-form α on M ($k \geq 1$) and for any vector fields $X_1, \ldots, X_k \in \mathfrak{X}(M)$, by (3.3.14), we have

$$(\mathcal{L}_X\alpha)(X_1, \ldots, X_k) = X(\alpha(X_1, \ldots, X_k)) - \sum_{i=1}^{k}\alpha(X_1, \ldots, [X, X_i], \ldots, X_k).$$

Obviously, $\mathcal{L}_X\alpha$ is also a differential k-form on M.

Corollary 3.3.28. *Let M be a manifold. For each $X \in \mathfrak{X}(M)$, the Lie derivative*

$$\mathcal{L}_X: \mathcal{A}(M) \to \mathcal{A}(M), \quad \alpha \mapsto \mathcal{L}_X\alpha$$

is a graded derivation of degree 0, and hence a derivation of $\mathcal{A}(M)$, i.e.,

$$\mathcal{L}_X(\alpha \wedge \beta) = (\mathcal{L}_X\alpha) \wedge \beta + \alpha \wedge \mathcal{L}_X\beta \text{ for all } \alpha, \beta \in \mathcal{A}(M). \tag{3.3.24}$$

Furthermore, we have

$$[\mathcal{L}_X, \mathcal{L}_Y] \overset{(1.3.10)}{:=} \mathcal{L}_X \circ \mathcal{L}_Y - \mathcal{L}_Y \circ \mathcal{L}_X = \mathcal{L}_{[X,Y]}; \quad X, Y \in \mathfrak{X}(M). \tag{3.3.25}$$

Proof. Relation (3.3.24) is an immediate consequence of the Leibnizian product rule for a tensor derivation (Definition 3.3.9(i)) and the definition of the wedge product, while relation (3.3.25) is just a special case of (3.3.13). □

Remark 3.3.29. As in 1.2.2 in an abstract setting, now we associate to each vector field X on M the *substitution operator*

$$i_X \colon \mathcal{A}(M) \to \mathcal{A}(M), \ \alpha \in \mathcal{A}_k(M) \mapsto i_X \alpha \in \mathcal{A}_{k-1}(M)$$

given by

$$(i_X \alpha)(X_1, \dots, X_{k-1}) := \alpha(X, X_1, \dots, X_{k-1}) \text{ if } k \geq 1;$$
$$i_X \alpha := 0 \text{ if } \alpha \in \mathcal{A}_0(M) = C^\infty(M). \tag{3.3.26}$$

Then, in particular,

$$i_X df = df(X) = Xf \text{ for all } f \in C^\infty(M). \tag{3.3.27}$$

Example 3.3.30. Let $(\mathcal{U}, (u^i)_{i=1}^n)$ be a chart on M, and consider the n-form $\omega := du^1 \wedge \cdots \wedge du^n$ over $\mathcal{U}$. If $X \in \mathfrak{X}(M)$, $X \upharpoonright \mathcal{U} = \sum_{i=1}^n X^i \frac{\partial}{\partial u^i}$, then

$$i_X \omega = \sum_{k=1}^n (-1)^{k-1} X^k du^1 \wedge \cdots \wedge \breve{du}^k \wedge \cdots \wedge du^n, \tag{3.3.28}$$

where the breve $\breve{}$ means that the term is omitted.

Indeed, by Proposition 1.3.15 the $(n-1)$-form $i_X \omega$ can uniquely be represented as

$$i_X \omega = \sum_{k=1}^n i_X \omega \left(\frac{\partial}{\partial u^1}, \dots, \frac{\breve{\partial}}{\partial u^k}, \dots, \frac{\partial}{\partial u^n} \right) du^1 \wedge \cdots \wedge \breve{du}^k \wedge \cdots \wedge du^n.$$

Since

$$i_X \omega \left(\frac{\partial}{\partial u^1}, \dots, \frac{\breve{\partial}}{\partial u^k}, \dots, \frac{\partial}{\partial u^n} \right) = \omega \left(\sum_{i=1}^n X^i \frac{\partial}{\partial u^i}, \frac{\partial}{\partial u^1}, \dots, \frac{\breve{\partial}}{\partial u^k}, \dots, \frac{\partial}{\partial u^n} \right)$$

$$= (-1)^{k-1} \omega \left(\frac{\partial}{\partial u^1}, \dots, \frac{\partial}{\partial u^{k-1}}, \sum_{i=1}^n X^i \frac{\partial}{\partial u^i}, \frac{\partial}{\partial u^{k+1}}, \dots, \frac{\partial}{\partial u^n} \right)$$

$$= (-1)^{k-1} X^k \omega \left(\frac{\partial}{\partial u^1}, \dots, \frac{\partial}{\partial u^n} \right) = (-1)^{k-1} X^k,$$

we obtain relation (3.3.28).

Lemma 3.3.31. *Let M be a manifold. For any vector field X on M, the substitution operator i_X is a graded derivation of degree -1 of the Grassmann algebra $\mathcal{A}(M)$, so we have*

$$i_X(\alpha \wedge \beta) = (i_X \alpha) \wedge \beta + (-1)^k \alpha \wedge i_X \beta; \quad \alpha \in \mathcal{A}_k(M), \ \beta \in \mathcal{A}(M). \tag{3.3.29}$$

For a proof, see, e.g., [76], p. 117.

Lemma 3.3.32. *For any vector fields* X, Y *on a manifold* M,

$$[i_X, i_Y] \overset{(1.3.10)}{=} i_X \circ i_Y + i_Y \circ i_X = 0, \tag{3.3.30}$$

$$[\mathcal{L}_X, i_Y] \overset{(1.3.10)}{=} \mathcal{L}_X \circ i_Y - i_Y \circ \mathcal{L}_X = i_{[X,Y]}. \tag{3.3.31}$$

Proof. It is sufficient to show that the effect of the two sides of these relations coincides on smooth functions and on their differentials. It is clear from the definition of the substitution operator that

$$[i_X, i_Y](f) = 0, \quad [i_X, i_Y](df) = 0 \quad \text{and} \quad [\mathcal{L}_X, i_Y](f) = i_{[X,Y]}(f) = 0$$

for all $f \in C^\infty(M)$. It remains to check that the effect of both sides of (3.3.31) is the same on differentials:

$$[\mathcal{L}_X, i_Y](df) \overset{(3.3.27)}{=} \mathcal{L}_X(Yf) - i_Y(\mathcal{L}_X(df)) = X(Yf) - \mathcal{L}_X(df)(Y)$$

$$\overset{(3.3.14)}{=} X(Yf) - X(Yf) + df[X,Y] = i_{[X,Y]}(df),$$

as was to be shown. $\qquad \square$

Definition 3.3.33. Let α be a differential k-form on a manifold M. By the *exterior derivative* of α we mean the mapping $d\alpha \colon (\mathfrak{X}(M))^{k+1} \to C^\infty(M)$ given by

$$d\alpha(X_0, \ldots, X_k) := \sum_{i=0}^{k}(-1)^i X_i \alpha\big(X_0, \ldots, \check{X}_i, \ldots X_k\big)$$

$$+ \sum_{0 \le i < j \le k} (-1)^{i+j} \alpha\big([X_i, X_j], X_0, \ldots, \check{X}_i, \ldots, \check{X}_j, \ldots, X_k\big) \tag{3.3.32}$$

if $k \ge 1$ and by $d\alpha(X) := X\alpha$ if $\alpha \in \mathcal{A}_0(M) = C^\infty(M)$. $(X, X_0, \ldots, X_k$ are vector fields on M, and the notation $\check{X}_i$ means that the argument X_i is deleted.)

Remark 3.3.34. It is easy to see that $d\alpha$ is an alternating $C^\infty(M)$-linear mapping, so $d\alpha \in \mathcal{A}_{k+1}(M)$ for all $k \in J_n$ $(n = \dim M)$. The operator

$$d \colon \mathcal{A}(M) \to \mathcal{A}(M), \ \alpha \mapsto d\alpha$$

so defined is called the *exterior derivative* in $\mathcal{A}(M)$, or on M.

Proposition 3.3.35 (Cartan's Formulae). *The exterior derivative in* $\mathcal{A}(M)$ *has the following properties:*

(i) $\mathcal{L}_X = i_X \circ d + d \circ i_X$, *for any vector field X on M.*

(ii) $d(\alpha \wedge \beta) = (d\alpha) \wedge \beta + (-1)^k \alpha \wedge d\beta$; $\alpha \in \mathcal{A}_k(M)$, $\beta \in \mathcal{A}(M)$, *hence d is a graded derivation of $\mathcal{A}(M)$ of degree 1.*

(iii) $d^2 = 0$.

(iv) $\mathcal{L}_X \circ d = d \circ \mathcal{L}_X$, *for all $X \in \mathfrak{X}(M)$.*

(v) *If $\varphi \colon M \to N$ is a smooth mapping, then $\varphi^* \circ d = d \circ \varphi^*$.*

Proof. (i) This relation may be obtained by a straightforward application of the definitions. For details, we refer to [76], p. 120.

(ii) Applying (i), the second formula may be proved by induction on $k + l$, where l is the degree of β. For details, see again [76], p. 120.

(iii) Since d is a graded derivation of degree 1, $d^2 = \frac{1}{2}[d, d]$ is a derivation of degree 2, as we have seen in 1.3.4. So it is sufficient to show that

$$d^2(f) = 0 \text{ and } d^2(df) = 0 \text{ for all } f \in C^\infty(M).$$

Since $d^2 f = d(df) \in \mathcal{A}_2(M)$, and for any vector fields X, Y on M,

$$d(df)(X, Y) \overset{(3.3.32)}{=} X(df(Y)) - Y(df(X)) - df([X, Y])$$
$$= X(Yf) - Y(Xf) - [X, Y]f = 0,$$

we have $d^2 f = 0$. This implies that $d^2 d(f) = d(d^2 f) = 0$.

(iv) Using the relations just obtained,

$$d \circ \mathcal{L}_X \overset{(i)}{=} d \circ i_X \circ d + d^2 \circ i_X \overset{(iii)}{=} d \circ i_X \circ d \overset{(i)}{=} (\mathcal{L}_X - i_X \circ d) \circ d \overset{(iii)}{=} \mathcal{L}_X \circ d.$$

(v) See [76], p. 121. $\qquad\square$

Remark 3.3.36. In terms of the graded commutator defined by (1.3.10), the formulas (i), (iii) and (iv) of the preceding lemma can be written as follows:

$$\mathcal{L}_X = [i_X, d], \ X \in \mathfrak{X}(M); \qquad (3.3.33)$$

$$\frac{1}{2}[d, d] = 0, \qquad (3.3.34)$$

$$[\mathcal{L}_X, d] = 0. \qquad (3.3.35)$$

Formula (3.3.33) relates the three fundamental graded derivations of the Grassmann algebra of a manifold; it is quoted as *H. Cartan's 'magic' formula.*

Definition 3.3.37. We say that a k-form $\alpha \in \mathcal{A}_k(M)$ is *closed* if $d\alpha = 0$. If $\alpha = d\beta$ for some $(k - 1)$-form $\beta \in \mathcal{A}_{k-1}(M)$, then α is called an *exact* form.

Since $d^2 = 0$, every exact form is closed. Conversely, locally every closed form is exact. More precisely, we have the following important result:

Lemma 3.3.38 (Poincaré Lemma). *If α is a closed differential form of degree $k \geq 1$ on a manifold M, then for each point $p \in M$ there is an open neighbourhood $\mathcal{U}$ of p for which $\alpha \upharpoonright \mathcal{U} \in \mathcal{A}_k(\mathcal{U})$ is exact.*

For a proof, see e.g. [2], pp. 435–436.

3.3.6   The Frölicher – Nijenhuis Theorem

According to Definition 3.3.20, by a TM-valued k-form on a manifold M we mean an element of the $C^\infty(M)$-module

$$\mathcal{A}_k(TM, TM) \cong A_k\big(\mathfrak{X}(M), \mathfrak{X}(M)\big), \quad k \in \{0, \ldots, \dim M\}.$$

In the following, for simplicity, we shall write

$$\mathcal{A}_k^1(M) := \mathcal{A}_k(TM, TM).$$

By convention, $\mathcal{A}_0^1(M) = \mathfrak{X}(M)$. Thus a TM-valued k-form on M can be regarded as a smooth mapping

$$K \colon M \to \bigcup_{p \in M} A_k(T_p M, T_p M)$$

such that for each $p \in M$,

$$K_p := K(p) \colon T_p M \times \cdots \times T_p M \to T_p M$$

is an alternating k-linear mapping, or, equivalently, as a $C^\infty(M)$-multilinear mapping $(\mathfrak{X}(M))^k \to \mathfrak{X}(M)$.

Proposition 3.3.39. *Let K be a TM-valued k-form on M. Define a mapping*

$$i_K \colon \mathcal{A}(M) \to \mathcal{A}(M), \quad \alpha \mapsto i_K \alpha$$

as follows:
   (i) *If $\alpha \in \mathcal{A}_l(M)$ and $l \geq 1$, then*

$$(i_K \alpha)(X_1, \ldots, X_{k+l-1})$$
$$:= \frac{1}{k!(l-1)!} \sum_{\sigma \in S_{k+l-1}} \varepsilon(\sigma) \alpha\big(K(X_{\sigma(1)}, \ldots, X_{\sigma(k)}), X_{\sigma(k+1)}, \ldots, X_{\sigma(k+l-1)}\big)$$

$$(3.3.36)$$

for all $X_1, \ldots, X_{k+l-1} \in \mathfrak{X}(M)$.

(ii) $i_K \alpha := 0$ *if* $\alpha \in \mathcal{A}_0(M) = C^\infty(M)$.

Then i_K *is an algebraic graded derivation of degree* $k-1$ *of* $\mathcal{A}(M)$, *and every algebraic graded derivation of the Grassmann algebra of a manifold is of this form.*

For a proof the reader is referred to [76], p. 192.

Remark 3.3.40. We keep the notation introduced in the preceding proposition.

(a) If $K := X \in \mathfrak{X}(M) = \mathcal{A}_0^1(M)$, then

$$i_X \alpha(X_1, \ldots, X_{l-1}) := \frac{1}{(l-1)!} \sum_{\sigma \in S_{l-1}} \varepsilon(\sigma) \alpha\big(X, X_{\sigma(1)}, \ldots, X_{\sigma(l-1)}\big)$$

$$= \alpha(X, X_1, \ldots, X_{l-1}),$$

so in this case i_K reduces to the classical substitution operator. We call, therefore, the graded derivation i_K defined by (3.3.36) the *substitution operator induced by the* TM-*valued* k-*form* K.

(b) Now consider the special case when $\alpha = df \in \mathcal{A}_1(M)$, $f \in C^\infty(M)$. Then

$$(i_K df)(X_1, \ldots, X_k) \overset{(3.3.36)}{=} \frac{1}{k!} \sum_{\sigma \in S_k} \varepsilon(\sigma) df\big(K(X_{\sigma(1)}, \ldots, X_{\sigma(k)})\big)$$

$$= df\big(K(X_1, \ldots, X_k)\big).$$

Since a graded derivation of $\mathcal{A}(M)$ is determined by its action on the smooth functions on M and their differentials, we may equivalently define the substitution operator i_K as the graded derivation of degree $\deg(K) - 1$ of $\mathcal{A}(M)$ such that

$$i_K f = 0 \quad \text{and} \quad i_K df = df \circ K \quad \text{for all } f \in C^\infty(M). \tag{3.3.37}$$

(c) If $K \in \mathcal{A}_1^1(M) \cong \mathrm{End}(\mathfrak{X}(M))$, then the general formula (3.3.36) reduces to

$$(i_K \alpha)(X_1, \ldots, X_l) = \sum_{i=1}^{l} \alpha\big(X_1, \ldots, K(X_i), \ldots, X_l\big). \tag{3.3.38}$$

In particular,

$$i_{1_{\mathfrak{X}(M)}} \alpha = l\alpha, \quad \alpha \in \mathcal{A}_l(M). \tag{3.3.39}$$

If α is a one-form on M, then

$$K^* \alpha = i_K \alpha \tag{3.3.40}$$

where the left-hand side is the pull-back of α under K defined by (1.2.6).

Definition 3.3.41. Let $K \in \mathcal{A}_k^1(M)$. By the *Lie derivative associated to* K we mean the graded derivation

$$\mathcal{L}_K := [i_K, d] = i_K \circ d - (-1)^{k-1} d \circ i_K \qquad (3.3.41)$$

of degree k of the Grassmann algebra $\mathcal{A}(M)$.

Remark 3.3.42. (a) If $K := X \in \mathfrak{X}(M) = \mathcal{A}_0^1(M)$, then (3.3.41) reduces to (3.3.33), so we obtain the classical Lie derivative with respect to a vector field. This justifies the terminology in some extent. However, *the Lie derivative associated to the identity transformation* $1_{\mathfrak{X}(M)}$ *is the exterior derivative:*

$$\mathcal{L}_{1_{\mathfrak{X}(M)}} = d. \qquad (3.3.42)$$

Indeed, for any l-form α on M,

$$\mathcal{L}_{1_{\mathfrak{X}(M)}} \alpha \overset{(3.3.41)}{=} i_{1_{\mathfrak{X}(M)}} d\alpha - d i_{1_{\mathfrak{X}(M)}} \alpha \overset{(3.3.39)}{=} (l+1) d\alpha - d(l\alpha) = d\alpha,$$

whence our claim.

(b) Since $i_K \upharpoonright C^\infty(M) = 0$ we have

$$\mathcal{L}_K f = i_K df \overset{(3.3.37)}{=} df \circ K \text{ for all } f \in C^\infty(M). \qquad (3.3.43)$$

(c) If $\mathcal{L}_{K_1} = \mathcal{L}_{K_2}$ *for some* TM-*valued forms* K_1, K_2 *on* M, *then* $K_1 = K_2$. To show this, notice first that $\mathcal{L}_{K_1} = \mathcal{L}_{K_2}$ implies that K_1 and K_2 are of the same degree; let $k := \deg(K_1) = \deg(K_2)$. Then

$$\mathcal{L}_{K_1} = i_{K_1} \circ d - (-1)^{k-1} d \circ i_{K_1}, \ \mathcal{L}_{K_2} = i_{K_2} \circ d - (-1)^{k-1} d \circ i_{K_2}.$$

Using (3.3.43) we find

$$df\big(K_1(X_1, \ldots, X_k)\big) = df(K_2(X_1, \ldots, X_k)),$$

or, equivalently,

$$K_1(X_1, \ldots, X_k)(f) = K_2(X_1, \ldots, X_k)(f),$$

for any smooth function f and vector fields $X_1, \ldots, X_k$ on M. This implies that $K_1 = K_2$. Our reasoning also shows that

$$\text{if } i_{K_1} = i_{K_2}, \text{ then } K_1 = K_2.$$

Lemma 3.3.43. *If a graded derivation* $\mathcal{D}$ *of* $\mathcal{A}(M)$ *satisfies* $[\mathcal{D}, d] = 0$, *then it is determined by its action on the smooth functions on* M.

Proof. Let $k := \deg(\mathcal{D})$. Then by our condition $[\mathcal{D}, d] = 0$,

$$\mathcal{D} \circ d = (-1)^k d \circ \mathcal{D}. \tag{$*$}$$

In view of Lemma 3.3.23, it is sufficient to show that the action of $\mathcal{D}$ on the differentials of smooth functions is determined by its action on $C^\infty(M)$. If $f \in C^\infty(M)$, then we find that

$$\mathcal{D}(df) \overset{(*)}{=} (-1)^k d(\mathcal{D}f),$$

and hence our claim follows. □

Proposition 3.3.44. *A graded derivation $\mathcal{D}$ of $\mathcal{A}(M)$ is the Lie derivative associated to a TM-valued form on M if, and only if, $[\mathcal{D}, d] = 0$.*

Proof. Suppose first that

$$\mathcal{D} = \mathcal{L}_K = i_K \circ d - (-1)^{k-1} d \circ i_K, \quad K \in \mathcal{A}_k^1(M).$$

Then

$$[\mathcal{D}, d] = \mathcal{L}_K \circ d - (-1)^k d \circ \mathcal{L}_K \overset{(3.3.34)}{=} -(-1)^{k-1} d \circ i_K \circ d$$
$$- (-1)^k d \circ i_K \circ d = 0,$$

which proves the necessity.

Conversely, let $\mathcal{D}$ be a graded derivation of degree k of $\mathcal{A}(M)$ satisfying the condition $[\mathcal{D}, d] = 0$. Define a mapping

$$K \colon (\mathfrak{X}(M))^k \to \mathfrak{X}(M), \quad (X_1, \ldots, X_k) \mapsto K(X_1, \ldots, X_k)$$

by

$$K(X_1, \ldots, X_k)(f) := (\mathcal{D}f)(X_1, \ldots, X_k), \quad f \in C^\infty(M).$$

Since $K(X_1, \ldots, X_k)(f) = df \circ K(X_1, \ldots, X_k)$, it follows that

$$\mathcal{L}_K f \overset{(3.3.43)}{=} df \circ K = \mathcal{D}f, \quad \text{for all } f \in C^\infty(M).$$

By Lemma 3.3.43, this implies that $\mathcal{D} = \mathcal{L}_K$. □

Corollary 3.3.45 (Frölicher – Nijenhuis Theorem). *Every graded derivation $\mathcal{D}$ of degree k of the Grassmann algebra $\mathcal{A}(M)$ can be written uniquely as*

$$\mathcal{D} = \mathcal{L}_K + i_L; \quad K \in \mathcal{A}_k^1(M), \ L \in \mathcal{A}_{k+1}^1(M).$$

Proof. In the light of the preceding lemmas, define a TM-valued form $K \in \mathcal{A}_k^1(M)$ by the condition that $\mathcal{L}_K$ *coincides with* $\mathcal{D}$ *on* $C^\infty(M)$. Then $(\mathcal{D} - \mathcal{L}_K) \restriction C^\infty(M) = 0$, so by Proposition 3.3.39 there exists a TM-valued form $L \in \mathcal{A}_{k+1}^1(M)$ such that $\mathcal{D} - \mathcal{L}_K = i_L$, therefore $\mathcal{D} = \mathcal{L}_K + i_L$. The uniqueness of K and L is clear from Remark 3.3.42(c). $\qquad\square$

Corollary 3.3.46. *Let* $K \in \mathcal{A}_k^1(M)$, $L \in \mathcal{A}_l^1(M)$. *There exists a unique* TM-*valued* $(k+l)$-*form* $[K, L]$ *on* M *such that*

$$[\mathcal{L}_K, \mathcal{L}_L] = \mathcal{L}_{[K,L]}. \tag{3.3.44}$$

Proof. By the graded Jacobi identity (1.3.8) we have

$$(-1)^k[[\mathcal{L}_K, \mathcal{L}_L], d] + (-1)^{lk}[[\mathcal{L}_L, d], \mathcal{L}_K] + (-1)^l[[d, \mathcal{L}_K], \mathcal{L}_L] = 0.$$

The second and third term vanish at the left-hand side by Proposition 3.3.44, hence $[[\mathcal{L}_K, \mathcal{L}_L], d] = 0$. This implies, again by the quoted proposition, the existence of a unique TM-valued form $[K, L] \in \mathcal{A}_{k+l}^1(M)$ satisfying (3.3.44). $\qquad\square$

Remark 3.3.47. The TM-valued form $[K, L]$ defined by (3.3.44) is said to be the *Frölicher – Nijenhuis bracket* of K and L. It is not difficult to verify that it reduces to

(i) the Lie bracket given by (3.1.38) if $K, L \in \mathcal{A}_0^1(M) = \mathfrak{X}(M)$;
(ii) the Frölicher – Nijenhuis bracket given by (3.3.16) if

$$K \in \mathcal{A}_1^1(M) = \mathrm{End}(\mathfrak{X}(M)), \quad L \in \mathcal{A}_0^1(M);$$

(iii) the Frölicher – Nijenhuis bracket given by (3.3.17) if $K, L \in \mathcal{A}_1^1(M)$.

Assertion (ii) will be proved below.

Corollary 3.3.48. (i) *For any* TM-*valued* k-*form* $K \in \mathcal{A}_k^1(M)$ *we have*

$$[K, 1_{\mathfrak{X}(M)}] = 0. \tag{3.3.45}$$

(ii) *The Frölicher – Nijenhuis bracket of* TM-*valued forms on* M *is graded anticommutative and satisfies the graded Jacobi identity, i.e., if* $K_i \in \mathcal{A}_{k_i}^1(M)$, $i \in \{1, 2, 3\}$, *then*

$$[K_1, K_2] = -(-1)^{k_1 k_2}[K_2, K_1], \tag{3.3.46}$$

$$(-1)^{k_1 k_3}[K_1, [K_2, K_3]] + (-1)^{k_2 k_1}[K_2, [K_3, K_1]] + (-1)^{k_3 k_2}[K_3, [K_1, K_2]] = 0. \tag{3.3.47}$$

Proof. (i) We have

$$\mathcal{L}_{[K,1_{\mathfrak{X}(M)}]} \overset{(3.3.44)}{=} [\mathcal{L}_K, \mathcal{L}_{1_{\mathfrak{X}}(M)}] \overset{(3.3.42)}{=} [\mathcal{L}_K, d] \overset{\text{Proposition 3.3.44}}{=} 0,$$

so $[K, 1_{\mathfrak{X}(M)}] = 0$ by Remark 3.3.42(c).

(ii) We verify only the graded anticommutativity of the Frölicher–Nijenhuis bracket, the graded Jacobi identity can be shown similarly. We have

$$\mathcal{L}_{[K_1,K_2]} \overset{(3.3.44)}{=} [\mathcal{L}_{K_1}, \mathcal{L}_{K_2}] \overset{(1.3.10)}{:=} \mathcal{L}_{K_1} \circ \mathcal{L}_{K_2} - (-1)^{k_1 k_2} \mathcal{L}_{K_2} \circ \mathcal{L}_{K_1}$$

$$= -(-1)^{k_1 k_2} (\mathcal{L}_{K_2} \circ \mathcal{L}_{K_1} - (-1)^{k_1 k_2} \mathcal{L}_{K_1} \circ \mathcal{L}_{K_2})$$

$$= -(-1)^{k_1 k_2} [\mathcal{L}_{K_2}, \mathcal{L}_{K_1}] = \mathcal{L}_{-(-1)^{k_1 k_2} [K_2, K_1]},$$

whence (3.3.46). □

Proposition 3.3.49 (Generalized Cartan Formulae). *Let K be a TM-valued one-form on M, and let X be a vector field on M. We have the following relations:*

$$[i_X, i_K] = i_X \circ i_K - i_K \circ i_X = i_{KX}, \tag{3.3.48}$$

$$[i_K, \mathcal{L}_X] = i_K \circ \mathcal{L}_X - \mathcal{L}_X \circ i_K = i_{[K,X]}, \tag{3.3.49}$$

$$[i_X, \mathcal{L}_K] = i_X \circ \mathcal{L}_K + \mathcal{L}_K \circ i_X = \mathcal{L}_{KX} + i_{[K,X]}, \tag{3.3.50}$$

$$[i_K, d] = i_K \circ d - d \circ i_K = \mathcal{L}_K, \tag{3.3.51}$$

$$[d, \mathcal{L}_K] = d \circ \mathcal{L}_K + \mathcal{L}_K \circ d = 0, \tag{3.3.52}$$

$$[\mathcal{L}_K, \mathcal{L}_X] = \mathcal{L}_K \circ \mathcal{L}_X - \mathcal{L}_X \circ \mathcal{L}_K = \mathcal{L}_{[K,X]}. \tag{3.3.53}$$

Proof. (1) Obviously, both $[i_X, i_K]$ and i_{KX} are algebraic graded derivations. For any function $f \in C^\infty(M)$,

$$[i_X, i_K] df = i_X(i_K df) - i_K(i_X df)$$

$$= (i_K df)(X) - i_K(Xf) = df(KX) = i_{KX} df;$$

which proves (3.3.48).

(2) The bracket $[i_K, \mathcal{L}_X]$ is also an algebraic graded derivation, so it is determined by its action on the differentials of smooth functions on M. For all $f \in C^\infty(M)$, $Y \in \mathfrak{X}(M)$ we have

$$([i_K, \mathcal{L}_X] df)(Y) = i_K(\mathcal{L}_X df)(Y) - \mathcal{L}_X(i_K df)(Y)$$

$$\overset{(3.3.38),(3.3.14)}{=} \mathcal{L}_X df(KY) - X(i_K df(Y)) + (i_K df)[X, Y]$$

$$= X(df(KY)) - df[X, KY] - X(df(KY)) + df(K[X, Y])$$

$$= df([KY, X] - K[Y, X]) \overset{(3.3.16)}{=} df([K, X](Y)) = i_{[K,X]} df(Y),$$

whence (3.3.49).

(3) Manipulating the left-hand side of relation (3.3.50), we find

$$[i_X, \mathcal{L}_K] = i_X \circ \mathcal{L}_K + \mathcal{L}_K \circ i_X \stackrel{(3.3.41)}{=} i_X \circ i_K \circ d - i_X \circ d \circ i_K$$

$$+ i_K \circ d \circ i_X - d \circ i_K \circ i_X \stackrel{(3.3.48)}{=} i_K \circ i_X \circ d + i_{KX} \circ d$$

$$- i_X \circ d \circ i_K + i_K \circ d \circ i_X - d \circ i_X \circ i_K + d \circ i_{KX}$$

$$= i_K \circ (i_X \circ d + d \circ i_X) - (i_X \circ d + d \circ i_X) \circ i_K$$

$$+ i_{KX} \circ d + d \circ i_{KX} \stackrel{(3.3.33)}{=} i_K \circ \mathcal{L}_X - \mathcal{L}_X \circ i_K + \mathcal{L}_{KX}$$

$$\stackrel{(3.3.49)}{=} \mathcal{L}_{KX} + i_{[K,X]},$$

as desired.

(4) Relation (3.3.51) is just a repetition of the definition of $\mathcal{L}_K$ when K is of degree 1. By Proposition 3.3.44, equality (3.3.52) is also evident.

(5) We turn to show (3.3.53), where $[K, X]$ is defined by (3.3.16), thus we also present the promised proof of Remark 3.3.47(ii). Since $[[\mathcal{L}_K, \mathcal{L}_X], d] = 0$ (see the proof of Corollary 3.3.46), $[\mathcal{L}_K, \mathcal{L}_X]$ is determined by its action on the smooth functions on M (Lemma 3.3.43). Let $f \in C^\infty(M)$. Then for all $Y \in \mathfrak{X}(M)$,

$$([\mathcal{L}_K, \mathcal{L}_X]f)(Y) = \mathcal{L}_K(\mathcal{L}_X f)(Y) - \mathcal{L}_X(\mathcal{L}_K f)(Y)$$

$$\stackrel{(3.3.43)}{=} d(Xf)(KY) - X(df)(KY) + df(K[X, Y])$$

$$= (KY)Xf - X((KY)f) + df(K[X, Y])$$

$$= [K(Y), X]f - K[Y, X]f \stackrel{(3.3.16)}{=} ([K, X]Y)f$$

$$= df \circ [K, X](Y) \stackrel{(3.3.43)}{=} (\mathcal{L}_{[K,X]}f)(Y),$$

which completes the proof.   □

3.4 Integration on Manifolds

Beyond the mountain are mountains.

Korean proverb

3.4.1 *Orientable Manifolds*

Definition 3.4.1. A manifold is said to be *orientable* if its tangent bundle is orientable.

Remark 3.4.2. (a) According to Definition 2.2.49, an n-dimensional manifold M is orientable if there exists an n-form $\omega \in \mathcal{A}_n(M)$ such that

$$\omega_p \neq 0 \text{ for all } p \in M.$$

Then we say that ω is a *volume form* on M. An *orientation* on M is an equivalence class of volume forms on M under the equivalence relation

$\omega_1 \sim \omega_2$ if there is a smooth function f on M such that $\omega_2 = f\omega_1$ and $f(p) > 0$ for all $p \in M$.

An *oriented manifold* is an orientable manifold endowed with an orientation.

(b) Suppose that M is an oriented manifold and let $\omega \in \mathcal{A}_n(M)$ represent the orientation of M. Then ω is called a *positive volume form* on M. A basis $(v_i)_{i=1}^n$ of a tangent space T_pM is said to be *positive* if $\omega_p(v_1, \ldots, v_n) > 0$. Obviously, this relation does not depend on the choice of the positive volume form ω.

(c) The friend of Proposition 2.2.51 in this context is the following result:

A connected manifold M is orientable if, and only if, there exists an atlas $\mathcal{A}$ of M such that for any two charts $(\mathcal{U}, x)$, $(\mathcal{V}, y)$ in $\mathcal{A}$,

$$\det\left((x \circ y^{-1})'(y(p))\right) > 0 \text{ for all } p \in \mathcal{U} \cap \mathcal{V}.$$

For a proof see, for example, [2], 449–450.

Lemma 3.4.3. *An orientable manifold is connected if, and only if, it has exactly two orientations.*

Proof. Suppose that M is an oriented manifold, and let ω be a positive volume form on M.

If M is connected, then by Remark 2.2.50(d) there are exactly two orientations on M. To prove the converse statement, suppose that M has exactly two orientations, but is not connected. Let $\mathcal{U} \notin \{\emptyset, M\}$ be a subset of M which is both open and closed. Define a section μ of $A_n(TM)$ as follows:

$$\mu_p := \omega_p \text{ if } p \in \mathcal{U}, \ \mu_p := -\omega_p \text{ if } p \in M \setminus \mathcal{U}.$$

Then μ is a volume form on M such that $\mu \nsim \omega$ and $\mu \nsim -\omega$. This is a contradiction. $\qquad\square$

Lemma and Definition 3.4.4. *Let M and N be oriented manifolds with positive volume forms ω_M and ω_N, respectively. If $\varphi \colon M \to N$ is a smooth local diffeomorphism (Definition 2.1.21), then $\varphi^* \omega_N = f \omega_M$, where f is a nowhere vanishing smooth function on M. The mapping φ is called*

orientation preserving   *if* $f(p) > 0$ *for all* $p \in M$,

orientation reversing   *if* $f(p) < 0$ *for all* $p \in M$.

If M is connected, then φ either preserves or reverses orientations.

The proof is an easy exercise. Now we translate Lemma 1.3.21 to manifolds.

Definition 3.4.5. Let M be a manifold. A two-form $\omega \in \mathcal{A}_2(M)$ is called *non-degenerate* if the alternating two-tensor

$$\omega_p \in A_2(T_pM) \subset T_2^0(T_pM)$$

is non-degenerate for each $p \in M$.

Corollary 3.4.6. *Let M be a manifold. A two-form $\omega \in \mathcal{A}_2(M)$ is non-degenerate if, and only if, M is even-dimensional, say $\dim M = 2n$, and $\omega^n = \omega \wedge \cdots \wedge \omega$ is a volume form on M. Thus, if $\omega \in \mathcal{A}_2(M)$ is non-degenerate, then M is orientable.* $\qquad\square$

3.4.2 Integration of Top Forms

In the following, M will be an n-dimensional manifold, where $n > 0$. By a top from on M we mean a differential form whose degree is n.

Definition 3.4.7. Let α be a differential form on M. The subset

$$\mathrm{supp}(\alpha) := \mathrm{cl}\{p \in M \mid \alpha_p \neq 0 \in A(T_pM)\}$$

of M is called the *support* of α. We say that α is *compactly supported* (*supported in* $\mathcal{U}$) if the set $\mathrm{supp}(\alpha)$ is compact (contained in a subset $\mathcal{U}$ of M, respectively). Similarly, if $X \in \mathfrak{X}(M)$, then

$$\mathrm{supp}(X) := \mathrm{cl}\{p \in M \mid X_p \neq 0 \in T_pM\}$$

is the support of X.

Remark 3.4.8. (a) We denote by $\mathcal{A}^c_k(M)$ and $\mathcal{A}^c(M)$ the set of compactly supported k-forms and compactly supported forms on M, respectively. Then, in particular, $\mathcal{A}^c_0(M) =: C^\infty_c(M)$ is the set of all compactly supported smooth functions on M (see Definition 2.1.28).

(b) The following relations can easily be checked:

(i) $\mathrm{supp}(\alpha + \beta) \subset \mathrm{supp}(\alpha) \cup \mathrm{supp}(\beta)$; $\alpha, \beta \in \mathcal{A}_k(M)$;
(ii) $\mathrm{supp}(\alpha \wedge \beta) \subset \mathrm{supp}(\alpha) \cap \mathrm{supp}(\beta)$; $\alpha, \beta \in \mathcal{A}(M)$;
(iii) $\mathrm{supp}([X,Y]) \subset \mathrm{supp}(X) \cap \mathrm{supp}(Y)$; $X, Y \in \mathfrak{X}(M)$;
(iv) $\mathrm{supp}(i_X\alpha) \subset \mathrm{supp}(X) \cap \mathrm{supp}(\alpha)$,
     $\mathrm{supp}(\mathcal{L}_X\alpha) \subset \mathrm{supp}(X) \cap \mathrm{supp}(\alpha)$; $X \in \mathfrak{X}(M)$, $\alpha \in \mathcal{A}(M)$;
(v) $\mathrm{supp}(d\alpha) \subset \mathrm{supp}(\alpha)$, $\alpha \in \mathcal{A}(M)$.

Thus each $\mathcal{A}^c_k(M)$ ($k \in J_n$) is a $C^\infty(M)$-module, and $\mathcal{A}^c(M)$ is a graded algebra over $C^\infty(M)$. In particular, $\mathcal{A}^c(M)$ is a graded $\mathbb{R}$-algebra, and this is more than enough for our purposes.

Remark 3.4.9. Until now, we have used the term 'volume form' in three different situations.

(i) In 1.3.7, we defined a volume form on an n-dimensional real vector space V as a nonzero alternating n-linear form on V, i.e., as a nonzero element of the one-dimensional real vector space $A_n(V)$.

(ii) In 2.2.5, we considered a vector bundle $\pi\colon E \to M$ of rank r, and introduced the following notion: a volume form on π is a smooth mapping

$$\omega\colon M \to A_r(E) := \bigcup_{p \in M} A_r(E_p)$$

such that $\omega_p \in A_r(E_p) \setminus \{0\}$ for every $p \in M$.

(iii) In 3.4.1, we defined a volume form on an (orientable) manifold M as a volume form on its tangent bundle $\tau\colon TM \to M$. Thus a volume form on M is just a top form $\mu \in \mathcal{A}_n(M)$ such that $\mu_p \neq 0$ for every $p \in M$.

When we work in an n-dimensional real vector space V, we have to be careful with the use of the term 'volume form' because V is also endowed with a smooth structure. If we need only the linear structure of V, then a volume form on V means an element of $A_n(V) \setminus \{0\}$, as in (i). If we regard V as a manifold with its canonical smooth structure (Example 2.1.9), then a volume form on V is a nowhere zero element of $\mathcal{A}_n(V)$, i.e., we use the term 'volume form' in the sense of (iii).

Every volume form $\mu \in A_n(V)$ canonically induces a volume form $\mu^\uparrow \in \mathcal{A}_n(V)$ as follows. We identify the tangent bundle of V with the trivial bundle $V \times V$ as in Example C.2.2, and define the mapping

$$\mu^\uparrow : p \in V \mapsto (\mu^\uparrow)_p \in A_n(T_pV)$$

by

$$(\mu^\uparrow)_p((p,v_1),\ldots,(p,v_n)) := \mu(v_1,\ldots,v_n), \tag{3.4.1}$$

where $v_1, \ldots, v_n \in V$.

Lemma 3.4.10. *Let $(V, \mathcal{O})$ be an n-dimensional oriented real vector space ($n \geq 1$). Choose a positive volume form $\mu \in \mathcal{O}$. Let $(b_i)_{i=1}^n$ be a basis of V such that $\mu(b_1,\ldots,b_n) = 1$, and let $(e_i)_{i=1}^n$ be the canonical basis of $\mathbb{R}^n$. Consider the linear isomorphism $\kappa_B : V \to \mathbb{R}^n$ determined by $\kappa_B(b_i) = e_i$ ($i \in J_n$). Given a nonempty open subset $\mathcal{U}$ of V, define a function*

$$\int_\mathcal{U} : \mathcal{A}_n^c(V) \to \mathbb{R}, \quad \omega \mapsto \int_\mathcal{U}(\omega) =: \int_\mathcal{U} \omega$$

by

$$\int_\mathcal{U} \omega := \int_{\kappa_B(\mathcal{U})} f \circ (\kappa_B)^{-1} \text{ (Riemann integral) if } \omega = f\mu^\uparrow, \quad f \in C_c^\infty(V),$$

where $\mu^\uparrow \in \mathcal{A}_n(V)$ is given by (3.4.1). Then the function $\int_\mathcal{U}$ is $\mathbb{R}$-linear and well-defined: it depends only on the orientation of V.

Proof. The linearity is clear. To show the well-definedness of $\int_\mathcal{U}$, let $\bar{\mu} \in \mathcal{O}$ be a further volume form, $(\bar{b}_i)_{i=1}^n$ a basis of V such that $\bar{\mu}(\bar{b}_1,\ldots,\bar{b}_n) = 1$, and let $\kappa_{\overline{B}} : V \to \mathbb{R}^n$ be the linear isomorphism determined by $\kappa_{\overline{B}}(\bar{b}_i) = e_i$, $i \in J_n$. Then $\omega \in \mathcal{A}_n^c(V)$ can also be written as

$$\omega = \bar{f}\bar{\mu}^\uparrow, \quad \bar{f} \in C_c^\infty(V).$$

Observe that

$$\bar{f}(p) = f(p)\mu(\bar{b}_1,\ldots,\bar{b}_n), \quad p \in V. \tag{$*$}$$

Indeed,

$$\bar{f}(p) = \bar{f}(p)\bar{\mu}(\bar{b}_1,\ldots,\bar{b}_n) = \bar{f}(p)(\bar{\mu}^\uparrow)((p,\bar{b}_1),\ldots,(p,\bar{b}_n))$$
$$= \omega_p(\bar{b}_1,\ldots,\bar{b}_n) = f(p)\mu(\bar{b}_1,\ldots,\bar{b}_n).$$

Let $(\alpha_j^i) \in M_n(\mathbb{R})$ be the transition matrix for the change of basis from $(b_i)_{i=1}^n$ to $(\bar{b}_i)_{i=1}^n$. Then

$$\bar{b}_j = \alpha_j^i b_i, \quad j \in J_n \qquad (**)$$

(see (1.1.6)). Since both bases are positive, $\det(\alpha_j^i) > 0$. Using $(**)$, relation $(*)$ can be rewritten as

$$\bar{f} = \det(\alpha_j^i)f.$$

Notice that from $(**)$ we also have $A(e_j) := \kappa_B \circ (\kappa_{\overline{B}})^{-1}(e_j) = \alpha_j^i e_i$ for all $j \in J_n$. Thus, by the change of variables formula for multiple integrals (see, e.g., [64], Ch. XX, Theorem 4),

$$\int_{\kappa_B(\mathcal{U})} f \circ (\kappa_B)^{-1} = \int_{A^{-1}(\kappa_B(\mathcal{U}))} (f \circ (\kappa_B)^{-1} \circ A) \cdot \det(A)$$

$$= \int_{\kappa_{\overline{B}}(\mathcal{U})} \det(\alpha_j^i)(f \circ (\kappa_{\overline{B}})^{-1}) = \int_{\kappa_{\overline{B}}(\mathcal{U})} \bar{f} \circ (\kappa_{\overline{B}})^{-1},$$

as was to be shown. $\qquad \square$

We say that $\int_{\mathcal{U}} \omega$ is the *integral* of the (compactly supported) top form $\omega \in \mathcal{A}_n^c(V)$. The following results can easily be proved.

Lemma 3.4.11. *Let $\omega \in \mathcal{A}_n^c(V)$ be supported in an open subset $\mathcal{V}$ of V. Then, for every nonempty open subset $\mathcal{U}$ of V,*

$$\int_{\mathcal{U}} \omega = \int_{\mathcal{U} \cap \mathcal{V}} \omega.$$

Lemma 3.4.12. *Suppose that V and W are oriented n-dimensional real vector spaces. Let $\mathcal{V} \subset V$, $\mathcal{W} \subset W$ be open subsets, and let $\varphi \colon \mathcal{V} \to \mathcal{W}$ be a diffeomorphism. If $\omega \in \mathcal{A}_n^c(\mathcal{W}) \subset \mathcal{A}_n^c(W)$, then $\varphi^*\omega \in \mathcal{A}_n^c(V)$, and for any nonempty open subset $\mathcal{U}$ of $\mathcal{V}$,*

$$\int_{\mathcal{U}} \varphi^*\omega = \varepsilon \int_{\varphi(\mathcal{U})} \omega, \qquad (3.4.2)$$

where $\varepsilon = 1$ if φ is orientation preserving, and $\varepsilon = -1$ if φ is orientation reversing. $\qquad \triangle$

Observe that (3.4.2) corresponds to the classical change of variables formula cited above.

Proposition 3.4.13. *Let M be an oriented n-dimensional manifold. There exists a unique $\mathbb{R}$-linear function*

$$\int_M : \mathcal{A}_n^c(M) \to \mathbb{R},$$

called integral, *satisfying the following conditions:*

(Int_1) *If $\varphi \in \text{Diff}(M)$ is an orientation preserving diffeomorphism and $\mathcal{U} \subset M$ is a (nonempty) open subset, then*

$$\int_{\mathcal{U}} \varphi^* \omega = \int_{\varphi(\mathcal{U})} \omega, \quad \omega \in \mathcal{A}_n^c(M)$$

(naturality with respect to orientation preserving diffeomorphisms).

(Int_2) *If $\mathcal{U} \subset M$ is an open subset and $\omega \in \mathcal{A}_n^c(M)$ is supported in $\mathcal{U}$, then*

$$\int_M \omega = \int_{\mathcal{U}} \omega.$$

(Int_3) *If $(V, [\mu])$ is an oriented n-dimensional real vector space, $(b_i)_{i=1}^n$ is a positive basis of V such that $\mu(b_1, \ldots, b_n) = 1$, and $(b^i)_{i=1}^n$ is the dual of $(b_i)_{i=1}^n$, then with the notation of Lemma 3.4.10,*

$$\int_V f(db^1 \wedge \cdots \wedge db^n)^\uparrow = \int_{\mathbb{R}^n} f \circ (\kappa_B)^{-1} \quad \text{(Riemann integral)},$$

for every $f \in C_c^\infty(V)$.

We omit the proof, but sketch the main steps of the construction of $\int_M$. For a detailed and careful treatment of integration on manifolds see, e.g., [13], [66] or [96]. We follow here the reasoning of [52], 4.13.

Given an open subset A of M, we define an $\mathbb{R}$-linear function

$$\int_A : \mathcal{A}_n^c(M) \to \mathbb{R}.$$

Step 1. Let $\omega \in \mathcal{A}_n^c(M)$, and suppose first that there exists a chart $(\mathcal{U}, u)$ for M such that $u(\mathcal{U}) = \mathbb{R}^n$ and $\text{supp}(\omega) \subset \mathcal{U}$. Here the first condition does not mean any restriction because every point of a topological n-manifold has an open neighbourhood homeomorphic to $\mathbb{R}^n$ (prove this yourself, or consult with [68], Lemma 2.13). Let us endow $\mathbb{R}^n$ with the orientation which makes $u \colon \mathcal{U} \to \mathbb{R}^n$ into an orientation preserving diffeomorphism. Define the integral of ω over A by

$$\int_A \omega := \int_{u(A \cap \mathcal{U})} (u^{-1})^* \omega. \tag{3.4.3}$$

We have to show that the right-hand side of (3.4.3) is independent of the choice of chart. Assume that $(\bar{\mathcal{U}}, \bar{u})$ is another chart such that $\bar{u}(\bar{\mathcal{U}}) = \mathbb{R}^n$ and $\text{supp}(\omega) \subset \bar{\mathcal{U}}$, and let $\bar{\mathbb{R}}^n$ denote $\mathbb{R}^n$ endowed with the orientation induced from M via $\bar{u}$. Let $\mathcal{V} := \mathcal{U} \cap \bar{\mathcal{U}}$. Then $u \circ \bar{u}^{-1}$ is an orientation preserving diffeomorphism from $\bar{u}(\mathcal{V}) \subset \bar{\mathbb{R}}^n$ to $u(\mathcal{V}) \subset \mathbb{R}^n$, and hence

$$\int_{u(A \cap \mathcal{U})} (u^{-1})^* \omega \overset{\text{Lemma 3.4.11}}{=} \int_{u(A \cap \mathcal{V})} (u^{-1})^* \omega$$

$$\overset{\text{Lemma 3.4.12}}{=} \int_{\bar{u}(A \cap \mathcal{V})} (u \circ \bar{u}^{-1})^* (u^{-1})^* \omega$$

$$\overset{(3.3.8)}{=} \int_{\bar{u}(A \cap \mathcal{V})} (\bar{u}^{-1})^* \omega \overset{\text{Lemma 3.4.11}}{=} \int_{\bar{u}(A \cap \bar{\mathcal{U}})} (\bar{u}^{-1})^* \omega,$$

as was to be shown.

Step 2. Now we turn to the general case. Let $\omega \in \mathcal{A}_n^c(M)$ be an arbitrary compactly supported top form on M. Then, by the compactness of $\text{supp}(\omega)$, there is a finite family $(\mathcal{U}_i, u_i)_{i=1}^k$ of charts of M such that

$$u_i(\mathcal{U}_i) = \mathbb{R}^n \quad (i \in J_k) \text{ and } \text{supp}(\omega) \subset \bigcup_{i=1}^k \mathcal{U}_i.$$

Let $\mathcal{U}_0 := M \setminus \text{supp}(\omega)$. Then $(\mathcal{U}_i)_{i=0}^k$ is an open covering of M; let $(f_i)_{i=0}^k$ be a subordinate partition of unity. Observe that $f_0 \omega = 0$, therefore

$$\omega = \sum_{i=1}^k f_i \omega \quad \text{and} \quad f_i \omega \in \mathcal{A}_n^c(\mathcal{U}_i), \quad i \in J_k.$$

Define the integral of ω over an open subset A of M by

$$\int_A \omega := \sum_{i=1}^k \int_A f_i \omega, \tag{3.4.4}$$

where $\int_A f_i \omega$ is given by (3.4.3). Again we have to check that $\int_A \omega$ is well-defined: the right-hand side of (3.4.4) does not depend on the choices we made. Accordingly, suppose that $(\bar{\mathcal{U}}_j, \bar{u}_j)_{j=1}^l$ is another family of charts such that

$$\bar{u}_j(\bar{\mathcal{U}}_j) = \mathbb{R}^n, \ j \in J_l; \ \text{supp}(\omega) \subset \bigcup_{j=1}^l \bar{\mathcal{U}}_j; \ \bar{\mathcal{U}}_0 := M \setminus \text{supp}(\omega),$$

and let $(\bar{f}_j)_{j=0}^l$ be a partition of unity subordinate to the open covering $(\bar{\mathcal{U}}_j)_{j=0}^l$ of M. Since $f_0 \omega = \bar{f}_0 \omega = 0$, we have

$$f_i \omega = \sum_{j=1}^l \bar{f}_j f_i \omega, \quad i \in J_k.$$

Therefore,

$$\sum_{i=1}^k \int_A f_i \omega = \sum_{i=1}^k \sum_{j=1}^l \int_A \bar{f}_j f_i \omega = \sum_{j=1}^l \int_A \bar{f}_j \omega,$$

which proves that the integral of ω is well-defined.

Now we list some basic results concerning integration on manifolds.

Lemma 3.4.14. *Let $(M, [\mu])$ be an oriented manifold, and let $-M$ denote the same manifold with the opposite orientation $[-\mu]$. Then*

$$\int_{-M} \omega = -\int_M \omega,$$

that is, $\int_M \omega$ changes sign when the orientation of M is reversed.

For a proof, see [96], Proposition 23.9.

Lemma 3.4.15. *Let $(M, [\mu])$ be an oriented manifold and $\omega := f\mu$ where $f \in C_c^\infty(M)$. Suppose that $f(p) \geq 0$ for every $p \in M$, and $\omega \neq 0$. Then $\int_M \omega > 0$.*

Idea of proof. With the help of a suitable partition of unity, reduce the problem to the case $M = \mathbb{R}^n$.

Lemma 3.4.16. *Let M be an oriented n-dimensional manifold, and let $\omega \in \mathcal{A}_{n-1}^c(M)$. Then $\int_M d\omega = 0$.*

For a proof we refer to [52], 4.13, Proposition XIV.

3.4.3   *Stokes' Theorem*

> Stokes' theorem shares three important attributes with many fully evolved major theorems:
>
> 1. It is trivial.
> 2. It is trivial because the terms appearing in it have been properly defined.
> 3. It has significant consequences.
>
> Michael Spivak [89], p. 104

In this subsection first we collect the definition of the remaining terms which we need to the statement of Stokes' theorem. The main new concept is a *regular domain* in a manifold, which we briefly discuss. After these preparations, we formulate the main result.

Remark 3.4.17. Suppose that n is a positive integer.

(a) We let $\mathbb{R}_-^n$ and $\mathbb{R}_+^n$ denote the half-spaces

$$\{(\alpha^1, \ldots, \alpha^n) \in \mathbb{R}^n \mid \alpha^1 \leq 0\} \quad \text{and} \quad \{(\alpha^1, \ldots, \alpha^n) \in \mathbb{R}^n \mid \alpha^1 \geq 0\},$$

respectively. They have the same *boundary*

$$\partial\mathbb{R}_-^n = \partial\mathbb{R}_+^n = \{(\alpha^1,\ldots,\alpha^n) \in \mathbb{R}^n \mid \alpha^1 = 0\} \cong \{0\} \times \mathbb{R}^{n-1}.$$

Observe that we write $\partial\mathbb{R}_-^n$ rather than $\mathrm{bd}(\mathbb{R}_-^n)$ introduced in Definition B.1.1(a). The reason will be clear in a moment.

(b) We assume that $\mathbb{R}_-^n$ and $\mathbb{R}_+^n$ are endowed with the subspace topology inherited from $\mathbb{R}^n$. If, for example, $\mathcal{U}$ is an open subset of $\mathbb{R}_-^n$, then

$$\partial\mathcal{U} := \mathcal{U} \cap \partial\mathbb{R}_-^n$$

is called the *manifold boundary* of $\mathcal{U}$, and the elements of $\partial\mathcal{U}$ are said to be the *boundary points* of $\mathcal{U}$.

In general, the manifold boundary does not coincide with the topological boundary. As an illustration, let

$$\mathcal{U} := \{a \in \mathbb{R}_-^2 \mid \|a\| < 1\},$$

where $\|\ \|$ is the Euclidean norm of $\mathbb{R}^2$. Then

$$\mathrm{bd}(\mathcal{U}) = \{a \in \mathbb{R}_-^2 \mid \|a\| = 1\} \cup \{(0,\tau) \in \mathbb{R}^2 \mid -1 < \tau < 1\}$$

($\mathcal{U}$ considered as a subset of $\mathbb{R}^2$), while $\partial\mathcal{U} = \{(0,\tau) \in \mathbb{R}^2 \mid -1 < \tau < 1\}$. By abuse of language, we also say that $\partial\mathcal{U}$ is the boundary of $\mathcal{U}$.

(c) Let $\mathcal{U}$ be an open subset of $\mathbb{R}_-^n$. A mapping $\varphi\colon \mathcal{U} \to \mathbb{R}^k$ is called *differentiable* if there exist an open subset $\widetilde{\mathcal{U}}$ of $\mathbb{R}^n$ and a differentiable mapping $\widetilde{\varphi}\colon \widetilde{\mathcal{U}} \to \mathbb{R}^k$ such that $\mathcal{U} = \widetilde{\mathcal{U}} \cap \mathbb{R}_-^n$ and $\varphi = \widetilde{\varphi} \restriction \mathcal{U}$; cf. Definition 2.1.14(c). Then φ has a well-defined derivative at every point p of $\mathcal{U}$ given by

$$\varphi'(p) := \widetilde{\varphi}'(p) \in L(\mathbb{R}^n, \mathbb{R}^k).$$

This is evident if $p \notin \partial\mathbb{R}_-^n$: no extension is needed. For points in $\partial\mathbb{R}_-^n$ the well-definedness takes some explaining, which we leave to the reader.

Lemma 3.4.18. *Let $\mathcal{U}$ and $\mathcal{V}$ be open subsets of $\mathbb{R}_-^n$, and let*

$$\varphi = \begin{pmatrix} \varphi^1 \\ \vdots \\ \varphi^n \end{pmatrix} \colon \mathcal{U} \to \mathcal{V}$$

be a diffeomorphism. Then

(i) *$\varphi(\partial\mathcal{U}) = \partial\mathcal{V}$, therefore $\varphi \restriction \partial\mathcal{U}\colon \partial\mathcal{U} \to \partial\mathcal{V}$ is a diffeomorphism between open subsets of $\{0\} \times \mathbb{R}^{n-1}$.*

(ii) *At every point p of $\partial\mathcal{U}$, the derivative $\varphi'(p) \in \operatorname{End}(\mathbb{R}^n)$ maps the sets $\{0\} \times \mathbb{R}^{n-1}$, $\mathbb{R}^n_-$ and $\mathbb{R}^n_+$ onto themselves. Thus the Jacobian matrix of φ at $p \in \partial\mathcal{U}$ has the form*

$$\begin{pmatrix} \begin{array}{c|c} D_1\varphi^1(p) & 0 \\ \hline D_1\varphi^2(p) & \\ \vdots & J_{f\restriction\{0\}\times\mathbb{R}^{n-1}}(p) \\ D_1\varphi^n(p) & \end{array} \end{pmatrix},$$

where $D_1\varphi^1(p) > 0$.

For a proof we refer to [61], 6.3.

Definition 3.4.19. Let M be an n-dimensional manifold.

(a) A closed subset $N \subset M$ is called a *regular domain* or a *domain with smooth boundary* if for every point $p \in N$ there exists a chart $(\mathcal{U}, u)$ of M around p such that

$$u(\mathcal{U} \cap N) = u(\mathcal{U}) \cap \mathbb{R}^n_-.$$

Then we say that $(\mathcal{U}, u)$ is a chart *adapted* to N.

(b) Suppose that N is a regular domain of M. A point $p \in M$ is called a *boundary point* of N if for some adapted chart $(\mathcal{U}, u)$ around p, we have $u(p) \in \partial\mathbb{R}^n_-$. The set

$$\partial N := \{p \in N \mid p \text{ is a boundary point}\}$$

is said to be the *boundary* of N. A tangent vector $v \in T_pM$ at a boundary point $p \in \partial N$ is called *outward-pointing* if there exists an adapted chart $(\mathcal{U}, u)$ around p such that the vector $(u_*)_p(v) \in T_{u(p)}\mathbb{R}^n$ has positive first coordinate.

Remark 3.4.20. (a) Lemma 3.4.18 guarantees that the boundary points and the outward-pointing tangent vectors are well-defined: the relevant properties are independent of the chosen adapted charts.

(b) Observe that above we defined the 'manifold boundary' of N. However, by the closedness of N, in this case $\partial N = \operatorname{bd}(N)$.

Lemma 3.4.21. *Let M be an n-dimensional manifold.*

(i) *If N is a regular domain of M, then ∂N is an $(n-1)$-dimensional submanifold of M.*

(ii) *If $n \geq 2$ and M is oriented, then the boundary of a regular domain N of M has an* induced orientation *with the following property: if $p \in \partial N$ and $v_1 \in T_p M$ is an outward-pointing tangent vector, then a basis $(v_2, \ldots, v_n)$ of $T_p \partial N$ is positive if, and only if, the basis $(v_1, v_2, \ldots, v_n)$ of $T_p M$ is positive.*

For a proof, see [70], Lemma 10.6.

Remark 3.4.22. If $\dim M = 1$, part (ii) of the lemma holds in the following modified form. The boundary ∂N of N is a zero-dimensional manifold, so an orientation of ∂N is just a function

$$p \in \partial N \mapsto \text{sign}(p) \in \{-1, 1\}.$$

Let $v_1 \in T_p M$ be an outward-pointing tangent vector at $p \in \partial N$. Then $\text{sign}(p) := 1$ if (v_1) is a positive basis of $T_p M$, otherwise $\text{sign}(p) := -1$. A 0-form on ∂N is a function $f \colon \partial N \to \mathbb{R}$. If $\text{supp}(f)$ is compact, we define

$$\int_{\partial N} f := \sum_{p \in \partial N} \text{sign}(p) f(p).$$

These conventions are used below when $\dim M = 1$.

Theorem 3.4.23 (Stokes' Theorem). *Suppose that M is an n-dimensional oriented manifold. Let N be a regular domain of M, and let ∂N have the induced orientation. Assume further that $\text{int}(N)$ is oriented as an open submanifold of M. Then for every compactly supported $(n-1)$-form $\omega \in \mathcal{A}_{n-1}^c(M)$ we have*

$$\int_{\text{int}(N)} d\omega = \int_{\partial N} i^* \omega, \tag{3.4.5}$$

where $i \colon \partial N \to M$ is the canonical inclusion.

Very careful proofs of this fundamental result can be found, among others, in references [13], [61], [69], [96]. In the excellent text of Madsen and Tornehave [70] we can also find a proof of Stokes' theorem which fits the approach sketched in this section well. We note that the proof itself is less laborious than the formulation of the necessary conceptual framework.

Observe finally that taking $N = M$ in (3.4.5), we obtain Lemma 3.4.16.

Chapter 4

Structures on Tangent Bundles

The human being follows the earth.
Earth follows heaven.
Heaven follows the Tao.
Tao follows what is natural.

Lao Tsu (Tao Te Ching [65], Ch. 25)

4.1 Vector Bundles on TM

4.1.1 *Finsler Bundles and Finsler Tensor Fields*

Remark 4.1.1. Let M be a manifold and consider its tangent bundle $\tau\colon TM \to M$.

(a) In our considerations a distinguished role will be played by the pull-back bundle

$$\pi\colon TM \times_M TM \to TM$$

of τ by its own projection. So it will prove convenient to introduce some special terminology and notation concerning this bundle. Usually we shall call the bundle $\pi\colon TM \times_M TM \to TM$ the *Finsler bundle over TM*.

The general pull-back construction has already been summarized in Example 2.2.10 and Example 2.2.37. In our case

$$TM \times_M TM = \big\{(u,v) \in TM \times TM \mid \tau(u) = \tau(v)\big\},$$
$$\pi = \mathrm{pr}_1 \restriction TM \times_M TM.$$

The fibre of π over $u \in TM$ is

$$\pi^{-1}(u) = \big\{(u,v) \in TM \times TM \mid v \in T_{\tau(u)}M\big\} = \{u\} \times T_{\tau(u)}M.$$

179

The real vector space structure of the fibres is given by

$$(u, v_1) + (u, v_2) := (u, v_1 + v_2), \quad \lambda(u, v) := (u, \lambda v),$$

where $u \in TM$; $v, v_1, v_2 \in T_{\tau(u)}M$, $\lambda \in \mathbb{R}$ (cf. Example 2.2.37). Thus the fibre of π over $u \in TM$ is canonically isomorphic to the tangent space $T_{\tau(u)}M$.

(b) As in general, we denote by $\Gamma(\pi)$ the $C^\infty(TM)$-module of the smooth sections of a Finsler bundle π. The elements of this module will be mentioned as *Finsler vector fields* on TM. According to Definition 2.2.11, $\Gamma_\tau(TM)$ stands for the $C^\infty(TM)$-module of sections of τ along τ, i.e.,

$$\Gamma_\tau(TM) := \{ \underline{X} \in C^\infty(TM, TM) \mid \tau \circ \underline{X} = \tau \}.$$

By Lemma 2.2.12, the modules $\Gamma(\pi)$ and $\Gamma_\tau(TM)$ are canonically isomorphic via the mapping

$$\widetilde{X} \in \Gamma(\pi) \mapsto \underline{X} := \pi_2 \circ \widetilde{X} \in \Gamma_\tau(TM),$$

where $\pi_2 := \mathrm{pr}_2 \restriction TM \times_M TM$. Thus we have the following commutative diagram:

$$
\begin{array}{ccc}
TM \times_M TM & \xrightarrow{\pi_2} & TM \\
\pi \big\uparrow\big\downarrow \, {}^{\widetilde{X}} \quad \nearrow {}_{\underline{X}} & & \big\downarrow {}^{\tau} \\
TM & \xrightarrow[\tau]{} & M
\end{array}
\qquad (4.1.1)
$$

Hence the Finsler vector fields on TM are of the form

$$\widetilde{X} \colon v \in TM \mapsto \widetilde{X}(v) = (v, \underline{X}(v)) \in TM \times_M TM, \qquad (4.1.2)$$

briefly $\widetilde{X} = (1_{TM}, \underline{X})$, where $\underline{X} \in C^\infty(TM, TM)$. We say that $\underline{X}$ is the *principal part* of the Finsler vector field $\widetilde{X}$. The Finsler vector field whose principal part is also the identity transformation of TM is called the *canonical Finsler vector field* on TM or the *canonical section* of π. It will be denoted by $\widetilde{\delta}$. Thus

$$\widetilde{\delta} := (1_{TM}, 1_{TM}) \colon v \in TM \mapsto \widetilde{\delta}(v) = (v, v) \in TM \times_M TM. \qquad (4.1.3)$$

Notice that if $\widetilde{Y} = (1_{TM}, \underline{Y}) \in \Gamma(\pi)$ and $X \in \mathfrak{X}(M)$, then $\underline{Y} \circ X \in \mathfrak{X}(M)$. Indeed,

$$\tau \circ (\underline{Y} \circ X) = (\tau \circ \underline{Y}) \circ X = \tau \circ X = 1_M.$$

(c) Let X be a vector field on M. Since

$$\tau \circ (X \circ \tau) = (\tau \circ X) \circ \tau = 1_M \circ \tau = \tau,$$

it follows that

$$\widehat{X} := (1_{TM}, X \circ \tau) \tag{4.1.4}$$

is a Finsler vector field on TM (cf. (2.2.7)). Finsler vector fields of this form are called *basic*. We also say that $\widehat{X}$ is the *lift* of $X \in \mathfrak{X}(M)$ into $\Gamma(\pi)$.

(d) **Coordinate description.** Choose a chart $(\mathcal{U}, (u^i)_{i=1}^n)$ on M. For each $v \in \tau^{-1}(\mathcal{U})$, the family

$$\left(\widehat{\frac{\partial}{\partial u^i}}(v) \right)_{i \in J_n} = \left(v, \left(\frac{\partial}{\partial u^i} \right)_{\tau(v)} \right)_{i \in J_n} \tag{4.1.5}$$

is a basis for the fibre $\pi^{-1}(v) = \{v\} \times T_{\tau(v)}M$, hence $\left(\widehat{\frac{\partial}{\partial u^i}} \right)_{i=1}^n$ *is a local frame of $TM \times_M TM$ over* $\tau^{-1}(\mathcal{U})$. Thus any Finsler vector field $\widetilde{X} \in \Gamma(\pi)$ has a local expression of the form

$$\widetilde{X} \upharpoonright \tau^{-1}(\mathcal{U}) = \sum_{i=1}^n \widetilde{X}^i \widehat{\frac{\partial}{\partial u^i}}, \tag{4.1.6}$$

where $\widetilde{X}^i \in C^\infty(\tau^{-1}(\mathcal{U}))$, $i \in J_n$. These functions are related to the principal part of $\widetilde{X}$ by

$$\widetilde{X}^i(v) = \underline{X}(v)(u^i), \quad v \in \tau^{-1}(\mathcal{U});$$

cf. (3.1.16). Instead of (4.1.6) we also write $\widetilde{X} \underset{(\mathcal{U})}{=} \sum_{i=1}^n \widetilde{X}^i \widehat{\frac{\partial}{\partial u^i}}$, or, less pedantically, $\widetilde{X} = \sum_{i=1}^n \widetilde{X}^i \widehat{\frac{\partial}{\partial u^i}}$.

Since for each $v \in \tau^{-1}(\mathcal{U})$,

$$\widetilde{\delta}(v) := (v, v) = \left(v, \sum_{i=1}^n y^i(v) \left(\frac{\partial}{\partial u^i} \right)_{\tau(v)} \right) = \sum_{i=1}^n y^i(v) \left(v, \left(\frac{\partial}{\partial u^i} \right)_{\tau(v)} \right)$$

$$= \sum_{i=1}^n y^i(v) \left(\widehat{\frac{\partial}{\partial u^i}} \right)_v = \left(\sum_{i=1}^n y^i \widehat{\frac{\partial}{\partial u^i}} \right)(v)$$

(where the functions y^i are defined by (3.1.14)), it follows that in terms of the local frame field $\left(\widehat{\frac{\partial}{\partial u^i}} \right)_{i=1}^n$ the canonical Finsler vector field has the coordinate expression

$$\widetilde{\delta} \underset{(\mathcal{U})}{=} \sum_{i=1}^n y^i \widehat{\frac{\partial}{\partial u^i}}. \tag{4.1.7}$$

If $\widehat{X}$ is a basic vector field and $X \restriction \mathcal{U} = \sum_{i=1}^{n} X^i \frac{\partial}{\partial u^i}$, then

$$\widehat{X} \underset{(\mathcal{U})}{=} \sum_{i=1}^{n} (X^i \circ \tau) \widehat{\frac{\partial}{\partial u^i}}. \tag{4.1.8}$$

In calculations, the following observation is extremely useful: *if* $(X_i)_{i=1}^{n}$ *is a frame field of* TM *over an open subset* $\mathcal{U}$ *of* M, *then* $(\widehat{X}_i)_{i=1}^{n}$ *is a frame field of* $TM \times_M TM$ *over* $\tau^{-1}(\mathcal{U})$. Thus *the* $C^\infty(TM)$-*module* $\Gamma(\pi)$ *is locally generated by the basic Finsler vector fields.*

(e) By a *Finsler one-form* on TM we mean an element of the dual module $(\Gamma(\pi))^*$ of the module $\Gamma(\pi)$ of Finsler vector fields on TM. Thus, on the analogy of (4.1.2), a Finsler one-form $\widetilde{\alpha}$ on TM can be regarded as a mapping

$$v \in TM \mapsto \widetilde{\alpha}(v) = (v, \underline{\alpha}(v)) \in TM \times_M T^*M,$$

where $\underline{\alpha} \in C^\infty(TM, T^*M)$. We say that $\underline{\alpha}$ is the *principal part* of $\widetilde{\alpha}$. The mapping

$$\widetilde{\alpha} \in (\Gamma(\pi))^* \mapsto \underline{\alpha} \in C^\infty(TM, T^*M), \ \widehat{\tau} \circ \underline{\alpha} = \widetilde{\alpha},$$

where $\widehat{\tau}$ is the projection of the cotangent bundle of M, is a canonical isomorphism of $C^\infty(TM)$-modules. As an element of the module $(\Gamma(\pi))^*$, a Finsler one-form $\widetilde{\alpha}$ acts on a Finsler vector field $\widetilde{X}$ to give a real-valued function $\widetilde{\alpha}(\widetilde{X})$ on TM by the rule

$$\widetilde{\alpha}(\widetilde{X})(v) := \underline{\alpha}(v)(\underline{X}(v)), \quad v \in TM.$$

If α is a one-form on M, then

$$\widehat{\alpha} := (1_{TM}, \alpha \circ \tau) \colon v \in TM \mapsto \widehat{\alpha}(v) = \bigl(v, \alpha(\tau(v))\bigr) \in TM \times_M T^*M$$

is a Finsler one-form with principal part $\alpha \circ \tau$, called the *lift* of α into $(\Gamma(\pi))^*$ or a *basic (Finsler) one-form*. Locally, the basic one-forms generate the $C^\infty(TM)$-module $(\Gamma(\pi))^*$.

If $(\mathcal{U}, (u^i)_{i=1}^{n})$ is a chart of M, then $(\widehat{du^i})_{i=1}^{n}$ is the dual coframe of the local frame field $\left(\widehat{\frac{\partial}{\partial u^i}} \right)_{i=1}^{n}$, i.e., we have

$$\widehat{du^i}\left(\widehat{\frac{\partial}{\partial u^j}} \right) = \delta_j^i; \quad i, j \in J_n.$$

In terms of this dual coframe, any Finsler one-form $\widetilde{\alpha} \in (\Gamma(\pi))^*$ can be locally expressed as follows:

$$\widetilde{\alpha} \underset{(\mathcal{U})}{=} \sum_{i=1}^{n} \widetilde{\alpha}_i \widehat{du^i}, \quad \widetilde{\alpha}_i \in C^\infty(\tau^{-1}(\mathcal{U})).$$

(f) In the case of a Finsler bundle $\pi\colon TM \times_M TM \to TM$ the general construction of π-tensors and π-tensor fields presented in 2.2.4 leads to the *Finsler tensors* and *Finsler tensor fields*.

For $(k,l) \in \mathbb{N} \times \mathbb{N}$, $(k,l) \neq (0,0)$, we form the disjoint union of the spaces of the type (k,l) tensors on the fibres $\pi^{-1}(v)$ ($v \in TM$):

$$F_l^k(TM) := T_l^k(TM \times_M TM) := \bigcup_{v \in TM} T_l^k(\pi^{-1}(v))$$

$$= \bigcup_{v \in TM} \left(\{v\} \times T_l^k(T_{\tau(v)}M) \right)$$

$$= \left\{ (v,A) \in TM \times T_l^k(TM) \mid \tau(v) = \tau_l^k(A) \right\}$$

$$= TM \times_M T_l^k(TM) = \tau^*(T_l^k(TM)).$$

Thus

$$F_l^k(TM) := T_l^k(TM \times_M TM) = \tau^*(T_l^k(TM)) = TM \times_M T_l^k(TM). \quad (4.1.9)$$

If

$$\pi_l^k\colon F_l^k(TM) \to TM, \quad (v,A) \mapsto v,$$

then π_l^k is a vector bundle over TM, called the bundle of *type (k,l) Finsler tensors* on TM. The $C^\infty(TM)$-module

$$\mathcal{F}_l^k(TM) := \Gamma(\pi_l^k) = \Gamma(F_l^k(TM))$$

of the smooth sections of π_l^k is said to be the module of *type (k,l) Finsler tensor fields* on TM. By an abuse of language, a Finsler tensor field will also be mentioned as a Finsler tensor, or simply a tensor.

We denote by $\Gamma_\tau(T_l^k(TM))$ the $C^\infty(TM)$-module of (smooth) sections of $T_l^k(TM)$ along τ. Then (cf. Definition 2.2.11)

$$\Gamma_\tau(T_l^k(TM)) = \left\{ \underline{A} \in C^\infty(TM, T_l^k(TM)) \mid \tau_l^k \circ \underline{A} = \tau \right\}.$$

Lemma 2.2.12 assures that the modules $\mathcal{F}_l^k(TM)$ and $\Gamma_\tau(T_l^k(TM))$ are canonically isomorphic by the mapping

$$\widetilde{A} \in \mathcal{F}_l^k(TM) \mapsto \underline{A} := \left(\pi_l^k\right)_2 \circ \widetilde{A} \in \Gamma_\tau(T_l^k(TM)),$$

where $(\pi_l^k)_2 := \mathrm{pr}_2 \upharpoonright TM \times_M T_l^k(TM)$. Thus, on the analogy of (4.1.1), we have the following commutative diagram:

$$\begin{array}{ccc}
F_l^k(TM) = TM \times_M T_l^k(TM) & \xrightarrow{\ (\pi_l^k)_2\ } & T_l^k(TM) \\[2mm]
\pi_l^k \Big\downarrow\Big\uparrow \widetilde{A} \qquad \qquad \searrow {\scriptstyle\underline{A}} & & \Big\downarrow \tau_l^k \\[2mm]
TM & \xrightarrow[\ \tau\ ]{} & M
\end{array} \qquad (4.1.10)$$

Thus any Finsler tensor field of type (k, l) is of the form

$$\widetilde{A} = (1_{TM}, \underline{A}), \quad \underline{A} \in \Gamma_\tau(T_l^k(TM)). \tag{4.1.11}$$

We say that $\underline{A}$ is the *principal part* of the tensor field $\widetilde{A} \in F_l^k(TM)$. Conversely, any smooth mapping

$$\underline{A} \colon TM \to T_l^k(TM) \text{ such that } \tau_l^k \circ \underline{A} = \tau$$

determines a Finsler tensor field with principal part $\underline{A}$.

Notice finally that the tensoriality criterion in Remark 2.2.47 yields the following interpretation of Finsler tensor fields:

$$\mathcal{F}_l^k(TM) \cong T_l^k(\Gamma(\pi)). \tag{4.1.12}$$

So any Finsler tensor field of type (k, l) on TM may be freely regarded as a $C^\infty(TM)$-multilinear mapping

$$\widetilde{A} \colon \underbrace{(\Gamma(\pi))^* \times \cdots \times (\Gamma(\pi))^*}_{k} \times \underbrace{\Gamma(\pi) \times \cdots \times \Gamma(\pi)}_{l} \to C^\infty(TM).$$

We also have the canonical module isomorphisms

$$\mathcal{F}_l^1(TM) \cong T_l^1(\Gamma(\pi)) \cong T_l^0\big(\Gamma(\pi), \Gamma(\pi)\big) \tag{4.1.13}$$

and

$$\mathcal{F}_1^1(TM) \cong \mathrm{End}(\Gamma(\pi)), \tag{4.1.14}$$

cf. (3.3.5) and (3.3.6). By convention, $\mathcal{F}_0^0(TM) = C^\infty(TM)$.

Coordinate description. Let $(\mathcal{U}, (u^i)_{i=1}^n)$ be a chart on M. In terms of the local frame field $\left(\widehat{\dfrac{\partial}{\partial u^i}}\right)_{i=1}^n$ and its dual coframe $(\widehat{du^i})_{i=1}^n$, any Finsler tensor field $\widetilde{A} \in \mathcal{F}_l^k(TM)$ has the coordinate expression

$$\widetilde{A}_{(\mathcal{U})} = \widetilde{A}_{j_1,\ldots,j_l}^{i_1,\ldots,i_k} \widehat{\frac{\partial}{\partial u^{i_1}}} \otimes \cdots \otimes \widehat{\frac{\partial}{\partial u^{i_k}}} \otimes \widehat{du^{j_1}} \otimes \cdots \otimes \widehat{du^{j_l}}, \tag{4.1.15}$$

where the *components* $\widetilde{A}_{j_1,\ldots,j_l}^{i_1,\ldots,i_k}$ of $\widetilde{A}$ with respect to $\left(\widehat{\dfrac{\partial}{\partial u^i}}\right)_{i=1}^n$ are smooth functions on $\tau^{-1}(\mathcal{U})$, and we use the summation convention; cf. (2.2.13) and (3.3.9).

Remark 4.1.2 (The Slit Finsler Bundle). We denote by $\mathring{T}M$ the subset of TM which consists of the non-zero tangent vectors to M. This is an open subset of TM, thus, by Example 2.1.7, it is a manifold on its own right. Furthermore, we let $\mathring{\tau} := \tau \restriction \mathring{T}M : \mathring{T}M \to M$.

We shall often encounter objects on $\mathring{T}M$ with homogeneity of various degrees, which, in general, cannot be extended smoothly, or even continuously, to the entire TM. Therefore, the pull-back bundle of τ by $\mathring{\tau}$ will also play a distinguished role. The special terminology and notation used in connection with this bundle is largely analogous to that connected with the Finsler bundle presented in the previous remark. However, for the readers' convenience, we also outline the special features encountered in the study of this so-called slit bundle.

(a) The pull-back bundle of τ by the restricted projection $\mathring{\tau}$ will be denoted by

$$\mathring{\pi} : \mathring{T}M \times_M TM \to TM$$

and called the *slit Finsler bundle* over $\mathring{T}M$. Thus, its total space is

$$\mathring{T}M \times_M TM := \{(u,v) \in \mathring{T}M \times TM \mid \mathring{\tau}(u) = \tau(v)\},$$

and its projection is $\mathring{\pi} := \mathrm{pr}_1 \restriction \mathring{T}M \times_M TM$. Its fibre over $u \in \mathring{T}M$ is $\mathring{\pi}^{-1}(u) = \{u\} \times T_{\mathring{\tau}(u)}M$, whose vector space structure is defined in the same way as in the case of the Finsler bundle.

(b) The smooth sections of $\mathring{\pi}$ form a $C^\infty(\mathring{T}M)$-module, denoted by $\Gamma(\mathring{\pi})$. These sections will be mentioned as *Finsler vector fields on* $\mathring{T}M$. Definition 2.2.11 now gives

$$\Gamma_{\mathring{\tau}}(TM) = \{\underline{X} \in C^\infty(\mathring{T}M, TM) \mid \tau \circ \underline{X} = \mathring{\tau}\}$$

for the $C^\infty(\mathring{T}M)$-module of the sections of τ along $\mathring{\tau}$. By Lemma 2.2.12 again, the mapping

$$\widetilde{X} \in \Gamma(\mathring{\pi}) \mapsto \underline{X} := \pi_2 \circ \widetilde{X} \in \Gamma_{\mathring{\tau}}(TM),$$

where $\pi_2 := \mathrm{pr}_2 \restriction \mathring{T}M \times_M TM$, is a canonical isomorphism between the modules $\Gamma(\mathring{\pi})$ and $\Gamma_{\mathring{\tau}}(TM)$. Hence every Finsler vector field on $\mathring{T}M$ is of the form

$$\widetilde{X} : v \in \mathring{T}M \mapsto \widetilde{X}(v) := (v, \underline{X}(v)) \in \mathring{T}M \times_M TM, \tag{4.1.2'}$$

where $\underline{X} \in C^\infty(\mathring{T}M, TM)$, the *principal part* of $\widetilde{X}$.

If we restrict the canonical Finsler vector field $\widetilde{\delta}$ introduced in Remark 4.1.1(b) to $\mathring{T}M$, then we obtain a Finsler vector field on $\mathring{T}M$, which we also denote by $\widetilde{\delta}$. The meaning of $\widetilde{\delta}$ is always clear from the context, thus, introducing a new symbol for this restriction would be superfluous pedantry.

(c) Similarly, a basic vector field $\widehat{X}$ defined in Remark 4.1.1(c) may also be restricted to $\mathring{T}M$ and denoted by the same symbol.

(d) **Coordinate description.** If $(\mathcal{U}, (u^i)_{i=1}^n)$ is a chart on M, then for each $v \in \mathring{\tau}^{-1}(\mathcal{U})$ the family

$$\left(\widehat{\frac{\partial}{\partial u^i}}(v) \right)_{i \in J_n} = \left(v, \left(\frac{\partial}{\partial u^i} \right)_{\mathring{\tau}(v)} \right)_{i \in J_n} \tag{4.1.5'}$$

is a basis for the fibre $\mathring{\pi}^{-1}(v) = \{v\} \times T_{\mathring{\tau}(v)}M$, hence $\left(\widehat{\frac{\partial}{\partial u^i}} \right)_{i=1}^n$ is a local frame of $\mathring{T}M \times_M TM$ over $\mathring{\tau}^{-1}(\mathcal{U})$ as well. Thus, similarly to (4.1.6), the local expression of a Finsler vector field $\widetilde{X} \in \Gamma(\mathring{\pi})$ is

$$\widetilde{X} \upharpoonright \mathring{\tau}^{-1}(\mathcal{U}) = \sum_{i=1}^n \widetilde{X}^i \widehat{\frac{\partial}{\partial u^i}}, \tag{4.1.6'}$$

where the smooth functions $\widetilde{X}^i$ are, of course, only defined on $\mathring{\tau}^{-1}(\mathcal{U})$ in this case. We use the shorthands $\widetilde{X} \underset{(\mathcal{U})}{=} \sum_{i=1}^n \widetilde{X}^i \widehat{\frac{\partial}{\partial u^i}}$ and $\widetilde{X} = \sum_{i=1}^n \widetilde{X}^i \widehat{\frac{\partial}{\partial u^i}}$ not only for (4.1.6), but for (4.1.6') as well.

The coordinate expressions

$$\widetilde{\delta} \underset{(\mathcal{U})}{=} \sum_{i=1}^n y^i \widehat{\frac{\partial}{\partial u^i}} \tag{4.1.7'}$$

and

$$\widehat{X} \underset{(\mathcal{U})}{=} \sum_{i=1}^n (X^i \circ \mathring{\tau}) \widehat{\frac{\partial}{\partial u^i}} \tag{4.1.8'}$$

are, of course, also valid for the canonical Finsler vector field on $\mathring{T}M$ and for basic vector fields on $\mathring{T}M$.

(e) A *Finsler one-form on $\mathring{T}M$* is an element of the dual module $(\Gamma(\mathring{\pi}))^*$ of the $C^\infty(\mathring{T}M)$-module $\Gamma(\mathring{\pi})$. Thus, analogously to (4.1.2'), a Finsler one-form $\widetilde{\alpha}$ on $\mathring{T}M$ can be regarded as a mapping

$$v \in \mathring{T}M \mapsto \widetilde{\alpha}(v) = (v, \underline{\alpha}(v)) \in \mathring{T}M \times_M T^*M,$$

where $\underline{\alpha} \in C^\infty(\mathring{T}M, T^*M)$.

Similarly to a basic vector field, a basic one-form $\widehat{\alpha}$ can also be restricted to $\mathring{T}M$, and we also denote the restricted basic one-form with the same symbol. The dual coframe of the local frame field $\left(\widehat{\frac{\partial}{\partial u^i}} \right)_{i=1}^n$ is $(\widehat{du^i})_{i=1}^n$ in this case as well.

(f) In the case of the slit Finsler bundle the general construction of π-tensors and π-tensor fields leads to Finsler tensors and Finsler tensor fields on $\mathring{T}M$ in the same way as in Remark 4.1.1(f). Thus, if $(k, l) \in \mathbb{N} \times \mathbb{N}$

and $(k, l) \neq (0, 0)$, then the total space of the type (k, l) Finsler tensor bundle over $\overset{\circ}{T}M$ is the disjoint union

$$F_l^k(\overset{\circ}{T}M) := T_l^k(\overset{\circ}{T}M \times_M TM) := \bigcup_{v \in \overset{\circ}{T}M} T_l^k(\overset{\circ}{\pi}^{-1}(v))$$

$$= \bigcup_{v \in \overset{\circ}{T}M} (\{v\} \times T_l^k(T_{\overset{\circ}{\tau}(v)}M)) \qquad (4.1.9')$$

$$= \{(v, A) \in \overset{\circ}{T}M \times T_l^k(TM) \mid \overset{\circ}{\tau}(v) = \tau_l^k(A)\}$$

$$= \overset{\circ}{T}M \times_M T_l^k(TM) = \overset{\circ}{\tau}^*(T_l^k(TM)),$$

its projection is

$$\overset{\circ}{\pi}_l^k : F_l^k(\overset{\circ}{T}M) \to \overset{\circ}{T}M, \quad (v, A) \mapsto v,$$

and the $C^\infty(\overset{\circ}{T}M)$-module of its smooth sections is

$$\mathcal{F}_l^k(\overset{\circ}{T}M) := \Gamma(\overset{\circ}{\pi}_l^k) = \Gamma(F_l^k(\overset{\circ}{T}M)).$$

This module, as we have already seen many times, is canonically isomorphic to the module

$$\Gamma_{\overset{\circ}{\tau}}(T_l^k(TM)) = \{\underline{A} \in C^\infty(\overset{\circ}{T}M, T_l^k(TM)) \mid \tau_l^k \circ \underline{A} = \overset{\circ}{\tau}\}$$

of sections of $T_l^k(TM)$ along $\overset{\circ}{\tau}$, and the isomorphism is given by

$$\widetilde{A} \in \mathcal{F}_l^k(\overset{\circ}{T}M) \mapsto \underline{A} := (\overset{\circ}{\pi}_l^k)_2 \circ \widetilde{A} \in \Gamma_{\overset{\circ}{\tau}}(T_l^k(TM)),$$

where $(\overset{\circ}{\pi}_l^k)_2 := \mathrm{pr}_2 \restriction \overset{\circ}{T}M \times_M T_l^k(TM)$.

Similarly as in the case of Finsler tensor fields on TM, a Finsler tensor field of type (k, l) on $\overset{\circ}{T}M$ may be regarded as a $C^\infty(\overset{\circ}{T}M)$-multilinear mapping

$$\widetilde{A} : \underbrace{(\Gamma(\overset{\circ}{\pi}))^* \times \cdots \times (\Gamma(\overset{\circ}{\pi}))^*}_{k} \times \underbrace{\Gamma(\overset{\circ}{\pi}) \times \cdots \times \Gamma(\overset{\circ}{\pi})}_{l} \to C^\infty(\overset{\circ}{T}M),$$

and the analogies of the module isomorphisms (4.1.13) and (4.1.14) are also valid. The coordinate description (4.1.15) of a Finsler tensor field also holds for Finsler tensor fields on $\overset{\circ}{T}M$ without any change.

(g) Finally, we note that any Finsler vector field on TM can be restricted to $\overset{\circ}{T}M$, and, since $\overset{\circ}{T}M$ is dense in TM, the original Finsler vector field may be uniquely reconstructed from this restriction. Thus we can identify each Finsler vector field on TM with its restriction to $\overset{\circ}{T}M$. After this identification we have $\Gamma(\pi) \subset \Gamma(\overset{\circ}{\pi})$. This identification is in harmony with our convention that we make no notational difference between the two meanings of the strictly speaking ambiguous symbols $\widetilde{\delta}$ and $\widehat{X}$. Of course, however, not all Finsler vector fields on $\overset{\circ}{T}M$ can be obtained in this way. Thus, on the one hand, we have the proper inclusion $TM \supset \overset{\circ}{T}M$, and on the other hand the proper inclusion $\Gamma(\pi) \subset \Gamma(\overset{\circ}{\pi})$. A similar remark applies also for the modules of Finsler one-forms and Finsler tensor fields of type (k, l) on TM and $\overset{\circ}{T}M$.

4.1.2   *The Vector Bundle Structure of $\tau_* : TTM \to TM$*

Definition 4.1.3. Let M be a manifold, $f \in C^\infty(M)$. By the *vertical lift* of f we mean the function

$$f^v := f \circ \tau \in C^\infty(TM).$$

The *complete lift* of f is the function

$$f^c : TM \to \mathbb{R}, \; v \mapsto f^c(v) := v(f).$$

Lemma 4.1.4. *Let $\varphi : M \to N$ be a smooth mapping between manifolds. Then for every $h \in C^\infty(N)$,*

$$(h \circ \varphi)^c = h^c \circ \varphi_*. \tag{4.1.16}$$

Proof. For any $v \in TM$,

$$(h \circ \varphi)^c(v) := v(h \circ \varphi) \overset{(3.1.4)}{=:} \varphi_*(v)(h) =: h^c(\varphi_*(v)) = h^c \circ \varphi_*(v),$$

whence our claim. $\square$

Lemma 4.1.5. *Let M be a manifold; $f, h \in C^\infty(M)$. Then*

$$(fh)^v = f^v h^v, \; (fh)^c = f^c h^v + f^v h^c.$$

If $(\mathcal{U}, (u^i)_{i=1}^n)$ is a chart of M, $\left(\tau^{-1}(\mathcal{U}), ((x^i), (y^i))\right)$ is the induced chart on TM, then

$$f^c \upharpoonright \tau^{-1}(\mathcal{U}) = \sum_{i=1}^n y^i \left(\frac{\partial f}{\partial u^i} \circ \tau \right), \tag{4.1.17}$$

therefore $f^c \in C^\infty(TM)$. In particular,

$$(u^i)^v = x^i, \; (u^i)^c = y^i; \quad i \in J_n. \tag{4.1.18}$$

Proof. A routine verification, which we leave to the reader. $\square$

Remark 4.1.6. With the notation already introduced, let

$$w := \sum_{i=1}^n \left(w^i \left(\frac{\partial}{\partial x^i} \right)_v + w^{n+i} \left(\frac{\partial}{\partial y^i} \right)_v \right); \; v = \sum_{i=1}^n v^i \left(\frac{\partial}{\partial u^i} \right)_p, \quad p \in \mathcal{U}. \tag{4.1.19}$$

Then

$$w(f^c) = \sum_{i=1}^n w^i \left(\frac{\partial}{\partial x^i} \right)_v \left(\sum_{j=1}^n y^j \left(\frac{\partial f}{\partial u^j} \circ \tau \right) \right)$$

$$+ \sum_{i=1}^{n} w^{n+i} \left(\frac{\partial}{\partial y^i} \right)_v \left(\sum_{j=1}^{n} y^j \left(\frac{\partial f}{\partial u^j} \circ \tau \right) \right)$$

$$= \sum_{i,j=1}^{n} w^i v^j \frac{\partial}{\partial x^i} \left(\frac{\partial f}{\partial u^j} \circ \tau \right)(v)$$

$$+ \sum_{i,j=1}^{n} w^{n+i} \left(\delta_i^j \frac{\partial f}{\partial u^j}(p) + v^j \frac{\partial}{\partial y^i} \left(\frac{\partial f}{\partial u^j} \circ \tau \right)(v) \right)$$

$$\overset{(*)}{=} \sum_{i,j=1}^{n} w^i v^j \sum_{k=1}^{n} \frac{\partial^2 f}{\partial u^k \partial u^j}(p) \frac{\partial(u^k \circ \tau)}{\partial x^i}(v) + \sum_{i=1}^{n} w^{n+i} \frac{\partial f}{\partial u^i}(p)$$

$$+ \sum_{i,j=1}^{n} w^{n+i} v^j \sum_{k=1}^{n} \frac{\partial^2 f}{\partial u^k \partial u^j}(p) \frac{\partial(u^k \circ \tau)}{\partial y^i}(v)$$

$$= \sum_{i,j=1}^{n} w^i v^j \frac{\partial^2 f}{\partial u^i \partial u^j}(p) + \sum_{i=1}^{n} w^{n+i} \frac{\partial f}{\partial u^i}(p),$$

applying at step $(*)$ the chain rule formulated in Lemma 2.1.19(iv).

Lemma 4.1.7. *Let w_1 and w_2 be tangent vectors in TTM. If*

(i) $\tau_{TM}(w_1) = \tau_{TM}(w_2)$,
(ii) $\tau_*(w_1) = \tau_*(w_2)$,
(iii) $w_1(f^c) = w_2(f^c)$ *for every* $f \in C^\infty(M)$,

then $w_1 = w_2$.

Proof. First we note that *a tangent vector $w \in T_v TM$ is uniquely determined by its action on the vertical and the complete lifts of smooth functions on M.* To see this we use the same notation as above. Let $w \in T_v TM$. Then, by the basis theorem (Proposition 3.1.8),

$$w = \sum_{i=1}^{n} \left(w(x^i) \left(\frac{\partial}{\partial x^i} \right)_v + w(y^i) \left(\frac{\partial}{\partial y^i} \right)_v \right).$$

Here, by (4.1.18), $x^i = (u^i)^{\vee}$, $y^i = (u^i)^c$, so our claim follows.

Now we can easily conclude that (i)–(iii) imply $w_1 = w_2$. Condition (i) assures that w_1 and w_2 are in the same tangent space of TM. For every smooth function f on M,

$$w_1(f^{\vee}) = w_1(f \circ \tau) \overset{(3.1.4)}{=} \tau_*(w_1)(f) \overset{(ii)}{=} \tau_*(w_2)(f) = w_2(f^{\vee}).$$

Thus, from (ii) and from our preliminary remark it follows that $w_1 = w_2$. $\square$

Proposition 4.1.8. *There exists one and only one diffeomorphism κ of TTM onto itself satisfying the following conditions:*

(i) $\tau_* \circ \kappa = \tau_{TM}$,

(ii) $\tau_{TM} \circ \kappa = \tau_*$,

(iii) $\kappa(w)(f^c) = w(f^c)$ *for all $w \in TTM$, $f \in C^\infty(M)$.*

Then κ is an involution of TTM, i.e., $\kappa^2 = 1_{TTM}$, and hence $\kappa^{-1} = \kappa$.

Proof. By the preceding lemma, κ is uniquely determined by conditions (i)–(iii). To show the existence, we may argue locally. With the same notation as before, define a mapping

$$\kappa \colon T(\tau^{-1}(\mathcal{U})) \to T(\tau^{-1}(\mathcal{U}))$$

by

$$\begin{cases} \kappa(w) := \sum_{i=1}^{n} \left(v^i \left(\frac{\partial}{\partial x^i} \right)_{\tau_*(w)} + w^{n+i} \left(\frac{\partial}{\partial y^i} \right)_{\tau_*(w)} \right) \\ \text{if } w \in T_v TM, \text{ where } v \in T_p M, \ p \in \mathcal{U}, \text{ is given by (4.1.19).} \end{cases} \tag{4.1.20}$$

Then, clearly, $\tau_{TM}(\kappa(w)) = \tau_*(w)$, so condition (ii) holds automatically. Since

$$\tau_*(\kappa(w)) \overset{(3.1.27)}{=} \sum_{i=1}^{n} v^i \left(\frac{\partial}{\partial u^i} \right)_{\tau(\tau_*(w))} \overset{(3.1.29)}{=} \sum_{i=1}^{n} v^i \left(\frac{\partial}{\partial u^i} \right)_p = v = \tau_{TM}(w),$$

condition (i) is also satisfied. To check (iii), let $f \in C^\infty(M)$. Then, by (4.1.17), $f^c \restriction \tau^{-1}(\mathcal{U}) = \sum_{i=1}^{n} y^i \left(\frac{\partial f}{\partial u^j} \circ \tau \right)$, and we obtain as in Remark 4.1.6 that

$$\kappa(w)(f^c) = \sum_{i,j=1}^{n} v^i w^j \frac{\partial^2 f}{\partial u^i \partial u^j}(p) + \sum_{i=1}^{n} w^{n+i} \frac{\partial f}{\partial u^i}(p)$$

$$= \sum_{i,j=1}^{n} w^i v^j \frac{\partial^2 f}{\partial u^i \partial u^j}(p) + \sum_{i=1}^{n} w^{n+i} \frac{\partial f}{\partial u^i}(p) = w(f^c),$$

taking into account that the matrix $\left(\frac{\partial^2 f}{\partial u^i \partial u^j}(p) \right)$ is symmetric. So κ satisfies (i)–(iii). The smoothness of κ is clear from the definition, and it can also be immediately seen that $\kappa^2 = 1_{T(\tau^{-1}(\mathcal{U}))}$. $\qquad \square$

Remark 4.1.9. The involutory diffeomorphism κ of TTM is determined by the *intrinsic* properties (i)–(iii), so it is called the *canonical involution* of TTM.

Proposition 4.1.10. *For each $v \in TM$, there exists a real vector space structure on the fibre $\tau_*^{-1}(v) \subset TTM$ such that $\tau_* : TTM \to TM$ becomes a vector bundle and the canonical involution $\kappa : TTM \to TTM$ is an isomorphism of the vector bundle $\tau_* : TTM \to TM$ onto the double tangent bundle $\tau_{TM} : TTM \to TM$.*

Proof. First we describe the fibre $(\tau_*)^{-1}(v)$ with the help of a chart $(\mathcal{U}, (u^i)_{i=1}^n)$ of M and the induced chart $(\tau^{-1}(\mathcal{U}), ((x^i), (y^i)))$ on TM. Let $v = \sum_{i=1}^n v^i \left(\frac{\partial}{\partial u^i}\right)_{\tau(v)}$. If

$$a = \sum_{i=1}^n \left(\alpha^i \left(\frac{\partial}{\partial x^i} \right)_w + \alpha^{n+i} \left(\frac{\partial}{\partial y^i} \right)_w \right) \in (\tau_*)^{-1}(v)$$

then we obtain

$$\tau_*(a) = \sum_{i=1}^n \alpha^i \left(\frac{\partial}{\partial u^i} \right)_{\tau(w)} = v = \sum_{i=1}^n v^i \left(\frac{\partial}{\partial u^i} \right)_{\tau(v)},$$

therefore

$$(\tau_*)^{-1}(v)$$
$$= \left\{ v^i \left(\frac{\partial}{\partial x^i} \right)_w + \alpha^{n+i} \left(\frac{\partial}{\partial y^i} \right)_w \in TTM \;\middle|\; \tau(w) = \tau(v), \alpha^{n+i} \in \mathbb{R}, i \in J_n \right\}$$

(summation convention in force).

Now let $a, b \in (\tau_*)^{-1}(v)$; $a \in T_{w_1}TM$, $b \in T_{w_2}TM$. Then

$$\tau(w_1) = \tau(w_2), \quad \tau_*(a) = \tau_*(b),$$

so $\kappa(a), \kappa(b) \in T_{\tau_*(a)}TM$ by (4.1.20). Thus we may form the sum

$$\kappa(a) + \kappa(b) \in T_{\tau_*(a)}TM.$$

Then, applying Proposition 4.1.8(i) repeatedly,

$$\kappa(\kappa(a) + \kappa(b)) \in T_{\tau_*(\kappa(a)+\kappa(b))}TM = T_{\tau_{TM}(a)+\tau_{TM}(b)}TM = T_{w_1+w_2}TM$$

and

$$\tau_*\big(\kappa(\kappa(a) + \kappa(b))\big) = \tau_{TM}(\kappa(a) + \kappa(b)) = \tau_*(a) = v,$$

whence $\kappa(\kappa(a) + \kappa(b)) \in (\tau_*)^{-1}(v)$.

After these preparations, we define addition and scalar multiplication on $(\tau_*)^{-1}(v)$ by

$$a \boxplus b := \kappa(\kappa(a) + \kappa(b)), \quad \lambda \boxdot a := \kappa(\lambda \kappa(a)) \quad (\lambda \in \mathbb{R}). \qquad (4.1.21)$$

Under these operations, $(\tau_*)^{-1}(v)$ is a $2n$-dimensional real vector space. Then $\tau_*\colon TTM \to TM$, with the so defined vector space structures on the fibres, becomes a vector bundle. Since $\kappa^2 = 1_{TTM}$, we see from (4.1.21) that the canonical involution of TTM is a bundle map between the vector bundles $\tau_*\colon TTM \to TM$ and $\tau_{TM}\colon TTM \to TM$. More precisely, κ restricts to linear isomorphisms on the fibres, and the induced mapping $\kappa_B\colon TM \to TM$ is the identical transformation, so Proposition 2.2.19 assures that κ is a (strong) isomorphism of vector bundles.   □

Remark 4.1.11. Using the charts introduced in the previous proof, the linear operations (4.1.21) can be given explicitly as follows: if

$$v = \sum_{i=1}^{n} v^i \left(\frac{\partial}{\partial u^i}\right)_{\tau(v)},$$

$$a = \sum_{i=1}^{n} \left(v^i \left(\frac{\partial}{\partial x^i}\right)_{w_1} + \alpha^{n+i} \left(\frac{\partial}{\partial y^i}\right)_{w_1}\right) \in (\tau_*)^{-1}(v),$$

$$b = \sum_{i=1}^{n} \left(v^i \left(\frac{\partial}{\partial x^i}\right)_{w_2} + \beta^{n+i} \left(\frac{\partial}{\partial y^i}\right)_{w_2}\right) \in (\tau_*)^{-1}(v),$$

then

$$a \boxplus b = \sum_{i=1}^{n} \left(v^i \left(\frac{\partial}{\partial x^i}\right)_{w_1+w_2} + (\alpha^{n+i} + \beta^{n+i}) \left(\frac{\partial}{\partial y^i}\right)_{w_1+w_2}\right), \quad (4.1.22)$$

$$\lambda \boxdot a = \sum_{i=1}^{n} \left(v^i \left(\frac{\partial}{\partial x^i}\right)_{\lambda w_1} + \lambda \alpha^{n+i} \left(\frac{\partial}{\partial y^i}\right)_{\lambda w_1}\right). \quad (4.1.23)$$

Comparing the second formula with the local expression of $(\mu_\lambda)_*$ obtained in Example 3.1.38(a), we see that

$$(\mu_\lambda)_*(a) = \lambda \boxdot a. \quad (4.1.24)$$

Lemma 4.1.12. *Let $\varphi\colon N \to M$ be a smooth mapping between manifolds. Suppose that $X, Y \in C^\infty(N, TM)$ are sections of the tangent bundle $\tau_M\colon TM \to M$ along φ, i.e.,*

$$\tau_M \circ X = \tau_M \circ Y = \varphi. \quad (*)$$

Then we have

$$(X + Y)_* = X_* \boxplus Y_* \quad (4.1.25)$$

and

$$(\lambda X)_* = \lambda \boxdot X_*, \quad \lambda \in \mathbb{R}, \quad (4.1.26)$$

where the right-hand sides are defined pointwise by (4.1.21).

Proof. (a) Relation (4.1.26) is an immediate consequence of (4.1.24) because $(\lambda X)_* = (\mu_\lambda \circ X)_* = (\mu_\lambda)_* \circ X_* = \lambda \boxdot X_*$.

(b) The verification of (4.1.25) is less straightforward. Let $\tau_N : TN \to N$ be the tangent bundle of N. Choose an arbitrary vector v in TN and let

$$w_1 := (X + Y)_*(v), \quad w_2 := (X_* \boxplus Y_*)(v).$$

We are going to show that w_1 and w_2 satisfy the conditions of Lemma 4.1.7, and hence $w_1 = w_2$. To do this note first that $X_* : TN \to TTM$ and $Y_* : TN \to TTM$ are bundle maps which induce X and Y, respectively, between the base manifolds (see Lemma 3.1.33). Thus we have

$$\tau_{TM} \circ X_* = X \circ \tau_N, \quad \tau_{TM} \circ Y_* = Y \circ \tau_N. \qquad (**)$$

Verification of condition (i). We have on the one hand

$$\tau_{TM} \circ (X + Y)_* \overset{(**)}{=} (X + Y) \circ \tau_N = X \circ \tau_N + Y \circ \tau_N.$$

On the other hand

$$
\begin{aligned}
\tau_{TM} \circ (X_* \boxplus Y_*) &\overset{(4.1.21)}{=} \tau_{TM} \circ \kappa \circ (\kappa \circ X_* + \kappa \circ Y_*) \\
&\overset{\text{Proposition } 4.1.8}{=} (\tau_M)_* \circ (\kappa \circ X_* + \kappa \circ Y_*) \\
&= \tau_{TM} \circ X_* + \tau_{TM} \circ Y_* \overset{(**)}{=} X \circ \tau_N + Y \circ \tau_N,
\end{aligned}
$$

whence $\tau_{TM}(w_1) = \tau_{TM}(w_2)$.

Verification of condition (ii). On the one hand,

$$(\tau_M)_*(w_1) = (\tau_M)_* \circ (X + Y)_*(v) = (\tau \circ (X + Y))_*(v) \overset{(*)}{=} \varphi_*(v).$$

On the other hand,

$$(\tau_M)_* \circ X_*(v) = (\tau_M \circ X)_*(v) = \varphi_*(v)$$

and, similarly, $(\tau_M)_* \circ Y_*(v) = \varphi_*(v)$. Since $X_*(v)$, $Y_*(v)$ and $X_*(v) \boxplus Y_*(v)$ are in the same fibre of $(\tau_M)_*$, we have

$$(\tau_M)_*(w_2) = (\tau_M)_* \circ (X_* \boxplus Y_*)(v) = (\tau_M)_*(X_*(v) \boxplus Y_*(v)) = \varphi_*(v),$$

therefore $(\tau_M)_*(w_1) = \varphi_*(v) = (\tau_M)_*(w_2)$.

Verification of condition (iii). Let f be a smooth function on M. Taking into account that f^c is fibrewise linear, we find, on the one hand, that

$$
\begin{aligned}
(X + Y)_*(v)(f^c) &\overset{(3.1.4)}{=} v(f^c \circ (X + Y)) = v(f^c \circ X + f^c \circ Y) \\
&= v(f^c \circ X) + v(f^c \circ Y) = X_*(v)(f^c) + Y_*(v)(f^c).
\end{aligned}
$$

On the other hand,

$$
\begin{aligned}
(X_* \boxplus Y_*)(v)(f^c) &\overset{(4.1.21)}{=} \kappa(\kappa \circ X_*(v) + \kappa \circ Y_*(v))(f^c) \\
&\overset{\text{Proposition } 4.1.8}{=} \kappa(X_*(v))(f^c) + \kappa(Y_*(v))(f^c) = X_*(v)(f^c) + Y_*(v)(f^c),
\end{aligned}
$$

so we also have $w_1(f^c) = w_2(f^c)$ for every $f \in C^\infty(M)$.

This concludes the proof. $\qquad\qquad\square$

Lemma 4.1.13. *Let $\varphi\colon M \to N$ be a smooth mapping between manifolds. Let κ_M and κ_N be the canonical involutions of TTM and TTN, respectively. Then we have*

$$\varphi_{**} \circ \kappa_M = \kappa_N \circ \varphi_{**}. \tag{4.1.27}$$

Proof. Let τ_M and τ_N be the tangent bundle projections $TM \to M$ and $TN \to N$, respectively. Then, by Lemma 3.1.33,

$$\tau_N \circ \varphi_* = \varphi \circ \tau_M, \quad \tau_{TN} \circ \varphi_{**} = \varphi_* \circ \tau_{TM}. \tag{$*$}$$

Next, we use the same argument as in part (b) of the preceding proof: we show that for every $w \in TTM$, the vectors $\varphi_{**}(\kappa_M(w))$ and $\kappa_N(\varphi_{**}(w))$ satisfy the conditions (i)–(iii) of Lemma 4.1.7.

Proof of (i). Using relation $(*)$ and Proposition 4.1.8, we get on the one hand that

$$\tau_{TN} \circ \varphi_{**} \circ \kappa_M = \varphi_* \circ \tau_{TM} \circ \kappa_M = \varphi_* \circ (\tau_M)_* = (\varphi \circ \tau_M)_*.$$

On the other hand, similarly, we find that

$$\tau_{TN} \circ \kappa_N \circ \varphi_{**} = (\tau_N)_* \circ \varphi_{**} = (\tau_N \circ \varphi_*)_* = (\varphi \circ \tau_M)_*,$$

as desired.

Proof of (ii). In the same way as above,

$$(\tau_N)_* \circ \varphi_{**} \circ \kappa_M = (\tau_N \circ \varphi_*)_* \circ \kappa_M \overset{(*)}{=} \varphi_* \circ (\tau_M)_* \circ \kappa_M$$
$$\overset{\text{Proposition 4.1.8}}{=} \varphi_* \circ \tau_{TM};$$
$$(\tau_N)_* \circ \kappa_N \circ \varphi_{**} = \tau_{TN} \circ \varphi_{**} = \varphi_* \circ \tau_{TM},$$

as wanted.

Proof of (iii). Let h be a smooth function on N. Then, on the one hand,

$$\varphi_{**}(\kappa_M(w))(h^{\mathsf{c}}) \overset{(3.1.4)}{=} \kappa_M(w)(h^{\mathsf{c}} \circ \varphi_*) \overset{(4.1.16)}{=} \kappa_M(w)(h \circ \varphi)^{\mathsf{c}}$$
$$\overset{\text{Proposition 4.1.8(iii)}}{=} w(h \circ \varphi)^{\mathsf{c}} \overset{(4.1.16)}{=} w(h^{\mathsf{c}} \circ \varphi_*) = \varphi_{**}(w)(h^{\mathsf{c}}).$$

On the other hand, again by Proposition 4.1.8(iii),

$$\kappa_N(\varphi_{**}(w))(h^{\mathsf{c}}) = \varphi_{**}(w)(h^{\mathsf{c}}).$$

This concludes the proof. $\square$

4.1.3   The Vertical Subbundle of TTM

Definition 4.1.14. Let M be a manifold. Consider its tangent bundle $\tau \colon TM \to M$ and the derivative $\tau_* \colon TTM \to TM$ of the tangent bundle projection. For each $v \in TM$, the subspace

$$V_v TM := \mathrm{Ker}(\tau_*)_v \subset T_v TM$$

is said to be the *vertical subspace* of $T_v TM$. The vectors of $T_v TM$ are called *vertical vectors* at v.

Remark 4.1.15. If $\pi \colon E \to M$ is an arbitrary vector bundle, then the vertical subspace $V_z E$ of $T_z E$ $(z \in E)$ is defined in the same way:

$$V_z E := \mathrm{Ker}(\pi_*)_z \subset T_z E.$$

Lemma 4.1.16. *Let M be an n-dimensional manifold.*

(i) *For each $v \in TM$, the vertical subspace $V_v TM$ is an n-dimensional subspace of $T_v TM$. If $v \in T_p M$, $(\mathcal{U}, (u^i)_{i=1}^n)$ is a chart of M at p and $\left(\tau^{-1}(\mathcal{U}), ((x^i)_{i=1}^n, (y^i)_{i=1}^n)\right)$ is the induced chart on TM, then*

$$\left(\left(\frac{\partial}{\partial y^i}\right)_v\right)_{i=1}^n \quad \text{is a basis of } V_v TM. \tag{4.1.28}$$

(ii) *Given a point $p \in M$, consider the canonical inclusion*

$$i_p \colon T_p M \to TM.$$

If $v \in T_p M$, then

$$V_v TM = \mathrm{Im}((i_p)_*)_v. \tag{4.1.29}$$

(iii) *Let $\iota_v \colon T_p M \to T_v T_p M$ be the canonical identification defined in Lemma 3.1.19. Then the mapping*

$$\begin{cases} \ell_v^\uparrow := ((i_p)_*)_v \circ \iota_v \colon T_p M \to T_v T_p M \to V_v TM \\ w \mapsto \ell_v^\uparrow(w) =: w^\uparrow(v) \end{cases} \tag{4.1.30}$$

is a linear isomorphism. In the basis (4.1.28),

$$w^\uparrow(v) = \sum_{i=1}^n y^i(w) \left(\frac{\partial}{\partial y^i}\right)_v. \tag{4.1.31}$$

Proof. (i) By (3.1.27), the linear mapping $(\tau_*)_v \colon T_v TM \to T_p M$ carries the basis $\left(\left(\frac{\partial}{\partial x^i}\right)_v, \left(\frac{\partial}{\partial y^i}\right)_v\right)_{i=1}^n$ of $T_v TM$ to the basis $\left(\left(\frac{\partial}{\partial u^i}\right)_p\right)_{i=1}^n$ of $T_p M$. Hence $(\tau_*)_v$ is surjective, and from the rank + nullity = dimension theorem we obtain that $\dim V_v TM = \dim \mathrm{Ker}(\tau_*)_v = 2n - n = n$. By the second relation in (3.1.27), the family $\left(\left(\frac{\partial}{\partial y^i}\right)_v\right)_{i=1}^n$ consists of vertical vectors, so it is a basis of $V_v TM$.

(ii) Since $T_p M$ is a submanifold of TM, the inclusion $i_p \colon T_p M \to TM$ is a smooth mapping. The composite mapping $\tau \circ i_p$ is clearly constant, so by the chain rule (Lemma 3.1.13), $\tau_* \circ (i_p)_*$ is the zero mapping. Hence for each $v \in T_p M$,

$$\mathrm{Im}((i_p)_*)_v \subset \mathrm{Ker}(\tau_*)_v = V_v TM.$$

On the other hand, the linear mapping

$$((i_p)_*)_v \colon T_v T_p M \to T_v TM$$

is injective, therefore

$$\dim \mathrm{Im}((i_p)_*)_v = \dim T_v T_p M = \dim T_p M = n.$$

Thus $\mathrm{Im}((i_p)_*)_v$ is an n-dimensional subspace of the n-dimensional vector space $V_v TM$ and hence $\mathrm{Im}((i_p)_*)_v = V_v TM$.

(iii) The mapping $\ell_v^\uparrow$, being a composition of linear isomorphisms, is also an isomorphism. We verify the coordinate expression (4.1.31). By the basis theorem (Proposition 3.1.8) and the definition of the function y^i ((3.1.14)), the family $(y^i \restriction T_p M)_{i=1}^n$ is the dual of the basis $\left(\left(\frac{\partial}{\partial u^i}\right)_p\right)_{i=1}^n$ of $T_p M$. To simplify the notation, we also write y^i for $y^i \restriction T_p M$. Then the mapping

$$(y^1, \ldots, y^n) \colon T_p M \to \mathbb{R}^n, \quad v \mapsto \sum_{i=1}^n y^i(v) e_i$$

is a linear coordinate system of $T_p M$. Applying Lemma 3.1.19 for $V := T_p M$, it follows that

$$\iota_v(w) = \sum_{i=1}^n y^i(w) \left(\frac{\partial}{\partial y^i}\right)_v, \quad w \in T_p M. \tag{4.1.32}$$

We show finally that the derivative of the harmless inclusion mapping i_p does not change this expression. Indeed, for each $j \in J_n$,

$$((i_p)_*)_v \left(\frac{\partial}{\partial y^j}\right)_v \overset{(3.1.5)}{=} \sum_{k=1}^n \frac{\partial(x^k \circ i_p)}{\partial y^j}(v) \left(\frac{\partial}{\partial x^k}\right)_v$$

$$+ \sum_{k=1}^n \frac{\partial(y^k \circ i_p)}{\partial y^j}(v) \left(\frac{\partial}{\partial y^k}\right)_v = \left(\frac{\partial}{\partial y^j}\right)_v,$$

hence

$$w^\uparrow(v) = ((i_p)_*)_v(\iota_v(w)) \overset{(4.1.32)}{=} ((i_p)_*)_v \left(\sum_{i=1}^n y^i(w) \left(\frac{\partial}{\partial y^i} \right)_v \right)$$

$$= \sum_{i=1}^n y^i(w) \left(\frac{\partial}{\partial y^i} \right)_v .$$

□

Remark 4.1.17. The vector $w^\uparrow(v) \in T_v TM$ is called the *vertical lift* of $w \in T_p M$ to $v \in T_p M$. It is also denoted by $w^v(v)$.

Let $\alpha \colon \mathbb{R} \to T_p M$ be the smooth curve given by

$$\alpha(t) := v + tw, \quad t \in \mathbb{R}.$$

Then the vertical lift $w^\uparrow(v)$ can be obtained as the velocity vector at 0 of the curve

$$i_p \circ \alpha \colon \mathbb{R} \to T_p M \to TM, \ t \mapsto v + tw.$$

Indeed,

$$w^\uparrow(v) \overset{(4.1.30)}{=} (i_p)_* \circ \iota_v(w) \overset{\text{Lemma } 3.1.19}{=} (i_p)_*(\dot{\alpha}(0)) \overset{(3.1.11)}{=} (i_p \circ \alpha)^{\cdot}(0).$$

Corollary 4.1.18. *Let M be a manifold, and F a smooth function on TM. Given a point p of M, let $F_p := F \upharpoonright T_p M$. Then for each $v, w \in T_p M$ we have*

$$w^\uparrow(v)(F) = (F_p)'(v)(w). \tag{4.1.33}$$

In particular, if $(\mathcal{U}, (u^i)_{i=1}^n)$ is a chart of M at p and $(\tau^{-1}(\mathcal{U}), (x^i), (y^i))$ is the induced chart on TM, then

$$\frac{\partial F}{\partial y^i}(v) = (F_p)'(v) \left(\left(\frac{\partial}{\partial u^i} \right)_p \right), \quad i \in J_n. \tag{4.1.34}$$

Proof. With the notation of the preceding remark,

$$w^\uparrow(v)(F) = (i_p \circ \alpha)^{\cdot}(0)(F) \overset{(3.1.9)}{=} (F \circ i_p \circ \alpha)'(0)$$

$$= \lim_{t \to 0} \frac{F \circ i_p \circ \alpha(t) - F \circ i_p \circ \alpha(0)}{t}$$

$$= \lim_{t \to 0} \frac{F_p(v + tw) - F_p(v)}{t} \overset{(C.1.3)}{=} (F_p)'(v)(w).$$

Since

$$\left(\left(\frac{\partial}{\partial u^i} \right)_p \right)^\uparrow (v) \overset{(4.1.31)}{=} \left(\frac{\partial}{\partial y^i} \right)_v \quad (i \in J_n),$$

from (4.1.33) we immediately obtain (4.1.34). □

Lemma 4.1.19. *Let M be a manifold. The mapping*

$$\mathbf{j} := (\tau_{TM}, \tau_*) \colon TTM \to TM \times_M TM, \ w \mapsto (\tau_{TM}(w), \tau_*(w)) \quad (4.1.35)$$

is a canonical strong bundle map from the double tangent bundle $\tau_{TM} \colon TTM \to TM$ onto the Finsler bundle $\pi \colon TM \times_M TM \to TM$. So we have the following commutative diagram:

$$
\begin{array}{ccc}
TTM & \xrightarrow{\ \mathbf{j}\ } & TM \times_M TM \\
{\scriptstyle \tau_{TM}}\downarrow & & \downarrow{\scriptstyle \pi} \\
TM & \xrightarrow[\ 1_{TM}\ ]{} & TM
\end{array}
\quad . \qquad (4.1.36)
$$

Proof. By Example 2.1.17(d), the mapping $\mathbf{j}$ is smooth. For each $v \in TM$, $w \in T_v TM$,

$$\pi \circ \mathbf{j}(w) = \mathrm{pr}_1(v, (\tau_*)_v(w)) = v = \tau_{TM}(w) = 1_{TM} \circ \tau_{TM}(w),$$

hence the diagram (4.1.36) commutes. Let $\mathbf{j}_v = \mathbf{j} \restriction T_v TM$, $v \in TM$. Then for any vector $w \in T_v TM$,

$$\mathbf{j}_v(w) = (v, (\tau_*)_v(w)) \in \{v\} \times T_{\tau(v)} M.$$

Since $(\tau_*)_v$ is a linear surjection (see Example 3.1.38(b)), it follows that $\mathbf{j}_v$ is a surjective linear mapping from $T_v TM$ onto $\{v\} \times T_{\tau(v)} M$. In light of Lemma 2.2.4, this concludes the proof. $\qquad\square$

Lemma 4.1.20. *Let $\gamma \colon I \to M$ be a smooth curve. Then*

$$\mathbf{j} \circ \ddot{\gamma} = \widetilde{\delta} \circ \dot{\gamma}. \qquad (4.1.37)$$

Proof. Since $\ddot{\gamma}$ is the velocity of the smooth curve $\dot{\gamma}$ in TM, $\ddot{\gamma}$ is a vector field along $\dot{\gamma}$ (see Example 3.1.39), that is, $\tau_{TM} \circ \ddot{\gamma} = \dot{\gamma}$. Furthermore, $\tau_* \circ \ddot{\gamma} = \dot{\gamma}$ by (3.1.32). Thus, for every $t \in I$,

$$\mathbf{j} \circ \ddot{\gamma}(t) \overset{(4.1.35)}{=} \left(\tau_{TM}(\ddot{\gamma}(t)), \tau_*(\ddot{\gamma}(t))\right) = (\dot{\gamma}(t), \dot{\gamma}(t)) = \widetilde{\delta} \circ (\dot{\gamma}(t)),$$

as claimed. $\qquad\square$

Corollary 4.1.21. *Let M be an n-dimensional manifold. If*

$$VTM := \bigcup_{v \in TM} V_v TM, \ \tau^{\mathrm{v}}_{TM} := \tau_{TM} \restriction VTM,$$

then $(VTM, \tau^{\mathrm{v}}_{TM}, TM)$ is a vector bundle with typical fibre $\mathbb{R}^n$, namely

$$VTM = \mathrm{Ker}(\mathbf{j}) = \mathrm{Ker}(\tau_{TM}, \tau_*).$$

Proof. Since $\mathbf{j}\colon TTM \to TM \times_M TM$ is a strong bundle map of constant rank n and

$$VTM = \bigcup_{v \in TM} \mathrm{Ker}(\tau_*)_v = \bigcup_{v \in TM} \mathrm{Ker}(\mathbf{j}_v) = \mathrm{Ker}(\mathbf{j}),$$

our assertion is a consequence of Example 2.2.35(ii). $\qquad\square$

Definition 4.1.22. The vector bundle $(VTM, \tau^{\mathsf{v}}_{TM}, TM)$ is said to be the *vertical subbundle* of the double tangent bundle (TTM, τ_{TM}, TM) or the *vertical bundle* of TTM.

Remark 4.1.23. More generally, consider the vector bundle $\pi\colon E \to M$. The *vertical bundle for* π, more precisely, the *vertical subbundle* of the tangent bundle $\tau_E\colon TE \to E$ (or simply, the vertical bundle of TE) is the triple

$$(VE, \tau^{\mathsf{v}}_E, E), \quad \tau^{\mathsf{v}}_E := \tau_E \upharpoonright VE,$$

where $VE := \bigcup_{z \in E} V_z E$ is the disjoint union of the vertical subspaces $V_z E$ defined in Remark 4.1.15. Since $\pi_*\colon TE \to TM$ maps every tangent space $T_z E$ linearly onto the tangent space $T_{\pi(z)} M$, it induces a surjective strong bundle map

$$(\tau_E, \pi_*)\colon TE \to E \times_M TM, \quad w \mapsto (\tau_E(w), \pi_*(w))$$

from the tangent bundle $\tau_E\colon TE \to E$ onto the pull-back bundle

$$\pi_1\colon E \times_M TM \to E, \quad \pi_1 := \mathrm{pr}_1 \upharpoonright E \times_M TM.$$

Thus, analogously to (4.1.36), we have the commutative diagram:

$$
\begin{array}{ccc}
TE & \xrightarrow{(\tau_E,\, \pi_*)} & E \times_M TM \\
{\scriptstyle \tau_E} \downarrow & & \downarrow {\scriptstyle \pi_1} \\
E & \xrightarrow[\; 1_E \;]{} & E
\end{array}\quad.
$$

The kernel of (τ_E, π_*) is just VE, which is therefore indeed a vector subbundle of TE.

Lemma 4.1.24. *The mapping*

$$\mathbf{i}\colon TM \times_M TM \to VTM, \quad (u, v) \mapsto \mathbf{i}(u, v) := v^{\uparrow}(u) \qquad (4.1.38)$$

is a strong bundle isomorphism of the Finsler bundle

$$\pi\colon TM \times_M TM \to TM$$

onto the vertical bundle $\tau^v_{TM} \colon VTM \to TM$, *making the following diagram commutative:*

$$
\begin{array}{ccc}
TM \times_M TM & \xrightarrow{\ i\ } & VTM \\
{\scriptstyle \pi}\downarrow & & \downarrow{\scriptstyle \tau^v_{TM}} \\
TM & \xrightarrow[\ 1_{TM}\ ]{} & TM
\end{array}\ \ .
$$

Proof. It is clear from the coordinate expression (4.1.31) that i is a smooth mapping. For each $u \in TM$,

$$
i_u := i \restriction \pi^{-1}(u) \colon \{u\} \times T_{\tau(u)}M \to V_u TM, \ v \mapsto i_u(v) := v^\uparrow(u)
$$

is a linear isomorphism by Lemma 4.1.16(iii). Finally, i induces the identity transformation of the common base manifold TM of the vector bundles π and τ^v_{TM}. Thus, by Proposition 2.2.19, i is a strong bundle isomorphism. $\square$

Corollary 4.1.25. *With the notation already introduced and the conventions of Example 2.2.41, the sequence*

$$
0 \longrightarrow TM \times_M TM \xrightarrow{\ i\ } TTM \xrightarrow{\ j\ } TM \times_M TM \longrightarrow 0 \qquad (4.1.39)
$$

is a short exact sequence of strong bundle maps. $\square$

Remark 4.1.26. (a) We say that the sequence (4.1.39) is the *canonical exact sequence* constructed from the tangent bundle $\tau \colon TM \to M$. Strictly speaking, the second mapping is the composite mapping

$$
TM \times_M TM \xrightarrow{\ i\ } VTM \xrightarrow{\text{inclusion}} TTM.
$$

Since $\mathrm{Im}(i) = \mathrm{Ker}(j)$, it follows that

$$
j \circ i = 0. \qquad (4.1.40)
$$

We shall see soon that the composition in reversed order yields a further important canonical object.

(b) By Proposition 2.2.31, the strong bundle maps i and j may be interpreted as tensorial mappings

$$
\Gamma(\pi) \longrightarrow \mathfrak{X}(TM), \ \widetilde{X} \longmapsto i\widetilde{X} := i \circ \widetilde{X}
$$

and

$$
\mathfrak{X}(TM) \longrightarrow \Gamma(\pi), \ \xi \longmapsto j\xi := j \circ \xi.
$$

So we obtain the canonical exact sequence

$$
0 \longrightarrow \Gamma(\pi) \xrightarrow{\ i\ } \mathfrak{X}(TM) \xrightarrow{\ j\ } \Gamma(\pi) \longrightarrow 0 \qquad (4.1.41)
$$

of $C^\infty(TM)$-homomorphisms, which contains the same information as the sequence (4.1.39).

Definition 4.1.27. Let M be a manifold. By a *vertical vector field* on TM we mean a smooth section of the vertical bundle of TTM.

Lemma 4.1.28. *For a vector field ξ on TM, the following conditions are equivalent:*

(i) ξ *is a vertical vector field, i.e.,* $\xi \in \Gamma(\tau_{TM}^{\mathsf{v}})$.
(ii) $\xi \underset{\tau}{\sim} o$*, where $o \in \mathfrak{X}(M)$ is the zero vector field.*
(iii) $\xi(\tilde{f}^{\mathsf{v}}) = 0$ *for all $f \in C^{\infty}(M)$.*

Proof. Let $\xi \in \mathfrak{X}(TM)$. The following sequence of equivalent statements proves the equivalence of (i) and (ii):

ξ is vertical;
$\xi(v) \in V_v TM$ for all $v \in TM$;
$(\tau_*)_v(\xi(v)) = 0_{\tau(v)} \in T_{\tau(v)}M$ for all $v \in TM$;
$\tau_* \circ \xi = o \circ \tau$;
$\xi \underset{\tau}{\sim} o$.

The equivalence of (ii) and (iii) is an immediate consequence of Lemma 3.1.49. □

Corollary 4.1.29. *The vertical vector fields on TM form a subalgebra of the (real) Lie algebra of vector fields on TM.*

Proof. We apply the related vector field lemma (Lemma 3.1.51) and our above observation

ξ is vertical if, and only if, $\xi \underset{\tau}{\sim} o$.

For example, if $\xi_1, \xi_2 \in \Gamma(\tau_{TM}^{\mathsf{v}})$, then $\xi_1 \underset{\tau}{\sim} o$, $\xi_2 \underset{\tau}{\sim} o$, which imply $[\xi_1, \xi_2] \underset{\tau}{\sim} o$, whence $[\xi_1, \xi_2] \in \Gamma(\tau_{TM}^{\mathsf{v}})$. □

Remark 4.1.30. Instead of $\Gamma(\tau_{TM}^{\mathsf{v}})$, we shall usually denote by $\mathfrak{X}^{\mathsf{v}}(TM)$ the $C^{\infty}(TM)$-module (and real Lie algebra) of the vertical vector fields on TM. Since $\mathbf{i}(TM \times_M TM) = VTM$, it follows by Remark 4.1.26(b) that

$$\mathfrak{X}^{\mathsf{v}}(TM) = \mathbf{i}(\Gamma(\pi)). \tag{4.1.42}$$

Definition 4.1.31. The *Liouville vector field* (or the *radial vector field*) on TM is

$$C := \mathbf{i}\tilde{\delta} \colon v \in TM \mapsto C(v) := \mathbf{i}(v, v) = v^{\uparrow}(v) \in TTM. \tag{4.1.43}$$

The *vertical lift* of a vector field X on M is the vertical vector field

$$X^{\mathsf{v}} := \mathbf{i}\hat{X} \colon v \in TM \mapsto X^{\mathsf{v}}(v) := \mathbf{i}(v, X(\tau(v))) = (X(\tau(v)))^{\uparrow}(v) \in TTM. \tag{4.1.44}$$

Lemma 4.1.32. *Let M be a manifold. For any vector fields X, Y on M and any function $f \in C^\infty(M)$,*

$$(X + Y)^\mathsf{v} = X^\mathsf{v} + Y^\mathsf{v}, \quad (fX)^\mathsf{v} = f^\mathsf{v} X^\mathsf{v}. \tag{4.1.45}$$

Proof. This is an easy consequence of the linearity of the mapping

$$\ell_v^\uparrow : w \in T_{\tau(v)}M \mapsto \ell_v^\uparrow(w) = w^\uparrow(v) \in V_v TM.$$

Indeed, for example,

$$(fX)^\mathsf{v}(v) := \big((fX)(\tau(v))\big)^\uparrow(v) = \big(f(\tau(v))X(\tau(v))\big)^\uparrow(v)$$

$$= f(\tau(v))\big(X(\tau(v))\big)^\uparrow(v) = f(\tau(v))X^\mathsf{v}(v) = (f^\mathsf{v} X^\mathsf{v})(v),$$

for all $v \in TM$, and hence $(fX)^\mathsf{v} = f^\mathsf{v} X^\mathsf{v}$. $\quad \cdot \Box$

Remark 4.1.33 (Coordinate Description). Let $(\mathcal{U}, (u^i)_{i=1}^n)$ be a chart on M, and consider the induced chart $\big(\tau^{-1}(\mathcal{U}), ((x^i)_{i=1}^n, (y^i)_{i=1}^n)\big)$ on TM.

(a) If ξ is a *vertical* vector field on TM and

$$\xi \underset{(\mathcal{U})}{=} \sum_{i=1}^n \left(\xi^i \frac{\partial}{\partial x^i} + \xi^{n+i} \frac{\partial}{\partial y^i} \right),$$

where ξ^i and ξ^{n+i} are smooth functions on $\tau^{-1}(\mathcal{U})$, then

$$0 = \xi((u^i)^\mathsf{v}) \overset{(4.1.18)}{=} \xi(x^i) = \xi^i \quad (i \in J_n)$$

by Lemma 4.1.28, whence

$$\xi \underset{(\mathcal{U})}{=} \sum_{i=1}^n \xi^{n+i} \frac{\partial}{\partial y^i}, \quad \xi^{n+i} \in C^\infty(\tau^{-1}(\mathcal{U})). \tag{4.1.46}$$

(b) We have, for any $v \in \tau^{-1}(\mathcal{U})$,

$$C(v) = v^\uparrow(v) \overset{(4.1.31)}{=} \sum_{i=1}^n y^i(v) \left(\frac{\partial}{\partial y^i} \right)_v = \left(\sum_{i=1}^n y^i \frac{\partial}{\partial y^i} \right)(v),$$

therefore

$$C \underset{(\mathcal{U})}{=} \sum_{i=1}^n y^i \frac{\partial}{\partial y^i}. \tag{4.1.47}$$

(c) If $X \in \mathfrak{X}(M)$, $X \upharpoonright \mathcal{U} = \sum_{i=1}^n X^i \frac{\partial}{\partial u^i}$, then for every $v \in \tau^{-1}(\mathcal{U})$,

$$X^\mathsf{v}(v) = X(\tau(v))^\uparrow(v) \overset{(4.1.31)}{=} \sum_{i=1}^n y^i\big(X(\tau(v))\big) \left(\frac{\partial}{\partial y^i} \right)_v$$

$$= \left(\sum_{i=1}^n (X^i \circ \tau) \frac{\partial}{\partial y^i} \right)(v),$$

whence

$$X^\mathsf{v} \underset{(\mathcal{U})}{=} \sum_{i=1}^n (X^i \circ \tau) \frac{\partial}{\partial y^i} = \sum_{i=1}^n (X^i)^\mathsf{v} \frac{\partial}{\partial y^i}. \tag{4.1.48}$$

Lemma 4.1.34. *The Liouville vector field C on TM has the following properties:*

(i) *For all $t \in \mathbb{R}$,*

$$C \underset{\mu_t}{\sim} C, \ \text{i.e.,} \ (\mu_t)_* \circ C = C \circ \mu_t. \tag{4.1.49}$$

(For the definition of μ_t, see Example 3.1.38(a).)
(ii) *For any smooth function f on M,*

$$Cf^c = f^c. \tag{4.1.50}$$

Furthermore, for any vector field X on M and any $f \in C^\infty(M)$,

$$X^v f^c = (Xf)^v. \tag{4.1.51}$$

Proof. We choose an induced chart $\left(\tau^{-1}(\mathcal{U}), ((x^i)_{i=1}^n, (y^i)_{i=1}^n)\right)$ on TM and calculate by using the local expression of C, $(\mu_t)_*$, f^c and X^v.

For every $v \in \tau^{-1}(\mathcal{U})$,

$$(\mu_t)_*(C(v)) \overset{(4.1.47)}{=} (\mu_t)_* \left(\sum_{i=1}^n y^i(v) \left(\frac{\partial}{\partial y^i} \right)_v \right) \overset{(3.1.25)}{=} \sum_{i=1}^n t y^i(v) \left(\frac{\partial}{\partial y^i} \right)_{tv}$$

$$= C(tv) = C \circ \mu_t(v),$$

whence (4.1.49).

$$Cf^c \overset{(4.1.47)}{\underset{(\mathcal{U})}{=}} \left(\sum_{i=1}^n y^i \frac{\partial}{\partial y^i} \right) f^c \overset{(4.1.17)}{=} \left(\sum_{i=1}^n y^i \frac{\partial}{\partial y^i} \right) \left(\sum_{k=1}^n y^k \left(\frac{\partial f}{\partial u^k} \right)^v \right)$$

$$= \sum_{i,k=1}^n y^i \delta_i^k \left(\frac{\partial f}{\partial u^k} \right)^v = \sum_{k=1}^n y^k \left(\frac{\partial f}{\partial u^k} \circ \tau \right) \underset{(\mathcal{U})}{=} f^c,$$

which proves (4.1.50). The remaining relation (4.1.51) may be checked in the same way. $\qquad \square$

Lemma 4.1.35. (i) *The mapping*

$$\varphi \colon \mathbb{R} \times TM \to TM, \ (t, v) \mapsto \varphi(t, v) := e^t v$$

is a global flow on TM whose velocity vector field is the Liouville vector field.

(ii) *Let X be a vector field on M. The mapping*

$$\psi \colon \mathbb{R} \times TM \to TM, \ (t, v) \mapsto \psi(t, v) := v + tX(\tau(v))$$

is a global flow on TM whose velocity field is the vertical lift of X.

Proof. It is clear that both φ and ψ are smooth mappings satisfying (Fl$_1$) and (Fl$_2$).

(i) Consider the flow φ. For any given $t \in TM$, the flow line of v is the smooth curve

$$\alpha_v \colon \mathbb{R} \to TM, \ t \mapsto \alpha_v(t) := e^t v.$$

Using an induced chart on TM as above,

$$\dot\alpha_v(0) \overset{(3.1.12)}{=} \sum_{i=1}^n \left((x^i \circ \alpha_v)'(0) \left(\frac{\partial}{\partial x^i} \right)_v + (y^i \circ \alpha_v)'(0) \left(\frac{\partial}{\partial y^i} \right)_v \right).$$

Since for each $t \in \mathbb{R}$,

$$x^i \circ \alpha_v(t) = u^i \circ \tau(e^t v) = u^i(p) \quad \text{if } v \in T_p M$$

and

$$y^i \circ \alpha_v(t) = y^i(e^t v) = e^t y^i(v),$$

the functions $x^i \circ \alpha_v$ are constant, while $(y^i \circ \alpha_v)'(t) = e^t y^i(v)$ whence

$$(y^i \circ \alpha_v)'(0) = y^i(v).$$

So we obtain

$$\dot\alpha_v(0) = \sum_{i=1}^n y^i(v) \left(\frac{\partial}{\partial y^i} \right)_v \overset{(4.1.47)}{=} C(v),$$

thus proving our first assertion.

(ii) Similarly, the flow line of $v \in TM$ under ψ is the curve

$$\beta_v \colon \mathbb{R} \to TM, \ t \mapsto \beta_v(t) := v + tX(\tau(v)),$$

whose velocity at 0 is

$$\dot\beta_v(0) = \sum_{i=1}^n \left((x^i \circ \beta_v)'(0) \left(\frac{\partial}{\partial x^i} \right)_v + (y^i \circ \beta_v)'(0) \left(\frac{\partial}{\partial y^i} \right)_v \right)$$

$$= \sum_{i=1}^n (y^i \circ X)(\tau(v)) \left(\frac{\partial}{\partial y^i} \right)_v = \left(\sum_{i=1}^n (X^i \circ \tau) \frac{\partial}{\partial y^i} \right)(v) \overset{(4.1.48)}{=} X^{\mathrm{v}}(v),$$

thus the velocity vector field of ψ is X^{v}. $\square$

Lemma 4.1.36. *A smooth function F on TM is the vertical lift of a smooth function on M if, and only if,*

$$X^{\mathrm{v}} F = 0 \quad \text{for all} \quad X \in \mathfrak{X}(M). \tag{$*$}$$

Proof. The condition is necessary by Lemma 4.1.28(iii).

Conversely, suppose that $(*)$ is satisfied. We shall show that in this case $F = (F \circ Z)^{\vee}$, where Z is a vector field on M.

Let p be any fixed point in M, and let u, v be tangent vectors to M at p. Choose a vector field X on M such that $X(p) = v - u$. (The existence of such a vector field is guaranteed by Lemma 2.2.24(i).) We claim that the function

$$h \colon \mathbb{R} \to \mathbb{R}, \ t \mapsto h(t) := F(u + t(v - u))$$

is constant. To see this, consider the (global) flow ψ of $X^{\vee}$ (Lemma 4.1.35(ii)). Then, for every $w \in T_p M$,

$$\psi(t, w) = w + tX(p), \quad t \in \mathbb{R}.$$

At each $t_0 \in \mathbb{R}$,

$$
\begin{aligned}
h'(t_0) &= \lim_{t \to 0} \frac{h(t_0 + t) - h(t_0)}{t} \\
&= \lim_{t \to 0} \frac{F(u + (t_0 + t)(v - u)) - F(u + t_0(v - u))}{t} \\
&= \lim_{t \to 0} \frac{F \circ \psi_t(u + t_0 X(p)) - F(u + t_0 X(p))}{t} \\
&\overset{(3.2.8)}{=} (X^{\vee}F)(u + t_0(v - u)) \overset{(*)}{=} 0.
\end{aligned}
$$

Thus $h' = 0$ and hence h is constant. Consequently, we have

$$F(u) = h(0) = h(1) = F(v).$$

This holds for any pair of tangent vectors u, v at any point of M. Therefore, there exists a vector field Z on M such that for every $u \in TM$,

$$F(u) = F(Z(\tau(u))) = (F \circ Z)^{\vee}(u).$$

This is what was to be shown. $\qquad \square$

Lemma 4.1.37. *A vector field on TM is uniquely determined by its action on the complete lifts of smooth functions on M.*

Proof. Let $\xi \in \mathfrak{X}(TM)$. It is sufficient to show that if

$$\xi f^{c} = 0 \text{ for all } f \in C^{\infty}(M), \tag{$*$}$$

then $\xi = 0$.

Observe first that condition $(*)$ implies that

$$0 = \frac{1}{2} \xi (f^2)^{c} \overset{\text{Lemma 4.1.5}}{=} \xi(f^c f^{\vee}) = (\xi f^c) f^{\vee} + f^c (\xi f^{\vee}) = f^c(\xi f^{\vee}).$$

Now choose a chart $(\mathcal{U}, (u^i)_{i=1}^n)$ on M. If $\left(\tau^{-1}(\mathcal{U}), ((x^i)_{i=1}^n, (y^i)_{i=1}^n)\right)$ is the induced chart on TM, then $x^i = (u^i)^{\vee}$, $y^i = (u^i)^{c}$ by (4.1.18), and with the choice $f := u^i$ we find that

$$y^i(\xi x^i) = 0 \text{ for all } i \in J_n.$$

So it follows that

$$\xi x^i \upharpoonright \tau^{-1}(\mathcal{U}) \setminus \mathrm{Ker}(y^i) = 0, \quad i \in J_n.$$

Since $\mathrm{Ker}(y^i)$ has empty interior, by continuity we have

$$\xi x^i \upharpoonright \tau^{-1}(\mathcal{U}) = 0, \quad i \in J_n. \tag{$**$}$$

Thus

$$\xi \underset{(\mathcal{U})}{=} \sum_{i=1}^n \left((\xi x^i)\frac{\partial}{\partial x^i} + (\xi y^i)\frac{\partial}{\partial y^i} \right) \overset{(*),(**)}{=} 0.$$

$\square$

Lemma 4.1.38. *Let X and Y be vector fields on a manifold M. Then*

$$[X^{\vee}, Y^{\vee}] = 0, \tag{4.1.52}$$
$$[C, X^{\vee}] = -X^{\vee}. \tag{4.1.53}$$

Proof. Since we have

$$[X^{\vee}, Y^{\vee}]f^{c} = X^{\vee}(Y^{\vee}f^{c}) - Y^{\vee}(X^{\vee}f^{c})$$
$$\overset{(4.1.51)}{=} X^{\vee}(Yf)^{\vee} - Y^{\vee}(Xf)^{\vee} \overset{\text{Lemma } 4.1.28}{=} 0$$

and

$$[C, X^{\vee}]f^{c} = C(Xf)^{\vee} - X^{\vee}(Cf^{c}) = -X^{\vee}(Cf^{c}) \overset{(4.1.50)}{=} -X^{\vee}f^{c}$$

for all $f \in C^{\infty}(M)$, both relations follow from the preceding lemma. $\square$

Lemma 4.1.39. *Given a vector field X on a manifold M, the mapping*

$$\ell_X^{\uparrow} : TM \to VTM, \quad v \mapsto \ell_X^{\uparrow}(v) := v^{\uparrow}(X(\tau(v)))$$

is an injective bundle map from the tangent bundle $\tau : TM \to M$ into the vertical bundle $\tau_{TM}^{\vee} : VTM \to TM$ which induces the mapping X between the base manifolds.

Conversely, define a mapping $\ell^{\downarrow} : VTM \to TM$ by

$$\ell^{\downarrow}(w) = a \in T_{\tau(v)}M \text{ if } \ell_v^{\uparrow}(a) = w, \text{ i.e., by } \ell^{\downarrow} \upharpoonright V_vTM := (\ell_v^{\uparrow})^{-1}.$$

Then $\ell^{\downarrow}$ is a surjective bundle map from the vertical bundle of TTM onto the tangent bundle TM inducing τ as the mapping between the base manifolds. So we have the following commutative diagrams:

$$
\begin{array}{ccc}
TM \xrightarrow{\ell_X^{\uparrow}} VTM & \qquad & VTM \xrightarrow{\ell^{\downarrow}} TM \\
\tau \downarrow \qquad \downarrow \tau_{TM}^{\vee} & , & \tau_{TM}^{\vee} \downarrow \qquad \downarrow \tau \\
M \xrightarrow[\;X\;]{} TM & \qquad & TM \xrightarrow[\;\tau\;]{} M
\end{array}
\qquad (4.1.54)
$$

For any vector field X on M, the mapping $\ell_X^{\uparrow}$ is a right inverse of $\ell^{\downarrow}$, i.e.,

$$
\ell^{\downarrow} \circ \ell_X^{\uparrow} = 1_{TM}. \qquad (4.1.55)
$$

We also have

$$
\ell^{\downarrow} \circ X^{\vee} = X \circ \tau \text{ for all } X \in \mathfrak{X}(M) \qquad (4.1.56)
$$

and

$$
\ell^{\downarrow} \circ C = 1_{TM}. \qquad (4.1.57)
$$

Proof. At each point $p \in M$,

$$
(\ell_X^{\uparrow})_p := \ell_X^{\uparrow} \upharpoonright T_pM = \ell_{X(p)}^{\uparrow} \colon T_pM \to T_{X(p)}TM
$$

is a linear isomorphism by Lemma 4.1.16, so $\ell_X^{\uparrow}$ restricts to linear isomorphisms on the fibres and clearly satisfies

$$
\tau_{TM} \circ \ell_X^{\uparrow} = X \circ \tau.
$$

If $\left(\tau^{-1}(\mathcal{U}), ((x^i)_{i=1}^n, (y^i)_{i=1}^n) \right)$ is an induced chart on TM, then

$$
\ell_X^{\uparrow}(v) := v^{\uparrow}\big(X(\tau(v))\big) \overset{(4.1.31)}{=} \sum_{i=1}^n y^i(v) \left(\frac{\partial}{\partial y^i} \right)_{X(\tau(v))}
$$

for all $v \in \tau^{-1}(\mathcal{U})$, therefore

$$
\ell_X^{\uparrow} \underset{(\mathcal{U})}{=} \sum_{i=1}^n y^i \left(\frac{\partial}{\partial y^i} \circ X \circ \tau \right),
$$

which proves the smoothness of $\ell_X^{\uparrow}$. From these, applying Lemma 2.2.4, we conclude that $\ell_X^{\uparrow}$ is an (injective) bundle map and $(\ell_X^{\uparrow})_B = X$.

Similarly,

$$
\ell^{\downarrow} \upharpoonright V_vTM := (\ell_v^{\uparrow})^{-1} \colon V_vTM \to T_{\tau(v)}M \text{ for all } v \in TM,
$$

so $\ell^{\downarrow}$ also restricts to linear isomorphisms on the fibres, and we have

$$
\tau \circ \ell^{\downarrow} = \tau \circ \tau_{TM}^{\vee}.
$$

In the induced chart chosen above

$$\ell^\downarrow \underset{(\mathcal{U})}{=} \sum_{i=1}^{n} Y^i \left(\frac{\partial}{\partial u^i} \circ \tau \circ \tau^{\vee}_{TM} \right),$$

where

$$Y^i \colon (\tau^{\vee}_{TM})^{-1}(\tau^{-1}(\mathcal{U})) \to \mathbb{R}, \ w \mapsto Y^i(w) := w(y^i), \quad i \in J_n.$$

Thus $\ell^\downarrow$ is also smooth, and Lemma 2.2.4 leads again to the desired conclusion.

Relations (4.1.54)–(4.1.57) are easy consequences of the definitions. For example,

$$\ell^\downarrow(X^{\vee}(v)) = \ell^\downarrow\big(X(\tau(v))^\uparrow(v)\big) = (\ell^\uparrow_v)^{-1}\big(\ell^\uparrow_v(X(\tau(v)))\big) = X \circ \tau(v)$$

for all $v \in TM$, whence $\ell^\downarrow \circ X^{\vee} = X \circ \tau$. $\qquad\square$

Remark 4.1.40. The bundle map $\ell^\downarrow \colon VTM \to TM$ is called the *canonical surjection of VTM onto TM*. Observe that both $\ell^\downarrow$ and $\ell^\uparrow_X$ ($X \in \mathfrak{X}(M)$) restrict to linear isomorphisms on the fibres, but they are not bundle isomorphisms: $\ell^\downarrow$ has infinitely many right inverses (see also Proposition 2.2.19).

Lemma 4.1.41. *For every real number λ,*

$$\ell^\downarrow \circ ((\mu_\lambda)_* \upharpoonright VTM) = \mu_\lambda \circ \ell^\downarrow. \tag{4.1.58}$$

Proof. With the notation as above, let $w = \sum_{i=1}^{n} w^i \left(\frac{\partial}{\partial y^i} \right)_v \in T_v TM$. If

$$\ell^\downarrow(w) = a = \sum_{i=1}^{n} \alpha^i \left(\frac{\partial}{\partial u^i} \right)_{\tau(v)} \in T_{\tau(v)} M,$$

then

$$w = a^\uparrow(w) \overset{(4.1.31)}{=} \sum_{i=1}^{n} y^i(a) \left(\frac{\partial}{\partial y^i} \right)_v = \sum_{i=1}^{n} \alpha^i \left(\frac{\partial}{\partial y^i} \right)_v.$$

Hence $\alpha^i = w^i$ ($i \in J_n$) and

$$\ell^\downarrow(w) = \sum_{i=1}^{n} w^i \left(\frac{\partial}{\partial u^i} \right)_{\tau(v)}.$$

Thus, on the one hand,

$$\ell^\downarrow((\mu_\lambda)_*(w)) \overset{(3.1.25)}{=} \ell^\downarrow \left(\sum_{i=1}^{n} \lambda w^i \left(\frac{\partial}{\partial y^i} \right)_{\lambda v} \right) = \sum_{i=1}^{n} \lambda w^i \left(\frac{\partial}{\partial u^i} \right)_{\tau(v)}.$$

On the other hand, $\mu_\lambda \circ \ell^\downarrow(w) = \sum_{i=1}^{n} \lambda w^i \left(\frac{\partial}{\partial u^i} \right)_{\tau(v)}$, whence our claim. $\quad\square$

4.1.4 Acceleration and Reparametrizations

Lemma 4.1.42. *Let M be a manifold, I a nonempty open interval and $\gamma \colon I \to M$ a smooth curve. Consider a reparametrization of γ by a smooth function $\theta \colon \bar{I} \to I$ with nowhere vanishing derivative. Then for every $t \in \bar{I}$,*

$$(\gamma \circ \theta)\dot{}(t) = \mu_{\theta'(t)} \circ \dot{\gamma} \circ \theta(t), \tag{4.1.59}$$

$$(\gamma \circ \theta)\ddot{}(t) = \tilde{\mu}_{\theta'(t)} \circ (\mu_{\theta'(t)})_* \circ \ddot{\gamma} \circ \theta(t) + \frac{\theta''(t)}{\theta'(t)}(C \circ (\gamma \circ \theta)\dot{})(t), \tag{4.1.60}$$

where $\tilde{\mu}_{\theta'(t)}$ and $\mu_{\theta'(t)}$ are the dilations defined in Remark 3.1.38(a), and C is the Liouville vector field.

If, in addition, the reparametrization is positive, then we have

$$(\gamma \circ \theta)\dot{}(t) = \mu^+_{\ln(\theta'(t))} \circ \dot{\gamma} \circ \theta(t), \tag{4.1.61}$$

$$(\gamma \circ \theta)\ddot{}(t) = \tilde{\mu}^+_{\ln(\theta'(t))} \circ (\mu^+_{\ln(\theta'(t))})_* \circ \ddot{\gamma} \circ \theta(t) + \frac{\theta''(t)}{\theta'(t)}(C \circ (\gamma \circ \theta)\dot{})(t), \tag{4.1.62}$$

where $\mu^+_t := \mu_{e^t}$, $\tilde{\mu}^+_t := \tilde{\mu}_{e^t}$.

Proof. Formula (4.1.59) is just a rewriting of (3.1.10). To prove (4.1.60), let, for short, $\bar{\gamma} := \gamma \circ \theta$. Choose a chart $(\mathcal{U}, (u^i)_{i=1}^n)$ on M with induced chart $(\tau^{-1}(\mathcal{U}), ((x^i)_{i=1}^n, (y^i)_{i=1}^n))$ on TM. Then the components of $\bar{\gamma}$ are $\bar{\gamma}^i := u^i \circ \bar{\gamma} = u^i \circ \gamma \circ \theta =: \gamma^i \circ \theta$, and so

$$(\bar{\gamma}^i)' = \theta'((\gamma^i)' \circ \theta), \quad (\bar{\gamma}^i)'' = \theta''((\gamma^i)' \circ \theta) + (\theta')^2((\gamma^i)'' \circ \theta).$$

Thus

$$\ddot{\bar{\gamma}}(t) \overset{(3.1.35)}{=} (\bar{\gamma}^i)'(t)\left(\frac{\partial}{\partial x^i}\right)_{\dot{\bar{\gamma}}(t)} + (\bar{\gamma}^i)''(t)\left(\frac{\partial}{\partial y^i}\right)_{\dot{\bar{\gamma}}(t)}$$

$$= \theta'(t)(\gamma^i)'(\theta(t))\left(\frac{\partial}{\partial x^i}\right)_{\theta'(t)\dot{\gamma}(\theta(t))}$$

$$+ \left((\theta'(t))^2(\gamma^i)''(\theta(t)) + \theta''(t)(\gamma^i)'(\theta(t))\right)\left(\frac{\partial}{\partial y^i}\right)_{\theta'(t)\dot{\gamma}(\theta(t))}$$

$$\overset{(3.1.25)}{=} \tilde{\mu}_{\theta'(t)} \circ (\mu_{\theta'(t)})_* \circ \left((\gamma^i)'\left(\frac{\partial}{\partial x^i} \circ \dot{\gamma}\right) + (\gamma^i)''\left(\frac{\partial}{\partial y^i} \circ \dot{\gamma}\right)\right)(\theta(t))$$

$$+ \frac{\theta''(t)}{\theta'(t)}(\bar{\gamma}^i)'(t)\left(\frac{\partial}{\partial y^i}\right)_{\dot{\bar{\gamma}}(t)}$$

$$= \tilde{\mu}_{\theta'(t)} \circ (\mu_{\theta'(t)})_* \circ \ddot{\gamma} \circ \theta(t) + \frac{\theta''(t)}{\theta'(t)}C(\dot{\bar{\gamma}}(t))$$

for all $t \in \bar{I}$. This proves (4.1.60). The rest of the lemma is immediate. $\square$

Remark 4.1.43. (a) Using (4.1.24), relation (4.1.60) can also be written in the variable-free form

$$(\gamma \circ \theta)^{\cdot\cdot} = \theta'(\theta' \boxdot (\ddot{\gamma} \circ \theta)) + \frac{\theta''}{\theta'}(C \circ (\gamma \circ \theta)^{\cdot}).$$

(b) For its technical interest, we present here an independent, coordinate-free proof of (4.1.62). Suppose therefore that the reparametrization given by θ is positive, i.e., $\theta'(t) > 0$ for all $t \in \bar{I}$. We shall need the flow of the Liouville vector field. By Lemma 4.1.35(i), it is given by

$$\mu^+ \colon \mathbb{R} \times TM \to TM, \ (t,v) \mapsto \mu^+(t,v) := e^t v. \tag{1}$$

We calculate the derivative of μ^+ at

$$\left(\lambda \left(\frac{d}{dr} \right)_t, w \right) \in T_t\mathbb{R} \oplus T_v TM \cong T_{(t,v)}(\mathbb{R} \times TM).$$

Using (3.1.23) we find that

$$(\mu^+)_* \left(\lambda \left(\frac{d}{dr} \right)_t, w \right) = (\mu^+)_* \left((j_v)_* \left(\lambda \left(\frac{d}{dr} \right)_t \right) + (_t j)_*(w) \right)$$

$$= \lambda(\mu^+ \circ j_v)_* \left(\frac{d}{dr} \right)_t + (\mu^+ \circ {}_t j)_*(w)$$

$$= \lambda(C \circ \mu^+)(t,v) + (\mu_t^+)_*(w)$$

$$= \lambda(C \circ \mu_t^+)(v) + (\mu_t^+)_*(w).$$

Formula (4.1.61) can be rewritten as

$$(\gamma \circ \theta)^{\cdot} = \mu^+ \circ (\ln \circ \theta', \dot{\gamma} \circ \theta) \tag{2}$$

(where μ^+ is given by (1)). Before proceeding the calculation, we note that for any smooth function $h \colon I \to \mathbb{R}$,

$$h_* \circ \frac{d}{dr} = h' \left(\frac{d}{dr} \circ h \right). \tag{3}$$

Thus we have

$$(\gamma \circ \theta)^{\cdot\cdot}(t) \overset{(3.1.8)}{=} ((\gamma \circ \theta)^{\cdot})_* \left(\frac{d}{dr} \right)_t \overset{(2)}{=} (\mu^+ \circ (\ln \circ \theta', \dot{\gamma} \circ \theta))_* \left(\frac{d}{dr} \right)_t$$

$$= \mu_*^+ \circ ((\ln \circ \theta')_*, (\dot{\gamma})_* \circ \theta_*) \left(\frac{d}{dr} \right)_t$$

$$\overset{(3)}{=} \mu_*^+ \left((\ln \circ \theta')'(t) \left(\frac{d}{dr} \right)_{\ln(\theta'(t))}, (\dot{\gamma})_* \left(\theta'(t) \left(\frac{d}{dr} \right)_{\theta(t)} \right) \right)$$

$$= \mu_*^+ \left(\frac{\theta''(t)}{\theta'(t)} \left(\frac{d}{dr} \right)_{\ln(\theta'(t))}, \theta'(t)\ddot{\tilde{\gamma}}(\theta(t)) \right)$$

$$= \frac{\theta''(t)}{\theta'(t)} (C \circ \mu_{\ln(\theta'(t))}^+)(\dot{\gamma} \circ \theta(t)) + (\mu_{\ln(\theta(t))}^+)_*(\theta'(t)\ddot{\tilde{\gamma}}(\theta(t)))$$

$$= \frac{\theta''(t)}{\theta'(t)} (C \circ (\gamma \circ \theta)^{\cdot})(t) + \tilde{\mu}_{\ln(\theta'(t))}^+ \circ (\mu_{\ln(\theta'(t))}^+)_* \circ \ddot{\tilde{\gamma}} \circ \theta(t),$$

for all $t \in \bar{I}$. This is what was to be shown. $\blacksquare$

4.1.5 The Complete Lift of a Vector Field

Definition 4.1.44. By the *complete lift* (or *tangent lift*) of a vector field X on M we mean the mapping

$$X^c := \kappa \circ X_* \colon TM \to TTM, \tag{4.1.63}$$

where κ is the canonical involution of TTM.

Lemma 4.1.45. *Let X be a vector field on a manifold M.*

(i) *The complete lift X^c of X is a vector field on TM.*

(ii) *The vector field X^c is τ-related to X, i.e.,*

$$\tau_* \circ X^c = X \circ \tau. \tag{4.1.64}$$

(iii) $\mathbf{j}X^c = \widehat{X}$.

(iv) *For every smooth function f on M,*

$$X^c f^v = (Xf)^v, \tag{4.1.65}$$

$$X^c f^c = (Xf)^c. \tag{4.1.66}$$

The second relation characterizes the complete lift of a vector field.

(v) *If $(\mathcal{U}, (u^i)_{i=1}^n)$ is a chart on M, $\left(\tau^{-1}(\mathcal{U}), ((x^i)_{i=1}^n, (y^i)_{i=1}^n)\right)$ is the induced chart on TM and $X \restriction \mathcal{U} = \sum_{i=1}^n X^i \frac{\partial}{\partial u^i}$, then*

$$X^c \underset{(\mathcal{U})}{=} \sum_{i=1}^n \left((X^i \circ \tau) \frac{\partial}{\partial x^i} + \sum_{j=1}^n y^j \left(\frac{\partial X^i}{\partial u^j} \circ \tau \right) \frac{\partial}{\partial y^i} \right)$$

$$= \sum_{i=1}^n \left((X^i)^v \frac{\partial}{\partial x^i} + (X^i)^c \frac{\partial}{\partial y^i} \right). \tag{4.1.67}$$

Proof. (i) Obviously $X^{\mathrm{c}} := \kappa \circ X_*$ is a smooth mapping. Applying Proposition 4.1.8(ii) we find that

$$\tau_{TM} \circ X^{\mathrm{c}} = \tau_{TM} \circ \kappa \circ X_* = \tau_* \circ X_* = (\tau \circ X)_* = (1_M)_* = 1_{TM},$$

hence X^{c} is a section of the tangent bundle $\tau_{TM} \colon TTM \to TM$.

(ii) This comes from Proposition 4.1.8(i):

$$\tau_* \circ X^{\mathrm{c}} = \tau_* \circ \kappa \circ X_* = \tau_{TM} \circ X_* = X \circ \tau,$$

since the diagram

$$
\begin{array}{ccc}
TM & \xrightarrow{\ X_*\ } & TTM \\
{\scriptstyle \tau}\downarrow & & \downarrow{\scriptstyle \tau_{TM}} \\
M & \xrightarrow{\ X\ } & TM
\end{array}
$$

commutes by Lemma 3.1.33.

(iii) At each point $v \in TM$,

$$(\mathbf{j}X^{\mathrm{c}})(v) = \mathbf{j}(X^{\mathrm{c}}(v)) \overset{(4.1.35)}{=} \big(v, (\tau_*)_v(X^{\mathrm{c}}(v))\big) = \big(v, (\tau_* \circ X^{\mathrm{c}})(v)\big)$$

$$\overset{(ii)}{=} \big(v, X \circ \tau(v)\big) \overset{(4.1.4)}{=} \widehat{X}(v).$$

(iv) By Lemma 3.1.49, $X^{\mathrm{c}} \underset{\tau}{\sim} X$ if, and only if,

$$X^{\mathrm{c}}(f \circ \tau) = (Xf) \circ \tau, \text{ i.e., } X^{\mathrm{c}}f^{\mathrm{v}} = (Xf)^{\mathrm{v}} \text{ for all } f \in C^{\infty}(M).$$

Thus (4.1.65) is just another expression of the τ-relatedness of X^{c} and X.

To prove (4.1.66), we apply Proposition 4.1.8(iii). Choose a vector v in TM, and let $p := \tau(v)$. Then

$$(X^{\mathrm{c}}f^{\mathrm{c}})(v) := X^{\mathrm{c}}(v)(f^{\mathrm{c}}) \overset{(4.1.63)}{=} (\kappa \circ X_*)_v(f^{\mathrm{c}}) = \kappa((X_*)_p(v))(f^{\mathrm{c}})$$

$$= (X_*)_p(v)(f^{\mathrm{c}}) \overset{(3.1.4)}{=} v(f^{\mathrm{c}} \circ X) = v(Xf) = (Xf)^{\mathrm{c}}(v),$$

whence $X^{\mathrm{c}}f^{\mathrm{c}} = (Xf)^{\mathrm{c}}$. This relation uniquely determines X^{c} by Lemma 4.1.37.

(v) Suppose that

$$X^{\mathrm{c}} \underset{(\mathcal{U})}{=} \sum_{i=1}^{n} \left(\xi^i \frac{\partial}{\partial x^i} + \xi^{n+i} \frac{\partial}{\partial y^i} \right); \quad \xi^i, \xi^{n+i} \in C^{\infty}(\tau^{-1}(\mathcal{U})).$$

Then

$$\tau_* \circ X^{\mathrm{c}} \underset{(\mathcal{U})}{\overset{(3.1.28)}{=}} \sum_{i=1}^{n} \xi^i \left(\frac{\partial}{\partial u^i} \circ \tau \right), \quad X \circ \tau \underset{(\mathcal{U})}{=} \sum_{i=1}^{n} (X^i)^{\mathrm{v}} \left(\frac{\partial}{\partial u^i} \circ \tau \right),$$

so by the τ-relatedness of X^{c} and X we obtain $\xi^i = (X^i)^{\mathrm{v}}$, $i \in J_n$. Applying X^{c} to y^i, we get

$$\xi^{n+i} = X^{\mathrm{c}}y^i = X^{\mathrm{c}}(u^i)^{\mathrm{c}} \overset{(4.1.66)}{=} (Xu^i)^{\mathrm{c}} = (X^i)^{\mathrm{c}}, \quad i \in J_n.$$

This concludes the proof. $\qquad\qquad\qquad\qquad\qquad\qquad\qquad\qquad\qquad\square$

Lemma 4.1.46. *A vector field ξ on TM is projectable, i.e., τ-related to a vector field on M (see Remark 3.1.52) if, and only if, there exists a vector field X on M such that $\xi - X^c$ is vertical.*

Proof. For a vector field X on M, the following assertions are equivalent:

$$\xi \underset{\tau}{\sim} X;$$

$$\tau_* \circ \xi = X \circ \tau;$$

$$\tau_* \circ \xi = \tau_* \circ X^c \text{ (by (4.1.64))};$$

$$\tau_* \circ (\xi - X^c) = 0;$$

$$\xi - X^c \text{ is a vertical vector field.}$$

This proves the lemma. $\qquad\qquad\qquad\qquad\qquad\qquad\qquad\qquad\qquad \square$

Remark 4.1.47. It may be checked by a direct calculation that a vector field ξ on TM is projectable if, and only if, in an induced chart $\left(\tau^{-1}(\mathcal{U}), ((x^i)_{i=1}^n, (y^i)_{i=1}^n)\right)$ it has the following coordinate expression:

$$\xi_{(\mathcal{U})} = \sum_{i=1}^n \left((X^i \circ \tau) \frac{\partial}{\partial x^i} + \xi^{n+i} \frac{\partial}{\partial y^i} \right),$$

$$X^i \in C^\infty(\mathcal{U}), \ \xi^{n+i} \in C^\infty(\tau^{-1}(\mathcal{U})); \quad i \in J_n.$$

Lemma 4.1.48. *Let $\varphi \colon M \to N$ be a smooth mapping between manifolds, and let X be a vector field on M, Y a vector field on N. Then $X \underset{\varphi}{\sim} Y$ implies that $X^c \underset{\varphi_*}{\sim} Y^c$. If, in particular, $\varphi \in \mathrm{Diff}(M)$, then*

$$(\varphi_*)_\# X^c = (\varphi_\# X)^c. \tag{4.1.68}$$

Proof. For every $h \in C^\infty(N)$,

$$X^c(h^c \circ \varphi_*) \overset{(4.1.16)}{=} X^c(h \circ \varphi)^c \overset{(4.1.66)}{=} (X(h \circ \varphi))^c$$

$$\overset{\text{Lemma } 3.1.49}{=} (Yh \circ \varphi)^c \overset{(4.1.16)}{=} (Yh)^c \circ \varphi_*,$$

so by Lemmas 3.1.49 and 4.1.37, $X^c \underset{\varphi_*}{\sim} Y^c$.

Now suppose that $\varphi \in \mathrm{Diff}(M)$. Then, by Remark 3.1.52(b), $X \underset{\varphi}{\sim} \varphi_\# X$, therefore $X^c \underset{\varphi_*}{\sim} (\varphi_\# X)^c$, and hence $(\varphi_*)_\# X^c = (\varphi_\# X)^c$. $\qquad \square$

Lemma 4.1.49. *Let X and Y be vector fields on a manifold M, and let κ be the canonical involution of TTM. Then*

$$Y_* \circ X - \kappa \circ X_* \circ Y = [X, Y]^v \circ Y, \tag{4.1.69}$$

therefore $[X, Y]$ vanishes at a point $p \in M$ if, and only if,

$$Y_* \circ X(p) = \kappa \circ X_* \circ Y(p).$$

Proof. Let p be a point in M. Obviously,

$$[X, Y]^{\mathsf{v}} \circ Y(p) \in T_{Y(p)}TM.$$

First we check that the left-hand side of (4.1.69) also results a tangent vector in $T_{Y(p)}TM$, and so the substraction is legitimate. This can be seen by an immediate calculation. On the one hand,

$$\tau_{TM} \circ Y_* \circ X(p) \overset{\text{Lemma } 3.1.33}{=} Y \circ \tau \circ X(p) = Y(p);$$

on the other hand

$$\tau_{TM} \circ \kappa \circ X_* \circ Y(p) \overset{\text{Proposition } 4.1.8}{=} (\tau \circ X)_* \circ Y(p) = Y(p),$$

as wanted.

Now we use the same argument we used twice before (see lemmas 4.1.12, 4.1.13): we show that the vectors

$$(Y_* \circ X - \kappa \circ X_* \circ Y)(p) \quad \text{and} \quad [X, Y]^{\mathsf{v}} \circ Y(p) \quad (p \in M)$$

satisfy conditions (i)–(iii) of Lemma 4.1.7, and hence they are equal. The first of these conditions holds by the calculation above.

Proof of (ii). Obviously, $\tau_*([X, Y]^{\mathsf{v}}(Y(p))) = 0_p$. Since

$$\tau_* \big(Y_*(X(p)) - \kappa \circ X_*(Y(p)) \big)$$

$$\overset{\text{Proposition } 4.1.8}{=} (\tau \circ Y)_*(X(p)) - \tau_{TM} \circ X_*(Y(p))$$

$$\overset{\text{Lemma } 3.1.33}{=} X(p) - X \circ \tau \circ Y(p) = X(p) - X(p) = 0_p,$$

we have the desired equality.

Proof of (iii). Let f be a smooth function on M. Then, on the one hand,

$$(Y_* \circ X - \kappa \circ X_* \circ Y)(p)(f^{\mathsf{c}}) = Y_*(X(p))(f^{\mathsf{c}}) - X^{\mathsf{c}}(Y(p))(f^{\mathsf{c}})$$

$$\overset{(3.1.4)}{=} X(p)(f^{\mathsf{c}} \circ Y) - X^{\mathsf{c}} f^{\mathsf{c}}(Y(p)) \overset{(4.1.66)}{=} X(p)(Yf) - (Xf)^{\mathsf{c}}(Y(p))$$

$$= X(Yf)(p) - Y(Xf)(p) = ([X, Y]f)(p).$$

On the other hand,

$$([X, Y]^{\mathsf{v}} \circ Y)(p)(f^{\mathsf{c}}) = [X, Y]^{\mathsf{v}}(Y(p))(f^{\mathsf{c}}) = ([X, Y]^{\mathsf{v}} f^{\mathsf{c}})(Y(p))$$

$$\overset{(4.1.51)}{=} ([X, Y]f)^{\mathsf{v}}(Y(p)) = ([X, Y]f) \circ \tau \circ Y(p) = ([X, Y]f(p)),$$

as desired.

This concludes the proof. $\qquad\qquad\qquad\qquad\qquad\qquad\qquad\qquad\qquad \square$

Lemma 4.1.50 (Symmetry Lemma). *Let M be a manifold, $\mathcal{U}$ an open subset of $\mathbb{R}^m$, and let $\varphi\colon \mathcal{U} \to M$ be a smooth mapping. Then the second partial derivatives $\ddot{\varphi}_{ij}$ obey the rule*

$$\ddot{\varphi}_{ij} = \kappa \circ \ddot{\varphi}_{ji}; \quad i,j \in J_m.$$

Proof. Since $\left[\frac{\partial}{\partial e^i}, \frac{\partial}{\partial e^j}\right] = 0$, by the preceding lemma we have

$$\left(\frac{\partial}{\partial e^i}\right)_* \circ \frac{\partial}{\partial e^j} = \kappa \circ \left(\frac{\partial}{\partial e^j}\right)_* \circ \frac{\partial}{\partial e^i}. \tag{$*$}$$

Then

$$\ddot{\varphi}_{ij} := (\dot{\varphi}_i)\dot{_j} \overset{(3.1.11)}{=} \left(\varphi_* \circ \frac{\partial}{\partial e^i}\right)_* \circ \frac{\partial}{\partial e^j} = \varphi_{**} \circ \left(\frac{\partial}{\partial e^i}\right)_* \circ \frac{\partial}{\partial e^j}$$

$$\overset{(*)}{=} \varphi_{**} \circ \kappa \circ \left(\frac{\partial}{\partial e^j}\right)_* \circ \frac{\partial}{\partial e^i} \overset{(4.1.27)}{=} \kappa \circ \varphi_{**} \circ \left(\frac{\partial}{\partial e^j}\right)_* \circ \frac{\partial}{\partial e^i} = \kappa \circ \ddot{\varphi}_{ji},$$

as was to be shown. $\qquad\square$

Lemma 4.1.51. *Let X be vector field on M, and consider its flow $\varphi^X\colon \mathcal{D}_X \to M$. If*

$$\widetilde{\mathcal{D}}_X := \big\{(t,v) \in \mathbb{R} \times TM \mid (t, \tau(v)) \in \mathcal{D}_X\big\},$$

then the mapping

$$\widetilde{\varphi}^X\colon \widetilde{\mathcal{D}}_X \to TM, \quad (t,v) \mapsto \widetilde{\varphi}^X(t,v) := (\varphi_t^X)_*(v)$$

is a local flow on TM whose velocity field is the complete lift of X.

Proof. Fix a point $(t,v) \in \widetilde{\mathcal{D}}_X$ and let $p := \tau(v)$. Choose a vector field Y on M such that $Y(p) = v$. (This is possible by Lemma 2.2.24(i).) Let $\frac{d}{dr} \in \mathfrak{X}(\mathbb{R})$ be the standard coordinate vector field on $\mathbb{R}$. Consider the vector fields

$$i_{\mathbb{R}}\frac{d}{dr}\colon \mathbb{R} \times M \to T(\mathbb{R} \times M), \quad (s,q) \mapsto (j_q)_*\left(\frac{d}{dr}\right)_s$$

and

$$i_M Y\colon \mathbb{R} \times M \to T(\mathbb{R} \times M), \quad (s,q) \mapsto (_s j)_* Y(q)$$

introduced in Example 3.1.53(b). Then

$$\widetilde{\varphi}_v^X(t) = \widetilde{\varphi}^X(t,v) = (\varphi_t^X)_*(v) = (\varphi^X)_* \circ (_t j)_*(v)$$

$$\overset{(3.1.49)}{=} (\varphi^X)_* \circ i_M Y(t,p) = (\varphi^X)_* \circ i_M Y \circ j_p(t).$$

Thus

$$(\widetilde{\varphi}_v^X)'(t) = \left((\varphi^X)_* \circ i_M Y \circ j_p\right)_* \left(\frac{d}{dr}\right)_t = (\varphi^X)_{**} \circ (i_M Y)_* \circ (j_p)_* \left(\frac{d}{dr}\right)_t$$

$$= (\varphi^X)_{**} \circ (i_M Y)_* \circ i_\mathbb{R} \frac{d}{dr}(t,p) \overset{(*)}{=} (\varphi^X)_{**} \circ \kappa \circ \left(i_\mathbb{R} \frac{d}{dr}\right)_* \circ i_M Y(t,p)$$

$$\overset{(4.1.27)}{=} \kappa \circ \left((\varphi^X)_* \circ i_\mathbb{R} \frac{d}{dr}\right)_* \circ i_M Y(t,p) \overset{(3.2.7)}{=} \kappa \circ \left(X \circ \varphi^X\right)_* \circ i_M Y(t,p)$$

$$= \kappa \circ X_* \circ \varphi_*^X \circ i_M Y \circ j_p(t) = X^c \circ \widetilde{\varphi}_v^X(t).$$

In step $(*)$ we applied Lemma 4.1.49 and the vanishing of $\left[i_M Y, i_\mathbb{R} \frac{d}{dr}\right]$. The latter can be seen as follows. From (3.1.48) and (3.1.50) we obtain

$$i_M Y \underset{\mathrm{pr}_1}{\sim} 0, \quad i_M Y \underset{\mathrm{pr}_2}{\sim} Y, \quad i_\mathbb{R} \frac{d}{dr} \underset{\mathrm{pr}_1}{\sim} \frac{d}{dr}, \quad i_\mathbb{R} \frac{d}{dr} \underset{\mathrm{pr}_2}{\sim} 0.$$

So, by Lemma 3.1.51(iii),

$$(\mathrm{pr}_1)_* \left[i_M Y, i_\mathbb{R} \frac{d}{dr}\right] = \left[0, \frac{d}{dr}\right] \circ \mathrm{pr}_1 = 0,$$

$$(\mathrm{pr}_2)_* \left[i_M Y, i_\mathbb{R} \frac{d}{dr}\right] = [Y, 0] \circ \mathrm{pr}_2 = 0.$$

This proves that $\left[i_M Y, i_\mathbb{R} \frac{d}{dr}\right] = 0$ and concludes the proof of Lemma 4.1.51. $\quad\square$

Lemma 4.1.52. *For any vector fields X, Y on M and any $f \in C^\infty(M)$,*

$$(X + Y)^c = X^c + Y^c, \quad (fX)^c = f^v X^c + f^c X^v; \tag{4.1.70}$$

$$[X^v, Y^c] = [X, Y]^v, \quad [X^c, Y^c] = [X, Y]^c; \tag{4.1.71}$$

$$[C, X^c] = 0. \tag{4.1.72}$$

Proof. In view of Lemma 4.1.37, it is sufficient to show that both sides of these relations act in the same way on the complete lifts of smooth functions on M. This is a routine verification. To illustrate how the calculus works, we check the second relation of (4.1.70) and (4.1.71).

Let h be a smooth function on M. Then

$$(fX)^c h^c \overset{(4.1.66)}{=} \left((fX)h\right)^c = \left(f(Xh)\right)^c \overset{\text{Lemma } 4.1.5}{=} f^c(Xh)^v + f^v(Xh)^c$$

$$\overset{(4.1.51),(4.1.66)}{=} f^c(X^v h^c) + f^v(X^c h^c) = (f^c X^v + f^v X^c)h^c,$$

whence $(fX)^c = f^c X^v + f^v X^c$. To prove the other relation, we use (4.1.66) repeatedly as follows:

$$[X^c, Y^c](h^c) = X^c(Y^c h^c) - Y^c(X^c h^c) = X^c(Yh)^c - Y^c(Xh)^c$$

$$= \left(X(Yh) - Y(Xh)\right)^c = ([X,Y]h)^c = [X,Y]^c(h^c),$$

so $[X^c, Y^c] = [X, Y]^c$. $\quad\square$

Lemma 4.1.53 (Local Frame Principle I). *Let M be an n-dimensional manifold. If $(X_i)_{i=1}^n$ is a frame field of TM over an open subset $\mathcal{U}$ of M, then $\big((X_i^{\mathrm{v}})_{i=1}^n, (X_i^{\mathrm{c}})_{i=1}^n\big)$ is a frame field of $\tau_{TM}\colon TTM \to TM$ over $\tau^{-1}(\mathcal{U})$.*

Proof. We have only to show that the family $\big((X_i^{\mathrm{v}}(v))_{i=1}^n, (X_i^{\mathrm{c}}(v))_{i=1}^n\big)$ is linearly independent at every point v of $\tau^{-1}(\mathcal{U})$. Suppose that

$$\sum_{i=1}^n \big(\lambda^i X_i^{\mathrm{v}}(v) + \mu^i X_i^{\mathrm{c}}(v)\big) = 0; \quad \lambda^i, \mu^i \in \mathbb{R}. \tag{$*$}$$

Then, for any smooth function f on M,

$$0 = \left(\sum_{i=1}^n \big(\lambda^i X_i^{\mathrm{v}}(v) + \mu^i X_i^{\mathrm{c}}(v)\big)\right)(f^{\mathrm{v}}) = \sum_{i=1}^n \big(\lambda^i (X_i^{\mathrm{v}} f^{\mathrm{v}})(v) + \mu^i (X_i^{\mathrm{c}} f^{\mathrm{v}})(v)\big)$$

$$\overset{\text{Lemma } 4.1.28}{=} \sum_{i=1}^n \mu^i (X_i^{\mathrm{c}} f^{\mathrm{v}})(v) \overset{4.1.65}{=} \sum_{i=1}^n \mu^i (X_i f)^{\mathrm{v}}(v)$$

$$= \sum_{i=1}^n \mu^i (X_i f)(\tau(v)) = \left(\sum_{i=1}^n \mu^i X_i(\tau(v))\right)(f).$$

Hence $\sum_{i=1}^n \mu^i X_i(\tau(v)) = 0$, which implies $\mu^1 = \cdots = \mu^n = 0$ by the linear independence of $\big(X_i(\tau(v))\big)_{i=1}^n$. So $(*)$ reduces to $\sum_{i=1}^n \lambda^i X_i^{\mathrm{v}}(v) = 0$. Applying both sides of this equality to f^{c}, we obtain

$$0 = \left(\sum_{i=1}^n \lambda^i X_i^{\mathrm{v}}(v)\right)(f^{\mathrm{c}}) = \sum_{i=1}^n \lambda^i (X_i^{\mathrm{v}} f^{\mathrm{c}})(v) \overset{(4.1.51)}{=} \sum_{i=1}^n \lambda^i (X_i f)^{\mathrm{v}}(v)$$

$$= \sum_{i=1}^n \lambda^i (X_i f)(\tau(v)) = \left(\sum_{i=1}^n \lambda^i X_i(\tau(v))\right)(f),$$

whence $\lambda^1 = \cdots = \lambda^n = 0$. This concludes the proof of the lemma. $\qquad\square$

Remark 4.1.54. (a) If, in particular, $(\mathcal{U}, (u^i)_{i=1}^n)$ is a chart on M with induced chart $\big(\tau^{-1}(\mathcal{U}), ((x^i)_{i=1}^n, (y^i)_{i=1}^n)\big)$ on TM, then by (4.1.67), (4.1.17) and (4.1.48),

$$\frac{\partial}{\partial x^i} = \left(\frac{\partial}{\partial u^i}\right)^{\mathrm{c}}, \quad \frac{\partial}{\partial y^i} = \left(\frac{\partial}{\partial u^i}\right)^{\mathrm{v}}. \tag{4.1.73}$$

So we obtain the induced local frame

$$\left(\left(\frac{\partial}{\partial u^i}\right)^{\mathrm{c}}_{i \in J_n}, \left(\frac{\partial}{\partial u^i}\right)^{\mathrm{v}}_{i \in J_n}\right) = \left(\left(\frac{\partial}{\partial x^i}\right)_{i \in J_n}, \left(\frac{\partial}{\partial y^i}\right)_{i \in J_n}\right)$$

on TM, consisting of coordinate vector fields.

For further references, we highlight the formulae

$$\mathbf{i}\left(\widehat{\frac{\partial}{\partial u^i}}\right) = \frac{\partial}{\partial y^i}; \ \mathbf{j}\left(\frac{\partial}{\partial x^i}\right) = \widehat{\frac{\partial}{\partial u^i}}, \ \mathbf{j}\left(\frac{\partial}{\partial y^i}\right) = 0 \quad (i \in J_n). \tag{4.1.74}$$

Here the first relation is immediate from (4.1.44); Lemma 4.1.45(iii) and (4.1.73) imply the second relation, while the third relation follows from the fact that $\mathrm{Ker}(\mathbf{j}) = \mathfrak{X}^\mathsf{v}(TM)$ by Corollary 4.1.21.

(b) The preceding lemma provides a simple and efficient method for some tensorial constructions on TM. Namely, in order to define a covariant or a type $(1, s)$ $(s \geq 1)$ tensor field on TM, it is sufficient to specify its action on the vertical and complete lifts of vector fields on M.

Example 4.1.55. (a) **Vertical lift of one-forms.** The *vertical lift* of a one-form $\alpha \in \mathcal{A}_1(M)$ is the one-form $\alpha^\mathsf{v} \in \mathcal{A}_1(TM)$ such that

$$\alpha^\mathsf{v}(X^\mathsf{v}) = 0, \ \alpha^\mathsf{v}(X^\mathsf{c}) = (\alpha(X))^\mathsf{v}; \quad X \in \mathfrak{X}(M). \tag{4.1.75}$$

We note that α^v is just the pull-back of α by the tangent bundle projection τ:

$$\alpha^\mathsf{v} = \tau^*\alpha. \tag{4.1.76}$$

Indeed, for any vector field X on M, and for every $v \in TM$,

$$\tau^*\alpha(X^\mathsf{v})(v) = (\tau^*\alpha)_v(X^\mathsf{v}(v)) \overset{(3.3.7)}{=} \alpha_{\tau(v)}(\tau_*(X^\mathsf{v}(v))) = 0 = \alpha^\mathsf{v}(X^\mathsf{v})(v),$$

$$\tau^*\alpha(X^\mathsf{c})(v) = (\tau^*\alpha)_v(X^\mathsf{c}(v)) = \alpha_{\tau(v)}(\tau_* \circ X^\mathsf{c}(v)) \overset{(4.1.64)}{=} \alpha_{\tau(v)}(X(\tau(v)))$$
$$= \alpha(X) \circ \tau(v) = (\alpha(X))^\mathsf{v}(v),$$

whence our claim.

For a coordinate description, choose a chart $(\mathcal{U}, (u^i)_{i=1}^n)$ on M and consider the induced chart $\left(\tau^{-1}(\mathcal{U}), ((x^i)_{i=1}^n, (y^i)_{i=1}^n)\right)$ on TM. Suppose that

$$\alpha \restriction \mathcal{U} = \sum_{i=1}^n \alpha_i du^i, \ \alpha^\mathsf{v} \restriction \tau^{-1}(\mathcal{U}) = \sum_{i=1}^n \left(\widetilde{\alpha}_i dx^i + \widetilde{\alpha}_{n+i} dy^i\right).$$

Then

$$0 =: \alpha^\mathsf{v}\left(\frac{\partial}{\partial u^i}\right)^\mathsf{v} \overset{(4.1.73)}{=} \alpha^\mathsf{v}\left(\frac{\partial}{\partial y^i}\right) = \widetilde{\alpha}_{n+i},$$

$$(\alpha_i)^\mathsf{v} = \left(\alpha\left(\frac{\partial}{\partial u^i}\right)\right)^\mathsf{v} =: \alpha^\mathsf{v}\left(\frac{\partial}{\partial u^i}\right)^\mathsf{c} \overset{(4.1.73)}{=} \alpha^\mathsf{v}\left(\frac{\partial}{\partial x^i}\right) = \widetilde{\alpha}_i,$$

therefore

$$\alpha^\mathsf{v} \underset{(\mathcal{U})}{=} \sum_{i=1}^n (\alpha_i)^\mathsf{v} dx^i. \tag{4.1.77}$$

This calculation also shows that the one-form α^v satisfying (4.1.75) indeed exists.

(b) **Vertical lift of type $(0,2)$ tensors.** By the *vertical lift* of a type $(0,2)$ tensor $B \in \mathcal{T}_2^0(M)$ we mean the type $(0,2)$ tensor $B^v \in \mathcal{T}_2^0(TM)$ given by

$$
\begin{cases}
B^v(\xi, \eta) := 0 & \text{if } \xi \text{ or } \eta \text{ is vertical;} \\
B^v(X^c, Y^c) := (B(X,Y))^v; & X, Y \in \mathfrak{X}(M).
\end{cases}
\tag{4.1.78}
$$

If, locally, $B \underset{(\mathcal{U})}{=} \sum_{i,j=1}^n B_{ij} du^i \otimes du^j$, then

$$
B^v \underset{(\mathcal{U})}{=} \sum_{i,j=1}^n (B_{ij} \circ \tau) dx^i \otimes dx^j.
\tag{4.1.79}
$$

Indeed,

$$
B^v \left(\frac{\partial}{\partial x^i}, \frac{\partial}{\partial y^j} \right) = B^v \left(\frac{\partial}{\partial y^i}, \frac{\partial}{\partial x^j} \right) = B^v \left(\frac{\partial}{\partial y^i}, \frac{\partial}{\partial y^j} \right) := 0,
$$

$$
B^v \left(\frac{\partial}{\partial x^i}, \frac{\partial}{\partial x^j} \right) = B^v \left(\left(\frac{\partial}{\partial u^i} \right)^c, \left(\frac{\partial}{\partial u^j} \right)^c \right) = \left(B \left(\frac{\partial}{\partial u^i}, \frac{\partial}{\partial u^j} \right) \right)^v = B_{ij} \circ \tau.
$$

Now we show that for every one-form $\alpha \in \mathcal{A}_1(M)$,

$$
d(\alpha^v) = (d\alpha)^v.
\tag{4.1.80}
$$

Note first that $d(\alpha^v)$ also vanishes if one of its arguments is vertical. Indeed, for example

$$
d(\alpha^v)(X^v, Y^c) = X^v \alpha^v(Y^c) - Y^c \alpha^v(X^v) - \alpha^v([X^v, Y^c])
$$

$$
\overset{(4.1.71)}{=} X^v(\alpha(Y))^v - \alpha^v([X,Y]^v) = 0.
$$

Thus $d(\alpha^v)$ is determined by its action on pairs of the form (X^c, Y^c), where X and Y are vector fields on M. Since

$$
d(\alpha^v)(X^c, Y^c) = X^c \alpha^v(Y^c) - Y^c \alpha^v(X^c) - \alpha^v([X^c, Y^c])
$$

$$
\overset{(4.1.71)}{=} X^c(\alpha(Y))^v - Y^c(\alpha(X))^v - \alpha^v([X,Y]^c)
$$

$$
\overset{(4.1.65)}{=} \left(X\alpha(Y) - Y\alpha(X) - \alpha([X,Y]) \right)^v
$$

$$
= (d\alpha(X,Y))^v = (d\alpha)^v(X^c, Y^c),
$$

we indeed have (4.1.80).

(c) **Vertical lift of type $(1,1)$ tensors.** Let

$$
A \in \mathcal{T}_1^1(M) \cong \text{End}(\mathfrak{X}(M)).
$$

The *vertical lift* $A^v \in \mathcal{T}_1^1(TM) \cong \text{End}(\mathfrak{X}(TM))$ of A is defined by

$$A^v(X^v) = 0, \ A^v(X^c) = (A(X))^v, \ X \in \mathfrak{X}(M). \tag{4.1.81}$$

The components of A and A^v with respect to the charts chosen above are given by

$$A\left(\frac{\partial}{\partial u^j}\right) = \sum_{i=1}^{n} A_j^i \frac{\partial}{\partial u^i},$$

and

$$A^v\left(\frac{\partial}{\partial x^j}\right) = \sum_{i=1}^{n} \left(\widetilde{A}_j^i \frac{\partial}{\partial x^i} + \widetilde{A}_j^{n+i} \frac{\partial}{\partial y^i}\right),$$

$$A^v\left(\frac{\partial}{\partial y^j}\right) = \sum_{i=1}^{n} \left(\widetilde{B}_j^i \frac{\partial}{\partial x^i} + \widetilde{B}_j^{n+i} \frac{\partial}{\partial y^i}\right), \ j \in J_n.$$

Then $A^v\left(\frac{\partial}{\partial u^j}\right)^v = A^v\left(\frac{\partial}{\partial y^j}\right) := 0$ implies that $\widetilde{B}_j^i = \widetilde{B}_j^{n+i} = 0 \ (j \in J_n)$. By the second condition in (4.1.81),

$$A^v\left(\frac{\partial}{\partial x^j}\right) = A^v\left(\frac{\partial}{\partial u^j}\right)^c = \left(A\left(\frac{\partial}{\partial u^j}\right)\right)^v = \left(\sum_{i=1}^{n} A_j^i \frac{\partial}{\partial u^i}\right)^v$$

$$\overset{\text{Lemma } 4.1.32}{=} \sum_{i=1}^{n} (A_j^i)^v \frac{\partial}{\partial y^i},$$

therefore $\widetilde{A}_j^i = 0$, $\widetilde{A}_j^{n+i} = (A_j^i)^v \ (i,j \in J_n)$ and hence

$$A^v\left(\frac{\partial}{\partial x^j}\right) = \sum_{i=1}^{n} (A_j^i \circ \tau)\left(\frac{\partial}{\partial y^i}\right), \ A^v\left(\frac{\partial}{\partial y^j}\right) = 0 \ (j \in J_n).$$

In particular, the vertical lift of the identity transformation $1_{\mathfrak{X}(M)}$ in $\text{End}(\mathfrak{X}(M))$ is the tensor field $(1_{\mathfrak{X}(M)})^v \in \text{End}(\mathfrak{X}(TM))$ given by

$$(1_{\mathfrak{X}(M)})^v(X^v) = 0, \ (1_{\mathfrak{X}(M)})^v(X^c) = X^v, \ X \in \mathfrak{X}(M). \tag{4.1.82}$$

4.1.6 The Vertical Endomorphism of TTM

Definition 4.1.56. By the *vertical endomorphism* of TTM we mean the strong bundle map

$$\mathbf{J} := \mathbf{i} \circ \mathbf{j} \colon TTM \to TM \times_M TM \to TTM. \tag{4.1.83}$$

Remark 4.1.57. We collect some elementary properties of the vertical endomorphism which come immediately from the definition and from our previous results.

(a) Corollary 4.1.21 and Lemma 4.1.24 imply that

$$\mathrm{Im}(\mathbf{J}) = \mathrm{Im}(\mathbf{i}) = \mathrm{Ker}(\mathbf{j}) = \mathrm{Ker}(\mathbf{J}) = VTM. \tag{4.1.84}$$

Since $\mathbf{j} \circ \mathbf{i} \stackrel{(4.1.40)}{=} 0$, we have

$$\mathbf{J}^2 = 0. \tag{4.1.85}$$

(b) In the same way as the strong bundle maps $\mathbf{i}$ and $\mathbf{j}$, the vertical endomorphism may also be interpreted as a $C^\infty(TM)$-linear transformation of $\mathfrak{X}(TM)$:

$$\mathbf{J} \in T_1^0\big(\mathfrak{X}(TM), \mathfrak{X}(TM)\big) = \mathrm{End}(\mathfrak{X}(TM)) \cong T_1^1(TM).$$

In this interpretation (4.1.84) takes the form

$$\mathrm{Im}(\mathbf{J}) = \mathrm{Im}(\mathbf{i}) = \mathrm{Ker}(\mathbf{j}) = \mathrm{Ker}(\mathbf{J}) = \mathfrak{X}^\mathrm{v}(TM). \tag{4.1.86}$$

From this it follows that

$$\xi \in \mathfrak{X}(TM) \ \textit{is vertical if, and only if,} \ \mathbf{J}\xi = 0.$$

Since for any vector field X on M, $X^\mathrm{v} \stackrel{(4.1.44)}{=} \mathbf{i}\widehat{X}$, and $\mathbf{j}X^\mathrm{c} = \widehat{X}$ by Lemma 4.1.45(iii), we obtain

$$\mathbf{J}X^\mathrm{v} = \mathbf{i} \circ \mathbf{j} \circ \mathbf{i}(\widehat{X}) = 0, \ \mathbf{J}X^\mathrm{c} = \mathbf{i} \circ \mathbf{j}(X^\mathrm{c}) = \mathbf{i}\widehat{X} = X^\mathrm{v},$$

i.e.,

$$\mathbf{J}X^\mathrm{v} = 0, \ \mathbf{J}X^\mathrm{c} = X^\mathrm{v} \ \text{for all} \ X \in \mathfrak{X}(M). \tag{4.1.87}$$

Comparing (4.1.87) and (4.1.82), we conclude that *the vertical endomorphism of TTM is just the vertical lift of the identity transformation* $1_{\mathfrak{X}(M)} \in \mathrm{End}(\mathfrak{X}(M))$, that is, $\mathbf{J} = (1_{\mathfrak{X}(M)})^\mathrm{v}$.

(c) For every smooth curve γ in M,

$$\mathbf{J} \circ \ddot{\gamma} = C \circ \dot{\gamma}. \tag{4.1.88}$$

Indeed, this is an immediate consequence of (4.1.37).

Lemma 4.1.58. *For any vector field X on M,*

$$[\mathbf{J}, X^\mathrm{v}] = [\mathbf{J}, X^\mathrm{c}] = 0. \tag{4.1.89}$$

Furthermore we have

$$[\mathbf{J}, C] = \mathbf{J}. \tag{4.1.90}$$

Proof. By Remark 4.1.54(b), it is sufficient to check that both sides of these relations take the same values on the vertical and complete lifts of vector fields on M. We leave the verification of (4.1.89) as an easy exercise for the reader. To show (4.1.90), let X be any vector field on M. Then

$$[\mathbf{J}, C]X^c \overset{(3.3.16)}{=} [\mathbf{J}X^c, C] - \mathbf{J}[X^c, C] \overset{(4.1.87),(4.1.72)}{=} [X^v, C] \overset{(4.1.53)}{=} X^v = \mathbf{J}X^c.$$

Similarly,

$$[\mathbf{J}, C]X^v = [\mathbf{J}X^v, C] - \mathbf{J}[X^v, C] = -\mathbf{J}X^v = 0 = \mathbf{J}X^v,$$

thus (4.1.90) is proved. $\qquad\qquad\square$

Lemma 4.1.59. *The Nijenhuis tensor of the vertical endomorphism vanishes.*

Proof. Since $N_{\mathbf{J}}$ is skew-symmetric, it is enough to check that

$$N_{\mathbf{J}}(X^v, Y^v) = N_{\mathbf{J}}(X^v, Y^c) = N_{\mathbf{J}}(X^c, Y^c) = 0$$

for any vector fields X, Y on M. This can be done by a straightforward calculation. For example,

$$N_{\mathbf{J}}(X^c, Y^c) \overset{(3.3.18)}{=} [\mathbf{J}X^c, \mathbf{J}Y^c] + \mathbf{J}^2[X^c, Y^c] - \mathbf{J}[\mathbf{J}X^c, Y^c] - \mathbf{J}[X^c, \mathbf{J}Y^c]$$

$$\overset{(4.1.85),(4.1.87)}{=} [X^v, Y^v] - \mathbf{J}[X^v, Y^c] - \mathbf{J}[X^c, Y^v]$$

$$\overset{(4.1.52),(4.1.71)}{=} -\mathbf{J}[X, Y]^v - \mathbf{J}[X, Y]^v \overset{(4.1.87)}{=} 0. \qquad\qquad\square$$

Lemma 4.1.60. *A vector field ξ on TM is*

(i) *the* vertical lift *of a vector field on M if, and only if,*

$$\mathbf{J}\xi = 0 \ \text{and} \ [\mathbf{J}, \xi] = 0;$$

(ii) *the* complete lift *of a vector field on M if, and only if,*

$$[C, \xi] = 0 \ \text{and} \ [\mathbf{J}, \xi] = 0.$$

Proof. The *necessity* of the conditions is clear from (4.1.87), (4.1.89) and (4.1.72). Since the question is local, we will show the *sufficiency* of the conditions on the domain of an induced chart of TM. So we choose a chart $(\mathcal{U}, (u^i)_{i=1}^n)$ on M, and consider the chart $(\tau^{-1}(\mathcal{U}), ((x^i)_{i=1}^n, (y^i)_{i=1}^n))$ given by (3.1.14) on TM. To shorten our formulae, in the forthcoming calculations we will use the summation convention introduced in Remark 1.1.12(d). Let

$$\xi := \xi^i \frac{\partial}{\partial x^i} + \xi^{n+i} \frac{\partial}{\partial y^i} \in \mathfrak{X}(\tau^{-1}(\mathcal{U})).$$

First we determine the coordinate expression of the vector field $[C, \xi]$ and the components of the type $(1,1)$ tensor field $[\mathbf{J}, \xi]$:

$$[C, \xi] \overset{(4.1.73)}{=} \left[C, \xi^i \left(\frac{\partial}{\partial u^i} \right)^{\mathrm{c}} \right] + \left[C, \xi^{n+i} \left(\frac{\partial}{\partial u^i} \right)^{\mathrm{v}} \right]$$

$$\overset{(4.1.53),(4.1.72)}{=} (C\xi^i) \left(\frac{\partial}{\partial u^i} \right)^{\mathrm{c}} + (C\xi^{n+i}) \left(\frac{\partial}{\partial u^i} \right)^{\mathrm{v}} - \xi^{n+i} \frac{\partial}{\partial y^i}$$

$$= (C\xi^i) \frac{\partial}{\partial x^i} + (-\xi^{n+i} + C\xi^{n+i}) \frac{\partial}{\partial y^i},$$

$$[\mathbf{J}, \xi] \left(\frac{\partial}{\partial x^j} \right) = \left[\mathbf{J} \frac{\partial}{\partial x^j}, \xi \right] - \mathbf{J} \left[\frac{\partial}{\partial x^j}, \xi \right] \overset{(4.1.74)}{=} \left[\frac{\partial}{\partial y^j}, \xi \right]$$

$$- \mathbf{J} \left[\frac{\partial}{\partial x^j}, \xi^i \frac{\partial}{\partial x^i} + \xi^{n+i} \frac{\partial}{\partial y^i} \right] = \frac{\partial \xi^i}{\partial y^j} \frac{\partial}{\partial x^i} + \left(\frac{\partial \xi^{n+i}}{\partial y^j} - \frac{\partial \xi^i}{\partial x^j} \right) \frac{\partial}{\partial y^i},$$

$$[\mathbf{J}, \xi] \left(\frac{\partial}{\partial y^j} \right) = -\mathbf{J} \left[\frac{\partial}{\partial y^j}, \xi^i \frac{\partial}{\partial x^i} + \xi^{n+i} \frac{\partial}{\partial y^i} \right] = -\frac{\partial \xi^i}{\partial y^j} \frac{\partial}{\partial y^i}.$$

Summing up, we have the following:

$$[C, \xi] = (C\xi^i) \frac{\partial}{\partial x^i} + (-\xi^{n+i} + C\xi^{n+i}) \frac{\partial}{\partial y^i}, \tag{4.1.91}$$

$$[\mathbf{J}, \xi] \left(\frac{\partial}{\partial x^j} \right) = \frac{\partial \xi^i}{\partial y^j} \frac{\partial}{\partial x^i} + \left(\frac{\partial \xi^{n+i}}{\partial y^j} - \frac{\partial \xi^i}{\partial x^j} \right) \frac{\partial}{\partial y^i}, \tag{4.1.92}$$

$$[\mathbf{J}, \xi] \left(\frac{\partial}{\partial y^j} \right) = -\frac{\partial \xi^i}{\partial y^j} \frac{\partial}{\partial y^i} \quad (j \in J_n). \tag{4.1.93}$$

Now suppose that $\mathbf{J}\xi = \xi^i \frac{\partial}{\partial y^i} = 0$ and $[\mathbf{J}, \xi] = 0$. Then, taking into account (4.1.92),

$$\xi^i = 0 \text{ and } \frac{\partial \xi^{n+i}}{\partial y^j} = 0 \text{ for all } i, j \in J_n.$$

From the second relation we conclude that

$$\xi^{n+i} = X^i \circ \tau; \quad X^i \in C^\infty(\mathcal{U}), \ i \in J_n.$$

Thus $\xi = X^{\mathrm{v}}$ if $X := X^i \frac{\partial}{\partial u^i}$. This proves the sufficiency in assertion (i).

Next, we turn to assertion (ii). We assume that $[C, \xi] = 0$ and $[\mathbf{J}, \xi] = 0$. Then (4.1.93) implies that $\xi^i = X^i \circ \tau = (X^i)^{\mathrm{v}}$; $X^i \in C^\infty(\mathcal{U})$, $i \in J_n$. Hence $\frac{\partial \xi^i}{\partial y^j} = 0$, $\frac{\partial \xi^i}{\partial x^j} = \frac{\partial X^i}{\partial u^j} \circ \tau$, therefore

$$\xi^{n+i} \overset{(4.1.91)}{=} y^j \frac{\partial \xi^{n+i}}{\partial y^j} \overset{(4.1.92)}{=} y^j \frac{\partial \xi^i}{\partial x^j} = y^j \left(\frac{\partial X^i}{\partial u^j} \circ \tau \right) \overset{(4.1.17)}{=} (X^i)^{\mathrm{c}}.$$

If $X := X^i \frac{\partial}{\partial u^i}$, then we find

$$\xi = (X^i)^{\mathrm{v}} \frac{\partial}{\partial x^i} + (X^i)^{\mathrm{c}} \frac{\partial}{\partial y^i} \overset{(4.1.67)}{=} X^{\mathrm{c}},$$

as was to be shown. $\qquad \square$

Corollary 4.1.61. *For a vector field ξ on TM, the following are equivalent:*

(i) ξ *is projectable.*

(ii) $\mathbf{J}\xi$ *is the vertical lift of a vector field on M.*

(iii) $[\mathbf{J}, \mathbf{J}\xi] = 0$.

Proof. The following assertions are equivalent:

(i) ξ is projectable.

(i_1) There is a vector field X on M such that $\xi - X^c \in \mathfrak{X}^v(TM)$ (Lemma 4.1.46).

(i_2) $\mathbf{J}(\xi - X^c) = 0$ (by (4.1.86)).

(ii) $\mathbf{J}\xi = X^v$ (by (4.1.87)), i.e., $\mathbf{J}\xi$ is the vertical lift of a vector field on M.

If $\mathbf{J}\xi$ is a vertical lift, then $[\mathbf{J}, \mathbf{J}\xi] = 0$ by (4.1.89), so we have (iii). Conversely, if $[\mathbf{J}, \mathbf{J}\xi] = 0$, then $\mathbf{J}\xi$ is a vertical lift by Lemma 4.1.60(i), since $\mathbf{J}(\mathbf{J}\xi) \overset{(4.1.85)}{=} 0$. So (iii) implies (ii), which completes the proof. $\quad\square$

Example 4.1.62. In applications the graded derivations $i_{\mathbf{J}}$ and $\mathcal{L}_{\mathbf{J}}$ of the Grassmann algebra $\mathcal{A}(TM)$ associated to the vertical endomorphism $\mathbf{J} \in \mathcal{A}_1^1(TM) = \mathrm{End}(\mathfrak{X}(TM))$ play an important role.

(a) By Remark 3.3.40, $i_{\mathbf{J}}$ is the unique graded derivation of degree 0 of $\mathcal{A}(TM)$ such that

$$i_{\mathbf{J}}F = 0 \text{ and } i_{\mathbf{J}}dF = dF \circ \mathbf{J} \text{ for all } F \in C^\infty(TM). \qquad (4.1.94)$$

If $\omega \in \mathcal{A}_k(TM)$, then by (3.3.38) we have

$$i_{\mathbf{J}}\omega(\xi_1, \dots, \xi_k) = \sum_{i=1}^{k} \omega(\xi_1, \dots, \mathbf{J}\xi_i, \dots, \xi_k), \qquad (4.1.95)$$

where $\xi_1, \dots, \xi_k \in \mathfrak{X}(TM)$.

(b) By Definition 3.3.41,

$$\mathcal{L}_{\mathbf{J}} = [i_{\mathbf{J}}, d] = i_{\mathbf{J}} \circ d - d \circ i_{\mathbf{J}}. \qquad (4.1.96)$$

Thus for any smooth function F on TM,

$$\mathcal{L}_{\mathbf{J}}F = dF \circ \mathbf{J} \in \mathcal{A}_1(TM), \qquad (4.1.97)$$

and $\mathcal{L}_{\mathbf{J}}$ is determined by this relation (cf. (3.3.43) and see Lemma 3.3.43). If $\left(\tau^{-1}(\mathcal{U}), ((x^i)_{i=1}^n, (y^i)_{i=1}^n)\right)$ is an induced chart on TM, then

$$\mathcal{L}_{\mathbf{J}}F\left(\frac{\partial}{\partial x^i}\right) \overset{(4.1.97)}{=} dF\left(\mathbf{J}\frac{\partial}{\partial x^i}\right) = dF\left(\frac{\partial}{\partial y^i}\right) = \frac{\partial F}{\partial y^i},$$

$$\mathcal{L}_{\mathbf{J}}F\left(\frac{\partial}{\partial y^i}\right) = dF\left(\mathbf{J}\frac{\partial}{\partial y^i}\right) = 0 \quad (i \in J_n),$$

so we have

$$\mathcal{L}_{\mathbf{J}}F \underset{(\mathcal{U})}{=} \sum_{i=1}^{n} \frac{\partial F}{\partial y^i} dx^i. \qquad (4.1.98)$$

(c) Let α be a one-form on M, and consider its vertical lift $\alpha^{\vee}$ on TM given by (4.1.78). Then

$$i_{\mathbf{J}}\alpha^{\vee} = 0, \quad \mathcal{L}_{\mathbf{J}}\alpha^{\vee} = 0. \qquad (4.1.99)$$

Indeed, the first relation is clear from the definition $\alpha^{\vee}$ (see (4.1.78)) and from (4.1.95). The second relation can also be readily seen:

$$\mathcal{L}_{\mathbf{J}}\alpha^{\vee} \overset{(4.1.96)}{=} i_{\mathbf{J}}d\alpha^{\vee} - di_{\mathbf{J}}\alpha^{\vee} = i_{\mathbf{J}}d\alpha^{\vee} \overset{(4.1.80)}{=} i_{\mathbf{J}}(d\alpha)^{\vee} \overset{(4.1.78),(4.1.95)}{=} 0.$$

Corollary 4.1.63. *We have the following relations:*

$$[i_C, i_{\mathbf{J}}] := i_C \circ i_{\mathbf{J}} - i_{\mathbf{J}} \circ i_C = 0, \qquad (4.1.100)$$

$$[i_{\mathbf{J}}, \mathcal{L}_C] := i_{\mathbf{J}} \circ \mathcal{L}_C - \mathcal{L}_C \circ i_{\mathbf{J}} = i_{\mathbf{J}}, \qquad (4.1.101)$$

$$[i_C, \mathcal{L}_{\mathbf{J}}] := i_C \circ \mathcal{L}_{\mathbf{J}} + \mathcal{L}_{\mathbf{J}} \circ i_C = i_{\mathbf{J}}, \qquad (4.1.102)$$

$$[\mathcal{L}_{\mathbf{J}}, \mathcal{L}_C] := \mathcal{L}_{\mathbf{J}} \circ \mathcal{L}_C - \mathcal{L}_C \circ \mathcal{L}_{\mathbf{J}} = \mathcal{L}_{\mathbf{J}}, \qquad (4.1.103)$$

$$\mathcal{L}_{\mathbf{J}}^2 = \frac{1}{2}[\mathcal{L}_{\mathbf{J}}, \mathcal{L}_{\mathbf{J}}] = 0, \qquad (4.1.104)$$

$$[i_{\mathbf{J}}, \mathcal{L}_{\mathbf{J}}] = i_{\mathbf{J}} \circ \mathcal{L}_{\mathbf{J}} - \mathcal{L}_{\mathbf{J}} \circ i_{\mathbf{J}} = 0. \qquad (4.1.105)$$

Proof. Taking into account that $\mathbf{J}C = 0$ and $[\mathbf{J}, C] \overset{(4.1.90)}{=} \mathbf{J}$, the first four relations are immediate consequences of (3.3.48), (3.3.49), (3.3.50) and (3.3.53), respectively. Applying Corollary 3.3.46 and Lemma 4.1.59, we find that $[\mathcal{L}_{\mathbf{J}}, \mathcal{L}_{\mathbf{J}}] = \mathcal{L}_{[\mathbf{J}, \mathbf{J}]} = 0$ whence (4.1.104).

To prove (4.1.105), note first that $[[i_{\mathbf{J}}, \mathcal{L}_{\mathbf{J}}], d] = 0$. Indeed, by the graded Jacobi identity (1.3.8),

$$0 = (-1)^{1\cdot1}[d, [i_{\mathbf{J}}, \mathcal{L}_{\mathbf{J}}]] + (-1)^{0\cdot1}[i_{\mathbf{J}}, [\mathcal{L}_{\mathbf{J}}, d]] + (-1)^{1\cdot0}[\mathcal{L}_{\mathbf{J}}, [d, i_{\mathbf{J}}]],$$

whence

$$[d, [i_{\mathbf{J}}, \mathcal{L}_{\mathbf{J}}]] = [i_{\mathbf{J}}, [\mathcal{L}_{\mathbf{J}}, d]] + [\mathcal{L}_{\mathbf{J}}, [d, i_{\mathbf{J}}]] \overset{(3.3.52),(3.3.51)}{=} -[\mathcal{L}_{\mathbf{J}}, \mathcal{L}_{\mathbf{J}}] \overset{(4.1.104)}{=} 0.$$

Thus, by Lemma 3.3.43, $[i_{\mathbf{J}}, \mathcal{L}_{\mathbf{J}}]$ is determined by its action on the smooth functions on TM. We show that

$$[i_{\mathbf{J}}, \mathcal{L}_{\mathbf{J}}](F) = 0 \quad \text{for all} \quad F \in C^{\infty}(TM).$$

Since $i_{\mathbf{J}}F = 0$ by (3.3.37), we have $[i_{\mathbf{J}}, \mathcal{L}_{\mathbf{J}}](F) = i_{\mathbf{J}}(\mathcal{L}_{\mathbf{J}}F)$. For every vector field ξ on TM,

$$i_{\mathbf{J}}(\mathcal{L}_{\mathbf{J}}F)(\xi) \overset{(3.3.38)}{=} \mathcal{L}_{\mathbf{J}}F(\mathbf{J}\xi) \overset{(3.3.43)}{=} dF(\mathbf{J}^2\xi) \overset{(4.1.85)}{=} 0,$$

therefore $[i_{\mathbf{J}}, \mathcal{L}_{\mathbf{J}}](F) = 0$. This concludes the proof of (4.1.105), and hence, of the corollary. $\qquad\square$

4.1.7 Push-forwards

Lemma 4.1.64. *If φ is a smooth transformation of M, then*

$$\varphi_{**} \circ \mathbf{i} = \mathbf{i} \circ (\varphi_* \times \varphi_*), \tag{4.1.106}$$

$$(\varphi_* \circ \varphi_*) \circ \mathbf{j} = \mathbf{j} \circ \varphi_{**}, \tag{4.1.107}$$

$$\varphi_{**} \circ \mathbf{J} = \mathbf{J} \circ \varphi_{**}. \tag{4.1.108}$$

Proof. To prove (4.1.106), let $(u,v) \in TM \times_M TM$, and consider the curve $\alpha \colon t \in \mathbb{R} \mapsto u + tv \in TM$. Then, for all $t \in \mathbb{R}$,

$$(\varphi_* \circ \alpha)(t) = \varphi_*(u + tv) = \varphi_*(u) + t\varphi_*(v).$$

By Remark 4.1.17,

$$(\varphi_*(v))^\uparrow(\varphi_*(u)) = (\varphi_* \circ \alpha)^{\cdot}(0), \quad v^\uparrow(u) = \dot{\alpha}(0),$$

therefore

$$\varphi_{**} \circ \mathbf{i}(u,v) \overset{(4.1.38)}{=} \varphi_{**}(v^\uparrow(u)) = \varphi_{**}(\dot{\alpha}(0)) \overset{(3.1.11)}{=} (\varphi_* \circ \alpha)^{\cdot}(0)$$

$$= (\varphi_*(v))^\uparrow(\varphi_*(u)) = \mathbf{i} \circ (\varphi_* \times \varphi_*)(u,v).$$

This is what was to be shown.

The second equality can be obtained by a straightforward calculation:

$$(\varphi_* \times \varphi_*) \circ \mathbf{j} \overset{(4.1.35)}{=} (\varphi_* \circ \tau_{TM}, \varphi_* \circ \tau_*) = (\varphi_* \circ \tau_{TM}, (\varphi \circ \tau)_*)$$

$$\overset{\text{Lemma 3.1.33}}{=} (\tau_{TM} \circ \varphi_{**}, (\tau \circ \varphi_*)_*) = (\tau_{TM}, \tau_*) \circ \varphi_{**} = \mathbf{j} \circ \varphi_{**}.$$

The third equality follows from the first two:

$$\varphi_{**} \circ \mathbf{J} = \varphi_{**} \circ \mathbf{i} \circ \mathbf{j} \overset{(4.1.106)}{=} \mathbf{i} \circ (\varphi_* \times \varphi_*) \circ \mathbf{j} \overset{(4.1.107)}{=} \mathbf{i} \circ \mathbf{j} \circ \varphi_{**} = \mathbf{J} \circ \varphi_{**}. \;\square$$

Corollary and Definition 4.1.65. *Let φ be a diffeomorphism of M. Then the mapping*

$$\varphi_* \times \varphi_* \colon TM \times_M TM \to TM \times_M TM, \;\; (u,v) \mapsto (\varphi_*(u), \varphi_*(v))$$

is an automorphism of the Finsler bundle π which induces the mapping φ_* on the base manifold:

$$
\begin{array}{ccc}
TM \times_M TM & \xrightarrow{\;\varphi_* \times \varphi_*\;} & TM \times_M TM \\
\pi \downarrow & & \downarrow \pi \\
TM & \xrightarrow{\quad\varphi_*\quad} & TM
\end{array}
.
$$

If $\widetilde{X}$ is a section of π, then

$$\varphi_\# \widetilde{X} := (\varphi_* \times \varphi_*) \circ \widetilde{X} \circ \varphi_*^{-1} \tag{4.1.109}$$

is also a section of π, called the push-forward *of $\widetilde{X}$ by φ.*

Proof. The assertions follow from Proposition 2.2.19 and Remark 2.2.32(b). □

We note that the precise notation and terminology would be $(\varphi_* \times \varphi_*)_\# \widetilde{X}$ and 'push-forward of $\widetilde{X}$ by $\varphi_* \times \varphi_*$'.

Lemma 4.1.66. *If $\varphi \in \mathrm{Diff}(M)$, then for every vector field X on M,*

$$\varphi_\# \widehat{X} = \widehat{\varphi_\# X}. \tag{4.1.110}$$

Further, we have

$$\varphi_\# \widetilde{\delta} = \widetilde{\delta}. \tag{4.1.111}$$

Proof. These follow immediately from the definitions. For example,

$$\varphi_\# \widehat{X} := (\varphi_* \times \varphi_*) \circ \widehat{X} \circ \varphi_*^{-1} \overset{(4.1.4)}{=} (\varphi_* \times \varphi_*) \circ (1_{TM}, X \circ \tau) \circ \varphi_*^{-1}$$

$$= (\varphi_*, \varphi_* \circ X \circ \tau) \circ \varphi_*^{-1} = (1_{TM}, \varphi_* \circ X \circ \tau \circ \varphi_*^{-1})$$

$$\overset{\text{Lemma 3.1.33}}{=} (1_{TM}, \varphi_* \circ X \circ \varphi^{-1} \circ \tau) \overset{(3.1.43)}{=} (1_{TM}, \varphi_\# X \circ \tau) =: \widehat{\varphi_\# X}. \quad \square$$

Corollary 4.1.67. *If $\varphi \in \mathrm{Diff}(M)$, then*

$$(\varphi_*)_\# C = C; \tag{4.1.112}$$

$$(\varphi_*)_\# X^{\mathsf{v}} = (\varphi_\# X)^{\mathsf{v}}, \quad X \in \mathfrak{X}(M); \tag{4.1.113}$$

$$(\varphi_*)_\# \circ \mathbf{i} = \mathbf{i} \circ \varphi_\#; \tag{4.1.114}$$

$$\varphi_\# \circ \mathbf{j} = \mathbf{j} \circ (\varphi_*)_\#; \tag{4.1.115}$$

$$(\varphi_*)_\# \circ \mathbf{J} = \mathbf{J} \circ (\varphi_*)_\#. \tag{4.1.116}$$

Proof. First we verify (4.1.114) and (4.1.115). Let $\widetilde{X} \in \Gamma(\pi)$, and let ξ be a vector field on TM. Then

$$(\varphi_*)_\# \mathbf{i} \widetilde{X} \overset{(3.1.43)}{=} \varphi_{**} \circ \mathbf{i} \circ \widetilde{X} \circ \varphi_*^{-1} \overset{(4.1.106)}{=} \mathbf{i} \circ (\varphi_* \times \varphi_*) \circ \widetilde{X} \circ \varphi_*^{-1}$$

$$:= \mathbf{i}(\varphi_\# \widetilde{X}),$$

whence $(\varphi_*)_\# \circ \mathbf{i} = \mathbf{i} \circ \varphi_\#$. Similarly,

$$\varphi_\# (\mathbf{j}\xi) \overset{(4.1.109)}{=} (\varphi_* \times \varphi_*) \circ \mathbf{j} \circ \xi \circ \varphi_*^{-1} \overset{(4.1.108)}{=} \mathbf{j} \circ \varphi_{**} \circ \xi \circ \varphi_*^{-1}$$

$$\overset{(3.1.43)}{=} \mathbf{j} \circ (\varphi_*)_\#(\xi),$$

so we have (4.1.115). Relation (4.1.116) follows immediately from (4.1.114) and (4.1.115). Finally,

$$(\varphi_*)_\# C = (\varphi_*)_\#(\mathbf{i} \circ \widetilde{\delta}) \overset{(4.1.114)}{=} \mathbf{i} \circ \varphi_\# \widetilde{\delta} \overset{(4.1.111)}{=} \mathbf{i} \circ \widetilde{\delta} = C;$$

$$(\varphi_*)_\# X^{\mathsf{v}} = (\varphi_*)_\#(\mathbf{i} \circ \widehat{X}) \overset{(4.1.114)}{=} \mathbf{i} \circ \varphi_\# \widehat{X} \overset{(4.1.110)}{=} \mathbf{i} \widehat{\varphi_\# X} = (\varphi_\# X)^{\mathsf{v}}. \quad \square$$

4.2   Homogeneity

4.2.1   *Homogeneous Mappings of Vector Spaces*

Throughout this subsection V and W are non-trivial finite-dimensional real vector spaces.

Remark 4.2.1. Given a real number t, we denote by μ_t^+ the positive dilation

$$v \in V \mapsto \mu_t^+(v) := e^t v \in V;$$

cf. Example 3.1.38(a). We say that a subset A of V is *conic* if $\mu_t^+(A) \subset A$ for all $t \in \mathbb{R}$. Obviously, the vector space V itself and $V \setminus \{0\}$ are conic subsets of V.

Definition 4.2.2. Let A be a conic subset of the vector space V. A mapping $f: A \to W$ is called *positive-homogeneous of degree* σ, where σ is any real number, if

$$f \circ \mu_t^+ = e^{\sigma t} f \quad \text{for all } t \in \mathbb{R}. \tag{4.2.1}$$

Then, for brevity, we also say that f is σ^+-*homogeneous*.

Lemma 4.2.3 (Euler's Theorem on Homogeneous Mappings). *Let $\mathcal{U}$ be an open conic subset of V. A differentiable mapping $f: \mathcal{U} \to W$ is positive-homogeneous of degree σ if, and only if, for each point p of $\mathcal{U}$*

$$f'(p)(p) = \sigma f(p). \tag{4.2.2}$$

Proof.   Suppose first that f is positive-homogeneous of degree σ. Then, for every $p \in \mathcal{U}$,

$$f'(p)(p) \overset{(\text{C.1.3})}{=} \lim_{t \to 0} \frac{f(p + tp) - f(p)}{t} = \lim_{t \to 0} \frac{f(e^{\ln(1+t)}p) - f(p)}{t}$$

$$\overset{(4.2.1)}{=} \left(\lim_{t \to 0} \frac{e^{\sigma \ln(1+t)} - 1}{t} \right) f(p) = (t \mapsto e^{\sigma \ln(1+t)})'(0) f(p) = \sigma f(p),$$

as desired.

Conversely, suppose that (4.2.2) holds for each $p \in \mathcal{U}$. For some fixed point $p_0 \in \mathcal{U}$, consider the initial value problem

$$x' = \sigma x, \quad x(0) = f(p_0) \tag{$*$}$$

in W. We obtain from elementary calculus that the unique solution to $(*)$ is the curve

$$\alpha \colon \mathbb{R} \to W, \ t \mapsto \alpha(t) := e^{\sigma t} f(p_0).$$

However, the curve

$$\beta \colon \mathbb{R} \to W, \ t \mapsto \beta(t) := f(e^t p_0)$$

is also a solution to $(*)$: $\beta(0) = f(p_0)$, and for every $t \in \mathbb{R}$,

$$\beta'(t) = \lim_{s \to 0} \frac{\beta(t+s) - \beta(t)}{s} = \lim_{s \to 0} \frac{f(e^{t+s}p_0) - f(e^t p_0)}{s}$$

$$= (s \mapsto f(e^s p_0))'(t) = f'(e^t p_0)(e^t p_0) \overset{(4.2.2)}{=} \sigma f(e^t p_0) = \sigma \beta(t).$$

Thus we conclude that $\beta = \alpha$, i.e.,

$$f(e^t p) = e^{\sigma t} f(p) \quad \text{for all } (t,p) \in \mathbb{R} \times \mathcal{U},$$

as required. $\qquad\square$

Lemma 4.2.4. *Let $\mathcal{U}$ be an open conic subset of V, and let $f \colon \mathcal{U} \to W$ be a differentiable mapping. If f is positive-homogeneous of degree σ, then its derivative $f' \colon \mathcal{U} \to L(V, W)$ is positive-homogeneous of degree $\sigma - 1$.*

Proof. With a fixed $t \in \mathbb{R}$, take the derivative of both sides of (4.2.1). Since μ_t^+ is linear, we have

$$(\mu_t^+)'(p) = \mu_t^+, \quad p \in \mathcal{U}$$

(Example C.1.4). Thus, by the chain rule, we get

$$f'(\mu_t^+(p)) \circ \mu_t^+ = e^{\sigma t} f'(p), \quad p \in \mathcal{U}.$$

Compose both sides of this equality with μ_{-t}^+ on the right. Using the linearity of $f'(p)$, we find that

$$f' \circ \mu_t^+ = e^{(\sigma-1)t} f',$$

as was to be shown. $\qquad\square$

Lemma 4.2.5. *Let a mapping $f \colon V \to W$ be given.*

(i) *If f is continuous at the origin and positive-homogeneous of degree 0, then f is a constant mapping.*

(ii) *If f is of class C^1 and positive-homogeneous of degree 1, then f is a linear mapping, namely $f = f'(0)$.*

(iii) *Suppose in particular that $W = \mathbb{R}$. If $f \colon V \to \mathbb{R}$ is of class C^2 and 2^+-homogeneous, then it is a quadratic form on V.*

Proof.   (i) Since f is 0^+-homogeneous on V, for every $(t, v) \in \mathbb{R} \times V$ we have

$$f(e^t v) = f(v). \qquad (*)$$

Thus, by the continuity of f at $0 \in V$,

$$f(0) = \lim_{t \to -\infty} f(e^t v) \overset{(*)}{=} \lim_{t \to -\infty} f(v) = f(v),$$

for all $v \in V$. This proves that f is constant.

(ii) Since f is 1^+-homogeneous and of class C^1, Euler's theorem applies and gives

$$f(p) = f'(p)(p), \quad p \in V. \qquad (**)$$

The mapping $f' \colon V \to L(V, W)$ is continuous and, by the preceding lemma, it is 0^+-homogeneous. Thus, by part (i), f' is constant. Hence

$$f(p) \overset{(**)}{=} f'(p)(p) = f'(0)(p) \quad \text{for all } p \in V,$$

which proves our second assertion.

(iii) Arguing as above,

$f' \colon V \to L(V, \mathbb{R}) = V^*$ is a 1^+-homogeneous C^1 mapping,
$f'' \colon V \to L(V, L(V, \mathbb{R})) \cong L^2(V, \mathbb{R})$ is a 0^+-homogeneous continuous mapping.

Thus, by part (i), the mapping f'' is constant, namely $f''(p) = f''(0)$ for all $p \in V$. By Euler's theorem,

$$f'(p)(p) = f''(p)(p, p), \quad 2f(p) = f'(p)(p), \quad p \in V.$$

Putting these together, we find that

$$f(p) = \frac{1}{2} f'(p)(p) = \frac{1}{2} f''(p)(p, p) = \frac{1}{2} f''(0)(p, p),$$

which concludes the proof. $\square$

Lemma 4.2.6. *Let $f \colon V \setminus \{0\} \to W$ be a mapping of class C^k. Suppose that f is σ^+-homogeneous, where $\sigma \geq k + 1$. Then the extension*

$$\bar{f} \colon V \to W, \ p \mapsto \bar{f}(p) := \begin{cases} f(p) \in W & \text{if } p \in V \setminus \{0\} \\ 0 \in W & \text{if } p = 0 \end{cases} \qquad (*)$$

of f onto V is of class C^k.

Proof. We equip V with a norm $\|\ \|$. To prove the lemma, we use induction on $k \in \mathbb{N}$.

Step 1. $k = 0$. Then f is continuous and σ^+-homogeneous, where $\sigma \geq 1$. It is sufficient to verify that the extension $\bar{f}$ of f is continuous at $0 \in V$. To see this, choose a sequence $(a_n)_{n\in\mathbb{N}}$ of nonzero points in V such that $\lim_{n\to\infty} a_n = 0$. Then

$$\lim_{n\to\infty} \bar{f}(a_n) = \lim_{n\to\infty} f\left(\frac{1}{\|a_n\|}\|a_n\|a_n\right) = \lim_{n\to\infty}\left(\|a_n\|^\sigma f\left(\frac{1}{\|a_n\|}a_n\right)\right) = 0,$$

because $\lim_{n\to\infty} \|a_n\|^\sigma = 0$, and f is continuous and hence bounded on the compact subset $\{v \in V \mid \|v\| = 1\}$ of V. This proves the continuity of $\bar{f}$ at 0.

Step 2. Let $l \in \mathbb{N}^*$, and suppose that the assertion is true for any positive integer $k < l$. Let $f \colon V \setminus \{0\} \to W$ be of class C^l and σ^+-homogeneous with $\sigma \geq l + 1$. Then the extension $\bar{f}$ of f is differentiable at 0 and $\bar{f}'(0) = 0$. Indeed,

$$\lim_{v\to 0} \frac{\bar{f}(v) - \bar{f}(0)}{\|v\|} = \lim_{v\to 0} \frac{f(v)}{\|v\|} = \lim_{v\to 0} \frac{1}{\|v\|} f\left(\|v\|\frac{1}{\|v\|}v\right)$$

$$= \lim_{v\to 0} \|v\|^{\sigma - 1} f\left(\frac{1}{\|v\|}v\right) = 0,$$

by the same reason as above. Thus the extension of f' onto V according to $(*)$ is just the derivative of $\bar{f}$. Since f' is $(\sigma - 1)^+$-homogeneous by Lemma 4.2.4, and $\sigma - 1 \geq l$, it follows by the induction hypothesis that the extension of f' onto V is of class $l - 1$, and hence $\bar{f}$ is of class C^l.

This concludes the proof. $\qquad\qquad\qquad\qquad\qquad\qquad\qquad\qquad\square$

4.2.2 Homogeneous Functions on TM

Remark 4.2.7. Let M be a manifold. Consider the tangent manifold TM of M, and let $\mathcal{V} \subset TM$ be a subset such that $\tau(\mathcal{V}) = M$.

(a) We say that $\mathcal{V}$ is *dilation invariant* if $\mu_\lambda(\mathcal{V}) = \mathcal{V}$ for any nonzero real number λ, i.e.,

if $v \in \mathcal{V}$ then $\lambda v \in \mathcal{V}$ for all $\lambda \in \mathbb{R}^*$.

The set $\mathring{T}M$ of nonzero tangent vectors to M is an immediate example of a dilation invariant subset of TM.

(b) A dilation of $\mathcal{V}$ by a positive factor can always be written in the form μ_{e^t}, $t \in \mathbb{R}$. We continue to use the notation

$$\mu_t^+ := \mu_{e^t}, \quad t \in \mathbb{R} \tag{4.2.3}$$

introduced in Remark 4.2.1 for positive dilations in TM as well. We say that $\mathcal{V}$ is *conic* if it is invariant under positive dilations, i.e., $\mu_t^+(\mathcal{V}) \subset \mathcal{V}$ for all $t \in \mathbb{R}$.

Definition 4.2.8. Let F be a real-valued function with domain $\mathcal{V} \subset TM$ such that $\tau(\mathcal{V}) = M$.

(a) Suppose that $\mathcal{V}$ is dilation invariant, and let k be an integer. Then F is called *projectively homogeneous of degree k* if

$$F \circ \mu_t = t^k F \quad \text{for all } t \in \mathbb{R}^*. \tag{4.2.4}$$

(b) If $\mathcal{V}$ is a conic subset of TM, then the function $F\colon \mathcal{V} \to \mathbb{R}$ is said to be *positive-homogeneous of degree σ*, where σ is any real number, if

$$F \circ \mu_t^+ = e^{\sigma t} F \quad \text{for all } t \in \mathbb{R}. \tag{4.2.5}$$

Lemma 4.2.9 (Euler's Theorem on the Tangent Bundle). *Let $\mathcal{V}$ be an open conic subset of TM. A C^1 function $F\colon \mathcal{V} \to \mathbb{R}$ is positive-homogeneous of degree σ if, and only if, $CF = \sigma F$, where C is the Liouville vector field.*

Proof. By Lemma 4.1.35(i), the Liouville vector field is generated by the global flow

$$\mu^+\colon \mathbb{R} \times TM \to TM, \quad (t, v) \mapsto \mu^+(t, v) := \mu_t^+(v) := e^t v.$$

Choose and fix a vector $v \in \mathcal{V}$, and let

$$\alpha_v\colon \mathbb{R} \to \mathcal{V}, \quad t \mapsto \alpha_v(t) := e^t v$$

be its flow line. Write, for brevity, $F_{\tau(v)} := F \restriction T_{\tau(v)} M$. Then

$$CF(v) \overset{(3.2.8)}{=} \lim_{t \to 0} \frac{F \circ \mu_t^+(v) - F(v)}{t} = \lim_{t \to 0} \frac{F_{\tau(v)}(\alpha_v(t)) - F_{\tau(v)}(\alpha_v(0))}{t}$$

$$= (F_{\tau(v)} \circ \alpha_v)'(0) \overset{(C.1.4)}{=} (F_{\tau(v)})'(\alpha_v(0))(\alpha_v'(0)) = F_{\tau(v)}'(v)(v)$$

and so the assertion follows from the classical Euler theorem (Lemma 4.2.3). $\qquad \square$

Lemma 4.2.10. *Let F be a real-valued function on the total manifold of the tangent bundle $\tau\colon TM \to M$.*

(i) *If F is continuous on $o(M) = \{0_p \in T_p M \mid p \in M\}$ and positive-homogeneous of degree 0, then F is fibrewise constant.*

(ii) *If F is of class C^1 and positive-homogeneous of degree 1, then F is fibrewise linear, i.e., $F \restriction T_p M \in (T_p M)^*$ for all $p \in M$.*

(iii) *If F is of class C^2 and positive-homogeneous of degree 2, then F restricts to a quadratic form on every fibre.*

This result is merely a rephrasing of Lemma 4.2.5 in the context of a tangent bundle.

Lemma 4.2.11. *Let $F: \mathring{T}M \to \mathbb{R}$ be a σ^+-homogeneous continuous function, where $\sigma \geq 1$. Then the extension*

$$\bar{F}: TM \to \mathbb{R}, \ v \mapsto \bar{F}(v) := \begin{cases} F(v) & \text{if } v \in \mathring{T}M \\ 0 & \text{if } v \in o(M) \end{cases} \tag{$*$}$$

of F onto TM is continuous. If, in particular, F is of class C^1 on $\mathring{T}M$ and $\sigma \geq 2$, then $\bar{F}$ is also of class C^1 on TM.

Proof. Choose a chart $(\mathcal{U}, u)$ on M with induced chart $(\tau^{-1}(\mathcal{U}), (x, y))$ on TM. Then the mapping

$$(x, y): \tau^{-1}(\mathcal{U}) \to u(\mathcal{U}) \times \mathbb{R}^n \subset \mathbb{R}^n \times \mathbb{R}^n$$

is a vector bundle isomorphism, so the function

$$F \circ (x, y)^{-1}: u(\mathcal{U}) \times \mathbb{R}^n \to \tau^{-1}(\mathcal{U}) \to \mathbb{R}$$

is σ^+-homogeneous in its second variable. Thus we may reduce the problem to the case where

(i) $M \subset \mathbb{R}^n$ is an open subset, $TM = M \times \mathbb{R}^n$;
(ii) $F: \mathring{T}M = M \times (\mathbb{R}^n \setminus \{0\}) \to \mathbb{R}$ is a continuous function which is σ^+-homogeneous in its second variable.

Let p be any point in M. We show that $\bar{F}$ is continuous at $(p, 0)$. To do this, choose a norm $\| \ \|$ on $\mathbb{R}^n$, and let $S := \{v \in \mathbb{R}^n \mid \|v\| = 1\}$ be the unit sphere with respect to this norm. Suppose that $\mathcal{U}$ is a compact neighbourhood of p. Let $(p_n, v_n)_{n \in \mathbb{N}}$ be a sequence such that

$$(p_n, v_n) \in \mathcal{U} \times (\mathbb{R}^n \setminus \{0\}) \quad \text{for all } n \in \mathbb{N}$$

and $\lim_{n \to \infty} p_n = p$, $\lim_{n \to \infty} v_n = 0$. Then

$$\lim_{n \to \infty} \bar{F}(p_n, v_n) = \lim_{n \to \infty} F\left(p_n, \|v_n\| \frac{1}{\|v_n\|} v_n\right)$$

$$= \lim_{n \to \infty} \|v_n\|^{\sigma} F\left(p_n, \frac{1}{\|v_n\|} v_n\right) = 0$$

because $\lim_{n \to \infty} \|v_n\|^{\sigma} = 0$ and F is continuous and hence bounded on the compact set $\mathcal{U} \times S$. This proves that $\bar{F}$ is continuous at $(p, 0)$.

Now suppose that F is of class C^1 and σ^+-homogeneous with $\sigma \geq 2$. We show that $\bar{F}$ has continuous partial derivatives; then it is of class C^1 by Proposition C.1.17.

For every $i \in J_n$, the function $D_i F$ is σ^+-homogeneous, while the function $D_{n+i} F$ is $(\sigma - 1)^+$-homogeneous (cf. Lemma 4.2.4). Thus, by the previous part, $D_i F$ and $D_{n+i} F$ have continuous extensions $\overline{D_i F}$ and $\overline{D_{n+i} F}$, given by $(*)$. To finish the proof, we check that

$$\overline{D_i F} = D_i \bar{F}, \quad \overline{D_{n+i} F} = D_{n+i} \bar{F}, \quad i \in J_n.$$

These equalities evidently hold at the points of $M \times (\mathbb{R}^n \setminus \{0\})$. Furthermore, we find that

$$D_i \bar{F}(p, 0) = \lim_{t \to 0} \frac{\bar{F}(p + te_i, 0) - \bar{F}(p, 0)}{t} \overset{(*)}{=} 0 = \overline{D_i F}(p, 0),$$

$$D_{n+i} \bar{F}(p, 0) = \lim_{t \to 0} \frac{\bar{F}(p, te_i) - \bar{F}(p, 0)}{t} = \lim_{t \to 0^+} \frac{t^\sigma F(p, e_i)}{t}$$

$$= 0 = \overline{D_{n+i} F}(p, 0),$$

as desired. $\square$

4.2.3   Homogeneous Vector Fields on TM

Definition 4.2.12. Let $\mathcal{V}$ be an open subset of TM such that $\tau(\mathcal{V}) = M$, and let ξ be a vector field on $\mathcal{V}$.

(a) Suppose that $\mathcal{V}$ is dilation invariant, and let k be an integer. We say that ξ is *projectively homogeneous of degree* k if

$$\xi \circ \mu_t = \widetilde{\mu}_{t^{k-1}} \circ (\mu_t)_* \circ \xi \quad \text{for all } t \in \mathbb{R}^*. \tag{4.2.6}$$

(b) If $\mathcal{V}$ is a conic subset of TM, then the vector field $\xi \colon \mathcal{V} \to T\mathcal{V}$ is said to be *positive-homogeneous of degree* σ, where σ is any real number, if

$$\xi \circ \mu_t^+ = \widetilde{\mu}_{(\sigma-1)t}^+ \circ (\mu_t^+)_* \circ \xi \quad \text{for all } t \in \mathbb{R}. \tag{4.2.7}$$

Remark 4.2.13. With the help of the push-forward of ξ by μ_t $(t \in \mathbb{R}^*)$, resp. μ_t^+ $(t \in \mathbb{R})$, relations (4.2.6) and (4.2.7) can be written more concisely in the form

$$(\mu_t)_\# \xi = \widetilde{\mu}_{t^{1-k}} \circ \xi, \quad \text{resp.} \quad (\mu_t^+)_\# \xi = \widetilde{\mu}_{(1-\sigma)t}^+ \circ \xi. \tag{4.2.8}$$

Lemma 4.2.14 (Euler's Theorem on Homogeneous Vector Fields). *In order that a vector field $\xi \colon \mathring{T}M \to T\mathring{T}M$ is positive-homogeneous of degree σ $(\sigma \in \mathbb{R})$ is necessary and sufficient that $[C, \xi] = (\sigma - 1)\xi$.*

Proof. Suppose first that ξ has the homogeneity property (4.2.7). Then, applying Lemma 4.1.35(i) and (3.2.11), we find

$$[C, \xi] = \lim_{t \to 0} \frac{1}{t}((\mu^+_{-t})_\# \xi - \xi) \overset{(4.2.8)}{=} \left(\lim_{t \to 0} \frac{e^{(\sigma-1)t} - 1}{t} \right) \xi = (\sigma - 1)\xi,$$

as was to be shown.

Conversely, suppose that $[C, \xi] = (\sigma - 1)\xi$. Arguing as in the proof of Lemma 4.2.9, let $v \in \overset{\circ}{T}M$ be an arbitrarily fixed vector, and consider the initial value problem

$$x' = (1 - \sigma)x, \quad x(0) = \xi(v) \tag{$*$}$$

on the vector space $T_v \overset{\circ}{T} M$. It is obvious that the curve

$$f \colon \mathbb{R} \to T_v \overset{\circ}{T} M, \ t \mapsto f(t) := e^{(1-\sigma)t}\xi(v)$$

is a solution of $(*)$. On the other hand, consider the curve

$$h \colon \mathbb{R} \to T_v \overset{\circ}{T} M, \ t \mapsto h(t) := (\mu^+_t)_\# \xi(v).$$

Then $h(0) = \xi(v)$ and, for each $t \in \mathbb{R}$,

$$h'(t) = \lim_{s \to 0} \frac{1}{s}(h(t + s) - h(t)) = \lim_{s \to 0} \frac{1}{s}((\mu^+_{t+s})_\# \xi(v) - (\mu^+_t)_\# \xi(v))$$

$$= \lim_{s \to 0} \frac{1}{s}((\mu^+_t)_\#(\mu^+_s)_\# \xi(v) - (\mu^+_t)_\# \xi(v))$$

$$= -(\mu^+_t)_\# \lim_{s \to 0} \frac{1}{s}(\xi - (\mu^+_s)_\# \xi)(v)$$

$$\overset{(3.2.11)}{=} -(\mu^+_t)_\# [C, \xi](v) \overset{\text{cond.}}{=} (1 - \sigma)(\mu^+_t)_\# \xi(v) = (1 - \sigma)h(t)$$

so h is also a solution of the initial value problem $(*)$. Thus the uniqueness of solutions implies $f = h$, and hence the second relation in (4.2.8), which was to be proved. $\qquad\square$

Corollary 4.2.15. *If ξ and η are positive-homogeneous vector fields on $\overset{\circ}{T}M$ of degree σ and ϱ, respectively, then their Lie bracket is positive-homogeneous of degree $\sigma + \varrho - 1$.*

Proof. Applying the Jacobi identity and the above lemma, we obtain that

$$[C, [\xi, \eta]] = [[C, \xi], \eta] + [\xi, [C, \eta]] = (\sigma - 1)[\xi, \eta] + (\varrho - 1)[\xi, \eta]$$

$$= (\sigma + \varrho - 2)[\xi, \eta],$$

whence our assertion. $\qquad\square$

Example 4.2.16. (a) Let X be a vector field on M. Since

$$[C, X^{\vee}] \overset{(4.1.53)}{=} -X^{\vee}, \quad [C, X^{\mathsf{c}}] \overset{(4.1.72)}{=} 0,$$

the vertical lift of X is positive-homogeneous of degree 0, the complete lift of X is positive-homogeneous of degree 1. Evidently, the Liouville vector field itself is positive-homogeneous of degree 1.

(b) Let $(\mathcal{U}, (u^i)_{i=1}^n)$ be a chart on M, and consider the induced chart $(\tau^{-1}(\mathcal{U}), ((x^i)_{i=1}^n, (y^i)_{i=1}^n))$ on TM. Since

$$Cx^i = C(u^i)^{\vee} = 0, \quad Cy^i = \left(\sum_{j=1}^n y^j \frac{\partial}{\partial y^j} \right) y^i = y^i,$$

the coordinate functions

$$x^i \text{ are positive-homogeneous of degree } 0,$$
$$y^i \text{ are positive-homogeneous of degree } 1$$

by Lemma 4.2.9. In view of (4.1.73), $\frac{\partial}{\partial x^i} = \left(\frac{\partial}{\partial u^i} \right)^{\mathsf{c}}$, $\frac{\partial}{\partial y^i} = \left(\frac{\partial}{\partial u^i} \right)^{\vee}$, so it follows that the coordinate vector fields

$$\frac{\partial}{\partial x^i} \text{ are positive-homogeneous of degree } 1,$$
$$\frac{\partial}{\partial y^i} \text{ are positive-homogeneous of degree } 0.$$

Now consider a vector field ξ on TM and let

$$\xi \underset{(\mathcal{U})}{=} \xi^i \frac{\partial}{\partial x^i} + \xi^{n+i} \frac{\partial}{\partial y^i}$$

(summation convention is in force). If ξ is positive-homogeneous of degree σ, then the component functions

$$\left. \begin{array}{l} \xi^i \text{ are positive-homogeneous of degree } \sigma - 1 \\ \xi^{n+i} \text{ are positive-homogeneous of degree } \sigma \end{array} \right\}, \qquad (*)$$

and the converse is also true: if the component functions of ξ with respect to any induced chart on TM have the homogeneity properties $(*)$, then ξ is positive-homogeneous of degree σ. Indeed,

$$[C, \xi] \underset{(\mathcal{U})}{=} (C\xi^i) \frac{\partial}{\partial x^i} + \xi^i \left[C, \frac{\partial}{\partial x^i} \right] + (C\xi^{n+i}) \frac{\partial}{\partial y^i} + \xi^{n+i} \left[C, \frac{\partial}{\partial y^i} \right]$$

$$= (C\xi^i) \frac{\partial}{\partial x^i} + (C\xi^{n+i} - \xi^{n+i}) \frac{\partial}{\partial y^i},$$

so $[C, \xi] = (\sigma - 1)\xi$ if, and only if, locally we have

$$C\xi^i = (\sigma - 1)\xi^i, \quad C\xi^{n+i} = \sigma \xi^{n+i}, \quad i \in J_n.$$

Chapter 5

Sprays and Lagrangians

5.1 Sprays and the Exponential Map

5.1.1 Second-order Vector Fields and Some of Their Mutants

Definition 5.1.1. A section S of the tangent bundle $\tau_{TM} \colon TTM \to TM$ is said to be a *second-order vector field of class C^r over M* (where $r \in \mathbb{N}^*$) if

$$S \in C^r(TM, TTM) \text{ and } \mathbf{J}S = C \text{ (or, equivalently, } \mathbf{j}S = \widetilde{\delta}).$$

If $S \in \mathfrak{X}(TM)$ and $\mathbf{J}S = C$, we simply say that S is a *second-order vector field* over M. By a *second-order vector field on $\mathring{T}M$* we mean a vector field $\mathring{S} \in \mathfrak{X}(\mathring{T}M)$ satisfying $\mathbf{J}\mathring{S} = C \restriction \mathring{T}M$.

Corollary 5.1.2. *If S is a second-order vector field on $\mathring{T}M$, then*

$$[i_S, i_\mathbf{J}] = i_S \circ i_\mathbf{J} - i_\mathbf{J} \circ i_S = i_C, \tag{5.1.1}$$

$$[i_S, \mathcal{L}_\mathbf{J}] = i_S \circ \mathcal{L}_\mathbf{J} + \mathcal{L}_\mathbf{J} \circ i_S = \mathcal{L}_C + i_{[\mathbf{J},S]}. \tag{5.1.2}$$

Proof. Since $\mathbf{J}S = C$, these relations are immediate consequences of (3.3.48) and (3.3.50), respectively. $\qquad\square$

Lemma 5.1.3. *A vector field of class C^r ($r \geq 1$) on TM is a second-order vector field if, and only if, it is a section also of the vector bundle $\tau_* \colon TTM \to TM$.*

Proof. Let $\xi \colon TM \to TTM$ be a vector field of class C^r ($r \geq 1$). The following assertions are equivalent:

(1) ξ is a second-order vector field, i.e., $\mathbf{J}\xi = C$.

(2) $\mathbf{j}\xi = \widetilde{\delta}$ (since $\mathbf{J} = \mathbf{i} \circ \mathbf{j}$, $C = \mathbf{i}\widetilde{\delta}$ and $\mathbf{i}$ is injective).

(3) For any vector $v \in TM$, $\mathbf{j}(\xi(v)) = (v, v)$.

(4) For any vector $v \in TM$, $\big(\tau_{TM}(\xi(v)), \tau_*(\xi(v))\big) = (v, v)$ (by (4.1.35)).

(5) For any vector $v \in TM$, $(v, \tau_*\xi(v)) = (v, v)$ (since ξ is a section of $\tau_{TM}\colon TTM \to TM$).

(6) $\tau_* \circ \xi = 1_{TM}$, i.e., ξ is a section of the bundle $\tau_*\colon TTM \to TM$. $\qquad\square$

Proposition 5.1.4. *A vector field on TM is a second-order vector field if, and only if, each of its integral curves is the canonical lift of a curve in M.*

Proof. Let $S \in \mathfrak{X}(TM)$, and suppose first that S is second-order. If $c\colon I \to TM$ is an integral curve of S, then by the previous lemma

$$\tau_* \circ \dot{c} = \tau_* \circ S \circ c = c.$$

Hence, by Lemma 3.1.40, c is a canonical lift, namely, it is the canonical lift of $\tau \circ c$.

Conversely, suppose that every integral curve c of S is a canonical lift. Then, taking into account Lemma 3.1.40 again,

$$c = \tau_* \circ \dot{c} = \tau_* \circ S \circ c. \qquad (*)$$

By Theorem 3.2.13, for each $v \in TM$ there exists a (unique, maximal) integral curve c_v of S such that $c_v(0) = v$. So we have

$$v = c_v(0) \overset{(*)}{=} \tau_* \circ S \circ c_v(0) = \tau_* \circ S(v), \quad v \in TM,$$

whence $\tau_* \circ S = 1_{TM}$, thus concluding the proof. $\qquad\square$

Proposition 5.1.5. *A vector field $S \in \mathfrak{X}(TM)$ is a second-order vector field if, and only if, $\kappa \circ S = S$, where κ is the canonical involution of TTM.*

Proof. Suppose first that $\kappa \circ S = S$. Then, applying Proposition 4.1.8(i),

$$\tau_* \circ S = \tau_* \circ \kappa \circ S = \tau_{TM} \circ S = 1_{TM},$$

so S is a second-order vector field by Lemma 5.1.3.

Conversely, let S be a second-order vector field on TM. Then $\kappa \circ S$ is a vector field on TM, since

$$\tau_{TM} \circ \kappa \circ S \overset{\text{Proposition 4.1.8(ii)}}{=} \tau_* \circ S = 1_{TM}.$$

For any smooth function f on M and any vector $v \in TM$,

$$[(\kappa \circ S)f^c](v) = (\kappa \circ S(v))f^c = \kappa(S(v))(f^c)$$

$$\overset{\text{Proposition 4.1.8(iii)}}{=} S(v)(f^c) = (Sf^c)(v),$$

therefore $(\kappa \circ S)(f^c) = S(f^c)$, $f \in C^\infty(M)$. By Lemma 4.1.37 this implies that $\kappa \circ S = S$. $\qquad\square$

Corollary 5.1.6. *If* $S\colon TM \to TTM$ *is a second-order vector field and* $\varphi \in \mathrm{Diff}(M)$, *then*

$$(\varphi_*)_\# S = \varphi_{**} \circ S \circ \varphi_*^{-1}$$

is also a second-order vector field.

Proof. According to the previous proposition, we only have to show that $\kappa \circ ((\varphi_*)_\# S) = (\varphi_*)_\# S$. However, this is true, since

$$\kappa \circ ((\varphi_*)_\# S) = \kappa \circ \varphi_{**} \circ S \circ \varphi_*^{-1} \overset{\text{Lemma 4.1.13}}{=} \varphi_{**} \circ \kappa \circ S \circ \varphi_*^{-1}$$
$$= \varphi_{**} \circ S \circ \varphi_*^{-1} = (\varphi_*)_\# S. \qquad \square$$

Lemma 5.1.7. *Let* $S \in \mathfrak{X}(TM)$ *be a second-order vector field. Then*

$$S f^{\mathsf{v}} = f^{\mathsf{c}} \text{ for all } f \in C^\infty(M).$$

Proof. By Lemma 5.1.3, we have $\tau_* \circ S = 1_{TM}$. Thus, for every $v \in TM$,

$$f^{\mathsf{c}}(v) = v(f) = \tau_* S(v)(f) \overset{(3.1.4)}{=} S(v)(f \circ \tau) = (S f^{\mathsf{v}})(v),$$

whence $S f^{\mathsf{v}} = f^{\mathsf{c}}$. $\qquad \square$

Remark 5.1.8. Let S be a second-order vector field of class C^r $(r \geq 1)$ over M. The coordinate expression of S with respect to an induced chart $\left(\tau^{-1}(\mathcal{U}), ((x^i)_{i=1}^n, (y^i)_{i=1}^n)\right)$ on TM is of the form

$$S \underset{(\mathcal{U})}{=} \sum_{i=1}^n \left(y^i \frac{\partial}{\partial x^i} - 2G^i \frac{\partial}{\partial y^i} \right), \tag{5.1.3}$$

where $G^i \in C^r(\tau^{-1}(\mathcal{U}))$, $i \in J_n$. Indeed, since S is a vector field of class C^r on TM, we have the local expression

$$S \underset{(\mathcal{U})}{=} \sum_{i=1}^n \left(S^i \frac{\partial}{\partial x^i} + S^{n+i} \frac{\partial}{\partial y^i} \right),$$

with some C^r functions S^i, S^{n+i} $(i \in J_n)$ defined on $\tau^{-1}(\mathcal{U})$. Now the condition $\mathbf{J}S = C$ implies that $S^i = y^i$ $(i \in J_n)$, and introducing the notation

$$G^i := -\frac{1}{2} S^{n+i},$$

which is traditional in Finsler geometry, we obtain (5.1.3). The functions G^i are called the *coefficients* of S or the *forces* defined by S with respect to the chart given on M.

Lemma 5.1.9 (Grifone's Identity). *If S is a second-order vector field over a manifold M, then we have*

$$\mathbf{J}[\mathbf{J}\xi, S] = \mathbf{J}\xi \text{ for all } \xi \in \mathfrak{X}(TM). \tag{5.1.4}$$

Proof. To show (5.1.4), note first that the mapping

$$\xi \in \mathfrak{X}(TM) \mapsto \mathbf{J}[\mathbf{J}\xi, S] \in \mathfrak{X}(TM)$$

is tensorial. Indeed, for any smooth function F on TM we have

$$\mathbf{J}[\mathbf{J}(F\xi), S] = \mathbf{J}[F(\mathbf{J}\xi), S] = \mathbf{J}(F[\mathbf{J}\xi, S] - (SF)\mathbf{J}\xi)$$

$$= F\mathbf{J}[\mathbf{J}\xi, S] - (SF)\mathbf{J}^2\xi \overset{(4.1.85)}{=} F\mathbf{J}[\mathbf{J}\xi, S],$$

and the additivity is obvious. So, by Remark 4.1.54(b), it is sufficient to check that (5.1.4) is true if $\xi \in \{X^v, X^c\}$, $X \in \mathfrak{X}(M)$. Since $\mathbf{J}X^v = 0$, the equality automatically holds for vertical lifts. Observe, furthermore, that

$$0 \overset{(4.1.89)}{=} [\mathbf{J}, X^v]S = [\mathbf{J}S, X^v] - \mathbf{J}[S, X^v]$$

$$= [C, X^v] - \mathbf{J}[S, X^v] \overset{(4.1.53)}{=} \mathbf{J}[X^v, S] - X^v.$$

Hence

$$\mathbf{J}[\mathbf{J}X^c, S] = \mathbf{J}[X^v, S] = X^v = \mathbf{J}X^c,$$

which is precisely what we want. $\qquad\square$

Corollary 5.1.10. *Let S be a second-order vector field over M, and X a vector field on M. Then*

$$\mathbf{J}[X^v, S] = X^v, \tag{5.1.5}$$

$$\mathbf{J}[X^c, S] = 0. \tag{5.1.6}$$

Proof. With the choice $\xi := X^c$, (5.1.4) leads immediately to (5.1.5). The second relation can be obtained as follows:

$$\mathbf{J}[X^c, S] \overset{(4.1.72)}{=} [C, X^c] + \mathbf{J}[X^c, S] = [\mathbf{J}S, X^c] - \mathbf{J}[S, X^c]$$

$$\overset{(3.3.16)}{=} [\mathbf{J}, X^c]S \overset{(4.1.89)}{=} 0. \qquad\square$$

Remark 5.1.11. Relation (5.1.5) can also be expressed as

$$[X^v, S] \underset{\tau}{\sim} X. \tag{5.1.7}$$

Indeed, (5.1.5) can equivalently be written in the form $\mathbf{j}[X^v, S] = \widehat{X}$. This holds if, and only if, for each $v \in TM$ we have

$$(v, X(\tau(v))) = \mathbf{j}([X^v, S](v)) \overset{(4.1.35)}{=} (v, \tau_*([X^v, S](v))).$$

Thus relation (5.1.5) is equivalent to $\tau_* \circ [X^v, S] = X \circ \tau$. The last equality means that $[X^v, S] \underset{\tau}{\sim} X$, concluding the proof.

It will be instructive to present two further proofs for (5.1.7), which are independent of Grifone's identity.

Proof 1. Let f be a smooth function on M. Applying Lemma 5.1.7, Lemma 4.1.28(iii) and (4.1.51), we find that

$$[X^{\mathsf{v}}, S](f \circ \tau) = [X^{\mathsf{v}}, S]f^{\mathsf{v}} = X^{\mathsf{v}}(Sf^{\mathsf{v}}) - S(X^{\mathsf{v}}f^{\mathsf{v}})$$
$$= X^{\mathsf{v}}f^{\mathsf{c}} = (Xf)^{\mathsf{v}} = (Xf) \circ \tau,$$

so $[X^{\mathsf{v}}, S] \underset{\tau}{\sim} X$ by Lemma 3.1.49.

Proof 2. By Lemma 4.1.35(ii), X^{v} is the velocity field of the global flow

$$\psi \colon \mathbb{R} \times TM \to TM, \quad (t, v) \mapsto \psi(t, v) := v + tX(\tau(v)).$$

Then, clearly, $\tau \circ \psi_t = \tau$ for all $t \in \mathbb{R}$. Now, applying Proposition 3.2.30,

$$[X^{\mathsf{v}}, S](v) = \lim_{t \to 0} \frac{1}{t}\big((\psi_{-t})_*(S(\psi_t(v))) - S(v)\big), \quad v \in TM.$$

Acting on both sides by τ_*, we find

$$\tau_*([X^{\mathsf{v}}, S](v)) = \lim_{t \to 0} \frac{1}{t}\big((\tau \circ \psi_{-t})_*(S(\psi_t(v))) - \tau_*(S(v))\big)$$
$$= \lim_{t \to 0} \frac{1}{t}\big(\tau_*(S(\psi_t(v))) - \tau_*(S(v))\big)$$
$$\overset{\text{Lemma 5.1.3}}{=} \lim_{t \to 0} \frac{1}{t}(\psi_t(v) - v) = \lim_{t \to 0} \frac{1}{t}(tX(\tau(v))) = X \circ \tau(v),$$

thus proving that $[X^{\mathsf{v}}, S] \underset{\tau}{\sim} X$. This nice reasoning is due to M. Crampin [30].

Corollary and Definition 5.1.12. *If S is a second-order vector field over M, then*

$$S^* := [C, S] - S \tag{5.1.8}$$

is a vertical vector field, called the deviation *of S.*

Proof. We have $\mathbf{J}S^* = \mathbf{J}[\mathbf{J}S, S] \overset{(5.1.4)}{=} \mathbf{J}S - \mathbf{J}S = 0$, whence our claim. $\qquad\qquad\square$

Corollary 5.1.13. *Let $S \colon TM \to TTM$ be an arbitrarily fixed second-order vector field. A vector field ξ on TM is vertical if, and only if,*

$$[\mathbf{J}, S]\xi = -\xi. \tag{5.1.9}$$

Proof. Suppose first that ξ is vertical. Then $\mathbf{J}\xi = 0$ and ξ can be written in the form $\xi = \mathbf{J}\eta$ $(\eta \in \mathfrak{X}(TM))$ because $\mathrm{Im}(\mathbf{J}) = \mathrm{Ker}(\mathbf{J}) = \mathfrak{X}^v(TM)$ by (4.1.86). Thus

$$[\mathbf{J}, S]\xi = [\mathbf{J}\xi, S] - \mathbf{J}[\xi, S] = -\mathbf{J}[\mathbf{J}\eta, S] \overset{(5.1.4)}{=} -\mathbf{J}\eta = -\xi,$$

as was to be shown.

Conversely, suppose that (5.1.9) is satisfied. Then

$$-\mathbf{J}\xi = \mathbf{J}([\mathbf{J}, S]\xi) = \mathbf{J}[\mathbf{J}\xi, S] - \mathbf{J}^2[\xi, S] \overset{(4.1.85),(5.1.4)}{=} \mathbf{J}\xi,$$

whence $\mathbf{J}\xi = 0$, and hence ξ is vertical. $\square$

Corollary 5.1.14. *For any second-order vector field S on TM, the mapping*

$$[\mathbf{J}, S]^2 := [\mathbf{J}, S] \circ [\mathbf{J}, S] \colon \mathfrak{X}(TM) \to \mathfrak{X}(TM)$$

is the identity transformation, i.e.,

$$[\mathbf{J}, S]^2 = 1_{\mathrm{End}(\mathfrak{X}(TM))}. \tag{5.1.10}$$

Proof. If $\xi \in \mathfrak{X}(TM)$ is vertical, then (5.1.9) implies that $[\mathbf{J}, S]^2\xi = \xi$. To complete the proof it suffices to show that

$$[\mathbf{J}, S]^2 X^c = X^c \text{ for all } X \in \mathfrak{X}(M).$$

Observe first that

$$[\mathbf{J}, S]X^c = [\mathbf{J}X^c, S] - \mathbf{J}[X^c, S] \overset{(4.1.87),(5.1.6)}{=} [X^v, S],$$

whence

$$\mathbf{J}(X^c - [\mathbf{J}, S]X^c) = \mathbf{J}(X^c - [X^v, S]) = X^v - \mathbf{J}[X^v, S] \overset{(5.1.5)}{=} X^v - X^v = 0.$$

Thus $X^c - [\mathbf{J}, S]X^c$ is vertical, so by Corollary 5.1.13,

$$[\mathbf{J}, S](X^c - [\mathbf{J}, S]X^c) = -X^c + [\mathbf{J}, S]X^c,$$

which leads immediately to the desired relation $[\mathbf{J}, S]^2 X^c = X^c$. $\square$

Definition 5.1.15. A second-order vector field of class C^2 is said to be an *affine spray* if it is positive-homogeneous of degree 2.

Lemma 5.1.16. *Let $S \in C^2(TM, TTM)$ be a second-order vector field. Then S is an affine spray if, and only if, the coefficients of S are positive-homogeneous of degree 2 with respect to any chart of M.*

Proof. This is an immediate consequence of assertion $(*)$ in Example 4.2.16(b). $\square$

Corollary 5.1.17. *A second-order vector field $S \in C^2(TM, TTM)$ is an affine spray if, and only if, the coordinate expression of S in any induced chart $(\tau^{-1}(\mathcal{U}), ((x^i)_{i=1}^n, (y^i)_{i=1}^n))$ is of the form*

$$S \underset{(\mathcal{U})}{=} y^i \frac{\partial}{\partial x^i} - y^j y^k (\Gamma_{jk}^i \circ \tau) \frac{\partial}{\partial y^i}, \tag{5.1.11}$$

where $\Gamma_{jk}^i \colon \mathcal{U} \to \mathbb{R}$ are smooth functions related to the spray coefficients G^i of S by

$$\Gamma_{jk}^i \circ \tau = \frac{\partial^2 G^i}{\partial y^j \partial y^k}; \quad i, j, k \in J_n \tag{5.1.12}$$

(summation convention in force).

Proof. Suppose that S is an affine spray. Then its coefficients G^i are positive-homogeneous of degree 2, so Lemma 4.2.10(iii) implies that they restrict to quadratic forms on each fibre $T_p M$, $p \in \mathcal{U}$. Thus there exist unique functions $\Gamma_{jk}^i \colon \mathcal{U} \to \mathbb{R}$ such that

$$G^i = \frac{1}{2} y^j y^k (\Gamma_{jk}^i \circ \tau) \text{ and } \Gamma_{jk}^i = \Gamma_{kj}^i.$$

From this we obtain that

$$\frac{\partial G^i}{\partial y^k} = \frac{1}{2} \frac{\partial}{\partial y^k} (y^r y^s (\Gamma_{rs}^i \circ \tau)) = \frac{1}{2} \delta_k^r y^s (\Gamma_{rs}^i \circ \tau) + \frac{1}{2} \delta_k^s y^r (\Gamma_{rs}^i \circ \tau)$$

$$= \frac{1}{2} (y^s (\Gamma_{ks}^i \circ \tau) + y^r (\Gamma_{rk}^i \circ \tau)),$$

$$\frac{\partial^2 G^i}{\partial y^j \partial y^k} = \frac{1}{2} (\Gamma_{kj}^i \circ \tau + \Gamma_{jk}^i \circ \tau) = \Gamma_{jk}^i \circ \tau,$$

which proves (5.1.12) and verifies the smoothness of the functions Γ_{jk}^i.

Conversely, if the local expression of S is of the form (5.1.11), then

$$[C, S] \underset{(\mathcal{U})}{=} \left[C, y^i \frac{\partial}{\partial x^i} \right] - \left[C, y^j y^k (\Gamma_{jk}^i \circ \tau) \frac{\partial}{\partial y^i} \right]$$

$$= y^i \frac{\partial}{\partial x^i} - C(y^j y^k)(\Gamma_{jk}^i \circ \tau) \frac{\partial}{\partial y^i} + y^j y^k (\Gamma_{jk}^i \circ \tau) \frac{\partial}{\partial y^i}$$

$$= y^i \frac{\partial}{\partial x^i} - y^j y^k (\Gamma_{jk}^i \circ \tau) \frac{\partial}{\partial y^i} \underset{(\mathcal{U})}{=} S,$$

therefore S is positive-homogeneous of degree 2, and hence it is an affine spray. $\qquad \square$

Now we give a seemingly slight, but important modification of the notion of an affine spray introduced by F. Brickell [19] and, probably independently, by P. Dazord [38].

Definition 5.1.18. A mapping $S\colon TM \to TTM$ is called a *spray* over M (or for M) if it satisfies the following conditions:

(S_1) S is a section of the tangent bundle $\tau_{TM}\colon TTM \to TM$, that is, $\tau_{TM} \circ S = 1_{TM}$.

(S_2) S is a section of the vector bundle $\tau_*\colon TTM \to TM$, i.e., $\tau_* \circ S = 1_{TM}$.

(S_3) S is smooth on $\overset{\circ}{T}M$.

(S_4) S is positive-homogeneous of degree 2.

(S_5) $S \circ o = o_{TM} \circ o$, where $o \in \mathfrak{X}(M)$ and $o_{TM} \in \mathfrak{X}(TM)$ are the zero vector fields.

A *spray manifold* is a pair (M, S) where M is a manifold and S is a spray for M.

Remark 5.1.19. (a) Having the concept of a spray, we can say that *an affine spray is a spray over M which is of class C^2 on TM*. Of course, condition (S_2) is just one of the equivalent possibilities discussed above. Condition (S_4) can also be expressed by saying that the deflection $S^* := [C, S] - S$ of S vanishes.

(b) As we have just learnt, the local expression of a spray S over M with respect to an induced chart $(\tau^{-1}(\mathcal{U}), ((x^i)_{i=1}^n, (y^i)_{i=1}^n))$ on TM is of the form

$$S \underset{(\mathcal{U})}{=} y^i \frac{\partial}{\partial x^i} - 2G^i \frac{\partial}{\partial y^i}, \qquad (*)$$

where the *spray coefficients* $G^i\colon \tau^{-1}(\mathcal{U}) \to \mathbb{R}$ are positive-homogeneous of degree 2. However, without assuming C^2 differentiability, this homogeneity property no longer implies that the restrictions $G^i \upharpoonright T_pM$ ($p \in \mathcal{U}$) are quadratic forms.

(c) Let $p \in \mathcal{U}$, and let 0_p denote, as usual, the zero vector of T_pM. From (S_5) it follows that $G_i(0_p) = 0$. Since the functions G_i are positive-homogeneous of degree 2, they are of class C^1 by Lemma 4.2.11. So, we arrive at the following important conclusion:

Corollary 5.1.20. *A spray over a manifold M is of class C^1 on TM.* $\square$

Lemma and Definition 5.1.21. *If S is a spray over M and*

$$\varrho\colon TM \to TM, \ v \mapsto \varrho(v) := -v$$

is the reflection map of TM *(Example 3.1.38(a)), then*

$$S^{\downarrow} := -\varrho_{\#}S = -\varrho_* \circ S \circ \varrho = \widetilde{\mu}_{-1} \circ \varrho_* \circ S \circ \varrho \tag{5.1.13}$$

is also a spray, called the reverse *of* S. *If* S *is equal to its reverse, and hence*

$$\varrho_*(S(-v)) = -S(v) \text{ for all } v \in TM, \tag{5.1.14}$$

then we say that S *is* reversible.

Proof. The push-forward $\varrho_{\#}S$ of S is a vector field on TM, and so is $-\varrho_{\#}S$, hence (S_1) holds for $S^{\downarrow}$. Taking into account that

$$\tau \circ \varrho = \tau, \quad \tau_* \circ \widetilde{\mu}_{-1} = \varrho \circ \tau_*,$$

we have

$$\tau_* \circ S^{\downarrow} = \tau_* \circ \widetilde{\mu}_{-1} \circ \varrho_* \circ S \circ \varrho = \varrho \circ (\tau \circ \varrho)_* \circ S \circ \varrho$$

$$= \varrho \circ \tau_* \circ S \circ \varrho \overset{(S_2)}{=} \varrho \circ 1_{TM} \circ \varrho = 1_{TM}.$$

Thus $S^{\downarrow}$ satisfies (S_2).

Condition (S_3) clearly holds for $S^{\downarrow}$. We check the 2^+-homogeneity of the reverse of S:

$$[C, S^{\downarrow}] = -[C, \varrho_{\#}S] \overset{(4.1.49)}{=} -[\varrho_{\#}C, \varrho_{\#}S] \overset{(3.1.44)}{=} -\varrho_{\#}[C, S] \overset{(S_4)}{=} -\varrho_{\#}S = S^{\downarrow},$$

as desired. Finally,

$$S^{\downarrow} \circ o = -\varrho_* \circ S \circ \varrho \circ o = -\varrho_* \circ S \circ o = -\varrho_* \circ o_{TM} \circ o = o_{TM} \circ o,$$

thus $S^{\downarrow}$ satisfies also (S_5), hence it is a spray. $\qquad \square$

Remark 5.1.22. (a) Keeping the notation of Remark 5.1.19(b), we express the condition of reversibility of a spray S in terms of the spray coefficients G^i. We have, on the one hand,

$$\varrho_*(S(v)) = \varrho_* \left(y^i(v) \left(\frac{\partial}{\partial x^i} \right)_v - 2G^i(v) \left(\frac{\partial}{\partial y^i} \right)_v \right)$$

$$\overset{(3.1.25)}{=} y^i(v) \left(\frac{\partial}{\partial x^i} \right)_{-v} + 2G^i(v) \left(\frac{\partial}{\partial y^i} \right)_{-v},$$

and, on the other hand,

$$-S(-v) = y^i(v) \left(\frac{\partial}{\partial x^i} \right)_{-v} + 2G^i(-v) \left(\frac{\partial}{\partial y^i} \right)_{-v}.$$

Therefore (5.1.14) holds, and hence S *is reversible if, and only if, the spray coefficients of S have the symmetry property*

$$G^i(-v) = G^i(v) \text{ for all } v \in \tau^{-1}(\mathcal{U}).$$

From this and from Remark 5.1.19(c) we conclude that *the coefficients of a reversible spray are completely homogeneous of degree* 2 in the sense that

$$G^i \circ \mu_t = t^2 G^i \text{ for all } t \in \mathbb{R}.$$

Hence *a reversible spray is completely homogeneous of degree* 2 in the same sense of the word 'completeness': we have

$$\widetilde{\mu}_t \circ (\mu_t)_* \circ S = S \circ \mu_t \text{ for all } t \in \mathbb{R}.$$

(Using (5.1.14), it can be shown also by a simple, direct calculation that a reversible spray is projectively homogeneous of degree 2.)

(b) *If S is a spray over M, and $S^{\downarrow}$ is its reverse, then $\bar{S} := \frac{1}{2}(S + S^{\downarrow})$ is a reversible spray.* Indeed, first $\bar{S}$ is clearly a spray. Since relation $\varrho \circ \varrho = 1_{TM}$ implies that $\varrho_* \circ \varrho_* = 1_{TTM}$, we have

$$\varrho_{\#} S^{\downarrow} = -\varrho_* \circ (\varrho_* \circ S \circ \varrho) \circ \varrho = -S,$$

therefore

$$\varrho_{\#} \bar{S} = \frac{1}{2}(\varrho_{\#} S + \varrho_{\#} S^{\downarrow}) = \frac{1}{2}(-S^{\downarrow} + (-S)) = -\frac{1}{2}(S + S^{\downarrow}) = -\bar{S},$$

as wanted.

The following (substantial) weakening of the notion of a spray will be useful in several applications.

Definition 5.1.23. A mapping $S \colon TM \to TTM$ is called a *semispray over M* if it satisfies (S_1), (S_2), (S_3) and (S_5), i.e., if it is a section both of the double tangent bundle $\tau_{TM} \colon TTM \to TM$ and the vector bundle $\tau_* \colon TTM \to TM$, smooth on $\mathring{T}M$, and sends zero vectors to zero vectors: $S(0_p) = 0_{0_p} \in T_{0_p} TM$ for all $p \in M$.

Remark 5.1.24. (a) Clearly, any second-order vector field on $\mathring{T}M$ can be extended to a semispray by defining it to be zero on zero vectors.

(b) Proposition 5.1.4 is true in a slightly more general form:

Let $S \colon TM \to TTM$ be a mapping satisfying (S_1), (S_3) and (S_5). Then S is a semispray over M if, and only if, each of its integral curves is the canonical lift of a curve in M.

Lemma 5.1.25. *If S is a semispray over M and $\varphi \in \mathrm{Diff}(M)$, then*

$$(\varphi_*)_\# S = \varphi_{**} \circ S \circ \varphi_*^{-1}$$

is also a semispray. If, in addition, S is a spray, then $(\varphi_*)_\# S$ is a spray as well.

Proof. Taking into account Corollary 5.1.6, we have immediately that $(\varphi_*)_\# S$ is a semispray. If S is a spray, then

$$[C, (\varphi_*)_\# S] \overset{(4.1.112)}{=} [(\varphi_*)_\# C, (\varphi_*)_\# S] \overset{(3.1.44)}{=} (\varphi_*)_\# [C, S] = (\varphi_*)_\# S,$$

so $(\varphi_*)_\# S$ is a spray. $\qquad\square$

Definition and Lemma 5.1.26. Let S be a semispray over M. A diffeomorphism $\varphi \in \mathrm{Diff}(M)$ is called an *automorphism of S* if S is invariant under its derivative, i.e., $(\varphi_*)_\# S = S$. The automorphisms of S form a group under composition called the *automorphism group* of S and denoted by $\mathrm{Aut}(S)$.

We leave the reader to check that $\mathrm{Aut}(S)$ is indeed a group.

5.1.2 Geodesics of a Semispray

Definition 5.1.27. Let M be a manifold, and let $S\colon TM \to TTM$ be a semispray over M.

(a) A smooth curve $\gamma\colon I \to M$ is called a *geodesic* of S if

$$S \circ \dot{\gamma} = \ddot{\gamma}. \tag{5.1.15}$$

If a smooth curve has a positive reparametrization as a geodesic of S, we call it a *pregeodesic*.

(b) Let $\gamma\colon I \to M$ be a geodesic of S. If $\tilde{I}$ is an open interval containing I and $\tilde{\gamma}\colon \tilde{I} \to M$ is a geodesic of S such that $\tilde{\gamma} \restriction I = \gamma$, then we say that $\tilde{\gamma}$ is an *extension of γ* and γ is a *restriction of $\tilde{\gamma}$*. A geodesic is *maximal* if it does not admit any proper extension (cf. Definition 3.2.11), or, equivalently, if it is not a proper restriction of any geodesic.

(c) A curve segment $\gamma\colon [a, b] \to M$ is a *geodesic segment* of S if it is a restriction of a geodesic. Then we also say that γ is a geodesic segment from $\gamma(a)$ to $\gamma(b)$.

Remark 5.1.28. Let S be a semispray over M. If $c\colon I \to TM$ is an integral curve of S, then, by Remark 5.1.24(b), c is the canonical lift of the curve $\gamma := \tau \circ c\colon I \to M$, i.e., $c = \dot{\gamma}$. In this case γ is a geodesic of S because

$$\ddot{\gamma} = \dot{c} = S \circ c = S \circ \dot{\gamma}.$$

Conversely, if $\gamma\colon I \to M$ is a geodesic of S, then its canonical lift $c := \dot{\gamma}$ is an integral curve of S.

Thus a smooth curve $\gamma\colon I \to M$ is

a *geodesic of* S if, and only if, $\dot{\gamma}$ is an integral curve of S;

a *maximal geodesic of* S if, and only if, $\dot{\gamma}$ is a maximal integral curve of S.

Proposition and Definition 5.1.29. *Let S be a semispray of class C^1 over a manifold M. For each point $p \in M$ and $v \in T_pM$, there exists a unique maximal geodesic $\gamma_v\colon I_v \to M$ such that $\dot{\gamma}_v(0) = v$. We say that γ_v is the maximal geodesic of S starting at p with initial velocity v.*

Proof.   Since S is a vector field of class C^1 on TM, we may apply the existence and uniqueness of integral curves of vector fields of class C^1 as described in Remark 3.2.14(b). Thus, for any $v \in TM$ there is a unique maximal integral curve $c\colon I \to TM$ of S such that $c(0) = v$. By the preceding remark, c is the canonical lift of the maximal geodesic $\gamma := \tau \circ c$, and we have $\dot{\gamma}(0) = c(0) = v$.

If $\tilde{\gamma}\colon \tilde{I} \to M$ is another maximal geodesic with $\dot{\tilde{\gamma}}(0) = v$, then $\tilde{c} := \dot{\tilde{\gamma}}$ is also an integral curve of S satisfying $\tilde{c}(0) = v$, hence $\tilde{c} = c$. $\square$

Corollary 5.1.30. *A geodesic of a semispray of class C^1 is either constant, or regular (i.e., an immersion).*

Proof.   Let $S \in C^1(TM, TTM)$ be a semispray, and let $\gamma\colon I \to M$ be a geodesic of S. If $\dot{\gamma}(t_0) = 0$ for some $t_0 \in I$, then consider the constant curve $\tilde{\gamma}\colon I \to M$, $t \mapsto \tilde{\gamma}(t) := \gamma(t_0)$. Then $\tilde{\gamma}$ is a geodesic of S starting at $\gamma(t_0)$ with zero initial velocity. Thus, by the uniqueness statement for geodesics, $\tilde{\gamma} = \gamma$. This proves our claim. $\square$

Definition and Lemma 5.1.31. Let S be a semispray over M. A diffeomorphism $\varphi \in \mathrm{Diff}(M)$ is called an *affine transformation* or an *affinity* of S if it preserves the geodesics, i.e., for every geodesic $\gamma\colon I \to M$ of S, the curve $\varphi \circ \gamma\colon I \to M$ is also a geodesic of S. The affine transformations of S form a group under composition called the *group of affinities* of S and denoted by $\mathrm{Aff}(S)$. $\triangle$

Proposition 5.1.32. *The automorphism group and the group of affinities of a semispray of class C^1 coincide.*

Proof. Let S be a semispray of class C^1 over M. Suppose first that $\varphi \in \mathrm{Aut}(S)$, and let $\gamma \colon I \to M$ be a geodesic of S. Then $\varphi_{**} \circ S = S \circ \varphi_*$, and so we obtain that

$$S \circ (\varphi \circ \gamma)^{\cdot} \overset{(3.1.11)}{=} S \circ \varphi_* \circ \dot\gamma \overset{\text{cond.}}{=} \varphi_{**} \circ S \circ \dot\gamma \overset{\text{cond.}}{=} \varphi_{**} \circ \ddot\gamma = \varphi_{**} \circ \dot\gamma_* \circ \frac{d}{dr}$$

$$= (\varphi_* \circ \dot\gamma)_* \circ \frac{d}{dr} = ((\varphi \circ \gamma)^{\cdot})_* \circ \frac{d}{dr} = (\varphi \circ \gamma)^{\cdot\cdot}.$$

This proves that $\varphi \circ \gamma$ is a geodesic of S, therefore $\mathrm{Aut}(S) \subset \mathrm{Aff}(S)$.

Conversely, suppose that φ is an affinity of S. Then our task is to show that $S \circ \varphi_* = \varphi_{**} \circ S$. Given any vector v in TM, by Proposition and Definition 5.1.29 there exists a unique (maximal) geodesic γ of S such that $\dot\gamma(0) = v$. Then, by condition, $\varphi \circ \gamma$ is also a geodesic of S. So, calculating as above, we find that

$$S \circ \varphi_*(v) = S \circ \varphi_* \circ \dot\gamma(0) = S \circ (\varphi \circ \gamma)^{\cdot}(0) = (\varphi \circ \gamma)^{\cdot\cdot}(0)$$

$$= \varphi_{**} \circ \ddot\gamma(0) = \varphi_{**} \circ S \circ \dot\gamma(0) = \varphi_{**} \circ S(v).$$

Thus $S \circ \varphi_* = \varphi_{**} \circ S$, as was to be proved. $\qquad\square$

Lemma 5.1.33. *Let I be a nonempty open interval and $h \colon I \to \mathbb{R}$ a smooth function. Then the second-order differential equation*

$$x'' + (h \circ x)(x')^2 = 0 \qquad\qquad (*)$$

has a strictly increasing smooth solution.

Proof. Let $t_0 \in I$ be a fixed point, and introduce two auxiliary functions $k, \ell \colon I \to \mathbb{R}$ by

$$k \colon t \in I \mapsto k(t) := \int_{t_0}^{t} h, \quad \ell \colon t \in I \mapsto \ell(t) := \int_{t_0}^{t} \exp \circ k.$$

Then ℓ is obviously smooth and strictly increasing, thus it has an inverse $\theta := \ell^{-1} \colon \widetilde{I} \to I$, where $\widetilde{I} \subset \mathbb{R}$ is another open interval. We show that θ is a solution of $(*)$. First, by the rule of differentiation of inverse functions, we get

$$\theta'(s) = \frac{1}{\ell'(\theta(s))} = \frac{1}{e^{k(\theta(s))}} = e^{-k(\theta(s))}, \quad s \in \widetilde{I}.$$

Now we substitute this into the left-hand side of $(*)$ to obtain

$$\theta''(s) + h(\theta(s))\theta'(s)^2 = e^{-k(\theta(s))}(-k'(\theta(s)))\theta'(s) + h(\theta(s))\theta'(s)^2$$

$$= -h(\theta(s))\theta'(s)^2 + h(\theta(s))\theta'(s)^2 = 0,$$

which shows that θ is indeed a solution of $(*)$. This concludes the proof of the lemma. $\qquad\square$

Lemma 5.1.34. *A curve* $\gamma\colon I \to M$ *is a pregeodesic of a spray* S *if, and only if, there exists a smooth function* $h\colon I \to \mathbb{R}$ *such that*

$$\ddot{\gamma} = S \circ \dot{\gamma} + h(C \circ \dot{\gamma}), \tag{5.1.16}$$

where C *is the Liouville vector field. In particular, if* $\bar{\gamma} := \gamma \circ \theta$ *is a positive reparametrization of a geodesic* γ, *then*

$$\ddot{\bar{\gamma}} = S \circ \dot{\bar{\gamma}} + \frac{\theta''}{\theta'}(C \circ \dot{\bar{\gamma}}). \tag{5.1.17}$$

Proof. First we prove (5.1.17), then it follows immediately that any pregeodesic satisfies (5.1.16). Let $\gamma\colon I \to M$ be a geodesic of S and let $\bar{\gamma} := \gamma \circ \theta \colon \bar{I} \to M$ be a positive reparametrization of γ. Taking into account the 2^+-homogeneity of S, for any $t \in \bar{I}$ we obtain

$$\ddot{\bar{\gamma}}(t) \overset{(4.1.62)}{=} \widetilde{\mu}^+_{\ln(\theta'(t))} \circ (\mu^+_{\ln(\theta'(t))})_* \circ \ddot{\gamma} \circ \theta(t) + \frac{\theta''(t)}{\theta'(t)}(C \circ \dot{\bar{\gamma}})(t)$$

$$\overset{(5.1.15)}{=} \widetilde{\mu}^+_{\ln(\theta'(t))} \circ (\mu^+_{\ln(\theta'(t))})_* \circ S \circ \dot{\gamma} \circ \theta(t) + \frac{\theta''(t)}{\theta'(t)}(C \circ \dot{\bar{\gamma}})(t)$$

$$\overset{(4.2.7)}{=} S \circ \mu^+_{\ln(\theta'(t))} \circ \dot{\gamma} \circ \theta(t) + \frac{\theta''(t)}{\theta'(t)}(C \circ \dot{\bar{\gamma}})(t)$$

$$\overset{(4.1.61)}{=} S \circ \dot{\bar{\gamma}}(t) + \frac{\theta''(t)}{\theta'(t)}(C \circ \dot{\bar{\gamma}})(t),$$

as desired.

Now let $\gamma\colon I \to M$ be a smooth curve satisfying (5.1.16), and $\bar{\gamma} := \gamma \circ \theta$ a positive reparametrization. Then, using the same tricks as in the previous calculation, and taking into account the 1^+-homogeneity of C, we get

$$\ddot{\bar{\gamma}}(t) = \widetilde{\mu}^+_{\ln(\theta'(t))} \circ (\mu^+_{\ln(\theta'(t))})_* \circ (S \circ \dot{\gamma} + h(C \circ \dot{\gamma})) \circ \theta(t) + \frac{\theta''(t)}{\theta'(t)}(C \circ \dot{\bar{\gamma}})(t)$$

$$= S \circ \dot{\bar{\gamma}}(t) + h(\theta(t))\widetilde{\mu}^+_{\ln(\theta'(t))} \circ C \circ \mu^+_{\ln(\theta'(t))} \circ \dot{\gamma} \circ \theta(t) + \frac{\theta''(t)}{\theta'(t)}(C \circ \dot{\bar{\gamma}})(t)$$

$$= S \circ \dot{\bar{\gamma}}(t) + \left(h(\theta(t))\theta'(t) + \frac{\theta''(t)}{\theta'(t)}\right)(C \circ \dot{\bar{\gamma}})(t).$$

Thus $\bar{\gamma}$ is a geodesic if, and only if, θ is a solution of the differential equation

$$x'' + (h \circ x)(x')^2 = 0.$$

This equation has a strictly increasing smooth solution by the preceding lemma, therefore γ is a pregeodesic. $\qquad\square$

Remark 5.1.35. Let S be a semispray over M. Using the coordinate expressions (5.1.3), (3.1.12) and (3.1.35) of S, $\dot{\gamma}$ and $\ddot{\gamma}$, respectively, we get the *local geodesic equations*

$$(\gamma^i)'' + 2(G^i \circ \dot{\gamma}) = 0, \quad i \in J_n \tag{5.1.18}$$

and the *local pregeodesic equations*

$$(\gamma^i)'' + 2(G^i \circ \dot{\gamma}) = h(\gamma^i)', \quad i \in J_n \tag{5.1.19}$$

for S, respectively.

Lemma 5.1.36. *Let S be a spray over M and γ a geodesic of S. Then*

$$\gamma^{\downarrow} \colon t \in -I \mapsto \gamma(-t) \in M$$

is a geodesic of the reverse $S^{\downarrow} := -\varrho_{\#} S$ of S.

Proof. For every $t \in I$,

$$(\gamma^{\downarrow})^{\cdot\cdot}(t) \overset{(4.1.60)}{=} \tilde{\mu}_{-1} \circ \varrho_* \circ \ddot{\gamma}(-t) \overset{(5.1.15)}{=} \tilde{\mu}_{-1} \circ \varrho_* \circ S \circ \dot{\gamma}(-t)$$

$$\overset{(4.1.59)}{=} \tilde{\mu}_{-1} \circ \varrho_* \circ S \circ \varrho \circ (\gamma^{\downarrow})^{\cdot}(t) = -\varrho_{\#} S \circ (\gamma^{\downarrow})^{\cdot}(t) = S^{\downarrow} \circ (\gamma^{\downarrow})^{\cdot}(t),$$

as was to be shown. $\qquad\qquad\square$

Lemma 5.1.37. *A semispray of class C^1 is a spray (resp. a reversible spray) if, and only if, its geodesics remain geodesics under positive affine reparametrizations (resp. affine reparametrizations).*

Proof. (a) Suppose first that S is a spray over M, and let $\gamma \colon I \to M$ be a geodesic of S. Consider a positive affine reparametrization

$$\theta \colon J \to I, \ t \mapsto \theta(t) := e^a t + b$$

of γ, where a, b are fixed real numbers. We show that $\gamma \circ \theta$ is also a geodesic of S. Indeed,

$$(\gamma \circ \theta)^{\cdot\cdot} \overset{(4.1.62)}{=} \tilde{\mu}_a^+ \circ (\mu_a^+)_* \circ S \circ \dot{\gamma} \circ \theta \overset{(4.2.7)}{=} S \circ \mu_a^+ \circ \dot{\gamma} \circ \theta \overset{(4.1.61)}{=} S \circ (\gamma \circ \theta)^{\cdot}$$

as desired.

Conversely, let S be a semispray of class C^1 over M whose geodesics remain geodesics under positive affine reparametrizations. Let v be a tangent vector in $\mathring{T}M$, and γ_v the unique maximal geodesic with initial velocity v. Then, for any real number a, the curve $\gamma_{e^a v} \colon t \mapsto \gamma_v(e^a t) \in M$ is also a geodesic of S. Thus we obtain that

$$S \circ \mu_a^+(v) = S(e^a v) = S \circ \dot{\gamma}_{e^a v}(0) = \ddot{\gamma}_{e^a v}(0) = \tilde{\mu}_a^+ \circ (\mu_a^+)_* \circ \ddot{\gamma}_v(0)$$

$$= \tilde{\mu}_a^+ \circ (\mu_a^+)_* \circ S \circ \dot{\gamma}_v(0) = \tilde{\mu}_a^+ \circ (\mu_a^+)_* \circ S(v),$$

which proves that S is positive-homogeneous of degree 2.

(b) The second assertion can be seen in an analogous way. We recall that a spray S is reversible if, and only if,

$$\tilde{\mu}_a \circ (\mu_a)_* \circ S = S \circ \mu_a \text{ for all } a \in \mathbb{R} \qquad (*)$$

(see Remark 5.1.22(a)). Applying (4.1.59) and (4.1.60) instead of (4.1.61) and (4.1.62), we obtain by the same calculations as in part (a) that $(*)$ holds if, and only if, the geodesics of S remain geodesics under affine reparametrizations. $\qquad \square$

Corollary 5.1.38. *A spray S over M is reversible if, and only if, for any geodesic $\gamma\colon I \to M$, the reverse curve*

$$\gamma^\downarrow\colon -I \to M, \ t \mapsto \gamma^\downarrow(t) := \gamma(-t)$$

is also a geodesic of S. $\qquad \square$

Proof. If the spray S is reversible, then any affine reparametrization of a geodesic of S is still a geodesic by the previous lemma. Thus, in particular, if γ is a geodesic, so is $\gamma^\downarrow$.

Conversely, suppose that the reverse of any geodesic is a geodesic. Consider a tangent vector $v \in TM$ and a geodesic $\gamma\colon I \to M$ with initial velocity $-v$. Then $\gamma^\downarrow$ is a geodesic of S, and by Lemma 5.1.36, it is also a geodesic of $S^\downarrow$, thus

$$S(v) = S \circ (\gamma^\downarrow)^{\cdot}(0) = (\gamma^\downarrow)^{\cdot\cdot}(0) = S^\downarrow \circ (\gamma^\downarrow)^{\cdot}(0) = S^\downarrow(v).$$

From the arbitrariness of v, we obtain that $S = S^\downarrow$, thus S is reversible. $\square$

Corollary 5.1.39. *Let (M, S) be a spray manifold, and let $\gamma_v\colon I_v \to M$ be the maximal geodesic of S with initial velocity v. Then, for any positive numbers r and s such that $rs \in I_v$, we have $r \in I_{sv}$, and $\gamma_v(rs) = \gamma_{sv}(r)$.*

If S is reversible, the property above holds for any real numbers r and s such that $rs \in I_v$.

Proof. Suppose that r and s are positive numbers such that $rs \in I_v$. Then from Lemma 5.1.37 it follows that

$$\tilde{\gamma}\colon t \in \frac{1}{s}I \mapsto \tilde{\gamma}(t) := \gamma_v(ts) \in M$$

is also a geodesic of S. Its initial velocity is obviously sv, and its domain contains r. By the uniqueness of geodesics, we have $\tilde{\gamma} = \gamma_{sv}$, hence $\gamma_{sv}(r) = \tilde{\gamma}(r) = \gamma_v(rs)$.

If the spray is reversible, the curve $\tilde{\gamma}$ defined above is a geodesic even if s is negative, and we can apply the same argument to prove the second assertion. $\qquad \square$

5.1.3   The Exponential Map

Definition 5.1.40. Let (M, S) be a spray manifold. Consider the set
$$\widetilde{TM} := \{v \in TM \mid \gamma_v(1) \text{ is defined}\}.$$
(Here, as above, γ_v is the maximal geodesic of S with initial velocity v.) Then the mapping
$$\exp \colon \widetilde{TM} \to \mathbb{R}, \ v \mapsto \exp(v) := \gamma_v(1)$$
is called the *exponential map* determined by S (or, less pedantically, the exponential map of S). The exponential map of S at a point $p \in M$ is $\exp_p := \exp \upharpoonright \widetilde{T_p M}, \ \widetilde{T_p M} := T_p M \cap \widetilde{TM}.$

Lemma 5.1.41. *Let S be a spray over M and $o \colon M \to TM$ the zero vector field. Then $\widetilde{TM}$ is an open neighbourhood of $o(M)$ in TM, the exponential map is of class C^1, and its restriction to $\widetilde{TM} \cap \mathring{T}M$ is smooth.*

Proof.   In Corollary 5.1.20 we saw that a spray is a vector field of class C^1. Then, according to Remark 3.2.26, S generates a unique maximal flow $\varphi^S \colon \mathcal{D}_S \to TM$ of class C^1, and its domain $\mathcal{D}_S$ is open. We can express the exponential map generated by S with the help of its flow by
$$\exp(v) = \tau \circ \varphi^S(1, v), \quad v \in \widetilde{TM}. \tag{5.1.20}$$
Indeed, let v be a tangent vector to M and γ_v the maximal geodesic of S with initial velocity v. Then, by the definition of geodesics, $\dot{\gamma}_v$ is an integral curve of S starting from v. If γ_v is defined at 1, then so is $\dot{\gamma}_v$, and hence $(1, v) \in \mathcal{D}_S$ and, since $\tau \circ \dot{\gamma}_v = \gamma_v$, the desired formula holds. This observation has two immediate consequences: the exponential map is of class C^1, and $\widetilde{TM}$ is open.

Now we show that $\exp \upharpoonright \widetilde{TM} \cap \mathring{T}M$ is smooth. Notice that the maximal flow of $\mathring{S} := S \upharpoonright \mathring{T}M$ is just $\varphi^S \upharpoonright (\mathcal{D}_S \cap (\mathbb{R} \times \mathring{T}M))$. This follows from Corollary 5.1.30 taking into account that the integral curves of S are the canonical lifts of its geodesics. Then, since the maximal flow of $\mathring{S}$ is smooth, so is the restriction of $\exp$ to $\widetilde{TM} \cap \mathring{T}M$. $\square$

Lemma 5.1.42. *Let (M, S) be a spray manifold with exponential map $\exp_p \colon \widetilde{TM} \to M$. If $v \in TM$ and $\gamma_v \colon I_v \to M$ is the maximal geodesic of S with initial velocity v, then for any positive $t \in I_v$, tv is in the domain of the exponential map, and*
$$\exp(tv) = \gamma_v(t). \tag{5.1.21}$$
If the spray is reversible, then (5.1.21) holds for every $t \in I_v$.

Proof. By Corollary 5.1.39, if $t \in I_v$ is positive, then 1 is in the domain of γ_{tv}, and $\gamma_{tv}(1) = \gamma_v(t)$. Thus we have $\exp(tv) := \gamma_{tv}(1) = \gamma_v(t)$, as claimed. In the reversible case, the above argument works for any $t \in I_v$. $\square$

Definition 5.1.43. A spray is said to be *positively complete* if the domains of its maximal geodesics are not bounded from above, and *negatively complete* if these domains are not bounded from below. A spray is said to be *complete* if it is both positively and negatively complete.

Proposition 5.1.44. *A spray is affine if, and only if, it determines an exponential map of class C^2.*

Proof. An affine spray is smooth on the whole TM, hence its flow is smooth. Then, from (5.1.20) it follows that the exponential map determined by an affine spray is smooth.

Conversely, suppose that S is a spray over a manifold M whose exponential map is of class C^2. Fix a point $p \in M$ and a chart $(\mathcal{U}, (u^i)_{i=1}^n)$ on M around p with induced chart $(\tau^{-1}(\mathcal{U}), ((x^i)_{i=1}^n, (y^i)_{i=1}^n))$ on TM. Define the component functions of exp by $\exp^i := u^i \circ \exp \upharpoonright \tau^{-1}(\mathcal{U})$. Let v be an arbitrary vector in T_pM and γ_v the maximal geodesic of S with initial velocity v. Define the parametrized straight line

$$c_v \colon \mathbb{R} \to TM, \ t \mapsto c_v(t) := tv.$$

Then, for all $i \in J_n$, $t \in \mathbb{R}$ we have $x^i \circ c_v(t) = u^i(\tau(v))$, $y^i \circ c_v(t) = ty^i(v)$, hence

$$(x^i \circ c_v)'(t) = 0, \quad (y^i \circ c_v)'(t) = ty^i(v). \tag{$*$}$$

From (5.1.21), $\exp^i \circ c_v = u^i \circ \gamma_v =: \gamma_v^i$, therefore

$$(\gamma_v^i)'(t) \overset{(2.1.1)}{=} (\exp^i \circ c_v)'(t) \overset{(2.1.1)}{=} \frac{\partial \exp^i}{\partial x^j}(tv)(x^j \circ c_v)'(t)$$

$$+ \frac{\partial \exp^i}{\partial y^j}(tv)(y^j \circ c_v)'(t) \overset{(*)}{=} y^j(v)\frac{\partial \exp^i}{\partial y^j}(tv) = y^j(v)\left(\frac{\partial \exp^i}{\partial y^j} \circ c_v\right)(t),$$

and, similarly,

$$(\gamma_v^i)''(t) = y^k(v)y^j(v)\frac{\partial^2 \exp^i}{\partial y^k \partial y^j}(tv).$$

In particular,

$$(\gamma_v^i)''(0) = y^k(v)y^j(v)\frac{\partial^2 \exp^i}{\partial y^k \partial y^j}(0). \tag{$**$}$$

Now let $(G^i)_{i=1}^n$ be the family of the coefficients of S with respect to the chosen chart. Then, for every $i \in J_n$,

$$2G^i(v) = 2G^i(\dot\gamma_v(0)) \overset{(5.1.18)}{=} -(\gamma_v^i)''(0) \overset{(**)}{=} y^j(v)y^k(v)\frac{\partial^2 \exp^i}{\partial y^j \partial y^k}(0).$$

Thus the functions G^i are quadratic forms, therefore, by Lemma 5.1.16, S is an affine spray. $\qquad\square$

Lemma 5.1.45. *Let (M,S) be a spray manifold, and let $\exp$ be the exponential map determined by S. Then for any $p \in M$, the mapping*

$$((\exp_p)_*)_{0_p} : T_{0_p}T_pM \to T_pM$$

is a linear isomorphism, namely, it is the inverse of the canonical isomorphism $\iota_{0_p} : T_pM \to T_{0_p}T_pM$ described in Lemma 3.1.19.

Proof. Given a tangent vector $v \in T_pM$, consider the parametrized straight line $c_v \colon \mathbb{R} \to T_pM$, $t \mapsto c_v(t) := tv$. Then

$$((\exp_p)_*)_{0_p} \circ \iota_{0_p}(v) \overset{\text{Lemma } 3.1.19}{=} ((\exp_p)_*)_{0_p} \circ \dot c_v(0)$$
$$\overset{(3.1.11)}{=} (\exp_p \circ c_v)\dot{}(0) \overset{(5.1.21)}{=} \dot\gamma_v(0) = v,$$

therefore $((\exp_p)_*)_{0_p} \circ \iota_{0_p} = 1_{T_pM}$, whence our assertion. $\qquad\square$

Corollary 5.1.46 (Normal Neighbourhood Lemma). *Let (M,S) be a spray manifold with exponential map $\exp \colon \widetilde{TM} \to M$. Then for any point $p \in M$, there exists an open neighbourhood $\widetilde{\mathcal{U}}$ of 0_p in T_pM and an open neighbourhood $\mathcal{U}$ of p in M such that $(\exp_p) \upharpoonright \widetilde{\mathcal{U}} \colon \widetilde{\mathcal{U}} \to \mathcal{U}$ is a diffeomorphism.*

Proof. The assertion is an immediate consequence of the preceding lemma and the inverse mapping theorem (Corollary 3.1.21). $\qquad\square$

Definition 5.1.47. (a) A subset H of a real vector space V is called *star-shaped about 0* (or *star-shaped* for short) if $v \in H$ implies $tv \in H$ for all $t \in [0,1]$.

(b) Let (M,S) be a spray manifold and $\exp \colon \widetilde{TM} \to M$ its exponential map. Given a point $p \in M$, an open neighbourhood of p is called a *normal neighbourhood* of p if it is the diffeomorphic image of a star-shaped open neighbourhood of $0_p \in T_pM$ under $\exp_p$. An open subset of M is said to be *totally normal* if it is a normal neighbourhood of each of its points.

Lemma 5.1.48. *Let (M, S) be a spray manifold. If $\mathcal{U}$ is a normal neighbourhood of a point $p \in M$, then for each point $q \in \mathcal{U}$ there exists a geodesic segment from p to q contained in $\mathcal{U}$, and all such geodesic segments differ only in a positive affine reparametrization.*

Proof. Let $\widetilde{\mathcal{U}}$ be a star-shaped open neighbourhood of $0_p \in T_pM$ such that $\exp_p(\widetilde{\mathcal{U}}) = \mathcal{U}$. Fix a point $q \in \mathcal{U}$ and set $v := (\exp_p \restriction \widetilde{\mathcal{U}})^{-1}(q)$. Then

$$\widetilde{\gamma} \colon t \in [0, 1] \mapsto \exp_p(tv)$$

is a geodesic segment from p to q. This geodesic segment is contained in $\mathcal{U}$, since $\widetilde{\mathcal{U}}$ is star-shaped, and hence $tv \in \widetilde{\mathcal{U}}$ if $t \in [0, 1]$.

To prove the uniqueness, let γ be a geodesic segment from p to q, contained in $\mathcal{U}$. After a positive affine reparametrization we may suppose that $\gamma(0) = p$ and $\gamma(1) = q$.

Consider the ray $\ell := \{t\,\dot{\gamma}(0) \in T_pM \mid t \in \mathbb{R}_+\}$ in T_pM. By (5.1.21),

$$\exp(t\,\dot{\gamma}(0)) = \gamma(t) \quad \text{for all } t \in [0, 1]. \tag{$*$}$$

Let $J := (\exp_p \restriction \widetilde{\mathcal{U}})^{-1}(\operatorname{Im}(\gamma))$. Since $\operatorname{Im}(\gamma) \subset \mathcal{U}$, and $\exp_p \restriction \widetilde{\mathcal{U}}$ is a diffeomorphism, J is a connected closed subset of ℓ, and hence it is a line segment contained also in $\widetilde{\mathcal{U}}$. Its endpoints are necessarily 0_p and $\delta\,\dot{\gamma}(0)$ for some positive real number δ.

Suppose that $\delta < 1$. Since $\widetilde{\mathcal{U}}$ is open, there exists a real number $\widetilde{\delta} \in \,]\delta, 1]$ such that $\widetilde{\delta}\,\dot{\gamma}(0)$ is still an element of $\widetilde{\mathcal{U}}$. Then, because $\exp_p$ is bijective on $\ell \cap \widetilde{\mathcal{U}}$, and $\exp_p(J) = \operatorname{Im}(\gamma)$, the point $\exp_p(\widetilde{\delta}\,\dot{\gamma}(0))$ cannot be in $\operatorname{Im}(\gamma)$. This is a contradiction, since the image of the line segment connecting 0_p and $\dot{\gamma}(0)$ under $\exp_p$ is $\operatorname{Im}(\gamma)$ by $(*)$. Thus $\delta \geq 1$, and hence $\dot{\gamma}(0) \in \widetilde{\mathcal{U}}$.

Now we can complete the proof. From $(*)$ we have

$$\exp_p(\dot{\gamma}(0)) = \gamma(1) = q$$

and hence $\dot{\gamma}(0) = (\exp_p \restriction \widetilde{\mathcal{U}})^{-1} = v$. Then $(*)$ implies that $\gamma = \widetilde{\gamma}$. $\qquad \square$

Remark 5.1.49. Let S be an affine spray over M, $\mathcal{U}$ a normal neighbourhood of a fixed point $p \in M$, and $\widetilde{\mathcal{U}}$ a star-shaped open neighbourhood of $0_p \in T_pM$ such that $\exp_p(\widetilde{\mathcal{U}}) = \mathcal{U}$. Since the exponential map of an affine spray is smooth, $\exp_p \restriction \widetilde{\mathcal{U}}$ is a smooth diffeomorphism. Thus, given a linear isomorphism $\varphi \colon T_pM \to \mathbb{R}^n$ ($n := \dim M$), the composite mapping $u := \varphi \circ (\exp_p \restriction \widetilde{\mathcal{U}})^{-1}$ yields a chart $(\mathcal{U}, u)$ of M around p. In view of Lemma 5.1.42, this chart has the convenient property that if γ is a geodesic starting from p, then $u \circ \gamma$ is a straight line through $0 \in \mathbb{R}^n$, i.e., $u \circ \gamma(t) = tv$ for some $v \in \mathbb{R}^n$.

A chart around p constructed in this manner is called a *normal chart* at this point.

Definition 5.1.50. Let (M, S) be a spray manifold. An open subset $\mathcal{U}$ of M is called

(i) *simple* if for each points $p, q \in \mathcal{U}$, all geodesic segments from p to q contained in $\mathcal{U}$ have the same image;

(ii) *convex* if for each point $p, q \in \mathcal{U}$, there is at least one geodesic segment from p to q contained in $\mathcal{U}$.

Remark 5.1.51. The following simple properties are immediate consequences of the definitions.

(a) An open subset of a simple set is also simple.

(b) A convex neighbourhood of a point contained in a normal neighbourhood of the point is also a normal neighbourhood of the point.

(c) A totally normal set is both simple and convex.

5.1.4 The Theorem of Whitehead

Our considerations in this subsection will be of purely local nature, so we assume that our base manifold M is an open subset (and hence an open submanifold) of $\mathbb{R}^n$. Obviously, we can do this without loss of generality. We identify the tangent bundle of M with the trivial $\mathbb{R}^n$-bundle $M \times \mathbb{R}^n$ as in Example C.2.2. We assume furthermore that $\mathbb{R}^n$ is equipped with the canonical scalar product $\langle\ ,\ \rangle$; $\|\ \|$ stands for the corresponding Euclidean norm. We use some basic concepts of convexity in a real vector space; these are explained (as we have already indicated before the statement of Hadamard's lemma) in some detail in subsection 9.1.1.

Proposition 5.1.52 (R. E. Traber, [95]). *In a spray manifold any point has a neighbourhood which is contained in a normal neighbourhood of each of its points.*

Proof. Let (M, S) be a spray manifold, and let p be a fixed point in M. Consider the mapping

$$(\exp, \mathrm{pr}_1)\colon \widetilde{TM} \to M \times M, \quad (q, v) \mapsto (\exp(q, v), q).$$

We calculate its derivative at $(p, 0)$. Applying (C.2.11), the formula $((\exp_p)_*)_{0_p} = \iota_{0_p}^{-1}$ obtained in Lemma 5.1.45 translates into $\exp_p'(0) = 1_{\mathbb{R}^n}$. Thus, taking into account that $\exp(q, 0) = q$ for all $q \in M$, we get

$$\exp'(p, 0)(v, w) = \exp'(p, 0)(v, 0) + \exp'(p, 0)(0, w)$$
$$= (1_{\mathbb{R}^n})'(p)(v) + \exp_p'(0)(w) = v + w.$$

Consequently,

$$(\exp, \mathrm{pr}_1)'(p, 0)(v, w) = \big(\exp'(p, 0)(v, w), (\mathrm{pr}_1)'(p, 0)(v, w)\big) = (v + w, v).$$

Then $(\exp, \mathrm{pr}_1)'(p, 0)$ is clearly a linear isomorphism, so there are open neighbourhoods $W \subset \widetilde{TM}$ of $(p, 0)$ and $\mathcal{V} \subset M \times M$ of (p, p) such that $(\exp, \mathrm{pr}_1) \upharpoonright W$ is a diffeomorphism onto $\mathcal{V}$.

If $\mathcal{U}$ is an open subset of M such that $\mathcal{U} \times \mathcal{U} \subset \mathcal{V}$, then for each $q \in \mathcal{U}$, the set $\exp_q^{-1}(\mathcal{U})$ is an open neighbourhood of $(q, 0)$, and the mapping $\exp_q \upharpoonright \exp_q^{-1}(\mathcal{U})$ is a diffeomorphism. Indeed, $\{q\} \times \exp_q^{-1}(\mathcal{U})$ is a subset of W containing $(q, 0)$ and $(\exp, \mathrm{pr}_1) \upharpoonright \{q\} \times \exp_q^{-1}(\mathcal{U})$ is just the mapping $(q, v) \mapsto (\exp_q(v), q)$ and hence is a diffeomorphism. Thus to prove the proposition, we need to find an open neighbourhood $\mathcal{U}$ of p such that $\mathcal{U} \times \mathcal{U} \subset \mathcal{V}$, and for each $q \in \mathcal{U}$ the set $\exp_q^{-1}(\mathcal{U})$ is contained in a star-shaped neighbourhood of 0_q, which is also a subset of $W \cap (\{q\} \times \mathbb{R}^n)$.

Let $(\mathcal{U}_m)_{m \in \mathbb{N}}$ be a sequence of open neighbourhoods of p in M satisfying the following conditions:

$$\mathcal{U}_m \times \mathcal{U}_m \subset \mathcal{V} \text{ and } \mathcal{U}_{m+1} \subset \mathcal{U}_m \text{ for all } m \in \mathbb{N}, \tag{$*$}$$

$$\lim_{m \to \infty} \mathrm{diam}(\mathcal{U}_m) = 0. \tag{$**$}$$

Define the sets

$$R_m := \big((\exp, \mathrm{pr}_1) \upharpoonright W\big)^{-1}(\mathcal{U}_m \times \mathcal{U}_m), \quad m \in \mathbb{N},$$

and let

$$r_m := \sup\big\{\|v\| \mid (q, v) \in R_m\big\} \in \ ]0, \infty], \quad m \in \mathbb{N}.$$

Then $R_m \subset \mathcal{U}_m \times B_{r_m}(0)$ for all $m \in \mathbb{N}$. From the continuity of the mapping $(\exp, \mathrm{pr}_1)$ and from condition $(**)$ it follows that

$$\lim_{m \to \infty} \mathrm{diam}\, R_m = 0 \text{ and } \lim_{m \to \infty} r_m = 0.$$

Thus we can find an index $N \in \mathbb{N}$ such that

$$\mathcal{U}_m \times B_{r_m}(0) \subset W \text{ whenever } m \geq N.$$

Now, by condition $(*)$, $\mathcal{U}_N \times \mathcal{U}_N \subset \mathcal{V}$, and for each $q \in \mathcal{U}_N$ we have

$$\{q\} \times \exp_q^{-1}(\mathcal{U}_N) \subset (\{p\} \times \mathbb{R}^n) \cap R_N \subset \{p\} \times B_{r_N}(0) \subset (\{p\} \times \mathbb{R}^n) \cap W.$$

The open ball $B_{r_N}(0)$ is star-shaped, so the open neighbourhood $\mathcal{U}_N$ of p satisfies our requirements. $\qquad \square$

Remark 5.1.53. Let U be an open subset of a spray manifold (M, S) which is contained in a normal neighbourhood of each of its points. Define the mapping

$$G\colon U \times U \times [0,1] \to M, \quad (a,b,t) \mapsto G(a,b,t) := \exp_a\left(t \exp_a^{-1}(b)\right).$$

Then G is of class C^1, and for any $a, b \in U$ the mapping

$$t \in [0,1] \mapsto G(a,b,t) \in M$$

is a geodesic segment from a to b by Lemma 5.1.42.

Theorem 5.1.54 (J. H. C. Whitehead [99]**).** *Each point of a spray manifold has a totally normal neighbourhood.*

Proof. Let (M, S) be a spray manifold, and let p be any point of M. Suppose that U is an open neighbourhood of p such that it is contained in a normal neighbourhood of each of its points. By Remark 5.1.51(b), we have only to show that U contains a convex neighbourhood of p.

Step 1. *There is a positive real number r, such that if the image of a geodesic γ is in the Euclidean ball $B_r(p)$, then the real function $\|\gamma\|^2$ is convex.* According to Lemma C.2.10, in our current setting the spray S is a mapping of the form

$$S\colon (p,v) \in M \times \mathbb{R}^n \mapsto S(p,v) := (p,v,v,S_2(p,v)),$$

where $S_2\colon M \times \mathbb{R}^n \to \mathbb{R}^n$ is a C^1 mapping, and it is 2^+-homogeneous in its second variable. The geodesic equation (5.1.15) translates into

$$\gamma'' = S_2 \circ (\gamma, \gamma').$$

Then

$$
\begin{aligned}
(\|\gamma\|^2)'' &= 2\langle \gamma', \gamma' \rangle + 2\langle \gamma, \gamma'' \rangle = 2\|\gamma'\|^2 + 2\langle \gamma, S_2 \circ (\gamma, \gamma') \rangle \\
&= 2\|\gamma'\|^2 (1 + \langle \gamma, S_2 \circ (\gamma, \|\gamma'\|^{-1}\gamma') \rangle).
\end{aligned}
\tag{$*$}
$$

Let R be a positive number such that $B_R(p) \subset U$. The mapping S_2 is continuous, so there is a positive number K such that $K \geq \|S(q,v)\|$ for all $(q,v) \in \mathrm{cl}(B_R(p)) \times \mathrm{bd}(B_1(0))$. If r is a positive number such that $r < \min\left\{\frac{1}{K}, R\right\}$, then for any $(q,v) \in B_r(p) \times \mathrm{bd}(B_1(0))$ we have

$$|\langle q, S_2(q,v) \rangle| \overset{(1.4.4)}{\leq} \|q\| \|S_2(q,v)\| < rK < 1.$$

If $\gamma(t) \in B_r(p)$, then $(\gamma(t), \|\gamma(t)\|^{-1}\gamma'(t)) \in B_r(p) \times \mathrm{bd}(B_1(0))$, hence

$$\langle \gamma(t), S_2(\gamma(t), \|\gamma'(t)\|^{-1}\gamma'(t)) \rangle \in \,]{-1}, 1[.$$

Comparing this with $(*)$, we see that $(\|\gamma\|^2)''(t) > 0$.

Step 2. *The open ball $B_r(p)$ is convex.* To see this, let

$$G\colon U \times U \times [0,1] \to M$$

be the mapping defined in Remark 5.1.53.

Suppose that for some $a, b \in B_r(p)$ with $\|a\| \geq \|b\|$ the geodesic $G(a, b, \cdot)$ is not contained in $B_r(p)$, i.e., there is a real number $t_1 \in [0,1]$ such that $\|G(a, b, t_1)\| \geq r$. Consider the parameterized line segment

$$\alpha\colon s \in [0,1] \mapsto (1-s)a + sb \in B_r(p).$$

Then for all $t \in [0,1]$,

$$G(a, \alpha(1), t) = G(a, b, t),$$
$$G(a, \alpha(0), t) = G(a, a, t) = a.$$

Since $\|a\| < r$, the continuity of G implies that for some $s_0, t_0 \in [0,1]$ we have

(i) $G(a, \alpha(s_0), t) \in B_r(p)$ for all $t \in [0,1]$;

(ii) $\|G(a, \alpha(s_0), t_0)\| > \|a\|$.

By Step 1, (i) yields that the function $f\colon t \in [0,1] \mapsto \|G(a, \alpha(s_0), t)\|^2 \in M$ must be convex. However,

$$f(0) = \|G(a, \alpha(s_0), 0)\|^2 = \|a\|^2,$$
$$f(1) = \|G(a, \alpha(s_0), 1)\|^2 = \|\alpha(s_0)\|^2 = \|(1-s_0)a + s_0 b\|^2 \leq \|a\|^2,$$

and (ii) gives $f(t_0) > \|a\|^2$. This means that $f(t_0) > \max\{f(0), f(1)\}$ which contradicts the convexity of f. □

Corollary 5.1.55. *In a spray manifold any neighbourhood of a point contains a totally normal neighbourhood of the point.*

Proof. This is obvious by step 2 of the proof of the preceding theorem. □

Proposition 5.1.56. *Let (M, S) be a spray manifold. Given a point $p \in M$ and a nonzero tangent vector $v \in T_pM$, there exist a point $q \neq p$ and a nonzero tangent vector $w \in T_qM$ satisfying the following conditions:*

(i) *p is in a normal neighbourhood of q;*

(ii) *the maximal geodesic*

$$\gamma_w\colon t \in I_w \mapsto \gamma_w(t) \overset{(5.1.21)}{=} \exp_q(tw)$$

runs through the point p and its velocity at p is λv for some positive λ.

Proof. Consider the maximal geodesic γ_v starting at p with initial velocity v. It is defined on an open interval which contains a closed interval $[-\delta, \delta]$ for some positive δ. Since γ_v is continuous, we may suppose that $\gamma_v([-\delta, \delta])$ is contained in a totally normal neighbourhood $\mathcal{U}$ of p. So if we set $q := \gamma_v(-\delta)$ and take into account that $\mathcal{U}$ is a normal neighbourhood of each of its points, we obtain assertion (i).

To prove (ii), suppose that $\mathcal{U} = \exp_q(\widetilde{\mathcal{U}})$, where $\widetilde{\mathcal{U}} \subset T_q M$ is a star-shaped open set. If $w_0 := \dot{\gamma}_v(-\delta)$, then we can find a positive number λ such that

$$w := \lambda w_0 \in \widetilde{\mathcal{U}} \subset T_q M.$$

Consider the maximal geodesic

$$\gamma_w \colon I_w \to M, \ t \mapsto \gamma_w(t) \overset{(5.1.21)}{=} \exp_q(tw).$$

Observe that γ_w is a positive affine reparametrization of γ_v, namely

$$\gamma_w(t) = \gamma_v(\lambda t - \delta) \text{ for all } t \in I_w.$$

Indeed, if $\widetilde{\gamma} := \gamma_v \circ \theta$, where $\theta(t) := \lambda t - \delta$, then $\widetilde{\gamma}$ is a positive affine reparametrization of γ_v, and hence it is also a geodesic by Lemma 5.1.37. The initial velocity of $\widetilde{\gamma}$ is

$$\dot{\widetilde{\gamma}}(0) = (\gamma_v \circ \theta)^{\cdot}(0) \overset{(3.1.10)}{=} \lambda \dot{\gamma}_v(-\delta) = w = \dot{\gamma}_w(0),$$

therefore γ_w and $\widetilde{\gamma}$ coincide. Thus

$$\gamma_w\left(\frac{\delta}{\lambda}\right) = \gamma_v(0) = p, \quad \dot{\gamma}_w\left(\frac{\delta}{\lambda}\right) \overset{(3.1.10)}{=} \lambda \dot{\gamma}_v(0) = \lambda v,$$

which proves that γ_w has the desired properties. $\qquad\square$

Remark 5.1.57. For convenient reference, we call the point q described by the proposition an *emanating point* for the given tangent vector $v \in T_p M$.

5.2   Lagrange Functions

5.2.1   *Regularity and Global Dynamics*

Definition 5.2.1. Let M be a manifold. A smooth function L on TM (or on $\mathring{T}M$) is sometimes called a *Lagrange function* or *Lagrangian*. Then we say that

$E_L := CL - L$ is the *energy function*,

$\theta_L := \mathcal{L}_J L$ is the *Lagrange one-form*,

$\omega_L := d\theta_L = d\mathcal{L}_J L$ is the *Lagrange two-form*

attached to L. A Lagrangian L is called *regular* if the Lagrange two-form ω_L is non-degenerate (in the sense of Definition 3.4.5).

Remark 5.2.2. Let $(\mathcal{U}, (u^i)_{i=1}^n)$ be a chart on M, and consider the induced chart $(\tau^{-1}(\mathcal{U}), ((x^i)_{i=1}^n, (y^i)_{i=1}^n))$ on TM. If $L \in C^\infty(TM)$ is a Lagrangian, then by (4.1.98) we have

$$\theta_L = \mathcal{L}_J L \underset{(\mathcal{U})}{=} \frac{\partial L}{\partial y^i} dx^i$$

(summation convention in force). Thus

$$\omega_L = d\theta_L \underset{(\mathcal{U})}{=} d\left(\frac{\partial L}{\partial y^j} dx^j\right) \overset{\text{Proposition 3.3.35}}{=} \left(d\frac{\partial L}{\partial y^j}\right) \wedge dx^j$$

$$\overset{(3.3.3)}{=} \frac{\partial^2 L}{\partial x^i \partial y^j} dx^i \wedge dx^j + \frac{\partial^2 L}{\partial y^i \partial y^j} dy^i \wedge dx^j,$$

and so, locally,

$$\omega_L \underset{(\mathcal{U})}{=} \frac{\partial^2 L}{\partial x^i \partial y^j} dx^i \wedge dx^j + \frac{\partial^2 L}{\partial y^i \partial y^j} dy^i \wedge dx^j. \tag{5.2.1}$$

Lemma 5.2.3. *A Lagrangian $L \in C^\infty(TM)$ is regular if, and only if, for any induced chart $(\tau^{-1}(\mathcal{U}), ((x^i)_{i=1}^n, (y^i)_{i=1}^n))$ on TM and any $v \in \tau^{-1}(\mathcal{U})$,*

$$\det\left(\frac{\partial^2 L}{\partial y^i \partial y^j}(v)\right) \neq 0.$$

Proof. By Corollary 3.4.6, ω_L is non-degenerate if, and only if, the $2n$-form $\omega_L^n \in \mathcal{A}_{2n}(TM)$ is a volume form. Since (5.2.1) implies that

$$\omega_L^n \underset{(\mathcal{U})}{=} \pm n! \det\left(\frac{\partial^2 L}{\partial y^i \partial y^j}\right) dy^1 \wedge \cdots \wedge dy^n \wedge dx^1 \wedge \cdots \wedge dx^n,$$

this holds if, and only if, $\det\left(\frac{\partial^2 L}{\partial y^i \partial y^j}\right)$ vanishes nowhere.   □

Lemma 5.2.4. *If ω_L is the Lagrange two-form attached to a Lagrangian $L \in C^\infty(TM)$, then for any vector fields ξ, η on TM,*

$$\omega_L(\xi, \eta) = \xi(\mathbf{J}\eta(L)) - \eta(\mathbf{J}\xi(L)) - \mathbf{J}[\xi, \eta](L), \qquad (5.2.2)$$

$$\omega_L(\mathbf{J}\xi, \eta) + \omega_L(\xi, \mathbf{J}\eta) = 0. \qquad (5.2.3)$$

Proof. From (3.3.32) with $k = 1$ we get

$$\omega_L(\xi, \eta) = d\mathcal{L}_\mathbf{J}L(\xi, \eta) = \xi\mathcal{L}_\mathbf{J}L(\eta) - \eta\mathcal{L}_\mathbf{J}L(\xi) - \mathcal{L}_\mathbf{J}L([\xi, \eta])$$
$$\overset{(4.1.97)}{=} \xi(dL(\mathbf{J}\eta)) - \eta(dL(\mathbf{J}\xi)) - dL(\mathbf{J}[\xi, \eta])$$
$$= \xi(\mathbf{J}\eta(L)) - \eta(\mathbf{J}\xi(L)) - \mathbf{J}[\xi, \eta](L),$$

as we claimed. From (5.2.2), taking into account that $\mathbf{J}^2 = 0$,

$$\omega_L(\mathbf{J}\xi, \eta) = \mathbf{J}\xi(\mathbf{J}\eta(L)) - \mathbf{J}[\mathbf{J}\xi, \eta](L),$$
$$\omega_L(\xi, \mathbf{J}\eta) = -\mathbf{J}\eta(\mathbf{J}\xi(L)) - \mathbf{J}[\xi, \mathbf{J}\eta](L).$$

Adding these two equalities, we find that

$$\omega_L(\mathbf{J}\xi, \eta) + \omega_L(\xi, \mathbf{J}\eta) = ([\mathbf{J}\xi, \mathbf{J}\eta] - \mathbf{J}[\mathbf{J}\xi, \eta] - \mathbf{J}[\xi, \mathbf{J}\eta])L$$
$$\overset{(3.3.18),(4.1.85)}{=} N_\mathbf{J}(\xi, \eta)(L) \overset{\text{Lemma } 4.1.59}{=} 0,$$

as was to be shown. $\qquad\square$

Lemma 5.2.5. *Let L be a Lagrangian on TM with attached energy E_L and Lagrange two-form ω_L. Then*

$$i_C\omega_L = \mathcal{L}_\mathbf{J}E_L, \qquad (5.2.4)$$

$$i_\mathbf{J}\omega_L = 0. \qquad (5.2.5)$$

Proof. We have

$$i_C\omega_L = i_C d\mathcal{L}_\mathbf{J}L \overset{(3.3.33)}{=} \mathcal{L}_C\mathcal{L}_\mathbf{J}L - d i_C\mathcal{L}_\mathbf{J}L$$
$$\overset{(4.1.102)}{=} \mathcal{L}_C\mathcal{L}_\mathbf{J}L - d(i_\mathbf{J}L - \mathcal{L}_\mathbf{J}i_C L) = \mathcal{L}_C\mathcal{L}_\mathbf{J}L$$
$$\overset{(4.1.103)}{=} \mathcal{L}_\mathbf{J}\mathcal{L}_C L - \mathcal{L}_\mathbf{J}L = \mathcal{L}_\mathbf{J}(CL - L) = \mathcal{L}_\mathbf{J}E_L,$$

thus proving (5.2.4).

Similarly,

$$i_\mathbf{J}\omega_L = i_\mathbf{J}d\mathcal{L}_\mathbf{J}L \overset{(3.3.52)}{=} -i_\mathbf{J}\mathcal{L}_\mathbf{J}dL \overset{(4.1.105)}{=} -\mathcal{L}_\mathbf{J}i_\mathbf{J}dL \overset{(3.3.43)}{=} -\mathcal{L}_\mathbf{J}^2 L \overset{(4.1.104)}{=} 0,$$

which completes the proof. $\qquad\square$

Lemma 5.2.6. *We keep the notation used above.*

(i) *If $S \in \mathfrak{X}(TM)$ is a second-order vector field, then*

$$i_{\mathbf{J}}(i_S\omega_L + dE_L) = 0. \tag{5.2.6}$$

(ii) *If L is a regular Lagrangian, and a vector field ξ on TM satisfies (5.2.6), then ξ is a second-order vector field.*

Proof. (i) If $S \in \mathfrak{X}(TM)$ is a second-order vector field, then

$$i_{\mathbf{J}}(i_S\omega_L + dE_L) \stackrel{(3.3.43)}{=} i_{\mathbf{J}}i_S\omega_L + \mathcal{L}_{\mathbf{J}}E_L \stackrel{(5.1.1)}{=} i_S i_{\mathbf{J}}\omega_L - i_C\omega_L + \mathcal{L}_{\mathbf{J}}E_L$$

$$\stackrel{(5.2.4),(5.2.5)}{=} -\mathcal{L}_{\mathbf{J}}E_L + \mathcal{L}_{\mathbf{J}}E_L = 0.$$

(ii) Now suppose that L is a regular Lagrangian, and let ξ be a vector field on TM such that $i_{\mathbf{J}}(i_\xi\omega_L + dE_L) = 0$, and hence $i_{\mathbf{J}}i_\xi\omega_L = -i_{\mathbf{J}}dE_L$. We manipulate both sides of this equality. On the one hand,

$$i_{\mathbf{J}}i_\xi\omega_L \stackrel{(3.3.48)}{=} i_\xi i_{\mathbf{J}}\omega_L - i_{\mathbf{J}\xi}\omega_L \stackrel{(5.2.5)}{=} -i_{\mathbf{J}\xi}\omega_L.$$

On the other hand,

$$-i_{\mathbf{J}}dE_L \stackrel{(3.3.43)}{=} -\mathcal{L}_{\mathbf{J}}E_L \stackrel{(5.2.4)}{=} -i_C\omega_L,$$

therefore $i_{\mathbf{J}\xi}\omega_L = i_C\omega_L$. Equivalently, $i_{\mathbf{J}\xi-C}\omega_L = 0$, which implies, by the non-degeneracy of ω_L, that $\mathbf{J}\xi = C$. Thus ξ is a second-order vector field, as was to be shown. $\square$

Definition 5.2.7. Let L be a Lagrangian on TM with attached energy E_L and Lagrange two-form ω_L. We say that L *admits global dynamics* if the set

$$\mathrm{Dyn}(L) := \{\xi \in \mathfrak{X}(TM) \mid i_\xi\omega_L = -dE_L\} \tag{5.2.7}$$

is nonempty.

Lemma 5.2.8. *We use the notation of the previous definition. If ξ belongs to $\mathrm{Dyn}(L)$, then*

(i) $\mathcal{L}_\xi\omega_L = 0$,

(ii) E_L *is a first integral for ξ.*

Proof. Both assertions can immediately be checked. As to the first, we have

$$\mathcal{L}_\xi\omega_L \stackrel{(3.3.33)}{=} (i_\xi \circ d + d \circ i_\xi)\omega_L = i_\xi dd\theta_L + di_\xi\omega_L = di_\xi\omega_L \stackrel{(5.2.7)}{=} 0,$$

as desired. As to the second,

$$\xi E_L = dE_L(\xi) \stackrel{(5.2.7)}{=} -i_\xi\omega_L(\xi) = -\omega_L(\xi,\xi) = 0,$$

which proves by Lemma 3.2.22 that E_L is a first integral of ξ. $\square$

Corollary and Definition 5.2.9. *Every regular Lagrangian admits a global dynamics. More precisely, if $L \in C^\infty(TM)$ is a regular Lagrangian, then there exists a unique second-order vector field $S \in \mathfrak{X}(TM)$ such that*

$$i_S \omega_L = -dE_L. \tag{5.2.8}$$

This second-order vector field is called the Lagrange vector field *(or Euler – Lagrange vector field) for L.*

Proof. By the non-degeneracy of ω_L, the mapping

$$(\omega_L)^\flat \colon \mathfrak{X}(TM) \to \mathfrak{X}^*(TM), \ \xi \mapsto (\omega_L)^\flat(\xi) := i_\xi \omega_L$$

is a $C^\infty(TM)$-linear isomorphism (cf. Corollary 1.4.1, and see also [1], 3.1.7 Proposition). Thus there exists a unique vector field S on TM such that $i_S \omega_L = -dE_L$. Then S clearly satisfies (5.2.6), therefore it is a second-order vector field. $\qquad\square$

Lemma 5.2.10. *If L is a 2^+-homogeneous Lagrangian, then $\mathcal{L}_C \omega_L = \omega_L$.*

Proof. Since $CL = 2L$, we have $E_L = L$. Thus

$$\mathcal{L}_C \omega_L = \mathcal{L}_C d\mathcal{L}_\mathbf{J} L \overset{(3.3.33)}{=} i_C dd\mathcal{L}_\mathbf{J} L + di_C d\mathcal{L}_\mathbf{J} L$$

$$= di_C \omega_L \overset{(5.2.4)}{=} d\mathcal{L}_\mathbf{J} E_L = d\mathcal{L}_\mathbf{J} L = \omega_L. \qquad\square$$

Corollary 5.2.11. *The Lagrange vector field for a 2^+-homogeneous regular Lagrangian is an affine spray.*

Proof. Let $L \in C^\infty(TM)$ be a 2^+-homogeneous regular Lagrangian. Then, as we just have seen, $E_L = L$, so the Lagrange vector field for L is the unique semispray S such that

$$i_S \omega_L = i_S d\mathcal{L}_\mathbf{J} L = -dL. \tag{5.2.9}$$

Then

$$i_{[C,S]} \omega_L = i_{[C,S]} d\mathcal{L}_\mathbf{J} L \overset{(3.3.31)}{=} \mathcal{L}_C i_S d\mathcal{L}_\mathbf{J} L - i_S \mathcal{L}_C d\mathcal{L}_\mathbf{J} L$$

$$\overset{(5.2.9)}{=} -\mathcal{L}_C dL - i_S \mathcal{L}_C \omega_L \overset{(3.3.35),\text{Lemma } 5.2.10}{=} -d\mathcal{L}_C L - i_S \omega_L$$

$$= -2dL + dL = -dL = i_S \omega_L,$$

therefore $i_{[C,S]} \omega_L = i_S \omega_L$. So, by the non-degeneracy of ω_L, it follows that $[C, S] = S$. $\qquad\square$

Corollary 5.2.12 (The Law of Conservation of Energy). *Let L be a regular Lagrangian on TM. If γ is a geodesic of the Lagrange vector field for L, then the energy E_L is constant along $\dot\gamma$, therefore $dE_L \circ \ddot\gamma = 0$.*

Proof.   This is immediate from Lemma 5.2.8(ii), (3.1.9) and (3.3.2).   □

Lemma 5.2.13. *Let $L \in C^\infty(TM)$ be a regular Lagrangian. In an induced chart $(\tau^{-1}(\mathcal{U}), ((x^i)_{i=1}^n, (y^i)_{i=1}^n))$ on TM, the Lagrange vector field S for L has the local expression*

$$S \underset{(\mathcal{U})}{=} y^i \frac{\partial}{\partial x^i} - 2G^i \frac{\partial}{\partial y^i}, \qquad (*)$$

where the forces G^i are given by

$$G^i = \frac{1}{2} L^{ij} \left(\frac{\partial^2 L}{\partial y^j \partial x^k} y^k - \frac{\partial L}{\partial x^j} \right), \ (L^{ij}) := \left(\frac{\partial^2 L}{\partial y^i \partial y^j} \right)^{-1} \qquad (5.2.10)$$

(summation convention is applied).

Proof.   Since S is a second-order vector field, its coordinate expression is of the form $(*)$ by (5.1.3). Therefore, our only task is to express the functions G^i in terms of the partial derivatives of L.

In view of (5.2.1),

$$i_S \omega_L \underset{(\mathcal{U})}{=} i_{y^k \frac{\partial}{\partial x^k} - 2G^k \frac{\partial}{\partial y^k}} \left(\frac{\partial^2 L}{\partial x^i \partial y^j} dx^i \wedge dx^j + \frac{\partial^2 L}{\partial y^i \partial y^j} dy^i \wedge dx^j \right)$$

$$\overset{(3.3.29)}{=} \frac{\partial^2 L}{\partial x^i \partial y^j} y^k \delta_k^i dx^j - \frac{\partial^2 L}{\partial x^i \partial y^j} y^k \delta_k^j dx^i - 2G^k \frac{\partial^2 L}{\partial y^i \partial y^j} \delta_k^i dx^j$$

$$- \frac{\partial^2 L}{\partial y^i \partial y^j} y^k \delta_k^j dy^i = \left(y^i \frac{\partial^2 L}{\partial x^i \partial y^j} - 2G^i \frac{\partial^2 L}{\partial y^i \partial y^j} \right) dx^j$$

$$- y^j \frac{\partial^2 L}{\partial x^i \partial y^j} dx^i - y^j \frac{\partial^2 L}{\partial y^i \partial y^j} dy^i.$$

On the other hand,

$$-dE_L = dL - d(CL) \underset{(\mathcal{U})}{=} \frac{\partial L}{\partial x^j} dx^j + \frac{\partial L}{\partial y^j} dy^j - d\left(y^j \frac{\partial L}{\partial y^j} \right)$$

$$= \frac{\partial L}{\partial x^j} dx^j - y^j \frac{\partial^2 L}{\partial x^i \partial y^j} dx^i - y^j \frac{\partial^2 L}{\partial y^i \partial y^j} dy^i.$$

Thus relation $i_S \omega_L = -dE_L$ implies that

$$y^i \frac{\partial^2 L}{\partial x^i \partial y^j} - 2G^i \frac{\partial^2 L}{\partial y^i \partial y^j} = \frac{\partial L}{\partial x^j}, \quad j \in J_n. \qquad (**)$$

By Lemma 5.2.3, the matrix $\left(\frac{\partial^2 L}{\partial y^i \partial y^j} \right) =: (L_{ij})$ is invertible at every point of $\tau^{-1}(\mathcal{U})$; let $(L^{ij}) := (L_{ij})^{-1}$. Then from $(**)$ we get

$$2G^i L_{ij} = \frac{\partial^2 L}{\partial y^j \partial x^k} y^k - \frac{\partial L}{\partial x^j},$$

from which
$$2G^l L_{lj} L^{ji} = \left(\frac{\partial^2 L}{\partial y^j \partial x^k} y^k - \frac{\partial L}{\partial x^j} \right) L^{ji}.$$
Thus, taking into account the symmetry of (L^{ij}),
$$G^i = \frac{1}{2} L^{ij} \left(\frac{\partial^2 L}{\partial y^j \partial x^k} y^k - \frac{\partial L}{\partial x^j} \right),$$
as was to be shown. $\qquad\square$

Lemma 5.2.14. *Let L be a regular Lagrangian on TM. Suppose that $\gamma: I \to M$ is a smooth curve such that $\mathrm{Im}(\gamma)$ is contained in the domain of a chart $(\mathcal{U}, (u^i)_{i=1}^n)$ of M. Then γ is a geodesic of the Lagrange vector field determined by L if, and only if,*
$$\left(\frac{\partial L}{\partial y^i} \circ \dot{\gamma} \right)' - \frac{\partial L}{\partial x^i} \circ \dot{\gamma} = 0, \quad i \in J_n, \tag{5.2.11}$$
where x^i, y^i are the coordinate functions of the chart induced by $(\mathcal{U}, (u^i)_{i=1}^n)$.

Proof. Let $S \in \mathrm{Dyn}(L)$ be the Lagrange vector field of L. In view of (5.1.18), (5.2.10) and (3.1.36), γ is a geodesic of S if, and only if, its component functions $\gamma^i = u^i \circ \gamma$ satisfy
$$(\gamma^i)'' + (L^{ij} \circ \dot{\gamma}) \left((\gamma^k)' \left(\frac{\partial^2 L}{\partial x^k \partial y^j} \circ \dot{\gamma} \right) - \frac{\partial L}{\partial x^j} \circ \dot{\gamma} \right) = 0, \quad i \in J_n. \tag{$*$}$$
Since
$$\left(\frac{\partial L}{\partial y^j} \circ \dot{\gamma} \right)' = \left(\frac{\partial^2 L}{\partial x^k \partial y^j} \circ \dot{\gamma} \right) (x^k \circ \dot{\gamma})' + \left(\frac{\partial^2 L}{\partial y^k \partial y^j} \circ \dot{\gamma} \right) (y^k \circ \dot{\gamma})'$$
$$\overset{(3.1.36)}{=} \left(\frac{\partial^2 L}{\partial x^k \partial y^j} \circ \dot{\gamma} \right) (\gamma^k)' + (L_{jk} \circ \dot{\gamma})(\gamma^k)'',$$
it follows that
$$(L^{ij} \circ \dot{\gamma}) \left(\frac{\partial^2 L}{\partial x^k \partial y^j} \circ \dot{\gamma} \right) (\gamma^k)' = (L^{ij} \circ \dot{\gamma}) \left(\frac{\partial L}{\partial y^j} \circ \dot{\gamma} \right)'$$
$$- ((L^{ij} L_{jk}) \circ \dot{\gamma})(\gamma^k)'' = (L^{ij} \circ \dot{\gamma}) \left(\frac{\partial L}{\partial y^j} \circ \dot{\gamma} \right)' - (\gamma^i)''.$$
Thus $(*)$ takes the form
$$(L^{ij} \circ \dot{\gamma}) \left(\left(\frac{\partial L}{\partial y^j} \circ \dot{\gamma} \right)' - \frac{\partial L}{\partial x^j} \circ \dot{\gamma} \right) = 0,$$
which is equivalent to (5.2.11). $\qquad\square$

In the situation described by the lemma we also say that the *Euler–Lagrange equations*
$$\frac{d}{dt} \frac{\partial L}{\partial y^i} - \frac{\partial L}{\partial x^i} = 0, \quad i \in J_n \tag{5.2.12}$$
are satisfied along $\dot{\gamma}$.

5.2.2   First Variation

In this subsection we present a brief introduction to calculus of variations in the context of manifolds. Our exposition is based on Section F in Ch. 1 of Besse's monograph [17].

Let M be a manifold. Throughout in the following, L denotes a Lagrangian on TM. Recall that $E_L := CL - L$ is the energy function, and $\omega_L := d\mathcal{L}_{\mathbf{J}} L$ is the Lagrange two-form attached to L. The results obtained in this setting are also valid for Lagrangians defined on $\mathring{T}M$.

We shall denote by I and J nonempty intervals in $\mathbb{R}$. In particular, we assume that $0 \in J$. The canonical coordinate system on $J \times I \subset \mathbb{R}^2$ is (e^1, e^2). If φ is a smooth mapping from $J \times I$ to M, then its partial derivatives are

$$\dot{\varphi}_i \overset{(3.2.14)}{=} \varphi_* \circ \frac{\partial}{\partial e^i}, \quad i \in \{1, 2\}.$$

Keeping these in mind, we begin with some simple observations which will be used later.

Definition and Lemma 5.2.15. Let $\gamma \colon I \to M$ be a smooth curve and let X be a vector field along γ. The *complete lift* of X is the mapping

$$X^c := \kappa \circ \dot{X} \colon I \to TTM \qquad (5.2.13)$$

(cf. (4.1.63)). Then

$$\tau_{TM} \circ X^c = \dot{\gamma}, \qquad (5.2.14)$$

therefore X^c is a vector field along $\dot{\gamma}$. We have

$$\tau_* \circ X^c = X, \qquad (5.2.15)$$
$$\mathbf{J} \circ X^c = \mathbf{i} \circ (\dot{\gamma}, X). \qquad (5.2.16)$$

Proof. By immediate calculations,

$$\tau_{TM} \circ X^c = \tau_{TM} \circ \kappa \circ \dot{X} \overset{\text{Proposition 4.1.8}}{=} \tau_* \circ \dot{X} \overset{(3.1.8)}{=} \tau_* \circ X_* \circ \frac{d}{dr}$$

$$= (\tau \circ X)_* \circ \frac{d}{dr} = \gamma_* \circ \frac{d}{dr} = \dot{\gamma},$$

$$\tau_* \circ X^c = \tau_* \circ \kappa \circ \dot{X} \overset{\text{Proposition 4.1.8}}{=} \tau_{TM} \circ \dot{X} = X,$$

so we have (5.2.14) and (5.2.15). Then

$$\mathbf{J} \circ X^c = \mathbf{i} \circ \mathbf{j} \circ X^c \overset{(4.1.35)}{=} \mathbf{i} \circ (\tau_{TM}, \tau_*) \circ X^c \overset{(5.2.14),(5.2.15)}{=} \mathbf{i} \circ (\dot{\gamma}, X),$$

which proves (5.2.16). $\qquad\qquad\qquad\qquad\qquad\qquad\qquad\qquad\qquad \square$

Lemma and Definition 5.2.16. *Let* $\gamma\colon I \to M$ *be a smooth curve. Then the mapping*

$$\begin{cases} \mathcal{E}\colon I \to T^*TM, \ t \mapsto \mathcal{E}_t, \\ \mathcal{E}_t(w) := (dE_L)_{\dot{\gamma}(t)}(w) + (\omega_L)_{\dot{\gamma}(t)}(\ddot{\gamma}(t), w), \ w \in T_{\dot{\gamma}(t)}TM \end{cases}$$

is a one-form along $\dot{\gamma}$*, i.e.,* $\mathcal{E} \in \mathfrak{X}^*_{\dot{\gamma}}(TM)$*. We write*

$$\mathcal{E} = (dE_L)_{\dot{\gamma}} + i_{\ddot{\gamma}}(\omega_L)_{\dot{\gamma}}, \qquad (5.2.17)$$

and we call $\mathcal{E}$ *the* Euler–Lagrange *one-form along* $\dot{\gamma}$*. If* $\mathcal{E} = 0$*, we say that the acceleration of* γ *satisfies the Euler–Lagrange equation along* $\dot{\gamma}$*.*

The assertion is obvious.

Lemma 5.2.17. *Let* $\gamma\colon I \to M$ *be a smooth curve, and let* $\mathcal{E} \in \mathfrak{X}^*_{\dot{\gamma}}(TM)$ *be the Euler–Lagrange one-form along* $\dot{\gamma}$*. Define a mapping* $\eta\colon I \to T^*M$ *as follows:*

Given a tangent vector $v \in T_{\gamma(t)}M$, choose a vector field Y along γ such that $Y(t) = v$, and let $\eta_t(v) := \mathcal{E}_t(Y^c(t))$. $\qquad$ (5.2.18)

Then:

(i) η *is well-defined;*
(ii) η *is a one-form along* γ*, i.e.,* $\eta \in \mathfrak{X}^*_{\gamma}(M)$*;*
(iii) $\mathcal{E} = \tau^*\eta$*.*

Proof. **Step 1.** Suppose that $\xi_1, \xi_2 \in \mathfrak{X}_{\dot{\gamma}}(TM)$ are such that $\tau_* \circ \xi_1 = \tau_* \circ \xi_2$. Then $\mathcal{E}(\xi_1) = \mathcal{E}(\xi_2)$.

Indeed, $\tau_* \circ \xi_1 = \tau_* \circ \xi_2$ implies that the values of $\xi_1 - \xi_2$ are vertical. On the other hand, by (5.2.6), $\mathcal{E}$ vanishes on vertical vectors. Thus, for every $t \in I$,

$$0 = \mathcal{E}_t((\xi_1 - \xi_2)(t)) = (\mathcal{E}_t(\xi_1) - \mathcal{E}_t(\xi_2))(t),$$

whence our claim.

Step 2. We show that the mapping given by (5.2.18) is well-defined.

Let $t \in I$, and let $Y_1, Y_2 \in \mathfrak{X}_{\gamma}(M)$ such that $Y_1(t) = Y_2(t)$. Since $\tau_* \circ Y_1^c = Y_1$ and $\tau_* \circ Y_2^c = Y_2$ by (5.2.15), it follows that $Y_1^c(t) - Y_2^c(t)$ is a vertical vector. Hence, as above, $\mathcal{E}_t(Y_1^c(t)) = \mathcal{E}_t(Y_2^c(t))$. This is what was to be shown.

Step 3. We prove that $\eta \in \mathfrak{X}^*_{\gamma}(M)$.

The smoothness of the mapping $\eta\colon I \to TM$ is clear because $\eta(X) = \mathcal{E}(X^c)$ for every $X \in \mathfrak{X}_{\gamma}(M)$. It remains to check that $\eta_t \in T^*_{\gamma(t)}M$ for all $t \in I$.

Let $v_1, v_2 \in T_{\gamma(t)}M$, and let $Y_1, Y_2 \in \mathfrak{X}_\gamma(M)$ such that $Y_1(t) = v_1$, $Y_2(t) = v_2$. Then

$$\eta_t(v_1 + v_2) := \mathcal{E}_t((Y_1 + Y_2)^c(t)).$$

Here

$$(Y_1 + Y_2)^c(t) := \kappa(Y_1 + Y_2)^{\cdot}(t) \overset{(3.1.8)}{=} \kappa\left(((Y_1 + Y_2)_*)_t\left(\frac{d}{dr}\right)_t\right)$$

$$\overset{(4.1.25)}{=} \kappa\left((Y_1)_*\left(\frac{d}{dr}\right)_t \boxplus (Y_2)_*\left(\frac{d}{dr}\right)_t\right)$$

$$\overset{(4.1.21)}{=} \kappa\left((Y_1)_*\left(\frac{d}{dr}\right)_t\right) + \kappa\left((Y_2)_*\left(\frac{d}{dr}\right)_t\right)$$

$$= \kappa(\dot{Y}_1(t)) + \kappa(\dot{Y}_2(t)) \overset{(5.2.13)}{=:} Y_1^c(t) + Y_2^c(t).$$

Thus

$$\eta_t(v_1 + v_2) = \mathcal{E}_t(Y_1^c(t) + Y_2^c(t)) = \mathcal{E}_t(Y_1^c(t)) + \mathcal{E}_t(Y_2^c(t)) = \eta_t(v_1) + \eta_t(v_2),$$

so the function $\eta_t \colon T_{\gamma(t)}M \to \mathbb{R}$ is additive. It can similarly be shown that

$$\eta_t(\lambda v) = \lambda \eta_t(v); \quad t \in I, \ \lambda \in \mathbb{R}, \ v \in T_{\gamma(t)}M.$$

Step 4. We prove finally that $\mathcal{E} = \tau^* \eta$.

Let $\xi \in \mathfrak{X}_{\dot\gamma}(TM)$, and let $X := \tau_* \circ \xi$. Then

$$\tau_* \circ X^c \overset{(5.2.15)}{=} X := \tau_* \circ \xi,$$

therefore

$$\mathcal{E}(\xi) \overset{\text{Step 1}}{=} \mathcal{E}(X^c) = \eta(X) = \eta(\tau_* \circ \xi),$$

as was to be shown. $\qquad\qquad\qquad\qquad\qquad\qquad\qquad\qquad\qquad\qquad\quad \square$

Definition 5.2.18. Let $\gamma \colon I \to M$ be a smooth curve.

(a) A *variation* of γ is a smooth mapping $\varphi \colon J \times I \to M$ such that

$$\varphi(0, t) = \gamma(t) \quad \text{for all } t \in I. \tag{5.2.19}$$

For any fixed $s \in J$, the smooth curve $\gamma_s \colon I \to M$, $t \mapsto \gamma_s(t) := \varphi(s, t)$ is called a *longitudinal curve* of the variation. We say, in particular, that $\gamma_0 = \gamma$ is the *base curve* of φ. Analogously, the *transverse curves* of the variation are $\gamma^t \colon J \to M$, $s \mapsto \gamma^t(s) := \varphi(s, t)$, for any fixed $t \in I$.

(b) Let $a, b \in I$, $a < b$. A variation $\varphi \colon J \times I \to M$ of γ is a *proper* or *fixed-endpoint variation over* $[a, b]$ if

$$\varphi(s, a) = \gamma(a) \quad \text{and} \quad \varphi(s, b) = \gamma(b) \quad \text{for all } s \in J. \tag{5.2.20}$$

(c) The *variation field* of a variation $\varphi \colon J \times I \to M$ of γ is the mapping

$$X \colon I \to TM, \ t \mapsto X(t) := \dot\varphi_1(0, t). \tag{5.2.21}$$

Lemma 5.2.19. *We use the same notation as in the preceding definition. The variation field X of φ is a vector field along γ. For every $t \in I$, the vector $X(t)$ is the initial velocity of the transverse curve γ^t, i.e.,*

$$X(t) = (\gamma^t)\dot{\ }(0). \tag{5.2.22}$$

If φ is a proper variation of γ over $[a,b]$, then $X(a) = X(b) = 0$.

Proof. It is clear that X is a smooth mapping. Let $\tau_{J \times I}$ be the tangent bundle projection $T(J \times I) \to J \times I$. Then, for every $t \in I$,

$$\tau \circ X(t) := \tau \circ \dot{\varphi}_1(0,t) = \tau \circ \varphi_* \circ \frac{\partial}{\partial e^1}(0,t)$$

$$\overset{\text{Lemma 3.1.33}}{=} \varphi \circ \tau_{J \times I} \circ \frac{\partial}{\partial e^1}(0,t) \overset{(5.2.19)}{=} \gamma(t),$$

whence $X \in \mathfrak{X}_\gamma(M)$.

Now suppose that $t \in I$ is fixed. Define the mapping

$$\alpha^t \colon J \to J \times I, \quad s \mapsto \alpha^t(s) := (s,t) = (0,t) + se_1$$

(cf. Remark 3.2.34). Then $\gamma^t = \varphi \circ \alpha^t$, and

$$(\gamma^t)\dot{\ }(0) \overset{(3.2.15)}{=} \dot{\varphi}_1(\alpha^t(0)) = \dot{\varphi}_1(0,t) =: X(t).$$

If φ is a proper variation over $[a,b]$, then $X(a) = (\gamma^a)\dot{\ }(0) = 0$, since γ^a is constant by (5.2.20). Similarly, $X(b) = (\gamma^b)\dot{\ }(0) = 0$. $\qquad\square$

Lemma 5.2.20. *Let $\varphi \colon J \times I \to M$ be a variation of a smooth curve $\gamma \colon I \to M$, and let X be the variation field of φ. If $\Phi := \dot{\varphi}_2$ then*

$$\dot{\gamma}_s(t) = \Phi(s,t), \quad \ddot{\gamma}_s(t) = \dot{\Phi}_2(s,t), \quad X^c(t) = \dot{\Phi}_1(0,t) \tag{5.2.23}$$

for all $(s,t) \in J \times I$. For the pull-back of the one-form $\mathcal{L}_{\mathbf{J}}L$ by Φ we have

$$(\Phi^* \mathcal{L}_{\mathbf{J}}L)\left(\frac{\partial}{\partial e^2}\right) = L \circ \Phi + E_L \circ \Phi. \tag{5.2.24}$$

Proof. For any fixed $s \in J$, define the mapping

$$\alpha_s \colon I \to J \times I, \quad t \mapsto \alpha_s(t) := (s,t) = (s,0) + te_2.$$

Then $\gamma_s = \varphi \circ \alpha_s$, $X = \dot{\varphi}_1 \circ \alpha_0$, and we find that

$$\dot{\gamma}_s(t) \overset{(3.2.15)}{=} \dot{\varphi}_2(\alpha_s(t)) = \dot{\varphi}_2(s,t) =: \Phi(s,t),$$

$$\ddot{\gamma}_s(t) \overset{(3.2.15)}{=} \ddot{\varphi}_{22}(s,t) = ((\dot{\varphi}_2)\dot{\ })_2(s,t) = \dot{\Phi}_2(s,t),$$

$$X^c(t) := \kappa(\dot{X}(t)) \overset{(5.2.21)}{=} \kappa((\dot{\varphi}_1 \circ \alpha_0)\dot{\ }(t)) \overset{(3.2.15)}{=} \kappa(((\dot{\varphi}_1)\dot{\ })_2(0,t))$$

$$=: \kappa(\ddot{\varphi}_{12}(0,t)) \overset{\text{Lemma 4.1.50}}{=} \ddot{\varphi}_{21}(0,t) := ((\dot{\varphi}_2)\dot{\ })_1(0,t) = \dot{\Phi}_1(0,t),$$

thus proving (5.2.23).

Finally, for every $(s,t) \in J \times I$,

$$
(\Phi^* \mathcal{L}_{\mathbf{J}} L) \left(\frac{\partial}{\partial e^2} \right) (s,t) = (\Phi^* \mathcal{L}_{\mathbf{J}} L)_{(s,t)} \left(\frac{\partial}{\partial e^2} \right)_{(s,t)}
$$

$$
= (\mathcal{L}_{\mathbf{J}} L)_{\Phi(s,t)} \left((\Phi_*)_{(s,t)} \left(\frac{\partial}{\partial e^2} \right)_{(s,t)} \right) = (\mathcal{L}_{\mathbf{J}} L)_{\Phi(s,t)} (\dot{\Phi}_2(s,t))
$$

$$
\overset{(5.2.23)}{=} (\mathcal{L}_{\mathbf{J}} L)_{\dot{\gamma}_s(t)} (\ddot{\gamma}_s(t)) \overset{(4.1.88)}{=} \mathbf{J}(\ddot{\gamma}_s(t)) L \overset{(4.1.88)}{=} C(\dot{\gamma}_s(t)) L
$$

$$
\overset{(5.2.23)}{=} CL(\Phi(s,t)) = (E_L \circ \Phi + L \circ \Phi)(s,t),
$$

whence (5.2.24). This concludes the proof. $\qquad \square$

Lemma 5.2.21. *Let $I \subset \mathbb{R}$ be a compact interval, $\gamma \colon I \to M$ a smooth curve and X a vector field along γ. Then there is a variation $\varphi \colon J \times I \to M$ whose variation vector field is X. If $X(a) = X(b) = 0$ for some $a, b \in I$, $a < b$, then the variation can be taken to be proper over $[a,b]$.*

Proof. We show that given an affine spray S over M with exponential map $\exp \colon \widetilde{TM} \to M$, there is an open interval J containing 0 such that

$$
\varphi \colon (s,t) \in J \times I \mapsto \exp(sX(t)) \in M
$$

is a suitable variation.

Since I is compact, from Remark 3.2.14 it follows that there is an open interval J containing zero such that for each $t \in J$ the maximal integral curve of S starting from $X(t)$ is defined on J. Equivalently, the maximal geodesic of S with initial velocity $X(t)$ is defined on J (see Remark 5.1.28). From Corollary 5.1.39, we infer that for all $(s,t) \in J \times I$ the maximal geodesic with initial velocity $sX(t)$ is defined at 1, thus $sX(t) \in \widetilde{TM}$. Therefore, the mapping φ is well-defined. It is also smooth, since the exponential map of an affine spray is smooth (see the proof of Proposition 5.1.44). The variation vector field of φ is indeed X, because the initial velocity of the curve

$$
\gamma^t \colon s \in J \mapsto \varphi(s,t) = \exp(sX(t)) \in M
$$

is just $X(t)$ by Lemma 5.1.42. It is clear that if $X(a) = X(b) = 0$, then γ^a and γ^b are constant, thus φ is proper over $[a,b]$.

The above argument proves the lemma only if the existence of an affine spray over M is guaranteed. This will be clear later, from Proposition 6.1.7 and Proposition and Definition 7.5.13. $\qquad \square$

Lemma 5.2.22 (Fundamental Lemma of Calculus of Variations).
Let $\gamma\colon I \to M$ be a smooth curve and $\theta\colon I \to T^*M$ a one-form along γ. Suppose that $a, b \in I$, $a < b$. If

$$\int_a^b \theta(X) = 0$$

for any vector field X along γ such that $X(a) = X(b) = 0$, then θ vanishes on $[a, b]$.

Proof. Consider the pull-back bundles γ^*T^*M and γ^*TM. Their base manifold I is contractible, so by Example 2.2.33(c) they are trivializable:

$$\gamma^*T^*M \cong I \times (\mathbb{R}^n)^*, \quad \gamma^*TM \cong I \times \mathbb{R}^n \quad (n := \dim M).$$

Thus θ can be regarded as a smooth mapping from I to $(\mathbb{R}^n)^*$, and so it can be written in the form

$$\theta = \theta_i e^i, \quad \theta_i \in C^\infty(I),$$

where $(e^i)_{i=1}^n$ is the dual of the canonical basis $(e_i)_{i=1}^n$ of $\mathbb{R}^n$. Similarly, if $X \in \mathfrak{X}_\gamma(M)$, then we can write

$$X = X^i e_i, \quad X^i \in C^\infty(I).$$

Using this interpretation, our lemma can be stated as follows:

If $\theta = \theta_i e^i \in C^\infty(I, (\mathbb{R}^n)^*)$ satisfies

$$\sum_{i=1}^n \int_a^b \theta_i X^i = 0 \tag{1}$$

for every family $(X^i)_{i=1}^n$ of smooth functions on I such that

$$X^i(a) = X^i(b) = 0, \quad i \in J_n, \tag{2}$$

then $\theta \restriction [a, b] = 0$.

To prove this, choose a smooth function f on $[a, b]$ such that

$$f(a) = f(b) = 0; \quad f(t) > 0 \quad \text{if } t \in {]}a, b{[}.$$

Let $j \in J_n$ be a fixed index and define the functions $X^i \in C^\infty([a, b])$ as follows:

$$X^i := \begin{cases} 0 & \text{if } i \neq j \\ f\theta_j & \text{if } i = j \end{cases}.$$

Then condition (2) holds, and (1) takes the form

$$\int_a^b (\theta_j)^2 f = 0.$$

Since f is positive on ${]}a, b{[}$, from this we infer that $\theta_j \restriction [a, b] = 0$. Hence, by the arbitrariness of the index j, it follows that $\theta \restriction [a, b] = 0$. $\qquad\square$

Proposition 5.2.23 (First Variation Formula). *Let* $\gamma\colon I \to M$ *be a smooth curve, and let* $a, b \in I$, $a < b$. *Suppose that* $\varphi\colon J \times I \to M$ *is a proper variation of* γ *over* $[a, b]$, *and let* $X \in \mathfrak{X}_\gamma(M)$ *be the variation field of* φ. *Define the function*

$$L_\varphi\colon J \to \mathbb{R}, \ s \mapsto L_\varphi(s) := \int_a^b L \circ \dot\gamma_s \stackrel{(5.2.23)}{=} \int_a^b L \circ \Phi_s, \qquad (5.2.25)$$

where γ_s *is the longitudinal curve belonging to the parameter* s, *and* $\Phi_s(t) := \Phi(s, t)$ *for all* $t \in I$. *Then*

$$(L_\varphi)'(0) = \int_a^b \mathcal{E}(X^c). \qquad (5.2.26)$$

Proof.   (a) Using (5.2.24), the integrand in (5.2.25) can be expressed as

$$L \circ \Phi_s = (\Phi^* \mathcal{L}_\mathbf{J} L)\left(\frac{\partial}{\partial e^2}\right) \circ \alpha_s - E_L \circ \Phi \circ \alpha_s,$$

where, as above, $\alpha_s\colon t \in I \mapsto \alpha_s(t) := (s, t) \in J \times I$. Thus, applying Lemma C.1.23,

$$(L_\varphi)'(0) = \int_a^b \frac{\partial}{\partial e^1}\left((\Phi^* \mathcal{L}_\mathbf{J} L)\left(\frac{\partial}{\partial e^2}\right) - E_L \circ \Phi\right) \circ \alpha_0. \qquad (1)$$

(b) We calculate the partial derivatives in (1):

$$\frac{\partial}{\partial e^1}\left(\Phi^*(\mathcal{L}_\mathbf{J} L)\left(\frac{\partial}{\partial e^2}\right)\right) = \frac{\partial}{\partial e^1}(i_{\frac{\partial}{\partial e^2}}(\Phi^*(\mathcal{L}_\mathbf{J} L)))$$

$$\stackrel{(3.3.2)}{=} di_{\frac{\partial}{\partial e^2}}\Phi^*(\mathcal{L}_\mathbf{J} L)\left(\frac{\partial}{\partial e^1}\right)$$

$$\stackrel{(3.3.33)}{=} \mathcal{L}_{\frac{\partial}{\partial e^2}}\Phi^*(\mathcal{L}_\mathbf{J} L)\left(\frac{\partial}{\partial e^1}\right) - i_{\frac{\partial}{\partial e^2}}d(\Phi^*(\mathcal{L}_\mathbf{J} L))\left(\frac{\partial}{\partial e^1}\right)$$

$$\stackrel{\text{Proposition } 3.3.35(v)}{=} \frac{\partial}{\partial e^2}\left(\Phi^*(\mathcal{L}_\mathbf{J} L)\left(\frac{\partial}{\partial e^1}\right)\right) - \Phi^* d\mathcal{L}_\mathbf{J} L\left(\frac{\partial}{\partial e^2}, \frac{\partial}{\partial e^1}\right).$$

$$= \frac{\partial}{\partial e^2}\left(\Phi^*(\mathcal{L}_\mathbf{J} L)\left(\frac{\partial}{\partial e^1}\right)\right) - \Phi^* \omega_L\left(\frac{\partial}{\partial e^2}, \frac{\partial}{\partial e^1}\right);$$

$$\frac{\partial}{\partial e^1}(E_L \circ \Phi) \stackrel{(3.3.2)}{=} d(E_L \circ \Phi)\left(\frac{\partial}{\partial e^1}\right) = d(\Phi^* E_L)\left(\frac{\partial}{\partial e^1}\right)$$

$$= \Phi^*(dE_L)\left(\frac{\partial}{\partial e^1}\right).$$

Thus (1) takes the form

$$(L_\varphi)'(0) = \int_a^b \frac{\partial}{\partial e^2}\left(\Phi^*(\mathcal{L}_\mathbf{J} L)\left(\frac{\partial}{\partial e^1}\right)\right) \circ \alpha_0$$

$$- \int_a^b \left(\Phi^* \omega_L\left(\frac{\partial}{\partial e^2}, \frac{\partial}{\partial e^1}\right) + \Phi^*(dE_L)\left(\frac{\partial}{\partial e^1}\right)\right) \circ \alpha_0. \qquad (2)$$

(c) We show that the first integral at the right-hand side of (2) is zero. Indeed,

$$\int_a^b \frac{\partial}{\partial e^2}\left(\Phi^*(\mathcal{L}_{\mathbf{J}}L)\left(\frac{\partial}{\partial e^1}\right)\right) \circ \alpha_0 = \Phi^*(\mathcal{L}_{\mathbf{J}}L)\left(\frac{\partial}{\partial e^1}\right)(0,b)$$

$$-\Phi^*(\mathcal{L}_{\mathbf{J}}L)\left(\frac{\partial}{\partial e^1}\right)(0,a) = \mathcal{L}_{\mathbf{J}}L(\dot{\Phi}_1(0,b)) - \mathcal{L}_{\mathbf{J}}L(\dot{\Phi}_1(0,a))$$

$$\overset{(5.2.23)}{=} \mathcal{L}_{\mathbf{J}}L(X^c(b)) - \mathcal{L}_{\mathbf{J}}L(X^c(a)) = dL(\mathbf{J} \circ X^c(b)) - dL(\mathbf{J} \circ X^c(a))$$

$$\overset{(5.2.16)}{=} dL\big(\mathbf{i}(\dot{\gamma}(b), X(b))\big) - dL\big(\mathbf{i}(\dot{\gamma}(a), X(a))\big)$$

$$\overset{\text{Lemma } 5.2.19}{=} dL\big(\mathbf{i}(\dot{\gamma}(b),0)\big) - dL\big(\mathbf{i}(\dot{\gamma}(a),0)\big) = 0.$$

(d) The second integrand in (2) can be manipulated as follows: for every $t \in I$,

$$\left(\Phi^*\omega_L\left(\frac{\partial}{\partial e^2},\frac{\partial}{\partial e^1}\right) + \Phi^*(dE_L)\left(\frac{\partial}{\partial e^1}\right)\right)(\alpha_0(t))$$

$$= (\omega_L)_{\Phi(0,t)}\left((\Phi_*)_{(0,t)}\left(\frac{\partial}{\partial e^2}\right)_{(0,t)}, (\Phi_*)_{(0,t)}\left(\frac{\partial}{\partial e^1}\right)_{(0,t)}\right)$$

$$+ (dE_L)_{\Phi(0,t)}\left((\Phi_*)_{(0,t)}\left(\frac{\partial}{\partial e^1}\right)_{(0,t)}\right)$$

$$\overset{(5.2.23)}{=} (\omega_L)_{\dot{\gamma}(t)}(\ddot{\gamma}(t), X^c(t)) + (dE_L)_{\dot{\gamma}(t)}(X^c(t))$$

$$\overset{(5.2.17)}{=} \mathcal{E}_t(X^c(t)) = \mathcal{E}(X^c)(t).$$

(e) From (c), (d) and (2) we conclude that

$$(L_\varphi)'(0) = \int_a^b \mathcal{E}(X^c),$$

as was to be shown. $\qquad\square$

Definition 5.2.24. Let $\gamma: I \to M$ be a smooth curve. We say that γ is an *extremal of the Lagrangian L* (or briefly an *L-extremal*) if for any closed interval $[a,b] \subset I$ with different endpoints and any proper variation $\varphi: J \times I \to M$ of γ over $[a,b]$, $(L_\varphi)'(0) = 0$.

Theorem 5.2.25. *A smooth curve* $\gamma: I \to M$ *is an L-extremal if, and only if, its acceleration satisfies the Euler – Lagrange equation along* $\dot{\gamma}$.

Proof. If $\ddot{\gamma}$ satisfies the Euler–Lagrange equation along $\dot{\gamma}$, i.e., $\mathcal{E} = 0$, then γ is an L-extremal by the first variation formula (5.2.26).

Conversely, suppose that γ is an L-extremal. Let $a, b \in I$, $a < b$. Consider a proper variation $\varphi \colon J \times I \to M$ of γ over $[a, b]$, and let X be its variation vector field. Then

$$0 = (L_\varphi)'(0) \qquad \text{since } \gamma \text{ is an } L\text{-extremal,}$$

$$= \int_a^b \mathcal{E}(X^c) \qquad \text{by (5.2.26),}$$

$$= \int_a^b \eta(X) \qquad \text{by Lemma 5.2.17.}$$

Since, according to Lemma 5.2.21, every vector field $X \in \mathfrak{X}_{\gamma \restriction [a,b]}(M)$ with $X(a) = X(b) = 0$ is the variation field of a proper variation of γ over $[a, b]$, the fundamental lemma of calculus of variations implies that η vanishes over $[a, b]$. Because the interval $[a, b] \subset I$ can be chosen arbitrarily, it follows that $\eta = 0$. However, by Lemma 5.2.17, $\mathcal{E} = \tau^* \eta$, so we also have $\mathcal{E} = 0$. $\square$

Corollary 5.2.26. *Suppose that $L \colon TM \to \mathbb{R}$ is a regular Lagrangian, and let S be the Lagrange vector field for L. Then the geodesics of S coincide with the L-extremals.*

Proof. Consider a smooth curve $\gamma \colon I \to M$. Then, by (5.2.8),

$$(\omega_L)_{\dot{\gamma}(t)}(S(\dot{\gamma}(t)), w) + (dE_L)_{\dot{\gamma}(t)}(w) = 0, \qquad (*)$$

for all $t \in I$, $w \in T_{\dot{\gamma}(t)} TM$.

If γ is a geodesic of S, then $S(\dot{\gamma}(t)) = \ddot{\gamma}(t)$, so $(*)$ yields

$$0 = (\omega_L)_{\dot{\gamma}(t)}(\ddot{\gamma}(t), w) + (dE_L)_{\dot{\gamma}(t)}(w) \overset{(5.2.17)}{=} \mathcal{E}_t(w),$$

therefore $\mathcal{E} = 0$. Thus, by the previous theorem, γ is an L-extremal.

Conversely, suppose that γ is an L-extremal. Then, again by Theorem 5.2.25, $\mathcal{E} = 0$. Therefore,

$$(\omega_L)_{\dot{\gamma}(t)}(\ddot{\gamma}(t), w) + (dE_L)_{\dot{\gamma}(t)}(w) = 0, \qquad (**)$$

for all $t \in I$, $w \in T_{\dot{\gamma}(t)} TM$. From $(*)$ and $(**)$ it follows that

$$(\omega_L)_{\dot{\gamma}(t)}(S(\dot{\gamma}(t)) - \ddot{\gamma}(t), w) = 0.$$

By the non-degeneracy of ω_L, this implies that $S(\dot{\gamma}(t)) = \ddot{\gamma}(t)$ for all $t \in I$. Thus γ is a geodesic of S, as was to be shown. $\square$

Chapter 6

Covariant Derivatives

6.1 Differentiation in Vector Bundles

> ...the manifold structure provides no *intrinsic* way of comparing tangent vectors at two distinct points ... In order to be able to make such comparisons, it is necessary to endow the manifold with an additional structure defined by what is called a *linear connection*.
>
> Jean Dieudonné ([43], p. 232)

6.1.1 *Covariant Derivative on a Vector Bundle*

Definition 6.1.1. (a) A *covariant derivative operator* (briefly a *covariant derivative*) on a vector bundle $\pi\colon E \to M$ is an $\mathbb{R}$-bilinear mapping

$$\nabla\colon \mathfrak{X}(M) \times \Gamma(\pi) \to \Gamma(\pi), \ (X, s) \mapsto \nabla_X s$$

such that

(i) ∇ is *tensorial* in its first variable, i.e.,

$$\nabla_{fX} s = f\nabla_X s, \tag{6.1.1}$$

(ii) ∇ is a *derivation* in its second variable, i.e.,

$$\nabla_X fs = (Xf)s + f\nabla_X s, \tag{6.1.2}$$

for all $f \in C^\infty(M)$, $X \in \mathfrak{X}(M)$ and $s \in \Gamma(\pi)$.

For a fixed vector field X on M, the mapping

$$\nabla_X\colon \Gamma(\pi) \to \Gamma(\pi), \ s \mapsto \nabla_X s$$

is called the *covariant derivative with respect to X*; the section $\nabla_X s$ is the *covariant derivative of s in the direction of X*.

Given a section s of $\Gamma(\pi)$, the mapping

$$\nabla s\colon \mathfrak{X}(M) \to \Gamma(\pi), \ X \mapsto (\nabla s)(X) := \nabla_X s \qquad (6.1.3)$$

is said to be the *covariant differential of s*. If $\nabla s = 0$, the section s is called *parallel* (with respect to ∇).

(b) A covariant derivative on the tangent bundle $\tau\colon TM \to M$, i.e., an $\mathbb{R}$-bilinear mapping

$$D\colon \mathfrak{X}(M) \times \mathfrak{X}(M) \to \mathfrak{X}(M), \ (X,Y) \mapsto D_X Y$$

satisfying conditions (6.1.1) and (6.1.2) (replacing the sections $s \in \Gamma(\pi)$ by vector fields $Y \in \mathfrak{X}(M)$) is also said to be (by abuse of language) a *covariant derivative on the manifold M*.

A pair (M, D) consisting of a manifold and a covariant derivative on its tangent bundle will be called an *affinely connected manifold*.

Remark 6.1.2. Let $\nabla\colon \mathfrak{X}(M) \times \Gamma(\pi) \to \Gamma(\pi)$ be a covariant derivative on the vector bundle $\pi\colon E \to M$.

(a) The tensoriality of the mapping

$$X \in \mathfrak{X}(M) \mapsto \nabla_X s \in \Gamma(\pi) \quad (s \in \Gamma(\pi) \text{ is fixed})$$

implies that at each point $p \in M$, $(\nabla_X s)(p)$ depends only on the value of X at p. (See Proposition 2.2.31 and its proof.) So we may define for any individual vector $v \in T_p M$ the vector $\nabla_v s \in E_p$ as follows:

$$\nabla_v s := (\nabla_X s)(p) \text{ if } X \in \mathfrak{X}(M) \text{ with } X(p) = v.$$

Then the mapping

$$T_p M \times \Gamma(\pi) \to E_p, \ (v, s) \mapsto \nabla_v s$$

is $\mathbb{R}$-bilinear and satisfies the Leibniz rule

$$\nabla_v f s = v(f)s(p) + f(p)\nabla_v s, \quad f \in C^\infty(M).$$

(b) Since the mapping

$$s \in \Gamma(\pi) \mapsto \nabla_X s \in \Gamma(\pi) \quad (X \in \mathfrak{X}(M) \text{ is fixed})$$

is a derivation, it follows (as in Lemma 3.3.11 or Lemma 3.3.22(ii)) that for every open subset $\mathcal{U}$ of M there exists a unique covariant derivative

$$\nabla^{\mathcal{U}}\colon \mathfrak{X}(\mathcal{U}) \times \Gamma_{\mathcal{U}}(\pi) \to \Gamma_{\mathcal{U}}(\pi)$$

such that

$$\nabla^{\mathcal{U}}_{X\restriction\mathcal{U}}(s \restriction \mathcal{U}) = (\nabla_X s) \restriction \mathcal{U} \text{ for all } X \in \mathfrak{X}(M), \ s \in \Gamma(\pi).$$

We say that $\nabla^{\mathcal{U}}$ is the *induced covariant derivative* on the restriction of π over $\mathcal{U}$ (Example 2.2.34).

(c) Let $(\mathcal{U}, (u^i)_{i=1}^n)$ be a chart on M. Suppose that $E_{\mathcal{U}} := \pi^{-1}(\mathcal{U})$ also has a trivialization. Then, by Lemma 2.2.26, there exists a local frame field $(s_\alpha)_{\alpha=1}^k$ of E over $\mathcal{U}$ (where k is the rank of π). So any local section of $\Gamma_{\mathcal{U}}(\pi)$ can uniquely be written as a $C^\infty(\mathcal{U})$-linear combination of $(s_\alpha)_{\alpha=1}^k$. We have, in particular,

$$\nabla^{\mathcal{U}}_{\frac{\partial}{\partial u^j}} s_\beta = \sum_{\alpha=1}^k \Gamma^\alpha_{j\beta} s_\alpha, \quad (j, \beta) \in J_n \times J_k \tag{6.1.4}$$

for some smooth functions $\Gamma^\alpha_{j\beta} \in C^\infty(\mathcal{U})$. These functions are called the *Christoffel symbols* of the covariant derivative ∇ *with respect to the chart* $(\mathcal{U}, (u^i)_{i=1}^n)$ *and the local frame field* $(s_\alpha)_{\alpha=1}^k$.

By a slight abuse of notation, we shall simply write ∇ instead of $\nabla^{\mathcal{U}}$ in the following.

In the special case when $D \colon \mathfrak{X}(M) \times \mathfrak{X}(M) \to \mathfrak{X}(M)$ is a covariant derivative on the manifold M (more precisely, on the tangent bundle $\tau \colon TM \to M$) the functions $\Gamma^i_{jk} \in C^\infty(\mathcal{U})$ defined by

$$D_{\frac{\partial}{\partial u^j}} \frac{\partial}{\partial u^k} = \sum_{i=1}^n \Gamma^i_{jk} \frac{\partial}{\partial u^i}; \quad j, k \in J_n \tag{6.1.5}$$

are called the *Christoffel symbols of D with respect to the chart* $(\mathcal{U}, (u^i)_{i=1}^n)$.

(d) The mapping

$$s \in \Gamma(\pi) \mapsto \nabla s \in \mathrm{Hom}(\mathfrak{X}(M), \Gamma(\pi)) \cong \mathcal{A}_1(TM, E),$$

where ∇s is given by (6.1.3), has the following property:

$$\nabla(fs) = df \otimes s + f \nabla s; \quad f \in C^\infty(M), \ s \in \Gamma(\pi). \tag{6.1.6}$$

Indeed, for any vector field X on M we have

$$\nabla(fs)(X) \overset{(6.1.3)}{=} \nabla_X(fs) \overset{(6.1.2)}{=} (Xf)s + f\nabla_X s = (df \otimes s + f\nabla s)(X).$$

Conversely, if an $\mathbb{R}$-linear mapping

$$\Gamma(\pi) \to \mathcal{A}_1(TM, E) \cong \mathrm{Hom}(\mathfrak{X}(M), \Gamma(\pi)), \ s \mapsto \nabla s$$

satisfies the 'Leibniz rule' (6.1.6), then the mapping

$$\mathfrak{X}(M) \times \Gamma(\pi) \to \Gamma(\pi), \ (X, s) \mapsto \nabla_X s := (\nabla s)(X)$$

is $\mathbb{R}$-bilinear and satisfies the conditions (6.1.1) and (6.1.2). So *a covariant derivative on a vector bundle* $\pi \colon E \to M$ *can be viewed equivalently as an* $\mathbb{R}$*-linear mapping*

$$\nabla \colon \Gamma(\pi) \to \mathcal{A}_1(TM, E), \ s \mapsto \nabla s$$

which satisfies the Leibniz rule (6.1.6). In the following we shall freely use this elegant interpretation. Then, given a chart $(\mathcal{U}, (u^i)_{i=1}^n)$ of M and a local frame $(s_\alpha)_{\alpha=1}^k$ of E over $\mathcal{U}$ as in (c), we define the *Christoffel symbols* $\Gamma_{j\beta}^\alpha$ of ∇ by

$$\nabla s_\beta = \sum_{j=1}^n \sum_{\alpha=1}^k \Gamma_{j\beta}^\alpha du^j \otimes s_\alpha.$$

If $s = \sum_{\alpha=1}^k \sigma^\alpha s_\alpha \in \Gamma_\mathcal{U}(\pi)$ is an arbitrary smooth section of E over $\mathcal{U}$, we find that

$$\nabla s = \nabla \left(\sum_{\alpha=1}^k \sigma^\alpha s_\alpha \right) = \sum_{\alpha=1}^k \nabla(\sigma^\alpha s_\alpha) \overset{(6.1.6)}{=} \sum_{\alpha=1}^k \left((d\sigma^\alpha) \otimes s_\alpha + \sigma^\alpha \nabla s_\alpha \right)$$

$$= \sum_{i=1}^n \sum_{\alpha=1}^k \left(\frac{\partial \sigma^\alpha}{\partial u^i} + \Gamma_{i\beta}^\alpha \sigma^\beta \right) du^i \otimes s_\alpha.$$

Hence

$$\nabla_{\frac{\partial}{\partial u^j}} s = (\nabla s) \left(\frac{\partial}{\partial u^j} \right) = \sum_{\alpha=1}^k \left(\frac{\partial \sigma^\alpha}{\partial u^j} + \Gamma_{j\beta}^\alpha \sigma^\beta \right) s_\alpha, \quad j \in J_n. \qquad (6.1.7)$$

Lemma 6.1.3. *Let* $\nabla \colon \mathfrak{X}(M) \times \Gamma(\pi) \to \Gamma(\pi)$ *be a covariant derivative on the vector bundle* $\pi \colon E \to M$. *If*

$$\omega \in \mathcal{A}_1\big(TM, \operatorname{End}(E)\big) \cong \operatorname{Hom}\big(\mathfrak{X}(M), \operatorname{End}(\Gamma(\pi)))\big)$$

and

$$\widetilde{\nabla}_X s := \nabla_X s + \omega_X(s); \quad X \in \mathfrak{X}(M), \ s \in \Gamma(\pi), \qquad (6.1.8)$$

then $\widetilde{\nabla}$ *is also a covariant derivative on* π. *Conversely, if* ∇ *and* $\widetilde{\nabla}$ *are two covariant derivatives on* π, *then the mapping*

$$\begin{cases} \omega \colon \mathfrak{X}(M) \to \operatorname{End}(\Gamma(\pi)), \ X \mapsto \omega_X, \\ \omega_X(s) := \widetilde{\nabla}_X s - \nabla_X s \text{ for all } s \in \Gamma(\pi) \end{cases} \qquad (6.1.9)$$

is $C^\infty(M)$-*linear, whence* ω *is an* $\operatorname{End}(E)$-*valued 1-form on* M.

We leave the easy verification to the reader.

Remark 6.1.4. If ∇ and $\widetilde{\nabla}$ are covariant derivatives on a vector bundle $\pi \colon E \to M$, then the $\operatorname{End}(E)$-valued 1-form ω given by (6.1.9) is called the *difference tensor* of $\widetilde{\nabla}$ and ∇. It may also be considered as a $C^\infty(M)$-bilinear mapping

$$\mathfrak{X}(M) \times \Gamma(E) \to \Gamma(E), \ (X, s) \mapsto \omega(X, s) := \omega_X s.$$

This interpretation will be freely used in the sequel.

Example 6.1.5. (a) Let M be an n-dimensional manifold, V a k-dimensional real vector space, and consider the trivial vector bundle $\mathrm{pr}_1\colon M \times V \to M$. Given a basis $(b_i)_{i=1}^k$ of V, the smooth sections

$$s_i\colon p \in M \mapsto s_i(p) := (p, b_i) \in M \times V \quad (i \in J_k)$$

form a (global) frame field of $M \times V$ (Example 2.2.27). So for any section $s \in \Gamma(M \times V)$ we have $s = \sum_{i=1}^k \sigma^i s_i$, $\sigma^i \in C^\infty(M)$. Now define a mapping

$$\nabla\colon \mathfrak{X}(M) \times \Gamma(M \times V) \to \Gamma(M \times V)$$

by

$$\nabla_X s = \nabla_X \left(\sum_{i=1}^k \sigma^i s_i \right) := \sum_{i=1}^k (X\sigma^i) s_i. \qquad (6.1.10)$$

Then ∇ is clearly $\mathbb{R}$-bilinear, tensorial in its first variable, and

$$\nabla_X f s := \sum_{i=1}^k \left(X(f\sigma^i) s_i \right) = \sum_{i=1}^k \left((Xf)\sigma^i + f(X\sigma^i) \right) s_i$$

$$= (Xf) \left(\sum_{i=1}^k \sigma^i s_i \right) + f \sum_{i=1}^k (X\sigma^i) s_i = (Xf)s + f\nabla_X s,$$

for all $f \in C^\infty(M)$. Thus ∇ is a derivation in its second variable, therefore it is a covariant derivative on the trivial bundle $M \times V$, called the *standard covariant derivative*. It can also be characterized as *the unique covariant derivative on $M \times V \to M$ such that the constant sections are parallel with respect to ∇*.

(b) If $\nabla\colon \Gamma(\pi) \to \mathcal{A}_1(TM, E)$ is a covariant derivative operator and $f \in C^\infty(M)$, then we define the mapping $f\nabla\colon \Gamma(\pi) \to \mathcal{A}_1(TM, E)$ by

$$(f\nabla)(s) := f\nabla(s), \text{ for all } s \in \Gamma(\pi).$$

It clearly satisfies

$$(f\nabla)(s)(X) = f(\nabla s)(X); \quad s \in \Gamma(\pi), \ X \in \mathfrak{X}(M).$$

If we regard a covariant derivative as a mapping

$$\mathfrak{X}(M) \times \Gamma(\pi) \to \Gamma(\pi), \ (X, s) \mapsto \nabla_X s$$

then

$$(f\nabla)(X, s) := f\nabla_X s; \quad f \in C^\infty(M).$$

Obviously, $f\nabla$ is no longer a covariant derivative. However, *any affine $C^\infty(M)$-linear combination of two covariant derivatives on a vector bundle $\pi\colon E \to M$ is a covariant derivative*. More precisely, we have the following:

Lemma 6.1.6. *If ∇_1 and ∇_2 are two covariant derivatives on a vector bundle $\pi\colon E \to M$ and $f_1, f_2 \in C^\infty(M)$ are two functions such that*
$$f_1(p) + f_2(p) = 1 \text{ for all } p \in M,$$
then $f_1\nabla_1 + f_2\nabla_2$ is a covariant derivative.

Proof. The mapping $f_1\nabla_1 + f_2\nabla_2\colon \Gamma(\pi) \to \mathcal{A}_1(TM, E)$ is clearly $\mathbb{R}$-linear. For any section $s \in \Gamma(\pi)$ and function $f \in C^\infty(M)$ we have

$$
\begin{aligned}
(f_1\nabla_1 + f_2\nabla_2)(fs) &= f_1\nabla_1(fs) + f_2\nabla_2(fs) = f_1(df \otimes s + f\nabla_1 s) \\
&\quad + f_2(df \otimes s + f\nabla_2 s) = (f_1 + f_2)(df \otimes s) \\
&\quad + f(f_1\nabla_1 + f_2\nabla_2)(s) = df \otimes s + f(f_1\nabla_1 + f_2\nabla_2)(s),
\end{aligned}
$$

which shows that $f_1\nabla_1 + f_2\nabla_2$ satisfies (6.1.6), hence it is a covariant derivative operator on π. □

Proposition 6.1.7. *Every vector bundle admits a covariant derivative.*

Proof. Let a vector bundle $\pi\colon E \to M$ with typical fibre V be given, and suppose that $(\mathcal{U}_i, \psi_i)_{i \in I}$ is a trivializing covering for π. Then for each $i \in I$, $\psi_i\colon \mathcal{U}_i \times V \to \pi^{-1}(\mathcal{U}_i)$ is a diffeomorphism such that the mappings
$$(\psi_i)_p\colon V \to E_p, \quad v \mapsto (\psi_i)_p(v) := \psi_i(p, v), \quad p \in \mathcal{U}_i$$
are linear isomorphisms (see Definition 2.2.16).

Observe that if $s\colon \mathcal{U}_i \to \pi^{-1}(\mathcal{U}_i) \subset E$ is a local section of π, then
$$(\psi_i)^{-1} \circ s\colon \mathcal{U}_i \to \mathcal{U}_i \times V$$
is a smooth section of the trivial vector bundle $\mathrm{pr}_1\colon \mathcal{U}_i \times V \to \mathcal{U}_i$, since by condition (LT) in Definition 2.2.1,
$$\mathrm{pr}_1 \circ (\psi_i)^{-1} \circ s = \pi \circ s = 1_{\mathcal{U}_i}.$$
Similarly, if $\sigma\colon \mathcal{U}_i \to \mathcal{U}_i \times V$ is a smooth section of the trivial bundle $\mathcal{U}_i \times V$, then $\psi_i \circ \sigma\colon \mathcal{U}_i \to \pi^{-1}(\mathcal{U}_i)$ is a local section of π over $\mathcal{U}_i$.

By Example 6.1.5(a), for each $i \in I$, we have the trivial covariant derivative ∇_i on the trivial vector bundle $\mathrm{pr}_1\colon \mathcal{U}_i \times V \to \mathcal{U}_i$. Then on the restricted vector bundle $\pi^{-1}(\mathcal{U}_i)$ over $\mathcal{U}_i$ (see Example 2.2.34) the mapping
$$\nabla_{\psi_i}\colon s \in \Gamma(\pi^{-1}(\mathcal{U}_i)) \mapsto \nabla_{\psi_i} s := \psi_i \circ \nabla_i(\psi_i^{-1} \circ s) \in \mathcal{A}_1\big(T\mathcal{U}_i, \pi^{-1}(\mathcal{U}_i)\big)$$
is a covariant derivative.

Let $(f_i)_{i \in I}$ be a smooth partition of unity on M, subordinate to the open covering $(\mathcal{U}_i)_{i \in I}$ (Definition 2.1.29, Theorem 2.1.30). Then the mapping
$$\nabla\colon s \in \Gamma(\pi) \mapsto \nabla s := \sum_{i \in I} f_i \nabla_{\psi_i}(s \restriction \mathcal{U}_i) \in \mathcal{A}_1(TM, E)$$
is well-defined, and, taking into account Lemma 6.1.6, it is a covariant derivative on π. □

Example 6.1.8 (New Covariant Derivatives from Old Ones).

(a) Let $\nabla_i \colon \Gamma(\pi_i) \to \mathcal{A}_1(TM, E_i)$ be covariant derivatives on the vector bundles $\pi_i \colon E_i \to M$, $i \in \{1, 2\}$. Identifying $\Gamma(E_1 \oplus E_2)$ with $\Gamma(E_1) \oplus \Gamma(E_2)$ (Lemma 2.2.43), the mapping

$$\begin{cases} \nabla_1 \oplus \nabla_2 \colon \Gamma(E_1 \oplus E_2) \to \mathcal{A}_1(TM, E_1 \oplus E_2), \\ \nabla_1 \oplus \nabla_2(s) := (\nabla_1 s_1, \nabla_2 s_2) \text{ if } s = (s_1, s_2) \end{cases}$$

is a covariant derivative on the Whitney sum $\pi_1 \oplus \pi_2$, called the *direct sum* of ∇_1 and ∇_2.

(b) Let $\pi \colon E \to M$ be a vector bundle, and consider its dual bundle $\widehat{\pi} \colon E^* \to M$. We recall that $\Gamma(\widehat{\pi}) \cong (\Gamma(\pi))^*$ (Proposition 2.2.44). If $\nabla \colon \mathfrak{X}(M) \times \Gamma(\pi) \to \Gamma(\pi)$ is a covariant derivative on π, then the mapping

$$\nabla^* \colon \mathfrak{X}(M) \times \Gamma(\widehat{\pi}) \to \Gamma(\widehat{\pi}), \quad (X, s^*) \mapsto \nabla^*_X s^*$$

given by

$$(\nabla^*_X s^*)(s) = X(s^*(s)) - s^* \nabla_X s, \quad s \in \Gamma(\pi) \tag{6.1.11}$$

is a covariant derivative on the dual bundle $\widehat{\pi}$.

Indeed, ∇^* is clearly $\mathbb{R}$-bilinear and tensorial in its first variable. To check that ∇^* satisfies also the derivation rule (6.1.2), let f be a smooth function on M. Then

$$\begin{aligned} (\nabla^*_X f s^*)(s) &:= X((f s^*)(s)) - (f s^*)(\nabla_X s) \\ &= X(f s^*(s)) + f\big((\nabla^*_X s^*)(s) - X(s^*(s))\big) \\ &= \big((Xf)s^* + f\nabla^*_X s^*\big)(s), \end{aligned}$$

whence $\nabla^*_X f s^* = (Xf)s^* + f\nabla^*_X s^*$, as was to be shown.

For simplicity, we shall denote the induced covariant derivative ∇^* on $\widehat{\pi}$ also by ∇.

(c) Let $\pi_1 \colon E_1 \to M$ and $\pi_2 \colon E_2 \to M$ be vector bundles, and consider the vector bundle $\mathrm{Lin}_M(\pi_1, \pi_2)$ whose fibre over a point $p \in M$ is the real vector space $L((E_1)_p, (E_2)_p)$ of the linear mappings $(E_1)_p \to (E_2)_p$ (Example 2.2.40). We recall that $\Gamma(\mathrm{Lin}_M(\pi_1, \pi_2)) \cong \mathrm{Hom}_M(\Gamma(\pi_1), \Gamma(\pi_2))$ (Proposition 2.2.44). Suppose that ∇_1 and ∇_2 are covariant derivatives of π_1 and π_2, respectively. Then the mapping

$$\nabla \colon \mathfrak{X}(M) \times \Gamma(\mathrm{Lin}_M(\pi_1, \pi_2)) \to \Gamma(\mathrm{Lin}_M(\pi_1, \pi_2)), \quad (X, \lambda) \mapsto \nabla_X \lambda$$

defined by

$$(\nabla_X \lambda)(s) := (\nabla_2)_X(\lambda(s)) - \lambda((\nabla_1)_X s), \quad s \in \Gamma(\pi_1) \tag{6.1.12}$$

is a covariant derivative on $\mathrm{Lin}_M(\pi_1, \pi_2)$. The $\mathbb{R}$-bilinearity and the tensoriality in X are again obvious. For any smooth function f on M,

$$
\begin{aligned}
(\nabla_X f\lambda)(s) &:= (\nabla_2)_X((f\lambda)(s)) - (f\lambda)((\nabla_1)_X(s)) \\
&= (\nabla_2)_X\big(f(\lambda(s))\big) - f\big(\lambda((\nabla_1)_X(s))\big) \\
&= (Xf)(\lambda(s)) + f\big((\nabla_2)_X(\lambda(s)) - \lambda(\nabla_1)_X(s)\big) \\
&\overset{(6.1.12)}{=} \big((Xf)\lambda + f\nabla_X\lambda\big)(s),
\end{aligned}
$$

so ∇ is a derivation in λ.

In particular, a covariant derivative ∇ on a vector bundle $\pi\colon E \to M$ induces a covariant derivative $\widehat{\nabla}$ on the bundle $\mathrm{End}(E)$ such that

$$
(\widehat{\nabla}_X\lambda)(s) = \nabla_X(\lambda(s)) - \lambda(\nabla_X s) \tag{6.1.13}
$$

for all $X \in \mathfrak{X}(M)$, $\lambda \in \Gamma(\mathrm{End}(E)) \cong \mathrm{End}(\Gamma(E))$, $s \in \Gamma(E)$.

In the sequel we sometimes write, by a slight abuse of notation, ∇ also for the induced covariant derivative $\widehat{\nabla}$ on $\mathrm{End}(E)$.

(d) Using the product rule for tensor derivations (Proposition 3.3.14) as a guiding principle, we extend the covariant derivatives $\nabla_X\colon \Gamma(\pi) \to \Gamma(\pi)$ ($X \in \mathfrak{X}(M)$) to any π-tensor fields on M as follows:

(i) We agree that $\nabla_X f := Xf$ if $f \in C^\infty(M) = T^0_0(\Gamma(\pi))$.

(ii) For $A \in \Gamma(T^k_l(E)) \cong T^k_l(\Gamma(E))$, $(k, l) \neq (0, 0)$, we define $\nabla_X A$ by

$$
\begin{aligned}
\nabla_X A(\sigma^1, \ldots, \sigma^k, s_1, \ldots, s_l) &:= X A(\sigma^1, \ldots, \sigma^k, s_1, \ldots, s_l) \\
&\quad - \sum_{i=1}^k A(\sigma^1, \ldots, \nabla_X\sigma^i, \ldots, \sigma^k, s_1, \ldots, s_l) \\
&\quad - \sum_{j=1}^l A(\sigma^1, \ldots, \sigma^k, s_1, \ldots, \nabla_X s_j, \ldots, s_l).
\end{aligned} \tag{6.1.14}
$$

It is easy to check that $\nabla_X A \in \Gamma(T^k_l(E))$ if $A \in \Gamma(T^k_l(E))$. The *covariant differential*

$$
\nabla A\colon \mathfrak{X}(M) \times T^k_l(\Gamma(E)) \to T^k_l(\Gamma(E))
$$

of A 'collects all the covariant derivatives of A': it is defined by

$$
i_X \circ \nabla A = \nabla_X A \quad \text{for all } X \in \mathfrak{X}(M), \tag{6.1.15}
$$

i.e., by

$$
\nabla A(X, \sigma^1, \ldots, \sigma^k, s_1, \ldots, s_l) := (\nabla_X A)(\sigma^1, \ldots, \sigma^k, s_1, \ldots, s_l), \tag{6.1.16}
$$

where $\sigma^i \in (\Gamma(E))^*$, $i \in J_k$; $s_j \in \Gamma(E)$, $j \in J_l$.

Observe that

$$\nabla f = df \text{ if } f \in T_0^0(\Gamma(E)) = C^\infty(M). \qquad (6.1.17)$$

Indeed, if $X \in \mathfrak{X}(M)$, then

$$i_X \circ \nabla f := \nabla_X f := Xf = df(X) = i_X \circ df.$$

We say that a π-tensor field $A \in T_l^k(\Gamma(E))$ is *parallel* (with respect to the covariant derivative ∇) if $\nabla A = 0$.

In the special case when D is a covariant derivative on a manifold M, the covariant differential of a type (k, l) tensor field on M is the type $(k, l+1)$ tensor field DA on M such that

$$DA(\theta^1, \ldots, \theta^k, X, X_1, \ldots, X_l) := (D_X A)(\theta^1, \ldots, \theta^k, X_1, \ldots, X_l) \quad (6.1.18)$$

for all $X, X_1, \ldots, X_l \in \mathfrak{X}(M)$; $\theta^1, \ldots, \theta^k \in \mathfrak{X}^*(M)$.

(e) Suppose that $\pi \colon E \to M$ and $\widetilde{\pi} \colon \widetilde{E} \to \widetilde{M}$ are isomorphic vector bundles, and let $\varphi \colon E \to \widetilde{E}$ be an isomorphism. The mapping

$$\varphi_\# \colon \Gamma(E) \to \Gamma(\widetilde{E}), \quad \sigma \mapsto \varphi_\# \sigma = \varphi \circ \sigma \circ \varphi_B^{-1}$$

introduced in Remark 2.2.32(b) is an isomorphism of modules if we identify the rings $C^\infty(M)$ and $C^\infty(\widetilde{M})$ via the ring isomorphism

$$\widetilde{f} \in C^\infty(\widetilde{M}) \mapsto f := \widetilde{f} \circ \varphi_B \in C^\infty(M).$$

If $\sigma \in \Gamma(E)$, $\widetilde{\sigma} \in \Gamma(\widetilde{E})$, we write

$$\widetilde{f} \cdot \sigma := (\widetilde{f} \circ \varphi_B)\sigma \quad \text{if} \quad \widetilde{f} \in C^\infty(\widetilde{M});$$

$$f \cdot \widetilde{\sigma} := (f \circ \varphi_B^{-1})\widetilde{\sigma} \quad \text{if} \quad f \in C^\infty(M).$$

Now let a covariant derivative ∇ on π be given. Since any section of $\widetilde{\pi}$ is of the form $\varphi_\#(\sigma)$, $\sigma \in \Gamma(E)$, and any vector field on $\widetilde{M}$ can be written as

$$(\varphi_B)_\# X := (\varphi_B)_* \circ X \circ \varphi_B^{-1}, \quad X \in \mathfrak{X}(M)$$

(see Remark 3.1.52(b)), we may define a mapping

$$\widetilde{\nabla} \colon \mathfrak{X}(\widetilde{M}) \times \Gamma(\widetilde{E}) \to \Gamma(\widetilde{E})$$

by

$$\widetilde{\nabla}_{(\varphi_B)_\# X}(\varphi_\# \sigma) := \varphi_\#(\nabla_X \sigma); \quad X \in \mathfrak{X}(M), \ \sigma \in \Gamma(E). \qquad (6.1.19)$$

We show that $\widetilde{\nabla}$ *is a covariant derivative on* $\widetilde{\pi}$. Since $\widetilde{\nabla}$ is clearly $\mathbb{R}$-linear, our only task is to verify (6.1.1) and (6.1.2).

Let $\widetilde{f} \in C^\infty(\widetilde{M}) \cong C^\infty(M)$. Then

$$\widetilde{\nabla}_{\widetilde{f}(\varphi_B)_\# X}(\varphi_\# \sigma) = \widetilde{\nabla}_{(\varphi_B)_\#(\widetilde{f} \cdot X)}(\varphi_\# \sigma) \overset{(6.1.19)}{:=} \varphi_\#(\nabla_{\widetilde{f} \cdot X} \sigma)$$

$$\overset{(6.1.1)}{=} \varphi_\#(\widetilde{f} \cdot \nabla_X \sigma) = \widetilde{f}\varphi_\#(\nabla_X \sigma) \overset{(6.1.19)}{=} \widetilde{f}\widetilde{\nabla}_{(\varphi_B)_\# X}(\varphi_\# \sigma),$$

so $\widetilde{\nabla}$ is tensorial in its first variable.

Next we check that $\widetilde{\nabla}$ is a derivation in its second variable. Applying the identifications mentioned above, we find

$$\widetilde{\nabla}_{(\varphi_B)_\# X}(\widetilde{f}(\varphi_\# \sigma)) = \widetilde{\nabla}_{(\varphi_B)_\# X}(\varphi_\#(\widetilde{f} \cdot \sigma)) := \varphi_\#(\nabla_X(\widetilde{f} \cdot \sigma))$$

$$\overset{(6.1.2)}{=} \varphi_\#((X\widetilde{f})\sigma + \widetilde{f} \cdot \nabla_X \sigma)$$

$$= (X(\widetilde{f} \circ \varphi_B) \circ \varphi_B^{-1})\varphi_\# \sigma + \widetilde{f}\varphi_\#(\nabla_X \sigma)$$

$$\overset{\text{Lemma 3.1.49}}{=} ((\varphi_B)_\# X)(\widetilde{f})\varphi_\# \sigma + \widetilde{f}\widetilde{\nabla}_{(\varphi_B)_\# X}(\varphi_\# \sigma),$$

as was to be shown.

Conversely, if $\widetilde{\nabla}$ is a covariant derivative on $\widetilde{\pi}$, and $\varphi^* \widetilde{\nabla}$ is defined by

$$\varphi_\#((\varphi^* \widetilde{\nabla})_X \sigma) := \widetilde{\nabla}_{(\varphi_B)_\# X}(\varphi_\# \sigma); \quad X \in \mathfrak{X}(M),\ \sigma \in \Gamma(\pi), \qquad (6.1.20)$$

then $\varphi^* \widetilde{\nabla}$ is a covariant derivative on π, called also the *pull-back* of $\widetilde{\nabla}$ via the bundle isomorphism φ.

(f) Consider, in particular, an affinely connected manifold (M, D), and let a diffeomorphism $\varphi \colon M \to \widetilde{M}$ be given. Its derivative $\varphi_* \colon TM \to T\widetilde{M}$ is an isomorphism of vector bundles, inducing φ between the base manifolds. The construction described in (e) yields a covariant derivative $\widetilde{D}$ on $\widetilde{M}$ such that

$$\widetilde{D}_{\varphi_\# X}(\varphi_\# Y) = \varphi_\#(D_X Y) \quad \text{for all } X, Y \in \mathfrak{X}(M). \qquad (6.1.21)$$

Conversely, the pull-back $\varphi^* \widetilde{D}$ of a covariant derivative $\widetilde{D}$ on $\widetilde{M}$ is given by

$$\varphi_\#((\varphi^* \widetilde{D})_X Y) := \widetilde{D}_{\varphi_\# X}(\varphi_\# Y); \quad X, Y \in \mathfrak{X}(M). \qquad (6.1.22)$$

Definition 6.1.9. (a) Let $\pi \colon E \to M$ and $\widetilde{\pi} \colon \widetilde{E} \to \widetilde{M}$ be isomorphic vector bundles with covariant derivatives ∇ and $\widetilde{\nabla}$, respectively. A vector bundle isomorphism $\varphi \colon E \to \widetilde{E}$ is called *covariant derivative preserving* if

$$\varphi_\#(\nabla_X \sigma) = \widetilde{\nabla}_{(\varphi_B)_\# X}(\varphi_\# \sigma) \quad \text{for all } X \in \mathfrak{X}(M),\ \sigma \in \Gamma(\pi).$$

(Here we identify the rings $C^\infty(M)$ and $C^\infty(\widetilde{M})$ with the help of the mapping $f \in C^\infty(M) \mapsto f \circ \varphi_B^{-1} \in C^\infty(\widetilde{M})$.)

(b) Two affinely connected manifolds (M_1, D_1) and (M_2, D_2) are said to be *isomorphic* if there exists a diffeomorphism $\varphi \colon M_1 \to M_2$ such that $D_1 = \varphi^* D_2$. In particular, an *automorphism* of an affinely connected manifold (M, D) is a diffeomorphism φ of M such that $\varphi^* D = D$.

Remark 6.1.10. It is easy to verify that the automorphisms of an affinely connected manifold (M, D) form a group under composition of mappings. We denote this group by $\mathrm{Aut}(M, D)$, or simply $\mathrm{Aut}(D)$, and we call it the *automorphism group of* (M, D).

6.1.2 The Second Covariant Differential

Lemma 6.1.11. *Let* ∇ *be a covariant derivative on a vector bundle* $\pi\colon E \to M$, *and suppose that a covariant derivative* D *on the base manifold* M *is also given. If* s *is a smooth section of* E, *then the mapping*

$$\nabla^2 s\colon \mathfrak{X}(M) \times \mathfrak{X}(M) \to \Gamma(E),$$

$$(X, Y) \mapsto \nabla^2 s(X, Y) := \nabla^2_{X,Y} s := \nabla_X \nabla_Y s - \nabla_{D_X Y} s \qquad (6.1.23)$$

is $C^\infty(M)$*-bilinear.*

Proof. The $C^\infty(M)$-linearity in the first slot and the additivity in the second slot are clear, so we have only to check $C^\infty(M)$-homogeneity in Y. For any function $f \in C^\infty(M)$, we have

$$\nabla^2 s(X, fY) = \nabla_X(\nabla_{fY} s) - \nabla_{D_X fY} s = \nabla_X(f \nabla_Y s) - \nabla_{(Xf)Y + f D_X Y} s$$
$$= (Xf)\nabla_Y s + f \nabla_X \nabla_Y s - (Xf)\nabla_Y s - f \nabla_{D_X Y} s$$
$$= f(\nabla_X \nabla_Y s - \nabla_{D_X Y} s) = f \nabla^2 s(X, Y),$$

as desired. $\qquad\square$

Next we turn our attention to affinely connected manifolds. Then we can define the second covariant differential of a tensor field without an auxiliary covariant derivative operator (which we need in (6.1.23)). We begin with the two simplest cases, these are the second covariant differential of a smooth function and a vector field.

For the rest of this subsection we assume that an affinely connected manifold (M, D) *is given.*

Definition 6.1.12. By the *D-Hessian*, or simply the *Hessian* of a function $f \in C^\infty(M)$ we mean the type $(0, 2)$ tensor field

$$\mathrm{Hess}(f) := D^2 f := D(Df) \overset{(6.1.17)}{=} D(df) \qquad (6.1.24)$$

on M.

Lemma 6.1.13. *For any vector fields* X, Y *on* M,

$$\mathrm{Hess}(f)(X, Y) = X(Yf) - (D_X Y)f. \qquad (6.1.25)$$

Proof. $\text{Hess}(f)(X,Y) := D^2 f(X,Y) = D(df)(X,Y)$

$$\overset{(6.1.18)}{=} (D_X df)(Y) \overset{(6.1.14)}{=} X(df(Y)) - df(D_X Y) = X(Yf) - (D_X Y)f. \qquad \square$$

Remark 6.1.14. In the sequel we shall usually write $D^2_{X,Y} f$ for the function $D^2 f(X,Y)$. The components of the tensor $\text{Hess}(f) \in \mathcal{T}^0_2(M)$ with respect to a chart $(\mathcal{U}, (u^i)^n_{i=1})$ on M are the functions

$$\text{Hess}(f)\left(\frac{\partial}{\partial u^i}, \frac{\partial}{\partial u^j}\right) = \frac{\partial^2 f}{\partial u^i \partial u^j} - \sum_{k=1}^n \Gamma^k_{ij} \frac{\partial f}{\partial u^k}; \quad i,j \in J_n,$$

where the functions Γ^k_{ij} are the Christoffel symbols of D defined by (6.1.5).

Definition 6.1.15. By the *second covariant differential* of a vector field Z on M we mean the type $(0,2)$ tensor field $D^2 Z := D(DZ) \in \mathcal{T}^1_2(M)$.

Remark 6.1.16. For any vector fields X, Y on M,

$$D^2(X,Y)Z := D^2_{X,Y}Z := D(DZ)(X,Y) \overset{(6.1.16)}{=} (D_X(DZ))(Y)$$

$$\overset{(6.1.14)}{=} D_X((DZ)(Y)) - (DZ)(D_X Y) \overset{(6.1.3)}{=} D_X D_Y Z - D_{D_X Y} Z.$$

Observe that the formula

$$D^2_{X,Y}Z = D_X D_Y Z - D_{D_X Y} Z \tag{6.1.26}$$

so obtained is a special case of (6.1.23).

Now we turn to the general case.

Definition 6.1.17. Let $A \in \mathcal{T}^0_l(M)$ or $A \in \mathcal{T}^1_l(M)$, $l \geq 1$. The second covariant differential of A is the tensor field $D^2 A \in \mathcal{T}^0_{l+2}(M) \cup \mathcal{T}^1_{l+2}(M)$ given by

$$D^2 A(X,Y,Z_1,\dots,Z_l) := (D^2_{X,Y}A)(Z_1,\dots,Z_l)$$

$$:= (D(DA))(X,Y,Z_1,\dots,Z_l) \overset{(6.1.16)}{=} (D_X(DA))(Y,Z_1,\dots,Z_l), \tag{6.1.27}$$

where $X, Y, Z_1, \dots, Z_l$ are vector fields on M.

Lemma 6.1.18. *With the notation of the preceding definition,*

$$(D^2_{X,Y}A)(Z_1,\dots,Z_l)$$
$$= (D_X(D_Y A))(Z_1,\dots,Z_l) - (D_{D_X Y}A)(Z_1,\dots,Z_l). \tag{6.1.28}$$

Proof. $(D^2_{X,Y}A)(Z_1, \ldots, Z_l) := (D_X(DA))(Y, Z_1, \ldots, Z_l)$

$\overset{(6.1.14)}{=} D_X\big((DA)(Y, Z_1, \ldots, Z_l)\big) - DA(D_X Y, Z_1, \ldots, Z_l)$

$\qquad - \sum_{i=1}^{l} DA(Y, Z_1, \ldots, D_X Z_i, \ldots, Z_l) \overset{(6.1.16)}{=} D_X\big((D_Y A)(Z_1, \ldots, Z_l)\big)$

$\qquad - (D_{D_X Y}A)(Z_1, \ldots, Z_l) - \sum_{i=1}^{l}(D_Y A)(Z_1, \ldots, D_X Z_i, \ldots, Z_l)$

$\overset{(6.1.14)}{=} (D_X(D_Y A))(Z_1, \ldots, Z_l) + \sum_{i=1}^{l}(D_Y A)(Z_1, \ldots, D_X Z_i, \ldots, Z_l)$

$\qquad - (D_{D_X Y}A)(Z_1, \ldots, Z_l) - \sum_{i=1}^{l}(D_Y A)(Z_1, \ldots, D_X Z_i, \ldots, Z_l)$

$\qquad = (D_X(D_Y A))(Z_1, \ldots, Z_l) - (D_{D_X Y}A)(Z_1, \ldots, Z_l). \qquad\qquad \square$

6.1.3 Exterior Covariant Derivative

In this subsection, we extend a covariant derivative operator

$$\nabla \colon \mathcal{A}_0(TM, E) = \Gamma(E) \to \mathcal{A}_1(TM, E)$$

to an $\mathbb{R}$-linear mapping

$$d^\nabla \colon \mathcal{A}(TM, E) \to \mathcal{A}(TM, E)$$

that satisfies a suitable derivation rule.

Definition 6.1.19. Let ∇ be a covariant derivative on a vector bundle $\pi \colon E \to M$, and let A be an E-valued k-form on M. By the *exterior covariant derivative* of A we mean the mapping $d^\nabla A \colon (\mathfrak{X}(M))^{k+1} \to \Gamma(E)$ given by

$$d^\nabla A := \nabla A \quad \text{if } k = 0, \text{ i.e., } A \in \Gamma(E),$$

and by

$$d^\nabla A(X_0, \ldots, X_k) := \sum_{i=0}^{k}(-1)^i \nabla_{X_i}(A(X_0, \ldots, \check{X}_i, \ldots, X_k))$$

$$+ \sum_{0 \le i < j \le k}(-1)^{i+j} A([X_i, X_j], X_0, \ldots, \check{X}_i, \ldots, \check{X}_j, \ldots, X_k) \quad (6.1.29)$$

if $1 \le k \le n = \dim M$ $(X_0, \ldots, X_k \in \mathfrak{X}(M))$.

Remark 6.1.20. A straightforward calculation shows that $d^\nabla A$ is an alternating $C^\infty(M)$-linear mapping, thus $d^\nabla A \in \mathcal{A}_{k+1}(TM, E)$. The operator

$$d^\nabla : \mathcal{A}(TM, E) \to \mathcal{A}(TM, E), \quad A \mapsto d^\nabla A$$

defined in this way is called the *exterior covariant derivative* with respect to the covariant derivative ∇.

The defining formula (6.1.29) should be compared with the analogous formula (3.3.32) for the exterior derivative of a k-form on M. In particular, if A is an E-valued one-form, then, for any vector fields $X, Y \in \mathfrak{X}(M)$, we obtain

$$(d^\nabla A)(X, Y) = \nabla_X(A(Y)) - \nabla_Y(A(X)) - A([X, Y]). \tag{6.1.30}$$

Proposition 6.1.21. *If $\alpha \in \mathcal{A}(M)$, and $B \in \mathcal{A}(TM, E)$, then we have*

$$d^\nabla(\alpha \wedge B) = d\alpha \wedge B + (-1)^{\deg(\alpha)}\alpha \wedge d^\nabla B.$$

(The wedge product $\wedge$ is defined by (3.3.21)).

Proof. Suppose that $\alpha \in \mathcal{A}_k(M)$ and $B \in \mathcal{A}_l(TM, E)$. Since every E-valued l-form can be locally written as the sum of tensor products of (ordinary) l-forms and sections of E, it is sufficient to show the assertion for $B = \beta \otimes s = \beta \wedge s$ with $\beta \in \mathcal{A}_l(M)$ and $s \in \Gamma(E)$. First we prove that

$$d^\nabla(\alpha \wedge s) = d\alpha \wedge s + (-1)^k \alpha \wedge \nabla s. \tag{6.1.31}$$

If $X_0, \ldots, X_k \in \mathfrak{X}(M)$, then we obtain

$$d^\nabla(\alpha \wedge s)(X_0, \ldots, X_k) \overset{(6.1.29)}{=} \sum_{i=0}^{k}(-1)^i \nabla_{X_i}(\alpha \wedge s(X_0, \ldots, \check{X}_i, \ldots, X_k))$$

$$+ \sum_{0 \le i < j \le k}(-1)^{i+j}\alpha \wedge s([X_i, X_j], X_0, \ldots, \check{X}_i, \ldots, \check{X}_j, \ldots, X_k)$$

$$= \sum_{i=0}^{k}(-1)^i \nabla_{X_i}(\alpha(X_0, \ldots, \check{X}_i, \ldots, X_k)s)$$

$$+ \sum_{0 \le i < j \le k}(-1)^{i+j}\alpha([X_i, X_j], X_0, \ldots, \check{X}_i, \ldots, \check{X}_j, \ldots, X_k)s$$

$$= \sum_{i=0}^{k}(-1)^i(X_i(\alpha(X_0, \ldots, \check{X}_i, \ldots, X_k))s + \alpha(X_0, \ldots, \check{X}_i, \ldots, X_k)\nabla_{X_i}s)$$

$$+ \sum_{0 \le i < j \le k}(-1)^{i+j}\alpha([X_i, X_j], X_0, \ldots, \check{X}_i, \ldots, \check{X}_j, \ldots, X_k)s$$

$$= (d\alpha \wedge s + (-1)^k \alpha \wedge \nabla s)(X_0, \ldots, X_k),$$

which shows (6.1.31). Now we turn to the proof of the original assertion:

$$d^\nabla(\alpha \wedge B) = d^\nabla(\alpha \wedge (\beta \wedge s)) = d^\nabla((\alpha \wedge \beta) \wedge s)$$

$$\overset{(6.1.31)}{=} d(\alpha \wedge \beta) \wedge s + (-1)^{k+l}(\alpha \wedge \beta) \wedge \nabla s$$

$$\overset{\text{Prop. 3.3.35(ii)}}{=} (d\alpha \wedge \beta + (-1)^k \alpha \wedge d\beta) \wedge s + (-1)^k \alpha \wedge ((-1)^l \beta \wedge \nabla s)$$

$$= d\alpha \wedge (\beta \wedge s) + (-1)^k \alpha \wedge (d\beta \wedge s + (-1)^l \beta \wedge \nabla s)$$

$$\overset{(6.1.31)}{=} d\alpha \wedge (\beta \wedge s) + (-1)^k \alpha \wedge d^\nabla(\beta \wedge s) = d\alpha \wedge B + (-1)^k \alpha \wedge d^\nabla B.$$

This concludes the proof. □

6.1.4 Metric Derivatives

Definition 6.1.22. Let $\pi\colon E \to M$ be a vector bundle with scalar product g (Definition 2.2.52). A covariant derivative ∇ on π is said to be *compatible with the scalar product* g or a *metric derivative* if g is parallel with respect to ∇, i.e., $\nabla g = 0$.

Remark 6.1.23. Since

$$\nabla g(X, s_1, s_2) \overset{(6.1.16)}{=} (\nabla_X g)(s_1, s_2)$$

$$\overset{(6.1.14)}{=} X g(s_1, s_2) - g(\nabla_X s_1, s_2) - g(s_1, \nabla_X s_2),$$

it follows that a covariant derivative ∇ on π is compatible with a scalar product $g \in T_2^0(\Gamma(E)) \cong \Gamma(T_2^0(\pi))$ if, and only if,

$$X g(s_1, s_2) = g(\nabla_X s_1, s_2) + g(s_1, \nabla_X s_2), \tag{6.1.32}$$

for all $X \in \mathfrak{X}(M)$ and $s_1, s_2 \in \Gamma(E)$.

Proposition 6.1.24. *On every semi-Euclidean vector bundle there exists a covariant derivative which is compatible with the scalar product.*

Proof. Let $\pi\colon E \to M$ be a vector bundle equipped with a non-degenerate scalar product g. Choose a covariant derivative $\overset{\circ}{\nabla}$ on π (its existence is guaranteed by Proposition 6.1.7). Define the mapping A as the covariant differential of g:

$$A := \overset{\circ}{\nabla} g : \mathfrak{X}(M) \times \Gamma(E) \times \Gamma(E) \to C^\infty(M). \tag{6.1.33}$$

Observe first that A is symmetric in its last two variables. Indeed, for any vector field X on M and any sections $s_1, s_2 \in \Gamma(E)$ we have

$$A(X, s_1, s_2) := \overset{\circ}{\nabla} g(X, s_1, s_2) = (\overset{\circ}{\nabla}_X g)(s_1, s_2)$$

$$= Xg(s_1, s_2) - g(\overset{\circ}{\nabla}_X s_1, s_2) - g(s_1, \overset{\circ}{\nabla}_X s_2)$$

$$= Xg(s_2, s_1) - g(\overset{\circ}{\nabla}_X s_2, s_1) - g(s_2, \overset{\circ}{\nabla}_X s_1)$$

$$= \overset{\circ}{\nabla}_X g(s_2, s_1) = A(X, s_2, s_1),$$

by the symmetry of g. The non-degeneracy of g assures that there exists a unique $\mathrm{End}(E)$-valued 1-form

$$\omega \in \mathcal{A}_1(TM, \mathrm{End}(E)) \cong \mathrm{Hom}\left(\mathfrak{X}(M), \mathrm{End}(\Gamma(E))\right)$$

such that for all $X \in \mathfrak{X}(M)$ and $s_1, s_2 \in \Gamma(\pi)$,

$$g(\omega(X)(s_1), s_2) = A(X, s_1, s_2). \tag{6.1.34}$$

Then the formula

$$\nabla_X s := \overset{\circ}{\nabla}_X s + \frac{1}{2}\omega(X)(s); \quad X \in \mathfrak{X}(M), \ s \in \Gamma(\pi) \tag{6.1.35}$$

defines a new covariant derivative on π by Lemma 6.1.3. We show that ∇ is compatible with the scalar product g.

Let $X \in \mathfrak{X}(M)$; $s_1, s_2 \in \Gamma(\pi)$. Then we get

$$\nabla g(X, s_1, s_2) \overset{(6.1.14)}{=} Xg(s_1, s_2) - g(\nabla_X s_1, s_2) - g(s_1, \nabla_X s_2)$$

$$\overset{(6.1.35)}{=} Xg(s_1, s_2) - g(\overset{\circ}{\nabla}_X s_1, s_2) - g(s_1, \overset{\circ}{\nabla}_X s_2)$$

$$- \frac{1}{2}g(\omega(X)(s_2), s_1) - \frac{1}{2}g(\omega(X)(s_1), s_2)$$

$$\overset{(6.1.14),(6.1.34)}{=} \overset{\circ}{\nabla} g(X, s_1, s_2) - \frac{1}{2}A(X, s_1, s_2) - \frac{1}{2}A(X, s_2, s_1)$$

$$\overset{\text{symmetry}}{=} \overset{\circ}{\nabla} g(X, s_1, s_2) - A(X, s_1, s_2) \overset{(6.1.33)}{=} 0$$

as desired.   □

Remark 6.1.25. Let $\pi\colon E \to M$ be a semi-Euclidean vector bundle with scalar product g. If ∇ and $\widetilde{\nabla}$ are metric covariant derivatives on π, and

$$\omega \in \mathcal{A}_1(TM, \mathrm{End}(E)) \cong \mathrm{Hom}\left(\mathfrak{X}(M), \mathrm{End}(\Gamma(E))\right)$$

is the difference tensor of $\widetilde{\nabla}$ and ∇ given by (6.1.9), then the endomorphism $\omega_X \in \mathrm{End}(\Gamma(\pi))$ is skew-symmetric with respect to g for all $X \in \mathfrak{X}(M)$, i.e.,

$$g(\omega_X(s_1), s_2) + g(s_1, \omega_X(s_2)) = 0; \quad s_1, s_2 \in \Gamma(\pi). \tag{6.1.36}$$

Indeed,

$$
\begin{aligned}
0 = (\nabla_X g)(s_1, s_2) &= X g(s_1, s_2) - g(\nabla_X s_1, s_2) - g(s_1, \nabla_X s_2) \\
&\overset{(6.1.9)}{=} X g(s_1, s_2) - g(\widetilde{\nabla}_X s_1, s_2) - g(s_1, \widetilde{\nabla}_X s_2) \\
&\quad + g(\omega_X(s_1), s_2) + g(s_1, \omega_X(s_2)) \\
&= \widetilde{\nabla}_X g(s_1, s_2) + g(\omega_X(s_1), s_2) + g(s_1, \omega_X(s_2)) \\
&= g(\omega_X(s_1), s_2) + g(s_1, \omega_X(s_2)).
\end{aligned}
$$

Conversely, it can be seen by the same calculation that if a one-form $\omega \in \mathcal{A}_1(TM, \mathrm{End}(E))$ has the skew-symmetry property (6.1.36), and

$$
\widetilde{\nabla}_X s := \nabla_X s + \omega_X(s); \quad X \in \mathfrak{X}(M), \ s \in \Gamma(\pi),
$$

where ∇ is a metric covariant derivative on π, then $\widetilde{\nabla}$ is also a metric derivative.

6.1.5 *Curvature and Torsion*

Lemma and Definition 6.1.26. *Let ∇ be a covariant derivative on a vector bundle $\pi \colon E \to M$. The mapping*

$$
R \colon \mathfrak{X}(M) \times \mathfrak{X}(M) \times \Gamma(\pi) \to \Gamma(\pi), \ (X, Y, s) \mapsto R(X, Y)s
$$

given by

$$
R(X, Y)s := \nabla_X \nabla_Y s - \nabla_Y \nabla_X s - \nabla_{[X,Y]} s \tag{6.1.37}
$$

is tensorial in all three arguments, and skew-symmetric in the first two arguments, i.e.,

$$
R(X, Y)s = -R(Y, X)s.
$$

We say that R is the curvature tensor *(or simply the* curvature*) of ∇. A covariant derivative on a vector bundle is called flat if its curvature is zero.*

Proof. Additivity in all three arguments is immediate from the additivity of covariant derivative and Lie bracket. Skew-symmetry in X and Y is also clear from the definition of R and the skew-symmetry of Lie bracket. So it remains only to check the homogeneity of R over $C^\infty(M)$ in the first and the third slot. This is just a simple calculation. Namely, given a smooth function f on M, we find that

$$R(X,Y)(fs) = \nabla_X \nabla_Y(fs) - \nabla_Y \nabla_X(fs) - \nabla_{[X,Y]}(fs)$$
$$\overset{(6.1.2)}{=} \nabla_X\big((Yf)s + f\nabla_Y s\big) - \nabla_Y\big((Xf)s + f\nabla_X s\big)$$
$$- ([X,Y]f)s - f\nabla_{[X,Y]}s = \big(X(Yf)\big)s + (Yf)\nabla_X s$$
$$+ (Xf)\nabla_Y s + f\nabla_X \nabla_Y s - \big(Y(Xf)\big)s - (Xf)\nabla_Y s$$
$$- (Yf)\nabla_X s - f\nabla_Y \nabla_X s - ([X,Y]f)s - f\nabla_{[X,Y]}s$$
$$= fR(X,Y)s.$$

Thus R is $C^\infty(M)$-homogeneous in its third argument. The proof of the $C^\infty(M)$-homogeneity in the first slot is simpler and is left to the reader. $\square$

Remark 6.1.27. Let ∇ be a covariant derivative on $\pi\colon E \to M$ with curvature R.

(a) By the preceding lemma, for any fixed vector fields X and Y on M, the mapping

$$R(X,Y)\colon \Gamma(\pi) \to \Gamma(\pi), \quad s \mapsto R(X,Y)s$$

is tensorial, therefore

$$R(X,Y) \in \mathrm{End}(\Gamma(\pi)) \cong \Gamma(\mathrm{End}(\pi)).$$

We also say that $R(X,Y)$ is the *curvature operator* defined for the vector fields $X, Y \in \mathfrak{X}(M)$.

Thus the curvature tensor of ∇ may also be regarded as an $\mathrm{End}(\pi)$-*valued 2-form on M*, denoted also by R, i.e., as an element of

$$\mathcal{A}_2(TM, \mathrm{End}(\pi)) \cong A_2\big(\mathfrak{X}(M), \mathrm{End}(\Gamma(\pi))\big)$$

such that for all $X, Y \in \mathfrak{X}(M)$, $R(X,Y) \in \Gamma(\mathrm{End}(\pi))$ is given by (6.1.37). Then R is also mentioned as the *curvature form* of ∇. Observe that

$$R \wedge s(X,Y) = R(X,Y)(s); \quad X, Y \in \mathfrak{X}(M), \ s \in \Gamma(\pi)$$

by (3.3.23).

(b) Given a section $s \in \Gamma(\pi)$, define an E-valued two-form

$$R(s) \in \mathcal{A}_2(TM, E) \cong A_2\big(\mathfrak{X}(M), \Gamma(\pi)\big)$$

on M by $R(s)(X,Y) := R(X,Y)(s)$. Then

$$R(s) = d^\nabla \circ \nabla(s). \tag{6.1.38}$$

Indeed, for any vector fields X, Y on M,

$$d^\nabla(\nabla s)(X,Y) \overset{(6.1.30)}{=} \nabla_X((\nabla s)Y) - \nabla_Y((\nabla s)X) - \nabla s([X,Y])$$
$$= \nabla_X \nabla_Y s - \nabla_Y \nabla_X s - \nabla_{[X,Y]}s = R(X,Y)s = R(s)(X,Y).$$

So there is also a third possibility to introduce the curvature of a covariant derivative ∇ on a vector bundle $\pi\colon E \to M$: it may be defined as the composite mapping of ∇ followed by d^∇, i.e., by the sequence

$$\mathcal{A}_0(TM, E) = \Gamma(\pi) \xrightarrow{\nabla} \mathcal{A}_1(TM, E) \xrightarrow{d^\nabla} \mathcal{A}_2(TM, E). \tag{6.1.39}$$

$$\underbrace{\phantom{\mathcal{A}_0(TM, E) = \Gamma(\pi) \xrightarrow{\nabla} \mathcal{A}_1(TM, E)}}_{R}$$

(c) Let $(\mathcal{U}, (u^i)_{i=1}^n)$ be a chart on M such that there exists a local frame $(s_\alpha)_{\alpha=1}^k$ of E over $\mathcal{U}$ (see Remark 6.1.2(c)). We define the *components* $R_{ij\beta}^\alpha \in C^\infty(\mathcal{U})$ of R with respect to this chart and frame by

$$R\left(\frac{\partial}{\partial u^i}, \frac{\partial}{\partial u^j}\right) s_\beta = \sum_{\alpha=1}^k R_{ij\beta}^\alpha s_\alpha; \quad i, j \in J_n; \ \beta \in J_k.$$

An immediate calculation shows that they can be expressed in terms of the Christoffel symbols $\Gamma_{i\beta}^\alpha$ of ∇ given by (6.1.4) as follows:

$$R_{ij\beta}^\alpha = \frac{\partial \Gamma_{j\beta}^\alpha}{\partial u^i} - \frac{\partial \Gamma_{i\beta}^\alpha}{\partial u^j} + \sum_{\gamma=1}^k (\Gamma_{i\gamma}^\alpha \Gamma_{j\beta}^\gamma - \Gamma_{j\gamma}^\alpha \Gamma_{i\beta}^\gamma). \tag{6.1.40}$$

If, in particular, D is a covariant derivative on M with Christoffel symbols Γ_{jk}^i given by (6.1.5), then we define the components R_{jkl}^i of R so that

$$R\left(\frac{\partial}{\partial u^j}, \frac{\partial}{\partial u^k}\right) \frac{\partial}{\partial u^l} = R_{jkl}^i \frac{\partial}{\partial u^i}; \quad j, k, l \in J_n;$$

summation convention in force. In this case (6.1.40) takes the form

$$R_{jkl}^i = \frac{\partial \Gamma_{kl}^i}{\partial u^j} - \frac{\partial \Gamma_{jl}^i}{\partial u^k} + \Gamma_{jm}^i \Gamma_{kl}^m - \Gamma_{km}^i \Gamma_{jl}^m. \tag{6.1.41}$$

Lemma 6.1.28. *Let ∇ and $\widetilde{\nabla}$ be covariant derivatives on a vector bundle $\pi\colon E \to M$. If $\omega \in \mathcal{A}_1(TM, \mathrm{End}(\pi))$ is the difference tensor of $\widetilde{\nabla}$ and ∇, then their curvatures $\widetilde{R}$ and R are related by*

$$\widetilde{R} = R + d^\nabla \omega + [\omega, \omega], \tag{6.1.42}$$

where $[\omega, \omega] \in \mathcal{A}_2(TM, \mathrm{End}(\pi))$ is defined by

$$[\omega, \omega](X, Y) := \omega_X \circ \omega_Y - \omega_Y \circ \omega_X; \quad X, Y \in \mathfrak{X}(M). \tag{6.1.43}$$

Proof.   Let $X, Y \in \mathfrak{X}(M)$ and $s \in \Gamma(\pi)$. Then

$$\widetilde{R}(X,Y)s := \widetilde{\nabla}_X \widetilde{\nabla}_Y s - \widetilde{\nabla}_Y \widetilde{\nabla}_X s - \widetilde{\nabla}_{[X,Y]} s$$

$$\overset{(6.1.9)}{=} \nabla_X(\nabla_Y s + \omega_Y(s)) + \omega_X(\nabla_Y s + \omega_Y(s))$$
$$- \nabla_Y(\nabla_X s + \omega_X(s)) - \omega_Y(\nabla_X s + \omega_X(s)) - \nabla_{[X,Y]}s - \omega_{[X,Y]}(s)$$

$$= R(X,Y)s + \big(\nabla_X(\omega_Y(s)) - \omega_Y(\nabla_X s)\big)$$
$$- \big(\nabla_Y(\omega_X(s)) - \omega_X(\nabla_Y s)\big) - \omega_{[X,Y]}(s) + [\omega, \omega](X,Y)(s)$$

$$\overset{(6.1.13)}{=} R(X,Y)s + (\nabla_X \omega_Y)(s) - (\nabla_Y \omega_X)(s)$$
$$- \omega_{[X,Y]}(s) + [\omega, \omega](X,Y)(s)$$

$$\overset{(6.1.30)}{=} R(X,Y)(s) + (d^{\nabla}\omega)(X,Y)(s) + [\omega, \omega](X,Y)(s),$$

thus proving (6.1.42).   □

Lemma 6.1.29. *Let $\pi \colon E \to M$ be a semi-Euclidean vector bundle with scalar product g. If ∇ is a metric covariant derivative on π, then for all $X, Y \in \mathfrak{X}(M)$ the corresponding curvature operator $R(X,Y) \in \mathrm{End}(\Gamma(\pi))$ is skew-symmetric with respect to g, i.e.,*

$$g(R(X,Y)s_1, s_2) = -g(s_1, R(X,Y)s_2); \quad s_1, s_2 \in \Gamma(\pi).$$

Proof.   It is sufficient to check that $g(R(X,Y)s, s) = 0$ for every $s \in \Gamma(\pi)$. We calculate the left-hand side:

$$g(R(X,Y)s, s) = g(\nabla_X \nabla_Y s, s) - g(\nabla_Y \nabla_X s, s) - g(\nabla_{[X,Y]}s, s)$$

$$\overset{(6.1.32)}{=} Xg(\nabla_Y s, s) - Yg(\nabla_X s, s) - \frac{1}{2}[X,Y]g(s,s)$$

$$= \frac{1}{2}\big(X(Yg(s,s)) - Y(Xg(s,s)) - [X,Y]g(s,s)\big)$$

$$= \frac{1}{2}\big([X,Y]g(s,s) - [X,Y]g(s,s)\big) = 0.$$

This proves our assertion.   □

Proposition 6.1.30 (Differential Bianchi Identity). *If ∇ is a covariant derivative on a vector bundle $\pi \colon E \to M$, then the exterior covariant derivative of its curvature form $R \in \mathcal{A}_2(TM, \mathrm{End}(\pi))$ vanishes:*

$$d^{\nabla} R = 0. \tag{6.1.44}$$

Proof.   For better clarity, we use a more pedantic notation: we write $\widehat{\nabla}$ for the induced covariant derivative on $\mathrm{End}(\pi)$ and $d^{\widehat{\nabla}}$ for the corresponding exterior derivative in $\mathcal{A}(TM, \text{End}(\pi))$. We have to verify that $d^{\widehat{\nabla}} R = 0$, where $d^{\widehat{\nabla}} R \in \mathcal{A}_3(TM, \text{End}(\pi))$. Let $X, Y, Z \in \mathfrak{X}(M)$, $s \in \Gamma(\pi)$. Then

$$
\begin{aligned}
d^{\widehat{\nabla}} R(X, Y, Z)(s) \overset{(6.1.29)}{=} & \left(\widehat{\nabla}_X R(Y, Z)\right)(s) - \left(\widehat{\nabla}_Y R(X, Z)\right)(s) \\
& + \left(\widehat{\nabla}_Z R(X, Y)\right)(s) - R([X, Y], Z)(s) \\
& + R([X, Z], Y)(s) - R([Y, Z], X)(s) \\
\overset{(6.1.13)}{=} & \nabla_X(R(Y, Z)s) - \nabla_Y(R(X, Z)s) + \nabla_Z(R(X, Y)s) \\
& - R(Y, Z)\nabla_X s + R(X, Z)\nabla_Y s - R(X, Y)\nabla_Z s \\
& - R([X, Y], Z)(s) + R([X, Z], Y)(s) - R([Y, Z], X)(s) \\
\overset{(6.1.37)}{=} & \nabla_X \nabla_Y \nabla_Z s - \nabla_X \nabla_Z \nabla_Y s - \nabla_X \nabla_{[Y,Z]} s \\
& - \nabla_Y \nabla_X \nabla_Z s + \nabla_Y \nabla_Z \nabla_X s + \nabla_Y \nabla_{[X,Z]} s \\
& + \nabla_Z \nabla_X \nabla_Y s - \nabla_Z \nabla_Y \nabla_X s - \nabla_Z \nabla_{[X,Y]} s \\
& - \nabla_Y \nabla_Z \nabla_X s + \nabla_Z \nabla_Y \nabla_X s + \nabla_{[Y,Z]} \nabla_X s \\
& + \nabla_X \nabla_Z \nabla_Y s - \nabla_Z \nabla_X \nabla_Y s - \nabla_{[X,Z]} \nabla_Y s \\
& - \nabla_X \nabla_Y \nabla_Z s + \nabla_Y \nabla_X \nabla_Z s + \nabla_{[X,Y]} \nabla_Z s \\
& - \nabla_{[X,Y]} \nabla_Z s + \nabla_Z \nabla_{[X,Y]} s + \nabla_{[[X,Y],Z]} s \\
& + \nabla_{[X,Z]} \nabla_Y s - \nabla_Y \nabla_{[X,Z]} s - \nabla_{[[X,Z],Y]} s \\
& - \nabla_{[Y,Z]} \nabla_X s + \nabla_X \nabla_{[Y,Z]} s + \nabla_{[[Y,Z],X]} s \\
= & \nabla_{[[X,Y],Z]+[[Y,Z],X]+[[Z,X],Y]} s = 0,
\end{aligned}
$$

as desired. $\qquad\qquad\qquad\qquad\qquad\qquad\qquad\qquad\qquad\qquad\qquad\qquad\square$

From the last step of our quite long (but immediate) calculations it turns out that the differential Bianchi identity has a purely algebraic origin: it comes from the Jacobi identity for the Lie bracket of vector fields. In the second volume of the Greub–Halperin–Vanstone monograph the proof is one (short) line ([53], p. 327). The price that one has to pay there for the ease of the proof is an exhausting elaboration of the calculus of bundle-valued differential forms.

Definition 6.1.31. Let $\pi \colon E \to M$ be a vector bundle, endowed with a covariant derivative $\nabla \colon \Gamma(\pi) \to \mathcal{A}_1(TM, E)$. If $\varphi \colon TM \to E$ is a strong bundle map, i.e.,

$$
\varphi \in \text{Bun}_M(\tau, \pi) \cong \text{Hom}(\mathfrak{X}(M), \Gamma(\pi)) \cong \mathcal{A}_1(TM, E),
$$

then the E-valued 2-form

$$
T^\varphi := d^\nabla \varphi \in \mathcal{A}_2(TM, E) \cong A_2(\mathfrak{X}(M), \Gamma(\pi))
$$

is also called the φ-*torsion of* ∇. If, in particular, D is a covariant derivative on M, then the 1_{TM}-torsion

$$T := T^{1_{TM}} := d^D 1_{TM}$$

is called the *torsion of* D. If $T = 0$, we say that D is *torsion-free*, or *symmetric*.

Remark 6.1.32. For any vector fields X, Y on M,

$$T^\varphi(X,Y) := d^\nabla \varphi(X,Y) \overset{(6.1.30)}{=} \nabla_X \varphi(Y) - \nabla_Y \varphi(X) - \varphi([X,Y]). \quad (6.1.45)$$

If $E = TM$, D is a covariant derivative on M and $\varphi = 1_{TM}$, we obtain the classical expression

$$T(X,Y) = D_X Y - D_Y X - [X,Y] \qquad (6.1.46)$$

for the torsion of D.

Example 6.1.33. Let (M, D) be an affinely connected manifold. The *D-Hessian of an arbitrary smooth function f on M is symmetric if, and only if, D is torsion-free.*
    Indeed, for all $X, Y \in \mathfrak{X}(M)$ we have

$$\mathrm{Hess}(f)(X,Y) - \mathrm{Hess}(f)(Y,X) \overset{(6.1.25)}{=} X(Yf) - (D_X Y)f$$
$$- Y(Xf) + (D_Y X)f = (D_Y X - D_X Y - [Y,X])f = T(Y,X)f,$$

whence our claim.

Lemma 6.1.34. *If D and $\widetilde{D}$ are two torsion-free covariant derivatives on a manifold M, then their difference tensor is symmetric.*

Proof. By Lemma 6.1.3, the difference tensor of $\widetilde{D}$ and D is given by

$$\omega(X,Y) := \omega_X(Y) := \widetilde{D}_X Y - D_X Y; \quad X, Y \in \mathfrak{X}(M).$$

Since $\widetilde{D}$ and D are torsion-free,

$$\widetilde{D}_X Y - \widetilde{D}_Y X = D_X Y - D_Y X = [X,Y]$$

for all $X, Y \in \mathfrak{X}(M)$, whence

$$\omega(X,Y) = \widetilde{D}_X Y - D_X Y = \widetilde{D}_Y X - D_Y X = \omega(Y,X),$$

as was to be shown. $\square$

Lemma 6.1.35 (Ricci Formula). *Let ∇ be a covariant derivative on a vector bundle $\pi\colon E \to M$, and let D be a covariant derivative on the base manifold M. Then for all $X, Y \in \mathfrak{X}(M)$ and $s \in \Gamma(\pi)$,*
$$\nabla^2_{X,Y}s - \nabla^2_{Y,X}s = R(X,Y)s - \nabla_{T(X,Y)}s, \qquad (6.1.47)$$
where R is the curvature of ∇, and T is the torsion of D.

Proof. By (6.1.23),
$$
\begin{aligned}
\nabla^2_{X,Y}s - \nabla^2_{Y,X}s &= \nabla_X\nabla_Y s - \nabla_Y\nabla_X s - \left(\nabla_{D_X Y}s - \nabla_{D_Y X}s\right) \\
&\overset{(6.1.46)}{=} \nabla_X\nabla_Y s - \nabla_Y\nabla_X s - \left(\nabla_{T(X,Y)+[X,Y]}s\right) \\
&= R(X,Y)s - \nabla_{T(X,Y)}s. \qquad \square
\end{aligned}
$$

Corollary 6.1.36. *If D is a torsion-free covariant derivative on a manifold M, and R is the curvature of D, then*
$$D^2_{X,Y}Z - D^2_{Y,X}Z = R(X,Y)Z; \quad X, Y, Z \in \mathfrak{X}(M). \qquad \square$$

Remark 6.1.37. There are several Ricci formulae for iterated covariant derivatives of tensor fields. As an instructive example, we deduce here an identity concerning the second covariant differential of a 1-form.

Let D be a covariant derivative on a manifold M with curvature R and torsion T. If α is a 1-form on M, then for vector fields X, Y on M we have
$$D^2_{X,Y}\alpha - D^2_{Y,X}\alpha = \alpha \circ R(Y,X) + D_{T(Y,X)}\alpha. \qquad (6.1.48)$$
Indeed, by Lemma 6.1.18, for all $Z \in \mathfrak{X}(M)$,
$$
\begin{aligned}
(D^2_{X,Y}\alpha)(Z) &= (D_X(D_Y\alpha))(Z) - (D_{D_X Y}\alpha)(Z) \\
&\overset{(6.1.14)}{=} X((D_Y\alpha)(Z)) - (D_Y\alpha)(D_X Z) - (D_{D_X Y}\alpha)(Z) \\
&\overset{(6.1.14)}{=} X((D_Y\alpha)(Z)) - Y(\alpha(D_X Z)) + \alpha(D_Y D_X Z) \\
&\quad - (D_X Y)(\alpha(Z)) + \alpha(D_{D_X Y}Z).
\end{aligned}
$$
Interchanging X and Y, and subtracting the resulting identity we find
$$
\begin{aligned}
(D^2_{X,Y}\alpha - D^2_{Y,X}\alpha)(Z) &= X((D_Y\alpha)(Z)) - Y((D_X\alpha)(Z)) \\
&\quad - Y(\alpha(D_X Z)) + X(\alpha(D_Y Z)) + \alpha(D_Y D_X Z - D_X D_Y Z) \\
&\quad + (D_Y X - D_X Y)(\alpha(Z)) + \alpha(D_{D_X Y - D_Y X}Z) \\
&= X(Y(\alpha(Z)) - \alpha(D_Y Z)) - Y(X(\alpha(Z)) - \alpha(D_X Z)) \\
&\quad - Y(\alpha(D_X Z)) + X(\alpha(D_Y Z)) + \alpha(D_Y D_X Z - D_X D_Y Z - D_{[Y,X]}Z) \\
&\quad - \alpha(D_{T(Y,X)}Z) + [Y,X](\alpha(Z)) + T(Y,X)(\alpha(Z)) \\
&= [X,Y](\alpha(Z)) + [Y,X](\alpha(Z)) + \alpha(R(Y,X)Z) + T(Y,X)(\alpha(Z)) \\
&\quad - \alpha(D_{T(Y,X)}Z) = \alpha(R(Y,X)Z) + (D_{T(Y,X)}\alpha)(Z),
\end{aligned}
$$
whence our claim.

Proposition 6.1.38. *Let D be a covariant derivative on a manifold M with curvature R and torsion T. Then the differential Bianchi identity takes the form*

$$\sum_{\text{cyc}} \left((D_X R)(Y, Z) + R(T(X, Y), Z)\right) = 0, \qquad (6.1.49)$$

where X, Y, Z are vector fields on M. If D is torsion-free, we have

$$\sum_{\text{cyc}} (D_X R)(Y, Z) = 0; \quad X, Y, Z \in \mathfrak{X}(M), \qquad (6.1.50)$$

where $\sum_{\text{cyc}}$ denotes the cyclic sum over X, Y, Z (see A.3.8).

Proof. By Proposition 6.1.30,

$$0 = d^{\widehat{D}} R(X, Y, Z) \overset{(6.1.29)}{=} \widehat{D}_X(R(Y, Z)) - \widehat{D}_Y(R(X, Z)) + \widehat{D}_Z(R(X, Y))$$
$$- R([X, Y], Z) + R([X, Z], Y) - R([Y, Z], X)$$
$$= \widehat{D}_X(R(Y, Z)) + \widehat{D}_Y(R(Z, X)) + \widehat{D}_Z(R(X, Y))$$
$$- \left(R([X, Y], Z) + R([Y, Z], X) + R([Z, X], Y)\right).$$

Here, for every $U \in \mathfrak{X}(M)$,

$$\left(\widehat{D}_X(R(Y, Z))\right)(U) \overset{(6.1.13)}{=} D_X(R(Y, Z)U) - R(Y, Z)D_X U$$
$$\overset{(6.1.14)}{=} (D_X R)(Y, Z)U + R(D_X Y, Z)U + R(Y, D_X Z)U$$
$$+ R(Y, Z)D_X U - R(Y, Z)D_X U,$$

therefore

$$\widehat{D}_X(R(Y, Z)) = (D_X R)(Y, Z) + R(D_X Y, Z) + R(Y, D_X Z).$$

Similarly,

$$\widehat{D}_Y(R(Z, X)) = (D_Y R)(Z, X) + R(D_Y Z, X) + R(Z, D_Y X),$$
$$\widehat{D}_Z(R(X, Y)) = (D_Z R)(X, Y) + R(D_Z X, Y) + R(X, D_Z Y).$$

Combining these results we find that

$$0 = \sum_{\text{cyc}} \left((D_X R)(Y, Z) + R(D_X Y - D_Y X - [X, Y], Z)\right)$$
$$= \sum_{\text{cyc}} \left((D_X R)(Y, Z) + R(T(X, Y), Z)\right),$$

completing the proof. $\square$

Proposition 6.1.39 (Algebraic Bianchi Identity). *Let D be a covariant derivative on a manifold M with curvature R and torsion T. Then*

$$\sum_{\text{cyc}} R(X,Y)Z = (d^D T)(X,Y,Z) = \sum_{\text{cyc}} \left((D_X T)(Y,Z) + T(T(X,Y),Z) \right),$$

(6.1.51)

where $X, Y, Z \in \mathfrak{X}(M)$. In particular, if D is torsion-free we have

$$\sum_{\text{cyc}} R(X,Y)Z = 0.$$
(6.1.52)

Proof. We evaluate the exterior covariant derivative $d^D T$ at (X,Y,Z) in two ways. First,

$$d^D T(X,Y,Z) \overset{(6.1.29)}{=} D_X(T(Y,Z)) - D_Y(T(X,Z)) + D_Z(T(X,Y))$$
$$- T([X,Y],Z) + T([X,Z],Y) - T([Y,Z],X)$$
$$= \sum_{\text{cyc}} \left(D_X(T(Y,Z)) - T([X,Y],Z) \right)$$
$$= D_X D_Y Z - D_X D_Z Y - D_X[Y,Z] - D_{[X,Y]}Z + D_Z[X,Y] + [[X,Y],Z]$$
$$+ D_Y D_Z X - D_Y D_X Z - D_Y[Z,X] - D_{[Y,Z]}X + D_X[Y,Z] + [[Y,Z],X]$$
$$+ D_Z D_X Y - D_Z D_Y X - D_Z[X,Y] - D_{[Z,X]}Y + D_Y[Z,X] + [[Z,X],Y]$$
$$= \sum_{\text{cyc}} \left(R(X,Y)Z + [[X,Y],Z] \right) = \sum_{\text{cyc}} R(X,Y)Z,$$

which proves the first half of (6.1.51).

Next, we calculate the covariant derivatives $D_X(T(Y,Z))$, $D_Y(T(Z,X))$ and $D_Z(T(X,Y))$ by using (6.1.14). We have

$$D_X(T(Y,Z)) = (D_X T)(Y,Z) + T(D_X Y, Z) + T(Y, D_X Z),$$

and the other two formulas can be obtained by cyclic permutations of (X,Y,Z). Thus

$$(d^D T)(X,Y,Z) = \left(\sum_{\text{cyc}} (D_X T)(Y,Z) \right) + T(D_X Y, Z) + T(Y, D_X Z)$$
$$+ T(D_Y Z, X) + T(Z, D_Y X) + T(D_Z X, Y) + T(X, D_Z Y)$$
$$- T([X,Y],Z) - T([Y,Z],X) - T([Z,X],Y)$$
$$= \sum_{\text{cyc}} \left((D_X T)(Y,Z) + T(D_X Y - D_Y X - [X,Y], Z) \right)$$
$$= \sum_{\text{cyc}} \left((D_X T)(Y,Z) + T(T(X,Y),Z) \right),$$

which concludes the proof. □

Example 6.1.40. The standard covariant derivative ∇ on a trivial vector bundle $\mathrm{pr}_1 \colon M \times V \to M$ has zero curvature. Indeed, with the notation of Example 6.1.5(a), if $s = \sum_{i=1}^{k} \sigma^i s_i \in \Gamma(M \times V)$ and $X, Y \in \mathfrak{X}(M)$, then

$$R(X,Y)s = \nabla_X \nabla_Y s - \nabla_Y \nabla_X s - \nabla_{[X,Y]} s$$

$$\overset{(6.1.10)}{=} \nabla_X \left(\sum_{i=1}^{k} (Y\sigma^i) s_i \right) - \nabla_Y \left(\sum_{i=1}^{k} (X\sigma^i) s_i \right) - \sum_{i=1}^{k} ([X,Y]\sigma^i) s_i$$

$$= \sum_{i=1}^{k} \left(X(Y\sigma^i) - Y(X\sigma^i) - [X,Y]\sigma^i \right) s_i = 0.$$

However, if $\omega \in \mathcal{A}_1(TM, \mathrm{End}(M \times V)) \cong \mathcal{A}_1(\mathfrak{X}(M), \mathrm{End}(\Gamma(M \times V)))$ and we define a new covariant derivative $\widetilde{\nabla}$ on $M \times V$ by (6.1.8), i.e., by the rule

$$\widetilde{\nabla}_X s := \nabla_X s + \omega_X(s); \quad X \in \mathfrak{X}(M), \ s \in \Gamma(M \times V),$$

then, by Lemma 6.1.28, the curvature of $\widetilde{\nabla}$ is

$$\widetilde{R} = d^\nabla \omega + [\omega, \omega],$$

which does not vanish in general.

6.1.6   The Levi-Civita Derivative

Definition 6.1.41. A manifold is said to be a *semi-Riemannian*, resp. a *Riemannian manifold*, if its tangent bundle is a semi-Euclidean, resp. a Euclidean vector bundle.

Remark 6.1.42. (a) In view of Definition 2.2.52, a semi-Riemannian manifold is a pair (M, g) consisting of a manifold M and a symmetric tensor field g on M such that $g_p \colon T_p M \times T_p M \to \mathbb{R}$ is a non-degenerate scalar product, and hence $(T_p M, g_p)$ is a semi-Euclidean vector space for every $p \in M$. In the Riemannian case the scalar products g_p $(p \in M)$ are positive definite. If (M, g) is a semi-Riemannian manifold, we also say that g is a *metric tensor* or a *semi-Riemannian metric* on M.

(b) Suppose that (M, g) is a Riemannian manifold, and let $N \subset M$ be a submanifold of M with the canonical inclusion $j \colon N \to M$. Then the pull-back $h := j^* g$ of g is a metric tensor, called the *induced metric* on N. Thus (N, h) is a Riemannian manifold, called a *Riemannian submanifold* of (M, g). The scalar product of two tangent vectors $u, v \in T_p N$ is given by

$$h_p(u, v) = g_p((j_*)_p(u), (j_*)_p(v)), \quad p \in N.$$

Since every tangent space T_pN can be regarded as a subspace of T_pM (Remark 3.1.24), we can simply write

$$h_p(u,v) = g_p(u,v); \quad u,v \in T_pN.$$

We note that if g is an indefinite metric on M, then j^*g need not be a metric tensor on N.

(c) Let (M,g) be a semi-Riemannian manifold. By the *speed* of a smooth curve $\gamma\colon I \to M$ we mean the function

$$\|\dot\gamma\|_g\colon I \to \mathbb{R}, \ t \mapsto \|\dot\gamma\|_g(t) := \|\dot\gamma(t)\|_g := |g_{\gamma(t)}(\dot\gamma(t),\dot\gamma(t))|^{\frac{1}{2}}. \quad (6.1.53)$$

(In the Riemannian case $\|\dot\gamma(t)\|_g = \left(g_{\gamma(t)}(\dot\gamma(t),\dot\gamma(t))\right)^{\frac{1}{2}}$.) We say that γ is of *constant speed* if the function $\|\dot\gamma\|_g$ is constant. In particular, γ has *unit speed* if $\|\dot\gamma(t)\|_g = 1$ for all $t \in I$.

(d) We continue to assume that (M,g) is a semi-Riemannian manifold. Then the musical isomorphisms $\flat$ and $\sharp$ introduced in Remark 2.2.55 take the form

$$\begin{cases} \flat\colon \mathfrak{X}(M) \to \mathfrak{X}^*(M) \cong \mathcal{A}_1(M), \ X \mapsto X^\flat, \\ X^\flat(Y) = g(X,Y) \text{ for all } Y \in \mathfrak{X}(M) \end{cases} \quad (6.1.54)$$

and

$$\begin{cases} \sharp\colon \mathfrak{X}^*(M) \to \mathfrak{X}(M), \ \alpha \mapsto \alpha^\sharp, \\ g(\alpha^\sharp,Y) = \alpha(Y) \text{ for all } Y \in \mathfrak{X}(M). \end{cases} \quad (6.1.55)$$

Choose a chart $(\mathcal{U}, (u^i)_{i=1}^n)$ on M, and let

$$g_{ij} := g\left(\frac{\partial}{\partial u^i}, \frac{\partial}{\partial u^j}\right); \quad i,j \in J_n$$

be the components of g with respect to $(\mathcal{U}, (u^i)_{i=1}^n)$. Since g_p is non-degenerate at each point p of $\mathcal{U}$, the matrices $(g_{ij}(p))$, $p \in \mathcal{U}$ are invertible by Corollary 1.4.1. We denote the inverse matrix of $(g_{ij}(p))$ by $(g^{ij}(p))$. Then the functions

$$g^{ij}\colon \mathcal{U} \to \mathbb{R}, \ p \mapsto g^{ij}(p); \quad i,j \in J_n$$

are smooth and we also have $(g^{ij}) = (g_{ij})^{-1}$.

If $X \in \mathfrak{X}(M)$, $X \restriction \mathcal{U} = \sum_{j=1}^n X^j \frac{\partial}{\partial u^j}$, then

$$X^\flat \restriction \mathcal{U} = \sum_{i=1}^n X_i du^i \text{ where } X_i = \sum_{j=1}^n g_{ij}X^j. \quad (6.1.56)$$

Indeed, if $X^\flat \restriction \mathcal{U} = \sum_{j=1}^n X_j du^j$, then for all $i \in J_n$,

$$X_i \overset{(1.2.11)}{=} X^\flat \left(\frac{\partial}{\partial u^i} \right) \overset{(6.1.54)}{=} g \left(\sum_{j=1}^n X^j \frac{\partial}{\partial u^j}, \frac{\partial}{\partial u^i} \right)$$

$$= \sum_{j=1}^n X^j g_{ji} = \sum_{j=1}^n g_{ij} X^j.$$

Similarly, if $\alpha \in \mathfrak{X}^*(M)$, $\alpha \restriction \mathcal{U} = \sum_{j=1}^n \alpha_j du^j$, then

$$\alpha^\sharp \restriction \mathcal{U} = \sum_{i=1}^n \alpha^i \frac{\partial}{\partial u^i}, \text{ where } \alpha^i = \sum_{j=1}^n g^{ij} \alpha_j. \tag{6.1.57}$$

This can also be seen by an immediate calculation. Over $\mathcal{U}$, for all $j \in J_n$ we have

$$\alpha_j = \alpha \left(\frac{\partial}{\partial u^j} \right) \overset{(6.1.55)}{=} g \left(\alpha^\sharp, \frac{\partial}{\partial u^j} \right) = g \left(\sum_{k=1}^n \alpha^k \frac{\partial}{\partial u^k}, \frac{\partial}{\partial u^j} \right) = \sum_{k=1}^n \alpha^k g_{kj},$$

therefore

$$\sum_{j=1}^n g^{ij} \alpha_j = \sum_{k=1}^n \sum_{j=1}^n \alpha^k g^{ij} g_{jk} = \sum_{k=1}^n \alpha^k \delta_k^i = \alpha^i,$$

as we claimed.

Thus the flat operator $\flat$ indeed 'lowers', the sharp operator $\sharp$ 'raises' an index, as we have already announced in Remark 2.2.55.

In particular, if f is a smooth function on M, then the vector field 'metrically equivalent' to the 1-form $df \in \mathfrak{X}^*(M)$ is called the *gradient* of f and is denoted by $\operatorname{grad} f$. Thus $\operatorname{grad} f := (df)^\sharp$, i.e., $\operatorname{grad} f$ is the unique vector field on M such that

$$g(\operatorname{grad} f, X) = df(X) = Xf \quad \text{for all} \quad X \in \mathfrak{X}(M). \tag{6.1.58}$$

From (3.3.3) and (6.1.57) we obtain the following coordinate expression for the gradient of f:

$$\operatorname{grad} f \underset{(\mathcal{U})}{=} \sum_{i,j=1}^n g^{ij} \frac{\partial f}{\partial u^i} \frac{\partial}{\partial u^j}.$$

Proposition 6.1.43 (Ricci Lemma). *Let (M, g) be a semi-Riemannian manifold, and let a TM-valued 2-form*

$$A \in \mathcal{A}_2(TM, TM) \cong A_2(\mathfrak{X}(M), \mathfrak{X}(M))$$

be given on M. Then there exists a unique metric covariant derivative on M with torsion A.

Proof. Start with an arbitrary metric covariant derivative $\mathring{D}$ on M. Then, by Remark 6.1.25, every metric covariant derivative D on M acts by the rule

$$D_X Y = \mathring{D}_X Y + \psi_X(Y); \quad X, Y \in \mathfrak{X}(M), \qquad (*)$$

where the tensor

$$\psi \colon \mathfrak{X}(M) \times \mathfrak{X}(M) \to \mathfrak{X}(M), \ (X, Y) \mapsto \psi_X(Y)$$

is skew-symmetric with respect to g. Let T and $\mathring{T}$ be the torsion of D and $\mathring{D}$, respectively. Then for any vector fields X, Y on M,

$$T(X, Y) \overset{(6.1.46)}{=} D_X Y - D_Y X - [X, Y] \overset{(*)}{=} \mathring{D}_X Y - \mathring{D}_Y X - [X, Y]$$
$$+ \psi_X(Y) - \psi_Y(X),$$

so we have

$$T(X, Y) = \mathring{T}(X, Y) + \psi_X(Y) - \psi_Y(X); \quad X, Y \in \mathfrak{X}(M). \qquad (**)$$

Define a mapping $\beta \colon \mathfrak{X}(M) \times \mathfrak{X}(M) \to \mathfrak{X}(M)$ by

$$\beta(X, Y) := A(X, Y) - \mathring{T}(X, Y). \qquad (***)$$

Obviously, β is $C^\infty(M)$-bilinear and alternating.

Now we apply Proposition 1.4.11 with the cast as follows:

$\mathsf{R} := C^\infty(M), V := \mathfrak{X}(M),$

$g :=$ the given semi-Riemannian metric on M.

Then the role of the isomorphism $f \colon V \to V^*$ in Lemma 1.4.9 is played by the musical isomorphism $\flat \colon \mathfrak{X}(M) \to \mathfrak{X}^*(M)$.

By the quoted proposition, there is a unique $C^\infty(M)$-linear mapping

$$\omega \colon \mathfrak{X}(M) \to \mathrm{End}(\mathfrak{X}(M)), \ X \mapsto \omega_X$$

such that

(i) ω_X is skew-symmetric with respect to g, i.e.,

$$g(\omega_X(Y), Z) + g(Y, \omega_X(Z)) = 0 \text{ for all } Y, Z \in \mathfrak{X}(M);$$

(ii) $\beta(X, Y) = \omega_X(Y) - \omega_Y(X).$

Let $\psi := \omega$ in relation $(*)$. The so obtained covariant derivative D remains metric, and its torsion T is the prescribed tensor A, since

$$T(X, Y) \overset{(**)}{=} \mathring{T}(X, Y) + \omega_X(Y) - \omega_Y(X) \overset{(ii)}{=} \mathring{T}(X, Y) + \beta(X, Y)$$
$$\overset{(***)}{=} A(X, Y)$$

for all $X, Y \in \mathfrak{X}(M)$. $\qquad \square$

Definition 6.1.44. The unique metric covariant derivative on a semi-Riemannian manifold with torsion equal to zero is called the *Levi-Civita derivative*.

Proposition 6.1.45. *Let (M, g) be a semi-Riemannian manifold and D the Levi-Civita derivative on M. Then for all $X, Y, Z \in \mathfrak{X}(M)$ we have the Koszul formula*

$$2g(D_X Y, Z) = Xg(Y, Z) + Yg(Z, X) - Zg(X, Y) \\ - g(X, [Y, Z]) + g(|Y, [Z, X]) + g(Z, [X, Y]). \tag{6.1.59}$$

Proof. We use the so-called Christoffel's trick. For any $X, Y, Z \in \mathfrak{X}(M)$,

$$Xg(Y, Z) = g(D_X Y, Z) + g(Y, D_X Z),$$

$$Yg(Z, X) = g(D_Y Z, X) + g(Z, D_Y X),$$

$$-Zg(X, Y) = -g(D_Z X, Y) - g(X, D_Z Y),$$

since the covariant derivative is compatible with the metric tensor. Adding up the three equalities and using the fact that the covariant derivative is torsion-free, we get

$$Xg(Y, Z) + Yg(Z, X) - Zg(X, Y)$$

$$= g(D_X Y + D_Y X, Z) + g(Y, D_X Z - D_Z X) + g(X, D_Y Z - D_Z Y)$$

$$= 2g(D_X Y, Z) - g(Z, [X, Y]) - g(Y, [Z, X]) + g(X, [Y, Z]),$$

which proves (6.1.59). $\qquad\qquad\qquad\qquad\qquad\qquad\qquad\qquad\qquad\qquad\square$

Remark 6.1.46. We keep the notation of the preceding proposition.

(a) Since g is non-degenerate and Z is arbitrary in (6.1.59), the Koszul formula determines $D_X Y$ uniquely. Thus we have obtained a further proof of the uniqueness of the Levi-Civita derivative. Conversely, if we define $g(D_X Y, Z)$ and hence $D_X Y$ by the Koszul formula, then it can be shown by a straightforward, but tedious calculation that D is a torsion-free, metric covariant derivative on M.

(b) Choose a chart $(\mathcal{U}, (u^i)_{i=1}^n)$ on M. The Christoffel symbols of D with respect to the chart are defined by

$$D_{\frac{\partial}{\partial u^j}} \frac{\partial}{\partial u^k} = \sum_{m=1}^n \Gamma_{jk}^m \frac{\partial}{\partial u^m}; \quad j, k \in J_n$$

(see (6.1.5)). Let $j, k, l \in J_n$. Plugging $X := \frac{\partial}{\partial u^j}$, $Y := \frac{\partial}{\partial u^k}$, $Z := \frac{\partial}{\partial u^l}$ into (6.1.59) we find

$$2g\left(D_{\frac{\partial}{\partial u^j}} \frac{\partial}{\partial u^k}, \frac{\partial}{\partial u^l}\right) = \frac{\partial}{\partial u^j} g_{kl} + \frac{\partial}{\partial u^k} g_{lj} - \frac{\partial}{\partial u^l} g_{jk},$$

since the Lie brackets of coordinate vector fields vanish. The left-hand side can be transformed as follows:

$$g\left(D_{\frac{\partial}{\partial u^j}}\frac{\partial}{\partial u^k},\frac{\partial}{\partial u^l}\right) = g\left(\sum_{m=1}^{n}\Gamma_{jk}^{m}\frac{\partial}{\partial u^m},\frac{\partial}{\partial u^l}\right) = \sum_{m=1}^{n}\Gamma_{jk}^{m}g_{ml}.$$

Thus

$$2\sum_{m=1}^{n}\Gamma_{jk}^{m}g_{ml} = \frac{\partial}{\partial u^j}g_{kl} + \frac{\partial}{\partial u^k}g_{lj} - \frac{\partial}{\partial u^l}g_{jk}.$$

Now we multiply both sides of this equality by g^{il} and sum over l. Since

$$\sum_{l,m=1}^{n}\Gamma_{jk}^{m}g^{il}g_{ml} = \sum_{m=1}^{n}\Gamma_{jk}^{m}\delta_{m}^{i} = \Gamma_{jk}^{i},$$

we get

$$\Gamma_{jk}^{i} = \frac{1}{2}\sum_{l=1}^{n}g^{il}\left(\frac{\partial}{\partial u^j}g_{kl} + \frac{\partial}{\partial u^k}g_{lj} - \frac{\partial}{\partial u^l}g_{jk}\right). \tag{6.1.60}$$

This shows that *the Christoffel symbols of the Levi-Civita derivative depend only on the metric tensor and the first partial derivatives* (with respect to the coordinate functions of the chosen chart) *of its components.*

6.1.7   Covariant Derivative Along a Curve

Let $\pi\colon E \to M$ be a vector bundle of rank k, and let a smooth curve $\gamma\colon I \to M$ be given, where I is a (nonempty) open interval. Then I is an open submanifold of $\mathbb{R}$, and hence, in its own right, it is a manifold. Consider the $C^\infty(I)$-module $\Gamma_\gamma(E)$ of sections of E along γ. This consists of the smooth mappings $s : I \to E$ such that $\pi \circ s = \gamma$ (see Definition 2.2.11).

If $\sigma \in \Gamma(E)$ is a section of E, then $\sigma \circ \gamma$ is obviously a section of E along γ. Moreover, if $(\sigma_\alpha)_{\alpha=1}^{k}$ is a local frame field of E over an open subset $\mathcal{U}$ of M such that $\gamma(I) \subset \mathcal{U}$, then $(\sigma_\alpha \circ \gamma)_{\alpha=1}^{k}$ is a basis of the $C^\infty(I)$-module $\Gamma_\gamma(E)$.

Definition 6.1.47. Let $\pi\colon E \to M$ be a vector bundle, and let a smooth curve $\gamma\colon I \to M$ be given. A *covariant derivative along γ* is an $\mathbb{R}$-linear mapping

$$\nabla^\gamma\colon \Gamma_\gamma(E) \to \Gamma_\gamma(E)$$

such that for all $f \in C^\infty(I)$ and $s \in \Gamma_\gamma(E)$ we have

$$\nabla^\gamma(fs) = f's + f\nabla^\gamma s. \tag{6.1.61}$$

Remark 6.1.48. Covariant derivatives along a smooth curve $\gamma\colon I \to M$ are also local operators: for any section $s \in \Gamma_\gamma(E)$ and any point $t_0 \in I$, the value of $\nabla^\gamma s$ at t_0 depends only on the values of s 'near t_0', i.e., its values in any interval $]t_0 - \varepsilon, t_0 + \varepsilon[\subset I$, where ε is an arbitrarily small positive real number. This can be seen by an argument similar to that used in the proof of Lemma 3.3.11.

Proposition 6.1.49. *Given a covariant derivative* ∇ *on a vector bundle* $\pi\colon E \to M$ *and a smooth curve* $\gamma\colon I \to M$, *there exists a unique covariant derivative* ∇^γ *along* γ *such that*

$$\nabla^\gamma(\sigma \circ \gamma) = (\nabla\sigma) \circ \dot\gamma \text{ for all } \sigma \in \Gamma(E). \tag{6.1.62}$$

Proof. *Uniqueness.* Suppose that $\nabla^\gamma\colon \Gamma_\gamma(E) \to \Gamma_\gamma(E)$ is a covariant derivative along γ satisfying (6.1.62). By the local character of ∇^γ mentioned in the previous remark, we may assume that $\mathrm{Im}(\gamma)$ is contained in an open subset $\mathcal{U}$ of M such that $E_\mathcal{U} = \pi^{-1}(\mathcal{U})$ has a trivialization. Then there exists a local frame field $(\sigma_\alpha)_{\alpha=1}^k$ ($k := \mathrm{rank}$ of π) of E over $\mathcal{U}$, and the family $(\sigma_\alpha \circ \gamma)_{\alpha=1}^k$ is a basis of $\Gamma_\gamma(E)$. So any section s along γ can uniquely be written in the form

$$s = \sum_{\alpha=1}^k s^\alpha(\sigma_\alpha \circ \gamma), \quad s^\alpha \in C^\infty(I).$$

Then we obtain

$$\nabla^\gamma s = \sum_{\alpha=1}^k \nabla^\gamma(s^\alpha(\sigma_\alpha \circ \gamma)) \overset{(6.1.61)}{=} \sum_{\alpha=1}^k \left((s^\alpha)'(\sigma_\alpha \circ \gamma) + s^\alpha\nabla^\gamma(\sigma_\alpha \circ \gamma)\right)$$

$$\overset{(6.1.62)}{=} \sum_{\alpha=1}^k \left((s^\alpha)'(\sigma_\alpha \circ \gamma) + s^\alpha(\nabla\sigma_\alpha) \circ \dot\gamma\right).$$

Therefore $\nabla^\gamma s$ is completely determined by ∇.

Existence. If $\mathrm{Im}(\gamma) \subset \mathcal{U}$, where $\mathcal{U}$ is an open subset of M such that $E_\mathcal{U}$ is trivializable, then we define ∇^γ by the formula

$$\nabla^\gamma s := \sum_{\alpha=1}^k \left((s^\alpha)'(\sigma_\alpha \circ \gamma) + s^\alpha((\nabla\sigma_\alpha) \circ \dot\gamma)\right) \text{ if } s = \sum_{\alpha=1}^k s^\alpha(\sigma_\alpha \circ \gamma) \tag{6.1.63}$$

obtained above. Then it can be checked by a straightforward calculation that ∇^γ is $\mathbb{R}$-linear and has the properties (6.1.61) and (6.1.62).

In the general case we cover $\mathrm{Im}(\gamma)$ by a family of $(\mathcal{U}_j)_{j\in J}$ of open subsets of M, where $E_{\mathcal{U}_j}$ has a trivialization for all $j \in J$, and define ∇^γ by the formula (6.1.63) on $\Gamma_\gamma(E_{\mathcal{U}_j})$ ($j \in J$). If $s \in \Gamma_\gamma(E)$ and $\mathcal{U}_i \cap \mathcal{U}_j \neq \emptyset$ ($i,j \in J$), then the values of $\nabla^\gamma s$ obtained from (6.1.63) coincide on $\gamma^{-1}(\mathcal{U}_i \cap \mathcal{U}_j)$ by the uniqueness, thus these local definitions give rise to a single section $\nabla^\gamma s$.

$\square$

Remark 6.1.50. We keep the notation introduced above.

(a) The covariant derivative ∇^γ described by Proposition 6.1.49 is called the covariant derivative along γ *induced* by ∇. Let us now give a coordinate description of ∇^γ. Choose a chart $(\mathcal{U}, (u^i)_{i=1}^n)$ on M with the property that $E_\mathcal{U} = \pi^{-1}(\mathcal{U})$ has a trivialization, and let $(\sigma_\alpha)_{\alpha=1}^k$ be a local frame field of E on $\mathcal{U}$. The Christoffel symbols $\Gamma_{j\beta}^\alpha$ of ∇ with respect to this chart and frame field are defined by

$$\nabla_{\frac{\partial}{\partial u^j}} \sigma_\beta = \sum_{\alpha=1}^k \Gamma_{j\beta}^\alpha \sigma_\alpha, \quad (j, \beta) \in J_n \times J_k$$

(see Remark 6.1.2(c)). If

$$s = \sum_{\alpha=1}^k s^\alpha(\sigma_\alpha \circ \gamma) \in \Gamma_\gamma(E) \text{ and } \dot{\gamma}(t) = \sum_{j=1}^n (\gamma^j)'(t) \left(\frac{\partial}{\partial u^j} \right)_{\gamma(t)}, \quad t \in I,$$

then

$$(\nabla^\gamma s)(t) \overset{(6.1.63)}{=} \sum_{\alpha=1}^k \left((s^\alpha)'(t)\sigma_\alpha(\gamma(t)) + s^\alpha(t)((\nabla\sigma_\alpha)(\dot{\gamma}(t))) \right)$$

$$= \sum_{\alpha=1}^k \left((s^\alpha)'(t)\sigma_\alpha(\gamma(t)) + s^\alpha(t)(\nabla_{\sum_{j=1}^n (\gamma^j)'(t)(\frac{\partial}{\partial u^j})_{\gamma(t)}} \sigma_\alpha) \right)$$

$$= \sum_{\alpha=1}^k \left((s^\alpha)'(t)\sigma_\alpha(\gamma(t)) \right) + \sum_{\beta=1}^k \sum_{j=1}^n s^\beta(t)(\gamma^j)'(t)\nabla_{(\frac{\partial}{\partial u^j})_{\gamma(t)}} \sigma_\beta$$

$$= \sum_{\alpha=1}^k \left((s^\alpha)'(t) + \sum_{\beta=1}^k \sum_{j=1}^n s^\beta(t)(\gamma^j)'(t)\Gamma_{j\beta}^\alpha(\gamma(t)) \right) \sigma_\alpha(\gamma(t)),$$

therefore

$$\nabla^\gamma s = \sum_{\alpha=1}^k \left((s^\alpha)' + \sum_{\beta=1}^k \sum_{j=1}^n (\gamma^j)'s^\beta(\Gamma_{j\beta}^\alpha \circ \gamma) \right)(\sigma_\alpha \circ \gamma). \tag{6.1.64}$$

(b) Suppose, in particular, that $E = TM$, and let D be a covariant derivative on M. If $(\mathcal{U}, (u^i)_{i=1}^n)$ is a chart on M, $X \in \mathfrak{X}_\gamma(M)$ and $X \upharpoonright \gamma^{-1}(\mathcal{U}) = \sum_{i=1}^n X^i \left(\frac{\partial}{\partial u^i} \circ \gamma \right)$, then (6.1.64) reduces to

$$D^\gamma X \underset{(\mathcal{U})}{=} \sum_{i=1}^n \left((X^i)' + \sum_{j,k=1}^n (\gamma^j)'X^k(\Gamma_{jk}^i \circ \gamma) \right) \left(\frac{\partial}{\partial u^i} \circ \gamma \right). \tag{6.1.65}$$

Lemma 6.1.51. *Let* $\pi\colon E \to M$ *and* $\widetilde{\pi}\colon \widetilde{E} \to \widetilde{M}$ *be vector bundles with covariant derivatives* ∇ *and* $\widetilde{\nabla}$, *respectively. Suppose that* $\varphi\colon E \to \widetilde{E}$ *is a covariant derivative preserving bundle isomorphism. Then, for any smooth curve* $\gamma\colon I \to M$, φ *preserves the covariant derivative along* γ *induced by* ∇. *More precisely, for every section* $s \in \Gamma_\gamma(E)$,

$$\varphi_\#^\gamma(\nabla^\gamma s) = \widetilde{\nabla}^{\varphi_B \circ \gamma}(\varphi_\#^\gamma(s)), \tag{6.1.66}$$

where $\varphi_B\colon M \to \widetilde{M}$ *is the mapping induced by* φ *(see Definition 2.2.5), and*

$$\varphi_\#^\gamma\colon \Gamma_\gamma(E) \to \Gamma_{\varphi_B \circ \gamma}(\widetilde{E}), \quad s \mapsto \varphi_\#^\gamma(s) := \varphi \circ s. \tag{$*$}$$

Proof.   Let, for short, $\widetilde{\gamma} := \varphi_B \circ \gamma$. Define a mapping

$$\bar{\nabla}\colon \Gamma_{\widetilde{\gamma}}(\widetilde{E}) \to \Gamma_{\widetilde{\gamma}}(\widetilde{E})$$

by the prescription

$$\bar{\nabla}(\varphi_\#^\gamma(s)) = \varphi_\#^\gamma(\nabla^\gamma s). \tag{$**$}$$

Then $\bar{\nabla}$ is well-defined because the mapping $\varphi_\#^\gamma$ is a module isomorphism if we identify the rings $C^\infty(M)$ and $C^\infty(\widetilde{M})$ as in Example 6.1.8(e). It can also easily be checked that $\bar{\nabla}$ is a covariant derivative along $\widetilde{\gamma}$. Next we show that

$$(\widetilde{\nabla}\widetilde{\sigma}) \circ \dot{\widetilde{\gamma}} = \bar{\nabla}(\widetilde{\sigma} \circ \widetilde{\gamma}) \quad \text{for all} \quad \widetilde{\sigma} \in \Gamma(\widetilde{E}).$$

Then we can conclude from Proposition 6.1.49 that $\bar{\nabla} = \widetilde{\nabla}^{\widetilde{\gamma}}$, and so we obtain (6.1.66).

Given a section $\widetilde{\sigma} \in \Gamma(\widetilde{E})$, there is a unique section $\sigma \in \Gamma(E)$ such that

$$\varphi_\#(\sigma) \stackrel{(2.2.10)}{=} \varphi \circ \sigma \circ \varphi_B^{-1} = \widetilde{\sigma},$$

because φ is a bundle isomorphism. Thus

$$\widetilde{\sigma} \circ \widetilde{\gamma} = \varphi_\#(\sigma) \circ \widetilde{\gamma} = \varphi \circ \sigma \circ \varphi_B^{-1} \circ \varphi_B \circ \gamma = \varphi \circ \sigma \circ \gamma =: \varphi_\#^\gamma(\sigma \circ \gamma). \tag{$***$}$$

For every $t \in I$,

$$(\widetilde{\nabla}\widetilde{\sigma}) \circ \dot{\widetilde{\gamma}}(t) = (\widetilde{\nabla}\varphi_\#(\sigma)) \circ (\varphi_B \circ \gamma)\dot{}(t) \stackrel{(3.1.11)}{=} (\widetilde{\nabla}\varphi_\#(\sigma))((\varphi_B)_*(\dot{\gamma}(t)))$$

$$= \widetilde{\nabla}_{(\varphi_B)_*(\dot{\gamma}(t))}\varphi_\#(\sigma) \stackrel{(6.1.19)}{=} \varphi(\nabla_{\dot{\gamma}(t)}\sigma) = \varphi((\nabla\sigma) \circ \dot{\gamma})(t)$$

$$\stackrel{(6.1.62)}{=} \varphi(\nabla^\gamma(\sigma \circ \gamma))(t) \stackrel{(*)}{=} (\varphi_\#^\gamma(\nabla^\gamma(\sigma \circ \gamma)))(t)$$

$$\stackrel{(**)}{=} \bar{\nabla}(\varphi_\#^\gamma(\sigma \circ \gamma))(t) \stackrel{(***)}{=} \bar{\nabla}(\widetilde{\sigma} \circ \widetilde{\gamma})(t),$$

whence $(\widetilde{\nabla}\widetilde{\sigma}) \circ \dot{\widetilde{\gamma}} = \bar{\nabla}(\widetilde{\sigma} \circ \widetilde{\gamma})$. This completes the proof. $\qquad\square$

Lemma 6.1.52. *Let ∇ be a covariant derivative on a vector bundle $\pi\colon E \to M$. Let $\gamma\colon I \to M$ be a smooth curve, $\theta\colon \widetilde{I} \to I$ a smooth function (where $\widetilde{I} \subset \mathbb{R}$ is also a nonempty open interval), and consider the reparametrization $\widetilde{\gamma} := \gamma \circ \theta\colon \widetilde{I} \to M$ of γ. Then the covariant derivatives induced by ∇ along γ and $\widetilde{\gamma}$ related by*

$$\nabla^{\widetilde{\gamma}}(s \circ \theta) = \theta'((\nabla^{\gamma}s) \circ \theta), \quad s \in \Gamma_{\gamma}(E). \tag{6.1.67}$$

Proof. We suppose that s is of the form $s = \sigma \circ \gamma$, where $\sigma \in \Gamma(E)$. The general case is left to the reader; Step 2 in the proof of Lemma 6.1.61 below indicates a possible method of calculation.

In our case, at each $t \in \widetilde{I}$, we have

$$(\nabla^{\widetilde{\gamma}}(s \circ \theta))(t) = (\nabla^{\widetilde{\gamma}}(\sigma \circ \widetilde{\gamma}))(t) \overset{(6.1.62)}{=} ((\nabla\sigma) \circ \dot{\widetilde{\gamma}})(t)$$

$$= \nabla_{\dot{\widetilde{\gamma}}(t)}\sigma \overset{(3.1.10)}{=} \theta'(t)\nabla_{\dot{\gamma}(\theta(t))}\sigma$$

$$= \theta'(t)(\nabla\sigma)(\dot{\gamma}(\theta(t))) = \theta'(t)(\nabla\sigma \circ \dot{\gamma})(\theta(t))$$

$$\overset{(6.1.62)}{=} \big(\theta'((\nabla^{\gamma}(\sigma \circ \gamma)) \circ \theta)\big)(t) = \big(\theta'((\nabla^{\gamma}s) \circ \theta)\big)(t),$$

as desired. $\qquad\square$

Corollary 6.1.53. *Let D be a covariant derivative on a manifold M. Consider a smooth curve $\gamma\colon I \to M$ and a reparametrization*

$$\widetilde{\gamma} := \gamma \circ \theta\colon \widetilde{I} \to I \to M$$

of γ. Then

$$D^{\widetilde{\gamma}}(\dot{\gamma} \circ \theta) = \theta'((D^{\gamma}\dot{\gamma}) \circ \theta), \tag{6.1.68}$$

$$D^{\widetilde{\gamma}}\dot{\widetilde{\gamma}} = \theta''(\dot{\gamma} \circ \theta) + (\theta')^2((D^{\gamma}\dot{\gamma}) \circ \theta). \tag{6.1.69}$$

Proof. Obviously, $\dot{\gamma} \in \mathfrak{X}_{\gamma}(M)$, $\dot{\widetilde{\gamma}} = \theta'(\dot{\gamma} \circ \theta) \in \mathfrak{X}_{\widetilde{\gamma}}(M)$, and (6.1.67) leads immediately to (6.1.68). Then

$$D^{\widetilde{\gamma}}\dot{\widetilde{\gamma}} = D^{\widetilde{\gamma}}(\theta'(\dot{\gamma} \circ \theta)) \overset{(6.1.61)}{=} \theta''(\dot{\gamma} \circ \theta) + \theta' D^{\widetilde{\gamma}}(\dot{\gamma} \circ \theta)$$

$$\overset{(6.1.68)}{=} \theta''(\dot{\gamma} \circ \theta) + (\theta')^2((D^{\gamma}\dot{\gamma}) \circ \theta),$$

as was to be shown. $\qquad\square$

6.1.8 *Parallel Translation with Respect to a Covariant Derivative*

Definition 6.1.54. Let $\pi\colon E \to M$ be a vector bundle with a covariant derivative ∇. Given a smooth curve $\gamma\colon I \to M$, consider the induced covariant derivative ∇^{γ} along γ. We say that

(i) a section $s \in \Gamma_\gamma(E)$ along γ is *parallel* if $\nabla^\gamma s = 0$,

(ii) a section $\sigma \in \Gamma(E)$ is *parallel along* γ if $\sigma \circ \gamma \in \Gamma_\gamma(E)$ is parallel.

Corollary 6.1.55. *The property that a section along γ is parallel is invariant under the reparametrizations of γ.*

Proof. This is clear from (6.1.67). $\qquad\qquad\qquad\qquad\qquad\qquad\qquad\square$

Lemma 6.1.56. *Let $f \colon I \times \mathbb{R}^n \to \mathbb{R}^n$ be a continuous mapping. Suppose that f is positive-homogeneous of degree 1 in its second variable and is of class C^1 on $I \times (\mathbb{R}^n \setminus \{0\})$. Then, for any $t_0 \in I$ and $v \in \mathbb{R}^n$, the initial value problem $x'(t) = f(t, x(t))$, $x(t_0) = v$ has a unique global solution.*

Proof. First, let J be a relatively compact subinterval of I, i.e., a bounded subinterval such that $\mathrm{cl}(J) \subset I$. We show that f satisfies the condition of Theorem 3.2.9 with $J \times \mathbb{R}^n$. For a fixed $t \in I$, consider the mapping $f_t \colon q \in \mathcal{U} \mapsto f_t(q) := f(t, q) \in \mathbb{R}$, and let

$$L := \max\{\|(f_t)'(v)\| \mid t \in \mathrm{cl}(J), v \in \mathbb{S}^{n-1} \subset \mathbb{R}^n\}.$$

Here, as in Remark 3.2.4, $\|(f_t)'(v)\|$ is the operator norm of the linear endomorphism $(f_t)'(v) \in \mathrm{End}(\mathbb{R}^n)$. By Lemma 4.2.4, $(f_t)'$ is 0^+-homogeneous, thus

$$\|(f_t)'(v)\| \leq L \quad \text{for any } t \in \mathrm{cl}(J) \text{ and } v \in \mathbb{R}^n \setminus \{0\}.$$

Now let $t \in J$, and let $v, w \in \mathbb{R}^n \setminus \{0\}$ be linearly independent vectors. Then the line segment $\overline{vw}$ does not contain the origin, therefore the mean value inequality (Lemma C.1.21) gives

$$\|f(t, w) - f(t, v)\| \leq \|w - v\| \sup_{\theta \in [0,1]} \|(f_t)'(v + \theta(w - v))\| \leq L\|w - v\|.$$

Thus the inequality required in the condition of Theorem 3.2.9 holds for linearly independent vectors $v, w \in \mathbb{R}^n$. Then, however, it follows by continuity that the inequality also holds for linearly dependent vectors v, w. So we may conclude that our initial value problem has a unique global solution on every relatively compact subset of I. Since I is the union of such intervals, it follows that the global solution exists on the whole of the given interval I. $\qquad\qquad\qquad\qquad\qquad\qquad\qquad\qquad\qquad\qquad\square$

Proposition 6.1.57. *Use the same notation and hypotheses as in Definition 6.1.54. For any parameter $t_0 \in I$ and any vector $z \in E_{\gamma(t_0)}$, there exists a unique parallel section $s \in \Gamma_\gamma(E)$ such that $s(t_0) = z$.*

Proof. Without loss of generality, we may assume that $\mathrm{Im}(\gamma)$ is contained in the domain of a chart $(\mathcal{U}, (u^i)_{i=1}^n)$, where $E_{\mathcal{U}} = \pi^{-1}(\mathcal{U})$ has a trivialization. Choose a local frame field $(\sigma_\alpha)_{\alpha=1}^k$ on $\mathcal{U}$, and let

$$s = \sum_{\alpha=1}^k s^\alpha(\sigma_\alpha \circ \gamma), \quad s^\alpha \in C^\infty(I).$$

Then, by (6.1.64), $\nabla^\gamma s = 0$ and $s(t_0) = z$ holds if, and only if,

$$(s^\alpha)' = -\sum_{j=1}^n \sum_{\beta=1}^k (\gamma^j)' s^\beta (\Gamma_{j\beta}^\alpha \circ \gamma), \quad s^\alpha(t_0) = z^\alpha; \quad \alpha \in J_k,$$

where $\sum_{\alpha=1}^k z^\alpha \sigma_\alpha(\gamma(t_0)) = z$. Otherwise stated, $s \in \Gamma_\gamma(E)$ is a parallel section satisfying $s(t_0) = z$ if, and only if, the mapping

$$\begin{pmatrix} s^1 \\ \vdots \\ s^k \end{pmatrix} : I \to \mathbb{R}^k$$

solves the initial value problem $x' = Ax$, $x(t_0) = z$, where

$$A \colon I \to M_k(\mathbb{R}), \ t \mapsto A(t) := -\left(\sum_{j=1}^n (\gamma^j)'(t) \Gamma_{j\beta}^\alpha(\gamma(t)) \right).$$

Now our assertion follows from the preceding lemma. $\qquad\square$

Lemma and Definition 6.1.58. *Keep notation and hypotheses as above. Let $t_0, t \in I$. Given a vector $z \in E_{\gamma(t_0)}$, let s be the unique parallel section along γ such that $s(t_0) = z$. Then the mapping*

$$P_{t_0}^t(\gamma) \colon E_{\gamma(t_0)} \to E_{\gamma(t)}, \ z \mapsto s(t)$$

is a linear isomorphism, called the parallel translation *along γ from $\gamma(t_0)$ to $\gamma(t)$.*

Proof. For simplicity, write $P := P_{t_0}^t(\gamma)$. Given two vectors z_1, z_2 in $E_{\gamma(t_0)}$, let s_1 and s_2 be the parallel sections along γ such that $s_1(t_0) = z_1$, $s_2(t_0) = z_2$. Since $\mathbb{R}$-linear combinations of parallel sections remain parallel, for any $\lambda_1, \lambda_2 \in \mathbb{R}$ we get

$$P(\lambda_1 z_1 + \lambda_2 z_2) = (\lambda_1 s_1 + \lambda_2 s_2)(t) = \lambda_1 s_1(t) + \lambda_2 s_2(t)$$
$$= \lambda_1 P(z_1) + \lambda_2 P(z_2).$$

Thus P is linear.

If $P(z) = 0 \in E_{\gamma(t)}$, then by the uniqueness statement of Proposition 6.1.57, $P(z) = o(t)$, where o is the zero section along γ. Hence $z = o(t_0) = 0$, therefore P is injective. This implies that P is also surjective, since the fibres $E_{\gamma(t_0)}$ and $E_{\gamma(t)}$ have the same finite dimension. $\quad\square$

Proposition 6.1.59. *Let* $\pi\colon E \to M$ *be a vector bundle with a covariant derivative* ∇. *Let* $p \in M$ *and* $v \in T_pM$. *Choose a smooth curve* $\gamma\colon I \to M$ *such that* $0 \in I$ *and* $\dot{\gamma}(0) = v$. *Then for any section* $\sigma \in \Gamma(E)$ *we have*

$$\nabla_v\sigma = \lim_{t \to 0, t \in I} \frac{(P_0^t(\gamma))^{-1}\big(\sigma(\gamma(t))\big) - \sigma(\gamma(0))}{t}. \tag{6.1.70}$$

Proof. Let $(z_i)_{i=1}^k$ be a basis of $E_{\gamma(0)}$. By Proposition 6.1.57, there exists a (unique) family $(s_i)_{i=1}^k$ of parallel sections along γ such that $s_i(0) = z_i$ $(i \in J_k)$. Then $\sigma \circ \gamma \in \Gamma_\gamma(E)$ can be written as

$$\sigma \circ \gamma = \sum_{\alpha=1}^k \sigma^\alpha s_\alpha, \quad \sigma^\alpha \in C^\infty(I).$$

Since the parallel translations along a curve are linear isomorphisms, for each $t \in I$ we have

$$(P_0^t(\gamma))^{-1}\big(\sigma(\gamma(t))\big) = (P_0^t(\gamma))^{-1}\left(\sum_{\alpha=1}^k \sigma^\alpha(t)s_\alpha(t)\right)$$

$$= \sum_{\alpha=1}^k \sigma^\alpha(t)(P_0^t(\gamma))^{-1}(s_\alpha(t)) = \sum_{\alpha=1}^k \sigma^\alpha(t)s_\alpha(0),$$

hence

$$\lim_{t \to 0, t \in I} \frac{(P_0^t(\gamma))^{-1}\big(\sigma(\gamma(t))\big) - \sigma(\gamma(0))}{t}$$

$$= \lim_{t \to 0, t \in I} \frac{\sum_{\alpha=1}^k \sigma^\alpha(t)s_\alpha(0) - \sum_{\alpha=1}^k \sigma^\alpha(0)s_\alpha(0)}{t}$$

$$= \left(\sum_{\alpha=1}^k \lim_{t \to 0, t \in I} \frac{\sigma^\alpha(t) - \sigma^\alpha(0)}{t}\right)s_\alpha(0) = \sum_{\alpha=1}^k (\sigma^\alpha)'(0)s_\alpha(0).$$

On the other hand, taking into account that $\nabla^\gamma s_\alpha = 0$ for all $\alpha \in J_k$,

$$\nabla_v\sigma = \nabla_{\dot{\gamma}(0)}\sigma = (\nabla\sigma)(\dot{\gamma}(0)) \overset{(6.1.62)}{=} \nabla^\gamma(\sigma \circ \gamma)(0)$$

$$= \nabla^\gamma\left(\sum_{\alpha=1}^k \sigma^\alpha s_\alpha\right)(0) = \sum_{\alpha=1}^k ((\sigma^\alpha)'s_\alpha + \sigma^\alpha\nabla^\gamma s_\alpha)(0)$$

$$= \sum_{\alpha=1}^k (\sigma^\alpha)'(0)s_\alpha(0).$$

This finishes the proof. $\qquad\qquad\qquad\qquad\qquad\qquad\qquad\qquad\square$

Proposition 6.1.60. *Let $\pi\colon E \to M$ and $\widetilde{\pi}\colon \widetilde{E} \to \widetilde{M}$ be vector bundles with covariant derivatives ∇ and $\widetilde{\nabla}$, respectively. Let $\varphi\colon E \to \widetilde{E}$ be a vector bundle isomorphism with induced mapping $\varphi_B\colon M \to \widetilde{M}$ between the base manifolds. Then the following are equivalent:*

(i) *The mapping φ is covariant derivative preserving.*

(ii) *For any smooth curve $\gamma\colon I \to M$ and any ∇^γ-parallel section s along γ, the section*

$$\varphi_\#^\gamma(s) := \varphi \circ s \in \Gamma_{\varphi_B \circ \gamma}(\widetilde{E})$$

is parallel with respect to $\widetilde{\nabla}^{\varphi_B \circ \gamma}$.

(iii) *For any smooth curve $\gamma\colon I \to M$, any points $t_0, t \in I$ and any vector $z \in E_{\gamma(t_0)}$,*

$$\varphi(P_{t_0}^t(\gamma)(z)) = P_{t_0}^t(\varphi_B \circ \gamma)(\varphi(z)).$$

Proof. $(i) \Rightarrow (ii)$: This is an immediate consequence of Lemma 6.1.51.

$(ii) \Rightarrow (iii)$: Let $\gamma\colon I \to M$ be a smooth curve and $s \in \Gamma_\gamma(E)$ a ∇^γ-parallel section. Pick two points t_0, t in I, and let $z := s(t_0)$. Then, by our assumption, $\varphi_\#^\gamma(s)$ is parallel along $\varphi_B \circ \gamma$, and $\varphi_\#^\gamma(s)(t_0) = \varphi(z)$. Thus

$$\varphi(P_{t_0}^t(\gamma)(z)) = \varphi(s(t)) = \varphi_\#^\gamma s(t) = P_{t_0}^t(\varphi_B \circ \gamma)(\varphi(z)),$$

as we claimed.

$(iii) \Rightarrow (i)$: Let $\sigma \in \Gamma(\pi)$ be any section, and let $v \in TM$ be any tangent vector to M. Choose a smooth curve $\gamma\colon I \to M$ such that $\dot{\gamma}(0) = v$ (Lemma 3.1.18). Then $\widetilde{\gamma} := \varphi_B \circ \gamma$ is a smooth curve in $\widetilde{M}$ satisfying

$$\dot{\widetilde{\gamma}}(0) = (\varphi_B \circ \gamma)^\cdot(0) \overset{(3.1.11)}{=} (\varphi_B)_*(\dot{\gamma}(0)) = (\varphi_B)_*(v).$$

Thus

$$
\begin{aligned}
\widetilde{\nabla}_{(\varphi_B)_*(v)} \varphi_\# \sigma &\overset{(6.1.70)}{=} \lim_{t \to 0, t \in I} \frac{(P_0^t(\widetilde{\gamma}))^{-1}(\varphi_\# \sigma(\widetilde{\gamma}(t))) - \varphi_\# \sigma(\widetilde{\gamma}(0))}{t} \\
&= \lim_{t \to 0, t \in I} \frac{(P_0^t(\widetilde{\gamma}))^{-1}(\varphi \circ \sigma(\gamma(t))) - \varphi \circ \sigma(\gamma(0))}{t} \\
&\overset{(iii)}{=} \lim_{t \to 0, t \in I} \frac{\varphi(P_0^t(\gamma))^{-1}(\sigma(\gamma(t))) - \varphi \circ \sigma(\gamma(0))}{t} \\
&= \varphi\left(\lim_{t \to 0, t \in I} \frac{(P_0^t(\gamma))^{-1}(\sigma(\gamma(t))) - \sigma(\gamma(0))}{t} \right) = \varphi(\nabla_v \sigma).
\end{aligned}
$$

This justifies that φ is covariant derivative preserving, and finishes the proof. $\qquad\square$

Lemma 6.1.61. *Let* $\pi\colon E \to M$ *be a vector bundle with scalar product* g, *and let* ∇ *be a covariant derivative on* π. *Then* ∇ *is compatible with* g *if, and only if, for any smooth curve* $\gamma\colon I \to M$ *and any sections* s_1, s_2 *in* $\Gamma_\gamma(E)$,

$$(g(s_1, s_2))' = g(\nabla^\gamma s_1, s_2) + g(s_1, \nabla^\gamma s_2). \tag{6.1.71}$$

Here the function $g(s_1, s_2)\colon I \to \mathbb{R}$ *is defined by*

$$g(s_1, s_2)(t) := g_{\gamma(t)}(s_1(t), s_2(t)), \quad t \in I,$$

and ∇^γ *is the covariant derivative along* γ *induced by* ∇.

Proof. **Step 1.** We show that (6.1.71) holds if, and only if, for every $\sigma_1, \sigma_2 \in \Gamma(E)$,

$$(g(\sigma_1 \circ \gamma, \sigma_2 \circ \gamma))' = g(\nabla^\gamma(\sigma_1 \circ \gamma), \sigma_2 \circ \gamma) + g(\sigma_1 \circ \gamma, \nabla^\gamma(\sigma_2 \circ \gamma)). \tag{6.1.72}$$

Indeed, (6.1.71) evidently implies (6.1.72). To see the converse, let s_1 and s_2 be sections of π along γ. Since the problem is local, we may assume that π is trivial, i.e., $E = M \times V$, where V is a k-dimensional real vector space and $\pi = \mathrm{pr}_1$. Then E has a global frame field $(\sigma_\alpha)_{\alpha=1}^k$ (Example 2.2.27), so the sections s_1, s_2 can be represented in the form

$$s_i = \sum_{\alpha=1}^k s_i^\alpha(\sigma_\alpha \circ \gamma) =: s_i^\alpha(\sigma_\alpha \circ \gamma), \quad i \in \{1, 2\}$$

with some smooth functions $s_i^\alpha\colon I \to \mathbb{R}$. Then, using the summation convention on Greek indices,

$$\begin{aligned}
(g(s_1, s_2))' &= \big(g(s_1^\alpha(\sigma_\alpha \circ \gamma), s_2^\beta(\sigma_\beta \circ \gamma))\big)' = \big(s_1^\alpha s_2^\beta g(\sigma_\alpha \circ \gamma, \sigma_\beta \circ \gamma)\big)' \\
&= ((s_1^\alpha)' s_2^\beta + s_1^\alpha (s_2^\beta)') g(\sigma_\alpha \circ \gamma, \sigma_\beta \circ \gamma) + s_1^\alpha s_2^\beta \big(g(\sigma_\alpha \circ \gamma, \sigma_\beta \circ \gamma)\big)' \\
&= g((s_1^\alpha)'(\sigma_\alpha \circ \gamma), s_2) + g(s_1, (s_2^\beta)'(\sigma_\beta \circ \gamma)) + s_1^\alpha s_2^\beta \big(g(\sigma_\alpha \circ \gamma, \sigma_\beta \circ \gamma)\big)' \\
&\stackrel{(6.1.72)}{=} g\big((s_1^\alpha)'(\sigma_\alpha \circ \gamma) + s_1^\alpha \nabla^\gamma(\sigma_\alpha \circ \gamma), s_2\big) \\
&\quad + g\big(s_1, (s_2^\beta)'(\sigma_\beta \circ \gamma) + s_2^\beta \nabla^\gamma(\sigma_\beta \circ \gamma)\big) \\
&\stackrel{(6.1.62),(6.1.63)}{=} g(\nabla^\gamma s_1, s_2) + g(s_1, \nabla^\gamma s_2),
\end{aligned}$$

as desired.

Step 2. Let $\sigma_1, \sigma_2 \in \Gamma(\pi)$. Observe that for every smooth curve $\gamma\colon I \to M$ and any fixed point $t \in I$,

$$\begin{aligned}
(\nabla_X g)(\sigma_1, \sigma_2)(\gamma(t)) &= (g(\sigma_1 \circ \gamma, \sigma_2 \circ \gamma))'(t) \\
&\quad - g(\nabla^\gamma(\sigma_1 \circ \gamma), \sigma_2 \circ \gamma)(t) - g(\sigma_1 \circ \gamma, \nabla^\gamma(\sigma_2 \circ \gamma))(t), \quad (*)
\end{aligned}$$

where X is a vector field on M such that $X(\gamma(t)) = \dot\gamma(t)$. Indeed,

$$(\nabla_X g)(\sigma_1, \sigma_2)(\gamma(t))$$

$$= \big(Xg(\sigma_1, \sigma_2) - g(\nabla_X \sigma_1, \sigma_2) - g(\sigma_1, \nabla_X \sigma_2)\big)(\gamma(t))$$

$$= \dot\gamma(t)g(\sigma_1, \sigma_2) - g_{\gamma(t)}(\nabla_{\dot\gamma(t)}\sigma_1, \sigma_2(\gamma(t))) - g_{\gamma(t)}(\sigma_1(\gamma(t)), \nabla_{\dot\gamma(t)}\sigma_2)$$

$$\overset{(3.1.9),(6.1.62)}{=} (g(\sigma_1, \sigma_2) \circ \gamma)'(t)$$

$$- g(\nabla^\gamma(\sigma_1 \circ \gamma), \sigma_2 \circ \gamma)(t) - g(\sigma_1 \circ \gamma, \nabla^\gamma(\sigma_2 \circ \gamma))(t)$$

as we claimed. Since every tangent vector to M can be realized as the velocity of a smooth curve on M, $(*)$ implies by Step 1 that $\nabla g = 0$ if, and only if, (6.1.71) is satisfied. This is what was to be shown. $\qquad\square$

Proposition 6.1.62. *Let* $\pi\colon E \to M$ *be a semi-Euclidean vector bundle with scalar product g. A covariant derivative ∇ on π is compatible with g if, and only if, the parallel translations induced by ∇ are orthogonal mappings.*

Proof. Assume first that ∇ is compatible with g. Let $\gamma\colon I \to M$ be a smooth curve. Pick two points t_0, t_1 in I, and let, for short, $P := P_{t_0}^{t_1}(\gamma)$. If $z_1, z_2 \in E_{\gamma(t_0)}$, then $P(z_1) = s_1(t_1)$, $P(z_2) = s_2(t_1)$, where $s_1, s_2 \in \Gamma_\gamma(E)$ are parallel sections such that $s_1(t_0) = z_1$, $s_2(t_0) = z_2$. By the preceding lemma, for every $t \in I$,

$$(g(s_1, s_2))'(t) = g_{\gamma(t)}(\nabla^\gamma s_1(t), s_2(t)) + g_{\gamma(t)}(s_1(t), \nabla^\gamma s_2(t))$$

$$= g_{\gamma(t)}(0, s_2(t)) + g_{\gamma(t)}(s_1(t), 0) = 0.$$

Thus the function $g(s_1, s_2)$ is constant. Therefore,

$$g_{\gamma(t_0)}(z_1, z_2) = (g(s_1, s_2))(t_0) = (g(s_1, s_2))(t_1) = g_{\gamma(t_1)}(s_1(t_1), s_2(t_1))$$

$$= g_{\gamma(t_1)}(P(z_1), P(z_2)).$$

This means that the mapping $P\colon E_{\gamma(t_0)} \to E_{\gamma(t_1)}$ preserves the scalar product. Since P is also linear (Lemma and Definition 6.1.58), it is an orthogonal mapping, as required.

Conversely, suppose that the parallel translations induced by ∇ are orthogonal mappings. Consider a smooth curve $\gamma\colon I \to M$ with $\gamma(0) = p$, $\dot\gamma(0) = v$. Let $(z_\alpha)_{\alpha=1}^k$ be an orthonormal basis of E_p. By Proposition 6.1.57, there is a unique family $(Z_\alpha)_{\alpha=1}^k$ of parallel sections along γ such that $Z_\alpha(0) = z_\alpha$ for $\alpha \in J_k$. Then, by our condition, $(Z_\alpha(t))_{\alpha=1}^k$ is an orthonormal basis of $E_{\gamma(t)}$ for every $t \in I$.

Now let σ_1 and σ_2 be any sections in $\Gamma(\pi)$. Using Fourier expansion, we can write

$$\sigma_i \circ \gamma = \sigma_i^\alpha Z_\alpha, \quad \sigma_i^\alpha := g(\sigma_i \circ \gamma, Z_\alpha); \quad i \in \{1, 2\},$$

with summation convention on repeated Geek indices. Thus,

$$g(\nabla_v \sigma_1, \sigma_2(p)) = g(\nabla_{\dot\gamma(0)}\sigma_1, \sigma_2(p)) \overset{(6.1.62)}{=} g(\nabla^\gamma(\sigma_1 \circ \gamma)(0), \sigma_2(p))$$

$$= g(\nabla^\gamma(\sigma_1^\alpha Z_\alpha)(0), \sigma_2(p)) \overset{(6.1.61)}{=} g((\sigma_1^\alpha)'(0)Z_\alpha(0), \sigma_2(p))$$

$$= g((\sigma_1^\alpha)'(0)Z_\alpha(0), \sigma_2^\beta(0)Z_\beta(0)) = \sum_{\alpha=1}^k (\sigma_1^\alpha)'(0)\sigma_2^\alpha(0).$$

Similarly, $g(\sigma_1(p), \nabla_v \sigma_2) = \sum_{\alpha=1}^k \sigma_1^\alpha(0)(\sigma_2^\alpha)'(0)$. Then

$$vg(\sigma_1, \sigma_2) = \dot\gamma(0)g(\sigma_1, \sigma_2) \overset{(3.1.9)}{=} (g(\sigma_1, \sigma_2) \circ \gamma)'(0) = \left(g(\sigma_1^\alpha Z_\alpha, \sigma_2^\beta Z_\beta)\right)'(0)$$

$$= \left(\sum \sigma_1^\alpha \sigma_2^\alpha\right)'(0) = g(\nabla_v \sigma_1, \sigma_2(p)) + g(\sigma_1(p), \nabla_v \sigma_2),$$

and so ∇ is compatible with g.　　　　　　　　　　　　　　　　$\square$

6.1.9　Geodesics of an Affinely Connected Manifold

In subsection 5.1.2, in the context of semisprays, we have already defined geodesics (and pregeodesics), so some parts of the following definition will be familiar to the reader.

Definition 6.1.63. Let M be a manifold equipped with a covariant derivative D. Suppose that $\gamma \colon I \to M$ is a smooth curve whose domain is a nonempty open interval in $\mathbb{R}$.

(a) We say that γ is a *geodesic* of (M, D) (or of D) if the velocity vector field of γ is parallel along γ, i.e., $D^\gamma \dot\gamma = 0$. A geodesic $\gamma \colon I \to M$ is called *maximal* if there is no other geodesic $\tilde\gamma \colon \tilde I \to M$ such that I is a proper subset of $\tilde I$ (i.e., $I \subset \tilde I$ but $I \neq \tilde I$) and $\tilde\gamma \upharpoonright I = \gamma$. An affinely connected manifold is said to be *geodesically complete* if the domain of every maximal geodesic of D is all of $\mathbb{R}$.

(b) Let $a, b \in \mathbb{R}$, $a < b$. A curve segment $\alpha \colon [a, b] \to M$ is called a *geodesic segment* if there is a geodesic $\gamma \colon I \to M$ such that $[a, b] \subset I$ and $\gamma \upharpoonright [a, b] = \alpha$. A curve $\beta \colon [a, \infty[\to M$ (or $\beta \colon]-\infty, a] \to M$) is a positive (resp. negative) *geodesic ray* if it is the restriction of a geodesic in the above sense. A continuous curve $\alpha \colon [a, b] \to M$ is called a *broken geodesic segment* if it is a piecewise smooth curve (see Definition 2.1.24) whose smooth segments are geodesic segments.

(c) The curve $\gamma \colon I \to M$ is said to be a *pregeodesic* of (M, D) if there is an open interval $\tilde I \subset \mathbb{R}$ and a diffeomorphism $\theta \colon \tilde I \to I$ such that the reparametrization $\gamma \circ \theta$ is a geodesic.

(d) By a geodesic (resp. a pregeodesic) of a semi-Riemannian manifold we mean a geodesic (resp. pregeodesic) of the Levi-Civita derivative on the manifold.

Remark 6.1.64. (a) Let (M, D) be an affinely connected manifold. Obviously, every constant curve in M is a geodesic of D. If $\gamma \colon I \to M$ is a geodesic such that $\dot{\gamma}(t) \neq 0$ for some $t \in I$ then $\dot{\gamma}$ never vanishes, because $\dot{\gamma}$ is parallel along γ and the parallel translations along a curve are injective mappings. Thus, as Barrett O'Neill says, 'a geodesic cannot slow down and stop'.

(b) Now suppose that D is the Levi-Civita derivative determined by a semi-Riemannian metric g on M, and let γ be a geodesic of (M, g). Then the function

$$g(\dot{\gamma}, \dot{\gamma}) \colon I \to \mathbb{R}, \ t \mapsto g(\dot{\gamma}, \dot{\gamma})(t) := g_{\gamma(t)}(\dot{\gamma}(t), \dot{\gamma}(t))$$

is constant, as we have already seen in the proof of Proposition 6.1.62. This implies the following important facts:

(i) If one of the tangent vectors of γ is spacelike (resp. isotropic, or timelike) then all of them are of this 'causal character'. So in a semi-Riemannian manifold we may speak of *spacelike, isotropic* or *timelike geodesics*, depending on the causal character of a single tangent vector of the curve.

(ii) The geodesics of a semi-Riemannian manifold have constant speed.

Corollary 6.1.65. *Let* $\gamma \colon I \to M$ *be a nonconstant geodesic of an affinely connected manifold* (M, D). *A reparametrization* $\widetilde{\gamma} := \gamma \circ \theta \colon \widetilde{I} \to I \to M$ *of* γ *is a geodesic if, and only if, there exist real numbers* a, b *such that*

$$\theta(t) = at + b \quad \text{for all } t \in \widetilde{I}.$$

Proof. Since $D^\gamma \dot{\gamma} = 0$, from (6.1.69) we obtain that $D^{\widetilde{\gamma}} \dot{\widetilde{\gamma}} = \theta''(\dot{\gamma} \circ \theta)$. By the remark above, the non-constancy of γ implies that $\dot{\gamma}$ is never zero. Thus $\widetilde{\gamma}$ is a geodesic if, and only if, $\theta'' = 0$. The latter relation is equivalent to the condition that θ has the given form. $\square$

Lemma 6.1.66. *Let* (M, D) *be an affinely connected manifold. A regular curve* $\gamma \colon I \to M$ *is a pregeodesic of* (M, D) *if, and only if, there is a smooth function* $h \colon I \to \mathbb{R}$ *such that*

$$D^\gamma \dot{\gamma} = h\dot{\gamma}. \tag{6.1.73}$$

Proof. Suppose first that $\gamma\colon I \to M$ is a pregeodesic. Let $\theta\colon \widetilde{I} \to I$ be a diffeomorphism such that $\widetilde{\gamma} := \gamma \circ \theta$ is a geodesic. Then $D^{\widetilde{\gamma}}\dot{\widetilde{\gamma}} = 0$, hence we have

$$\theta''(\dot{\gamma} \circ \theta) + (\theta')^2((D^{\gamma}\dot{\gamma}) \circ \theta) = 0, \qquad (*)$$

by (6.1.69). The derivative of θ never vanishes, and θ has a smooth inverse because it is a diffeomorphism. So we obtain from $(*)$ that

$$D^{\gamma}\dot{\gamma} = -\left(\frac{\theta''}{(\theta')^2} \circ \theta^{-1}\right)\dot{\gamma} =: h\dot{\gamma}.$$

This is what was to be shown.

Assume conversely that γ satisfies (6.1.73). We want to find a diffeomorphism $\theta\colon \widetilde{I} \to I$ such that $\widetilde{\gamma} := \gamma \circ \theta$ is a geodesic, i.e., $D^{\widetilde{\gamma}}\dot{\widetilde{\gamma}} = 0$. In view of $(*)$ and our condition (6.1.73), this requirement leads to the second-order differential equation $x'' + (h \circ x)(x')^2 = 0$ for θ. According to Lemma 5.1.33, this equation has a strictly increasing smooth solution, which concludes the proof of the lemma. $\qquad\square$

Lemma 6.1.67 (Local Geodesic Equations). *Let (M, D) be an affinely connected manifold. Suppose that $\gamma\colon I \to M$ is a smooth curve, and $\mathrm{Im}(\gamma)$ is contained in the domain of a chart $(\mathcal{U}, (u^i)_{i=1}^n)$ of M. Then γ is a geodesic of (M, D) if, and only if,*

$$(\gamma^i)'' + \sum_{j,k=1}^n (\gamma^j)'(\gamma^k)'(\Gamma^i_{jk} \circ \gamma) = 0, \quad i \in J_n, \qquad (6.1.74)$$

where $\gamma^i := u^i \circ \gamma$, and the functions Γ^i_{jk} are the Christoffel symbols of D with respect to the chosen chart.

Proof. With the choice $X := \dot{\gamma} = \sum_{i=1}^n (\gamma^i)'(\frac{\partial}{\partial u^i} \circ \gamma)$, from (6.1.65) we obtain that

$$D^{\gamma}\dot{\gamma} = \sum_{i=1}^n \left((\gamma^i)'' + \sum_{j,k=1}^n (\gamma^j)'(\gamma^k)'(\Gamma^i_{jk} \circ \gamma)\right)\left(\frac{\partial}{\partial u^i} \circ \gamma\right).$$

Therefore, $D^{\gamma}\dot{\gamma} = 0$ if, and only if, (6.1.74) is satisfied. $\qquad\square$

Remark 6.1.68. Assume that (M, D) is an affinely connected manifold.

(a) Let $(\mathcal{U}, (u^i)_{i=1}^n)$ be a chart on M. By the preceding lemma, a smooth curve $\gamma\colon I \to \mathcal{U} \subset M$ is a geodesic of D if, and only if, its component functions $\gamma^i := u^i \circ \gamma$ satisfy the system of n second-order ordinary differential equations

$$(x^i)'' + \sum_{j,k=1}^n (x^j)'(x^k)'(\Gamma^i_{jk} \circ \gamma) = 0, \quad i \in J_n, \qquad (6.1.75)$$

called the *(local) geodesic equations*. In the classical literature they are often abbreviated as

$$\frac{d^2x^i}{dt^2} + \sum_{j,k=1}^{n} \Gamma^i_{jk} \frac{dx^j}{dt} \frac{dx^k}{dt} = 0, \quad i \in J_n.$$

(b) More generally, consider a smooth curve $\gamma \colon I \to M$ whose image is not contained in the domain of a single chart on M. Then for each $t \in I$, there exists a positive real number ε such that $\mathrm{Im}(\gamma \upharpoonright\,]t - \varepsilon, t + \varepsilon[) \subset \mathcal{U}$, where $\mathcal{U}$ is a chart domain. The curve γ is a geodesic of D if, and only if, all of the restrictions $\gamma \upharpoonright\,]t - \varepsilon, t + \varepsilon[$ satisfy the corresponding local geodesic equations of the form (6.1.75).

(c) In subsection 7.5.2 we shall show that a covariant derivative D on M gives rise canonically to an affine spray over M which has the same geodesics as D. Then we can conclude immediately that *for any $v \in TM$, there is a unique maximal geodesic γ_v of D with $\dot\gamma(0) = v$.*

6.1.10 *Hypersurfaces in a Riemannian Manifold*

Throughout this subsection $(\bar M, \bar g)$ will be an $(n+1)$-dimensional Riemannian manifold with $n \geq 1$, and M a hypersurface of $\bar M$. We suppose that M is equipped with the induced metric tensor $g := j^* \bar g$, where $j \colon M \to \bar M$ is the canonical inclusion. Sometimes we shall restrict ourselves to hypersurfaces of the form $M := f^{-1}(\lambda)$, where $f \in C^\infty(\bar M)$ and $\lambda \in \mathrm{Im}(f)$ is a regular value of f. To distinguish geometric objects on M and $\bar M$, we systematically use overbars.

Definition 6.1.69. A vector field U defined on an open neighbourhood of a hypersurface M of $(\bar M, \bar g)$ is called a *unit normal vector field on M* if

$$\bar g(U, U) = 1 \quad \text{and} \quad \bar g(U, \bar X) = 0 \text{ whenever } \bar X \in \mathfrak{X}(\bar M) \text{ is tangent to } M.$$

Remark 6.1.70. Suppose that U is a unit normal vector field on M, and let p be a point of M. It can easily be seen that

$$\begin{cases} \text{a vector } v \text{ in } T_p\bar M \text{ is a tangent vector to } M \text{ at } p \\ \text{if, and only if, } \bar g_p(v, U(p)) = 0. \end{cases} \tag{6.1.76}$$

Lemma 6.1.71. *Consider the hypersurface $M = f^{-1}(\lambda)$ of $(\bar M, \bar g)$. If*

$$U := -\frac{1}{\|\mathrm{grad} f\|} \overline{\mathrm{grad}} f \tag{6.1.77}$$

where $\overline{\mathrm{grad}}f$ *is the gradient of* f *with respect to the metric tensor* $\bar{g}$ *(see (6.1.58)), and* $\|\overline{\mathrm{grad}}f\| := (\bar{g}(\overline{\mathrm{grad}}f, \overline{\mathrm{grad}}f))^{\frac{1}{2}}$, *then* U *and* $-U$ *are unit normal vector fields on* M.

Proof. Condition $\bar{g}(U, U) = 1$ is clearly satisfied. To check the second relation, let $p \in M$, $v \in T_pM$. Then, by Corollary 3.3.6, $(df)_p(v) = 0$. Therefore,

$$\bar{g}_p(U(p), v) = -\frac{1}{\|\overline{\mathrm{grad}}f\|(p)}\bar{g}_p(\overline{\mathrm{grad}}f(p), v) = -\frac{1}{\|\overline{\mathrm{grad}}f\|(p)}(df)_p(v) = 0.$$

Obviously, the same are true for $-U$. $\qquad\qquad\qquad\qquad\qquad\qquad\qquad\square$

Proposition 6.1.72. *Consider a hypersurface* M *of* $(\bar{M}, \bar{g})$. *Let* U *be a unit normal vector field on* M, *and let* $\bar{D}$ *be the Levi-Civita derivative on* $\bar{M}$. *If* $X, Y \in \mathfrak{X}(M)$, *define a mapping* $p \in M \mapsto D_X Y(p) \in T_p\bar{M}$ *by*

$$D_X Y(p) := \bar{D}^{\gamma_p}(Y \circ \gamma_p)(0) - \bar{g}_p(\bar{D}^{\gamma_p}(Y \circ \gamma_p)(0), U(p))U(p), \qquad (6.1.78)$$

where γ_p *is an integral curve of* X *starting at* p *and* $\bar{D}^{\gamma_p}$ *is the covariant derivative along* γ_p *induced by* $\bar{D}$. *Then* $D_X Y$ *is a vector field on* M, *and the mapping*

$$D: (X, Y) \in \mathfrak{X}(M) \times \mathfrak{X}(M) \mapsto D_X Y \in \mathfrak{X}(M)$$

is the Levi-Civita derivative on (M, g).

Proof. It is clear from (6.1.78) that

$$\bar{g}_p(D_X Y(p), U(p)) = 0, \quad p \in M.$$

This implies by (6.1.76) that $D_X Y(p) \in T_pM$, so $D_X Y$ is a section of TM. **Step 1.** We show that D *is a derivation in its second variable.* To do this, let $X, Y \in \mathfrak{X}(M)$, $f \in C^\infty(M)$. Then for every $p \in M$,

$$\bar{D}^{\gamma_p}((fY) \circ \gamma_p)(0)$$

$$\overset{(6.1.61)}{=} (f \circ \gamma_p)'(0)(Y \circ \gamma_p)(0) + (f \circ \gamma_p)(0)\bar{D}^{\gamma_p}(Y \circ \gamma_p)(0)$$

$$\overset{(3.1.9)}{=} \dot{\gamma}_p(0)(f)Y(p) + f(p)\bar{D}^{\gamma_p}(Y \circ \gamma_p)(0)$$

$$= ((Xf)Y)(p) + f(p)\bar{D}^{\gamma_p}(Y \circ \gamma_p)(0),$$

from which we immediately obtain the desired relation

$$D_X(fY)(p) = ((Xf)Y)(p) + f(p)D_X Y(p).$$

Step 2. We verify that D *is a covariant derivative on* M. We begin with a useful technical observation.

Let $\bar{X}, \bar{Y} \in \mathfrak{X}(\bar{M})$, and consider the vector fields $X := \bar{X} \upharpoonright M$, $Y := \bar{Y} \upharpoonright M$ (see Example 3.1.53(c)). Then we have
$$D_X Y = \left(\bar{D}_{\bar{X}}\bar{Y} - \bar{g}(\bar{D}_{\bar{X}}\bar{Y}, U)U\right) \upharpoonright M, \tag{$*$}$$
as it can be seen by a direct calculation, using (6.1.62).

Now let p be an arbitrarily fixed point in M. By the Gram–Schmidt orthogonalization process (see Lemma 1.4.3) we can construct a local orthonormal frame field $(\bar{E}_i)_{i=1}^{n+1}$ of $T\bar{M}$ over an open neighbourhood $\bar{\mathcal{U}}$ of p such that $\bar{E}_{n+1} = U \upharpoonright \bar{\mathcal{U}}$. Let $\mathcal{U} := \bar{\mathcal{U}} \cap M$ and $E_\alpha := \bar{E}_\alpha \upharpoonright \mathcal{U}$, $\alpha \in J_n$. Then, by (6.1.76), $(E_\alpha)_{\alpha=1}^n$ is a local orthonormal frame field of TM over $\mathcal{U}$. Given any two vector fields X, Y on M, they can be represented as
$$X = X^\alpha E_\alpha, \quad Y = Y^\alpha E_\alpha; \quad X^\alpha, Y^\alpha \in C^\infty(\mathcal{U}), \ \alpha \in J_n$$
(with summation on Greek indices). Thus, at every point p of $\mathcal{U}$,
$$D_X Y(p) = D_{X(p)}(Y^\beta E_\beta) \stackrel{\text{Step } 1}{=} X(Y^\beta)(p)E_\beta(p) + Y^\beta(p)D_{X(p)}E_\beta$$
$$= \left(X(Y^\beta)E_\beta + X^\alpha Y^\beta D_{E_\alpha}E_\beta\right)(p).$$
This shows that D is $C^\infty(M)$-linear in its first variable and additive in its second variable. Furthermore, the sections $D_{E_\alpha}E_\beta$ $(\alpha, \beta \in J_n)$ are smooth by $(*)$, so it follows that the section $D_X Y$ is also smooth, i.e., it is a vector field on M. This concludes the proof that D is a covariant derivative.

Step 3. We show that D *is torsion-free and compatible with g.*

For any indices $i, j \in J_n$ we have
$$D_{E_i}E_j - D_{E_j}E_i - [E_i, E_j]$$
$$\stackrel{(3.1.53),(*)}{=} \left(\bar{D}_{\bar{E}_i}\bar{E}_j - \bar{D}_{\bar{E}_j}\bar{E}_i - [\bar{E}_i, \bar{E}_j] - \bar{g}(\bar{D}_{\bar{E}_i}\bar{E}_j - \bar{D}_{\bar{E}_j}\bar{E}_i, U)U\right) \upharpoonright M$$
$$= -\bar{g}([\bar{E}_i, \bar{E}_j], U)U \upharpoonright M,$$
because $\bar{D}$ is torsion-free. In the expression so obtained $[\bar{E}_i, \bar{E}_j]$ is tangent to M, so by (6.1.76) it follows that $\bar{g}([\bar{E}_i, \bar{E}_j], U) \upharpoonright M = 0$. Since $\mathfrak{X}(M)$ is locally generated by $(E_i)_{i=1}^n$, we conclude that the torsion of D vanishes.

Similarly, taking into account the definition of the induced metric g (Remark 6.1.42(b)),
$$Dg(E_i, E_j, E_k) = E_i(g(E_j, E_k)) - g(D_{E_i}E_j, E_k) - g(E_j, D_{E_i}E_k)$$
$$\stackrel{(*)}{=} E_i(\bar{g}(\bar{E}_j, \bar{E}_k) \upharpoonright M) - g\left((\bar{D}_{\bar{E}_i}\bar{E}_j - \bar{g}(\bar{D}_{\bar{E}_i}\bar{E}_j, U)U) \upharpoonright M, E_k\right)$$
$$- g\left(E_j, (\bar{D}_{\bar{E}_i}\bar{E}_k - \bar{g}(\bar{D}_{\bar{E}_i}\bar{E}_k, U)U) \upharpoonright M\right)$$
$$\stackrel{(6.1.76)}{=} E_i(\bar{g}(\bar{E}_j, \bar{E}_k) \upharpoonright M) - g((\bar{D}_{\bar{E}_i}\bar{E}_j) \upharpoonright M, E_k) - g(E_j, (\bar{D}_{\bar{E}_i}\bar{E}_k) \upharpoonright M)$$
$$= \left(\bar{E}_i(\bar{g}(\bar{E}_j, \bar{E}_k)) - \bar{g}(\bar{D}_{\bar{E}_i}\bar{E}_j, \bar{E}_k) - \bar{g}(\bar{E}_j, \bar{D}_{\bar{E}_i}\bar{E}_k)\right) \upharpoonright M$$
$$= \bar{D}\bar{g}(\bar{E}_i, \bar{E}_j, \bar{E}_k) \upharpoonright M \stackrel{\bar{D} \text{ is metric}}{=} 0,$$
for all $i, j, k \in J_n$. This proves that D is compatible with g. $\qquad\square$

Lemma and Definition 6.1.73. *Let M be a hypersurface of the Riemannian manifold $(\bar{M}, \bar{g})$. Let $\bar{D}$ be the Levi-Civita derivative on $\bar{M}$, and suppose that U is a unit normal vector field on M.*

(i) *For any point $p \in M$ and any tangent vector $v \in T_pM$,*

$$W_p(v) := \bar{D}_v U \quad \text{is a tangent vector to } M \text{ at } p.$$

(ii) *The mapping $W_p \colon T_pM \to T_pM$, $v \mapsto W_p(v) = \bar{D}_v U$ is linear, and it is called the* Weingarten operator *or* shape operator *of M at p.*

(iii) *The mapping*

$$W \colon p \in M \mapsto W_p \in \mathrm{End}(T_pM) \cong T^1_1(T_pM)$$

is a type $(1,1)$ tensor field on M, called the Weingarten tensor *or* shape tensor *of M derived from U.*

Proof. We check only that $W_p(v) \in T_pM$, the other assertions are clear.
Since $\bar{g}(U, U)$ is the constant function of value 1, and $v \in T_pM$ is a tangent vector also to $\bar{M}$ (Remark 3.1.24), we find that

$$0 = v\,\bar{g}(U, U) = \bar{g}_p(\bar{D}_v U, U(p)) + \bar{g}_p(U(p), \bar{D}_v U) = 2\,\bar{g}_p(\bar{D}_v U, U(p)),$$

so $\bar{D}_v U \in T_pM$ by (6.1.76). $\qquad\square$

Proposition 6.1.74. *Preserving the above notation, suppose that the hypersurface in $\bar{M}$ is of the form $M = f^{-1}(\lambda)$, and consider the Weingarten tensor W of M derived from the unit normal vector field U given by (6.1.77). Then for any point p of M and for any tangent vectors $v, w \in T_pM$,*

$$g_p(W_p(v), w) = -\frac{(\overline{\mathrm{Hess}}(f))_p(v, w)}{\|\overline{\mathrm{grad}}f\|(p)} \overset{(6.1.24)}{:=} -\frac{(\bar{D}(\bar{D}f))_p(v, w)}{\|\overline{\mathrm{grad}}f\|(p)}, \quad (6.1.79)$$

therefore W_p is a self-adjoint linear transformation of T_pM.

Proof. Using the definition of W_p and U we obtain

$$W_p(v) := -\bar{D}_v\left(\frac{1}{\|\overline{\mathrm{grad}}f\|}\overline{\mathrm{grad}}f\right)$$

$$= -v\left(\frac{1}{\|\overline{\mathrm{grad}}f\|}\right)\overline{\mathrm{grad}}f(p) - \frac{1}{\|\overline{\mathrm{grad}}f\|(p)}\bar{D}_v\overline{\mathrm{grad}}f.$$

Since $\bar{g}_p(\overline{\mathrm{grad}}f(p), w) \overset{(6.1.58)}{=} (df)_p(w) \overset{\text{Corollary 3.3.6}}{=} 0$, it follows that

$$\bar{g}_p(W_p(v), w) = -\frac{\bar{g}_p(\bar{D}_v\overline{\mathrm{grad}}f, w)}{\|\overline{\mathrm{grad}}f\|(p)}.$$

Now let $\bar{X}, \bar{Y} \in \mathfrak{X}(\bar{M})$ be tangent vector fields to M such that $\bar{X}(p) = v$ and $\bar{Y}(p) = w$. Then

$$0 = \bar{D}\bar{g}(\bar{X}, \overline{\mathrm{grad}}f, \bar{Y}) = (\bar{D}_{\bar{X}}\bar{g})(\overline{\mathrm{grad}}f, \bar{Y})$$
$$= \bar{X}(\bar{g}(\overline{\mathrm{grad}}f, \bar{Y})) - \bar{g}(\bar{D}_{\bar{X}}\overline{\mathrm{grad}}f, \bar{Y}) - \bar{g}(\overline{\mathrm{grad}}f, \bar{D}_{\bar{X}}\bar{Y}),$$

whence

$$\bar{g}(\bar{D}_{\bar{X}}\overline{\mathrm{grad}}f, \bar{Y}) = \bar{X}(\bar{g}(\overline{\mathrm{grad}}f, \bar{Y})) - \bar{g}(\overline{\mathrm{grad}}f, \bar{D}_{\bar{X}}\bar{Y})$$

$$\overset{(6.1.58)}{=} \bar{X}(df(\bar{Y})) - df(\bar{D}_{\bar{X}}\bar{Y}) = \bar{D}(df)(\bar{X}, \bar{Y}) =: \overline{\mathrm{Hess}}(f)(\bar{X}, \bar{Y}).$$

Evaluating both sides at p, we obtain that

$$\bar{g}_p(\bar{D}_v\overline{\mathrm{grad}}f, w) = (\overline{\mathrm{Hess}}(f))_p(v, w),$$

therefore

$$\bar{g}_p(W_p(v), w) = -\frac{(\overline{\mathrm{Hess}}(f))_p(v, w)}{\|\overline{\mathrm{grad}}f\|(p)}.$$

Since $W_p(v)$ and w are tangent vectors to M at p, we have $\bar{g}_p(W_p(v), w) = g_p(W_p(v), w)$ by Remark 6.1.42(b).

This proves (6.1.79). Finally, the Levi-Civita derivative $\bar{D}$ is torsion-free, hence $\overline{\mathrm{Hess}}(f)$ is symmetric by Example 6.1.33, and therefore W_p is self-adjoint. $\qquad\square$

Example 6.1.75. Let V be a finite-dimensional real vector space with $\dim V \geq 2$.

(a) Suppose that f is a smooth function defined on an open subset $\mathcal{U}$ of V, and $\lambda \in \mathrm{Im}(f)$ is a regular value of f in the sense that

$$f'(p) \neq 0 \quad \text{for all} \quad p \in M := f^{-1}(\lambda).$$

Then M *is a hypersurface of* V, *and for every* $p \in M$, *we have the canonical isomorphisms*

$$T_pM \cong \{p\} \times \mathrm{Ker}(f'(p)) \cong \mathrm{Ker}(f'(p)). \qquad (*)$$

To see this, recall first that T_pV is canonically isomorphic to $\{p\} \times V$ by Example C.2.2. Thus any tangent vector $v \in T_pM$ can be represented as a pair $(p, \underline{v}) \in \{p\} \times V$, and so we can write

$$(df)_p(v) = v(f) = (p, \underline{v})(f) \overset{(C.2.1)}{=} f'(p)(\underline{v}).$$

From this we infer that $(df)_p \neq 0$ if, and only if, $f'(p) \neq 0$. Now, by Lemma 3.3.5, it follows that M is a hypersurface in V, and relations $(*)$ are immediate consequences of Corollary 3.3.6.

(b) Suppose that V is equipped with a positive definite scalar product β, and let $\bar{g}$ be the Riemannian metric on V determined by β. Then for every $p \in V$,

$$\bar{g}_p((p,\underline{v}),(p,\underline{w})) := \beta(\underline{v},\underline{w}); \quad \underline{v},\underline{w} \in V,$$

and the Levi-Civita derivative on $(V,\bar{g})$ is the standard covariant derivative $\bar{D}$ on V given by (C.3.1); see also Proposition C.3.3. With the same notation and hypotheses as in (a), the gradient of $f \in C^\infty(\mathcal{U})$ is given by

$$\bar{g}_p((\overline{\operatorname{grad}f})(p),v) = (df)_p(v) = v(f); \quad p \in \mathcal{U}, \ v \in T_pV.$$

Let W be the Weingarten tensor of M derived from the unit normal vector field $-\frac{1}{\|\operatorname{grad}f\|}\overline{\operatorname{grad}}f$. Then for any $v = (p,\underline{v})$, $w = (p,\underline{w}) \in T_pM$,

$$\bar{g}_p(W_p(v),w) \stackrel{(6.1.79)}{=} -\frac{(\bar{D}(\bar{D}f))_p(v,w)}{\|\operatorname{grad}f\|(p)} \stackrel{(C.3.2)}{=} -\frac{f''(p)(\underline{v},\underline{w})}{\|\operatorname{grad}f\|(p)}. \tag{6.1.80}$$

The function

$$K\colon M \to \mathbb{R}, \quad p \mapsto K(p) := \det(W_p)$$

is called the *Gauss–Kronecker curvature* of M, and $K(p)$ is the Gauss–Kronecker curvature of M at p. When $\dim V = 3$, we simply speak of *Gaussian curvature*. The function K depends, of course, on the choice of the Euclidean structure on V. If we wish to emphasize this, we say that K is the *Gauss–Kronecker curvature of M with respect to the scalar product b*.

6.2   Covariant Derivatives on a Finsler Bundle

In this section π will stand for the Finsler bundle

$$TM \times_M TM \to TM$$

over TM introduced in Remark 4.1.1. For the $C^\infty(TM)$-module of $TM \times_M TM$-valued k-forms on TM we use the shorter notation $\mathcal{A}_k(\tau_{TM},\pi)$ instead of $\mathcal{A}_k(TTM,TM \times_M TM)$.

6.2.1   *Curvature and Torsion*

First we recall two canonical strong bundle maps, the

$$\text{canonical surjection} \quad \mathbf{j} := (\tau_{TM},\tau_*)\colon TTM \to TM \times_M TM$$

(see Lemma 4.1.19), and the

$$\text{canonical injection} \quad \mathbf{i}\colon TM \times_M TM \to TTM, \ (u,v) \mapsto v^\uparrow(u)$$

(see (4.1.30) and Lemma 4.1.24). By Remark 4.1.26(b), we may also write

$$\mathbf{j} \in \mathrm{Hom}(\mathfrak{X}(TM), \Gamma(\pi)) = \mathcal{A}_1(\tau_{TM}, \pi) \text{ and } \mathbf{i} \in \mathrm{Hom}(\Gamma(\pi), \mathfrak{X}(TM)).$$

According to our general Definition 6.1.1(a), a covariant derivative on π is an $\mathbb{R}$-linear mapping

$$D \colon \mathfrak{X}(TM) \times \Gamma(\pi) \to \Gamma(\pi), \quad (\xi, \widetilde{Y}) \mapsto D_\xi \widetilde{Y}$$

which is tensorial in its first variable and a derivation in its second variable. The latter condition means that

$$D_\xi F \widetilde{Y} = (\xi F)\widetilde{Y} + F D_\xi \widetilde{Y} \text{ for all } F \in C^\infty(TM).$$

The curvature tensor

$$R^D \colon \mathfrak{X}(TM) \times \mathfrak{X}(TM) \times \Gamma(\pi) \to \Gamma(\pi)$$

of D is defined by (6.1.37), which in our case gives

$$R^D(\xi, \eta)\widetilde{Z} = D_\xi D_\eta \widetilde{Z} - D_\eta D_\xi \widetilde{Z} - D_{[\xi, \eta]}\widetilde{Z}$$

for all $\xi, \eta \in \mathfrak{X}(TM)$, $\widetilde{Z} \in \Gamma(\pi)$. As we have explained in Remark 6.1.27(a), R^D may also be regarded as an element of

$$\mathcal{A}_2(\tau_{TM}, \mathrm{End}(\pi)) \cong A_2(\mathfrak{X}(TM), \mathrm{End}(\Gamma(\pi)))$$

such that for all $\xi, \eta \in \mathfrak{X}(TM)$,

$$R^D(\xi, \eta) \colon \widetilde{Z} \in \Gamma(\pi) \mapsto R^D(\xi, \eta)(\widetilde{Z}) := R^D(\xi, \eta)\widetilde{Z} \in \Gamma(\pi).$$

Thus we may consider the wedge product

$$R^D \wedge \mathbf{j} \in \mathcal{A}_3(TM, E)$$

given by (3.3.23):

$$(R^D \wedge \mathbf{j})(\xi_1, \xi_2, \xi_3) := \frac{1}{2!} \sum_{\sigma \in S_3} \varepsilon(\sigma) R^D(\xi_{\sigma(1)}, \xi_{\sigma(2)})\mathbf{j}(\xi_{\sigma(3)}).$$

Taking into account the skew-symmetry of R^D (Lemma 6.1.26), we obtain immediately that for all $\xi_1, \xi_2, \xi_3 \in \mathfrak{X}(TM)$,

$$(R^D \wedge \mathbf{j})(\xi_1, \xi_2, \xi_3) = \sum_{\mathrm{cyc}} R^D(\xi_1, \xi_2)\mathbf{j}(\xi_3). \tag{6.2.1}$$

We can attach to any covariant derivative D on π a 'natural' torsion, the $\mathbf{j}$-torsion $T^{\mathbf{j}} := d^D\mathbf{j}$ of D (see Definition 6.1.31). We denote it by T^D and call it the *torsion of D*. Thus

$$T^D := T^{\mathbf{j}} := d^D\mathbf{j}, \tag{6.2.2}$$

so for all $\xi, \eta \in \mathfrak{X}(TM)$, we have

$$T^D(\xi, \eta) \overset{(6.1.45)}{=} D_\xi \mathbf{j}\eta - D_\eta \mathbf{j}\xi - \mathbf{j}[\xi, \eta]. \tag{6.2.3}$$

Corollary and Definition 6.2.1. *Let D be a covariant derivative on the Finsler bundle $\pi\colon TM \times_M TM \to TM$. Then the mapping*

$$\mathbf{Q}\colon \Gamma(\pi) \times \Gamma(\pi) \times \Gamma(\pi) \to \Gamma(\pi), \quad (\widetilde{X}, \widetilde{Y}, \widetilde{Z}) \mapsto \mathbf{Q}(\widetilde{X}, \widetilde{Y})\widetilde{Z}$$

given by

$$\mathbf{Q}(\widetilde{X}, \widetilde{Y})\widetilde{Z} := R^D(\mathbf{i}\widetilde{X}, \mathbf{i}\widetilde{Y})\widetilde{Z} = D_{\mathbf{i}\widetilde{X}} D_{\mathbf{i}\widetilde{Y}} \widetilde{Z} - D_{\mathbf{i}\widetilde{Y}} D_{\mathbf{i}\widetilde{X}} \widetilde{Z} - D_{[\mathbf{i}\widetilde{X}, \mathbf{i}\widetilde{Y}]} \widetilde{Z} \quad (6.2.4)$$

is a type $(1,3)$ tensor field, i.e., $\mathbf{Q} \in \mathcal{F}_3^1(TM)$. This tensor field is called the vertical curvature *of the covariant derivative D. It is skew-symmetric in the first two variables.* $\quad\square$

Lemma and Definition 6.2.2. *Let D be a covariant derivative on the Finsler bundle $\pi\colon TM \times_M TM \to TM$. The mapping*

$$\mathfrak{Q}\colon \Gamma(\pi) \times \Gamma(\pi) \to \Gamma(\pi)$$

given by

$$\mathfrak{Q}(\widetilde{X}, \widetilde{Y}) := D_{\mathbf{i}\widetilde{X}} \widetilde{Y} - D_{\mathbf{i}\widetilde{Y}} \widetilde{X} - \mathbf{i}^{-1}[\mathbf{i}\widetilde{X}, \mathbf{i}\widetilde{Y}]; \quad \widetilde{X}, \widetilde{Y} \in \Gamma(\pi) \qquad (6.2.5)$$

is skew-symmetric and $C^\infty(TM)$-bilinear. It is called the vertical torsion *of D.*

Proof. The skew-symmetry of $\mathfrak{Q}$ is clear. So it is sufficient to check that $\mathfrak{Q}$ is tensorial in its first variable. For any function $F \in C^\infty(TM)$, we find that

$$
\begin{aligned}
\mathfrak{Q}(F\widetilde{X}, \widetilde{Y}) :&= D_{\mathbf{i}(F\widetilde{X})} \widetilde{Y} - D_{\mathbf{i}\widetilde{Y}} F\widetilde{X} - \mathbf{i}^{-1}[\mathbf{i}(F\widetilde{X}), \mathbf{i}\widetilde{Y}] \\
&= D_{F\mathbf{i}(\widetilde{X})} \widetilde{Y} - ((\mathbf{i}\widetilde{Y})F)\widetilde{X} - F D_{\mathbf{i}\widetilde{Y}} \widetilde{X} - \mathbf{i}^{-1}[F(\mathbf{i}\widetilde{X}), \mathbf{i}\widetilde{Y}] \\
&= F\big(D_{\mathbf{i}\widetilde{X}} \widetilde{Y} - D_{\mathbf{i}\widetilde{Y}} \widetilde{X} - \mathbf{i}^{-1}[\mathbf{i}\widetilde{X}, \mathbf{i}\widetilde{Y}]\big) \\
&\quad - ((\mathbf{i}\widetilde{Y})F)\widetilde{X} + \mathbf{i}^{-1}\big(((\mathbf{i}\widetilde{Y})F)\mathbf{i}\widetilde{X}\big) = F\mathfrak{Q}(\widetilde{X}, \widetilde{Y}),
\end{aligned}
$$

as was to be shown. $\quad\square$

Remark 6.2.3. Let D be a covariant derivative on the Finsler bundle $TM \times_M TM \to TM$. Choose a chart $(\mathcal{U}, (u^i)_{i=1}^n)$ on M, and consider the induced chart $(\tau^{-1}(\mathcal{U}), ((x^i)_{i=1}^n, (y^i)_{i=1}^n))$ on TM. The family $\left(\dfrac{\partial}{\partial u^i}\right)_{i=1}^n$ defined by $(4.1.5)$ is a local frame of $TM \times_M TM$ on $\tau^{-1}(\mathcal{U})$. The *Christoffel symbols* of D with respect to the chart $(\mathcal{U}, (u^i)_{i=1}^n)$ are the functions

$$\Gamma_{jk}^i, \; C_{jk}^i \in C^\infty(\tau^{-1}(\mathcal{U})); \quad i, j, k \in J_n,$$

given by

$$D_{\frac{\partial}{\partial x^j}}\widehat{\frac{\partial}{\partial u^k}} = \sum_{i=1}^{n}\Gamma^i_{jk}\widehat{\frac{\partial}{\partial u^i}}, \quad D_{\frac{\partial}{\partial y^j}}\widehat{\frac{\partial}{\partial u^k}} = \sum_{i=1}^{n}C^i_{jk}\widehat{\frac{\partial}{\partial u^i}} \tag{6.2.6}$$

(cf. (6.1.4)). Now we determine the coordinate representation of T^D with respect to the local frames $\left(\left(\frac{\partial}{\partial x^i}\right)_{i=1}^{n}, \left(\frac{\partial}{\partial y^i}\right)_{i=1}^{n}\right)$ and $\left(\widehat{\frac{\partial}{\partial u^i}}\right)_{i=1}^{n}$. Taking into account (4.1.74), we find that

$$T^D\left(\frac{\partial}{\partial x^j}, \frac{\partial}{\partial x^k}\right) \overset{(6.2.3)}{=} D_{\frac{\partial}{\partial x^j}}\widehat{\frac{\partial}{\partial u^k}} - D_{\frac{\partial}{\partial x^k}}\widehat{\frac{\partial}{\partial u^j}} \overset{(6.2.6)}{=} \sum_{i=1}^{n}(\Gamma^i_{jk}-\Gamma^i_{kj})\widehat{\frac{\partial}{\partial u^i}},$$

$$T^D\left(\frac{\partial}{\partial x^j}, \frac{\partial}{\partial y^k}\right) = -D_{\frac{\partial}{\partial y^k}}\widehat{\frac{\partial}{\partial u^j}} = \sum_{i=1}^{n}-C^i_{kj}\widehat{\frac{\partial}{\partial u^i}}, \quad T^D\left(\frac{\partial}{\partial y^j}, \frac{\partial}{\partial y^k}\right) = 0,$$

$j, k \in J_n$. Thus

$$T^D \underset{(\mathcal{U})}{=} \sum_{i,j,k=1}^{n}(\Gamma^i_{jk}-\Gamma^i_{kj})\widehat{\frac{\partial}{\partial u^i}}\otimes dx^j \otimes dx^k - \sum_{i,j,k=1}^{n}C^i_{kj}\widehat{\frac{\partial}{\partial u^i}}\otimes dx^j \wedge dy^k. \tag{6.2.7}$$

Finally, the components of the vertical torsion of D with respect to the local frame field $\left(\widehat{\frac{\partial}{\partial u^i}}\right)_{i=1}^{n}$ are the functions

$$\mathcal{Q}\left(\widehat{\frac{\partial}{\partial u^j}}, \widehat{\frac{\partial}{\partial u^k}}\right) = D_{\mathbf{i}\widehat{\frac{\partial}{\partial u^j}}}\widehat{\frac{\partial}{\partial u^k}} - D_{\mathbf{i}\widehat{\frac{\partial}{\partial u^k}}}\widehat{\frac{\partial}{\partial u^j}}$$

$$\overset{(4.1.73)}{=} D_{\frac{\partial}{\partial y^j}}\widehat{\frac{\partial}{\partial u^k}} - D_{\frac{\partial}{\partial y^k}}\widehat{\frac{\partial}{\partial u^j}} \overset{(6.2.6)}{=} \sum_{i=1}^{n}(C^i_{jk}-C^i_{kj})\widehat{\frac{\partial}{\partial u^i}}.$$

For later references, we repeat the result:

$$\mathcal{Q}\left(\widehat{\frac{\partial}{\partial u^j}}, \widehat{\frac{\partial}{\partial u^k}}\right) = \sum_{i=1}^{n}(C^i_{jk}-C^i_{kj})\widehat{\frac{\partial}{\partial u^i}}; \quad j, k \in J_n. \tag{6.2.8}$$

Proposition 6.2.4 (Algebraic Bianchi Identity). *Let D be a covariant derivative on the Finsler bundle $\pi\colon TM \times_M TM \to TM$ with curvature tensor R^D and torsion tensor T^D. Then*

$$d^D d^D\mathbf{j} = d^D T^D = R^D \wedge \mathbf{j}, \tag{6.2.9}$$

or, equivalently,

$$d^D T^D(\xi, \eta, \zeta) = \sum_{\text{cyc}} R^D(\xi, \eta)\mathbf{j}\zeta; \quad \xi, \eta, \zeta \in \mathfrak{X}(TM). \tag{6.2.10}$$

Proof. The equivalence of (6.2.9) and (6.2.10) is evident from (6.2.1). We verify (6.2.10), calculating in the same way as in the first part of the proof of Proposition 6.1.39. We find

$$
\begin{aligned}
(d^D T^D)(\xi, \eta, \zeta) &= D_\xi(T^D(\eta, \zeta)) - D_\eta(T^D(\xi, \zeta)) + D_\zeta(T^D(\xi, \eta)) \\
&\quad - T^D([\xi, \eta], \zeta) + T^D([\xi, \zeta], \eta) - T^D([\eta, \zeta], \xi) \\
&= D_\xi D_\eta \mathbf{j}\zeta - D_\xi D_\zeta \mathbf{j}\eta - D_\xi \mathbf{j}[\eta, \zeta] - D_{[\xi, \eta]}\mathbf{j}\zeta + D_\zeta \mathbf{j}[\xi, \eta] + \mathbf{j}[[\xi, \eta], \zeta] \\
&\quad - D_\eta D_\xi \mathbf{j}\zeta + D_\eta D_\zeta \mathbf{j}\xi + D_\eta \mathbf{j}[\xi, \zeta] + D_{[\xi, \zeta]}\mathbf{j}\eta - D_\eta \mathbf{j}[\xi, \zeta] - \mathbf{j}[[\xi, \zeta], \eta] \\
&\quad + D_\zeta D_\xi \mathbf{j}\eta - D_\zeta D_\eta \mathbf{j}\xi - D_\zeta \mathbf{j}[\xi, \eta] - D_{[\eta, \zeta]}\mathbf{j}\xi + D_\xi \mathbf{j}[\eta, \zeta] + \mathbf{j}[[\eta, \zeta], \xi] \\
&= \left(D_\xi D_\eta \mathbf{j}\zeta - D_\eta D_\xi \mathbf{j}\zeta - D_{[\xi, \eta]}\mathbf{j}\zeta \right) + \left(D_\eta D_\zeta \mathbf{j}\xi - D_\zeta D_\eta \mathbf{j}\xi - D_{[\eta, \zeta]}\mathbf{j}\xi \right) \\
&\quad + \left(D_\zeta D_\xi \mathbf{j}\eta - D_\xi D_\zeta \mathbf{j}\eta - D_{[\zeta, \xi]}\mathbf{j}\eta \right) + \mathbf{j}\left([[\xi, \eta], \zeta] + [[\eta, \zeta], \xi] + [[\zeta, \xi], \eta] \right) \\
&= \sum_{\text{cyc}} R^D(\xi, \eta)\mathbf{j}\zeta.
\end{aligned}
$$

$\square$

6.2.2   Deflection and Regularities

Definition 6.2.5. Let D be a covariant derivative on the Finsler bundle $\pi \colon TM \times_M TM \to TM$. By the *deflection* of D we mean the covariant differential of the canonical section $\widetilde{\delta} = (1_{TM}, 1_{TM})$ of π, i.e., the mapping

$$
\mu := D\widetilde{\delta} \colon \xi \in \mathfrak{X}(TM) \mapsto \mu(\xi) := (D\widetilde{\delta})(\xi) = D_\xi \widetilde{\delta} \in \Gamma(\pi). \tag{6.2.11}
$$

The *vertical deflection* of D is

$$
\mu^{\mathsf{v}} := \mu \circ \mathbf{i} \colon \Gamma(\pi) \to \Gamma(\pi), \ \ \widetilde{X} \mapsto \mu^{\mathsf{v}}(\widetilde{X}) := \mu(\mathbf{i}\widetilde{X}) = D_{\mathbf{i}\widetilde{X}}\widetilde{\delta}. \tag{6.2.12}
$$

Remark 6.2.6. Hypotheses and notation as in the above definition.

Since a covariant derivative $D \colon \mathfrak{X}(TM) \times \Gamma(\pi) \to \Gamma(\pi)$ is $C^\infty(TM)$-linear in its first variable, it follows that both the deflection and the vertical deflection of D are tensorial. Thus, by Proposition 2.2.31, μ and μ^{v} come from a strong bundle map, namely there exist (uniquely determined) strong bundle maps

$$
\underline{\mu} \colon TTM \to TM \times_M TM \ \text{and} \ \underline{\mu}^{\mathsf{v}} \colon TM \times_M TM \to TM \times_M TM
$$

such that

$$
\mu(\xi) = \underline{\mu} \circ \xi, \ \xi \in \mathfrak{X}(TM) \ \text{and} \ \mu^{\mathsf{v}}(\widetilde{X}) = \underline{\mu}^{\mathsf{v}} \circ \widetilde{X}, \ \widetilde{X} \in \Gamma(\pi).
$$

Then for every $u \in TM$,

$$
\underline{\mu}(\xi(u)) = \mu(\xi)(u) = (D_\xi \widetilde{\delta})(u) = D_{\xi(u)}\widetilde{\delta} \in \{u\} \times T_{\tau(u)}M
$$

and

$$\underline{\mu}^{\vee}(\widetilde{X}(u)) = \mu(\widetilde{X})(u) = (D_{\mathbf{i}\widetilde{X}}\widetilde{\delta})(u) = D_{\mathbf{i}(u,\underline{X}(u))}\widetilde{\delta}$$
$$\overset{(4.1.38)}{=} D_{(\underline{X}(u))^{\uparrow}(u)}\widetilde{\delta} \in \{u\} \times T_{\tau(u)}M,$$

where $\widetilde{X} = (1_{TM}, \underline{X})$, $\underline{X} \in C^{\infty}(TM, TM)$ (see Remark 4.1.1(b)). We say that

$$\underline{\mu}_u := \underline{\mu} \upharpoonright T_u TM : T_u TM \to \{u\} \times T_{\tau(u)}M$$

and

$$(\underline{\mu}^{\vee})_u := \underline{\mu}^{\vee} \upharpoonright \{u\} \times T_{\tau(u)}M : \{u\} \times T_{\tau(u)}M \to \{u\} \times T_{\tau(u)}M$$

are the restrictions of $\underline{\mu}$ and $\underline{\mu}^{\vee}$ onto the fibre $T_u TM$ and $\{u\} \times T_{\tau(u)}M$, respectively.

Coordinate description. Let the notation be as in Remark 6.2.3. Then we find by an immediate calculation that for every $j \in J_n$,

$$\mu\left(\frac{\partial}{\partial x^j}\right) = y^k \Gamma^i_{jk} \widehat{\frac{\partial}{\partial u^i}}, \tag{6.2.13}$$

$$\mu\left(\frac{\partial}{\partial y^j}\right) = \mu^{\vee}\left(\widehat{\frac{\partial}{\partial u^j}}\right) = (\delta^i_j + y^k C^i_{jk}) \widehat{\frac{\partial}{\partial u^i}}, \tag{6.2.14}$$

where we used the summation convention formulated in Remak 1.1.12(d).

Definition 6.2.7. A covariant derivative D on the Finsler bundle $\pi \colon TM \times_M TM \to TM$ is said to be *regular* if its vertical deflection $\mu^{\vee} := (D\widetilde{\delta}) \circ \mathbf{i}$ is an automorphism of the $C^{\infty}(TM)$-module $\Gamma(\pi)$; *strongly regular* if $\mu^{\vee} = 1_{\Gamma(\pi)}$.

Remark 6.2.8. (a) Since the $C^{\infty}(TM)$-module $\mathrm{Hom}_{TM}(\Gamma(\pi), \Gamma(\pi))$ of the tensorial mappings of $\Gamma(\pi)$ into itself is canonically isomorphic to the module $\mathrm{Bun}_{TM}(\pi, \pi)$ of strong bundle transformations of π (see Remark 2.2.32(a)), it follows that *a covariant derivative on a Finsler bundle is regular if, and only if, its vertical deflection restricts to a linear automorphism on each fibre.* (As to the converse statement, it is sufficient to require only the fibrewise injectivity or surjectivity of $\mu^{\vee}$.)

(b) With the notation as above, (6.2.14) implies immediately that a covariant derivative $D \colon \mathfrak{X}(TM) \times \Gamma(\pi) \to \Gamma(\pi)$ is

regular if, and only if, $\det(\delta^i_j + y^k(v)C^i_{jk}(v)) \neq 0$,
strongly regular if, and only if $y^k(v)C^i_{jk}(v) = 0$

for all $v \in \tau^{-1}(\mathcal{U})$, where $(\mathcal{U}, (u^i)_{i=1}^n)$ is an arbitrarily chosen chart of M.

Next, we prove a more conceptual characterization of the regularity of a covariant derivative on a Finsler bundle.

Proposition 6.2.9. *A covariant derivative D on the Finsler bundle $\pi \colon TM \times_M TM \to TM$ is regular if, and only if, at each $u \in TM$ we have*

$$T_u TM = \operatorname{Ker}(\underline{\mu}_u) \oplus V_u TM, \tag{$*$}$$

where $\underline{\mu} \colon TTM \to TM \times_M TM$ is the strong bundle map associated to the deflection $\mu \in \operatorname{Hom}_{TM}(\mathfrak{X}(TM), \Gamma(\pi))$ of D.

Proof. Suppose first that D is regular. Then the vertical deflection μ^{v} of D is 'fibrewise a linear isomorphism', i.e., by Remark 6.2.6,

$$(\underline{\mu}^{\mathsf{v}})_u \in \operatorname{GL}(\{u\} \times T_{\tau(u)}M), \quad u \in TM.$$

The restriction of $\underline{\mu}_u$ onto $V_u TM$ is a linear isomorphism of $V_u TM$ onto $\{u\} \times T_{\tau(u)}M$, because $\dim(V_u TM) = \dim(\{u\} \times T_{\tau(u)}M) = n := \dim M$ and

$$\underline{\mu}_u(V_u TM) \overset{\text{Lemma 4.1.24}}{=} \underline{\mu}_u \circ \mathbf{i}_u(\{u\} \times T_{\tau(u)}M)$$
$$= \underline{\mu}_u^{\mathsf{v}}(\{u\} \times T_{\tau(u)}M) = \{u\} \times T_{\tau(u)}M,$$

whence $\underline{\mu}_u \restriction V_u TM$ is surjective, and hence also injective. This implies that

$$\operatorname{Ker}(\underline{\mu}_u) \cap V_u TM = \{0\}. \tag{$**$}$$

By the rank + nullity = dimension theorem, $\dim(\operatorname{Ker}(\underline{\mu}_u)) = n$, so $\operatorname{Ker}(\underline{\mu}_u)$ and $V_u TM$ are n-dimensional subspaces of the $2n$-dimensional vector space $T_u TM$ satisfying $(**)$. This proves relation $(*)$.

Conversely, suppose that $(*)$ is satisfied. If $\underline{\mu}_u^{\mathsf{v}}(v_1) = \underline{\mu}_u^{\mathsf{v}}(v_2)$ for some $v_1, v_2 \in T_{\tau(u)}M$, then

$$\underline{\mu}_u^{\mathsf{v}}(v_1 - v_2) = \underline{\mu}_u(\mathbf{i}_u(v_1 - v_2)) = 0,$$

whence $\mathbf{i}_u(v_1 - v_2) \in \operatorname{Ker}(\underline{\mu}_u)$. However, $\mathbf{i}_u(v_1 - v_2) \in V_u TM$, so our condition implies that $\mathbf{i}_u(v_1 - v_2) = 0$, from which we conclude that $v_1 = v_2$. Thus $\underline{\mu}_u^{\mathsf{v}}$ is injective and therefore bijective, as was to be shown. $\qquad\square$

Corollary 6.2.10. *Let D be a regular covariant derivative on the Finsler bundle $\pi \colon TM \times_M TM \to TM$. Then the kernel of the deflection of D is a complementary submodule of $\mathfrak{X}^{\mathsf{v}}(TM)$ in $\mathfrak{X}(TM)$, i.e.,*

$$\operatorname{Ker}(\mu) \oplus \mathfrak{X}^{\mathsf{v}}(TM) = \mathfrak{X}(TM). \qquad\square$$

Lemma 6.2.11. *If D is a strongly regular covariant derivative on π, then the vertical curvature and the vertical torsion of D are related by*

$$\mathbf{Q}(\widetilde{X}, \widetilde{Y})\widetilde{\delta} = \mathfrak{Q}(\widetilde{X}, \widetilde{Y}); \quad \widetilde{X}, \widetilde{Y} \in \Gamma(\pi). \tag{6.2.15}$$

Proof. By the strong regularity of D, for every Finsler vector field $\widetilde{X}$ we have $D_{\mathbf{i}\widetilde{X}}\widetilde{\delta} = \mu^{\mathsf{v}}(\widetilde{X}) = \widetilde{X}$. Thus

$$
\begin{aligned}
\mathbf{Q}(\widetilde{X}, \widetilde{Y})\widetilde{\delta} &\overset{(6.2.4)}{=} D_{\mathbf{i}\widetilde{X}}D_{\mathbf{i}\widetilde{Y}}\widetilde{\delta} - D_{\mathbf{i}\widetilde{Y}}D_{\mathbf{i}\widetilde{X}}\widetilde{\delta} - D_{[\mathbf{i}\widetilde{X},\mathbf{i}\widetilde{Y}]}\widetilde{\delta} \\
&= D_{\mathbf{i}\widetilde{X}}\widetilde{Y} - D_{\mathbf{i}\widetilde{Y}}\widetilde{X} - \mu([\mathbf{i}\widetilde{X},\mathbf{i}\widetilde{Y}]) \\
&= D_{\mathbf{i}\widetilde{X}}\widetilde{Y} - D_{\mathbf{i}\widetilde{Y}}\widetilde{X} - \mu^{\mathsf{v}}(\mathbf{i}^{-1}[\mathbf{i}\widetilde{X},\mathbf{i}\widetilde{Y}]) \\
&\overset{\text{cond.}}{=} D_{\mathbf{i}\widetilde{X}}\widetilde{Y} - D_{\mathbf{i}\widetilde{Y}}\widetilde{X} - \mathbf{i}^{-1}([\mathbf{i}\widetilde{X},\mathbf{i}\widetilde{Y}]) \overset{(6.2.5)}{=} \mathfrak{Q}(\widetilde{X}, \widetilde{Y}). \qquad \square
\end{aligned}
$$

6.2.3 Vertical Covariant Derivative Operators

Definition 6.2.12. A *vertical covariant derivative operator* (briefly a *vertical covariant derivative* or a *vertical derivative*) on the Finsler bundle $\pi\colon TM \times_M TM \to TM$ is an $\mathbb{R}$-bilinear mapping

$$D^{\mathsf{v}}\colon \Gamma(\pi) \times \Gamma(\pi) \to \Gamma(\pi), \quad (\widetilde{X}, \widetilde{Y}) \mapsto D^{\mathsf{v}}_{\widetilde{X}}\widetilde{Y}$$

such that D^{v} is tensorial (i.e., $C^\infty(TM)$-linear) in its first variable, and satisfies the derivation rule

$$D^{\mathsf{v}}_{\widetilde{X}}F\widetilde{Y} = ((\mathbf{i}\widetilde{X})F)\widetilde{Y} + FD^{\mathsf{v}}_{\widetilde{X}}\widetilde{Y}, \quad F \in C^\infty(TM). \tag{6.2.16}$$

Lemma and Definition 6.2.13. *Let D^{v} be a vertical covariant derivative on the Finsler bundle π. Define the mappings*

$$R(D^{\mathsf{v}})\colon \Gamma(\pi) \times \Gamma(\pi) \times \Gamma(\pi) \to \Gamma(\pi), \quad (\widetilde{X}, \widetilde{Y}, \widetilde{Z}) \mapsto R(D^{\mathsf{v}})(\widetilde{X}, \widetilde{Y})\widetilde{Z}$$

and

$$T(D^{\mathsf{v}})\colon \Gamma(\pi) \times \Gamma(\pi) \to \Gamma(\pi), \quad (\widetilde{X}, \widetilde{Y}) \mapsto T(D^{\mathsf{v}})(\widetilde{X}, \widetilde{Y})$$

by

$$R(D^{\mathsf{v}})(\widetilde{X}, \widetilde{Y})\widetilde{Z} := D^{\mathsf{v}}_{\widetilde{X}}D^{\mathsf{v}}_{\widetilde{Y}}\widetilde{Z} - D^{\mathsf{v}}_{\widetilde{Y}}D^{\mathsf{v}}_{\widetilde{X}}\widetilde{Z} - D^{\mathsf{v}}_{\mathbf{i}^{-1}[\mathbf{i}\widetilde{X},\mathbf{i}\widetilde{Y}]}\widetilde{Z} \tag{6.2.17}$$

and

$$T(D^{\mathsf{v}})(\widetilde{X}, \widetilde{Y}) := D^{\mathsf{v}}_{\widetilde{X}}\widetilde{Y} - D^{\mathsf{v}}_{\widetilde{Y}}\widetilde{X} - \mathbf{i}^{-1}[\mathbf{i}\widetilde{X}, \mathbf{i}\widetilde{Y}], \tag{6.2.18}$$

respectively. Then $R(D^{\mathsf{v}})$ is a type $(1, 3)$ Finsler tensor field, called the curvature of D^{v}; $T(D^{\mathsf{v}})$ is a type $(1, 2)$ Finsler tensor field, called the torsion of D^{v}. The curvature $R(D^{\mathsf{v}})$ is skew-symmetric in the first two arguments and the torsion $T(D^{\mathsf{v}})$ is skew-symmetric.

Proof. The skew-symmetry properties are clear. We show that $T(D^v)$ is $C^\infty(TM)$-linear in its first argument, then this implies that it is a type $(1,2)$ Finsler tensor field on TM. For every function $F \in C^\infty(TM)$, we have

$$
\begin{aligned}
T(D^v)(F\widetilde{X}, \widetilde{Y}) &= D^v_{F\widetilde{X}}\widetilde{Y} - D^v_{\widetilde{Y}}F\widetilde{X} - \mathbf{i}^{-1}[\mathbf{i}(F\widetilde{X}), \mathbf{i}\widetilde{Y}] \\
&= FD^v_{\widetilde{X}}\widetilde{Y} - ((\mathbf{i}\widetilde{Y})F)\widetilde{X} - FD^v_{\widetilde{Y}}\widetilde{X} - F\mathbf{i}^{-1}[\mathbf{i}\widetilde{X}, \mathbf{i}\widetilde{Y}] \\
&\quad + \mathbf{i}^{-1}\big(((\mathbf{i}\widetilde{Y})F)\mathbf{i}\widetilde{X}\big) \\
&= F(D^v_{\widetilde{X}}\widetilde{Y} - D^v_{\widetilde{Y}}\widetilde{X} - \mathbf{i}^{-1}[\mathbf{i}\widetilde{X}, \mathbf{i}\widetilde{Y}]) = FT(D^v)(\widetilde{X}, \widetilde{Y}).
\end{aligned}
$$

The remainder is also straightforward and left to the reader. $\qquad\Box$

Example 6.2.14. Any covariant derivative D on π induces a vertical covariant derivative D^v given by

$$
D^v_{\widetilde{X}}\widetilde{Y} := D_{\mathbf{i}\widetilde{X}}\widetilde{Y}; \quad \widetilde{X}, \widetilde{Y} \in \Gamma(\pi). \tag{6.2.19}
$$

It follows immediately that in this case $R(D^v) = \mathbf{Q}$, $T(D^v) = \mathcal{Q}$.

Remark 6.2.15 (Extension of D^v to other Finsler Tensor Fields).
Let a vertical derivative $D^v \colon \Gamma(\pi) \times \Gamma(\pi) \to \Gamma(\pi)$ be given. Following the scheme of Example 6.1.8(d), we define the vertical derivative for any Finsler tensor field $A \in \mathcal{F}^k_l(TM)$, $(k, l) \in \mathbb{N} \times \mathbb{N}$.

Let $\widetilde{X}$ be a fixed Finsler vector field on TM.

First step. We agree that

$$
D^v_{\widetilde{X}}F := (\mathbf{i}\widetilde{X})F \text{ if } F \in \mathcal{F}^0_0(TM) = C^\infty(TM). \tag{6.2.20}
$$

Second step. If $\widetilde{\alpha} \in (\Gamma(\pi))^* = \mathcal{F}^0_1(TM)$, we define $D^v_{\widetilde{X}}\widetilde{\alpha}$ on the analogy of (6.1.11):

$$
(D^v_{\widetilde{X}}\widetilde{\alpha})(\widetilde{Y}) := \mathbf{i}\widetilde{X}(\widetilde{\alpha}(\widetilde{Y})) - \widetilde{\alpha}(D^v_{\widetilde{X}}\widetilde{Y}), \quad \widetilde{Y} \in \Gamma(\pi). \tag{6.2.21}
$$

Third step. Now we turn to the general case. Let $A \in \mathcal{F}^k_l(TM)$ where $k + l \geq 2$. Then, to make sure that the 'product rule' (cf. (6.1.14)) holds, we define $D^v_{\widetilde{X}}A$ by

$$
\begin{aligned}
(D^v_{\widetilde{X}}A)(\widetilde{\alpha}^1, \ldots, \widetilde{\alpha}^k, \widetilde{X}_1, \ldots, \widetilde{X}_l) &:= (\mathbf{i}\widetilde{X})A(\widetilde{\alpha}^1, \ldots, \widetilde{\alpha}^k, \widetilde{X}_1, \ldots, \widetilde{X}_l) \\
&\quad - \sum_{i=1}^k A(\widetilde{\alpha}^1, \ldots, D^v_{\widetilde{X}}\widetilde{\alpha}^i, \ldots, \widetilde{\alpha}^k, \widetilde{X}_1, \ldots, \widetilde{X}_l) \\
&\quad - \sum_{j=1}^l A(\widetilde{\alpha}^1, \ldots, \widetilde{\alpha}^k, \widetilde{X}_1, \ldots, D^v_{\widetilde{X}}\widetilde{X}_j, \ldots \widetilde{X}_l),
\end{aligned} \tag{6.2.22}
$$

where $\widetilde{\alpha}^i \in (\Gamma(\pi))^*$, $i \in J_k$ and $\widetilde{X}_j \in \Gamma(\pi)$, $j \in J_l$.

Now the *vertical differential* of $A \in \mathcal{F}_l^k(TM)$ $((k,l) \in \mathbb{N} \times \mathbb{N})$ is the Finsler tensor field $D^v A \in \mathcal{F}_{l+1}^k(TM)$ given by

$$D^v A(\widetilde{\alpha}^1, \ldots, \widetilde{\alpha}^k, \widetilde{X}, \widetilde{X}_1, \ldots, \widetilde{X}_l) := (D_{\widetilde{X}}^v A)(\widetilde{\alpha}^1, \ldots, \widetilde{\alpha}^k, \widetilde{X}_1, \ldots, \widetilde{X}_l),$$
$$(6.2.23)$$

cf. (6.1.16). In particular,

$$D^v F = dF \circ \mathbf{i}, \quad F \in C^\infty(TM). \tag{6.2.24}$$

Lemma and Definition 6.2.16. *The mapping*

$$\nabla^v : \Gamma(\pi) \times \Gamma(\pi) \to \Gamma(\pi), \quad (\widetilde{X}, \widetilde{Y}) \mapsto \nabla_{\widetilde{X}}^v \widetilde{Y} := \mathbf{j}[\mathbf{i}\widetilde{X}, \eta], \tag{6.2.25}$$

where

$$\eta \in \mathfrak{X}(TM) \text{ is such that } \mathbf{j}\eta = \widetilde{Y}, \tag{6.2.26}$$

is a vertical derivative, called the canonical vertical derivative *on the Finsler bundle* π.

Proof. First we check that $\nabla_{\widetilde{X}}^v \widetilde{Y}$ is well-defined, i.e., does not depend on the choice of the vector field η satisfying (6.2.26). Indeed, if $\mathbf{j}\eta_1 = \mathbf{j}\eta_2 = \widetilde{Y}$, then $\mathbf{j}(\eta_1 - \eta_2) = 0$, therefore $\eta_1 - \eta_2$ is vertical, so $[\mathbf{i}\widetilde{X}, \eta_1 - \eta_2]$ is also vertical (Corollary 4.1.29), whence $\mathbf{j}[\mathbf{i}\widetilde{X}, \eta_1] = \mathbf{j}[\mathbf{i}\widetilde{X}, \eta_2]$.

Obviously, ∇^v is $\mathbb{R}$-bilinear. Now let $F \in C^\infty(TM)$. Then

$$\nabla_{F\widetilde{X}}^v \widetilde{Y} := \mathbf{j}[\mathbf{i}(F\widetilde{X}), \eta] = \mathbf{j}[F(\mathbf{i}\widetilde{X}), \eta] = F\nabla_{\widetilde{X}}^v \widetilde{Y} - \mathbf{j}((\eta F)\mathbf{i}\widetilde{X}) \overset{(4.1.40)}{=} F\nabla_{\widetilde{X}}^v \widetilde{Y},$$

which proves that ∇^v is tensorial in its first argument. Finally, we verify (6.2.16):

$$\nabla_{\widetilde{X}}^v F\widetilde{Y} := \mathbf{j}[\mathbf{i}\widetilde{X}, F\eta] = \mathbf{j}\big(F[\mathbf{i}\widetilde{X}, \eta] + ((\mathbf{i}\widetilde{X})F)\eta\big)$$
$$= F\mathbf{j}[\mathbf{i}\widetilde{X}, \eta] + ((\mathbf{i}\widetilde{X})F)\mathbf{j}\eta = ((\mathbf{i}\widetilde{X})F)\widetilde{Y} + F\nabla_{\widetilde{X}}^v \widetilde{Y}. \qquad \square$$

Remark 6.2.17. According to Remark 6.2.15, the canonical vertical derivative and the canonical vertical differential of an arbitrary Finsler tensor field is also defined. If $F \in C^\infty(TM)$, then we have

$$\nabla^v F = dF \circ \mathbf{i} \tag{6.2.27}$$

as in the general case (see (6.2.24)). Evaluating both sides of (6.2.27) at a basic section $\widehat{X} \in \Gamma(\pi)$, we find that

$$\nabla^v F(\widehat{X}) = X^v F, \tag{6.2.28}$$

because $dF \circ i(\widehat{X}) \overset{(4.1.44)}{=} dF(X^v) = X^v F$. Observe that the vertical differential of F and its Lie derivative with respect to the vertical endomorphism (see (4.1.97)) are related by

$$\mathcal{L}_{\mathbf{J}} F = (\nabla^v F) \circ \mathbf{j}. \tag{6.2.29}$$

Indeed, for any vector field ξ on TM, we have

$$\mathcal{L}_{\mathbf{J}} F(\xi) \overset{(4.1.97)}{=} dF(\mathbf{J}\xi) = (\mathbf{J}\xi)F = (i(\mathbf{j}\xi))F = \nabla^v F(\mathbf{j}\xi) = (\nabla^v F) \circ \mathbf{j}(\xi).$$

Coordinate description. Choose a chart $(\mathcal{U}, (u^i)_{i=1}^n)$ on M, and consider the induced chart $(\tau^{-1}(\mathcal{U}), ((x^i)_{i=1}^n, (y^i)_{i=1}^n))$ on TM. Let $(\widehat{du^i})_{i=1}^n$ be the dual coframe of the local frame $\left(\dfrac{\partial}{\partial u^i}\right)_{i=1}^n$ of $TM \times_M TM$ over $\tau^{-1}(\mathcal{U})$.

(a) If $F \in C^\infty(TM)$, then

$$\nabla^v F \underset{(\mathcal{U})}{=} \sum_{i=1}^n \frac{\partial F}{\partial y^i} \widehat{du^i}. \tag{6.2.30}$$

Indeed, $\nabla^v F \left(\dfrac{\partial}{\partial u^i}\right) \overset{(6.2.28)}{=} \left(\dfrac{\partial}{\partial u^i}\right)^v F \overset{(4.1.73)}{=} \dfrac{\partial F}{\partial y^i}$ for every $i \in J_n$, so we have (6.2.30).

(b) Let $\widetilde{X}, \widetilde{Y} \in \Gamma(\pi)$. Working locally, and using the summation convention, we write $\widetilde{X} = \widetilde{X}^i \dfrac{\partial}{\partial u^i}$, $\widetilde{Y} = \widetilde{Y}^i \dfrac{\partial}{\partial u^i}$ $(\widetilde{X}^i, \widetilde{Y}^i \in C^\infty(\tau^{-1}(\mathcal{U})))$. We claim that

$$\nabla^v_{\widetilde{X}} \widetilde{Y} = \widetilde{X}^i \frac{\partial \widetilde{Y}^j}{\partial y^i} \frac{\partial}{\partial u^j}. \tag{6.2.31}$$

To see this, let $\eta := \widetilde{Y}^i \dfrac{\partial}{\partial x^i}$. Then $\eta \in \mathfrak{X}(\tau^{-1}(\mathcal{U}))$, and

$$\mathbf{j}\eta = \widetilde{Y}^i \mathbf{j} \left(\frac{\partial}{\partial x^i}\right) \overset{(4.1.74)}{=} \widetilde{Y}^i \frac{\widehat{\partial}}{\partial u^i} = \widetilde{Y}.$$

Since, similarly, $i\widetilde{X} = \widetilde{X}^i \dfrac{\partial}{\partial y^i}$, we obtain

$$\nabla^v_{\widetilde{X}} \widetilde{Y} := \mathbf{j}[i\widetilde{X}, \eta] = \mathbf{j}\left[\widetilde{X}^i \frac{\partial}{\partial y^i}, \eta\right] = \widetilde{X}^i \mathbf{j}\left[\frac{\partial}{\partial y^i}, \eta\right] - \eta(\widetilde{X}^i)\mathbf{j}\left(\frac{\partial}{\partial y^i}\right)$$

$$\overset{(4.1.74)}{=} \widetilde{X}^i \mathbf{j}\left[\frac{\partial}{\partial y^i}, \widetilde{Y}^k \frac{\partial}{\partial x^k}\right] = \widetilde{X}^i \frac{\partial \widetilde{Y}^k}{\partial y^i} \frac{\widehat{\partial}}{\partial u^k},$$

as desired.

Corollary 6.2.18. (i) *For any Finsler vector field* $\widetilde{X} \in \Gamma(\pi)$ *and any vector field* Y *on* M,

$$\nabla^v_{\widetilde{X}} \widehat{Y} = 0. \tag{6.2.32}$$

Equivalently, $\nabla^{\mathrm{v}}\widehat{Y} = 0$.

(ii) *For every* $\widetilde{X} \in \Gamma(\pi)$,

$$\nabla^{\mathrm{v}}_{\widetilde{X}}\widetilde{\delta} = \widetilde{X}, \tag{6.2.33}$$

whence $\nabla^{\mathrm{v}}\widetilde{\delta} = 1_{\Gamma(\pi)}$.

Proof. Locally, matters are manifest. In the setting as above, let $Y = Y^j \frac{\partial}{\partial u^j} \in \mathfrak{X}(\mathcal{U})$. Then $\widehat{Y} = (Y^j \circ \tau)\widehat{\frac{\partial}{\partial u^j}}$, and we have

$$\nabla^{\mathrm{v}}_{\widetilde{X}}\widehat{Y} \stackrel{(6.2.31)}{=} \widetilde{X}^i \frac{\partial(Y^j \circ \tau)}{\partial y^i} \widehat{\frac{\partial}{\partial u^j}} = 0,$$

since the functions $Y^j \circ \tau$ are constant on each tangent space. Similarly,

$$\nabla^{\mathrm{v}}_{\widetilde{X}}\widetilde{\delta} \stackrel{(4.1.7)}{=} \nabla^{\mathrm{v}}_{\widetilde{X}}\left(y^j \widehat{\frac{\partial}{\partial u^j}}\right) \stackrel{(6.2.31)}{=} \widetilde{X}^i \frac{\partial y^j}{\partial y^i} \widehat{\frac{\partial}{\partial u^j}} = \widetilde{X}^i \delta^j_i \widehat{\frac{\partial}{\partial u^j}} = \widetilde{X}^i \widehat{\frac{\partial}{\partial u^i}} = \widetilde{X}. \qquad \square$$

Remark 6.2.19. We present a more conceptual, coordinate-free proof of the corollary.

(i) Since $\mathbf{j}Y^{\mathrm{c}} = \widehat{Y}$ by Lemma 4.1.45(iii), and the Finsler vector field $\widetilde{X}$ can be written in the form $\widetilde{X} = \mathbf{j}\xi$, where $\xi \in \mathfrak{X}(TM)$, we have

$$\nabla^{\mathrm{v}}_{\widetilde{X}}\widehat{Y} = \mathbf{j}[\mathbf{J}\xi, Y^{\mathrm{c}}] \stackrel{(*)}{=} \mathbf{j} \circ \mathbf{i} \circ \mathbf{j}[\xi, Y^{\mathrm{c}}] \stackrel{(4.1.40)}{=} 0.$$

At step $(*)$ we used the following trick:

$$0 \stackrel{(4.1.89)}{=} [\mathbf{J}, Y^{\mathrm{c}}]\xi = [\mathbf{J}\xi, Y^{\mathrm{c}}] - \mathbf{J}[\xi, Y^{\mathrm{c}}],$$

so $[\mathbf{J}\xi, Y^{\mathrm{c}}]$ can be changed to $\mathbf{J}[\xi, Y^{\mathrm{c}}]$.

(ii) Let $\widetilde{X} = \mathbf{j}\xi$, as above. If $S \in \mathfrak{X}(TM)$ is a second-order vector field, then $\mathbf{j}S = \widetilde{\delta}$, so we have

$$\mathbf{i}\nabla^{\mathrm{v}}_{\widetilde{X}}\widetilde{\delta} = \mathbf{J}[\mathbf{J}\xi, S] \stackrel{(5.1.4)}{=} \mathbf{J}\xi = \mathbf{i}\widetilde{X},$$

whence $\nabla^{\mathrm{v}}_{\widetilde{X}}\widetilde{\delta} = \widetilde{X}$.

Definition 6.2.20. Let A be a type $(0, s)$ or of type $(1, s)$ Finsler tensor field on TM or on $\mathring{T}M$. We say that A is *positive-homogeneous of degree k* or, briefly, k^+-*homogeneous*, where k is an integer, if $\nabla^{\mathrm{v}}_{\widetilde{\delta}}A = kA$.

Corollary 6.2.21. *A Finsler tensor field is k^+-homogeneous if, and only if, its components with respect to any chart are positive-homogeneous of degree k.*

Proof. For clarity, consider a type $(0,2)$ Finsler tensor field g on TM; the general case does not cause any extra difficulty.

Choose a chart $(\mathcal{U}, (u^i)_{i=1}^n)$ on M. Then, locally

$$g \underset{(\mathcal{U})}{=} g_{ij} \widehat{du^i} \otimes \widehat{du^j}, \quad g_{ij} = g\left(\frac{\widehat{\partial}}{\partial u^i}, \frac{\widehat{\partial}}{\partial u^j}\right) \in C^\infty(\tau^{-1}(\mathcal{U}))$$

(cf. (3.3.9)), and

$$(\nabla^{\mathsf{v}}_{\widetilde{\delta}} g)\left(\frac{\widehat{\partial}}{\partial u^i}, \frac{\widehat{\partial}}{\partial u^j}\right) \overset{(6.2.22)}{=} Cg_{ij} - g\left(\nabla^{\mathsf{v}}_{\widetilde{\delta}}\frac{\widehat{\partial}}{\partial u^i}, \frac{\widehat{\partial}}{\partial u^j}\right) - g\left(\frac{\widehat{\partial}}{\partial u^i}, \nabla^{\mathsf{v}}_{\widetilde{\delta}}\frac{\widehat{\partial}}{\partial u^j}\right)$$

$$\overset{(6.2.32)}{=} Cg_{ij}.$$

Thus g is k^+-homogeneous if, and only if, $Cg_{ij} = y^l \frac{\partial g_{ij}}{\partial y^l} = kg_{ij}$ for all $i, j \in J_n$, so Lemma 4.2.9 implies our assertion. □

Lemma 6.2.22. *If a Finsler tensor field $A \in \mathcal{F}^0_s(\overset{\circ}{T}M) \cup \mathcal{F}^1_s(\overset{\circ}{T}M)$ is k^+-homogeneous, then $\nabla^{\mathsf{v}} A$ is $(k-1)^+$-homogeneous.*

Proof. Applying Lemma 4.2.4, this can be deduced as a consequence of the preceding lemma. However, we present here a coordinate-free proof, which does not rely on Lemma 4.2.4.

Let, for definiteness, $A \in \mathcal{F}^1_s(\overset{\circ}{T}M)$. Then $\nabla^{\mathsf{v}} A \in \mathcal{F}^1_{s+1}(\overset{\circ}{T}M)$, and for any vector fields $X_0, X_1, \ldots, X_s$ on M,

$$(\nabla^{\mathsf{v}}_{\widetilde{\delta}}(\nabla^{\mathsf{v}} A))(\widehat{X}_0, \widehat{X}_1, \ldots, \widehat{X}_s) \overset{(6.2.32)}{=} \nabla^{\mathsf{v}}_{\widetilde{\delta}}((\nabla^{\mathsf{v}} A)(\widehat{X}_0, \widehat{X}_1, \ldots, \widehat{X}_s))$$

$$= \nabla^{\mathsf{v}}_{\widetilde{\delta}}(\nabla^{\mathsf{v}}_{\widehat{X}_0}(A(\widehat{X}_1, \ldots, \widehat{X}_s))) \overset{(6.2.25)}{=} \nabla^{\mathsf{v}}_{\widetilde{\delta}}(\mathbf{j}[X_0^{\mathsf{v}}, \eta]) = \mathbf{j}[C, [X_0^{\mathsf{v}}, \eta]],$$

where $\eta \in \mathfrak{X}(\overset{\circ}{T}M)$ is such that $\mathbf{j}\eta = A(\widehat{X}_1, \ldots, \widehat{X}_s)$. By the Jacobi identity (L2) in 1.3.3, we have

$$\mathbf{j}[C, [X_0^{\mathsf{v}}, \eta]] = -\mathbf{j}[X_0^{\mathsf{v}}, [\eta, C]] - \mathbf{j}[\eta, [C, X_0^{\mathsf{v}}]] \overset{(4.1.53)}{=} \mathbf{j}[X_0^{\mathsf{v}}, [C, \eta]] - \mathbf{j}[X_0^{\mathsf{v}}, \eta].$$

Here

$$\mathbf{j}[X_0^{\mathsf{v}}, \eta] = \nabla^{\mathsf{v}}_{\widehat{X}_0}(A(\widehat{X}_1, \ldots, \widehat{X}_s)) \overset{(6.2.32)}{=} (\nabla^{\mathsf{v}}_{\widehat{X}_0} A)(\widehat{X}_1, \ldots, \widehat{X}_s)$$

$$= \nabla^{\mathsf{v}} A(\widehat{X}_0, \widehat{X}_1, \ldots, \widehat{X}_s),$$

$$\mathbf{j}[C, \eta] = \nabla^{\mathsf{v}}_{\widetilde{\delta}}(A(\widehat{X}_1, \ldots, \widehat{X}_s)) = (\nabla^{\mathsf{v}}_{\widetilde{\delta}} A)(\widehat{X}_1, \ldots, \widehat{X}_s) = kA(\widehat{X}_1, \ldots, \widehat{X}_s),$$

so

$$\mathbf{j}[X_0^{\mathsf{v}}, [C, \eta]] = k\nabla^{\mathsf{v}}_{\widehat{X}_0}(A(\widehat{X}_1, \ldots, \widehat{X}_s)) = k\nabla^{\mathsf{v}} A(\widehat{X}_0, \widehat{X}_1, \ldots, \widehat{X}_s),$$

therefore

$$(\nabla^{\mathsf{v}}_{\tilde{\delta}}(\nabla^{\mathsf{v}} A))(\widehat{X}_0, \widehat{X}_1, \ldots, \widehat{X}_s) = k\nabla^{\mathsf{v}} A(\widehat{X}_0, \widehat{X}_1, \ldots, \widehat{X}_s)$$
$$- \nabla^{\mathsf{v}} A(\widehat{X}_0, \widehat{X}_1, \ldots, \widehat{X}_s) = (k-1)\nabla^{\mathsf{v}} A(\widehat{X}_0, \widehat{X}_1, \ldots, \widehat{X}_s),$$

as was to be shown. $\qquad\square$

Lemma 6.2.23. *The canonical vertical derivative has vanishing curvature and torsion:*

$$R(\nabla^{\mathsf{v}}) = 0, \quad T(\nabla^{\mathsf{v}}) = 0. \tag{6.2.34}$$

Proof. Since both $R(\nabla^{\mathsf{v}})$ and $T(\nabla^{\mathsf{v}})$ are tensor fields, it is sufficient to check (6.2.34) for basic Finsler vector fields. Then (6.2.17), (6.2.18) and (6.2.32) immediately give the result. $\qquad\square$

Lemma and Definition 6.2.24. *Let D be a covariant derivative on the Finsler bundle π, and consider its torsion T^D. Define a mapping*

$$\mathcal{S}\colon \Gamma(\pi) \times \Gamma(\pi) \to \Gamma(\pi), \ (\widetilde{X}, \widetilde{Y}) \mapsto \mathcal{S}(\widetilde{X}, \widetilde{Y})$$

by

$$\mathcal{S}(\widetilde{X}, \widetilde{Y}) := T^D(\xi, \mathbf{i}\widetilde{Y}) \ \textit{if} \ \xi \in \mathfrak{X}(TM) \ \textit{is such that} \ \mathbf{j}\xi = \widetilde{X}. \tag{6.2.35}$$

Then $\mathcal{S}$ is well-defined, and it can be obtained by the formula

$$\mathcal{S}(\widetilde{X}, \widetilde{Y}) = \nabla^{\mathsf{v}}_{\widetilde{Y}}\widetilde{X} - D^{\mathsf{v}}_{\widetilde{Y}}\widetilde{X}. \tag{6.2.36}$$

We say that $\mathcal{S}$ is the vertical deviation *of D. If $\mathcal{S} = 0$ (i.e., $D^{\mathsf{v}} = \nabla^{\mathsf{v}}$), then D is called* vertically natural.

Proof. It is sufficient to verify (6.2.36), then it follows automatically that $\mathcal{S}$ does not depend on the choice of ξ in (6.2.35). From the definitions,

$$\mathcal{S}(\widetilde{X}, \widetilde{Y}) = T^D(\xi, \mathbf{i}\widetilde{Y}) = D_\xi \mathbf{j}(\mathbf{i}\widetilde{Y}) - D_{\mathbf{i}\widetilde{Y}}\mathbf{j}\xi - \mathbf{j}[\xi, \mathbf{i}\widetilde{Y}]$$
$$\overset{(4.1.40)}{=} \mathbf{j}[\mathbf{i}\widetilde{Y}, \xi] - D^{\mathsf{v}}_{\widetilde{Y}}\mathbf{j}\xi \overset{(6.2.25)}{=} \nabla^{\mathsf{v}}_{\widetilde{Y}}\widetilde{X} - D^{\mathsf{v}}_{\widetilde{Y}}\widetilde{X}. \qquad\square$$

Lemma 6.2.25. *Let D be a covariant derivative on the Finsler bundle π. The vertical deflection and the vertical deviation of D are related by*

$$\mu^{\mathsf{v}} = 1_{\Gamma(\pi)} - i_{\tilde{\delta}}\mathcal{S}. \tag{6.2.37}$$

Consequently, D is strongly regular if, and only if, $i_{\tilde{\delta}}\mathcal{S} = 0$. (Here $i_{\tilde{\delta}}$ is the substitution operator associated to $\tilde{\delta}$; see 1.2.2.)

Proof.   For any Finsler vector field $\widetilde{X} \in \Gamma(\pi)$,

$$i_{\widetilde{\delta}}\mathcal{S}(\widetilde{X}) = \mathcal{S}(\widetilde{\delta}, \widetilde{X}) \stackrel{(6.2.36)}{=} \nabla^{\mathsf{v}}_{\widetilde{X}}\widetilde{\delta} - D^{\mathsf{v}}_{\widetilde{X}}\widetilde{\delta}$$

$$\stackrel{(6.2.33)}{=} \widetilde{X} - D_{\mathbf{i}\widetilde{X}}\widetilde{\delta} \stackrel{(6.2.12)}{=} \widetilde{X} - \mu^{\mathsf{v}}(\widetilde{X}) = (1_{\Gamma(\pi)} - \mu^{\mathsf{v}})(\widetilde{X}),$$

whence our claim.   □

Remark 6.2.26. With the notations of Remark 6.2.3, the components of the vertical deviation $\mathcal{S}$ with respect to the local frame field $\left(\widehat{\frac{\partial}{\partial u^i}}\right)^n_{i=1}$ are the negatives of the functions C^i_{jk} introduced in (6.2.6). More precisely,

$$\text{if } \mathcal{S}\left(\widehat{\frac{\partial}{\partial u^j}}, \widehat{\frac{\partial}{\partial u^k}}\right) = \sum_{i=1}^n S^i_{jk}\,\widehat{\frac{\partial}{\partial u^i}}, \text{ then } S^i_{jk} = -C^i_{kj}; \quad i,j,k \in J_n. \quad (6.2.38)$$

Indeed,

$$\mathcal{S}\left(\widehat{\frac{\partial}{\partial u^j}}, \widehat{\frac{\partial}{\partial u^k}}\right) \stackrel{(6.2.36)}{=} \nabla^{\mathsf{v}}_{\widehat{\frac{\partial}{\partial u^k}}}\widehat{\frac{\partial}{\partial u^j}} - D^{\mathsf{v}}_{\widehat{\frac{\partial}{\partial u^k}}}\widehat{\frac{\partial}{\partial u^j}} \stackrel{(6.2.32)}{=} -D^{\mathsf{v}}_{\widehat{\frac{\partial}{\partial u^k}}}\widehat{\frac{\partial}{\partial u^j}}$$

$$= -D_{\frac{\partial}{\partial v^k}}\widehat{\frac{\partial}{\partial u^j}} \stackrel{(6.2.6)}{=} -\sum_{i=1}^n C^i_{kj}\,\widehat{\frac{\partial}{\partial u^i}}.$$

The moral is that *the Christoffel symbols of an induced vertical covariant derivative are tensor components.*

6.2.4   *The Vertical Hessian of a Lagrangian*

Definition 6.2.27. Let M be a manifold and F a smooth function on TM or on $\overset{\circ}{T}M$. The type $(0,2)$ Finsler tensor field $g_F := \nabla^{\mathsf{v}}\nabla^{\mathsf{v}}F$ is called the *vertical Hessian* of F.

Lemma 6.2.28. *Let $F \in C^\infty(TM)$ be a Lagrangian.*

(i) *The vertical Hessian of F and the Lagrange two-form attached to F are related by*

$$g_F(\mathbf{j}\xi, \mathbf{j}\eta) = \omega_F(\mathbf{J}\xi, \eta); \quad \xi, \eta \in \mathfrak{X}(TM). \quad (6.2.39)$$

(ii) *The vertical Hessian $g_F \in \mathcal{F}^0_2(TM)$ is a symmetric bilinear Finsler tensor field.*

(iii) *For any vector fields X, Y on M,*

$$g_F(\widehat{X}, \widehat{Y}) = \omega_F(X^{\mathsf{v}}, Y^{\mathsf{c}}) = X^{\mathsf{v}}(Y^{\mathsf{v}}F). \quad (6.2.40)$$

(iv) *Let* $(\mathcal{U}, (u^i)_{i=1}^n)$ *be a chart on* M, *and consider the induced chart* $(\tau^{-1}(\mathcal{U}), ((x^i)_{i=1}^n, (y^i)_{i=1}^n))$ *on* TM. *The components of* g_F *with respect to the local frame* $\left(\frac{\partial}{\partial u^i}\right)_{i=1}^n$ *of* $TM \times_M TM$ *are the functions*

$$g_{ij} := g_F\left(\widehat{\frac{\partial}{\partial u^i}}, \widehat{\frac{\partial}{\partial u^j}}\right) = \frac{\partial^2 F}{\partial y^i \partial y^j}; \quad i, j \in J_n. \tag{6.2.41}$$

Proof. We have seen in the proof of Lemma 5.2.4 that

$$\omega_F(\mathbf{J}\xi, \eta) = \mathbf{J}\xi(\mathbf{J}\eta(F)) - \mathbf{J}[\mathbf{J}\xi, \eta](F). \tag{6.2.42}$$

We show that the left-hand side of (6.2.39) gives the same expression. Using the general rules (6.2.22) and (6.2.23),

$$g_F(\mathbf{j}\xi, \mathbf{j}\eta) = \nabla^\mathrm{v}\nabla^\mathrm{v}F(\mathbf{j}\xi, \mathbf{j}\eta) = (\nabla^\mathrm{v}_{\mathbf{j}\xi}(\nabla^\mathrm{v}F))(\mathbf{j}\eta)$$

$$= \mathbf{J}\xi(\mathbf{J}\eta(F)) - \nabla^\mathrm{v}F(\nabla^\mathrm{v}_{\mathbf{j}\xi}\mathbf{j}\eta) \overset{(6.2.25)}{=} \mathbf{J}\xi(\mathbf{J}\eta(F)) - \nabla^\mathrm{v}F(\mathbf{j}[\mathbf{J}\xi, \eta])$$

$$= \mathbf{J}\xi(\mathbf{J}\eta(F)) - \mathbf{J}[\mathbf{J}\xi, \eta](F),$$

as was to be shown. Having this result, the symmetry of g_F can easily be deduced:

$$g_F(\mathbf{j}\xi, \mathbf{j}\eta) = \omega_F(\mathbf{J}\xi, \eta) \overset{(5.2.3)}{=} -\omega_F(\xi, \mathbf{J}\eta) = \omega_F(\mathbf{J}\eta, \xi) = g(\mathbf{j}\eta, \mathbf{j}\xi).$$

Using (6.2.39), (6.2.42) we find that

$$g_F(\widehat{X}, \widehat{Y}) = g_F(\mathbf{j}X^\mathrm{c}, \mathbf{j}Y^\mathrm{c}) = \omega_F(\mathbf{J}X^\mathrm{c}, Y^\mathrm{c}) = \omega_F(X^\mathrm{v}, Y^\mathrm{c})$$

$$= X^\mathrm{v}(Y^\mathrm{v}F) - \mathbf{J}[X^\mathrm{v}, Y^\mathrm{c}](F) \overset{(4.1.71)}{=} X^\mathrm{v}(Y^\mathrm{v}F) - \mathbf{J}[X, Y]^\mathrm{v}(F) = X^\mathrm{v}(Y^\mathrm{v}F),$$

thus proving (6.2.40). Finally, (6.2.41) is an immediate consequence of (6.2.40), and so the proof is complete. $\square$

Remark 6.2.29. (a) In the course of the previous proof, as a by-product, we obtained the following useful formula:

$$g_F(\mathbf{j}\xi, \mathbf{j}\eta) = \mathbf{J}\xi(\mathbf{J}\eta(F)) - \mathbf{J}[\mathbf{J}\xi, \eta](F); \quad \xi, \eta \in \mathfrak{X}(TM). \tag{6.2.43}$$

(b) Let, for all $\xi, \eta \in \mathfrak{X}(TM)$,

$$\bar{g}_F(\mathbf{J}\xi, \mathbf{J}\eta) := g_F(\mathbf{j}\xi, \mathbf{j}\eta) \overset{(6.2.39)}{=} \omega_F(\mathbf{J}\xi, \eta). \tag{6.2.44}$$

Then $\bar{g}_F$ is a scalar product on the vertical bundle VTM, called the *scalar product induced (or determined) by* F. Since

$$\bar{g}_F\left(\frac{\partial}{\partial y^i}, \frac{\partial}{\partial y^j}\right) = \bar{g}_F\left(\mathbf{J}\frac{\partial}{\partial x^i}, \mathbf{J}\frac{\partial}{\partial x^j}\right) := g_F\left(\mathbf{j}\frac{\partial}{\partial x^i}, \mathbf{j}\frac{\partial}{\partial x^j}\right)$$

$$\overset{(4.1.74)}{=} g_F\left(\widehat{\frac{\partial}{\partial u^i}}, \widehat{\frac{\partial}{\partial u^j}}\right) \overset{(6.2.41)}{=} \frac{\partial^2 F}{\partial y^i \partial y^j} \quad (i, j \in J_n),$$

the scalar product $\bar{g}_F$ and the vertical Hessian g_F of F have the same component functions.

Corollary 6.2.30. *For a Lagrangian* $F \in C^\infty(TM)$, *the following are equivalent:*

(i) F *is regular, i.e., the Lagrangian two-form* $\omega_F = d\mathcal{L}_J F$ *is non-degenerate.*

(ii) *The vertical Hessian* $g_F = \nabla^v \nabla^v F$ *is non-degenerate.*

(iii) *The scalar product* $\bar{g}_F$ *on* VTM *is non-degenerate.*

(iv) *For any induced chart* $(\tau^{-1}(\mathcal{U}), ((x^i)_{i=1}^n, (y^i)_{i=1}^n))$ *on* TM *and for every* $v \in \tau^{-1}(\mathcal{U})$, $\det\left(\frac{\partial^2 F}{\partial y^i \partial y^j}(v)\right) \neq 0$.

Proof. By Lemma 5.2.3, Lemma 6.2.28(iv) and Remark 6.2.29(b), assertions (i)–(iii) are equivalent to (iv). $\qquad\qquad\square$

Our next result shows that the canonical vertical differential and the vertical Hessian of a smooth function on TM have a natural pointwise meaning.

Lemma 6.2.31. *Let* $F: TM \to \mathbb{R}$ *be a smooth function, and let* $F_p := F \upharpoonright T_pM$ *for every* $p \in M$. *Given a vector* $u \in TM$, *define the functions* $(\nabla^v F)_u : \{u\} \times T_{\tau(u)}M \to \mathbb{R}$ *and*

$$(\nabla^v \nabla^v F)_u : (\{u\} \times T_{\tau(u)}M) \times (\{u\} \times T_{\tau(u)}M) \to \mathbb{R}$$

by $(\nabla^v F)_u(u, v) := (\nabla^v F)(\widehat{X})(u)$ *and*

$$(\nabla^v \nabla^v F)_u((u, v), (u, w)) := (\nabla^v \nabla^v F)(\widehat{X}, \widehat{Y})(u)$$

respectively, where $X, Y \in \mathfrak{X}(M)$ *are so chosen that* $X(\tau(u)) = v$ *and* $Y(\tau(u)) = w$. *Then*

$$(\nabla^v F)_u(u, v) = (F_{\tau(u)})'(u)(v), \tag{6.2.45}$$

$$(\nabla^v \nabla^v F)_u((u, v), (u, w)) = (F_{\tau(u)})''(u)(v, w) \tag{6.2.46}$$

(therefore $(\nabla^v F)_u$ *and* $(\nabla^v \nabla^v F)_u$ *are well-defined).*

Proof. Relation (6.2.45) is an easy consequence of the definition of $(\nabla^v F)_u$ and Corollary 4.1.18. Indeed,

$$(\nabla^v F)_u(u, v) := (\nabla^v F)(\widehat{X})(u) = (\nabla^v_{\widehat{X}} F)(u) = (X^v F)(u) = X^v(u)(F)$$

$$\overset{(4.1.44)}{=} \left(X(\tau(u))\right)^{\uparrow}(u)(F) = v^{\uparrow}(u)(F) \overset{(4.1.33)}{=} (F_p)'(u)(v).$$

We now turn to showing relation (6.2.46). Starting with the definition of $(\nabla^v \nabla^v F)_u$,

$$(\nabla^v \nabla^v F)_u((u, v), (u, w)) = (\nabla^v \nabla^v F)(\widehat{X}, \widehat{Y})(u) = (\nabla^v_{\widehat{X}}(\nabla^v F))(\widehat{Y})(u)$$

$$= (X^v(\nabla^v_{\widehat{Y}} F))(u) = v^{\uparrow}(u)(\nabla^v_{\widehat{Y}} F) \overset{(6.2.45)}{=} (\nabla^v_{\widehat{Y}} F)'_{\tau(u)}(u)(v).$$

Applying Proposition C.1.6,

$$(\nabla^v_{\widehat{Y}} F)'_{\tau(u)}(u)(v) = \lim_{t \to 0} \frac{1}{t}\big((\nabla^v_{\widehat{Y}} F)_{\tau(u)}(u + tv) - (\nabla^v_{\widehat{Y}} F)_{\tau(u)}(u)\big).$$

Here

$$\begin{aligned}
(\nabla^v_{\widehat{Y}} F)_{\tau(u)}(u + tv) &= (\nabla^v F)(\widehat{Y})(u + tv)\\
&=: (\nabla^v F)_{u+tv}(u + tv, Y \circ \tau(u + tv))\\
&= (\nabla^v F)_{u+tv}(u + tv, w) \overset{(6.2.45)}{=} (F_{\tau(u)})'(u + tv)(w).
\end{aligned}$$

Similarly,

$$(\nabla^v_{\widehat{Y}} F)_{\tau(u)}(u) = (\nabla^v F)(\widehat{Y})(u) =: (\nabla^v F)_u(u, w) \overset{(6.2.45)}{=} (F_{\tau(u)})'(u)(w).$$

Thus

$$\begin{aligned}
(\nabla^v \nabla^v F)_u((u, v), (u, w)) &= \lim_{t \to 0} \frac{(F_{\tau(u)})'(u + tv) - (F_{\tau(u)})'(u)}{t}(w)\\
&= (F_{\tau(u)})''(v)(w) \overset{\text{Remark C.1.3}}{=} (F_{\tau(u)})''(v, w),
\end{aligned}$$

as was to be shown. $\qquad\qquad\qquad\qquad\qquad\qquad\qquad\qquad\qquad\qquad\qquad\square$

6.2.5   *Parallelism and Geodesics*

Let $I \subset \mathbb{R}$ be a nonempty open interval, and let a smooth curve $\gamma\colon I \to M$ be given.

In this subsection for the elements of the $C^\infty(I)$-module $\Gamma_\gamma(TM) = \mathfrak{X}_\gamma(M)$ introduced in Remark 3.1.30(d) we shall use the terminology 'lift of γ to TM' mentioned in Definition 2.2.11. We recall that an element of $\Gamma_\gamma(TM)$ is a smooth curve $c\colon I \to TM$ such that $\tau \circ c = \gamma$. Notice that $\Gamma_\gamma(TM)$ is surely nonempty: the canonical lift $\dot{\gamma}$ of γ (see Example 3.1.39) belongs to $\Gamma_\gamma(TM)$.

We say that $c \in \Gamma_\gamma(TM)$ is *extendible* if there exists a vector field X defined on an open neighbourhood of $\mathrm{Im}(\gamma)$ such that $c = X \circ \gamma$. Then X is also called an *extension* of c. In general, not all elements of $\Gamma_\gamma(TM)$ can be extended: if $\gamma(t_1) = \gamma(t_2)$ ($t_1 \neq t_2 \in I$) but $\dot{\gamma}(t_1) \neq \dot{\gamma}(t_2)$, then $\dot{\gamma}$ is not extendible. However, if $\dot{\gamma}(t_0) \neq 0$ for some $t_0 \in I$, then there exists a vector field X on an open neighbourhood of $\gamma(t_0)$ such that $c(t) = X(\gamma(t))$ for t near t_0.

If $\widetilde{X} \in \Gamma(\pi)$ and $c \in \Gamma_\gamma(TM)$, then, obviously,

$$\widetilde{X} \circ c\colon I \to TM \to TM \times_M TM$$

is a section of π along c, i.e., $\widetilde{X} \circ c \in \Gamma_c(TM \times_M TM) = \Gamma_c(\pi)$. In particular, $\widehat{X} \circ c \in \Gamma_c(\pi)$ for all $X \in \mathfrak{X}(M)$ and $\widetilde{\delta} \circ c = (c, c) \in \Gamma_c(\pi)$.

Now let a covariant derivative operator D on π be given, and suppose that $c \colon I \to TM$ is an arbitrary smooth curve. By Proposition 6.1.49, there exists a unique $\mathbb{R}$-linear mapping $D^c \colon \Gamma_c(\pi) \to \Gamma_c(\pi)$ such that

$$D^c(fs) = f's + fD^c s \text{ for all } f \in C^\infty(I) \text{ and } s \in \Gamma_c(\pi)$$

and

$$D^c(\widetilde{X} \circ c) = (D\widetilde{X}) \circ \dot{c} \text{ for all } \widetilde{X} \in \Gamma(\pi).$$

If, in particular, $c \in \Gamma_\gamma(TM)$, then we write $D^\gamma(\widetilde{X} \circ c)$ instead of $D^c(\widetilde{X} \circ c)$, and we say that $D^\gamma(\widetilde{X} \circ c)$ *is the covariant derivative of* $\widetilde{X}$ *along* γ *with respect to the lift* $c \in \Gamma_\gamma(TM)$ *of* γ.

Definition 6.2.32. We use the same notation and hypotheses as above.

(a) A section $\widetilde{X} \in \Gamma(\pi)$ is said to be *D-parallel* or, if there is no risk of confusion, simply *parallel along* γ *with respect to a lift* c *of* γ if

$$D^\gamma(\widetilde{X} \circ c) = (D\widetilde{X}) \circ \dot{c} = 0. \tag{6.2.47}$$

If, in particular,

$$D^\gamma(\widetilde{X} \circ \dot{\gamma}) = (D\widetilde{X}) \circ \ddot{\gamma} = 0, \tag{6.2.48}$$

then X is called *D-parallel* (or in short *parallel*) along γ.

Similarly, a vector field $X \in \mathfrak{X}(M)$ is *D-parallel* (or *parallel*) *along* γ *with respect to a lift* c *of* γ if

$$D^\gamma(\widehat{X} \circ c) = (D\widehat{X}) \circ \dot{c} = 0, \tag{6.2.49}$$

and *D-parallel* (or *parallel*) along γ if

$$D^\gamma(\widehat{X} \circ \dot{\gamma}) = (D\widehat{X}) \circ \ddot{\gamma} = 0. \tag{6.2.50}$$

(b) The curve γ is said to be a *geodesic* of D if the canonical section of π is parallel along γ, i.e., if

$$D^\gamma(\widetilde{\delta} \circ \dot{\gamma}) \overset{(4.1.37)}{=} D^\gamma(\mathbf{j} \circ \ddot{\gamma}) = (D\widetilde{\delta}) \circ \ddot{\gamma} = 0. \tag{6.2.51}$$

Lemma 6.2.33. *We use the notation of Remark 6.2.3. A vector field X on M is D-parallel along a smooth curve $\gamma \colon I \to M$ with respect to a lift $c \in \Gamma_\gamma(TM)$ if, and only if, locally we have*

$$(X^i \circ \gamma)' + (\Gamma^i_{jk} \circ c)(\gamma^j)'(X^k \circ \gamma) + (C^i_{jk} \circ c)(c^j)'(X^k \circ \gamma) = 0, \tag{6.2.52}$$

where $X \underset{(\mathcal{U})}{=} X^j \frac{\partial}{\partial u^j}$; $\gamma^i = u^i \circ \gamma$, $c^i = y^i \circ c$ $(i \in J_n)$. *In particular, X is parallel along γ if, and only if, locally*

$$(X^i \circ \gamma)' + (\Gamma^i_{jk} \circ \dot{\gamma})(\gamma^j)'(X^k \circ \gamma) + (C^i_{jk} \circ \dot{\gamma})(\gamma^j)''(X^k \circ \gamma) = 0 \tag{6.2.53}$$

holds. (Summation convention in force.)

Proof. By (3.1.12),

$$\dot{c}(t) = (x^i \circ c)'(t) \left(\frac{\partial}{\partial x^i}\right)_{c(t)} + (y^i \circ c)'(t) \left(\frac{\partial}{\partial y^i}\right)_{c(t)}, \quad t \in I.$$

Since $c \in \Gamma_\gamma(TM)$, $x^i \circ c = u^i \circ \tau \circ c = u^i \circ \gamma = \gamma^i$, so we have

$$\dot{c}(t) = (\gamma^i)'(t) \left(\frac{\partial}{\partial x^i}\right)_{c(t)} + (c^i)'(t) \left(\frac{\partial}{\partial y^i}\right)_{c(t)}, \quad t \in I; \tag{6.2.54}$$

cf. (3.1.35). Now, at each $t \in I$,

$$D^\gamma(\widehat{X} \circ c)(t) \stackrel{(6.2.49)}{=} (D\widehat{X}) \circ \dot{c}(t) = D_{\dot{c}(t)}\widehat{X} = D_{\dot{c}(t)}\left((X^k \circ \tau)\widehat{\frac{\partial}{\partial u^k}}\right)$$

$$= \dot{c}(t)(X^k \circ \tau)\left(\widehat{\frac{\partial}{\partial u^k}}\right)_{c(t)} + (X^k \circ \tau)(c(t))D_{\dot{c}(t)}\widehat{\frac{\partial}{\partial u^k}}.$$

Here in the first term of the right-hand side

$$\dot{c}(t)(X^k \circ \tau) \stackrel{(6.2.54)}{=} (\gamma^i)'(t)\frac{\partial(X^k \circ \tau)}{\partial x^i}(c(t)) = (\gamma^i)'(t)\frac{\partial X^k}{\partial u^i}(\tau(c(t)))$$

$$= \left[(\gamma^i)'\left(\frac{\partial X^k}{\partial u^i} \circ \gamma\right)\right](t) = (X^k \circ \gamma)'(t).$$

In the second term the covariant derivative can be manipulated as follows:

$$D_{\dot{c}(t)}\widehat{\frac{\partial}{\partial u^k}} \stackrel{(6.2.54)}{=} (\gamma^j)'(t)\left(D_{\frac{\partial}{\partial x^j}}\widehat{\frac{\partial}{\partial u^k}}\right)(c(t)) + (c^j)'(t)\left(D_{\frac{\partial}{\partial y^j}}\widehat{\frac{\partial}{\partial u^k}}\right)(c(t))$$

$$\stackrel{(6.2.6)}{=} ((\gamma^j)'(t)\Gamma^i_{jk}(c(t)) + (c^j)'(t)C^i_{jk}(c(t)))\left(\widehat{\frac{\partial}{\partial u^i}}\right)_{c(t)}.$$

Thus

$$D^\gamma(\widehat{X} \circ c) = ((X^i \circ \gamma)' + (\Gamma^i_{jk} \circ c)(\gamma^j)'(X^k \circ \gamma)$$

$$+ (C^i_{jk} \circ c)(c^j)'(X^k \circ \gamma))\left(\widehat{\frac{\partial}{\partial u^i}} \circ c\right),$$

whence our first assertion. If, in particular, $c = \dot{\gamma}$, then $c^j = y^j \circ c = y^j \circ \dot{\gamma} = (\gamma^j)'$, and we obtain the second assertion. $\square$

Corollary 6.2.34. *We keep the notation of the preceding lemma. Let* $X, Y \in \mathfrak{X}(M)$, *and suppose that* $X \circ \gamma = Y \circ \gamma$. *Then*

$$D^\gamma(\widehat{X} \circ \dot{\gamma}) = D^\gamma(\widehat{Y} \circ \dot{\gamma}).$$

Proof. Since $X \circ \gamma = Y \circ \gamma$ implies that

$$X^i \circ \gamma = y^i \circ X \circ \gamma = y^i \circ Y \circ \gamma = Y^i \circ \gamma, \quad i \in J_n,$$

our assertion is an immediate consequence of the local expression of $D^\gamma(\widehat{X} \circ \dot\gamma)$ obtained above.   □

Lemma 6.2.35. *Let D be a covariant derivative on the Finsler bundle π. Suppose that $\gamma \colon I \to M$ is a regular curve whose canonical lift $\dot\gamma$ is extendible. Then γ is a geodesic of D if, and only if, for some (and hence all) extension X of $\dot\gamma$ we have*

$$D^\gamma(\widehat{X} \circ \dot\gamma) = 0. \tag{6.2.55}$$

Proof. The preceding corollary guarantees that (6.2.55) does not depend on the chosen extension of $\dot\gamma$. Since

$$\widehat{X} \circ \dot\gamma \overset{(4.1.4)}{=} (1_{TM}, X \circ \tau) \circ \dot\gamma = (\dot\gamma, X \circ \tau \circ \dot\gamma) = (\dot\gamma, X \circ \gamma) = (\dot\gamma, \dot\gamma) = \widetilde{\delta} \circ \dot\gamma,$$

relation (6.2.55) holds if, and only if, $D^\gamma(\widetilde{\delta} \circ \dot\gamma) = 0$, i.e., γ is a geodesic of D.   □

Corollary 6.2.36. *With the notation of Lemma 6.2.33, a regular curve $\gamma \colon I \to M$ is a geodesic of a covariant derivative D on π if, and only if, locally we have*

$$(\gamma^i)'' + (\Gamma^i_{jk} \circ \dot\gamma)(\gamma^j)'(\gamma^k)' + (C^i_{jk} \circ \dot\gamma)(\gamma^j)''(\gamma^k)' = 0. \tag{6.2.56}$$

If, in particular, D is strongly regular, then (6.2.56) reduces to

$$(\gamma^i)'' + (\Gamma^i_{jk} \circ \dot\gamma)(\gamma^j)'(\gamma^k)' = 0. \tag{6.2.57}$$

Proof. Since the question is local, we may suppose that over the chart domain $\mathcal{U}$ we have

$$\dot\gamma = X \circ \gamma, \quad \text{where } X = X^i \frac{\partial}{\partial u^i} \in \mathfrak{X}(\mathcal{U}).$$

Then

$$X^i \circ \gamma = y^i \circ X \circ \gamma = y^i \circ \dot\gamma = (\gamma^i)', \quad (X^i \circ \gamma)' = (\gamma^i)'',$$

so (6.2.53) leads to (6.2.56). If D is strongly regular, then $C^i_{jk} y^k = 0$ by Remark 6.2.8(b), so the third term in (6.2.56) vanishes.   □

6.2.6 Metric v-covariant Derivatives

Definition 6.2.37. Let g be a semi-Euclidean scalar product on the Finsler bundle π, i.e., a symmetric type $(0,2)$ Finsler tensor field

$$(\widetilde{X}, \widetilde{Y}) \in \Gamma(\pi) \times \Gamma(\pi) \mapsto g(\widetilde{X}, \widetilde{Y}) \in C^\infty(TM),$$

such that for each $u \in TM$, the symmetric bilinear form

$$g_u \colon (\{u\} \times T_{\tau(u)}M) \times (\{u\} \times T_{\tau(u)}M) \to \mathbb{R}$$

is non-degenerate.

(a) By the *Cartan tensor* of g we mean the canonical vertical differential

$$\mathcal{C}_\flat := \nabla^v g \in \mathcal{F}_3^0(TM) \tag{6.2.58}$$

of g. The *vector-valued Cartan tensor* (called also simply Cartan tensor) of g is the type $(1,2)$ Finsler tensor field $\mathcal{C}$ 'metrically equivalent' to $\mathcal{C}_\flat$, i.e., given by the condition

$$g(\mathcal{C}(\widetilde{X}, \widetilde{Y}), \widetilde{Z}) = \mathcal{C}_\flat(\widetilde{X}, \widetilde{Y}, \widetilde{Z}) \text{ for all } \widetilde{X}, \widetilde{Y}, \widetilde{Z} \in \Gamma(\pi). \tag{6.2.59}$$

(b) The *Christoffel – Cartan tensor* of g is the type $(0,3)$ Finsler tensor field $\overset{\circ}{\mathcal{C}}_\flat$ constructed from $\mathcal{C}_\flat$ by the 'Christoffel process':

$$\overset{\circ}{\mathcal{C}}_\flat(\widetilde{X}, \widetilde{Y}, \widetilde{Z}) := \mathcal{C}_\flat(\widetilde{X}, \widetilde{Y}, \widetilde{Z}) + \mathcal{C}_\flat(\widetilde{Y}, \widetilde{Z}, \widetilde{X}) - \mathcal{C}_\flat(\widetilde{Z}, \widetilde{X}, \widetilde{Y}). \tag{6.2.60}$$

The type $(1,2)$ Christoffel – Cartan tensor $\overset{\circ}{\mathcal{C}}$ is defined by

$$g(\overset{\circ}{\mathcal{C}}(\widetilde{X}, \widetilde{Y}), \widetilde{Z}) = \overset{\circ}{\mathcal{C}}_\flat(\widetilde{X}, \widetilde{Y}, \widetilde{Z}). \tag{6.2.61}$$

Remark 6.2.38. (a) The musical isomorphism $\sharp$ established in Remark 2.2.55 guarantees that both $\mathcal{C}$ and $\overset{\circ}{\mathcal{C}}$ exist and are uniquely determined.

(b) The Cartan tensor $\mathcal{C}_\flat$ is symmetric in its last two arguments, and the Christoffel – Cartan tensor $\overset{\circ}{\mathcal{C}} \in \mathcal{F}_2^1(TM)$ is symmetric:

$$\mathcal{C}_\flat(\widetilde{X}, \widetilde{Y}, \widetilde{Z}) = \mathcal{C}_\flat(\widetilde{X}, \widetilde{Z}, \widetilde{Y}), \quad \overset{\circ}{\mathcal{C}}_\flat(\widetilde{X}, \widetilde{Y}) = \overset{\circ}{\mathcal{C}}_\flat(\widetilde{Y}, \widetilde{X}) \tag{6.2.62}$$

for all $\widetilde{X}, \widetilde{Y}, \widetilde{Z} \in \Gamma(\pi)$. Indeed,

$$\mathcal{C}_\flat(\widetilde{X}, \widetilde{Y}, \widetilde{Z}) := (\nabla^v g)(\widetilde{X}, \widetilde{Y}, \widetilde{Z}) = (\nabla^v_{\widetilde{X}} g)(\widetilde{Y}, \widetilde{Z})$$

$$= (i\widetilde{X})g(\widetilde{Y}, \widetilde{Z}) - g(\nabla^v_{\widetilde{X}} \widetilde{Y}, \widetilde{Z}) - g(\widetilde{Y}, \nabla^v_{\widetilde{X}} \widetilde{Z}),$$

and by the symmetry of g we obtain the first relation of (6.2.62). This implies the symmetry of $\overset{\circ}{\mathcal{C}}$:

$$g(\overset{\circ}{\mathcal{C}}(\widetilde{X}, \widetilde{Y}), \widetilde{Z}) = \mathcal{C}_\flat(\widetilde{X}, \widetilde{Y}, \widetilde{Z}) + \mathcal{C}_\flat(\widetilde{Y}, \widetilde{Z}, \widetilde{X}) - \mathcal{C}_\flat(\widetilde{Z}, \widetilde{X}, \widetilde{Y})$$

$$= \mathcal{C}_\flat(\widetilde{X}, \widetilde{Z}, \widetilde{Y}) + \mathcal{C}_\flat(\widetilde{Y}, \widetilde{X}, \widetilde{Z}) - \mathcal{C}_\flat(\widetilde{Z}, \widetilde{Y}, \widetilde{X})$$

$$= \overset{\circ}{\mathcal{C}}_\flat(\widetilde{Y}, \widetilde{X}, \widetilde{Z}) = g(\overset{\circ}{\mathcal{C}}(\widetilde{Y}, \widetilde{X}), \widetilde{Z}),$$

whence, by the non-degeneracy of g, $\overset{\circ}{\mathcal{C}}(\widetilde{X}, \widetilde{Y}) = \overset{\circ}{\mathcal{C}}(\widetilde{Y}, \widetilde{X})$.

Definition 6.2.39. Let g be a semi-Euclidean scalar product in the Finsler bundle π. A *vertical* covariant derivative D^{v} on π is said to be *metric* if $D^{\mathsf{v}}g = 0$, i.e., see (6.2.22),

$$(\mathrm{i}\widetilde{X})g(\widetilde{Y},\widetilde{Z}) = g(D^{\mathsf{v}}_{\widetilde{X}}\widetilde{Y},\widetilde{Z}) + g(\widetilde{Y}, D^{\mathsf{v}}_{\widetilde{X}}\widetilde{Z}) \text{ for all } \widetilde{X},\widetilde{Y},\widetilde{Z} \in \Gamma(\pi). \quad (6.2.63)$$

Proposition 6.2.40. *Given a semi-Euclidean scalar product on the Finsler bundle π, there exists a unique vertical covariant derivative on π which is metric and whose torsion vanishes.*

Proof. Let g be the given semi-Euclidean scalar product on π.

Existence. Using the notation introduced above, define a mapping D^{v} by

$$D^{\mathsf{v}}_{\widetilde{X}}\widetilde{Y} := \nabla^{\mathsf{v}}_{\widetilde{X}}\widetilde{Y} + \frac{1}{2}\mathring{\mathcal{C}}(\widetilde{X},\widetilde{Y}) \quad (6.2.64)$$

(cf. (6.1.35) in the proof of the analogous Proposition 6.1.24). Then, obviously, D^{v} is a vertical covariant derivative on π. We show that D^{v} is metric and $T(D^{\mathsf{v}}) = 0$.

Let $\widetilde{X},\widetilde{Y},\widetilde{Z}$ be Finsler vector fields on TM. Then

$$(D^{\mathsf{v}}g)(\widetilde{X},\widetilde{Y},\widetilde{Z}) = (\mathrm{i}\widetilde{X})g(\widetilde{Y},\widetilde{Z}) - g(D^{\mathsf{v}}_{\widetilde{X}}\widetilde{Y},\widetilde{Z}) - g(\widetilde{Y}, D^{\mathsf{v}}_{\widetilde{X}}\widetilde{Z})$$

$$= (\mathrm{i}\widetilde{X})g(\widetilde{Y},\widetilde{Z}) - g(\nabla^{\mathsf{v}}_{\widetilde{X}}\widetilde{Y},\widetilde{Z}) - g(\widetilde{Y}, \nabla^{\mathsf{v}}_{\widetilde{X}}\widetilde{Z})$$

$$- \frac{1}{2}\left(\mathring{\mathcal{C}}_\flat(\widetilde{X},\widetilde{Y},\widetilde{Z}) + \mathring{\mathcal{C}}_\flat(\widetilde{X},\widetilde{Z},\widetilde{Y})\right) = \mathcal{C}_\flat(\widetilde{X},\widetilde{Y},\widetilde{Z})$$

$$- \frac{1}{2}\left(\mathcal{C}_\flat(\widetilde{X},\widetilde{Y},\widetilde{Z}) + \mathcal{C}_\flat(\widetilde{Y},\widetilde{Z},\widetilde{X}) - \mathcal{C}_\flat(\widetilde{Z},\widetilde{X},\widetilde{Y})\right)$$

$$- \frac{1}{2}\left(\mathcal{C}_\flat(\widetilde{X},\widetilde{Z},\widetilde{Y}) + \mathcal{C}_\flat(\widetilde{Z},\widetilde{Y},\widetilde{X}) - \mathcal{C}_\flat(\widetilde{Y},\widetilde{X},\widetilde{Z})\right) \overset{(6.2.62)}{=} 0,$$

$$T(D^{\mathsf{v}})(\widetilde{X},\widetilde{Y}) \overset{(6.2.18)}{:=} D^{\mathsf{v}}_{\widetilde{X}}\widetilde{Y} - D^{\mathsf{v}}_{\widetilde{Y}}\widetilde{X} - \mathrm{i}^{-1}[\mathrm{i}\widetilde{X},\mathrm{i}\widetilde{Y}] = T(\nabla^{\mathsf{v}})(\widetilde{X},\widetilde{Y})$$

$$+ \frac{1}{2}\mathring{\mathcal{C}}(\widetilde{X},\widetilde{Y}) - \frac{1}{2}\mathring{\mathcal{C}}(\widetilde{Y},\widetilde{X}) \overset{(6.2.34),(6.2.62)}{=} 0,$$

as desired.

Uniqueness. Suppose that $\widetilde{D}^{\mathsf{v}}$ is another metric vertical derivative on π with vanishing torsion. Define the *difference tensor* ψ^{v} of D^{v} and $\widetilde{D}^{\mathsf{v}}$ by

$$\psi^{\mathsf{v}}(\widetilde{X},\widetilde{Y}) := D^{\mathsf{v}}_{\widetilde{X}}\widetilde{Y} - \widetilde{D}^{\mathsf{v}}_{\widetilde{X}}\widetilde{Y}; \quad \widetilde{X},\widetilde{Y} \in \Gamma(\pi) \quad (6.2.65)$$

(cf. (6.1.9)). Since D^{v} and $\widetilde{D}^{\mathsf{v}}$ are both metric derivatives, we have, for any Finsler vector fields $\widetilde{X},\widetilde{Y},\widetilde{Z}$,

$$0 = (\widetilde{D}^{\mathsf{v}}_{\widetilde{X}}g)(\widetilde{Y},\widetilde{Z}) = (\mathrm{i}\widetilde{X})g(\widetilde{Y},\widetilde{Z}) - g(\widetilde{D}^{\mathsf{v}}_{\widetilde{X}}\widetilde{Y},\widetilde{Z}) - g(\widetilde{Y}, \widetilde{D}^{\mathsf{v}}_{\widetilde{X}}\widetilde{Z})$$

$$= g(D^{\mathsf{v}}_{\widetilde{X}}\widetilde{Y},\widetilde{Z}) + g(\widetilde{Y}, D^{\mathsf{v}}_{\widetilde{X}}\widetilde{Z}) - g(\widetilde{D}^{\mathsf{v}}_{\widetilde{X}}\widetilde{Y},\widetilde{Z}) - g(\widetilde{Y}, \widetilde{D}^{\mathsf{v}}_{\widetilde{X}}\widetilde{Z})$$

$$= g(\psi^{\mathsf{v}}(\widetilde{X},\widetilde{Y}),\widetilde{Z}) + g(\widetilde{Y}, \psi^{\mathsf{v}}(\widetilde{X},\widetilde{Z})).$$

On the other hand, by the vanishing of $T(D^{\mathsf{v}})$ and $T(\widetilde{D}^{\mathsf{v}})$,

$$0 = D^{\mathsf{v}}_{\widetilde{X}}\widetilde{Y} - D^{\mathsf{v}}_{\widetilde{Y}}\widetilde{X} - \mathbf{i}^{-1}[\mathbf{i}\widetilde{X}, \mathbf{i}\widetilde{Y}] \overset{(6.2.65)}{=} \widetilde{D}^{\mathsf{v}}_{\widetilde{X}}\widetilde{Y} - \widetilde{D}^{\mathsf{v}}_{\widetilde{Y}}\widetilde{X} - \mathbf{i}^{-1}[\mathbf{i}\widetilde{X}, \mathbf{i}\widetilde{Y}]$$
$$+ \psi^{\mathsf{v}}(\widetilde{X}, \widetilde{Y}) - \psi^{\mathsf{v}}(\widetilde{Y}, \widetilde{X}) = \psi^{\mathsf{v}}(\widetilde{X}, \widetilde{Y}) - \psi^{\mathsf{v}}(\widetilde{Y}, \widetilde{X}).$$

Summing up, we find the following:

(i) For any fixed $\widetilde{X} \in \Gamma(\pi)$, the endomorphism $\psi^{\mathsf{v}}_{\widetilde{X}}$ defined by $\psi^{\mathsf{v}}_{\widetilde{X}}(\widetilde{Y}) := \psi^{\mathsf{v}}(\widetilde{X}, \widetilde{Y})$ is skew-symmetric with respect to the scalar product g, i.e.,

$$g(\psi^{\mathsf{v}}_{\widetilde{X}}(\widetilde{Y}), \widetilde{Z}) = g(\widetilde{Y}, \psi^{\mathsf{v}}_{\widetilde{X}}(\widetilde{Z})); \quad \widetilde{Y}, \widetilde{Z} \in \Gamma(\pi).$$

(ii) $\psi^{\mathsf{v}}_{\widetilde{X}}(\widetilde{Y}) - \psi^{\mathsf{v}}_{\widetilde{Y}}(\widetilde{X}) = 0;\ \widetilde{X}, \widetilde{Y} \in \Gamma(\pi).$

Now let

$$\mathsf{R} := C^{\infty}(TM),\ V := \Gamma(\pi) \text{ and } \beta := 0$$

in Proposition 1.4.11. Then $\omega := \psi^{\mathsf{v}}$ satisfies conditions (i) and (ii) of the quoted proposition, therefore, as we have seen, $\psi^{\mathsf{v}} = 0$. Hence $\widetilde{D}^{\mathsf{v}} = D^{\mathsf{v}}$, as was to be shown. $\qquad\square$

Chapter 7

Theory of Ehresmann Connections

7.1 Horizontal Subbundles

Definition 7.1.1. Let M be a manifold. Consider the tangent bundle $\tau \colon TM \to M$ of M and the tangent bundle $\tau_{TM} \colon TTM \to TM$ of TM. A subbundle

$$\tau_{TM}^{\mathsf{h}} \colon HTM \to TM$$

of τ_{TM} (or of TTM) is said to be *horizontal* if

$$\tau_{TM} = \tau_{TM}^{\mathsf{h}} \oplus \tau_{TM}^{\mathsf{v}} \quad (\text{or } TTM = HTM \oplus VTM) \tag{7.1.1}$$

(Whitney sum, see Example 2.2.38). The fibres $H_v TM$ ($v \in TM$) of a horizontal subbundle are called the *horizontal subspaces* (with respect to the choice of HTM). A smooth vector field ξ on TM is said to be *horizontal* if $\xi(v) \in H_v TM$ for all $v \in TM$, i.e., if ξ is a smooth section of τ_{TM}^{h}.

Proposition 7.1.2. *The double tangent bundle* $\tau_{TM} \colon TTM \to TM$ *admits a horizontal subbundle.*

Idea of proof. Choose a positive definite scalar product $g \in \mathcal{T}_2^0(TM)$. (The existence of g is guaranteed by Lemma 2.2.56.) At each $v \in TM$ take the orthogonal complement $(V_v TM)^\perp$ of $V_v TM$ in $T_v TM$ with respect to the scalar product $g_v \in T_2^0(T_v TM)$. If $H_v TM := (V_v TM)^\perp$ and $HTM :=$ $\bigcup_{v \in TM} H_v TM$, then HTM is a horizontal subbundle of TTM. $\triangle$

Remark 7.1.3. Let a horizontal subbundle HTM of TTM be specified. (We emphasize that the vertical bundle VTM, in general, does not have a distinguished complement of 'horizontal vectors' in TTM.)

(a) For any $v \in TM$ we have

$$T_v TM = H_v TM \oplus V_v TM,$$

which implies

$$\dim H_v TM = \dim T_v TM - \dim V_v TM = 2n - n = n := \dim M,$$

therefore the horizontal subbundle HTM of TTM is of rank n. (As a manifold, the dimension of HTM is $2n + n = 3n$.) Thus, taking into account that $\mathrm{Ker}(\tau_*)_v = V_v TM$, we deduce that the restriction of the linear mapping

$$(\tau_*)_v : T_v TM = H_v TM \oplus V_v TM \to T_{\tau(v)} M$$

to $H_v TM$ is a linear isomorphism onto $T_{\tau(v)} M$.

Since $\mathbf{j} = (\tau_{TM}, \tau_*)$, it follows that the restriction of $\mathbf{j}$ to HTM (denoted by the same symbol) is a strong bundle isomorphism of HTM onto the Finsler bundle $TM \times_M TM$:

$$
\begin{array}{ccc}
HTM & \xrightarrow{\ \mathbf{j}\ } & TM \times_M TM \\
{\scriptstyle \tau^h_{TM}}\downarrow & & \downarrow{\scriptstyle \pi} \\
TM & \xrightarrow[\ 1_{TM}\ ]{} & TM
\end{array}
.
$$

(b) The horizontal vector fields on TM form a submodule of the $C^\infty(TM)$-module $\mathfrak{X}(TM)$, denoted by $\mathfrak{X}^h(TM)$. However, in general, $\mathfrak{X}^h(TM)$ is not a subalgebra of the Lie algebra $\mathfrak{X}(TM)$.

Every vector field ξ on TM can be uniquely decomposed as

$$\xi = \xi_H + \xi_V; \quad \xi_H \in \mathfrak{X}^h(TM),\ \xi_V \in \mathfrak{X}^v(TM).$$

Thus the Whitney sum decomposition (7.1.1) induces a direct decomposition

$$\mathfrak{X}(TM) = \mathfrak{X}^h(TM) \oplus \mathfrak{X}^v(TM) \tag{7.1.2}$$

of the $C^\infty(TM)$-module $\mathfrak{X}(TM)$ into two complementary submodules (where one of them is canonical). Define the transformations $\mathbf{h}$ and $\mathbf{v}$ of $\mathfrak{X}(TM)$ by

$$\mathbf{h}\xi := \xi_H,\ \mathbf{v}\xi := \xi_V \ \text{if}\ \xi = \xi_H + \xi_V,\ \xi_H \in \mathfrak{X}^h(TM),\ \xi_V \in \mathfrak{X}^v(TM).$$

Then $\mathbf{h}$ and $\mathbf{v}$ are complementary projection operators of $\mathfrak{X}(TM)$, that is, $\mathbf{h}, \mathbf{v} \in \mathrm{End}_{C^\infty(TM)}(\mathfrak{X}(TM))$ satisfying

$$\mathbf{h}^2 = \mathbf{h},\ \mathbf{v} = 1_{\mathfrak{X}(TM)} - \mathbf{h},\ \mathbf{h} \circ \mathbf{v} = \mathbf{v} \circ \mathbf{h} = 0. \tag{7.1.3}$$

By Proposition 2.2.31, $\mathbf{h}$ and $\mathbf{v}$ have pointwise interpretations, i.e., they may be regarded also as strong bundle endomorphisms of TTM. We say that $\mathbf{h}$ and $\mathbf{v}$ are the *horizontal* and *vertical projections* determined by (or associated with) the horizontal subbundle HTM.

Lemma and Definition 7.1.4. *Let a horizontal subbundle* HTM *of* TTM *be specified. For any vector field* X *on* M *there exists a unique horizontal vector field* X^h *on* TM *such that* $X^h \underset{\tau}{\sim} X$, *namely*

$$X^h = \mathbf{h}X^c, \qquad (7.1.4)$$

where $\mathbf{h}$ *is the horizontal projection determined by* HTM. *We say that* X^h *is the* horizontal lift *of* X *with respect to the chosen horizontal subbundle.*

Proof. The vector field $\mathbf{h}X^c$ is evidently horizontal. Since, by Lemma 4.1.45(ii), $\mathbf{h}X^c = X^c - \mathbf{v}X^c$, where $X^c \underset{\tau}{\sim} X$ and, furthermore, $\mathbf{v}X^c \underset{\tau}{\sim} 0$ because $\mathbf{v}X^c$ is vertical, Lemma 3.1.51(i) implies that $X^h = \mathbf{h}X^c \underset{\tau}{\sim} X$. To show the uniqueness of the horizontal lift, suppose that $\eta_1 \in \mathfrak{X}^h(TM)$ and $\eta_2 \in \mathfrak{X}^h(TM)$ are both τ-related to X. Then $\eta_1 - \eta_2 \underset{\tau}{\sim} 0$, whence $\eta_1 - \eta_2 \in \mathfrak{X}^v(TM)$. On the other hand, $\eta_1 - \eta_2 \in \mathfrak{X}^h(TM)$. Hence, by (7.1.2), $\eta_1 - \eta_2 = 0$, as was to be shown. $\qquad \square$

Lemma 7.1.5. *Let a horizontal subbundle* HTM *of* TTM *be given, and let* $\mathbf{h}$ *and* $\mathbf{v}$ *be the horizontal and vertical projections associated with* HTM.

(i) *For any vector field* $X \in \mathfrak{X}(M)$ *and function* $f \in C^\infty(M)$,

$$(fX)^h = (f \circ \tau)X^h = f^v X^h, \qquad (7.1.5)$$
$$X^h(f \circ \tau) = X^h f^v = (Xf)^v. \qquad (7.1.6)$$

(ii) *The mapping*

$$\mathfrak{X}(M) \times \mathfrak{X}(M) \to \mathfrak{X}(TM), \ (X, Y) \mapsto [X^h, Y^v]$$

has the following properties:

(a) $[X^h, Y^v]$ *is vertical,*
(b) *it is* $\mathbb{R}$-*bilinear,*
(c) $[(fX)^h, Y^v] = f^v[X^h, Y^v]$,
(d) $[X^h, (fY)^v] = (Xf)^v Y^v + f^v[X^h, Y^v]$.

(iii) *For any vector fields* X, Y *on* M,

$$\mathbf{h}[X^h, Y^h] = [X, Y]^h. \qquad (7.1.7)$$

Proof. (i) Since $X^h \underset{\tau}{\sim} X$, by Lemma 3.1.51(ii) we also have $(f \circ \tau)X^h \underset{\tau}{\sim} fX$. On the other hand, $(fX)^h \underset{\tau}{\sim} fX$, therefore the uniqueness of the horizontal lift implies that $(f \circ \tau)X^h = (fX)^h$. By Lemma 3.1.49, relation $X^h \underset{\tau}{\sim} X$ is an immediate consequence of (7.1.6).

(ii) (a) Since $X^h \underset{\tau}{\sim} X$, $Y^v \underset{\tau}{\sim} 0$, from Lemma 3.1.51(iii) we infer that $[X^h, Y^v] \underset{\tau}{\sim} [X, 0] = 0$, and $[X^h, Y^v]$ is therefore vertical.

(b) This is clear, because the mappings $X \mapsto X^h$ and $Y \mapsto Y^v$ are both $\mathbb{R}$-linear. (c) and (d) can be obtained by direct calculation:

$$[(fX)^h, Y^v] \overset{(7.1.5)}{=} [f^v X^h, Y^v]$$

$$\overset{(3.1.39)}{=} f^v[X^h, Y^v] - (Y^v f^v) X^h \overset{\text{Lemma 4.1.28}}{=} f^v[X^h, Y^v],$$

$$[X^h, (fY)^v] \overset{(4.1.45)}{=} [X^h, f^v Y^v]$$

$$\overset{(3.1.39)}{=} f^v[X^h, Y^v] + (X^h f^v) Y^v \overset{(7.1.6)}{=} (Xf)^v Y^v + f^v[X^h, Y^v].$$

(iii) We apply again the related vector field lemma: relations $X^h \underset{\tau}{\sim} X$ and $Y^h \underset{\tau}{\sim} Y$ imply that $[X^h, Y^h] \underset{\tau}{\sim} [X, Y]$. Since $[X^h, Y^h] = \mathbf{h}[X^h, Y^h] + \mathbf{v}[X^h, Y^h]$ and $\mathbf{v}[X^h, Y^h] \underset{\tau}{\sim} 0$, then it follows that $\mathbf{h}[X^h, Y^h] \underset{\tau}{\sim} [X, Y]$. On the other hand, $[X, Y]^h \underset{\tau}{\sim} [X, Y]$, so we conclude as above that $\mathbf{h}[X^h, Y^h] = [X, Y]^h$. □

Remark 7.1.6. With the notation as above, let a horizontal subbundle HTM of TTM be specified.

(a) Let us call a mapping

$$\widehat{A} \colon \underbrace{\mathfrak{X}(M) \times \cdots \times \mathfrak{X}(M)}_{k} \to \mathfrak{X}^v(TM),$$

where k is a positive integer, $C^\infty(M)$-multilinear if it is additive in each arguments and

$$\widehat{A}(X_1, \ldots, fX_i, \ldots, X_k) = f^v \widehat{A}(X_1, \ldots, X_k)$$

for all $X_1, \ldots, X_k \in \mathfrak{X}(M)$, $f \in C^\infty(M)$, $i \in J_k$.

Now define two mappings $\widehat{R}$ and $\widehat{T}$ from $\mathfrak{X}(M) \times \mathfrak{X}(M)$ to $\mathfrak{X}^v(TM)$ by

$$\widehat{R}(X, Y) := [X^h, Y^h] - [X, Y]^h \tag{7.1.8}$$

and

$$\widehat{T}(X, Y) := [X^h, Y^v] - [Y^h, X^v] - [X, Y]^v, \tag{7.1.9}$$

respectively. Then $\widehat{R}$ and $\widehat{T}$ are both skew-symmetric $C^\infty(M)$-bilinear mappings. Indeed, the skew-symmetry is obvious. We show that $\widehat{R}$ is $C^\infty(M)$-linear in its first argument, the same property of $\widehat{T}$ can be verified similarly. The additivity can be seen by inspection. Furthermore, for any smooth function f on M, we have

$$\widehat{R}(fX,Y) := [(fX)^{\mathsf{h}}, Y^{\mathsf{h}}] - [fX,Y]^{\mathsf{h}} \stackrel{(7.1.5)}{=} [f^{\mathsf{v}}X^{\mathsf{h}}, Y^{\mathsf{h}}]$$

$$- (f[X,Y])^{\mathsf{h}} + ((Yf)X)^{\mathsf{h}} = f^{\mathsf{v}}[X^{\mathsf{h}}, Y^{\mathsf{h}}] - f^{\mathsf{v}}[X,Y]^{\mathsf{h}}$$

$$- (Y^{\mathsf{h}}f^{\mathsf{v}})X^{\mathsf{h}} + (Yf)^{\mathsf{v}}X^{\mathsf{h}} \stackrel{(7.1.6)}{=} f^{\mathsf{v}}[X^{\mathsf{h}}, Y^{\mathsf{h}}] - f^{\mathsf{v}}[X,Y]^{\mathsf{h}}$$

$$= f^{\mathsf{v}}\widehat{R}(X,Y),$$

so the functions can be factored out in the desired form. By the skew-symmetry, the $C^{\infty}(M)$-linearity in the second slot holds automatically.

(b) Properties (a)–(d) in part (ii) of the lemma resemble the axioms of a covariant derivative operator, while the mappings $\widehat{R}$ and $\widehat{T}$ are strongly analogous to the curvature and the torsion of a covariant derivative. These analogies, among others, prepare us for the introduction of an important covariant derivative on a Finsler bundle (in the presence of the horizontal subbundle HTM). At this point, we make only some preliminary observations, a systematic treatment will be presented in the next sections (in a somewhat different setting).

Define a mapping $\nabla \colon \mathfrak{X}(TM) \times \Gamma(\pi) \to \Gamma(\pi)$, $(\xi, \widetilde{Y}) \mapsto \nabla_{\xi}\widetilde{Y}$ by

$$\mathbf{i}(\nabla_{\xi}\widetilde{Y}) := \mathbf{v}[\mathbf{h}\xi, \mathbf{J}\eta] + \mathbf{J}[\mathbf{v}\xi, \eta], \qquad (7.1.10)$$

where $\eta \in \mathfrak{X}(TM)$ is such that $\mathbf{j}\eta = \widetilde{Y}$. It is easy to see that ∇ is well-defined (cf. also Lemma and Definition 6.2.16). The vector field $\xi \in \mathfrak{X}(TM)$ can be written in the form

$$\xi = \mathbf{h}\xi + \mathbf{v}\xi = \mathbf{h}\xi + \mathbf{i}\widetilde{X}, \quad \widetilde{X} \in \Gamma(\pi).$$

If ξ is vertical and hence is of the form $\xi = \mathbf{i}\widetilde{X}$, then relation (7.1.10) yields

$$\mathbf{i}(\nabla_{\mathbf{i}\widetilde{X}}\widetilde{Y}) = \mathbf{J}[\mathbf{i}\widetilde{X}, \eta]$$

which is equivalent to

$$\nabla_{\mathbf{i}\widetilde{X}}\widetilde{Y} = \mathbf{j}[\mathbf{i}\widetilde{X}, \eta], \quad \mathbf{j}\eta = \widetilde{Y}.$$

Thus, by the prescription (6.2.19), the mapping ∇ induces the canonical vertical derivative ∇^{v} as a vertical covariant derivative.

Next, suppose that ξ is a horizontal lift, namely $\xi = \mathbf{h}X^{\mathsf{c}} = X^{\mathsf{h}}$ where $X \in \mathfrak{X}(M)$, and let $\widetilde{Y} := \widehat{Y}$, $Y \in \mathfrak{X}(M)$. If $\eta := Y^{\mathsf{c}}$ then $\mathbf{J}\eta \stackrel{(4.1.87)}{=} Y^{\mathsf{v}}$, and (7.1.10) reduces to

$$\mathbf{i}(\nabla_{X^{\mathsf{h}}}\widehat{Y}) = \mathbf{v}[X^{\mathsf{h}}, Y^{\mathsf{v}}] = [X^{\mathsf{h}}, Y^{\mathsf{v}}], \qquad (7.1.11)$$

so we arrive at the covariant-derivative-like mapping introduced in Lemma 7.1.5.

These suggest that the mapping ∇ is in fact a covariant derivative on the Finsler bundle π. We show that this is true.

Proposition and Definition 7.1.7. *Let a horizontal subbundle HTM of TTM be specified. The mapping*

$$\nabla \colon \mathfrak{X}(TM) \times \Gamma(\pi) \to \Gamma(\pi)$$

given by (7.1.10) *is a covariant derivative on the Finsler bundle $\pi \colon TM \times_M TM \to TM$, called the* Berwald derivative *determined by the horizontal subbundle HTM.*

Proof. The $\mathbb{R}$-bilinearity of ∇ is clear. We show that ∇ is tensorial in its first argument and satisfies the derivation rule in its second argument. Choose a smooth function F on TM. Then

$$\mathbf{i}(\nabla_{F\xi}\widetilde{Y}) := \mathbf{v}[\mathbf{h}(F\xi), \mathbf{J}\eta] + \mathbf{J}[\mathbf{v}(F\xi), \eta] = \mathbf{v}[F(\mathbf{h}\xi), \mathbf{J}\eta] + \mathbf{J}[F(\mathbf{v}\xi), \eta]$$

$$= \mathbf{i}(F\nabla_{\xi}\widetilde{Y}) - \mathbf{v}((\mathbf{J}\eta)(F)\mathbf{h}\xi) - \mathbf{J}(\eta(F)\mathbf{v}\xi) = \mathbf{i}(F\nabla_{\xi}\widetilde{Y}),$$

because $\mathbf{v}\circ\mathbf{h} = 0$ and $\mathbf{J}\circ\mathbf{v} = 0$. Thus $\nabla_{F\xi}\widetilde{Y} = F\nabla_{\xi}\widetilde{Y}$, as desired. Similarly,

$$\mathbf{i}(\nabla_{\xi}F\widetilde{Y}) := \mathbf{v}[\mathbf{h}\xi, \mathbf{J}(F\eta)] + \mathbf{J}[\mathbf{v}\xi, F\eta] = \mathbf{i}(F\nabla_{\xi}\widetilde{Y}) + \mathbf{v}(\mathbf{h}\xi(F)\mathbf{J}\eta)$$

$$+ \mathbf{J}(\mathbf{v}\xi(F)\eta) = \mathbf{i}(F\nabla_{\xi}\widetilde{Y}) + (\mathbf{h}\xi(F) + \mathbf{v}\xi(F))\mathbf{J}\eta$$

$$= \mathbf{i}((\xi F)\mathbf{j}\eta + F\nabla_{\xi}\widetilde{Y}) = \mathbf{i}((\xi F)\widetilde{Y} + F\nabla_{\xi}\widetilde{Y}),$$

taking into account that $\mathbf{v}\circ\mathbf{J} = \mathbf{J}$. So we have

$$\nabla_{\xi}F\widetilde{Y} = (\xi F)\widetilde{Y} + F\nabla_{\xi}\widetilde{Y},$$

which concludes the proof. $\qquad\qquad\square$

Remark 7.1.8. Lemma 7.1.5 suggests another important idea. Let us say that a horizontal subbundle HTM of TTM is *induced by a covariant derivative D on M* if

$$[X^{\mathbf{h}}, Y^{\mathbf{v}}] = (D_X Y)^{\mathbf{v}} \quad \text{for all } X, Y \in \mathfrak{X}(M). \tag{7.1.12}$$

In terms of the Berwald derivative determined by HTM, this condition can be written equivalently in the form

$$\mathbf{i}(\nabla_{X^{\mathbf{h}}}\widehat{Y}) = (D_X Y)^{\mathbf{v}} \quad \text{for all } X, Y \in \mathfrak{X}(M). \tag{7.1.13}$$

We also say then that the Berwald derivative ∇ is *horizontally basic*. The curvature R^D and the torsion T^D of D are related to the mappings $\widehat{R}$ and $\widehat{T}$ defined in (7.1.8) and (7.1.9) as follows: for all $X, Y, Z \in \mathfrak{X}(M)$,

$$[\widehat{R}(X, Y), Z^{\mathbf{v}}] = (R^D(X, Y)Z)^{\mathbf{v}}, \tag{7.1.14}$$

$$\widehat{T}(X, Y) = (T^D(X, Y))^{\mathbf{v}}. \tag{7.1.15}$$

The proof is an immediate calculation. Using the Jacobi identity, we find

$$[\hat{R}(X,Y), Z^v] \overset{(7.1.8)}{=} [[X^h, Y^h], Z^v] - [[X,Y]^h, Z^v] = -[[Y^h, Z^v], X^h]$$
$$- [[Z^v, X^h], Y^h] - [[X,Y]^h, Z^v] = [X^h, [Y^h, Z^v]]$$
$$- [Y^h, [X^h, Z^v]] - [[X,Y]^h, Z^v] \overset{(7.1.12)}{=} (D_X D_Y Z)^v$$
$$- (D_Y D_X Z)^v - (D_{[X,Y]} Z)^v = (R^D(X,Y)Z)^v,$$

which proves (7.1.14). The second relation can also be easily verified.

7.2 Ehresmann Connections and Associated Objects

Definition 7.2.1. Let M be a manifold. Consider its tangent bundle $\tau\colon TM \to M$, the double tangent bundle $\tau_{TM}\colon TTM \to TM$, and the Finsler bundle $\pi\colon TM \times_M TM \to TM$. By an *Ehresmann connection* in TM we mean a mapping

$$\mathcal{H}\colon TM \times_M TM \to TTM, \ (u,v) \mapsto \mathcal{H}(u,v),$$

which satisfies the following conditions:

(C_1) $\mathcal{H}$ is a fibre-preserving and fibrewise linear mapping between the vector bundles π and τ_{TM} which induces the identity in their common base manifold, i.e., for each $u \in TM$, $v, v_1, v_2 \in T_{\tau(u)}M$ and $\lambda_1, \lambda_2 \in \mathbb{R}$ we have

$$\mathcal{H}(u,v) \in T_u TM \tag{7.2.1}$$

and

$$\mathcal{H}(u, \lambda_1 v_1 + \lambda_2 v_2) = \lambda_1 \mathcal{H}(u, v_1) + \lambda_2 \mathcal{H}(u, v_2). \tag{7.2.2}$$

(C_2) $\mathcal{H}$ is a right inverse for the canonical mapping $\mathbf{j} = (\tau_{TM}, \tau_*)$, i.e., for all $(u,v) \in TM \times_M TM$ we have

$$\mathbf{j}_u(\mathcal{H}(u,v)) = (u,v), \ \text{briefly } \mathbf{j} \circ \mathcal{H} = 1_{TM \times_M TM}. \tag{7.2.3}$$

(C_3) $\mathcal{H}$ is smooth over $\mathring{T}M \times_M TM$, where $\mathring{T}M = TM \setminus o(M)$, and o is the zero vector field on M.

(C_4) For each $p \in M$ and $v \in T_pM$,

$$\mathcal{H}(o(p), v) = (o_*)_p(v). \tag{7.2.4}$$

Remark 7.2.2. Keep the notation and hypotheses as in Definition 7.2.1.

(a) For a fixed $u \in TM$, denote by $\mathcal{H}_u$ the restriction of $\mathcal{H}$ to the vector space $\{u\} \times T_{\tau(u)}M$. Identifying $\{u\} \times T_{\tau(u)}M$ with $T_{\tau(u)}M$, the mapping

$$\mathcal{H}_u \colon T_{\tau(u)}M \to T_u TM, \quad v \mapsto \mathcal{H}_u(v) := \mathcal{H}(u,v) \tag{7.2.5}$$

is linear by condition (C_1). In view of conditions (C_1) and (C_3), the diagram

$$
\begin{array}{ccc}
TM \times_M TM & \xrightarrow{\;\mathcal{H}\;} & TTM \\
\pi \downarrow & & \downarrow \tau_{TM} \\
TM & \xrightarrow[\;1_{TM}\;]{} & TM
\end{array}
\tag{7.2.6}
$$

is commutative, and, apart from the lack of differentiability on $o(M) \times_M TM$, displays a strong bundle map between π and τ_{TM}.

(b) Condition (C_4) is compatible with conditions (C_1) and (C_2). Indeed,

$$(o_*)_p \colon T_pM \to T_{o(p)}TM$$

is a linear mapping for all $p \in M$. Furthermore, if we define $\mathcal{H}_{o(p)}$ by (C_4), then

$$\mathbf{j}_{o(p)}(\mathcal{H}(o(p),v)) = \mathbf{j}_{o(p)}((o_*)_p(v)) = \big(o(p), (\tau_*)_{o(p)}((o_*)_p(v))\big)$$

$$= (o(p), ((\tau \circ o)_*)_p(v)) = (o(p), v),$$

hence the property

$$\mathbf{j} \circ \mathcal{H} \restriction o(M) \times_M TM = 1_{o(M) \times_M TM}$$

required by (C_2) is indeed satisfied.

(c) Since

$$\mathbf{j}_u(\mathcal{H}(u,v)) = \big(u, (\tau_*)_u(\mathcal{H}(u,v))\big), \quad (u,v) \in TM \times_M TM,$$

condition (C_2) can be written equivalently in the form

$$(\tau_*)_u(\mathcal{H}(u,v)) = v, \quad (u,v) \in TM \times_M TM. \tag{7.2.7}$$

From this it follows that for any fixed $v \in TM$,

$$\mathcal{H}(u,v) \in (\tau_*)^{-1}(v) \text{ if } u \in T_{\tau(v)}M. \tag{7.2.8}$$

Now let $\pi_2 := \mathrm{pr}_2 \restriction TM \times_M TM$. Then $\pi_2 \colon TM \times_M TM \to TM$ also carries a natural vector bundle structure with fibres

$$(\pi_2)^{-1}(v) = T_{\tau(v)}M \times \{v\}, \quad v \in TM,$$

endowed with the linear operations

$$\lambda_1(u_1,v) + \lambda_2(u_2,v) := (\lambda_1 u_1 + \lambda_2 u_2, v); \quad u_1, u_2 \in T_{\tau(v)}M; \; \lambda_1, \lambda_2 \in \mathbb{R}$$

(cf. Remark 4.1.1(a)). In terms of the mapping π_2, (7.2.7) takes the form

$$\tau_* \circ \mathcal{H} = \pi_2. \tag{7.2.9}$$

Thus condition (C_2) can also be expressed as follows:

$\mathcal{H}$ is a fibre-preserving mapping between the vector bundles $\pi_2 \colon TM \times_M TM \to TM$ and $\tau_* \colon TTM \to TM$ which induces the identity in the common base manifold TM, i.e., the diagram

$$
\begin{array}{ccc}
TM \times_M TM & \xrightarrow{\ \mathcal{H}\ } & TTM \\
\pi_2 \downarrow & & \downarrow \tau_* \\
TM & \xrightarrow[\ 1_{TM}\ ]{} & TM
\end{array}
\qquad (7.2.10)
$$

is commutative.

Lemma 7.2.3. *If* $\mathcal{H} \colon TM \times_M TM \to TTM$ *is an Ehresmann connection in* TM, *then*

$$
T_u TM = \mathrm{Im}(\mathcal{H}_u) \oplus V_u TM, \quad u \in TM. \qquad (7.2.11)
$$

Proof. Suppose that M is an n-dimensional manifold. Then TM is $2n$-dimensional, whence $\dim T_u TM = 2n$, $\dim V_u TM = n$. By condition (C_2) the linear mapping $\mathcal{H}_u$ has a left inverse. Hence it is injective, therefore $\mathrm{Im}(\mathcal{H}_u)$ is an n-dimensional subspace of $T_u TM$. If $w \in \mathrm{Im}(\mathcal{H}_u) \cap V_u TM$, then, on the one hand, $w = \mathcal{H}_u(v)$ for a unique $v \in T_{\tau(u)} M$. On the other hand, $\tau_*(w) = 0$, so we find $0 = \tau_*(w) = \tau_*(\mathcal{H}_u(v)) \overset{(7.2.7)}{=} v$. Then w is also zero by the linearity of $\mathcal{H}_u$, which shows that $\mathrm{Im}(\mathcal{H}_u) \cap V_u TM = \{0\}$. Since both $\mathrm{Im}(\mathcal{H}_u)$ and $V_u TM$ are n-dimensional subspaces in $T_u TM$, this implies (7.2.11). $\qquad \square$

Remark 7.2.4. Let us use the same notation and hypotheses as above.

(a) We say that $H_u TM := \mathrm{Im}(\mathcal{H}_u)$ is the *horizontal subspace* of $T_u TM$ with respect to the Ehresmann connection $\mathcal{H}$. (If the Ehresmann connection is clear from the context, then we speak simply of the horizontal subspace at u.) An individual element $w \in T_u TM$ is called *horizontal* if $w \in H_u TM$.

(b) Now consider the disjoint union

$$
HTM := \mathrm{Im}(\mathcal{H}) := \bigcup_{u \in TM} H_u TM \subset TTM
$$

of all horizontal subspaces. Then, by Lemma 7.2.3, we can write

$$
TTM = HTM \oplus VTM
$$

(cf. also Example 2.2.38). On the analogy of the situation described in Definition 7.1.1, we say that HTM is the *horizontal subbundle of* TTM *determined by the Ehresmann connection* $\mathcal{H}$, although, by the lack of differentiability of $\mathcal{H}$ over $o(M) \times_M TM$, in general, HTM *is not a vector subbundle of* TTM in the strict sense of the word. We denote the natural projection of HTM onto TM by τ_{TM}^{h}, as above. Also, by a slight abuse of language, we shall use the terminology 'horizontal vector fields', 'horizontal projection', etc., introduced in 7.1, in the more general context of Ehresmann connections.

(i) We say that a mapping $\xi\colon TM \to TTM$ is a *horizontal vector field* with respect to the Ehresmann connection $\mathcal{H}$, or $\mathcal{H}$-*horizontal*, if it is a section of $\tau_{TM}^{\mathrm{h}}\colon HTM \to TM$, i.e.,

$$\xi(v) \in H_v TM = \mathrm{Im}(\mathcal{H}_v) \text{ for all } v \in TM,$$

and it is smooth over $\mathring{T}M = TM \setminus o(M)$. As in 7.1, we denote by $\mathfrak{X}^{\mathrm{h}}(TM)$ the set of $\mathcal{H}$-horizontal vector fields on TM (although the notation $\mathfrak{X}^{\mathcal{H}}(TM)$ would be more precise). We equip $\mathfrak{X}^{\mathrm{h}}(TM)$ with the usual $C^\infty(TM)$-module operations. Then we can also write

$$\mathfrak{X}(TM) = \mathfrak{X}^{\mathrm{h}}(TM) \oplus \mathfrak{X}^{\mathrm{v}}(TM),$$

and we can use (with some caution) everything that we have explained so far in 7.1.

(ii) We can define the *horizontal projection* determined by $\mathrm{Im}(\mathcal{H})$ immediately in terms of $\mathcal{H}$ as the composite mapping

$$\mathbf{h} := \mathcal{H} \circ \mathbf{j}\colon TTM \to TM \times_M TM \to TTM. \qquad (7.2.12)$$

Then, obviously, $\mathrm{Im}(\mathbf{h}) = \mathrm{Im}(\mathcal{H}) = HTM$. Since

$$\mathbf{h}^2 = \mathcal{H} \circ \mathbf{j} \circ \mathcal{H} \circ \mathbf{j} \overset{(C_2)}{=} \mathcal{H} \circ \mathbf{j} = \mathbf{h},$$

it follows that $\mathbf{h}$ is indeed a projection operator.

(c) Before continuing we make a general remark. Consider the tangent bundle $\tau_{TM}\colon TTM \to TM$ and suppose that $\Phi\colon TTM \to TTM$ is a fibre-preserving, fibrewise linear mapping which induces the identity on TM (cf. Lemma 2.2.4). Then Φ can be interpreted naturally as a mapping of the form $v \in TM \mapsto \Phi_v \in \mathrm{End}(T_v TM)$. If, in addition, Φ is smooth, then

$$\Phi \in \mathrm{End}(TTM) \cong \mathcal{T}_1^1(TM) \cong \mathrm{End}(\mathfrak{X}(TM))$$

(cf. Remark 3.3.8). We note that Φ is smooth if, and only if, for any vector field $\xi \in \mathfrak{X}(TM)$, the mapping

$$\Phi(\xi)\colon TM \to TTM, \ v \mapsto \Phi(\xi)(v) := \Phi_v(\xi(v))$$

is smooth, i.e., if $\xi \in \mathfrak{X}(TM)$ implies $\Phi(\xi) \in \mathfrak{X}(TM)$.

Similarly, let $\overset{\circ}{\Phi}\colon T\overset{\circ}{T}M \to T\overset{\circ}{T}M$ be a fibre-preserving, fibrewise linear mapping. Then it can also be regarded as a mapping of the form

$$v \in \overset{\circ}{T}M \mapsto \overset{\circ}{\Phi}(v) \in \mathrm{End}(T_v\overset{\circ}{T}M).$$

If $\overset{\circ}{\Phi}$ is smooth, then it is a $T\overset{\circ}{T}M$-valued tensor field on $\overset{\circ}{T}M$, or equivalently,

$$\overset{\circ}{\Phi} \in \mathcal{T}_1^1(\overset{\circ}{T}M) \cong \mathrm{End}(\mathfrak{X}(\overset{\circ}{T}M)).$$

Evidently, what we have just said is valid for the horizontal projection **h** and the vertical projection **v**. Finally, we mention that the fibrewise action of **h** is given by

$$\mathbf{h}_u(w) = \mathcal{H}(u, (\tau_*)_u(w)), \quad w \in T_u TM \tag{7.2.13}$$

for all $u \in TM$.

(d) Although (in general) an Ehresmann connection $\mathcal{H}$ is not a bundle map in the strict sense of the word, it induces a $C^\infty(\overset{\circ}{T}M)$-linear homomorphism

$$\widetilde{X} \in \Gamma(\pi) \mapsto \mathcal{H}(\widetilde{X}) := \mathcal{H} \circ \widetilde{X} \in \mathfrak{X}(\overset{\circ}{T}M), \tag{7.2.14}$$

cf. Proposition 2.2.31. More precisely, one has to write

$$\mathcal{H} \circ \widetilde{X} \restriction \overset{\circ}{T}M \in \mathfrak{X}(\overset{\circ}{T}M).$$

However, the slight abuse of notation in (7.2.14) is tolerable, and makes things more convenient. If $\underline{X}$ is the principal part of $\widetilde{X}$, then

$$\mathcal{H}(\widetilde{X})(v) = \mathcal{H}(v, \underline{X}(v)) = \mathcal{H}_v(\underline{X}(v)), \quad v \in TM. \tag{7.2.15}$$

In particular, the vector field

$$X^{\mathsf{h}} := \mathcal{H}(\widehat{X}), \quad X \in \mathfrak{X}(M) \tag{7.2.16}$$

is the *horizontal lift* of X (with respect to $\mathcal{H}$) in the sense of Lemma and Definition 7.1.4. Indeed, for every $v \in \overset{\circ}{T}M$, we have

$$\mathcal{H}(\widehat{X})(v) \overset{(7.2.15)}{:=} \mathcal{H}_v(X(\tau(v))) \in H_v\overset{\circ}{T}M,$$

so $\mathcal{H}(\widehat{X})$ is a horizontal vector field. Moreover, $X^{\mathsf{h}} := \mathcal{H}(\widehat{X})$ is τ-related to X, because

$$\tau_* \circ X^{\mathsf{h}} = \tau_* \circ \mathcal{H} \circ \widehat{X} \overset{(7.2.9)}{=} \pi_2 \circ \widehat{X} = X \circ \tau.$$

The pointwise action of X^{h} is given by

$$X^{\mathsf{h}}(v) = \mathcal{H}(v, X(\tau(v))) = \mathcal{H}_v(X(\tau(v))), \quad v \in TM. \tag{7.2.17}$$

(e) Sometimes it is more convenient to restrict our considerations to the slit tangent bundle $\overset{\circ}{T}M$. By an *Ehresmann connection in* $\overset{\circ}{T}M$ we mean a smooth mapping $\overset{\circ}{\mathcal{H}}\colon \overset{\circ}{T}M \times_M TM \to T\overset{\circ}{T}M$ satisfying (mutatis mutandis) our above conditions (C_1) and (C_2). To be more accurate, they take the following form:

$(\mathring{C}_1)$ For each $u \in \mathring{T}M$, $v, v_1, v_2 \in T_{\mathring{\tau}(u)}M$ and $\lambda_1, \lambda_2 \in \mathbb{R}$ we have

$$\mathring{\mathcal{H}}(u, v) \in T_u \mathring{T}M \text{ and } \mathring{\mathcal{H}}(u, \lambda_1 v_1 + \lambda_2 v_2) = \lambda_1 \mathring{\mathcal{H}}(u, v_1) + \lambda_2 \mathring{\mathcal{H}}(u, v_2).$$

$(\mathring{C}_2)$ For all $(u, v) \in \mathring{T}M \times_M TM$, $\mathbf{j}_u(\mathring{\mathcal{H}}(u, v)) = (u, v)$.

At the level of sections, the definition can be formulated as follows: *an Ehresmann connection in $\mathring{T}M$ is a $C^\infty(\mathring{T}M)$-linear mapping*

$$\mathring{\mathcal{H}} \colon \Gamma(\mathring{\pi}) \to \mathfrak{X}(\mathring{T}M) \quad \text{such that} \quad \mathbf{j} \circ \mathring{\mathcal{H}} = 1_{\Gamma(\mathring{\pi})}.$$

The associated objects to an Ehresmann connection in $\mathring{T}M$ (horizontal lifting, projections, etc.) can be defined as above (and as below, see Definition 7.2.7) in TM. Clearly, an Ehresmann connection $\mathring{\mathcal{H}}$ in $\mathring{T}M$ can be extended to an Ehresmann connection $\mathcal{H}$ in TM such that

$$\mathcal{H} \restriction \mathring{T}M \times_M TM = \mathring{\mathcal{H}} \text{ and } \mathcal{H}(0_p, v) := (o_*)_p(v) \quad (p \in M, \ v \in T_pM).$$

Lemma 7.2.5. *Given a manifold M, there exists an Ehresmann connection in TM.*

Proof. By Proposition 7.1.2, we can choose a horizontal subbundle HTM in TTM. Then

$$TTM = HTM \oplus VTM \text{ (Whitney sum)}.$$

The canonical surjection $\mathbf{j} \colon TTM \to TM \times_M TM$ restricts to a strong bundle isomorphism $HTM \to TM \times_M TM$; let $\mathcal{H} := (\mathbf{j} \restriction HTM)^{-1}$ (cf. proof of Lemma 1.1.19). Then it can be checked immediately that $\mathcal{H}$ is an Ehresmann connection in TM. (This Ehresmann connection, in fact, is smooth on its whole domain.) $\qquad \square$

Lemma 7.2.6 (Local Frame Principle II). *Let M be an n-dimensional manifold equipped with an Ehresmann connection $\mathcal{H}$ in TM. If $(X_i)_{i=1}^n$ is a frame field of TM over an open subset $\mathcal{U}$ of M, then $((X_i^{\mathrm{v}})_{i=1}^n, (X_i^{\mathrm{h}})_{i=1}^n)$ (where $X_i^{\mathrm{h}} = \mathcal{H}(\widehat{X}_i)$) is a frame field of $\tau_{TM} \colon TTM \to TM$ over $\tau^{-1}(\mathcal{U})$.*

This can be proved essentially in the same way as the analogous Lemma 4.1.53.

Definition 7.2.7. Let $\mathcal{H} \colon TM \times_M TM \to TTM$ be an Ehresmann connection, and consider the vertical projection

$$\mathbf{v} = 1_{TTM} - \mathbf{h} = 1_{TTM} - \mathcal{H} \circ \mathbf{j}$$

determined by $\mathcal{H}$.

(a) We say that

$$\mathcal{V} := \mathbf{i}^{-1} \circ \mathbf{v} \colon TTM \to VTM \to TM \times_M TM \tag{7.2.18}$$

is the *vertical mapping* associated to $\mathcal{H}$.

(b) The mapping

$$\mathbf{F} := \mathcal{H} \circ \mathcal{V} - \mathbf{i} \circ \mathbf{j} \colon TTM \to TTM \tag{7.2.19}$$

is called the *almost complex structure* determined by $\mathcal{H}$.

(c) The mapping

$$K := \ell^{\downarrow} \circ \mathbf{v} \colon TTM \to VTM \to TM \tag{7.2.20}$$

is said to be the *connector* associated to the Ehresmann connection $\mathcal{H}$.

Lemma 7.2.8. *If $\mathcal{V}$ is the vertical mapping associated with an Ehresmann connection $\mathcal{H}$ in TM, then*

$$\mathrm{Ker}(\mathcal{V}) = \mathrm{Im}(\mathcal{H}) \text{ whence } \mathcal{V} \circ \mathcal{H} = 0 \tag{7.2.21}$$

and

$$\mathcal{V} \circ \mathbf{i} = 1_{TM \times_M TM}. \tag{7.2.22}$$

Proof. Both assertions are quite evident. If $\mathbf{h}$ and $\mathbf{v}$ are the horizontal and the vertical projection associated with $\mathcal{H}$, then

$$\mathrm{Ker}(\mathcal{V}) = \mathrm{Ker}(\mathbf{i}^{-1} \circ \mathbf{v}) = \mathrm{Ker}(\mathbf{v}) = \mathrm{Im}(\mathbf{h}) = \mathrm{Im}(\mathcal{H}),$$

and

$$\mathcal{V} \circ \mathbf{i} = \mathbf{i}^{-1} \circ \mathbf{v} \circ \mathbf{i} = \mathbf{i}^{-1} \circ (1_{TTM} - \mathcal{H} \circ \mathbf{j}) \circ \mathbf{i} \overset{(4.1.40)}{=} \mathbf{i}^{-1} \circ \mathbf{i} = 1_{TM \times_M TM}.$$

$\square$

Remark 7.2.9. Thus, specifying an Ehresmann connection in TM, we obtain a split short exact sequence (Example 2.2.41)

$$0 \rightleftarrows \overset{\circ}{T}M \times_M TM \underset{\mathcal{V}}{\overset{\mathbf{i}}{\rightleftarrows}} T\overset{\circ}{T}M \underset{\mathcal{H}}{\overset{\mathbf{j}}{\rightleftarrows}} \overset{\circ}{T}M \times_M TM \rightleftarrows 0 \tag{7.2.23}$$

of strong bundle maps, which induces a similar short exact sequence

$$0 \rightleftarrows \Gamma(\overset{\circ}{\pi}) \underset{\mathcal{V}}{\overset{\mathbf{i}}{\rightleftarrows}} \mathfrak{X}(\overset{\circ}{T}M) \underset{\mathcal{H}}{\overset{\mathbf{j}}{\rightleftarrows}} \Gamma(\overset{\circ}{\pi}) \rightleftarrows 0 \tag{7.2.24}$$

of $C^{\infty}(\overset{\circ}{T}M)$-homomorphisms. We recall that the property 'splitting' means here that

$$\mathcal{V} \circ \mathbf{i} = \mathbf{j} \circ \mathcal{H} = 1_{\overset{\circ}{T}M \times_M TM} \text{ (or } 1_{\Gamma(\overset{\circ}{\pi})}).$$

Since $C \overset{(4.1.43)}{=} \mathbf{i}\widetilde{\delta}$, it follows that

$$\mathcal{V}(C) = \widetilde{\delta}. \tag{7.2.25}$$

Now we collect some basic relations between the vertical endomorphism **J** (as a canonical object) and the geometric data obtained from an Ehresmann connection above.

Lemma 7.2.10. *Let* **h**, **v**, **F** *and* K *be the horizontal projection, the vertical projection, the almost complex structure and the connector associated to an Ehresmann connection* $\mathcal{H}$ *in* TM. *Then the following formulae are valid:*

$$\mathbf{h}^2 = \mathbf{h}, \quad \mathbf{v}^2 = \mathbf{v}, \quad \mathbf{F}^2 = -1_{TTM}; \tag{7.2.26}$$

$$\mathbf{J} \circ \mathbf{h} = \mathbf{J}, \quad \mathbf{h} \circ \mathbf{J} = 0; \tag{7.2.27}$$

$$\mathbf{J} \circ \mathbf{v} = 0, \quad \mathbf{v} \circ \mathbf{J} = \mathbf{J}; \tag{7.2.28}$$

$$\mathbf{J} \circ \mathbf{F} = \mathbf{v}, \quad \mathbf{F} \circ \mathbf{J} = \mathbf{h}; \tag{7.2.29}$$

$$\mathbf{F} \circ \mathbf{h} = \mathbf{v} \circ \mathbf{F} = -\mathbf{J}, \quad \mathbf{h} \circ \mathbf{F} = \mathbf{F} \circ \mathbf{v} = \mathbf{J} + \mathbf{F}; \tag{7.2.30}$$

$$\tau \circ K = \tau \circ \tau_{TM}; \tag{7.2.31}$$

$$K \circ X^{\mathbf{v}} = X \circ \tau; \tag{7.2.32}$$

$$X^{\mathbf{h}} = \mathbf{h}X^{c}, \quad \mathbf{J}X^{\mathbf{h}} = X^{\mathbf{v}}; \tag{7.2.33}$$

$$\mathbf{F}X^{\mathbf{h}} = -X^{\mathbf{v}}, \quad \mathbf{F}X^{\mathbf{v}} = X^{\mathbf{h}}; \tag{7.2.34}$$

$$\mathbf{h}[X^{\mathbf{h}}, Y^{\mathbf{h}}] = [X, Y]^{\mathbf{h}}, \quad \mathbf{J}[X^{\mathbf{h}}, Y^{\mathbf{h}}] = [X, Y]^{\mathbf{v}}. \tag{7.2.35}$$

In (7.2.32)–(7.2.35) X *and* Y *are arbitrary vector fields on* M.

Proof. (a) The first relation of (7.2.26) has already been shown. From this it follows immediately that $\mathbf{v}^2 = \mathbf{v}$. Next we show that $\mathbf{F}^2 = -1_{TTM}$. Applying (7.2.21) and (7.2.22), we find

$$\mathbf{F}^2 = (\mathcal{H} \circ \mathcal{V} - \mathbf{i} \circ \mathbf{j}) \circ (\mathcal{H} \circ \mathcal{V} - \mathbf{i} \circ \mathbf{j}) = -\mathbf{i} \circ \mathcal{V} - \mathcal{H} \circ \mathbf{j} + \mathbf{J}^2$$
$$= -\mathbf{v} - \mathbf{h} = -1_{TTM},$$

as desired. (This result explains why **F** is called an almost complex structure.)

(b) Now we verify (7.2.27)–(7.2.29):

$$\mathbf{J} \circ \mathbf{h} = \mathbf{i} \circ \mathbf{j} \circ \mathcal{H} \circ \mathbf{j} \overset{(C_2)}{=} \mathbf{i} \circ \mathbf{j} = \mathbf{J}, \qquad \mathbf{h} \circ \mathbf{J} = \mathcal{H} \circ \mathbf{j} \circ \mathbf{i} \circ \mathbf{j} \overset{(4.1.40)}{=} 0,$$

$$\mathbf{J} \circ \mathbf{v} = \mathbf{J} - \mathbf{J} \circ \mathbf{h} \overset{(7.2.27)}{=} \mathbf{J} - \mathbf{J} = 0, \qquad \mathbf{v} \circ \mathbf{J} = \mathbf{J} - \mathbf{h} \circ \mathbf{J} \overset{(7.2.27)}{=} \mathbf{J},$$

$$\mathbf{J} \circ \mathbf{F} = \mathbf{i} \circ \mathbf{j} \circ \mathcal{H} \circ \mathcal{V} - \mathbf{J}^2 = \mathbf{i} \circ \mathcal{V} \overset{(7.2.18)}{=} \mathbf{v},$$

$$\mathbf{F} \circ \mathbf{J} = \mathcal{H} \circ \mathcal{V} \circ \mathbf{J} - \mathbf{J}^2 = \mathcal{H} \circ \mathbf{i}^{-1} \circ \mathbf{v} \circ \mathbf{J} \overset{(7.2.28)}{=} \mathcal{H} \circ \mathbf{i}^{-1} \circ \mathbf{J} = \mathcal{H} \circ \mathbf{j} = \mathbf{h}.$$

(c) The proof of (7.2.30) is similarly easy, and we leave it to the reader.

(d) By the second diagram in (4.1.54),

$$\tau \circ K = \tau \circ \ell^{\downarrow} \circ \mathbf{v} = \tau \circ \tau^{\mathsf{v}}_{TM} \circ \mathbf{v} = \tau \circ \tau_{TM},$$

which proves (7.2.31). If $X \in \mathfrak{X}(M)$,

$$K \circ X^{\mathsf{v}} = \ell^{\downarrow} \circ \mathbf{v} \circ X^{\mathsf{v}} = \ell^{\downarrow} \circ X^{\mathsf{v}} \overset{(4.1.56)}{=} X \circ \tau,$$

as we asserted.

(e) We have already shown that $X^{\mathsf{h}} = \mathbf{h}X^{\mathsf{c}}$ (see (7.1.4)). In the context of Ehresmann connections it can be seen as follows:

$$X^{\mathsf{h}} := \mathcal{H}(\widehat{X}) = \mathcal{H}(\mathbf{j}X^{\mathsf{c}}) = \mathcal{H} \circ \mathbf{j}(X^{\mathsf{c}}) = \mathbf{h}X^{\mathsf{c}}.$$

As to the second formula in (7.2.33),

$$\mathbf{J}X^{\mathsf{h}} = \mathbf{i} \circ \mathbf{j} \circ \mathcal{H}(\widehat{X}) = \mathbf{i}\widehat{X} = X^{\mathsf{v}},$$

as we claimed.

(f) Also an easy calculation shows that (7.2.34) holds:

$$\mathbf{F}X^{\mathsf{h}} = \mathbf{F} \circ \mathbf{h}(X^{\mathsf{c}}) \overset{(7.2.30)}{=} -\mathbf{J}X^{\mathsf{c}} = -X^{\mathsf{v}},$$

$$\mathbf{F}X^{\mathsf{v}} = \mathbf{F} \circ \mathbf{J}(X^{\mathsf{c}}) \overset{(7.2.29)}{=} \mathbf{h}X^{\mathsf{c}} = X^{\mathsf{h}}.$$

(Notice that we obtain $\mathbf{F}^2 = -1_{TTM}$ also from these relations.)

(g) The first relation of (7.2.35) has been asserted and proved in Lemma 7.1.5. From this the second relation can be obtained easily:

$$\mathbf{J}[X^{\mathsf{h}}, Y^{\mathsf{h}}] \overset{(7.2.27)}{=} \mathbf{J} \circ \mathbf{h}[X^{\mathsf{h}}, Y^{\mathsf{h}}] = \mathbf{J}[X, Y]^{\mathsf{h}} = [X, Y]^{\mathsf{v}}.$$

This concludes the proof of the lemma. $\qquad\square$

Lemma and Definition 7.2.11. *Let $\mathcal{H}_1$ and $\mathcal{H}_2$ be Ehresmann connections in TM. Then there exists a unique endomorphism $\mathcal{A}$ of $\Gamma(\pi)$ such that $\mathcal{H}_1 - \mathcal{H}_2 = \mathbf{i} \circ \mathcal{A}$. We say that $\mathcal{A}$ is the* difference tensor *of $\mathcal{H}_1$ and $\mathcal{H}_2$. We have $\mathcal{A} = \mathcal{V}_2 \circ \mathcal{H}_1$ or, equivalently,*

$$\mathbf{i} \circ \mathcal{A} = \mathbf{v}_2 \circ \mathcal{H}_1. \tag{7.2.36}$$

Proof. Since $\mathbf{J} \circ (\mathcal{H}_1 - \mathcal{H}_2) = \mathbf{i} \circ (\mathbf{j} \circ \mathcal{H}_1 - \mathbf{j} \circ \mathcal{H}_2) \overset{(C_2)}{=} 0$, $\mathcal{H}_1 - \mathcal{H}_2$ is vertical-valued. Thus it can be written as

$$\mathcal{H}_1 - \mathcal{H}_2 = \mathbf{i} \circ \mathcal{A}, \quad \mathcal{A} \in \mathrm{End}(\Gamma(\pi)).$$

Clearly, the endomorphism $\mathcal{A}$ is unique. Moreover, we have

$$\mathbf{i} \circ \mathcal{A} = \mathcal{H}_1 - \mathcal{H}_2 \circ (\mathbf{j} \circ \mathcal{H}_1) = \mathcal{H}_1 - \mathbf{h}_2 \circ \mathcal{H}_1$$

$$= (1_{\mathfrak{X}(TM)} - \mathbf{h}_2) \circ \mathcal{H}_1 = \mathbf{v}_2 \circ \mathcal{H}_1,$$

as was to be shown. $\qquad\square$

Remark 7.2.12 (Christoffel Symbols). Let an Ehresmann connection $\mathcal{H}$ be specified in TM. Choose a chart $(\mathcal{U}, (u^i)_{i=1}^n)$ on M and consider the induced chart $(\tau^{-1}(\mathcal{U}), ((x^i)_{i=1}^n, (y^i)_{i=1}^n))$ on TM. In the local frame field $\left(\left(\frac{\partial}{\partial x^i} \right)_{i=1}^n, \left(\frac{\partial}{\partial y^i} \right)_{i=1}^n \right)$ on $\tau^{-1}(\mathcal{U})$, the $\mathcal{H}$-horizontal vector fields $\mathcal{H} \left(\widehat{\frac{\partial}{\partial u^j}} \right)$ $(j \in J_n)$ can be expressed uniquely in the form

$$\mathcal{H} \left(\widehat{\frac{\partial}{\partial u^j}} \right) = A^i_j \frac{\partial}{\partial x^i} + B^i_j \frac{\partial}{\partial y^i}, \tag{$*$}$$

where A^i_j, B^i_j are real-valued functions defined on $\tau^{-1}(\mathcal{U})$ and smooth on $\tau^{-1}(\mathcal{U}) \cap \overset{\circ}{T}M$. (Here and later, we use the summation convention.) Since $\mathbf{j} \circ \mathcal{H} = 1_{TM \times_M TM}$, from $(*)$ we obtain

$$\widehat{\frac{\partial}{\partial u^j}} = A^i_j \mathbf{j} \left(\frac{\partial}{\partial x^i} \right) + B^i_j \mathbf{j} \left(\frac{\partial}{\partial y^i} \right) \overset{(4.1.74)}{=} A^i_j \widehat{\frac{\partial}{\partial u^i}},$$

whence $(A^i_j) = (\delta^i_j)$. Following the usual sign convention, let

$$N^i_j := -B^i_j; \quad i,j \in J_n.$$

Then we have

$$\mathcal{H} \left(\widehat{\frac{\partial}{\partial u^j}} \right) = \left(\frac{\partial}{\partial u^j} \right)^{\mathrm{h}} = \frac{\partial}{\partial x^j} - N^i_j \frac{\partial}{\partial y^i}, \quad j \in J_n. \tag{7.2.37}$$

We say that the functions

$$N^i_j : \tau^{-1}(\mathcal{U}) \to \mathbb{R}, \quad N^i_j \in C^\infty(\tau^{-1}(\mathcal{U}) \cap \overset{\circ}{T}M)$$

are the *Christoffel symbols* of the Ehresmann connection $\mathcal{H}$ with respect to the chart $(\mathcal{U}, (u^i)_{i=1}^n)$. Using them, we find by a straightforward calculation that $\mathbf{h}$, $\mathbf{v}$ and $\mathbf{F}$ have the following matrix representations with respect to the local frame field $\left(\left(\frac{\partial}{\partial x^i} \right)_{i=1}^n, \left(\frac{\partial}{\partial y^i} \right)_{i=1}^n \right)$:

$$\mathbf{h} \underset{(\mathcal{U})}{=} \begin{pmatrix} (\delta^i_j) & 0 \\ -(N^i_j) & 0 \end{pmatrix}, \quad \mathbf{v} \underset{(\mathcal{U})}{=} \begin{pmatrix} 0 & 0 \\ (N^i_j) & (\delta^i_j) \end{pmatrix}; \tag{7.2.38}$$

$$\mathbf{F} \underset{(\mathcal{U})}{=} \begin{pmatrix} (N^i_j) & (\delta^i_j) \\ -(N^i_k N^k_j) - (\delta^i_j) & -(N^i_j) \end{pmatrix}. \tag{7.2.39}$$

For each $j \in J_n$,

$$\mathcal{V} \left(\frac{\partial}{\partial x^j} \right) \overset{(7.2.18)}{=} \mathbf{i}^{-1} \circ \mathbf{v} \left(\frac{\partial}{\partial x^j} \right) \overset{(7.2.38)}{=} \mathbf{i}^{-1} \left(N^i_j \frac{\partial}{\partial y^i} \right) \overset{(4.1.74)}{=} N^i_j \widehat{\frac{\partial}{\partial u^i}},$$

$$\mathcal{V} \left(\frac{\partial}{\partial y^j} \right) = \mathbf{i}^{-1} \circ \mathbf{v} \left(\frac{\partial}{\partial y^j} \right) = \mathbf{i}^{-1} \left(\frac{\partial}{\partial y^j} \right) = \widehat{\frac{\partial}{\partial u^j}},$$

so the vertical mapping associated to $\mathcal{H}$ has the local representation

$$\mathcal{V} = N_j^i \widehat{\frac{\partial}{\partial u^i}} \otimes dx^j + \widehat{\frac{\partial}{\partial u^i}} \otimes dy^i. \tag{7.2.40}$$

Finally,

$$K\left(\frac{\partial}{\partial x^j}\right) \overset{(7.2.20)}{=} \ell^{\downarrow} \circ \mathbf{v}\left(\frac{\partial}{\partial x^j}\right) \overset{(7.2.38)}{=} \ell^{\downarrow}\left(N_j^i \frac{\partial}{\partial y^i}\right)$$

$$= \ell^{\downarrow}\left(N_j^i \left(\frac{\partial}{\partial u^i}\right)^{\mathsf{v}}\right) \overset{(4.1.56)}{=} N_j^i \left(\frac{\partial}{\partial u^i} \circ \tau\right)$$

and

$$K\left(\frac{\partial}{\partial y^j}\right) = \ell^{\downarrow} \circ \mathbf{v}\left(\frac{\partial}{\partial y^j}\right) = \ell^{\downarrow}\left(\frac{\partial}{\partial y^j}\right) = \frac{\partial}{\partial u^j} \circ \tau,$$

therefore, at every $v \in \tau^{-1}(\mathcal{U})$, the connector

$$K_v := K \restriction T_v TM = \ell^{\downarrow} \circ \mathbf{v}_v \colon T_v TM \to T_{\tau(v)} M$$

acts by the rule

$$K_v(w) = (w^{n+i} + w^j N_j^i(v)) \left(\frac{\partial}{\partial u^i}\right)_{\tau(v)} \tag{7.2.41}$$

where $w = w^i \left(\frac{\partial}{\partial x^i}\right)_v + w^{n+i} \left(\frac{\partial}{\partial y^i}\right)_v$.

Lemma and Definition 7.2.13. *If $\mathcal{H}$ is an Ehresmann connection in TM, then $S_{\mathcal{H}} := \mathcal{H}\widetilde{\delta}$ is a semispray over M, called the semispray associated to $\mathcal{H}$. Then*

$$\mathbf{h}[C, S_{\mathcal{H}}] = S_{\mathcal{H}}, \tag{7.2.42}$$

where $\mathbf{h}$ is the horizontal projection determined by $\mathcal{H}$. If the Christoffel symbols of $\mathcal{H}$ with respect to a chart $(\mathcal{U}, (u^i)_{i=1}^n)$ of M are the functions N_j^i, then the coefficients of $S_{\mathcal{H}}$ with respect to the same chart are the functions

$$\frac{1}{2} y^j N_j^i, \quad i \in J_n. \tag{7.2.43}$$

Proof. The mapping $S_{\mathcal{H}}$ obviously satisfies conditions (S_1) and (S_2). It is also clear that $S_{\mathcal{H}}$ is smooth on $\overset{\circ}{T}M$. Finally, for each $p \in M$ we have

$$S_{\mathcal{H}} \circ o(p) = \mathcal{H}(o(p), o(p)) \overset{(C_4)}{=} (o_*)_p(o_p) = o_{TM} \circ o(p),$$

so $S_{\mathcal{H}}$ has the property (S_5). Thus $S_{\mathcal{H}}$ is indeed a semispray over TM. Locally,

$$S_{\mathcal{H}} \underset{(\mathcal{U})}{=} \mathcal{H}\left(y^j \widehat{\frac{\partial}{\partial u^j}}\right) \overset{(C_1)}{=} y^j \mathcal{H}\left(\widehat{\frac{\partial}{\partial u^j}}\right) \overset{(7.2.37)}{=} y^j \frac{\partial}{\partial x^j} - y^j N_j^i \frac{\partial}{\partial y^i},$$

hence the coefficients of $S_{\mathcal{H}}$ with respect to the chosen chart are $y^j N^i_j$, $i \in J_n$ (cf. Remark (5.1.8)).

To show (7.2.42), consider the deviation $S^*_{\mathcal{H}}$ of $S_{\mathcal{H}}$ (defined, tacitly, as that of a second-order vector field, see Corollary and Definition 5.1.12). Since $S^*_{\mathcal{H}}$ is vertical,

$$0 = \mathbf{h} \, S^*_{\mathcal{H}} = \mathbf{h}[C, S_{\mathcal{H}}] - \mathbf{h} \, S_{\mathcal{H}} = \mathbf{h}[C, S_{\mathcal{H}}] - \mathcal{H} \circ \mathbf{j} \circ S_{\mathcal{H}}$$
$$= \mathbf{h}[C, S_{\mathcal{H}}] - \mathcal{H}\widetilde{\delta} = \mathbf{h}[C, S_{\mathcal{H}}] - S_{\mathcal{H}},$$

whence our assertion.   □

If $\overset{\circ}{\mathcal{H}}$ is an Ehresmann connection in $\overset{\circ}{T}M$, then $S_{\overset{\circ}{\mathcal{H}}} := \overset{\circ}{\mathcal{H}}(\widetilde{\delta} \restriction \overset{\circ}{T}M)$ is a second-order vector field on $\overset{\circ}{T}M$, called also *associated to* $\overset{\circ}{\mathcal{H}}$. In the following, less pedantically, we shall simply write $S_{\overset{\circ}{\mathcal{H}}} = \overset{\circ}{\mathcal{H}}\widetilde{\delta}$.

7.3 Constructions of Ehresmann Connections

7.3.1 *Ehresmann Connections and Projection Operators*

Lemma 7.3.1. *Consider the double tangent bundle* $\tau_{TM} \colon TTM \to TM$. *Let* $\mathbf{h} \colon TTM \to TTM$ *be a mapping such that*

(i) *for each* $v \in TM$, $\mathbf{h}_v := \mathbf{h} \restriction T_v TM \in \mathrm{End}(T_v TM)$;
(ii) $\mathbf{h}$ *is smooth on* $T\overset{\circ}{T}M$;
(iii) $\mathbf{h}^2 = \mathbf{h}$, $\mathrm{Ker}(\mathbf{h}) = VTM$;
(iv) $\mathbf{h} \circ o_* = o_*$ *(* $o \in \mathfrak{X}(M)$ *is the zero vector field).*

Then there exists a unique Ehresmann connection in TM with associated projection mapping $\mathbf{h}$. *This Ehresmann connection can be given in the form* $\mathcal{H} = \mathbf{h} \circ \mathcal{H}_0$, *where $\mathcal{H}_0$ is an arbitrarily chosen Ehresmann connection in* TM.

Proof. First we show the uniqueness. If $\mathcal{H}_1$ and $\mathcal{H}_2$ are Ehresmann connections in TM with the same horizontal projection $\mathbf{h}$, then $\mathcal{H}_1 \circ \mathbf{j} = \mathcal{H}_2 \circ \mathbf{j}$, whence $(\mathcal{H}_1 - \mathcal{H}_2) \circ \mathbf{j} = 0$. This implies that $\mathcal{H}_1 - \mathcal{H}_2 = 0$, because $\mathbf{j}$ is surjective, and hence has a right inverse.

To prove the existence, let $\mathcal{H} := \mathbf{h} \circ \mathcal{H}_0$, where $\mathcal{H}_0$ is an arbitrarily chosen Ehresmann connection in TM (Lemma 7.2.5). Then $\mathcal{H} \circ \mathbf{j} = \mathbf{h} \circ \mathcal{H}_0 \circ \mathbf{j} = \mathbf{h}$, and conditions (C_1) and (C_3) also hold automatically.

Let $\mathbf{v} := 1_{TTM} - \mathbf{h}$. Then

$$\mathrm{Im}(\mathbf{v}) = VTM(\overset{\text{(iii)}}{=} \mathrm{Ker}(\mathbf{h})). \qquad (*)$$

Indeed, for each $w \in TTM$,

$$\mathbf{h}(\mathbf{v}(w)) = \mathbf{h}(w - \mathbf{h}(w)) \overset{(iii)}{=} \mathbf{h}(w) - \mathbf{h}(w) = 0,$$

whence $\mathrm{Im}(\mathbf{v}) \subset VTM$. Conversely, if $w \in VTM$, then

$$w = \mathbf{h}(w) + \mathbf{v}(w) \overset{(iii)}{=} \mathbf{v}(w) \in \mathrm{Im}(\mathbf{v}),$$

so we also have $VTM \subset \mathrm{Im}(\mathbf{v})$. Since $\mathrm{Ker}(\mathbf{j}) = VTM$ by Corollary 4.1.21, from $(*)$ it follows that $\mathbf{j} \circ \mathbf{v} = 0$. Thus

$$\mathbf{j} \circ \mathcal{H} = \mathbf{j} \circ \mathbf{h} \circ \mathcal{H}_0 = \mathbf{j} \circ (1_{TTM} - \mathbf{v}) \circ \mathcal{H}_0 = \mathbf{j} \circ \mathcal{H}_0 = 1_{TM \times_M TM}.$$

Finally, we check (C_4). For each $p \in M$ and $v \in T_p M$,

$$\mathcal{H}(o(p), v) := \mathbf{h} \circ \mathcal{H}_0(o(p), v) \overset{(7.2.4)}{=} \mathbf{h}((o_*)_p(v)) \overset{(iv)}{=} (o_*)_p(v),$$

as was to be shown. $\qquad\qquad\qquad\qquad\qquad\qquad\qquad\qquad\qquad\qquad\square$

Remark 7.3.2. By the lemma, an Ehresmann connection in TM can be defined equivalently as a projection operator on TM, satisfying the above conditions (i)–(iv). Technically, this idea can also be carried out as follows:

We start with a mapping $\overset{\circ}{\mathbf{h}} \colon T\overset{\circ}{T}M \to T\overset{\circ}{T}M$ satisfying requirements (i)–(iii) of the lemma. Then there exists a unique Ehresmann connection $\mathcal{H} \colon TM \times_M TM \to TTM$ such that its restriction $\overset{\circ}{\mathcal{H}} := \mathcal{H} \upharpoonright \overset{\circ}{T}M \times_M TM$ induces the given mapping $\overset{\circ}{\mathbf{h}}$, i.e., $\overset{\circ}{\mathcal{H}} \circ \mathbf{j} = \overset{\circ}{\mathbf{h}}$. Here, equivalently, we can regard $\overset{\circ}{\mathbf{h}}$ as a $C^\infty(\overset{\circ}{T}M)$-linear endomorphism of $\mathfrak{X}(\overset{\circ}{T}M)$ satisfying $(\overset{\circ}{\mathbf{h}})^2 = \overset{\circ}{\mathbf{h}}$ and $\mathrm{Ker}(\overset{\circ}{\mathbf{h}}) = \mathfrak{X}^\mathrm{v}(\overset{\circ}{T}M)$.

From now on, to simplify notation, we shall omit the circle from the objects associated to an Ehresmann connection defined only on $\overset{\circ}{T}M$, i.e., we shall simply write $\mathcal{H}$ and $\mathbf{h}$ instead of $\overset{\circ}{\mathcal{H}}$ and $\overset{\circ}{\mathbf{h}}$, etc.

7.3.2 Ehresmann Connections from Regular Covariant Derivatives

Example 7.3.3. Let D be a covariant derivative on the Finsler bundle $\pi \colon TM \times_M TM \to TM$. Consider its deflection

$$\mu := D\widetilde{\delta} \in \mathrm{Hom}_{C^\infty(TM)}(\mathfrak{X}(TM), \Gamma(\pi))$$

and vertical deflection $\mu^\mathrm{v} := \mu \circ \mathbf{i} \in \mathrm{End}(\Gamma(\pi))$. We recall that μ and μ^v can also be interpreted as strong bundle maps

$$\underline{\mu} \colon TTM \to TM \times_M TM \quad \text{and} \quad \underline{\mu}^\mathrm{v} \colon TM \times_M TM \to TM \times_M TM,$$

related by $\underline{\mu}^{\mathsf{v}} = \underline{\mu} \circ \mathbf{i}$, see Remark 6.2.6.

(a) Now suppose that D *is regular*. Then the vertical deflection $\underline{\mu}^{\mathsf{v}}$ of D restricts to a linear automorphism on each fibre of π (Remark 6.2.8(a)), and by Proposition 6.2.9, $\text{Ker}(\underline{\mu}) = \bigcup_{v \in TM} \text{Ker}(\underline{\mu}_v)$ is a horizontal subbundle of TTM. We show that the mapping

$$\mathbf{h} := 1_{TTM} - \mathbf{i} \circ (\underline{\mu}^{\mathsf{v}})^{-1} \circ \underline{\mu} \qquad (7.3.1)$$

satisfies requirements (i)–(iv) of Lemma 7.3.1, and $\text{Im}(\mathbf{h}) = \text{Ker}(\underline{\mu})$.

Obviously, $\mathbf{h}$ restricts to a linear endomorphism in each fibre, and it is smooth on its entire domain. An immediate calculation shows that $\mathbf{h}^2 = \mathbf{h}$ and $\mathbf{h} \circ o_* = o_*$. (We suggest that the reader should verify it.) We check that $\text{Ker}(\mathbf{h}) = VTM$.

If $\mathbf{h}(w) = 0$, then by (7.3.1)

$$w = \mathbf{i} \circ (\underline{\mu}^{\mathsf{v}})^{-1} \circ \underline{\mu}(w) \in \text{Im}(\mathbf{i}) = VTM,$$

therefore $\text{Ker}(\mathbf{h}) \subset VTM$. Conversely, if $w \in VTM$, then $w = \mathbf{J}(z) = \mathbf{i} \circ \mathbf{j}(z)$ for some $z \in TTM$, and

$$\mathbf{h}(w) \overset{(7.3.1)}{=} w - \mathbf{i} \circ (\underline{\mu}^{\mathsf{v}})^{-1} \circ \underline{\mu} \circ \mathbf{i}(\mathbf{j}(z)) = w - \mathbf{i}(\mathbf{j}(z)) = w - w = 0,$$

which proves the inclusion $VTM \subset \text{Ker}(\mathbf{h})$.

It remains to show that $\text{Im}(\mathbf{h}) = \text{Ker}(\underline{\mu})$. For each $w \in TTM$,

$$\underline{\mu}(\mathbf{h}(w)) = \underline{\mu}(w) - \underline{\mu} \circ \mathbf{i} \circ (\underline{\mu}^{\mathsf{v}})^{-1} \circ \underline{\mu}(w) = \underline{\mu}(w) - \underline{\mu}(w) = 0,$$

so $\text{Im}(\mathbf{h}) \subset \text{Ker}(\underline{\mu})$. To verify the converse inclusion, consider the vertical projection $\mathbf{v} := 1_{TTM} - \mathbf{h}$. Then any vector $w \in TTM$ can be written uniquely in the form $w = \mathbf{h}(w) + \mathbf{v}(w)$. Now we suppose that $w \in \text{Ker}(\underline{\mu})$. In this case, taking into account that $\underline{\mu} \circ \mathbf{h} = 0$ by our above calculation, we find that

$$0 = \underline{\mu}(w) = \underline{\mu} \circ \mathbf{h}(w) + \underline{\mu} \circ \mathbf{v}(w) = \underline{\mu}^{\mathsf{v}} \circ \mathbf{i}^{-1}(\mathbf{v}(w)).$$

Since $\underline{\mu}^{\mathsf{v}}$ is a linear automorphism on every fibre, it follows that $\mathbf{i}^{-1}(\mathbf{v}(w)) = 0$, whence $\mathbf{v}(w) = 0$ and $w = \mathbf{h}(w) \in \text{Im}(\mathbf{h})$, as was to be shown.

The Ehresmann connection $\mathcal{H}_D$ determined by $\mathbf{h}$ according to Lemma 7.3.1 is called the *Ehresmann connection induced by the regular covariant derivative* D. If D is strongly regular, (7.3.1) takes the simpler form $\mathbf{h} = 1_{TTM} - \mathbf{i} \circ \underline{\mu}$. As we have learnt, $\mathbf{h}$ can be freely regarded as a projection operator of the $C^\infty(TM)$-module $\mathfrak{X}(TM)$. Then the horizontal lift of a vector field $X \in \mathfrak{X}(M)$ with respect to $\mathcal{H}_D$ is

$$X^{\mathsf{h}} := \mathcal{H}_D(\widehat{X}) = \mathbf{h}X^{\mathsf{c}} = X^{\mathsf{c}} - \mathbf{i} \circ (\mu^{\mathsf{v}})^{-1} D_{X^{\mathsf{c}}} \widetilde{\delta}. \qquad (7.3.2)$$

In the strongly regular case

$$X^{\mathsf{h}} = X^{\mathsf{c}} - \mathbf{i}\, D_{X^{\mathsf{c}}}\widetilde{\delta}. \tag{7.3.3}$$

(b) We continue to assume that D is a regular covariant derivative on π. Using the notation of Remark 6.2.3, we determine the Christoffel symbols of the Ehresmann connection $\mathcal{H}_D$. The summation convention will be in force. For each $j \in J_n$,

$$
\mathcal{H}_D\left(\widehat{\frac{\partial}{\partial u^j}}\right) \overset{(7.3.2)}{=} \frac{\partial}{\partial x^j} - \mathbf{i} \circ (\mu^{\mathsf{v}})^{-1} D_{\frac{\partial}{\partial x^j}} y^k \widehat{\frac{\partial}{\partial u^k}}
$$

$$
\overset{(6.2.6)}{=} \frac{\partial}{\partial x^j} - y^k \mathbf{i} \circ (\mu^{\mathsf{v}})^{-1}\left(\Gamma^s_{jk}\widehat{\frac{\partial}{\partial u^s}}\right)
$$

$$
= \frac{\partial}{\partial x^j} - y^k \Gamma^s_{jk}\mathbf{i}\left(M^r_s\widehat{\frac{\partial}{\partial u^r}}\right) = \frac{\partial}{\partial x^j} - y^k\Gamma^s_{jk}M^r_s\frac{\partial}{\partial y^r},
$$

where, taking into account (6.2.14),

$$(M^r_s) := (L^r_s)^{-1}, \quad (L^r_s) := (\delta^r_s + y^k C^r_{sk}).$$

Thus, if (N^i_j) is the family of the Christoffel symbols of $\mathcal{H}_D$, then for each $i, j \in J_n$ we have $N^i_j = y^k\Gamma^s_{jk}M^i_s$, whence

$$N^i_j L^r_i = y^k\Gamma^s_{jk}M^i_s L^r_i = y^k\Gamma^s_{jk}\delta^r_s = y^k\Gamma^r_{jk}.$$

Therefore the Christoffel symbols of $\mathcal{H}_D$ are uniquely determined by the relations

$$N^i_j(\delta^r_i + y^k C^r_{ik}) = y^k\Gamma^r_{jk}; \quad j, r \in J_n. \tag{7.3.4}$$

If, in particular, D is strongly regular, then

$$N^i_j = y^k\Gamma^i_{jk}; \quad i, j \in J_n. \tag{7.3.5}$$

7.3.3 The Crampin – Grifone Construction

Proposition 7.3.4 (M. Crampin and J. Grifone). *Let $S \in \mathfrak{X}(\mathring{T}M)$ be a second-order vector field. If*

$$\mathbf{h} := \frac{1}{2}(1_{\mathfrak{X}(\mathring{T}M)} + [\mathbf{J}, S]), \tag{7.3.6}$$

then $\mathbf{h} \in \mathrm{End}(\mathfrak{X}(\mathring{T}M))$ satisfying $\mathbf{h}^2 = \mathbf{h}$ and $\mathrm{Ker}(\mathbf{h}) = \mathfrak{X}^{\mathsf{v}}(\mathring{T}M)$. Therefore, there exists a unique Ehresmann connection $\mathcal{H}$ in $\mathring{T}M$ such that

$\mathcal{H} \circ \mathbf{j} = \mathbf{h}$. *The horizontal lift of a vector field $X \in \mathfrak{X}(M)$ with respect to $\mathcal{H}$ is given by*

$$X^{\mathbf{h}} := \mathcal{H}(\widehat{X}) = \frac{1}{2}(X^{\mathbf{c}} + [X^{\mathbf{v}}, S]). \tag{7.3.7}$$

The second-order vector field associated to $\mathcal{H}$ is

$$S_{\mathcal{H}} := \mathcal{H}\widetilde{\delta} = S + \frac{1}{2}S^* = \frac{1}{2}(S + [C, S]). \tag{7.3.8}$$

If S is a spray over M, then

$$\mathbf{h} := \frac{1}{2}(1_{\mathfrak{X}(TM)} + [\mathbf{J}, S])$$

(regarded as a bundle endomorphism of τ_{TM}) satisfies conditions (i)–(iv) of Lemma 7.3.1, therefore there exists a unique Ehresmann connection $\mathcal{H}$ in TM with horizontal projection $\mathbf{h}$. The semispray associated to $\mathcal{H}$ is the starting spray S.

Proof.   Obviously, our results obtained for second-order vector fields over M (including 5.1.9 and its consequences) are valid without any change also for second-order vector fields on $\mathring{T}M$. Taking this into account,

$$\mathbf{h}^2 = \frac{1}{4}(1_{\mathfrak{X}(\mathring{T}M)} + 2[\mathbf{J}, S] + [\mathbf{J}, S]^2) \overset{(5.1.10)}{=} \mathbf{h}.$$

Now we verify that $\mathrm{Ker}(\mathbf{h}) = \mathfrak{X}^{\mathbf{v}}(\mathring{T}M)$.

If $\mathbf{h}(\xi) = 0$ for some $\xi \in \mathfrak{X}(\mathring{T}M)$, then (7.3.6) gives

$$\xi = -[\mathbf{J}, S]\xi \overset{(3.3.16)}{=} -[\mathbf{J}\xi, S] + \mathbf{J}[\xi, S].$$

Hence

$$\mathbf{J}\xi = -\mathbf{J}[\mathbf{J}\xi, S] \overset{(5.1.4)}{=} -\mathbf{J}\xi,$$

and therefore $\mathbf{J}\xi = 0$. This proves that $\mathrm{Ker}(\mathbf{h}) \subset \mathfrak{X}^{\mathbf{v}}(\mathring{T}M)$. On the other hand, for all $\xi \in \mathfrak{X}(\mathring{T}M)$,

$$2\mathbf{h}(\mathbf{J}\xi) \overset{(7.3.6)}{=} \mathbf{J}\xi + [\mathbf{J}, S]\mathbf{J}\xi = \mathbf{J}\xi - \mathbf{J}[\mathbf{J}\xi, S] \overset{(5.1.4)}{=} \mathbf{J}\xi - \mathbf{J}\xi = 0,$$

so the converse inclusion $\mathfrak{X}^{\mathbf{v}}(\mathring{T}M) \subset \mathrm{Ker}(\mathbf{h})$ is also true. From these, by Remark 7.3.2, we conclude the first assertion of the proposition.

For any vector field X on M,

$$\mathcal{H}(\widehat{X}) \overset{(7.2.33)}{=} \mathbf{h}X^{\mathbf{c}} \overset{(7.3.6)}{=} \frac{1}{2}(X^{\mathbf{c}} + [\mathbf{J}, S]X^{\mathbf{c}})$$

$$= \frac{1}{2}(X^{\mathbf{c}} + [X^{\mathbf{v}}, S] - \mathbf{J}[X^{\mathbf{c}}, S]) \overset{(5.1.6)}{=} \frac{1}{2}(X^{\mathbf{c}} + [X^{\mathbf{v}}, S]),$$

so we have (7.3.7).

To determine the semispray associated to $\mathcal{H}$ choose an arbitrary second-order vector field $\widetilde{S} \in \mathfrak{X}(\mathring{T}M)$. Then

$$S_{\mathcal{H}} := \mathcal{H}\widetilde{\delta} = \mathcal{H} \circ \mathbf{j} \circ \widetilde{S} = \mathbf{h}\widetilde{S} = \frac{1}{2}(\widetilde{S} + [\mathbf{J}, S]\widetilde{S}) = \frac{1}{2}(\widetilde{S} + [C, S] - \mathbf{J}[\widetilde{S}, S]).$$

Since the vector field $\widetilde{S} - S$ is vertical, the Grifone identity gives

$$\widetilde{S} - S = \mathbf{J}[\widetilde{S} - S, S] = \mathbf{J}[\widetilde{S}, S].$$

Hence

$$S_{\mathcal{H}} = \frac{1}{2}(\widetilde{S} + [C, S] - \widetilde{S} + S) = \frac{1}{2}(S + [C, S]),$$

as wanted.

If S is a spray over M, then our above argument shows, without modification, that conditions (i)–(iii) of Lemma 7.3.1 hold. Since $[C, S] = S$, we have $S_{\mathcal{H}} = S$. So it remains to verify that $\mathbf{h} \circ o_* = o_*$, where $o \in \mathfrak{X}(M)$ is the zero vector field.

Let p be a point of M. Choose a chart $(\mathcal{U}, (u^i)_{i=1}^n)$ of M at p, and consider the induced chart $(\tau^{-1}(\mathcal{U}), ((x^i)_{i=1}^n, (y^i)_{i=1}^n))$ in TM. Let the functions G^i be the coefficients of S with respect to the chosen chart. Then for each $j \in J_n$,

$$(o_*)_p \left(\frac{\partial}{\partial u^j}\right)_p \overset{(3.1.5)}{=} \frac{\partial(x^k \circ o)}{\partial u^j}(p) \left(\frac{\partial}{\partial x^k}\right)_{o(p)} + \frac{\partial(y^k \circ o)}{\partial u^j}(p) \left(\frac{\partial}{\partial y^k}\right)_{o(p)}$$

$$= \frac{\partial u^k}{\partial u^j}(p) \left(\frac{\partial}{\partial x^k}\right)_{o(p)} = \left(\frac{\partial}{\partial x^j}\right)_{o(p)}.$$

Thus, on the other hand,

$$\mathbf{h} \circ o_* \left(\frac{\partial}{\partial u^j}\right)_p = \mathbf{h}\left(\frac{\partial}{\partial x^j}\right)_{o(p)} = \mathcal{H} \circ \mathbf{j}\left(\frac{\partial}{\partial x^j}\right)_{o(p)} = \mathcal{H}\left(o(p), \left(\frac{\partial}{\partial u^j}\right)_p\right).$$

Since

$$\mathcal{H}\left(\widehat{\frac{\partial}{\partial u^j}}\right) \overset{(7.3.7)}{=} \frac{1}{2}\left(\frac{\partial}{\partial x^j} + \left[\frac{\partial}{\partial y^j}, y^i \frac{\partial}{\partial x^i} - 2G^i \frac{\partial}{\partial y^i}\right]\right)$$

$$= \frac{1}{2}\left(2\frac{\partial}{\partial x^j} - 2\frac{\partial G^i}{\partial y^j}\frac{\partial}{\partial y^i}\right) = \frac{\partial}{\partial x^j} - \frac{\partial G^i}{\partial y^j}\frac{\partial}{\partial y^i},$$

the Christoffel symbols of $\mathcal{H}$ with respect to the chart $(\mathcal{U}, (u^i)_{i=1}^n)$ are the functions

$$G^i_j := \frac{\partial G^i}{\partial y^j} : \tau^{-1}(\mathcal{U}) \to \mathbb{R}; \quad i, j \in J_n.$$

Because S is a spray, its coefficients G^i are positive-homogeneous of degree 2 (see Remark 5.1.19(b)(c)), and hence the functions $\frac{\partial G^i}{\partial y^j}$ are 1^+-homogeneous. This implies that $\frac{\partial G^i}{\partial y^j}(0_p) = 0$. Hence

$$\mathbf{h} \circ o_* \left(\frac{\partial}{\partial u^j} \right)_p = \left(\frac{\partial}{\partial x^j} \right)_{0_p} - \frac{\partial G^i}{\partial y^j}(0_p) \left(\frac{\partial}{\partial y^i} \right)_{0_p} = \left(\frac{\partial}{\partial x^j} \right)_{0_p},$$

which concludes the proof of the relation $\mathbf{h} \circ o_* = o_*$, and hence that of the proposition. $\qquad\square$

Remark 7.3.5. We say that the Ehresmann connection $\mathcal{H}$ (and the horizontal projection $\mathbf{h}$ given by (7.3.6)) in the first part of Proposition 7.3.4 is *generated by* the second-order vector field S. Our above calculation shows that *the Christoffel symbols of $\mathcal{H}$ with respect to a chart $(\mathcal{U}, (u^i)_{i=1}^n)$ are*

$$G_j^i := \frac{\partial G^i}{\partial y^j} \text{ over } \tau^{-1}(\mathcal{U}) \cap \mathring{T}M, \tag{7.3.9}$$

where the functions G^i are the coefficients of S, while condition (C_4) forces that

$$G_j^i(0_p) = 0 \text{ for all } p \in \mathcal{U}.$$

Corollary 7.3.6. *A second-order vector field on $\mathring{T}M$ is horizontal with respect to the Ehresmann connection generated by itself if, and only if, it is positive-homogeneous of degree two.*

Proof. Suppose that $S \in \mathfrak{X}(\mathring{T}M)$ is a second-order vector field. Consider the Ehresmann connection $\mathcal{H}$ generated by S, and let $\mathbf{h}$ be the horizontal projection determined by $\mathcal{H}$. Then

$$\mathbf{h}S = \mathcal{H} \circ \mathbf{j}(S) = \mathcal{H}\widetilde{\delta} \stackrel{(7.3.8)}{=} \frac{1}{2}(S + [C, S]),$$

therefore $\mathbf{h}S = S$ if, and only if, $[C, S] = S$, as was to be shown. $\qquad\square$

7.4   Some Useful Technicalities

Let $\mathcal{H}$ be an Ehresmann connection in $\mathring{T}M$, and consider the horizontal projection determined by $\mathcal{H}$. Then $\mathbf{h} \in \text{End}(\mathfrak{X}(\mathring{T}M)) = \mathcal{A}_1^1(\mathring{T}M)$, so by the Frölicher–Nijenhuis theory sketched in 3.3.6 two graded derivations $i_\mathbf{h}$ and $\mathcal{L}_\mathbf{h}$ of $\mathcal{A}(\mathring{T}M)$ (of degree 0 and 1, resp.) are associated to $\mathbf{h}$. We recall that

$$i_\mathbf{h}F = 0, \quad i_\mathbf{h}dF = dF \circ \mathbf{h} \quad (F \in C^\infty(\mathring{T}M)), \tag{7.4.1}$$

and i_h is determined by these relations; and, moreover,

$$\mathcal{L}_\mathsf{h} := [i_\mathsf{h}, d] = i_\mathsf{h} \circ d - d \circ i_\mathsf{h}. \tag{7.4.2}$$

Then for any smooth function F on $\mathring{T}M$,

$$\mathcal{L}_\mathsf{h} F = i_\mathsf{h} dF = dF \circ \mathsf{h}. \tag{7.4.3}$$

Locally, for each $j \in J_n$, we have

$$\mathcal{L}_\mathsf{h} F\left(\frac{\partial}{\partial x^j}\right) = dF\left(\mathsf{h}\left(\frac{\partial}{\partial u^j}\right)^\mathsf{c}\right) = dF\left(\frac{\partial}{\partial u^j}\right)^\mathsf{h}$$

$$\overset{(7.2.37)}{=} \left(\frac{\partial}{\partial x^j} - N_j^i \frac{\partial}{\partial y^i}\right) F = \frac{\partial F}{\partial x^j} - N_j^i \frac{\partial F}{\partial y^i},$$

$$\mathcal{L}_\mathsf{h} F\left(\frac{\partial}{\partial y^j}\right) = dF\left(\mathsf{h}\frac{\partial}{\partial y^j}\right) = 0,$$

therefore

$$\mathcal{L}_\mathsf{h} F \underset{(\mathcal{U})}{=} \left(\frac{\partial F}{\partial x^j} - N_j^i \frac{\partial F}{\partial y^i}\right) dx^j. \tag{7.4.4}$$

Lemma 7.4.1. *Let $\mathcal{H}$ be an Ehresmann connection in $\mathring{T}M$ with horizontal projection h. Then, for every smooth function F on $\mathring{T}M$,*

$$\mathcal{L}_\mathsf{h} \mathcal{L}_\mathsf{J} F = \mathsf{h}^* d\mathcal{L}_\mathsf{J} F. \tag{7.4.5}$$

Proof. In view of (7.4.2), $\mathcal{L}_\mathsf{h} \mathcal{L}_\mathsf{J} F = i_\mathsf{h} d\mathcal{L}_\mathsf{J} F - d i_\mathsf{h} \mathcal{L}_\mathsf{J} F$. Here, by (3.3.40), $i_\mathsf{h} \mathcal{L}_\mathsf{J} F = \mathsf{h}^* \mathcal{L}_\mathsf{J} F$. Since for every $X \in \mathfrak{X}(M)$,

$$\mathsf{h}^* \mathcal{L}_\mathsf{J} F(X^\mathsf{v}) = 0 = \mathcal{L}_\mathsf{J} F(X^\mathsf{v}), \quad \mathsf{h}^* \mathcal{L}_\mathsf{J} F(X^\mathsf{h}) = \mathcal{L}_\mathsf{J} F(X^\mathsf{h}),$$

it follows by Lemma 7.2.6 that $\mathsf{h}^* \mathcal{L}_\mathsf{J} F = \mathcal{L}_\mathsf{J} F$, and hence

$$\mathcal{L}_\mathsf{h} \mathcal{L}_\mathsf{J} F = i_\mathsf{h} d\mathcal{L}_\mathsf{J} F - d\mathcal{L}_\mathsf{J} F. \tag{$*$}$$

It can immediately be checked that the right-hand side of ($*$) vanishes on all pairs of the form

$$(X^\mathsf{v}, Y^\mathsf{v}), \ (X^\mathsf{v}, Y^\mathsf{h}), \ (X^\mathsf{h}, Y^\mathsf{v}); \quad X, Y \in \mathfrak{X}(M).$$

Since, furthermore,

$$(i_\mathsf{h} d\mathcal{L}_\mathsf{J} F - d\mathcal{L}_\mathsf{J} F)(X^\mathsf{h}, Y^\mathsf{h}) \overset{(3.3.38)}{=} 2d\mathcal{L}_\mathsf{J} F(X^\mathsf{h}, Y^\mathsf{h}) - d\mathcal{L}_\mathsf{J} F(X^\mathsf{h}, Y^\mathsf{h})$$

$$= d\mathcal{L}_\mathsf{J} F(X^\mathsf{h}, Y^\mathsf{h}) = \mathsf{h}^* d\mathcal{L}_\mathsf{J} F(X^\mathsf{h}, Y^\mathsf{h}),$$

the desired relation follows. $\qquad\square$

Lemma and Definition 7.4.2. *Let β be a one-form on a manifold M, and consider its vertical lift $\beta^{\vee}$ given by (4.1.75). If*

$$\bar{\beta} := i_S \beta^{\vee}, \tag{7.4.6}$$

where S is an arbitrary semispray, then $\bar{\beta}$ is a well-defined smooth function on TM, called the function determined by the one-form β. This function is 1^+-homogeneous and satisfies the identity

$$\mathcal{L}_{\mathbf{J}}\bar{\beta} = \beta^{\vee}. \tag{7.4.7}$$

Proof. Since the difference of two semisprays is a vertical vector field, and $\beta^{\vee}$ kills vertical vector fields, the function $\bar{\beta}$ is independent of the choice of semispray. To show the 1^+-homogeneity of $\bar{\beta}$, we first express the function $C\bar{\beta}$ in terms of $\beta^{\vee}$:

$$C\bar{\beta} = \mathcal{L}_C i_S \beta^{\vee} \stackrel{(3.3.31)}{=} i_S \mathcal{L}_C \beta^{\vee} + i_{[C,S]}\beta^{\vee}.$$

Here, for every vector field X on M,

$$(\mathcal{L}_C \beta^{\vee})(X^{\vee}) = C(\beta^{\vee}(X^{\vee})) - \beta^{\vee}([C, X^{\vee}])$$
$$\stackrel{(4.1.53)}{=} C(\beta^{\vee}(X^{\vee})) + \beta^{\vee}(X^{\vee}) \stackrel{(4.1.75)}{=} 0,$$

$$(\mathcal{L}_C \beta^{\vee})(X^c) = C(\beta^{\vee}(X^c)) - \beta^{\vee}([C, X^c]) \stackrel{(4.1.75),(4.1.72)}{=} C(\beta(X))^{\vee} = 0,$$

therefore $\mathcal{L}_C \beta^{\vee} = 0$. Since S can be chosen to be a spray, $i_{[C,S]}\beta^{\vee} = i_S \beta^{\vee} =: \bar{\beta}$. So we get $C\bar{\beta} = \bar{\beta}$, as desired.

Now we turn to prove (7.4.7). For every $X \in \mathfrak{X}(M)$,

$$0 \stackrel{(4.1.99)}{=} \mathcal{L}_{\mathbf{J}}\beta^{\vee}(S, X^c) \stackrel{(4.1.96)}{=} i_{\mathbf{J}}d\beta^{\vee}(S, X^c) - di_{\mathbf{J}}\beta^{\vee}(S, X^c)$$
$$\stackrel{(4.1.95),(4.1.99)}{=} d\beta^{\vee}(C, X^c) + d\beta^{\vee}(S, X^{\vee}) \stackrel{(4.1.80)}{=} (d\beta)^{\vee}(C, X^c)$$
$$+ S\beta^{\vee}(X^{\vee}) - X^{\vee}\beta^{\vee}(S) - \beta^{\vee}([S, X^{\vee}]) \stackrel{(4.1.78)}{=} -X^{\vee}\bar{\beta} - \beta^{\vee}([S, X^{\vee}]),$$

whence $(\mathcal{L}_{\mathbf{J}}\bar{\beta})(X^c) = -\beta^{\vee}([S, X^{\vee}])$. Here

$$[S, X^{\vee}] \stackrel{(7.3.7)}{=} X^c - 2X^{\mathsf{h}} = X^c - 2\mathbf{h}X^c = X^c - 2(X^c - \mathbf{v}X^c) = -X^c + 2\mathbf{v}X^c,$$

therefore

$$(\mathcal{L}_{\mathbf{J}}\bar{\beta})(X^c) = \beta^{\vee}(X^c - 2\mathbf{v}X^c) = \beta^{\vee}(X^c),$$

as was to be shown. $\qquad\square$

Lemma 7.4.3. *If β is a one-form on M, then for every $v \in TM$,*

$$\bar{\beta}(v) = \beta_{\tau(v)}(v). \tag{7.4.8}$$

In particular, the differential of a function $f \in C^\infty(M)$ induces the complete lift of f on TM:

$$\overline{df} = f^c. \tag{7.4.9}$$

Proof. Starting with the definition of $\bar{\beta}$, we have

$$\bar{\beta}(v) = (i_S \beta^\vee)(v) = (\beta^\vee)_v(S(v)) \overset{(4.1.76)}{=} (\tau^*\beta)_v(S(v))$$
$$\overset{(3.3.7)}{=} \beta_{\tau(v)}(\tau_*(S(v))) \overset{\text{Lemma } 5.1.3}{=} \beta_{\tau(v)}(v),$$

as desired. Thus, in particular,

$$\overline{df}(v) = (df)_{\tau(v)}(v) \overset{(3.3.1)}{=} v(f) =: f^c(v),$$

whence (7.4.9). $\qquad\square$

We note that the function $\bar{\beta}$ is usually defined by (7.4.8).

Lemma 7.4.4. *For every one-form β on M,*

$$(d\beta)^\vee = \frac{1}{2}\mathcal{L}_J i_S (d\beta)^\vee, \tag{7.4.10}$$

where S is an arbitrary semispray over M.

Proof. The two-form $(d\beta)^\vee$ can be manipulated as follows:

$$(d\beta)^\vee \overset{(4.1.80)}{=} d\beta^\vee \overset{(7.4.7)}{=} d\mathcal{L}_J\bar{\beta} \overset{(3.3.52)}{=} -\mathcal{L}_J d\bar{\beta}$$
$$= -\mathcal{L}_J di_S\beta^\vee \overset{(3.3.33)}{=} -\mathcal{L}_J\mathcal{L}_S\beta^\vee + \mathcal{L}_J i_S d\beta^\vee.$$

Here

$$\mathcal{L}_J\mathcal{L}_S\beta^\vee \overset{(3.3.53)}{=} \mathcal{L}_S\mathcal{L}_J\beta^\vee + \mathcal{L}_{[J,S]}\beta^\vee \overset{(4.1.99)}{=} \mathcal{L}_{[J,S]}\beta^\vee \overset{(7.3.6)}{=} 2\mathcal{L}_h\beta^\vee - \mathcal{L}_1\beta^\vee$$
$$\overset{(3.3.42)}{=} 2\mathcal{L}_h\beta^\vee - d\beta^\vee \overset{(7.4.7)}{=} 2\mathcal{L}_h\mathcal{L}_J\bar{\beta} - d\beta^\vee \overset{(7.4.5)}{=} 2h^*d\mathcal{L}_J\bar{\beta} - d\beta^\vee$$
$$\overset{(7.4.7)}{=} 2h^*d\beta^\vee - d\beta^\vee \overset{\text{Lemma } 7.2.6}{=} 2d\beta^\vee - d\beta^\vee = d\beta^\vee = (d\beta)^\vee,$$

therefore $(d\beta)^\vee = -(d\beta)^\vee + \mathcal{L}_J i_S(d\beta)^\vee$, so we have the desired relation. $\qquad\square$

7.5 Homogeneity and Linearity

7.5.1 *Homogeneity Conditions*

In the following, the symbols $\boxplus$ and $\boxdot$ stand for the linear operations defined by (4.1.21).

Definition 7.5.1. Let $\mathcal{H}$ be an Ehresmann connection in TM. We say that

(i) $\mathcal{H}$ is *homogeneous* if for all $(u, v) \in TM \times_M TM$, $\lambda \in \mathbb{R}$ we have

$$\mathcal{H}(\lambda u, v) = \lambda \boxdot \mathcal{H}(u, v); \qquad (7.5.1)$$

(ii) $\mathcal{H}$ is *positive-homogeneous* if relation (7.5.1) holds for each element of $TM \times_M TM$, and for any positive number λ;

(iii) $\mathcal{H}$ is *linear* if

$$\mathcal{H}(\lambda_1 u_1 + \lambda_2 u_2, v) = \lambda_1 \boxdot \mathcal{H}(u_1, v) \boxplus \lambda_2 \boxdot \mathcal{H}(u_2, v) \qquad (7.5.2)$$

for all $(u_1, v), (u_2, v) \in TM \times_M TM$ and $\lambda_1, \lambda_2 \in \mathbb{R}$.

Lemma 7.5.2. *Let the notation be as in Remark 7.2.12. An Ehresmann connection $\mathcal{H}$ in TM is*

homogeneous, positive-homogeneous *or* linear, *respectively,*

if, and only if, the Christoffel symbols of $\mathcal{H}$ with respect to any chart $(\mathcal{U}, (u^i)_{i=1}^n)$ *of M are*

homogeneous, positive-homogeneous *or* linear

functions on $\tau^{-1}(\mathcal{U})$, respectively, i.e., we have:

$N_j^i(\lambda v) = \lambda N_j^i(v)$ – homogeneous case,
$N_j^i(e^\lambda v) = e^\lambda N_j^i(v)$ – positive-homogeneous case,
$N_j^i(\lambda_1 v_1 + \lambda_2 v_2) = \lambda_1 N_j^i(v_1) + \lambda_2 N_j^i(v_2)$ – linear case,

where $\lambda, \lambda_1, \lambda_2 \in \mathbb{R}$, $v, v_1, v_2 \in \tau^{-1}(\mathcal{U})$ and $\tau(v_1) = \tau(v_2)$.

Proof. According to (7.2.37),

$$\mathcal{H}\left(v, \left(\frac{\partial}{\partial u^j}\right)_{\tau(v)}\right) = \left(\frac{\partial}{\partial x^j}\right)_v - N_j^i(v)\left(\frac{\partial}{\partial y^i}\right)_v,$$

$$\mathcal{H}\left(\lambda v, \left(\frac{\partial}{\partial u^j}\right)_{\tau(v)}\right) = \left(\frac{\partial}{\partial x^j}\right)_{\lambda v} - N_j^i(\lambda v)\left(\frac{\partial}{\partial y^i}\right)_{\lambda v},$$

for every $j \in J_n$. On the other hand,

$$\lambda \boxdot \mathcal{H}\left(v, \left(\frac{\partial}{\partial u^j}\right)_{\tau(v)}\right) \overset{(4.1.23)}{=} \left(\frac{\partial}{\partial x^j}\right)_{\lambda v} - \lambda N_j^i(v) \left(\frac{\partial}{\partial y^i}\right)_{\lambda v},$$

therefore

$$\begin{cases} \mathcal{H}\left(\lambda v, \left(\frac{\partial}{\partial u^j}\right)_{\tau(v)}\right) = \lambda \boxdot \mathcal{H}\left(v, \left(\frac{\partial}{\partial u^j}\right)_{\tau(v)}\right) \text{ if, and only if,} \\ N_j^i(\lambda v) = \lambda N_j^i(v) \text{ for all } i, j \in J_n. \end{cases} \quad (*)$$

Similarly,

$$\mathcal{H}\left(v_1 + v_2, \left(\frac{\partial}{\partial u^j}\right)_p\right) = \left(\frac{\partial}{\partial x^j}\right)_{v_1+v_2} - N_j^i(v_1 + v_2)\left(\frac{\partial}{\partial y^i}\right)_{v_1+v_2},$$

$$\mathcal{H}\left(v_1, \left(\frac{\partial}{\partial u^j}\right)_p\right) \boxplus \mathcal{H}\left(v_2, \left(\frac{\partial}{\partial u^j}\right)_p\right)$$

$$\overset{(4.1.22)}{=} \left(\frac{\partial}{\partial x^j}\right)_{v_1+v_2} - (N_j^i(v_1) + N_j^i(v_2))\left(\frac{\partial}{\partial y^i}\right)_{v_1+v_2},$$

so

$$\begin{cases} \mathcal{H}\left(v_1 + v_2, \left(\frac{\partial}{\partial u^j}\right)_p\right) = \mathcal{H}\left(v_1, \left(\frac{\partial}{\partial u^j}\right)_p\right) \boxplus \mathcal{H}\left(v_2, \left(\frac{\partial}{\partial u^j}\right)_p\right) \\ \text{if, and only if, } N_j^i(v_1 + v_2) = N_j^i(v_1) + N_j^i(v_2) \text{ for all } i, j \in J_n. \end{cases} \quad (**)$$

From $(*)$ and $(**)$ all our assertions follow at once. $\qquad \square$

Corollary 7.5.3. *If an Ehresmann connection is linear, then it is smooth on all of its domain.*

Proof. Suppose that $\mathcal{H}$ is a linear Ehresmann connection in TM. Given any chart $(\mathcal{U}, u)$ on M, by the previous lemma the Christoffel symbols N_j^i of $\mathcal{H}$ with respect to $(\mathcal{U}, u)$ are fibrewise linear functions. Thus, for any fixed $(i, j) \in J_n \times J_n$ and for any $p \in \mathcal{U}$, there exists a unique family $(\Gamma_{jk}^i(p))_{k=1}^n$ of real numbers such that

$$N_j^i \upharpoonright T_pM : v \in T_pM \mapsto N_j^i(v) = \Gamma_{jk}^i(p)y^k(v) \in \mathbb{R}.$$

With the help of the functions $\Gamma_{jk}^i : \mathcal{U} \to \mathbb{R}$, $p \mapsto \Gamma_{jk}^i(p)$, $k \in J_n$, the Christoffel symbols N_j^i can be written in the form

$$N_j^i = (\Gamma_{jl}^i \circ \tau)y^l. \quad (7.5.3)$$

Composing both sides of this relation with a coordinate vector field $\frac{\partial}{\partial u^k}$, we find

$$N_j^i \circ \frac{\partial}{\partial u^k} = \left(\Gamma_{jl}^i \circ \tau \circ \frac{\partial}{\partial u^k}\right)\left(y^l \circ \frac{\partial}{\partial u^k}\right) = \Gamma_{jl}^i \delta_k^l = \Gamma_{jk}^i.$$

Since the functions N_j^i are smooth on $\tau^{-1}(\mathcal{U}) \cap \overset{\circ}{T}M$ by (C_3) and the coordinate vector fields are nowhere vanishing, it follows that the functions Γ_{jk}^i are smooth. So the same is true for the Christoffel symbols N_j^i by (7.5.3). $\square$

Remark 7.5.4. We recall that in Example 3.1.38 we defined the dilation

$$\mu_\lambda : TM \to TM, \ v \mapsto \mu_\lambda(v) := \lambda v$$

for any fixed $\lambda \in \mathbb{R}$. In Remark 4.1.11 we saw that $(\mu_\lambda)_*(w) = \lambda \boxdot w$. For our next purposes, it will also be useful to introduce the mapping

$$\mu_\lambda^1 : TM \times_M TM \to TM \times_M TM, \ (v, w) \mapsto (\lambda v, w). \tag{7.5.4}$$

Now we immediately obtain the following result.

Corollary 7.5.5. *An Ehresmann connection $\mathcal{H}$ in TM is homogeneous (positive-homogeneous) if, and only if,*

$$(\mu_\lambda)_* \circ \mathcal{H} = \mathcal{H} \circ \mu_\lambda^1 \tag{7.5.5}$$

holds for all $\lambda \in \mathbb{R}$ (for all $\lambda \in \mathbb{R}_+^*$). $\square$

Corollary 7.5.6. *The associated semispray of a positive-homogeneous Ehresmann connection is a spray.*

Proof. Let $\mathcal{H}$ be a positive-homogeneous Ehresmann connection in $\overset{\circ}{T}M$ with associated semispray $S := \mathcal{H}\widetilde{\delta}$. Then, for any positive number λ and any vector $u \in \overset{\circ}{T}M$,

$$\widetilde{\mu}_\lambda \circ (\mu_\lambda)_* \circ S(v) = \widetilde{\mu}_\lambda \circ (\mu_\lambda)_* \circ \mathcal{H}(v, v) \overset{(7.5.5)}{=} \lambda \mathcal{H}(\lambda v, v)$$
$$= \mathcal{H}(\lambda v, \lambda v) = S(\lambda v) = S \circ \mu_\lambda(v).$$

This proves that S is 2^+-homogeneous (see (4.2.7)). $\square$

Proposition 7.5.7. *Let $\mathcal{H}$ be an Ehresmann connection in TM, and consider the horizontal projection $\mathbf{h}$, the vertical projection $\mathbf{v}$ and the connector K associated with $\mathcal{H}$. Then $\mathcal{H}$ is homogeneous (positive-homogeneous) if, and only if, one, and hence all, of the following equivalent relations holds for all $\lambda \in \mathbb{R}$ (for all $\lambda \in \mathbb{R}_+^*$):*

$$\mathbf{h} \circ (\mu_\lambda)_* = (\mu_\lambda)_* \circ \mathbf{h}, \tag{7.5.6}$$

$$\mathbf{v} \circ (\mu_\lambda)_* = (\mu_\lambda)_* \circ \mathbf{v}, \tag{7.5.7}$$

$$K \circ (\mu_\lambda)_* = \mu_\lambda \circ K. \tag{7.5.8}$$

Proof. First we show that (7.5.6) is equivalent to (7.5.5). Indeed, (7.5.6) holds if, and only if,

$$\mathbf{h}_{\lambda v} \circ ((\mu_\lambda)_*)_v = ((\mu_\lambda)_*)_v \circ \mathbf{h}_v, \ \text{for all } v \in TM.$$

If $w \in T_v TM$, then

$$\mathbf{h}_{\lambda v} \circ ((\mu_\lambda)_*)_v(w) \overset{(7.2.13)}{=} \mathcal{H}\big(\lambda v, \tau_*((\mu_\lambda)_*(w))\big) = \mathcal{H}(\lambda v, (\tau \circ \mu_\lambda)_*(w))$$

$$= \mathcal{H}(\lambda v, \tau_*(w)) \overset{(7.5.4)}{=} \mathcal{H} \circ \mu_\lambda^1(v, \tau_*(w)),$$

and $((\mu_\lambda)_*)_v \circ \mathbf{h}_v(w) = (\mu_\lambda)_* \circ \mathcal{H}(v, \tau_*(w))$, whence our assertion. Thus, by Corollary 7.5.5, (7.5.6) is equivalent to the homogeneity (or the positive-homogeneity) of $\mathcal{H}$.

Next, obviously, (7.5.7) is equivalent to (7.5.6). Finally, since

$$K \circ (\mu_\lambda)_* = \ell^\downarrow \circ \mathbf{v} \circ (\mu_\lambda)_* \text{ and } \mu_\lambda \circ K = \mu_\lambda \circ \ell^\downarrow \circ \mathbf{v} \overset{(4.1.58)}{=} \ell^\downarrow \circ (\mu_\lambda)_* \circ \mathbf{v},$$

relations (7.5.7) and (7.5.8) are also equivalent. $\qquad\square$

Proposition 7.5.8. *An Ehresmann connection is positive-homogeneous if, and only if, the horizontal lift of every vector field is positive-homogeneous of degree 1.*

Proof. First we recall that the Liouville vector field is generated by the flow

$$\mu^+ \colon \mathbb{R} \times TM \to TM, \quad (t, v) \mapsto \mu^+(t, v) = e^t v.$$

Let X be a vector field on M. Then $[C, X^c] \overset{(4.1.72)}{=} 0$, $[C, X^v] \overset{(4.1.53)}{=} -X^v$, so by Definition 4.2.12(b) and Lemma 4.2.14 we have the following relations:

$$(\mu_t^+)_* \circ X^c = X^c \circ \mu_t^+, \quad t \in \mathbb{R}; \tag{$*$}$$

$$(\mu_t^+)_* \circ X^v = \widetilde{\mu}_t^+ \circ X^v \circ \mu_t^+, \quad t \in \mathbb{R}. \tag{$**$}$$

Now let $\mathcal{H}$ be an Ehresmann connection in TM. If $\mathcal{H}$ is positive-homogeneous, then

$$(\mu_t^+)_* \circ X^h = (\mu_t^+)_* \circ \mathbf{h} \circ X^c \overset{(7.5.6)}{=} \mathbf{h} \circ (\mu_t^+)_* \circ X^c$$

$$\overset{(*)}{=} \mathbf{h} \circ X^c \circ \mu_t^+ = X^h \circ \mu_t^+,$$

so X^h is positive-homogeneous of degree 1.

Conversely, if every horizontal lift X^h has the homogeneity property in question, then, by a similar calculation as above, we obtain

$$(\mu_t^+)_* \circ \mathbf{h} \circ X^c = (\mu_t^+)_* \circ X^h \overset{(4.2.7)}{=} X^h \circ \mu_t^+ = \mathbf{h} \circ X^c \circ \mu_t^+ \overset{(*)}{=} \mathbf{h} \circ (\mu_t^+)_* \circ X^c.$$

On the other hand, we clearly have $(\mu_t^+)_* \circ \mathbf{h} \circ X^v = 0$, and $\mathbf{h} \circ (\mu_t^+)_* \circ X^v = 0$ also holds, because

$$\mathbf{h} \circ (\mu_t^+)_* \circ X^v \overset{(**)}{=} \mathbf{h} \circ \widetilde{\mu}_t^+ \circ X^v \circ \mu_t^+ = \widetilde{\mu}_t^+ \circ \mathbf{h} \circ X^v \circ \mu_t^+.$$

Thus $(\mu_t^+)_* \circ \mathbf{h}$ and $\mathbf{h} \circ (\mu_t^+)_*$ act in the same way on the complete and on the vertical lifts of vector fields on M. By the local frame field principle I this implies that $\mathbf{h} \circ (\mu_t^+)_* = (\mu_t^+)_* \circ \mathbf{h}$ for all $t \in \mathbb{R}$. $\qquad\square$

Lemma and Definition 7.5.9. *Let $\mathcal{H}$ be an Ehresmann connection in TM, and let $\mathcal{V}$ be the vertical mapping associated with $\mathcal{H}$. The mapping*

$$\mathbf{t}\colon \Gamma(\mathring{\pi}) \to \Gamma(\mathring{\pi}), \quad \widetilde{X} \mapsto \mathbf{t}(\widetilde{X}) := \mathcal{V}[\mathcal{H}\widetilde{X}, C] \tag{7.5.9}$$

is a type $(1,1)$ Finsler tensor on $\mathring{T}M$, i.e., $\mathbf{t} \in \mathcal{F}_1^1(\mathring{T}M) \cong \mathrm{End}(\Gamma(\mathring{\pi}))$.

We say that $\mathbf{t}$ is the tension *of the Ehresmann connection $\mathcal{H}$. For any basic section $\widehat{X} \in \Gamma(\mathring{\pi})$ we have*

$$\mathbf{i}\,\mathbf{t}(\widehat{X}) = [X^{\mathsf{h}}, C]. \tag{7.5.10}$$

If $(\mathcal{U}, (u^i)_{i=1}^n)$ is a chart on M, then

$$\mathbf{t} \underset{(\mathcal{U})}{=} \left(y^k \frac{\partial N_j^i}{\partial y^k} - N_j^i \right) \widehat{\frac{\partial}{\partial u^i}} \otimes \widehat{du^j}, \tag{7.5.11}$$

where (N_j^i) is the family of the Christoffel symbols of $\mathcal{H}$ with respect to the chosen chart, and the summation convention is employed.

Proof.   Let F be a smooth function on TM. Then

$$\mathbf{t}(F\widetilde{X}) := \mathcal{V}[\mathcal{H}(F\widetilde{X}), C] = \mathcal{V}[F\mathcal{H}(\widetilde{X}), C]$$

$$= F\,\mathbf{t}(\widetilde{X}) - (CF)\mathcal{V} \circ \mathcal{H}(\widetilde{X}) \overset{(7.2.21)}{=} F\,\mathbf{t}(\widetilde{X}).$$

Since $\mathbf{t}$ is clearly additive, it follows that the tension of $\mathcal{H}$ is indeed a type $(1,1)$ Finsler tensor. As $[X^{\mathsf{h}}, C]$ is obviously vertical, formula (7.5.10) is an immediate consequence of the definition. Finally, for each $j \in J_n$,

$$\mathbf{t}\left(\widehat{\frac{\partial}{\partial u^j}} \right) \overset{(7.2.37)}{=} \mathcal{V}\left[\frac{\partial}{\partial x^j} - N_j^i \frac{\partial}{\partial y^i}, C \right] =$$

$$= \mathcal{V}\left[\left(\frac{\partial}{\partial u^j} \right)^{\mathsf{c}}, C \right] + \mathcal{V}\left((CN_j^i)\frac{\partial}{\partial y^i} + N_j^i \left[C, \left(\frac{\partial}{\partial u^i} \right)^{\mathsf{v}} \right] \right)$$

$$\overset{(4.1.72),(4.1.53)}{=} \left(y^k \frac{\partial N_j^i}{\partial y^k} - N_j^i \right) \mathcal{V}\left(\frac{\partial}{\partial y^i} \right) \overset{(7.2.40)}{=} \left(y^k \frac{\partial N_j^i}{\partial y^k} - N_j^i \right) \widehat{\frac{\partial}{\partial u^i}},$$

whence (7.5.11). $\square$

Corollary 7.5.10. *Let $\mathcal{H}$ be an Ehresmann connection in TM with horizontal projection $\mathbf{h}$. The following properties are equivalent:*

(i) *$\mathcal{H}$ is positive-homogeneous.*
(ii) *The tension of $\mathcal{H}$ vanishes.*
(iii) *For every $X \in \mathfrak{X}(M)$, $[X^{\mathsf{h}}, C] = 0$.*
(iv) *$[\mathbf{h}, C] = 0$.*

Proof. The equivalence of assertions (i), (ii) and (iii) is clear from Lemma 4.2.14, Proposition 7.5.8 and (7.5.10). Since for every $X \in \mathfrak{X}(M)$ we have

$$[\mathbf{h}, C]X^{\mathsf{c}} \overset{(3.3.16)}{=} [X^{\mathsf{h}}, C] - \mathbf{h}[X^{\mathsf{c}}, C] \overset{(4.1.72)}{=} [X^{\mathsf{h}}, C]$$

and

$$[\mathbf{h}, C]X^{\mathsf{v}} = -\mathbf{h}[X^{\mathsf{v}}, C] \overset{(4.1.53)}{=} -\mathbf{h}X^{\mathsf{v}} = 0,$$

it follows that (iv) is equivalent to (iii). $\qquad\square$

7.5.2 The Ehresmann Connection of an Affinely Connected Manifold

Proposition 7.5.11. *Let M be a manifold and D a covariant derivative on M. Define a mapping*

$$\mathcal{H} \colon TM \times_M TM \to TTM, \quad (v, w) \mapsto \mathcal{H}(v, w)$$

by

$$\mathcal{H}(v, w) := Y_*(w) - (D_w Y)^{\uparrow}(v) \text{ if } Y \in \mathfrak{X}(M), \ Y(\tau(v)) = v. \quad (7.5.12)$$

Then:

(i) *$\mathcal{H}$ is well-defined.*

(ii) *$\mathcal{H}$ is a linear Ehresmann connection in TM.*

(iii) *For any vector fields X, Y on M we have*

$$[X^{\mathsf{h}}, Y^{\mathsf{v}}] = (D_X Y)^{\mathsf{v}}, \quad (7.5.13)$$

where $X^{\mathsf{h}} := \mathcal{H}(\widehat{X})$ is the horizontal lift of X with respect to $\mathcal{H}$.

(iv) *The covariant derivative D is torsion-free if, and only if,*

$$\widehat{T}(X, Y) \overset{(7.1.9)}{:=} [X^{\mathsf{h}}, Y^{\mathsf{v}}] - [Y^{\mathsf{h}}, X^{\mathsf{v}}] - [X, Y]^{\mathsf{v}} = 0$$

for all $X, Y \in \mathfrak{X}(M)$.

Locally, given a chart $(\mathcal{U}, (u^i)_{i=1}^n)$ on M and the induced chart $(\tau^{-1}(\mathcal{U}), ((x^i)_{i=1}^n, (y^i)_{i=1}^n))$ on TM,

$$\mathcal{H}(v, w) = w^j \left(\left(\frac{\partial}{\partial x^j} \right)_v - v^k (\Gamma_{jk}^i \circ \tau)(v) \left(\frac{\partial}{\partial y^i} \right)_v \right), \quad (7.5.14)$$

if $v = v^i \left(\frac{\partial}{\partial u^i} \right)_p$, $w = w^i \left(\frac{\partial}{\partial u^i} \right)_p$ $(p = \tau(v) = \tau(w))$, and (Γ_{jk}^i) is the family of the Christoffel symbols of D with respect to the chart $(\mathcal{U}, (u^i)_{i=1}^n)$. (The summation convention is in force.)

Conversely, every linear Ehresmann connection $\mathcal{H}$ determines a covariant derivative D on M given by

$$D_X Y := K \circ Y_* \circ X; \quad X, Y \in \mathfrak{X}(M), \tag{7.5.15}$$

where K is the connector associated with $\mathcal{H}$. This covariant derivative operator induces the starting Ehresmann connection by (7.5.12)

Proof. First we deduce (7.5.14) from (7.5.12), then assertions (i)–(iv) can be obtained as easy consequences.

Let $Y \underset{(\mathcal{U})}{=} Y^i \frac{\partial}{\partial u^i}$. Since $Y(p) = Y(\tau(v)) = v = v^i \left(\frac{\partial}{\partial u^i} \right)_p$, it follows that $Y^i \circ \tau(v) = Y^i(p) = v^i$ for all $i \in J_n$. Now, applying (3.1.20), we find that

$$Y_*(w) = (Y_*)_p(w) = w^j \left(\left(\frac{\partial}{\partial x^j} \right)_v + \frac{\partial Y^i}{\partial u^j}(p) \left(\frac{\partial}{\partial y^i} \right)_v \right).$$

We have, furthermore

$$(D_w Y)^\uparrow(v) = \left(w^j D_{\left(\frac{\partial}{\partial u^j} \right)_p} \left(Y^k \frac{\partial}{\partial u^k} \right) \right)^\uparrow(v)$$

$$= w^j \left(\frac{\partial Y^i}{\partial u^j}(p) \left(\frac{\partial}{\partial u^i} \right)_p + Y^k(p) \Gamma^i_{jk}(p) \left(\frac{\partial}{\partial u^i} \right)_p \right)^\uparrow(v)$$

$$\overset{(4.1.31)}{=} w^j \left(\frac{\partial Y^i}{\partial u^j}(p) + v^k \Gamma^i_{jk}(p) \right) \left(\frac{\partial}{\partial y^i} \right)_v.$$

Hence

$$\mathcal{H}(v, w) = Y_*(w) - (D_w Y)^\uparrow(v) = w^j \left(\left(\frac{\partial}{\partial x^j} \right)_v - v^k \Gamma^i_{jk}(p) \left(\frac{\partial}{\partial y^i} \right)_v \right),$$

which proves (7.5.14). From this formula it can be seen that $\mathcal{H}$ is well-defined (does not depend on the choice of Y), and $\mathcal{H}$ satisfies conditions (C_1), (C_3) and (C_4). Since

$$\mathbf{j}_v(\mathcal{H}(v, w)) := \left(\tau_{TM}(\mathcal{H}(v, w)), (\tau_*)_v(\mathcal{H}(v, w)) \right)$$

$$= \left(v, (\tau_*)_v((Y_*)_p(w) - (D_w Y)^\uparrow(v)) \right) = (v, (\tau \circ Y)_*(w)) = (v, w),$$

condition (C_2) is also satisfied. Thus $\mathcal{H}$ is an Ehresmann connection. This Ehresmann connection is linear: for any $(v_1, w), (v_2, w) \in TM \times_M TM$ and $\lambda_1, \lambda_2 \in \mathbb{R}$ we have

$$\mathcal{H}(\lambda_1 v_1 + \lambda_2 v_2, w) \overset{(7.5.14)}{=} w^j \left(\frac{\partial}{\partial x^j} \right)_{\lambda_1 v_1 + \lambda_2 v_2}$$

$$- w^j (\lambda_1 v_1^k + \lambda_2 v_2^k)(\Gamma^i_{jk} \circ \tau)(v_1) \left(\frac{\partial}{\partial y^i} \right)_{\lambda_1 v_1 + \lambda_2 v_2}$$

$$\overset{(4.1.22),(4.1.23)}{=} \lambda_1 \boxdot \mathcal{H}(v_1, w) \boxplus \lambda_2 \boxdot \mathcal{H}(v_2, w).$$

Now we show that (7.5.13) holds. We see from (7.5.14) that the Christoffel symbols of $\mathcal{H}$ with respect to the chosen chart are the functions $y^k(\Gamma^i_{jk} \circ \tau)$; $i, j \in J_n$. So the $\mathcal{H}$-horizontal lift of a vector field $X \in \mathfrak{X}(M)$, $X \underset{(\mathcal{U})}{=} X^i \frac{\partial}{\partial u^i}$ is

$$X^{\mathrm{h}} = \left(X^j \frac{\partial}{\partial u^j}\right)^{\mathrm{h}} \overset{(7.1.5)}{=} (X^j \circ \tau)\left(\frac{\partial}{\partial u^j}\right)^{\mathrm{h}}$$

$$\overset{(7.2.37)}{=} (X^j \circ \tau)\left(\frac{\partial}{\partial x^j} - y^k(\Gamma^i_{jk} \circ \tau)\frac{\partial}{\partial y^i}\right).$$

Let $Y \in \mathfrak{X}(M)$, $Y \underset{(\mathcal{U})}{=} Y^i \frac{\partial}{\partial u^i}$ be another vector field. Then

$$[X^{\mathrm{h}}, Y^{\mathrm{v}}] \underset{(\mathcal{U})}{=} \left[(X^j \circ \tau)\frac{\partial}{\partial x^j}, Y^{\mathrm{v}}\right] - \left[y^k(X^j \circ \tau)(\Gamma^i_{jk} \circ \tau)\frac{\partial}{\partial y^i}, Y^{\mathrm{v}}\right]$$

$$= (X^j \circ \tau)\left[\left(\frac{\partial}{\partial u^j}\right)^{\mathrm{c}}, Y^{\mathrm{v}}\right] + (Y^{\mathrm{v}}y^k)(X^j \circ \tau)(\Gamma^i_{jk} \circ \tau)\frac{\partial}{\partial y^i}$$

$$= (X^j \circ \tau)\left[\left(\frac{\partial}{\partial u^j}\right), Y\right]^{\mathrm{v}} + (Yu^k)^{\mathrm{v}}(X^j \circ \tau)(\Gamma^i_{jk} \circ \tau)\frac{\partial}{\partial y^i}$$

$$= (X^j \circ \tau)\left(\frac{\partial Y^i}{\partial u^j} \circ \tau + (Y^k\Gamma^i_{jk}) \circ \tau\right)\frac{\partial}{\partial y^i}$$

$$= \left(X^j\left(\frac{\partial Y^i}{\partial u^j} + Y^k\Gamma^i_{jk}\right)\frac{\partial}{\partial u^i}\right)^{\mathrm{v}} \overset{(6.1.7)}{\underset{(\mathcal{U})}{=}} (D_XY)^{\mathrm{v}}$$

as was to be shown. After this has been proved, (iv) is a consequence of Remark 7.1.8, see mainly (7.1.15).

Next we turn to the proof of the converse statement. Let $\mathcal{H}$ be a linear Ehresmann connection in TM with horizontal projection $\mathbf{h}$, vertical projection $\mathbf{v}$, connector K and Christoffel symbols N^i_j ($i, j \in J_n$). Using the same notation as above, we derive a coordinate expression for D_XY defined by (7.5.15).

At any point p in $\mathcal{U}$,

$$D_XY(p) := K \circ Y_*(X(p)) = K_{Y(p)}\big((Y_*)_p(X(p))\big)$$

$$\overset{(3.1.20)}{=} K_{Y(p)}\left(X^j(p)\left(\frac{\partial}{\partial x^j}\right)_{Y(p)} + X^j(p)\frac{\partial Y^i}{\partial u^j}(p)\left(\frac{\partial}{\partial y^i}\right)_{Y(p)}\right)$$

$$\overset{(7.2.41)}{=} \left(X^j(p)\frac{\partial Y^i}{\partial u^j}(p) + X^j(p)N^i_j(Y(p))\right)\left(\frac{\partial}{\partial u^i}\right)_p.$$

Since $\mathcal{H}$ is linear, by (7.5.3) its Christoffel symbols are of the form $N^i_j = y^k(\Gamma^i_{jk} \circ \tau)$, where $\Gamma^i_{jk} \colon \mathcal{U} \to \mathbb{R}$ are smooth functions. So we obtain the local expression

$$D_X Y \underset{(\mathcal{U})}{=} X^j \left(\frac{\partial Y^i}{\partial u^j} + Y^k \Gamma^i_{jk} \right) \frac{\partial}{\partial u^i},$$

from which it is clear that D is a covariant derivative on M.

Let $\mathcal{H}^D$ be the Ehresmann connection determined by D. Then for each $X, Y \in \mathfrak{X}(M)$, $p \in M$,

$$
\begin{aligned}
\mathcal{H}^D(Y(p), X(p)) &\overset{(7.5.12)}{:=} (Y_*)_p(X(p)) - (D_{X(p)}Y)^\uparrow(Y(p)) \\
&\overset{(7.5.15)}{=} (Y_*)_p(X(p)) - \big(K_{Y(p)}((Y_*)_p(X(p)))\big)^\uparrow(Y(p)) \\
&\overset{(7.2.20)}{=} (Y_*)_p(X(p)) - \big(\ell^\downarrow(\mathbf{v}((Y_*)_p(X(p))))\big)^\uparrow(Y(p)) \\
&= (Y_*)_p(X(p)) - \mathbf{v}((Y_*)_p(X(p))) \\
&= \mathbf{h}((Y_*)_p(X(p))) = \mathcal{H} \circ \mathbf{j}((Y_*)_p(X(p))) \\
&= \mathcal{H}\big(Y(p), (\tau_*)_{Y(p)} \circ (Y_*)_p(X(p))\big) \\
&= \mathcal{H}\big(Y(p), ((\tau \circ Y)_*)_p(X(p))\big) = \mathcal{H}(Y(p), X(p)),
\end{aligned}
$$

therefore $\mathcal{H}^D = \mathcal{H}$. This concludes the proof. $\qquad\square$

Remark 7.5.12. The last conclusion in the previous proof can also be deduced as follows.

If the Christoffel symbols of $\mathcal{H}$ are the functions N^i_j, then the Christoffel symbols Γ^i_{jk} of the induced covariant derivative D are determined by $N^i_j = y^l(\Gamma^i_{jl} \circ \tau)$, whence

$$\frac{\partial N^i_j}{\partial y^k} = \delta^l_k(\Gamma^i_{jl} \circ \tau) = \Gamma^i_{jk} \circ \tau.$$

Hence the Christoffel symbols of $\mathcal{H}^D$ are

$$y^k(\Gamma^i_{jk} \circ \tau) = y^k \frac{\partial N^i_j}{\partial y^k} = N^i_j,$$

taking into account the homogeneity of the functions N^i_j and applying Lemma 4.2.9.

Proposition and Definition 7.5.13. *Let D be a covariant derivative on a manifold M, and let $\mathcal{H}$ be the Ehresmann connection arising from D by (7.5.12). Then its associated semispray*

$$S := \mathcal{H}\widetilde{\delta} \colon TM \to TTM, \quad v \mapsto S(v) := \mathcal{H}(v, v) \tag{7.5.16}$$

is an affine spray over M*, which has the same geodesics as* D*. This spray is called the* geodesic spray *of* D*. Locally, with the notation of Proposition 7.5.11,*

$$S \underset{(\mathcal{U})}{=} y^i \frac{\partial}{\partial x^i} - (\Gamma^i_{jk} \circ \tau) y^j y^k \frac{\partial}{\partial y^i}, \qquad (7.5.17)$$

where the functions $\Gamma^i_{jk} \in C^\infty(\mathcal{U})$ *are the Christoffel symbols of* D *with respect to the chosen chart.*

Proof. From the coordinate expression (7.5.14) of $\mathcal{H}$, we obtain immediately (7.5.17). Thus, taking into account Corollary 5.1.17, it follows that S is an affine spray with spray coefficients

$$G^i = \frac{1}{2}(\Gamma^i_{jk} \circ \tau) y^j y^k.$$

Thus the local geodesic equations (5.1.18) for S take the form

$$(\gamma^i)'' + (\Gamma^i_{jk} \circ \gamma)(\gamma^j)'(\gamma^k)' = 0, \quad i \in J_n.$$

Comparing this to (6.1.74), we conclude that S and D have the same geodesics. $\qquad \square$

Corollary 7.5.14. *Let* (M, D) *be an affinely connected manifold. Given a tangent vector* $v \in TM$*, there is a unique maximal geodesic* γ_v *starting at* $\tau(v)$ *with initial velocity* v*, i.e., with* $\dot{\gamma}_v(0) = v$*.*

Proof. Let S be the geodesic spray of D. Then, by the preceding proposition, D has the same geodesics as S. Thus the assertion follows from Proposition and Definition 5.1.29. $\qquad \square$

7.5.3 The Linear Deviation

Lemma 7.5.15. *Let* $\mathcal{H}$ *be an Ehresmann connection in* TM*. The mapping*

$$\begin{aligned} \mathsf{N} \colon TM \times_M TM &\to TTM, \\ (v, w) &\mapsto \mathsf{N}(v, w) := \mathcal{H}(v, w) - \kappa\, \mathcal{H}(w, v), \end{aligned} \qquad (7.5.18)$$

where $\kappa \colon TTM \to TTM$ *is the canonical involution, is a well-defined vertical-valued mapping.*

Proof. The vectors $\mathcal{H}(v, w)$ and $\mathcal{H}(w, v)$ are in $T_v TM$ and $T_w TM$, respectively. Then

$$\kappa\, \mathcal{H}(w, v) \in T_{\tau_* \mathcal{H}(w,v)} TM \overset{(7.2.7)}{=} T_v TM,$$

by condition (ii) in Proposition 4.1.8, so the difference $\mathcal{H}(v,w) - \kappa\,\mathcal{H}(w,v)$ makes sense. Since

$$\tau_*\mathsf{N}(v,w) = \tau_*(\mathcal{H}(v,w)) - \tau_*\,\kappa(\mathcal{H}(w,v))$$

$$\overset{\text{(7.2.7), Proposition 4.1.8}}{=} w - \tau_{TM}(\mathcal{H}(w,v)) = w - w = 0,$$

it follows that $\mathsf{N}(v,w) \in VTM$. $\square$

Remark 7.5.16. We call the mapping N defined by (7.5.18) the *linear deviation* of $\mathcal{H}$. It was introduced by Z. Shen [88] under the name *N-curvature*. However, it has no curvature meaning at all. We shall see soon that the vanishing of N characterizes the linear Ehresmann connections which are torsion-free in the sense that the tensor $\widehat{T}$ given by (7.1.9) is zero.

Coordinate description. Choose a chart $(\mathcal{U}, (u^i)_{i=1}^n)$ on M and consider the induced chart $(\tau^{-1}(\mathcal{U}), ((x^i)_{i=1}^n, (y^i)_{i=1}^n))$ on TM. Let (N_j^i) be the family of the Christoffel symbols of the Ehresmann connection $\mathcal{H}$ with respect to the chart $(\mathcal{U}, (u^i)_{i=1}^n)$. If $v = v^i \left(\frac{\partial}{\partial u^i}\right)_p$, $w = w^i \left(\frac{\partial}{\partial u^i}\right)_p$ ($p \in \mathcal{U}$; summation convention is in force), then

$$\mathcal{H}(v,w) = \mathcal{H}_v\left(w^j \left(\frac{\partial}{\partial u^j}\right)_p\right) \overset{(C_1)}{=} w^j \mathcal{H}_v\left(\left(\frac{\partial}{\partial u^j}\right)_p\right)$$

$$\overset{(7.2.37)}{=} w^j \left(\left(\frac{\partial}{\partial x^j}\right)_v - N_j^i(v)\left(\frac{\partial}{\partial y^i}\right)_v\right),$$

$$\kappa\,\mathcal{H}(w,v) = \kappa\left(v^j \left(\frac{\partial}{\partial x^j}\right)_w - v^j N_j^i(w)\left(\frac{\partial}{\partial y^i}\right)_w\right)$$

$$\overset{(4.1.20)}{=} w^i \left(\frac{\partial}{\partial x^i}\right)_{\tau_*\mathcal{H}(w,v)} - v^j N_j^i(w)\left(\frac{\partial}{\partial y^i}\right)_{\tau_*\mathcal{H}(w,v)}$$

$$\overset{(7.2.7)}{=} w^i \left(\frac{\partial}{\partial x^i}\right)_v - v^j N_j^i(w)\left(\frac{\partial}{\partial y^i}\right)_v,$$

therefore

$$\mathsf{N}(v,w) = (v^j N_j^i(w) - w^j N_j^i(v))\left(\frac{\partial}{\partial y^i}\right)_v. \qquad (7.5.19)$$

Proposition 7.5.17. *An Ehresmann connection $\mathcal{H}$ in TM has vanishing linear deviation if, and only if, it is a linear connection and the covariant derivative determined by $\mathcal{H}$ is torsion-free.*

Proof. Suppose first that the linear deviation of $\mathcal{H}$ vanishes, i.e.,

$$\mathsf{N}(v, w) = \mathcal{H}(v, w) - \kappa \, \mathcal{H}(w, v) = 0 \text{ for all } (v, w) \in TM \times_M TM.$$

Then for each $v_1, v_2 \in T_{\tau(w)}M$; $\lambda_1, \lambda_2 \in \mathbb{R}$,

$$\mathcal{H}(\lambda_1 v_1 + \lambda_2 v_2, w) = \kappa \, \mathcal{H}(w, \lambda_1 v_1 + \lambda_2 v_2) \overset{(C_1)}{=} \kappa(\lambda_1 \mathcal{H}(w, v_1) + \lambda_2 \mathcal{H}(w, v_2))$$
$$= \kappa(\lambda_1 \kappa \, \mathcal{H}(v_1, w) + \lambda_2 \kappa \, \mathcal{H}(v_2, w)) = \lambda_1 \boxdot \mathcal{H}(v_1, w) \boxplus \lambda_2 \boxdot \mathcal{H}(v_2, w),$$

which proves that the Ehresmann connection $\mathcal{H}$ is linear.

We proceed with coordinate calculations, keeping our above notation and conventions. Since $\mathcal{H}$ is linear, its Christoffel symbols are of the form $N_j^i = y^k(\Gamma_{jk}^i \circ \tau)$, $\Gamma_{jk}^i \in C^\infty(\mathcal{U})$. Then for each $p \in \mathcal{U}$, $v = v^i \left(\frac{\partial}{\partial u^i}\right)_p$, $w = w^i \left(\frac{\partial}{\partial u^i}\right)_p$ we have

$$0 = \mathsf{N}(v, w) \overset{(7.5.19)}{=} \left(v^j(y^k(\Gamma_{jk}^i \circ \tau)(w)) - w^j(y^k(\Gamma_{jk}^i \circ \tau)(v))\right) \left(\frac{\partial}{\partial y^i}\right)_v$$
$$= (v^j w^k \Gamma_{jk}^i(p) - w^j v^k \Gamma_{jk}^i(p)) \left(\frac{\partial}{\partial y^i}\right)_v = v^j w^k (\Gamma_{jk}^i - \Gamma_{kj}^i)(p) \left(\frac{\partial}{\partial y^i}\right)_v.$$

This implies that over $\mathcal{U}$, $\Gamma_{jk}^i = \Gamma_{kj}^i$; $i, j, k \in J_n$. From this it follows that the covariant derivative D determined by $\mathcal{H}$ is torsion-free, because, as it turns out from the proof of Proposition 7.5.11, the Christoffel symbols of D with respect to the chosen chart are just the functions Γ_{jk}^i.

Conversely, if $\mathcal{H}$ is a linear Ehresmann connection with torsion-free induced covariant derivative on M, then its Christoffel symbols with respect to a chart $(\mathcal{U}, (u^i)_{i=1}^n)$ of M are of the form

$$N_j^i = y^k(\Gamma_{jk}^i \circ \tau), \quad \Gamma_{jk}^i = \Gamma_{kj}^i; \quad i, j, k \in J_n,$$

and our above calculation shows that the linear deviation of $\mathcal{H}$ vanishes. $\square$

7.6 Parallel Translation with Respect to an Ehresmann Connection

If $I \subset \mathbb{R}$ is an open interval, and $\gamma \colon I \to M$ is a smooth curve, then, as we saw in Remark 3.1.30(d), a vector field $X \in \Gamma_\gamma(TM) = \mathfrak{X}_\gamma(M)$ is a section of TM along γ, i.e., a smooth mapping $X \colon I \to TM$ such that $\tau \circ X = \gamma$. Thus, we may consider X as a smooth curve in TM as well, and we can speak about its tangent vector field $\dot{X}$, which is a vector field along X, i.e., $\dot{X} \in \mathfrak{X}_X(TM)$.

Definition 7.6.1. Let $\mathcal{H}$ be a positive-homogeneous Ehresmann connection in TM, $I \subset \mathbb{R}$ an open interval and $\gamma\colon I \to M$ a smooth curve. A vector field $X \in \mathfrak{X}_\gamma(M)$ is said to be *parallel along γ with respect to $\mathcal{H}$* (briefly $\mathcal{H}$-*parallel*, or simply *parallel*) if

$$\dot{X}(t) = \mathcal{H}(X(t), \dot{\gamma}(t)) \quad \text{for all } t \in I, \tag{7.6.1}$$

briefly if $\dot{X} = \mathcal{H} \circ (X, \dot{\gamma})$.

Remark 7.6.2. If we look at X as a curve in TM, then the fact that X is parallel may be expressed by saying that X is a *horizontal lift* of γ (with respect to $\mathcal{H}$).

The next lemma gives a geometric characterization of parallel vector fields along a curve.

Lemma 7.6.3. *With the notation of the previous definition, a vector field X along γ is parallel if, and only if, $\dot{X}(t) \in H_{X(t)}TM$ for each $t \in I$.*

Proof. From the definition it is obvious that each $\dot{X}(t)$ is a horizontal vector if the vector field X is parallel. Conversely, suppose that we have $\dot{X}(t) \in H_{X(t)}TM$ for each $t \in I$. Then

$$\dot{X}(t) = \mathbf{h}\dot{X}(t) = \mathcal{H}(\mathbf{j}\dot{X}(t)) = \mathcal{H}(\tau_{TM}\dot{X}(t), \tau_*\dot{X}(t))$$

$$= \mathcal{H}\left(X(t), \tau_*\left(X_*\left(\frac{d}{dr}\right)_t\right)\right) = \mathcal{H}\left(X(t), (\tau \circ X)_*\left(\frac{d}{dr}\right)_t\right)$$

$$= \mathcal{H}\left(X(t), \gamma_*\left(\frac{d}{dr}\right)_t\right) = \mathcal{H}(X(t), \dot{\gamma}(t)),$$

thus X is indeed a parallel vector field along γ. $\qquad\square$

Lemma and Definition 7.6.4. *Let $\mathcal{H}$ be a positive-homogeneous Ehresmann connection in TM, $I \subset \mathbb{R}$ an open interval such that $0 \in I$ and $\gamma\colon I \to M$ a smooth curve.*

(i) *If $v \in T_{\gamma(0)}M$, then there exists a unique parallel vector field X along γ such that $X(0) = v$.*

(ii) *Fix a point $t \in I$. Given a vector $v \in T_{\gamma(0)}M$, let X be the unique parallel vector field along γ such that $X(0) = v$. Then the mapping*

$$P_0^t(\gamma)\colon T_{\gamma(0)}M \to T_{\gamma(t)}M, \quad v \mapsto P_0^t(\gamma)(v) := X(t)$$

is called the parallel translation along γ *from $\gamma(0)$ to $\gamma(t)$ with respect to the Ehresmann connection $\mathcal{H}$ (or, briefly, the $\mathcal{H}$-parallel translation).*

Proof. Choose a chart $(\mathcal{U}, (u^i)_{i=1}^n)$ on M with induced chart $(\tau^{-1}(\mathcal{U}), (x^i, y^i)_{i=1}^n)$ on TM, and let N_j^i $(i, j \in J_n)$ be the Christoffel symbols of $\mathcal{H}$ with respect to the given chart. Suppose first that $\gamma(I) \subset \mathcal{U}$. By Lemma 7.6.3, a vector field X along γ is parallel if, and only if, $\dot{X}(t) = \mathbf{h}\dot{X}(t)$. Now we express this condition in local coordinates. Let $\gamma^i := u^i \circ \gamma$ for each $i \in J_n$ and $X(t) = X^i(t) \left(\frac{\partial}{\partial u^i}\right)_{\gamma(t)}$ for $t \in I$. Then γ^i and X^i are real-valued smooth functions on I. By (3.1.12), we have

$$\dot{X}(t) = (x^i \circ X)'(t) \left(\frac{\partial}{\partial x^i}\right)_{X(t)} + (y^i \circ X)'(t) \left(\frac{\partial}{\partial y^i}\right)_{X(t)}$$

$$= (\gamma^i)'(t) \left(\frac{\partial}{\partial x^i}\right)_{X(t)} + (X^i)'(t) \left(\frac{\partial}{\partial y^i}\right)_{X(t)}.$$

We act on both sides by $\mathbf{h}$ and we apply (7.2.38) to obtain

$$\mathbf{h}(\dot{X}(t)) = (\gamma^j)'(t) \left(\delta_j^i \left(\frac{\partial}{\partial x^i}\right)_{X(t)} - N_j^i(X(t)) \left(\frac{\partial}{\partial y^i}\right)_{X(t)}\right)$$

$$= (\gamma^i)'(t) \left(\frac{\partial}{\partial x^i}\right)_{X(t)} - (\gamma^j)'(t) N_j^i(X(t)) \left(\frac{\partial}{\partial y^i}\right)_{X(t)}.$$

Thus, condition $\dot{X}(t) = \mathbf{h}\dot{X}(t)$ is equivalent to

$$(X^i)'(t) = -(\gamma^j)'(t) N_j^i(X(t)); \quad i \in J_n, \ t \in I. \tag{7.6.2}$$

This is a system of time-dependent ordinary differential equations on I with homogeneous coefficients for the functions X^i, thus Lemma 6.1.56 guarantees the existence and uniqueness of the parallel vector field X along γ with $X(0) = v$.

If the image of γ is not contained in the domain of a single chart, then, for a fixed $t \in I$, the set $\gamma([0, t])$ (or $\gamma([t, 0])$) is a compact subset of M, thus it can be covered by a finite number of coordinate neighbourhoods, and it follows by the previous paragraph that the required unique parallel vector field exists on $[0, t]$ (or on $[t, 0]$). Since t was arbitrary, X can be uniquely extended to the whole of I. $\qquad\Box$

Corollary 7.6.5. *Under the conditions of the previous lemma and definition, the restriction of the parallel translation $P_0^t(\gamma)$ to $\mathring{T}_{\gamma(0)}M$ is a positive-homogeneous diffeomorphism between $\mathring{T}_{\gamma(0)}M$ and $\mathring{T}_{\gamma(t)}M$.*

Proof. If $v \in T_{\gamma(0)}M$, and X is the unique parallel vector field along γ such that $X(0) = v$, then $X(t) = P_0^t(\gamma)(v)$ by the definition of $P_0^t(\gamma)$, thus X is the unique parallel vector field along γ such that $X(t) = P_0^t(\gamma)(v)$.

It follows that $P_t^0(\gamma)(P_0^t(\gamma)(v)) = v$ and $P_t^0(\gamma) \circ P_0^t(\gamma) = 1_{T_{\gamma(0)}M}$. Thus $P_0^t(\gamma)$ is a bijective mapping between $T_{\gamma(0)}M$ and $T_{\gamma(t)}M$, and, since obviously $P_0^t(\gamma)(0) = 0$, it is a bijection of $\overset{\circ}{T}_{\gamma(0)}M$ onto $\overset{\circ}{T}_{\gamma(t)}M$ as well. The smoothness of $P_0^t(\gamma)$ on $\overset{\circ}{T}_{\gamma(0)}M$ and its inverse follows from the theorem on the smooth dependence on the initial condition (Theorem 3.2.7). Finally, by the homogeneity of $\mathcal{H}$, its Christoffel symbols N_j^i with respect to an arbitrary chart are also positive-homogeneous functions of degree 1, thus the component functions of X and λX satisfy the same differential equation for an arbitrary positive number λ, therefore $P_0^t(\gamma)(\lambda v) = \lambda P_0^t(\gamma)(v)$ if $v \in \overset{\circ}{T}_{\gamma(0)}M$, which proves the positive-homogeneity of $P_0^t(\gamma)$. $\qquad\square$

Corollary 7.6.6. *Let $\mathcal{H}$ be an Ehresmann connection in TM induced by a covariant derivative D on M. Then a vector field along a smooth curve in M is $\mathcal{H}$-parallel if, and only if, it is parallel with respect to D.*

Proof. We argue locally. Chose a chart $(\mathcal{U}, (u^i)_{i=1}^n)$ on M. Let $\gamma: I \to \mathcal{U}$ be a smooth curve, and let $X \in \mathfrak{X}_\gamma(M)$. Let (N_j^i) and (Γ_{jk}^i) be the families of the Christoffel symbols of $\mathcal{H}$ and D, respectively, with respect to the chosen chart. Then, as we have seen in Remark 7.5.12,

$$N_j^i = y^k(\Gamma_{jk}^i \circ \tau); \quad i, j \in J_n. \tag{$*$}$$

The following assertions are equivalent:

(i) X is $\mathcal{H}$-parallel, i.e., $\dot{X} = \mathcal{H} \circ (X, \dot\gamma)$.

(ii) For every $i \in J_n$, $(X^i)' = -(\gamma^j)'(N_j^i \circ X)$, by (7.6.2).

(iii) For every $i \in J_n$, $(X^i)' = -(\gamma^j)'(y^k(\Gamma_{jk}^i \circ \tau)) \circ X = -(\gamma^j)'X^k(\Gamma_{jk}^i \circ \gamma)$, by ($*$).

(iv) $D^\gamma X = 0$, i.e., X is parallel with respect to D, by (6.1.65).

Thus (i) and (iv) are equivalent, as was to be shown. $\qquad\square$

Remark 7.6.7. With the same hypotheses and notation as above, suppose that the vector field along γ is of the form $X = Z \circ \gamma$, $Z \in \mathfrak{X}(M)$. Then we can prove the corollary by a simple coordinate-free argument as follows:

In view of the definition of the induced Ehresmann connection,

$$\mathcal{H}(X(t), \dot\gamma(t)) = \mathcal{H}(Z(\gamma(t)), \dot\gamma(t)) \overset{(7.5.12)}{:=} Z_*(\dot\gamma(t)) - (D_{\dot\gamma(t)}Z)^\uparrow(Z(\gamma(t)))$$

$$\overset{(3.1.11)}{=} \dot{X}(t) - (D_{\dot\gamma(t)}Z)^\uparrow(Z(\gamma(t))), \quad \text{for all } t \in I.$$

Thus $X = Z \circ \gamma$ is $\mathcal{H}$-parallel if, and only if,

$$(D_{\dot\gamma(t)}Z)^\uparrow(Z(\gamma(t))) = 0, \quad t \in I. \tag{$*$}$$

Since the vertical lifting is an injective linear mapping by Lemma 4.1.16(iii), $(*)$ holds if, and only if,

$$0 = D_{\dot{\gamma}(t)}Z \overset{(6.1.62)}{=} D^{\gamma}(Z \circ \gamma)(t) = (D^{\gamma}X)(t) \quad \text{for all } t \in I,$$

i.e., X is parallel with respect to D.

Lemma 7.6.8. *Let $\mathcal{H}$ be a positive-homogeneous Ehresmann connection in TM. Then $\mathcal{H}$ is linear if, and only if, the $\mathcal{H}$-parallel translations are linear mappings.*

Proof. Suppose that $\mathcal{H}$ is linear. Take a smooth curve $\gamma \colon I \to M$. Then for any $\mathcal{H}$-parallel vector fields $X, Y \in \mathfrak{X}_{\gamma}(M)$ and for any $\lambda \in \mathbb{R}$,

$$(X+Y)^{\cdot} := (X+Y)_* \circ \frac{d}{dr} \overset{(4.1.25)}{=} (X_* \boxplus Y_*) \circ \frac{d}{dr} = \dot{X} \boxplus \dot{Y}$$

$$= \mathcal{H} \circ (X, \dot{\gamma}) \boxplus \mathcal{H} \circ (Y, \dot{\gamma}) \overset{(7.5.2)}{=} \mathcal{H} \circ (X+Y, \dot{\gamma}),$$

$$(\lambda X)^{\cdot} = (\lambda X)_* \circ \frac{d}{dr} \overset{(4.1.26)}{=} \lambda \boxdot \dot{X} = \lambda \boxdot \mathcal{H} \circ (X, \dot{\gamma}) = \mathcal{H} \circ (\lambda X, \dot{\gamma}),$$

therefore $X + Y$ and λX are parallel along γ. From this it follows immediately that the $\mathcal{H}$-parallel translations are linear mappings.

Every element of $TM \times_M TM$ can be represented in the form $(X(0), \dot{\gamma}(0))$, where $\gamma \colon I \to M$ is a smooth curve and $X \in \mathfrak{X}_{\gamma}(M)$ is an $\mathcal{H}$-parallel vector field (Lemma and Definition 7.6.4), so the converse statement can be shown similarly. $\qquad \square$

Definition 7.6.9. If $\gamma_1, \gamma_2 \colon [0,1] \to M$ are piecewise smooth curve segments (Definition 2.1.24(c)) such that $\gamma_1(1) = \gamma_2(0)$, then their *product* is the piecewise smooth curve segment $\gamma_1 * \gamma_2 \colon [0,1] \to M$ given by

$$\gamma_1 * \gamma_2(t) := \begin{cases} \gamma_1(2t) & \text{if } 0 \le t \le \frac{1}{2}, \\ \gamma_2(2t-1) & \text{if } \frac{1}{2} \le t \le 1. \end{cases}$$

Remark 7.6.10. If $\gamma \colon [a,b] \to M$ is a smooth curve segment, then the parallel translation introduced in Lemma and Definition 7.6.4 can be easily generalized to a mapping $P_a^b(\gamma) \colon T_{\gamma(a)}M \to T_{\gamma(b)}M$. Namely, let $I \subset \mathbb{R}$ be an open interval containing $[a,b]$, $\tilde{\gamma} \colon I \to M$ a smooth curve such that $\tilde{\gamma} \upharpoonright [a,b] = \gamma$ and

$$\begin{cases} P_a^b(\gamma) \colon T_{\gamma(a)}M \to T_{\gamma(b)}M, \quad v \mapsto P_a^b(\gamma)(v) := X(b) \\ \text{if } X \text{ is the unique parallel vector field along } \tilde{\gamma} \text{ such that } X(a) = v. \end{cases}$$

Then $P_a^b(\gamma)$ is independent of the extension $\tilde{\gamma}$.

Moreover, if $\gamma \colon [a, b] \to M$ is a piecewise smooth curve segment and $a = t_0 < t_1 < \cdots < t_k = b$ is a partition of $[a, b]$ as in Definition 2.1.24(c), then we define parallel translation along γ by

$$P_a^b(\gamma) := P_{t_{k-1}}^b(\gamma) \circ P_{t_{k-2}}^{t_{k-1}}(\gamma) \circ \cdots \circ P_{t_1}^{t_2}(\gamma) \circ P_a^{t_1}(\gamma) \colon$$

$$T_{\gamma(a)}M \to T_{\gamma(t_1)}M \to \cdots \to T_{\gamma(t_{k-1})}M \to T_{\gamma(b)}M.$$

Equivalently, if $v \in T_{\gamma(a)}M$, then $P_a^b(\gamma)(v) = X(b)$, where X is the unique 'piecewise smooth' parallel vector field along γ with $X(a) = v$.

Lemma and Definition 7.6.11. *Let $\mathcal{H}$ be a positive-homogeneous Ehresmann connection in TM and fix a point $p \in M$. Let $C(p)$ denote the set of all piecewise smooth curves $\gamma \colon [0, 1] \to M$ with $\gamma(0) = \gamma(1) = p$. Then*

$$\mathrm{Hol}(p) := \left\{ P_0^1(\gamma) \in \mathrm{Map}(T_pM, T_pM) \mid \gamma \in C(p) \right\}$$

is a group under the composition of mappings, called the holonomy group *at p associated to $\mathcal{H}$.*

Proof. First we show that $P_0^1(\gamma_1 * \gamma_2) = P_0^1(\gamma_2) \circ P_0^1(\gamma_1)$ whenever $\gamma_1, \gamma_2 \in C(p)$. For simplicity, we may assume that γ_1 and γ_2 are smooth. Let $v \in T_pM$, let X_1 be the unique parallel vector field along γ_1 such that $X_1(0) = v$ and X_2 the unique parallel vector field along γ_2 such that $X_2(0) = X_1(1)$. We define a 'piecewise smooth vector field' X along $\gamma := \gamma_1 * \gamma_2$ by

$$X(t) := \begin{cases} X_1(2t) & \text{if } 0 \le t \le \frac{1}{2}, \\ X_2(2t - 1) & \text{if } \frac{1}{2} \le t \le 1. \end{cases}$$

Then, from Remark 3.1.17(b) we obtain

$$\dot{X}(t) := \begin{cases} 2\dot{X}_1(2t) & \text{if } 0 \le t < \frac{1}{2}, \\ 2\dot{X}_2(2t - 1) & \text{if } \frac{1}{2} < t \le 1, \end{cases}$$

thus each tangent vector of X is horizontal, which, by Lemma 7.6.3, implies that X is a piecewise smooth parallel vector field along γ. In fact, it is the unique piecewise smooth parallel vector field along γ such that $X(0) = v$, therefore

$$P_0^1(\gamma_1 * \gamma_2)(v) = X(1) = X_2(1) = P_0^1(\gamma_2)\left(P_0^1(\gamma_1)(v)\right),$$

and hence $P_0^1(\gamma_1 * \gamma_2) = P_0^1(\gamma_2) \circ P_0^1(\gamma_1)$. This immediately implies that $\mathrm{Hol}(p)$ is closed under the composition of mappings.

Property (G1) in A.3.1 is an obvious consequence of the associativity of composition. If γ is the constant curve with value p, then $1_{T_pM} = P_0^1(\gamma)$ belongs to $\mathrm{Hol}(p)$, so (G2) is also satisfied. Finally, if $\gamma \in C(p)$, and $\hat{\gamma}(t) := \gamma(1 - t)$ ($t \in [0, 1]$), then $P_0^1(\gamma) \circ P_0^1(\hat{\gamma}) = 1_{T_pM}$, which proves (G3). This concludes the proof that $\mathrm{Hol}(p)$ is a group. $\qquad\square$

7.7 Geodesics of an Ehresmann Connection

Throughout this subsection, M is an n-dimensional manifold, $\tau \colon TM \to M$ is the tangent bundle of M, and I denotes an open interval in $\mathbb{R}$.

Definition 7.7.1. Let $\mathcal{H}$ be an Ehresmann connection in TM with vertical mapping $\mathcal{V}$. A smooth curve $\gamma \colon I \to M$ is called a *geodesic* of $\mathcal{H}$ if $\mathcal{V} \circ \ddot{\gamma} = 0$.

Remark 7.7.2. Equivalently speaking, $\gamma \colon I \to M$ is a geodesic of $\mathcal{H}$ if, and only if,

$$\ddot{\gamma}(t) \in H_{\dot{\gamma}(t)} TM \text{ for every } t \in I, \text{ or } \mathbf{h} \circ \ddot{\gamma} = \ddot{\gamma} \tag{7.7.1}$$

($\mathbf{h} := \mathcal{H} \circ \mathbf{j}$), i.e., if *the acceleration of γ is a horizontal vector field along $\dot{\gamma}$.*

Lemma 7.7.3. *The geodesics of an Ehresmann connection coincide with the geodesics of its associated semispray.*

Proof. Let $\mathcal{H}$ be an Ehresmann connection in TM with associated vertical mapping $\mathcal{V}$ and horizontal projection $\mathbf{h}$. Consider the semispray $S_{\mathcal{H}} := \mathcal{H}\widetilde{\delta}$ associated to $\mathcal{H}$, and let $\gamma \colon I \to M$ be a smooth curve.

If γ is a geodesic of $S_{\mathcal{H}}$, then $\ddot{\gamma} = S_{\mathcal{H}} \circ \dot{\gamma}$, and hence

$$\mathcal{V} \circ \ddot{\gamma} = \mathcal{V} \circ S_{\mathcal{H}} \circ \dot{\gamma} = \mathcal{V} \circ \mathcal{H} \circ \widetilde{\delta} \circ \dot{\gamma} \overset{(7.2.21)}{=} 0,$$

so γ is also a geodesic of $\mathcal{H}$.

Conversely, suppose that γ is a geodesic of $\mathcal{H}$. Then

$$\ddot{\gamma} \overset{(7.7.1)}{=} \mathbf{h} \circ \ddot{\gamma} = \mathcal{H} \circ \mathbf{j} \circ \ddot{\gamma} \overset{(4.1.37)}{=} \mathcal{H} \circ \widetilde{\delta} \circ \dot{\gamma} = S_{\mathcal{H}} \circ \dot{\gamma},$$

which proves that γ is at the same time a geodesic of $S_{\mathcal{H}}$. $\qquad\square$

Remark 7.7.4. With the same notation as above, let (N^i_j) be the family of Christoffel symbols of $\mathcal{H}$ with respect to the usual local frame field $\left(\widehat{\frac{\partial}{\partial u^i}} \right)^n_{i=1}$ over $\mathcal{U}$. Then, by (7.2.43), the coefficients of $S_{\mathcal{H}}$ are the functions $\frac{1}{2} y^j N^i_j$ so by Lemma 7.7.3 and (5.1.18) it follows that the *local geodesic equations for $\mathcal{H}$* are

$$(\gamma^i)'' + (N^i_j \circ \dot{\gamma})(\gamma^j)' = 0, \quad i \in J_n. \tag{7.7.2}$$

Corollary 7.7.5. *Suppose that $\mathcal{H}$ is a positive-homogeneous Ehresmann connection in TM. A smooth curve $\gamma \colon I \to M$ is a geodesic of $\mathcal{H}$ if, and only if, its velocity vector field is parallel along γ.*

Proof. This is an immediate consequence of Lemma 7.6.3 and Remark 7.7.2. ☐

Corollary 7.7.6. *If $\mathcal{H}$ is an Ehresmann connection in TM, and $v \in TM$, then there exists a geodesic $\gamma\colon I \to M$ of $\mathcal{H}$ such that $0 \in I$ and $\dot{\gamma}(0) = v$. If I 'is sufficiently small', this geodesic is uniquely determined.*

Proof. The local geodesic equations (7.7.2) for $\mathcal{H}$ form a system of time-independent second-order differential equations. If $v = 0$, then the constant curve with value $\tau(v)$ is obviously a solution of the problem. If $v \neq 0$, then the assertion follows from the local existence and uniqueness theorem (Theorem 3.2.3). ☐

7.8   Curvature and Torsion

Relations (7.1.14) and (7.1.15) suggest that the mappings $\widehat{R}$ and $\widehat{T}$ given by (7.1.8) and (7.1.9) are good candidates for the role of curvature and torsion of an Ehresmann connection. We adopt the suggestion, albeit technically in a somewhat different form.

Lemma and Definition 7.8.1. *Keeping the notation introduced above, let $\mathcal{H}$ be an Ehresmann connection in $\overset{\circ}{T}M$. The mappings $\mathcal{R}$ and $\mathbf{T}$ given by*

$$\mathcal{R}(\widetilde{X}, \widetilde{Y}) := \mathcal{V}[\mathcal{H}\widetilde{X}, \mathcal{H}\widetilde{Y}] \qquad (7.8.1)$$

and

$$\mathbf{T}(\widetilde{X}, \widetilde{Y}) := \mathcal{V}[\mathcal{H}\widetilde{X}, \mathbf{i}\widetilde{Y}] - \mathcal{V}[\mathcal{H}\widetilde{Y}, \mathbf{i}\widetilde{X}] - \mathbf{j}[\mathcal{H}\widetilde{X}, \mathcal{H}\widetilde{Y}], \qquad (7.8.2)$$

where $\widetilde{X}, \widetilde{Y} \in \Gamma(\overset{\circ}{\pi})$, are skew-symmetric Finsler tensor fields of type $(1,2)$, called the curvature *and the* torsion *of $\mathcal{H}$, respectively. Evaluating $\mathcal{R}$ and $\mathbf{T}$ on basic sections $\widehat{X}$ and $\widehat{Y}$, we obtain that*

$$\mathbf{i}\,\mathcal{R}(\widehat{X}, \widehat{Y}) = \mathbf{v}[X^{\mathsf{h}}, Y^{\mathsf{h}}] = [X^{\mathsf{h}}, Y^{\mathsf{h}}] - [X, Y]^{\mathsf{h}} = \widehat{R}(X, Y), \qquad (7.8.3)$$

$$\mathbf{i}\,\mathbf{T}(\widehat{X}, \widehat{Y}) = [X^{\mathsf{h}}, Y^{\mathsf{v}}] - [Y^{\mathsf{h}}, X^{\mathsf{v}}] - [X, Y]^{\mathsf{v}} = \widehat{T}(X, Y). \qquad (7.8.4)$$

The type $(1,1)$ Finsler tensor field

$$\mathbf{T}^{s} = \mathbf{t} + i_{\widetilde{\delta}}\mathbf{T}, \qquad (7.8.5)$$

where $\mathbf{t}$ is the tension, is called the strong torsion *of $\mathcal{H}$.*

Proof. The additivity and skew-symmetry of $\mathcal{R}$ and $\mathbf{T}$ are clear at first sight, so to show that they are tensors, it is sufficient to verify their linearity in the first slot. Let $F \in C^\infty(\mathring{T}M)$. Then

$$\mathcal{R}(F\widetilde{X}, \widetilde{Y}) := \mathcal{V}[\mathcal{H}(F\widetilde{X}), \mathcal{H}\widetilde{Y}] = F\mathcal{V}[\mathcal{H}\widetilde{X}, \mathcal{H}\widetilde{Y}] - (\mathcal{H}\widetilde{Y})F\mathcal{V}(\mathcal{H}\widetilde{X})$$

$$\overset{(7.2.21)}{=} F\mathcal{R}(\widetilde{X}, \widetilde{Y}),$$

$$\mathbf{T}(F\widetilde{X}, \widetilde{Y}) := \mathcal{V}[\mathcal{H}(F\widetilde{X}), \mathbf{i}\widetilde{Y}] - \mathcal{V}[\mathcal{H}\widetilde{Y}, \mathbf{i}(F\widetilde{X})] - \mathbf{j}[\mathcal{H}(F\widetilde{X}), \mathcal{H}\widetilde{Y}]$$

$$= F\mathbf{T}(\widetilde{X}, \widetilde{Y}) - (\mathbf{i}\widetilde{Y})F\mathcal{V}(\mathcal{H}\widetilde{X}) - (\mathcal{H}\widetilde{Y})F\mathcal{V}(\mathbf{i}\widetilde{X}) + (\mathcal{H}\widetilde{Y})F\mathbf{j}(\mathcal{H}\widetilde{X})$$

$$\overset{(7.2.21),(7.2.22),(7.2.3)}{=} F\mathbf{T}(\widetilde{X}, \widetilde{Y}) - (\mathcal{H}\widetilde{Y})F\widetilde{X} + (\mathcal{H}\widetilde{Y})F\widetilde{X} = F\mathbf{T}(\widetilde{X}, \widetilde{Y}),$$

as desired.

Now let $X, Y \in \mathfrak{X}(M)$. Evaluating $\mathcal{R}$ and $\mathbf{T}$ at $(\widehat{X}, \widehat{Y})$, we find that

$$\mathbf{i}\,\mathcal{R}(\widehat{X}, \widehat{Y}) = \mathbf{v}[X^\mathsf{h}, Y^\mathsf{h}] = [X^\mathsf{h}, Y^\mathsf{h}] - \mathbf{h}[X^\mathsf{h}, Y^\mathsf{h}]$$

$$\overset{(7.1.7)}{=} [X^\mathsf{h}, Y^\mathsf{h}] - [X, Y]^\mathsf{h} \overset{(7.1.8)}{=} \widehat{R}(X, Y),$$

$$\mathbf{i}\,\mathbf{T}(\widehat{X}, \widehat{Y}) = \mathbf{v}[X^\mathsf{h}, Y^\mathsf{v}] - \mathbf{v}[Y^\mathsf{h}, X^\mathsf{v}] - \mathbf{J}[X^\mathsf{h}, Y^\mathsf{h}]$$

$$\overset{(7.2.35)}{=} [X^\mathsf{h}, Y^\mathsf{v}] - [Y^\mathsf{h}, X^\mathsf{v}] - [X, Y]^\mathsf{v} \overset{(7.1.9)}{=} \widehat{T}(X, Y),$$

which concludes the proof. $\qquad\square$

Remark 7.8.2. Let an Ehresmann connection $\mathcal{H}$ be specified in $\mathring{T}M$.

(a) Consider the Nijenhuis tensor $N_\mathbf{h} := \frac{1}{2}[\mathbf{h}, \mathbf{h}]$ of the horizontal projection determined by $\mathcal{H}$. Since $\mathbf{h}^2 = \mathbf{h}$, (3.3.18) gives

$$N_\mathbf{h}(\xi, \eta) = [\mathbf{h}\xi, \mathbf{h}\eta] + \mathbf{h}[\xi, \eta] - \mathbf{h}[\mathbf{h}\xi, \eta] - \mathbf{h}[\xi, \mathbf{h}\eta]$$

for all $\xi, \eta \in \mathfrak{X}(TM)$. In particular,

$$N_\mathbf{h}(\mathbf{J}\xi, \eta) = \mathbf{h}[\mathbf{J}\xi, \eta] - \mathbf{h}[\mathbf{J}\xi, \mathbf{h}\eta] = \mathbf{h}[\mathbf{J}\xi, \mathbf{v}\eta] = 0,$$

therefore $N_\mathbf{h}$ vanishes whenever one of its arguments is vertical. Thus $N_\mathbf{h}$ is completely determined by its action on the pairs $(X^\mathsf{c}, Y^\mathsf{c})$, where $X, Y \in \mathfrak{X}(M)$. So, using (7.2.35), and taking into account that $X^\mathsf{h} \underset{\tau}{\sim} X$ and $\mathbf{v}Y^\mathsf{c} \underset{\tau}{\sim} 0$ imply $[X^\mathsf{h}, \mathbf{v}Y^\mathsf{c}] \in \mathfrak{X}^\mathsf{v}(\mathring{T}M)$, we obtain that

$$N_\mathbf{h}(X^\mathsf{c}, Y^\mathsf{c}) = [X^\mathsf{h}, Y^\mathsf{h}] + \mathbf{h}[X^\mathsf{c}, Y^\mathsf{c}] - \mathbf{h}[X^\mathsf{h}, Y^\mathsf{c}] - \mathbf{h}[X^\mathsf{c}, Y^\mathsf{h}]$$

$$= [X^\mathsf{h}, Y^\mathsf{h}] + \mathbf{h}[X, Y]^\mathsf{c} - \mathbf{h}[X^\mathsf{h}, Y^\mathsf{h} + \mathbf{v}Y^\mathsf{c}] - \mathbf{h}[X^\mathsf{h} + \mathbf{v}X^\mathsf{c}, Y^\mathsf{h}]$$

$$= [X^\mathsf{h}, Y^\mathsf{h}] + [X, Y]^\mathsf{h} - 2[X, Y]^\mathsf{h} = [X^\mathsf{h}, Y^\mathsf{h}] - [X, Y]^\mathsf{h} = \mathbf{v}[X^\mathsf{h}, Y^\mathsf{h}].$$

Therefore, the Nijenhuis tensor $N_\mathbf{h}$ and the curvature of $\mathcal{H}$ are related by

$$N_\mathbf{h}(X^\mathsf{c}, Y^\mathsf{c}) = \mathbf{i}\,\mathcal{R}(\widehat{X}, \widehat{Y}); \quad X, Y \in \mathfrak{X}(M). \tag{7.8.6}$$

We may state that *the Nijenhuis tensor $N_{\mathbf{h}}$ carries the same information as the curvature of $\mathcal{H}$* defined by (7.8.1). However, $N_{\mathbf{h}}$ lives on the tangent bundle of $\mathring{T}M$, while $\mathcal{R}$ lives on the Finsler bundle $\mathring{\pi} \colon \mathring{T}M \times_M TM \to \mathring{T}M$.

(b) Next we consider the Frölicher–Nijenhuis bracket of the vertical endomorphism $\mathbf{J}$ and the horizontal projection $\mathbf{h}$. Since $\mathbf{J} \circ \mathbf{h} + \mathbf{h} \circ \mathbf{J} = \mathbf{J}$ by (7.2.27), (3.3.17) gives

$$[\mathbf{J}, \mathbf{h}](\xi, \eta) = [\mathbf{J}\xi, \mathbf{h}\eta] + [\mathbf{h}\xi, \mathbf{J}\eta] + \mathbf{J}[\xi, \eta] - \mathbf{J}[\mathbf{h}\xi, \eta]$$
$$- \mathbf{J}[\xi, \mathbf{h}\eta] - \mathbf{h}[\mathbf{J}\xi, \eta] - \mathbf{h}[\xi, \mathbf{J}\eta],$$

for all $\xi, \eta \in \mathfrak{X}(\mathring{T}M)$. Replacing ξ by $\mathbf{J}\xi$, we find that

$$[\mathbf{J}, \mathbf{h}](\mathbf{J}\xi, \eta) = \mathbf{J}[\mathbf{J}\xi, \eta] - \mathbf{J}[\mathbf{J}\xi, \mathbf{h}\eta] = \mathbf{J}[\mathbf{J}\xi, \mathbf{v}\eta] = 0.$$

So, as $N_{\mathbf{h}} = \frac{1}{2}[\mathbf{h}, \mathbf{h}]$ above, $[\mathbf{J}, \mathbf{h}]$ is also completely determined by its action on the pairs (X^c, Y^c), $(X, Y) \in \mathfrak{X}(M) \times \mathfrak{X}(M)$, and then

$$[\mathbf{J}, \mathbf{h}](X^c, Y^c) = [X^v, Y^h] + [X^h, Y^v] + \mathbf{J}[X^c, Y^c] - \mathbf{J}[X^h, Y^c]$$
$$- \mathbf{J}[X^c, Y^h] - \mathbf{h}[X^v, Y^c] - \mathbf{h}[X^c, Y^v].$$

On the right-hand side, the last two terms vanish because $[X^v, Y^c]$ and $[X^c, Y^v]$ are vertical vector fields. Since

$$0 \overset{(4.1.89)}{=} [\mathbf{J}, Y^c]X^h = [X^v, Y^c] - \mathbf{J}[X^h, Y^c],$$

we have $\mathbf{J}[X^h, Y^c] = [X^v, Y^c] \overset{(4.1.71)}{=} [X, Y]^v$. Similarly,

$$\mathbf{J}[X^c, Y^h] = -\mathbf{J}[Y^h, X^c] = -[Y, X]^v = [X, Y]^v.$$

Furthermore, $\mathbf{J}[X^c, Y^c] \overset{(4.1.71)}{=} \mathbf{J}[X, Y]^c = [X, Y]^v$. Thus

$$[\mathbf{J}, \mathbf{h}](X^c, Y^c) = [X^h, Y^v] - [Y^h, X^v] - [X, Y]^v,$$

therefore the type $(1, 2)$ tensor $[\mathbf{J}, \mathbf{h}]$ on $\mathring{T}M$ and the torsion of $\mathcal{H}$ are related by

$$[\mathbf{J}, \mathbf{h}](X^c, Y^c) = \mathbf{i}\,\mathbf{T}(\widehat{X}, \widehat{Y}); \quad X, Y \in \mathfrak{X}(M). \tag{7.8.7}$$

So the Frölicher–Nijenhuis bracket $[\mathbf{J}, \mathbf{h}]$ contains the same information as the torsion of $\mathcal{H}$.

(c) Let $S_{\mathcal{H}}$ be the second-order vector field associated to $\mathcal{H}$. Then, for every vector field X on M, we have $\mathbf{T}^s(\widehat{X}) = \mathcal{V}[S_{\mathcal{H}}, X^v] - \mathbf{j}[S_{\mathcal{H}}, X^h]$, or, equivalently,

$$\mathbf{i}\mathbf{T}^s(\widehat{X}) = \mathbf{v}[S_{\mathcal{H}}, X^v] - \mathbf{J}[S_{\mathcal{H}}, X^h]. \tag{7.8.8}$$

Indeed,

$$\mathbf{T}^s(\widehat{X}) := \mathbf{t}(\widehat{X}) + \mathbf{T}(\widetilde{\delta}, \widehat{X}) := \mathcal{V}[X^h, C] + \mathcal{V}[S_{\mathcal{H}}, X^v]$$
$$- \mathcal{V}[X^h, C] - \mathbf{j}[S_{\mathcal{H}}, X^h] = \mathcal{V}[S_{\mathcal{H}}, X^v] - \mathbf{j}[S_{\mathcal{H}}, X^h].$$

Remark 7.8.3. Choose a chart $(\mathcal{U}, u) = (\mathcal{U}, (u^i)_{i=1}^n)$ on M, and consider the induced chart $(\tau^{-1}(\mathcal{U}), ((x^i)_{i=1}^n, (y^i)_{i=1}^n))$ on TM. Let $\mathcal{H}$ be an Ehresmann connection in TM and let (N_j^i) be the family of Christoffel symbols of $\mathcal{H}$ with respect to $(\mathcal{U}, u)$. Then the components of the curvature, torsion and strong torsion of $\mathcal{H}$ with respect to $(\mathcal{U}, u)$ are the functions

$$R_{jk}^i = \frac{\partial N_j^i}{\partial x^k} - \frac{\partial N_k^i}{\partial x^j} + N_j^l \frac{\partial N_k^i}{\partial y^l} - N_k^l \frac{\partial N_j^i}{\partial y^l}, \qquad (7.8.9)$$

$$T_{jk}^i = \frac{\partial N_j^i}{\partial y^k} - \frac{\partial N_k^i}{\partial y^j}, \qquad (7.8.10)$$

and

$$(T^s)_j^i = y^k \frac{\partial N_k^i}{\partial y^j} - N_j^i, \qquad (7.8.11)$$

defined and smooth on $\tau^{-1}(\mathcal{U}) \cap \mathring{T}M$. (Indices run over J_n, summation convention in force.) Indeed,

$$\begin{aligned}
i\mathcal{R}\left(\frac{\partial}{\partial u^j}, \frac{\partial}{\partial u^k}\right) &\overset{(7.8.3)}{=} \mathbf{v}\left[\left(\frac{\partial}{\partial u^j}\right)^{\mathsf{h}}, \left(\frac{\partial}{\partial u^k}\right)^{\mathsf{h}}\right] \\
&\overset{(7.2.37)}{=} \mathbf{v}\left[\frac{\partial}{\partial x^j} - N_j^i \frac{\partial}{\partial y^i}, \frac{\partial}{\partial x^k} - N_k^l \frac{\partial}{\partial y^l}\right] \\
&= \frac{\partial N_j^i}{\partial x^k}\frac{\partial}{\partial y^i} - \frac{\partial N_k^i}{\partial x^j}\frac{\partial}{\partial y^i} - \frac{\partial N_j^i}{\partial y^l}N_k^l\frac{\partial}{\partial y^i} + \frac{\partial N_k^l}{\partial y^i}N_j^i\frac{\partial}{\partial y^l} \\
&= \left(\frac{\partial N_j^i}{\partial x^k} - \frac{\partial N_k^i}{\partial x^j} + N_j^l\frac{\partial N_k^i}{\partial y^l} - N_k^l\frac{\partial N_j^i}{\partial y^l}\right)\frac{\partial}{\partial y^i},
\end{aligned}$$

which proves (7.8.9). The two other tensors can be handled similarly.

Lemma 7.8.4. *If an Ehresmann connection is positive-homogeneous, then its curvature is 1^+-homogeneous.*

Proof. With the same notation as above, we have to show that $\nabla_{\overset{\mathrm{v}}{\delta}}\mathcal{R} = \mathcal{R}$, or equivalently, that

$$(\nabla_{\overset{\mathrm{v}}{\delta}}\mathcal{R})(\widehat{X}, \widehat{Y}) = \mathcal{R}(\widehat{X}, \widehat{Y}) \quad \text{for all } X, Y \in \mathfrak{X}(M).$$

Manipulating the left-hand side of this relation, we obtain

$$(\nabla_{\overset{\mathrm{v}}{\delta}}\mathcal{R})(\widehat{X}, \widehat{Y}) \overset{(6.2.22)}{=} \nabla_{\overset{\mathrm{v}}{\delta}}(\mathcal{R}(\widehat{X}, \widehat{Y})) - \mathcal{R}(\nabla_{\overset{\mathrm{v}}{\delta}}\widehat{X}, \widehat{Y}) - \mathcal{R}(\widehat{X}, \nabla_{\overset{\mathrm{v}}{\delta}}\widehat{Y})$$

$$\overset{(6.2.32)}{=} \nabla_{\overset{\mathrm{v}}{\delta}}(\mathcal{R}(\widehat{X}, \widehat{Y})) \overset{(6.2.25)}{=} \mathbf{j}[C, \mathcal{H}\mathcal{R}(\widehat{X}, \widehat{Y})] = \mathbf{j}[C, \mathcal{H} \circ \mathcal{V}[X^{\mathsf{h}}, Y^{\mathsf{h}}]],$$

taking into account that $\mathbf{j}(\mathcal{H}\,\mathcal{R}(\widehat{X},\widehat{Y})) = \mathcal{R}(\widehat{X},\widehat{Y})$. Thus

$$\mathbf{i}(\nabla^{\mathrm{v}}_{\widehat{\delta}}\mathcal{R})(\widehat{X},\widehat{Y}) = \mathbf{J}[C,\mathcal{H}\circ\mathcal{V}[X^{\mathsf{h}},Y^{\mathsf{h}}]] \overset{(7.2.19)}{=} \mathbf{J}[C,\mathbf{F}[X^{\mathsf{h}},Y^{\mathsf{h}}]]$$
$$+ \mathbf{J}[C,\mathbf{J}[X^{\mathsf{h}},Y^{\mathsf{h}}]] = \mathbf{J}[C,\mathbf{F}[X^{\mathsf{h}},Y^{\mathsf{h}}]].$$

On the other hand,

$$\mathbf{i}\,\mathcal{R}(\widehat{X},\widehat{Y}) = \mathbf{v}[X^{\mathsf{h}},Y^{\mathsf{h}}] \overset{(7.2.29)}{=} \mathbf{J}\,\mathbf{F}[X^{\mathsf{h}},Y^{\mathsf{h}}] \overset{(4.1.90)}{=} [\mathbf{J},C]\mathbf{F}[X^{\mathsf{h}},Y^{\mathsf{h}}]$$
$$\overset{(3.3.16)}{=} [\mathbf{v}[X^{\mathsf{h}},Y^{\mathsf{h}}],C] - \mathbf{J}[\mathbf{F}[X^{\mathsf{h}},Y^{\mathsf{h}}],C] = \mathbf{J}[C,\mathbf{F}[X^{\mathsf{h}},Y^{\mathsf{h}}]],$$

since

$$[\mathbf{v}[X^{\mathsf{h}},Y^{\mathsf{h}}],C] = [[X^{\mathsf{h}},Y^{\mathsf{h}}],C] - [\mathbf{h}[X^{\mathsf{h}},Y^{\mathsf{h}}],C]$$
$$= -[[Y^{\mathsf{h}},C],X^{\mathsf{h}}] - [[C,X^{\mathsf{h}}],Y^{\mathsf{h}}] - [[X,Y]^{\mathsf{h}},C] = 0$$

by the positive-homogeneity of $\mathcal{H}$ (see, e.g., Proposition 7.5.8). □

Remark 7.8.5. The above proof is conceptual, but quite tricky and lengthy. We have to admit that using the components of $\mathcal{R}$ given by (7.8.9) and taking into account the positive-homogeneity of the Christoffel symbols N^i_j, it follows immediately that

$$CR^i_{jk} = y^l \frac{\partial R^i_{jk}}{\partial y^l} = R^i_{jk}, \tag{7.8.12}$$

which implies by Corollary 6.2.21 the 1^+-homogeneity of $\mathcal{R}$.

So, although we definitely prefer coordinate-free methods, sometimes they also have certain disadvantages.

Proposition 7.8.6. *If $S \in \mathfrak{X}(\mathring{T}M)$ is a second-order vector field, then the torsion of the Ehresmann connection generated by S in $\mathring{T}M$ vanishes.*

First proof. By Remark 7.8.2(b), it is sufficient to show that $[\mathbf{J},\mathbf{h}] = 0$, where $\mathbf{h}$ is given by (7.3.6). We have

$$[\mathbf{J},\mathbf{h}] = \frac{1}{2}\big([\mathbf{J},1_{\mathfrak{X}(\mathring{T}M)}] + [\mathbf{J},[\mathbf{J},S]]\big) \overset{(3.3.45)}{=} \frac{1}{2}[\mathbf{J},[\mathbf{J},S]].$$

Using the graded Jacobi identity (3.3.47), we find

$$0 = (-1)^{1\cdot 0}[\mathbf{J},[\mathbf{J},S]] + (-1)^{1\cdot 1}[\mathbf{J},[S,\mathbf{J}]] + (-1)^{0\cdot 1}[S,[\mathbf{J},\mathbf{J}]]$$
$$\overset{\text{Lemma } 4.1.59}{=} [\mathbf{J},[\mathbf{J},S]] - [\mathbf{J},[S,\mathbf{J}]] \overset{(3.3.46)}{=} 2[\mathbf{J},[\mathbf{J},S]],$$

thereby proving the proposition.

Second proof. We now apply (7.3.7) to show that

$$\mathbf{i}\,\mathbf{T}(\widehat{X},\widehat{Y}) = [X^{\mathsf{h}},Y^{\mathsf{v}}] - [Y^{\mathsf{h}},X^{\mathsf{v}}] - [X,Y]^{\mathsf{v}} = 0 \text{ for all } X,Y \in \mathfrak{X}(M).$$

We have

$$\mathbf{i}\,\mathbf{T}(\widehat{X},\widehat{Y}) = \frac{1}{2}[[X^{\mathsf{v}},S],Y^{\mathsf{v}}] - \frac{1}{2}[[Y^{\mathsf{v}},S],X^{\mathsf{v}}] + \frac{1}{2}[X^{\mathsf{c}},Y^{\mathsf{v}}]$$

$$- \frac{1}{2}[Y^{\mathsf{c}},X^{\mathsf{v}}] - [X,Y]^{\mathsf{v}} \overset{(4.1.71)}{=} \frac{1}{2}\big([[X^{\mathsf{v}},S],Y^{\mathsf{v}}] + [[S,Y^{\mathsf{v}}],X^{\mathsf{v}}]\big)$$

$$\overset{(\mathrm{L2})}{=} -\frac{1}{2}[[Y^{\mathsf{v}},X^{\mathsf{v}}],S] \overset{(4.1.52)}{=} 0,$$

which gives the result. $\square$

The next important theorem shows that the converse of Proposition 7.8.6 is also true, under some refinement of the conditions. We consider a mapping $\mathcal{H}\colon TM \times_M TM \to TTM$ which is 'essentially' an Ehresmann connection, however, we drop condition (C_4), and require smoothness everywhere (although differentiability of class C^2 would be sufficient). We define the torsion of $\mathcal{H}$ also by (7.8.2). In the formulation of the theorem below, we regard $\mathcal{H}$ as a $C^\infty(TM)$-linear mapping from $\Gamma(\pi)$ into $\mathfrak{X}(TM)$ satisfying $\mathbf{j}\circ\mathcal{H} = 1_{\Gamma(\pi)}$, cf. Remark 7.2.4(d).

Theorem 7.8.7 (M. Crampin). *Let $\mathcal{H}\colon \Gamma(\pi) \to \mathfrak{X}(TM)$ be a $C^\infty(TM)$-linear mapping satisfying $\mathbf{j}\circ\mathcal{H} = 1_{\Gamma(\pi)}$. If the torsion of $\mathcal{H}$ vanishes, then $\mathcal{H}$ is generated by a second-order vector field over M, i.e., there exists a second-order vector field $S \in \mathfrak{X}(TM)$ such that*

$$\mathbf{h} = \mathcal{H}\circ\mathbf{j} = \frac{1}{2}(1_{\mathfrak{X}(TM)} + [\mathbf{J},S]),$$

or, equivalently,

$$X^{\mathsf{h}} := \mathcal{H}(\widehat{X}) = \frac{1}{2}(X^{\mathsf{c}} + [X^{\mathsf{v}},S]) \text{ for all } X \in \mathfrak{X}(M).$$

Proof. **Step 1.** Let $\widetilde{S}$ be an arbitrary second-order vector field on TM. Then, apart from condition (C_4), $\widetilde{S}$ generates a (smooth) Ehresmann connection in TM such that the $\widetilde{\mathcal{H}}$-horizontal lift of a vector field $X \in \mathfrak{X}(M)$ is given by

$$X^{\widetilde{\mathsf{h}}} := \widetilde{\mathcal{H}}(\widehat{X}) = \frac{1}{2}(X^{\mathsf{c}} + [X^{\mathsf{v}},\widetilde{S}]).$$

By Proposition 7.8.6, the torsion of $\widetilde{\mathcal{H}}$ vanishes, so for any vector fields X,Y on M we have

$$[X^{\widetilde{\mathsf{h}}},Y^{\mathsf{v}}] - [Y^{\widetilde{\mathsf{h}}},X^{\mathsf{v}}] = [X,Y]^{\mathsf{v}}. \tag{1}$$

Step 2. Define a mapping $\lambda\colon \mathfrak{X}(M) \to \mathfrak{X}^{\mathsf{v}}(TM)$ by

$$\lambda(X) := X^{\mathsf{h}} - X^{\tilde{\mathsf{h}}} = X^{\mathsf{h}} - \frac{1}{2}(X^{\mathsf{c}} + [X^{\mathsf{v}}, \widetilde{S}]), \quad X \in \mathfrak{X}(M), \qquad (2)$$

where X^{h} is the $\mathcal{H}$-horizontal lift of X. Applying (7.2.33), we find that $\mathbf{J}(\lambda(X)) = \mathbf{J}X^{\mathsf{h}} - \mathbf{J}X^{\tilde{\mathsf{h}}} = X^{\mathsf{v}} - X^{\mathsf{v}} = 0$, therefore λ is indeed vertical-valued. It is also linear in the sense of Remark 7.1.6(a). Indeed, the additivity is clear, and we have

$$\lambda(fX) = (fX)^{\mathsf{h}} - (fX)^{\tilde{\mathsf{h}}} \overset{(7.1.5)}{=} f^{\mathsf{v}}X^{\mathsf{h}} - f^{\mathsf{v}}X^{\tilde{\mathsf{h}}} = f^{\mathsf{v}}\lambda(X)$$

for all $f \in C^{\infty}(M)$.

Step 3. Since the torsion of $\mathcal{H}$ vanishes by our condition, for any vector fields X, Y on M,

$$0 = [X^{\mathsf{h}}, Y^{\mathsf{v}}] - [Y^{\mathsf{h}}, X^{\mathsf{v}}] - [X, Y]^{\mathsf{v}} \overset{(2)}{=} [\lambda(X) + X^{\tilde{\mathsf{h}}}, Y^{\mathsf{v}}]$$

$$- [\lambda(Y) + Y^{\tilde{\mathsf{h}}}, X^{\mathsf{v}}] - [X, Y]^{\mathsf{v}} = [\lambda(X), Y^{\mathsf{v}}] - [\lambda(Y), X^{\mathsf{v}}] + [X^{\tilde{\mathsf{h}}}, Y^{\mathsf{v}}]$$

$$- [Y^{\tilde{\mathsf{h}}}, X^{\mathsf{v}}] - [X, Y]^{\mathsf{v}} \overset{(1)}{=} [\lambda(X), Y^{\mathsf{v}}] - [\lambda(Y), X^{\mathsf{v}}].$$

Thus we have

$$[\lambda(X), Y^{\mathsf{v}}] = [\lambda(Y), X^{\mathsf{v}}] \text{ for all } X, Y \in \mathfrak{X}(M). \qquad (3)$$

Step 4. Now we choose a chart $(\mathcal{U}, (u^i)_{i=1}^n)$ on M, and consider the induced chart $(\tau^{-1}(\mathcal{U}), ((x^i)_{i=1}^n, (y^i)_{i=1}^n))$ on TM. By the $C^{\infty}(M)$-linearity of λ, there exist unique smooth functions $\lambda_j^i \in C^{\infty}(\tau^{-1}(\mathcal{U}))$ $(i, j \in J_n)$ such that

$$\lambda\left(\frac{\partial}{\partial u^j}\right) = \lambda_j^i \frac{\partial}{\partial y^i}, \quad j \in J_n$$

(summation convention in force). Then (3) gives that

$$\left[\lambda_j^i \frac{\partial}{\partial y^i}, \frac{\partial}{\partial y^k}\right] = \left[\lambda_k^i \frac{\partial}{\partial y^i}, \frac{\partial}{\partial y^j}\right]; \quad j, k \in J_n,$$

whence

$$\frac{\partial \lambda_j^i}{\partial y^k} = \frac{\partial \lambda_k^i}{\partial y^j}; \quad i, j, k \in J_n. \qquad (4)$$

Define the functions $Z^i\colon \tau^{-1}(\mathcal{U}) \to \mathbb{R}$ $(i \in J_n)$ by

$$v \mapsto Z^i(v) := \int_0^1 (\lambda_j^i \circ \mu_v) y^j(v),$$

where $\mu_v\colon [0, 1] \to \tau^{-1}(\mathcal{U})$, $t \mapsto \mu_v(t) := tv$. Using more traditional notation,

$$Z^i(v) := \int_0^1 \lambda_j^i(tv) y^j(v)\, dt.$$

Then, e.g., by Theorem 8.1 in [63], these functions are smooth, and their partial derivatives can be obtained by differentiation under the integral sign. Thus we find that

$$\frac{\partial Z^i}{\partial y^k}(v) = \int_0^1 \left(\left(\frac{\partial \lambda_j^i}{\partial y^k} \circ \mu_v \right) y^j(v) 1_{[0,1]} + \lambda_k^i \circ \mu_v \right)$$

$$\overset{(4)}{=} \int_0^1 \left(\left(\frac{\partial \lambda_k^i}{\partial y^j} \circ \mu_v \right) y^j(v) 1_{[0,1]} + \lambda_k^i \circ \mu_v \right).$$

In traditional notation,

$$\frac{\partial Z^i}{\partial y^k}(v) = \int_0^1 \left(\frac{\partial \lambda_j^i}{\partial y^k}(tv) t y^j(v) + \lambda_k^i(tv) \right) dt.$$

Observe that

$$(\lambda_k^i \circ \mu_v)' = \left(\frac{\partial \lambda_k^i}{\partial y^j} \circ \mu_v \right) (y^j \circ \mu_v)' = \left(\frac{\partial \lambda_k^i}{\partial y^j} \circ \mu_v \right) y^j(v).$$

So, integrating by parts,

$$\int_0^1 \left(\frac{\partial \lambda_j^i}{\partial y^k} \circ \mu_v \right) y^j(v) 1_{[0,1]} = \int_0^1 (\lambda_k^i \circ \mu_v)' 1_{[0,1]}$$

$$= (\lambda_k^i \circ \mu_v) 1_{[0,1]} \Big|_0^1 - \int_0^1 \lambda_k^i \circ \mu_v = \lambda_k^i(v) - \int_0^1 \lambda_k^i \circ \mu_v,$$

therefore

$$\frac{\partial Z^i}{\partial y^k}(v) = \lambda_k^i(v); \quad v \in \tau^{-1}(\mathcal{U}); \quad i, k \in J_n. \tag{5}$$

Now let $Z_{\mathcal{U}} := Z^i \frac{\partial}{\partial y^i}$. If X is an arbitrary vector field on M such that $X \underset{(\mathcal{U})}{=} X^k \frac{\partial}{\partial u^k}$, then, using also the $C^\infty(M)$-linearity of λ, we find that

$$[X^v, Z_{\mathcal{U}}] \underset{(\mathcal{U})}{=} \left[(X^k)^v \frac{\partial}{\partial y^k}, Z^i \frac{\partial}{\partial y^i} \right] = (X^k)^v \frac{\partial Z^i}{\partial y^k} \frac{\partial}{\partial y^i}$$

$$\overset{(5)}{=} (X^k)^v \lambda_k^i \frac{\partial}{\partial y^i} = (X^k)^v \lambda \left(\frac{\partial}{\partial u^k} \right) = \lambda \left(X^k \frac{\partial}{\partial u^k} \right) = \lambda(X).$$

Step 5. Owing to the preceding argument, it follows that there exists a *vertical* vector field Z on TM such that $\lambda(X) = [X^v, Z]$ for all $X \in \mathfrak{X}(M)$. Let $S := \widetilde{S} + 2Z$. Then S is also a second-order vector field on TM, and for any vector field X on M, we have

$$X^h = \mathcal{H}(\widehat{X}) \overset{(2)}{=} X^{\widetilde{h}} + \lambda(X) = \frac{1}{2}(X^c + [X^v, \widetilde{S}] + 2[X^v, Z])$$

$$= \frac{1}{2}(X^c + [X^v, \widetilde{S} + 2Z]) = \frac{1}{2}(X^c + [X^v, S]),$$

so $\mathcal{H}$ is generated by S. $\qquad\square$

Proposition 7.8.8. *Let $\mathcal{H}$ be an Ehresmann connection in $\mathring{T}M$. If $S_{\mathcal{H}}$ is the semispray associated to $\mathcal{H}$ and $\mathcal{H}_{S_{\mathcal{H}}}$ is the Ehresmann connection generated by $S_{\mathcal{H}}$, then*

$$\mathcal{H} - \mathcal{H}_{S_{\mathcal{H}}} = \frac{1}{2}\mathbf{i}\,\mathbf{T}^s, \qquad (7.8.13)$$

where $\mathbf{T}^s$ is the strong torsion of $\mathcal{H}$.

Proof. For any vector field X on M,

$$\mathbf{i}\,\mathbf{T}^s(\widehat{X}) \overset{(7.8.8)}{=} \mathbf{v}[S_{\mathcal{H}}, X^{\mathsf{v}}] - \mathbf{J}[S_{\mathcal{H}}, X^{\mathsf{h}}] \overset{(7.3.7)}{=} \mathbf{v}(X^{\mathsf{c}} - 2\mathcal{H}_{S_{\mathcal{H}}}(\widehat{X})) - \mathbf{J}[S_{\mathcal{H}}, X^{\mathsf{h}}]$$

$$= \mathbf{v}X^{\mathsf{c}} - 2\,\mathbf{v}\,\mathcal{H}_{S_{\mathcal{H}}}(\widehat{X}) + \mathbf{J}[X^{\mathsf{h}}, S_{\mathcal{H}}]$$

$$\overset{(7.2.36)}{=} \mathbf{v}X^{\mathsf{c}} - 2(\mathcal{H}_{S_{\mathcal{H}}} - \mathcal{H})(\widehat{X}) + \mathbf{J}[X^{\mathsf{c}}, S_{\mathcal{H}}] - \mathbf{J}[\mathbf{v}X^{\mathsf{c}}, S_{\mathcal{H}}]$$

$$\overset{(5.1.4)}{=} 2(\mathcal{H} - \mathcal{H}_{S_{\mathcal{H}}})(\widehat{X}) + \mathbf{J}[X^{\mathsf{c}}, S_{\mathcal{H}}] \overset{(5.1.6)}{=} 2(\mathcal{H} - \mathcal{H}_{S_{\mathcal{H}}})(\widehat{X}),$$

whence our assertion. $\qquad\square$

Corollary 7.8.9. *The strong torsion of an Ehresmann connection vanishes if, and only if, its torsion and tension vanish.*

Proof. Let $\mathcal{H}$ be an Ehresmann connection in $\mathring{T}M$ with torsion, strong torsion and tension $\mathbf{T}$, $\mathbf{T}^s$ and $\mathbf{t}$, respectively. It is clear from the definition (see (7.8.5)) that $\mathbf{T} = 0$ and $\mathbf{t} = 0$ imply $\mathbf{T}^s = 0$.

Conversely, suppose that $\mathbf{T}^s = 0$. Then $\mathcal{H} = \mathcal{H}_{S_{\mathcal{H}}}$ from (7.8.13). Since the torsion of $\mathcal{H}_{S_{\mathcal{H}}}$ vanishes by Proposition 7.8.6, it follows that $\mathbf{T} = 0$. So in (7.8.5) $\mathbf{T} = 0$ and $\mathbf{T}^s = 0$, therefore the tension $\mathbf{t}$ also vanishes. $\qquad\square$

Corollary 7.8.10 (A Uniqueness Theorem for Ehresmann Connections). *An Ehresmann connection in $\mathring{T}M$ is uniquely determined by its associated semispray and strong torsion.*

Proof. Let $\mathcal{H}_1$ and $\mathcal{H}_2$ be Ehresmann connection in $\mathring{T}M$. Suppose that they have the same associated semispray $S = \mathcal{H}_1\widetilde{\delta} = \mathcal{H}_2\widetilde{\delta}$ and the same strong torsion $\mathbf{T}^s$. Then, by (7.8.13),

$$\frac{1}{2}\mathbf{i}\,\mathbf{T}^s = \mathcal{H}_1 - \mathcal{H}_S \text{ and } \frac{1}{2}\mathbf{i}\,\mathbf{T}^s = \mathcal{H}_2 - \mathcal{H}_S,$$

whence $\mathcal{H}_1 = \mathcal{H}_2$. $\qquad\square$

Remark 7.8.11. The last two results were discovered by Joseph Grifone, see section 10 of his seminal paper [54]. The deduction presented here is different (and simpler) from that of Grifone.

7.9 Ehresmann Connections and Covariant Derivatives

Let D be a covariant derivative on the slit Finsler bundle

$$\mathring{\pi}\colon \mathring{T}M \times_M TM \to \mathring{T}M.$$

As in subsection 6.2.1, we define the *torsion* T^D, the *vertical torsion* $\mathfrak{Q}$, the *vertical deviation* $\mathcal{S}$ and the *vertical curvature* $\mathbf{Q}$ of D by the formulae

$$T^D(\xi,\eta) := T^{\mathbf{j}}(\xi,\eta) \overset{(6.2.3)}{=} D_\xi \mathbf{j}\eta - D_\eta \mathbf{j}\xi - \mathbf{j}[\xi,\eta];$$

$$\mathfrak{Q}(\widetilde{X},\widetilde{Y}) \overset{(6.2.5)}{:=} D_{\mathbf{i}\widetilde{X}}\widetilde{Y} - D_{\mathbf{i}\widetilde{Y}}\widetilde{X} - \mathbf{i}^{-1}[\mathbf{i}\widetilde{X},\mathbf{i}\widetilde{Y}];$$

$$\mathcal{S}(\widetilde{X},\widetilde{Y}) := T^D(\xi,\mathbf{i}\widetilde{Y}) = \mathbf{j}[\mathbf{i}\widetilde{Y},\xi] - D_{\mathbf{i}\widetilde{Y}}\widetilde{X}, \quad \mathbf{j}\xi = \widetilde{X};$$

$$\mathbf{Q}(\widetilde{X},\widetilde{Y})\widetilde{Z} \overset{(6.2.4)}{:=} R^D(\mathbf{i}\widetilde{X},\mathbf{i}\widetilde{Y})\widetilde{Z} = D_{\mathbf{i}\widetilde{X}}D_{\mathbf{i}\widetilde{Y}}\widetilde{Z} - D_{\mathbf{i}\widetilde{Y}}D_{\mathbf{i}\widetilde{X}}\widetilde{Z} - D_{[\mathbf{i}\widetilde{X},\mathbf{i}\widetilde{Y}]}\widetilde{Z};$$

where $\xi,\eta \in \mathfrak{X}(\mathring{T}M)$; $\widetilde{X},\widetilde{Y} \in \Gamma(\mathring{\pi})$.

In the presence of an Ehresmann connection $\mathcal{H}$ in $\mathring{T}M$, we can speak of the $\mathcal{V}$-torsion of D, where $\mathcal{V}\colon \mathfrak{X}(\mathring{T}M) \to \Gamma(\mathring{\pi})$ is the vertical mapping associated to $\mathcal{H}$ (the general notion of φ-torsion was formulated in Definition 6.1.31). It becomes also possible to introduce some further useful 'partial torsions' and 'partial curvatures'.

Definition and Lemma 7.9.1. Let $D\colon \mathfrak{X}(\mathring{T}M) \times \Gamma(\mathring{\pi}) \to \Gamma(\mathring{\pi})$ be a covariant derivative on $\mathring{\pi}$, and let an Ehresmann connection $\mathcal{H}$ be specified in $\mathring{T}M$.

(a) The $\mathcal{H}$-*horizontal torsion* (or, if there is no danger of confusion, the *horizontal torsion*) of D is the type $(1,2)$ Finsler tensor field $\mathfrak{T} \in \mathcal{F}^1_2(\mathring{T}M)$ given by

$$\begin{aligned} \mathfrak{T}(\widetilde{X},\widetilde{Y}) &:= T^D(\mathcal{H}\widetilde{X},\mathcal{H}\widetilde{Y}) \\ &= D_{\mathcal{H}\widetilde{X}}\widetilde{Y} - D_{\mathcal{H}\widetilde{Y}}\widetilde{X} - \mathbf{j}[\mathcal{H}\widetilde{X},\mathcal{H}\widetilde{Y}]; \quad \widetilde{X},\widetilde{Y} \in \Gamma(\mathring{\pi}). \end{aligned} \tag{7.9.1}$$

(b) The $\mathcal{H}$-*horizontal* (or simply *horizontal*) *curvature* $\mathbf{R}$ and the $\mathcal{H}$-*mixed* (briefly *mixed*) *curvature* $\mathbf{P}$ of D are defined by

$$\begin{aligned} \mathbf{R}(\widetilde{X},\widetilde{Y})\widetilde{Z} &:= R^D(\mathcal{H}\widetilde{X},\mathcal{H}\widetilde{Y})\widetilde{Z} \\ &= D_{\mathcal{H}\widetilde{X}}D_{\mathcal{H}\widetilde{Y}}\widetilde{Z} - D_{\mathcal{H}\widetilde{Y}}D_{\mathcal{H}\widetilde{X}}\widetilde{Z} - D_{[\mathcal{H}\widetilde{X},\mathcal{H}\widetilde{Y}]}\widetilde{Z} \end{aligned} \tag{7.9.2}$$

and

$$\begin{aligned} \mathbf{P}(\widetilde{X},\widetilde{Y})\widetilde{Z} &:= R^D(\mathbf{i}\widetilde{X},\mathcal{H}\widetilde{Y})\widetilde{Z} \\ &= D_{\mathbf{i}\widetilde{X}}D_{\mathcal{H}\widetilde{Y}}\widetilde{Z} - D_{\mathcal{H}\widetilde{Y}}D_{\mathbf{i}\widetilde{X}}\widetilde{Z} - D_{[\mathbf{i}\widetilde{X},\mathcal{H}\widetilde{Y}]}\widetilde{Z}, \end{aligned} \tag{7.9.3}$$

respectively.

(c) Let $\mathcal{V}$ be the vertical mapping associated to $\mathcal{H}$. Then the $\mathcal{V}$-*torsion* $T^{\mathcal{V}}(D)$ of D is given by

$$T^{\mathcal{V}}(D)(\xi, \eta) := T^{\mathcal{V}}(\xi, \eta) \overset{(6.1.45)}{=} D_{\xi}\mathcal{V}\eta - D_{\eta}\mathcal{V}\xi - \mathcal{V}[\xi, \eta], \qquad (7.9.4)$$

where $\xi, \eta \in \mathfrak{X}(\mathring{T}M)$. We define the $\mathcal{H}$-*deviation* (or *horizontal deviation*) $\mathcal{P} \in \mathcal{F}_2^1(\mathring{T}M)$ of D by

$$\mathcal{P}(\widetilde{X}, \widetilde{Y}) := T^{\mathcal{V}}(D)(\mathrm{i}\widetilde{X}, \mathcal{H}\widetilde{Y}) = -D_{\mathcal{H}\widetilde{Y}}\widetilde{X} - \mathcal{V}[\mathrm{i}\widetilde{X}, \mathcal{H}\widetilde{Y}]. \qquad (7.9.5)$$

For any sections $\widetilde{X}, \widetilde{Y} \in \Gamma(\mathring{\pi})$, we have

$$T^{\mathcal{V}}(D)(\mathcal{H}\widetilde{X}, \mathcal{H}\widetilde{Y}) = -\mathcal{R}(\widetilde{X}, \widetilde{Y}) \text{ and } T^{\mathcal{V}}(D)(\mathrm{i}\widetilde{X}, \mathrm{i}\widetilde{Y}) = \mathcal{Q}(\widetilde{X}, \widetilde{Y}). \qquad (7.9.6)$$

Proof. All assertions can be verified either by inspection or by simple direct calculation. As an example, we show that $\mathcal{P}$ is indeed a Finsler tensor field, i.e., it is $C^{\infty}(\mathring{T}M)$-bilinear.

Obviously, $\mathcal{P}$ is additive in both slots. To check the $C^{\infty}(\mathring{T}M)$-homogeneity, let F be a smooth function on $\mathring{T}M$. For any sections $\widetilde{X}, \widetilde{Y} \in \Gamma(\mathring{\pi})$,

$$\mathcal{P}(F\widetilde{X}, \widetilde{Y}) = -D_{\mathcal{H}\widetilde{Y}}F\widetilde{X} - \mathcal{V}[\mathrm{i}(F\widetilde{X}), \mathcal{H}\widetilde{Y}] = -FD_{\mathcal{H}\widetilde{Y}}\widetilde{X} - ((\mathcal{H}\widetilde{Y})F)\widetilde{X}$$

$$- F\mathcal{V}[\mathrm{i}\widetilde{X}, \mathcal{H}\widetilde{Y}] + \mathcal{V}((\mathcal{H}\widetilde{Y})F)\mathrm{i}\widetilde{X} \overset{(7.2.22)}{=} F\mathcal{P}(\widetilde{X}, \widetilde{Y});$$

$$\mathcal{P}(\widetilde{X}, F\widetilde{Y}) = -D_{\mathcal{H}(F\widetilde{Y})}\widetilde{X} - \mathcal{V}[\mathrm{i}\widetilde{X}, \mathcal{H}(F\widetilde{Y})] = -FD_{\mathcal{H}\widetilde{Y}}\widetilde{X}$$

$$- F\mathcal{V}[\mathrm{i}\widetilde{X}, \mathcal{H}\widetilde{Y}] - \mathcal{V}((\mathrm{i}\widetilde{X})F)\mathcal{H}\widetilde{Y} \overset{(7.2.21)}{=} F\mathcal{P}(\widetilde{X}, \widetilde{Y}),$$

as was to be shown. $\qquad\qquad\qquad\qquad\qquad\qquad\qquad\qquad\qquad\qquad\qquad\qquad\quad\square$

Remark 7.9.2. (a) Using an Ehresmann connection $\mathcal{H}$ in $\mathring{T}M$, the vertical deviation of a covariant derivative D on $\mathring{\pi}$ can also be expressed by the formula

$$\mathcal{S}(\widetilde{X}, \widetilde{Y}) = T^D(\mathcal{H}\widetilde{X}, \mathrm{i}\widetilde{Y}); \quad \widetilde{X}, \widetilde{Y} \in \Gamma(\mathring{\pi}). \qquad (7.9.7)$$

However, of course, the vertical deviation of a covariant derivative on $\mathring{\pi}$ does not depend on any additional structure. Similarly, the vertical torsion of D can be obtained from the $\mathcal{V}$-torsion of D by the second formula in (7.9.6).

(b) *The torsion of a covariant derivative D on $\mathring{\pi}$ is completely determined by its horizontal torsion and vertical deviation.* Explicitly, for any vector fields ξ, η on $\mathring{T}M$ we have

$$T^D(\xi, \eta) = \mathcal{T}(\mathbf{j}\xi, \mathbf{j}\eta) + \mathcal{S}(\mathbf{j}\xi, \mathcal{V}\eta) - \mathcal{S}(\mathbf{j}\eta, \mathcal{V}\xi). \qquad (7.9.8)$$

Indeed, if $\mathcal{H}$ is an Ehresmann connection in $\mathring{T}M$ with associated projections **h** and **v** and vertical mapping $\mathcal{V}$, then by (7.9.1), (7.9.7) and (6.2.3),

$$T^D(\xi, \eta) = T^D(\mathbf{h}\xi + \mathbf{v}\xi, \mathbf{h}\eta + \mathbf{v}\eta) = T^D(\mathcal{H}(\mathbf{j}\xi), \mathcal{H}(\mathbf{j}\eta))$$
$$+ T^D(\mathbf{i}(\mathcal{V}\xi), \mathcal{H}(\mathbf{j}\eta)) + T^D(\mathcal{H}(\mathbf{j}\xi), \mathbf{i}(\mathcal{V}\eta)) + T^D(\mathbf{i}(\mathcal{V}\xi), \mathbf{i}(\mathcal{V}\eta))$$
$$= \mathcal{T}(\mathbf{j}\xi, \mathbf{j}\eta) - \mathcal{S}(\mathbf{j}\eta, \mathcal{V}\xi) + \mathcal{S}(\mathbf{j}\xi, \mathcal{V}\eta),$$

as was to be shown.

Definition 7.9.3. Let D be a covariant derivative on the slit Finsler bundle $\mathring{\pi}$, and let an Ehresmann connection $\mathcal{H}$ be specified in $\mathring{T}M$.

(a) We say that

D is *associated* to $\mathcal{H}$ if $\mathrm{Ker}(\underline{\mu}) = \mathrm{Im}(\mathcal{H})$,
D is *strongly associated* to $\mathcal{H}$ if $\mu = \mathcal{V}$.

Here $\underline{\mu} \colon T\mathring{T}M \to \mathring{T}M \times_M TM$ is the strong bundle map determined by the deflection $\mu := D\mathring{\delta}$ of D (see Definition 6.2.5), and $\mathcal{V}$ is the vertical mapping belonging to $\mathcal{H}$. The mapping

$$\mu^{\mathcal{H}} := \mu \circ \mathcal{H} \colon \Gamma(\mathring{\pi}) \to \Gamma(\mathring{\pi}), \ \widetilde{X} \mapsto \mu^{\mathcal{H}}(\widetilde{X}) = D_{\mathcal{H}\widetilde{X}}\widetilde{\delta} \tag{7.9.9}$$

is called the $\mathcal{H}$-*deflection* or *horizontal deflection* of D.

Remark 7.9.4. Clearly, the $\mathcal{H}$-deflection $\mu^{\mathcal{H}}$ of D is tensorial. To describe it locally, choose a chart $(\mathcal{U}, (u^i)_{i=1}^n)$ on M. Let (Γ^i_{jk}) and (C^i_{jk}) be the families of the Christoffel symbols of D with respect to the chosen chart (see, mutatis mutandis, Remark 6.2.3), and let (N^i_j) be the family of Christoffel symbols of $\mathcal{H}$ with respect to the same chart. Then, by (7.2.37) and (6.2.6),

$$\mu^{\mathcal{H}}\left(\widehat{\frac{\partial}{\partial u^j}}\right) = D_{\left(\frac{\partial}{\partial u^j}\right)^{\mathbf{h}}}\widetilde{\delta} = D_{\frac{\partial}{\partial x^j} - N^i_j \frac{\partial}{\partial y^i}} y^k \widehat{\frac{\partial}{\partial u^k}}$$
$$= y^k \Gamma^i_{jk} \widehat{\frac{\partial}{\partial u^i}} - N^i_j \left(\delta^k_i \widehat{\frac{\partial}{\partial u^k}} + y^k C^l_{ik} \widehat{\frac{\partial}{\partial u^l}}\right).$$

After some index manipulation on the right-hand side, we find that

$$\mu^{\mathcal{H}}\left(\widehat{\frac{\partial}{\partial u^j}}\right) = \left(y^k \Gamma^i_{jk} - N^k_j(\delta^i_k + y^l C^i_{kl})\right)\widehat{\frac{\partial}{\partial u^i}}, \quad j \in J_n. \tag{7.9.10}$$

If, in particular, D is strongly regular, then $y^l C^i_{kl} = 0$ by Remark 6.2.8(b), and we obtain

$$\mu^{\mathcal{H}}\left(\widehat{\frac{\partial}{\partial u^j}}\right) = (y^k \Gamma^i_{jk} - N^i_j)\widehat{\frac{\partial}{\partial u^i}}, \quad j \in J_n. \tag{7.9.11}$$

Lemma 7.9.5. *Let $\mathcal{H}$ be an Ehresmann connection in $\overset{\circ}{T}M$. A covariant derivative D on $\overset{\circ}{\pi}$ is*

(i) *associated to $\mathcal{H}$ if, and only if, D is regular and its $\mathcal{H}$-deflection vanishes;*

(ii) *strongly associated to $\mathcal{H}$ if, and only if, D is strongly regular and its $\mathcal{H}$-deflection vanishes.*

Proof. (i) Suppose that D is associated to $\mathcal{H}$. Then $\mathrm{Ker}(\mu) = \mathrm{Im}(\mathcal{H})$ implies by Lemma 7.2.3 that $T\overset{\circ}{T}M = \mathrm{Ker}(\underline{\mu}) \oplus V\overset{\circ}{T}M$. So D is regular by Proposition 6.2.9, while $\mu^{\mathcal{H}} = \mu \circ \mathcal{H} = 0$ holds automatically.

Conversely, let D be a regular covariant derivative with vanishing $\mathcal{H}$-deflection. Using the projection operators associated to $\mathcal{H}$, any vector field ξ on $\overset{\circ}{T}M$ can uniquely be written as

$$\xi = \mathbf{h}\xi + \mathbf{v}\xi = \mathcal{H}(\mathbf{j}\xi) + \mathbf{i}(\mathcal{V}\xi).$$

If $\xi \in \mathrm{Ker}(\mu)$, then $\mu^{\mathcal{H}} = 0$ implies that

$$0 = \mu(\xi) = \mu \circ \mathbf{i}(\mathcal{V}\xi) = \mu^{\mathsf{v}}(\mathcal{V}\xi).$$

Since μ^{v} is injective by the regularity of D, from this it follows that $\mathcal{V}\xi = 0$. Hence $\xi = \mathbf{h}\xi \in \mathrm{Im}(\mathcal{H})$, therefore $\mathrm{Ker}(\mu) \subset \mathrm{Im}(\mathcal{H})$. To show the converse, let $\xi \in \mathrm{Im}(\mathcal{H})$. Then $\mu(\xi) = \mu(\mathbf{h}\xi) = \mu \circ \mathcal{H}(\mathbf{j}\xi) = \mu^{\mathcal{H}}(\mathbf{j}\xi) = 0$, so we have $\xi \in \mathrm{Ker}(\mu)$ and hence $\mathrm{Im}(\mathcal{H}) = \mathrm{Ker}(\mu)$.

(ii) Suppose that D is strongly associated to $\mathcal{H}$. Then

$$\mu^{\mathsf{v}} := (D\widetilde{\delta}) \circ \mathbf{i} = \mu \circ \mathbf{i} \overset{\text{cond.}}{=} \mathcal{V} \circ \mathbf{i} \overset{(7.2.22)}{=} 1_{\Gamma(\overset{\circ}{\pi})},$$

so D is strongly regular. The $\mathcal{H}$-deflection of D also vanishes, since

$$\mu^{\mathcal{H}} = \mu \circ \mathcal{H} = \mathcal{V} \circ \mathcal{H} \overset{(7.2.21)}{=} 0.$$

Conversely, if D is strongly regular, i.e., $\mu^{\mathsf{v}} = 1_{\gamma(\overset{\circ}{\pi})}$, and $\mu^{\mathcal{H}} = 0$, then for any vector field ξ on $\overset{\circ}{T}M$,

$$\mu(\xi) = \mu(\mathcal{H}(\mathbf{j}\xi)) + \mu(\mathbf{i}(\mathcal{V}\xi)) = \mu^{\mathcal{H}}(\mathbf{j}\xi) + \mu^{\mathsf{v}}(\mathcal{V}\xi) = \mathcal{V}\xi,$$

whence $\mu = \mathcal{V}$, as was to be shown. $\square$

Corollary 7.9.6. *If a covariant derivative D on $\overset{\circ}{\pi}$ is associated to an Ehresmann connection $\mathcal{H}$ in $\overset{\circ}{T}M$, then $\mathcal{H} = \mathcal{H}_D$, where $\mathcal{H}_D$ is the Ehresmann connection induced by D according to Example 7.3.3(a).*

First proof – *using local coordinates.* Choose a chart $(\mathcal{U}, (u^i)_{i=1}^n)$ on M. Let (N_j^i) and $(\widetilde{N}_j^i)$ be the families of Christoffel symbols of $\mathcal{H}$ and $\mathcal{H}_D$, respectively, with respect to the chosen chart. Since, in particular, D is regular by Lemma 7.9.5(i), from (7.3.4), after changing two summation indices on the left-hand side, we obtain $\widetilde{N}_j^k(\delta_k^r + y^l C_{kl}^r) = y^k \Gamma_{jk}^r$; $i, j \in J_n$. On the other hand, by the same lemma, $\mu^{\mathcal{H}} = 0$, so from (7.9.10) we get $N_j^k(\delta_k^i + y^l C_{kl}^i) = y^k \Gamma_{jk}^i$; $i, j \in J_n$. By the regularity of D, the matrix $(\delta_k^i + y^l C_{kl}^i)$ is invertible at each point, thus $\mathcal{H}_D$ and $\mathcal{H}$ have the same Christoffel symbols with respect to the same chart, therefore they are equal.

Second proof – *in a coordinate-free manner.* Since D is associated to $\mathcal{H}$, $\mu^{\mathcal{H}} = 0$ by the previous lemma. Thus for any vector field X on M,

$$
\begin{aligned}
\mathcal{H}_D(\widehat{X}) &\overset{(7.3.2)}{=} X^c - i(\mu^v)^{-1} D_{X^c}\widetilde{\delta} = X^c - i(\mu^v)^{-1} D_{hX^c + vX^c}\widetilde{\delta} \\
&= X^c - i(\mu^v)^{-1} D\widetilde{\delta}(\mathcal{H}(jX^c)) - i(\mu^v)^{-1} D\widetilde{\delta}(i(\mathcal{V}X^c)) \\
&= X^c - i(\mu^v)^{-1}\mu^{\mathcal{H}}(jX^c) - i(\mu^v)^{-1}\mu^v(\mathcal{V}X^c) \\
&= X^c - i(\mathcal{V}X^c) = X^c - vX^c = \mathcal{H}(\widehat{X}),
\end{aligned}
$$

whence $\mathcal{H}_D = \mathcal{H}$. $\qquad\square$

Lemma 7.9.7. *Let D be a covariant derivative on the Finsler bundle $\mathring{\pi}$, and let $\mathcal{H}$ be an Ehresmann connection in $\mathring{T}M$. If D is strongly associated to $\mathcal{H}$, then*

$$
i_{\widetilde{\delta}}\mathcal{S} = 0 \quad and \quad i_{\widetilde{\delta}}\mathcal{P} = \mathbf{t}, \tag{7.9.12}
$$

where $i_{\widetilde{\delta}}$ is the substitution operator associated to $\widetilde{\delta}$. Moreover, for any sections $\widetilde{X}, \widetilde{Y} \in \Gamma(\mathring{\pi})$, we have $\mathbf{Q}(\widetilde{X}, \widetilde{Y})\widetilde{\delta} = \mathcal{Q}(\widetilde{X}, \widetilde{Y})$,

$$
\mathbf{R}(\widetilde{X}, \widetilde{Y})\widetilde{\delta} = -\mathcal{R}(\widetilde{X}, \widetilde{Y}), \tag{7.9.13}
$$

$$
\mathbf{P}(\widetilde{X}, \widetilde{Y})\widetilde{\delta} = \mathcal{P}(\widetilde{X}, \widetilde{Y}). \tag{7.9.14}
$$

Proof. Since D is strongly associated to $\mathcal{H}$, it is strongly regular ($\mu^v = 1_{\Gamma(\mathring{\pi})}$) and its $\mathcal{H}$-deflection vanishes ($\mu^{\mathcal{H}} = 0$). Thus $i_{\widetilde{\delta}}\mathcal{S} = 0$ and $\mathbf{Q}(\widetilde{X}, \widetilde{Y})\widetilde{\delta} = \mathcal{Q}(\widetilde{X}, \widetilde{Y})$ follow from Lemma 6.2.25 and Lemma 6.2.11, respectively. The verification of the other equalities is just a calculation. For any section $\widetilde{X} \in \Gamma(\mathring{\pi})$,

$$
i_{\widetilde{\delta}}\mathcal{P}(\widetilde{X}) = \mathcal{P}(\widetilde{\delta}, \widetilde{X}) \overset{(7.9.5)}{=} -D_{\mathcal{H}\widetilde{X}}\widetilde{\delta} - \mathcal{V}[C, \mathcal{H}\widetilde{X}] = -\mu^{\mathcal{H}}(\widetilde{X}) + \mathbf{t}(\widetilde{X}) = \mathbf{t}(\widetilde{X}),
$$

so $i_{\widetilde{\delta}}\mathcal{P} = \mathbf{t}$. Starting from the definitions of $\mathbf{R}$ and $\mathbf{P}$,

$$\mathbf{R}(\widetilde{X},\widetilde{Y})(\widetilde{\delta}) = D_{\mathcal{H}\widetilde{X}}D_{\mathcal{H}\widetilde{Y}}\widetilde{\delta} - D_{\mathcal{H}\widetilde{Y}}D_{\mathcal{H}\widetilde{X}}\widetilde{\delta} - D_{[\mathcal{H}\widetilde{X},\mathcal{H}\widetilde{Y}]}\widetilde{\delta} = -D_{[\mathcal{H}\widetilde{X},\mathcal{H}\widetilde{Y}]}\widetilde{\delta}$$

$$= -D_{\mathcal{H}(\mathbf{j}[\mathcal{H}\widetilde{X},\mathcal{H}\widetilde{Y}])}\widetilde{\delta} - D_{\mathbf{i}(\mathcal{V}[\mathcal{H}\widetilde{X},\mathcal{H}\widetilde{Y}])}\widetilde{\delta} = -\mu^{\mathcal{H}}(\mathbf{j}[\mathcal{H}\widetilde{X},\mathcal{H}\widetilde{Y}])$$

$$- \mu^{\mathcal{V}}(\mathcal{V}[\mathcal{H}\widetilde{X},\mathcal{H}\widetilde{Y}]) = -\mathcal{V}[\mathcal{H}\widetilde{X},\mathcal{H}\widetilde{Y}] = -\mathcal{R}(\widetilde{X},\widetilde{Y}),$$

$$\mathbf{P}(\widetilde{X},\widetilde{Y})\widetilde{\delta} = D_{\mathbf{i}\widetilde{X}}D_{\mathcal{H}\widetilde{Y}}\widetilde{\delta} - D_{\mathcal{H}\widetilde{Y}}D_{\mathbf{i}\widetilde{X}}\widetilde{\delta} - D_{[\mathbf{i}\widetilde{X},\mathcal{H}\widetilde{Y}]}\widetilde{\delta} = -D_{\mathcal{H}\widetilde{Y}}\widetilde{X}$$

$$- D_{\mathbf{i}(\mathcal{V}[\mathbf{i}\widetilde{X},\mathcal{H}\widetilde{Y}])}\widetilde{\delta} = -D_{\mathcal{H}\widetilde{Y}}\widetilde{X} - \mathcal{V}[\mathbf{i}\widetilde{X},\mathcal{H}\widetilde{Y}] \overset{(7.9.5)}{=} \mathcal{P}(\widetilde{X},\widetilde{Y}),$$

thus we find that (7.9.13) and (7.9.14) are also valid. $\square$

Proposition 7.9.8. *Let D be a covariant derivative on $\mathring{\pi}$ and let $\mathcal{H}$ be an Ehresmann connection in $\mathring{T}M$. Suppose that D is strongly associated to $\mathcal{H}$.*

(i) *If the vertical deviation of D is symmetric, then the vertical torsion of D vanishes, so we have*

$$[\mathbf{i}\widetilde{X}, \mathbf{i}\widetilde{Y}] = \mathbf{i}(D_{\mathbf{i}\widetilde{X}}\widetilde{Y} - D_{\mathbf{i}\widetilde{Y}}\widetilde{X}). \tag{7.9.15}$$

(ii) *If the horizontal torsion of D vanishes, then*

$$[\mathbf{i}\widetilde{X}, \mathcal{H}\widetilde{Y}] = \mathcal{H}(D_{\mathbf{i}\widetilde{X}}\widetilde{Y} + \mathcal{S}(\widetilde{Y},\widetilde{X})) - \mathbf{i}(D_{\mathcal{H}\widetilde{Y}}\widetilde{X} + \mathcal{P}(\widetilde{X},\widetilde{Y})); \tag{7.9.16}$$

$$[\mathcal{H}\widetilde{X}, \mathcal{H}\widetilde{Y}] = \mathbf{i}\,\mathcal{R}(\widetilde{X},\widetilde{Y}) + \mathcal{H}(D_{\mathcal{H}\widetilde{X}}\widetilde{Y} - D_{\mathcal{H}\widetilde{Y}}\widetilde{X}). \tag{7.9.17}$$

Here $\widetilde{X}$ and $\widetilde{Y}$ are arbitrary sections in $\Gamma(\mathring{\pi})$.

Proof. (i) By the symmetry of $\mathcal{S}$, we have

$$0 = \mathcal{S}(\widetilde{X},\widetilde{Y}) - \mathcal{S}(\widetilde{Y},\widetilde{X}) = D_{\mathbf{i}\widetilde{X}}\widetilde{Y} - D_{\mathbf{i}\widetilde{Y}}\widetilde{X} + \mathbf{j}[\mathbf{i}\widetilde{Y},\mathcal{H}\widetilde{X}] - \mathbf{j}[\mathbf{i}\widetilde{X},\mathcal{H}\widetilde{Y}].$$

Thus

$$\mathcal{Q}(\widetilde{X},\widetilde{Y}) \overset{(6.2.5)}{=} D_{\mathbf{i}\widetilde{X}}\widetilde{Y} - D_{\mathbf{i}\widetilde{Y}}\widetilde{X} - \mathbf{i}^{-1}[\mathbf{i}\widetilde{X},\mathbf{i}\widetilde{Y}]$$

$$= \mathbf{j}[\mathcal{H}\widetilde{X},\mathbf{i}\widetilde{Y}] - \mathbf{j}[\mathcal{H}\widetilde{Y},\mathbf{i}\widetilde{X}] - \mathbf{i}^{-1}[\mathbf{i}\widetilde{X},\mathbf{i}\widetilde{Y}].$$

Representing $\widetilde{X}$ and $\widetilde{Y}$ in the form $\widetilde{X} = \mathbf{j}\xi$ and $\widetilde{Y} = \mathbf{j}\eta$ ($\xi, \eta \in \mathfrak{X}(\mathring{T}M)$), we find that

$$\mathbf{i}\,\mathcal{Q}(\widetilde{X},\widetilde{Y}) = \mathbf{J}[\mathbf{h}\xi,\mathbf{J}\eta] - \mathbf{J}[\mathbf{h}\eta,\mathbf{J}\xi] - [\mathbf{J}\xi,\mathbf{J}\eta]$$

$$= \mathbf{J}[\xi,\mathbf{J}\eta] + \mathbf{J}[\mathbf{J}\xi,\eta] - [\mathbf{J}\xi,\mathbf{J}\eta] \overset{(3.3.18)}{=} -N_{\mathbf{J}}(\xi,\eta) \overset{\text{Lemma 4.1.59}}{=} 0.$$

Relation (7.9.15) is just another expression of the vanishing of $\mathcal{Q}$.

(ii) If $\mathcal{T} = 0$, from (7.9.8) we obtain that

$$D_{\xi}\mathbf{j}\eta - D_{\eta}\mathbf{j}\xi - \mathbf{j}[\xi,\eta] = \mathcal{S}(\mathbf{j}\xi,\mathcal{V}\eta) - \mathcal{S}(\mathbf{j}\eta,\mathcal{V}\xi); \quad \xi, \eta \in \mathfrak{X}(\mathring{T}M). \tag{$*$}$$

With the choice $\xi := \mathrm{i}\widetilde{X}$, $\eta = \mathcal{H}\widetilde{Y}$ we find that $D_{\mathrm{i}\widetilde{X}}\widetilde{Y} - \mathbf{j}[\mathrm{i}\widetilde{X}, \mathcal{H}\widetilde{Y}] = -\mathcal{S}(\widetilde{Y}, \widetilde{X})$, whence

$$\mathbf{h}[\mathrm{i}\widetilde{X}, \mathcal{H}\widetilde{Y}] = \mathcal{H}(D_{\mathrm{i}\widetilde{X}}\widetilde{Y} + \mathcal{S}(\widetilde{Y}, \widetilde{X})).$$

On the other hand $\mathcal{P}(\widetilde{X}, \widetilde{Y}) \overset{(7.9.5)}{=} -D_{\mathcal{H}\widetilde{Y}}\widetilde{X} - \mathcal{V}[\mathrm{i}\widetilde{X}, \mathcal{H}\widetilde{Y}]$, from which we obtain that

$$\mathbf{v}[\mathrm{i}\widetilde{X}, \mathcal{H}\widetilde{Y}] = -\mathrm{i}(D_{\mathcal{H}\widetilde{Y}}\widetilde{X} + \mathcal{P}(\widetilde{X}, \widetilde{Y})).$$

Adding the last two equalities, we get (7.9.16).

Now let $\xi := \mathcal{H}\widetilde{X}$, $\eta := \mathcal{H}\widetilde{Y}$ in (∗). These substitutions yield

$$D_{\mathcal{H}\widetilde{X}}\widetilde{Y} - D_{\mathcal{H}\widetilde{Y}}\widetilde{X} - \mathbf{j}[\mathcal{H}\widetilde{X}, \mathcal{H}\widetilde{Y}] = 0,$$

whence $\mathbf{h}[\mathcal{H}\widetilde{X}, \mathcal{H}\widetilde{Y}] = \mathcal{H}(D_{\mathcal{H}\widetilde{X}}\widetilde{Y} - D_{\mathcal{H}\widetilde{Y}}\widetilde{X})$. From the definition of the curvature of $\mathcal{H}$, $\mathbf{v}[\mathcal{H}\widetilde{X}, \mathcal{H}\widetilde{Y}] = \mathrm{i}\mathcal{R}(\widetilde{X}, \widetilde{Y})$. Adding the last two equalities again, we obtain (7.9.17). □

7.10   The Induced Berwald Derivative

Lemma and Definition 7.10.1. *Let $\mathcal{H}$ be an Ehresmann connection in $\mathring{T}M$. There exists a unique covariant derivative on the slit Finsler bundle $\mathring{\pi}: \mathring{T}M \times_M TM \to \mathring{T}M$ with vanishing vertical and $\mathcal{H}$-horizontal deviation. This covariant derivative is called the* Berwald derivative *induced by the Ehresmann connection $\mathcal{H}$.*

Proof.   Suppose that a covariant derivative $\nabla: \mathfrak{X}(\mathring{T}M) \times \Gamma(\mathring{\pi}) \to \Gamma(\mathring{\pi})$ has vanishing vertical and $\mathcal{H}$-horizontal deviation. Then for any sections $\widetilde{X}, \widetilde{Y} \in \Gamma(\mathring{\pi})$,

$$\nabla_{\mathrm{i}\widetilde{X}}\widetilde{Y} \overset{(6.2.36),(6.2.25)}{=} \mathbf{j}[\mathrm{i}\widetilde{X}, \mathcal{H}\widetilde{Y}] \qquad (7.10.1)$$

(since $\mathbf{j}(\mathcal{H}\widetilde{Y}) = \widetilde{Y}$), and

$$\nabla_{\mathcal{H}\widetilde{X}}\widetilde{Y} \overset{(7.9.5)}{=} \mathcal{V}[\mathcal{H}\widetilde{X}, \mathrm{i}\widetilde{Y}]. \qquad (7.10.2)$$

If ξ is any vector field on $\mathring{T}M$, then

$$\nabla_{\xi}\widetilde{Y} = \nabla_{\mathbf{v}\xi + \mathbf{h}\xi}\widetilde{Y} = \nabla_{\mathrm{i}(\mathcal{V}\xi)}\widetilde{Y} + \nabla_{\mathcal{H}(\mathbf{j}\xi)}\widetilde{Y},$$

therefore

$$\nabla_{\xi}\widetilde{Y} = \mathbf{j}[\mathbf{v}\xi, \mathcal{H}\widetilde{Y}] + \mathcal{V}[\mathbf{h}\xi, \mathrm{i}\widetilde{Y}]. \qquad (7.10.3)$$

Thus ∇ is uniquely determined by the conditions $\mathcal{S} = 0$ and $\mathcal{P} = 0$.

If we write $\widetilde{Y} = \mathbf{j}\eta$ $(\eta \in \mathfrak{X}(\mathring{T}M))$ in (7.10.3), and compose both sides by $\mathbf{i}$, we obtain

$$\mathbf{i}(\nabla_\xi \widetilde{Y}) = \mathbf{J}[\mathbf{v}\xi, \eta] + \mathbf{v}[\mathbf{h}\xi, \mathbf{J}\eta],$$

which is just relation (7.1.10), if we restrict ourselves to the slit Finsler bundle. Thus, by Proposition and Definition 7.1.7, the mapping given by (7.10.3) is a covariant derivative on $\mathring{\pi}$, which proves the existence assertion of the lemma. $\qquad\square$

Corollary 7.10.2. *If $\mathcal{H}$ is an Ehresmann connection in $\mathring{T}M$, and ∇ is the Berwald derivative induced by $\mathcal{H}$, then*

$$\nabla_{X^\mathrm{v}}\widehat{Y} = \nabla^\mathrm{v}_{\widehat{X}}\widehat{Y} = 0, \tag{7.10.4}$$

$$\nabla_{X^\mathrm{h}}\widehat{Y} = \mathcal{V}[X^\mathrm{h}, Y^\mathrm{v}], \quad \mathbf{i}\nabla_{X^\mathrm{h}}\widehat{Y} = [X^\mathrm{h}, Y^\mathrm{v}], \tag{7.10.5}$$

for any vector fields X, Y on M.

Example 7.10.3. Let an Ehresmann connection $\mathcal{H}$ be specified in $\mathring{T}M$, and let ∇ be the Berwald derivative induced by $\mathcal{H}$.

(a) We say that the mapping

$$\nabla^\mathrm{h} \colon \Gamma(\mathring{\pi}) \times \Gamma(\mathring{\pi}) \to \Gamma(\mathring{\pi}),$$

$$(\widetilde{X}, \widetilde{Y}) \mapsto \nabla^\mathrm{h}_{\widetilde{X}}\widetilde{Y} := \nabla_{\mathcal{H}\widetilde{X}}\widetilde{Y} = \mathcal{V}[\mathcal{H}\widetilde{X}, \mathbf{i}\widetilde{Y}] \tag{7.10.6}$$

is the *horizontal covariant derivative* (*horizontal derivative* or *h-Berwald derivative* for short) determined by ∇, or induced by $\mathcal{H}$. Then (cf. Definition 6.2.12) ∇^h is $\mathbb{R}$-bilinear, $C^\infty(\mathring{T}M)$-linear in its first argument, and satisfies the derivation rule

$$\nabla^\mathrm{h}_{\widetilde{X}}F\widetilde{Y} = ((\mathcal{H}\widetilde{X})F)\widetilde{Y} + F\nabla^\mathrm{h}_{\widetilde{X}}\widetilde{Y}, \quad F \in C^\infty(\mathring{T}M). \tag{7.10.7}$$

Observe that the vertical and the horizontal deviation of a covariant derivative D on $\mathring{\pi}$ are just the difference tensors given by

$$\mathcal{S} \colon (\widetilde{X}, \widetilde{Y}) \in \Gamma(\mathring{\pi}) \times \Gamma(\mathring{\pi}) \mapsto \mathcal{S}(\widetilde{X}, \widetilde{Y}) = \nabla^\mathrm{v}_{\widetilde{Y}}\widetilde{X} - D^\mathrm{v}_{\widetilde{Y}}\widetilde{X}$$

and

$$\mathcal{P} \colon (\widetilde{X}, \widetilde{Y}) \in \Gamma(\mathring{\pi}) \times \Gamma(\mathring{\pi}) \mapsto \mathcal{P}(\widetilde{X}, \widetilde{Y}) = \nabla^\mathrm{h}_{\widetilde{Y}}\widetilde{X} - D^\mathrm{h}_{\widetilde{Y}}\widetilde{X},$$

where $D^\mathrm{h}_{\widetilde{X}}\widetilde{Y} := D_{\mathcal{H}\widetilde{X}}\widetilde{Y}$.

(b) Let $\widetilde{X}$ be a fixed section in $\Gamma(\mathring{\pi})$. The 'horizontal derivation'

$$\nabla^\mathrm{h}_{\widetilde{X}} \colon \Gamma(\mathring{\pi}) \to \Gamma(\mathring{\pi}), \ \widetilde{Y} \mapsto \nabla^\mathrm{h}_{\widetilde{X}}\widetilde{Y}$$

can be naturally extended to any Finsler tensor field on $\mathring{T}M$. We agree that

$$\nabla^{\mathsf{h}}_{\widetilde{X}} F := (\mathcal{H}\widetilde{X})F, \ F \in C^\infty(\mathring{T}M). \tag{7.10.8}$$

If $\widetilde{A} \in \mathcal{F}^k_l(\mathring{T}M)$, with $(k,l) \neq (0,0)$, let

$$\nabla^{\mathsf{h}}_{\widetilde{X}} \widetilde{A} := \nabla_{\mathcal{H}\widetilde{X}} \widetilde{A}, \tag{7.10.9}$$

where the right-hand side is defined by (6.1.14). Thus, in particular, if $\widetilde{\alpha} \in (\Gamma(\mathring{\pi}))^* = \mathcal{F}^0_1(\mathring{T}M)$, then for any $\widetilde{Y} \in \Gamma(\mathring{\pi})$ we have

$$(\nabla^{\mathsf{h}}_{\widetilde{X}} \widetilde{\alpha})(\widetilde{Y}) = \mathcal{H}\widetilde{X}(\widetilde{\alpha}(\widetilde{Y})) - \widetilde{\alpha}(\nabla^{\mathsf{h}}_{\widetilde{X}} \widetilde{Y}), \tag{7.10.10}$$

cf. (6.1.11). In general, if $\widetilde{A} \in \mathcal{F}^k_l(\mathring{T}M)$, $(k,l) \neq (0,0)$, then

$$\begin{aligned}
(\nabla^{\mathsf{h}}_{\widetilde{X}} \widetilde{A})(\widetilde{\alpha}^1,\dots,\widetilde{\alpha}^k,\widetilde{Y}_1,\dots,\widetilde{Y}_l) = {} & (\mathcal{H}\widetilde{X})\widetilde{A}(\widetilde{\alpha}^1,\dots,\widetilde{\alpha}^k,\widetilde{Y}_1,\dots,\widetilde{Y}_l) \\
& - \sum_{i=1}^k \widetilde{A}(\widetilde{\alpha}^1,\dots,\nabla^{\mathsf{h}}_{\widetilde{X}}\widetilde{\alpha}^i,\dots,\widetilde{\alpha}^k,\widetilde{Y}_1,\dots,\widetilde{Y}_l) \\
& - \sum_{j=1}^l \widetilde{A}(\widetilde{\alpha}^1,\dots,\widetilde{\alpha}^k,\widetilde{Y}_1,\dots,\nabla^{\mathsf{h}}_{\widetilde{X}}\widetilde{Y}_j,\dots,\widetilde{Y}_l),
\end{aligned} \tag{7.10.11}$$

where $\widetilde{\alpha}_i \in (\Gamma(\mathring{\pi}))^*$, $i \in J_k$; $\widetilde{Y}_j \in \Gamma(\mathring{\pi})$, $j \in J_l$ (cf. (6.2.22)).

The $\mathcal{H}$-*horizontal differential* (*horizontal differential, h-Berwald differential*) of $\widetilde{A}$ is the type $(k, l+1)$ Finsler tensor field $\nabla^{\mathsf{h}}\widetilde{A}$ on $\mathring{T}M$ given by

$$\nabla^{\mathsf{h}}\widetilde{A}(\widetilde{\alpha}^1,\dots,\widetilde{\alpha}^k,\widetilde{X},\widetilde{Y}_1,\dots,\widetilde{Y}_l) := (\nabla^{\mathsf{h}}_{\widetilde{X}}\widetilde{A})(\widetilde{\alpha}^1,\dots,\widetilde{\alpha}^k,\widetilde{Y}_1,\dots,\widetilde{Y}_l), \tag{7.10.12}$$

cf. (6.1.18) and (6.2.23). Then, in particular,

$$\nabla^{\mathsf{h}}F = dF \circ \mathcal{H}, \tag{7.10.13}$$

and, therefore,

$$\nabla^{\mathsf{h}}F \circ \mathbf{j} = dF \circ \mathbf{h} \overset{(7.4.3)}{=} \mathcal{L}_{\mathbf{h}}F, \tag{7.10.14}$$

for every smooth function F on $\mathring{T}M$.

Corollary 7.10.4. *Let ∇ be the Berwald derivative induced by an Ehresmann connection $\mathcal{H}$ in $\mathring{T}M$. Then for any sections $\widetilde{X},\widetilde{Y} \in \Gamma(\mathring{\pi})$ we have*

$$[\mathbf{i}\widetilde{X}, \mathbf{i}\widetilde{Y}] = \mathbf{i}(\nabla_{\mathbf{i}\widetilde{X}}\widetilde{Y} - \nabla_{\mathbf{i}\widetilde{Y}}\widetilde{X}), \tag{7.10.15}$$

$$[\mathbf{i}\widetilde{X}, \mathcal{H}\widetilde{Y}] = \mathcal{H}\nabla_{\mathbf{i}\widetilde{X}}\widetilde{Y} - \mathbf{i}\nabla_{\mathcal{H}\widetilde{Y}}\widetilde{X}. \tag{7.10.16}$$

Proof. Since the vertical deviation of ∇ vanishes, and hence it is automatically symmetric, (7.10.15) is an immediate consequence of Proposition 7.9.8(i). The second relation comes from the definition of ∇, because

$$[i\widetilde{X}, \mathcal{H}\widetilde{Y}] = \mathcal{H}(j[i\widetilde{X}, \mathcal{H}\widetilde{Y}]) - i(\mathcal{V}[\mathcal{H}\widetilde{Y}, i\widetilde{X}]) \overset{(7.10.1),(7.10.2)}{=} \mathcal{H}\nabla_{i\widetilde{X}}\widetilde{Y} - i\nabla_{\mathcal{H}\widetilde{Y}}\widetilde{X}.$$

$\square$

Lemma 7.10.5. *The vertical curvature of the Berwald derivative induced by an Ehresmann connection vanishes.*

Proof. Let $\mathbf{Q}^\nabla$ denote the vertical curvature of ∇. Then, by (6.2.4), for any sections $\widetilde{X}, \widetilde{Y}, \widetilde{Z}$ in $\Gamma(\mathring{\pi})$,

$$\mathbf{Q}^\nabla(\widetilde{X}, \widetilde{Y})\widetilde{Z} = \nabla_{i\widetilde{X}}\nabla_{i\widetilde{Y}}\widetilde{Z} - \nabla_{i\widetilde{Y}}\nabla_{i\widetilde{X}}\widetilde{Z} - \nabla_{[i\widetilde{X}, i\widetilde{Y}]}\widetilde{Z}$$

$$\overset{(7.10.1)}{=} j\big([i\widetilde{X}, h[i\widetilde{Y}, \mathcal{H}\widetilde{Z}]] - [i\widetilde{Y}, h[i\widetilde{X}, \mathcal{H}\widetilde{Z}]] - [[i\widetilde{X}, i\widetilde{Y}], \mathcal{H}\widetilde{Z}]\big)$$

$$\overset{(4.1.40)}{=} j\big([i\widetilde{X}, [i\widetilde{Y}, \mathcal{H}\widetilde{Z}]] + [i\widetilde{Y}, [\mathcal{H}\widetilde{Z}, i\widetilde{X}]] + [\mathcal{H}\widetilde{Z}, [i\widetilde{X}, i\widetilde{Y}]]\big) \overset{(L2)}{=} 0. \square$$

The preceding proof illustrates again the crucial role of the Jacobi identity (L2) concerning curvature identities. However, the vanishing of $\mathbf{Q}^\nabla$ can also be seen by inspection if we evaluate it at basic sections.

Lemma 7.10.6 (Mixed Ricci Formula for Functions). *Let F be a smooth function on $\mathring{T}M$. For any sections $\widetilde{X}, \widetilde{Y} \in \Gamma(\mathring{\pi})$ we have*

$$\nabla^v\nabla^h F(\widetilde{X}, \widetilde{Y}) - \nabla^h\nabla^v F(\widetilde{Y}, \widetilde{X}) = 0. \tag{7.10.17}$$

Proof. The first term on the left-hand side of (7.10.17) can be manipulated as follows:

$$\nabla^v\nabla^h F(\widetilde{X}, \widetilde{Y}) = \nabla_{i\widetilde{X}}(\nabla^h F)(\widetilde{Y}) = \nabla_{i\widetilde{X}}\nabla_{\mathcal{H}\widetilde{Y}}F - \nabla^h F(\nabla_{i\widetilde{X}}\widetilde{Y})$$

$$= i\widetilde{X}(\mathcal{H}\widetilde{Y}(F)) - \mathcal{H}(\nabla_{i\widetilde{X}}\widetilde{Y})F.$$

Similarly,

$$\nabla^h\nabla^v F(\widetilde{Y}, \widetilde{X}) = \mathcal{H}\widetilde{Y}(i\widetilde{X}(F)) - i(\nabla_{\mathcal{H}\widetilde{Y}}\widetilde{X})F.$$

Thus

$$\nabla^v\nabla^h(\widetilde{X}, \widetilde{Y}) - \nabla^h\nabla^v F(\widetilde{Y}, \widetilde{X}) = [i\widetilde{X}, \mathcal{H}\widetilde{Y}]F + (i\nabla_{\mathcal{H}\widetilde{Y}}\widetilde{X} - \mathcal{H}\nabla_{i\widetilde{X}}\widetilde{Y})F$$

$$\overset{(7.10.16)}{=} [i\widetilde{X}, \mathcal{H}\widetilde{Y}]F + [\mathcal{H}\widetilde{Y}, i\widetilde{X}]F = 0,$$

as was to be shown. $\square$

Lemma 7.10.7. *Suppose that* $\mathcal{H}$ *is a torsion-free Ehresmann connection in* $\mathring{T}M$ *with horizontal projection* $\mathbf{h}$, *and let* F *be a smooth function on* $\mathring{T}M$. *Then*

$$\mathcal{L}_{\mathbf{h}}\mathcal{L}_{\mathbf{J}}F(X^{\mathsf{h}},Y^{\mathsf{h}}) = \nabla^{\mathsf{h}}\nabla^{\mathsf{v}}F(\widehat{X},\widehat{Y}) - \nabla^{\mathsf{h}}\nabla^{\mathsf{v}}F(\widehat{Y},\widehat{X})$$

for all $X, Y \in \mathfrak{X}(M)$.

Proof. We have, on the one hand,

$$\mathcal{L}_{\mathbf{h}}\mathcal{L}_{\mathbf{J}}F(X^{\mathsf{h}},Y^{\mathsf{h}}) \overset{(7.4.5)}{=} \mathbf{h}^{*}d\mathcal{L}_{\mathbf{J}}F(X^{\mathsf{h}},Y^{\mathsf{h}}) \overset{(1.2.6)}{=} d\mathcal{L}_{\mathbf{J}}F(X^{\mathsf{h}},Y^{\mathsf{h}})$$
$$= X^{\mathsf{h}}(\mathcal{L}_{\mathbf{J}}F(Y^{\mathsf{h}})) - Y^{\mathsf{h}}(\mathcal{L}_{\mathbf{J}}F(X^{\mathsf{h}})) - \mathcal{L}_{\mathbf{J}}F([X^{\mathsf{h}},Y^{\mathsf{h}}])$$
$$\overset{(7.2.33),(4.1.97)}{=} X^{\mathsf{h}}(Y^{\mathsf{v}}F) - Y^{\mathsf{h}}(X^{\mathsf{v}}F) - \mathbf{J}[X^{\mathsf{h}},Y^{\mathsf{h}}](F).$$

On the other hand,

$$\nabla^{\mathsf{h}}\nabla^{\mathsf{v}}F(\widehat{X},\widehat{Y}) = (\nabla_{X^{\mathsf{h}}}\nabla^{\mathsf{v}}F)(\widehat{Y}) = X^{\mathsf{h}}(Y^{\mathsf{v}}F) - \nabla^{\mathsf{v}}F(\nabla_{X^{\mathsf{h}}}\widehat{Y})$$
$$\overset{(7.10.5)}{=} X^{\mathsf{h}}(Y^{\mathsf{v}}F) - \nabla^{\mathsf{v}}F(\mathcal{V}[X^{\mathsf{h}},Y^{\mathsf{v}}]).$$

Therefore, taking into account that $\mathbf{T} = 0$,

$$\nabla^{\mathsf{h}}\nabla^{\mathsf{v}}F(\widehat{X},\widehat{Y}) - \nabla^{\mathsf{h}}\nabla^{\mathsf{v}}F(\widehat{Y},\widehat{X}) = X^{\mathsf{h}}(Y^{\mathsf{v}}F) - Y^{\mathsf{h}}(X^{\mathsf{h}}F)$$
$$- \nabla^{\mathsf{v}}F(\mathcal{V}[X^{\mathsf{h}},Y^{\mathsf{v}}] - \mathcal{V}[Y^{\mathsf{h}},X^{\mathsf{v}}]) \overset{(7.8.2)}{=} X^{\mathsf{h}}(Y^{\mathsf{v}}F) - Y^{\mathsf{h}}(X^{\mathsf{v}}F)$$
$$- \nabla^{\mathsf{v}}F(\mathbf{j}[X^{\mathsf{h}},Y^{\mathsf{h}}]) = X^{\mathsf{h}}(Y^{\mathsf{v}}F) - Y^{\mathsf{h}}(X^{\mathsf{v}}F) - \mathbf{J}[X^{\mathsf{h}},Y^{\mathsf{h}}](F),$$

whence our claim. $\qquad\square$

Proposition 7.10.8. *If two positive-homogeneous Ehresmann connections induce the same Berwald derivative, then the Ehresmann connections are identical.*

Proof. Let $\mathcal{H}$ and $\bar{\mathcal{H}}$ be positive-homogeneous Ehresmann connections in $\mathring{T}M$; the geometric data arising from them will be distinguished by a bar. Suppose that $\mathcal{H}$ and $\bar{\mathcal{H}}$ induce the same Berwald derivative ∇. Then for any vector fields X, Y on M we have

$$0 = \mathbf{i}(\nabla_{X^{\mathsf{h}}}\widehat{Y} - \nabla_{X^{\mathsf{h}}}\widehat{Y}) \overset{(7.10.5)}{=} [X^{\mathsf{h}},Y^{\mathsf{v}}] - [X^{\bar{\mathsf{h}}},Y^{\mathsf{v}}] = [X^{\mathsf{h}} - X^{\bar{\mathsf{h}}},Y^{\mathsf{v}}].$$

Since

$$\mathbf{J}(X^{\mathsf{h}} - X^{\bar{\mathsf{h}}}) = \mathbf{J}\circ\mathbf{h}(X^{\mathsf{c}}) - \mathbf{J}\circ\bar{\mathbf{h}}(X^{\mathsf{c}}) \overset{(7.2.27)}{=} \mathbf{J}X^{\mathsf{c}} - \mathbf{J}X^{\mathsf{c}} = 0,$$

the vector field $X^{\mathsf{h}} - X^{\bar{\mathsf{h}}}$ is vertical. Furthermore, $[\mathbf{J}, X^{\mathsf{h}} - X^{\bar{\mathsf{h}}}]Y^{\mathsf{c}} = [Y^{\mathsf{v}}, X^{\mathsf{h}} - X^{\bar{\mathsf{h}}}] - \mathbf{J}[Y^{\mathsf{c}}, X^{\mathsf{h}} - X^{\bar{\mathsf{h}}}] = 0$, and, clearly, $[\mathbf{J}, X^{\mathsf{h}} - X^{\bar{\mathsf{h}}}]Y^{\mathsf{v}} = 0$, therefore $[\mathbf{J}, X^{\mathsf{h}} - X^{\bar{\mathsf{h}}}] = 0$. Thus, by Lemma 4.1.60(i), $X^{\mathsf{h}} - X^{\bar{\mathsf{h}}}$ is the vertical lift of a vector field on M. Then, taking into account (4.1.53) and Lemma 4.2.14, $X^{\mathsf{h}} - X^{\bar{\mathsf{h}}}$ is positive-homogeneous of degree 0. On the other hand, since both Ehresmann connections are positive-homogeneous, we have $[C, X^{\mathsf{h}} - X^{\bar{\mathsf{h}}}] = 0$. This implies that $X^{\mathsf{h}} - X^{\bar{\mathsf{h}}}$ is 1^{+}-homogeneous as well. Therefore $X^{\mathsf{h}} - X^{\bar{\mathsf{h}}} = 0$, and hence $\mathcal{H} = \bar{\mathcal{H}}$. $\square$

7.11   The Debauch of Indices

Throughout this section, $\mathcal{H}$ is an Ehresmann connection in $\mathring{T}M$, ∇ is the Berwald derivative induced by $\mathcal{H}$ and $(\mathcal{U}, (u^i)_{i=1}^n)$ is a chart on M with induced chart $(\tau^{-1}(\mathcal{U}), ((x^i)_{i=1}^n, (y^i)_{i=1}^n))$ on TM. Summation convention in force.

Example 7.11.1 (The Christoffel Symbols). Let (N^i_j) be the family of the Christoffel symbols of $\mathcal{H}$ with respect to the chosen chart. Then for all $i, j \in J_n$, $N^i_j \colon \mathring{\tau}^{-1}(\mathcal{U}) \to \mathbb{R}$ are smooth functions such that

$$\left(\frac{\partial}{\partial u^j}\right)^{\mathsf{h}} = \frac{\partial}{\partial x^j} - N^i_j \frac{\partial}{\partial y^i},$$

see Remark 7.2.12. For each $j, k \in J_n$,

$$\nabla_{\frac{\partial}{\partial x^j}} \widehat{\frac{\partial}{\partial u^k}} = \nabla_{\mathsf{h}(\frac{\partial}{\partial u^j})^{\mathsf{c}} + \mathsf{v}(\frac{\partial}{\partial u^j})^{\mathsf{c}}} \widehat{\frac{\partial}{\partial u^k}} = \nabla_{(\frac{\partial}{\partial u^j})^{\mathsf{h}}} \widehat{\frac{\partial}{\partial u^k}} + \nabla^{\mathsf{v}}_{(\frac{\partial}{\partial u^j})^{\mathsf{c}}} \widehat{\frac{\partial}{\partial u^k}}$$

$$\overset{(6.2.32)}{=} \nabla_{(\frac{\partial}{\partial u^j})^{\mathsf{h}}} \widehat{\frac{\partial}{\partial u^k}} = \mathcal{V}\left[\left(\frac{\partial}{\partial u^j}\right)^{\mathsf{h}}, \frac{\partial}{\partial y^k}\right]$$

$$= \mathcal{V}\left[\frac{\partial}{\partial x^j} - N^i_j \frac{\partial}{\partial y^i}, \frac{\partial}{\partial y^k}\right] = \frac{\partial N^i_j}{\partial y^k} \mathcal{V}\left(\frac{\partial}{\partial y^i}\right) \overset{(7.2.40)}{=} \frac{\partial N^i_j}{\partial y^k} \widehat{\frac{\partial}{\partial u^i}};$$

$$\nabla_{\frac{\partial}{\partial y^j}} \widehat{\frac{\partial}{\partial u^k}} = \nabla_{(\frac{\partial}{\partial u^j})^{\mathsf{v}}} \widehat{\frac{\partial}{\partial u^k}} \overset{(7.10.4)}{=} 0.$$

Thus, the non-vanishing Christoffel symbols of ∇ with respect to the chart $(\mathcal{U}, (u^i)_{i=1}^n)$ are the functions

$$N^i_{jk} := \frac{\partial N^i_j}{\partial y^k} \colon \mathring{\tau}^{-1}(\mathcal{U}) \to \mathbb{R}; \quad i, j, k \in J_n, \tag{7.11.1}$$

cf. (6.2.6). We say for short that (N^i_{jk}) *is the family of the Christoffel symbols of the Berwald derivative induced by $\mathcal{H}$ with Christoffel symbols N^i_j.*

Using these new ingredients,

$$\nabla_{\frac{\partial}{\partial x^j}} \widehat{\frac{\partial}{\partial u^k}} = \nabla_{(\frac{\partial}{\partial u^j})^{\mathsf{h}}} \widehat{\frac{\partial}{\partial u^k}} = N^i_{jk} \widehat{\frac{\partial}{\partial u^i}}; \quad j, k \in J_n. \tag{7.11.2}$$

Remark 7.11.2 (The Comma Operator). Given a type (k, l) Finsler tensor field $\widetilde{A}$ on $\overset{\circ}{T}M$ and an index $j \in J_n$, we define the *comma operator* $, j$ by the following rule:

If $\widetilde{A}^{i_1 \ldots i_k}_{j_1 \ldots j_l}$ are the components of $\widetilde{A}$ with respect to the chosen chart, then the functions

$$\widetilde{A}^{i_1 \ldots i_k}_{j_1 \ldots j_l, j}$$

are the components of $\nabla^{\mathrm{v}} \widetilde{A} \in \mathcal{F}^k_{l+1}(\overset{\circ}{T}M)$ with respect to the same chart.

We show that the *comma operator* $, j$ *acts as partial differentiation with respect to* y^j.

(a) To begin with the simplest case, let $F \in \mathcal{F}^0_0(\overset{\circ}{T}M) := C^\infty(\overset{\circ}{T}M)$. Then $\nabla^{\mathrm{v}} F \overset{(6.2.30)}{\underset{(\mathcal{U})}{=}} \frac{\partial F}{\partial y^i} \widehat{du^i}$, therefore

$$F_{,j} = \frac{\partial F}{\partial y^j}. \tag{7.11.3}$$

(b) Next, consider a Finsler one-form $\widetilde{\alpha}$ on $\overset{\circ}{T}M$ represented locally as $\widetilde{\alpha} \underset{(\mathcal{U})}{=} \widetilde{\alpha}_i \widehat{du^i}$. Then for each $j, k \in J_n$,

$$\nabla^{\mathrm{v}} \widetilde{\alpha} \left(\widehat{\frac{\partial}{\partial u^j}}, \widehat{\frac{\partial}{\partial u^k}} \right) \overset{(6.2.23)}{=} (\nabla^{\mathrm{v}}_{\frac{\partial}{\partial u^j}} \widetilde{\alpha}) \left(\widehat{\frac{\partial}{\partial u^k}} \right)$$

$$\overset{(6.2.22)}{=} \frac{\partial}{\partial y^j} \left(\widetilde{\alpha} \left(\widehat{\frac{\partial}{\partial u^k}} \right) \right) - \widetilde{\alpha} \left(\nabla^{\mathrm{v}}_{\frac{\partial}{\partial u^j}} \widehat{\frac{\partial}{\partial u^k}} \right) \overset{(7.10.4)}{=} \frac{\partial \widetilde{\alpha}_k}{\partial y^j},$$

therefore

$$\nabla^{\mathrm{v}} \widetilde{\alpha} \underset{(\mathcal{U})}{=} \frac{\partial \widetilde{\alpha}_k}{\partial y^j} \widehat{du^j} \otimes \widehat{du^k}. \tag{7.11.4}$$

Note that, in particular,

$$\nabla^{\mathrm{v}} \widehat{du^i} = 0, \quad i \in J_n. \tag{7.11.5}$$

So in this case

$$\widetilde{\alpha}_{i,j} = \frac{\partial \widetilde{\alpha}_i}{\partial y^j}. \tag{7.11.6}$$

(c) Finally, we turn to the general case. Let $\widetilde{A} \in \mathcal{F}^k_l(\overset{\circ}{T}M)$ with components

$$\widetilde{A}^{i_1 \ldots i_k}_{j_1 \ldots j_l} = \widetilde{A} \left(\widehat{du^{i_1}}, \ldots, \widehat{du^{i_k}}, \widehat{\frac{\partial}{\partial u^{j_1}}}, \ldots, \widehat{\frac{\partial}{\partial u^{j_l}}} \right),$$

where all indices run from 1 to n. Then

$$\nabla^v \widetilde{A}\left(\widehat{du^{i_1}}, \ldots, \widehat{du^{i_k}}, \widehat{\frac{\partial}{\partial u^j}}, \widehat{\frac{\partial}{\partial u^{j_1}}}, \ldots, \widehat{\frac{\partial}{\partial u^{j_l}}}\right)$$

$$:= \left(\nabla^v_{\widehat{\frac{\partial}{\partial u^j}}} \widetilde{A}\right)\left(\widehat{du^{i_1}}, \ldots, \widehat{du^{i_k}}, \widehat{\frac{\partial}{\partial u^{j_1}}}, \ldots, \widehat{\frac{\partial}{\partial u^{j_l}}}\right) \overset{(7.10.4),(7.11.5)}{=} \frac{\partial \widetilde{A}^{i_1 \ldots i_k}_{j_1 \ldots j_l}}{\partial y^j}.$$

Thus

$$\widetilde{A}^{i_1 \ldots i_k}_{j_1 \ldots j_l,j} = \frac{\partial \widetilde{A}^{i_1 \ldots i_k}_{j_1 \ldots j_l}}{\partial y^j}. \tag{7.11.7}$$

At this point it is worth emphasizing that in a coordinate description *the canonical vertical differential reduces to partial differentiation with respect to the 'natural fibre coordinates' y^i.*

Example 7.11.3 (The Semicolon Operator). Let $\widetilde{A} \in \mathcal{F}^k_l(\mathring{T}M)$. We define for every $j \in J_n$ the *semicolon operator* $;j$ by the following prescription:

If $\widetilde{A}^{i_1 \ldots i_k}_{j_1 \ldots j_l}$ are the components of $\widetilde{A}$ with respect to the chosen chart, then the functions

$$\widetilde{A}^{i_1 \ldots i_k}_{j_1 \ldots j_l;j}$$

are the components of $\nabla^h \widetilde{A} \in \mathcal{F}^k_{l+1}(\mathring{T}M)$ with respect to the same chart.

Consider the following examples.

(a) $F \in \mathcal{F}^0_0(\mathring{T}M) = C^\infty(\mathring{T}M)$.
    Then

$$\nabla^h F\left(\widehat{\frac{\partial}{\partial u^j}}\right) \overset{(7.10.8)}{=} \left(\frac{\partial}{\partial u^j}\right)^h F \overset{(7.2.37)}{=} \frac{\partial F}{\partial x^j} - N^i_j \frac{\partial F}{\partial y^i},$$

so $\nabla^h F \underset{(\mathcal{U})}{=} \left(\frac{\partial F}{\partial x^j} - N^i_j \frac{\partial F}{\partial y^i}\right)\widehat{du^j}$, therefore

$$F_{;j} = \frac{\partial F}{\partial x^j} - N^i_j \frac{\partial F}{\partial y^i}. \tag{7.11.8}$$

(b) $\widetilde{X} \in \mathcal{F}^1_0(\mathring{T}M) = \Gamma(\mathring{\pi})$, $\widetilde{X} \underset{(\mathcal{U})}{=} \widetilde{X}^i \widehat{\frac{\partial}{\partial u^i}}$.

Now $\nabla^h \widetilde{X}$ is a type $(1,1)$ tensor field, whose components $\widetilde{X}^i_{;j} = (\nabla^h \widetilde{X})^i_j$ are defined by

$$\nabla^h \widetilde{X}\left(\widehat{\frac{\partial}{\partial u^j}}\right) = \widetilde{X}^i_{;j} \widehat{\frac{\partial}{\partial u^i}}; \quad j \in J_n.$$

Since

$$\nabla^{\mathsf h}\widetilde{X}\left(\widehat{\frac{\partial}{\partial u^j}}\right) := \nabla_{(\frac{\partial}{\partial u^j})^{\mathsf h}}\widetilde{X} = \nabla_{(\frac{\partial}{\partial u^j})^{\mathsf h}}\left(\widetilde{X}^i\widehat{\frac{\partial}{\partial u^i}}\right) = \left(\left(\frac{\partial}{\partial u^j}\right)^{\mathsf h}\widetilde{X}^i\right)\widehat{\frac{\partial}{\partial u^i}}$$

$$+ \widetilde{X}^k\nabla_{(\frac{\partial}{\partial u^j})^{\mathsf h}}\widehat{\frac{\partial}{\partial u^k}} \overset{(7.11.2)}{=} \left(\left(\frac{\partial}{\partial u^j}\right)^{\mathsf h}\widetilde{X}^i\right)\widehat{\frac{\partial}{\partial u^i}} + \widetilde{X}^k N^i_{jk}\widehat{\frac{\partial}{\partial u^i}}$$

$$= \left(\frac{\partial\widetilde{X}^i}{\partial x^j} - N^k_j\frac{\partial\widetilde{X}^i}{\partial y^k} + N^i_{jk}\widetilde{X}^k\right)\widehat{\frac{\partial}{\partial u^i}},$$

it follows that

$$\widetilde{X}^i_{;j} = \frac{\partial\widetilde{X}^i}{\partial x^j} - N^k_j\frac{\partial\widetilde{X}^i}{\partial y^k} + N^i_{jk}\widetilde{X}^k. \tag{7.11.9}$$

If, in particular, $\widetilde{X} = \widehat{\frac{\partial}{\partial u^k}}$, where $k \in J_n$ is an arbitrarily fixed index, then $\widehat{\frac{\partial}{\partial u^k}} = \delta^i_k\widehat{\frac{\partial}{\partial u^i}}$, $\left(\nabla^{\mathsf h}\widehat{\frac{\partial}{\partial u^k}}\right)\widehat{\frac{\partial}{\partial u^j}} = \nabla_{(\frac{\partial}{\partial u^j})^{\mathsf h}}\widehat{\frac{\partial}{\partial u^k}} \overset{(7.11.2)}{=} N^i_{jk}\widehat{\frac{\partial}{\partial u^i}}$, therefore

$$\delta^i_{k;j} = N^i_{jk} \text{ for any fixed } k \in J_n. \tag{7.11.10}$$

As a second (and more important) special case, consider the canonical section $\widetilde{\delta}$ of $\mathring{\pi}$. By (4.1.7), $\widetilde{\delta} \underset{(\mathcal{U})}{=} y^i\widehat{\frac{\partial}{\partial u^i}}$, so relation (7.11.9) leads to

$$y^i_{;j} = y^k N^i_{jk} - N^i_j. \tag{7.11.11}$$

Thus, by (7.5.11), *the functions* $y^i_{;j}$ *are just the components of the tension of* $\mathcal{H}$, therefore $\mathbf{t} = \nabla^{\mathsf h}\widetilde{\delta}$. For a coordinate-free proof, see Proposition 7.12.1.

(c) $\widetilde{\alpha} \in \mathcal{F}^0_1(\mathring{T}M) = (\Gamma(\mathring{\pi}))^*$, $\widetilde{\alpha} \underset{(\mathcal{U})}{=} \widetilde{\alpha}_i\widehat{du^i}$.

Then for each $j, k \in J_n$,

$$(\nabla^{\mathsf h}\widetilde{\alpha})\left(\widehat{\frac{\partial}{\partial u^j}}, \widehat{\frac{\partial}{\partial u^k}}\right) = (\nabla_{(\frac{\partial}{\partial u^j})^{\mathsf h}}\widetilde{\alpha})\left(\widehat{\frac{\partial}{\partial u^k}}\right) = \left(\frac{\partial}{\partial u^j}\right)^{\mathsf h}\widetilde{\alpha}_k - \widetilde{\alpha}\left(N^i_{jk}\widehat{\frac{\partial}{\partial u^i}}\right)$$

$$= \frac{\partial\widetilde{\alpha}_k}{\partial x^j} - N^i_j\frac{\partial\widetilde{\alpha}_k}{\partial y^i} - N^i_{jk}\widetilde{\alpha}_i,$$

therefore

$$\widetilde{\alpha}_{k;j} = \frac{\partial\widetilde{\alpha}_k}{\partial x^j} - N^i_j\frac{\partial\widetilde{\alpha}_k}{\partial y^i} - N^i_{jk}\widetilde{\alpha}_i. \tag{7.11.12}$$

If, in particular, $\widetilde{\alpha} := \widehat{du^i} = \delta^i_k\widehat{du^k}$ for a fixed $i \in J_n$, then

$$(\nabla^{\mathsf h}\widehat{du^i})\left(\widehat{\frac{\partial}{\partial u^j}}, \widehat{\frac{\partial}{\partial u^k}}\right) = \left(\nabla_{(\frac{\partial}{\partial u^j})^{\mathsf h}}\widehat{du^i}\right)\widehat{\frac{\partial}{\partial u^k}}$$

$$= \left(\frac{\partial}{\partial u^j}\right)^{\mathsf h}\delta^i_k - \widehat{du^i}\left(N^l_{jk}\widehat{\frac{\partial}{\partial u^l}}\right) = -N^i_{jk},$$

thus

$$\delta^i_{k;j} = -N^i_{jk} \text{ for any fixed } i \in J_n. \tag{7.11.13}$$

(d) *For every smooth function F on $\mathring{T}M$,*

$$F_{;j,k} - F_{,k;j} = 0; \quad j,k \in J_n. \tag{7.11.14}$$

This is just the mixed Ricci formula (7.10.17) expressed in terms of the tensor components. It can be obtained directly as follows:

$$F_{;j,k} \overset{(7.11.8),(7.11.3)}{=} \frac{\partial}{\partial y^k}\left(\frac{\partial F}{\partial x^j} - N^i_j \frac{\partial F}{\partial y^i}\right) = \frac{\partial^2 F}{\partial y^k \partial x^j} - N^i_{jk}\frac{\partial F}{\partial y^i} - N^i_j \frac{\partial^2 F}{\partial y^k \partial y^i}$$

$$= \frac{\partial}{\partial x^j}\left(\frac{\partial F}{\partial y^k}\right) - N^i_j \frac{\partial}{\partial y^i}\left(\frac{\partial F}{\partial y^k}\right) - N^i_{jk}\frac{\partial F}{\partial y^i} \overset{(7.11.12)}{=} F_{,k;j}.$$

(e) $\widetilde{A} \in \mathcal{F}^1_1(\mathring{T}M)$, $\widetilde{A} \underset{(\mathcal{U})}{=} \widetilde{A}^i_j \widehat{\frac{\mathring{\partial}}{\partial u^i}} \otimes \widehat{du^j}$.

Then $\nabla^h \widetilde{A} \in \mathcal{F}^1_2(\mathring{T}M)$; we calculate its components:

$$\nabla^h \widetilde{A}\left(\widehat{du^i}, \widehat{\frac{\partial}{\partial u^j}}, \widehat{\frac{\partial}{\partial u^k}}\right) = \left(\nabla_{(\frac{\partial}{\partial u^j})^h}\widetilde{A}\right)\left(\widehat{du^i}, \widehat{\frac{\partial}{\partial u^k}}\right)$$

$$= \left(\frac{\partial}{\partial u^j}\right)^h \widetilde{A}^i_k - \widetilde{A}\left(\nabla_{(\frac{\partial}{\partial u^j})^h}\widehat{du^i}, \widehat{\frac{\partial}{\partial u^k}}\right) - \widetilde{A}\left(\widehat{du^i}, \nabla_{(\frac{\partial}{\partial u^j})^h} \widehat{\frac{\partial}{\partial u^k}}\right)$$

$$\overset{(b),(c)}{=} \left(\frac{\partial}{\partial u^j}\right)^h \widetilde{A}^i_k + \widetilde{A}\left(N^i_{jl}\widehat{du^l}, \widehat{\frac{\partial}{\partial u^k}}\right) - \widetilde{A}\left(\widehat{du^i}, N^l_{jk}\widehat{\frac{\partial}{\partial u^l}}\right)$$

$$= \frac{\partial \widetilde{A}^i_k}{\partial x^j} - N^l_j \frac{\partial \widetilde{A}^i_k}{\partial y^l} + N^i_{jl}\widetilde{A}^l_k - N^l_{jk}\widetilde{A}^i_l.$$

So in this case

$$\widetilde{A}^i_{k;j} = \frac{\partial \widetilde{A}^i_k}{\partial x^j} - N^l_j \frac{\partial \widetilde{A}^i_k}{\partial y^l} + N^i_{jl}\widetilde{A}^l_k - N^l_{jk}\widetilde{A}^i_l. \tag{7.11.15}$$

(f) $g \in \mathcal{F}^0_2(\mathring{T}M)$, $g \underset{(\mathcal{U})}{=} g_{kl}\widehat{du^k} \otimes \widehat{du^l}$.

Then $\nabla^h g \in \mathcal{F}_3^0(\mathring{T}M)$, and its components are

$$\nabla^h g \left(\widehat{\frac{\partial}{\partial u^j}}, \widehat{\frac{\partial}{\partial u^k}}, \widehat{\frac{\partial}{\partial u^l}} \right) = \left(\nabla_{\left(\frac{\partial}{\partial u^j} \right)^h} g \right) \left(\widehat{\frac{\partial}{\partial u^k}}, \widehat{\frac{\partial}{\partial u^l}} \right)$$

$$= \left(\frac{\partial}{\partial u^j} \right)^h g_{kl} - g \left(N_{jk}^i \widehat{\frac{\partial}{\partial u^i}}, \widehat{\frac{\partial}{\partial u^l}} \right) - g \left(\widehat{\frac{\partial}{\partial u^k}}, N_{jl}^i \widehat{\frac{\partial}{\partial u^i}} \right)$$

$$= \left(\frac{\partial}{\partial u^j} \right)^h g_{kl} - N_{jk}^i g_{il} - N_{jl}^i g_{ki}.$$

Therefore,

$$g_{kl;j} = \frac{\partial g_{kl}}{\partial x^j} - N_j^i \frac{\partial g_{kl}}{\partial y^i} - N_{jk}^i g_{il} - N_{jl}^i g_{ki}. \tag{7.11.16}$$

Remark 7.11.4. (a) Formulae containing the comma operator $, j$ and the semicolon operator $; j$ are typical in traditional tensor calculus ([72, 74], and see also [8]). We note that the index conventions of Matsumoto and his school are somewhat different from ours. For example, they introduce the Christoffel symbols N_{jk}^i of ∇ by

$$\nabla^h_{\frac{\partial}{\partial u^k}} \widehat{\frac{\partial}{\partial u^j}} = N_{jk}^i \widehat{\frac{\partial}{\partial u^i}}.$$

This innocent interchange of two indices, and some other similarly small differences in the conventions generate considerable (but not significant) changes in the coordinate expressions.

(b) Using the traditional formalism, the components of the curvature and the torsion of $\mathcal{H}$ given by (7.8.9) and (7.8.10) can be written as follows:

$$R_{jk}^i = N_{j;k}^i - N_{k;j}^i, \tag{7.11.17}$$

$$T_{jk}^i = N_{j,k}^i - N_{k,j}^i. \tag{7.11.18}$$

7.12 Tension, Torsion, Curvature and Geodesics Again

Proposition 7.12.1. *The tension, torsion and the curvature of an Ehresmann connection $\mathcal{H}$ can be expressed in terms of the Berwald derivative induced by $\mathcal{H}$. Namely,*

$$\mathbf{t} = \nabla^h \widetilde{\delta}, \tag{7.12.1}$$

$$\mathbf{T} = d^\nabla \mathbf{j} \circ (\mathcal{H} \times \mathcal{H}) = T^\nabla \circ (\mathcal{H} \times \mathcal{H}), \tag{7.12.2}$$

$$\mathcal{R} = -d^\nabla \mathcal{V} \circ (\mathcal{H} \times \mathcal{H}) = -T^\mathcal{V}(\nabla) \circ (\mathcal{H} \times \mathcal{H}), \tag{7.12.3}$$

where $\mathcal{H} \times \mathcal{H}$ is defined by (A.2.2).

Proof. Using the first relation in (7.9.6), the curvature of an Ehresmann connection can be expressed in terms of the $\mathcal{V}$-torsion of any covariant derivative on $\overset{\circ}{\pi}$, so (7.12.3) does not require verification. For any section $\widetilde{X}$ in $\Gamma(\overset{\circ}{\pi})$,

$$\nabla^{\mathsf{h}}\widetilde{\delta}(\widetilde{X}) = \nabla_{\mathcal{H}\widetilde{X}}\widetilde{\delta} \overset{(7.10.6)}{=} \mathcal{V}[\mathcal{H}\widetilde{X}, \mathbf{i}\widetilde{\delta}] = \mathcal{V}[\mathcal{H}\widetilde{X}, C] \overset{(7.5.9)}{=:} \mathbf{t}(\widetilde{X}),$$

which proves (7.12.1). Finally, for all $\widetilde{X}, \widetilde{Y} \in \Gamma(\overset{\circ}{\pi})$,

$$T^{\nabla}(\mathcal{H}\widetilde{X}, \mathcal{H}\widetilde{Y}) \overset{(7.9.1)}{=} \nabla_{\mathcal{H}\widetilde{X}}\widetilde{Y} - \nabla_{\mathcal{H}\widetilde{Y}}\widetilde{X} - \mathbf{j}[\mathcal{H}\widetilde{X}, \mathcal{H}\widetilde{Y}]$$

$$\overset{(7.10.6)}{=} \mathcal{V}[\mathcal{H}\widetilde{X}, \mathbf{i}\widetilde{Y}] - \mathcal{V}[\mathcal{H}\widetilde{Y}, \mathbf{i}\widetilde{X}] - \mathbf{j}[\mathcal{H}\widetilde{X}, \mathcal{H}\widetilde{Y}]$$

$$\overset{(7.8.2)}{=:} \mathbf{T}(\widetilde{X}, \widetilde{Y}),$$

as was to be shown.   □

Lemma 7.12.2. *If ∇ is the Berwald derivative induced by an Ehresmann connection $\mathcal{H}$ in $\overset{\circ}{T}M$, then*

$$\nabla\widetilde{\delta} = \mathbf{t} \circ \mathbf{j} + \mathcal{V}, \tag{7.12.4}$$

therefore ∇ is strongly associated to $\mathcal{H}$ if, and only if, $\mathcal{H}$ is positive-homogeneous.

Proof. For any vector field ξ on $\overset{\circ}{T}M$,

$$\nabla\widetilde{\delta}(\xi) = \nabla_\xi\widetilde{\delta} = \nabla_{\mathsf{v}\xi+\mathsf{h}\xi}\widetilde{\delta} = \nabla_{\mathbf{i}(\mathcal{V}\xi)}\widetilde{\delta} + \nabla_{\mathcal{H}(\mathbf{j}\xi)}\widetilde{\delta}$$

$$= \nabla^{\mathsf{v}}_{\mathcal{V}\xi}\widetilde{\delta} + \nabla^{\mathsf{h}}_{\mathbf{j}\xi}\widetilde{\delta} \overset{(6.2.33),(7.12.1)}{=} \mathcal{V}\xi + \mathbf{t}(\mathbf{j}\xi),$$

which proves (7.12.4). Since $\mathbf{j} \colon \mathfrak{X}(\overset{\circ}{T}M) \to \Gamma(\overset{\circ}{\pi})$ is surjective, from this it follows that ∇ is strongly associated to $\mathcal{H}$, i.e., $\nabla\widetilde{\delta} = \mathcal{V}$ if, and only if, $\mathbf{t} = 0$. Taking into account Corollary 7.5.10, this concludes the proof.   □

Lemma 7.12.3. *The geodesics of a positive-homogeneous Ehresmann connection coincide with the geodesics of the induced Berwald derivative.*

Proof. Let $\mathcal{H}$ be a positive-homogeneous Ehresmann connection in $\overset{\circ}{T}M$. Let $\mathcal{V}$ and ∇ be the vertical mapping associated to $\mathcal{H}$ and the Berwald derivative induced by $\mathcal{H}$, respectively. Consider a smooth curve $\gamma \colon I \to M$. By (6.2.51), γ is a geodesic of ∇ if, and only if,

$$0 = \nabla^\gamma(\widetilde{\delta} \circ \dot{\gamma}) = \nabla^\gamma(\mathbf{j} \circ \ddot{\gamma}) = (\nabla\widetilde{\delta}) \circ \ddot{\gamma}.$$

Since $\mathcal{H}$ is positive-homogeneous, by the previous lemma $\nabla\widetilde{\delta} = \mathcal{V}$. Thus

$$\nabla^\gamma(\widetilde{\delta} \circ \dot{\gamma}) = \mathcal{V} \circ \ddot{\gamma},$$

therefore (see also Definition 7.7.1) the geodesics of ∇ and $\mathcal{H}$ coincide.   □

Proposition 7.12.4. *The geodesics of an Ehresmann connection coincide with the geodesics of the induced Berwald derivative if, and only if, the semispray associated to the Ehresmann connection is a spray.*

Proof. Using the same notation as just above,

$$\nabla^\gamma(\widetilde{\delta} \circ \dot\gamma) = (\nabla\widetilde{\delta}) \circ \dot\gamma \overset{(7.12.4)}{=} \mathcal{V} \circ \ddot\gamma + \mathbf{t} \circ \mathbf{j} \circ \ddot\gamma \overset{(4.1.37)}{=} \mathcal{V} \circ \ddot\gamma + \mathbf{t} \circ \widetilde{\delta} \circ \dot\gamma.$$

Here

$$\mathbf{t}(\widetilde{\delta}) \overset{(7.5.9)}{=} \mathcal{V}[\mathcal{H}\widetilde{\delta}, C] = \mathcal{V}[S_\mathcal{H}, C] = -\mathcal{V}[C, S_\mathcal{H}],$$

where $S_\mathcal{H}$ is the semispray associated to $\mathcal{H}$. If $S_\mathcal{H}$ is a spray, then

$$\mathcal{V}[C, S_\mathcal{H}] = \mathcal{V}S_\mathcal{H} = \mathcal{V} \circ \mathcal{H} \circ \widetilde{\delta} \overset{(7.2.21)}{=} 0,$$

therefore $\nabla^\gamma(\widetilde{\delta} \circ \dot\gamma) = \mathcal{V} \circ \ddot\gamma$, and so the geodesics of ∇ and $\mathcal{H}$ coincide.

Conversely, the last property implies that $\nabla^\gamma(\widetilde{\delta} \circ \dot\gamma) = \mathcal{V} \circ \ddot\gamma$. Then $\mathbf{t}(\widetilde{\delta}) = 0$, whence

$$0 = \mathbf{v}[C, S_\mathcal{H}] = [C, S_\mathcal{H}] - \mathbf{h}[C, S_\mathcal{H}] \overset{(7.2.42)}{=} [C, S_\mathcal{H}] - S_\mathcal{H}.$$

So $S_\mathcal{H}$ is a spray, as was to be shown. $\square$

Proposition 7.12.5 (An Interpretation of the Tension). *Let $\mathcal{H}$ be and Ehresmann connection in $\overset{\circ}{T}M$. If ∇ is the Berwald derivative induced by $\mathcal{H}$, and $\mathcal{H}_\nabla$ is the Ehresmann connection induced by ∇ according to Example 7.3.3(a), then*

$$\mathcal{H} - \mathcal{H}_\nabla = \mathbf{i} \circ \mathbf{t}. \tag{7.12.5}$$

Proof. The Berwald derivative ∇ is strongly regular because $(\nabla\widetilde{\delta}) \circ \mathbf{i} = 1_{\Gamma(\hat\pi)}$ by (6.2.33). Thus the horizontal lift with respect to $\mathcal{H}_\nabla$ is given by (7.3.3), i.e.,

$$\mathcal{H}_\nabla(\widehat{X}) = X^c - \mathbf{i}\nabla_{X^c}\widetilde{\delta}, \quad X \in \mathfrak{X}(M).$$

So we obtain

$$\begin{aligned}
(\mathcal{H} - \mathcal{H}_\nabla)(\widehat{X}) &= X^h - X^c + \mathbf{i}\nabla_{X^c}\widetilde{\delta} = X^h - X^c + \mathbf{i}\nabla_{\mathbf{v}X^c}\widetilde{\delta} + \mathbf{i}\nabla_{X^h}\widetilde{\delta} \\
&= X^h - X^c + \mathbf{i}(\nabla\widetilde{\delta} \circ \mathbf{i}(\mathcal{V}X^c)) + \mathbf{i}(\nabla^h\widetilde{\delta}(\widehat{X})) \\
&= \mathbf{h}X^c - X^c + \mathbf{v}X^c + \mathbf{i} \circ \mathbf{t}(\widehat{X}) = \mathbf{i} \circ \mathbf{t}(\widehat{X}),
\end{aligned}$$

as desired. $\square$

Lemma 7.12.6 (Horizontal Ricci Formula for Functions). *Let $\mathcal{H}$ be an Ehresmann connection in $\mathring{T}M$, and let ∇^h be the horizontal derivative determined by $\mathcal{H}$. Then for any function $F \in C^\infty(\mathring{T}M)$ and any sections $\widetilde{X}, \widetilde{Y} \in \Gamma(\mathring{\pi})$ we have*

$$\nabla^h\nabla^h F(\widetilde{X}, \widetilde{Y}) - \nabla^h\nabla^h F(\widetilde{Y}, \widetilde{X}) = \big(\mathbf{i}\,\mathcal{R}(\widetilde{X}, \widetilde{Y}) - \mathcal{H}\mathbf{T}(\widetilde{X}, \widetilde{Y})\big)F$$
$$= (\nabla^v F \circ \mathcal{R} - \nabla^h F \circ \mathbf{T})(\widetilde{X}, \widetilde{Y}), \quad (7.12.6)$$

where $\mathcal{R}$ is the curvature, $\mathbf{T}$ is the torsion of $\mathcal{H}$.

Proof. By tensoriality, it is enough to check this relation for basic sections $\widehat{X}, \widehat{Y}$; $X, Y \in \mathfrak{X}(M)$. Note first that we have

$$\mathcal{H}\mathbf{T}(\widehat{X}, \widehat{Y}) \overset{(7.12.2)}{=} \mathcal{H}\mathbf{T}^\nabla(X^h, Y^h) \overset{(7.9.1)}{=} \mathcal{H}(\nabla_{X^h}\widehat{Y} - \nabla_{Y^h}\widehat{X}) - \mathbf{h}[X^h, Y^h]$$
$$\overset{(7.1.7)}{=} \mathcal{H}(\nabla_{X^h}\widehat{Y} - \nabla_{Y^h}\widehat{X}) - [X, Y]^h,$$

and

$$\mathbf{i}\,\mathcal{R}(\widehat{X}, \widehat{Y}) \overset{(7.8.3)}{=} [X^h, Y^h] - [X, Y]^h.$$

Since

$$\nabla^h\nabla^h F(\widehat{X}, \widehat{Y}) = (\nabla_{X^h}(\nabla^h F))(\widehat{Y}) \overset{(7.10.11)}{=} X^h(Y^h F) - \nabla^h F(\nabla_{X^h}\widehat{Y})$$
$$= X^h(Y^h F) - \mathcal{H}(\nabla_{X^h}\widehat{Y})F,$$

thus we obtain

$$\nabla^h\nabla^h F(\widehat{X}, \widehat{Y}) - \nabla^h\nabla^h F(\widehat{Y}, \widehat{X}) = [X^h, Y^h]F - \mathcal{H}(\nabla_{X^h}\widehat{Y} - \nabla_{Y^h}\widehat{X})F$$
$$= ([X^h, Y^h] - [X, Y]^h)F - \mathcal{H}\mathbf{T}(\widehat{X}, \widehat{Y})F = \mathbf{i}\,\mathcal{R}(\widehat{X}, \widehat{Y})F - \mathcal{H}\mathbf{T}(\widehat{X}, \widehat{Y})F,$$

as was to be shown. $\qquad\square$

Remark 7.12.7. If the Ehresmann connection is *torsion-free*, then relation (7.12.6) reduces to

$$\nabla^h\nabla^h F(\widetilde{X}, \widetilde{Y}) - \nabla^h\nabla^h F(\widetilde{Y}, \widetilde{X}) = \mathbf{i}\mathcal{R}(\widetilde{X}, \widetilde{Y})F.$$

Translating this into the language of traditional tensor calculus, we obtain

$$F_{;k;j} - F_{;j;k} = R^i_{jk}F_{,i} \quad (j, k \in J_n), \quad (7.12.7)$$

where R^i_{jk} is given by (7.8.9). We invite the reader to elaborate an independent proof of (7.12.7).

To conclude this subsection we now derive the counterpart of Proposition 6.1.38 for Ehresmann connections.

Proposition 7.12.8 (h-differential Bianchi Identity for Ehresmann Connections). *Let $\mathcal{H}$ be an Ehresmann connection in $\overset{\circ}{T}M$, and let ∇^h be the horizontal derivative determined by $\mathcal{H}$. Then for any sections $\widetilde{X}, \widetilde{Y}, \widetilde{Z}$ in $\Gamma(\overset{\circ}{\pi})$,*

$$\sum_{\text{cyc}} ((\nabla^h \mathcal{R})(\widetilde{X}, \widetilde{Y}, \widetilde{Z}) + \mathcal{R}(\mathbf{T}(\widetilde{X}, \widetilde{Y}), \widetilde{Z})) = 0. \tag{7.12.8}$$

Proof. Again, it is enough to verify the relation for basic sections $\widehat{X}, \widehat{Y}, \widehat{Z}$. We start with the definition of $(\nabla^h \mathcal{R})(\widehat{X}, \widehat{Y}, \widehat{Z})$, and repeat it twice, by permuting the letters cyclically:

$$(\nabla^h \mathcal{R})(\widehat{X}, \widehat{Y}, \widehat{Z}) = \nabla_{X^h}(\mathcal{R}(\widehat{Y}, \widehat{Z})) - \mathcal{R}(\nabla_{X^h}\widehat{Y}, \widehat{Z}) - \mathcal{R}(\widehat{Y}, \nabla_{X^h}\widehat{Z}),$$

$$(\nabla^h \mathcal{R})(\widehat{Y}, \widehat{Z}, \widehat{X}) = \nabla_{Y^h}(\mathcal{R}(\widehat{Z}, \widehat{X})) - \mathcal{R}(\nabla_{Y^h}\widehat{Z}, \widehat{X}) - \mathcal{R}(\widehat{Z}, \nabla_{Y^h}\widehat{X}),$$

$$(\nabla^h \mathcal{R})(\widehat{Z}, \widehat{X}, \widehat{Y}) = \nabla_{Z^h}(\mathcal{R}(\widehat{X}, \widehat{Y})) - \mathcal{R}(\nabla_{Z^h}\widehat{X}, \widehat{Y}) - \mathcal{R}(\widehat{X}, \nabla_{Z^h}\widehat{Y}).$$

In the next step we add these equalities. Taking into account that

$$\mathbf{T}(\widehat{X}, \widehat{Y}) = \nabla_{X^h}\widehat{Y} - \nabla_{Y^h}\widehat{X} - \mathbf{j}[X^h, Y^h],$$

we find that

$$\sum_{\text{cyc}} (\nabla^h \mathcal{R})(\widehat{X}, \widehat{Y}, \widehat{Z}) = \sum_{\text{cyc}} \nabla^h_{\widehat{X}}(\mathcal{R}(\widehat{Y}, \widehat{Z})) - \mathcal{R}(\nabla_{X^h}\widehat{Y} - \nabla_{Y^h}\widehat{X}, \widehat{Z})$$

$$- \mathcal{R}(\nabla_{Y^h}\widehat{Z} - \nabla_{Z^h}\widehat{Y}, \widehat{X}) - \mathcal{R}(\nabla_{Z^h}\widehat{X} - \nabla_{X^h}\widehat{Z}, \widehat{Y})$$

$$= \sum_{\text{cyc}} \nabla^h_{\widehat{X}}(\mathcal{R}(\widehat{Y}, \widehat{Z})) - \sum_{\text{cyc}} \mathcal{R}(\mathbf{T}(\widehat{X}, \widehat{Y}), \widehat{Z})$$

$$- (\mathcal{R}(\mathbf{j}[X^h, Y^h], \widehat{Z}) + \mathcal{R}(\mathbf{j}[Y^h, Z^h], \widehat{X}) + \mathcal{R}(\mathbf{j}[Z^h, X^h], \widehat{Y}))$$

$$\overset{(7.10.2),(7.8.1)}{=} \mathcal{V}[X^h, \mathbf{i}\mathcal{R}(\widehat{Y}, \widehat{Z})] + \mathcal{V}[Y^h, \mathbf{i}\mathcal{R}(\widehat{Z}, \widehat{X})] + \mathcal{V}[Z^h, \mathbf{i}\mathcal{R}(\widehat{X}, \widehat{Y})]$$

$$- \sum_{\text{cyc}} \mathcal{R}(\mathbf{T}(\widehat{X}, \widehat{Y}), \widehat{Z}) - \sum_{\text{cyc}} \mathcal{V}[\mathbf{h}[X^h, Y^h], Z^h]$$

$$\overset{(7.1.7)}{=} - \sum_{\text{cyc}} \mathcal{R}(\mathbf{T}(\widehat{X}, \widehat{Y}), \widehat{Z}) + \mathcal{V}[X^h, \mathbf{i}\mathcal{R}(\widehat{Y}, \widehat{Z}) + [Y, Z]^h]$$

$$+ \mathcal{V}[Y^h, \mathbf{i}\mathcal{R}(\widehat{Z}, \widehat{X}) + [Z, X]^h] + \mathcal{V}[Z^h, \mathbf{i}\mathcal{R}(\widehat{X}, \widehat{Y}) + [X, Y]^h]$$

$$\overset{(7.8.3)}{=} - \sum_{\text{cyc}} \mathcal{R}(\mathbf{T}(\widehat{X}, \widehat{Y}), \widehat{Z}) + \mathcal{V} \sum_{\text{cyc}} [X^h, [Y^h, Z^h]]$$

$$\overset{\text{Jacobi}}{=} - \sum_{\text{cyc}} \mathcal{R}(\mathbf{T}(\widehat{X}, \widehat{Y}), \widehat{Z}),$$

which concludes the proof. $\qquad\square$

7.13 The Berwald Curvature

Throughout this section, we assume that an Ehresmann connection $\mathcal{H}$ is specified in $\overset{\circ}{T}M$. As above, ∇ and ∇^h stand for the Berwald derivative and the horizontal Berwald derivative induced by $\mathcal{H}$, respectively.

Remark 7.13.1. For the $\mathcal{H}$-mixed curvature of ∇, defined in general by (7.9.3), we use a special terminology and notation. We call it the *Berwald curvature* of the Ehresmann connection $\mathcal{H}$, and denote it, following Z. Shen's monograph [88], by **B**. Thus for any sections $\widetilde{X}, \widetilde{Y}, \widetilde{Z} \in \Gamma(\overset{\circ}{\pi})$,

$$\mathbf{B}(\widetilde{X}, \widetilde{Y})\widetilde{Z} := R^\nabla(i\widetilde{X}, \mathcal{H}\widetilde{Y})\widetilde{Z}$$
$$= \nabla_{i\widetilde{X}}\nabla_{\mathcal{H}\widetilde{Y}}\widetilde{Z} - \nabla_{\mathcal{H}\widetilde{Y}}\nabla_{i\widetilde{X}}\widetilde{Z} - \nabla_{[i\widetilde{X}, \mathcal{H}\widetilde{Y}]}\widetilde{Z}. \tag{7.13.1}$$

Lemma 7.13.2. *With the notation introduced above, we have, for any vector fields X, Y, Z on M,*

$$\mathbf{B}(\widehat{X}, \widehat{Y})\widehat{Z} = \mathbf{j}[X^v, \mathbf{F}[Y^h, Z^v]] = (\nabla^v\nabla^h\widehat{Z})(\widehat{X}, \widehat{Y}), \tag{7.13.2}$$

where $\mathbf{F}$ is the almost complex structure determined by $\mathcal{H}$.

Proof. Using (7.13.1), (7.10.1) and (7.10.2), we find that

$$\mathbf{B}(\widehat{X}, \widehat{Y})\widehat{Z} = \nabla_{X^v}\nabla_{Y^h}\widehat{Z} - \nabla_{Y^h}\nabla_{X^v}\widehat{Z} - \nabla_{[X^v, Y^h]}\widehat{Z} = \nabla_{X^v}(\mathcal{V}[Y^h, Z^v])$$
$$= \mathbf{j}[X^v, \mathcal{H} \circ \mathcal{V}[Y^h, Z^v]].$$

Here $\mathcal{H} \circ \mathcal{V}[Y^h, Z^v] \overset{(7.2.19)}{=} \mathbf{F}[Y^h, Z^v] + \mathbf{J}[Y^h, Z^v] = \mathbf{F}[Y^h, Z^v]$, so we have $\mathbf{B}(\widehat{X}, \widehat{Y})\widehat{Z} = \mathbf{j}[X^v, \mathbf{F}[Y^h, Z^v]]$. On the other hand

$$(\nabla^v\nabla^h\widehat{Z})(\widehat{X}, \widehat{Y}) = \nabla_{X^v}(\nabla^h\widehat{Z})(\widehat{Y}) = \nabla_{X^v}\nabla_{Y^h}\widehat{Z} = \mathbf{j}[X^v, \mathcal{H} \circ \mathcal{V}[Y^h, Z^v]],$$

which completes the proof of (7.13.2). □

Corollary 7.13.3. *If $\mathbf{B}$ is the Berwald curvature of the Ehresmann connection $\mathcal{H}$, then for any vector fields X, Y, Z on M,*

$$\mathbf{i}\,\mathbf{B}(\widehat{X}, \widehat{Y})\widehat{Z} = [X^v, [Y^h, Z^v]] = [[X^v, Y^h], Z^v], \tag{7.13.3}$$

therefore the Berwald curvature is symmetric in its first and third arguments.

Proof. To obtain (7.13.3), observe that

$$0 \overset{(4.1.89)}{=} [\mathbf{J}, X^v](\mathbf{F}[Y^h, Z^v]) = [\mathbf{J}\,\mathbf{F}[Y^h, Z^v], X^v] - \mathbf{J}[\mathbf{F}[Y^h, Z^v], X^v]$$
$$\overset{(7.2.29)}{=} [\mathbf{v}[Y^h, Z^v], X^v] + \mathbf{J}[X^v, \mathbf{F}[Y^h, Z^v]].$$

Since $[Y^h, Z^v]$ is vertical, it follows that

$$\mathbf{i}\, \mathbf{B}(\widehat{X}, \widehat{Y})\widehat{Z} \overset{(7.13.2)}{=} \mathbf{J}[X^v, \mathbf{F}[Y^h, Z^v]] = [X^v, [Y^h, Z^v]].$$

By using the Jacobi identity for the Lie bracket of vector fields, we also have

$$\mathbf{i}\, \mathbf{B}(\widehat{X}, \widehat{Y})\widehat{Z} = -[Y^h, [Z^v, X^v]] - [Z^v, [X^v, Y^h]] \overset{(4.1.52)}{=} [[X^v, Y^h], Z^v],$$

which concludes the proof of (7.13.3). Thus, finally,

$$\mathbf{i}\, \mathbf{B}(\widehat{Z}, \widehat{Y})\widehat{X} = [Z^v, [Y^h, X^v]] = [[X^v, Y^h], Z^v] = \mathbf{i}\, \mathbf{B}(\widehat{X}, \widehat{Y})\widehat{Z},$$

so **B** has the desired symmetry property. □

Corollary 7.13.4. *If the torsion of an Ehresmann connection vanishes, then its Berwald curvature is symmetric also in its first two arguments, and hence it is totally symmetric.*

Proof. Let X, Y and Z be vector fields on M. By the vanishing of the torsion of $\mathcal{H}$, $[X^h, Y^v] - [Y^h, X^v] \overset{(7.8.4)}{=} [X, Y]^v$. Thus

$$\mathbf{i}\, \mathbf{B}(\widehat{X}, \widehat{Y})\widehat{Z} - \mathbf{i}\, \mathbf{B}(\widehat{Y}, \widehat{X})\widehat{Z} \overset{(7.13.3)}{=} [[X^v, Y^h], Z^v] - [[Y^v, X^h], Z^v]$$

$$= [[X^h, Y^v] - [Y^h, X^v], Z^v] = [[X, Y]^v, Z^v] \overset{(4.1.52)}{=} 0. \;\square$$

Remark 7.13.5 (Coordinate Description). Choose a chart $(\mathcal{U}, (u^i)_{i=1}^n)$ on M, and consider the induced chart $(\tau^{-1}(\mathcal{U}), ((x^i)_{i=1}^n, (y^i)_{i=1}^n))$ on TM. Let (N_j^i) be the family of the Christoffel symbols of $\mathcal{H}$ with respect to the chosen charts. Then by (7.11.1) the Christoffel symbols of the induced Berwald derivative are the functions $N_{jk}^i = \frac{\partial N_j^i}{\partial y^k}$ $(i, j, k \in J_n)$, defined and smooth on $\mathring{\tau}^{-1}(\mathcal{U})$. The components B_{jkl}^i of the Berwald curvature of $\mathcal{H}$ are given by

$$B_{jkl}^i \widehat{\frac{\partial}{\partial u^i}} = \mathbf{B}\left(\widehat{\frac{\partial}{\partial u^j}}, \widehat{\frac{\partial}{\partial u^k}}, \widehat{\frac{\partial}{\partial u^l}}\right) \overset{(7.13.3)}{=} \mathcal{V}\left[\left[\left(\frac{\partial}{\partial u^j}\right)^v, \left(\frac{\partial}{\partial u^k}\right)^h\right], \left(\frac{\partial}{\partial u^l}\right)^v\right]$$

$$= \mathcal{V}\left[\left[\frac{\partial}{\partial y^j}, \frac{\partial}{\partial x^k} - N_k^i \frac{\partial}{\partial y^i}\right], \frac{\partial}{\partial y^l}\right] = \mathcal{V}\left[-\frac{\partial N_k^i}{\partial y^j} \frac{\partial}{\partial y^i}, \frac{\partial}{\partial y^l}\right]$$

$$= \mathcal{V}\left(\frac{\partial^2 N_k^i}{\partial y^l \partial y^j} \widehat{\frac{\partial}{\partial y^i}}\right) = \frac{\partial^2 N_k^i}{\partial y^l \partial y^j} \widehat{\frac{\partial}{\partial u^i}},$$

that is, we have

$$B_{jkl}^i = \frac{\partial^2 N_k^i}{\partial y^l \partial y^j} = \frac{\partial N_{kj}^i}{\partial y^l}; \quad i, j, k, l \in J_n. \tag{7.13.4}$$

From this we see at once that $B^i_{jkl} = B^i_{lkj}$, which also implies the symmetry of $\mathbf{B}$ in its first and third arguments. If the torsion of $\mathcal{H}$ vanishes, then $N^i_{kj} = N^i_{jk}$ by (7.8.10), and we obtain (without tricky calculations) that $\mathbf{B}$ is symmetric in its first two arguments.

Lemma 7.13.6. *The Berwald curvature and the tension of an Ehresmann connection are related by*

$$\mathbf{B}(\widetilde{X}, \widetilde{Y})\widetilde{\delta} = \nabla^{\mathsf{v}} \mathbf{t}(\widetilde{X}, \widetilde{Y}); \quad \widetilde{X}, \widetilde{Y} \in \Gamma(\overset{\circ}{\pi}). \tag{7.13.5}$$

Proof. By direct calculation we get

$$\mathbf{B}(\widetilde{X}, \widetilde{Y})\widetilde{\delta} \overset{(7.13.1)}{=} \nabla_{\mathbf{i}\widetilde{X}}\nabla_{\mathcal{H}\widetilde{Y}}\widetilde{\delta} - \nabla_{\mathcal{H}\widetilde{Y}}\nabla_{\mathbf{i}\widetilde{X}}\widetilde{\delta} - \nabla_{[\mathbf{i}\widetilde{X},\mathcal{H}\widetilde{Y}]}\widetilde{\delta}$$

$$\overset{(7.12.1),(6.2.33)}{=} \nabla_{\mathbf{i}\widetilde{X}}(\mathbf{t}(\widetilde{Y})) - \nabla_{\mathcal{H}\widetilde{Y}}\widetilde{X} - \nabla_{[\mathbf{i}\widetilde{X},\mathcal{H}\widetilde{Y}]}\widetilde{\delta}$$

$$= \nabla_{\mathbf{i}\widetilde{X}}(\mathbf{t}(\widetilde{Y})) - \nabla_{\mathcal{H}\widetilde{Y}}\widetilde{X} - \nabla_{\mathcal{H}\nabla_{\mathbf{i}\widetilde{X}}\widetilde{Y}}\widetilde{\delta} + \nabla_{\mathbf{i}\nabla_{\mathcal{H}\widetilde{Y}}\widetilde{X}}\widetilde{\delta}$$

$$\overset{(7.12.1),(6.2.33)}{=} \nabla_{\mathbf{i}\widetilde{X}}(\mathbf{t}(\widetilde{Y})) - \nabla_{\mathcal{H}\widetilde{Y}}\widetilde{X} - \mathbf{t}(\nabla_{\mathbf{i}\widetilde{X}}\widetilde{Y}) + \nabla_{\mathcal{H}\widetilde{Y}}\widetilde{X}$$

$$= \nabla^{\mathsf{v}}_{\widetilde{X}}(\mathbf{t}(\widetilde{Y})) - \mathbf{t}(\nabla^{\mathsf{v}}_{\widetilde{X}}\widetilde{Y}) = (\nabla^{\mathsf{v}}_{\widetilde{X}}\mathbf{t})(\widetilde{Y}) = \nabla^{\mathsf{v}}\mathbf{t}(\widetilde{X}, \widetilde{Y}),$$

as was to be shown. $\square$

Corollary 7.13.7. *If the torsion and the vertical differential of the tension of an Ehresmann connection vanish, then*

$$\delta \in \{\widetilde{X}, \widetilde{Y}, \widetilde{Z}\} \text{ implies } \mathbf{B}(\widetilde{X}, \widetilde{Y})\widetilde{Z} = 0. \tag{7.13.6}$$

$\square$

Lemma 7.13.8. *The Berwald curvature of a positive-homogeneous Ehresmann connection is positive-homogeneous of degree -1, i.e.,*

$$\nabla^{\mathsf{v}}_{\widetilde{\delta}}\mathbf{B} = \nabla_C \mathbf{B} = -\mathbf{B}. \tag{7.13.7}$$

Proof. We recall that $\mathbf{J}\xi \overset{(4.1.90)}{=} [\mathbf{J}, C]\xi = [\mathbf{J}\xi, C] - \mathbf{J}[\xi, C]$, whence

$$\mathbf{J}[C, \xi] = \mathbf{J}\xi - [\mathbf{J}\xi, C], \quad \xi \in \mathfrak{X}(\overset{\circ}{T}M). \tag{$*$}$$

We have also learnt (see (7.5.10) and Corollary 7.5.10) that the positive-homogeneity of $\mathcal{H}$ is equivalent to the condition

$$[X^{\mathsf{h}}, C] = 0 \text{ for all } X \in \mathfrak{X}(M). \tag{$**$}$$

Taking these into account, we have for any vector fields X, Y, Z on M,

$$\mathbf{i}((\nabla_C \mathbf{B})(\widehat{X}, \widehat{Y})\widehat{Z}) = \mathbf{i}\nabla_C(\mathbf{B}(\widehat{X}, \widehat{Y})\widehat{Z}) \overset{(7.10.1)}{=} \mathbf{J}[C, \mathcal{H}\mathbf{B}(\widehat{X}, \widehat{Y})\widehat{Z}]$$

$$\overset{(*)}{=} \mathbf{i}\mathbf{B}(\widehat{X}, \widehat{Y})\widehat{Z} - [\mathbf{i}\mathbf{B}(\widehat{X}, \widehat{Y})\widehat{Z}, C].$$

We show that the second term on the right-hand side is equal to $2i\mathbf{B}(\widehat{X}, \widehat{Y})\widehat{Z}$:

$$[i\mathbf{B}(\widehat{X}, \widehat{Y})\widehat{Z}, C] \stackrel{(7.13.3)}{=} [[[X^{\mathsf{v}}, Y^{\mathsf{h}}], Z^{\mathsf{v}}], C] \stackrel{(L2)}{=} -[[Z^{\mathsf{v}}, C], [X^{\mathsf{v}}, Y^{\mathsf{h}}]]$$

$$- [[C, [X^{\mathsf{v}}, Y^{\mathsf{h}}]], Z^{\mathsf{v}}] \stackrel{(4.1.53),(L2)}{=} -[Z^{\mathsf{v}}, [X^{\mathsf{v}}, Y^{\mathsf{h}}]] + [[X^{\mathsf{v}}, [Y^{\mathsf{h}}, C]], Z^{\mathsf{v}}]$$

$$+ [[Y^{\mathsf{h}}, [C, X^{\mathsf{v}}]], Z^{\mathsf{v}}] \stackrel{(**),(4.1.53)}{=} -[Z^{\mathsf{v}}, [X^{\mathsf{v}}, Y^{\mathsf{h}}]] - [[Y^{\mathsf{h}}, X^{\mathsf{v}}], Z^{\mathsf{v}}]$$

$$= 2[[X^{\mathsf{v}}, Y^{\mathsf{h}}], Z^{\mathsf{v}}] \stackrel{(7.13.3)}{=} 2i\mathbf{B}(\widehat{X}, \widehat{Y})\widehat{Z}.$$

This concludes the proof. $\qquad\square$

By using local coordinates, a simpler and more straightforward proof can be given, cf. Remark 7.8.5. As an instructive exercise, we leave it to the reader to work it out.

Lemma 7.13.9 (Mixed Ricci Formula for Sections). *Let a section* $\widetilde{Z} \in \Gamma(\overset{\circ}{\pi})$ *be given. Then for any sections* $\widetilde{X}, \widetilde{Y} \in \Gamma(\overset{\circ}{\pi})$,

$$\nabla^{\mathsf{v}}\nabla^{\mathsf{h}}\widetilde{Z}(\widetilde{X}, \widetilde{Y}) - \nabla^{\mathsf{h}}\nabla^{\mathsf{v}}\widetilde{Z}(\widetilde{Y}, \widetilde{X}) = \mathbf{B}(\widetilde{X}, \widetilde{Y})\widetilde{Z}. \qquad (7.13.8)$$

Proof. By a direct calculation, we obtain

$$\nabla^{\mathsf{v}}\nabla^{\mathsf{h}}\widetilde{Z}(\widetilde{X}, \widetilde{Y}) - \nabla^{\mathsf{h}}\nabla^{\mathsf{v}}\widetilde{Z}(\widetilde{Y}, \widetilde{X})$$

$$= \nabla_{\mathsf{i}\widetilde{X}}\nabla_{\mathcal{H}\widetilde{Y}}\widetilde{Z} - \nabla_{\mathcal{H}\widetilde{Y}}\nabla_{\mathsf{i}\widetilde{X}}\widetilde{Z} - \nabla_{\mathcal{H}\nabla_{\mathsf{i}\widetilde{X}}\widetilde{Y}}\widetilde{Z} + \nabla_{\mathsf{i}\nabla_{\mathcal{H}\widetilde{Y}}\widetilde{X}}\widetilde{Z}$$

$$\stackrel{(7.10.16)}{=} \nabla_{\mathsf{i}\widetilde{X}}\nabla_{\mathcal{H}\widetilde{Y}}\widetilde{Z} - \nabla_{\mathcal{H}\widetilde{Y}}\nabla_{\mathsf{i}\widetilde{X}}\widetilde{Z} - \nabla_{[\mathsf{i}\widetilde{X}, \mathcal{H}\widetilde{Y}]}\widetilde{Z} \stackrel{(7.13.1)}{=} \mathbf{B}(\widetilde{X}, \widetilde{Y})\widetilde{Z},$$

as was to be shown. $\qquad\square$

Lemma 7.13.10 (Mixed Ricci Formula for $(0, 2)$-tensors). *Let g be a type $(0, 2)$ Finsler tensor field on $\overset{\circ}{T}M$. Then for any sections* $\widetilde{X}, \widetilde{Y}, \widetilde{Z}_1, \widetilde{Z}_2$ *of* $\overset{\circ}{\pi}$,

$$(\nabla^{\mathsf{v}}\nabla^{\mathsf{h}}g)(\widetilde{X}, \widetilde{Y}, \widetilde{Z}_1, \widetilde{Z}_2) - (\nabla^{\mathsf{h}}\nabla^{\mathsf{v}}g)(\widetilde{Y}, \widetilde{X}, \widetilde{Z}_1, \widetilde{Z}_2)$$

$$= -g(\mathbf{B}(\widetilde{X}, \widetilde{Y})\widetilde{Z}_1, \widetilde{Z}_2) - g(\widetilde{Z}_1, \mathbf{B}(\widetilde{X}, \widetilde{Y})\widetilde{Z}_2). \quad (7.13.9)$$

The proof is left to the reader (or see [7], Lemma 3.7).

Remark 7.13.11. In the language of traditional tensor calculus, (7.13.8) takes the form

$$\widetilde{Z}^i_{;j,k} - \widetilde{Z}^i_{,k;j} = B^i_{kjl}\widetilde{Z}^l; \quad i, j, k \in J_n, \qquad (7.13.10)$$

if $\widetilde{Z} \underset{(\mathcal{U})}{=} \widetilde{Z}^i \widehat{\frac{\partial}{\partial u^i}}$, and B^i_{kjl} is given by (7.13.4). Now we prove this formula also directly, using the comma and the semicolon operators. We have, on the one hand,

$$\widetilde{Z}^i_{;j,k} \overset{(7.11.9)}{=} \frac{\partial}{\partial y^k}\left(\frac{\partial \widetilde{Z}^i}{\partial x^j} - N^l_j \frac{\partial \widetilde{Z}^i}{\partial y^l} + N^i_{jl}\widetilde{Z}^l\right)$$

$$= \frac{\partial^2 \widetilde{Z}^i}{\partial y^k \partial x^j} - N^l_{jk}\frac{\partial \widetilde{Z}^i}{\partial y^l} - N^l_j \frac{\partial^2 \widetilde{Z}^i}{\partial y^k \partial y^l} + \frac{\partial N^i_{jl}}{\partial y^k}\widetilde{Z}^l + N^i_{jl}\frac{\partial \widetilde{Z}^l}{\partial y^k}.$$

On the other hand,

$$\widetilde{Z}^i_{,k;j} = \left(\frac{\partial \widetilde{Z}^i}{\partial y^k}\right)_{;j} \overset{(7.11.15)}{=} \frac{\partial^2 \widetilde{Z}^i}{\partial x^j \partial y^k} - N^l_j \frac{\partial^2 \widetilde{Z}^i}{\partial y^l \partial y^k} + N^i_{jl}\frac{\partial \widetilde{Z}^l}{\partial y^k} - N^l_{jk}\frac{\partial \widetilde{Z}^i}{\partial y^l}.$$

Thus, taking into account Corollary 7.13.3,

$$\widetilde{Z}^i_{;j,k} - \widetilde{Z}^i_{,k;j} = \frac{\partial N^i_{jl}}{\partial y^k}\widetilde{Z}^l \overset{(7.13.4)}{=} B^i_{ljk}\widetilde{Z}^l = B^i_{kjl}\widetilde{Z}^l,$$

so we have the desired equality.

Proposition 7.13.12 (Differential Bianchi Identity for the Berwald Curvature). *For any sections* $\widetilde{X}, \widetilde{Y}, \widetilde{Z}, \widetilde{U}$ *in* $\Gamma(\mathring{\pi})$,

$$(\nabla_{\mathbf{i}\widetilde{X}}\mathbf{B})(\widetilde{Y}, \widetilde{Z}, \widetilde{U}) = (\nabla_{\mathbf{i}\widetilde{Y}}\mathbf{B})(\widetilde{X}, \widetilde{Z}, \widetilde{U}). \tag{7.13.11}$$

Proof. It is enough to verify (7.13.11) for basic sections $\widehat{X}, \widehat{Y}, \widehat{Z}, \widehat{U}$. We start from the Bianchi identity $d^\nabla R^\nabla = 0$ (see (6.1.44)), and evaluate $d^\nabla R^\nabla$ at $(X^\mathsf{v}, Y^\mathsf{v}, Z^\mathsf{h}, \widehat{U})$. Taking into account the second step of the proof of Proposition 6.1.30, we find that

$$0 = \nabla_{X^\mathsf{v}}(R^\nabla(Y^\mathsf{v}, Z^\mathsf{h})\widehat{U}) - \nabla_{Y^\mathsf{v}}(R^\nabla(X^\mathsf{v}, Z^\mathsf{h})\widehat{U}) + \nabla_{Z^\mathsf{h}}(R^\nabla(X^\mathsf{v}, Y^\mathsf{v})\widehat{U})$$
$$- R^\nabla(Y^\mathsf{v}, Z^\mathsf{h})\nabla_{X^\mathsf{v}}\widehat{U} + R^\nabla(X^\mathsf{v}, Z^\mathsf{h})\nabla_{Y^\mathsf{v}}\widehat{U} - R^\nabla(X^\mathsf{v}, Y^\mathsf{v})\nabla_{Z^\mathsf{h}}\widehat{U}$$
$$- R^\nabla([X^\mathsf{v}, Y^\mathsf{v}], Z^\mathsf{h})\widehat{U} + R^\nabla([X^\mathsf{v}, Z^\mathsf{h}], Y^\mathsf{v})\widehat{U} - R^\nabla([Y^\mathsf{v}, Z^\mathsf{h}], X^\mathsf{v})\widehat{U}.$$

Here

$$\nabla_{X^\mathsf{v}}(R^\nabla(Y^\mathsf{v}, Z^\mathsf{h})\widehat{U}) = \nabla_{X^\mathsf{v}}(\mathbf{B}(\widehat{Y}, \widehat{Z})\widehat{U}) \overset{(7.10.4)}{=} (\nabla_{X^\mathsf{v}}\mathbf{B})(\widehat{Y}, \widehat{Z}, \widehat{U}),$$

and, similarly,

$$\nabla_{Y^\mathsf{v}}(R^\nabla(X^\mathsf{v}, Z^\mathsf{h})\widehat{U}) = (\nabla_{Y^\mathsf{v}}\mathbf{B})(\widehat{X}, \widehat{Z}, \widehat{U}).$$

The other terms on the right-hand side of the above relation vanish, as it can be seen essentially by inspection, so we have the desired relation. $\square$

Remark 7.13.13. In terms of the components of $\mathbf{B}$, (7.13.11) takes the form

$$B^i_{klm,j} = B^i_{jlm,k}; \quad i, j, k, l, m \in J_n. \tag{7.13.12}$$

7.14   The Affine Curvature

We continue to assume that an Ehresmann connection $\mathcal{H}$ is specified in $\mathring{T}M$. For the Berwald derivative and the horizontal derivative induced by $\mathcal{H}$ we also keep the notation introduced above.

For the $\mathcal{H}$-horizontal curvature of ∇ defined in general by (7.9.2), we also use a special terminology and notation. We call it the *affine curvature* of the Ehresmann connection $\mathcal{H}$, and denote it by $\mathbf{H}$. Thus, for any sections $\widetilde{X}, \widetilde{Y}, \widetilde{Z}$ in $\Gamma(\mathring{\pi})$,

$$\begin{aligned}
\mathbf{H}(\widetilde{X}, \widetilde{Y})\widetilde{Z} &:= R^{\nabla}(\mathcal{H}\widetilde{X}, \mathcal{H}\widetilde{Y})\widetilde{Z} \\
&= \nabla_{\mathcal{H}\widetilde{X}}\nabla_{\mathcal{H}\widetilde{Y}}\widetilde{Z} - \nabla_{\mathcal{H}\widetilde{Y}}\nabla_{\mathcal{H}\widetilde{X}}\widetilde{Z} - \nabla_{[\mathcal{H}\widetilde{X},\mathcal{H}\widetilde{Y}]}\widetilde{Z}.
\end{aligned} \tag{7.14.1}$$

This tensor was essentially introduced by L. Berwald in his epoch-making posthumous paper [16]. Berwald, as we shall briefly explain below, built his theory in the more specific context of Ehresmann connections generated by sprays, and (of course) using the language of classical tensor calculus. We felt appropriate to preserve his terminology and notation, as well as, what is more important, to keep the main ideas of his exposition.

First, using our tools, we formulate and prove some basic relations found by Berwald. We begin by showing that, roughly speaking, 'the affine curvature is essentially the vertical differential of the curvature of the Ehresmann connection'. More precisely, we have the following result.

Lemma 7.14.1. *For any sections $\widetilde{X}, \widetilde{Y}, \widetilde{Z}$ in $\Gamma(\mathring{\pi})$,*

$$\mathbf{H}(\widetilde{X}, \widetilde{Y})\widetilde{Z} = -\nabla^{\mathsf{v}}\mathcal{R}(\widetilde{Z}, \widetilde{X}, \widetilde{Y}). \tag{7.14.2}$$

Proof. By tensoriality, it is enough to check this equality for basic sections $\widehat{X}, \widehat{Y}, \widehat{Z}$. Starting with the definition of $\mathbf{H}$, we obtain

$$\begin{aligned}
\mathbf{H}(\widehat{X}, \widehat{Y})\widehat{Z} &= \nabla_{X^{\mathsf{h}}}\nabla_{Y^{\mathsf{h}}}\widehat{Z} - \nabla_{Y^{\mathsf{h}}}\nabla_{X^{\mathsf{h}}}\widehat{Z} - \nabla_{[X^{\mathsf{h}},Y^{\mathsf{h}}]}\widehat{Z} \\
&\overset{(7.10.2),(6.2.32)}{=} \nabla_{X^{\mathsf{h}}}\mathcal{V}[Y^{\mathsf{h}}, Z^{\mathsf{v}}] - \nabla_{Y^{\mathsf{h}}}\mathcal{V}[X^{\mathsf{h}}, Z^{\mathsf{v}}] - \nabla_{\mathbf{h}[X^{\mathsf{h}},Y^{\mathsf{h}}]}\widehat{Z} \\
&\overset{(7.10.2),(7.1.7)}{=} \mathcal{V}([X^{\mathsf{h}}, \mathbf{v}[Y^{\mathsf{h}}, Z^{\mathsf{v}}]] - [Y^{\mathsf{h}}, \mathbf{v}[X^{\mathsf{h}}, Z^{\mathsf{v}}]] - [[X,Y]^{\mathsf{h}}, Z^{\mathsf{v}}]) \\
&= \mathcal{V}([X^{\mathsf{h}}, [Y^{\mathsf{h}}, Z^{\mathsf{v}}]] + [Y^{\mathsf{h}}, [Z^{\mathsf{v}}, X^{\mathsf{h}}]] - [[X,Y]^{\mathsf{h}}, Z^{\mathsf{v}}]) \\
&\overset{(\mathrm{L2})}{=} \mathcal{V}([[X^{\mathsf{h}}, Y^{\mathsf{h}}] - [X,Y]^{\mathsf{h}}, Z^{\mathsf{v}}]) \overset{(7.8.3)}{=} \mathcal{V}[\mathbf{i}\mathcal{R}(\widehat{X}, \widehat{Y}), Z^{\mathsf{v}}].
\end{aligned}$$

On the other hand,

$$\nabla^{\mathsf{v}}\mathcal{R}(\widehat{Z}, \widehat{X}, \widehat{Y}) = (\nabla^{\mathsf{v}}_{\widehat{Z}}\mathcal{R})(\widehat{X}, \widehat{Y}) \overset{(6.2.32)}{=} \nabla^{\mathsf{v}}_{\widehat{Z}}(\mathcal{R}(\widehat{X}, \widehat{Y})) \overset{(7.10.1)}{=} \mathbf{j}[Z^{\mathsf{v}}, \mathcal{H}\mathcal{R}(\widehat{X}, \widehat{Y})].$$

Applying the first equality of (4.1.89),

$$0 = [\mathbf{J}, Z^{\mathsf{v}}]\mathcal{H}\mathcal{R}(\widehat{X}, \widehat{Y}) = [\mathrm{i}\mathcal{R}(\widehat{X}, \widehat{Y}), Z^{\mathsf{v}}] - \mathbf{J}[\mathcal{H}\mathcal{R}(\widehat{X}, \widehat{Y}), Z^{\mathsf{v}}],$$

whence

$$\mathbf{H}(\widehat{X}, \widehat{Y})\widehat{Z} = \mathcal{V}[\mathrm{i}\mathcal{R}(\widehat{X}, \widehat{Y}), Z^{\mathsf{v}}] = \mathbf{j}[\mathcal{H}\mathcal{R}(\widehat{X}, \widehat{Y}), Z^{\mathsf{v}}] = -\nabla^{\mathsf{v}}\mathcal{R}(\widehat{Z}, \widehat{X}, \widehat{Y}),$$

as was to be shown. $\qquad\qquad\qquad\qquad\qquad\qquad\qquad\qquad\qquad\qquad\square$

Remark 7.14.2. Since $\mathbf{H}$ is skew-symmetric in its first two arguments, (7.14.2) implies that $\nabla^{\mathsf{v}}\mathcal{R}$ is skew-symmetric in its second and third arguments. So we also have

$$\mathbf{H}(\widetilde{X}, \widetilde{Y})\widetilde{Z} = \nabla^{\mathsf{v}}\mathcal{R}(\widetilde{Z}, \widetilde{Y}, \widetilde{X}). \qquad (7.14.3)$$

Lemma 7.14.3. *The traces of the affine curvature and the curvature of an Ehresmann connection are related by*

$$(\mathrm{tr}\,\mathbf{H})(\widetilde{X}, \widetilde{Y}) = -(\nabla^{\mathsf{v}}\,\mathrm{tr}\,\mathcal{R})(\widetilde{Y}, \widetilde{X}), \qquad (7.14.4)$$

where $\widetilde{X}, \widetilde{Y} \in \Gamma(\mathring{\pi})$.

Proof. Using Lemma 1.2.6 and the skew-symmetry of $\nabla_{\mathbf{i}\widetilde{Y}}\mathcal{R}$, we find that

$$(\mathrm{tr}\,\mathbf{H})(\widetilde{X}, \widetilde{Y}) = \mathrm{tr}(\widetilde{Z} \mapsto \mathbf{H}(\widetilde{Z}, \widetilde{X})\widetilde{Y}) \stackrel{(7.14.3)}{=} \mathrm{tr}(\widetilde{Z} \mapsto \nabla^{\mathsf{v}}\mathcal{R}(\widetilde{Y}, \widetilde{X}, \widetilde{Z}))$$
$$= \mathrm{tr}(\widetilde{Z} \mapsto (\nabla_{\mathbf{i}\widetilde{Y}}\mathcal{R})(\widetilde{X}, \widetilde{Z})) = -\,\mathrm{tr}(\widetilde{Z} \mapsto (\nabla_{\mathbf{i}\widetilde{Y}}\mathcal{R})(\widetilde{Z}, \widetilde{X}))$$
$$= -(\mathrm{tr}\,\nabla_{\mathbf{i}\widetilde{Y}}\mathcal{R})(\widetilde{X}).$$

Since the covariant derivative with respect to a vector field is a tensor derivation (Example 6.1.8(d)), and tensor derivations commute with contractions (Definition 3.3.9), $\mathrm{tr}\,\nabla_{\mathbf{i}\widetilde{Y}}\mathcal{R} = \nabla_{\mathbf{i}\widetilde{Y}}\,\mathrm{tr}\,\mathcal{R}$. Thus we obtain the desired equality. $\qquad\qquad\qquad\qquad\qquad\qquad\qquad\qquad\qquad\square$

Remark 7.14.4 (Coordinate Description). We use the same notation as in Remark 7.13.5. The components of $\mathbf{H}$ with respect to the chart $(\mathcal{U}, (u^i)_{i=1}^n)$ are the functions H^i_{jkl} determined by

$$H^i_{jkl}\,\widehat{\frac{\partial}{\partial u^i}} = \mathbf{H}\left(\widehat{\frac{\partial}{\partial u^j}}, \widehat{\frac{\partial}{\partial u^k}}\right)\widehat{\frac{\partial}{\partial u^l}}; \quad j, k, l \in J_n. \qquad (7.14.5)$$

Since, by (7.8.3),

$$\left[\left(\frac{\partial}{\partial u^j}\right)^{\mathsf{h}}, \left(\frac{\partial}{\partial u^k}\right)^{\mathsf{h}}\right] = \mathrm{i}\mathcal{R}\left(\widehat{\frac{\partial}{\partial u^j}}, \widehat{\frac{\partial}{\partial u^k}}\right),$$

and

$$\nabla_{\mathbf{i}\mathcal{R}(\widehat{X},\widehat{Y})}\widehat{Z} = \nabla^{\mathsf{v}}_{\mathcal{R}(\widehat{X},\widehat{Y})}\widehat{Z} \overset{(6.2.32)}{=} 0 \text{ for all } X, Y, Z \in \mathfrak{X}(M),$$

we find that

$$\mathbf{H}\left(\widehat{\frac{\partial}{\partial u^j}}, \widehat{\frac{\partial}{\partial u^k}}\right)\widehat{\frac{\partial}{\partial u^l}} \overset{(7.14.1)}{=} \nabla_{(\frac{\partial}{\partial u^j})^{\mathsf{h}}}\nabla_{(\frac{\partial}{\partial u^k})^{\mathsf{h}}}\widehat{\frac{\partial}{\partial u^l}} - \nabla_{(\frac{\partial}{\partial u^k})^{\mathsf{h}}}\nabla_{(\frac{\partial}{\partial u^j})^{\mathsf{h}}}\widehat{\frac{\partial}{\partial u^l}}$$

$$- \nabla_{[(\frac{\partial}{\partial u^j})^{\mathsf{h}},(\frac{\partial}{\partial u^k})^{\mathsf{h}}]}\widehat{\frac{\partial}{\partial u^l}} = \nabla_{(\frac{\partial}{\partial u^j})^{\mathsf{h}}}N^i_{kl}\widehat{\frac{\partial}{\partial u^i}} - \nabla_{(\frac{\partial}{\partial u^k})^{\mathsf{h}}}N^i_{jl}\widehat{\frac{\partial}{\partial u^i}}$$

$$= N^i_{kl;j}\widehat{\frac{\partial}{\partial u^i}} + N^r_{kl}\nabla_{(\frac{\partial}{\partial u^j})^{\mathsf{h}}}\widehat{\frac{\partial}{\partial u^r}} - N^i_{jl;k}\widehat{\frac{\partial}{\partial u^i}} - N^r_{jl}\nabla_{(\frac{\partial}{\partial u^k})^{\mathsf{h}}}\widehat{\frac{\partial}{\partial u^r}}$$

$$= \left(N^i_{kl;j} - N^i_{jl;k} + N^r_{kl}N^i_{jr} - N^r_{jl}N^i_{kr}\right)\widehat{\frac{\partial}{\partial u^i}}.$$

Thus

$$H^i_{jkl} = N^i_{kl;j} - N^i_{jl;k} + N^i_{jr}N^r_{kl} - N^i_{kr}N^r_{jl}, \tag{7.14.6}$$

where the semicolon operator $; k$ acts by (7.11.8). Taking this into account, we get

$$H^i_{jkl} = \frac{\partial N^i_{kl}}{\partial x^j} - \frac{\partial N^i_{jl}}{\partial x^k} + N^i_{jr}N^r_{kl} - N^i_{kr}N^r_{jl} + N^r_k N^i_{jlr} - N^r_j N^i_{klr}, \tag{7.14.7}$$

where

$$N^i_{klr} := N^i_{kl,r} = \frac{\partial N^i_{kl}}{\partial y^r}, \qquad N^i_{jlr} := N^i_{jl,r} = \frac{\partial N^i_{jl}}{\partial y^r}.$$

Now we deduce the coordinate expression of (7.14.2):

$$R^i_{jk,l} \overset{(7.8.9)}{=} \frac{\partial}{\partial y^l}\left(\frac{\partial N^i_j}{\partial x^k} - \frac{\partial N^i_k}{\partial x^j} + N^r_j N^i_{kr} - N^r_k N^i_{jr}\right)$$

$$= \frac{\partial N^i_{jl}}{\partial x^k} - \frac{\partial N^i_{kl}}{\partial x^j} + N^i_{kr}N^r_{jl} - N^i_{jr}N^r_{kl} + N^r_j N^i_{krl} - N^r_k N^i_{jrl} \overset{(7.14.7)}{=} -H^i_{jkl},$$

so we have

$$H^i_{jkl} = -R^i_{jk,l} = R^i_{kj,l}. \tag{7.14.8}$$

Lemma 7.14.5. *The curvature, the torsion, the tension and the affine curvature of an Ehresmann connection are related by*

$$\mathbf{H}(\widetilde{X},\widetilde{Y})\widetilde{\delta} + \mathcal{R}(\widetilde{X},\widetilde{Y}) = (\mathbf{t}\circ\mathbf{T} + \nabla^{\mathsf{h}}\mathbf{t})(\widetilde{X},\widetilde{Y}) - \nabla^{\mathsf{h}}\mathbf{t}(\widetilde{Y},\widetilde{X}), \tag{7.14.9}$$

where $\widetilde{X}, \widetilde{Y} \in \Gamma(\widetilde{\pi})$.

Proof.

$$\mathbf{H}(\widetilde{X}, \widetilde{Y})\widetilde{\delta} \overset{(7.14.1)}{=} \nabla_{\mathcal{H}\widetilde{X}}\nabla_{\mathcal{H}\widetilde{Y}}\widetilde{\delta} - \nabla_{\mathcal{H}\widetilde{Y}}\nabla_{\mathcal{H}\widetilde{X}}\widetilde{\delta} - \nabla_{[\mathcal{H}\widetilde{X},\mathcal{H}\widetilde{Y}]}\widetilde{\delta}$$

$$\overset{(7.12.1)}{=} \nabla_{\mathcal{H}\widetilde{X}}(\mathbf{t}(\widetilde{Y})) - \nabla_{\mathcal{H}\widetilde{Y}}(\mathbf{t}(\widetilde{X})) - \nabla_{\mathbf{h}[\mathcal{H}\widetilde{X},\mathcal{H}\widetilde{Y}]}\widetilde{\delta} - \nabla_{\mathbf{v}[\mathcal{H}\widetilde{X},\mathcal{H}\widetilde{Y}]}\widetilde{\delta}$$

$$\overset{(7.12.1),(6.2.33)}{=} \nabla_{\mathcal{H}\widetilde{X}}(\mathbf{t}(\widetilde{Y})) - \nabla_{\mathcal{H}\widetilde{Y}}(\mathbf{t}(\widetilde{X})) - \mathbf{t}(\mathbf{j}[\mathcal{H}\widetilde{X},\mathcal{H}\widetilde{Y}]) - \mathcal{V}[\mathcal{H}\widetilde{X},\mathcal{H}\widetilde{Y}],$$

whence

$$\mathbf{H}(\widetilde{X}, \widetilde{Y})\widetilde{\delta} + \mathcal{R}(\widetilde{X}, \widetilde{Y}) = (\nabla^{\mathbf{h}}\mathbf{t})(\widetilde{X}, \widetilde{Y}) - (\nabla^{\mathbf{h}}\mathbf{t})(\widetilde{Y}, \widetilde{X})$$

$$+ \mathbf{t}(\nabla_{\mathcal{H}\widetilde{X}}\widetilde{Y}) - \mathbf{t}(\nabla_{\mathcal{H}\widetilde{Y}}\widetilde{X}) - \mathbf{t}(\mathbf{j}[\mathcal{H}\widetilde{X},\mathcal{H}\widetilde{Y}])$$

$$\overset{(7.12.2)}{=} (\nabla^{\mathbf{h}}\mathbf{t})(\widetilde{X}, \widetilde{Y}) + \mathbf{t}(\mathbf{T}(\widetilde{X}, \widetilde{Y})) - (\nabla^{\mathbf{h}}\mathbf{t})(\widetilde{Y}, \widetilde{X}),$$

as was to be shown. $\qquad\square$

Corollary 7.14.6. *The affine curvature and the curvature of a* positive-homogeneous *Ehresmann connection are related by*

$$\mathbf{H}(\widetilde{X}, \widetilde{Y})\widetilde{\delta} = -\mathcal{R}(\widetilde{X}, \widetilde{Y}). \tag{7.14.10}$$

Written in terms of tensor components,

$$y^l H^i_{jkl} = -R^i_{jk} = R^i_{kj}; \quad i, j, k \in J_n. \tag{7.14.11}$$

Proof. In the homogeneous case $\mathbf{t} = 0$, so (7.14.9) implies immediately (7.14.10). We can also refer to (7.9.13), since (under homogeneity) ∇ is strongly associated to $\mathcal{H}$ by Lemma 7.12.2. $\qquad\square$

Remark 7.14.7. Applying (7.14.2) and (7.14.10), Lemma 7.8.4 can be shortly proved. Indeed, suppose that $\mathcal{H}$ is positive-homogeneous. Then, for any sections $\widetilde{X}, \widetilde{Y} \in \Gamma(\overset{\circ}{\pi})$ we have

$$(\nabla^{\mathbf{v}}_{\widetilde{\delta}}\mathcal{R})(\widetilde{X}, \widetilde{Y}) = \nabla^{\mathbf{v}}\mathcal{R}(\widetilde{\delta}, \widetilde{X}, \widetilde{Y}) = -\mathbf{H}(\widetilde{X}, \widetilde{Y})\widetilde{\delta} = \mathcal{R}(\widetilde{X}, \widetilde{Y}).$$

Thus $\nabla^{\mathbf{v}}_{\widetilde{\delta}}\mathcal{R} = \mathcal{R}$, therefore $\mathcal{R}$ is positive-homogeneous.

Corollary 7.14.8. *The affine curvature of a positive-homogeneous Ehresmann connection is 0^+-homogeneous.*

Proof. This is an immediate consequence of lemmas 7.8.4, 7.14.1 and 6.2.22. $\qquad\square$

Remark 7.14.9. Using tensor components as in Remark 7.14.4, we give a direct proof of the corollary:

$$\frac{\partial H^i_{jkl}}{\partial y^r} y^r \overset{(7.14.8)}{=} \frac{\partial R^i_{kj}}{\partial y^r \partial y^l} y^r = \frac{\partial}{\partial y^l}\left(\frac{\partial R^i_{kj}}{\partial y^r} y^r\right) - \frac{\partial R^i_{kj}}{\partial y^l}$$

$$\overset{(7.8.12)}{=} \frac{\partial R^i_{kj}}{\partial y^l} - \frac{\partial R^i_{kj}}{\partial y^l} = 0.$$

Thus, by Corollary 6.2.21 and Lemma 4.2.9, **H** is 0^+-homogeneous.

Proposition 7.14.10 (Algebraic Bianchi Identity for the Affine Curvature). *For any sections* $\widetilde{X}, \widetilde{Y}, \widetilde{Z} \in \Gamma(\mathring{\pi})$,

$$\sum_{\text{cyc}} \mathbf{H}(\widetilde{X}, \widetilde{Y})\widetilde{Z} = \sum_{\text{cyc}} \left((\nabla^h \mathbf{T})(\widetilde{X}, \widetilde{Y}, \widetilde{Z}) + \mathbf{T}(\mathbf{T}(\widetilde{X}, \widetilde{Y}), \widetilde{Z})\right), \qquad (7.14.12)$$

where **T** *is the torsion of the given Ehresmann connection.*

Proof. We apply (6.2.10) with the choice

$$D := \nabla, \quad (\xi, \eta, \zeta) := (\mathcal{H}\widetilde{X}, \mathcal{H}\widetilde{Y}, \mathcal{H}\widetilde{Z}).$$

Then, on the one hand,

$$\sum_{\text{cyc}} R^\nabla(\mathcal{H}\widetilde{X}, \mathcal{H}\widetilde{Y})\widetilde{Z} \overset{(7.14.1)}{=} \sum_{\text{cyc}} \mathbf{H}(\widetilde{X}, \widetilde{Y})\widetilde{Z}.$$

On the other hand, taking into account the first step of the proof of Proposition 6.2.4 and the vanishing of the vertical deviation of ∇,

$$d^\nabla T^\nabla(\mathcal{H}\widetilde{X}, \mathcal{H}\widetilde{Y}, \mathcal{H}\widetilde{Z})$$
$$= \nabla_{\mathcal{H}\widetilde{X}}(T^\nabla(\mathcal{H}\widetilde{Y}, \mathcal{H}\widetilde{Z})) - \nabla_{\mathcal{H}\widetilde{Y}}(T^\nabla(\mathcal{H}\widetilde{X}, \mathcal{H}\widetilde{Z})) + \nabla_{\mathcal{H}\widetilde{Z}}(T^\nabla(\mathcal{H}\widetilde{X}, \mathcal{H}\widetilde{Y}))$$
$$\quad - T^\nabla([\mathcal{H}\widetilde{X}, \mathcal{H}\widetilde{Y}], \mathcal{H}\widetilde{Z}) + T^\nabla([\mathcal{H}\widetilde{X}, \mathcal{H}\widetilde{Z}], \mathcal{H}\widetilde{Y}) - T^\nabla([\mathcal{H}\widetilde{Y}, \mathcal{H}\widetilde{Z}], \mathcal{H}\widetilde{X})$$
$$= \nabla_{\mathcal{H}\widetilde{X}}(\mathbf{T}(\widetilde{Y}, \widetilde{Z})) + \nabla_{\mathcal{H}\widetilde{Y}}(\mathbf{T}(\widetilde{Z}, \widetilde{X})) + \nabla_{\mathcal{H}\widetilde{Z}}(\mathbf{T}(\widetilde{X}, \widetilde{Y}))$$
$$\quad - T^\nabla(\mathcal{H}\mathbf{j}[\mathcal{H}\widetilde{X}, \mathcal{H}\widetilde{Y}], \mathcal{H}\widetilde{Z}) - T^\nabla(\mathcal{H}\mathbf{j}[\mathcal{H}\widetilde{Y}, \mathcal{H}\widetilde{Z}], \mathcal{H}\widetilde{X})$$
$$\quad - T^\nabla(\mathcal{H}\mathbf{j}[\mathcal{H}\widetilde{Z}, \mathcal{H}\widetilde{X}], \mathcal{H}\widetilde{Y})$$
$$= \sum_{\text{cyc}} \left(\nabla_{\mathcal{H}\widetilde{X}}(\mathbf{T}(\widetilde{Y}, \widetilde{Z})) - \mathbf{T}(\mathbf{j}[\mathcal{H}\widetilde{X}, \mathcal{H}\widetilde{Y}], \widetilde{Z})\right)$$
$$\overset{(7.9.1)}{=} \sum_{\text{cyc}} \left(\nabla_{\mathcal{H}\widetilde{X}}(\mathbf{T}(\widetilde{Y}, \widetilde{Z})) - \mathbf{T}(\nabla_{\mathcal{H}\widetilde{X}}\widetilde{Y}, \widetilde{Z})\right)$$
$$\quad + \sum_{\text{cyc}} \left(\mathbf{T}(\nabla_{\mathcal{H}\widetilde{Y}}\widetilde{X}, \widetilde{Z}) + \mathbf{T}(\mathbf{T}(\widetilde{X}, \widetilde{Y}), \widetilde{Z})\right)$$

$$\begin{aligned}
&= \nabla_{\mathcal{H}\widetilde{X}}(\mathbf{T}(\widetilde{Y},\widetilde{Z})) - \mathbf{T}(\nabla_{\mathcal{H}\widetilde{X}}\widetilde{Y},\widetilde{Z}) + \mathbf{T}(\nabla_{\mathcal{H}\widetilde{Y}}\widetilde{X},\widetilde{Z}) + \mathbf{T}(\mathbf{T}(\widetilde{X},\widetilde{Y}),\widetilde{Z}) \\
&\quad + \nabla_{\mathcal{H}\widetilde{Y}}(\mathbf{T}(\widetilde{Z},\widetilde{X})) - \mathbf{T}(\nabla_{\mathcal{H}\widetilde{Y}}\widetilde{Z},\widetilde{X}) + \mathbf{T}(\nabla_{\mathcal{H}\widetilde{Z}}\widetilde{Y},\widetilde{X}) + \mathbf{T}(\mathbf{T}(\widetilde{Y},\widetilde{Z}),\widetilde{X}) \\
&\quad + \nabla_{\mathcal{H}\widetilde{Z}}(\mathbf{T}(\widetilde{X},\widetilde{Y})) - \mathbf{T}(\nabla_{\mathcal{H}\widetilde{Z}}\widetilde{X},\widetilde{Y}) + \mathbf{T}(\nabla_{\mathcal{H}\widetilde{X}}\widetilde{Z},\widetilde{Y}) + \mathbf{T}(\mathbf{T}(\widetilde{Z},\widetilde{X}),\widetilde{Y}) \\
&= \sum_{\mathrm{cyc}}\left((\nabla^{\mathrm{h}}\mathbf{T})(\widetilde{X},\widetilde{Y},\widetilde{Z}) + \mathbf{T}(\mathbf{T}(\widetilde{X},\widetilde{Y}),\widetilde{Z})\right).
\end{aligned}$$

Thus (6.2.10) leads to the identity (7.14.12). $\qquad\square$

Proposition 7.14.11 (The Mixed Differential Bianchi Identity).
For any sections $\widetilde{X},\widetilde{Y},\widetilde{Z},\widetilde{U}$ *in* $\Gamma(\mathring{\pi})$, *we have*

$$\begin{aligned}
\nabla^{\mathrm{v}}\mathbf{H}(\widetilde{X},\widetilde{Y},\widetilde{Z},\widetilde{U}) - \nabla^{\mathrm{h}}\mathbf{B}(\widetilde{Y},\widetilde{X},\widetilde{Z},\widetilde{U}) + \nabla^{\mathrm{h}}\mathbf{B}(\widetilde{Z},\widetilde{X},\widetilde{Y},\widetilde{U}) \\
= \mathbf{B}(\widetilde{X},\mathbf{T}(\widetilde{Y},\widetilde{Z}))\widetilde{U}.
\end{aligned} \qquad (7.14.13)$$

Proof. We evaluate the left-hand side of the general Bianchi identity $d^{\nabla}R^{\nabla} = 0$ at $(\mathrm{i}\widetilde{X},\mathcal{H}\widetilde{Y},\mathcal{H}\widetilde{Z},\widetilde{U})$, applying the expression obtained in the second step of the proof of Proposition 6.1.30. In the forthcoming manipulations we shall use the following observations:

(i) $\mathbf{T}(\widetilde{X},\widetilde{Y}) = \nabla_{\mathcal{H}\widetilde{X}}\widetilde{Y} - \nabla_{\mathcal{H}\widetilde{Y}}\widetilde{X} - \mathbf{j}[\mathcal{H}\widetilde{X},\mathcal{H}\widetilde{Y}]$ (see (7.12.2));

(ii) the vertical curvature $\mathbf{Q}^{\nabla}$ defined by $\mathbf{Q}^{\nabla}(\widetilde{X},\widetilde{Y})\widetilde{Z} := R^{\nabla}(\mathrm{i}\widetilde{X},\mathrm{i}\widetilde{Y})\widetilde{Z}$ vanishes;

(iii) $R^{\nabla}(\mathrm{i}\widetilde{X},[\mathcal{H}\widetilde{Y},\mathcal{H}\widetilde{Z}])\widetilde{U} = R^{\nabla}(\mathrm{i}\widetilde{X},\mathcal{H}\mathbf{j}[\mathcal{H}\widetilde{Y},\mathcal{H}\widetilde{Z}])\widetilde{U}$

$$+ R^{\nabla}(\mathrm{i}\widetilde{X},\mathrm{i}\mathcal{V}[\mathcal{H}\widetilde{Y},\mathcal{H}\widetilde{Z}])\widetilde{U} \overset{\text{(ii)}}{=} \mathbf{B}(\widetilde{X},\mathbf{j}[\mathcal{H}\widetilde{Y},\mathcal{H}\widetilde{Z}])\widetilde{U}$$

$$\overset{\text{(i)}}{=} \mathbf{B}(\widetilde{X},\nabla_{\mathcal{H}\widetilde{Y}}\widetilde{Z})\widetilde{U} - \mathbf{B}(\widetilde{X},\nabla_{\mathcal{H}\widetilde{Z}}\widetilde{Y})\widetilde{U} - \mathbf{B}(\widetilde{X},\mathbf{T}(\widetilde{Y},\widetilde{Z}))\widetilde{U};$$

(iv) $R^{\nabla}([\mathrm{i}\widetilde{X},\mathcal{H}\widetilde{Y}],\mathcal{H}\widetilde{Z})\widetilde{U} \overset{(7.10.16)}{=} R^{\nabla}(\mathcal{H}\nabla_{\mathrm{i}\widetilde{X}}\widetilde{Y},\mathcal{H}\widetilde{Z})\widetilde{U}$

$$- R^{\nabla}(\mathrm{i}\nabla_{\mathcal{H}\widetilde{Y}}\widetilde{X},\mathcal{H}\widetilde{Z})\widetilde{U} = \mathbf{H}(\nabla_{\mathrm{i}\widetilde{X}}\widetilde{Y},\widetilde{Z})\widetilde{U} - \mathbf{B}(\nabla_{\mathcal{H}\widetilde{Y}}\widetilde{X},\widetilde{Z})\widetilde{U}.$$

Taking these into account, we find that

$$\begin{aligned}
0 &= d^{\nabla}R^{\nabla}(\mathrm{i}\widetilde{X},\mathcal{H}\widetilde{Y},\mathcal{H}\widetilde{Z})\widetilde{U} \\
&= \nabla_{\mathrm{i}\widetilde{X}}(R^{\nabla}(\mathcal{H}\widetilde{Y},\mathcal{H}\widetilde{Z})\widetilde{U}) - \nabla_{\mathcal{H}\widetilde{Y}}(R^{\nabla}(\mathrm{i}\widetilde{X},\mathcal{H}\widetilde{Z})\widetilde{U}) + \nabla_{\mathcal{H}\widetilde{Z}}(R^{\nabla}(\mathrm{i}\widetilde{X},\mathcal{H}\widetilde{Y})\widetilde{U}) \\
&\quad - R^{\nabla}(\mathcal{H}\widetilde{Y},\mathcal{H}\widetilde{Z})\nabla_{\mathrm{i}\widetilde{X}}\widetilde{U} + R^{\nabla}(\mathrm{i}\widetilde{X},\mathcal{H}\widetilde{Z})\nabla_{\mathcal{H}\widetilde{Y}}\widetilde{U} - R^{\nabla}(\mathrm{i}\widetilde{X},\mathcal{H}\widetilde{Y})\nabla_{\mathcal{H}\widetilde{Z}}\widetilde{U} \\
&\quad - R^{\nabla}([\mathrm{i}\widetilde{X},\mathcal{H}\widetilde{Y}],\mathcal{H}\widetilde{Z})\widetilde{U} + R^{\nabla}([\mathrm{i}\widetilde{X},\mathcal{H}\widetilde{Z}],\mathcal{H}\widetilde{Y})\widetilde{U} - R^{\nabla}([\mathcal{H}\widetilde{Y},\mathcal{H}\widetilde{Z}],\mathrm{i}\widetilde{X})\widetilde{U} \\
&= \nabla_{\mathrm{i}\widetilde{X}}(\mathbf{H}(\widetilde{Y},\widetilde{Z})\widetilde{U}) - \nabla_{\mathcal{H}\widetilde{Y}}(\mathbf{B}(\widetilde{X},\widetilde{Z})\widetilde{U}) + \nabla_{\mathcal{H}\widetilde{Z}}(\mathbf{B}(\widetilde{X},\widetilde{Y})\widetilde{U}) \\
&\quad - \mathbf{H}(\widetilde{Y},\widetilde{Z})\nabla_{\mathrm{i}\widetilde{X}}\widetilde{U} + \mathbf{B}(\widetilde{X},\widetilde{Z})\nabla_{\mathcal{H}\widetilde{Y}}\widetilde{U} - \mathbf{B}(\widetilde{X},\widetilde{Y})\nabla_{\mathcal{H}\widetilde{Z}}\widetilde{U}
\end{aligned}$$

$$- \mathbf{H}(\nabla_{i\widetilde{X}}\widetilde{Y}, \widetilde{Z})\widetilde{U} + \mathbf{B}(\nabla_{\mathcal{H}\widetilde{Y}}\widetilde{X}, \widetilde{Z})\widetilde{U} + \mathbf{H}(\nabla_{i\widetilde{X}}\widetilde{Z}, \widetilde{Y})\widetilde{U} - \mathbf{B}(\nabla_{\mathcal{H}\widetilde{Z}}\widetilde{X}, \widetilde{Y})\widetilde{U}$$

$$+ \mathbf{B}(\widetilde{X}, \nabla_{\mathcal{H}\widetilde{Y}}\widetilde{Z})\widetilde{U} - \mathbf{B}(\widetilde{X}, \nabla_{\mathcal{H}\widetilde{Z}}\widetilde{Y})\widetilde{U} - \mathbf{B}(\widetilde{X}, \mathbf{T}(\widetilde{Y}, \widetilde{Z}))\widetilde{U}$$

$$= (\nabla_{i\widetilde{X}}\mathbf{H})(\widetilde{Y}, \widetilde{Z})\widetilde{U} - (\nabla_{\mathcal{H}\widetilde{Y}}\mathbf{B})(\widetilde{X}, \widetilde{Z})\widetilde{U} + (\nabla_{\mathcal{H}\widetilde{Z}}\mathbf{B})(\widetilde{X}, \widetilde{Y})\widetilde{U}$$

$$- \mathbf{B}(\widetilde{X}, \mathbf{T}(\widetilde{Y}, \widetilde{Z}))\widetilde{U},$$

thus proving (7.14.13). $\square$

Proposition 7.14.12 (h-differential Bianchi Identity for the Affine Curvature). *For any sections* $\widetilde{X}, \widetilde{Y}, \widetilde{Z}, \widetilde{U}$ *in* $\Gamma(\mathring{\pi})$, *we have*

$$\sum_{\mathrm{cyc}(\widetilde{X}, \widetilde{Y}, \widetilde{Z})} (\nabla^{\mathbf{h}}\mathbf{H})(\widetilde{X}, \widetilde{Y}, \widetilde{Z}, \widetilde{U})$$

$$= \sum_{\mathrm{cyc}(\widetilde{X}, \widetilde{Y}, \widetilde{Z})} (\mathbf{B}(\mathcal{R}(\widetilde{X}, \widetilde{Y}), \widetilde{Z})\widetilde{U} - \mathbf{H}(\mathbf{T}(\widetilde{X}, \widetilde{Y}), \widetilde{Z})\widetilde{U}). \quad (7.14.14)$$

Proof. Observe first that

$$R^{\nabla}([\mathcal{H}\widetilde{X}, \mathcal{H}\widetilde{Y}], \mathcal{H}\widetilde{Z})\widetilde{U} = R^{\nabla}(\mathcal{H}\mathbf{j}[\mathcal{H}\widetilde{X}, \mathcal{H}\widetilde{Y}], \mathcal{H}\widetilde{Z})\widetilde{U}$$

$$+ R^{\nabla}(\mathbf{i}\mathcal{V}[\mathcal{H}\widetilde{X}, \mathcal{H}\widetilde{Y}], \mathcal{H}\widetilde{Z})\widetilde{U} = \mathbf{H}(\mathbf{j}[\mathcal{H}\widetilde{X}, \mathcal{H}\widetilde{Y}], \widetilde{Z})\widetilde{U} + \mathbf{B}(\mathcal{R}(\widetilde{X}, \widetilde{Y}), \widetilde{Z})\widetilde{U}$$

$$= \mathbf{H}(\nabla_{\mathcal{H}\widetilde{X}}\widetilde{Y}, \widetilde{Z})\widetilde{U} - \mathbf{H}(\nabla_{\mathcal{H}\widetilde{Y}}\widetilde{X}, \widetilde{Z})\widetilde{U} - \mathbf{H}(\mathbf{T}(\widetilde{X}, \widetilde{Y}), \widetilde{Z})\widetilde{U}$$

$$+ \mathbf{B}(\mathcal{R}(\widetilde{X}, \widetilde{Y}), \widetilde{Z})\widetilde{U}.$$

Thus

$$0 = d^{\nabla}R^{\nabla}(\mathcal{H}\widetilde{X}, \mathcal{H}\widetilde{Y}, \mathcal{H}\widetilde{Z})\widetilde{U} = \nabla_{\mathcal{H}\widetilde{X}}(R^{\nabla}(\mathcal{H}\widetilde{Y}, \mathcal{H}\widetilde{Z})\widetilde{U})$$

$$+ \nabla_{\mathcal{H}\widetilde{Y}}(R^{\nabla}(\mathcal{H}\widetilde{Z}, \mathcal{H}\widetilde{X})\widetilde{U}) + \nabla_{\mathcal{H}\widetilde{Z}}(R^{\nabla}(\mathcal{H}\widetilde{X}, \mathcal{H}\widetilde{Y})\widetilde{U})$$

$$- R^{\nabla}(\mathcal{H}\widetilde{Y}, \mathcal{H}\widetilde{Z})\nabla_{\mathcal{H}\widetilde{X}}\widetilde{U} - R^{\nabla}(\mathcal{H}\widetilde{Z}, \mathcal{H}\widetilde{X})\nabla_{\mathcal{H}\widetilde{Y}}\widetilde{U} - R^{\nabla}(\mathcal{H}\widetilde{X}, \mathcal{H}\widetilde{Y})\nabla_{\mathcal{H}\widetilde{Z}}\widetilde{U}$$

$$- R^{\nabla}([\mathcal{H}\widetilde{X}, \mathcal{H}\widetilde{Y}], \mathcal{H}\widetilde{Z})\widetilde{U} - R^{\nabla}([\mathcal{H}\widetilde{Y}, \mathcal{H}\widetilde{Z}], \mathcal{H}\widetilde{X})\widetilde{U} - R^{\nabla}([\mathcal{H}\widetilde{Z}, \mathcal{H}\widetilde{X}], \mathcal{H}\widetilde{Y})\widetilde{U}$$

$$= \nabla_{\mathcal{H}\widetilde{X}}(\mathbf{H}(\widetilde{Y}, \widetilde{Z})\widetilde{U}) + \nabla_{\mathcal{H}\widetilde{Y}}(\mathbf{H}(\widetilde{Z}, \widetilde{X})\widetilde{U}) + \nabla_{\mathcal{H}\widetilde{Z}}(\mathbf{H}(\widetilde{X}, \widetilde{Y})\widetilde{U})$$

$$- \mathbf{H}(\widetilde{Y}, \widetilde{Z})\nabla_{\mathcal{H}\widetilde{X}}\widetilde{U} - \mathbf{H}(\widetilde{Z}, \widetilde{X})\nabla_{\mathcal{H}\widetilde{Y}}\widetilde{U} - \mathbf{H}(\widetilde{X}, \widetilde{Y})\nabla_{\mathcal{H}\widetilde{Z}}\widetilde{U}$$

$$- \mathbf{H}(\nabla_{\mathcal{H}\widetilde{X}}\widetilde{Y}, \widetilde{Z})\widetilde{U} + \mathbf{H}(\nabla_{\mathcal{H}\widetilde{Y}}\widetilde{X}, \widetilde{Z})\widetilde{U} + \mathbf{H}(\mathbf{T}(\widetilde{X}, \widetilde{Y}), \widetilde{Z})\widetilde{U}$$

$$- \mathbf{H}(\nabla_{\mathcal{H}\widetilde{Y}}\widetilde{Z}, \widetilde{X})\widetilde{U} + \mathbf{H}(\nabla_{\mathcal{H}\widetilde{Z}}\widetilde{Y}, \widetilde{X})\widetilde{U} + \mathbf{H}(\mathbf{T}(\widetilde{Y}, \widetilde{Z}), \widetilde{X})\widetilde{U}$$

$$- \mathbf{H}(\nabla_{\mathcal{H}\widetilde{Z}}\widetilde{X}, \widetilde{Y})\widetilde{U} + \mathbf{H}(\nabla_{\mathcal{H}\widetilde{X}}\widetilde{Z}, \widetilde{Y})\widetilde{U} + \mathbf{H}(\mathbf{T}(\widetilde{Z}, \widetilde{X}), \widetilde{Y})\widetilde{U}$$

$$- \mathbf{B}(\mathcal{R}(\widetilde{X}, \widetilde{Y}), \widetilde{Z})\widetilde{U} - \mathbf{B}(\mathcal{R}(\widetilde{Y}, \widetilde{Z}), \widetilde{X})\widetilde{U} - \mathbf{B}(\mathcal{R}(\widetilde{Z}, \widetilde{X}), \widetilde{Y})\widetilde{U}$$

$$= \sum_{\mathrm{cyc}(\widetilde{X}, \widetilde{Y}, \widetilde{Z})} (\nabla^{\mathbf{h}}\mathbf{H}(\widetilde{X}, \widetilde{Y}, \widetilde{Z}, \widetilde{U}) - \mathbf{B}(\mathcal{R}(\widetilde{X}, \widetilde{Y}), \widetilde{Z})\widetilde{U} + \mathbf{H}(\mathbf{T}(\widetilde{X}, \widetilde{Y}), \widetilde{Z})\widetilde{U}),$$

as asserted. $\square$

Proposition 7.14.13 (Horizontal Ricci Formula for Sections). *Let* $\widetilde{Z}$ *be a section in* $\Gamma(\mathring{\pi})$. *Then for any sections* $\widetilde{X}, \widetilde{Y} \in \Gamma(\mathring{\pi})$ *we have*

$$\nabla^h \nabla^h \widetilde{Z}(\widetilde{X}, \widetilde{Y}) - \nabla^h \nabla^h \widetilde{Z}(\widetilde{Y}, \widetilde{X})$$
$$= \mathbf{H}(\widetilde{X}, \widetilde{Y})\widetilde{Z} + (\nabla^v \widetilde{Z} \circ \mathcal{R} - \nabla^h \widetilde{Z} \circ \mathbf{T})(\widetilde{X}, \widetilde{Y}). \quad (7.14.15)$$

Proof.   Note first that

$$\mathcal{H}(\nabla_{\mathcal{H}\widetilde{X}}\widetilde{Y} - \nabla_{\mathcal{H}\widetilde{Y}}\widetilde{X}) = \mathcal{H}(\mathbf{T}(\widetilde{X}, \widetilde{Y}) + \mathbf{j}[\mathcal{H}\widetilde{X}, \mathcal{H}\widetilde{Y}])$$
$$= \mathcal{H}\mathbf{T}(\widetilde{X}, \widetilde{Y}) + \mathbf{h}[\mathcal{H}\widetilde{X}, \mathcal{H}\widetilde{Y}] = \mathcal{H}\mathbf{T}(\widetilde{X}, \widetilde{Y}) + [\mathcal{H}\widetilde{X}, \mathcal{H}\widetilde{Y}] - \mathbf{v}[\mathcal{H}\widetilde{X}, \mathcal{H}\widetilde{Y}],$$

whence

$$\mathcal{H}(\nabla_{\mathcal{H}\widetilde{X}}\widetilde{Y} - \nabla_{\mathcal{H}\widetilde{Y}}\widetilde{X}) = \mathcal{H}\mathbf{T}(\widetilde{X}, \widetilde{Y}) - \mathbf{i}\mathcal{R}(\widetilde{X}, \widetilde{Y}) + [\mathcal{H}\widetilde{X}, \mathcal{H}\widetilde{Y}]. \quad (7.14.16)$$

Now, on the one hand,

$$\nabla^h \nabla^h \widetilde{Z}(\widetilde{X}, \widetilde{Y}) = (\nabla_{\mathcal{H}\widetilde{X}}(\nabla^h \widetilde{Z}))(\widetilde{Y}) = \nabla_{\mathcal{H}\widetilde{X}}\nabla_{\mathcal{H}\widetilde{Y}}\widetilde{Z} - \nabla^h \widetilde{Z}(\nabla_{\mathcal{H}\widetilde{X}}\widetilde{Y})$$
$$= \nabla_{\mathcal{H}\widetilde{X}}\nabla_{\mathcal{H}\widetilde{Y}}\widetilde{Z} - \nabla_{\mathcal{H}\nabla_{\mathcal{H}\widetilde{X}}\widetilde{Y}}\widetilde{Z}.$$

On the other hand, similarly,

$$\nabla^h \nabla^h \widetilde{Z}(\widetilde{Y}, \widetilde{X}) = \nabla_{\mathcal{H}\widetilde{Y}}\nabla_{\mathcal{H}\widetilde{X}}\widetilde{Z} - \nabla_{\mathcal{H}\nabla_{\mathcal{H}\widetilde{Y}}\widetilde{X}}\widetilde{Z}.$$

Therefore,

$$\nabla^h \nabla^h \widetilde{Z}(\widetilde{X}, \widetilde{Y}) - \nabla^h \nabla^h \widetilde{Z}(\widetilde{Y}, \widetilde{X}) = \nabla_{\mathcal{H}\widetilde{X}}\nabla_{\mathcal{H}\widetilde{Y}}\widetilde{Z} - \nabla_{\mathcal{H}\widetilde{Y}}\nabla_{\mathcal{H}\widetilde{X}}\widetilde{Z}$$
$$- \nabla_{\mathcal{H}(\nabla_{\mathcal{H}\widetilde{X}}\widetilde{Y} - \nabla_{\mathcal{H}\widetilde{Y}}\widetilde{X})}\widetilde{Z} \overset{(7.14.16)}{=} \nabla_{\mathcal{H}\widetilde{X}}\nabla_{\mathcal{H}\widetilde{Y}}\widetilde{Z} - \nabla_{\mathcal{H}\widetilde{Y}}\nabla_{\mathcal{H}\widetilde{X}}\widetilde{Z}$$
$$- \nabla_{[\mathcal{H}\widetilde{X}, \mathcal{H}\widetilde{Y}]}\widetilde{Z} + \nabla_{\mathbf{i}\mathcal{R}(\widetilde{X}, \widetilde{Y})}\widetilde{Z} - \nabla_{\mathcal{H}\mathbf{T}(\widetilde{X}, \widetilde{Y})}\widetilde{Z}$$
$$= \mathbf{H}(\widetilde{X}, \widetilde{Y})\widetilde{Z} + \nabla^v_{\mathcal{R}(\widetilde{X}, \widetilde{Y})}\widetilde{Z} - \nabla^h_{\mathbf{T}(\widetilde{X}, \widetilde{Y})}\widetilde{Z},$$

whence our assertion.   □

Proposition 7.14.14 (Horizontal Ricci Formula for Finsler One-forms). *Let* $\widetilde{\alpha}$ *be a Finsler one-form on* $\mathring{T}M$. *Then for any sections* $\widetilde{X}, \widetilde{Y}, \widetilde{Z} \in \Gamma(\mathring{\pi})$ *we have*

$$\nabla^h \nabla^h \widetilde{\alpha}(\widetilde{X}, \widetilde{Y}, \widetilde{Z}) - \nabla^h \nabla^h \widetilde{\alpha}(\widetilde{Y}, \widetilde{X}, \widetilde{Z})$$
$$= -\widetilde{\alpha}(\mathbf{H}(\widetilde{X}, \widetilde{Y})\widetilde{Z}) + (\nabla^v_{\mathcal{R}(\widetilde{X}, \widetilde{Y})}\widetilde{\alpha} - \nabla^h_{\mathbf{T}(\widetilde{X}, \widetilde{Y})}\widetilde{\alpha})(\widetilde{Z}). \quad (7.14.17)$$

Proof. First we evaluate $\nabla^h\nabla^h\widetilde{\alpha}$ at $(\widetilde{X}, \widetilde{Y}, \widetilde{Z})$. We find that

$$\nabla^h\nabla^h\widetilde{\alpha}(\widetilde{X}, \widetilde{Y}, \widetilde{Z}) = (\nabla_{\mathcal{H}\widetilde{X}}(\nabla^h\widetilde{\alpha}))(\widetilde{Y}, \widetilde{Z}) = \nabla_{\mathcal{H}\widetilde{X}}(\nabla^h\widetilde{\alpha}(\widetilde{Y}, \widetilde{Z}))$$
$$- \nabla^h\widetilde{\alpha}(\nabla_{\mathcal{H}\widetilde{X}}\widetilde{Y}, \widetilde{Z}) - \nabla^h\widetilde{\alpha}(\widetilde{Y}, \nabla_{\mathcal{H}\widetilde{X}}\widetilde{Z}) = \nabla_{\mathcal{H}\widetilde{X}}((\nabla_{\mathcal{H}\widetilde{Y}}\widetilde{\alpha})(\widetilde{Z}))$$
$$- (\nabla_{\mathcal{H}\nabla_{\mathcal{H}\widetilde{X}}\widetilde{Y}}\widetilde{\alpha})(\widetilde{Z}) - (\nabla_{\mathcal{H}\widetilde{Y}}\widetilde{\alpha})(\nabla_{\mathcal{H}\widetilde{X}}\widetilde{Z})$$
$$= \mathcal{H}\widetilde{X}(\mathcal{H}\widetilde{Y}(\widetilde{\alpha}(\widetilde{Z}))) - \mathcal{H}\widetilde{X}(\widetilde{\alpha}(\nabla_{\mathcal{H}\widetilde{Y}}\widetilde{Z})) - \mathcal{H}\nabla_{\mathcal{H}\widetilde{X}}\widetilde{Y}(\widetilde{\alpha}(\widetilde{Z}))$$
$$+ \widetilde{\alpha}(\nabla_{\mathcal{H}\nabla_{\mathcal{H}\widetilde{X}}\widetilde{Y}}\widetilde{Z}) - \mathcal{H}\widetilde{Y}(\widetilde{\alpha}(\nabla_{\mathcal{H}\widetilde{X}}\widetilde{Z})) + \widetilde{\alpha}(\nabla_{\mathcal{H}\widetilde{Y}}\nabla_{\mathcal{H}\widetilde{X}}\widetilde{Z}).$$

Exchanging $\widetilde{X}$ and $\widetilde{Y}$,

$$\nabla^h\nabla^h\widetilde{\alpha}(\widetilde{Y}, \widetilde{X}, \widetilde{Z}) = \mathcal{H}\widetilde{Y}(\mathcal{H}\widetilde{X}(\widetilde{\alpha}(\widetilde{Z}))) - \mathcal{H}\widetilde{Y}(\widetilde{\alpha}(\nabla_{\mathcal{H}\widetilde{X}}\widetilde{Z})) - \mathcal{H}\nabla_{\mathcal{H}\widetilde{Y}}\widetilde{X}(\widetilde{\alpha}(\widetilde{Z}))$$
$$+ \widetilde{\alpha}(\nabla_{\mathcal{H}\nabla_{\mathcal{H}\widetilde{Y}}\widetilde{X}}\widetilde{Z}) - \mathcal{H}\widetilde{X}(\widetilde{\alpha}(\nabla_{\mathcal{H}\widetilde{Y}}\widetilde{Z})) + \widetilde{\alpha}(\nabla_{\mathcal{H}\widetilde{X}}\nabla_{\mathcal{H}\widetilde{Y}}\widetilde{Z}).$$

Finally, taking the difference of the two equalities, we get

$$\nabla^h\nabla^h\widetilde{\alpha}(\widetilde{X}, \widetilde{Y}, \widetilde{Z}) - \nabla^h\nabla^h\widetilde{\alpha}(\widetilde{Y}, \widetilde{X}, \widetilde{Z}) = [\mathcal{H}\widetilde{X}, \mathcal{H}\widetilde{Y}]\widetilde{\alpha}(\widetilde{Z})$$
$$- (\mathcal{H}\nabla_{\mathcal{H}\widetilde{X}}\widetilde{Y} - \mathcal{H}\nabla_{\mathcal{H}\widetilde{Y}}\widetilde{X})\widetilde{\alpha}(\widetilde{Z}) + \widetilde{\alpha}(\nabla_{\mathcal{H}\nabla_{\mathcal{H}\widetilde{X}}\widetilde{Y}}\widetilde{Z} - \nabla_{\mathcal{H}\nabla_{\mathcal{H}\widetilde{Y}}\widetilde{X}}\widetilde{Z})$$
$$- \widetilde{\alpha}(\nabla_{\mathcal{H}\widetilde{X}}\nabla_{\mathcal{H}\widetilde{Y}}\widetilde{Z} - \nabla_{\mathcal{H}\widetilde{Y}}\nabla_{\mathcal{H}\widetilde{X}}\widetilde{Z})$$
$$= (\nabla_{[\mathcal{H}\widetilde{X}, \mathcal{H}\widetilde{Y}]}\widetilde{\alpha})(\widetilde{Z}) - \widetilde{\alpha}(\nabla_{\mathcal{H}\widetilde{X}}\nabla_{\mathcal{H}\widetilde{Y}}\widetilde{Z} - \nabla_{\mathcal{H}\widetilde{Y}}\nabla_{\mathcal{H}\widetilde{X}}\widetilde{Z} - \nabla_{[\mathcal{H}\widetilde{X}, \mathcal{H}\widetilde{Y}]}\widetilde{Z})$$
$$- (\nabla_{\mathcal{H}(\nabla_{\mathcal{H}\widetilde{X}}\widetilde{Y} - \nabla_{\mathcal{H}\widetilde{Y}}\widetilde{X})}\widetilde{\alpha})(\widetilde{Z})$$
$$\stackrel{(7.14.16)}{=} -\widetilde{\alpha}(\mathbf{H}(\widetilde{X}, \widetilde{Y})\widetilde{Z}) + (\nabla_{i\mathcal{R}(\widetilde{X}, \widetilde{Y}) - \mathcal{H}\mathbf{T}(\widetilde{X}, \widetilde{Y})}\widetilde{\alpha})(\widetilde{Z})$$
$$= -\widetilde{\alpha}(\mathbf{H}(\widetilde{X}, \widetilde{Y})\widetilde{Z}) + (\nabla^v_{\mathcal{R}(\widetilde{X}, \widetilde{Y})}\widetilde{\alpha} - \nabla^h_{\mathbf{T}(\widetilde{X}, \widetilde{Y})}\widetilde{\alpha})(\widetilde{Z}),$$

as was to be shown. $\qquad\square$

7.15 Linear Ehresmann Connections Revisited

In this section we summarize some important and useful characterizations of linear Ehresmann connections.

Proposition 7.15.1. *Let $\mathcal{H}$ be a positive-homogeneous Ehresmann connection in TM. Then the following assertions are equivalent:*

(i) *$\mathcal{H}$ is linear.*

(ii) *With respect to any chart of M, the Christoffel symbols of $\mathcal{H}$ are fibrewise linear functions.*

(iii) *With respect to any chart of M, the Christoffel symbols of the Berwald derivative induced by $\mathcal{H}$ are vertical lifts.*

(iv) *The Berwald curvature of $\mathcal{H}$ vanishes.*

(v) *For any vector fields X, Y on M, the Lie bracket $[X^{\mathsf{h}}, Y^{\mathsf{v}}] \in \mathfrak{X}(TM)$ is a vertical lift.*

(vi) *There exists a unique covariant derivative D on M such that*

$$(D_X Y)^{\mathsf{v}} = [X^{\mathsf{h}}, Y^{\mathsf{v}}] \overset{(7.10.5)}{=} \mathbf{i} \nabla_{X^{\mathsf{h}}} \widehat{Y}.$$

(vii) *The $\mathcal{H}$-parallel translations along any smooth curve are linear mappings.*

Proof. Note first that assertion (i)$\Leftrightarrow$(vii) is just a restatement of Lemma 7.6.8, while the equivalence of (i) and (vi) follows from Proposition 7.5.11.

For our next calculations in local coordinates, we choose a chart $(\mathcal{U}, (u^i)_{i=1}^n)$ in M with induced chart $(\tau^{-1}(\mathcal{U}), ((x^i)_{i=1}^n, (y^i)_{i=1}^n))$ on TM. Let N_j^i and N_{jk}^i be the Christoffel symbols of $\mathcal{H}$ and the induced Berwald derivative ∇ with respect to the chosen chart. Then the components of the Berwald curvature $\mathbf{B}$ of $\mathcal{H}$ are

$$B_{jkl}^i \overset{(7.13.4)}{=} \frac{\partial N_{jk}^i}{\partial y^l}; \quad i, j, k, l \in J_n.$$

If the Christoffel symbols N_{jk}^i are vertical lifts, then $B_{jkl}^i = \frac{\partial N_{jk}^i}{\partial y^l} = 0$ by Lemma 4.1.28(iii). Thus (iii) implies (iv).

Now suppose that $\mathbf{B} = 0$. Then, for all $X, Y, Z \in \mathfrak{X}(M)$,

$$0 = \mathbf{B}(\widehat{Z}, \widehat{X})\widehat{Y} \overset{(7.13.3)}{=} [Z^{\mathsf{v}}, [X^{\mathsf{h}}, Y^{\mathsf{v}}]]. \qquad (*)$$

Here $[X^{\mathsf{h}}, Y^{\mathsf{v}}]$ is a vertical vector field by (7.1.5)(ii),(c), thus $\mathbf{J}[X^{\mathsf{h}}, Y^{\mathsf{v}}] = 0$. We show that $[\mathbf{J}, [X^{\mathsf{h}}, Y^{\mathsf{v}}]] = 0$ also holds; then from Lemma 4.1.60(i) we infer that $[X^{\mathsf{h}}, Y^{\mathsf{v}}]$ is a vertical lift. To do this, we take an arbitrary vector field Z on M, and evaluate the endomorphism $[\mathbf{J}, [X^{\mathsf{h}}, Y^{\mathsf{v}}]]$ at Z^{v} and Z^{c}. We find that

$$[\mathbf{J}, [X^{\mathsf{h}}, Y^{\mathsf{v}}]]Z^{\mathsf{v}} \overset{(3.3.16)}{=} -\mathbf{J}[Z^{\mathsf{v}}, [X^{\mathsf{h}}, Y^{\mathsf{v}}]] \overset{(*)}{=} 0,$$

$$[\mathbf{J}, [X^{\mathsf{h}}, Y^{\mathsf{v}}]]Z^{\mathsf{c}} \overset{(3.3.16)}{=} [Z^{\mathsf{v}}, [X^{\mathsf{h}}, Y^{\mathsf{v}}]] - \mathbf{J}[Z^{\mathsf{c}}, [X^{\mathsf{h}}, Y^{\mathsf{v}}]]$$

$$\overset{(*)}{=} -\mathbf{J}[Z^{\mathsf{c}}, [X^{\mathsf{h}}, Y^{\mathsf{v}}]] = 0,$$

since $Z^{\mathsf{c}} \underset{\tau}{\sim} Z$ and $[X^{\mathsf{h}}, Y^{\mathsf{v}}] \underset{\tau}{\sim} o$ imply by the related vector field lemma that $[Z^{\mathsf{c}}, [X^{\mathsf{h}}, Y^{\mathsf{v}}]]$ is vertical. Thus $[X^{\mathsf{h}}, Y^{\mathsf{v}}]$ is indeed a vertical lift, so (iv) implies (v). To close a 'subcircle' of implications, we prove that (v) implies (iii).

If $[X^{\mathsf{h}}, Y^{\mathsf{v}}]$ is a vertical lift for all $X, Y \in \mathfrak{X}(M)$, then, in particular, the Lie brackets

$$\left[\left(\frac{\partial}{\partial u^j}\right)^{\mathsf{h}}, \left(\frac{\partial}{\partial u^k}\right)^{\mathsf{v}}\right] \overset{(7.10.5)}{=} \mathrm{i}\nabla_{\left(\frac{\partial}{\partial u^j}\right)^{\mathsf{h}}} \widehat{\frac{\partial}{\partial u^k}} \overset{(7.11.2)}{=} N^i_{jk} \frac{\partial}{\partial y^i}$$

are vertical lifts for all $j, k \in J_n$. Hence, taking into account (4.1.48), the Christoffel symbols N^i_{jk} of ∇ are vertical lifts. Thus (v) implies (iii). Summarizing this part of the proof, we have

$$\text{(iii)} \quad \Rightarrow \quad \text{(iv)}$$
$$\nwarrow \quad \nearrow \quad .$$
$$\text{(v)}$$

We know from Lemma 7.5.2 that the linearity of an Ehresmann connection manifests in the fibrewise linearity of its Christoffel symbols, so (ii) just rephrases (i). In the next step, we suppose first that (ii) is satisfied. Then, in particular, the Christoffel symbols N^i_j are 1^+-homogeneous, therefore, by Lemma 4.2.4, the Christoffel symbols

$$N^i_{jk} \overset{(7.11.1)}{=} \frac{\partial N^i_j}{\partial y^k}$$

of ∇ are 0^+-homogeneous. By the fibrewise linearity, the functions N^i_{jk} are smooth on their whole domain $\tau^{-1}(\mathcal{U})$. So, from Lemma 4.2.10(i), we may conclude that the N^i_{jk}'s are fibrewise constant and hence vertical lifts. Thus (ii) implies (iii). To complete the proof, we show that the converse implication (iii)$\Rightarrow$(ii) is also true.

Assuming (iii), we define an Ehresmann connection $\bar{\mathcal{H}}$ in $\tau^{-1}(\mathcal{U}) \subset TM$, by giving its Christoffel symbols as follows:

$$\bar{N}^i_j := y^k N^i_{jk}; \quad i, j \in J_n.$$

Then $\bar{\mathcal{H}}$ is clearly a linear Ehresmann connection, and the Christoffel symbols of the Berwald derivative induced by $\bar{\mathcal{H}}$ are

$$\bar{N}^i_{jk} = \frac{\partial \bar{N}^i_j}{\partial y^k} = N^i_{jk}; \quad i, j, k \in J_n.$$

Since both $\mathcal{H}$ and $\bar{\mathcal{H}}$ are positive homogeneous Ehresmann connections, the equality of the Christoffel symbols of the induced Berwald derivatives implies by Proposition 7.10.8 that $\mathcal{H}$ and $\bar{\mathcal{H}}$ are also equal over $\tau^{-1}(\mathcal{U})$. Therefore, the Christoffel symbols of $\mathcal{H}$ (with respect to the chosen chart) are fibrewise linear functions. Thus we have obtained the following closed chain of implications:

$$\text{(vii)} \Leftrightarrow \text{(i)} \Leftrightarrow \text{(ii)} \Leftrightarrow \text{(iii)} \quad \Rightarrow \quad \text{(iv)}$$
$$\Updownarrow \qquad\qquad \nwarrow \quad \nearrow \quad .$$
$$\text{(vi)} \qquad\qquad\quad \text{(v)} \qquad\qquad \square$$

Remark 7.15.2. Keeping the notation of the previous proposition, we recall that ∇ stands for the Berwald derivative induced by $\mathcal{H}$. It will be useful to note that conditions (iii) and (vii) can be reformulated as follows:

(iii') There exists a covariant derivative D on M such that the Christoffel symbols of ∇ with respect to a chart on M are the vertical lifts of the Christoffel symbols of D with respect to the same chart.

(vii') There exists a covariant derivative D on M such that the $\mathcal{H}$-parallel translations are precisely the parallel translations with respect to D.

Indeed, the equivalence of (iii') and (iii) is clear from Remark 7.5.12, while the equivalence of (vii') and (vii) can easily be seen applying Corollary 7.6.6.

Lemma 7.15.3. *Let* (M, D) *be an affinely connected manifold,* T *and* R *the torsion and curvature of* D*. If* $\mathcal{H}$ *is the linear Ehresmann connection induced by* D*, then the torsion* $\mathbf{T}$ *and the affine curvature* $\mathbf{H}$ *of* $\mathcal{H}$ *are related to* T *and* R *by*

$$\mathbf{iT}(\widehat{X}, \widehat{Y}) = (T(X, Y))^{\mathsf{v}}, \quad X, Y \in \mathfrak{X}(M); \tag{7.15.1}$$

$$\mathbf{iH}(\widehat{X}, \widehat{Y})\widehat{Z} = (R(X, Y)Z)^{\mathsf{v}}, \quad X, Y, Z \in \mathfrak{X}(M). \tag{7.15.2}$$

Proof. The first relation is an immediate consequence of (7.1.15) and (7.8.4). As for the second formula, we saw in the proof of Lemma 7.14.1 that $\mathbf{H}(\widehat{X}, \widehat{Y})\widehat{X} = \mathcal{V}[\mathbf{i}\mathfrak{R}(\widehat{X}, \widehat{Y}), Z^{\mathsf{v}}]$, from which we obtain

$$\mathbf{iH}(\widehat{X}, \widehat{Y})\widehat{Z} = \mathbf{v}[\mathbf{i}\mathfrak{R}(\widehat{X}, \widehat{Y}), Z^{\mathsf{v}}] = [\mathbf{i}\mathfrak{R}(\widehat{X}, \widehat{Y}), Z^{\mathsf{v}}]$$

$$\overset{(7.8.3)}{=} [\widehat{R}(X, Y), Z^{\mathsf{v}}] \overset{(7.1.14)}{=} (R(X, Y)Z)^{\mathsf{v}},$$

as claimed. □

Proposition 7.15.4. *Let* $\mathcal{H}$ *be an Ehresmann connection in* TM*. Then the following are equivalent:*

(i) $\mathcal{H}$ *is linear and torsion-free.*

(ii) *The linear deviation of* $\mathcal{H}$ *vanishes.*

(iii) $\mathcal{H}$ *is generated by an affine spray.*

(iv) $\mathcal{H}$ *is induced by a uniquely determined torsion-free covariant derivative on* M*.*

Proof. Taking into account (7.15.1), the equivalence (i)⇔(iv) is an immediate consequence of the previous proposition, while the equivalence (ii)⇔(iv) is just a reformulation of Proposition 7.5.17. Thus our only task is to show that (i) is equivalent to (iii).

Suppose first that $\mathcal{H}$ is linear and torsion-free. Then the tension of $\mathcal{H}$ automatically vanishes, so from Corollary 7.8.9 and Proposition 7.8.8 we conclude that $\mathcal{H}$ is generated by its own associated semispray $S := \mathcal{H}\widetilde{\delta}$. The semispray S is in fact an affine spray. Indeed, by Corollary 7.5.3, $\mathcal{H}$ is smooth on the whole TM, so S also has the same smoothness property. Moreover, by Corollary 7.5.6, S is 2^+-homogeneous. Thus we have shown that (i) implies (iii).

Conversely, suppose that $\mathcal{H}$ is generated by an affine spray S. Then, by Proposition 7.8.6, $\mathcal{H}$ is torsion-free. To show the linearity of $\mathcal{H}$, choose a chart $(\mathcal{U}, (u^i)_{i=1}^n)$ on M with induced chart $(\tau^{-1}(\mathcal{U}), (x^i), (y^i))$ on TM. From (5.1.11), the coefficients of S with respect to the chosen chart are of the form

$$G^i = \frac{1}{2} y^k y^l (\Gamma^i_{kl} \circ \tau), \quad i \in J_n,$$

where the Γ^i_{jk}'s are smooth functions on $\mathcal{U}$. In the proof of Corollary 5.1.17, we obtained that $\frac{\partial G^i}{\partial y^j} = y^k(\Gamma^i_{jk} \circ \tau)$. Then the Christoffel symbols of $\mathcal{H}$ are

$$G^i_j \overset{(7.3.9)}{:=} \frac{\partial G^i}{\partial y^j} = y^k(\Gamma^i_{jk} \circ \tau).$$

From this expression we see immediately that the functions G^i_j are fibrewise linear. Thus, by Lemma 7.5.2, the Ehresmann connection $\mathcal{H}$ is linear, and this concludes the proof. $\qquad\square$

Chapter 8

Geometry of Spray Manifolds

8.1 The Berwald Connection and Related Constructions

8.1.1 *The Berwald Connection*

We recall that a *spray manifold* is a manifold equipped with a spray, i.e., a pair (M, S) where M is a manifold and $S \colon TM \to TTM$ is a mapping satisfying conditions (S_1)–(S_5) of Definition 5.1.18. The Ehresmann connection $\mathcal{H}$ generated by S according to the Crampin–Grifone theorem (Proposition 7.3.4) is called the *Berwald connection* of (M, S). Thus $\mathcal{H}$ is the unique Ehresmann connection in $\mathring{T}M$ such that

$$\mathbf{h} = \mathcal{H} \circ \mathbf{j} = \frac{1}{2}(1_{\mathfrak{X}(\mathring{T}M)} + [\mathbf{J}, S]).$$

The Berwald connection in $\mathring{T}M$ is *torsion-free* by Proposition 7.8.6 and *positive-homogeneous*. The latter property was shown (at least implicitly) in the proof of the Crampin–Grifone theorem, using coordinate calculations. Now we present a simple, direct argument.

For any vector field X on M we have

$$2[X^{\mathsf{h}}, C] \overset{(7.3.7)}{=} [X^{\mathsf{c}}, C] + [[X^{\mathsf{v}}, S], C] \overset{(4.1.72)}{=} [[X^{\mathsf{v}}, S], C] \overset{\text{Jacobi}}{=} -[[S, C], X^{\mathsf{v}}]$$

$$- [[C, X^{\mathsf{v}}], S] = [S, X^{\mathsf{v}}] + [S, [C, X^{\mathsf{v}}]] \overset{(4.1.53)}{=} [S, X^{\mathsf{v}}] - [S, X^{\mathsf{v}}] = 0,$$

which proves, by Proposition 7.5.8, that $\mathcal{H}$ is positive-homogeneous.

In view of Corollary 7.3.6, *the spray S is horizontal with respect to the Berwald connection of (M, S)*, so we have

$$\mathbf{h}S = S, \quad \mathbf{v}S = 0, \quad \mathcal{V}S = 0, \quad \mathbf{F}S = -C. \tag{8.1.1}$$

To close this brief subsection, we mention that for the horizontal lift of a vector field with respect to the Berwald connection the following counterpart of Grifone's identity (5.1.5) holds:

$$\mathbf{J}[X^{\mathsf{h}}, S] = X^{\mathsf{h}} - X^{\mathsf{c}}, \quad X \in \mathfrak{X}(M). \tag{8.1.2}$$

To see this, observe first that

$$[\mathbf{J}, S]X^{\mathsf{h}} \overset{(3.3.16)}{=} [\mathbf{J}X^{\mathsf{h}}, S] - \mathbf{J}[X^{\mathsf{h}}, S] \overset{(7.2.33)}{=} [X^{\mathsf{v}}, S] - \mathbf{J}[X^{\mathsf{h}}, S].$$

Thus

$$\mathbf{J}[X^{\mathsf{h}}, S] = [X^{\mathsf{v}}, S] - [\mathbf{J}, S]X^{\mathsf{h}} \overset{(7.3.7), (7.3.6)}{=} 2X^{\mathsf{h}} - X^{\mathsf{c}} - (2X^{\mathsf{h}} - X^{\mathsf{h}})$$
$$= X^{\mathsf{h}} - X^{\mathsf{c}},$$

as we claimed.

8.1.2 *The Induced Berwald Derivative*

Suppose that (M, S) is a spray manifold, and let $\mathcal{H}$ be the Berwald connection of (M, S). As we agreed in section 7.10, we denote by ∇ the Berwald derivative induced by $\mathcal{H}$. We recall the following rules of calculation:

$$\nabla_{\mathbf{i}\widetilde{X}}\widetilde{Y} = \mathbf{j}[\mathbf{i}\widetilde{X}, \mathcal{H}\widetilde{Y}], \quad \nabla_{\mathcal{H}\widetilde{X}}\widetilde{Y} = \mathcal{V}[\mathcal{H}\widetilde{X}, \mathbf{i}\widetilde{Y}]; \quad \widetilde{X}, \widetilde{Y} \in \Gamma(\mathring{\pi}).$$

We have, in particular,

$$\mathbf{i}\nabla_S\widehat{X} = X^{\mathsf{c}} - X^{\mathsf{h}}, \text{ for all } X \in \mathfrak{X}(M). \tag{8.1.3}$$

Indeed, since S is horizontal,

$$\mathbf{i}\nabla_S\widehat{X} = \mathbf{v}[S, X^{\mathsf{v}}] \overset{(7.3.7)}{=} \mathbf{v}(X^{\mathsf{c}} - 2X^{\mathsf{h}}) = \mathbf{v}X^{\mathsf{c}} = X^{\mathsf{c}} - \mathbf{h}X^{\mathsf{c}} = X^{\mathsf{c}} - X^{\mathsf{h}}.$$

By the vanishing of the tension of $\mathcal{H}$, (7.12.1) implies that

$$\nabla_{\mathcal{H}\widetilde{X}}\widetilde{\delta} = 0 \text{ for all } \widetilde{X} \in \Gamma(\mathring{\pi}). \tag{8.1.4}$$

Since $\mathcal{H}$ is determined by S, we also say that ∇ is the *Berwald derivative induced by the spray* S. As the next simple observation shows, there is no other spray over M that induces the same covariant derivative on $\Gamma(\mathring{\pi})$.

Lemma 8.1.1. *If two sprays over a manifold induce the same Berwald derivative, then the sprays are identical.*

Proof. Let S and $\bar{S}$ be sprays of M, and suppose that the Berwald derivatives induced by them coincide. Then, by Proposition 7.10.8, the Berwald connections induced by S and $\bar{S}$ are identical, so $[\mathbf{J}, S - \bar{S}] = 0$. Since $S - \bar{S}$ is vertical, it follows from Lemma 4.1.60(i) that $S - \bar{S}$ is actually a vertical lift. Thus $S - \bar{S}$ is 0^+-homogeneous. On the other hand, since S and $\bar{S}$ are sprays, $[C, S - \bar{S}] = [C, S] - [C, \bar{S}] = S - \bar{S}$, so $S - \bar{S}$ is also 2^+-homogeneous. This is possible only if $S - \bar{S} = 0$. $\qquad\square$

Throughout the rest of this chapter, we assume that (M, S) is a spray manifold, $\mathcal{H}$ is the Berwald connection of (M, S), and ∇ is the Berwald derivative induced by S.

8.1.3   *Torsion and Curvature*

By Proposition 7.8.6, the torsion of the Berwald connection $\mathcal{H}$ vanishes. Thus (see, e.g., Remark 7.9.2(b)) the torsion T^∇ of the Berwald derivative ∇ also vanishes. So we have

(i) $T^\nabla(\xi, \eta) \overset{(6.2.3)}{=} \nabla_\xi \mathbf{j}\eta - \nabla_\eta \mathbf{j}\xi - \mathbf{j}[\xi, \eta] = 0$

or

(ii) $T^\nabla(\mathcal{H}\widetilde{X}, \mathcal{H}\widetilde{Y}) = \mathfrak{T}(\widetilde{X}, \widetilde{Y}) = \nabla_{\mathcal{H}\widetilde{X}}\widetilde{Y} - \nabla_{\mathcal{H}\widetilde{Y}}\widetilde{X} - \mathbf{j}[\mathcal{H}\widetilde{X}, \mathcal{H}\widetilde{Y}] = 0.$

The *Berwald curvature* $\mathbf{B}$ of ∇, defined by

$$\mathbf{B}(\widetilde{X}, \widetilde{Y})\widetilde{Z} := R^\nabla(\mathbf{i}\widetilde{X}, \mathcal{H}\widetilde{Y})\widetilde{Z} = \nabla_{\mathbf{i}\widetilde{X}}\nabla_{\mathcal{H}\widetilde{Y}}\widetilde{Z} - \nabla_{\mathcal{H}\widetilde{Y}}\nabla_{\mathbf{i}\widetilde{X}}\widetilde{Z} - \nabla_{[\mathbf{i}\widetilde{X}, \mathcal{H}\widetilde{Y}]}\widetilde{Z},$$

has the following properties:

(iii) it is *totally symmetric*, i.e., $\mathbf{B}(\widetilde{X}_{\sigma(1)}, \widetilde{X}_{\sigma(2)})\widetilde{X}_{\sigma(3)} = \mathbf{B}(\widetilde{X}_1, \widetilde{X}_2)\widetilde{X}_3$ for all $\sigma \in S_3$ (Corollary 7.13.4);
(iv) it is $(-1)^+$-homogeneous, i.e., $\nabla^{\mathsf{v}}_{\widetilde{\delta}}\mathbf{B} = \nabla_C \mathbf{B} = -\mathbf{B}$ (Lemma 7.13.8);
(v) $\widetilde{\delta} \in \{\widetilde{X}, \widetilde{Y}, \widetilde{Z}\}$ implies $\mathbf{B}(\widetilde{X}, \widetilde{Y})\widetilde{Z} = 0$ (Corollary 7.13.7).

The *affine curvature* $\mathbf{H}$ of ∇, defined by

$$\mathbf{H}(\widetilde{X}, \widetilde{Y})\widetilde{Z} := R^\nabla(\mathcal{H}\widetilde{X}, \mathcal{H}\widetilde{Y})\widetilde{Z} = \nabla_{\mathcal{H}\widetilde{X}}\nabla_{\mathcal{H}\widetilde{Y}}\widetilde{Z} - \nabla_{\mathcal{H}\widetilde{Y}}\nabla_{\mathcal{H}\widetilde{X}}\widetilde{Z} - \nabla_{[\mathcal{H}\widetilde{X}, \mathcal{H}\widetilde{Y}]}\widetilde{Z},$$

can also be obtained as the vertical differential of the curvature of $\mathcal{H}$, more precisely,

(vi) $\mathbf{H}(\widetilde{X}, \widetilde{Y})\widetilde{Z} = -\nabla^{\mathsf{v}}\mathfrak{R}(\widetilde{Z}, \widetilde{X}, \widetilde{Y})$ (Lemma 7.14.1),

and

(vii) $\mathbf{H}$ is 0^+-homogeneous, i.e., $\nabla^{\mathsf{v}}_{\widetilde{\delta}}\mathbf{H} = \nabla_C \mathbf{H} = 0$ (Corollary 7.14.8).

Owing to the homogeneity of $\mathcal{H}$, the curvature of the Berwald connection can be reconstructed form the affine curvature:

(viii) $\mathfrak{R}(\widetilde{X}, \widetilde{Y}) = -\mathbf{H}(\widetilde{X}, \widetilde{Y})\widetilde{\delta}$ (by Corollary 7.14.6).

In the following, we shall also say that $\mathbf{B}$, $\mathfrak{R}$ and $\mathbf{H}$ are the *Berwald curvature*, the *fundamental affine curvature* and the *affine curvature of the spray manifold* (M, S) *(or of the spray S)*, respectively. If the curvature tensor R^∇ of the Berwald derivative induced by S vanishes, then (M, S) (or S) is said to be *flat*. Obviously, this holds if, and only if, both the Berwald curvature and the affine curvature of (M, S) vanish.

The vanishing of the torsion of the Berwald connection implies that some of the Ricci formulae and the Bianchi identities deduced earlier take a simpler form. For convenience of reference we collect these important equalities below.

RICCI FORMULAE IN (M, S)

(cf. (7.12.6), (7.13.8), (7.14.15), (7.14.17))

(R1) $\nabla^h \nabla^h F(\widetilde{X}, \widetilde{Y}) - \nabla^h \nabla^h F(\widetilde{Y}, \widetilde{X}) = \mathbf{i}\, \mathcal{R}(\widetilde{X}, \widetilde{Y})F,$

(R2) $\nabla^v \nabla^h \widetilde{Z}(\widetilde{X}, \widetilde{Y}) - \nabla^h \nabla^v \widetilde{Z}(\widetilde{Y}, \widetilde{X}) = \mathbf{B}(\widetilde{X}, \widetilde{Y})\widetilde{Z},$

(R3) $\nabla^h \nabla^h \widetilde{Z}(\widetilde{X}, \widetilde{Y}) - \nabla^h \nabla^h \widetilde{Z}(\widetilde{Y}, \widetilde{X}) = \mathbf{H}(\widetilde{X}, \widetilde{Y})\widetilde{Z} + \nabla^v \widetilde{Z}(\mathcal{R}(\widetilde{X}, \widetilde{Y})),$

(R4) $\nabla^h \nabla^h \widetilde{\alpha}(\widetilde{X}, \widetilde{Y}, \widetilde{Z}) - \nabla^h \nabla^h \widetilde{\alpha}(\widetilde{Y}, \widetilde{X}, \widetilde{Z})$

$$= -\widetilde{\alpha}(\mathbf{H}(\widetilde{X}, \widetilde{Y})\widetilde{Z}) + (\nabla^v_{\mathcal{R}(\widetilde{X}, \widetilde{Y})} \widetilde{\alpha})(\widetilde{Z}).$$

BIANCHI IDENTITIES IN (M, S)

(cf. (7.13.11), (7.14.12), (7.14.13), (7.14.14))

(B1) $(\nabla_{\mathbf{i}\widetilde{X}} \mathbf{B})(\widetilde{Y}, \widetilde{Z}, \widetilde{U}) = (\nabla_{\mathbf{i}\widetilde{Y}} \mathbf{B})(\widetilde{X}, \widetilde{Z}, \widetilde{U}),$

(B2) $\sum_{\mathrm{cyc}} \mathbf{H}(\widetilde{X}, \widetilde{Y})\widetilde{Z} = 0,$

(B3) $\nabla^v \mathbf{H}(\widetilde{X}, \widetilde{Y}, \widetilde{Z}, \widetilde{U}) = \nabla^h \mathbf{B}(\widetilde{Y}, \widetilde{Z}, \widetilde{X}, \widetilde{U}) - \nabla^h \mathbf{B}(\widetilde{Z}, \widetilde{Y}, \widetilde{X}, \widetilde{U}),$

(B4) $\displaystyle\sum_{\mathrm{cyc}(\widetilde{X}, \widetilde{Y}, \widetilde{Z})} (\nabla^h \mathbf{H})(\widetilde{X}, \widetilde{Y}, \widetilde{Z}, \widetilde{U}) - \sum_{\mathrm{cyc}(\widetilde{X}, \widetilde{Y}, \widetilde{Z})} \mathbf{B}(\mathcal{R}(\widetilde{X}, \widetilde{Y}), \widetilde{Z})\widetilde{U} = 0.$

For the fundamental affine curvature of (M, S) we have the following *h-differential* and *v-differential Bianchi identities*:

(B5) $\sum_{\mathrm{cyc}} \nabla^h \mathcal{R}(\widetilde{X}, \widetilde{Y}, \widetilde{Z}) = 0,$

(B6) $\sum_{\mathrm{cyc}} \nabla^v \mathcal{R}(\widetilde{X}, \widetilde{Y}, \widetilde{Z}) = 0.$

Indeed, **(B5)** is an immediate consequence of (7.12.8) by the vanishing of the torsion, while **(B6)** follows from **(B2)** and (vi).

8.1.4 *Coordinate Description*

Choose a chart $(\mathcal{U}, (u^i)_{i=1}^n)$ on M, and consider, as always in similar situations, the induced chart $(\tau^{-1}(\mathcal{U}), ((x^i)_{i=1}^n, (y^i)_{i=1}^n))$ on TM.

(a) According to Remark 5.1.8, the spray S can locally be represented in the form

$$S \underset{(\mathcal{U})}{=} y^i \frac{\partial}{\partial x^i} - 2G^i \frac{\partial}{\partial y^i}.$$

(Here, and in the following, summation convention is assumed.) The *spray coefficients* G^i are smooth on $\overset{\circ}{\tau}{}^{-1}(\mathcal{U}) = \tau^{-1}(\mathcal{U}) \cap \overset{\circ}{T}M$ and of class C^1 on $\tau^{-1}(\mathcal{U})$. The local manifestation of the 2^+-homogeneity of S is the relation

$$y^j \frac{\partial G^i}{\partial y^j} = 2G^i, \quad i \in J_n. \tag{8.1.5}$$

From this it follows immediately that

$$G^i(0_p) = 0 \text{ for all } p \in \mathcal{U}.$$

(b) We introduce the functions

$$G^i_j := \frac{\partial G^i}{\partial y^j}, \quad G^i_{jk} := \frac{\partial G^i_j}{\partial y^k}, \quad G^i_{jkl} := \frac{\partial G^i_{jk}}{\partial y^l}, \dots. \tag{8.1.6}$$

The functions G^i_j are smooth on $\overset{\circ}{\tau}{}^{-1}(\mathcal{U})$, but they are merely continuous on $\tau^{-1}(\mathcal{U})$ in general. So the (smooth) functions $G^i_{jk}, G^i_{jkl}, \dots$ are defined only on $\overset{\circ}{\tau}{}^{-1}(\mathcal{U})$. Obviously, they are symmetric in the lower indices. Taking into account, for example, Lemma 6.2.22, it follows that $G^i_j, G^i_{jk}, G^i_{jkl}, \dots$ are positive-homogeneous of degree $1, 0, -1, \dots$ respectively. Therefore we have

$$y^k G^i_{jk} = G^i_j, \quad y^l G^i_{jkl} = 0, \quad y^m G^i_{jklm} = -G^i_{jkl}. \tag{8.1.7}$$

(c) We learnt in Remark 7.3.5 that the Christoffel symbols of the *Berwald connection* $\mathcal{H}$ with respect to the chosen chart are the functions G^i_j. The vanishing of the torsion of $\mathcal{H}$ is reflected in the symmetry $G^i_{jk} = G^i_{kj}$, cf. (7.8.10). The first relation in (8.1.7) can be regarded as the expression of the vanishing of the strong torsion of $\mathcal{H}$ in terms of the spray coefficients, cf. (7.8.11). For the components of the curvature of $\mathcal{H}$ we have

$$i\mathcal{R}\left(\frac{\widehat{\partial}}{\partial u^j}, \frac{\widehat{\partial}}{\partial u^k}\right) = R^i_{jk} \frac{\partial}{\partial y^i}, \quad R^i_{jk} = \frac{\partial G^i_j}{\partial x^k} - \frac{\partial G^i_k}{\partial x^j} + G^l_j G^i_{kl} - G^l_k G^i_{jl}. \tag{8.1.8}$$

The Ricci formula **(R1)** takes the local form

$$F_{;j;k} - F_{;k;j} = R^i_{kj} F_{,i}. \tag{8.1.9}$$

We recall that

$$F_{;j} \overset{(7.11.8)}{=} \frac{\partial F}{\partial x^j} - G^i_j \frac{\partial F}{\partial y^i} = \frac{\partial F}{\partial x^j} - \frac{\partial G^i}{\partial y^j} \frac{\partial F}{\partial y^i},$$

and

$$\widetilde{X}^i_{;j} \overset{(7.11.9)}{=} \frac{\partial \widetilde{X}^i}{\partial x^j} - G^k_j \frac{\partial \widetilde{X}^i}{\partial y^k} + G^i_{jk} \widetilde{X}^k,$$

if $\widetilde{X} \in \Gamma(\overset{\circ}{\pi})$, $\widetilde{X} \underset{(\mathcal{U})}{=} \widetilde{X}^i \widehat{\frac{\partial}{\partial u^i}}$; cf. formula (3.1) in Berwald's paper [16]. As
we saw in Example 7.11.1, (G^i_{jk}) is the family of Christoffel symbols of the
Berwald derivative ∇ with respect to the chart $(\mathcal{U}, (u^i)^n_{i=1})$.

(d) By Remark 7.13.5, the functions G^i_{jkl} are the components of the
Berwald curvature of (M, S). So we have

$$\mathbf{B}\left(\widehat{\frac{\partial}{\partial u^j}}, \widehat{\frac{\partial}{\partial u^k}}\right)\widehat{\frac{\partial}{\partial u^l}} = R^\nabla\left(\frac{\partial}{\partial y^j}, \left(\frac{\partial}{\partial u^k}\right)^{\mathsf{h}}\right)\widehat{\frac{\partial}{\partial u^l}} = G^i_{jkl}\widehat{\frac{\partial}{\partial u^i}}. \tag{8.1.10}$$

The second and the third equalities in (8.1.7) are just the coordinate expressions of (7.13.6) and (7.13.7), respectively. The identities **(R2)** and
(B1) take the forms

$$\widetilde{Z}^i_{;j,k} - \widetilde{Z}^i_{,k;j} = G^i_{jkl}\widetilde{Z}^l \tag{8.1.11}$$

and

$$G^i_{jkl,m} = G^i_{jkm,l}; \tag{8.1.12}$$

cf. (7.13.10) and (7.13.12), respectively.

(e) By (7.14.7), the components of the *affine curvature* of (M, S) are
given by

$$H^i_{jkl} = \frac{\partial G^i_{kl}}{\partial x^j} - \frac{\partial G^i_{jl}}{\partial x^k} + G^i_{jr}G^r_{kl} - G^i_{kr}G^r_{jl} + G^r_k G^i_{jlr} - G^r_j G^i_{klr}.$$

The local expression for the Ricci formula **(R3)** and the Bianchi identities
(B2)–**(B4)** are the following:

$$\widetilde{Z}^i_{;j,k} - \widetilde{Z}^i_{;k;j} = H^i_{kjl}\widetilde{Z}^l + R^l_{kj}\widetilde{Z}^i_{,l}; \tag{8.1.13}$$

$$H^i_{jkl} + H^i_{klj} + H^i_{ljk} = 0; \tag{8.1.14}$$

$$H^i_{jkl,m} = B^i_{klm;j} - B^i_{jlm;k}; \tag{8.1.15}$$

$$\sum_{\text{cyc}(j,k,l)} H^i_{klm;j} = \sum_{\text{cyc}(j,k,l)} R^r_{jk}B^i_{rlm}. \tag{8.1.16}$$

By the symmetry of **B**, (8.1.15) implies that

$$H^i_{jkl,m} = H^i_{jkm,l}, \tag{8.1.17}$$

but this also follows from (7.14.8), or directly from **(B3)**: the symmetry of
B leads to

$$\nabla^{\mathsf{v}}\mathbf{H}(\widetilde{X}, \widetilde{Y}, \widetilde{Z}, \widetilde{U}) = \nabla^{\mathsf{v}}\mathbf{H}(\widetilde{U}, \widetilde{Y}, \widetilde{Z}, \widetilde{X}), \tag{8.1.18}$$

and (8.1.17) is just the local expression of this relation.

8.2 Affine Deviation

8.2.1 *The Jacobi Endomorphism*

Definition 8.2.1. By the *Jacobi endomorphism* of a spray manifold (M, S) (or simply of a spray S) we mean the type $(1, 1)$ Finsler tensor field $\mathbf{K}$ on $\mathring{T}M$ given by

$$\mathbf{K}(\widetilde{X}) := \mathcal{V}[S, \mathcal{H}\widetilde{X}], \quad \widetilde{X} \in \Gamma(\mathring{\pi}), \tag{8.2.1}$$

where $\mathcal{H}$ is the Berwald connection of (M, S), and $\mathcal{V}$ is the vertical mapping associated to $\mathcal{H}$. If $n = \dim M \geq 2$, the function

$$K := \frac{1}{n-1} \operatorname{tr} \mathbf{K} \tag{8.2.2}$$

is called the *curvature function* of (M, S) (or of S).

Remark 8.2.2. (a) The Jacobi endomorphism was introduced by L. Berwald in [16] under the name *affiner Abweichungstensor*, in English translation *affine deviation tensor*. For the curvature function he used the term *affiner Krümmungsskalar*. We borrowed the name 'Jacobi endomorphism' from a paper of E. Martínez, J. F. Cariñena and W. Sarlet [71]. Z. Shen [88] uses the term *Riemann curvature* and the notation $\mathbf{R}$. Instead of curvature function, Shen speaks about *Ricci scalar* in this context ([88], p. 113). We agree that when we consider the curvature function of a spray manifold, then the base manifold is tacitly assumed to be at least two-dimensional.

(b) Since $\mathcal{H}\widetilde{\delta} = S$ by Corollary 7.3.6, we have

$$\mathcal{R}(\widetilde{\delta}, \widetilde{X}) = \mathcal{V}[S, \mathcal{H}\widetilde{X}] = \mathbf{K}(\widetilde{X}), \tag{8.2.3}$$

for all $\widetilde{X} \in \Gamma(\mathring{\pi})$. We shall frequently use the simple expression

$$\mathcal{R}(\widetilde{\delta}, \cdot\,) = i_{\widetilde{\delta}}\mathcal{R}$$

for the Jacobi endomorphism in the following. Since, by relation (viii) above,

$$\mathbf{H}(\widetilde{X}, \widetilde{Y})\widetilde{\delta} = -\mathcal{R}(\widetilde{X}, \widetilde{Y}) = \mathcal{R}(\widetilde{Y}, \widetilde{X}),$$

we also have

$$\mathbf{K}(\widetilde{X}) = \mathbf{H}(\widetilde{X}, \widetilde{\delta})\widetilde{\delta}, \quad \widetilde{X} \in \Gamma(\mathring{\pi}). \tag{8.2.4}$$

Lemma 8.2.3. *The (canonical) vertical differential of the Jacobi endomorphism is given by*

$$\nabla^{\mathrm{v}}\mathbf{K}(\widetilde{X}, \widetilde{Y}) = \mathcal{R}(\widetilde{X}, \widetilde{Y}) + \mathbf{H}(\widetilde{Y}, \widetilde{\delta})\widetilde{X}; \quad \widetilde{X}, \widetilde{Y} \in \Gamma(\mathring{\pi}). \tag{8.2.5}$$

Proof. $\nabla^{\mathrm{v}}\mathbf{K}(\widetilde{X}, \widetilde{Y}) = (\nabla_{\mathrm{i}\widetilde{X}}\mathbf{K})(\widetilde{Y}) = \nabla_{\mathrm{i}\widetilde{X}}(\mathbf{K}(\widetilde{Y})) - \mathbf{K}(\nabla_{\mathrm{i}\widetilde{X}}\widetilde{Y})$

$\overset{(8.2.3)}{=} \nabla_{\mathrm{i}\widetilde{X}}(\mathcal{R}(\widetilde{\delta}, \widetilde{Y})) - \mathcal{R}(\widetilde{\delta}, \nabla_{\mathrm{i}\widetilde{X}}\widetilde{Y}) = (\nabla_{\mathrm{i}\widetilde{X}}\mathcal{R})(\widetilde{\delta}, \widetilde{Y}) + \mathcal{R}(\nabla_{\mathrm{i}\widetilde{X}}\widetilde{\delta}, \widetilde{Y})$

$\overset{(6.2.33)}{=} \mathcal{R}(\widetilde{X}, \widetilde{Y}) + \nabla^{\mathrm{v}}\mathcal{R}(\widetilde{X}, \widetilde{\delta}, \widetilde{Y}) \overset{(7.14.3)}{=} \mathcal{R}(\widetilde{X}, \widetilde{Y}) + \mathbf{H}(\widetilde{Y}, \widetilde{\delta})\widetilde{X},$

as was to be shown. $\qquad\qquad\qquad\qquad\qquad\qquad\qquad\qquad\qquad\qquad\qquad\quad\square$

Corollary 8.2.4. *The Jacobi endomorphism and the curvature function of a spray manifold are positive-homogeneous of degree two, i.e., we have*

$$\nabla_C \mathbf{K} = 2\mathbf{K}, \quad CK = 2K. \qquad (8.2.6)$$

Proof. For any section $\widetilde{X} \in \Gamma(\mathring{\pi})$,

$$\nabla_C \mathbf{K}(\widetilde{X}) = \nabla^{\mathrm{v}}\mathbf{K}(\widetilde{\delta}, \widetilde{X}) \overset{(8.2.5)}{=} \mathcal{R}(\widetilde{\delta}, \widetilde{X}) + \mathbf{H}(\widetilde{X}, \widetilde{\delta})\widetilde{\delta} \overset{(8.2.3),(8.2.4)}{=} 2\mathbf{K}(\widetilde{X}),$$

which proves the 2^+-homogeneity of $\mathbf{K}$. Since the covariant derivative and the trace operator are interchangeable (see the argument in the proof of Lemma 7.14.3), this implies also the desired homogeneity property of the curvature function. $\qquad\qquad\qquad\qquad\qquad\qquad\qquad\qquad\qquad\quad\square$

Corollary 8.2.5. *For the trace of $\nabla^{\mathrm{v}}\mathbf{K}$ we have the identity*

$$(\mathrm{tr}\,\nabla^{\mathrm{v}}\mathbf{K})(\widetilde{\delta}) = -\,\mathrm{tr}\,\mathbf{K}. \qquad (8.2.7)$$

Proof. For any section $\widetilde{X} \in \Gamma(\mathring{\pi})$,

$$\nabla^{\mathrm{v}}\mathbf{K}(\widetilde{X}, \widetilde{\delta}) \overset{(8.2.5)}{=} \mathcal{R}(\widetilde{X}, \widetilde{\delta}) + \mathbf{H}(\widetilde{\delta}, \widetilde{\delta})\widetilde{X} = -\mathbf{K}(\widetilde{X}),$$

therefore

$$(\mathrm{tr}\,\nabla^{\mathrm{v}}\mathbf{K})(\widetilde{\delta}) = \mathrm{tr}(\widetilde{X} \mapsto \nabla^{\mathrm{v}}\mathbf{K}(\widetilde{X}, \widetilde{\delta})) = \mathrm{tr}(\widetilde{X} \mapsto -\mathbf{K}(\widetilde{X})) = -\,\mathrm{tr}\,\mathbf{K}. \quad\square$$

Corollary 8.2.6. *The fundamental affine curvature of a spray manifold can be expressed by the formula*

$$\mathcal{R}(\widetilde{X}, \widetilde{Y}) = \frac{1}{3}(\nabla^{\mathrm{v}}\mathbf{K}(\widetilde{X}, \widetilde{Y}) - \nabla^{\mathrm{v}}\mathbf{K}(\widetilde{Y}, \widetilde{X})), \qquad (8.2.8)$$

where $\widetilde{X}, \widetilde{Y} \in \Gamma(\mathring{\pi})$.

Proof. Applying (8.2.5), the Bianchi identity (**B2**) and (7.14.10), we obtain that

$$\nabla^{\mathrm{v}}\mathbf{K}(\widetilde{X}, \widetilde{Y}) - \nabla^{\mathrm{v}}\mathbf{K}(\widetilde{Y}, \widetilde{X}) = 2\mathcal{R}(\widetilde{X}, \widetilde{Y}) + \mathbf{H}(\widetilde{Y}, \widetilde{\delta})\widetilde{X} - \mathbf{H}(\widetilde{X}, \widetilde{\delta})\widetilde{Y}$$

$$= 2\mathcal{R}(\widetilde{X}, \widetilde{Y}) + \mathbf{H}(\widetilde{Y}, \widetilde{\delta})\widetilde{X} + \mathbf{H}(\widetilde{\delta}, \widetilde{X})\widetilde{Y} = 2\mathcal{R}(\widetilde{X}, \widetilde{Y}) - \mathbf{H}(\widetilde{X}, \widetilde{Y})\widetilde{\delta}$$

$$= 3\mathcal{R}(\widetilde{X}, \widetilde{Y}),$$

as was to be shown. $\qquad\qquad\qquad\qquad\qquad\qquad\qquad\qquad\qquad\qquad\qquad\quad\square$

Corollary and Definition 8.2.7. *In a spray manifold the equalities*
$$\mathbf{K} = 0, \quad \mathcal{R} = 0 \quad and \quad \mathbf{H} = 0$$
are equivalent. If one, and therefore all of these relations are satisfied, then the spray manifold (and the given spray also) is said to be horizontally flat *(or* h-flat *for short).*

Proof. Indeed, if $\mathbf{K} = 0$, then $\mathcal{R} = 0$ by (8.2.8), and $\mathcal{R} = 0$ implies that $\mathbf{H} = 0$ by (7.14.2). Conversely, if $\mathbf{H} = 0$, then $\mathcal{R} = 0$ by (7.14.10), and from $\mathcal{R} = 0$ if follows that $\mathbf{K} = 0$ by (8.2.3). $\qquad\square$

Corollary 8.2.8. *The trace of the fundamental affine curvature of a spray manifold can be expressed in terms of the Jacobi endomorphism by the formula*

$$\operatorname{tr} \mathcal{R} = \frac{1}{3}(\operatorname{tr} \nabla^{\mathsf{v}} \mathbf{K} - \nabla^{\mathsf{v}} \operatorname{tr} \mathbf{K}). \tag{8.2.9}$$

Proof. For any section $\widetilde{X}$ in $\Gamma(\mathring{\pi})$, we have
$$(\operatorname{tr} \mathcal{R})(\widetilde{X}) = \operatorname{tr}(\widetilde{Z} \mapsto \mathcal{R}(\widetilde{Z}, \widetilde{X}))$$
$$\overset{(8.2.8)}{=} \frac{1}{3} \operatorname{tr}\left(\widetilde{Z} \mapsto (\nabla^{\mathsf{v}}\mathbf{K}(\widetilde{Z}, \widetilde{X}) - \nabla^{\mathsf{v}}\mathbf{K}(\widetilde{X}, \widetilde{Z}))\right)$$
$$= \frac{1}{3}(\operatorname{tr} \nabla^{\mathsf{v}}\mathbf{K})(\widetilde{X}) - \frac{1}{3}\operatorname{tr}(\widetilde{Z} \mapsto (\nabla_{\mathsf{i}\widetilde{X}}\mathbf{K})(\widetilde{Z}))$$
$$= \frac{1}{3}(\operatorname{tr} \nabla^{\mathsf{v}}\mathbf{K})(\widetilde{X}) - \frac{1}{3}\operatorname{tr} \nabla_{\mathsf{i}\widetilde{X}}\mathbf{K} = \frac{1}{3}(\operatorname{tr} \nabla^{\mathsf{v}}\mathbf{K})(\widetilde{X}) - \frac{1}{3}\nabla_{\mathsf{i}\widetilde{X}}\operatorname{tr}\mathbf{K}$$
$$= \frac{1}{3}(\operatorname{tr} \nabla^{\mathsf{v}}\mathbf{K} - \nabla^{\mathsf{v}}\operatorname{tr}\mathbf{K})(\widetilde{X}),$$
as was to be shown. $\qquad\square$

Proposition 8.2.9 (h-differential Bianchi Identity for the Jacobi Endomorphism). *If $\mathcal{R}$ is the fundamental affine curvature and $\mathbf{K}$ is the Jacobi endomorphism of a spray manifold, then for any sections $\widetilde{X}, \widetilde{Y}$ in $\Gamma(\mathring{\pi})$ we have*

(B7) $\nabla^{\mathsf{h}}\mathbf{K}(\widetilde{X}, \widetilde{Y}) - \nabla^{\mathsf{h}}\mathbf{K}(\widetilde{Y}, \widetilde{X}) - \nabla^{\mathsf{h}}\mathcal{R}(\delta, \widetilde{X}, \widetilde{Y}) = 0.$

Proof. From the Bianchi identity (7.12.8),
$$\nabla^{\mathsf{h}}\mathcal{R}(\widetilde{X}, \delta, \widetilde{Y}) + \nabla^{\mathsf{h}}\mathcal{R}(\delta, \widetilde{Y}, \widetilde{X}) + \nabla^{\mathsf{h}}\mathcal{R}(\widetilde{Y}, \widetilde{X}, \delta) = 0.$$
Here
$$\nabla^{\mathsf{h}}\mathcal{R}(\widetilde{X}, \delta, \widetilde{Y}) = (\nabla_{\mathcal{H}\widetilde{X}}\mathcal{R})(\delta, \widetilde{Y}) = \nabla_{\mathcal{H}\widetilde{X}}(\mathbf{K}(\widetilde{Y})) - \mathcal{R}(\nabla_{\mathcal{H}\widetilde{X}}\delta, \widetilde{Y})$$
$$- \mathbf{K}(\nabla_{\mathcal{H}\widetilde{X}}\widetilde{Y}) \overset{(8.1.4)}{=} \nabla_{\mathcal{H}\widetilde{X}}(\mathbf{K}(\widetilde{Y})) - \mathbf{K}(\nabla_{\mathcal{H}\widetilde{X}}\widetilde{Y})$$
$$= (\nabla_{\mathcal{H}\widetilde{X}}\mathbf{K})(\widetilde{Y}) = \nabla^{\mathsf{h}}\mathbf{K}(\widetilde{X}, \widetilde{Y}).$$

Similarly, $\nabla^h \mathcal{R}(\widetilde{Y}, \widetilde{X}, \widetilde{\delta}) = -(\nabla^h \mathbf{K})(\widetilde{Y}, \widetilde{X})$. Since $\nabla^h \mathcal{R}$ is clearly skew-symmetric in its second and third arguments, thus we obtain the desired identity. $\qquad\square$

Lemma 8.2.10. *For any section $\widetilde{X} \in \Gamma(\mathring{\pi})$ we have*

$$(\mathrm{tr}\,\mathbf{H})(\widetilde{X}, \widetilde{\delta}) = -(\mathrm{tr}\,\mathcal{R})(\widetilde{X}), \tag{8.2.10}$$

$$(\mathrm{tr}\,\mathbf{H})(\widetilde{\delta}, \widetilde{X}) = ((n-1)\nabla^v K + \mathrm{tr}\,\mathcal{R})(\widetilde{X}). \tag{8.2.11}$$

In particular,

$$(\mathrm{tr}\,\mathbf{H})(\widetilde{\delta}, \widetilde{\delta}) = -(\mathrm{tr}\,\mathcal{R})(\widetilde{\delta}) = (n-1)K. \tag{8.2.12}$$

Proof. From relation (7.14.4),

$$(\mathrm{tr}\,\mathbf{H})(\widetilde{X}, \widetilde{\delta}) = -(\nabla^v \mathrm{tr}\,\mathcal{R})(\widetilde{\delta}, \widetilde{X}) = -(\nabla_C \mathrm{tr}\,\mathcal{R})(\widetilde{X})$$

$$= -(\mathrm{tr}\,\nabla_C \mathcal{R})(\widetilde{X}) \overset{\text{Lemma 7.8.4}}{=} -(\mathrm{tr}\,\mathcal{R})(\widetilde{X}),$$

as desired. Similarly,

$$(\mathrm{tr}\,\mathbf{H})(\widetilde{\delta}, \widetilde{X}) = -(\nabla^v \mathrm{tr}\,\mathcal{R})(\widetilde{X}, \widetilde{\delta}) = -(\nabla_{i\widetilde{X}} \mathrm{tr}\,\mathcal{R})(\widetilde{\delta})$$

$$= -\mathbf{i}\widetilde{X}(\mathrm{tr}\,\mathcal{R}(\widetilde{\delta})) + \mathrm{tr}\,\mathcal{R}(\nabla_{i\widetilde{X}}\widetilde{\delta}) \overset{(6.2.33)}{=} -\mathbf{i}\widetilde{X}(\mathrm{tr}\,\mathcal{R}(\widetilde{\delta})) + \mathrm{tr}\,\mathcal{R}(\widetilde{X}).$$

Since

$$\mathrm{tr}\,\mathcal{R}(\widetilde{\delta}) \overset{(1.2.23)}{=} \mathrm{tr}(\widetilde{Z} \mapsto \mathcal{R}(\widetilde{Z}, \widetilde{\delta})) = -\mathrm{tr}(\widetilde{Z} \mapsto \mathbf{K}(\widetilde{Z}))$$

$$= -\mathrm{tr}\,\mathbf{K} \overset{(8.2.2)}{=} -(n-1)K,$$

we obtain

$$(\mathrm{tr}\,\mathbf{H})(\widetilde{\delta}, \widetilde{X}) = (n-1)\mathbf{i}\widetilde{X}(K) + \mathrm{tr}\,\mathcal{R}(\widetilde{X}) = ((n-1)\nabla^v K + \mathrm{tr}\,\mathcal{R})(\widetilde{X})$$

and

$$(\mathrm{tr}\,\mathbf{H})(\widetilde{\delta}, \widetilde{\delta}) \overset{(8.2.10)}{=} -(\mathrm{tr}\,\mathcal{R})(\widetilde{\delta}) = (n-1)K,$$

as was to be shown. $\qquad\square$

Remark 8.2.11. With the help of the substitution operators $i_{\widetilde{\delta}}$ and $j_{\widetilde{\delta}}$ defined by (1.2.4) and (1.2.23), respectively, (8.2.10) and (8.2.11) can be written in the more compact forms

$$j_{\widetilde{\delta}} \mathrm{tr}\,\mathbf{H} = -\mathrm{tr}\,\mathcal{R} \quad \text{and} \quad i_{\widetilde{\delta}} \mathrm{tr}\,\mathbf{H} = (n-1)\nabla^v K + \mathrm{tr}\,\mathcal{R}. \tag{8.2.13}$$

Lemma 8.2.12. *For any sections $\widetilde{X}, \widetilde{Y} \in \Gamma(\mathring{\pi})$,*

$$(\mathrm{tr}\,\mathbf{H})(\widetilde{X}, \widetilde{Y}) - (\mathrm{tr}\,\mathbf{H})(\widetilde{Y}, \widetilde{X}) = (C_3^1 \mathbf{H})(\widetilde{Y}, \widetilde{X}), \tag{8.2.14}$$

where $(C_3^1 \mathbf{H})(\widetilde{Y}, \widetilde{X}) := \mathrm{tr}(\widetilde{Z} \mapsto \mathbf{H}(\widetilde{Y}, \widetilde{X})\widetilde{Z})$, cf. 1.2.5.

Proof. Straightforward calculation:

$$(\operatorname{tr}\mathbf{H})(\widetilde{X},\widetilde{Y}) - (\operatorname{tr}\mathbf{H})(\widetilde{Y},\widetilde{X}) \stackrel{(7.14.4)}{=} (\nabla^{\mathrm{v}}\operatorname{tr}\mathcal{R})(\widetilde{X},\widetilde{Y}) - (\nabla^{\mathrm{v}}\operatorname{tr}\mathcal{R})(\widetilde{Y},\widetilde{X})$$

$$= (\nabla_{\mathbf{i}\widetilde{X}}\operatorname{tr}\mathcal{R})(\widetilde{Y}) - (\nabla_{\mathbf{i}\widetilde{Y}}\operatorname{tr}\mathcal{R})(\widetilde{X}) = (\operatorname{tr}(\nabla_{\mathbf{i}\widetilde{X}}\mathcal{R}))(\widetilde{Y}) - (\operatorname{tr}(\nabla_{\mathbf{i}\widetilde{Y}}\mathcal{R}))(\widetilde{X})$$

$$= \operatorname{tr}\left(\widetilde{Z} \mapsto ((\nabla_{\mathbf{i}\widetilde{X}}\mathcal{R})(\widetilde{Z},\widetilde{Y}) - (\nabla_{\mathbf{i}\widetilde{Y}}\mathcal{R})(\widetilde{Z},\widetilde{X}))\right)$$

$$= \operatorname{tr}\left(\widetilde{Z} \mapsto (\nabla^{\mathrm{v}}\mathcal{R}(\widetilde{X},\widetilde{Z},\widetilde{Y}) - \nabla^{\mathrm{v}}\mathcal{R}(\widetilde{Y},\widetilde{Z},\widetilde{X}))\right)$$

$$= \operatorname{tr}\left(\widetilde{Z} \mapsto (\nabla^{\mathrm{v}}\mathcal{R}(\widetilde{Y},\widetilde{X},\widetilde{Z}) + \nabla^{\mathrm{v}}\mathcal{R}(\widetilde{X},\widetilde{Z},\widetilde{Y}))\right)$$

$$\stackrel{(\mathbf{B6})}{=} \operatorname{tr}(\widetilde{Z} \mapsto -\nabla^{\mathrm{v}}\mathcal{R}(\widetilde{Z},\widetilde{Y},\widetilde{X})) \stackrel{(7.14.2)}{=} \operatorname{tr}(\widetilde{Z} \mapsto \mathbf{H}(\widetilde{Y},\widetilde{X})\widetilde{Z})$$

$$= (C_3^1\mathbf{H})(\widetilde{Y},\widetilde{X}). \qquad \square$$

Lemma 8.2.13. *If the Jacobi endomorphism of a spray manifold is of the form*

$$\mathbf{K} = \lambda\,1_{\Gamma(\mathring{\pi})}, \quad \lambda \in C^\infty(\mathring{T}M), \tag{8.2.15}$$

then the spray manifold is horizontally flat.

Proof. By the skew-symmetry of $\mathcal{R}$,

$$0 = \mathcal{R}(\widetilde{\delta},\widetilde{\delta}) = \mathbf{K}(\widetilde{\delta}) \stackrel{(8.2.15)}{=} \lambda\widetilde{\delta}.$$

Since the section $\widetilde{\delta}$ never vanishes on $\mathring{T}M$, this implies that $\lambda = 0$, whence $\mathbf{K} = 0$. $\qquad \square$

Definition 8.2.14. A spray manifold (or a spray) is said to be *isotropic* if its Jacobi endomorphism is of the form

$$\mathbf{K} = \lambda\,1_{\Gamma(\mathring{\pi})} + \widetilde{\alpha} \otimes \widetilde{\delta}, \tag{8.2.16}$$

where $\lambda \in C^\infty(\mathring{T}M)$, $\widetilde{\alpha} \in \mathcal{F}_1^0(\mathring{T}M)$.

Lemma 8.2.15. *If a spray manifold is isotropic, then*

$$\mathbf{K} = K\,1_{\Gamma(\mathring{\pi})} + \widetilde{\alpha} \otimes \widetilde{\delta}, \tag{8.2.17}$$

where K is the curvature function. The Finsler 1-form $\widetilde{\alpha}$ is uniquely determined, it is positive-homogeneous of degree 1, and we have

$$\operatorname{tr}(\widetilde{\alpha} \otimes \widetilde{\delta}) = \widetilde{\alpha}(\widetilde{\delta}) = -K. \tag{8.2.18}$$

Proof. As in the proof of the preceding lemma,

$$0 = \mathcal{R}(\widetilde{\delta},\widetilde{\delta}) = \mathbf{K}(\widetilde{\delta}) \stackrel{(8.2.16)}{=} (\lambda + \widetilde{\alpha}(\widetilde{\delta}))\widetilde{\delta},$$

whence $\widetilde{\alpha}(\widetilde{\delta}) = -\lambda$. Taking this into account,

$$\operatorname{tr}\mathbf{K} = \lambda\operatorname{tr}(1_{\Gamma(\mathring{\pi})}) + \operatorname{tr}(\widetilde{\alpha}\otimes\widetilde{\delta}) \overset{(1.2.19),(1.2.16)}{=} \lambda n - \lambda$$

(where $n = \dim M$). Since $\operatorname{tr}\mathbf{K} \overset{(8.2.2)}{=} (n-1)K$, it follows that $\lambda = K$. These prove (8.2.17) and (8.2.18).

By (8.2.6), $\mathbf{K} - K\,1_{\Gamma(\mathring{\pi})}$ is 2^+-homogeneous, so the same is true for $\widetilde{\alpha}\otimes\widetilde{\delta}$. Thus

$$2(\widetilde{\alpha}\otimes\widetilde{\delta}) = \nabla_C(\widetilde{\alpha}\otimes\widetilde{\delta}) = (\nabla_C\widetilde{\alpha})\otimes\widetilde{\delta} + \widetilde{\alpha}\otimes\nabla_C\widetilde{\delta} \overset{(6.2.33)}{=} (\nabla_C\widetilde{\alpha})\otimes\widetilde{\delta} + \widetilde{\alpha}\otimes\widetilde{\delta},$$

whence $(\nabla_C\widetilde{\alpha})\otimes\widetilde{\delta} = \widetilde{\alpha}\otimes\widetilde{\delta}$. Therefore, for any section $\widetilde{X} \in \Gamma(\mathring{\pi})$,

$$(\nabla_C\widetilde{\alpha})(\widetilde{X})\widetilde{\delta} = \widetilde{\alpha}(\widetilde{X})\widetilde{\delta}.$$

This implies, as above, that we have the desired relation $\nabla_C\widetilde{\alpha} = \widetilde{\alpha}$.

If a Finsler 1-form $\widetilde{\beta}$ also satisfies (8.2.17), then for any section $\widetilde{X} \in \Gamma(\mathring{\pi})$ we have

$$(\mathbf{K} - K\,1_{\Gamma(\mathring{\pi})})(\widetilde{X}) = \widetilde{\alpha}(\widetilde{X})\widetilde{\delta} \quad\text{and}\quad (\mathbf{K} - K\,1_{\Gamma(\mathring{\pi})})(\widetilde{X}) = \widetilde{\beta}(\widetilde{X})\widetilde{\delta}.$$

Hence

$$(\widetilde{\alpha}(\widetilde{X}) - \widetilde{\beta}(\widetilde{X}))\widetilde{\delta} = 0 \quad\text{for all }\widetilde{X} \in \Gamma(\mathring{\pi}),$$

from which we conclude that $\widetilde{\alpha} = \widetilde{\beta}$. $\qquad\square$

Remark 8.2.16. Later, in the proof of Proposition 8.3.12, we shall give an explicit expression for the Finsler 1-form $\widetilde{\alpha}$ in terms of the Jacobi endomorphism.

Local computations. Choose a chart $(\mathcal{U}, (u^i)_{i=1}^n)$ on M. Consider the induced chart $(\mathring{\tau}^{-1}(\mathcal{U}), ((x^i)_{i=1}^n, (y^i)_{i=1}^n))$ on $\mathring{T}M$ and the local frame field $\left(\widehat{\dfrac{\partial}{\partial u^i}}\right)_{i=1}^n$ of $\mathring{\pi}$ over $\mathring{\tau}^{-1}(\mathcal{U})$. The components of the Jacobi endomorphism with respect to the chosen chart are the smooth functions K_j^i determined by

$$\mathbf{K}\left(\widehat{\frac{\partial}{\partial u^j}}\right) = K_j^i\widehat{\frac{\partial}{\partial u^i}}, \quad i \in J_n.$$

Then

$$K_j^i\widehat{\frac{\partial}{\partial u^i}} = \mathbf{K}\left(\widehat{\frac{\partial}{\partial u^j}}\right) \overset{(8.2.3)}{=} \mathcal{R}\left(\widetilde{\delta}, \widehat{\frac{\partial}{\partial u^j}}\right) = y^k R_{kj}^i\widehat{\frac{\partial}{\partial u^i}}$$

$$\overset{(8.1.8)}{=} y^k\left(\frac{\partial G_k^i}{\partial x^j} - \frac{\partial G_j^i}{\partial x^k} + G_k^r G_{jr}^i - G_j^r G_{kr}^i\right)\widehat{\frac{\partial}{\partial u^i}}$$

$$\overset{(8.1.5),(8.1.6)}{=} \left(2\frac{\partial G^i}{\partial x^j} - \frac{\partial G_j^i}{\partial x^k}y^k + 2G_{jk}^i G^k - G_r^i G_j^r\right)\widehat{\frac{\partial}{\partial u^i}},$$

therefore

$$K_j^i = 2\frac{\partial G^i}{\partial x^j} - \frac{\partial G_j^i}{\partial x^k}y^k + 2G_{jk}^i G^k - G_k^i G_j^k. \qquad (8.2.19)$$

This is just the formula (2.6) in Berwald's paper [16], obtained by him in a completely different manner. We shall present a possible modern reconstruction of Berwald's arguments in the next subsection.

Using (8.1.5) and (8.1.7), we obtain

$$K_j^i y^j = 0; \qquad (8.2.20)$$

this is formula (2.7) of the article [16]. In our setting, (8.2.20) is just the coordinate expression of the identity $\mathbf{K}(\widetilde{\delta}) = 0$. For the curvature function of (M, S) we obtain

$$K := \frac{1}{n-1}\operatorname*{tr}_{(\mathcal{U})}\mathbf{K} = \frac{1}{n-1}K_i^i$$
$$= \frac{1}{n-1}\left(2\frac{\partial G^i}{\partial x^i} - \frac{\partial G_i^i}{\partial x^k}y^k + 2G_{ik}^i G^k - G_k^i G_i^k\right) \quad (8.2.21)$$

(summation in force!).

By (8.2.8), the components of the fundamental affine curvature can be expressed in terms of the components of the Jacobi endomorphism. We have

$$R_{jk}^i \widehat{\frac{\partial}{\partial u^i}} = \mathcal{R}\left(\widehat{\frac{\partial}{\partial u^j}}, \widehat{\frac{\partial}{\partial u^k}}\right) = \frac{1}{3}\left(\nabla^{\mathrm{v}}\mathbf{K}\left(\widehat{\frac{\partial}{\partial u^j}}, \widehat{\frac{\partial}{\partial u^k}}\right) - \nabla^{\mathrm{v}}\mathbf{K}\left(\widehat{\frac{\partial}{\partial u^k}}, \widehat{\frac{\partial}{\partial u^j}}\right)\right)$$
$$= \frac{1}{3}\left(\nabla_{\frac{\partial}{\partial y^j}}\left(\mathbf{K}\left(\widehat{\frac{\partial}{\partial u^k}}\right)\right) - \nabla_{\frac{\partial}{\partial y^k}}\left(\mathbf{K}\left(\widehat{\frac{\partial}{\partial u^j}}\right)\right)\right)$$
$$= \frac{1}{3}\left(\frac{\partial K_k^i}{\partial y^j} - \frac{\partial K_j^i}{\partial y^k}\right)\widehat{\frac{\partial}{\partial u^i}},$$

whence

$$R_{jk}^i = \frac{1}{3}\left(\frac{\partial K_k^i}{\partial y^j} - \frac{\partial K_j^i}{\partial y^k}\right) = \frac{1}{3}(K_{k,j}^i - K_{j,k}^i). \qquad (8.2.22)$$

Berwald *defines* the fundamental affine curvature by this formula, writing K_{jk}^i instead of R_{jk}^i. The components of the affine curvature can be obtained as

$$H_{jkl}^i \overset{(7.14.8)}{=} -\frac{\partial R_{jk}^i}{\partial y^l} \overset{(8.2.22)}{=} \frac{1}{3}\left(\frac{\partial^2 K_j^i}{\partial y^l \partial y^k} - \frac{\partial^2 K_k^i}{\partial y^l \partial y^j}\right). \qquad (8.2.23)$$

Again, this formula serves as the *definition* of the affine curvature in Berwald's treatment. More precisely, he uses the notation K^i_{jkl}, and, what is more important, he uses another convention for the arrangement of the lower indices. The corresponding formula in [16] is the following:

$$K_h{}^i{}_{jk} = \frac{\partial K^i_{jk}}{\partial y^h} = \frac{1}{3}\left(\frac{\partial^2 K^i_k}{\partial y^h \partial y^j} - \frac{\partial^2 K^i_j}{\partial y^h \partial y^k} \right).$$

From (8.2.19), taking into account (8.1.5) and (8.1.7), we easily obtain that

$$\frac{\partial K^i_j}{\partial y^k} y^k = 2K^i_j, \tag{8.2.24}$$

which proves the 2^+-homogeneity of $\mathbf{K}$. (Compare this argument with the proof of Corollary 8.2.4.) Using (8.2.22) and (8.2.24),

$$\frac{\partial R^i_{jk}}{\partial y^l} y^l = \frac{1}{3}\left(\frac{\partial}{\partial y^j}\left(\frac{\partial K^i_k}{\partial y^l} y^l \right) - \frac{\partial K^i_k}{\partial y^j} - \frac{\partial}{\partial y^k}\left(\frac{\partial K^i_j}{\partial y^l} y^l \right) + \frac{\partial K^i_j}{\partial y^k} \right)$$

$$= \frac{1}{3}\left(2\frac{\partial K^i_k}{\partial y^j} - \frac{\partial K^i_k}{\partial y^j} - 2\frac{\partial K^i_j}{\partial y^k} + \frac{\partial K^i_j}{\partial y^k} \right) = R^i_{jk},$$

thus proving again that the fundamental affine curvature is positive-homogeneous of degree 1. A proof of this style for the 0^+-homogeneity of $\mathbf{H}$ was presented in Remark 7.14.9.

For later use, we determine the components of $\nabla^{\mathrm{v}}\mathbf{K} \in \mathcal{F}^1_2(\mathring{T}M)$ and $\operatorname{tr}\nabla^{\mathrm{v}}\mathbf{K} \in \mathcal{F}^0_1(\mathring{T}M) = (\Gamma(\mathring{\pi}))^*$ with respect to the chosen chart. Suppose that

$$\nabla^{\mathrm{v}}\mathbf{K} \underset{(\mathcal{U})}{=} K^i_{jk} \widehat{\frac{\partial}{\partial u^i}} \otimes \widehat{du^j} \otimes \widehat{du^k}.$$

Then

$$K^i_{jk}\widehat{\frac{\partial}{\partial u^i}} = \nabla^{\mathrm{v}}\mathbf{K}\left(\widehat{\frac{\partial}{\partial u^j}}, \widehat{\frac{\partial}{\partial u^k}} \right) = \left(\nabla_{\frac{\partial}{\partial y^j}}\mathbf{K} \right)\left(\widehat{\frac{\partial}{\partial u^k}} \right)$$

$$\overset{(6.2.32)}{=} \nabla_{\frac{\partial}{\partial y^j}}\left(\mathbf{K}\left(\widehat{\frac{\partial}{\partial u^k}} \right) \right) = \nabla_{\frac{\partial}{\partial y^j}}\left(K^i_k \widehat{\frac{\partial}{\partial u^i}} \right) = \frac{\partial K^i_k}{\partial y^j}\widehat{\frac{\partial}{\partial u^i}},$$

therefore

$$\nabla^{\mathrm{v}}\mathbf{K} \underset{(\mathcal{U})}{=} \frac{\partial K^i_k}{\partial y^j}\widehat{\frac{\partial}{\partial u^i}} \otimes \widehat{du^j} \otimes \widehat{du^k}, \tag{8.2.25}$$

and in the components $\frac{\partial K_k^i}{\partial y^j}$ the index j plays the role of the *first* lower index. Thus, by (1.2.20),

$$\operatorname{tr} \nabla^{\mathsf{v}} \mathbf{K} \underset{(\mathcal{U})}{=} \frac{\partial K_k^i}{\partial y^i} \widehat{du^k}. \tag{8.2.26}$$

As a further example of using traditional tensor calculus, we deduce (8.2.5):

$$H_{jkl}^i y^k \overset{(8.2.23)}{=} \frac{1}{3} \left(\frac{\partial^2 K_j^i}{\partial y^l \partial y^k} y^k - \frac{\partial^2 K_k^i}{\partial y^l \partial y^j} y^k \right)$$

$$= \frac{1}{3} \left(\frac{\partial}{\partial y^l} \left(\frac{\partial K_j^i}{\partial y^k} y^k \right) - \frac{\partial K_j^i}{\partial y^l} - \frac{\partial^2 K_k^i}{\partial y^l \partial y^j} y^k \right)$$

$$\overset{(8.2.24)}{=} \left(\frac{\partial K_j^i}{\partial y^l} - \frac{\partial^2 K_k^i}{\partial y^l \partial y^j} y^k \right).$$

On the other hand,

$$0 \overset{(8.2.20)}{=} \frac{\partial}{\partial y^l} \left(\frac{\partial}{\partial y^j} (K_k^i y^k) \right) = \frac{\partial}{\partial y^l} \left(\frac{\partial K_k^i}{\partial y^j} y^k + K_j^i \right)$$

$$= \frac{\partial^2 K_k^i}{\partial y^l \partial y^j} y^k + \frac{\partial K_l^i}{\partial y^j} + \frac{\partial K_j^i}{\partial y^l},$$

therefore

$$H_{jkl}^i y^k = \frac{2}{3} \frac{\partial K_j^i}{\partial y^l} + \frac{1}{3} \frac{\partial K_l^i}{\partial y^j} = \frac{\partial K_j^i}{\partial y^l} + \frac{1}{3} \left(\frac{\partial K_l^i}{\partial y^j} - \frac{\partial K_j^i}{\partial y^l} \right) = \frac{\partial K_j^i}{\partial y^l} + R_{jl}^i.$$

This is just the coordinate expression of (8.2.5).

Finally, we mention that in the isotropic case the components of the Jacobi endomorphism are of the form

$$K_j^i = K \delta_j^i + \widetilde{\alpha}_j y^i, \tag{8.2.27}$$

if $\widetilde{\alpha} \underset{(\mathcal{U})}{=} \widetilde{\alpha}_j \widehat{du^j}$. Then

$$(n-1)K = \operatorname{tr} \mathbf{K} \underset{(\mathcal{U})}{=} K_i^i = nK + \widetilde{\alpha}_i y^i,$$

whence $\widetilde{\alpha}_i y^i = -K \upharpoonright \overset{\circ}{\tau}^{-1}(\mathcal{U})$. This is the local form of relation (8.2.18).

8.2.2   Jacobi Fields

In this subsection we try to reconstruct in our own way the train of thought that may have led Berwald to the 'affiner Abweichungstensor'.

Proposition 8.2.17. *Let (M, S) be a spray manifold endowed with the Berwald derivative ∇ induced by S. Assume that X is a complete vector field on M, and let $(\varphi_t)_{t \in \mathbb{R}}$ be the one-parameter group of X. Then the following assertions are equivalent:*

  (i) $\varphi_t \in \mathrm{Aff}(S)$ *for all* $t \in \mathbb{R}$.
  (ii) $\varphi_t \in \mathrm{Aut}(S)$ *for all* $t \in \mathbb{R}$.
  (iii) $[X^c, S] = 0$.
  (iv) *X satisfies the partial differential equation*

$$\nabla_S \nabla_S \widehat{X} + \mathbf{K}(\widehat{X}) = 0. \tag{8.2.28}$$

Proof.   The equivalence of (i) and (ii) is clear from Proposition 5.1.32. By definition, $\varphi_t \in \mathrm{Aff}(S)$ if $((\varphi_t)_*)_\# S = S$. Since $((\varphi_t)_*)_{t \in \mathbb{R}}$ is the one-parameter group of X^c (Lemma 4.1.51), from Proposition 3.2.32 we infer that (ii) and (iii) are equivalent.

Since

$$\mathbf{i}\nabla_S \nabla_S \widehat{X} + \mathbf{i}\mathbf{K}(\widehat{X}) \overset{(7.10.2),(8.2.1)}{=} \mathbf{v}[S, \mathbf{i}\nabla_S \widehat{X}] + \mathbf{v}[S, X^{\mathsf{h}}]$$

$$\overset{(8.1.3)}{=} \mathbf{v}[S, X^c - X^{\mathsf{h}}] + \mathbf{v}[S, X^{\mathsf{h}}] = \mathbf{v}[S, X^c] \overset{(5.1.6)}{=} [S, X^c],$$

we also have the equivalence of (iii) and (iv). $\square$

Remark 8.2.18. If we introduce the concept of a *local affine transformation* of a spray manifold, the assumption of completeness can be dropped in the proposition above. It will be useful to sketch some details.

Let (M, S) be a spray manifold. We say that a diffeomorphism

$$\psi \colon \mathcal{U} \to \mathcal{V}$$

between two open subsets of M is a *local affine transformation* if for every geodesic $\gamma \colon I \to \mathcal{U}$ of S the curve $\psi \circ \gamma \colon I \to \mathcal{V}$ is also a geodesic of S. A vector field X on M is called an *affine vector field* if the local one-parameter group of X consists of local affine transformations. We note, finally, that a local diffeomorphism $\psi \colon \mathcal{U} \to \mathcal{V}$ is a *local automorphism* of S if

$$(\psi_*)_\#(S \restriction T\mathcal{U}) = S \restriction T\mathcal{V}.$$

Now we can reformulate Proposition 8.2.17 as follows.

Corollary 8.2.19. *Let (M, S) be a spray manifold. For a vector field X on M the following properties are equivalent:*

(i) X *is an affine vector field.*

(ii) *The one-parameter group of X consists of local automorphisms of S.*

(iii) $[X^c, S] = 0$.

(iv) $\nabla_S \nabla_S \widehat{X} + \mathbf{K}(\widehat{X}) = 0$.

Definition 8.2.20. Let (M, S) be a spray manifold, and let γ be a geodesic of S. A vector field X along γ is said to be a *Jacobi field* if

$$\nabla^{\dot\gamma}\nabla^{\dot\gamma}(\dot\gamma, X) + \mathbf{K} \circ (\dot\gamma, X) = 0. \tag{8.2.29}$$

Remark 8.2.21. Let I be the domain of γ. Then, for every $t \in I$,

$$(\dot\gamma, X)(t) \overset{(A.2.1)}{:=} (\dot\gamma(t), X(t)) \in \{\dot\gamma(t)\} \times T_{\gamma(t)}M,$$

so $(\dot\gamma, X)$ is a section of the Finsler bundle π along $\dot\gamma$, i.e., $(\dot\gamma, X) \in \Gamma_{\dot\gamma}(\pi)$. We regard the Jacobi endomorphism $\mathbf{K}$ in (8.2.29) as a section of the vector bundle

$$\mathring{T}M \times_M T_1^1(TM) \to \mathring{T}M.$$

In this interpretation, $\mathbf{K}$ is a mapping of the form

$$v \in \mathring{T}M \mapsto \mathbf{K}(v) = (v, \underline{\mathbf{K}}(v)) \in \{v\} \times T_1^1(T_{\dot\tau(v)}M) \cong \mathrm{End}(T_{\dot\tau(v)}M),$$

where $\underline{\mathbf{K}}$ is the principal part of $\mathbf{K}$, see (4.1.10) and (4.1.11).

In the construction of the 'affiner Abweichungstensor', Berwald may have been motivated by the following observation.

Proposition 8.2.22. *Let (M, S) be a spray manifold, and let $\gamma \colon I \to M$ be a geodesic of S. Suppose that X is a vector field on M such that for a sufficiently small open interval J containing $0 \in \mathbb{R}$, the smooth curves $\varphi_s^X \circ \gamma \colon I \to M$ are defined and are geodesics of S for all $s \in J$. Then $X \circ \gamma$ is a Jacobi field.*

Proof. Define the mapping

$$\psi \colon J \times I \to M, \quad (s, t) \mapsto \psi(s, t) := \varphi_s^X \circ \gamma(t).$$

Let (e^1, e^2) be the canonical coordinate system on $J \times I \subset \mathbb{R}^2$.
Step 1. We show that

$$\dot\psi_1 = X \circ \psi, \quad \dot\psi_2(s, t) = (\varphi_s^X \circ \gamma)^{\cdot}(t), \quad (s, t) \in J \times I \tag{$*$}$$

where $\dot\psi_1 := \psi_* \circ \frac{\partial}{\partial e^1}$, $\dot\psi_2 := \psi_* \circ \frac{\partial}{\partial e^2}$ (see (3.2.14)).

For any fixed $t \in I$,

$$\varphi_{\gamma(t)} \colon J \to M, \ s \mapsto \varphi_{\gamma(t)}(s) := \varphi^X(s, \gamma(t)) = \psi(s, t)$$

is an integral curve of X, so we have

$$X \circ \psi(s, t) = X \circ \varphi_{\gamma(t)}(s) = \dot\varphi_{\gamma(t)}(s) = ((\varphi_{\gamma(t)})_*)_s \left(\frac{d}{dr}\right)_s$$

$$= (\psi_*)_{(s,t)} \left(\frac{\partial}{\partial e^1}\right)_{(s,t)} = \dot\psi_1(s, t).$$

This proves the first relation in $(*)$. The second can also be obtained by an immediate calculation:

$$\dot\psi_2(s, t) = (\psi_*)_{(s,t)} \left(\frac{\partial}{\partial e^2}\right)_{(s,t)} = ((\varphi^X_s \circ \gamma)_*)_t \left(\frac{d}{dr}\right)_t = (\varphi^X_s \circ \gamma)\dot{}(t).$$

Step 2. We show that $\frac{\partial}{\partial e^1}$ and X^c are $\dot\psi_2$-related. For every $(s, t) \in J \times I$,

$$X^c \circ \dot\psi_2(s, t) = \kappa \circ X_* \circ \psi_* \circ \frac{\partial}{\partial e^2}(s, t) = \kappa \circ (X \circ \psi)_* \circ \frac{\partial}{\partial e^2}(s, t)$$

$$\overset{(*)}{=} \kappa \circ (\dot\psi_1)_* \circ \frac{\partial}{\partial e^2}(s, t) =: \kappa \circ \ddot\psi_{12}(s, t)$$

$$\overset{\text{Lemma 4.1.50}}{=} \ddot\psi_{21}(s, t) = (\dot\psi_2)_* \circ \frac{\partial}{\partial e^1}(s, t),$$

as desired.

Step 3. We show that $\frac{\partial}{\partial e^2}$ and S are $\dot\psi_2$-related.

Indeed, for any fixed $s \in J$, $\varphi^X_s \circ \gamma$ is a geodesic of S by our condition. Thus

$$(\dot\psi_2)_* \circ \frac{\partial}{\partial e^2}(s, t) =: \ddot\psi_{22}(s, t) \overset{(*)}{=} (\varphi^X_s \circ \gamma)\ddot{}(t)$$

$$= S \circ (\varphi^X_s \circ \gamma)\dot{}(t) \overset{(*)}{=} S \circ \dot\psi_2(s, t),$$

whence our claim.

Step 4. By Step 2 and Step 3, the related vector field lemma implies that

$$[X^c, S] \circ \dot\psi_2 = (\dot\psi_2)_* \circ \left[\frac{\partial}{\partial e^1}, \frac{\partial}{\partial e^2}\right] = 0.$$

Thus, for every $(s, t) \in J \times I$,

$$[X^c, S] \circ (\varphi^X_s \circ \gamma)\dot{}(t) \overset{(*)}{=} [X^c, S] \circ \dot\psi_2(s, t) = 0.$$

In particular, with the choice $s := 0$ it follows that $[X^c, S] \circ \dot{\gamma} = 0$. Then, by the same calculation as in the proof of Proposition 8.2.17, we obtain that

$$(\nabla_S \nabla_S \widehat{X} + \mathbf{K}(\widehat{X})) \circ \dot{\gamma} = 0. \qquad (**)$$

Here, for every $t \in I$,

$$(\nabla_S \nabla_S \widehat{X}) \circ \dot{\gamma}(t) = \nabla_{S(\dot{\gamma}(t))} \nabla_S \widehat{X} = \nabla_{\ddot{\gamma}(t)} \nabla_S \widehat{X} \overset{(6.1.62)}{=} (\nabla^{\dot{\gamma}}((\nabla_S \widehat{X}) \circ \dot{\gamma}))(t)$$

$$= (\nabla^{\dot{\gamma}} \nabla^{\dot{\gamma}} (\widehat{X} \circ \dot{\gamma}))(t) = (\nabla^{\dot{\gamma}} \nabla^{\dot{\gamma}} (\dot{\gamma}, X \circ \gamma))(t),$$

and, taking into account Remark 8.2.21,

$$\mathbf{K}(\widehat{X}) \circ \dot{\gamma}(t) = \mathbf{K} \circ (\dot{\gamma}, X \circ \gamma)(t).$$

Thus $(**)$ takes the form

$$\nabla^{\dot{\gamma}} \nabla^{\dot{\gamma}} (\dot{\gamma}, X \circ \gamma) + \mathbf{K} \circ (\dot{\gamma}, X \circ \gamma) = 0.$$

This concludes the proof. $\qquad\qquad\qquad\qquad\qquad\qquad\qquad\qquad\qquad \Box$

Remark 8.2.23. Observe that in the above proof the mapping ψ is a variation of γ, and every longitudinal curve of ψ is a geodesic. A variation with this property is called a *geodesic variation*. Thus the proposition can be restated as follows: *If γ is a geodesic of a spray manifold, then the variation vector field of a geodesic variation of γ is a Jacobi field.*

8.3 The Weyl Endomorphism

Definition 8.3.1. By the *Weyl endomorphism* of a spray manifold (M, S) (or of a spray S) we mean the type $(1, 1)$ Finsler tensor field

$$\mathbf{W}_1 := \mathbf{K} - K\mathbf{1} - \frac{1}{n+1}(\operatorname{tr} \nabla^v \mathbf{K} - \nabla^v K) \otimes \widetilde{\delta}, \qquad (8.3.1)$$

where $\mathbf{K}$ is the Jacobi endomorphism, K is the curvature function of (M, S), $\mathbf{1}$ abbreviates $\mathbf{1}_{\Gamma(\mathring{\pi})}$, and $n := \dim M$. The type $(1, 2)$ and $(1, 3)$ Finsler tensor fields $\mathbf{W}_2$ and $\mathbf{W}_3$, given by

$$\mathbf{W}_2(\widetilde{X}, \widetilde{Y}) := \frac{1}{3}(\nabla^v \mathbf{W}_1(\widetilde{X}, \widetilde{Y}) - \nabla^v \mathbf{W}_1(\widetilde{Y}, \widetilde{X})) \qquad (8.3.2)$$

and

$$\mathbf{W}_3(\widetilde{X}, \widetilde{Y})\widetilde{Z} := \nabla^v \mathbf{W}_2(\widetilde{Z}, \widetilde{X}, \widetilde{Y}), \qquad (8.3.3)$$

where $\widetilde{X}, \widetilde{Y}, \widetilde{Z} \in \Gamma(\mathring{\pi})$, are said to be the *fundamental projective curvature tensor* and the *projective curvature tensor* of (M, S) (or of S), respectively.

Remark 8.3.2. The reason for the introduction of the Weyl endomorphism will be clear in the next section. We note that Berwald uses the term *projective deviation tensor* ('projektiver Abweichungstensor' in German) for the Weyl endomorphism $\mathbf{W}_1$. We borrowed the names of $\mathbf{W}_2$ and $\mathbf{W}_3$ from Berwald; these are the English translations of his terms 'Grundtensor der Projektivkrümmung' and 'Projektivkrümmungstensor'. Observe that, by definition, the Weyl endomorphism bears the same relationship to the fundamental projective curvature and the projective curvature tensor as the Jacobi endomorphism does to the fundamental affine curvature and the affine curvature tensor. As we shall see soon, the analogies do not stop here.

Clearly, the tensor $\mathbf{W}_2$ is skew-symmetric. This implies that $\mathbf{W}_3$ is skew-symmetric in its first two arguments.

Coordinate description. We use the same notation as in Remark 8.2.16. First we calculate the components W^i_j of the Weyl endomorphism $\mathbf{W}_1$ with respect to the chosen chart $(\mathcal{U}, (u^i)^n_{i=1})$.

For any $j \in J_n$,

$$
W^i_j \widehat{\frac{\partial}{\partial u^i}} = \mathbf{W}_1 \left(\widehat{\frac{\partial}{\partial u^j}} \right) \stackrel{(8.3.1)}{=} \mathbf{K} \left(\widehat{\frac{\partial}{\partial u^j}} \right) - K \widehat{\frac{\partial}{\partial u^j}}
$$

$$
- \frac{1}{n+1} \left((\operatorname{tr} \nabla^{\mathrm{v}} \mathbf{K}) \left(\widehat{\frac{\partial}{\partial u^j}} \right) - \frac{\partial K}{\partial y^j} \right) y^i \widehat{\frac{\partial}{\partial u^i}}
$$

$$
\stackrel{(8.2.26)}{=} \left(K^i_j - K \delta^i_j - \frac{1}{n+1} \left(\frac{\partial K^r_j}{\partial y^r} - \frac{\partial K}{\partial y^j} \right) y^i \right) \widehat{\frac{\partial}{\partial u^i}},
$$

therefore

$$
W^i_j = K^i_j - K \delta^i_j - \frac{1}{n+1} \left(\frac{\partial K^r_j}{\partial y^r} - \frac{\partial K}{\partial y^j} \right) y^i. \tag{8.3.4}
$$

Then

$$
\mathbf{W}_2 \left(\widehat{\frac{\partial}{\partial u^j}}, \widehat{\frac{\partial}{\partial u^k}} \right) \stackrel{(8.3.2)}{:=} \frac{1}{3} \left(\nabla^{\mathrm{v}} \mathbf{W}_1 \left(\widehat{\frac{\partial}{\partial u^j}}, \widehat{\frac{\partial}{\partial u^k}} \right) - \nabla^{\mathrm{v}} \mathbf{W}_1 \left(\widehat{\frac{\partial}{\partial u^k}}, \widehat{\frac{\partial}{\partial u^j}} \right) \right)
$$

$$
= \frac{1}{3} \left(\nabla_{\frac{\partial}{\partial y^j}} \left(W^i_k \widehat{\frac{\partial}{\partial u^i}} \right) - \nabla_{\frac{\partial}{\partial y^k}} \left(W^i_j \widehat{\frac{\partial}{\partial u^i}} \right) \right)
$$

$$
= \frac{1}{3} \left(\frac{\partial W^i_k}{\partial y^j} - \frac{\partial W^i_j}{\partial y^k} \right) \widehat{\frac{\partial}{\partial u^i}},
$$

so the components of $\mathbf{W}_2$ are

$$
W^i_{jk} := \frac{1}{3} \left(\frac{\partial W^i_k}{\partial y^j} - \frac{\partial W^i_j}{\partial y^k} \right); \tag{8.3.5}
$$

cf. (8.2.22). Finally,

$$\mathbf{W}_3\left(\widehat{\frac{\partial}{\partial u^j}},\widehat{\frac{\partial}{\partial u^k}}\right)\widehat{\frac{\partial}{\partial u^l}} \overset{(8.3.3)}{:=} \nabla^v\mathbf{W}_2\left(\widehat{\frac{\partial}{\partial u^l}},\widehat{\frac{\partial}{\partial u^j}},\widehat{\frac{\partial}{\partial u^k}}\right) = \nabla_{\frac{\partial}{\partial y^l}}\left(W^i_{jk}\widehat{\frac{\partial}{\partial u^i}}\right)$$

$$= \frac{\partial W^i_{jk}}{\partial y^l}\widehat{\frac{\partial}{\partial u^i}},$$

thus the components of the projective curvature tensor $\mathbf{W}_3$ are

$$W^i_{jkl} := \frac{\partial W^i_{jk}}{\partial y^l} = \frac{1}{3}\left(\frac{\partial^2 W^i_k}{\partial y^l \partial y^j} - \frac{\partial^2 W^i_j}{\partial y^l \partial y^k}\right). \tag{8.3.6}$$

We note that Berwald writes $W_l{}^i{}_{jk} := \frac{\partial W^i_{jk}}{\partial y^l}$ instead of W^i_{jkl}. In his context, this notation (and typography) is more logical than ours.

Lemma 8.3.3. *For the Weyl endomorphism of a spray manifold we have the following identities:*

$$\mathbf{W}_1(\widetilde{\delta}) = 0, \quad \operatorname{tr}\mathbf{W}_1 = 0, \quad \operatorname{tr}\nabla^v\mathbf{W}_1 = 0. \tag{8.3.7}$$

In terms of the components of $\mathbf{W}_1$,

$$W^i_j y^j = 0, \quad W^i_i = 0, \quad \frac{\partial W^i_j}{\partial y^i} = 0. \tag{8.3.8}$$

Proof. Observe first that

$$(\operatorname{tr}\nabla^v\mathbf{K} - \nabla^v K)(\widetilde{\delta}) = \operatorname{tr}\nabla^v\mathbf{K}(\widetilde{\delta}) - \nabla_C K \overset{(8.2.6),(8.2.7)}{=} -\operatorname{tr}\mathbf{K} - 2K$$

$$\overset{(8.2.2)}{=} -(n+1)K.$$

Taking this into account,

$$\mathbf{W}_1(\widetilde{\delta}) = \mathbf{K}(\widetilde{\delta}) - K\widetilde{\delta} + K\widetilde{\delta} = 0,$$

$$\operatorname{tr}\mathbf{W}_1 = \operatorname{tr}\mathbf{K} - nK - \frac{1}{n+1}(\operatorname{tr}\nabla^v\mathbf{K} - \nabla^v K)(\widetilde{\delta})$$

$$= (n-1)K - nK + K = 0.$$

Thus we have proved the first two relations in (8.3.7). To show the third relation, we verify that $\frac{\partial W^i_j}{\partial y^i} = 0$. For any $k \in J_n$, we obtain from (8.3.4) that

$$\frac{\partial W^i_j}{\partial y^k} = \frac{\partial K^i_j}{\partial y^k} - \frac{\partial K}{\partial y^k}\delta^i_j - \frac{1}{n+1}\left(\frac{\partial^2 K^r_j}{\partial y^k \partial y^r} - \frac{\partial^2 K}{\partial y^k \partial y^j}\right)y^i$$

$$- \frac{1}{n+1}\left(\frac{\partial K^r_j}{\partial y^r} - \frac{\partial K}{\partial y^j}\right)\delta^i_k.$$

Thus, taking also into account that the functions K and K_j^i are 2^+-homogeneous, we obtain that

$$\frac{\partial W_j^i}{\partial y^i} = \frac{\partial K_j^i}{\partial y^i} - \frac{\partial K}{\partial y^j} - \frac{1}{n+1}\left(\frac{\partial K_j^r}{\partial y^r} - \frac{\partial K}{\partial y^j}\right) - \frac{n}{n+1}\left(\frac{\partial K_j^r}{\partial y^r} - \frac{\partial K}{\partial y^j}\right) = 0,$$

as desired. $\qquad\square$

Corollary 8.3.4. *The projective curvature tensors $\mathbf{W}_2$ and $\mathbf{W}_3$ are trace-free.*

Proof. Using the components W_{jk}^i and W_{jkl}^i of $\mathbf{W}_2$ and $\mathbf{W}_3$ introduced in Remark 8.3.2, we find that

$$(\operatorname{tr}\mathbf{W}_2)_k \overset{(1.2.20)}{=} W_{ik}^i \overset{(8.3.5)}{=} \frac{1}{3}\left(\frac{\partial W_k^i}{\partial y^i} - \frac{\partial W_i^i}{\partial y^k}\right) \overset{(8.3.8)}{=} 0.$$

Thus

$$(\operatorname{tr}\mathbf{W}_3)_{kl} = W_{ikl}^i \overset{(8.3.6)}{=} \frac{\partial W_{ik}^i}{\partial y^l} = 0,$$

whence our assertion. $\qquad\square$

Corollary 8.3.5. *The Weyl endomorphism, the fundamental projective curvature tensor and the projective curvature tensor of a spray manifold are positive-homogeneous of degree 2, 1 and 0, respectively.*

Proof. Since the Jacobi endomorphism $\mathbf{K}$ and the curvature function K are 2^+-homogeneous, $\nabla^v\mathbf{K}$ and $\nabla^v K$ are 1^+-homogeneous by Lemma 6.2.22. Thus, taking also into account that the operators ∇_C and tr are interchangeable, we have

$$\nabla_C\mathbf{W}_1 \overset{(8.3.1)}{=} \nabla_C\mathbf{K} - (\nabla_C K)\mathbf{1} - \frac{1}{n+1}(\operatorname{tr}\nabla_C\nabla^v\mathbf{K} - \nabla_C\nabla^v K)\otimes\widetilde{\delta}$$

$$- \frac{1}{n+1}(\operatorname{tr}\nabla^v\mathbf{K} - \nabla^v K)\otimes\nabla_C\widetilde{\delta}$$

$$= 2\left(\mathbf{K} - K\mathbf{1} - \frac{1}{n+1}(\operatorname{tr}\nabla^v\mathbf{K} - \nabla^v K)\otimes\widetilde{\delta}\right) = 2\mathbf{W}_1,$$

so $\mathbf{W}_1$ is 2^+-homogeneous. Again, by Lemma 6.2.22, this implies that $\mathbf{W}_2$ is 1^+-homogeneous, and hence $\mathbf{W}_3$ is 0^+-homogeneous. $\qquad\square$

Lemma 8.3.6. *For any Finsler vector fields $\widetilde{X}, \widetilde{Y}$ on $\overset{\circ}{T}M$,*

$$\mathbf{W}_2(\widetilde{\delta}, \widetilde{X}) = \mathbf{W}_1(\widetilde{X}), \quad \mathbf{W}_3(\widetilde{X}, \widetilde{Y})\widetilde{\delta} = \mathbf{W}_2(\widetilde{X}, \widetilde{Y}), \qquad (8.3.9)$$

$$\mathbf{W}_3(\widetilde{\delta}, \widetilde{X})\widetilde{\delta} = \mathbf{W}_1(\widetilde{X}), \quad \mathbf{W}_3(\widetilde{X}, \widetilde{\delta})\widetilde{Y} = -\nabla^v\mathbf{W}_1(\widetilde{Y}, \widetilde{X}) - \mathbf{W}_2(\widetilde{X}, \widetilde{Y}).$$

$$(8.3.10)$$

Proof. We have

$$\mathbf{W}_2(\tilde{\delta}, \tilde{X}) := \frac{1}{3}(\nabla^{\mathrm{v}}\mathbf{W}_1(\tilde{\delta}, \tilde{X}) - \nabla^{\mathrm{v}}\mathbf{W}_1(\tilde{X}, \tilde{\delta})) = \frac{1}{3}(\nabla_C\mathbf{W}_1)(\tilde{X})$$

$$- \frac{1}{3}(\nabla_{\mathrm{i}\tilde{X}}\mathbf{W}_1)(\tilde{\delta}) \overset{\text{Corollary 8.3.5}}{=} \frac{2}{3}\mathbf{W}_1(\tilde{X}) - \frac{1}{3}\nabla_{\mathrm{i}\tilde{X}}(\mathbf{W}_1(\tilde{\delta}))$$

$$+ \frac{1}{3}\mathbf{W}_1(\nabla_{\mathrm{i}\tilde{X}}\tilde{\delta}) \overset{(6.2.33),(8.3.7)}{=} \frac{2}{3}\mathbf{W}_1(\tilde{X}) + \frac{1}{3}\mathbf{W}_1(\tilde{X}) = \mathbf{W}_1(\tilde{X}),$$

which proves the first relation in (8.3.9). Further,

$$\mathbf{W}_3(\tilde{X}, \tilde{Y})\tilde{\delta} := (\nabla^{\mathrm{v}}\mathbf{W}_2)(\tilde{\delta}, \tilde{X}, \tilde{Y}) = (\nabla_C\mathbf{W}_2)(\tilde{X}, \tilde{Y})$$

$$\overset{\text{Corollary 8.3.5}}{=} \mathbf{W}_2(\tilde{X}, \tilde{Y}),$$

so the second identity in (8.3.9) is also true. The first relation in (8.3.10) is an immediate consequence of (8.3.9). Finally,

$$\mathbf{W}_3(\tilde{X}, \tilde{\delta})\tilde{Y} := (\nabla^{\mathrm{v}}\mathbf{W}_2)(\tilde{Y}, \tilde{X}, \tilde{\delta}) = \nabla_{\mathrm{i}\tilde{Y}}(\mathbf{W}_2(\tilde{X}, \tilde{\delta})) - \mathbf{W}_2(\nabla_{\mathrm{i}\tilde{Y}}\tilde{X}, \tilde{\delta})$$

$$- \mathbf{W}_2(\tilde{X}, \nabla_{\mathrm{i}\tilde{Y}}\tilde{\delta}) \overset{(8.3.9),(6.2.33)}{=} -\nabla_{\mathrm{i}\tilde{Y}}(\mathbf{W}_1(\tilde{X})) + \mathbf{W}_1(\nabla_{\mathrm{i}\tilde{Y}}\tilde{X})$$

$$- \mathbf{W}_2(\tilde{X}, \tilde{Y}) = -\nabla^{\mathrm{v}}\mathbf{W}_1(\tilde{Y}, \tilde{X}) - \mathbf{W}_2(\tilde{X}, \tilde{Y})$$

which concludes the proof. $\square$

Corollary and Definition 8.3.7. *In a spray manifold the equalities*

$$\mathbf{W}_1 = 0, \quad \mathbf{W}_2 = 0 \quad and \quad \mathbf{W}_3 = 0$$

are equivalent. If one, and therefore all of these relations are satisfied, then the spray manifold (and the given spray also) is said to be Weyl flat.

Proof. If is clear from the definition of $\mathbf{W}_2$ and $\mathbf{W}_3$ that $\mathbf{W}_1 = 0$ implies $\mathbf{W}_2 = 0$, and $\mathbf{W}_2 = 0$ implies $\mathbf{W}_3 = 0$. Conversely, if $\mathbf{W}_3 = 0$, then $\mathbf{W}_2 = 0$ by the second identity in (8.3.9), and $\mathbf{W}_2 = 0$ leads to $\mathbf{W}_1 = 0$ by the first identity in (8.3.9). $\square$

Proposition 8.3.8. *The fundamental projective curvature and the fundamental affine curvature tensor of a spray manifold are related by*

$$\mathbf{W}_2(\tilde{X}, \tilde{Y}) = \mathcal{R}(\tilde{X}, \tilde{Y}) + \frac{1}{n+1}(C_3^1\mathbf{H})(\tilde{X}, \tilde{Y})\tilde{\delta}$$

$$- \frac{1}{n^2 - 1}(n\operatorname{tr}\mathcal{R}(\tilde{Y}) - \operatorname{tr}\mathbf{H}(\tilde{\delta}, \tilde{Y}))\tilde{X} \qquad (8.3.11)$$

$$+ \frac{1}{n^2 - 1}(n\operatorname{tr}\mathcal{R}(\tilde{X}) - \operatorname{tr}\mathbf{H}(\tilde{\delta}, \tilde{X}))\tilde{Y}$$

where $\tilde{X}, \tilde{Y} \in \Gamma(\mathring{\pi})$.

Proof. By tensoriality, it is enough to check this relation for basic vector fields $\widehat{X}, \widehat{Y}$. Then

$$3\mathbf{W}_2(\widehat{X}, \widehat{Y}) := \nabla^{\mathsf{v}}\mathbf{W}_1(\widehat{X}, \widehat{Y}) - \nabla^{\mathsf{v}}\mathbf{W}_1(\widehat{Y}, \widehat{X})$$

$$= (\nabla_{X^{\mathsf{v}}}\mathbf{W}_1)(\widehat{Y}) - (\nabla_{Y^{\mathsf{v}}}\mathbf{W}_1)(\widehat{X})$$

$$\stackrel{(6.2.32)}{=} \nabla_{X^{\mathsf{v}}}(\mathbf{W}_1(\widehat{Y})) - \nabla_{Y^{\mathsf{v}}}(\mathbf{W}_1(\widehat{X}))$$

$$\stackrel{(8.3.1)}{=} \nabla_{X^{\mathsf{v}}}(\mathbf{K}(\widehat{Y})) - \nabla_{Y^{\mathsf{v}}}(\mathbf{K}(\widehat{X})) - ((X^{\mathsf{v}}K)\widehat{Y} - (Y^{\mathsf{v}}K)\widehat{X})$$

$$- \frac{1}{n+1}\left(X^{\mathsf{v}}(\mathrm{tr}\,\nabla^{\mathsf{v}}\mathbf{K}(\widehat{Y})) - Y^{\mathsf{v}}(\mathrm{tr}\,\nabla^{\mathsf{v}}\mathbf{K}(\widehat{X}))\right)\widetilde{\delta}$$

$$- \frac{1}{n+1}(\mathrm{tr}\,\nabla^{\mathsf{v}}\mathbf{K}(\widehat{Y})\widehat{X} - \mathrm{tr}\,\nabla^{\mathsf{v}}\mathbf{K}(\widehat{X})\widehat{Y})$$

$$- \frac{1}{n+1}((X^{\mathsf{v}}K)\widehat{Y} - (Y^{\mathsf{v}}K)\widehat{X})$$

$$\stackrel{(8.2.8)}{=} 3\mathcal{R}(\widehat{X}, \widehat{Y}) - \frac{1}{n+1}\left(\mathrm{tr}\,\nabla^{\mathsf{v}}\mathbf{K}(\widehat{Y}) - (n+2)\nabla^{\mathsf{v}}K(\widehat{Y})\right)\widehat{X}$$

$$+ \frac{1}{n+1}\left(\mathrm{tr}\,\nabla^{\mathsf{v}}\mathbf{K}(\widehat{X}) - (n+2)\nabla^{\mathsf{v}}K(\widehat{X})\right)\widehat{Y}$$

$$- \frac{1}{n+1}\left(X^{\mathsf{v}}(\mathrm{tr}\,\nabla^{\mathsf{v}}\mathbf{K}(\widehat{Y})) - Y^{\mathsf{v}}(\mathrm{tr}\,\nabla^{\mathsf{v}}\mathbf{K}(\widehat{X}))\right)\widetilde{\delta}.$$

Here, by (8.2.9),

$$\mathrm{tr}\,\nabla^{\mathsf{v}}\mathbf{K} - (n+2)\nabla^{\mathsf{v}}K = 3(\mathrm{tr}\,\mathcal{R} - \nabla^{\mathsf{v}}K), \tag{$*$}$$

and

$$X^{\mathsf{v}}(\mathrm{tr}\,\nabla^{\mathsf{v}}\mathbf{K}(\widehat{Y})) = (\nabla_{X^{\mathsf{v}}}\,\mathrm{tr}\,\nabla^{\mathsf{v}}\mathbf{K})(\widehat{Y}) = (\nabla_{X^{\mathsf{v}}}(3\,\mathrm{tr}\,\mathcal{R} + (n-1)\nabla^{\mathsf{v}}K))(\widehat{Y})$$

$$= 3(\nabla^{\mathsf{v}}\,\mathrm{tr}\,\mathcal{R})(\widehat{X}, \widehat{Y}) + (n-1)\nabla^{\mathsf{v}}\nabla^{\mathsf{v}}K(\widehat{X}, \widehat{Y}).$$

Thus, taking into account the symmetry of the vertical Hessian (Lemma 6.2.28(ii)),

$$X^{\mathsf{v}}(\mathrm{tr}\,\nabla^{\mathsf{v}}\mathbf{K}(\widehat{Y})) - Y^{\mathsf{v}}(\mathrm{tr}\,\nabla^{\mathsf{v}}\mathbf{K}(\widehat{X})) = 3((\nabla^{\mathsf{v}}\,\mathrm{tr}\,\mathcal{R})(\widehat{X}, \widehat{Y}) - (\nabla^{\mathsf{v}}\,\mathrm{tr}\,\mathcal{R})(\widehat{Y}, \widehat{X}))$$

$$\stackrel{(7.14.4)}{=} -3(\mathrm{tr}\,\mathbf{H}(\widehat{Y}, \widehat{X}) - \mathrm{tr}\,\mathbf{H}(\widehat{X}, \widehat{Y})) \stackrel{(8.2.14)}{=} -3(C_3^1\mathbf{H})(\widehat{X}, \widehat{Y}).$$

For the vertical differential of K we obtain from (8.2.11) that

$$\nabla^{\mathsf{v}}K(\widehat{Z}) = \frac{1}{n-1}(\mathrm{tr}\,\mathbf{H}(\widetilde{\delta}, \widehat{Z}) - \mathrm{tr}\,\mathcal{R}(\widehat{Z})), \quad Z \in \mathfrak{X}(M). \tag{$**$}$$

By relations $(*)$ and $(**)$,

$$\frac{1}{n+1}(\mathrm{tr}\,\nabla^{\mathsf{v}}\mathbf{K}(\widehat{Z}) - (n+2)\nabla^{\mathsf{v}}K(\widehat{Z}))$$

$$= \frac{3}{n+1}\left(\mathrm{tr}\,\mathcal{R}(\widehat{Z}) - \frac{1}{n-1}(\mathrm{tr}\,\mathbf{H}(\widetilde{\delta}, \widehat{Z}) - \mathrm{tr}\,\mathcal{R}(\widehat{Z}))\right)$$

$$= \frac{3}{n^2-1}(n\,\mathrm{tr}\,\mathcal{R}(\widehat{Z}) - \mathrm{tr}\,\mathbf{H}(\widetilde{\delta}, \widehat{Z})).$$

Putting these together, we obtain the desired equality (8.3.11). $\qquad\square$

Proposition 8.3.9. *The projective and the affine curvature tensor of a spray manifold are related by*

$$
\mathbf{W}_3(\widetilde{X},\widetilde{Y})\widetilde{Z} = -\mathbf{H}(\widetilde{X},\widetilde{Y})\widetilde{Z} + \frac{1}{n+1}(C_3^1\mathbf{H}(\widetilde{X},\widetilde{Y}))\widetilde{Z}
$$

$$
+ \frac{1}{n+1}\big((\nabla^{\mathrm{v}}\operatorname{tr}\mathbf{H})(\widetilde{Z},\widetilde{Y},\widetilde{X}) - (\nabla^{\mathrm{v}}\operatorname{tr}\mathbf{H})(\widetilde{Z},\widetilde{X},\widetilde{Y})\big)\widetilde{\delta}
$$

$$
+ \frac{1}{n^2-1}\big(n\operatorname{tr}\mathbf{H}(\widetilde{Y},\widetilde{Z}) + \operatorname{tr}\mathbf{H}(\widetilde{Z},\widetilde{Y}) + (\nabla^{\mathrm{v}}\operatorname{tr}\mathbf{H})(\widetilde{Z},\widetilde{\delta},\widetilde{Y})\big)\widetilde{X} \quad (8.3.12)
$$

$$
- \frac{1}{n^2-1}\big(n\operatorname{tr}\mathbf{H}(\widetilde{X},\widetilde{Z}) + \operatorname{tr}\mathbf{H}(\widetilde{Z},\widetilde{X}) + (\nabla^{\mathrm{v}}\operatorname{tr}\mathbf{H})(\widetilde{Z},\widetilde{\delta},\widetilde{X})\big)\widetilde{Y}
$$

where $\widetilde{X},\widetilde{Y},\widetilde{Z} \in \Gamma(\mathring{\pi})$.

Proof. Again, it is enough to check this equality for basic vector fields $\widehat{X},\widehat{Y},\widehat{Z}$. Then by definition, we have

$$
\mathbf{W}_3(\widehat{X},\widehat{Y})\widehat{Z} := \nabla_{\widehat{Z}^{\mathrm{v}}}\big(\mathbf{W}_2(\widehat{X},\widehat{Y})\big) \overset{(8.3.11)}{=} \underbrace{\nabla_{\widehat{Z}^{\mathrm{v}}}\big(\mathcal{R}(\widehat{X},\widehat{Y})\big)}_{(a)}
$$

$$
+ \frac{1}{n+1}\underbrace{\nabla_{\widehat{Z}^{\mathrm{v}}}\big(C_3^1\mathbf{H}(\widehat{X},\widehat{Y})\widehat{\delta}\big)}_{(b)} + \frac{1}{n^2-1}\underbrace{\nabla_{\widehat{Z}^{\mathrm{v}}}\big(n\operatorname{tr}\mathcal{R}(\widehat{X}) - \operatorname{tr}\mathbf{H}(\widehat{\delta},\widehat{X})\big)}_{(c)}\widehat{Y}
$$

$$
- \frac{1}{n^2-1}\underbrace{\nabla_{\widehat{Z}^{\mathrm{v}}}\big(n\operatorname{tr}\mathcal{R}(\widehat{Y}) - \operatorname{tr}\mathbf{H}(\widehat{\delta},\widehat{Y})\big)}_{(d)}\widehat{X}.
$$

We calculate $(a),(b),(c)$ and (d):

$$
(a) = \nabla^{\mathrm{v}}_{\widehat{Z}}\mathcal{R}(\widehat{X},\widehat{Y}) = \nabla^{\mathrm{v}}\mathcal{R}(\widehat{Z},\widehat{X},\widehat{Y}) \overset{(7.14.2)}{=} -\mathbf{H}(\widehat{X},\widehat{Y})\widehat{Z};
$$

$$
(b) = \nabla_{\widehat{Z}^{\mathrm{v}}}\big(C_3^1\mathbf{H}(\widehat{X},\widehat{Y})\big)\widehat{\delta} + (C_3^1\mathbf{H}(\widehat{X},\widehat{Y}))\widehat{Z}
$$

$$
\overset{(8.2.14)}{=} \big((\nabla^{\mathrm{v}}\operatorname{tr}\mathbf{H})(\widehat{Z},\widehat{Y},\widehat{X}) - (\nabla^{\mathrm{v}}\operatorname{tr}\mathbf{H})(\widehat{Z},\widehat{X},\widehat{Y})\big)\widehat{\delta} + (C_3^1\mathbf{H}(\widehat{X},\widehat{Y}))\widehat{Z};
$$

$$
(c) = \nabla_{\widehat{Z}^{\mathrm{v}}}\big(n\operatorname{tr}\mathcal{R}(\widehat{X}) - \operatorname{tr}\mathbf{H}(\widehat{\delta},\widehat{X})\big) = n(\nabla^{\mathrm{v}}\operatorname{tr}\mathcal{R})(\widehat{Z},\widehat{X})
$$

$$
- (\nabla^{\mathrm{v}}\operatorname{tr}\mathbf{H})(\widehat{Z},\widehat{\delta},\widehat{X}) - \operatorname{tr}\mathbf{H}(\widehat{Z},\widehat{X})
$$

$$
\overset{(7.14.4)}{=} -n\operatorname{tr}\mathbf{H}(\widehat{X},\widehat{Z}) - \operatorname{tr}\mathbf{H}(\widehat{Z},\widehat{X}) - (\nabla^{\mathrm{v}}\operatorname{tr}\mathbf{H})(\widehat{Z},\widehat{\delta},\widehat{X});
$$

$$
(d) = -n\operatorname{tr}\mathbf{H}(\widehat{Y},\widehat{Z}) - \operatorname{tr}\mathbf{H}(\widehat{Z},\widehat{Y}) - (\nabla^{\mathrm{v}}\operatorname{tr}\mathbf{H})(\widehat{Z},\widehat{\delta},\widehat{Y}).
$$

Thus

$$\mathbf{W}_3(\widehat{X}, \widehat{Y})\widehat{Z} = -\mathbf{H}(\widehat{X}, \widehat{Y})\widehat{Z} + \frac{1}{n+1}(C_3^1 \mathbf{H}(\widehat{X}, \widehat{Y}))\widehat{Z}$$

$$+ \frac{1}{n+1}((\nabla^{\mathrm{v}}\operatorname{tr}\mathbf{H})(\widehat{Z}, \widehat{Y}, \widehat{X}) - (\nabla^{\mathrm{v}}\operatorname{tr}\mathbf{H})(\widehat{Z}, \widehat{X}, \widehat{Y}))\widetilde{\delta}$$

$$+ \frac{1}{n^2 - 1}\left(n\operatorname{tr}\mathbf{H}(\widehat{Y}, \widehat{Z}) + \operatorname{tr}\mathbf{H}(\widehat{Z}, \widehat{Y}) + (\nabla^{\mathrm{v}}\operatorname{tr}\mathbf{H})(\widehat{Z}, \widetilde{\delta}, \widehat{Y})\right)\widehat{X}$$

$$- \frac{1}{n^2 - 1}\left(n\operatorname{tr}\mathbf{H}(\widehat{X}, \widehat{Z}) + \operatorname{tr}\mathbf{H}(\widehat{Z}, \widehat{X}) + (\nabla^{\mathrm{v}}\operatorname{tr}\mathbf{H})(\widehat{Z}, \widetilde{\delta}, \widehat{X})\right)\widehat{Y}$$

as was to be shown. $\qquad\square$

Proposition 8.3.10. *The Weyl endomorphism of a two-dimensional spray manifold vanishes.*

Proof. Let (M, S) be a spray manifold with $\dim M = 2$. We choose a chart $(\mathcal{U}, (u^1, u^2))$ on M, and argue locally.

With the notation of Remark 8.3.2, the components of the Weyl endomorphism with respect to the chosen chart are

$$W_j^i = K_j^i - K\delta_j^i - \frac{1}{3}\left(\frac{\partial K_j^r}{\partial y^r} - \frac{\partial K}{\partial y^j}\right)y^i; \quad i, j \in \{1, 2\}.$$

For the curvature function of (M, S) we simply have

$$K \underset{(\mathcal{U})}{=} K_1^1 + K_2^2, \tag{1}$$

whence

$$\frac{\partial K}{\partial y^k} = \frac{\partial K_1^1}{\partial y^k} + \frac{\partial K_2^2}{\partial y^k}, \quad k \in \{1, 2\}. \tag{2}$$

By (8.2.20), $K_1^i y^1 + K_2^i y^2 = 0$, so for each $i \in \{1, 2\}$ we have

$$\frac{\partial K_1^i}{\partial y^1}y^1 + K_1^i + \frac{\partial K_2^i}{\partial y^1}y^2 = 0, \quad \frac{\partial K_1^i}{\partial y^2}y^1 + \frac{\partial K_2^i}{\partial y^2}y^2 + K_2^i = 0. \tag{3}$$

Finally, by the 2^+-homogeneity of $\mathbf{K}$,

$$\frac{\partial K_j^i}{\partial y^1}y^1 + \frac{\partial K_j^i}{\partial y^2}y^2 = 2K_j^i; \quad i, j, \in \{1, 2\}. \tag{4}$$

Taking these into account,

$$W_1^1 = K_1^1 - K - \frac{1}{3}\left(\frac{\partial K_1^1}{\partial y^1}y^1 + \frac{\partial K_1^2}{\partial y^2}y^1 - \frac{\partial K}{\partial y^1}y^1\right)$$

$$\overset{(1),(2)}{=} -K_2^2 - \frac{1}{3}\left(\frac{\partial K_1^2}{\partial y^2}y^1 - \frac{\partial K_2^2}{\partial y^1}y^1\right)$$

$$\overset{(3)}{=} -K_2^2 - \frac{1}{3}\left(-\frac{\partial K_2^2}{\partial y^1}y^1 - \frac{\partial K_2^2}{\partial y^2}y^2 - K_2^2\right)$$

$$\overset{(4)}{=} -K_2^2 - \frac{1}{3}(-2K_2^2 - K_2^2) = -K_2^2 + K_2^2 = 0.$$

It can be shown in the same way that the components W_2^1, W_1^2 and W_2^2 of $\mathbf{W}_1$ also vanish, and we leave it to the reader to finish the proof. $\quad\square$

Corollary 8.3.11. *Every two-dimensional spray manifold is isotropic.*

Proof. By the vanishing of the Weyl endomorphism, the Jacobi endomorphism takes the form

$$\mathbf{K} = K\mathbf{1} + \frac{1}{3}(\operatorname{tr}\nabla^{\mathrm{v}}\mathbf{K} - \nabla^{\mathrm{v}}K) \otimes \widetilde{\delta},$$

so (8.2.17) holds with the choice $\widetilde{\alpha} := \frac{1}{3}(\operatorname{tr}\nabla^{\mathrm{v}}\mathbf{K} - \nabla^{\mathrm{v}}K)$. $\quad\square$

Proposition 8.3.12. *A spray manifold is isotropic if, and only if, its Weyl endomorphism vanishes.*

Proof. Let (M, S) be a spray manifold. If $\dim M = 2$, then by the previous proposition and corollary, the assertion is obvious. Now let $\dim M > 2$. As we argued above, if $\mathbf{W}_1 = 0$, then

$$\mathbf{K} = K\mathbf{1} + \frac{1}{n+1}(\operatorname{tr}\nabla^{\mathrm{v}}\mathbf{K} - \nabla^{\mathrm{v}}K) \otimes \widetilde{\delta},$$

so (M, S) is isotropic.

To prove the converse, suppose that (M, S) is isotropic, i.e., its Jacobi endomorphism is of the form

$$\mathbf{K} = K\mathbf{1} + \widetilde{\alpha} \otimes \widetilde{\delta}, \tag{$*$}$$

where $\widetilde{\alpha} \in \mathcal{F}_1^0(\mathring{T}M)$ and $\nabla_C\widetilde{\alpha} = \widetilde{\alpha}$ (see Lemma 8.2.15). We show that then

$$\widetilde{\alpha} = \frac{1}{n+1}(\operatorname{tr}\nabla^{\mathrm{v}}\mathbf{K} - \nabla^{\mathrm{v}}K),$$

which implies immediately the vanishing of $\mathbf{W}_1$.

First we calculate the (canonical) vertical differential of both sides of $(*)$. Since for any sections $\widetilde{X}, \widetilde{Y} \in \Gamma(\mathring{\pi})$,

$$\nabla^{\mathrm{v}}(\widetilde{\alpha} \otimes \widetilde{\delta})(\widetilde{X}, \widetilde{Y}) = \nabla_{\mathbf{i}\widetilde{X}}(\widetilde{\alpha} \otimes \widetilde{\delta})(\widetilde{Y}) = ((\nabla_{\mathbf{i}\widetilde{X}}\widetilde{\alpha}) \otimes \widetilde{\delta})(\widetilde{Y}) + (\widetilde{\alpha} \otimes \nabla_{\mathbf{i}\widetilde{X}}\widetilde{\delta})(\widetilde{Y})$$

$$= (\nabla_{\mathbf{i}\widetilde{X}}\widetilde{\alpha})(\widetilde{Y})\widetilde{\delta} + \widetilde{\alpha}(\widetilde{Y})\widetilde{X} = \nabla^{\mathrm{v}}\widetilde{\alpha}(\widetilde{X}, \widetilde{Y})\widetilde{\delta} + \mathbf{1} \otimes \widetilde{\alpha}(\widetilde{X}, \widetilde{Y}),$$

we have

$$\nabla^{\mathrm{v}}(\widetilde{\alpha} \otimes \widetilde{\delta}) = \nabla^{\mathrm{v}}\widetilde{\alpha} \otimes \widetilde{\delta} + \mathbf{1} \otimes \widetilde{\alpha},$$

therefore

$$\nabla^{\mathrm{v}}\mathbf{K} = \nabla^{\mathrm{v}}K \otimes \mathbf{1} + (\nabla^{\mathrm{v}}\widetilde{\alpha}) \otimes \widetilde{\delta} + \mathbf{1} \otimes \widetilde{\alpha}. \tag{$**$}$$

Next, we determine the traces of the three type $(1,2)$ Finsler tensor fields on the right-hand side of $(**)$. We use the inductive definition provided by $(1.2.22)$. In our present context it takes the following form: given a Finsler tensor field $A \in \mathcal{F}^1_{s+1}(\overset{\circ}{T}M)$, $i_{\widetilde{X}} \operatorname{tr} A := \operatorname{tr}(j_{\widetilde{X}} A)$ for all $\widetilde{X} \in \Gamma(\overset{\circ}{\pi})$.

We proceed step by step.

(i) Since
$$j_{\widetilde{X}}(\nabla^{\mathsf{v}} K \otimes \mathbf{1})(\widetilde{Y}) := \nabla^{\mathsf{v}} K \otimes \mathbf{1}(\widetilde{Y}, \widetilde{X}) = \nabla^{\mathsf{v}} K(\widetilde{Y})\widetilde{X} = (\nabla^{\mathsf{v}} K \otimes \widetilde{X})(\widetilde{Y}),$$
we have $j_{\widetilde{X}}(\nabla^{\mathsf{v}} K \otimes \mathbf{1}) = \nabla^{\mathsf{v}} K \otimes \widetilde{X}$. Thus
$$i_{\widetilde{X}} \operatorname{tr}(\nabla^{\mathsf{v}} K \otimes \mathbf{1}) = \operatorname{tr} j_{\widetilde{X}}(\nabla^{\mathsf{v}} K \otimes \mathbf{1}) = \operatorname{tr}(\nabla^{\mathsf{v}} K \otimes \widetilde{X}) = \nabla^{\mathsf{v}} K(\widetilde{X}),$$
whence
$$\operatorname{tr}(\nabla^{\mathsf{v}} K \otimes \mathbf{1}) = \nabla^{\mathsf{v}} K.$$

(ii) Similarly,
$$\begin{aligned}
i_{\widetilde{X}} \operatorname{tr}(\nabla^{\mathsf{v}} \widetilde{\alpha} \otimes \widetilde{\delta}) &= \operatorname{tr} j_{\widetilde{X}}(\nabla^{\mathsf{v}} \widetilde{\alpha} \otimes \widetilde{\delta}) = \operatorname{tr}((j_{\widetilde{X}} \nabla^{\mathsf{v}} \widetilde{\alpha}) \otimes \widetilde{\delta}) \\
&= (j_{\widetilde{X}} \nabla^{\mathsf{v}} \widetilde{\alpha})(\widetilde{\delta}) = \nabla^{\mathsf{v}} \widetilde{\alpha}(\widetilde{\delta}, \widetilde{X}) = (\nabla_C \widetilde{\alpha})(\widetilde{X}) = \widetilde{\alpha}(\widetilde{X}),
\end{aligned}$$
therefore
$$\operatorname{tr}(\nabla^{\mathsf{v}} \widetilde{\alpha} \otimes \widetilde{\delta}) = \widetilde{\alpha}.$$

(iii) By an easy calculation as in (i) above, we find that $j_{\widetilde{X}}(\mathbf{1} \otimes \widetilde{\alpha}) = \widetilde{\alpha}(\widetilde{X})\mathbf{1}$. So
$$i_{\widetilde{X}} \operatorname{tr}(\mathbf{1} \otimes \widetilde{\alpha}) = \operatorname{tr} j_{\widetilde{X}}(\mathbf{1} \otimes \widetilde{\alpha}) = \operatorname{tr}(\widetilde{\alpha}(\widetilde{X})\mathbf{1}) = \widetilde{\alpha}(\widetilde{X}) \operatorname{tr}(\mathbf{1}) = n\,\widetilde{\alpha}(\widetilde{X}),$$
whence
$$\operatorname{tr}(\mathbf{1} \otimes \widetilde{\alpha}) = n\,\widetilde{\alpha}.$$

From (i)–(iii) we conclude that $\operatorname{tr} \nabla^{\mathsf{v}} \mathbf{K} = \nabla^{\mathsf{v}} K + (n+1)\widetilde{\alpha}$, and so we obtain the desired expression for $\widetilde{\alpha}$. Thus we kept also our promise made in Remark 8.2.16. $\qquad\square$

8.4 Projective Changes

8.4.1 *Projectively Related Sprays*

Definition 8.4.1. Two sprays S and $\bar{S}$ over a manifold M are said to be (pointwise) *projectively related* or *projectively equivalent* if there exists a smooth function P on $\overset{\circ}{T}M$ such that over $\overset{\circ}{T}M$ we have
$$\bar{S} = S - 2PC, \tag{8.4.1}$$
where C is the Liouville vector field. Then the function P is called the *projective factor*. We say that a spray is *projectively reversible* if it is projectively equivalent to its reverse.

Lemma 8.4.2. *Let (M, S) be a spray manifold. A further spray $\bar{S}$ over M is projectively related to S if, and only if, $\bar{S}$ is contained by the submodule* $\mathrm{span}(C, S)$ *of the $C^\infty(\mathring{T}M)$-module $\mathfrak{X}(\mathring{T}M)$.*

Proof. The necessity of the condition is evident. Conversely, suppose that $\bar{S} \in \mathrm{span}(C, S)$. Then $\bar{S} = fC + hS$ holds for some smooth functions $f, h \in C^\infty(\mathring{T}M)$. Acting on both sides by $\mathbf{J}$, we find that

$$C = \mathbf{J}\bar{S} = \mathbf{J}(fC + hS) = h\mathbf{J}(S) = hC.$$

Hence h is the constant function of value 1, therefore $\bar{S} = S + fC$, as was to be shown. $\qquad\square$

Remark 8.4.3. Let, locally,

$$\underset{(\mathcal{U})}{S} = y^i \frac{\partial}{\partial x^i} - 2G^i \frac{\partial}{\partial y^i}, \quad \underset{(\mathcal{U})}{\bar{S}} = y^i \frac{\partial}{\partial x^i} - 2\bar{G}^i \frac{\partial}{\partial y^i}$$

(see 8.1.4(a)). Then $\underset{(\mathcal{U})}{\bar{S} - S} = 2(G^i - \bar{G}^i)\frac{\partial}{\partial y^i}$, so the coordinate expression of (8.4.1) is the following:

$$\bar{G}^i = G^i + Py^i, \quad i \in J_n. \tag{8.4.2}$$

Lemma 8.4.4. *If two sprays over a manifold M are projectively related by (8.4.1), then the projective factor is positive-homogeneous of degree 1.*

Proof. Since S and $\bar{S}$ are 2^+-homogeneous, we have

$$\bar{S} = [C, \bar{S}] = [C, S - 2PC] = [C, S] - 2(CP)C = S - 2(CP)C,$$

whence $CP = P$. $\qquad\square$

Remark 8.4.5. (a) By the preceding calculation it can also be seen that if S is a spray over M and $P \in C^\infty(\mathring{T}M)$ is a 1^+-homogeneous function, then $\bar{S} := S - 2PC$ is again a spray. We say that $\bar{S}$ has been obtained from S by a *projective change* with projective factor P. The set

$$[S] := \{S - 2PC \in \mathfrak{X}(\mathring{T}M) \mid P \in C^\infty(\mathring{T}M) \text{ and } CP = P\}$$

is called the *projective class* of S or simply a *projective spray*.

(b) If A is a Finsler tensor field constructed from a spray S, then the corresponding Finsler tensor field constructed from a spray $\bar{S} \in [S]$ will be denoted by $\bar{A}$. If $\bar{S} = S - 2PC$ and $\bar{A} = A$ then we say that A is *invariant under the projective change with projective factor P*. If $\bar{A} = A$ for each $\bar{S} \in [S]$, then A is said to be *invariant under the projective changes of S*, or a *projectively invariant tensor* on the spray manifold (M, S).

(c) *If S is a projectively reversible spray then its projective class contains a reversible spray.* Indeed, $\bar{S} := \frac{1}{2}(S + S^{\downarrow})$ is a reversible spray by Remark 5.1.22(b). Since $S^{\downarrow} \in [S]$ by condition, we have

$$\bar{S} = \frac{1}{2}(2S - 2PC) = S - PC,$$

where $P \in C^{\infty}(\mathring{T}M)$ satisfying $CP = P$. This proves our claim.

The useful observation made here is due to M. Crampin and his collaborators [35].

(d) If S and $\bar{S} = S - 2PC \in [S]$ are reversible sprays, then the projective factor has the property

$$P(-v) = -P(v), \quad v \in \mathring{T}M. \tag{8.4.3}$$

Indeed,

$$-\bar{S} \overset{(5.1.13)}{=} \varrho_{\#}\bar{S} = \varrho_{\#}(S - 2PC) \overset{(3.1.45)}{=} \varrho_{\#}S - 2(P \circ \varrho^{-1})\varrho_{\#}C$$

$$\overset{(4.1.49)}{=} \varrho_{\#}S - 2(P \circ \varrho)C \overset{(5.1.13)}{=} -S - 2(P \circ \varrho)C,$$

whence $\bar{S} = S + 2(P \circ \varrho)C$. Comparing the two expressions of $\bar{S}$, we see that $P \circ \varrho = -P$, as was to be shown.

Proposition 8.4.6. *Two sprays over a manifold M are projectively equivalent if, and only if, they have the same pregeodesics. In more detail, if S is a spray over M, then $\bar{S} \in [S]$ if, and only if, for each $v \in \mathring{T}M$ the maximal geodesics γ_v and $\bar{\gamma}_v$ of S and $\bar{S}$ with initial velocity v differ only by a positive reparametrization.*

Proof. Let S be a spray, and let $\bar{S} := S - 2PC$ for some smooth function P on $\mathring{T}M$. If γ is a geodesic of S, then

$$\bar{S} \circ \dot{\gamma} = S \circ \dot{\gamma} - 2(P \circ \dot{\gamma})(C \circ \dot{\gamma}) = \ddot{\gamma} - 2(P \circ \dot{\gamma})(C \circ \dot{\gamma}),$$

so γ satisfies the pregeodesic equation (5.1.16) of $\bar{S}$ with the function $h := -2(P \circ \dot{\gamma})$.

Now suppose that any geodesic of a spray $\bar{S}$ is a pregeodesic of another spray S. Choose an arbitrary point $p \in M$ and a tangent vector $v \in \mathring{T}_pM$. Let $\gamma_v : I \to M$ be the maximal geodesic of $\bar{S}$ starting at p with initial velocity v. Then γ_v is a pregeodesic of S, hence (5.1.16) yields

$$\bar{S}(v) = \bar{S} \circ \dot{\gamma}_v(0) = \ddot{\gamma}_v(0) = S \circ \dot{\gamma}_v(0) + h(0)(C \circ \dot{\gamma}_v)(0) = S(v) + h(0)C(v),$$

where $h(0)$ depends on the choice of the tangent vector $v \in \mathring{T}_pM$. Define a function P on $\mathring{T}M$ by $P(v) := -\frac{1}{2}h(0)$. Then $\bar{S} = S - 2PC$ holds on $\mathring{T}M$. The function P is necessarily smooth by the smoothness of S, $\bar{S}$ and C, therefore the sprays $\bar{S}$ and S are projectively equivalent with projective factor P. $\square$

8.4.2 Changes of Associated Objects

Lemma 8.4.7. *Let S be a spray over M. Under a projective change $\bar{S} := S - 2PC$ of S, the Berwald connection determined by S changes by the rule*

$$\bar{\mathcal{H}} = \mathcal{H} - P\mathbf{i} - \nabla^{\mathrm{v}}P \otimes C. \tag{8.4.4}$$

The horizontal projections and the vertical mappings associated to $\mathcal{H}$ and $\bar{\mathcal{H}}$ are related by

$$\bar{\mathbf{h}} = \mathbf{h} - P\mathbf{J} - (\nabla^{\mathrm{v}}P \circ \mathbf{j}) \otimes C \overset{(6.2.29)}{=} \mathbf{h} - P\mathbf{J} - \mathcal{L}_{\mathbf{J}}P \otimes C \tag{8.4.5}$$

and

$$\bar{\mathcal{V}} = \mathcal{V} + P\mathbf{j} + (\nabla^{\mathrm{v}}P \circ \mathbf{j}) \otimes \widetilde{\delta} = \mathcal{V} + P\mathbf{j} + \mathcal{L}_{\mathbf{J}}P \otimes \widetilde{\delta}. \tag{8.4.6}$$

Proof. For any basic section $\widehat{X}$ in $\Gamma(\mathring{\pi})$ we have

$$\bar{\mathcal{H}}(\widehat{X}) \overset{(7.3.7)}{=} \frac{1}{2}(X^{\mathrm{c}} + [X^{\mathrm{v}}, \bar{S}]) = \frac{1}{2}(X^{\mathrm{c}} + [X^{\mathrm{v}}, S] - 2[X^{\mathrm{v}}, PC])$$

$$= \mathcal{H}(\widehat{X}) - (X^{\mathrm{v}}P)C - PX^{\mathrm{v}} = (\mathcal{H} - P\mathbf{i} - \nabla^{\mathrm{v}}P \otimes C)(\widehat{X}),$$

thus proving (8.4.4). This easily implies the next two relations:

$$\bar{\mathbf{h}} := \bar{\mathcal{H}} \circ \mathbf{j} \overset{(8.4.4)}{=} \mathbf{h} - P\mathbf{J} - (\nabla^{\mathrm{v}}P \circ \mathbf{j}) \otimes C,$$

$$\bar{\mathcal{V}} := \mathbf{i}^{-1} \circ \bar{\mathbf{v}} = \mathbf{i}^{-1} \circ (1 - \bar{\mathbf{h}}) \overset{(8.4.5)}{=} \mathbf{i}^{-1} \circ (1 - \mathbf{h})$$

$$+ \mathbf{i}^{-1} \circ (P\mathbf{J} + (\nabla^{\mathrm{v}}P \circ \mathbf{j}) \otimes \mathbf{i}\widetilde{\delta}) = \mathcal{V} + P\mathbf{j} + (\nabla^{\mathrm{v}}P \circ \mathbf{j}) \otimes \widetilde{\delta}. \qquad \square$$

Remark 8.4.8. Using the notation of 8.1.4, the local form of relation (8.4.4) is

$$\bar{G}^i_j = G^i_j + P\delta^i_j + \frac{\partial P}{\partial y^j}y^i, \tag{8.4.7}$$

where the functions $\bar{G}^i_j$ and G^i_j are the Christoffel symbols of the Berwald connection of $(M, \bar{S})$ and of (M, S), respectively, with respect to the chosen chart. Indeed, for each $j \in J_n$ we have

$$\frac{\partial}{\partial x^j} - \bar{G}^i_j \frac{\partial}{\partial y^i} = \bar{\mathcal{H}}\left(\widehat{\frac{\partial}{\partial u^j}}\right) \overset{(8.4.4)}{=} \mathcal{H}\left(\widehat{\frac{\partial}{\partial u^j}}\right) - P\frac{\partial}{\partial y^j} - \frac{\partial P}{\partial y^j}y^i\frac{\partial}{\partial y^i}$$

$$= \frac{\partial}{\partial x^j} - \left(G^i_j + P\delta^i_j + \frac{\partial P}{\partial y^j}y^i\right)\frac{\partial}{\partial y^i},$$

whence (8.4.7). Observe that this can also be obtained immediately from (8.4.2) by partial differentiation with respect to y^i.

Proposition 8.4.9. *Let (M, S) be a spray manifold. Under a projective change $\bar{S} := S - 2PC$ of S, the Jacobi endomorphism of (M, S) changes by the rule*

$$\bar{\mathbf{K}} = \mathbf{K} + \lambda \mathbf{1} + \widetilde{\beta} \otimes \widetilde{\delta}, \tag{8.4.8}$$

where

$$\lambda := P^2 - SP \in C^\infty(\mathring{T}M), \quad \widetilde{\beta} := 3(\nabla^h P - P\nabla^v P) + \nabla^v \lambda \in (\Gamma(\mathring{\pi}))^*. \tag{8.4.9}$$

The function λ is 2^+-homogeneous, the one-form $\widetilde{\beta}$ is 1^+-homogeneous, and we have

$$\widetilde{\beta}(\widetilde{\delta}) = -\lambda. \tag{8.4.10}$$

Proof. Let $\widehat{X}$ be a basic section in $\Gamma(\mathring{\pi})$. Then

$$\bar{\mathbf{K}}(\widehat{X}) := \bar{\mathcal{V}}[\bar{S}, \bar{\mathcal{H}}(\widehat{X})] \overset{(8.4.6)}{=} \mathcal{V}[\bar{S}, \bar{\mathcal{H}}(\widehat{X})] + P\mathbf{j}[\bar{S}, \bar{H}(\widehat{X})] + (\mathbf{J}[\bar{S}, \bar{\mathcal{H}}(\widehat{X})]P)\widetilde{\delta}.$$

First we express the Lie bracket $[\bar{S}, \bar{\mathcal{H}}(\widehat{X})]$ in terms of the original spray S and the projective factor P. Taking into account that $\bar{\mathcal{H}}$ is also a positive-homogeneous Ehresmann connection, and P is a 1^+-homogeneous function, we find that

$$[\bar{S}, \bar{\mathcal{H}}(\widehat{X})] = [S, \bar{\mathcal{H}}(\widehat{X})] + 2(\bar{\mathcal{H}}(\widehat{X})P)C$$

$$\overset{(8.4.4)}{=} [S, X^h - PX^v - (X^vP)C] + 2(X^hP - 2P(X^vP))C$$

$$= [S, X^h] - (SP)X^v - P[S, X^v] + (X^vP)S - S(X^vP)C$$

$$+ 2(X^hP)C - 4P(X^vP)C.$$

From this, using (4.1.43), (4.1.44), (4.1.83), (7.2.22), (7.2.25) and (8.1.1), we obtain the following:

$$\mathbf{J}[\bar{S}, \bar{\mathcal{H}}(\widehat{X})] = \mathbf{J}[S, X^h] + PX^v + (X^vP)C,$$

$$\mathbf{j}[\bar{S}, \bar{\mathcal{H}}(\widehat{X})] = \mathbf{j}[S, X^h] + P\widehat{X} + (X^vP)\widetilde{\delta},$$

$$\mathcal{V}[\bar{S}, \bar{\mathcal{H}}(\widehat{X})] = \mathbf{K}(\widehat{X}) - P\mathcal{V}[S, X^v] - (SP)\widehat{X} - S(X^vP)\widetilde{\delta}$$

$$+ 2(X^hP)\widetilde{\delta} - 4P(X^vP)\widetilde{\delta}.$$

Thus

$$\bar{\mathbf{K}}(\widehat{X}) = \mathbf{K}(\widehat{X}) - P\mathcal{V}[S, X^v] - (SP)\widehat{X} - S(X^vP)\widetilde{\delta} + 2(X^hP)\widetilde{\delta}$$

$$- 4P(X^vP)\widetilde{\delta} + P\mathbf{j}[S, X^h] + P^2\widehat{X} + P(X^vP)\widetilde{\delta}$$

$$+ (\mathbf{J}[S, X^h])P\widetilde{\delta} + 2P(X^vP)\widetilde{\delta}$$

$$\overset{(8.4.9)}{=} \mathbf{K}(\widehat{X}) + \lambda\widehat{X} + \big(3(X^{\mathsf{h}}P - P(X^{\mathsf{v}}P)) + 2P(X^{\mathsf{v}}P) - X^{\mathsf{v}}(SP)\big)\widetilde{\delta}$$
$$- (X^{\mathsf{h}}P)\widetilde{\delta} + X^{\mathsf{v}}(SP)\widetilde{\delta} - S(X^{\mathsf{v}}P)\widetilde{\delta} - P\mathcal{V}[S, X^{\mathsf{v}}]$$
$$+ P\mathbf{j}[S, X^{\mathsf{h}}] + (\mathbf{J}[S, X^{\mathsf{h}}])P\widetilde{\delta}$$
$$\overset{(8.4.9)}{=} (\mathbf{K} + \lambda\mathbf{1} + (\widetilde{\beta} \otimes \widetilde{\delta}))(\widehat{X}) + E,$$

where

$$E := -(X^{\mathsf{h}}P)\widetilde{\delta} + ([X^{\mathsf{v}}, S]P)\widetilde{\delta} - P\mathcal{V}[S, X^{\mathsf{v}}] + P\mathbf{j}[S, X^{\mathsf{h}}] + (\mathbf{J}[S, X^{\mathsf{h}}])P\widetilde{\delta}.$$

We show that $E = 0$. To do this, observe that from (7.3.7) we get $[X^{\mathsf{v}}, S] = 2X^{\mathsf{h}} - X^{\mathsf{c}}$, whence $\mathcal{V}[X^{\mathsf{v}}, S] = -\mathcal{V}X^{\mathsf{c}}$; and $\mathbf{J}[X^{\mathsf{h}}, S] \overset{(8.1.2)}{=} X^{\mathsf{h}} - X^{\mathsf{c}}$, which implies that $\mathbf{j}[X^{\mathsf{h}}, S] = -\mathcal{V}X^{\mathsf{c}}$. Thus

$$E = -(X^{\mathsf{h}}P)\widetilde{\delta} + ([X^{\mathsf{v}}, S]P)\widetilde{\delta} + P\mathcal{V}[X^{\mathsf{v}}, S] - P\mathbf{j}[X^{\mathsf{h}}, S] - (\mathbf{J}[X^{\mathsf{h}}, S]P)\widetilde{\delta}$$
$$= -(X^{\mathsf{h}}P)\widetilde{\delta} + 2(X^{\mathsf{h}}P)\widetilde{\delta} - (X^{\mathsf{c}}P)\widetilde{\delta} - P\mathcal{V}X^{\mathsf{c}}$$
$$+ P\mathcal{V}X^{\mathsf{c}} - (X^{\mathsf{h}}P - X^{\mathsf{c}}P)\widetilde{\delta} = 0,$$

as desired. This concludes the proof of (8.4.8).

Since

$$C\lambda = C(P^2 - SP) = 2P(CP) - C(SP) = 2P^2 - [C, S]P - S(CP)$$
$$= 2(P^2 - SP) = 2\lambda,$$

the function λ is 2^+-homogeneous. Furthermore,

$$\widetilde{\beta}(\widetilde{\delta}) = 3(SP - P(CP)) + C\lambda = -3\lambda + 2\lambda = -\lambda,$$

so relation (8.4.10) is also valid. Note, finally, that

$$(\nabla_C \nabla^{\mathsf{h}} P)(\widehat{X}) = C(X^{\mathsf{h}}P) = [C, X^{\mathsf{h}}]P + X^{\mathsf{h}}(CP) = X^{\mathsf{h}}P = \nabla^{\mathsf{h}}P(\widehat{X})$$

for each $X \in \mathfrak{X}(M)$, so $\nabla^{\mathsf{h}}P$ is 1^+-homogeneous. Since the other two terms in the expression of $\widetilde{\beta}$ are clearly 1^+-homogeneous, $\widetilde{\beta}$ has the same homogeneity property. This finishes the proof of the proposition. $\qquad\square$

Corollary 8.4.10. *The Jacobi endomorphism of a spray manifold (M, S) is invariant under a projective change of S with projective factor P if, and only if, P satisfies the partial differential equation*

$$\nabla^{\mathsf{h}}P - P\nabla^{\mathsf{v}}P = 0. \tag{8.4.11}$$

Proof. The necessity of the condition is clear from (8.4.8). Conversely, suppose that (8.4.11) is satisfied. Then by (8.4.9), (8.4.10) and the 2^+-homogeneity of λ we obtain

$$-\lambda = \nabla^{\mathsf{v}}\lambda(\widetilde{\delta}) = C\lambda = 2\lambda,$$

whence $\lambda = 0$ and consequently $\bar{\mathbf{K}} = \mathbf{K}$. $\qquad\square$

Corollary 8.4.11. *Under a projective change* $\bar{S} = S - 2PC$ *of a spray* S, *the curvature function of* (M, S) *changes by*

$$\bar{K} = K + \lambda, \tag{8.4.12}$$

where $\lambda \overset{(8.4.9)}{:=} P^2 - SP$. *Thus the curvature function remains invariant under the projective change if, and only if, the projective factor satisfies the partial differential equation*

$$P^2 - SP = 0. \tag{8.4.13}$$

Proof. From (8.4.8) we obtain that

$$\operatorname{tr} \bar{\mathbf{K}} = \operatorname{tr} \mathbf{K} + n\lambda + \operatorname{tr}(\tilde{\beta} \otimes \tilde{\delta}) \overset{(8.4.10)}{=} \operatorname{tr} \mathbf{K} + (n-1)\lambda.$$

By (8.2.2), this implies (8.4.12). Obviously, $\bar{K} = K$ if, and only if, $\lambda = 0$, i.e., P satisfies (8.4.13). □

Remark 8.4.12. (a) In the language of traditional tensor calculus, (8.4.11) and (8.4.13) take the forms

$$P_{;j} - PP_{,j} = 0 \quad (j \in J_n) \quad \text{and} \quad P^2 - P_{;j}y^j = 0, \tag{8.4.14}$$

respectively. Indeed, the first transcription is clear by (7.11.3) and (7.11.8). To obtain the second, let

$$S \underset{(\mathcal{U})}{=} y^i \frac{\partial}{\partial x^i} - 2G^i \frac{\partial}{\partial y^i}$$

over a chart domain. Then the Christoffel symbols of the Berwald connection of (M, S) are the functions $G^i_j = \frac{\partial G^i}{\partial y^j}$, and

$$P_{;j}y^j \overset{(7.11.8)}{=} \left(\frac{\partial P}{\partial x^j} - G^i_j \frac{\partial P}{\partial y^i} \right) y^j \overset{(8.1.5)}{=} y^i \frac{\partial P}{\partial x^i} - 2G^i \frac{\partial P}{\partial y^i} \underset{(\mathcal{U})}{=} SP,$$

so the local form of (8.4.13) is indeed the second relation in (8.4.14).

(b) Let (M, S) be a spray manifold. A 1^+-homogeneous smooth function P on $\mathring{T}M$ is called by Z. Shen

    a *Funk function* if $\nabla^h P - P\nabla^v P = 0$,
    a *weak Funk function* if $P^2 - SP = 0$;

see [88], p. 177. Using these terms, the two preceding corollaries can be restated as follows: *the Jacobi endomorphism (resp. the curvature function) of a spray manifold is invariant under a projective change of the spray if, and only if, the projective factor is a Funk function (resp. a weak Funk function)*. We note that every Funk function P is also a weak Funk function, because

$$(\nabla^h P - P\nabla^v P)(\tilde{\delta}) = SP - P\nabla_C P = SP - P^2.$$

The following nice observations are also due to Z. Shen.

Proposition 8.4.13. *Let (M, S) be a spray manifold, and suppose that $P \in C^\infty(\mathring{T}M)$ is a weak Funk function. If $\gamma \colon I \to M$ is a geodesic of S starting at $p \in M$ with initial velocity $v \in \mathring{T}_p M$, then*

$$P \circ \dot\gamma(t) = \frac{P(v)}{1 - P(v)t} \text{ for all } t \in I. \tag{8.4.15}$$

If, in particular, S is complete, then $P = 0$.

Proof. Since $S \circ \dot\gamma = \ddot\gamma$, for all $t \in I$ we have

$$(P \circ \dot\gamma)'(t) \overset{(3.1.9)}{=} \ddot\gamma(t)P = S(\dot\gamma(t))P = SP(\dot\gamma(t))$$
$$\overset{\text{cond.}}{=} P^2(\dot\gamma(t)) = (P^2 \circ \dot\gamma)(t).$$

Thus the function $P \circ \dot\gamma \colon I \to \mathbb{R}$ solves the initial value problem

$$y' = y^2, \quad y(0) = P(v) \tag{$*$}$$

over I. The differential equation here is separable, therefore it can be integrated. We obtain the following explicit form for the solutions:

$$h(t) = \frac{1}{c - t}, \quad t \in I,$$

with $c \in \mathbb{R}$. The initial condition implies that $c = \frac{1}{P(v)}$, so the solution of $(*)$ is given by

$$h(t) = \frac{P(v)}{1 - P(v)t}, \quad t \in I.$$

This proves (8.4.15).

If S is complete, then the domain of every maximal geodesic of S is $\mathbb{R}$. So the maximal geodesic γ_v is defined, in particular, at $t := \frac{1}{P(v)}$. This, however, contradicts (8.4.15). Therefore, in the complete case, the initial value problem $(*)$ has only the trivial solution $h = 0$; hence $P = 0$. $\qquad\square$

Proposition 8.4.14. *Let S be a positively complete spray. If $\bar{S} := S - 2PC \in [S]$, and for the curvature functions of S and $\bar{S}$ we have the inequality $\bar{K} \le K$, then the projective factor P is non-positive. If, in addition, both S and $\bar{S}$ are reversible, then $\bar{S} = S$.*

For a proof see [88], pp. 178–179.

8.4.3 Projectively Related Covariant Derivatives

Lemma 8.4.15. *Let (M, S) be a spray manifold. Suppose that $\bar{S} \in \mathfrak{X}(\mathring{T}M)$ is projectively related to S with projective factor P, i.e., $\bar{S} = S - 2PC$. Let ∇ and $\bar{\nabla}$ be the Berwald derivatives induced by S and $\bar{S}$, respectively. Then, with the conventions of 8.4.2,*

$$\bar{\nabla}_{X^{\mathfrak{h}}}\widehat{Y} = \nabla_{X^{\mathfrak{h}}}\widehat{Y} + (X^{\mathsf{v}}P)\widehat{Y} + (Y^{\mathsf{v}}P)\widehat{X} + Y^{\mathsf{v}}(X^{\mathsf{v}}P)\widetilde{\delta}, \qquad (8.4.16)$$

for all $X, Y \in \mathfrak{X}(M)$.

Proof. This is an immediate calculation:

$$\bar{\nabla}_{X^{\mathfrak{h}}}\widehat{Y} = \bar{\nabla}_{\bar{\mathcal{H}}(\widehat{X})}\widehat{Y} \overset{(7.10.2)}{=} \bar{\mathcal{V}}[\bar{\mathcal{H}}(\widehat{X}), Y^{\mathsf{v}}] \overset{(8.4.4)}{=} \bar{\mathcal{V}}[X^{\mathfrak{h}} - PX^{\mathsf{v}} - (X^{\mathsf{v}}P)C, Y^{\mathsf{v}}]$$

$$= \bar{\mathcal{V}}[X^{\mathfrak{h}}, Y^{\mathsf{v}}] + (Y^{\mathsf{v}}P)\bar{\mathcal{V}}(X^{\mathsf{v}}) + Y^{\mathsf{v}}(X^{\mathsf{v}}P)\bar{\mathcal{V}}C + (X^{\mathsf{v}}P)\bar{\mathcal{V}}(Y^{\mathsf{v}})$$

$$\overset{(8.4.6)}{=} \mathcal{V}[X^{\mathfrak{h}}, Y^{\mathsf{v}}] + (X^{\mathsf{v}}P)\widehat{Y} + (Y^{\mathsf{v}}P)\widehat{X} + Y^{\mathsf{v}}(X^{\mathsf{v}}P)\widetilde{\delta}$$

$$= \nabla_{X^{\mathfrak{h}}}\widehat{Y} + (X^{\mathsf{v}}P)\widehat{Y} + (Y^{\mathsf{v}}P)\widehat{X} + Y^{\mathsf{v}}(X^{\mathsf{v}}P)\widetilde{\delta}.$$

In the fifth equality we also used the fact that $[X^{\mathfrak{h}}, Y^{\mathsf{v}}]$ is vertical by Lemma 7.1.5(ii),(a). $\qquad\square$

Theorem 8.4.16. *Let D and $\bar{D}$ be torsion-free covariant derivatives on a manifold M. Then the following are equivalent:*

(i) *D and $\bar{D}$ have the same pregeodesics.*

(ii) *There exists a one-form α on M such that*

$$\bar{D}_X Y = D_X Y + \alpha(X)Y + \alpha(Y)X, \qquad (8.4.17)$$

for all $X, Y \in \mathfrak{X}(M)$.

(iii) *If S and $\bar{S}$ are the geodesic sprays of D and $\bar{D}$, resp., then S and $\bar{S}$ are projectively related. Namely,*

$$\bar{S} = S - 2\bar{\alpha}C, \qquad (8.4.18)$$

where $\bar{\alpha} \in C^{\infty}(TM)$ is the function determined by a one-form α on M according to (7.4.6).

Proof. Note first that if two *affine* sprays S and $\bar{S}$ are projectively related, say $\bar{S} = S - 2PC$, then the projective factor P is necessarily of the form

$$P = i_{\xi}\alpha^{\mathsf{v}} =: \bar{\alpha}, \quad \alpha \in \mathfrak{X}^*(M),$$

where $\xi = \mathfrak{X}(TM)$ is a semispray. Indeed, since S and $\bar{S}$ are smooth on TM, the 1^+-homogeneous function P is also smooth on the whole TM. Then, by Lemma 4.2.10(ii), P is fibrewise linear, so there is a one-form α on M such that

$$P \upharpoonright T_{\tau(v)}M = \alpha_{\tau(v)} \in T^*_{\tau(v)}M, \quad v \in TM.$$

Hence, for every $v \in TM$,

$$P(v) = \alpha_{\tau(v)}(v) \overset{(7.4.8)}{=} \bar{\alpha}(v),$$

as we claimed.

Since a covariant derivative and its geodesic spray have the same geodesics, the equivalence of (i) and (iii) is clear from Proposition 8.4.6.

We show the equivalence of (ii) and (iii). Let $\mathcal{H}$ and $\bar{\mathcal{H}}$ be the (linear) Ehresmann connections determined by D and $\bar{D}$, respectively. Then by (7.5.12) and (7.5.16), the geodesic sprays of D and $\bar{D}$ are given by

$$v \mapsto S(v) := \mathcal{H}(v,v) = Y_*(v) - (D_vY)^\uparrow(v)$$

and

$$v \mapsto \bar{S}(v) := \bar{\mathcal{H}}(v,v) = Y_*(v) - (\bar{D}_vY)^\uparrow(v),$$

where $Y \in \mathfrak{X}(M)$ is such that $Y(\tau(v)) = v$. Thus

$$\bar{S}(v) = S(v) + (D_vY)^\uparrow(v) - (\bar{D}_vY)^\uparrow(v) = S(v) + (D_vY - \bar{D}_vY)^\uparrow(v)$$

$$\overset{(4.1.38)}{=} S(v) + \mathbf{i}(v, D_vY - \bar{D}_vY).$$

Now suppose that (ii) is satisfied. Then

$$\mathbf{i}(v, D_vY - \bar{D}_vY) = \mathbf{i}(v, (D_YY - \bar{D}_YY)(p))$$

$$\overset{(8.4.17)}{=} -\mathbf{i}(v, (2\alpha(Y)Y)(p)) = -\mathbf{i}(v, 2\alpha_{\tau(v)}(v)v)$$

$$= -2\alpha_{\tau(v)}(v)\mathbf{i}(v,v) = -2\alpha_{\tau(v)}(v)C(v) \overset{(7.4.8)}{=} -2(\bar{\alpha}C)(v),$$

therefore $\bar{S} = S - 2\bar{\alpha}C$, as was to be shown.

Conversely, we assume that condition (iii) holds. Let ∇ and $\bar{\nabla}$ be the Berwald derivatives induced by S and $\bar{S}$, respectively. Then, for any vector fields X, Y on M, we have

$$\mathbf{i}\nabla_{X^\mathsf{h}}\widehat{Y} \overset{(7.10.2)}{=} [X^\mathsf{h}, Y^\mathsf{v}] \overset{(7.5.13)}{=} (D_XY)^\mathsf{v}.$$

In the same way, $\mathbf{i}\bar{\nabla}_{X^\mathsf{h}}\widehat{Y} = (\bar{D}_XY)^\mathsf{v}$. Thus relation (8.4.16) (which is valid by the projective relatedness of S and $\bar{S}$) takes the following form:

$$(\bar{D}_XY)^\mathsf{v} = (D_XY)^\mathsf{v} + (X^\mathsf{v}\bar{\alpha})Y^\mathsf{v} + (Y^\mathsf{v}\bar{\alpha})X^\mathsf{v} + Y^\mathsf{v}(X^\mathsf{v}\bar{\alpha})C.$$

We saw in the proof of Proposition 8.4.23 that $X^{\vee}\bar{\alpha} = (\alpha(X))^{\vee}$, whence $Y^{\vee}(X^{\vee}\bar{\alpha}) = 0$, so we get
$$(\bar{D}_X Y)^{\vee} = (D_X Y + \alpha(X)Y + \alpha(Y)X)^{\vee},$$
which is equivalent to (8.4.17). This completes the proof. $\square$

Remark 8.4.17. Given a chart $(\mathcal{U}, (u^i)_{i=1}^n)$ on M, let (Γ^i_{jk}) and $(\bar{\Gamma}^i_{jk})$ be the families of the Christoffel symbols of D and $\bar{D}$ resp., with respect to the chosen chart. If $\underset{(\mathcal{U})}{\alpha} = \alpha_i du^i$, it can be shown by a simple calculation that (8.4.17) takes the local form
$$\bar{\Gamma}^i_{jk} = \Gamma^i_{jk} + \alpha_j \delta^i_k + \alpha_k \delta^i_j; \quad i, j, k \in J_n. \tag{8.4.19}$$
It is a classical result of Hermann Weyl [98] that two covariant derivatives D and $\bar{D}$ on a manifold M have the 'same geodesics with possibly different parametrization', i.e., have the same pregeodesics if, and only if, there exists a (unique) one-form α on M such that locally (8.4.19) is satisfied.

  Two covariant derivatives D and $\bar{D}$ are called *projectively equivalent* (or briefly *projective*) if one (and hence all) of the three conditions above holds.

8.4.4 *The Meaning of the Weyl Endomorphism*

Proposition 8.4.18. *The Weyl endomorphism, the fundamental projective curvature tensor and the projective curvature tensor of a spray manifold are projectively invariant.*

Proof. Let (M, S) be a spray manifold, and consider a projective change $\bar{S} := S - 2PC$ of S. We have only to show that $\bar{\mathbf{W}}_1 = \mathbf{W}_1$; then the projective invariance of $\mathbf{W}_2$ and $\mathbf{W}_3$ is an immediate consequence. Starting with the definition of the Weyl endomorphism of $(M, \bar{S})$, we find that

$$\bar{\mathbf{W}}_1 := \bar{\mathbf{K}} - \bar{K}\mathbf{1} - \frac{1}{n+1}(\operatorname{tr}\nabla^{\vee}\bar{\mathbf{K}} - \nabla^{\vee}\bar{K}) \otimes \widetilde{\delta}$$

$$\overset{(8.4.8),(8.4.12)}{=} \mathbf{K} + \lambda\mathbf{1} + \widetilde{\beta} \otimes \widetilde{\delta} - K\mathbf{1} - \lambda\mathbf{1}$$

$$\quad - \frac{1}{n+1}\Big(\operatorname{tr}(\nabla^{\vee}(\mathbf{K} + \lambda\mathbf{1} + \widetilde{\beta} \otimes \widetilde{\delta})) - \nabla^{\vee}K - \nabla^{\vee}\lambda\Big) \otimes \widetilde{\delta}$$

$$= \mathbf{W}_1 + \widetilde{\beta} \otimes \widetilde{\delta} - \frac{1}{n+1}\Big(\operatorname{tr}(\nabla^{\vee}(\lambda\mathbf{1} + \widetilde{\beta} \otimes \widetilde{\delta})) - \nabla^{\vee}\lambda\Big) \otimes \widetilde{\delta}.$$

Here $\operatorname{tr}(\nabla^{\vee}(\lambda\mathbf{1})) = \operatorname{tr}(\nabla^{\vee}\lambda \otimes \mathbf{1}) = \nabla^{\vee}\lambda$ (see part (i) in the proof of Proposition 8.3.12), so it remains only to check that $\operatorname{tr}\nabla^{\vee}(\widetilde{\beta} \otimes \widetilde{\delta}) = (n+1)\widetilde{\beta}$. To do this, note first that
$$\nabla^{\vee}(\widetilde{\beta} \otimes \widetilde{\delta}) = \nabla^{\vee}\widetilde{\beta} \otimes \widetilde{\delta} + \mathbf{1} \otimes \widetilde{\beta}.$$

Indeed, for any sections $\widehat{X}, \widehat{Y}$ in $\Gamma(\mathring{\pi})$,

$$\nabla^{\mathrm{v}}(\widetilde{\beta} \otimes \widetilde{\delta})(\widehat{X}, \widehat{Y}) = (\nabla_{X^{\mathrm{v}}}(\widetilde{\beta} \otimes \widetilde{\delta}))(\widehat{Y}) = (\nabla_{X^{\mathrm{v}}}\widetilde{\beta})(\widehat{Y})\widetilde{\delta} + (\widetilde{\beta} \otimes \widehat{X})(\widehat{Y})$$
$$= \nabla^{\mathrm{v}}\widetilde{\beta}(\widehat{X}, \widehat{Y})\widetilde{\delta} + \widetilde{\beta}(\widehat{Y})\widehat{X} = (\nabla^{\mathrm{v}}\widetilde{\beta} \otimes \widetilde{\delta} + \mathbf{1} \otimes \widetilde{\beta})(\widehat{X}, \widehat{Y}),$$

as claimed.

Thus, using the 1^{+}-homogeneity of $\widetilde{\beta}$ and the results of parts (ii) and (iii) of the proof quoted above, we have

$$\operatorname{tr} \nabla^{\mathrm{v}}(\widetilde{\beta} \otimes \widetilde{\delta}) = \operatorname{tr}(\nabla^{\mathrm{v}}\widetilde{\beta} \otimes \widetilde{\delta}) + \operatorname{tr}(\mathbf{1} \otimes \widetilde{\beta}) = \widetilde{\beta} + n\widetilde{\beta} = (n+1)\widetilde{\beta},$$

which concludes the proof of the present proposition. $\qquad\square$

Corollary 8.4.19. *The property of being isotropic is a projectively invariant property of a spray.*

Proof. The assertion follows from the preceding proposition and from Proposition 8.3.12. $\qquad\square$

Remark 8.4.20. Let (M, S) be a spray manifold with $\dim M > 2$. Suppose that there is a projective change $\bar{S} := S - 2PC$ of S such that $\bar{S}$ is horizontally flat. Then $\bar{\mathbf{K}} = 0$ (see Corollary and Definition 8.2.7), whence $\bar{\mathbf{W}}_1 = 0$. Since $\mathbf{W}_1 = \bar{\mathbf{W}}_1$ by Proposition 8.4.18, it follows that (M, S) is also Weyl flat, and hence isotropic. Thus *(for $\dim M > 2$) a spray over M is isotropic if it is projectively equivalent to a horizontally flat spray.* The converse is also true, at least locally.

Theorem 8.4.21 (M. Crampin). *Suppose that (M, S) is an at least 3-dimensional* isotropic *spray manifold. Then S is locally projectively equivalent to a horizontally flat spray, i.e., for each point $p \in M$ there is an open neighbourhood $\mathcal{U}$ of p and a 1^{+}-homogeneous smooth function $P \colon \mathring{\tau}^{-1}(\mathcal{U}) \to \mathbb{R}$ such that the spray $\bar{S} := S - 2PC$ is horizontally flat.*

For a proof we refer to Crampin's paper [32].

8.4.5 The Douglas Tensor

Lemma 8.4.22. *Let (M, S) be a spray manifold. Under a projective change $\bar{S} := S - 2PC$ of S, the Berwald curvature of (M, S) changes by the rule*

$$\bar{\mathbf{B}} = \mathbf{B} + (\nabla^{\mathrm{v}}\nabla^{\mathrm{v}}\nabla^{\mathrm{v}}P) \otimes \widetilde{\delta} + (\nabla^{\mathrm{v}}\nabla^{\mathrm{v}}P) \odot \mathbf{1}. \qquad (8.4.20)$$

The change of the trace of the Berwald curvature is given by

$$\operatorname{tr} \bar{\mathbf{B}} = \operatorname{tr} \mathbf{B} + (n+1)\nabla^{\mathrm{v}}\nabla^{\mathrm{v}}P. \qquad (8.4.21)$$

Proof. For any vector fields X, Y, Z on M we have

$$i\bar{\mathbf{B}}(\widehat{X}, \widehat{Y}, \widehat{Z}) \overset{(7.13.3)}{=} [X^{\mathsf{v}}, [\bar{\mathcal{H}}(\widehat{Y}), Z^{\mathsf{v}}]] \overset{(8.4.4)}{=} [X^{\mathsf{v}}, [Y^{\mathsf{h}} - PY^{\mathsf{v}} - (Y^{\mathsf{v}}P)C, Z^{\mathsf{v}}]]$$

$$= i\mathbf{B}(\widehat{X}, \widehat{Y})\widehat{Z} - [X^{\mathsf{v}}, [PY^{\mathsf{v}}, Z^{\mathsf{v}}]] - [X^{\mathsf{v}}, [(Y^{\mathsf{v}}P)C, Z^{\mathsf{v}}]]$$

$$\overset{(4.1.52),(4.1.53)}{=} i\mathbf{B}(\widehat{X}, \widehat{Y})\widehat{Z} + X^{\mathsf{v}}(Z^{\mathsf{v}}P)Y^{\mathsf{v}} + [X^{\mathsf{v}}, (Y^{\mathsf{v}}P)Z^{\mathsf{v}}]$$

$$+ [X^{\mathsf{v}}, Z^{\mathsf{v}}(Y^{\mathsf{v}}P)C] = i\mathbf{B}(\widehat{X}, \widehat{Y})\widehat{Z} + X^{\mathsf{v}}(Z^{\mathsf{v}}P)Y^{\mathsf{v}}$$

$$+ X^{\mathsf{v}}(Y^{\mathsf{v}}P)Z^{\mathsf{v}} + Z^{\mathsf{v}}(Y^{\mathsf{v}}P)X^{\mathsf{v}} + X^{\mathsf{v}}(Z^{\mathsf{v}}(Y^{\mathsf{v}}P))C$$

$$= i\mathbf{B}(\widehat{X}, \widehat{Y})\widehat{Z} + (\nabla^{\mathsf{v}}\nabla^{\mathsf{v}}P(\widehat{X}, \widehat{Z}))Y^{\mathsf{v}} + (\nabla^{\mathsf{v}}\nabla^{\mathsf{v}}P(\widehat{X}, \widehat{Y}))Z^{\mathsf{v}}$$

$$+ (\nabla^{\mathsf{v}}\nabla^{\mathsf{v}}P(\widehat{Z}, \widehat{Y}))X^{\mathsf{v}} + (\nabla^{\mathsf{v}}\nabla^{\mathsf{v}}\nabla^{\mathsf{v}}P(\widehat{X}, \widehat{Z}, \widehat{Y}))C$$

$$= (i\mathbf{B} + \nabla^{\mathsf{v}}\nabla^{\mathsf{v}}\nabla^{\mathsf{v}}P \otimes C + \nabla^{\mathsf{v}}\nabla^{\mathsf{v}}P \odot i)(\widehat{X}, \widehat{Y}, \widehat{Z}),$$

taking into account in the last step that the vertical Hessian $\nabla^{\mathsf{v}}\nabla^{\mathsf{v}}P$ of P is symmetric (see Lemma 6.2.28(ii)), and $\nabla^{\mathsf{v}}\nabla^{\mathsf{v}}\nabla^{\mathsf{v}}P$ is totally symmetric, i.e., $\nabla^{\mathsf{v}}\nabla^{\mathsf{v}}\nabla^{\mathsf{v}}P \in S_3(\Gamma(\mathring{\pi}))$. (The latter can be checked in the same way as the symmetry of $\nabla^{\mathsf{v}}\nabla^{\mathsf{v}}P$.) Thus

$$i\bar{\mathbf{B}} = i\mathbf{B} + \nabla^{\mathsf{v}}\nabla^{\mathsf{v}}\nabla^{\mathsf{v}}P \otimes C + \nabla^{\mathsf{v}}\nabla^{\mathsf{v}}P \odot i.$$

Acting on both sides by the vertical mapping $\mathcal{V}$, we obtain (8.4.20).

The easiest way to prove (8.4.21) is to use local coordinates. With the notation of 8.1.4, the components of $\mathbf{B}$ are $G^i_{jkl} = \frac{\partial^3 G^i}{\partial y^j \partial y^k \partial y^l}$, where the functions G^i are the spray coefficients of S. Then (8.4.20) takes the form

$$\bar{G}^i_{jkl} = G^i_{jkl} + \frac{\partial^3 P}{\partial y^j \partial y^k \partial y^l}y^i + \sum_{\mathrm{cyc}(j,k,l)} \frac{\partial^2 P}{\partial y^j \partial y^k}\delta^i_l$$

$$= G^i_{jkl} + \frac{\partial^3 P}{\partial y^j \partial y^k \partial y^l}y^i + \frac{\partial^2 P}{\partial y^j \partial y^k}\delta^i_l + \frac{\partial^2 P}{\partial y^k \partial y^l}\delta^i_j + \frac{\partial^2 P}{\partial y^l \partial y^j}\delta^i_k,$$

therefore the components of $\operatorname{tr}\bar{\mathbf{B}}$ are

$$(\operatorname{tr}\bar{\mathbf{B}})_{kl} \overset{(1.2.20)}{=} \bar{G}^i_{ikl} = G^i_{ikl} + \frac{\partial^3 P}{\partial y^i \partial y^k \partial y^l}y^i + \frac{\partial^2 P}{\partial y^l \partial y^k} + n\frac{\partial^2 P}{\partial y^k \partial y^l} + \frac{\partial^2 P}{\partial y^l \partial y^k}.$$

By the 1^+-homogeneity of P the functions $\frac{\partial^2 P}{\partial y^k \partial y^l}$ $(k, l \in J_n)$ are -1^+-homogeneous. Hence

$$\frac{\partial^3 P}{\partial y^i \partial y^k \partial y^l}y^i = -\frac{\partial^2 P}{\partial y^k \partial y^l},$$

from which we obtain that

$$(\operatorname{tr}\bar{\mathbf{B}})_{kl} = (\operatorname{tr}\mathbf{B})_{kl} + (n+1)\frac{\partial^2 P}{\partial y^k \partial y^l}; \quad k, l \in J_n.$$

This is just the local form of (8.4.21). $\square$

Proposition 8.4.23. *The Berwald curvature of a spray manifold remains invariant under a projective change with projective factor P if, and only if, $\nabla^v\nabla^v P = 0$. All positive-homogeneous solutions of this partial differential equation are of the form*

$$\bar{\beta} = i_\xi \beta^v, \quad \beta \in \mathfrak{X}^*(M), \tag{8.4.22}$$

where ξ is an arbitrary semispray over M.

Proof. If $\nabla^v\nabla^v P = 0$, then from (8.4.20) we see that $\bar{\mathbf{B}} = \mathbf{B}$. Conversely, if $\bar{\mathbf{B}} = \mathbf{B}$, then we also have $\operatorname{tr}\bar{\mathbf{B}} = \operatorname{tr}\mathbf{B}$, and (8.4.21) implies that $\nabla^v\nabla^v P = 0$.

Next we show that the functions given by (8.4.22) are solutions of the partial differential equation $\nabla^v\nabla^v P = 0$. Indeed, let X and Y be vector fields on M. Then

$$Y^v\bar{\beta} = d\bar{\beta}(Y^v) = d\bar{\beta}\circ\mathbf{J}(Y^c) \stackrel{(4.1.97)}{=} (\mathcal{L}_\mathbf{J}\bar{\beta})(Y^c) \stackrel{(7.4.7)}{=} \beta^v(Y^c) \stackrel{(4.1.75)}{=} (\beta(Y))^v, \tag{$*$}$$

so

$$(\nabla^v\nabla^v\bar{\beta})(\widehat{X},\widehat{Y}) \stackrel{(6.2.40)}{=} X^v(Y^v\bar{\beta}) \stackrel{(*)}{=} X^v(\beta(Y))^v = 0,$$

therefore $\nabla^v\nabla^v\bar{\beta} = 0$, as claimed.

Conversely, suppose that a positive-homogeneous function P in $C^\infty(\mathring{T}M)$ satisfies $\nabla^v\nabla^v P = 0$. Then, for any vector fields $X, Y \in \mathfrak{X}(M)$,

$$0 = (\nabla^v\nabla^v P)(\widehat{X},\widehat{Y}) \stackrel{(6.2.40)}{=} X^v(Y^v P),$$

therefore by Lemma 4.1.36, the function $Y^v P$ is the vertical lift of a smooth function on M. Thus there is a mapping $\beta\colon \mathfrak{X}(M) \to C^\infty(M)$ such that $Y^v P = (\beta(Y))^v$ for each $Y \in \mathfrak{X}(M)$. We have

$$(\beta(fY))^v = (fY)^v P = f^v(Y^v P) = (f\beta(Y))^v$$

for any $Y \in \mathfrak{X}(M)$ and $f \in C^\infty(M)$, thus the mapping β is in fact $C^\infty(M)$-linear, i.e., $\beta \in \mathfrak{X}^*(M)$, and

$$Y^v P = (\beta(Y))^v \stackrel{(*)}{=} Y^v\bar{\beta}.$$

Since Y was arbitrary, there is a function $f \in C^\infty(M)$ such that $P = \bar{\beta} + f^v$. However, both P and $\bar{\beta}$ are 1^+-homogeneous, which forces that f is identically zero. $\quad\square$

Corollary and Definition 8.4.24. *Let (M, S) be a spray manifold, and let $\mathbf{B}$ be the Berwald curvature of (M, S). Then the tensor*

$$\mathbf{D} := \mathbf{B} - \frac{1}{n+1}\big((\nabla^v \operatorname{tr}\mathbf{B}) \otimes \widetilde{\delta} + (\operatorname{tr}\mathbf{B}) \odot \mathbf{1}\big) \tag{8.4.23}$$

is projectively invariant. It is called the Douglas tensor *of the spray manifold.*

Proof. Consider a projective change of S with projective factor P. Then

$$\bar{\mathbf{D}} := \bar{\mathbf{B}} - \frac{1}{n+1}\big((\nabla^{\mathsf{v}} \operatorname{tr} \bar{\mathbf{B}}) \otimes \tilde{\delta} + (\operatorname{tr} \bar{\mathbf{B}}) \odot \mathbf{1}\big)$$

$$\overset{(8.4.20),(8.4.21)}{=} \mathbf{B} + (\nabla^{\mathsf{v}}\nabla^{\mathsf{v}}\nabla^{\mathsf{v}} P) \otimes \tilde{\delta} + (\nabla^{\mathsf{v}}\nabla^{\mathsf{v}} P) \odot \mathbf{1}$$

$$- \frac{1}{n+1}\big((\nabla^{\mathsf{v}} \operatorname{tr} \mathbf{B}) \otimes \tilde{\delta} + (n+1)(\nabla^{\mathsf{v}}\nabla^{\mathsf{v}}\nabla^{\mathsf{v}} P) \otimes \tilde{\delta}\big)$$

$$- \frac{1}{n+1}\big((\operatorname{tr} \mathbf{B}) \odot \mathbf{1} + (n+1)(\nabla^{\mathsf{v}}\nabla^{\mathsf{v}} P) \odot \mathbf{1}\big)$$

$$= \mathbf{B} - \frac{1}{n+1}\big((\nabla^{\mathsf{v}} \operatorname{tr} \mathbf{B}) \otimes \tilde{\delta} + (\operatorname{tr} \mathbf{B}) \odot \mathbf{1}\big) =: \mathbf{D},$$

as was to be shown. $\qquad\square$

Remark 8.4.25. With the notation of subsection 8.1.4, we determine the components of the Douglas tensor with respect to the chart $(\mathcal{U}, (u^i)_{i=1}^n)$ fixed there:

$$D^i_{jkl} \widehat{\frac{\partial}{\partial u^i}} := \mathbf{D}\left(\widehat{\frac{\partial}{\partial u^j}}, \widehat{\frac{\partial}{\partial u^k}}\right)\widehat{\frac{\partial}{\partial u^l}} \overset{(8.4.23)}{=} \mathbf{B}\left(\widehat{\frac{\partial}{\partial u^j}}, \widehat{\frac{\partial}{\partial u^k}}\right)\widehat{\frac{\partial}{\partial u^l}}$$

$$- \frac{1}{n+1}(\nabla_{\frac{\partial}{\partial y^j}} \operatorname{tr} \mathbf{B})\left(\widehat{\frac{\partial}{\partial u^k}}, \widehat{\frac{\partial}{\partial u^l}}\right)\tilde{\delta}$$

$$- \frac{1}{n+1}\sum_{\operatorname{cyc}(j,k,l)}(\operatorname{tr} \mathbf{B})\left(\widehat{\frac{\partial}{\partial u^j}}, \widehat{\frac{\partial}{\partial u^k}}\right)\widehat{\frac{\partial}{\partial u^l}}$$

$$= \left(G^i_{jkl} - \frac{1}{n+1}\left(\frac{\partial G_{kl}}{\partial y^j}y^i + \sum_{\operatorname{cyc}(j,k,l)}G_{jk}\delta^i_l\right)\right)\widehat{\frac{\partial}{\partial u^i}},$$

therefore

$$D^i_{jkl} = G^i_{jkl} - \frac{1}{n+1}\left(G_{kl,j}y^i + \sum_{\operatorname{cyc}(j,k,l)}G_{jk}\delta^i_l\right) \qquad (8.4.24)$$

$$= G^i_{jkl} - \frac{1}{n+1}(G_{kl,j}y^i + G_{jk}\delta^i_l + G_{kl}\delta^i_j + G_{lj}\delta^i_k),$$

where $G_{kl} := G^i_{ikl}$, $G_{kl,j} \overset{(7.11.7)}{=} \frac{\partial G_{kl}}{\partial y^j}$; cf. formula (5.11) in Douglas's fundamental paper [46].

Lemma 8.4.26. *The Douglas tensor of a spray manifold is symmetric, it has the property that*

$$\tilde{\delta} \in \{\tilde{X}, \tilde{Y}, \tilde{Z}\} \quad \text{implies} \quad \mathbf{D}(\tilde{X}, \tilde{Y})\tilde{Z} = 0, \qquad (8.4.25)$$

and its trace vanishes.

Proof. By Corollary 7.13.4, the Berwald curvature $\mathbf{B}$ is (totally) symmetric, hence $\operatorname{tr} \mathbf{B}$ is symmetric. The symmetry of $\mathbf{B}$ and the Bianchi identity (B1) (see 8.1.3 or (7.13.11)) imply the symmetry of $\nabla^{\mathsf{v}}\mathbf{B}$. Since for all $\widetilde{X}, \widetilde{Y}, \widetilde{Z} \in \Gamma(\mathring{\pi})$ we have

$$(\nabla^{\mathsf{v}}\operatorname{tr}\mathbf{B})(\widetilde{X}, \widetilde{Y}, \widetilde{Z}) = (\nabla_{\mathbf{i}\widetilde{X}}\operatorname{tr}\mathbf{B})(\widetilde{Y}, \widetilde{Z}) = (\operatorname{tr}\nabla_{\mathbf{i}\widetilde{X}}\mathbf{B})(\widetilde{Y}, \widetilde{Z}),$$

it follows that $\nabla^{\mathsf{v}}\operatorname{tr}\mathbf{B}$ is also symmetric, which proves the symmetry of $\mathbf{D}$:

$$\mathbf{D}(\widetilde{X}_{\sigma(1)}, \widetilde{X}_{\sigma(2)})\widetilde{X}_{\sigma(3)} = \mathbf{D}(\widetilde{X}_1, \widetilde{X}_2)\widetilde{X}_3 \text{ for all } \sigma \in S_3. \qquad (8.4.26)$$

Next we show that $i_{\widetilde{\delta}}\mathbf{D} = 0$; by (8.4.26) this implies (8.4.25). Let $\widetilde{Y}, \widetilde{Z} \in \Gamma(\mathring{\pi})$. Then

$$i_{\widetilde{\delta}}\mathbf{D}(\widetilde{Y}, \widetilde{Z}) = i_{\widetilde{\delta}}\mathbf{B}(\widetilde{Y}, \widetilde{Z}) - \frac{1}{n+1}(\nabla_C\operatorname{tr}\mathbf{B})(\widetilde{Y}, \widetilde{Z})\widetilde{\delta}$$

$$- \frac{1}{n+1}\left((\operatorname{tr}\mathbf{B}(\widetilde{\delta}, \widetilde{Y}))\widetilde{Z} + (\operatorname{tr}\mathbf{B}(\widetilde{Y}, \widetilde{Z}))\widetilde{\delta} + (\operatorname{tr}\mathbf{B}(\widetilde{Z}, \widetilde{\delta}))\widetilde{Y}\right)$$

$$\overset{(7.13.6)}{=} -\frac{1}{n+1}\left((\nabla_C\operatorname{tr}\mathbf{B})(\widetilde{Y}, \widetilde{Z})\widetilde{\delta} + (\operatorname{tr}\mathbf{B}(\widetilde{Y}, \widetilde{Z}))\widetilde{\delta}\right)$$

$$\overset{(7.13.7)}{=} -\frac{1}{n+1}\left(-(\operatorname{tr}\mathbf{B}(\widetilde{Y}, \widetilde{Z}))\widetilde{\delta} + (\operatorname{tr}\mathbf{B}(\widetilde{Y}, \widetilde{Z}))\widetilde{\delta}\right) = 0,$$

as desired.

To prove that $\mathbf{D}$ is trace-free, we use the coordinate expression (8.4.24):

$$(\operatorname{tr}\mathbf{D})_{kl} \overset{(1.2.20)}{=} D^i_{ikl} = G_{kl} - \frac{1}{n+1}\frac{\partial G_{kl}}{\partial y^i}y^i - \frac{1}{n+1}(G_{lk} + nG_{kl} + G_{lk})$$

$$= G_{kl} + \frac{1}{n+1}G_{kl} - \frac{n+2}{n+1}G_{kl} = 0,$$

taking into account that the functions G_{jk} are -1^+-homogeneous because, as we just have seen, $\nabla_C\operatorname{tr}\mathbf{B} = -\operatorname{tr}\mathbf{B}$. $\qquad\square$

8.4.6 The Meaning of the Douglas Tensor

Lemma 8.4.27. *A spray is affine if, and only if, its Berwald curvature vanishes.*

Proof. Let S be a spray over a manifold M, and let $\mathcal{H}$ be the Berwald connection of (M, S). Then, as we saw in 8.1.1, $\mathcal{H}$ is torsion-free. We recall that by the Berwald curvature of S we mean the Berwald curvature $\mathbf{B}$ of $\mathcal{H}$.

If $\mathbf{B} = 0$, then $\mathcal{H}$ is a torsion-free *linear* Ehresmann connection by Proposition 7.15.1, so from Proposition 7.15.4 it follows that $\mathcal{H}$ is generated by an *affine* spray $\bar{S}$. Then, in view of the last assertion of Proposition

7.3.4, for the associated semispray of $\mathcal{H}$ we have $\mathcal{H}\tilde{\delta} = S$ and $\mathcal{H}\tilde{\delta} = \bar{S}$. Thus $S = \bar{S}$, whence S is an affine spray.

Conversely, suppose that the spray S is affine. Then $\mathcal{H}$ is a torsion-free linear Ehresmann connection by the equivalence (i)⇔(iii) in Proposition 7.15.4, so the implication (i)⇒(iv) in Proposition 7.15.1 assures that the Berwald curvature of S vanishes. □

Corollary 8.4.28. *If a spray is projectively related to an affine spray, then its Douglas tensor vanishes.*

Proof. By the previous lemma, the Berwald curvature of an affine spray vanishes. Hence its Douglas tensor also vanishes by (8.4.23). Since the Douglas tensor is projectively invariant, our statement follows. □

We devote the rest of this subsection to the proof of the converse of Corollary 8.4.28: *if a spray has vanishing Douglas tensor then it is projectively related to an affine spray.* To do this we need some preparation. As a first step, we introduce the *divergence* of a vector field on a manifold with volume form, and establish some of its basic properties.

Lemma and Definition 8.4.29. *Suppose that M is an orientable manifold and let μ be a volume form on M.*

(i) *Given a vector field X on M, there exists a unique smooth function $\operatorname{div}_\mu X$ on M such that*

$$\mathcal{L}_X \mu = (\operatorname{div}_\mu X)\mu. \tag{8.4.27}$$

This function is called the divergence *of X with respect to the volume form μ.*

(ii) *For $f \in C^\infty(M)$,*

$$\operatorname{div}_\mu fX = f\operatorname{div}_\mu X + Xf. \tag{8.4.28}$$

(iii) *If $(\mathcal{U}, (u^i)_{i=1}^n)$ is a chart on M, and*

$$\mu \underset{(\mathcal{U})}{=} \sigma du^1 \wedge \cdots \wedge du^n, \quad X \underset{(\mathcal{U})}{=} \sum_{i=1}^n X^i \frac{\partial}{\partial u^i},$$

then

$$\operatorname{div}_\mu X = \sum_{k=1}^n \frac{1}{\sigma} \frac{\partial(\sigma X^k)}{\partial u^k}. \tag{8.4.29}$$

Proof. Since the volume form μ constitutes a basis for the 1-dimensional $C^\infty(M)$-module $\mathcal{A}_n(M)$, every n-form on M can uniquely be obtained as a multiple of μ. Thus the function $\mathrm{div}_\mu X \in C^\infty(M)$ exists and is uniquely determined by (8.4.27).

Secondly, we derive the coordinate expression (8.4.29):

$$\mathcal{L}_X\mu \overset{(3.3.33)}{=} i_X d\mu + d i_X \mu = d i_X \mu \underset{(\mathcal{U})}{=} d i_X(\sigma du^1 \wedge \cdots \wedge du^n)$$

$$= d(\sigma i_X(du^1 \wedge \cdots \wedge du^n))$$

$$\overset{(3.3.28)}{=} d\left(\sigma \sum_{k=1}^n (-1)^{k-1} X^k du^1 \wedge \cdots \wedge \overset{\vee}{du}^k \wedge \cdots \wedge du^n\right)$$

$$= \sum_{k=1}^n (-1)^{k-1}\left(\sum_{j=1}^n \frac{\partial(\sigma X^k)}{\partial u^j} du^j \wedge du^1 \wedge \cdots \wedge \overset{\vee}{du}^k \wedge \cdots \wedge du^n\right)$$

$$= \sum_{k=1}^n \left(\frac{1}{\sigma}\frac{\partial(\sigma X^k)}{\partial u^k}\right)\sigma du^1 \wedge \cdots \wedge du^n = \left(\sum_{k=1}^n \frac{1}{\sigma}\frac{\partial(\sigma X^k)}{\partial u^k}\right)\mu,$$

from which we conclude the desired expression for $\mathrm{div}_\mu X$. Using (8.4.29), we easily obtain (8.4.28). $\qquad\square$

Remark 8.4.30. (a) A coordinate-free proof of the rule (8.4.28) can be found, e.g., in [2], p. 456.

(b) Observe that if we replace the volume form μ with the opposite volume form $-\mu$, then in the coordinate expression of μ the function σ changes to $-\sigma$, therefore the coordinate expression (8.4.29) remains invariant. For more details, we refer to [66], p. 400.

Definition 8.4.31. Let M be a manifold. We say that a volume form ω on the tangent manifold TM is *vertically invariant* if $\mathcal{L}_{X^v}\omega = 0$ for all $X \in \mathfrak{X}(M)$.

Lemma 8.4.32. *Let M be a manifold.*

(i) *A volume form ω on TM is vertically invariant if, and only if, $\mathrm{div}_\omega X^v = 0$ for all $X \in \mathfrak{X}(M)$.*

(ii) *Suppose that ω is a vertically invariant volume form on TM, and let $\widetilde\omega$ be another volume form on TM. Then $\widetilde\omega$ is vertically invariant if, and only if, $\widetilde\omega = f^v\omega$ for some smooth function f on M.*

Proof. The first assertion is clear from the definition of divergence. As to the second claim, we surely have $\widetilde\omega = h\omega$ for some smooth function h on

TM. If $\widetilde{\omega}$ is vertically invariant, then for any vector field X on M,

$$0 = \mathcal{L}_{X^\vee}\widetilde{\omega} = \mathcal{L}_{X^\vee}(h\omega) = (X^\vee h)\omega + h\mathcal{L}_{X^\vee}\omega = (X^\vee h)\omega.$$

Thus $X^\vee h = 0$ for all $X \in \mathfrak{X}(M)$, hence h is a vertical lift by Lemma 4.1.36. The converse can be shown by a similar argument. □

Remark 8.4.33. Let $(\mathcal{U}, (u^i)_{i=1}^n)$ be a chart on M with induced chart $(\tau^{-1}(\mathcal{U}), ((x^i)_{i=1}^n, (y^i)_{i=1}^n))$ on TM. Then

$$\mu := dx^1 \wedge \cdots \wedge dx^n \wedge dy^1 \wedge \cdots \wedge dy^n$$

is a vertically invariant volume form on $\tau^{-1}(\mathcal{U})$.

Indeed, if $X = X^i \frac{\partial}{\partial u^i}$ is a vector field on $\mathcal{U}$, then its vertical lift is $X^\vee = (X^i)^\vee \frac{\partial}{\partial y^i} \in \mathfrak{X}(\tau^{-1}(\mathcal{U}))$, and

$$\mathrm{div}_\mu(X^\vee) \overset{(8.4.29)}{=} \sum_{i=1}^n \frac{\partial(X^i)^\vee}{\partial y^i} = 0,$$

so our claim follows by part (i) of the previous lemma. We say that μ is an *induced volume form* on $\tau^{-1}(\mathcal{U})$. In view of part (ii) of Lemma 8.4.32, every vertically invariant volume form on TM can locally be written as $f^\vee \mu$, where f is a smooth function on M.

Lemma 8.4.34. *Let M be an n-dimensional manifold. If ω is a vertically invariant volume form on TM and $C \in \mathfrak{X}(TM)$ is the Liouville vector field, then $\mathrm{div}_\omega C = n$.*

Proof. We argue locally. With the notation of the previous remark, let

$$\omega \underset{(\mathcal{U})}{=} f^\vee \mu, \quad C \underset{(\mathcal{U})}{=} y^i \frac{\partial}{\partial y^i}.$$

Then

$$\mathrm{div}_\omega C \overset{(8.4.29)}{\underset{(\mathcal{U})}{=}} \sum_{i=1}^n \frac{1}{f^\vee} \frac{\partial(f^\vee y^i)}{\partial y^i} = \sum_{i=1}^n \frac{\partial y^i}{\partial y^i} = n.$$

□

Proposition 8.4.35. *The tangent manifold of a manifold admits a vertically invariant volume form, therefore it is always orientable.*

Proof. Let M be an n-dimensional manifold and consider its $2n$-dimensional tangent manifold TM. Suppose that $(u^i)_{i=1}^n$ and $(\bar{u}^i)_{i=1}^n$ are local coordinate systems on M with the same domain $\mathcal{U}$, and consider the induced coordinate systems $((x^i)_{i=1}^n, (y^i)_{i=1}^n)$ and $((\bar{x}^i)_{i=1}^n, (\bar{y}^i)_{i=1}^n)$ on TM.

Step 1. We show that

$$d\bar{x}^i = (A^i_j)^{\vee}dx^j, \quad d\bar{y}^i = (A^i_j)^{c}dx^j + (A^i_j)^{\vee}dy^j, \quad i \in J_n, \qquad (8.4.30)$$

where $A^i_j := \frac{\partial \bar{u}^i}{\partial u^j}$; $i, j \in J_n$. This can be done by an immediate computation:

$$d\bar{x}^i \overset{(3.3.3)}{=} \frac{\partial \bar{x}^i}{\partial x^j}dx^j + \frac{\partial \bar{x}^i}{\partial y^j}dy^j \overset{(4.1.18),(4.1.73)}{=} \left(\left(\frac{\partial}{\partial u^j} \right)^{c} (\bar{u}^i)^{\vee} \right) dx^j$$

$$+ \left(\left(\frac{\partial}{\partial u^j} \right)^{\vee} (\bar{u}^i)^{\vee} \right) dy^j \overset{(4.1.65), \text{Lemma } 4.1.36}{=} \left(\frac{\partial \bar{u}^i}{\partial u^j} \right)^{\vee} dx^j = (A^i_j)^{\vee}dx^j;$$

$$d\bar{y}^i = \frac{\partial \bar{y}^i}{\partial x^j}dx^j + \frac{\partial \bar{y}^i}{\partial y^j}dy^j = \left(\left(\frac{\partial}{\partial u^j} \right)^{c} (\bar{u}^i)^{c} \right) dx^j + \left(\left(\frac{\partial}{\partial u^j} \right)^{\vee} (\bar{u}^i)^{c} \right) dy^j$$

$$\overset{(4.1.66),(4.1.51)}{=} \left(\frac{\partial \bar{u}^i}{\partial u^j} \right)^{c} dx^j + \left(\frac{\partial \bar{u}^i}{\partial u^j} \right)^{\vee} dy^j = (A^i_j)^{c}dx^j + (A^i_j)^{\vee}dy^j.$$

Step 2. Consider the induced volume forms

$$\mu = dx^1 \wedge \cdots \wedge dx^n \wedge dy^1 \wedge \cdots \wedge dy^n$$

and

$$\bar{\mu} = d\bar{x}^1 \wedge \cdots \wedge d\bar{x}^n \wedge d\bar{y}^1 \wedge \cdots \wedge d\bar{y}^n.$$

We claim that μ and $\bar{\mu}$ represent the same orientation of the vector bundle $\tau_{TM} \upharpoonright \tau^{-1}(\mathcal{U})$. To see this, we express the volume form $\bar{\mu}$ in terms of the coordinate differentials dx^i and dy^i. Since the wedge product is $C^{\infty}(TM)$-multilinear and $\alpha \wedge \alpha = 0$ for any one-form α, we find that

$$\bar{\mu}_1 := d\bar{x}^1 \wedge \cdots \wedge d\bar{x}^n \overset{(8.4.30)}{=} (A^1_{j_1})^{\vee}dx^{j_1} \wedge \cdots \wedge (A^n_{j_n})^{\vee}dx^{j_n}$$

$$= \sum_{\sigma \in S_n} (A^1_{\sigma(1)})^{\vee}dx^{\sigma(1)} \wedge \cdots \wedge (A^n_{\sigma(n)})^{\vee}dx^{\sigma(n)}$$

$$= \left(\sum_{\sigma \in S_n} \varepsilon(\sigma)(A^1_{\sigma(1)})^{\vee} \ldots (A^n_{\sigma(n)})^{\vee} \right) dx^1 \wedge \cdots \wedge dx^n$$

$$\overset{(1.3.18)}{=} \det((A^i_j)^{\vee})dx^1 \wedge \cdots \wedge dx^n = (\det(A^i_j))^{\vee}dx^1 \wedge \cdots \wedge dx^n.$$

Thus

$$\bar{\mu}_1 \wedge d\bar{y}^1 = (\det(A^i_j))^{\vee}dx^1 \wedge \cdots \wedge dx^n \wedge ((A^1_j)^{c}dx^j + (A^1_j)^{\vee}dy^j)$$

$$= (\det(A^i_j))^{\vee}dx^1 \wedge \cdots \wedge dx^n \wedge ((A^1_j)^{\vee}dy^j) = \bar{\mu}_1 \wedge ((A^1_j)^{\vee}dy^j).$$

Continuing in the same way, we obtain that

$$\bar{\mu} = \bar{\mu}_1 \wedge d\bar{y}^1 \wedge \cdots \wedge d\bar{y}^n = \bar{\mu}_1 \wedge (A^1_{j_1})^{\vee}dy^{j_1} \wedge \cdots \wedge (A^n_{j_n})^{\vee}dy^{j_n}$$

$$= \bar{\mu}_1 \wedge (\det(A^i_j))^{\vee} \wedge dy^1 \wedge \cdots \wedge dy^n = ((\det(A^i_j))^2)^{\vee}\mu.$$

Here the function $((\det(A^i_j))^2)^{\vee}$ is positive, so μ and $\bar{\mu}$ represent the same orientation.

Step 3. Let $\mathcal{A} = \{(\mathcal{U}_i, u_i) \mid i \in I\}$ be an atlas of M, and let $(f_i)_{i \in I}$ be a partition of unity on M subordinate to the covering $(\mathcal{U}_i)_{i \in I}$. Every chart $(\mathcal{U}_i, u_i)$ yields an induced volume form μ_i on $\tau^{-1}(\mathcal{U})$. Define a top form ω on TM by

$$v \in TM \mapsto \omega_v := \sum_{i \in I}(f_i)^{\vee}(v)\mu_i(v) \in A_{2n}(T_vTM).$$

For an arbitrarily fixed vector w in TM, let

$$I(w) := \{i \in I \mid \tau(w) \in \mathcal{U}_i, \ f_i(\tau(v)) \neq 0\} \quad \text{and} \quad \mathcal{U}_w := \bigcap_{j \in I(w)} \mathcal{U}_j.$$

Then, by Step 2,

$$\mu_j \underset{(\mathcal{U}_w)}{=} (h_j)^{\vee}\mu_{j_0}, \quad j \in I(w),$$

where $j_0 \in I(w)$ is a fixed index and h_j is a positive smooth function on $\mathcal{U}_w$. So it follows that

$$\omega \underset{(\mathcal{U}_w)}{=} \sum_{j \in I(w)} (f_j h_j)^{\vee}\mu_{j_0},$$

and here $\sum_{j \in I(w)}(f_j h_j)^{\vee} \in C^{\infty}(\tau^{-1}(\mathcal{U}_w))$ is a positive function. Thus, taking into account Remark 8.4.33, ω is a vertically invariant volume form on $\tau^{-1}(\mathcal{U}_w)$. Since the vector w can be chosen arbitrarily, this completes the proof. $\square$

Lemma 8.4.36. *Let (M, S) be a spray manifold. If ω is a vertically invariant volume form on TM, then*

$$\nabla^{\vee}\nabla^{\vee}(\mathrm{div}_{\omega} S) = -2\,\mathrm{tr}\,\mathbf{B}. \tag{8.4.31}$$

Proof. With the notation of Remark 8.4.33, we calculate locally. Then $\omega \underset{(\mathcal{U})}{=} f^{\vee}\mu$, where f is a smooth function on $\mathcal{U}$ and μ is the induced volume form on $\tau^{-1}(\mathcal{U})$. If $S \underset{(\mathcal{U})}{=} y^i\frac{\partial}{\partial x^i} - 2G^i\frac{\partial}{\partial y^i}$, then

$$\mathrm{div}_{\omega} S \underset{(\mathcal{U})}{\overset{(8.4.29)}{=}} \sum_{i=1}^{n}\left(\frac{1}{f^{\vee}}\frac{\partial(f^{\vee}y^i)}{\partial x^i} - \frac{2}{f^{\vee}}\frac{\partial(f^{\vee}G^i)}{\partial y^i}\right)$$

$$\overset{(4.1.65)}{=} \sum_{i=1}^{n}\left(\frac{y^i}{f^{\vee}}\left(\frac{\partial f}{\partial u^i}\right)^{\vee} - 2\frac{\partial G^i}{\partial y^i}\right).$$

Thus, taking into account (4.1.17),

$$\operatorname{div}_\omega S \underset{(\mathcal{U})}{=} \frac{f^c}{f^v} - 2\sum_{i=1}^n \frac{\partial G^i}{\partial y^i} = \frac{f^c}{f^v} - 2G^i_i. \qquad (8.4.32)$$

Now we turn the the proof of (8.4.31). For any $j,k \in J_n$,

$$\nabla^v\nabla^v(\operatorname{div}_\omega S)\left(\widehat{\frac{\partial}{\partial u^j}}, \widehat{\frac{\partial}{\partial u^k}}\right) \overset{(8.4.32)}{=} \nabla^v\nabla^v\left(\frac{f^c}{f^v} - 2G^i_i\right)\left(\widehat{\frac{\partial}{\partial u^j}}, \widehat{\frac{\partial}{\partial u^k}}\right)$$

$$\overset{(6.2.40)}{=} \frac{\partial}{\partial y^j}\left(\frac{\partial}{\partial y^k}\left(\frac{f^c}{f^v} - 2G^i_i\right)\right)$$

$$\overset{(4.1.51)}{=} \left(\frac{\partial}{\partial u^j}\right)^v\left(\left(\frac{\partial f}{\partial u^k}\right)^v\frac{1}{f^v}\right) - 2\frac{\partial^2 G^i_i}{\partial y^j\partial y^k} = -2\frac{\partial^2 G^i_i}{\partial y^j\partial y^k}$$

$$\overset{(7.13.4)}{=} -2G^i_{jik} \overset{8.1.3(\text{iii})}{=} -2G^i_{ijk} = -2\operatorname{tr}\mathbf{B}\left(\widehat{\frac{\partial}{\partial u^j}}, \widehat{\frac{\partial}{\partial u^k}}\right),$$

as was to be shown. □

Theorem 8.4.37 (J. Douglas). *If the Douglas tensor of a spray manifold vanishes, then the spray is projectively related to an affine spray.*

Proof. Let M be an n-dimensional manifold. Suppose that S is a spray over M such that $\mathbf{D} = 0$. We construct an affine spray over M which is projectively related to S. Choosing a vertically invariant volume form ω on TM, let

$$\bar{S} := S - 2PC \quad \text{where} \quad P := \frac{1}{2(n+1)}\operatorname{div}_\omega S. \qquad (*)$$

We claim that the function P is 1^+-homogeneous. Indeed,

$$CP = \frac{1}{2(n+1)}C\operatorname{div}_\omega S \overset{(8.4.32)}{\underset{(\mathcal{U})}{=}} \frac{1}{2(n+1)}\left(C\frac{f^c}{f^v} - C2G^i_i\right)$$

$$\overset{(4.1.50)}{=} \frac{1}{2(n+1)}\left(\frac{f^c}{f^v} - 2CG^i_i\right) = \frac{1}{2(n+1)}\left(\frac{f^c}{f^v} - 2G^i_i\right) \underset{(\mathcal{U})}{=} P,$$

taking into account that the Christoffel symbols G^i_j of the Berwald connection of (M,S) are also 1^+-homogeneous. Thus relation $(*)$ is a projective change of S. The vertical Hessian of P is

$$\nabla^v\nabla^v P = \frac{1}{2(n+1)}\nabla^v\nabla^v(\operatorname{div}_\omega S) \overset{(8.4.31)}{=} -\frac{1}{n+1}\operatorname{tr}\mathbf{B}. \qquad (**)$$

For the Berwald curvature of $\bar{S}$ we obtain that

$$\bar{\mathbf{B}} \overset{(8.4.20)}{=} \mathbf{B} + (\nabla^v\nabla^v\nabla^v P)\otimes\tilde{\delta} + (\nabla^v\nabla^v P)\odot\mathbf{1}$$

$$\overset{(**)}{=} \mathbf{B} - \frac{1}{n+1}\left((\nabla^v\operatorname{tr}\mathbf{B})\otimes\tilde{\delta} + (\operatorname{tr}\mathbf{B})\odot\mathbf{1}\right) \overset{(8.4.23)}{=:} \mathbf{D} = 0.$$

So, by Lemma 8.4.27, $\bar{S}$ is an affine spray. This concludes the proof. □

Remark 8.4.38. A local version of Theorem 8.4.37 was proved by J. Douglas [46]. A global version is due to Z. Shen [88]. Shen, however, assumed that the base manifold is endowed with a volume form. In our version this assumption is eliminated.

8.5 Integrability and Flatness

Definition 8.5.1. Let $\mathcal{H}$ be an (everywhere smooth) Ehresmann connection in TM. We say that $\mathcal{H}$ is *integrable* if for each $v_0 \in TM$ there exists an open neighbourhood $\mathcal{U}$ of $\tau(v_0)$ and a vector field $Z \in \mathfrak{X}(\mathcal{U})$ such that $Z(\tau(v_0)) = v_0$ and

$$Z_*(v) = \mathcal{H}(Z(\tau(v)), v) \text{ for all } v \in \tau^{-1}(\mathcal{U}). \tag{8.5.1}$$

Remark 8.5.2. Taking into account (7.2.17), condition (8.5.1) can be reformulated as follows: *for every vector field X on $\mathcal{U}$, X and X^{h} are Z-related, i.e.,*

$$Z_* \circ X = X^{\mathsf{h}} \circ Z. \tag{8.5.2}$$

To see the local meaning of the integrability condition (8.5.1) or (8.5.2), choose a chart $(\mathcal{U}, (u^i)_{i=1}^n)$ on M with induced chart $(\tau^{-1}(\mathcal{U}), ((x^i)_{i=1}^n, (y^i)_{i=1}^n))$ on TM. If

$$Z \underset{(\mathcal{U})}{=} Z^i \frac{\partial}{\partial u^i}, \quad \mathcal{H}\left(\widehat{\frac{\partial}{\partial u^j}}\right) = \frac{\partial}{\partial x^j} - N_j^i \frac{\partial}{\partial y^i} \quad (j \in J_n),$$

then applying (3.1.20) we find that the local expression of (8.5.1) and (8.5.2) is the following:

$$\frac{\partial Z^i}{\partial u^j} = -N_j^i \circ Z, \quad i, j \in J_n. \tag{8.5.3}$$

Thus, roughly speaking, *the Ehresmann connection $\mathcal{H}$ is integrable if, and only if, the system of partial differential equations* (8.5.3) *has a solution.*

Lemma 8.5.3. *We use the same notation as above. Given a connected open neighbourhood $\mathcal{W}$ of $\tau(v_0)$, there is at most one vector field Z defined on $\mathcal{W}$ which satisfies $Z(\tau(v_0)) = v_0$ and (8.5.1).*

Proof. Let q be an arbitrarily chosen point in $\mathcal{W}$. By the connectedness of $\mathcal{W}$, there is a smooth curve $\gamma : I \to \mathcal{W}$ such that $\gamma(t_0) = \tau(v_0)$ and $\gamma(t_1) = q$ for some $t_0, t_1 \in I$ (Proposition 2.1.26). Suppose that Z is a vector field on $\mathcal{W}$ satisfying $Z(\tau(v_0)) = v_0$ and (8.5.1). Then for every $t \in I$,

$$(Z \circ \gamma)^{\cdot}(t) \overset{(3.1.11)}{=} Z_*(\dot{\gamma}(t)) \overset{(8.5.1)}{=} \mathcal{H}\big(Z(\tau(\dot{\gamma}(t))), \dot{\gamma}(t)\big) = \mathcal{H}(Z \circ \gamma(t), \dot{\gamma}(t)).$$

Thus $Z \circ \gamma$ satisfies (7.6.1), and in this sense it is $\mathcal{H}$-parallel, although $\mathcal{H}$ has not been assumed to be positive-homogeneous (cf. Definition 7.6.1). Choose a local coordinate system $(u^i)_{i=1}^n$ on an open neighbourhood of $\tau(v_0)$, and let $X^i := u^i \circ Z \circ \gamma$ $(i \in J_n)$. Then we obtain the system of ordinary differential equations

$$(X^i)' = -(\gamma^j)'(N_j^i \circ X), \quad i \in J_n \tag{$*$}$$

$(\gamma^j := u^j \circ \gamma,\ X := Z \circ \gamma)$ for the functions X^i with initial condition

$$X(t_0) = Z(\gamma(t_0)) = Z(\tau(v_0)) = v_0;$$

cf. (7.6.2). By Remark 3.2.4, there is an open interval I_0 containing t_0 such that $(*)$ has a *unique* solution $X \colon I_0 \to T\mathcal{W}$. This implies the uniqueness statement of the lemma. $\qquad\square$

Proposition 8.5.4 (Integrability Theorem for Ehresmann Connections). *An Ehresmann connection is integrable if, and only if, its curvature vanishes.*

Proof. Let $\mathcal{H}$ be an Ehresmann connection in TM, and suppose first that $\mathcal{H}$ is integrable. Then, given a vector $v_0 \in TM$, there is a vector field Z, defined on an open neighbourhood $\mathcal{U}$ of $\tau(v_0)$, such that $Z(\tau(v_0)) = v_0$ and (8.5.1) is satisfied. Applying the equivalence of (8.5.1) and (8.5.2), we show that the curvature of $\mathcal{H}$ vanishes.

Let X and Y be any vector fields on M. By (8.5.2) and by the related vector field lemma, $X \underset{Z}{\sim} X^{\mathsf{h}}$, $Y \underset{Z}{\sim} Y^{\mathsf{h}}$, whence $[X, Y] \underset{Z}{\sim} [X^{\mathsf{h}}, Y^{\mathsf{h}}]$. Thus

$$[X^{\mathsf{h}}, Y^{\mathsf{h}}](v_0) = [X^{\mathsf{h}}, Y^{\mathsf{h}}] \circ Z(\tau(v_0)) = Z_* \circ [X, Y](\tau(v_0))$$

$$\overset{(8.5.2)}{=} [X, Y]^{\mathsf{h}} \circ Z(\tau(v_0)) = [X, Y]^{\mathsf{h}}(v_0).$$

Therefore, by the arbitrariness of v_0,

$$[X^{\mathsf{h}}, Y^{\mathsf{h}}] = [X, Y]^{\mathsf{h}},$$

and hence

$$i\mathfrak{R}(\widehat{X}, \widehat{Y}) \overset{(7.8.3)}{=} \mathbf{v}[X^{\mathsf{h}}, Y^{\mathsf{h}}] = \mathbf{v}[X, Y]^{\mathsf{h}} = 0,$$

as desired.

Conversely, suppose that the curvature of $\mathcal{H}$ is zero. Let v_0 be any fixed vector in TM, and choose a chart $(\mathcal{U}, u) = (\mathcal{U}, (u^i)_{i=1}^n)$ of M centred on $\tau(v_0) =: p_0$. We shall write

$$X_i := \frac{\partial}{\partial u^i}, \quad i \in J_n \tag{1}$$

for short. As in the proof of Proposition 3.2.35, define the mapping

$$\varphi \colon (\alpha^1, \ldots, \alpha^n) \in u(\mathcal{U}) \mapsto \varphi_{\alpha^1}^{X_1} \circ \cdots \circ \varphi_{\alpha^n}^{X_n}(p_0) \in M.$$

Then, by the quoted proposition and by Corollary 3.2.36,

$$\varphi = u^{-1} \quad \text{and} \quad \frac{\partial}{\partial e^i} \underset{\varphi}{\sim} X_i \quad (i \in J_n), \tag{2}$$

where (e^i) is the canonical coordinate system of $\mathbb{R}^n$. Since $\mathcal{R}(\widehat{X}_i, \widehat{X}_j) = 0$, from (7.8.3) we obtain that

$$[X_i^{\mathsf{h}}, X_j^{\mathsf{h}}] = [X_i, X_j]^{\mathsf{h}} \overset{(1)}{=} 0; \quad i, j \in J_n.$$

Thus, applying Proposition 3.2.35 to the family $(X_i^{\mathsf{h}})_{i=1}^n$, we see that there is an open neighbourhood $\mathcal{V}$ of $0 \in \mathbb{R}^n$ such that the mapping

$$\bar{Z} \colon (\alpha^1, \ldots, \alpha^n) \in \mathcal{V} \mapsto \varphi_{\alpha^1}^{X_1^{\mathsf{h}}} \circ \cdots \circ \varphi_{\alpha^n}^{X_n^{\mathsf{h}}}(v_0) \in TM$$

has the properties

$$\bar{Z}(0) = v_0 \quad \text{and} \quad \frac{\partial}{\partial e^i} \underset{\bar{Z}}{\sim} X_i^{\mathsf{h}} \quad (i \in J_n). \tag{3}$$

Let $\mathcal{W} := u^{-1}(\mathcal{V}) \subset M$ and $Z := \bar{Z} \circ u \upharpoonright \mathcal{W}$. Since $X_i^{\mathsf{h}} \underset{\tau}{\sim} X_i$ $(i \in J_n)$, we have $\varphi_{\alpha^i}^{X_i} \circ \tau = \tau \circ \varphi_{\alpha^i}^{X_i^{\mathsf{h}}}$ by (3.2.12). Thus, for every $p \in \mathcal{W}$,

$$\tau(Z(p)) = \tau(\bar{Z}(u(p))) = \tau \circ \varphi_{u^1(p)}^{X_1^{\mathsf{h}}} \circ \cdots \circ \varphi_{u^n(p)}^{X_n^{\mathsf{h}}}(v_0)$$

$$= \varphi_{u^1(p)}^{X_1} \circ \cdots \circ \varphi_{u^n(p)}^{X_n} \circ \tau(v_0) = \varphi_{u^1(p)}^{X_1} \circ \cdots \circ \varphi_{u^n(p)}^{X_n}(p_0)$$

$$= \varphi(u^1(p), \ldots, u^n(p)) = \varphi(u(p)) = p.$$

Therefore the mapping $Z \colon \mathcal{W} \subset M \to TM$ is a vector field on M such that $Z(p_0) = v_0$. Over the neighbourhood $\mathcal{W}$,

$$Z_* \circ X_i = \bar{Z}_* \circ u_* \circ X_i = \bar{Z}_* \circ \varphi_*^{-1} \circ X_i$$

$$\overset{(2)}{=} \bar{Z}_* \circ \frac{\partial}{\partial e^i} \circ \varphi^{-1} \overset{(3)}{=} X_i^{\mathsf{h}} \circ \bar{Z} \circ u = X_i^{\mathsf{h}} \circ Z,$$

i.e.,

$$X_i \underset{Z}{\sim} X_i^{\mathsf{h}} \quad \text{for all} \quad i \in J_n. \tag{4}$$

If $X \underset{(\mathcal{W})}{=} f^i X_i$ is an arbitrary vector field on $\mathcal{W}$, then $X^{\mathsf{h}} \overset{(7.1.5)}{=} (f^i \circ \tau) X_i^{\mathsf{h}}$, and (4) implies by Lemma 3.1.51(i),(ii) that $X \underset{Z}{\sim} X^{\mathsf{h}}$. Therefore Z satisfies (8.5.2) and hence (8.5.1). This concludes the proof. $\qquad\square$

Corollary 8.5.5. *An Ehresmann connection is integrable if, and only if, the Lie bracket of an two horizonal vector fields is horizontal.*

Proof. With the same notation as above,

$$i\mathcal{R}(\widetilde{X}, \widetilde{Y}) \overset{(7.8.1)}{=} \mathbf{v}[\mathcal{H}\widetilde{X}, \mathcal{H}\widetilde{Y}] = [\mathcal{H}\widetilde{X}, \mathcal{H}\widetilde{Y}] - \mathbf{h}[\mathcal{H}\widetilde{X}, \mathcal{H}\widetilde{Y}],$$

for all $\widetilde{X}, \widetilde{Y} \in \Gamma(\mathring{\pi})$. So $[\mathcal{H}\widetilde{X}, \mathcal{H}\widetilde{Y}] \in \mathfrak{X}^{\mathsf{h}}(TM)$ if, and only if, $\mathcal{R} = 0$, whence our claim. $\square$

Corollary 8.5.6 (The Local Frobenius Theorem). *Let $(\mathcal{U}, (u^i)_{i=1}^n)$ be a chart on a manifold M with induced chart $(\tau^{-1}(\mathcal{U}), ((x^i)_{i=1}^n, (y^i)_{i=1}^n))$ on TM, and let $(f_j^i)_{(i,j) \in J_n \times J_n}$ be a family of smooth functions on $\tau^{-1}(\mathcal{U})$. Then the following are equivalent:*

(i) *For every $v_0 \in \tau^{-1}(\mathcal{U})$, there is an open neighbourhood $\mathcal{W} \subset \mathcal{U}$ of $\tau(v_0)$ and a unique vector field $Z \in \mathfrak{X}(\mathcal{W})$ such that $Z(\tau(v_0)) = v_0$ and*

$$\frac{\partial Z^i}{\partial u^j} = f_j^i \circ Z; \quad i, j \in J_n.$$

(ii) *The functions f_j^i satisfy the relations*

$$\frac{\partial f_k^i}{\partial x^j} - \frac{\partial f_j^i}{\partial x^k} + \frac{\partial f_k^i}{\partial y^l} f_j^l - \frac{\partial f_j^i}{\partial y^l} f_k^l = 0; \quad i, j, k \in J_n.$$

(iii) *The Ehresmann connection in $\tau^{-1}(\mathcal{U})$ defined by*

$$\mathcal{H}\left(\widehat{\frac{\partial}{\partial u^j}}\right) := \frac{\partial}{\partial x^j} + f_j^i \frac{\partial}{\partial y^i}, \quad j \in J_n$$

(i.e., the Ehresmann connection with Christoffel symbols $N_j^i := -f_j^i$) has vanishing curvature.

Proof. The equivalence of (ii) and (iii) is clear from (7.8.9). By Remark 8.5.2, (i) is equivalent to the integrability of the Ehresmann connection $\mathcal{H}$. Therefore, the equivalence of (i) and (iii) follows from Proposition 8.5.4. The uniqueness of the vector field Z in (i) is a consequence of Lemma 8.5.3. $\square$

Corollary 8.5.7. *Let D be a covariant derivative on a manifold M.*

(i) *In order that D be flat is necessary and sufficient that for every point $p \in M$ and tangent vector $v \in T_pM$ there exist an open neighbourhood $\mathcal{V}$ of p and a parallel vector field $X: \mathcal{V} \to TM$ with $X(p) = v$.*

(ii) *A necessary and sufficient condition that D be a torsion-free flat covariant derivative is that there exists a chart at every point of M for which the Christoffel symbols of D are zero.*

Proof. Let $\mathcal{H}$ be the linear Ehresmann connection generated by D according to (7.5.12). Let R, $\mathcal{R}$ and $\mathbf{H}$ be the curvature of D, the curvature of $\mathcal{H}$ and the affine curvature of $\mathcal{H}$, respectively.

(i) For every vector field $Z \in \mathfrak{X}(\mathcal{V})$ and every vector $w \in \tau^{-1}(\mathcal{U})$,

$$(D_w Z)^{\uparrow}(Z(\tau(w))) \overset{(7.5.12)}{=} Z_*(w) - \mathcal{H}(Z(\tau(w)), w).$$

Comparing this to (8.5.1), we see that Z *is parallel, i.e.,* $DZ = 0$ *if, and only if, $\mathcal{H}$ is integrable.* On the other hand,

$\mathcal{H}$ is integrable if, and only if, $\mathcal{R} = 0$ (Proposition 8.5.4);
$\mathcal{R} = 0$ if, and only if, $\mathbf{H} = 0$ (Corollary and Definition 8.2.7);
$\mathbf{H} = 0$ if, and only if, $R = 0$ (Lemma 7.15.3),

and so assertion (i) follows.

(ii) The condition is clearly sufficient: when at each point of M there exists a chart with respect to which the Christoffel symbols of D vanish, then D is flat and torsion-free. Conversely, suppose that $R = 0$ and

$$T(X, Y) := D_X Y - D_Y X - [X, Y] = 0; \quad X, Y \in \mathfrak{X}(M).$$

Choose a point p in M and a basis $(v_i)_{i=1}^n$ in $T_p M$. By part (i), there exist an open neighbourhood $\mathcal{V}$ of p and a local frame field $(X_i)_{i=1}^n$ of TM over $\mathcal{V}$ such that

$$DX_i = 0 \text{ and } X_i(p) = v_i \text{ for all } i \in J_n.$$

Further, since $T = 0$, we have

$$[X_i, X_j] = D_{X_i} X_j - D_{X_j} X_i = 0; \quad i, j \in J_n.$$

Thus, by Corollary 3.2.36, there exists a chart $(\mathcal{U}, (u^i)_{i=1}^n)$ of M centred on p such that

$$X_i \upharpoonright \mathcal{U} = \frac{\partial}{\partial u^i}, \quad i \in J_n.$$

So we are done: the Christoffel symbols of D vanish with respect to the chart above. $\qquad\qquad\qquad\qquad\qquad\qquad\qquad\qquad\qquad\qquad\qquad\qquad\qquad\quad\square$

Corollary 8.5.8. *Let S be a spray over a manifold M. A necessary and sufficient condition that S be flat is that for every point p of M there exists a chart $(\mathcal{U}, (u^i)_{i=1}^n)$ of M at p with induced chart $(\tau^{-1}(\mathcal{U}), ((x^i)_{i=1}^n, (y^i)_{i=1}^n))$ on TM such that $S \underset{(\mathcal{U})}{=} y^i \frac{\partial}{\partial x^i}$.*

Proof. First we recall that, by definition, a spray S is flat if it is horizontally flat, i.e., one (and hence all) of the equivalent conditions $\mathbf{K} = 0$, $\mathcal{R} = 0$ and $\mathbf{H} = 0$ is satisfied, and the Berwald curvature $\mathbf{B}$ of S is also zero.

The given condition is sufficient. Indeed, if a chart $(\mathcal{U}, (u^i)_{i=1}^n)$ of M is such that in the induced chart $(\tau^{-1}(\mathcal{U}), ((x^i)_{i=1}^n, (y^i)_{i=1}^n))$ we have $S \underset{(\mathcal{U})}{=} y^i \frac{\partial}{\partial x^i}$, then the Christoffel symbols of the Berwald connection and the induced Berwald derivative of (M, S) are zero with respect to $(\mathcal{U}, (u^i)_{i=1}^n)$, hence, by (8.1.8) and (8.1.10), $\mathcal{R} = 0$ and $\mathbf{B} = 0$.

Now we show the *necessity* of the condition. Suppose that S is flat. Let $\mathcal{H}$ be the Berwald connection of (M, S). Then $\mathcal{H}$ is torsion-free and its Berwald curvature vanishes, therefore, by Proposition 7.15.1

$\mathcal{H}$ *is a torsion-free linear Ehresmann connection.*

Equivalently (see Proposition 7.15.4)

the linear deviation of $\mathcal{H}$, given by (7.5.18), vanishes.

Since we also have that $\mathcal{R} = 0$, $\mathcal{H}$ is integrable. Thus Proposition 8.5.4 assures that there exists a local frame $(X_i)_{i=1}^n$ of TM over an open neighbourhood $\mathcal{V}$ of p such that

$$(X_i)_*(v) = \mathcal{H}(X_i(\tau(v)), v) \text{ for all } v \in \tau^{-1}(\mathcal{V}), \quad i \in J_n.$$

Then

$$X_i^{\mathsf{c}}(v) = \kappa \circ (X_i)_*(v) = \kappa \circ \mathcal{H}(X_i(\tau(v)), v)$$
$$\overset{(7.5.18)}{=} \mathcal{H}(v, X_i(\tau(v))) =: X_i^{\mathsf{h}}(v), \tag{$*$}$$

therefore,

$$0 \overset{(7.8.4),(*)}{=} [X_i^{\mathsf{c}}, X_j^{\mathsf{v}}] - [X_j^{\mathsf{c}}, X_i^{\mathsf{v}}] - [X_i, X_j]^{\mathsf{v}} \overset{(4.1.71)}{=} [X_i, X_j]^{\mathsf{v}}, \quad i, j \in J_n.$$

We found in this way that the frame $(X_i)_{i=1}^n$ consists of mutually commuting vector fields. Thus, by Corollary 3.2.36, at every point of $\mathcal{V}$ there is chart $(\mathcal{U}, (u^i)_{i=1}^n)$ such that

$$\mathcal{U} \subset \mathcal{V} \quad \text{and} \quad X_i \upharpoonright \mathcal{U} = \frac{\partial}{\partial u^i}.$$

Let (G^i) and (G^i_j) be the families of the spray coefficients of S and the Christoffel symbols of $\mathcal{H}$ with respect to the chart $(\mathcal{U}, (u^i)_{i=1}^n)$, respectively. Then

$$\frac{\partial}{\partial x^j} = \left(\frac{\partial}{\partial u^j} \right)^{\mathsf{c}} \overset{(*)}{=} \left(\frac{\partial}{\partial u^j} \right)^{\mathsf{h}} = \frac{\partial}{\partial x^j} - G^i_j \frac{\partial}{\partial y^i}, \quad j \in J_n,$$

whence $G^i_j = 0$ and, therefore,

$$G^i \overset{(8.1.5)}{=} \frac{1}{2} y^j G^i_j = 0, \quad i \in J_n.$$

This concludes the proof.   □

Definition 8.5.9. A spray manifold (M, S), or a spray S is said to be *projectively flat* if it is projectively related to a flat spray. If, in particular, D is a torsion-free covariant derivative on M and its geodesic spray is projectively flat, then (M, D), or D is called projectively flat.

Corollary and Definition 8.5.10. *A spray manifold (M, S) is projectively flat if, and only if, there exists*

(i) *a 1^+-homogeneous function $P \in C^\infty(\mathring{T}M)$,*
(ii) *at every point of M a chart $(\mathcal{U}, (u^i)^n_{i=1})$ with induced chart $(\tau^{-1}(\mathcal{U}), ((x^i)^n_{i=1}, (y^i)^n_{i=1}))$*

such that

$$S \underset{(\mathcal{U})}{=} y^i \frac{\partial}{\partial x^i} - 2Py^i \frac{\partial}{\partial y^i}. \tag{8.5.4}$$

Then we say that $(\mathcal{U}, (u^i)^n_{i=1})$ is a rectilinear chart on M with induced rectilinear (local) coordinate system $((x^i)^n_{i=1}, (y^i)^n_{i=1})$ on TM.

Proof. Suppose first that S is a projectively flat spray over M. Then there exist a flat spray S_0 over M and a 1^+-homogeneous smooth function P on $\mathring{T}M$ such that $S = S_0 - 2PC$. By the flatness of S_0, Corollary 8.5.8 assures the existence of an induced chart $(\tau^{-1}(\mathcal{U}), ((x^i)^n_{i=1}, (y^i)^n_{i=1}))$ at every point of TM on which $S_0 = y^i \frac{\partial}{\partial x^i}$. Then

$$S \underset{(\mathcal{U})}{=} y^i \frac{\partial}{\partial x^i} - 2Py^i \frac{\partial}{\partial y^i},$$

as required.

Conversely, suppose that (i) and (ii) are satisfied, and consider the spray $S_0 := S + 2PC$. Then $S_0 \underset{(\mathcal{U})}{=} y^i \frac{\partial}{\partial x^i}$, so S_0 is flat by the previous proposition. Therefore, S is projectively related to a flat spray. This was to be shown. □

Remark 8.5.11. (a) Suppose, in particular, that D is a torsion-free covariant derivative on M. Then, taking into account Theorem 8.4.16 and Remark 8.4.17, D is projectively flat if, and only if, at every point of M

there is a chart $(\mathcal{U}, (u^i)_{i=1}^n)$ of M such that the Christoffel symbols of D with respect to this chart are of the form

$$\alpha_j \delta_k^i + \alpha_k \delta_j^i; \quad i, j, k \in J_n,$$

where the functions $\alpha_i \in C^\infty(\mathcal{U})$ are the components of a one-form with respect to the chosen chart.

(b) As a preparation to the following important result, we note that by the *Weyl tensors* (Weyl endomorphism $\mathbf{W}_1$; projective curvature tensors $\mathbf{W}_2$, $\mathbf{W}_3$) *of a covariant derivative* D on M we mean the corresponding data of the geodesic spray of D.

Theorem 8.5.12 (H. Weyl [98]). *A torsion-free covariant derivative on an at least 3-dimensional manifold is projectively flat if, and only if, one (and hence each) of its Weyl tensors vanishes.*

A proof, formulated in the language of traditional tensor calculus can be found in the quoted paper of Weyl, or in section 34 of Eisenhart's book 'Non-Riemannian Geometry' [47]. For recent index-free proofs we refer to [23] and [10]. Clearly, the theorem can also be stated as follows:

An affine spray over an at least 3-dimensional manifold is projectively flat if, and only if, its Weyl endomorphism vanishes.

Theorem 8.5.13. *An at least 3-dimensional spray manifold is projectively flat if, and only if, its Douglas tensor and Weyl endomorphism are zero.*

Proof. Let S be a spray over M, where $\dim M \geq 3$. Let $\mathbf{D}$ and $\mathbf{W}_1$ be the Douglas tensor and the Weyl endomorphism of S, respectively.

Suppose first that S is projectively related to a flat spray S_0. Since, obviously, the Douglas tensor and the Weyl endomorphism of S_0 are zero, and these tensors are projectively invariant, we have $\mathbf{D} = 0$ and $\mathbf{W}_1 = 0$.

Conversely, let $\mathbf{D} = 0$ and $\mathbf{W}_1 = 0$. Then, by Theorem 8.4.37, the vanishing of the Douglas tensor implies that S is projectively related to an affine spray S_0. Since $\mathbf{W}_1$ is projectively invariant, the Weyl endomorphism of S_0 is also zero. Thus, by the above quoted theorem of Weyl, S_0 is projectively flat, as was to be shown. $\square$

Chapter 9

Finsler Norms and Finsler Functions

9.1 Finsler Vector Spaces

> See simplicity in the complicated.
> Achieve greatness in small things.
>
> Lao Tsu (Tao Te Ching [65], Ch. 63)

> ... the study of Minkowskian geometry ought to be the first and main step, the passage from there to general Finsler spaces will be the second and simpler step.
>
> Herbert Busemann ([24], p. 11)

9.1.1 *Convexity*

Throughout Section 9.1, we work in finite-dimensional real vector spaces. We assume that they are endowed with their canonical topology (see B.4).

Definition 9.1.1. Let V be a finite-dimensional real vector space with $\dim V \geq 1$.

(a) If a and b are distinct points of V then the subset

$$\overline{ab} := \{(1-t)a + tb \in V \mid t \in [0,1]\}$$

of V is called the *line segment* joining a and b.

(b) A subset $\mathcal{K}$ in V is said to be *convex* if the line segment joining any two points of $\mathcal{K}$ is contained in $\mathcal{K}$, i.e.,

$$(1-t)a + tb \in \mathcal{K} \text{ for all } t \in [0,1]; \quad a, b \in \mathcal{K}, \ a \neq b.$$

The subset $\mathcal{K}$ is called *strictly convex* if

$$(1-t)a + tb \in \text{int}(\mathcal{K}) \text{ for all } t \in]0,1[; \quad a, b \in \mathcal{K}, \ a \neq b.$$

(Here 'int' stands for the interior, see Definition B.1.1(a)). A *convex body* in V is a compact convex subset of V whose interior is nonempty.

(c) Let $\mathcal{K}$ be a convex subset of V. A function $f\colon \mathcal{K} \to \mathbb{R}$ is said to be *convex* if

$$f((1-t)a + tb) \le (1-t)f(a) + tf(b) \qquad (9.1.1)$$

for all $a, b \in \mathcal{K}$ and for every $t \in [0,1]$. The function f is called *strictly convex* if for any distinct points a, b in $\mathcal{K}$ and every $t \in \,]0,1[$ we have

$$f((1-t)a + tb) < (1-t)f(a) + tf(b). \qquad (9.1.2)$$

Lemma 9.1.2. *Let $\mathcal{K}$ be a convex subset of the vector space V. A function $f\colon \mathcal{K} \to \mathbb{R}$ is convex if, and only if, its epigraph*

$$\operatorname{epi}(f) := \{(p,t) \in \mathcal{K} \times \mathbb{R} \mid f(p) \le t\}$$

is a convex subset of $V \times \mathbb{R}$.

We leave the proof to the reader or refer to [80], pp. 164–165.

Lemma 9.1.3. *Let V and W be real vector spaces, and let $\mathcal{K}$ be a convex subset of V. If $\psi\colon W \to V$ is an affine mapping, $\mathcal{K}_0$ is a convex subset of W such that $\psi(\mathcal{K}_0) \subset \mathcal{K}$ and $f\colon \mathcal{K} \to \mathbb{R}$ is a convex function, then $f \circ (\psi \restriction \mathcal{K}_0)\colon \mathcal{K}_0 \to \mathbb{R}$ is also a convex function.*

Proof. Let $\psi = t_v \circ \varphi$, where $\varphi \in L(W,V)$ and $v \in V$ is a fixed vector (see Definition 1.1.36). Then for all $a, b \in \mathcal{K}_0$ and for every $t \in [0,1]$,

$$\begin{aligned}
f \circ \psi((1-t)a + tb) &= f((1-t)\varphi(a) + t\varphi(b) + v) \\
&= f((1-t)(\varphi(a) + v) + t(\varphi(b) + v)) \\
&\le (1-t)f(\psi(a)) + tf(\psi(b)),
\end{aligned}$$

so $f \circ (\psi \restriction \mathcal{K}_0)$ is indeed a convex function. $\qquad\square$

Proposition 9.1.4. *Let $\mathcal{K}$ be a convex subset of V. If a function defined on $\mathcal{K}$ is convex, then it is continuous in the interior of $\mathcal{K}$. In particular, every convex function defined on the whole vector space V is continuous.*

For a proof we refer to [55], pp. 21–22. Here it is also shown that the continuity of a convex function on $\operatorname{int}(\mathcal{K})$ cannot be extended, in general, to the boundary of $\mathcal{K}$.

Corollary 9.1.5. *If $f\colon V \to \mathbb{R}$ is a convex function, then*

$$\mathcal{U} := \{v \in V \mid f(v) < 1\}$$

is an open subset of V.

Proof. By the preceding proposition f is a continuous function, so $f^{-1}(]-\infty, 1[) = \mathcal{U}$ is an open subset of V. $\qquad\square$

Definition 9.1.6. A real-valued function f on a real vector space V is said to be *subadditive* if

$$f(u + v) \leq f(u) + f(v) \qquad (9.1.3)$$

for all u and v in V.

Lemma 9.1.7. *A 1^+-homogeneous function $f \colon V \to \mathbb{R}$ is convex if, and only if, it is subadditive.*

Proof. Suppose first that f is convex. Then, by using the 1^+-homogeneity, for all $u, v \in V$ we have

$$\frac{1}{2} f(u + v) = f\left(\frac{1}{2}u + \frac{1}{2}v\right) \overset{(9.1.1)}{\leq} \frac{1}{2}f(u) + \frac{1}{2}f(v),$$

whence (9.1.3).

Conversely, if f is subadditive, then for all $u, v \in V$ and for every $t \in [0, 1]$,

$$f((1 - t)u + tv) \overset{(9.1.3)}{\leq} f((1 - t)u) + f(tv) = (1 - t)f(u) + tf(v),$$

as was to be shown. $\qquad\square$

Remark 9.1.8. *A 1^+-homogeneous function $f \colon V \to \mathbb{R}$ can never be strictly convex.* Indeed, if $a, b \in V$ and $b = \lambda a$ with $\lambda \in \mathbb{R}_+^* \setminus \{1\}$, then for every $t \in]0, 1[$ we have

$$f((1 - t)a + tb) = f((1 - t + \lambda t)a) = (1 - t)f(a) + \lambda t f(a)$$
$$= (1 - t)f(a) + tf(b),$$

so the strict inequality required in (9.1.2) does not hold. The best we can hope for in this situation is the *strong convexity* of f in the sense of Busemann ([25], p. 99):

Definition 9.1.9. A 1^+-homogeneous function $f \colon V \to \mathbb{R}$ is said to be *strongly convex* if it is convex and for $a, b \in V \setminus \{0\}$, $a \neq b$, $t \in]0, 1[$ the equality

$$f((1 - t)a + tb) = (1 - t)f(a) + tf(b)$$

holds if, and only if, $b = \lambda a$ with $\lambda \in \mathbb{R}_+^* \setminus \{1\}$.

Lemma 9.1.10. *If* $f \colon V \to \mathbb{R}$ *is a* 1^+*-homogeneous and continuous function, and*

$$A := \{v \in V \mid f(v) \leq 1\}, \quad \mathcal{U} := \{v \in V \mid f(v) < 1\},$$

then $\mathcal{U} = \operatorname{int}(A)$, $A = \operatorname{cl}(\mathcal{U})$.

Proof. If $v \in \mathcal{U}$, then $\mathcal{U}$ is an open neighbourhood of v contained in A, so v is in the interior of A.

Now let $v \in A \setminus \mathcal{U}$. Then $f(v) = 1$. Consider the sequence $(v_n)_{n \in \mathbb{N}^*}$ in V given by

$$v_n := \left(1 + \frac{1}{n}\right) v, \quad n \in \mathbb{N}^*.$$

Then, by the 1^+-homogeneity of f, $f(v_n) = \left(1 + \frac{1}{n}\right) f(v) = 1 + \frac{1}{n}$. Thus $v_n \in V \setminus A$ for all $n \in \mathbb{N}^*$. Since $\lim_{n \to \infty} v_n = v$, it follows that v cannot be an interior point of A; therefore $\operatorname{int}(A) = \mathcal{U}$.

It can be shown in a similar way that $A = \operatorname{cl}(\mathcal{U})$, we leave the details to the reader. $\qquad\square$

Remark 9.1.11. Neither the requirement of continuity nor the condition of 1^+-homogeneity of f can be omitted in the lemma. To see this, define first a function $f_1 \colon \mathbb{R}^2 \to \mathbb{R}$ by

$$v = (v^1, v^2) \mapsto f_1(v) := \begin{cases} \|v\| & \text{if } v^1 = 0 \text{ or } \frac{v^2}{v^1} \in \mathbb{Q}, \\ 2\|v\| & \text{if } \frac{v^2}{v^1} \in \mathbb{R} \setminus \mathbb{Q}, \end{cases}$$

where $\| \ \|$ stands for the Euclidean norm of $\mathbb{R}^2$. Then f_1 fails to be continuous; the subset $\mathcal{U}_1 := \{v \in \mathbb{R}^2 \mid f_1(v) < 1\}$ is not open, and the subset $A_1 := \{v \in \mathbb{R}^2 \mid f_1(v) \leq 1\}$ is not closed.

As a second counterexample, consider the function

$$f_2 \colon \mathbb{R}^2 \to \mathbb{R}, \ v \mapsto f_2(v) := \begin{cases} \|v\| & \text{if } \|v\| \leq 1, \\ 1 & \text{if } \|v\| > 1. \end{cases}$$

Then f_2 is continuous, but it is not homogeneous. The subset

$$\mathcal{U}_2 := \{v \in \mathbb{R}^2 \mid f_2(v) < 1\}$$

is the open Euclidean unit disc in $\mathbb{R}^2$, while $A_2 := \{v \in \mathbb{R}^2 \mid f_2(v) \leq 1\}$ is the whole $\mathbb{R}^2$.

Proposition 9.1.12. *Let* $f \colon V \to \mathbb{R}$ *be a* 1^+*-homogeneous function such that* $f(v) > 0$ *if* $v \neq 0$, *and let* $\mathcal{K} := \{v \in V \mid f(v) \leq 1\}$. *Then*

(i) *f is convex if, and only if, the subset $\mathcal{K}$ of V is convex.*

(ii) *f is strongly convex if, and only if, the subset $\mathcal{K}$ of V is strictly convex.*

Proof. (i) First, let f be convex. Then for all $a, b \in \mathcal{K}$ and for every $t \in [0, 1]$ we have

$$f((1-t)a + tb) \le (1-t)f(a) + tf(b) \le 1 - t + t = 1,$$

whence $(1-t)a + tb \in \mathcal{K}$. This proves that $\mathcal{K}$ is convex.

Conversely, suppose that $\mathcal{K}$ is a convex subset of V. We show that then f is subadditive, which is equivalent to the convexity of f by Lemma 9.1.7. Let $u, v \in V$. We may suppose that u and v are nonzero and $v \ne \lambda u$, $\lambda \in \mathbb{R}_+^*$. Then

$$f\left(\frac{u}{f(u)}\right) = f\left(\frac{v}{f(v)}\right) = 1,$$

therefore $\frac{u}{f(u)}$ and $\frac{v}{f(v)}$ are (distinct) points in $\mathcal{K}$. By the convexity of $\mathcal{K}$, the point

$$\frac{u+v}{f(u) + f(v)} = \frac{f(u)}{f(u) + f(v)}\frac{u}{f(u)} + \frac{f(v)}{f(u) + f(v)}\frac{v}{f(v)}$$

also lies in $\mathcal{K}$. Hence

$$\frac{f(u+v)}{f(u) + f(v)} = f\left(\frac{u+v}{f(u) + f(v)}\right) \le 1,$$

which proves that f is subadditive.

(ii) Now suppose that f is strongly convex. We have to show that for all $a, b \in \mathcal{K}$ such that $a \ne b$ and for every $t \in {]0, 1[}$ we have

$$(1-t)a + tb \in \text{int}(\mathcal{K}).$$

Note first that by Lemma 9.1.10, $\text{int}(\mathcal{K}) = \{v \in V \mid f(v) < 1\}$. We consider two cases.

(1) $b \ne \lambda a$, $\lambda \in \mathbb{R}_+^*$. Then by the strong convexity of f we have

$$f((1-t)a + tb) < (1-t)f(a) + tf(b) \le 1,$$

so $(1-t)a + tb \in \text{int}(\mathcal{K})$.

(2) $b = \lambda a$, $\lambda \in \mathbb{R}_+^*$. Observe that here $\lambda \ne 1$ because $a \ne b$. Now the strong convexity of f yields the equality

$$f((1-t)a + tb) = (1-t)f(a) + tf(b), \quad t \in {]0, 1[}.$$

We have $f(a) \le 1$ and $f(b) \le 1$ (because $a, b \in \mathcal{K}$), furthermore $f(a) \ne f(b)$ (otherwise we should have $f(b) = f(\lambda a) = \lambda f(a)$, whence $\lambda = 1$, which is a contradiction). So at least one of $f(a)$ and $f(b)$ is less than 1, therefore

$$(1-t)f(a) + tf(b) < 1 - t + t = 1.$$

Thus we conclude that $(1-t)a + tb \in \text{int}(\mathcal{K})$, as was to be shown.

Conversely, suppose that $\mathcal{K}$ is strictly convex. Then, by part (i), f is certainly convex; our task is to show that f is actually strongly convex. Let $u, v \in V$. We may assume again that

$$u, v \in V \setminus \{0\}; \quad v \neq tu, \ t \in \mathbb{R}_+^*.$$

As above, $\frac{u}{f(u)}$ and $\frac{v}{f(v)}$ are distinct points in $\mathcal{K}$. Let $\lambda \in \,]0,1[$. Then

$$\frac{(1-\lambda)u + \lambda v}{f((1-\lambda)u) + f(\lambda v)} = \frac{(1-\lambda)f(u)}{f((1-\lambda)u) + f(\lambda v)} \frac{u}{f(u)}$$
$$+ \frac{\lambda f(v)}{f((1-\lambda)u) + f(\lambda v)} \frac{v}{f(v)}.$$

If $t := \frac{\lambda f(v)}{f((1-\lambda)u) + f(\lambda v)}$, then $t \in \,]0,1[$ and $1 - t = \frac{(1-\lambda)f(u)}{f((1-\lambda)u) + f(\lambda v)}$, so the right-hand side of the above relation can be written in the form

$$(1-t)\frac{u}{f(u)} + t\frac{v}{f(v)}, \quad t \in \,]0,1[.$$

Since $\mathcal{K}$ is strictly convex, it follows that $(1-t)\frac{u}{f(u)} + t\frac{v}{f(v)} \in \text{int}(\mathcal{K})$. Thus

$$1 > f\left((1-t)\frac{u}{f(u)} + t\frac{v}{f(v)}\right) = f\left(\frac{(1-\lambda)u + \lambda v}{f((1-\lambda)u) + f(\lambda v)}\right)$$
$$= \frac{f((1-\lambda)u + \lambda v)}{(1-\lambda)f(u) + \lambda f(v)},$$

whence

$$f((1-\lambda)u + \lambda v) < (1-\lambda)f(u) + \lambda f(v), \quad \lambda \in \,]0,1[.$$

This concludes the proof. $\qquad\square$

In the rest of this subsection we recall some calculus tests for convexity of a function. Detailed expositions are available, e.g., in [48], [55] or [86].

Proposition 9.1.13. *Let $\mathcal{K}$ be an open convex subset of the vector space V, and let $f \colon \mathcal{K} \to \mathbb{R}$ be a differentiable function. Then*

(i) *f is convex if, and only if,*

$$f(b) \geq f(a) + f'(a)(b-a) \tag{9.1.4}$$

for all $a, b \in \mathcal{K}$;

(ii) *f is strictly convex if, and only if,*

$$f(b) > f(a) + f'(a)(b-a) \tag{9.1.5}$$

for all $a, b \in \mathcal{K}$ with $a \neq b$.

Proof. (i) If f is convex, then for all $a, b \in \mathcal{K}$ and for every $t \in \,]0, 1[$ we have (9.1.1), i.e., $f((1-t)a+tb) \leq (1-t)f(a) + tf(b)$. Let $v := b - a$. Then the inequality can be rewritten as

$$f(a + tv) - f(a) \leq t(f(a+v) - f(a)). \qquad (*)$$

Subtracting $tf'(a)(v)$ from both sides and dividing by t, we get

$$\frac{f(a+tv) - f(a)}{t} - f'(a)(v) \leq f(a+v) - f(a) - f'(a)(v).$$

Here the left-hand side tends to zero as $t \to 0^+$ by Proposition C.1.6, while the right-hand side remains constant. This proves (9.1.4).

Conversely, suppose that (9.1.4) holds throughout $\mathcal{K}$. Given $a_1, a_2 \in \mathcal{K}$, $a_1 \neq a_2$, and $t \in \,]0, 1[$, let $a := (1-t)a_1 + ta_2$, $v := a_2 - a$. Then a_1 can be expressed as $a_1 = a - \frac{t}{1-t}v$. With the choices $b := a_1$ and $b := a_2$ we obtain from (9.1.4) that

$$f(a_1) \geq f(a) + f'(a)\left(-\frac{t}{1-t}v\right) \quad \text{and} \quad f(a_2) \geq f(a) + f'(a)(v).$$

Multiplying the second inequality by $\frac{t}{1-t}$ and adding it to the first one, we find that

$$f(a_1) + \frac{t}{1-t}f(a_2) \geq \frac{1}{1-t}f(a),$$

whence

$$f(a) = f((1-t)a_1 + ta_2) \leq (1-t)f(a_1) + tf(a_2).$$

This is the desired inequality (9.1.1) under the assumption that $t \in \,]0, 1[$. However, if $t \in \{0, 1\}$, then (9.1.1) holds automatically. Therefore f is convex on $\mathcal{K}$.

(ii) Now let f be strictly convex on $\mathcal{K}$. Then, in particular, f is convex, so (9.1.4) holds for all $a, b \in \mathcal{K}$. Suppose that $a \neq b$. Let $v := b - a$, $t \in \,]0, 1[$. Replacing b in (9.1.4) by $a + tv$, we obtain

$$f'(a)(tv) \leq f(a + tv) - f(a).$$

Since f is strictly convex, relation $(*)$ gives

$$f(a + tv) - f(a) < t(f(a+v) - f(a)).$$

Thus it follows that

$$f'(a)(tv) < t(f(a+v) - f(a)).$$

Dividing both sides by t, we get

$$f(a + v) > f(a) + f'(a)(v),$$

as was to be shown.

If f satisfies (9.1.5), then the proof of the converse statement in part (i) works word for word, with strict inequality everywhere. $\qquad \square$

In the special case when V is the one-dimensional real vector space $\mathbb{R}$, we have some simple criteria for convexity or strict convexity of a differentiable function. To formulate them, we first recall the following notions.

Definition 9.1.14. Let $S \subset \mathbb{R}$. A function $f\colon S \to \mathbb{R}$ is said to be

     increasing if $s_1 < s_2$ implies $f(s_1) \le f(s_2)$,
     strictly increasing if $s_1 < s_2$ implies $f(s_1) < f(s_2)$,

where s_1 and s_2 are in S. *Decreasing* and *strictly decreasing* functions are defined similarly.

Proposition 9.1.15. *Let $I \subset \mathbb{R}$ be an interval and f a real-valued function on I. Suppose that f is continuous on I and differentiable on* $\mathrm{int}(I)$. *Then:*

(i) *f is convex on I if, and only if, f' is increasing on* $\mathrm{int}(I)$.
(ii) *f is strictly convex on I if, and only if, f' is strictly increasing on* $\mathrm{int}(I)$.

For a proof, see, e.g., [49], (5.10.2).

Corollary 9.1.16. *Let f be a real-valued function which is continuous on an interval I and twice differentiable on* $\mathrm{int}(I)$. *Then f is convex on I if, and only if, $f''(t) \ge 0$ for every $t \in \mathrm{int}(I)$.* $\triangle$

The next result provides a convenient criterion for convexity of C^2 functions.

Proposition 9.1.17 (Brunn and Hadamard). *Let $\mathcal{K}$ be an open convex subset of the real vector space V, and let $f\colon \mathcal{K} \to \mathbb{R}$ be a function of class C^2. Then f is convex on $\mathcal{K}$ if, and only if, the second derivative $f''(p)$ of f at p is positive semidefinite for every $p \in \mathcal{K}$. If $f''(p)$ is positive definite for each $p \in \mathcal{K}$, then f is strictly convex.*

Proof. **Step 1.** Since $\mathcal{K}$ is convex, we may use Taylor's formula (see Theorem C.1.22) with $r = 2$ and for any two points $a, b \in \mathcal{K}$. Thus

$$f(b) = f(a) + f'(a)(v) + \frac{1}{2} f''(a + sv)(v, v), \tag{9.1.6}$$

where $s \in \,]0, 1[$ and $v := b - a$.
Step 2. Suppose that $f''(p)$ *is positive definite* for every $p \in \mathcal{K}$. Then, in particular,

$$f''(a + sv)(v, v) > 0 \text{ if } v \ne 0,$$

and from (9.1.6) we conclude that

$$f(b) > f(a) + f'(a)(v).$$

Thus, by Proposition 9.1.13(ii), f is strictly convex.

Step 3. If $f''(p)$ *is positive semidefinite* for every $p \in \mathcal{K}$, then the same reasoning shows that $f(b) \geq f(a) + f'(a)(v)$, so, by Proposition 9.1.13(i), f is convex.

Step 4. Suppose that the function f *is convex.* Given two points $a, b \in \mathcal{K}$, let $v := b - a$, and consider the parametrized straight line

$$\gamma \colon \mathbb{R} \to V, \ t \mapsto \gamma(t) := a + tv = (1 - t)a + tb.$$

Then γ is an affine mapping (see Definition 1.1.39) and $\mathrm{Im}(\gamma \restriction [0, 1]) \subset \mathcal{K}$ by the convexity of $\mathcal{K}$, so in view of Lemma 9.1.3, the function

$$f \circ (\gamma \restriction [0, 1]) \colon t \in [0, 1] \mapsto f(\gamma(t)) = f(a + tv) \in \mathbb{R}$$

is convex. It is twice differentiable, and applying the chain rule we find that

$$(f \circ \gamma)'(t) = f'(\gamma(t))(\gamma'(t)) = f'(a + tv)(v),$$
$$(f \circ \gamma)''(t) = f''(a + tv)(v, v)$$

for all $t \in [0, 1]$. From Corollary 9.1.16 it follows that

$$(f \circ \gamma)''(t) \geq 0 \text{ for every } t \in [0, 1].$$

In particular, $(f \circ \gamma)''(0) \geq 0$, therefore $f''(a)(v, v) \geq 0$ for all $a \in \mathcal{K}$ and $v \in V$ such that $a + v \in \mathcal{K}$. Then, however, by the bilinearity of $f''(a)$, the desired inequality holds for all $v \in V$. This concludes the proof. $\qquad \square$

To close this subsection, we mention that for convex functions the necessary condition $f'(a) = 0$ for a minimum is also sufficient. More precisely, we have the following:

Lemma 9.1.18. *Let $\mathcal{K} \subset V$ be an open convex subset, and let $f \colon \mathcal{K} \to \mathbb{R}$ be a differentiable convex function. If $a \in \mathcal{K}$ is a critical point of f, i.e., $f'(a) = 0$, then f has an absolute minimum at a.*

Proof. Indeed, from (9.1.4) we obtain that $f(b) \geq f(a)$ for every $b \in \mathcal{K}$. $\qquad \square$

Remark 9.1.19. *A strictly convex differentiable function has at most one critical point.* To see this, let $\mathcal{K} \subset V$ be a convex subset, and let $f \colon \mathcal{K} \to \mathbb{R}$ be a strictly convex differentiable function. Suppose that f has two distinct critical points $a_1, a_2 \in \mathcal{K}$. Applying (9.1.5) with the castings $a_1 = a$, $a_2 = b$ and $a_1 = b$, $a_2 = a$, we obtain that $f(a_2) > f(a_1)$ and $f(a_1) > f(a_2)$, which is a contradiction.

9.1.2   Pre-Finsler Norms

Definition 9.1.20. Let V be a (non-trivial finite-dimensional real) vector space. A function $f \colon V \to \mathbb{R}$ is said to be a *pre-Finsler norm* on V if it has the following properties:

(FN$_1$) f is continuous on V and smooth on the pointed space $\overset{\circ}{V} := V \setminus \{0\}$.
(FN$_2$) f is positive-homogeneous of degree 1.

The function $\psi := \frac{1}{2} f^2$ is called the *energy function* (or *energy* for short) associated to the pre-Finsler norm f. A *pre-Finsler vector space* (V, f) is a vector space V equipped with a pre-Finsler norm f. A pre-Finsler norm f (and the pre-Finsler vector space (V, f)) is called *positive definite* if f satisfies

(FN$_3$) $f(v) > 0$ if $v \neq 0$.

Remark 9.1.21. Let (V, f) be a pre-Finsler vector space.

(a) The *unit ball* in (V, f) is the set

$$B := \{ v \in V \mid f(v) \leq 1 \},$$

and the *unit sphere* in (V, f) is the boundary

$$\mathrm{bd}(B) = \{ v \in V \mid f(v) = 1 \} = f^{-1}(1)$$

of the unit ball.

(b) Continuity and 1^+-homogeneity of f imply that $f(0) = 0$. Indeed, for any fixed $v \in V$ and for any sequence $(t_n)_{n \in \mathbb{N}^*}$ of positive real numbers such that $\lim_{n \to \infty} t_n = 0$, we have

$$f(0) = f\left(\lim_{n \to \infty} t_n v \right) = \lim_{n \to \infty} f(t_n v) = \left(\lim_{n \to \infty} t_n \right) f(v) = 0,$$

as we claimed.

(c) In view of Lemma 4.2.6, the energy function $\psi := \frac{1}{2} f^2$ of (V, f) *is of class C^1 on the whole vector space V*. The first and second derivative of ψ can be expressed as follows:

$$\psi' = f f', \quad \psi'' = f' \otimes f' + f f''. \tag{9.1.7}$$

Thus, in a less concise form, at every point p of $\overset{\circ}{V}$ we have

$$\psi'(p)(v) = f(p) f'(p)(v), \quad v \in V; \tag{9.1.8}$$

$$\psi''(p)(v, w) = f'(p)(v) f'(p)(w) + f(p) f''(p)(v, w); \quad v, w \in V. \tag{9.1.9}$$

The left-hand side of (9.1.8) is also defined at $p = 0$, and, as it turns out from the proof of Lemma 4.2.6, $\psi'(0) = 0 \in V^*$.

(d) By Schwarz's theorem (Theorem C.1.19), the second derivative $g_p := \psi''(p)$ of the energy function is a symmetric bilinear form at every point p of $\mathring{V}$, i.e., $g_p \in S_2(V)$ for all $p \in \mathring{V}$. If we identify the tangent space $T_p\mathring{V}$ with the vector space $\{p\} \times V$ via (C.2.4), we find that the mapping

$$\begin{cases} g \colon p \in \mathring{V} \mapsto \widetilde{g}_p \in T_2^0(T_p\mathring{V}), \\ \widetilde{g}_p((p,v),(p,w)) := g_p(v,w) = \psi''(p)(v,w) \end{cases} \tag{9.1.10}$$

is a symmetric type $(0,2)$ tensor field on $\mathring{V}$ as a manifold. The distinction between $\widetilde{g}_p$ and g_p can clearly be omitted. We say that $g \in T_2^0(\mathring{V})$ is the *fundamental tensor* of the pre-Finsler vector space (V, f).

Lemma 9.1.22. *Let (V, f) be a pre-Finsler vector space, and let $\psi := \frac{1}{2}f^2$ be the energy associated to f. Then for each $p \in \mathring{V}$; $v, w \in V$ we have*

$$f'(p)(p) = f(p), \quad f''(p)(p,v) = 0; \tag{9.1.11}$$

$$f''(p)(v, w + \lambda p) = f''(p)(v,w) \quad (\lambda \in \mathbb{R}), \tag{9.1.12}$$

$$g_p(p,v) = \psi'(p)(v) = f(p)f'(p)(v), \tag{9.1.13}$$

$$g_p(p,p) = f^2(p) = 2\psi(p), \tag{9.1.14}$$

$$g_{\lambda p} = g_p \quad (\lambda \in \mathbb{R}_+^*). \tag{9.1.15}$$

Proof. By the 1^+-homogeneity of f, (9.1.11) is an immediate consequence of Euler's theorem (Lemma 4.2.3) and Lemma 4.2.4. Since $f''(p)$ is bilinear, the second relation in (9.1.11) implies (9.1.12).

The energy function ψ is clearly 2^+-homogeneous, so, by Lemma 4.2.4, its derivative

$$\psi' \colon \mathring{V} \to V^* \text{ is } 1^+\text{-homogeneous}, \tag{$*$}$$

its second derivative

$$\psi'' \colon \mathring{V} \to T_2^0(V) \text{ is } 0^+\text{-homogeneous.} \tag{$**$}$$

Thus

$$g_p(p,v) := \psi''(p)(p,v) \overset{\text{Remark C.1.3}}{=} (\psi''(p)(p))(v)$$

$$= \left(\lim_{t \to 0^+} \frac{\psi'(p + tp) - \psi'(p)}{t} \right)(v)$$

$$\overset{(*)}{=} \left(\lim_{t \to 0^+} \frac{(1+t)\psi'(p) - \psi'(p)}{t} \right)(v) = \psi'(p)(v) \overset{(9.1.8)}{=} f(p)f'(p)(v),$$

which proves (9.1.13). Then, in particular,

$$g_p(p,p) = \psi'(p)(p) \overset{(4.2.2)}{=} 2\psi(p) = f^2(p),$$

so we have (9.1.14). Finally, for every positive real number λ,

$$g_{\lambda p} := \psi''(\lambda p) \overset{(**)}{=} \psi''(p) =: g_p.$$

This completes the proof. $\qquad\square$

Corollary 9.1.23. *If (V, f) is a pre-Finsler vector space with $\dim V \geq 2$, then the unit sphere $\mathrm{bd}(B)$ is a hypersurface in V.*

Proof. At every point $p \in \mathrm{bd}(B)$, $f'(p)(p) \overset{(9.1.11)}{=} f(p) = 1 \neq 0$. Hence $f'(p) \neq 0 \in V^*$, therefore $\mathrm{bd}(B) = f^{-1}(1)$ is indeed a hypersurface in V by Example 6.1.75(a). $\qquad\square$

Definition and Lemma 9.1.24. Let (V, f) be a pre-Finsler vector space.

(a) The function

$$\langle\ ,\ \rangle_F \colon \mathring{V} \times V \to \mathbb{R},\ (u, v) \mapsto \langle u, v \rangle_F := g_u(u, v) \tag{9.1.16}$$

is said to be the *Finslerian scalar product*, or, for short, the *F-scalar product* on V. It is

(FS$_1$) 1^+-homogeneous in its first argument,
(FS$_2$) $\mathbb{R}$-linear in its second argument.

(b) We say that a vector $u \in \mathring{V}$ is *F-orthogonal* to a vector $v \in V$ if $\langle u, v \rangle_F = 0$. Then we write $u \perp_F v$. If $f(u) \neq 0$, then

$$u \perp_F v \text{ if, and only if, } v \in \mathrm{Ker}(f'(u)). \tag{9.1.17}$$

Proof. For all $u \in \mathring{V}$, $v \in V$ and $\lambda \in \mathbb{R}_+^*$ we have

$$\langle \lambda u, v \rangle_F := g_{\lambda u}(\lambda u, v) \overset{(9.1.15)}{=} g_u(\lambda u, v) = \lambda g_u(u, v) =: \lambda \langle u, v \rangle_F,$$

which proves (FS$_1$). Property (FS$_2$) is evident by the bilinearity of the second derivative. Under the condition $f(u) \neq 0$, (9.1.17) is an immediate consequence of (9.1.13). $\qquad\square$

Lemma 9.1.25 (Projection onto $(\mathrm{span}(p))^{\perp_{g_p}}$). *Let (V, f) be a pre-Finsler vector space. If $p \in \mathring{V}$ and $f(p) \neq 0$, then the mapping*

$$P_p \colon V \to V,\ v \mapsto P_p(v) := v - \frac{g_p(p, v)}{g_p(p, p)}p \overset{(9.1.13)}{=} v - \frac{f'(p)(v)}{f(p)}p \tag{9.1.18}$$

is a linear transformation of V having the following properties:

(i) $\operatorname{Ker}(P_p) = \operatorname{span}(p);$

(ii) $P_p^2 = P_p$, *i.e.*, P_p *is a projection in* V;

(iii) $\operatorname{Im}(P_p) = \operatorname{Ker}(f'(p))$, *therefore* $p \perp_F P_p(v)$ *for all* $v \in V$;

(iv) $V = \operatorname{span}(p) \oplus \operatorname{Ker}(f'(p))$.

Proof. The linearity of P_p is obvious from the definition. It is also clear from (9.1.18) that $\operatorname{Ker}(P_p) = \operatorname{span}(p)$. For all $v \in V$ we have

$$P_p^2(v) = P_p \left(v - \frac{f'(p)(v)}{f(p)} p \right) \overset{\text{(i)}}{=} P_p(v),$$

and so $P_p^2 = P_p$. Since

$$f'(p)(P_p(v)) = f'(p)(v) - \frac{f'(p)(v)}{f(p)} f'(p)(p) \overset{(9.1.11)}{=} f'(p)(v) - f'(p)(v) = 0,$$

it follows that $\operatorname{Im}(P_p) \subset \operatorname{Ker}(f'(p))$. Conversely, if $v \in \operatorname{Ker}(f'(p))$, then

$$v = v - \frac{f'(p)(v)}{f(p)} p = P_p(v) \in \operatorname{Im}(P_p),$$

so we have (iii).

Finally, $P_p^2 = P_p$ implies by elementary linear algebra (see, e.g., [51], 2.19) that $V = \operatorname{Ker}(P_p) \oplus \operatorname{Im}(P_p)$, so (i) and (iii) lead to (iv). $\qquad\square$

Corollary and Definition 9.1.26. *Let* (V, f) *be a pre-Finsler vector space, and let* $\mathcal{U} := \{p \in V \mid f(p) \neq 0\}$. *Then the mapping*

$$P \colon p \in \mathcal{U} \mapsto P_p \in \operatorname{End}(V) \cong T_1^1(V),$$

where P_p *is given by* (9.1.18)*, is a type* $(1,1)$ *tensor field on* $\mathcal{U}$*, called the* projection tensor *associated to* f*. The type* $(0,2)$ *tensor field* $P_\flat \in \mathcal{T}_2^0(\mathcal{U})$ *defined by*

$$(P_\flat)_p(u, v) := g_p(P_p(u), v); \quad p \in \mathcal{U}; \ u, v, \in V \qquad (9.1.19)$$

is said to be the angular tensor *of* (V, f)*. The angular tensor has the following properties:*

$$(P_\flat)_p = g_p - f'(p) \otimes f'(p) = f(p) f''(p), \quad p \in \mathcal{U}; \qquad (9.1.20)$$

$$g_p(P_p(v), P_p(w)) = (P_\flat)_p(v, w); \quad p \in \mathcal{U}; \ v, w \in V. \qquad (9.1.21)$$

Proof. Given a point $p \in \mathcal{U}$, for all $v, w \in V$ we have

$$(P_\flat)_p(v, w) := g_p(P_p(v), w) \overset{(9.1.18)}{=} g_p \left(v - \frac{f'(p)(v)}{f(p)} p, w \right)$$

$$\overset{(9.1.13)}{=} g_p(v, w) - f'(p)(v) f'(p)(w)$$

$$= (g_p - f'(p) \otimes f'(p))(v, w) \overset{(9.1.9),(9.1.10)}{=} f(p) f''(p)(v, w),$$

thus proving (9.1.20). From this it is clear that $P_\flat$ is symmetric. Taking this into account, we find that

$$g_p(P_p(v), P_p(w)) =: (P_\flat)_p(v, P_p(w)) = (P_\flat)_p(P_p(w), v) := g_p(P_p^2(w), v)$$
$$= g_p(P_p(w), v) =: (P_\flat)_p(w, v) = (P_\flat)_p(v, w),$$

as was to be shown. $\qquad\qquad\square$

Remark 9.1.27. We can say that the angular tensor $P_\flat$ is obtained from the projection tensor P by *type-changing*, and it is *metrically equivalent* (more precisely, *g-equivalent*) to P. Classically, in the language of traditional tensor calculus, the transition from P to $P_\flat$ is called the *lowering of the contravariant index*, cf. Remark 6.1.42(d).

Corollary 9.1.28. *Let $f\colon V \to \mathbb{R}$ be a pre-Finsler norm, and let P be the projection tensor associated to f. Then for every $p \in \mathcal{U}$,*

$$f''(p)(v, w) = f''(p)(P_p(v), P_p(w)); \quad v, w \in V. \tag{9.1.22}$$

Proof. Indeed, we have

$$f''(p)(v, w) \overset{(9.1.20)}{=} \frac{1}{f(p)}(P_\flat)_p(v, w) \overset{(9.1.21)}{=} \frac{1}{f(p)}g_p(P_p(v), P_p(w))$$
$$= \frac{1}{f(p)}g_p\big(P_p(P_p(v)), P_p(P_p(w))\big)$$
$$\overset{(9.1.21)}{=} \frac{1}{f(p)}(P_\flat)_p(P_p(v), P_p(w)) \overset{(9.1.20)}{=} f''(p)(P_p(v), P_p(w)),$$

as desired. $\qquad\qquad\square$

Lemma and Definition 9.1.29. *Let (V, f) be a pre-Finsler vector space. Given a point $p \in \overset{\circ}{V}$, consider the quotient space $V/\operatorname{span}(p)$ of V with respect to the subspace generated by p. Then the function*

$$\begin{cases} h_p\colon V/\operatorname{span}(p) \times V/\operatorname{span}(p) \to \mathbb{R}, \\ ([v], [w]) \mapsto h_p([v], [w]) := f''(p)(v, w) \quad \text{if } v \in [v],\ w \in [w] \end{cases} \tag{9.1.23}$$

is a well-defined symmetric bilinear form. We say that the mapping

$$h\colon p \in \overset{\circ}{V} \mapsto h_p \in S_2(V/\operatorname{span}(p))$$

is the reduced fundamental tensor *of (V, f).*

Proof. The well-definedness of h_p is guaranteed by (9.1.12); its bilinearity and symmetry are clear. $\qquad\qquad\square$

Proposition 9.1.30. *Let (V, f) be a pre-Finsler vector space. If f is positive at a point $p \in \mathring{V}$, then g_p and h_p have the same index and nullity, and their ranks are related by $\mathrm{rank}(g_p) = \mathrm{rank}(h_p) + 1$.*

Proof. First we note that for all $v, w \in V$ we have

$$g_p(v, w) := \psi''(p)(v, w) \overset{(9.1.9)}{=} f'(p)(v)f'(p)(w) + f(p)f''(p)(v, w).$$

If, in particular, v and w belong to $\mathrm{Ker}(f'(p))$, then

$$g_p(v, w) = f(p)f''(p)(v, w) = h_p\big(\sqrt{f(p)}[v], \sqrt{f(p)}[w]\big). \qquad (*)$$

Since $V = \mathrm{span}(p) \oplus \mathrm{Ker}(f'(p))$ by Lemma 9.1.25(iv), the subspace $\mathrm{Ker}(f'(p))$ is $(n-1)$-dimensional ($n := \dim V$), and, by Lemma 1.4.3, it has a g_p-orthonormal basis $(b_i)_{i=1}^{n-1}$. Let $b_0 := \frac{1}{f(p)}p$. Then

$$g_p(b_0, b_0) = \frac{1}{(f(p))^2}g_p(p, p) \overset{(9.1.14)}{=} 1,$$

$$g_p(b_0, b_i) = \frac{1}{f(p)}g_p(p, b_i) \overset{(9.1.13)}{=} f'(p)(b_i) = 0 \quad (i \in J_{n-1});$$

therefore $(b_i)_{i=0}^{n-1}$ is a g_p-orthonormal basis of V, and, taking into account $(*)$, $(\sqrt{f(p)}[b_i])_{i=1}^{n-1}$ is an h_p-orthonormal basis of $V/\mathrm{span}(p)$. Now our assertions follow from Sylvester's law of inertia (see the end of subsection 1.4.3). $\qquad \square$

Corollary 9.1.31. *Let $f: V \to \mathbb{R}$ be a pre-Finsler norm. If $f(p) > 0$ at a point $p \in \mathring{V}$, then the scalar products g_p and h_p are positive definite, resp. non-degenerate at the same time.*

9.1.3 Finsler Norms and Some of Their Characterizations

Definition 9.1.32. A positive definite pre-Finsler norm $f: V \to \mathbb{R}$ is called a *Finsler norm* if it satisfies the following condition:

(FN$_4$) **Ellipticity.** The symmetric bilinear form

$$g_p := \psi''(p) := \frac{1}{2}(f^2)''(p)$$

is positive definite for all $p \in \mathring{V}$.

A *Finsler vector space* (V, f) is a (non-trivial, finite-dimensional) real vector space V equipped with a Finsler norm f. A Finsler norm f (and the Finsler vector space (V, f)) is called *reversible* if f also satisfies the condition

(FN$_5$) **Reversibility.** $f(-v) = f(v)$ for every $v \in \overset{\circ}{V}$.

Remark 9.1.33. (a) In the book of Bao, Chern and Shen [8], the term 'Minkowski norm' is used for a Finsler norm, and the property of ellipticity is mentioned as 'strong convexity'. However, in physicists' use a Minkowski norm has a different meaning. Furthermore, we prefer reserving the term 'strong convexity' for the property introduced in Definition 9.1.9, so we decided to apply a different terminology in both cases.

(b) The requirement of ellipticity can also be formulated as follows:

The fundamental tensor $g \in \mathcal{T}_2^0(\overset{\circ}{V})$ is a Riemannian metric on $\overset{\circ}{V}$.

(c) It follows immediately that a reversible Finsler norm is *absolutely homogeneous* in the sense that $f(\lambda v) = |\lambda| f(v)$ for any $\lambda \in \mathbb{R}$ and $v \in V$; cf. (N2) in Definition B.4.2.

Lemma 9.1.34. *If* $\dim V \geq 2$ *and a pre-Finsler norm* $f \colon V \to \mathbb{R}$ *is elliptic (i.e., satisfies* (FN$_4$)*), then* f *is either everywhere positive or everywhere negative on* $\overset{\circ}{V}$*, therefore either* (V, f) *of* $(V, -f)$ *is a Finsler vector space.*

Proof. By (9.1.14), $f^2(p) = g_p(p, p)$ for all $p \in \overset{\circ}{V}$. If f is elliptic, and hence g_p is positive definite, then it follows that f^2 cannot vanish outside the origin. Combining this observation with the continuity of f, we conclude that either f or $-f$ is everywhere positive. $\square$

Proposition 9.1.35 (M. Crampin [33]). *A positive definite pre-Finsler norm whose reduced fundamental tensor is (pointwise) non-degenerate is a Finsler norm.*

Proof. Let $f \colon V \to \mathbb{R}$ be a positive definite pre-Finsler norm such that the scalar product h_p given by (9.1.23) is non-degenerate for all $p \in \overset{\circ}{V}$. By Corollary 9.1.31 it is enough to show that h is pointwise positive definite. The continuity of f'' and the non-degeneracy of h imply that the signature of an individual scalar product h_p does not depend on the point $p \in \overset{\circ}{V}$. Therefore, to prove the positive definiteness of h, it is sufficient to check this property at a suitably chosen point of $\overset{\circ}{V}$.

Step 1. Choose a positive definite scalar product $\langle \, , \, \rangle$ on V. Write $\| \, \|$ and $\perp$ for the corresponding Euclidean norm and orthogonality; then the Euclidean unit sphere in V is $S_E := \{u \in V \mid \|u\| = 1\}$. Since S_E is a compact subset of V and f is continuous on S_E, the function $f \restriction S_E$ attains a minimum value, that is, *there is a point $e \in S_E$ such that $f(e) \leq f(u)$ for all $u \in S_E$.*

Step 2. We show that

$$\mathrm{Ker}(f'(e)) = (\mathrm{span}(e))^{\perp}.$$

Let v be an arbitrary nonzero vector in $(\mathrm{span}(e))^{\perp}$. Consider the parametrized straight line

$$\gamma \colon \mathbb{R} \to V, \ t \mapsto \gamma(t) := e + tv$$

in V, and define a spherical curve c in S_E by $c := \frac{1}{\|\gamma\|}\gamma$. Then $c(0) = e$,

$$c' = -\frac{\|\gamma\|'}{\|\gamma\|^2}\gamma + \frac{1}{\|\gamma\|}\gamma' = -\frac{\langle\gamma,\gamma'\rangle}{\|\gamma\|^3}\gamma + \frac{1}{\|\gamma\|}\gamma',$$

whence $c'(0) = v$. Thus the composite function

$$f \circ c \colon \mathbb{R} \to S_E \to \mathbb{R}$$

has a minimum at $0 \in \mathbb{R}$. Since $f \circ c$ is differentiable at 0, it follows that

$$0 = (f \circ c)'(0) = f'(c(0))(c'(0)) = f'(e)(v).$$

This shows that $v \in \mathrm{Ker}(f'(e))$, therefore

$$(\mathrm{span}(e))^{\perp} \subset \mathrm{Ker}(f'(e)).$$

However, both $(\mathrm{span}(e))^{\perp}$ and $\mathrm{Ker}(f'(e))$ have codimension 1, so the subset relation implies the desired equality.

Step 3. We claim that

$$f'(e)(c''(0)) = -\|v\|^2 f(e). \tag{$*$}$$

To see this, note first that the vector $c''(0)$ has a unique decomposition of the form

$$c''(0) = u + \langle c''(0), e\rangle e, \quad u \perp e.$$

Then $f'(e)(u) = 0$ by Step 2, hence

$$f'(e)(c''(0)) = \langle c''(0), e\rangle f'(e)(e) \overset{(9.1.11)}{=} \langle c''(0), e\rangle f(e).$$

On the other hand, the function $\langle c, c\rangle \colon \mathbb{R} \to \mathbb{R}$ is constant because $\mathrm{Im}(c)$ is contained in S_E, so we have

$$0 = \langle c, c\rangle' = 2\langle c, c'\rangle, \quad 0 = \frac{1}{2}\langle c, c\rangle'' = \langle c'', c\rangle + \|c'\|^2.$$

Evaluating the second relation at 0, we find that

$$\langle c''(0), e\rangle = \langle c'', c\rangle(0) = -\|c'(0)\|^2 = -\|v\|^2,$$

which finishes the proof of equality $(*)$.

Step 4. Now we turn again to the function $f \circ c$. Since it has a minimum at $0 \in \mathbb{R}$ and $f \circ c \in C^2(\mathbb{R})$ (in fact, it is smooth), the second derivative of $f \circ c$ at 0 is non-negative. So we have

$$0 \le (f \circ c)''(0) \stackrel{\text{(C.1.6)}}{=} f''(c(0))(c'(0), c'(0)) + f'(c(0))(c''(0))$$

$$= f''(e)(v, v) + f'(e)(c''(0)) \stackrel{(*)}{=} f''(e)(v, v) - \|v\|^2 f(e),$$

whence

$$h_e([v], [v]) := f''(e)(v, v) \ge \|v\|^2 f(e) > 0.$$

Thus h is positive definite at $e \in S_E$, which concludes the proof of the proposition. $\qquad\qquad\square$

Theorem 9.1.36. *For a positive definite pre-Finsler norm $f \colon V \to \mathbb{R}$ the following properties are equivalent:*

  (i) *The fundamental tensor of (V, f) is Euclidean, i.e., f is a Finsler norm.*

  (ii) *The fundamental tensor of (V, f) is semi-Euclidean.*

  (iii) *The reduced fundamental tensor of (V, f) is positive definite.*

  (iv) *The reduced fundamental tensor of (V, f) is non-degenerate.*

  (v) *At every point p of $\overset{\circ}{V}$, the symmetric bilinear form*

$$\widetilde{f''(p)} := f''(p) \restriction \operatorname{Ker}(f'(p)) \times \operatorname{Ker}(f'(p))$$

  is non-degenerate.

  (vi) *The nullity of the symmetric bilinear form $f''(p)$ is 1 at every point $p \in \overset{\circ}{V}$.*

  (vii) *If $p \in \overset{\circ}{V}$, $v \in V$, then $f''(p)(v, v) \ge 0$ and equality holds if, and only if, $v \in \operatorname{span}(p)$.*

  (viii) *If $p \in \overset{\circ}{V}$ and $f''(p)(v, w) = 0$ holds for all $w \in V$, then $v \in \operatorname{span}(p)$.*

Proof. By Corollary 9.1.31 assertion (i) is equivalent to assertion (iii), and assertion (ii) is equivalent to assertion (iv). Evidently, (i) implies (ii). Furthermore, by Proposition 9.1.35, (iv) implies (i). Thus properties (i)–(iv) are equivalent.

Now we show the equivalence of (iv) and (v). Observe first that

$$V/\operatorname{span}(p) \cong \operatorname{Ker}(f'(p)). \qquad\qquad (*)$$

Indeed, by Lemma 9.1.25, $\operatorname{span}(p) = \operatorname{Ker}(P_p)$ and $\operatorname{Im}(P_p) = \operatorname{Ker}(f'(p))$, so the first isomorphism theorem for vector spaces (Lemma and Definition 1.1.38(iii)) gives $(*)$. The canonical isomorphism of $\operatorname{Ker}(f'(p))$ onto $V/\operatorname{span}(p)$ is given by the mapping

$$\varphi_p \colon v \in \operatorname{Ker}(f'(p)) \mapsto \varphi_p(v) := [v] = v + \operatorname{span}(p) \in V/\operatorname{span}(p).$$

Using this, we have

$$\widetilde{f''(p)} = h_p \circ (\varphi_p \times \varphi_p),$$

which implies immediately that the non-degeneracy of h_p and $\widetilde{f''(p)}$ are equivalent properties.

Assertions (vii) and (viii) are just reformulations of (iii) and (iv), respectively. By the second relation of (9.1.11), the nullspace of $f''(p)$ ($p \in \overset{\circ}{V}$) contains the one-dimensional subspace $\mathrm{span}(p)$ of V, therefore assertions (vi) and (viii) are equivalent. Thus we have the following closed chain of implications:

$$
\begin{array}{ccccccc}
\text{(ii)} & \Leftarrow & \text{(i)} & \Leftrightarrow & \text{(iii)} & \Leftrightarrow & \text{(vii)} \\
 & \nwarrow & \Uparrow & & & & \\
\text{(v)} & \Leftrightarrow & \text{(iv)} & \Leftrightarrow & \text{(viii)} & \Leftrightarrow & \text{(vi)}.
\end{array}
$$

This concludes the proof of the theorem. $\qquad\qquad\square$

Proposition 9.1.37 (The Fundamental Inequality). *Let (V, f) be a Finsler vector space. Then for any $p \in \overset{\circ}{V}$ and $v \in V$ we have*

$$f'(p)(v) \leq f(v). \tag{9.1.24}$$

Equality holds if, and only if, v is a non-negative scalar multiple of p.

Proof. **Step 1.** If $v = 0$, then by the linearity of the derivative and by Remark 9.1.21(b), both sides of (9.1.24) are zero. Let, secondly, $v = \lambda p$, where λ is a positive real number. Then

$$f'(p)(v) = \lambda f'(p)(p) \overset{(9.1.11)}{=} \lambda f(p) = f(\lambda p) = f(v).$$

Thus, *if $v = \lambda p$, where $\lambda \in \mathbb{R}^*_+$, then equality holds in* (9.1.24). Third, if $v = \lambda p$, where λ is a negative real number, then

$$f'(p)(v) = \lambda f'(p)(p) = \lambda f(p) = -(-\lambda)f(p) = -f(-\lambda p) = -f(-v) < 0.$$

Hence *if $v = \lambda p$, where λ is negative, then strict inequality is valid in* (9.1.24).

Step 2. Suppose that p and v are linearly independent. From Taylor's formula (see Theorem C.1.22),

$$f(v) = f(p + v - p) = f(p) + f'(p)(v - p) + \frac{1}{2}f''(p + \vartheta(v - p))(v - p, v - p),$$

where $\vartheta \in \ ]0, 1[$. The linear independence of p and $v - p$ implies that

$$v - p \notin \mathrm{span}(p + \vartheta(v - p)).$$

Then, by property (vii) in Theorem 9.1.36,

$$f''(p + \vartheta(v - p))(v - p, v - p) > 0,$$

hence

$$f(v) > f(p) + f'(p)(v - p) \overset{(9.1.11)}{=} f'(p)(v).$$

So we conclude that *if p and v are linearly independent, then (9.1.24) holds as a strict inequality.*

This concludes the proof. $\qquad\qquad\qquad\qquad\qquad\qquad\qquad\qquad\square$

Corollary 9.1.38 (The Finslerian Schwarz Inequality). *Let (V, f) be a Finsler vector space. Then for every $u \in \overset{\circ}{V}$, $v \in V$,*

$$\langle u, v \rangle_F \le f(u)f(v). \qquad\qquad\qquad (9.1.25)$$

Equality holds if, and only if, v is a non-negative multiple of u.

Proof. Multiplying both sides of the fundamental inequality $f'(u)(v) \le f(v)$ by the positive real number $f(u)$ yields

$$f(u)f'(u)(v) \le f(u)f(v).$$

Here the left-hand side can be manipulated as follows:

$$f(u)f'(u)(v) \overset{(9.1.8)}{=} \psi'(u)(v) \overset{(4.2.2)}{=} \psi''(u)(u, v) =: g_u(u, v) =: \langle u, v \rangle_F,$$

where we used the fact that $\psi' \overset{(9.1.7)}{=} ff'$ is 1^+-homogeneous. Thus we proved inequality (9.1.25). Clearly, we have the same criterion of equality in (9.1.25) as in the fundamental inequality. $\qquad\qquad\qquad\square$

Corollary 9.1.39. *If $f \colon V \to \mathbb{R}$ is a Finsler norm, then f is a convex function.*

Proof. Let $a, b \in V$, and suppose first that the line segment $\overline{ab}$ does not contain the origin. Using the fundamental inequality in the first step, we find that

$$f(b) \ge f'(a)(b) = f(a) + f'(a)(b) - f(a)$$
$$\overset{(9.1.11)}{=} f(a) + f'(a)(b) - f'(a)(a) = f(a) + f'(a)(b - a).$$

It follows that (9.1.1) holds for all $t \in [0, 1]$ as in the proof of Proposition 9.1.13(i). If $\overline{ab}$ contains the origin, then (9.1.1) holds by continuity. This proves that f is convex on V. $\qquad\qquad\qquad\qquad\qquad\qquad\square$

Remark 9.1.40. It follows from the previous corollary and from Lemma 9.1.7 that *reversible Finsler vector spaces are (finite-dimensional) normed spaces in the usual sense* (see Definition B.4.2). Finite dimensional normed spaces are also called *Minkowski spaces*. At this point it becomes understandable why Busemann referred to 'Minkowskian geometry' in the quotation at the beginning of this section. An excellent book on this subject is 'Minkowski Geometry' by A. C. Thompson [93].

We close this subsection with a more geometric characterization of Finsler norms.

Theorem 9.1.41. *Let* (V, f) *be an at least two-dimensional pre-Finsler vector space with a positive definite pre-Finsler norm. Endow* V *with a positive definite scalar product. Then* f *is a Finsler norm if, and only if, the unit sphere of* (V, f) *has nowhere vanishing Gauss–Kronecker curvature with respect to the chosen Euclidean structure.*

Proof. By Corollary 9.1.23, the unit sphere $\mathrm{bd}(B) := f^{-1}(1)$ of (V, f) is indeed a hypersurface in V. In our forthcoming argument we shall use the notation of Example 6.1.75(b) with the choice $M := \mathrm{bd}(B)$.

At every point $p \in \mathrm{bd}(B)$, the function

$$b_p \colon T_p(\mathrm{bd}(B)) \times T_p(\mathrm{bd}(B)) \to \mathbb{R}, \quad (v, w) \mapsto b_p(v, w) := \bar{g}_p(W_p(v), w)$$

is a symmetric bilinear form, the *second fundamental form* of $\mathrm{bd}(B)$ at p (see, e.g., [94]). (Warning: in the above formula $\bar{g}$ is the Riemannian metric determined by the scalar product on V according to the quoted example or Proposition C.3.3, and not the fundamental tensor of the pre-Finsler vector space (V, f).) Since

$$T_p(\mathrm{bd}(B)) \cong \{p\} \times \mathrm{Ker}(f'(p)) \cong \mathrm{Ker}(f'(p)),$$

b_p can also be regarded as a symmetric bilinear form on $\mathrm{Ker}(f'(p))$. Thus, by relation (6.1.80) it follows that

$$\begin{aligned} &\underbrace{b_p \text{ is non-degenerate if, and only if,}}_{} \\ &\widetilde{f''(p)} \in S_2(\mathrm{Ker}(f'(p))) \text{ is non-degenerate.} \end{aligned} \tag{$*$}$$

Now choose a $\bar{g}_p$-orthonormal basis $(e_i)_{i=1}^{n-1}$ in $T_p(\mathrm{bd}(B))$, and let $(W_j^i(p))$ be the matrix of the Weingarten operator W_p with respect to this basis. Then

$$b_p(e_j, e_k) = \bar{g}_p(W_p(e_j), e_k) = \bar{g}_p(W_j^i(p)e_i, e_k) = W_j^i(p)\bar{g}_p(e_i, e_k) = W_j^k(p),$$

Connections, Sprays and Finsler Structures

whence

$$K(p) := \det W_p = \det(W^i_j(p)) = \det(b_p(e_i, e_j)).$$

Taking into account Corollary 1.4.1, from this it follows that

$$b_p \text{ is non-degenerate if, and only if, } K(p) \neq 0. \qquad (**)$$

Assertions $(*)$ and $(**)$, and the equivalence of (v) and (i) in Theorem 9.1.36 lead to the desired conclusion. $\qquad\square$

9.1.4 Reduction to Euclidean Vector Space

In the formulation of the next observation and definition we use the same canonical identifications

$$V \cong \{p\} \times V \cong T_p\mathring{V} \quad (p \in \mathring{V})$$

as in Remark 9.1.21(d), where we introduced the fundamental tensor of a pre-Finsler vector space.

Lemma and Definition 9.1.42. *Let (V, f) be a Finsler vector space with energy function $\psi := \frac{1}{2}f^2$.*

(i) *The mapping*

$$\mathcal{C}_\flat \colon p \in \mathring{V} \mapsto (\mathcal{C}_\flat)_p := \psi'''(p) \in T^0_3(V) \cong T^0_3(T_p\mathring{V})$$

is a symmetric type $(0,3)$ tensor field on $\mathring{V}$, called the Cartan tensor of the Finsler vector space (V, f).

(ii) *There exists a unique symmetric tensor field $\mathcal{C}$ of type $(1,2)$ on $\mathring{V}$ such that*

$$g_p(\mathcal{C}_p(u, v), w) = (\mathcal{C}_\flat)_p(u, v, w) \qquad (9.1.26)$$

for every $p \in \mathring{V}$ and for all $u, v, w \in V \cong T_pV$. We say that $\mathcal{C}$ is the vector-valued Cartan tensor of (V, f). (For short, $\mathcal{C}$ is also named the Cartan tensor of (V, f).)

Proof. Clearly, $(\mathcal{C}_\flat)_p$ depends smoothly on p, because ψ is a smooth function of $\mathring{V}$. The symmetry of $\mathcal{C}_\flat$ is guaranteed by the symmetry of ψ'''. To prove (ii), let $p \in \mathring{V}$ be any fixed point, and let $u, v \in V$ be fixed vectors. Then (V, g_p) is a Euclidean vector space, thus, in particular, g_p is a non-degenerate symmetric bilinear form on V. The function

$$w \in V \mapsto (\mathcal{C}_\flat)_p(u, v, w) \in \mathbb{R}$$

is a linear form on V, so by Corollary 1.4.1 there exists a unique vector $\mathcal{C}_p(u,v) \in V$ such that (9.1.26) is satisfied. Then the mapping

$$\mathcal{C}_p \colon V \times V \to V, \ (u,v) \mapsto \mathcal{C}_p(u,v)$$

is bilinear and symmetric, and it is also clear that $\mathcal{C}_p$ depends on $p \in \mathring{V}$ smoothly. This concludes the proof. $\qquad\qquad\qquad\qquad\qquad\square$

Lemma 9.1.43. *If $\mathcal{C}_\flat$ is the Cartan tensor of a Finsler vector space (V,f), then*

$$(\mathcal{C}_\flat)_p(p,u,v) = 0 \qquad\qquad (9.1.27)$$

for every $p \in \mathring{V}$ and for all $u,v \in V$.

Proof. Since the energy function ψ is 2^+-homogeneous, its second derivative ψ'' is 0^+-homogeneous by Lemma 4.2.4. Thus

$$(\mathcal{C}_\flat)_p(p,u,v) := \psi'''(p)(p,u,v) = \psi'''(p)(p)(u,v) \overset{(4.2.2)}{=} 0.$$

Notice that the last equality can be directly obtained as follows:

$$\psi'''(p)(p)(u,v) \overset{(\text{C.1.3})}{=} \left(\lim_{t \to 0^+} \frac{\psi''(p+tp) - \psi''(p)}{t} \right)(u,v)$$

$$\overset{(*)}{=} \left(\lim_{t \to 0^+} \frac{\psi''(p) - \psi''(p)}{t} \right)(u,v) = 0,$$

where we used the 0^+-homogeneity of ψ'' in step $(*)$. $\qquad\qquad\square$

Definition 9.1.44. We say that a Finsler vector space (V,f) *reduces to a Euclidean vector space* if there exists a positive definite scalar product b on V such that

$$f(v) = (b(v,v))^{\frac{1}{2}}, \quad v \in V, \qquad\qquad (9.1.28)$$

i.e., f is the norm associated with a positive definite scalar product on V. Then we also say that the Finsler norm f is *Euclidean*.

Proposition 9.1.45. *Let (V,f) be a Finsler vector space. The following conditions are equivalent:*

  (i) *The Finsler norm f is Euclidean.*
 (ii) *The energy function $\psi := \frac{1}{2}f^2$ is a quadratic form on V.*
(iii) *The derivative of the energy function is a linear mapping from V into V^*.*
 (iv) *The F-scalar product $\langle \, , \, \rangle_F$ is symmetric.*
  (v) *The energy function is of class C^2.*

(vi) *The fundamental tensor of (V, f) is constant.*
(vii) *The Cartan tensor of (V, f) vanishes.*

Proof.   **Step 1.** The equivalence of (i) and (ii) is evident. Since

$$g_p(u, v) := \psi''(p)(u, v), \quad (\mathcal{C}_b)_p(u, v, w) := \psi'''(p)(u, v, w),$$

where $p \in \overset{\circ}{V}$; $u, v, w \in V$, both assertion (vi) and assertion (vii) are equivalent to the constancy of ψ'', so they are equivalent to each other.
Step 2. Suppose that ψ *is a quadratic form,* i.e.,

$$\psi(v) = \frac{1}{2} f^2(v) = \frac{1}{2} b(v, v)$$

for some positive definite scalar product $b \in S_2(V)$. Then, by Example C.1.10, it can immediately be seen that ψ is smooth, so (v) holds automatically. Applying (C.1.5) to ψ, we have

$$\psi'(p)(v) = b(p, v); \quad p, v \in V.$$

Thus $\psi' : V \to V^*$ is a linear mapping, namely, with the notation of Lemma 1.2.1, $\psi' = j_b$. We also have

$$\langle u, v \rangle_F := g_u(u, v) \overset{(9.1.13)}{=} \psi'(u)(v) = b(u, v),$$

therefore the F-scalar product is symmetric. So we have shown that *(ii)* implies *(iii), (iv) and (v).*

Conversely, from Lemma 4.2.5(iii) we obtain immediately that (v) implies (ii).
Step 3. We verify the implications

$$(iv) \Rightarrow (iii) \Rightarrow (vi) \Rightarrow (ii).$$

To begin, suppose that $\langle \, , \, \rangle_F$ is symmetric. Then, for all $p, v \in \overset{\circ}{V}$,

$$\langle p, v \rangle_F := g_p(p, v) := \psi''(p)(p, v), \quad \langle v, p \rangle_F := g_v(v, p) := \psi''(v)(v, p),$$

whence

$$\psi''(p)(p, v) = \psi''(v)(v, p). \tag{$**$}$$

Thus, for every fixed $v \in \overset{\circ}{V}$,

$$\psi'(p)(v) \overset{(9.1.13)}{=} g_p(p, v) = \psi''(p)(p, v) \overset{(**)}{=} \psi''(v)(v, p),$$

therefore ψ' depends on p linearly.

To proceed, assume that $\psi' : V \to V^*$ is linear. Then, for every $p \in V$, $\psi''(p) = \psi'$ (Example C.1.4). Therefore, for all $u, v \in V$,

$$g_p(u, v) := \psi''(p)(u, v) = \psi'(u)(v).$$

This shows that the fundamental tensor g is constant.

To conclude this part, we turn to the implication (vi)$\Rightarrow$(ii). As we have just seen, condition (vi) is equivalent to the constancy of ψ''. Thus, for any fixed point $p \in \mathring{V}$ and for every $v \in \mathring{V}$,

$$2\psi(v) \overset{(9.1.14)}{=} g_v(v,v) = \psi''(v)(v,v) = \psi''(p)(v,v),$$

therefore ψ is the quadratic form associated with $\frac{1}{2}\psi''(p)$.

Summing up, we have the following closed chain of implications:

$$
\begin{array}{ccccc}
\text{(v)} & & \text{(vi)} & \Leftrightarrow & \text{(vii)} \\
 & \nwarrow & \Downarrow & \nwarrow & \\
\text{(i)} & \Leftrightarrow & \text{(ii)} & \Rightarrow & \text{(iii)} \\
 & & \Downarrow & \nearrow & \\
 & & \text{(iv)} & &
\end{array}
$$

$\square$

Remark 9.1.46. As we have already mentioned, every reversible Finsler vector space is a (finite-dimensional) normed space. There is a very extensive literature on the characterizations of orthogonal spaces with positive definite scalar product among (not necessarily finite-dimensional) normed spaces. For a comprehensive survey we refer to Amir [5].

9.1.5 *Averaged Scalar Product on a Gauge Vector Space*

Definition 9.1.47. (a) A convex and positive definite pre-Finsler norm is called a *gauge function*, or for short, a *gauge*. A *gauge vector space* is a (finite-dimensional, real) vector space equipped with a gauge function.

If f is a gauge function on V, then (as above) $\psi := \frac{1}{2}f^2$ is the *energy* associated to f. The sets

$$B := \{v \in V \mid f(v) \leq 1\} \quad \text{and} \quad \mathrm{bd}(B) = \{v \in V \mid f(v) = 1\} = f^{-1}(v)$$

are the *unit ball* and the *unit sphere* in the gauge vector space (V, f), respectively.

(b) Two gauge vector spaces (V_1, f_1) and (V_2, f_2) are called *linearly isometric* if there exists a *gauge-preserving* linear isomorphism from V_1 onto V_2, i.e., a linear isomorphism $\varphi \colon V_1 \to V_2$ such that $f_1 = f_2 \circ \varphi$. Then we also say that φ is a *linear isometry* from (V_1, f_1) onto (V_2, f_2). In the special case of Finsler vector spaces we use these terms in the same sense.

For the rest of this subsection we assume that $n := \dim V \geq 2$.

Remark 9.1.48. Let (V, f) be a gauge vector space with unit ball B. It can be shown that B is a regular domain in V. Thus, by Remark 3.4.20(b), $\partial B = \mathrm{bd}(B)$, i.e., *the manifold boundary of the unit ball coincides with the unit sphere.* By Corollary 9.1.23, ∂B is a hypersurface in V. In view of Example 6.1.75(a), for every $p \in M$, we have the canonical identification

$$T_p(\partial B) \cong \{p\} \times \mathrm{Ker}(f'(p)) \subset T_p V \cong \{p\} \times V.$$

Lemma and Definition 9.1.49. *Let (V, f) be an n-dimensional gauge vector space with unit ball B and let*

$$j := j_{\partial B} \colon \partial B \to V$$

be the canonical inclusion of the unit sphere of (V, f) into V. Define the radial vector field Z on V by

$$Z(p) := (p, p) \in \{p\} \times V \cong T_p V, \quad p \in V.$$

Suppose that $\omega \in A_n(V)$ is a volume form on the vector space V, and let $\omega^\uparrow \in \mathcal{A}_n(V)$ be the induced volume form on V as on a manifold defined by (3.4.1). Then the mapping

$$\begin{cases} \omega_{\partial B} \colon p \in \partial B \mapsto (\omega_{\partial B})_p \in A_{n-1}(T_p \partial B), \\ (\omega_{\partial B})_p((p, w_1), \dots, (p, w_{n-1})) := (\omega^\uparrow)_p((p, p), (p, w_1), \dots, (p, w_{n-1})) \end{cases}$$

is a volume form on ∂B, and we have

$$\omega_{\partial B} = j^* i_Z \omega^\uparrow. \tag{9.1.29}$$

The volume form $\omega_{\partial B}$ is called the induced volume form *on ∂B. The orientation of the unit ball determined by $\omega_{\partial B}$ is said to be the* induced orientation *on ∂B.*

Proof. To show that $\omega_{\partial B}$ is a volume form on ∂B, it is sufficient to check that $(\omega_{\partial B})_p \neq 0$ for all $p \in \partial B$. So let a point p in ∂B be given, and choose a basis $(p, w_i)_{i=1}^{n-1}$ for $T_p \partial B$. Then $(p, w_1, \dots, w_{n-1})$ is a basis of V by Lemma 9.1.25(iv), so $\omega(p, w_1, \dots, w_{n-1}) \neq 0$. This implies that

$$0 \neq \omega(p, w_1, \dots, w_{n-1}) =: (\omega_{\partial B})_p((p, w_1), \dots, (p, w_{n-1}))$$

as desired.

Relation (9.1.29) can be verified immediately. $\qquad\square$

Remark 9.1.50. We use the same notation as above.

(a) Since the mapping $\omega \in A_n(V) \mapsto \omega_{\partial B} \in \mathcal{A}_{n-1}(\partial B)$ is $\mathbb{R}$-linear, it follows that the induced orientation on the unit ball reverses if we reverse the orientation of V.

(b) For every $p \in \partial B$, $Z(p)$ is an outward pointing vector in $T_p V$. Therefore, the induced orientation on ∂B coincides with the orientation induced by the orientation $[\omega^\uparrow]$ of the *manifold* V according to Lemma 3.4.21(ii).

Proposition and Definition 9.1.51. *Let* (V, f) *be a gauge vector space with energy function* $\psi = \frac{1}{2} f^2$ *and unit ball* B. *Suppose that* $\omega \in A_n(V)$ *orients* V, *and let the unit sphere* ∂B *be equipped with the induced orientation* $[\omega_{\partial B}]$. *For every* $(u, v) \in V \times V$, *denote by* $\psi''(u, v)$ *the function*

$$V \setminus \{0\} \to \mathbb{R}, \ p \mapsto \psi''(u, v)(p) := \psi''(p)(u, v).$$

Then the function

$$\begin{cases} b_f : V \times V \to \mathbb{R}, \ (u, v) \mapsto b_f(u, v), \\ b_f(u, v) := \dfrac{1}{\int_{\partial B} \omega_{\partial B}} \displaystyle\int_{\partial B} \psi''(u, v) \omega_{\partial B} \end{cases} \tag{9.1.30}$$

is a well-defined positive definite scalar product, called the averaged scalar product *on* (V, f).

Proof. **Step 1.** We show that *the definition of* b_f *is independent of the choice of the volume form on* V. Indeed, let μ be another volume form on V. Then $\omega = \alpha \mu$ for some nonzero real number α, so by part (a) of the preceding remark, $\omega_{\partial B} = \alpha \mu_{\partial B}$. Thus, taking into account the linearity of the integral,

$$\frac{1}{\int_{\partial B} \omega_{\partial B}} \int_{\partial B} \psi''(u, v) \omega_{\partial B} = \frac{1}{\int_{\partial B} \alpha \mu_{\partial B}} \int_{\partial B} \alpha \psi''(u, v) \mu_{\partial B}$$

$$= \frac{\alpha}{\alpha \int_{\partial B} \mu_{\partial B}} \int_{\partial B} \psi''(u, v) \mu_{\partial B} = \frac{1}{\int_{\partial B} \mu_{\partial B}} \int_{\partial B} \psi''(u, v) \mu_{\partial B},$$

as desired.

Step 2. We prove that b_f *is a positive definite scalar product.*

Since the second derivative $\psi''(p) : V \times V \to \mathbb{R}$ is a symmetric bilinear form for every $p \in V \setminus \{0\}$ (Theorem C.1.19) and the integral is $\mathbb{R}$-linear, b_f is also a symmetric bilinear form on V.

By the convexity of f, $f''(p)$ is positive semidefinite for all $p \in V \setminus \{0\}$ (Proposition 9.1.17). This implies that $\psi''(p)$ is also positive semidefinite because

$$\psi''(p)(v, v) \overset{(9.1.9)}{=} (f'(p)(v))^2 + f(p) f''(p)(v, v) \geq 0$$

for all $v \in V$. Thus the function $\psi''(v, v)$ $(v \in V)$ is surely non-negative. Now suppose that $v \in V \setminus \{0\}$, and let $q := \frac{1}{f(v)}v$. Then $q \in \partial B$, and

$$\psi''(v, v)(q) := \psi''(q)(v, v) = \psi'' \left(\frac{1}{f(v)}v \right)(v, v)$$

$$\overset{(9.1.15)}{=} \psi''(v)(v, v) \overset{(9.1.14)}{=} (f(v))^2 > 0,$$

that is, $\psi''(v, v)$ is positive at the point q in ∂B. Then the integrals $\int_{\partial B} \psi''(v, v)\omega_{\partial B}$ and $\int_{\partial B} \omega_{\partial B}$ are both positive by Lemma 3.4.15, hence $b_f(v, v) > 0$. This proves the positive definiteness of b_f, and concludes the proof of the proposition. $\square$

Example 9.1.52. Assume that b is a positive definite scalar product on V, i.e., (V, b) is a Euclidean vector space. Consider the Euclidean norm

$$f: V \to \mathbb{R}, \quad v \mapsto f(v) := \sqrt{b(v, v)}$$

associated to b. Then, in particular, (V, f) is a gauge vector space. We claim that the averaged scalar product on (V, f) is just the given scalar product b, i.e., $b_f = b$.

Indeed, in this case the energy function $\psi := \frac{1}{2}f^2$ is the quadratic form associated with b. So, taking into account Example C.1.10, we obtain easily that $\psi''(p) = b$ for all $p \in V$. Therefore, for every $(u, v) \in V \times V$, the function

$$\psi''(u, v): V \to \mathbb{R}, \quad p \mapsto \psi''(u, v)(p) := \psi''(p)(u, v)$$

is the constant function of value $b(u, v)$, and hence (9.1.30) gives $b_f(u, v) = b(u, v)$.

Proposition 9.1.53. *Let (V_1, f_1) and (V_2, f_2) be gauge vector spaces. Let b_{f_1} and b_{f_2} be the averaged scalar products on (V_1, f_1) and (V_2, f_2), respectively. If $\varphi: V_1 \to V_2$ is a linear isometry from (V_1, f_1) onto (V_2, f_2), then φ is an orthogonal isomorphism between the Euclidean vector spaces (V_1, b_{f_1}) and (V_2, b_{f_2}).*

Proof.   (a) First we introduce the following notations:

$$\psi_i := \frac{1}{2}(f_i)^2, \quad \partial B_i = \mathrm{bd}(B_i) = f_i^{-1}(1),$$

$$j_i := j_{\partial B_i}: \partial B_i \to V_i, \quad i \in \{1, 2\};$$

$$\varphi_{\partial B_1} := \varphi \restriction \partial B_1.$$

Since φ is gauge-preserving, we have

$$\psi_1 = \psi_2 \circ \varphi. \tag{1}$$

The mapping $\varphi_{\partial B_1} : \partial B_1 \to V_2$ is a diffeomorphism from ∂B_1 onto ∂B_2, and makes the following diagram commutative:

$$
\begin{array}{ccc}
\partial B_1 & \xrightarrow{\varphi_{\partial B_1}} & \partial B_2 \\
{\scriptstyle j_1}\downarrow & & \downarrow{\scriptstyle j_2} \\
V_1 & \xrightarrow{\varphi} & V_2
\end{array} \qquad (2)
$$

We denote by Z_i the radial vector field on V_i; then

$$
Z_1(p) = (p, p), \quad p \in V_1; \qquad Z_2(q) = (q, q), \quad q \in V_2.
$$

It is easy to see that

$$
\varphi_* \circ Z_1 \circ \varphi^{-1} =: \varphi_{\#} Z_1 = Z_2 \qquad (3)
$$

and

$$
i_{Z_1} \circ \varphi^* = \varphi^* \circ i_{\varphi_{\#} Z_1} \overset{(3)}{=} \varphi^* \circ i_{Z_2}. \qquad (4)
$$

(b) Let ω be a volume form on the vector space V_2, and let

$$
\mu := \varphi^* \omega \in A_n(V_1)
$$

($n := \dim V_1 = \dim V_2$). Then, by (3.4.1), we also have $\mu^{\uparrow} = \varphi^* \omega^{\uparrow}$. We show that

$$
(\varphi_{\partial B_1})^* \omega_{\partial B_2} = \mu_{\partial B_1}. \qquad (5)
$$

Indeed,

$$
(\varphi_{\partial B_1})^* \omega_{\partial B_2} \overset{(9.1.29)}{=} (\varphi_{\partial B_1})^* \circ (j_2)^* (i_{Z_2} \omega^{\uparrow}) \overset{(3.3.8)}{=} (j_2 \circ \varphi_{\partial B_1})^* (i_{Z_2} \omega^{\uparrow})
$$

$$
\overset{(2)}{=} (\varphi \circ j_1)^* (i_{Z_2} \omega^{\uparrow}) = (j_1)^* \circ \varphi^* (i_{Z_2} \omega^{\uparrow})
$$

$$
\overset{(4)}{=} (j_1)^* (i_{Z_1}(\varphi^* \omega^{\uparrow})) = (j_1)^* (i_{Z_1} \mu^{\uparrow}) \overset{(9.1.29)}{=} \mu_{\partial B_1},
$$

as we claimed.

(c) From (1), using the chain rule and Example C.1.4, we obtain that for each $p \in V \setminus \{0\}$,

$$
\psi_1'(p) = \psi_2'(\varphi(p)) \circ \varphi, \quad \psi_1''(p) = \psi_2''(\varphi(p)) \circ (\varphi, \varphi).
$$

Therefore, for every $(u, v) \in V_1 \times V_1$,

$$
\psi_1''(u, v) = \psi_2''(\varphi(u), \varphi(v)) \circ \varphi. \qquad (6)
$$

(d) After these preparations, we prove that φ preserves the averaged scalar product. Let $(u, v) \in V_1 \times V_1$. Then

$$
b_{f_2}(\varphi(u), \varphi(v)) = \frac{1}{\int_{\partial B_2} \omega_{\partial B_2}} \int_{\partial B_2} \psi_2''(\varphi(u), \varphi(v)) \, \omega_{\partial B_2}
$$

$$= \frac{1}{\int_{\varphi_{\partial B_1}(\partial B_1)} \omega_{\partial B_2}} \int_{\varphi_{\partial B_1}(\partial B_1)} \psi_2''(\varphi(u), \varphi(v)) \, \omega_{\partial B_2}$$

$$\overset{(\text{Int}_1)}{=} \frac{1}{\int_{\partial B_1} (\varphi_{\partial B_1})^* \omega_{\partial B_2}} \int_{\partial B_1} (\varphi_{\partial B_1})^* (\psi_2''(\varphi(u), \varphi(v)) \, \omega_{\partial B_2})$$

$$\overset{(5)}{=} \frac{1}{\int_{\partial B_1} \mu_{\partial B_1}} \int_{\partial B_1} (\psi_2''(\varphi(u), \varphi(v)) \circ \varphi) \, \mu_{\partial B_1}$$

$$\overset{(6)}{=} \frac{1}{\int_{\partial B_1} \mu_{\partial B_1}} \int_{\partial B_1} \psi_1''(u, v) \, \mu_{\partial B_1} = b_{f_1}(u, v).$$

This completes the proof.   □

Remark 9.1.54. Averaged scalar product construction in Finsler geometry was first applied by Z. I. Szabó [90]. The treatment given here is partly based on Matveev et al. [75]. For a thorough discussion of averaging process on Finsler vector spaces we refer to Crampin [34].

9.2   Fundamentals on Finsler Functions

9.2.1   *Pre-Finsler Manifolds*

Definition 9.2.1. (a) Let M be a manifold. A function $F \colon TM \to \mathbb{R}$ is called a *pre-Finsler function* (for M) if it satisfies the following conditions:

(F_1) F is continuous on TM and smooth on $\mathring{T}M$.
(F_2) F is positive-homogeneous of degree 1.

A *pre-Finsler manifold* is a pair (M, F) consisting of a manifold M and a pre-Finsler function F on TM. We say that $E := \frac{1}{2} F^2$ is the *energy function* associated to F or the energy function (briefly *energy*) of (M, F). A pre-Finsler function F (and the pre-Finsler manifold (M, F)) is called *positive definite* if F also has the property

(F_3) $F(v) > 0$ if $v \neq 0$.

(b) If (M, F) is a pre-Finsler manifold, then the vertical Hessian

$$g := \nabla^v \nabla^v E = \frac{1}{2} \nabla^v \nabla^v F^2 \in \mathcal{F}_2^0(\mathring{T}M) \tag{9.2.1}$$

of E is called the *fundamental tensor* of (M, F).

Remark 9.2.2. Let (M, F) be a pre-Finsler manifold.

(a) Clearly, the energy function E of (M, F) is 2^+-homogeneous. This implies by Lemma 4.2.11 that E *is of class* C^1 *on* TM.

We note that the phrase 'energy function of F' may also refer to the energy of F as a Lagrangian on $\mathring{T}M$ described in Definition 5.2.1 and denoted by E_F. However, the difference in notation clearly distinguishes the two possible meanings.

(b) In view of Lemma 6.2.28(ii), the fundamental tensor of (M, F) is a symmetric type $(0, 2)$ Finsler tensor field on $\mathring{T}M$, and hence it is a scalar product on $\mathring{\pi}$ in the sense of Definition 2.2.52(a). By formula (6.2.43),

$$g(\mathbf{j}\xi, \mathbf{j}\eta) = \mathbf{J}\xi(\mathbf{J}\eta(E)) - \mathbf{J}[\mathbf{J}\xi, \eta](E), \tag{9.2.2}$$

for all $\xi, \eta \in \mathfrak{X}(\mathring{T}M)$. This can equivalently be written as

$$g(\widetilde{X}, \widetilde{Y}) = \mathbf{i}\widetilde{X}(\mathbf{i}\widetilde{Y}(E)) - \mathbf{J}[\mathbf{i}\widetilde{X}, \mathcal{H}\widetilde{Y}]E, \tag{9.2.3}$$

where $\widetilde{X}, \widetilde{Y} \in \Gamma(\mathring{\pi})$ and $\mathcal{H}$ is an arbitrarily chosen Ehresmann connection in TM. In particular, for any vector fields X, Y on M,

$$g(\widehat{X}, \widehat{Y}) = X^{\mathrm{v}}(Y^{\mathrm{v}}E) = \frac{1}{2}X^{\mathrm{v}}(Y^{\mathrm{v}}F^2), \tag{9.2.4}$$

cf. (6.2.40). With the same notation as in Lemma 6.2.28(iv), the components of g are the functions

$$g_{ij} := g\left(\widehat{\frac{\partial}{\partial u^i}}, \widehat{\frac{\partial}{\partial u^j}}\right) = \frac{\partial^2 E}{\partial y^i \partial y^j} = \frac{1}{2}\frac{\partial^2 F^2}{\partial y^i \partial y^j}; \quad i, j \in J_n, \tag{9.2.5}$$

defined and smooth on $\mathring{\tau}^{-1}(\mathcal{U}) = \tau^{-1}(\mathcal{U}) \cap \mathring{T}M$.

(c) By Remark 6.2.29(b), the energy function E of (M, F) induces a scalar product $\bar{g}$ on the vertical bundle $V\mathring{T}M$ such that

$$\bar{g}(\mathbf{J}\xi, \mathbf{J}\eta) := g(\mathbf{j}\xi, \mathbf{j}\eta) \overset{(6.2.39)}{=} \omega_E(\mathbf{J}\xi, \eta), \tag{9.2.6}$$

for all $\xi, \eta \in \mathfrak{X}(\mathring{T}M)$. Then

$$\bar{g}_{ij} := \bar{g}\left(\frac{\partial}{\partial y^i}, \frac{\partial}{\partial y^j}\right) = g\left(\widehat{\frac{\partial}{\partial u^i}}, \widehat{\frac{\partial}{\partial u^j}}\right) = g_{ij}; \quad i, j \in J_n,$$

thus $\bar{g}$ may also be called the fundamental tensor of (M, F). Although g and $\bar{g}$ live on different bundles, they carry the same information.

(d) Evidently, (T_pM, F_p), where $F_p := F \upharpoonright T_pM$, is a pre-Finsler vector space for every $p \in M$. We show that *the fundamental tensor* g *of* (M, F) *restricts to the fundamental tensor of* $(T_{\tau(u)}M, F_{\tau(u)})$ *for every* $u \in \mathring{T}M$.

First we recall (see Remark 4.1.2(f)) that g, as a section of the Finsler tensor bundle $F_2^0(\mathring{T}M)$, is a smooth mapping

$$g \colon \mathring{T}M \to \mathring{T}M \times_M T_2^0(TM)$$

of the form

$$u \in \mathring{T}M \mapsto g_u = (u, \underline{g}_u) \in \{u\} \times T_2^0(T_{\tau(u)}M),$$

and we can freely identify g_u with its principal part $\underline{g}_u$. In view of (6.2.45) and (6.2.46),

$$(\nabla^\mathrm{v} E)_u(u, v) = (E_{\tau(u)})'(u)(v),$$
$$(\nabla^\mathrm{v}\nabla^\mathrm{v} E)_u((u, v), (u, w)) = (E_{\tau(u)})''(u)(v, w),$$

therefore

$$g_u = (\nabla^\mathrm{v}\nabla^\mathrm{v} E)_u = (E_{\tau(u)})''(u) \tag{9.2.7}$$

for every $u \in \mathring{T}M$. Thus the mapping

$$v \in \mathring{T}_{\tau(u)}M \mapsto g_v = (\nabla^\mathrm{v}\nabla^\mathrm{v} E)_v \in T_2^0(T_{\tau(u)}M)$$

is just the fundamental tensor of the pre-Finsler vector space $(T_{\tau(u)}M, F_{\tau(u)})$.

These observations make it possible to translate the results of the preceding section into the context of pre-Finsler (and later of Finsler) manifolds. However, instead of mechanical translations, we shall usually give independent proofs, using the apparatus of Finsler tensor fields on $\mathring{T}M$.

Lemma 9.2.3. *Let (M, F) be a pre-Finsler manifold with energy function $E := \frac{1}{2} F^2$.*

(i) *The fundamental tensor $g := \nabla^\mathrm{v}\nabla^\mathrm{v} E$ of (M, F) is positive-homogeneous of degree 0, i.e.,*

$$\nabla^\mathrm{v}_{\widetilde{\delta}} g = 0. \tag{9.2.8}$$

(ii) *We have*

$$g(\widetilde{\delta}, \widetilde{\delta}) = \bar{g}(C, C) = 2E = F^2. \tag{9.2.9}$$

Proof. Let X and Y be arbitrary vector fields on M. Then

$$(\nabla^\mathrm{v}_{\widetilde{\delta}} g)(\widehat{X}, \widehat{Y}) \overset{(6.2.22)}{=} Cg(\widehat{X}, \widehat{Y}) - g(\nabla^\mathrm{v}_{\widetilde{\delta}}\widehat{X}, \widehat{Y}) - g(\widehat{X}, \nabla^\mathrm{v}_{\widetilde{\delta}}\widehat{Y})$$

$$\overset{(9.2.4),(6.2.32)}{=} C(X^\mathrm{v}(Y^\mathrm{v} E)) = [C, X^\mathrm{v}](Y^\mathrm{v} E) + X^\mathrm{v}(C(Y^\mathrm{v} E))$$

$$\overset{(4.1.53)}{=} -X^\mathrm{v}(Y^\mathrm{v} E) + X^\mathrm{v}([C, Y^\mathrm{v}]E) + X^\mathrm{v}(Y^\mathrm{v}(CE))$$

$$= -2X^\mathrm{v}(Y^\mathrm{v} E) + 2X^\mathrm{v}(Y^\mathrm{v} E) = 0,$$

which proves (9.2.8).

To see the first equality in (9.2.9), let $S \in \mathfrak{X}(TM)$ be an arbitrary second-order vector field. Then

$$\bar{g}(C, C) = \bar{g}(\mathbf{J}S, \mathbf{J}S) := g(\mathbf{j}S, \mathbf{j}S) = g(\widetilde{\delta}, \widetilde{\delta}),$$

as desired. Now, using (9.2.3),

$$g(\widetilde{\delta}, \widetilde{\delta}) = C(CE) - \mathbf{J}[C, S]E \overset{(5.1.4)}{=} C(CE) - CE = 4E - 2E = 2E,$$

which completes the proof. $\qquad\square$

Remark 9.2.4. The 0^+-homogeneity of the fundamental tensor $g = \nabla^{\mathsf{v}}\nabla^{\mathsf{v}}E$ can be deduced immediately from Lemma 6.2.22. However, it can be instructive to see also an independent proof for this result.

By Corollary 6.2.21, the 0^+-homogeneity of g in terms of its components can be expressed as follows:

$$y^k \frac{\partial g_{ij}}{\partial y^k} = 0; \quad i, j \in J_n. \tag{9.2.10}$$

Locally, relation (9.2.9) takes the form

$$g_{ij}y^i y^j = 2E = F^2. \tag{9.2.11}$$

Observe that (9.2.9) is the exact analogue of (9.1.14) on $\mathring{T}M$.

Lemma and Definition 9.2.5. *Let (M, F) be a pre-Finsler manifold with energy function E and fundamental tensor g.*

(i) *Define a Finsler one-form θ_g on $\mathring{T}M$ by*

$$\theta_g(\widetilde{X}) := g(\widetilde{X}, \widetilde{\delta}), \quad \widetilde{X} \in \Gamma(\mathring{\pi}). \tag{9.2.12}$$

Then

$$\theta_g = \nabla^{\mathsf{v}}E = F\nabla^{\mathsf{v}}F. \tag{9.2.13}$$

The one-form $\theta_g \in \mathcal{F}_1^0(\mathring{T}M)$ is related to the Lagrange one-form $\theta_E := \mathcal{L}_{\mathbf{J}}E \in \mathfrak{X}^*(\mathring{T}M)$ attached to E by

$$\theta_E = \theta_g \circ \mathbf{j}. \tag{9.2.14}$$

In terms of the usual local coordinates,

$$\theta_g \underset{(\mathcal{U})}{=} \frac{\partial E}{\partial y^i}\widehat{du^i}, \quad \theta_E \underset{(\mathcal{U})}{=} \frac{\partial E}{\partial y^i}dx^i. \tag{9.2.15}$$

We say that θ_g is the Lagrange one-form *of the pre-Finsler manifold (M, F).*

(ii) *Suppose that F is positive definite. Then we say that*

$$\ell := \frac{1}{F}\widetilde{\delta} \tag{9.2.16}$$

is the canonical unit section *of $\mathring{\pi}$. It has unit g-norm in the sense that*

$$g(\ell, \ell) = 1, \tag{9.2.17}$$

and it is 0^+-homogeneous, i.e.,

$$\nabla^{\mathrm{v}}_{\widetilde{\delta}}\ell = 0. \tag{9.2.18}$$

If

$$\ell_\flat := \frac{1}{F}\theta_g = \nabla^{\mathrm{v}}F, \tag{9.2.19}$$

then $\ell_\flat$ is the dual one-form to ℓ, so we have $\ell_\flat(\ell) = 1$.

(iii) *The* angular tensor field *of a positive definite pre-Finsler manifold (M, F) is*

$$\eta := g - \ell_\flat \otimes \ell_\flat = g - \nabla^{\mathrm{v}}F \otimes \nabla^{\mathrm{v}}F. \tag{9.2.20}$$

It is related to the vertical Hessian of F by

$$\frac{1}{F}\eta = \nabla^{\mathrm{v}}\nabla^{\mathrm{v}}F. \tag{9.2.21}$$

Proof. (i) Let $\widetilde{X} \in \Gamma(\mathring{\pi})$. Then, for some vector field ξ on $\mathring{T}M$, $\widetilde{X} = \mathbf{j}\xi$, and hence $\mathbf{i}\widetilde{X} = \mathbf{J}\xi$. So we obtain

$$\theta_g(\widetilde{X}) := g(\widetilde{X}, \widetilde{\delta}) \overset{(9.2.3)}{=} (\mathbf{i}\widetilde{X})(CE) - \mathbf{J}[\mathbf{J}\xi, \mathcal{H}\widetilde{\delta}]E$$
$$\overset{(5.1.4)}{=} 2(\mathbf{i}\widetilde{X})E - (\mathbf{J}\xi)E = (\mathbf{i}\widetilde{X})E \overset{(6.2.27)}{=} \nabla^{\mathrm{v}}E(\widetilde{X}),$$

whence (9.2.13). Relation (9.2.14) is an immediate consequence of (6.2.29). The coordinate expressions in (9.2.15) are already known, see (6.2.30) and (4.1.98).

(ii) From (9.2.9) we get (9.2.17). The 0^+-homogeneity of ℓ can be verified by a simple calculation:

$$\nabla^{\mathrm{v}}_{\widetilde{\delta}}\ell = \nabla^{\mathrm{v}}_{\widetilde{\delta}}\frac{1}{F}\widetilde{\delta} = \left(C\frac{1}{F}\right)\widetilde{\delta} + \frac{1}{F}\nabla^{\mathrm{v}}_{\widetilde{\delta}}\widetilde{\delta} \overset{(6.2.33)}{=} -\frac{CF}{F^2}\widetilde{\delta} + \frac{1}{F}\widetilde{\delta} \overset{(\mathrm{F2})}{=} -\frac{1}{F}\widetilde{\delta} + \frac{1}{F}\widetilde{\delta}.$$

Furthermore, acting by $\ell_\flat$ on ℓ, we find

$$\ell_\flat(\ell) = \frac{1}{F^2}\theta_g(\widetilde{\delta}) = \frac{1}{F^2}g(\widetilde{\delta}, \widetilde{\delta}) \overset{(9.2.9)}{=} 1,$$

as desired.

(iii) Finally, we check (9.2.21). Let X and Y be vector fields on M. Then

$$\frac{1}{F}\eta(\widehat{X},\widehat{Y}) := \frac{1}{F}\left(g(\widehat{X},\widehat{Y}) - \nabla^{\mathrm{v}}F(\widehat{X})\nabla^{\mathrm{v}}F(\widehat{Y})\right)$$

$$\overset{(9.2.4)}{=} \frac{1}{F}\left(\frac{1}{2}X^{\mathrm{v}}(Y^{\mathrm{v}}F^2) - (X^{\mathrm{v}}F)(Y^{\mathrm{v}}F)\right)$$

$$= \frac{1}{F}\left(X^{\mathrm{v}}((Y^{\mathrm{v}}F)F) - (X^{\mathrm{v}}F)(Y^{\mathrm{v}}F)\right)$$

$$= X^{\mathrm{v}}(Y^{\mathrm{v}}F) = \nabla^{\mathrm{v}}\nabla^{\mathrm{v}}F(\widehat{X},\widehat{Y}),$$

as was to be shown. $\qquad\square$

Lemma 9.2.6. *Let* (M,F) *be a pre-Finsler manifold. Then we have*

$$\mathcal{L}_C\mathcal{L}_{\mathbf{J}}F = 0, \quad \mathcal{L}_C\theta_E = \mathcal{L}_C\mathcal{L}_{\mathbf{J}}E = \mathcal{L}_{\mathbf{J}}E = \theta_E, \tag{9.2.22}$$

$$i_C d\mathcal{L}_{\mathbf{J}}F = 0. \tag{9.2.23}$$

Proof. By direct calculation,

$$\mathcal{L}_C\mathcal{L}_{\mathbf{J}}F \overset{(4.1.103)}{=} \mathcal{L}_{\mathbf{J}}\mathcal{L}_C F - \mathcal{L}_{\mathbf{J}}F = \mathcal{L}_{\mathbf{J}}F - \mathcal{L}_{\mathbf{J}}F = 0.$$

In the same way,

$$\mathcal{L}_C\theta_E = \mathcal{L}_C\mathcal{L}_{\mathbf{J}}E = 2\mathcal{L}_{\mathbf{J}}E - \mathcal{L}_{\mathbf{J}}E = \mathcal{L}_{\mathbf{J}}E = \theta_E.$$

Finally,

$$i_C d\mathcal{L}_{\mathbf{J}}F = \mathcal{L}_C\mathcal{L}_{\mathbf{J}}F - d i_C\mathcal{L}_{\mathbf{J}}F \overset{(9.2.22)}{=} -d i_C\mathcal{L}_{\mathbf{J}}F$$

$$\overset{(4.1.102)}{=} d\mathcal{L}_{\mathbf{J}}i_C F - d i_{\mathbf{J}}F \overset{(3.3.26),(4.1.94)}{=} 0,$$

which completes the proof. $\qquad\square$

Corollary 9.2.7. *If* (M,F) *is a pre-Finsler manifold, then the Lagrange two-form* $\omega_F := d\mathcal{L}_{\mathbf{J}}F$ *attached to* F *is always degenerate.*

Proof. Indeed, the nullspace of $d\mathcal{L}_{\mathbf{J}}F$ contains the Liouville vector field by (9.2.23). $\qquad\square$

Now we formulate and prove the analogue of Lemma 9.1.25.

Lemma 9.2.8. *Let* (M,F) *be a positive definite pre-Finsler manifold. Define a transformation* $\mathbf{p}$ *of* $\Gamma(\mathring{\pi})$ *by the rule*

$$\mathbf{p}(\widetilde{X}) := \widetilde{X} - \frac{g(\widetilde{X},\widetilde{\delta})}{g(\widetilde{\delta},\widetilde{\delta})}\widetilde{\delta}, \quad \widetilde{X} \in \Gamma(\mathring{\pi}). \tag{9.2.24}$$

Then $\mathbf{p}$ *is* $C^\infty(\mathring{T}M)$*-linear, whence*

$$\mathbf{p} \in \mathrm{End}(\Gamma(\mathring{\pi})) \cong \mathcal{F}_1^1(\mathring{T}M),$$

and has the following properties:

(i) $\mathrm{Ker}(\mathbf{p}) = \mathrm{span}(\widetilde{\delta})$;

(ii) $\mathbf{p}^2 = \mathbf{p}$, *i.e.*, $\mathbf{p}$ *is a projection in* $\Gamma(\mathring{\pi})$;

(iii) $\mathrm{Im}(\mathbf{p}) = \mathrm{Ker}(\theta_g) = \mathrm{Ker}(\nabla^\mathrm{v} E)$;

(iv) $\Gamma(\mathring{\pi}) = \mathrm{span}(\widetilde{\delta}) \oplus \mathrm{Ker}(\theta_g)$.

Proof. The $C^\infty(\mathring{T}M)$-linearity and property (i) of $\mathbf{p}$ are clear from the definition. For every $\widetilde{X} \in \Gamma(\mathring{\pi})$

$$\mathbf{p}^2(\widetilde{X}) = \mathbf{p}\left(\widetilde{X} - \frac{g(\widetilde{X}, \widetilde{\delta})}{g(\widetilde{\delta}, \widetilde{\delta})}\widetilde{\delta}\right) = \mathbf{p}(\widetilde{X}) - \frac{g(\widetilde{X}, \widetilde{\delta})}{g(\widetilde{\delta}, \widetilde{\delta})}\mathbf{p}(\widetilde{\delta}) \stackrel{(i)}{=} \mathbf{p}(\widetilde{X}),$$

therefore $\mathbf{p}^2 = \mathbf{p}$. Since

$$\theta_g(\mathbf{p}(\widetilde{X})) = g\left(\widetilde{X} - \frac{g(\widetilde{X}, \widetilde{\delta})}{g(\widetilde{\delta}, \widetilde{\delta})}\widetilde{\delta}, \widetilde{\delta}\right) = g(\widetilde{X}, \widetilde{\delta}) - g(\widetilde{X}, \widetilde{\delta}) = 0,$$

we have $\mathrm{Im}(\mathbf{p}) \subset \mathrm{Ker}(\theta_g)$. Conversely, if $\widetilde{X} \in \mathrm{Ker}(\theta_g)$, then

$$\widetilde{X} = \widetilde{X} - \frac{g(\widetilde{X}, \widetilde{\delta})}{g(\widetilde{\delta}, \widetilde{\delta})}\widetilde{\delta} = \mathbf{p}(\widetilde{X}) \in \mathrm{Im}(\mathbf{p}),$$

so the converse relation $\mathrm{Ker}(\theta_g) \subset \mathrm{Im}(\mathbf{p})$ also holds.

Properties (i)–(iii) imply (iv); we refer again to [51], 2.19. $\qquad\square$

Remark 9.2.9. We say that $\mathbf{p}$ is the *projection tensor* on $\Gamma(\mathring{\pi})$ onto $(\mathrm{span}(\widetilde{\delta}))^{\perp_g}$ determined by g. It is also referred to as the *projection tensor of* (M, F). In view of (9.2.9), (9.2.12) and (9.2.13), $\mathbf{p}$ can equivalently be written in the form

$$\mathbf{p} = 1 - \frac{1}{2E}\nabla^\mathrm{v} E \otimes \widetilde{\delta} = 1 - \frac{1}{F}\nabla^\mathrm{v} F \otimes \widetilde{\delta}, \qquad (9.2.25)$$

where $\mathbf{1}$ is an abbreviation for $1_{\Gamma(\mathring{\pi})}$.

Lemma 9.2.10. *If* (M, F) *is a positive definite n-dimensional pre-Finsler manifold, then the trace of its projection is* $n - 1$.

Proof. By an immediate calculation,

$$\mathrm{tr}(\mathbf{p}) \stackrel{(9.2.25)}{=} \mathrm{tr}\left(1 - \frac{1}{F}\nabla^\mathrm{v} F \otimes \widetilde{\delta}\right) = \mathrm{tr}\,\mathbf{1} - \frac{1}{F}\mathrm{tr}(\widetilde{\delta} \otimes \nabla^\mathrm{v} F)$$

$$\stackrel{(1.2.17)}{=} n - \frac{1}{F}\nabla^\mathrm{v} F(\widetilde{\delta}) = n - \frac{1}{F}CF = n - 1. \qquad\square$$

Definition 9.2.11. Let (M, F) be a positive definite pre-Finsler manifold; $A \in \mathcal{F}_k^0(\mathring{T}M)$, $B \in \mathcal{F}_k^1(\mathring{T}M)$ ($k \in \mathbb{N}^*$). The *projected tensors* of A and B are the tensors $\mathbf{p}A \in \mathcal{F}_k^0(\mathring{T}M)$ and $\mathbf{p}B \in \mathcal{F}_k^1(\mathring{T}M)$ given by

$$\mathbf{p}A(\widetilde{X}_1, \dots, \widetilde{X}_k) := A(\mathbf{p}(\widetilde{X}_1), \dots, \mathbf{p}(\widetilde{X}_k)) \qquad (9.2.26)$$

and

$$\mathbf{p}B(\widetilde{X}_1, \dots, \widetilde{X}_k) := \mathbf{p}\big(B(\mathbf{p}(\widetilde{X}_1), \dots, \mathbf{p}(\widetilde{X}_k))\big), \qquad (9.2.27)$$

respectively, where $\widetilde{X}_1, \dots, \widetilde{X}_k \in \Gamma(\mathring{\pi})$.

Lemma 9.2.12. *Keeping the notation and hypotheses just introduced, suppose that*

$$\widetilde{\delta} \in \{\widetilde{X}_1, \dots, \widetilde{X}_k\} \text{ implies } A(\widetilde{X}_1, \dots, \widetilde{X}_k) = 0 \text{ and } B(\widetilde{X}_1, \dots, \widetilde{X}_k) = 0.$$

Then $\mathbf{p}A = A$, $\mathbf{p}B = \mathbf{p} \circ B$.

Proof. Immediate from the definition and from Lemma 9.2.8(i). $\qquad\square$

Lemma 9.2.13. *Let (M, F) be a positive definite pre-Finsler manifold. Then the projected tensor of the fundamental tensor of (M, F) is the angular tensor, i.e., we have*

$$\mathbf{p}g = \eta. \qquad (9.2.28)$$

Proof. For any vector fields X, Y on M,

$$\mathbf{p}g(\widehat{X}, \widehat{Y}) := g(\mathbf{p}(\widehat{X}), \mathbf{p}(\widehat{Y})) \overset{(9.2.25)}{=} g\left(\widehat{X} - \frac{1}{2E}(X^{\mathsf{v}}E)\widetilde{\delta}, \widehat{Y} - \frac{1}{2E}(Y^{\mathsf{v}}E)\widetilde{\delta}\right)$$

$$\overset{(9.2.9)}{=} g(\widehat{X}, \widehat{Y}) - \frac{1}{2E}X^{\mathsf{v}}Eg(\widetilde{\delta}, \widehat{Y}) - \frac{1}{2E}Y^{\mathsf{v}}Eg(\widehat{X}, \widetilde{\delta})$$

$$+ \frac{1}{2E}(X^{\mathsf{v}}E)(Y^{\mathsf{v}}E) \overset{(9.2.12),(9.2.13)}{=} g(\widehat{X}, \widehat{Y}) - \frac{2}{2E}(X^{\mathsf{v}}E)(Y^{\mathsf{v}}E)$$

$$+ \frac{1}{2E}(X^{\mathsf{v}}E)(Y^{\mathsf{v}}E) = \left(g - \frac{1}{2E}\nabla^{\mathsf{v}}E \otimes \nabla^{\mathsf{v}}E\right)(\widehat{X}, \widehat{Y})$$

$$= (g - \nabla^{\mathsf{v}}F \otimes \nabla^{\mathsf{v}}F)(\widehat{X}, \widehat{Y}) =: \eta(\widehat{X}, \widehat{Y}),$$

as was to be shown. $\qquad\square$

9.2.2 Finsler Functions and the Canonical Spray

Definition 9.2.14. Let M be a manifold. A positive definite pre-Finsler function $F: TM \to \mathbb{R}$ is called a *Finsler function* if it satisfies the condition

(F_4) **Ellipticity.** The fundamental tensor

$$g := \nabla^v \nabla^v E = \frac{1}{2} \nabla^v \nabla^v F^2$$

of (M, F) is *Euclidean*, i.e., the slit Finsler bundle $\mathring{\pi}$ is a Euclidean vector bundle with the scalar product g.

A *Finsler manifold* (M, F) is a manifold M together with a Finsler function for M. A Finsler function F (and also the Finsler manifold (M, F)) is called *reversible* if F satisfies the condition

(F_5) **Reversibility.** $F(-v) = F(v)$ for every $v \in \mathring{T}M$.

Remark 9.2.15. By the *reverse* of a Finsler function F we mean the function $F^{\downarrow} := F \circ \varrho$, where ϱ is the reflection map defined in Example 3.1.38(a). Clearly, $F^{\downarrow}$ is also a Finsler function. The Finsler manifold $(M, F^{\downarrow})$ is called the reverse of (M, F). Using our new term, condition (F_5) can be rephrased as $F = F^{\downarrow}$.

Corollary 9.2.16. *Let* $F \colon TM \to \mathbb{R}$ *be a positive definite pre-Finsler function. Then the following properties are equivalent:*

 (i) *The fundamental tensor of* (M, F) *is Euclidean, i.e., F is a Finsler function.*
 (ii) *The fundamental tensor of* (M, F) *is semi-Euclidean, i.e., the slit Finsler bundle* $\mathring{\pi}$ *together with the scalar product g is a semi-Euclidean vector bundle.*
 (i') *The vertical bundle* $V\mathring{T}M$ *with the scalar product $\bar{g}$ given by (9.2.6) is Euclidean.*
 (ii') *The vertical bundle* $V\mathring{T}M$ *with the scalar product $\bar{g}$ is semi-Euclidean.*
 (iii) *At each point u of* $\mathring{T}M$, *the nullspace of the vertical Hessian of F is* $\mathrm{span}(\tilde{\delta}(u))$.
 (iv) *At each point of* $\mathring{T}M$, *the vertical Hessian of F has rank $n - 1$, where* $n := \dim M$.
 (v) *The Lagrange two-form* $\omega_E = d\mathcal{L}_{\mathbf{J}}E$ *of E is non-degenerate.*
 (vi) *At each* $u \in \mathring{T}M$, *the nullspace of the angular tensor field of* (M, F) *is* $\mathrm{span}(\tilde{\delta}(u))$.

Proof. Choose an arbitrary vector $u \in \mathring{T}M$, and let $p := \mathring{\tau}(u)$. By the pointwise interpretation of the vertical Hessian discussed in Lemma 6.2.31 and Remark 9.2.2(d), conditions (i)–(iv) take the following forms:

(1) The bilinear form $(E_p)''(u) \in T_2^0(T_pM)$ is positive definite.

(2) The bilinear form $(E_p)''(u) \in T_2^0(T_pM)$ is non-degenerate.

(3) The nullspace of the bilinear form $(F_p)''(u)$ is $\mathrm{span}(\widetilde{\underline{\delta}}(u)) = \mathrm{span}(u)$.

(4) The rank of the bilinear form $(F_p)''(u)$ is $n - 1$.

These are just conditions (i), (ii), (viii) and (vi) in Theorem 9.1.36, so they are equivalent. By Remark 9.2.2(c), the equivalences (i)$\Leftrightarrow$(i') and (ii)$\Leftrightarrow$(ii') are evident. Taking into account (9.2.21), (iii) is clearly equivalent to (vi). Finally the equivalence of (ii) and (v) follows from Corollary 6.2.30. $\qquad\square$

Lemma and Definition 9.2.17. *Let (M, F) be a Finsler manifold. There exists a unique spray S over M such that*

$$i_S \omega_E = -dE. \tag{9.2.29}$$

This spray is called the canonical spray *of the Finsler manifold (M, F) (or the canonical spray for the energy E).*

Proof. By the above corollary, the energy of (M, F) is a regular Lagrangian on $\overset{\circ}{T}M$. The energy associated to this Lagrangian in the sense of Definition 5.2.1 is $CE - E = E$. Thus, by the same argument as in the proof of Corollary and Definition 5.2.9, we obtain that there exists a unique second-order vector field $S \in \mathfrak{X}(\overset{\circ}{T}M)$ which satisfies (9.2.29). Since E is 2^+-homogeneous, it follows as in the proof of Corollary 5.2.11 that $[C, S] = S$. Extending S to the whole TM by the prescription $S \circ o_M := o_{TM} \circ o_M$, i.e., by requiring condition (S_5) of Definition 5.1.18 to hold, we obtain the desired spray over M. $\qquad\square$

Remark 9.2.18. Let (M, F) be a Finsler manifold. Choose a chart $(\mathcal{U}, (u^i)_{i=1}^n)$ on M, and let $(\tau^{-1}(\mathcal{U}), ((x^i)_{i=1}^n, (y^i)_{i=1}^n))$ be the induced chart on TM. By Lemma 5.2.13 it follows that the coefficients of the canonical spray of (M, F) with respect to the chosen chart are the functions

$$G^i = \frac{1}{2} g^{ij} \left(\frac{\partial^2 E}{\partial y^j \partial x^k} y^k - \frac{\partial E}{\partial x^j} \right), \quad i \in J_n, \tag{9.2.30}$$

where

$$(g^{ij}) := (g_{ij})^{-1}, \quad g_{ij} = \frac{\partial^2 E}{\partial y^i \partial y^j} = \frac{1}{2} \frac{\partial^2 F^2}{\partial y^i \partial y^j}.$$

The functions G^i are smooth on $\tau^{-1}(\mathcal{U}) \cap \overset{\circ}{T}M$ and of class C^1 on $\tau^{-1}(\mathcal{U})$.

Remark 9.2.19. If (M, F) is a Finsler manifold and S is its canonical spray, then the energy function E is a first integral for S by Lemma 5.2.8(ii). Thus

$$0 = SE = \frac{1}{2} SF^2 = F(SF),$$

whence $SF = 0$. Therefore, the Finsler function F is also a first integral for the canonical spray of (M, F). Under an additional condition, the converse is also true.

Lemma 9.2.20. *Let (M, F) be a Finsler manifold. A semispray S over M is the canonical spray of (M, F) if, and only if, the following are satisfied:*

(i) *F is a first integral for S, i.e., $SF = 0$.*
(ii) *$i_S \omega_F = 0$, where ω_F is the Lagrange two-form attached to F.*

Proof. First we express ω_E in terms of ω_F:

$$\omega_E := d\mathcal{L}_{\mathbf{J}} E = \frac{1}{2} d\mathcal{L}_{\mathbf{J}} F^2 = d(F\mathcal{L}_{\mathbf{J}} F)$$

$$\overset{\text{Proposition 3.3.35(ii)}}{=} dF \wedge \mathcal{L}_{\mathbf{J}} F + F d\mathcal{L}_{\mathbf{J}} F = dF \wedge \mathcal{L}_{\mathbf{J}} F + F\omega_F.$$

If S is an arbitrary semispray over M, then we find

$$i_S \omega_E \overset{(3.3.29)}{=} i_S dF \wedge \mathcal{L}_{\mathbf{J}} F - dF \wedge i_S \mathcal{L}_{\mathbf{J}} F + F i_S \omega_F$$

$$= (SF)\mathcal{L}_{\mathbf{J}} F - \mathbf{J}S(F)dF + F i_S \omega_F,$$

whence

$$i_S \omega_E = (SF)\mathcal{L}_{\mathbf{J}} F - dE + F i_S \omega_F. \tag{9.2.31}$$

To show the necessity of (i) and (ii), suppose that S is the canonical spray of (M, F). Then, by the preceding remark, $SF = 0$. On the other hand, $i_S \omega_E = -dE$, therefore (9.2.31) reduces to $F i_S \omega_F = 0$, from which we conclude that $i_S \omega_F = 0$.

The sufficiency of (i) and (ii) is clear: if $SF = 0$ and $i_S \omega_F = 0$, then from (9.2.31) we obtain $i_S \omega_E = -dE$. □

Proposition 9.2.21. *If (M, F) is an n-dimensional Finsler manifold, then the Lagrange two-form attached to F has rank $2(n-1)$. The nullspace of ω_F is spanned by the Liouville vector field and the canonical spray of (M, F).*

Proof. Let $N(\omega_F)$ denote the nullspace of ω_F as in 1.4.1. It is sufficient to show that $N(\omega_F) = \text{span}(C, S)$, where S is the canonical spray of (M, F). By Corollary 9.2.7 and Lemma 9.2.20 we clearly have $\text{span}(C, S) \subset N(\omega_F)$, so our task reduces to proving that

$$\xi \in \mathfrak{X}(\overset{\circ}{T}M) \text{ and } \xi \in N(\omega_F) \text{ implies } \xi \in \text{span}(C, S). \tag{$*$}$$

To do this, we make two preparatory steps.

Step 1. We show that $\xi \in \mathfrak{X}(\mathring{T}M)$ and $\mathbf{J}\xi \in N(\omega_F)$ imply that $\mathbf{J}\xi$ belongs to $\mathrm{span}(C)$. Indeed, then $i_{\mathbf{J}\xi}\omega_F = 0$, so for every $\zeta \in \mathfrak{X}(\mathring{T}M)$,

$$0 = \omega_F(\mathbf{J}\xi, \zeta) \overset{(6.2.39)}{=} g_F(\mathbf{j}\xi, \mathbf{j}\zeta).$$

Since $N(g_F) = \mathrm{span}(\widetilde{\delta})$ by Corollary 9.2.16(iii), from this it follows that $\mathbf{j}\xi = \varphi\widetilde{\delta}$ for some $\varphi \in C^\infty(\mathring{T}M)$, whence $\mathbf{J}\xi = \mathbf{i}(\mathbf{j}\xi) = \varphi(\mathbf{i}\widetilde{\delta}) = \varphi C$, as desired.

Step 2. We assert that $\xi \in N(\omega_F)$ implies $\mathbf{J}\xi \in N(\omega_F)$. This can be verified by a simple calculation:

$$i_{\mathbf{J}\xi}\omega_F \overset{(3.3.48)}{=} i_\xi i_{\mathbf{J}}\omega_F - i_{\mathbf{J}}i_\xi\omega_F \overset{(5.2.5)}{=} -i_{\mathbf{J}}i_\xi\omega_F \overset{\text{condition}}{=} 0.$$

Now we turn to the proof of ($*$). Let $\xi \in N(\omega_F)$. Then $\mathbf{J}\xi \in N(\omega_F)$ also holds by Step 2, therefore $\mathbf{J}\xi = \varphi C$ ($\varphi \in C^\infty(\mathring{T}M)$) by Step 1. Thus

$$0 = \mathbf{J}\xi - \varphi C = \mathbf{J}(\xi - \varphi S),$$

so $\xi - \varphi S$ is vertical. Since $\xi - \varphi S$ belongs to $N(\omega_F)$, we conclude, using again Step 1, that

$$\xi - \varphi S = \psi C \text{ for some } \psi \in C^\infty(\mathring{T}M).$$

So we have $\xi = \psi C + \varphi S \in \mathrm{span}(C, S)$, as was to be shown. $\qquad\square$

9.2.3 The Rapcsák Equations

Theorem 9.2.22. *Let (M, F) be a pre-Finsler manifold. Suppose that S is a spray over M, and (with the notation of 8.1.1) let $\mathcal{H}$ be the Berwald connection of (M, S). Then the following are equivalent:*

(R_1) $i_S\omega_F := i_S d\mathcal{L}_{\mathbf{J}}F = 0$;

(R_2) $S(X^{\mathsf{v}}F) - X^{\mathsf{c}}F = 0$, $\quad X \in \mathfrak{X}(M)$;

(R_3) $X^{\mathsf{v}}(SF) - 2X^{\mathsf{h}}F = 0$, $\quad X \in \mathfrak{X}(M)$;

(R_4) $\nabla_S\nabla^{\mathsf{v}}F = \nabla^{\mathsf{h}}F$;

(R_5) $i_{\widetilde{\delta}}\nabla^{\mathsf{h}}\nabla^{\mathsf{v}}F = \nabla^{\mathsf{h}}F$;

(R_6) $\mathcal{L}_{\mathbf{h}}\mathcal{L}_{\mathbf{J}}F = 0$;

(R_7) $\nabla^{\mathsf{h}}\nabla^{\mathsf{v}}F(\widehat{X}, \widehat{Y}) = \nabla^{\mathsf{h}}\nabla^{\mathsf{v}}F(\widehat{Y}, \widehat{X})$, $\quad X, Y \in \mathfrak{X}(M)$;

(R_8) $\nabla^{\mathsf{v}}\nabla^{\mathsf{h}}F(\widehat{X}, \widehat{Y}) = \nabla^{\mathsf{v}}\nabla^{\mathsf{h}}F(\widehat{Y}, \widehat{X})$, $\quad X, Y \in \mathfrak{X}(M)$.

Proof. **Step 1.** The equivalence of (R_2) and (R_3) is almost immediate since

$$2X^{\mathsf{h}}F \overset{(7.3.7)}{=} X^{\mathsf{c}}F + [X^{\mathsf{v}}, S]F = X^{\mathsf{c}}F + X^{\mathsf{v}}(SF) - S(X^{\mathsf{v}}F),$$

whence

$$S(X^v F) - X^c F = X^v(SF) - 2X^h F.$$

We show that both (R_1) and (R_4) are equivalent to (R_2). Observe first that $i_S \omega_F$ is identically zero on $\mathfrak{X}^v(\mathring{T}M)$, since $i_\mathbf{J}(i_S \omega_F + dE_F) \overset{(5.2.6)}{=} 0$ and $E_F := CF - F = 0$. On the other hand,

$$i_S \omega_F(X^c) = \omega_F(S, X^c) \overset{(5.2.2)}{=} S(\mathbf{J}X^c(F)) - X^c(\mathbf{J}S(F)) - \mathbf{J}[S, X^c](F)$$

$$= S(X^v F) - X^c F + \mathbf{J}[X^c, S](F) \overset{(5.1.6)}{=} S(X^v F) - X^c F,$$

therefore (R_1) and (R_2) are equivalent.

Evaluating now the Finsler one-form $\nabla_S \nabla^v F - \nabla^h F$ at a basic section $\widehat{X}$, we find that

$$(\nabla_S \nabla^v F - \nabla^h F)(\widehat{X}) = S(X^v F) - \nabla^v F(\nabla_S \widehat{X}) - X^h F$$

$$= S(X^v F) - X^h F - (i \nabla_S \widehat{X})F \overset{(8.1.3)}{=} S(X^v F) - X^c F.$$

This proves the equivalence of (R_4) and (R_2).

Thus we have

$$
\begin{array}{c}
(R_3) \\
\Updownarrow \\
(R_1) \Leftrightarrow (R_2) \Leftrightarrow (R_4)
\end{array}
\qquad (*)
$$

Step 2. We proceed with some simple observations. Since

$$(i_{\tilde{\delta}} \nabla^h \nabla^v F - \nabla^h F)(\widehat{X}) = \nabla^h \nabla^v F(\tilde{\delta}, \widehat{X}) - \nabla^h F(\widehat{X}) = (\nabla_S \nabla^v F - \nabla^h F)(\widehat{X}),$$

for every $X \in \mathfrak{X}(M)$, relations (R_4) and (R_5) are equivalent. The equivalence of (R_7) and (R_8) is clear from the mixed Ricci formula (7.10.17). Further, from Lemmas 7.4.1 and 7.10.7, it follows that (R_6) is equivalent to (R_7).

Next we show that (R_6) implies (R_2). Indeed, for every $X \in \mathfrak{X}(M)$,

$$\mathcal{L}_\mathbf{h} \mathcal{L}_\mathbf{J} F(S, X^h) \overset{(7.4.5)}{=} \mathbf{h}^* d\mathcal{L}_\mathbf{J} F(S, X^h) = d\mathcal{L}_\mathbf{J} F(S, X^h) = \omega_F(S, X^h)$$

$$\overset{(5.2.2)}{=} S(X^v F) - X^h F - \mathbf{J}[S, X^h](F) \overset{(8.1.2)}{=} S(X^v F) - X^c F,$$

as desired.

So diagram $(*)$ can be enlarged as follows:

$$
\begin{array}{cccc}
(R_7) \Leftrightarrow (R_6) & & (R_3) & (R_5) \\
\Updownarrow & \searrow & \Updownarrow & \Updownarrow \\
(R_8) & (R_1) & \Leftrightarrow (R_2) & \Leftrightarrow (R_4)
\end{array}
$$

Step 3. To complete the proof, we show that (R_1) implies (R_6).

Note first that the Ehresmann connection generated by S is positive-homogeneous, because S is a spray (see, e.g., 8.1.1). Thus $[\mathbf{h}, C] = 0$ by Corollary 7.5.10, so from (3.3.53) we infer that $\mathcal{L}_\mathbf{h} \circ \mathcal{L}_C = \mathcal{L}_C \circ \mathcal{L}_\mathbf{h}$. Taking this into account,

$$0 \overset{(9.2.22)}{=} \mathcal{L}_\mathbf{h}\mathcal{L}_C\mathcal{L}_\mathbf{J}F = \mathcal{L}_C\mathcal{L}_\mathbf{h}\mathcal{L}_\mathbf{J}F \overset{(5.1.2)}{=} [i_S, \mathcal{L}_\mathbf{J}]\mathcal{L}_\mathbf{h}\mathcal{L}_\mathbf{J}F - i_{[\mathbf{J},S]}\mathcal{L}_\mathbf{h}\mathcal{L}_\mathbf{J}F$$

$$\overset{(7.3.6)}{=} [i_S, \mathcal{L}_\mathbf{J}]\mathcal{L}_\mathbf{h}\mathcal{L}_\mathbf{J}F - 2i_\mathbf{h}\mathcal{L}_\mathbf{h}\mathcal{L}_\mathbf{J}F + i_1\mathcal{L}_\mathbf{h}\mathcal{L}_\mathbf{J}F$$

$$\overset{(7.4.5),(3.3.39)}{=} [i_S, \mathcal{L}_\mathbf{J}]\mathcal{L}_\mathbf{h}\mathcal{L}_\mathbf{J}F - 2\mathcal{L}_\mathbf{h}\mathcal{L}_\mathbf{J}F,$$

whence

$$2\mathcal{L}_\mathbf{h}\mathcal{L}_\mathbf{J}F = i_S\mathcal{L}_\mathbf{J}\mathcal{L}_\mathbf{h}\mathcal{L}_\mathbf{J}F + \mathcal{L}_\mathbf{J}i_S\mathcal{L}_\mathbf{h}\mathcal{L}_\mathbf{J}F. \qquad (**)$$

Since the torsion of $\mathcal{H}$ vanishes, from (7.8.7) we get that $[\mathbf{J}, \mathbf{h}] = 0$. By (3.3.44) this implies that $\mathcal{L}_\mathbf{J} \circ \mathcal{L}_\mathbf{h} = -\mathcal{L}_\mathbf{h} \circ \mathcal{L}_\mathbf{J}$. So in the first term in $(**)$,

$$\mathcal{L}_\mathbf{J} \circ \mathcal{L}_\mathbf{h} \circ \mathcal{L}_\mathbf{J} = -\mathcal{L}_\mathbf{h} \circ \mathcal{L}_\mathbf{J} \circ \mathcal{L}_\mathbf{J} = -\frac{1}{2}\mathcal{L}_\mathbf{h} \circ \mathcal{L}_{N_\mathbf{J}} \overset{\text{Lemma } 4.1.59}{=} 0,$$

thus $(**)$ takes the form

$$2\mathcal{L}_\mathbf{h}\mathcal{L}_\mathbf{J}F = \mathcal{L}_\mathbf{J}i_S\mathcal{L}_\mathbf{h}\mathcal{L}_\mathbf{J}F. \qquad (***)$$

Now suppose that (R_1) holds. If $\xi \in \mathfrak{X}(\mathring{T}M)$, then

$$i_S\mathcal{L}_\mathbf{h}\mathcal{L}_\mathbf{J}F(\xi) = \mathcal{L}_\mathbf{h}\mathcal{L}_\mathbf{J}F(S, \xi) \overset{(7.4.5)}{=} \mathbf{h}^*d\mathcal{L}_\mathbf{J}F(S, \xi)$$

$$= d\mathcal{L}_\mathbf{J}F(S, \mathbf{h}\xi) = i_Sd\mathcal{L}_\mathbf{J}F(\mathbf{h}\xi) \overset{(R_1)}{=} 0.$$

Substituting this into $(***)$ we obtain $\mathcal{L}_\mathbf{h}\mathcal{L}_\mathbf{J}F = 0$, which shows that (R_1) implies (R_6). This concludes the proof. $\qquad\square$

Remark 9.2.23. Choose a chart $(\mathcal{U}, (u^i)_{i=1}^n)$ on M together with the induced chart $(\tau^{-1}(\mathcal{U}), ((x^i)_{i=1}^n, (y^i)_{i=1}^n))$ on TM. Let, locally,

$$S \underset{(\mathcal{U})}{=} y^i\frac{\partial}{\partial x^i} - 2G^i\frac{\partial}{\partial y^i}.$$

Then the Christoffel symbols of the Berwald connection and the Berwald derivative induced by S are the functions

$$G_j^i := \frac{\partial G^i}{\partial y^j} \quad \text{and} \quad G_{jk}^i := \frac{\partial G_j^i}{\partial y^k} \quad (i, j, k \in J_n),$$

respectively (see, e.g., 8.1.4).

(a) **Coordinate expression of** (R_7) For all $j, k \in J_n$,

$$\nabla^h \nabla^v F \left(\widehat{\frac{\partial}{\partial u^j}}, \widehat{\frac{\partial}{\partial u^k}} \right) = \left(\frac{\partial}{\partial u^j} \right)^h \frac{\partial F}{\partial y^k} - \nabla^v F \left(\nabla_{(\frac{\partial}{\partial u^j})^h} \widehat{\frac{\partial}{\partial u^k}} \right)$$

$$\overset{(7.2.37),(7.11.2)}{=} \frac{\partial^2 F}{\partial x^j \partial y^k} - G^i_j \frac{\partial^2 F}{\partial y^i \partial y^k} - G^i_{jk} \frac{\partial F}{\partial y^i}.$$

By the vanishing of the torsion of the Berwald connection, $G^i_{jk} = G^i_{kj}$ for all $i, j, k \in J_n$, so in terms of the usual local coordinates (R_7) can be expressed as

$$\frac{\partial^2 F}{\partial x^j \partial y^k} - G^i_j \frac{\partial^2 F}{\partial y^i \partial y^k} = \frac{\partial^2 F}{\partial x^k \partial y^j} - G^i_k \frac{\partial^2 F}{\partial y^i \partial y^j}; \quad j, k \in J_n. \quad (9.2.32)$$

(b) **Coordinate expression of** (R_1) Taking into account that the functions $\frac{\partial F}{\partial x^i}$ are 1^+-homogeneous and the functions $\frac{\partial^2 F}{\partial y^i \partial y^j}$ are 0^+-homogeneous, from our calculation in the proof of Lemma 5.2.13 we get

$$i_S \omega_F \underset{(\mathcal{U})}{=} \left(y^i \frac{\partial^2 F}{\partial x^i \partial y^j} - 2G^i \frac{\partial^2 F}{\partial y^i \partial y^j} - \frac{\partial F}{\partial x^j} \right) dx^j.$$

Thus, locally, relation (R_1) takes the form

$$y^i \frac{\partial^2 F}{\partial x^i \partial y^j} - 2G^i \frac{\partial^2 F}{\partial y^i \partial y^j} = \frac{\partial F}{\partial x^j}, \quad j \in J_n. \quad (9.2.33)$$

(c) We show that $(9.2.33)$ can equivalently be written as

$$F_{;j} - \frac{\partial F_{;i}}{\partial y^j} y^i = 0 \quad (j \in J_n), \quad (9.2.34)$$

where ; is the semicolon operator given by $(7.11.8)$. Indeed,

$$\frac{\partial F_{;i}}{\partial y^j} y^i = \left(\frac{\partial}{\partial y^j} \left(\frac{\partial F}{\partial x^i} - G^r_i \frac{\partial F}{\partial y^r} \right) \right) y^i$$

$$= y^i \frac{\partial^2 F}{\partial y^j \partial x^i} - y^i \frac{\partial G^r_i}{\partial y^j} \frac{\partial F}{\partial y^r} - y^i G^r_i \frac{\partial^2 F}{\partial y^j \partial y^r}$$

$$\overset{(8.1.5),(8.1.7)}{=} y^i \frac{\partial^2 F}{\partial x^i \partial y^j} - G^r_j \frac{\partial F}{\partial y^r} - 2G^r \frac{\partial^2 F}{\partial y^j \partial y^r},$$

whence

$$F_{;j} - \frac{\partial F_{;i}}{\partial y^j} y^i = \frac{\partial F}{\partial x^j} + 2G^i \frac{\partial^2 F}{\partial y^i \partial y^j} - y^i \frac{\partial^2 F}{\partial x^i \partial y^j}.$$

Thus $(9.2.34)$ holds if, and only if, $(9.2.33)$ holds.

(d) In 1961 A. Rapcsák proved that *two Finsler functions F and $\bar{F}$ over the same base manifold M have common pregeodesics if, and only if,*

$$\bar{F}_{;j} - \frac{\partial \bar{F}_{;i}}{\partial y^j} y^i = 0 \quad (j \in J_n),$$

where the semicolon denotes the h-Berwald derivative induced by the canonical spray of (M, F); see [85]. So we call relations (R_1)–(R_8) (and (9.2.33), (9.2.34)) the *Rapcsák equations* for the (pre-Finsler) function F with respect to the spray S.

The aspects indicated by Rapcsák's quoted theorem will be briefly discussed in section 9.6.

9.2.4   Riemannian Finsler Functions

Example 9.2.24. (a) Let (M, g_M) be a Riemannian manifold. Define a function E on TM by

$$E(v) := \frac{1}{2}(g_M)_{\tau(v)}(v, v), \quad v \in TM; \qquad (9.2.35)$$

and let $F := \sqrt{2E}$. We show that F *is a reversible Finsler function*.

(1) Choose a chart $(\mathcal{U}, (u^i)_{i=1}^n)$ on M, and consider the induced chart $(\tau^{-1}(\mathcal{U}), ((x^i)_{i=1}^n, (y^i)_{i=1}^n))$ on TM. Let $(g_M)_{ij} := g_M\left(\frac{\partial}{\partial u^i}, \frac{\partial}{\partial u^j}\right)$ be the components of g_M with respect to $(\mathcal{U}, (u^i)_{i=1}^n)$. Then for every $p \in \mathcal{U}$ and $v = v^i \left(\frac{\partial}{\partial u^i}\right)_p \in T_pM$,

$$E(v) = \frac{1}{2}(g_M)_{ij}(p)v^i v^j = \frac{1}{2}\left(((g_M)_{ij} \circ \tau)y^i y^j\right)(v),$$

therefore

$$E \underset{(\mathcal{U})}{=} \frac{1}{2}((g_M)_{ij} \circ \tau)y^i y^j, \quad F \underset{(\mathcal{U})}{=} \sqrt{((g_M)_{ij} \circ \tau)y^i y^j}. \qquad (9.2.36)$$

It is clear from the obtained coordinate expression that E is smooth, and, since g_M is positive definite, E is positive on $\overset{\circ}{T}M$. Thus F is continuous, and smooth on $\overset{\circ}{T}M$, which verifies (F_1).

(2) For every $v \in TM$ and $\lambda \in \mathbb{R}$,

$$E(\lambda v) = \frac{1}{2}(g_M)_{\tau(v)}(\lambda v, \lambda v) = \lambda^2 E(v),$$

therefore $F(\lambda v) = |\lambda| F(v)$. This implies that F is 1^+-homogeneous and reversible. So F satisfies (F_2) and (F_5). Immediately from the definition, (F_3) also holds.

(3) Finally, by the first equality in (9.2.36), the component functions of $g := \nabla^v \nabla^v E$ are

$$g_{ij} = \frac{\partial^2 E}{\partial y^i \partial y^j} = (g_M)_{ij} \circ \tau,$$

so the positive definiteness of g_M implies that F satisfies (F_4).

We say that a Finsler function of the form $F = \sqrt{2E}$, where E arises from a Riemannian metric according to (9.2.35), is a *Riemannian Finsler function*.

(b) Keeping the preceding notation, we show that the fundamental tensor $g = \nabla^v \nabla^v E$ of (M, F) and the Riemannian metric g_M are related by

$$g(\widehat{X}, \widehat{Y}) = (g_M(X, Y))^v; \quad X, Y \in \mathfrak{X}(M). \tag{9.2.37}$$

Working locally, write $X \underset{(\mathcal{U})}{=} X^i \frac{\partial}{\partial u^i}$, $Y \underset{(\mathcal{U})}{=} Y^i \frac{\partial}{\partial u^i}$. Then

$$g(\widehat{X}, \widehat{Y}) \overset{(6.2.40)}{=} X^v(Y^v E) \underset{(\mathcal{U})}{\overset{(9.2.36)}{=}} \frac{1}{2}((g_M)_{ij} \circ \tau)(X^v(Y^v(y^i y^j))).$$

Here

$$Y^v(y^i y^j) = (Y^v y^i)y^j + y^i(Y^v y^j) \overset{(4.1.18)}{=} (Y^v(u^i)^c)y^j + y^i(Y^v(u^j)^c)$$
$$\overset{(4.1.51)}{=} (Yu^i)^v y^j + y^i(Yu^j)^v = (Y^i)^v(u^j)^c + (u^i)^c(Y^j)^v,$$

therefore

$$X^v(Y^v(y^i y^j)) = (Y^i X^j + X^i Y^j)^v.$$

So we obtain

$$g(\widehat{X}, \widehat{Y}) \underset{(\mathcal{U})}{=} \frac{1}{2}((g_M)_{ij}(X^i Y^j + X^j Y^i))^v = ((g_M)_{ij} X^i Y^j)^v \underset{(\mathcal{U})}{=} (g(X, Y))^v,$$

as was to be shown.

(c) We continue to assume that F is the Riemannian Finsler function arising from the Riemannian metric g_M. We show that the local expression of the canonical spray S of (M, F) is

$$S \underset{(\mathcal{U})}{=} y^i \frac{\partial}{\partial x^i} - (\Gamma^i_{jk} \circ \tau)y^j y^k \frac{\partial}{\partial y^i}, \tag{9.2.38}$$

where the functions Γ^i_{jk} are the Christoffel symbols of the Levi-Civita derivative of (M, g_M) with respect to the chart $(\mathcal{U}, (u^i)^n_{i=1})$.

By Remark 9.2.18, $S \underset{(\mathcal{U})}{=} y^i \frac{\partial}{\partial x^i} - 2G^i \frac{\partial}{\partial y^i}$, where the spray coefficients G^i are given by (9.2.30). From (9.2.36) we obtain

$$\frac{\partial E}{\partial x^j} = \frac{1}{2}\left(\frac{\partial(g_M)_{rs}}{\partial u^j} \circ \tau\right) y^r y^s,$$

whence

$$\frac{\partial^2 E}{\partial y^j \partial x^k} = \frac{1}{2}\left(\frac{\partial(g_M)_{rs}}{\partial u^k} \circ \tau\right)(\delta^r_j y^s + y^r \delta^s_j) = \left(\frac{\partial(g_M)_{jl}}{\partial u^k} \circ \tau\right) y^l.$$

Thus

$$G^i \overset{(9.2.30)}{=} \frac{1}{2}\left(\left((g_M)^{ij}\left(\frac{\partial(g_M)_{jl}}{\partial u^k} - \frac{1}{2}\frac{\partial(g_M)_{kl}}{\partial u^j}\right)\right) \circ \tau\right)y^k y^l.$$

On the other hand, the Christoffel symbols of the Levi-Civita derivative of (M, g_M) are

$$\Gamma^i_{kl} \overset{(6.1.60)}{=} \frac{1}{2}(g_M)^{ij}\left(\frac{\partial(g_M)_{lj}}{\partial u^k} + \frac{\partial(g_M)_{jk}}{\partial u^l} - \frac{\partial(g_M)_{kl}}{\partial u^j}\right),$$

from where we get that

$$G^i = \frac{1}{2}(\Gamma^i_{kl} \circ \tau)y^k y^l,$$

because

$$\left(\frac{\partial(g_M)_{lj}}{\partial u^k}\right)y^k y^l + \left(\frac{\partial(g_M)_{jk}}{\partial u^l}\right)y^k y^l$$

$$= \left(\frac{\partial(g_M)_{lj}}{\partial u^k}\right)y^k y^l + \left(\frac{\partial(g_M)_{jl}}{\partial u^k}\right)y^l y^k = 2\left(\frac{\partial(g_M)_{lj}}{\partial u^k} \circ \tau\right)y^k y^l.$$

This proves (9.2.38). From this it follows that *the canonical spray of the Riemannian Finsler manifold (M, F) has the same geodesics as the starting Riemannian manifold (M, g_M).*

Now the question arises: when does a Finsler function come from a Riemannian metric in the above manner? Our answer will be a partial transcription of Proposition 9.1.45 from the context of Finsler vector spaces. As a first step, we introduce the Cartan tensors in Finsler manifolds.

Definition and Lemma 9.2.25. Let (M, F) be a Finsler manifold with energy $E := \frac{1}{2}F^2$ and fundamental tensor $g = \nabla^v\nabla^v E$.

(i) The type $(0,3)$ Finsler tensor field

$$\mathcal{C}_\flat := \nabla^v g = \nabla^v\nabla^v\nabla^v E \in \mathcal{F}^0_3(\mathring{T}M) \tag{9.2.39}$$

is called the *Cartan tensor* of (M, F). It is symmetric, $(-1)^+$-homogeneous, and has the property

$$\tilde{\delta} \in \{\tilde{X}, \tilde{Y}, \tilde{Z}\} \text{ implies } \mathcal{C}_\flat(\tilde{X}, \tilde{Y}, \tilde{Z}) = 0. \tag{9.2.40}$$

(ii) There exists a unique type $(1,2)$ Finsler tensor field $\mathcal{C} \in \mathcal{F}^1_2(\mathring{T}M)$ such that

$$g(\mathcal{C}(\tilde{X}, \tilde{Y}), \tilde{Z}) = \mathcal{C}_\flat(\tilde{X}, \tilde{Y}, \tilde{Z}) \tag{9.2.41}$$

for all $\tilde{X}, \tilde{Y}, \tilde{Z} \in \Gamma(\mathring{\pi})$. We say that $\mathcal{C}$ is the *vector-valued Cartan tensor* of (M, F), but $\mathcal{C}$ is also mentioned as Cartan tensor for short. It is symmetric, $(-1)^+$-homogeneous, and has the property

$$\mathcal{C}(\tilde{X}, \tilde{\delta}) = \mathcal{C}(\tilde{\delta}, \tilde{X}) = 0, \quad \tilde{X} \in \Gamma(\mathring{\pi}). \tag{9.2.42}$$

Proof. We recall that we have already defined Cartan tensors in the generality of Finsler bundles with scalar product; see Definition 6.2.37. Then we saw that $C_\flat$ is symmetric in its last two arguments. In our case, by the immediate pointwise interpretation of $C_\flat$ (cf. Remark 9.2.2(d)) and the symmetry of the third derivative (see Corollary C.1.20), the tensor $C_\flat$ is symmetric, and hence the same is true for C.

Since the fundamental tensor g of (M, F) is non-degenerate, we obtain the existence and uniqueness of C by the reasoning applied in Remark 6.2.38(a).

The 0^+-homogeneity of g (Lemma 9.2.3(i)) implies by Lemma 6.2.22 that $C_\flat = \nabla^{\mathsf{v}} g$ is $(-1)^+$-homogeneous. We show that C shares the same property. Let $X, Y, Z \in \mathfrak{X}(M)$. Then, using (6.2.32) repeatedly,

$$g((\nabla^{\mathsf{v}}_{\widehat{\delta}} C)(\widehat{X}, \widehat{Y}), \widehat{Z}) = g(\nabla^{\mathsf{v}}_{\widehat{\delta}}(C(\widehat{X}, \widehat{Y})), \widehat{Z})$$

$$= C(g(C(\widehat{X}, \widehat{Y}), \widehat{Z})) - (\nabla^{\mathsf{v}}_{\widehat{\delta}} g)(C(\widehat{X}, \widehat{Y}), \widehat{Z}) \overset{(9.2.8)}{=} C(g(C(\widehat{X}, \widehat{Y}), \widehat{Z}))$$

$$= C(C_\flat(\widehat{X}, \widehat{Y}, \widehat{Z})) = (\nabla^{\mathsf{v}}_{\widehat{\delta}} C_\flat)(\widehat{X}, \widehat{Y}, \widehat{Z}) = -C_\flat(\widehat{X}, \widehat{Y}, \widehat{Z}) = g(-C(\widehat{X}, \widehat{Y}), \widehat{Z}),$$

whence $\nabla^{\mathsf{v}}_{\widehat{\delta}} C = -C$.

Finally, for any sections $\widetilde{Y}, \widetilde{Z} \in \Gamma(\overset{\circ}{\pi})$,

$$C_\flat(\widetilde{\delta}, \widetilde{Y}, \widetilde{Z}) := \nabla^{\mathsf{v}} g(\widetilde{\delta}, \widetilde{Y}, \widetilde{Z}) = (\nabla^{\mathsf{v}}_{\widetilde{\delta}} g)(\widetilde{Y}, \widetilde{Z}) \overset{(9.2.8)}{=} 0,$$

so we have (9.2.40). This implies immediately that (9.2.42) is also true. $\square$

Proposition 9.2.26 (Characterizations of Riemannian Finsler Functions). *For a Finsler manifold (M, F), the following conditions are equivalent:*

(i) *For every point p in M, the Finsler vector space $(T_p M, F_p)$ reduces to a Euclidean vector space.*

(ii) *The energy function $E = \frac{1}{2} F^2$ is smooth on the whole TM.*

(iii) *For any vector fields X, Y on M, the function $g(\widehat{X}, \widehat{Y}) \in C^\infty(\overset{\circ}{T}M)$ is a vertical lift.*

(iv) *The Cartan tensor of (M, F) vanishes.*

If one (and hence all) of these conditions holds, then there exists a Riemannian metric g_M on M such that

$$g(\widehat{X}, \widehat{Y}) = (g_M(X, Y))^{\mathsf{v}} \quad \text{for all } X, Y \in \mathfrak{X}(M),$$

therefore F is a Riemannian Finsler function arising from g_M.

Proof. We organize our reasoning according to the following scheme:

$$(i) \implies (iii) \iff (iv)$$
$$\nwarrow \quad \Downarrow$$
$$(ii)$$

(i)⇒(iii): Let p be any point in M. By Proposition 9.1.45, the mapping

$$(E_p)'' \colon \mathring{T}_p M \to T_2^0(T_p M)$$

is constant, therefore the scalar products

$$g_u((u,v),(u,w)) \overset{(9.2.7)}{=} (E_p)''(u)(v,w); \quad u \in \mathring{T}_p M; \; v,w \in T_p M$$

do not depend on the choice of the point u. This implies that the function $g(\widehat{X}, \widehat{Y})$ is a vertical lift in $C^\infty(\mathring{T}M)$, for all $X, Y \in \mathfrak{X}(M)$.

(iii)⇒(ii): Now for any given vector fields X, Y on M, the function $g(\widehat{X}, \widehat{Y}) \in C^\infty(\mathring{T}M)$ is constant on the tangent spaces. Thus, it can naturally be extended to a function

$$\overline{g(\widehat{X}, \widehat{Y})} \colon TM \to \mathbb{R}$$

by the following rule:

For every $u \in o(M)$, $\overline{g(\widehat{X}, \widehat{Y})}(u) := g(\widehat{X}, \widehat{Y})(v)$, where $v \in \mathring{T}_{\tau(u)}M$ is an arbitrarily chosen tangent vector.

We claim that the function $\overline{g(\widehat{X}, \widehat{Y})}$ is smooth. Indeed, choose an open subset $\mathcal{U}$ of M with a nowhere vanishing vector field $Z \in \mathfrak{X}(\mathcal{U})$. Then

$$\overline{g(\widehat{X}, \widehat{Y})} \restriction \tau^{-1}(\mathcal{U}) = (g(\widehat{X}, \widehat{Y}) \circ Z)^{\mathrm{v}} \in C^\infty(\tau^{-1}(\mathcal{U})),$$

which implies our claim by Proposition 2.1.16(ii).

In the next step, we smoothly extend the fundamental tensor g of (M, F) to TM as follows:

For any $X, Y \in \mathfrak{X}(M)$, let $\bar{g}(\widehat{X}, \widehat{Y}) := \overline{g(\widehat{X}, \widehat{Y})}$.

Then, as it can easily be seen, $\bar{g} \in \mathcal{F}_2^0(TM)$ and $\bar{g}_u = g_u$ for all $u \in \mathring{T}M$, so $\bar{g}$ is indeed a smooth extension of g.

By (9.2.9) we have

$$E \restriction \mathring{T}M = \frac{1}{2}\bar{g}(\widetilde{\delta}, \widetilde{\delta}) \restriction \mathring{T}M.$$

However, $\bar{g}(\widetilde{\delta}, \widetilde{\delta})(0_p) = 0$ for all $p \in M$, so the equality $E = \frac{1}{2}\bar{g}(\widetilde{\delta}, \widetilde{\delta})$ holds on the whole TM. Since $\bar{g}(\widetilde{\delta}, \widetilde{\delta})$ is a smooth function on TM, so is E.

(ii)⇒(i): For every $p \in M$, $E := E_p \restriction T_pM$ is of class C^2 (actually smooth). Thus, by Proposition 9.1.45, (T_pM, F_p) reduces to a Euclidean vector space.

(iii)⇔(iv): For any vector fields X, Y, Z on M,

$$C_b(\widehat{X}, \widehat{Y}, \widehat{Z}) := (\nabla^v g)(\widehat{X}, \widehat{Y}, \widehat{Z}) = (\nabla^v_{\widehat{X}} g)(\widehat{Y}, \widehat{Z}) \stackrel{(6.2.32)}{=} X^v g(\widehat{Y}, \widehat{Z}).$$

By Lemma 4.1.36, $g(\widehat{Y}, \widehat{Z})$ is a vertical lift if, and only if, $X^v g(\widehat{Y}, \widehat{Z}) = 0$ for all $X \in \mathfrak{X}(M)$. This proves the desired equivalence, and completes the proof of the proposition.   □

9.3   Notable Covariant Derivatives on a Finsler Manifold

9.3.1   *The Fundamental Lemma of Finsler Geometry*

To begin this subsection, we first derive a formula which will be of crucial importance in our forthcoming considerations.

Lemma 9.3.1 (J. Klein [62]). *Let M be a manifold. Suppose that S is a semispray over M, and let $\mathcal{H}$ be the Ehresmann connection generated by S according to the Crampin–Grifone theorem. Then for every C^1 function L on TM,*

$$2\mathcal{L}_{\mathbf{h}}L = d(L - CL) - i_S d\mathcal{L}_{\mathbf{J}}L + \mathcal{L}_{\mathbf{J}} i_S dL, \tag{9.3.1}$$

where $\mathbf{h}$ is the horizontal projection determined by $\mathcal{H}$.

Proof.   From (7.3.6), $2\mathbf{h} = 1_{\mathfrak{X}(TM)} + [\mathbf{J}, S]$. Thus

$$2\mathcal{L}_{\mathbf{h}}L = \mathcal{L}_{1_{\mathfrak{X}(TM)} + [\mathbf{J},S]}L \stackrel{(3.3.42)}{=} dL + \mathcal{L}_{[\mathbf{J},S]}L \stackrel{(3.3.53)}{=} dL + \mathcal{L}_{\mathbf{J}}\mathcal{L}_S L - \mathcal{L}_S \mathcal{L}_{\mathbf{J}}L$$

$$\stackrel{(3.3.33)}{=} dL + \mathcal{L}_{\mathbf{J}} i_S dL + \mathcal{L}_{\mathbf{J}} di_S L - i_S d\mathcal{L}_{\mathbf{J}}L - di_S \mathcal{L}_{\mathbf{J}}L$$

$$\stackrel{(3.3.26)}{=} dL - di_S \mathcal{L}_{\mathbf{J}}L - i_S d\mathcal{L}_{\mathbf{J}}L + \mathcal{L}_{\mathbf{J}} i_S dL.$$

Here

$$di_S \mathcal{L}_{\mathbf{J}}L = d(\mathcal{L}_{\mathbf{J}}L(S)) \stackrel{(3.3.43)}{=} d(dL \circ \mathbf{J}(S)) = d(dL(C)) = d(CL),$$

so we obtain (9.3.1).   □

Definition 9.3.2. Let $\mathcal{H}$ be an Ehresmann connection in TM with associated horizontal projection $\mathbf{h}$. We say that $\mathcal{H}$ is *compatible with a C^1 function* L on TM if $\mathcal{L}_{\mathbf{h}}L = 0$.

Remark 9.3.3. An Ehresmann connection $\mathcal{H}$ in TM is compatible with a function $L \in C^1(TM)$ if, and only if, $X^{\mathsf{h}}L = 0$ for all $X \in \mathfrak{X}(M)$.

Indeed,

$$\mathcal{L}_{\mathbf{h}}L(X^{\mathsf{c}}) \overset{(3.3.43)}{=} dL \circ \mathbf{h}(X^{\mathsf{c}}) = dL(X^{\mathsf{h}}) = X^{\mathsf{h}}L,$$

$$\mathcal{L}_{\mathbf{h}}L(X^{\mathsf{v}}) = dL \circ \mathbf{h}(X^{\mathsf{v}}) = 0,$$

whence our claim.

By (7.4.4), the local condition of the compatibility is

$$\frac{\partial L}{\partial x^j} - N_j^i \frac{\partial L}{\partial y^i} = 0, \quad j \in J_n,$$

where the functions N_j^i are the Christoffel symbols of $\mathcal{H}$ with respect to the chosen chart.

Lemma 9.3.4. *A positive-homogeneous Ehresmann connection $\mathcal{H}$ in TM is compatible with a function $L \in C^1(TM)$ if, and only if, for any $\mathcal{H}$-parallel vector field $X \colon I \to TM$ along a smooth curve $\gamma \colon I \to M$, the function*

$$L \circ X \colon I \to \mathbb{R}$$

is constant.

Proof. Let $X \colon I \to TM$ be an $\mathcal{H}$-parallel vector field along the smooth curve $\gamma \colon I \to M$. Then for every $t \in I$, $\dot{X}(t) \in \mathrm{Im}(\mathcal{H}_{X(t)})$, therefore $\mathbf{h} \circ \dot{X} = \dot{X}$. Thus

$$\mathcal{L}_{\mathbf{h}}L(\dot{X}(t)) = dL \circ \mathbf{h}(\dot{X}(t)) = dL(\dot{X}(t)) \overset{(3.3.1)}{=} \dot{X}(t)(L) \overset{(3.1.9)}{=} (L \circ X)'(t),$$

and our claim follows from Lemma and Definition 7.6.4. $\qquad\square$

Theorem 9.3.5 (The Fundamental Lemma of Finsler Geometry).
Let (M, F) be a Finsler manifold. There exists a unique Ehresmann connection $\mathcal{H}$ in TM satisfying the following conditions:

(CC$_1$) *$\mathcal{H}$ is positive-homogeneous.*
(CC$_2$) *The torsion of $\mathcal{H}$ vanishes.*
(CC$_3$) *$\mathcal{H}$ is compatible with the Finsler function F.*

This Ehresmann connection is generated by the canonical spray S of (M, F), i.e., it is the Berwald connection of the spray manifold (M, S).

Proof. *Existence.* Let S be the canonical spray of (M, F), and let $\mathcal{H}$ be the Berwald connection of (M, S). Then, as we discussed in 8.1.1, $\mathcal{H}$ is positive-homogeneous and torsion-free. To prove (CC$_3$), first we show that $\mathcal{H}$ is compatible with the energy $E = \frac{1}{2}F^2$ of (M, F). Indeed, since $CE = 2E$, formula (9.3.1) leads to

$$2\mathcal{L}_{\mathbf{h}}E = -dE - i_S d\mathcal{L}_{\mathbf{J}}E + \mathcal{L}_{\mathbf{J}}i_S dE \overset{(9.2.29)}{=} \mathcal{L}_{\mathbf{J}}i_S dE.$$

Obviously, $S \in \mathrm{Dyn}(E)$, so by Lemma 5.2.8(ii), $E_E := CE - E = E$ is a first integral for S. Therefore, by Lemma 3.2.22, $SE = i_S dE = 0$, whence $\mathcal{L}_{\mathbf{h}}E = 0$. Thus, for every vector field X on M,

$$0 = X^{\mathbf{h}}E = \frac{1}{2}X^{\mathbf{h}}(F^2) = F(X^{\mathbf{h}}F).$$

Hence, by the positive definiteness of F, $X^{\mathbf{h}}F = 0$ for all $X \in \mathfrak{X}(M)$. This completes the proof of (CC_3).

Uniqueness. Suppose that $\bar{\mathcal{H}}$ is an Ehresmann connection in TM, satisfying conditions (CC_1)–(CC_3). Let $\bar{S} := \bar{\mathcal{H}}\tilde{\delta}$ be the semispray associated to $\bar{\mathcal{H}}$. Then the positive-homogeneity of $\bar{\mathcal{H}}$ implies by Corollary 7.5.6 that $\bar{S}$ is a spray. This spray *generates the Ehresmann connection* $\bar{\mathcal{H}}$. Indeed, since $\bar{\mathcal{H}}$ is homogeneous and torsion-free, its strong torsion also vanishes by Corollary 7.8.9. Then, in view of Proposition 7.8.8, the Ehresmann connection generated by $\bar{S}$ is $\bar{\mathcal{H}}$.

To conclude the proof, we show that the spray $\bar{S}$ *coincides with the canonical spray of* (M, F). Consider the horizontal projection $\bar{\mathbf{h}}$ determined by $\bar{\mathcal{H}}$. Then

$$\bar{\mathbf{h}}\bar{S} = \bar{\mathbf{h}}(\bar{\mathcal{H}}\tilde{\delta}) = (\bar{\mathcal{H}} \circ \mathbf{j} \circ \bar{\mathcal{H}})(\tilde{\delta}) \overset{(7.2.3)}{=} \bar{\mathcal{H}}\tilde{\delta} = \bar{S}.$$

By condition (CC_3),

$$0 = F\mathcal{L}_{\bar{\mathbf{h}}}F = \frac{1}{2}\mathcal{L}_{\bar{\mathbf{h}}}F^2 = \mathcal{L}_{\bar{\mathbf{h}}}E,$$

thus $\bar{\mathcal{H}}$ is compatible also with E. This implies that

$$i_{\bar{S}}dE = \bar{S}E = \bar{\mathbf{h}}\bar{S}(E) = dE \circ \bar{\mathbf{h}}(\bar{S}) \overset{(3.3.43)}{=} (\mathcal{L}_{\bar{\mathbf{h}}}E)(\bar{S}) = 0.$$

Now apply formula (9.3.1) with $L := E$. We obtain that

$$i_{\bar{S}}d\mathcal{L}_{\mathbf{J}}E = -dE.$$

Therefore, by the uniqueness of the canonical spray,

$$\bar{S} = S = \text{the canonical spray of } (M, F),$$

and hence

$$\bar{\mathcal{H}} = \mathcal{H} = \text{the Berwald connection of } (M, S). \qquad \square$$

Remark 9.3.6. The Ehresmann connection characterized by conditions (CC_1)–(CC_3) is called the *canonical connection* on a Finsler manifold. The first intrinsic formulation and index-free proof of this truly fundamental result is due to J. Grifone [54]. Our proof is essentially different from that of Grifone.

Corollary 9.3.7. *Assume that F is a Riemannian Finsler function arising from a Riemannian metric g_M. Then the canonical connection on (M, F) is the linear Ehresmann connection induced by the Levi-Civita derivative of (M, g_M).*

Proof. Let D denote the Levi-Civita derivative of (M, g_M), and let $\mathcal{H}$ be the linear Ehresmann connection generated by D. Then, evidently, $\mathcal{H}$ is positive homogeneous, and Lemma 7.15.3 implies that it is also torsion-free.

It remains to show that $\mathcal{H}$ satisfies (CC_3). To do this, let $\gamma\colon I \to M$ be a smooth curve, and let $X \in \mathfrak{X}_\gamma(M)$ be a parallel vector field along γ with respect to $\mathcal{H}$. Then, by Corollary 7.6.6, X is parallel also with respect to D. In view of Proposition 6.1.62, the function $g_M \circ (X, X)\colon I \to \mathbb{R}$ is constant, and hence the function $F^2 \circ X \overset{(9.2.35)}{=} g_M \circ (X, X)$ is also constant. By Lemma 9.3.4 this implies that $\mathcal{H}$ is compatible with F. $\qquad\square$

9.3.2 The Finslerian Berwald Derivative

Proposition 9.3.8. *Let (M, F) be a Finsler manifold. There exists a unique Ehresmann connection $\widetilde{\mathcal{H}}$ in TM and a covariant derivative*

$$\widetilde{\nabla}\colon \mathfrak{X}(\mathring{T}M) \times \Gamma(\mathring{\pi}) \to \Gamma(\mathring{\pi})$$

on the slit Finsler bundle $\mathring{\pi}$ such that the following conditions are satisfied:

$(B\nabla_1)$ $\widetilde{\nabla}$ *is vertically natural, i.e., for all $\widetilde{X}, \widetilde{Y} \in \Gamma(\mathring{\pi})$,*

$$\widetilde{\nabla}^{\mathrm{v}}_{\widetilde{X}}\widetilde{Y} := \widetilde{\nabla}_{\mathbf{i}\widetilde{X}}\widetilde{Y} = \nabla^{\mathrm{v}}_{\widetilde{X}}\widetilde{Y} \overset{(6.2.25)}{=} \mathbf{j}[\mathbf{i}\widetilde{X}, \widetilde{\mathcal{H}}\widetilde{Y}].$$

$(B\nabla_2)$ *The $\widetilde{\mathcal{H}}$-deviation of $\widetilde{\nabla}$ vanishes, i.e., by (7.9.5),*

$$\widetilde{\nabla}_{\widetilde{\mathcal{H}}\widetilde{X}}\widetilde{Y} = \widetilde{\mathcal{V}}[\widetilde{\mathcal{H}}\widetilde{X}, \mathbf{i}\widetilde{Y}]; \quad \widetilde{X}, \widetilde{Y} \in \Gamma(\mathring{\pi}),$$

where $\widetilde{\mathcal{V}}$ is the vertical mapping associated to $\widetilde{\mathcal{H}}$.

$(B\nabla_3)$ $\widetilde{\nabla}$ *is strongly associated to $\widetilde{\mathcal{H}}$, i.e., $\widetilde{\nabla}\widetilde{\delta} = \widetilde{\mathcal{V}}$.*

$(B\nabla_4)$ *The $\widetilde{\mathcal{H}}$-torsion $\mathcal{T}$ of $\widetilde{\nabla}$, given by*

$$\mathcal{T}(\widetilde{X}, \widetilde{Y}) := T^{\widetilde{\nabla}}(\widetilde{\mathcal{H}}\widetilde{X}, \widetilde{\mathcal{H}}\widetilde{Y}) = \widetilde{\nabla}_{\widetilde{\mathcal{H}}\widetilde{X}}\widetilde{Y} - \widetilde{\nabla}_{\widetilde{\mathcal{H}}\widetilde{Y}}\widetilde{X} - \mathbf{j}[\widetilde{\mathcal{H}}\widetilde{X}, \widetilde{\mathcal{H}}\widetilde{Y}],$$

vanishes.

$(B\nabla_5)$ $\widetilde{\mathcal{H}}$ *is compatible with* F.

Then $\widetilde{\mathcal{H}}$ *is the canonical connection of* (M, F)*, and* $\widetilde{\nabla}$ *is the Berwald derivative induced by* $\widetilde{\mathcal{H}}$.

Proof.   *Existence.* Let $\mathcal{H}$ be the canonical connection of (M, F), and let ∇ be the induced Berwald derivative. Then $(B\nabla_1)$, $(B\nabla_2)$ and $(B\nabla_5)$ hold automatically. Since $\mathcal{H}$ is positive-homogeneous, from Lemma 7.12.2 it follows that ∇ is strongly associated to $\mathcal{H}$.

Finally, the torsion $\mathbf{T}$ of $\mathcal{H}$ coincides with the $\mathcal{H}$-torsion $\mathcal{T}$ of ∇ by 8.1.3(ii), so $\mathbf{T} = 0$ implies that $(B\nabla_4)$ also holds.

Uniqueness. Suppose that $(\widetilde{\mathcal{H}}, \widetilde{\nabla})$ satisfies $(B\nabla_1)$–$(B\nabla_5)$. Then, by $(B\nabla_1)$ and $(B\nabla_2)$, $\widetilde{\nabla}$ is the Berwald derivative induced by $\widetilde{\mathcal{H}}$. Again from Lemma 7.12.2, we conclude that $(B\nabla_3)$ implies the positive-homogeneity of $\widetilde{\mathcal{H}}$. As above, the torsion of $\widetilde{\mathcal{H}}$ coincides with the $\widetilde{\mathcal{H}}$-torsion of $\widetilde{\nabla}$ (see Proposition 7.12.1), so $(B\nabla_4)$ assures that $\widetilde{\mathcal{H}}$ is torsion-free. Thus, taking into account also condition $(B\nabla_5)$, it follows from the uniqueness statement of Theorem 9.3.5 that

$$\widetilde{\mathcal{H}} = \mathcal{H} = \text{the canonical connection of } (M, F).$$

Therefore

$$\widetilde{\nabla} = \nabla = \text{the Berwald derivative induced by } \mathcal{H}.$$

This concludes the proof. $\square$

Remark 9.3.9. (a) Axioms $(B\nabla_2)$–$(B\nabla_5)$, as a characterization of the Finslerian h-Berwald derivative, were first formulated by T. Okada [78] in terms of traditional tensor calculus. We now explain the approach employed in [78] in a simplified manner.

Suppose that $\widetilde{\mathcal{H}}$ is an Ehresmann connection in TM and D is a covariant derivative on $\mathring{\pi}$. Choose a chart $(\mathcal{U}, (u^i)_{i=1}^n)$ on M and consider the induced chart $(\tau^{-1}(\mathcal{U}), ((x^i)_{i=1}^n, (y^i)_{i=1}^n))$ on TM. Let (N_j^i), (Γ_{jk}^i) and (C_{jk}^i) be the families of the Christoffel symbols of $\widetilde{\mathcal{H}}$ and D with respect to the chosen chart. Then

$$\widetilde{\mathcal{H}}\left(\widehat{\frac{\partial}{\partial u^j}}\right) \overset{(7.2.37)}{=} \frac{\partial}{\partial x^j} - N_j^i \frac{\partial}{\partial y^i}$$

and, by (6.2.6),

$$D_{\frac{\partial}{\partial x^j}} \widehat{\frac{\partial}{\partial u^k}} = \Gamma_{jk}^i \widehat{\frac{\partial}{\partial u^i}}, \quad D_{\frac{\partial}{\partial y^j}} \widehat{\frac{\partial}{\partial u^k}} = C_{jk}^i \widehat{\frac{\partial}{\partial u^i}}.$$

We recall that

$$\nabla^v_{\frac{\partial}{\partial u^j}} \widehat{\frac{\partial}{\partial u^k}} = \nabla_{\frac{\partial}{\partial y^j}} \widehat{\frac{\partial}{\partial u^k}} \overset{(7.10.4)}{=} 0$$

and

$$\tilde{\mathcal{V}}\left[\tilde{\mathcal{H}}\left(\widehat{\frac{\partial}{\partial u^j}}\right), \frac{\partial}{\partial y^k}\right] = \frac{\partial N_j^i}{\partial y^k} \widehat{\frac{\partial}{\partial u^i}} \quad \text{by Example 7.11.1.}$$

Taking these into account, conditions $(B\nabla_1)$–$(B\nabla_5)$ take the following forms:

(i) $C_{jk}^i = 0$,   (ii) $\Gamma_{jk}^i = \dfrac{\partial N_j^i}{\partial y^k}$,   (iii) $y^k \Gamma_{jk}^i = N_j^i$,

$$(i, j, k \in J_n).$$

(iv) $\Gamma_{jk}^i = \Gamma_{kj}^i$, (v) $\dfrac{\partial F}{\partial x^k} - N_k^i \dfrac{\partial F}{\partial y^i} = 0$

Relations (iii) and (iv) will take some explaining. First,

$$D\tilde{\delta}\left(\frac{\partial}{\partial x^j}\right) = D_{\frac{\partial}{\partial x^j}} y^k \widehat{\frac{\partial}{\partial u^k}} = y^k \Gamma_{jk}^i \widehat{\frac{\partial}{\partial u^i}}, \quad \tilde{\mathcal{V}}\left(\frac{\partial}{\partial x^j}\right) \overset{(7.2.40)}{=} N_j^i \widehat{\frac{\partial}{\partial u^i}},$$

$$D\tilde{\delta}\left(\frac{\partial}{\partial y^j}\right) = D_{\frac{\partial}{\partial y^j}} y^k \widehat{\frac{\partial}{\partial u^k}} \overset{(i)}{=} \widehat{\frac{\partial}{\partial u^j}} \overset{(7.2.40)}{=} \tilde{\mathcal{V}}\left(\frac{\partial}{\partial y^j}\right),$$

therefore (iii) is indeed the local expression of $(B\nabla_3)$.

Second,

$$\mathcal{T}\left(\widehat{\frac{\partial}{\partial u^j}}, \widehat{\frac{\partial}{\partial u^k}}\right) =$$

$$= D_{\tilde{\mathcal{H}}\left(\frac{\partial}{\partial u^j}\right)} \widehat{\frac{\partial}{\partial u^k}} - D_{\tilde{\mathcal{H}}\left(\frac{\partial}{\partial u^k}\right)} \widehat{\frac{\partial}{\partial u^j}} - \mathbf{j}\left[\tilde{\mathcal{H}}\left(\widehat{\frac{\partial}{\partial u^j}}\right), \tilde{\mathcal{H}}\left(\widehat{\frac{\partial}{\partial u^k}}\right)\right]$$

$$\overset{(i),(7.2.35)}{=} D_{\frac{\partial}{\partial x^j}} \widehat{\frac{\partial}{\partial u^k}} - D_{\frac{\partial}{\partial x^k}} \widehat{\frac{\partial}{\partial u^j}} = (\Gamma_{jk}^i - \Gamma_{kj}^i) \widehat{\frac{\partial}{\partial u^i}},$$

thus (iv) is the local form of $(B\nabla_4)$.

By Remark 9.3.3, (v) is the coordinate expression of condition $(B\nabla_5)$. Notice that (v) is equivalent to

$$(\text{v'}) \quad \frac{\partial E}{\partial x^k} - N_k^i \frac{\partial E}{\partial y^i} = 0, \quad k \in J_n,$$

where $E = \frac{1}{2} F^2$.

We show that $\tilde{\mathcal{H}}$ *is the canonical connection of* (M, F); then, clearly, D is the Berwald derivative induced by the canonical connection. In this way we prove once again the uniqueness statement of the fundamental lemma of Finsler geometry.

Step 1. Conditions (ii) and (iii) imply that

$$y^k \frac{\partial N_j^i}{\partial y^k} = N_j^i; \quad i, j \in J_n; \tag{$*$}$$

therefore the Ehresmann connection $\widetilde{\mathcal{H}}$ *is positive-homogeneous.* Let

$$\widetilde{S} := \widetilde{\mathcal{H}}\widetilde{\delta} \underset{(\mathcal{U})}{=} y^i \frac{\partial}{\partial x^i} - 2\widetilde{S}^i \frac{\partial}{\partial y^i}$$

be the semispray associated to $\widetilde{\mathcal{H}}$. Then by Lemma and Definition 7.2.13, the coefficients of $\widetilde{S}$ with respect to the chosen chart are the functions $\widetilde{S}^i = \frac{1}{2} y^j N_j^i$, $i \in J_n$. These are 2^+-homogeneous because

$$y^k \frac{\partial \widetilde{S}^i}{\partial y^k} = \frac{1}{2}\left(y^k N_k^i + y^j y^k \frac{\partial N_j^i}{\partial y^k} \right) \overset{(*)}{=} y^j N_j^i = 2\widetilde{S}^i.$$

Therefore, $\widetilde{S}$ is a *spray* over M. By Remark 7.3.5, the Christoffel symbols of the Ehresmann connection generated by $\widetilde{S}$ are

$$\frac{\partial \widetilde{S}^i}{\partial y^j} = \frac{\partial}{\partial y^j}\left(\frac{1}{2} y^k N_k^i \right) = \frac{1}{2}\left(N_j^i + y^k \frac{\partial N_k^i}{\partial y^j} \right) \overset{(ii)}{=} \frac{1}{2}(N_j^i + y^k \Gamma_{kj}^i)$$

$$\overset{(iv)}{=} \frac{1}{2}(N_j^i + y^k \Gamma_{jk}^i) \overset{(iii)}{=} \frac{1}{2}(N_j^i + N_j^i) = N_j^i,$$

thus $\widetilde{S} := \widetilde{\mathcal{H}}\widetilde{\delta}$ generates $\widetilde{\mathcal{H}}$.

Step 2. From (v'), by differentiation with respect to y^j, we obtain that

$$0 = \frac{\partial^2 E}{\partial y^j \partial x^k} - \frac{\partial N_k^i}{\partial y^j} \frac{\partial E}{\partial y^i} - N_k^i \frac{\partial^2 E}{\partial y^i \partial y^j} \overset{(9.2.5)}{=} \frac{\partial^2 E}{\partial y^j \partial x^k} - \frac{\partial N_k^i}{\partial y^j} \frac{\partial E}{\partial y^i} - N_k^i g_{ij},$$

whence

$$N_k^i g_{ij} = \frac{\partial^2 E}{\partial y^j \partial x^k} - \frac{\partial N_k^i}{\partial y^j} \frac{\partial E}{\partial y^i}.$$

Multiplying both sides by y^k, and taking into account that by our above calculation $\frac{\partial N_k^i}{\partial y^j} y^k = N_j^i$, we obtain that

$$y^k N_k^i g_{ij} = \frac{\partial^2 E}{\partial y^j \partial x^k} y^k - N_j^i \frac{\partial E}{\partial y^i} \overset{(v')}{=} \frac{\partial^2 E}{\partial y^j \partial x^k} y^k - \frac{\partial E}{\partial x^j}.$$

From this we can express $\widetilde{S}^i = \frac{1}{2} y^k N_k^i$ as

$$\widetilde{S}^i = \frac{1}{2} g^{ij}\left(\frac{\partial^2 E}{\partial y^j \partial x^k} y^k - \frac{\partial E}{\partial x^j} \right) \overset{(9.2.30)}{=} G^i,$$

where the functions G^i are the coefficients of the canonical spray S of (M, F). Thus $\widetilde{S} = S$, $\widetilde{\mathcal{H}}$ is the canonical connection of (M, F), and D is the Berwald derivative induced by the canonical connection.

(b) The covariant derivative described by the axioms $(B\nabla_1)$–$(B\nabla_5)$ is called the *Finslerian Berwald derivative* in $\mathring{\pi}$ determined by the Finsler function F. As previously, we shall denote it by ∇. By a slight abuse of language, we also say that ∇ is *the Berwald derivative on the Finsler manifold* (M, F). If S is the canonical spray of (M, F), then ∇ is just the Berwald derivative of (M, S). Thus, everything we explained in Chapter 8 in the generality of spray manifolds remains valid for Finsler manifolds. In the following, by

the *Berwald curvature* $\mathbf{B}$
the *fundamental affine curvature* $\mathcal{R}$ $\left.\right\}$ of a Finsler manifold (M, F)
the *affine curvature* $\mathbf{H}$

we shall mean the corresponding curvature of the attached spray manifold (M, S). A Finsler manifold with vanishing Berwald and affine curvature is called *flat*.

Similarly, the *Jacobi endomorphism* of (M, F) is the Jacobi endomorphism $\mathbf{K}$ of (M, S) defined by (8.2.1). Also, we speak about the *Weyl tensors* $\mathbf{W}_1, \mathbf{W}_2, \mathbf{W}_3$ of (M, F) in this sense. We say that (M, F) is *h-flat* if $\mathbf{K} = 0$ (equivalently, if $\mathcal{R} = 0$ or $\mathbf{H} = 0$, see Corollary and Definition 8.2.7), and (M, F) is *isotropic* if its canonical spray is isotropic.

Finally, by the *Douglas tensor* of a Finsler manifold we mean the Douglas tensor of its attached spray manifold.

Of course, in the richer world of Finsler geometry we meet several new actors and new phenomena.

Proposition 9.3.10. *A Finsler manifold (M, F) is flat if, and only if, at every point of M there exists a chart $(\mathcal{U}, (u^i)_{i=1}^n)$ with induced chart $(\tau^{-1}(\mathcal{U}), ((x^i)_{i=1}^n, (y^i)_{i=1}^n))$ on TM such that $\frac{\partial F}{\partial x^i} = 0$ for all $i \in J_n$.*

Proof. Let S be the canonical spray of (M, F). If (M, F) is flat, then locally, by Corollary 8.5.8, $S \underset{(\mathcal{U})}{=} y^i \frac{\partial}{\partial x^i}$. Thus

$$\frac{\partial F}{\partial x^i} = \left(\frac{\partial}{\partial u^i}\right)^{\mathsf{h}} F \overset{(CC_3)}{=} 0, \quad i \in J_n,$$

so the charts provided by the quoted corollary have the desired property.

Conversely, under the condition of our proposition, we obtain from (9.2.30) that $S \underset{(\mathcal{U})}{=} y^i \frac{\partial}{\partial x^i}$. Thus, by Corollary 8.5.8, S is flat, and therefore, by definition, so is (M, F). $\qquad\square$

Remark 9.3.11. In the literature, flat Finsler manifolds are usually called *locally Minkowski manifolds*; see, e.g., [72], Definition 24.1 or [74], Definition 1.2.3.1.

Definition and Lemma 9.3.12. Let (M, F) be a Finsler manifold with fundamental tensor g. Let ∇^h be the h-Berwald derivative determined by the Berwald connection of (M, F). Then the type $(0, 3)$ Finsler tensor field

$$\mathcal{C}^h_\flat := \mathbf{L} := \nabla^h g \in \mathcal{F}^0_3(\mathring{T}M) \tag{9.3.2}$$

is called the *second Cartan tensor* or the *Landsberg tensor* of (M, F). There exists a unique type $(1, 2)$ Finsler tensor field $\mathcal{C}^h \in \mathcal{F}^1_2(\mathring{T}M)$ such that

$$g(\mathcal{C}^h(\widetilde{X}, \widetilde{Y}), \widetilde{Z}) = \mathcal{C}^h_\flat(\widetilde{X}, \widetilde{Y}, \widetilde{Z}), \tag{9.3.3}$$

for all $\widetilde{X}, \widetilde{Y}, \widetilde{Z} \in \Gamma(\mathring{\pi})$. We say that $\mathcal{C}^h$ is the *vector-valued second Cartan* (or *Landsberg*) *tensor* of (M, F), but it is also mentioned as the second Cartan (or Landsberg) tensor for short.

Proof. The existence and uniqueness of $\mathcal{C}^h$ can be shown, again, with the help of the musical isomorphism $\sharp$, see Remark 2.2.55. □

Proposition 9.3.13. *The Berwald curvature and the Landsberg tensor of a Finsler manifold are related by*

$$\nabla^v E \circ \mathbf{B} = \mathbf{L}, \tag{9.3.4}$$

where E is the energy function.

Proof. As in most cases, we denote the underlying manifold by M. For any vector fields X, Y, Z on M,

$$\mathbf{L}(\widehat{X}, \widehat{Y}, \widehat{Z}) := \nabla^h g(\widehat{X}, \widehat{Y}, \widehat{Z}) = (\nabla_{X^h} g)(\widehat{Y}, \widehat{Z})$$

$$= X^h(g(\widehat{Y}, \widehat{Z})) - g(\nabla_{X^h}\widehat{Y}, \widehat{Z}) - g(\widehat{Y}, \nabla_{X^h}\widehat{Z})$$

$$\overset{(9.2.3)}{=} X^h(Y^v(Z^v E)) - \mathbf{i}\nabla_{X^h}\widehat{Y}(Z^v E) + \mathbf{J}[\mathbf{i}\nabla_{X^h}\widehat{Y}, Z^h](E)$$

$$- Y^v(\mathbf{i}\nabla_{X^h}\widehat{Z}(E)) + \mathbf{J}[Y^v, \mathcal{H}\nabla_{X^h}\widehat{Z}](E).$$

Here

$$\mathbf{J}[\mathbf{i}\nabla_{X^h}\widehat{Y}, Z^h] \overset{(7.10.1)}{=} \mathbf{i}\nabla_{\mathbf{i}\nabla_{X^h}\widehat{Y}}\widehat{Z} = \mathbf{i}\nabla^v_{\nabla_{X^h}\widehat{Y}}\widehat{Z} \overset{(6.2.32)}{=} 0,$$

$$\mathbf{J}[Y^v, \mathcal{H}\nabla_{X^h}\widehat{Z}] \overset{(7.10.2)}{=} \mathbf{J}[Y^v, \mathcal{H}\mathcal{V}[X^h, Z^v]]$$

$$\overset{(7.2.19)}{=} \mathbf{J}[Y^v, \mathbf{F}[X^h, Z^v]] + \mathbf{J}[Y^v, \mathbf{J}[X^h, Z^v]]$$

$$\overset{[X^h, Z^v] \text{ is vertical}}{=} \mathbf{J}[Y^v, \mathbf{F}[X^h, Z^v]] \overset{(7.13.2)}{=} \mathbf{i}\mathbf{B}(\widehat{Y}, \widehat{X})\widehat{Z}$$

$$\overset{\text{Corollary 7.13.4}}{=} \mathbf{i}\mathbf{B}(\widehat{X},\widehat{Y})\widehat{Z}.$$

Thus we obtain that

$$\mathbf{L}(\widehat{X},\widehat{Y},\widehat{Z}) = X^{\mathsf{h}}(Y^{\mathsf{v}}(Z^{\mathsf{v}}E)) - \mathbf{i}\nabla_{X^{\mathsf{h}}}\widehat{Y}(Z^{\mathsf{v}}E) - Y^{\mathsf{v}}(\mathbf{i}\nabla_{X^{\mathsf{h}}}\widehat{Z}(E))$$
$$+ \mathbf{i}\mathbf{B}(\widehat{X},\widehat{Y})\widehat{Z}(E) \overset{(7.10.2)}{=} X^{\mathsf{h}}(Y^{\mathsf{v}}(Z^{\mathsf{v}}E)) - [X^{\mathsf{h}},Y^{\mathsf{v}}](Z^{\mathsf{v}}E)$$
$$- Y^{\mathsf{v}}([X^{\mathsf{h}},Z^{\mathsf{v}}](E)) + \mathbf{i}\mathbf{B}(\widehat{X},\widehat{Y})\widehat{Z}(E) = Y^{\mathsf{v}}(X^{\mathsf{h}}(Z^{\mathsf{v}}E))$$
$$- Y^{\mathsf{v}}(X^{\mathsf{h}}(Z^{\mathsf{v}}E)) + Y^{\mathsf{v}}(Z^{\mathsf{v}}(X^{\mathsf{h}}E)) + \mathbf{i}\mathbf{B}(\widehat{X},\widehat{Y})\widehat{Z}(E)$$
$$\overset{(\text{CC}_3)}{=} \mathbf{i}\mathbf{B}(\widehat{X},\widehat{Y})\widehat{Z}(E) \overset{(6.2.32)}{=} \nabla^{\mathsf{v}}E(\mathbf{B}(\widehat{X},\widehat{Y})\widehat{Z}),$$

whence our assertion. $\qquad\square$

Remark 9.3.14. Since

$$\nabla^{\mathsf{v}}g := \mathcal{C}_\flat, \quad \nabla^{\mathsf{h}}g \overset{(9.3.4)}{=} \nabla^{\mathsf{v}}E \circ \mathbf{B},$$

it follows that neither ∇^{v} nor ∇^{h} is a metric derivative in general, therefore *the Finslerian Berwald derivative is not a metric derivative on $\overset{\circ}{\pi}$.*

Corollary 9.3.15. *The Landsberg tensor of a Finsler manifold has the following properties:*

  (i) *It is symmetric in all three variables.*
 (ii) *$\widetilde{\delta} \in \{\widetilde{X},\widetilde{Y},\widetilde{Z}\}$ implies that $\mathbf{L}(\widetilde{X},\widetilde{Y},\widetilde{Z}) = 0$.*
(iii) *It is 0^+-homogeneous, i.e.,*

$$\nabla_C\mathbf{L} = 0. \tag{9.3.5}$$

Proof. Assertions (i) and (ii) are immediate consequences of (9.3.4) and the corresponding properties of the Berwald curvature (Corollary 7.13.4 and (7.13.6)). To show (9.3.5), let X, Y and Z be arbitrary vector fields on M. Then, taking into account (6.2.32),

$$\nabla_C\mathbf{L}(\widehat{X},\widehat{Y},\widehat{Z}) = C(\mathbf{L}(\widehat{X},\widehat{Y},\widehat{Z})) \overset{(9.3.4)}{=} C(\nabla^{\mathsf{v}}E(\mathbf{B}(\widehat{X},\widehat{Y})\widehat{Z}))$$
$$= (\nabla^{\mathsf{v}}_{\widetilde{\delta}}\nabla^{\mathsf{v}}E)(\mathbf{B}(\widehat{X},\widehat{Y})\widehat{Z}) + \nabla^{\mathsf{v}}E(\nabla^{\mathsf{v}}_{\widetilde{\delta}}(\mathbf{B}(\widehat{X},\widehat{Y})\widehat{Z}))$$
$$\overset{\text{Lemma 6.2.22, (7.13.7)}}{=} \nabla^{\mathsf{v}}E(\mathbf{B}(\widehat{X},\widehat{Y})\widehat{Z}) - \nabla^{\mathsf{v}}E(\mathbf{B}(\widehat{X},\widehat{Y})\widehat{Z}) = 0. \qquad\square$$

Corollary 9.3.16. *If g is the fundamental tensor and S is the canonical spray of a Finsler manifold, then $\nabla_S g = 0$.*

Proof. For any sections $\widetilde{X}, \widetilde{Y}$ of $\overset{\circ}{\pi}$,

$$(\nabla_S g)(\widetilde{X}, \widetilde{Y}) = (\nabla^{\mathsf{h}}_{\widetilde{\delta}} g)(\widetilde{X}, \widetilde{Y}) = \nabla^{\mathsf{h}} g(\widetilde{\delta}, \widetilde{X}, \widetilde{Y})$$
$$= \mathbf{L}(\widetilde{\delta}, \widetilde{X}, \widetilde{Y}) \overset{\text{Corollary } 9.3.15(\text{ii})}{=} 0,$$

as was to be shown.   □

Proposition 9.3.17. *The Cartan tensor and the second Cartan tensor of a Finsler manifold are related by*

$$\nabla_S \mathcal{C}_\flat = -\mathcal{C}^{\mathsf{h}}_\flat = -\mathbf{L}, \tag{9.3.6}$$

where S is the canonical spray.

Proof. For any vector fields X, Y, Z on M,

$$(\nabla_S \mathcal{C}_\flat)(\widehat{X}, \widehat{Y}, \widehat{Z}) = (\nabla^{\mathsf{h}}_{\widetilde{\delta}} \nabla^{\mathsf{v}} g)(\widehat{X}, \widehat{Y}, \widehat{Z}) = (\nabla^{\mathsf{h}} \nabla^{\mathsf{v}} g)(\widetilde{\delta}, \widehat{X}, \widehat{Y}, \widehat{Z})$$

$$\overset{(7.13.9)}{=} (\nabla^{\mathsf{v}} \nabla^{\mathsf{h}} g)(\widehat{X}, \widetilde{\delta}, \widehat{Y}, \widehat{Z}) + g(\mathbf{B}(\widehat{X}, \widetilde{\delta})\widehat{Y}, \widehat{Z})$$

$$+ g(\widehat{Y}, \mathbf{B}(\widehat{X}, \widetilde{\delta})\widehat{Z}) \overset{(7.13.6)}{=} (\nabla^{\mathsf{v}} \nabla^{\mathsf{h}} g)(\widehat{X}, \widetilde{\delta}, \widehat{Y}, \widehat{Z})$$

$$= (\nabla_{X^{\mathsf{v}}} \mathbf{L})(\widetilde{\delta}, \widehat{Y}, \widehat{Z}) = X^{\mathsf{v}}(\mathbf{L}(\widetilde{\delta}, \widehat{Y}, \widehat{Z})) - \mathbf{L}(\nabla_{X^{\mathsf{v}}} \widetilde{\delta}, \widehat{Y}, \widehat{Z})$$

$$\overset{\text{Cor. } 9.3.15(\text{ii}), (6.2.33)}{=} -\mathbf{L}(\widehat{X}, \widehat{Y}, \widehat{Z}) = -\mathcal{C}^{\mathsf{h}}_\flat(\widehat{X}, \widehat{Y}, \widehat{Z}),$$

as desired.   □

Lemma 9.3.18. *The projected tensor of the Berwald curvature of a Finsler manifold is*

$$\mathbf{pB} = \mathbf{B} - \frac{1}{2E} \mathbf{L} \otimes \widetilde{\delta}. \tag{9.3.7}$$

Proof. Let X, Y, Z be vector fields on M. Then, by Corollary 7.13.7 and Lemma 9.2.12,

$$(\mathbf{pB})(\widehat{X}, \widehat{Y}, \widehat{Z}) = \mathbf{p}(\mathbf{B}(\widehat{X}, \widehat{Y})\widehat{Z}) \overset{(9.2.25)}{=} \mathbf{B}(\widehat{X}, \widehat{Y})\widehat{Z} - \frac{1}{2E} \nabla^{\mathsf{v}} E(\mathbf{B}(\widehat{X}, \widehat{Y})\widehat{Z})\widetilde{\delta}$$

$$\overset{(9.3.4)}{=} \mathbf{B}(\widehat{X}, \widehat{Y})\widehat{Z} - \frac{1}{2E} \mathbf{L}(\widehat{X}, \widehat{Y}, \widehat{Z})\widetilde{\delta}$$

$$= \left(\mathbf{B} - \frac{1}{2E} \mathbf{L} \otimes \widetilde{\delta} \right)(\widehat{X}, \widehat{Y}, \widehat{Z}),$$

as wanted.   □

9.3.3   *The Cartan Derivative*

From a 'modern' point of view, the covariant derivative operators introduced in Finsler geometry by L. Berwald, É. Cartan, S.-S. Chern, H. Rund and M. Hashiguchi using classical tensor calculus, can be interpreted as covariant derivatives on the Finsler bundle $\mathring{\pi}$, specified by some characteristic properties (compatibility or 'semi-compatibility' with the metric tensor and the canonical connection, vanishing of some of the torsion tensors). We discussed the Finslerian Berwald derivative in the previous subsection. Now we continue with the covariant derivative constructed by Cartan in 1934 [28]. This is the analogue of the Levi-Civita derivative on a semi-Riemannian manifold (Definition 6.1.44) in the sense that it is the only covariant derivative on a Finsler manifold (more precisely, on the slit Finsler bundle) which is metric and whose vertical and $\mathcal{H}$-horizontal torsions vanish.

If (M, F) is a Finsler manifold with canonical connection $\mathcal{H}$, and

$$D\colon \mathfrak{X}(\mathring{T}M) \times \Gamma(\mathring{\pi}) \to \Gamma(\mathring{\pi})$$

is a covariant derivative operator, then D is uniquely determined by its *vertical part* and *horizontal part* given by

$$D^{\mathsf{v}}_{\widetilde{X}}\widetilde{Y} := D_{\mathbf{i}\widetilde{X}}\widetilde{Y}, \quad D^{\mathsf{h}}_{\widetilde{X}}\widetilde{Y} := D_{\mathcal{H}\widetilde{X}}\widetilde{Y}; \quad \widetilde{X}, \widetilde{Y} \in \Gamma(\mathring{\pi}),$$

respectively (cf. Example 6.2.14). We say that *the horizontal part is metric* if it satisfies a condition analogous to (6.2.63), namely, if

$$(\mathcal{H}\widetilde{X})g(\widetilde{Y}, \widetilde{Z}) = g(D^{\mathsf{h}}_{\widetilde{X}}\widetilde{Y}, \widetilde{Z}) + g(\widetilde{Y}, D^{\mathsf{h}}_{\widetilde{X}}\widetilde{Z}) \text{ for all } \widetilde{X}, \widetilde{Y}, \widetilde{Z} \in \Gamma(\mathring{\pi}).$$

Proposition and Definition 9.3.19. *Let (M, F) be a Finsler manifold and $\mathcal{H}$ its canonical connection. There exists a unique covariant derivative operator*

$$D\colon \mathfrak{X}(\mathring{T}M) \times \Gamma(\mathring{\pi}) \to \Gamma(\mathring{\pi})$$

on the slit Finsler bundle $\mathring{\pi}$ such that the following conditions are satisfied:

(CD$_1$) *The vertical part D^{v} of D is metric.*

(CD$_2$) *The vertical torsion of D given by*

$$\mathfrak{Q}(\widetilde{X}, \widetilde{Y}) := D_{\mathbf{i}\widetilde{X}}\widetilde{Y} - D_{\mathbf{i}\widetilde{Y}}\widetilde{X} - \mathbf{i}^{-1}[\mathbf{i}\widetilde{X}, \mathbf{i}\widetilde{Y}]$$

vanishes.

(CD$_3$) *The horizontal part D^{h} of D is metric.*

(CD$_4$) *The $\mathcal{H}$-horizontal torsion of D given by*

$$\mathfrak{T}(\widetilde{X}, \widetilde{Y}) := D_{\mathcal{H}\widetilde{X}}\widetilde{Y} - D_{\mathcal{H}\widetilde{Y}}\widetilde{X} - \mathbf{j}[\mathcal{H}\widetilde{X}, \mathcal{H}\widetilde{Y}]$$

vanishes.

The operator D is said to be Cartan's covariant derivative *on (M, F). Its action is given by*

$$D_{\mathbf{i}\widetilde{X}}\widetilde{Y} = \nabla_{\mathbf{i}\widetilde{X}}\widetilde{Y} + \frac{1}{2}\mathcal{C}(\widetilde{X}, \widetilde{Y}), \quad D_{\mathcal{H}\widetilde{X}}\widetilde{Y} = \nabla_{\mathcal{H}\widetilde{X}}\widetilde{Y} + \frac{1}{2}\mathcal{C}^{\mathsf{h}}(\widetilde{X}, \widetilde{Y}), \quad (9.3.8)$$

where ∇ is the Finslerian Berwald derivative and $\mathcal{C}$ and $\mathcal{C}^{\mathsf{h}}$ are the Cartan tensor and the second Cartan tensor of (M, F), respectively (see (9.2.41) and (9.3.3)).

Proof.   We apply Proposition 6.2.40 to the fundamental tensor g to see that there is a unique vertical covariant derivative D^{v} on $\mathring{\pi}$ which is metric and whose torsion vanishes. By (CD_1) and (CD_2), this vertical covariant derivative necessarily coincides with the vertical part of Cartan's derivative. Since the Cartan tensor of a Finsler manifold is symmetric, the tensors $\mathcal{C}$ and $\mathring{\mathcal{C}}$ introduced in (6.2.59) and (6.2.61) coincide in our case, thus (6.2.64) gives the first formula in (9.3.8).

The construction and the uniqueness of the horizontal part of Cartan's covariant derivative is completely analogous to the proof of Proposition 6.2.40, thus we leave the details to the reader. The calculation is made somewhat simpler by the fact that the second Cartan tensor $\mathcal{C}^{\mathsf{h}}$ is also symmetric according to Corollary 9.3.15(i). $\square$

9.3.4   *The Chern – Rund and the Hashiguchi Derivative*

Proposition and Definition 9.3.20. *Let (M, F) be a Finsler manifold and $\mathcal{H}$ its canonical connection. There exists a unique covariant derivative operator*

$$D \colon \mathfrak{X}(\mathring{T}M) \times \Gamma(\mathring{\pi}) \to \Gamma(\mathring{\pi})$$

on the slit Finsler bundle $\mathring{\pi}$ such that the following conditions are satisfied:

(CRD_1) *It is vertically natural, i.e.,*

$$D_{\widetilde{X}}^{\mathsf{v}}\widetilde{Y} = \nabla_{\widetilde{X}}^{\mathsf{v}}\widetilde{Y} = \mathbf{j}[\mathbf{i}\widetilde{X}, \mathcal{H}\widetilde{Y}]$$

for all $\widetilde{X}, \widetilde{Y} \in \Gamma(\mathring{\pi})$.

(CRD_2) *The horizontal part D^{h} of D is metric.*

(CRD_3) *The $\mathcal{H}$-horizontal torsion of D given by*

$$\mathcal{T}(\widetilde{X}, \widetilde{Y}) := D_{\mathcal{H}\widetilde{X}}\widetilde{Y} - D_{\mathcal{H}\widetilde{Y}}\widetilde{X} - \mathbf{j}[\mathcal{H}\widetilde{X}, \mathcal{H}\widetilde{Y}]$$

vanishes.

The operator D is said to be the Chern–Rund *covariant derivative on* (M, F). *Its action is given by*

$$D_{i\tilde{X}}\tilde{Y} = \nabla_{i\tilde{X}}\tilde{Y}, \quad D_{\mathcal{H}\tilde{X}}\tilde{Y} = \nabla_{\mathcal{H}\tilde{X}}\tilde{Y} + \frac{1}{2}\mathcal{C}^{h}(\tilde{X}, \tilde{Y}), \qquad (9.3.9)$$

where ∇ is the Finslerian Berwald derivative and $\mathcal{C}^h$ is the second Cartan tensor of (M, F).

Proof. Condition (CRD$_1$) uniquely determines the vertical part of the Chern–Rund derivative. By conditions (CRD$_2$) and (CRD$_3$), the horizontal part of the Chern–Rund derivative coincides with that of Cartan's derivative. All of our assertions follow from these two observations. □

Remark 9.3.21. This covariant derivative was constructed by S.-S. Chern in 1943 using Cartan's calculus of differential forms, and later, independently, by H. Rund in 1951 by means of tensor calculus. The fact that these two constructions are identical was only discovered in 1996 by M. Anastasiei [6].

Another construction of the Chern–Rund derivative can be found in a paper of H.-B. Rademacher [84]: it is given locally as a covariant derivative on the base manifold M (Definition 6.1.1(b)), 'parametrized' by a nowhere vanishing vector field on an open subset of M. Now we reconstruct the covariant derivative operator defined by Rademacher with the help of our Chern–Rund derivative introduced in Proposition and Definition 9.3.20.

Theorem 9.3.22. *Let (M, F) be a Finsler manifold with fundamental tensor g, $\mathcal{U} \subset M$ an open set and U a nowhere vanishing vector field on $\mathcal{U}$. If g_U is defined by*

$$g_U(X, Y) := g(\hat{X}, \hat{Y}) \circ U, \quad X, Y \in \mathfrak{X}(\mathcal{U}),$$

then g_U is a Riemannian metric on $\mathcal{U}$ (Remark 6.1.42(a)), and there exists a unique covariant derivative

$$D^U : \mathfrak{X}(\mathcal{U}) \times \mathfrak{X}(\mathcal{U}) \to \mathfrak{X}(\mathcal{U}), \quad (X, Y) \mapsto D^U_X Y$$

such that the torsion of D^U given by

$$T(X, Y) := D^U_X Y - D^U_Y X - [X, Y], \quad X, Y \in \mathfrak{X}(M)$$

vanishes (cf. (6.1.46)) and it is almost metric *in the sense that*

$$X g_U(Y, Z) = g_U(D^U_X Y, Z) + g_U(Y, D^U_X Z) + \mathcal{C}_\flat(\widehat{D^U_X U}, \hat{Y}, \hat{Z}) \circ U$$

for any vector fields X, Y, Z on $\mathcal{U}$. Furthermore, D^U is related to the Chern–Rund derivative D by

$$D^U_X Y = (D_{X^c}\hat{Y}) \circ U = (D_{X^h}\hat{Y}) \circ U, \quad X, Y \in \mathfrak{X}(\mathcal{U}). \qquad (9.3.10)$$

Proof.   To show the existence, we define D^U by the prescription (9.3.10). Then it is obvious that D^U is additive in both of its variables. To show the $C^\infty(\mathcal{U})$-linearity in the first variable and the derivation property in the second variable, we choose a smooth function f on $\mathcal{U}$. Using the properties of the Chern–Rund derivative, we obtain

$$D^U_{fX}Y = (D_{(fX)^c}\widehat{Y}) \circ U \overset{(4.1.70)}{=} (f^c D_{X^v}\widehat{Y} + f^v D_{X^c}\widehat{Y}) \circ U$$

$$= (f^v \circ U)(D_{X^c}\widehat{Y} \circ U) = f D^U_X Y,$$

$$D^U_X fY = (D_{X^c} f^v \widehat{Y}) \circ U = ((X^c f^v)\widehat{Y} + f^v D_{X^c}\widehat{Y}) \circ U = (Xf)Y + f D^U_X Y.$$

Now we determine the torsion of D^U:

$$T(X,Y) = D^U_X Y - D^U_Y X - [X,Y] = (D_{X^c}\widehat{Y} - D_{Y^c}\widehat{X} - \widehat{[X,Y]}) \circ U$$

$$= (D_{X^h}\widehat{Y} - D_{Y^h}\widehat{X} - \mathbf{j}[X^h,Y^h]) \circ U \overset{(7.9.1)}{=} \mathcal{T}(\widehat{X},\widehat{Y}) \circ U \overset{(CRD_3)}{=} 0,$$

where we also used the fact that D is vertically natural by (CRD_2), thus

$$D_{X^c}\widehat{Y} = D_{X^h}\widehat{Y} + D_{\mathbf{v}X^c}\widehat{Y} = D_{X^h}\widehat{Y}.$$

It remains to show that D^U is almost metric. Starting from the left-hand side, we get

$$X(g_U(Y,Z)) = X(g(\widehat{Y},\widehat{Z}) \circ U).$$

Now we apply Lemma 4.1.49 to the vector fields X and U to obtain

$$U_* \circ X - \kappa \circ X_* \circ U = U_* \circ X - X^c \circ U = [X,U]^v \circ U.$$

When evaluated at a point $p \in \mathcal{U}$, both sides of this equality are tangent vectors to $\mathring{T}\mathcal{U}$. If we apply them to the smooth function $g(\widehat{Y},\widehat{Z})$ on $\mathring{T}\mathcal{U}$, then we find

$$X_p(g(\widehat{Y},\widehat{Z}) \circ U) = (U_* \circ X)(p)(g(\widehat{Y},\widehat{Z}) \circ U)$$

$$= (X^c + [X,U]^v)(U(p))(g(\widehat{Y},\widehat{Z}) \circ U) = (X^c + [X,U]^v)(g(\widehat{Y},\widehat{Z}))(U(p)),$$

therefore

$$X(g(\widehat{Y},\widehat{Z}) \circ U) = ((X^c + [X,U]^v)g(\widehat{Y},\widehat{Z})) \circ U.$$

We turn back to the proof of the almost metric property. Applying (CRD_2) and the definition (9.2.39) of $\mathcal{C}_\flat$ we get

$$X(g_U(Y,Z)) = ((X^h + \mathbf{v}X^c + [X,U]^v)g(\widehat{Y},\widehat{Z})) \circ U$$

$$= g(D_{X^h}\widehat{Y},\widehat{Z}) \circ U + g(\widehat{Y}, D_{X^h}\widehat{Z}) \circ U + \mathcal{C}_\flat(\mathbf{v}X^c + \widehat{[X,U]},\widehat{Y},\widehat{Z}) \circ U$$

$$= g_U(D^U_X Y, Z) + g_U(Y, D^U_X Z) + \mathcal{C}_\flat(\mathbf{v}X^c + \widehat{[X,U]},\widehat{Y},\widehat{Z}) \circ U.$$

It remains to show that $(\mathcal{V}X^{\mathrm{c}} + \widehat{[X,U]}) \circ U = \widehat{D^U_X U} \circ U$. Observe that at each point $p \in \mathcal{U}$,

$$U^{\mathrm{h}}(U(p)) = \mathcal{H}\widehat{U}(U(p)) = \mathcal{H}(U(p), U(p)) = \mathcal{H}\delta(U(p)) = S(v(p)),$$

where S is the canonical spray of (M, F). Thus we have

$$
\begin{aligned}
(\mathcal{V}X^{\mathrm{c}} + \widehat{[X,U]}) \circ U \;&\overset{(7.2.35),(\mathrm{B}\nabla_4)}{=}\; (\mathcal{V}X^{\mathrm{c}} + \nabla_{X^{\mathrm{h}}}\widehat{U} - \nabla_{U^{\mathrm{h}}}\widehat{X}) \circ U \\
&= (\mathcal{V}X^{\mathrm{c}} + \nabla_{X^{\mathrm{h}}}\widehat{U} - \nabla_S \widehat{X}) \circ U \\
&\overset{(7.10.2),\,\text{Corollary }9.3.15(\mathrm{ii})}{=}\; \left(\mathcal{V}(X^{\mathrm{c}} - [S, X^{\mathrm{v}}]) + \nabla_{X^{\mathrm{h}}}\widehat{U} + \frac{1}{2}\mathcal{C}^{\mathrm{h}}(\widehat{X}, \delta)\right) \circ U \\
&\overset{(7.3.7)}{=}\; \left(\nabla_{X^{\mathrm{h}}}\widehat{U} + \frac{1}{2}\mathcal{C}^{\mathrm{h}}(\widehat{X}, U)\right) \circ U \\
&\overset{(9.3.9)}{=}\; (D_{X^{\mathrm{h}}}\widehat{U}) \circ U \overset{(9.3.10)}{=}\; \widehat{D^U_X X} = \widehat{D^U_X U} \circ U.
\end{aligned}
$$

This completes the proof of the existence assertion. As for uniqueness, it can be shown that if D^U is an almost metric, torsion-free covariant derivative operator on $\mathcal{U}$, then it obeys a Koszul-type formula (see (6.1.59)), and hence it is uniquely determined. For details see Rademacher's cited paper [84]. $\square$

Lemma and Definition 9.3.23. *Let (M, F) be a Finsler manifold and $\mathcal{H}$ its canonical connection. There exists a unique covariant derivative operator $D\colon \mathfrak{X}(\mathring{T}M) \times \Gamma(\mathring{\pi}) \to \Gamma(\mathring{\pi})$ on the slit Finsler bundle $\mathring{\pi}$ such that the following conditions are satisfied:*

(HD$_1$) *The vertical part D^{v} of D is metric.*

(HD$_2$) *The vertical torsion of D given by*

$$\mathcal{Q}(\widetilde{X}, \widetilde{Y}) := D_{\mathbf{i}\widetilde{X}}\widetilde{Y} - D_{\mathbf{i}\widetilde{Y}}\widetilde{X} - \mathbf{i}^{-1}[\mathbf{i}\widetilde{X}, \mathbf{i}\widetilde{Y}]$$

vanishes.

(HD$_3$) *The horizontal part D^{h} of D coincides with the horizontal part ∇^{h} of the Finslerian Berwald derivative.*

The operator D is said to be Hashiguchi's covariant derivative *on (M, F). Its action is given by*

$$D_{\mathbf{i}\widetilde{X}}\widetilde{Y} = \nabla_{\mathbf{i}\widetilde{X}}\widetilde{Y} + \frac{1}{2}\mathcal{C}(\widetilde{X}, \widetilde{Y}), \quad D_{\mathcal{H}\widetilde{X}}\widetilde{Y} = \nabla_{\mathcal{H}\widetilde{X}}\widetilde{Y}, \tag{9.3.11}$$

where ∇ is the Finslerian Berwald derivative and $\mathcal{C}$ is the Cartan tensor of (M, F).

Proof. Condition (HD$_3$) determines the horizontal part of Hashiguchi's derivative, and by conditions (HD$_1$) and (HD$_2$), the vertical part of Hashiguchi's derivative coincides with that of Cartan's derivative. These two observations imply the assertion of the lemma. □

Remark 9.3.24. We summarize the discussed notable covariant derivative operators in a tabular form:

name	Berwald	Cartan	Chern – Rund	Hashiguchi
vertical part	∇^{v}	$\nabla^{\mathsf{v}} + \dfrac{1}{2}\mathcal{C}$	∇^{v}	$\nabla^{\mathsf{v}} + \dfrac{1}{2}\mathcal{C}$
horizontal part	∇^{h}	$\nabla^{\mathsf{h}} + \dfrac{1}{2}\mathcal{C}^{\mathsf{h}}$	$\nabla^{\mathsf{h}} + \dfrac{1}{2}\mathcal{C}^{\mathsf{h}}$	∇^{h}
$D^{\mathsf{v}}g$	$\mathcal{C}_{\flat}$	0	$\mathcal{C}_{\flat}$	0
$D^{\mathsf{h}}g$	$\mathcal{C}_{\flat}^{\mathsf{h}}$	0	0	$\mathcal{C}_{\flat}^{\mathsf{h}}$
$\mathfrak{Q}$	0	0	0	0
$\mathfrak{T}$	0	0	0	0

9.4　Isotropic Finsler Manifolds

Throughout this section (M, F) is an n-dimensional Finsler manifold, where $n \geq 2$.

9.4.1　*Characterizations of Isotropy*

In the usual manner, let S be the canonical spray of (M, F), and let $\mathcal{H}$ be the Ehresmann connection generated by S. Then, as we learnt in Theorem 9.3.5, $\mathcal{H}$ is the canonical connection of (M, F). The Jacobi endomorphism $\mathbf{K}$ of (M, S) is given by

$$\mathbf{K}(\widetilde{X}) \overset{(8.2.1)}{:=} \mathcal{V}[S, \mathcal{H}\widetilde{X}], \quad \widetilde{X} \in \Gamma(\overset{\circ}{\pi}).$$

We agreed in Remark 9.3.9(b) that $\mathbf{K}$ is called also the *Jacobi endomorphism of* (M, F). To obtain a 0^+-homogeneous function, we modify (8.2.2), and define the *curvature function of* (M, F) by

$$K^0 := \frac{1}{2E} K \overset{(8.2.2)}{=} \frac{1}{2(n-1)E} \operatorname{tr} \mathbf{K}. \tag{9.4.1}$$

The Finsler manifold (M, F) is called *isotropic* if its canonical spray is isotropic in the sense of Definition 8.2.14. In this case the curvature function K^0 is also mentioned as the *curvature scalar* of (M, F). An isotropic Finsler manifold is said to be of *constant curvature* (resp. *zero curvature*) if the curvature scalar is a constant function (resp. K^0 is identically zero).

Lemma 9.4.1. *With the notation as above,*

$$\nabla^{\mathrm{v}} F \circ \mathbf{K} = \nabla^{\mathrm{v}} E \circ \mathbf{K} = 0. \tag{9.4.2}$$

Proof. For every vector field X on M,

$$\nabla^{\mathrm{v}} F(\mathbf{K}(\widehat{X})) = i\mathbf{K}(\widehat{X})(F) = \mathbf{v}[S, X^{\mathrm{h}}](F)$$
$$= [S, X^{\mathrm{h}}](F) - \mathbf{h}[S, X^{\mathrm{h}}](F) \overset{(\mathrm{CC}_3)}{=} 0,$$

taking into account that the canonical spray of (M, F) is also $\mathcal{H}$-horizontal. Since $\nabla^{\mathrm{v}} E = F \nabla^{\mathrm{v}} F$, the second equality $\nabla^{\mathrm{v}} E \circ \mathbf{K} = 0$ is an immediate consequence of the first. $\qquad \square$

Proposition 9.4.2. *The Jacobi endomorphism of a Finsler manifold is self-adjoint with respect to the fundamental tensor, i.e.,*

$$g(\mathbf{K}(\widetilde{X}), \widetilde{Y}) = g(\widetilde{X}, \mathbf{K}(\widetilde{Y})), \tag{9.4.3}$$

for all $\widetilde{X}, \widetilde{Y} \in \Gamma(\mathring{\pi})$.

Proof. It is sufficient to check (9.4.3) for basic sections $\widehat{X}, \widehat{Y}$. Then

$$g(\mathbf{K}(\widehat{X}), \widehat{Y}) \overset{(9.2.3)}{=} i\mathbf{K}(\widehat{X})(Y^{\mathrm{v}} E) - \mathbf{J}[i\mathbf{K}(\widehat{X}), Y^{\mathrm{h}}](E) = i\mathbf{K}(\widehat{X})(Y^{\mathrm{v}} E)$$

because $i\mathbf{K}(\widehat{X}) \underset{\tau}{\sim} 0$ and $Y^{\mathrm{h}} \underset{\tau}{\sim} Y$ imply by the related vector field lemma that $[i\mathbf{K}(\widehat{X}), Y^{\mathrm{h}}]$ is vertical. Taking into account (9.4.2), we can write

$$i\mathbf{K}(\widehat{X})(Y^{\mathrm{v}} E) = [i\mathbf{K}(\widehat{X}), Y^{\mathrm{v}}](E),$$

therefore

$$g(\mathbf{K}(\widehat{X}), \widehat{Y}) = [i\mathbf{K}(\widehat{X}), Y^{\mathrm{v}}](E). \tag{$*$}$$

Similarly,

$$g(\widehat{X}, \mathbf{K}(\widehat{Y})) \overset{(9.2.3)}{=} X^{\mathrm{v}}(i\mathbf{K}(\widehat{Y})(E)) - \mathbf{J}[X^{\mathrm{v}}, \mathcal{H}\mathbf{K}(\widehat{Y})](E)$$
$$\overset{(9.4.2)}{=} -\mathbf{J}[X^{\mathrm{v}}, \mathcal{H}\mathbf{K}(\widehat{Y})](E).$$

The right-hand side of $(*)$ can also be written in a similar form:

$$0 \overset{(4.1.89)}{=} [\mathbf{J}, Y^\mathsf{v}] \mathcal{H} \mathbf{K}(\widehat{X}) \overset{(3.3.16)}{=} [\mathbf{i}\mathbf{K}(\widehat{X}), Y^\mathsf{v}] - \mathbf{J}[\mathcal{H}\mathbf{K}(\widehat{X}), Y^\mathsf{v}],$$

whence $[\mathbf{i}\mathbf{K}(\widehat{X}), Y^\mathsf{v}] = \mathbf{J}[\mathcal{H}\mathbf{K}(\widehat{X}), Y^\mathsf{v}]$. Thus

$$g(\mathbf{K}(\widehat{X}), \widehat{Y}) - g(\widehat{X}, \mathbf{K}(\widehat{Y})) = (\mathbf{J}[X^\mathsf{v}, \mathcal{H}\mathbf{K}(\widehat{Y})] - \mathbf{J}[Y^\mathsf{v}, \mathcal{H}\mathbf{K}(\widehat{X})])(E).$$

On the other hand,

$$3\mathbf{i}\mathfrak{R}(\widehat{X}, \widehat{Y}) \overset{(8.2.8)}{=} \mathbf{i}\big(\nabla^\mathsf{v}\mathbf{K}(\widehat{X}, \widehat{Y}) - \nabla^\mathsf{v}\mathbf{K}(\widehat{Y}, \widehat{X})\big)$$

$$= \mathbf{i}\big(\nabla_{X^\mathsf{v}}(\mathbf{K}(\widehat{Y})) - \nabla_{Y^\mathsf{v}}(\mathbf{K}(\widehat{X}))\big)$$

$$\overset{(7.10.1)}{=} \mathbf{J}[X^\mathsf{v}, \mathcal{H}\mathbf{K}(\widehat{Y})] - \mathbf{J}[Y^\mathsf{v}, \mathcal{H}\mathbf{K}(\widehat{X})].$$

So it follows that

$$g(\mathbf{K}(\widehat{X}), \widehat{Y}) - g(\widehat{X}, \mathbf{K}(\widehat{Y})) = 3\mathbf{i}\mathfrak{R}(\widehat{X}, \widehat{Y})(E)$$

$$\overset{(7.8.3)}{=} 3([X^\mathsf{h}, Y^\mathsf{h}]E - [X, Y]^\mathsf{h}(E)) \overset{(\mathrm{CC}_3)}{=} 0,$$

as was to be shown. $\qquad\qquad\square$

Lemma 9.4.3. *With our previous notation, a Finsler manifold is isotropic if, and only if, its Jacobi endomorphism is of the form*

$$\mathbf{K} = K^0(2E\mathbf{1} - \nabla^\mathsf{v}E \otimes \widetilde{\delta}) \tag{9.4.4}$$

$$= K\left(\mathbf{1} - \frac{\nabla^\mathsf{v}F}{F} \otimes \widetilde{\delta}\right) = K(\mathbf{1} - \ell_\flat \otimes \ell). \tag{9.4.5}$$

Then, for every section $\widetilde{X} \in \Gamma(\overset{\circ}{\pi})$,

$$\mathbf{K}(\widetilde{X}) = K^0(g(\widetilde{\delta}, \widetilde{\delta})\widetilde{X} - g(\widetilde{\delta}, \widetilde{X})\widetilde{\delta}), \tag{9.4.6}$$

and, fibrewise,

$$\mathbf{K}_u(v) = K^0(u)(g_u(u, u)v - g_u(u, v)u) \tag{9.4.7}$$

$$= K^0(u)(\langle u, u \rangle_F v - \langle u, v \rangle_F u). \tag{9.4.8}$$

Proof. Taking into account Lemma 8.2.15, (M, F) is isotropic if, and only if, its Jacobi endomorphism is of the form

$$\mathbf{K} = K\mathbf{1} + \widetilde{\alpha} \otimes \widetilde{\delta}, \tag{$*$}$$

where $\widetilde{\alpha} \in \mathcal{F}_1^0(\overset{\circ}{T}M)$ is a uniquely determined Finsler 1-form. We show that $\widetilde{\alpha}$ can be expressed in terms of the Finsler function and the curvature scalar.

Indeed, for any vector field X on M,

$$0 \overset{(9.4.2)}{=} \nabla^v F(\mathbf{K}(\widehat{X})) \overset{(*)}{=} \nabla^v F(K\widehat{X}) + \nabla^v F(\widetilde{\alpha}(\widehat{X})\widetilde{\delta})$$
$$= K\nabla^v F(\widehat{X}) + \widetilde{\alpha}(\widehat{X})CF = K\nabla^v F(\widehat{X}) + F\widetilde{\alpha}(\widehat{X}),$$

whence

$$\widetilde{\alpha} = -K\frac{\nabla^v F}{F} \overset{(9.4.1)}{=} -2EK^0\frac{\nabla^v F}{F} = -K^0 \nabla^v E,$$

or

$$\widetilde{\alpha} = -K\frac{\nabla^v F}{F} \overset{(9.2.19)}{=} -\frac{K}{F}\ell_\flat.$$

Taking into account that $\ell := \frac{1}{F}\widetilde{\delta}$, thus we obtain (9.4.4) and (9.4.5).

Evaluating both sides of $(*)$ at $\widetilde{X}$, we find that

$$\mathbf{K}(\widetilde{X}) = K^0(2E\widetilde{X} - \nabla^v E(\widetilde{X})\widetilde{\delta}) \overset{(9.2.9),(9.2.13)}{=} K^0(g(\widetilde{\delta},\widetilde{\delta})\widetilde{X} - \theta_g(\widetilde{X})\widetilde{\delta})$$
$$\overset{(9.2.12)}{=} K^0(g(\widetilde{\delta},\widetilde{\delta})\widetilde{X} - g(\widetilde{\delta},\widetilde{X})\widetilde{\delta}),$$

as we claimed. From (9.4.6) we immediately obtain (9.4.7). By the definition of the F-scalar product (see (9.1.16)), (9.4.7) can also be written as (9.4.8).

Conversely, if $\mathbf{K}$ is of the form (9.4.4) or (9.4.5), then it can be immediately seen from the definition that the Finsler manifold is isotropic. $\qquad \square$

Corollary 9.4.4. *A Finsler manifold is isotropic if, and only if, its Jacobi endomorphism is a multiple of the projection tensor $\mathbf{p} \in \mathrm{End}(\Gamma(\mathring{\pi}))$ determined by the fundamental tensor, namely*

$$\mathbf{K} = K\mathbf{p} = 2EK^0\mathbf{p}. \tag{9.4.9}$$

Proof. This is an immediate consequence of the previous lemma and (9.2.25). $\qquad \square$

Corollary 9.4.5. *An isotropic Finsler manifold is of zero curvature if, and only if, it is horizontally flat.*

Proof. This is obvious from Corollary and Definition 8.2.7 and (9.4.9).
$\qquad \square$

Proposition 9.4.6. *A Finsler manifold is isotropic if, and only if, its fundamental affine curvature is of the form*

$$\mathcal{R} = \left(K^0 \nabla^v E + \frac{2}{3}E\nabla^v K^0\right) \wedge \mathbf{p}, \tag{9.4.10}$$

where K^0 is the curvature function given by (9.4.1) and $\mathbf{p}$ is the projection tensor.

Proof. *Necessity.* Suppose that the Finsler manifold (M, F) is isotropic. Then from (9.4.9),

$$\nabla^{\mathrm{v}}\mathbf{K} = \nabla^{\mathrm{v}}(K\mathbf{p}) = (\nabla^{\mathrm{v}}K) \otimes \mathbf{p} + K\nabla^{\mathrm{v}}\mathbf{p}$$

$$\overset{(9.2.25)}{=} (\nabla^{\mathrm{v}}K) \otimes \mathbf{p} + \frac{K}{F^2}\nabla^{\mathrm{v}}F \otimes \nabla^{\mathrm{v}}F \otimes \widetilde{\delta} - \frac{K}{F}\nabla^{\mathrm{v}}(\nabla^{\mathrm{v}}F \otimes \widetilde{\delta}).$$

For any vector fields X, Y on M,

$$\nabla^{\mathrm{v}}(\nabla^{\mathrm{v}}F \otimes \widetilde{\delta})(\widehat{X}, \widehat{Y}) = ((\nabla_{X^{\mathrm{v}}}\nabla^{\mathrm{v}}F) \otimes \widetilde{\delta} + \nabla^{\mathrm{v}}F \otimes \nabla_{X^{\mathrm{v}}}\widetilde{\delta})(\widehat{Y})$$

$$\overset{(6.2.33)}{=} \nabla^{\mathrm{v}}\nabla^{\mathrm{v}}F(\widehat{X}, \widehat{Y})\widetilde{\delta} + \nabla^{\mathrm{v}}F(\widehat{Y})\widehat{X} = ((\nabla^{\mathrm{v}}\nabla^{\mathrm{v}}F) \otimes \widetilde{\delta} + \mathbf{1} \otimes \nabla^{\mathrm{v}}F)(\widehat{X}, \widehat{Y}),$$

therefore

$$\nabla^{\mathrm{v}}\mathbf{K} = (\nabla^{\mathrm{v}}K) \otimes \mathbf{p} + \frac{K}{F^2}\nabla^{\mathrm{v}}F \otimes \nabla^{\mathrm{v}}F \otimes \widetilde{\delta} - \frac{K}{F}(\nabla^{\mathrm{v}}\nabla^{\mathrm{v}}F) \otimes \widetilde{\delta} - \frac{K}{F}\mathbf{1} \otimes \nabla^{\mathrm{v}}F.$$

Thus

$$3\mathfrak{R}(\widehat{X}, \widehat{Y}) \overset{(8.2.8)}{=} \nabla^{\mathrm{v}}\mathbf{K}(\widehat{X}, \widehat{Y}) - \nabla^{\mathrm{v}}\mathbf{K}(\widehat{Y}, \widehat{X})$$

$$= \left(\nabla^{\mathrm{v}}K \wedge \mathbf{p} + \frac{K}{F}(\nabla^{\mathrm{v}}F \wedge \mathbf{1})\right)(\widehat{X}, \widehat{Y})$$

$$\overset{(9.2.25)}{=} \left(\nabla^{\mathrm{v}}K + \frac{K}{F}\nabla^{\mathrm{v}}F\right) \wedge \mathbf{p}(\widehat{X}, \widehat{Y}).$$

Here

$$\nabla^{\mathrm{v}}K + \frac{K}{F}\nabla^{\mathrm{v}}F \overset{(9.4.1)}{=} 2\nabla^{\mathrm{v}}(EK^0) + K^0 F\nabla^{\mathrm{v}}F = 3K^0\nabla^{\mathrm{v}}E + 2E\nabla^{\mathrm{v}}K^0,$$

therefore

$$\mathfrak{R} = \left(K^0\nabla^{\mathrm{v}}E + \frac{2}{3}E\nabla^{\mathrm{v}}K^0\right) \wedge \mathbf{p},$$

as we claimed.

Sufficiency. Suppose that the fundamental affine curvature of (M, F) is of the form (9.4.10). Then for every $\widetilde{X} \in \Gamma(\mathring{\pi})$,

$$\mathbf{K}(\widetilde{X}) \overset{(8.2.3)}{=} \mathfrak{R}(\widetilde{\delta}, \widetilde{X}) \overset{(9.4.10)}{=} \left(2K^0 E + \frac{2}{3}ECK^0\right)\mathbf{p}(\widetilde{X}) = 2EK^0\mathbf{p}(\widetilde{X})$$

(taking into account that $\mathbf{p}(\widetilde{\delta}) = 0$ by Lemma 9.2.8(i)). Thus (M, F) is isotropic by Corollary 9.4.4. $\qquad\square$

9.4.2 The Flag Curvature

Definition and Lemma 9.4.7. Keeping the notation of the preceding subsection, let (M, F) be a Finsler manifold. Choose and fix a vector $u \in \mathring{T}M$, and let $v \in \mathring{T}_{\hat{\tau}(u)}M$ be a further vector such that u and v are linearly independent. Then we say that the pair $(u, \sigma) := (u, \text{span}(u, v))$ is a *flag* in $T_{\hat{\tau}(u)}M$ with *flagpole* u. The real number

$$\mathcal{K}(u, \sigma) := \frac{g_u(\mathbf{K}_u(v), v)}{g_u(u, u)g_u(v, v) - (g_u(u, v))^2} \tag{9.4.11}$$

is well-defined:

(i) the denominator in (9.4.11) is not zero,
(ii) $\mathcal{K}(u, \sigma)$ is independent of the choice of $v \in \sigma$.

The number $\mathcal{K}(u, \sigma)$ is called the *flag curvature* of (u, σ). We also have

(iii) $\mathcal{K}(\lambda u, \sigma) = \mathcal{K}(u, \sigma)$ for every $\lambda \in \mathbb{R}_+^*$,

i.e., 'the flag curvature does not depend on the length of the flagpole'.

Proof. (i) Since u and v are assumed to be linearly independent, and g_u is a positive definite scalar product, the Schwarz inequality (1.4.4) guarantees that the denominator in (9.4.11) never vanishes (see also the comment below (1.4.4)).

(ii) Let $w := \lambda u + \mu v$; $\lambda, \mu \in \mathbb{R}$, $\mu \neq 0$. We show that

$$\frac{g_u(\mathbf{K}_u(w), w)}{g_u(u, u)g_u(w, w) - (g_u(u, w))^2} = \mathcal{K}(u, \sigma).$$

As to the denominators,

$$g_u(u, u)g_u(w, w) - (g_u(u, w))^2 = \mu^2(g_u(u, u)g_u(v, v) - (g_u(u, v))^2).$$

Since

$$\mathbf{K}_u(u) = \mathbf{K}(\tilde{\delta}) \overset{(8.2.1)}{=} (\mathcal{V}[S, S])(u) = 0,$$

we have

$$\mathbf{K}_u(w) = \mathbf{K}_u(\lambda u + \mu v) = \mu \mathbf{K}_u(v).$$

Thus

$$g_u(\mathbf{K}_u(w), w) = g_u(\mu \mathbf{K}_u(v), \lambda u + \mu v) = \lambda \mu g_u(\mathbf{K}_u(v), u) + \mu^2 g_u(\mathbf{K}_u(v), v)$$

$$\overset{(9.4.3)}{=} \lambda \mu g_u(v, \mathbf{K}_u(u)) + \mu^2 g_u(\mathbf{K}_u(v), v) = \mu^2 g_u(\mathbf{K}_u(v), v),$$

therefore $\mathcal{K}(u, \sigma)$ remains invariant under the change $v \rightsquigarrow \lambda u + \mu v$.

(iii) The third assertion is an immediate consequence of (9.1.15) and the 2^+-homogeneity of the Jacobi endomorphism. $\square$

Remark 9.4.8. With the notation as above, choose a vector field X on M such that $X(\mathring{\tau}(u)) = v$ (Lemma 2.2.24(i)). Then $\widehat{X}(u) = (u, v)$, and (9.4.11) can be rewritten as follows:

$$\mathcal{K}(u, \sigma) = \frac{g(\mathbf{K}(\widehat{X}), \widehat{X})}{g(\widetilde{\delta}, \widetilde{\delta})g(\widehat{X}, \widehat{X}) - (g(\widetilde{\delta}, \widehat{X}))^2}(u). \tag{9.4.12}$$

Owing to the lemma, we may assume without loss of generality that the 'transverse edge' $v \in \sigma$ is g_u-orthogonal to u. Then $g(\widetilde{\delta}, \widehat{X})(u) = 0$, and (9.4.12) reduces to

$$\mathcal{K}(u, \sigma) = \frac{g(\mathbf{K}(\widehat{X}), \widehat{X})}{g(\widetilde{\delta}, \widetilde{\delta})g(\widehat{X}, \widehat{X})}(u) = \frac{g(\mathbf{K}(\widehat{X}), \widehat{X})}{2Eg(\widehat{X}, \widehat{X})}(u). \tag{9.4.13}$$

Proposition 9.4.9. *A Finsler manifold* (M, F) *is isotropic if, and only if, for every* $u \in \mathring{T}M$ *and for any two-dimensional subspaces* $\sigma_1, \sigma_2 \subset T_{\mathring{\tau}(u)}M$ *containing* u, *we have*

$$\mathcal{K}(u, \sigma_1) = \mathcal{K}(u, \sigma_2), \tag{$*$}$$

i.e., the flag curvature of (u, σ) *does not depend on the plane* σ *containing* u.

Proof. *Necessity.* Suppose that (M, F) is isotropic. Then

$$\mathbf{K}(\widehat{X}) \overset{(9.4.4)}{=} 2K^0 E\widehat{X} - K^0(X^{\vee}E)\widetilde{\delta}.$$

Let $u \in \mathring{T}M$, and choose a flag

$$(u, \sigma) = (u, \operatorname{span}(u, v)), \quad g_u(u, v) = 0$$

in $T_{\mathring{\tau}(u)}M$. If $X \in \mathfrak{X}(M)$ is such that $\widehat{X}(u) = (u, v)$, then $g(\widetilde{\delta}, \widehat{X})(u) = 0$ and

$$\mathcal{K}(u, \sigma) \overset{(9.4.13)}{=} \frac{g(\mathbf{K}(\widehat{X}), \widehat{X})}{2Eg(\widehat{X}, \widehat{X})}(u)$$

$$= \frac{2K^0 Eg(\widehat{X}, \widehat{X}) - K^0(X^{\vee}E)g(\widetilde{\delta}, \widehat{X})}{2Eg(\widehat{X}, \widehat{X})}(u) = K^0(u),$$

therefore $\mathcal{K}(u, \sigma)$ is independent of the plane $\sigma \subset T_{\mathring{\tau}(u)}M$ containing the flagpole.

Sufficiency. Suppose that $(*)$ is satisfied. Define, by a slight abuse of notation, the function

$$\mathcal{K}: \mathring{T}M \to \mathbb{R}, \quad u \mapsto \mathcal{K}(u) := \mathcal{K}(u, \sigma),$$

where $\sigma \subset T_{\dot{\tau}(u)}M$ is any two-dimensional subspace containing u. If v is a tangent vector in $T_{\dot{\tau}(u)}M$ such that u and v are linearly independent, then

$$\mathcal{K}(u) := \mathcal{K}(u, \sigma) \stackrel{(9.4.11)}{:=} \frac{g_u(\mathbf{K}_u(v), v)}{g_u(u, u)g_u(v, v) - (g_u(u, v))^2},$$

whence

$$g_u\big(\mathbf{K}_u(v) - \mathcal{K}(u)(g_u(u, u)v - g_u(u, v)u), v\big) = 0. \qquad (**)$$

This relation remains valid if $v \in \mathrm{span}(u)$ because $\mathbf{K}(\widetilde{\delta}) = 0$.

Now define a linear endomorphism φ of $T_{\dot{\tau}(u)}M$ by

$$\varphi(v) := \mathbf{K}_u(v) - \mathcal{K}(u)(g_u(u, u)v - g_u(u, v)u), \quad v \in T_{\dot{\tau}(u)}M.$$

It can be checked immediately that φ is self-adjoint with respect to g_u, so the bilinear form

$$(v, w) \in T_{\dot{\tau}(u)}M \times T_{\dot{\tau}(u)}M \mapsto g_u(\varphi(v), w) \in \mathbb{R}$$

is symmetric. Since

$$g_u(\varphi(v), v) \stackrel{(**)}{=} 0 \quad \text{for all } v \in T_{\dot{\tau}(u)}M,$$

it follows by polarization (see (1.4.2)) that

$$g_u(\varphi(v), w) = 0 \quad \text{for all } v, w \in T_{\dot{\tau}(u)}M.$$

From this we conclude by the non-degeneracy of g_u that $\varphi = 0$. Hence

$$\mathbf{K}_u(v) = \mathcal{K}(u)(g_u(u, u)v - g_u(u, v)u)$$

for any fixed $u \in \mathring{T}M$ and any $v \in T_{\dot{\tau}(u)}M$. Therefore, by Lemma 9.4.3, the Finsler manifold (M, F) is isotropic with curvature scalar $\mathcal{K}$. $\qquad \square$

9.4.3 *The Generalized Schur Theorem*

Lemma 9.4.10. *If (M, F) is a Finsler manifold, then*

$$\nabla^{\mathrm{h}}\nabla^{\mathrm{v}}F = 0, \quad \nabla^{\mathrm{h}}\nabla^{\mathrm{v}}E = 0; \qquad (9.4.14)$$

$$\nabla^{\mathrm{h}}\mathbf{p} = \nabla^{\mathrm{h}}\left(1 - \frac{1}{F}\nabla^{\mathrm{v}}F \otimes \widetilde{\delta}\right) = 0. \qquad (9.4.15)$$

Proof. Since $\nabla^{\mathrm{h}}F = 0$ by (CC$_3$), the first relation in (9.4.14) immediately follows from (7.10.17). The second relation in (9.4.14) is a consequence of the first because $\nabla^{\mathrm{v}}E = F\nabla^{\mathrm{v}}F$. Finally, (9.4.15) follows from (CC$_3$), (9.4.14) and (8.1.4). $\qquad \square$

Theorem 9.4.11 (Generalized Schur's Theorem). *Let (M, F) be an at least 3-dimensional connected isotropic Finsler manifold. If the curvature scalar of (M, F) is fibrewise constant, i.e., $\nabla^v K^0 = 0$, then (M, F) is of constant curvature.*

Proof. Since (M, F) is isotropic and $\nabla^v K^0 = 0$, from (9.4.10) we obtain that the fundamental affine curvature of (M, F) can be written in the form

$$\mathcal{R} = K^0(\nabla^v E \otimes \mathbf{p} - \mathbf{p} \otimes \nabla^v E).$$

Step 1. We calculate the h-Berwald differential of $\mathcal{R}$. To do this, let X, Y, Z be arbitrarily chosen vector fields on M. Then

$$\nabla^h(K^0\nabla^v E \otimes \mathbf{p})(\widehat{X}, \widehat{Y}, \widehat{Z}) = (\nabla_{X^h}(K^0\nabla^v E \otimes \mathbf{p}))(\widehat{Y}, \widehat{Z})$$

$$= (X^h K^0)(\nabla^v E \otimes \mathbf{p})(\widehat{Y}, \widehat{Z}) + K^0((\nabla_{X^h}\nabla^v E) \otimes \mathbf{p} + \nabla^v E \otimes \nabla_{X^h}\mathbf{p})(\widehat{Y}, \widehat{Z})$$

$$\overset{(9.4.14),(9.4.15)}{=} (\nabla^h K^0 \otimes \nabla^v E \otimes \mathbf{p})(\widehat{X}, \widehat{Y}, \widehat{Z}).$$

Thus

$$\nabla^h(K^0\nabla^v E \otimes \mathbf{p}) = \nabla^h K^0 \otimes \nabla^v E \otimes \mathbf{p},$$

and, in the same way,

$$\nabla^h(K^0\mathbf{p} \otimes \nabla^v E) = \nabla^h K^0 \otimes \mathbf{p} \otimes \nabla^v E,$$

therefore

$$\nabla^h\mathcal{R} = \nabla^h K^0 \otimes (\nabla^v E \otimes \mathbf{p} - \mathbf{p} \otimes \nabla^v E). \tag{1}$$

Step 2. Taking into account the Bianchi identity (**B5**) in 8.1.3, (1) leads to the following equality:

$$\nabla^h K^0 \otimes \nabla^v E \otimes \mathbf{p} + \nabla^v E \otimes \mathbf{p} \otimes \nabla^h K^0 + \mathbf{p} \otimes \nabla^h K^0 \otimes \nabla^v E$$
$$- \nabla^h K^0 \otimes \mathbf{p} \otimes \nabla^v E - \mathbf{p} \otimes \nabla^v E \otimes \nabla^h K^0 - \nabla^v E \otimes \nabla^h K^0 \otimes \mathbf{p} = 0. \tag{2}$$

Using (1.2.22) and (1.2.23), we calculate the traces of the six terms on the left-hand side separately. For this, let $X, Y \in \mathfrak{X}(M)$.

(a) $i_{\widehat{X}} \operatorname{tr}(\nabla^h K^0 \otimes \nabla^v E \otimes \mathbf{p}) = \operatorname{tr}(\nabla^h K^0 \otimes X^v E \otimes \mathbf{p})$

$$= \nabla^v E(\widehat{X}) \operatorname{tr}(\nabla^h K^0 \otimes \mathbf{p}),$$

therefore

$$\operatorname{tr}(\nabla^h K^0 \otimes \nabla^v E \otimes \mathbf{p}) = \nabla^v E \otimes \operatorname{tr}(\nabla^h K^0 \otimes \mathbf{p}).$$

Similarly,

$$i_{\widehat{Y}} \operatorname{tr}(\nabla^h K^0 \otimes \mathbf{p}) = \operatorname{tr}(\nabla^h K^0 \otimes \mathbf{p}(\widehat{Y})) = \nabla^h K^0(\mathbf{p}(\widehat{Y})) = \mathcal{H}(\mathbf{p}(\widehat{Y}))K^0$$

$$\overset{(9.2.25)}{=} \left(Y^h - \frac{1}{F}(Y^v F)S\right)K^0 = \left(\nabla^h K^0 - \frac{SK^0}{F}\nabla^v F\right)(\widehat{Y}).$$

Thus we get

$$\mathrm{tr}(\nabla^h K^0 \otimes \nabla^v E \otimes \mathbf{p}) = \nabla^v E \otimes \nabla^h K^0 - (SK^0)\nabla^v F \otimes \nabla^v F. \qquad (3)$$

(b) $i_{\widehat{X}} \,\mathrm{tr}(\nabla^v E \otimes \mathbf{p} \otimes \nabla^h K^0) = \mathrm{tr}(\nabla^v E \otimes \mathbf{p}(\widehat{X}) \otimes \nabla^h K^0)$

$$= \mathrm{tr}(\nabla^v E \otimes \nabla^h K^0 \otimes \mathbf{p}(\widehat{X})),$$

$i_{\widehat{Y}} \,\mathrm{tr}(\nabla^v E \otimes \nabla^h K^0 \otimes \mathbf{p}(\widehat{X})) = \mathrm{tr}(\nabla^v E \otimes \nabla^h K^0(\widehat{Y}) \otimes \mathbf{p}(\widehat{X}))$

$$= (Y^h K^0) \,\mathrm{tr}(\nabla^v E \otimes \mathbf{p}(\widehat{X})) = (Y^h K^0)\mathbf{i}\mathbf{p}(\widehat{X})E$$

$$= (Y^h K^0)\left(X^v E - \frac{1}{F}(X^v F)CE\right) = (Y^h K^0)(X^v E - FX^v F) = 0,$$

therefore

$$\mathrm{tr}(\nabla^v E \otimes \mathbf{p} \otimes \nabla^h K^0) = 0.$$

(c) $i_{\widehat{X}} \,\mathrm{tr}(\mathbf{p} \otimes \nabla^h K^0 \otimes \nabla^v E) = \nabla^h K^0(\widehat{X}) \,\mathrm{tr}(\mathbf{p} \otimes \nabla^v E),$

$i_{\widehat{Y}}(\mathbf{p} \otimes \nabla^v E) = \nabla^v E(\widehat{Y}) \,\mathrm{tr}(\mathbf{p}) \overset{\text{Lemma } 9.2.10}{=} (n-1)\nabla^v E(\widehat{Y}),$

whence

$$\mathrm{tr}(\mathbf{p} \otimes \nabla^h K^0 \otimes \nabla^v E) = (n-1)\nabla^h K^0 \otimes \nabla^v E. \qquad (4)$$

(d) The traces of the remaining three terms can be obtained in the same way. The results are as follows:

$$\mathrm{tr}(\nabla^h K^0 \otimes \mathbf{p} \otimes \nabla^v E) = \nabla^h K^0 \otimes \nabla^v E - (SK^0)\nabla^v F \otimes \nabla^v F, \qquad (5)$$

$$\mathrm{tr}(\mathbf{p} \otimes \nabla^v E \otimes \nabla^h K^0) = (n-1)\nabla^v E \otimes \nabla^h K^0, \qquad (6)$$

$$\mathrm{tr}(\nabla^v E \otimes \nabla^h K^0 \otimes \mathbf{p}) = 0.$$

Step 3. We subtract the sum of equalities (3) and (4) from the sum of equalities (5) and (6). Then we find that

$$0 = (n-2)(\nabla^v E \otimes \nabla^h K^0 - \nabla^h K^0 \otimes \nabla^v E) = (n-2)\nabla^v E \wedge \nabla^h K^0.$$

Here $n > 2$ by condition, so we conclude that $\nabla^v E \wedge \nabla^h K^0 = 0$. Evaluating the left-hand side at $(\widehat{X}, \widetilde{\delta})$, we get

$$(SK^0)\nabla^v E(\widehat{X}) - 2E\nabla^h K^0(\widehat{X}) = 0,$$

whence

$$\nabla^h K^0 = \frac{SK^0}{2E}\nabla^v E =: f\nabla^v E. \qquad (7)$$

Now we turn again to our condition $\nabla^v K^0 = 0$. Taking the h-Berwald differential of this relation, we find that

$$0 = \nabla^h \nabla^v K^0(\widehat{Y}, \widehat{X}) \stackrel{(7.10.17)}{=} \nabla^v \nabla^h K^0(\widehat{X}, \widehat{Y}) \stackrel{(7)}{=} \nabla^v (f \nabla^v E)(\widehat{X}, \widehat{Y})$$
$$= \nabla_{X^v}(f \nabla^v E)(\widehat{Y}) = ((X^v f) \nabla^v E + f \nabla_{X^v} \nabla^v E)(\widehat{Y})$$
$$= (\nabla^v f \otimes \nabla^v E + f \nabla^v \nabla^v E)(\widehat{X}, \widehat{Y}),$$

and therefore

$$f \nabla^v \nabla^v E = -\nabla^v f \otimes \nabla^v E. \tag{8}$$

By the ellipticity condition, the fundamental tensor $g = \nabla^v \nabla^v E$ of (M, F) is, in particular (fibrewise) non-degenerate, while, as it can easily be checked, $\nabla^v f \otimes \nabla^v E$ is everywhere degenerate. Thus (8) implies that the function f is identically zero. Hence, from (7), we conclude that $\nabla^h K^0 = 0$. Then the differential of K^0 also vanishes, because, for every $\xi \in \mathfrak{X}(\overset{\circ}{T}M)$,

$$dK^0(\xi) = \xi K^0 = (\mathbf{h}\xi)K^0 + (\mathbf{v}\xi)K^0 = (\nabla^h K^0)(\xi) + (\nabla^v K^0)(\xi) = 0.$$

Since M is assumed to be connected, from $dK^0 = 0$ it follows (see, e.g., 1.24 Theorem in [97]) that K^0 is a constant function. This concludes the proof of the theorem. $\qquad\square$

Remark 9.4.12. The generalization of Schur's classical theorem (see, e.g., [81], pp. 39–40) from Riemannian geometry to Finsler manifolds is due to Berwald [16]. Berwald's formulation and proof uses, of course, the language and technique of traditional tensor calculus. An elegant, index- and variable-free formulation and proof of the generalized Schur theorem was discovered by L. del Castillo [39]. A nice, index-free treatment can also be found in Diaz's Thèse [41].

Discussion 9.4.13 (The Riemannian Case). (A) Assume that (M, g_M) is an n-dimensional Riemannian manifold with $n \geq 2$. Let D be the Levi-Civita derivative on M, and let R be the curvature tensor of D.

(a) The type $(0, 4)$ tensor field $R^\flat$ defined by

$$R^\flat(X, Y, Z, U) := g_M(R(X, Y)Z, U); \quad X, Y, Z, U \in \mathfrak{X}(M) \tag{9.4.16}$$

is called the *Riemannian curvature tensor* of (M, g_M). It has the following symmetry properties:

(i) $R^\flat$ is skew-symmetric in the first two and in the last two entries;
(ii) $R^\flat$ is symmetric between the first two and last two entries, i.e.,

$$R^\flat(X, Y, Z, U) = R^\flat(Z, U, X, Y);$$

(iii) $R^\flat$ satisfies the *algebraic Bianchi identity*

$$\sum_{\text{cyc}(X,Y,Z)} R^\flat(X,Y,Z,U) = 0 \qquad \text{(cf. (6.1.52))};$$

(iv) $R^\flat$ satisfies the *differential Bianchi identity*

$$\sum_{\text{cyc}(X,Y,Z)} (D_X R^\flat)(Y,Z,U_1,U_2) = 0 \qquad \text{(cf. (6.1.50))}.$$

The only novelty here is property (ii). For a proof of this, see, e.g., [81], p. 34.

(b) Choose a point p of M, and denote by $\text{Gr}_2(T_pM)$ the set of all two-dimensional subspaces of T_pM. Further, let $\text{Gr}_2(M) := \bigcup_{p \in M} T_pM$. The set $\text{Gr}_2(T_pM)$ has a natural smooth structure which makes it a smooth manifold of dimension $2(n-2)$ (see, e.g., [69], Example 1.24), and $\text{Gr}_2(M)$ is a fibre bundle over M, but these facts are not necessary for the present discussion.

Given a plane $\sigma \in \text{Gr}_2(T_pM)$ and basis (u,v) of σ, the number

$$\mathcal{K}(\sigma) := \frac{R^\flat(u,v,v,u)}{g_M(u,u)g_M(v,v) - (g_M(u,v))^2} \qquad (9.4.17)$$

is independent of the choice of basis (u,v) for σ, and is called the *sectional curvature* of σ. If, in particular, (u,v) is a $(g_M)_p$-orthonormal basis, then (9.4.17) reduces to $\mathcal{K}(\sigma) = R^\flat(u,v,v,u)$. If the function

$$\mathcal{K} \colon \text{Gr}_2(M) \to \mathbb{R}, \ \sigma \mapsto \mathcal{K}(\sigma)$$

is constant, say $\mathcal{K}(\sigma) = k$ for every $\sigma \in \text{Gr}_2(M)$, then the Riemannian manifold is called *of constant curvature k*. The following result goes back to Riemann:

A Riemannian manifold has constant curvature k if, and only if,

$$R_p(u,v)w = k(g_M(v,w)u - g_M(w,u)v)$$

for every $p \in M$ and for all $u, v, w \in T_pM$.

Now we are in a position to formulate the classical Schur theorem (usually quoted as *Schur lemma*):

Let (M, g_M) be an at least three-dimensional, connected Riemannian manifold. If there exists a function $f \colon M \to \mathbb{R}$ such that

$$\mathcal{K}(\sigma) = f(p) \text{ for all } \sigma \in \text{Gr}_2(T_pM),$$

then f is constant, therefore (M, g_M) has constant curvature.

(**B**) Suppose that $F: TM \to \mathbb{R}$ is a Riemannian Finsler function arising from (M, g_M). We show that then the flag curvature $\mathcal{K}(u, \sigma)$ defined by (9.4.11) reduces to the sectional curvature $\mathcal{K}(\sigma)$ given by (9.4.17), and the generalized Schur theorem yields the classical Schur lemma stated above.

To see this, first we recall that by Lemma 7.15.3,

$$\mathbf{iH}(\widehat{X}, \widehat{Y})\widehat{Z} = (R(X, Y)Z)^{\mathrm{v}} \quad \text{for all } X, Y, Z \in \mathfrak{X}(M).$$

Thus, for any point p in M and any tangent vectors $u, v \in T_pM$,

$$\mathbf{K}_u(v) \stackrel{(8.2.4)}{=} \mathbf{H}_u(v, u, u) = R_p(v, u)u.$$

Therefore,

$$g_u(\mathbf{K}_u(v), v) = g_u(R_p(v, u)u, v) \stackrel{(9.2.37)}{=} (g_M)_p(R_p(v, u)u, v)$$

$$\stackrel{(9.4.16)}{=} R^\flat(v, u, u, v) = R^\flat(u, v, v, u),$$

where in the last step we used property (ii) above.

Now, comparing (9.4.11) with (9.4.17), we conclude that for any plane $\sigma \in \mathrm{Gr}_2(T_pM)$ and any tangent vector $u \in \sigma$, *the flag curvature of (u, σ) coincides with the sectional curvature of σ.* Thus, if there is a function $f: M \to \mathbb{R}$ such that

$$\mathcal{K}(\sigma) = f(p) \quad \text{for all } p \in M, \ \sigma \in \mathrm{Gr}_2(T_pM),$$

then F is isotropic by Proposition 9.4.9, and the flag curvature does not depend on the flagpole. Since in the isotropic case the curvature scalar of (M, F) can be expressed in terms of the flag curvature (see the necessity part in the proof of Proposition 9.4.9), it follows from the generalized Schur theorem that the function f is constant.

Thus for Riemannian Finsler functions the generalized Schur theorem indeed reduces to the classical Schur theorem.

9.5   Geodesics and Distance

9.5.1   *Finslerian Geodesics and Isometries*

Definition 9.5.1. A smooth curve is called a *geodesic* of a Finsler manifold if it is a geodesic of its canonical spray. Then we also use the phrase 'Finslerian geodesic'. By a *Finslerian pregeodesic* we mean a pregeodesic of the canonical spray of the considered Finsler manifold.

Remark 9.5.2. Let (M, F) be a Finsler manifold with canonical spray S, canonical connection $\mathcal{H}$ and induced Berwald derivative ∇.

(a) For a smooth curve $\gamma \colon I \to M$ the following are equivalent:

(i) γ is a Finslerian geodesic, i.e., $\ddot{\gamma} = S \circ \dot{\gamma}$.

(ii) γ is a geodesic of $\mathcal{H}$, i.e., $\mathcal{V} \circ \ddot{\gamma} = 0$, where $\mathcal{V}$ is the vertical mapping associated to $\mathcal{H}$.

(iii) The velocity field of γ is parallel along γ with respect to the canonical connection.

(iv) γ is a geodesic of the Finslerian Berwald derivative ∇, i.e.,

$$\nabla^{\dot{\gamma}}(\widetilde{\delta} \circ \dot{\gamma}) = \nabla^{\dot{\gamma}}(\mathbf{j} \circ \ddot{\gamma}) = 0.$$

Indeed, *(i) is equivalent to (ii)* by Lemma 7.7.3, since $S_{\mathcal{H}} := \mathcal{H}\widetilde{\delta} = S$ by the 2^+-homogeneity of S; *(ii) is equivalent to (iii)* by Corollary 7.7.5, since $\mathcal{H}$ is positive-homogeneous; finally, *(ii) is equivalent to (iv)* by Lemma 7.12.3.

In the next subsection we shall describe the Finsler geodesics as the extremals of a variational problem.

(b) *Given a tangent vector v in TM, there exists a unique maximal Finslerian geodesic $\gamma_v \colon I \to M$ with initial velocity v, i.e., satisfying $\dot{\gamma}_v(0) = v$;* see Proposition and Definition 5.1.29. The local geodesic equations take the familiar form

$$(\gamma^i)'' + 2(G^i \circ \dot{\gamma}) = 0, \quad i \in J_n$$

in the usual local coordinates (cf. (5.1.18)), where the functions G^i are given by (9.2.30).

(c) *Finslerian geodesics have constant Finslerian speed:* if $\gamma \colon I \to M$ is a geodesic of (M, F) then the *Finslerian speed function*

$$F \circ \dot{\gamma} \colon I \to \mathbb{R}, \quad t \mapsto F(\dot{\gamma}(t))$$

is constant. Indeed, F is a first integral for S by Remark 9.2.19.

(d) We show that *a smooth curve $\gamma \colon I \to M$ of nonzero constant Finslerian speed is a Finsler geodesic if, and only if, it satisfies the Euler–Lagrange equations*

$$\frac{d}{dt}\frac{\partial F}{\partial y^i} - \frac{\partial F}{\partial x^i} = 0, \quad i \in J_n$$

along $\dot{\gamma}$, i.e.,

$$\left(\frac{\partial F}{\partial y^i} \circ \dot{\gamma}\right)' - \frac{\partial F}{\partial x^i} \circ \dot{\gamma} = 0, \quad i \in J_n. \tag{9.5.1}$$

Indeed, by Lemma 5.2.14, a smooth curve $\gamma\colon I \to M$ is a Finslerian geodesic if, and only if, it satisfies

$$\left(\frac{\partial E}{\partial y^i} \circ \dot{\gamma}\right)' - \frac{\partial E}{\partial x^i} \circ \dot{\gamma} = 0, \quad i \in J_n, \tag{$*$}$$

where E is the energy of the Finsler manifold. Since

$$\frac{\partial E}{\partial x^i} \circ \dot{\gamma} = \frac{1}{2}\frac{\partial F^2}{\partial x^i} \circ \dot{\gamma} = F \circ \dot{\gamma}\left(\frac{\partial F}{\partial x^i} \circ \dot{\gamma}\right),$$

$$\left(\frac{\partial E}{\partial y^i} \circ \dot{\gamma}\right)' = \left(F \circ \dot{\gamma}\left(\frac{\partial F}{\partial y^i} \circ \dot{\gamma}\right)\right)'$$

$$= (F \circ \dot{\gamma})'\left(\frac{\partial F}{\partial y^i} \circ \dot{\gamma}\right) + F \circ \dot{\gamma}\left(\frac{\partial F}{\partial y^i} \circ \dot{\gamma}\right)',$$

$(*)$ can be written in the form

$$(F \circ \dot{\gamma})'\left(\frac{\partial F}{\partial y^i} \circ \dot{\gamma}\right) + F \circ \dot{\gamma}\left(\left(\frac{\partial F}{\partial y^i} \circ \dot{\gamma}\right)' - \frac{\partial F}{\partial x^i} \circ \dot{\gamma}\right) = 0,$$

so for smooth curves of constant nonzero Finsler speed relations (9.5.1) and $(*)$ are equivalent.

Definition 9.5.3. Let (M, F) and $(N, \bar{F})$ be Finsler manifolds of the same dimension, and let $\varphi\colon M \to N$ be a smooth mapping. We say that φ is a *local Finslerian isometry* if its derivative preserves the Finslerian norms of the tangent vectors, i.e.,

$$F(v) = \bar{F}((\varphi_*)_p(v)) \text{ for all } p \in M,\ v \in T_pM,$$

briefly, if $F = \bar{F} \circ \varphi_*$. A local Finslerian isometry is said be a *Finslerian isometry* (or simply an *isometry*) if it is a diffeomorphism.

Remark 9.5.4. (a) Since the derivatives of a local Finslerian isometry φ preserve the Finslerian norms, they are linear isomorphisms. Thus, by the inverse mapping theorem (Corollary 3.1.21), for each point $p \in M$ there exist open neighbourhoods $\mathcal{U}$ of p and $\mathcal{V}$ of $\varphi(p)$ such that $\varphi \restriction \mathcal{U}\colon \mathcal{U} \to \mathcal{V}$ is a Finslerian isometry. This justifies the term 'local isometry'.

(b) The Finslerian isometries of a Finsler manifold (M, F) onto itself form a group under composition. This group is called the *isometry group* of (M, F), and denoted by $\mathrm{Iso}(M, F)$ or simply $\mathrm{Iso}(F)$. A Finsler manifold is said to be *homogeneous* if its isometry group acts transitively, i.e., for each pair of points p, q in M, there is a $\varphi \in \mathrm{Iso}(F)$ such that $\varphi(p) = q$ (cf. A.3.6).

Proposition 9.5.5. *The isometries of a Finsler manifold are affine transformations: if (M, F) is a Finsler manifold and $\varphi \in \mathrm{Iso}(F)$, then for every Finslerian geodesic γ, the curve $\varphi \circ \gamma$ is also a Finslerian geodesic. Thus*

$$\mathrm{Iso}(F) \text{ is a subgroup of } \mathrm{Aff}(S), \tag{9.5.2}$$

where S is the canonical spray of (M, F).

Proof. By Proposition 5.1.32, (9.5.2) is equivalent to

$$\mathrm{Iso}(F) \text{ is a subgroup of } \mathrm{Aut}(S). \tag{9.5.3}$$

We prove assertion (9.5.3), i.e., that

$$(\varphi_*)_\# S := \varphi_{**} \circ S \circ \varphi_*^{-1} = S$$

for every $\varphi \in \mathrm{Iso}(F)$. For brevity, let $\bar{S} := (\varphi_*)_\# S$. From Lemma 5.1.25 we know that $\bar{S}$ is a spray. We show that

$$\bar{S}F = 0 \text{ and } \bar{S}(Y^\mathsf{v}F) - Y^\mathsf{c}F = 0 \text{ for all } Y \in \mathfrak{X}(M);$$

then Lemma 9.2.20 and the equivalence of (R_1) and (R_2) (Theorem 9.2.22) imply that $\bar{S}$ is the canonical spray of (M, F), and hence $\bar{S} = S$.

(a) *Proof of $\bar{S}F = 0$.* Let $v \in \overset{\circ}{T}M$. Then

$$(\bar{S}F)(v) = \bar{S}(v)(F) = (\varphi_{**} \circ S \circ \varphi_*^{-1})(v)(F) = \varphi_{**}(S_{\varphi_*^{-1}(v)})(F)$$

$$\overset{(3.1.4)}{=} S_{\varphi_*^{-1}(v)}(F \circ \varphi_*) \overset{\varphi \in \mathrm{Iso}(F)}{=} S_{\varphi_*^{-1}(v)}(F)$$

$$= (SF)(\varphi_*^{-1}(v)) \overset{\text{Lemma } 9.2.20}{=} 0.$$

(b) *Proof of '$\bar{S}(Y^\mathsf{v}F) - Y^\mathsf{c}F = 0$ for all $Y \in \mathfrak{X}(M)$'.* Let, for convenience, $X := \varphi_\#^{-1}Y$. Then $Y = \varphi_\# X$,

$$Y^\mathsf{c} = (\varphi_\# X)^\mathsf{c} \overset{(4.1.68)}{=} (\varphi_*)_\# X^\mathsf{c}, \quad Y^\mathsf{v} = (\varphi_\# X)^\mathsf{v} \overset{(4.1.113)}{=} (\varphi_*)_\# X^\mathsf{v},$$

and hence

$$X^\mathsf{c} \underset{\varphi_*}{\sim} Y^\mathsf{c}, \quad X^\mathsf{v} \underset{\varphi_*}{\sim} Y^\mathsf{v}$$

(see Remark 3.1.52(b)). Thus, by Lemma 3.1.49, we have

$$X^\mathsf{c}F = X^\mathsf{c}(F \circ \varphi_*) = (Y^\mathsf{c}F) \circ \varphi_*, \quad X^\mathsf{v}F = X^\mathsf{v}(F \circ \varphi_*) = (Y^\mathsf{v}F) \circ \varphi_*. \tag{$*$}$$

Now we are ready to calculate $\bar{S}(Y^\mathsf{v}F) - Y^\mathsf{c}F$. Replacing F with $Y^\mathsf{v}F$ in (a), we find that

$$\bar{S}(Y^\mathsf{v}F)(v) = S_{\varphi_*^{-1}(v)}(Y^\mathsf{v}F \circ \varphi_*) \overset{(*)}{=} S_{\varphi_*^{-1}(v)}X^\mathsf{v}F = S(X^\mathsf{v}F)(\varphi_*^{-1}(v)).$$

Therefore, for every $v \in \overset{\circ}{T}M$,

$$(\bar{S}(Y^\mathsf{v}F) - Y^\mathsf{c}F)(v) = S(X^\mathsf{v}F)(\varphi_*^{-1}(v)) - Y^\mathsf{c}F(v)$$

$$\overset{(*)}{=} (S(X^\mathsf{v}F) - X^\mathsf{c}F)(\varphi_*^{-1}(v)) = 0,$$

applying again Lemma 9.2.20 and the equivalence of (R_1) and (R_2) in the last step. This completes the proof of the proposition. $\qquad\square$

Remark 9.5.6. We note that the preceding proposition also holds for local Finslerian isometries in the following form:

If (M, F) is a Finsler manifold with canonical spray S, then for any local isometry $\varphi \colon M \to M$ we have $S \underset{\varphi_*}{\sim} S$.

The proof of this slightly more general result does not need any new idea and we leave it to the reader.

Proposition 9.5.7. *Let (M, F) be a Finsler manifold with canonical spray S, and let $\exp$ be the exponential map determined by S. If $\varphi \colon M \to M$ is a local Finslerian isometry, then*

(i) *φ maps geodesics into geodesics;*
(ii) *at any point $p \in M$,*

$$\exp_{\varphi(p)} \circ (\varphi_*)_p = \varphi \circ \exp_p, \tag{9.5.4}$$

i.e., we have the commutative diagram

$$
\begin{array}{ccc}
\widetilde{T_pM} & \xrightarrow{\;\exp_p\;} & M \\
{\scriptstyle (\varphi_*)_p} \downarrow & & \downarrow {\scriptstyle \varphi} \\
\widetilde{T_{\varphi(p)}M} & \xrightarrow[\exp_{\varphi(p)}]{} & M
\end{array}\;,
$$

where $\widetilde{T_pM} := T_pM \cap \widetilde{TM}$, $\widetilde{T_{\varphi(p)}M} := T_{\varphi(p)}M \cap \widetilde{TM}$.

Proof. (i) Let $\gamma \colon I \to M$ be a geodesic of (M, F). Since φ is a local isometry, by the preceding remark we have $\varphi_{**} \circ S = S \circ \varphi_*$. Thus

$$(\varphi \circ \gamma)^{\cdot\cdot} \overset{(3.1.11)}{=} \varphi_{**} \circ \ddot{\gamma} = \varphi_{**} \circ S \circ \dot{\gamma} = S \circ \varphi_* \circ \dot{\gamma} = S \circ (\varphi \circ \gamma)^{\cdot},$$

therefore $\varphi \circ \gamma$ is also a geodesic of (M, F).

(ii) Let $v \in T_pM$, and let γ_v be the maximal geodesic starting at p with initial velocity v. Then, by part (i), $\varphi \circ \gamma_v$ is also a geodesic, starting at $\varphi(p)$ with initial velocity

$$(\varphi \circ \gamma_v)^{\cdot}(0) = (\varphi_*)_{\gamma_v(0)}(\dot{\gamma}_v(0)) = (\varphi_*)_p(v).$$

Thus we have

$$\varphi \circ \exp_p(v) \overset{(5.1.21)}{=} \varphi \circ \gamma_v(1) \overset{(5.1.21)}{=} \exp_{\varphi(p)} \circ (\varphi_*)_p(v),$$

as desired. $\qquad\qquad\qquad\qquad\qquad\qquad\qquad\qquad\qquad\qquad\qquad\square$

Proposition 9.5.8 (Uniqueness of Finslerian Isometries). *Let* (M, F) *be a connected Finsler manifold. If* $\varphi, \psi \in C^{\infty}(M, M)$ *are local Finslerian isometries and there is a point p in M such that $(\varphi_*)_p = (\psi_*)_p$, then $\varphi = \psi$.*

Proof. Let $A := \{a \in M \mid (\varphi_*)_a = (\psi_*)_a\}$. Then A is non-empty because $p \in A$, and A is closed by the the continuity of φ_* and ψ_* (see the proof of Theorem 3.2.5). So it suffices to show that A is open.

Let a be any point in A, and let $\mathcal{U}$ be a normal neighbourhood of a. Then $\exp_a$ is a diffeomorphism of a star-shaped open subset $\widetilde{\mathcal{U}}$ of $T_a M$ onto $\mathcal{U}$. So we can write

$$\varphi \upharpoonright \mathcal{U} \overset{(9.5.4)}{=} \exp_{\varphi(a)} \circ (\varphi_*)_a \circ (\exp_a \upharpoonright \widetilde{\mathcal{U}})^{-1}$$

$$\overset{a \in A}{=} \exp_{\psi(a)} \circ (\psi_*)_a \circ (\exp_a \upharpoonright \widetilde{\mathcal{U}})^{-1} \overset{(9.5.4)}{=} \psi \upharpoonright \mathcal{U}.$$

From this it follows that $(\varphi_*)_q = (\psi_*)_q$ for all $q \in \mathcal{U}$. Thus $\mathcal{U}$ is a subset of A, and hence A is open. $\qquad\square$

9.5.2 The Finslerian Distance

Throughout this subsection, M is an at least two-dimensional *connected* manifold and $F : TM \to \mathbb{R}$ is a Finsler function for M. If $(p, q) \in M \times M$ and $\gamma : [a, b] \to M$ is a piecewise smooth curve segment such that $\gamma(a) = p$, $\gamma(b) = q$, then we say that γ is a *piecewise smooth curve segment from p to q*. The connectedness of M guarantees that from any point p to any point q there is such a curve segment (cf. Proposition 2.1.26).

Definition 9.5.9. Let $\gamma : [a, b] \to M$ be a smooth curve segment. The *Finslerian length* of γ is

$$\ell_F(\gamma) = \int_a^b F \circ \dot{\gamma}.$$

If γ is a piecewise smooth curve segment as in Definition 2.1.24(c), then its Finslerian length is

$$\ell_F(\gamma) := \sum_{i=1}^k \ell_F(\gamma \upharpoonright [t_{i-1}, t_i]).$$

Lemma 9.5.10. *The Finslerian length of a curve segment is invariant under positive reparametrizations.*

Proof. Let $\gamma\colon [a,b] \to M$ be a smooth curve segment, and let θ be a strictly increasing diffeomorphism of $[c,d]$ onto $[a,b]$. Consider the reparametrization $\widetilde{\gamma} := \gamma \circ \theta$ of γ by θ. Then

$$\ell_F(\widetilde{\gamma}) := \int_c^d F \circ \dot{\widetilde{\gamma}} \stackrel{(3.1.10)}{=} \int_c^d F \circ (\theta'(\dot{\gamma} \circ \theta)) \stackrel{(F_2)}{=} \int_c^d \theta'((F \circ \dot{\gamma}) \circ \theta)$$

$$= \int_{\theta(c)}^{\theta(d)} F \circ \dot{\gamma} = \int_a^b F \circ \dot{\gamma} =: \ell_F(\gamma),$$

as was to be shown. $\qquad\qquad\qquad\qquad\qquad\qquad\qquad\qquad\qquad\qquad\quad\square$

Remark 9.5.11. We can extend in a straightforward way the definition of reparametrization from smooth curve segments to piecewise smooth curve segments (see, e.g., [67], p. 93). Then it remains true that *the Finslerian length of a piecewise smooth curve segment does not change under positive reparametrizations.*

Definition and Proposition 9.5.12. Given an ordered pair (p,q) of points in M, let $\Omega(p,q)$ be the set of all piecewise smooth curve segments in M from p to q. The *Finslerian distance* $\varrho(p,q)$ from p to q is the greatest lower bound of $\{\ell_F(\gamma) \in \mathbb{R} \mid \gamma \in \Omega(p,q)\}$, i.e.,

$$\varrho(p,q) := \inf\{\ell_F(\gamma) \in \mathbb{R} \mid \gamma \in \Omega(p,q)\}.$$

Then the function

$$\varrho\colon M \times M \to \mathbb{R}, \ (p,q) \mapsto \varrho(p,q)$$

is a quasi-distance on M, i.e., satisfies conditions (M_1)–(M_3) in Definition and Lemma B.2.1. This quasi-distance is called the *Finslerian distance* on M determined by the Finsler function F.

For a proof we refer to [8], Section 6.2.

Definition 9.5.13. A smooth curve $\gamma\colon I \to M$ is called a *length-minimizing* (or simply a *minimizing*) curve if

$$\varrho(\gamma(a),\gamma(b)) = \ell_F(\gamma \upharpoonright [a,b]) \quad \text{for all } a,b \in I, \ a < b.$$

Proposition 9.5.14. *In a Finsler manifold the geodesics are precisely the extremals of the energy function associated to the Finsler function. Furthermore, a regular smooth curve is a pregeodesic if, and only if, it is an extremal of the Finsler function.*

Proof. The first assertion is immediate from Corollary 5.2.26. To prove the second, consider a regular smooth curve $\gamma\colon I \to M$. Since $E_F :=$ $CF - F = 0$, the Euler–Lagrange one-form along $\dot\gamma$ is given by

$$\mathcal{E}_t(w) = (\omega_F)_{\dot\gamma(t)}(\ddot\gamma(t), w); \quad t \in I, \; w \in T_{\dot\gamma(t)}TM$$

(cf. (5.2.17)). Thus, by Theorem 5.2.25, γ is an F-extremal if, and only if,

$$(\omega_F)_{\dot\gamma(t)}(\ddot\gamma(t), w) = 0 \quad \text{for all } t \in I, \; w \in T_{\dot\gamma(t)}TM.$$

The latter condition can be equivalently stated as follows:

$\ddot\gamma(t)$ *belongs to the nullspace of* $(\omega_F)_{\dot\gamma(t)}$ *for all* $t \in I.$

By Proposition 9.2.21, the nullspace of ω_F is spanned by the Liouville vector field and the canonical spray S of (M, F). Thus

$$\begin{cases} \ddot\gamma \in N(\omega_F)_{\dot\gamma} \text{ if, and only if, there exist functions } f, h \text{ on } I \\ \text{such that } \ddot\gamma = f(S \circ \dot\gamma) + h(C \circ \dot\gamma). \end{cases} \tag{$*$}$$

We show that f is the constant function of value 1 and h is smooth, whenever $\ddot\gamma$ can be expressed in the above form. Observe that

$$C \circ \dot\gamma \overset{(4.1.88)}{=} \mathbf{J} \circ \ddot\gamma \overset{(*)}{=} \mathbf{J} \circ (f(S \circ \dot\gamma) + h(C \circ \dot\gamma)) = f(C \circ \dot\gamma).$$

By the regularity of γ, $\dot\gamma(t) \in \mathring{T}M$ for every $t \in I$, and hence $C(\dot\gamma(t)) \neq 0$. Thus the above relation implies our first claim. We have, furthermore,

$$dF \circ \ddot\gamma \overset{(*)}{=} dF(S) \circ \dot\gamma + h\, dF(C) \circ \dot\gamma \overset{\text{Lemma } 9.2.20}{=} h\, F \circ \dot\gamma.$$

Since the Finsler function is positive and smooth on $\mathring{T}M$, from this we obtain that

$$h = \frac{dF \circ \ddot\gamma}{F \circ \dot\gamma} \in C^\infty(I).$$

Therefore,

$$\begin{cases} \gamma \text{ is an } F\text{-extremal if, and only if, } \ddot\gamma = S \circ \dot\gamma + h(C \circ \dot\gamma) \text{ for} \\ \text{some } h \in C^\infty(I). \end{cases}$$

By Lemma 5.1.34, this finishes the proof. $\qquad\square$

Corollary 9.5.15. *In a Finsler manifold the regular length-minimizing curves are pregeodesics.*

Proof. Let $\gamma\colon I \to M$ be a regular length-minimizing curve. We show that γ is an F-extremal, then the previous proposition implies the claim.

Let $a, b \in I$, $a < b$; and let $\varphi\colon J \times I \to M$ be a proper variation of γ over $[a, b]$. Since γ is length-minimizing, for any longitudinal curve γ_s $(s \in J)$ we have

$$\ell_F(\gamma \restriction [a, b]) \le \ell_F(\gamma_s \restriction [a, b]).$$

Thus the function F_φ defined by (5.2.25) has a minimum at 0, therefore $(F_\varphi)'(0) = 0$. By the arbitrariness of the closed interval $[a, b] \subset I$ and the variation φ, this proves that γ is an F-extremal. $\qquad\square$

Remark 9.5.16. (a) Locally, the converse of Corollary 9.5.15 is also true. Namely, if $\gamma\colon I \to M$ is a pregeodesic of (M, F), then for each $t \in I$ we can find a positive number ε such that

$$\varrho(\gamma(t), \gamma(t + \varepsilon)) = \ell_F(\gamma \restriction [t, t + \varepsilon]).$$

Actually, we have a much stronger result. To formulate it, let exp be the exponential map of the canonical spray of (M, F), and choose a point p in M. Let $\mathcal{U}$ be a normal neighbourhood of p and $\widetilde{\mathcal{U}}$ a star-shaped open neighbourhood of 0_p in $T_p M$ such that $\exp_p \restriction \widetilde{\mathcal{U}}\colon \widetilde{\mathcal{U}} \to \mathcal{U}$ is a diffeomorphism. Then

$$\varrho(p, \exp_p(v)) = F(v) \quad \text{for all } v \in \widetilde{\mathcal{U}}. \tag{9.5.5}$$

The reader may consult with the reference [8] for details.

(b) By Lemma 5.1.41, the exponential map of the canonical spray of (M, F) is of class C^1, and it is smooth outside the zeros. Thus it follows from (9.5.5) that for any fixed point $p \in M$ and any normal neighbourhood $\mathcal{U}$ of p, the function

$$\varrho_p\colon q \in \mathcal{U} \mapsto \varrho_p(q) := \varrho(p, q) \in \mathbb{R}$$

is of class C^1 on $\mathcal{U}$ and smooth on $\mathcal{U} \setminus \{p\}$. We say that ϱ_p is the *distance function from* p.

A further consequence of (9.5.5) is the following result due to Busemann and Mayer [27].

Proposition 9.5.17 (Busemann–Mayer Theorem). *Let ϱ be the Finslerian distance on M determined by the Finsler function F. Given a point $p \in M$, a tangent vector v to M at p and a smooth curve γ in M such that $\dot{\gamma}(0) = v$, we have*

$$F(v) = \lim_{t \to 0^+} \frac{\varrho(p, \gamma(t))}{t}.$$

Proof. Let $\mathcal{U}$ be a normal neighbourhood of p and $\widetilde{\mathcal{U}}$ a star-shaped open neighbourhood of $0_p \in T_pM$ such that $\exp_p \upharpoonright \widetilde{\mathcal{U}} \colon \widetilde{\mathcal{U}} \to \mathcal{U}$ is a diffeomorphism. Since F is positive-homogeneous, it suffices to prove the proposition in the case when v is in $\widetilde{\mathcal{U}}$. Let $\alpha := (\exp_p \upharpoonright \widetilde{\mathcal{U}})^{-1} \circ \gamma$. Then α is a smooth curve in the vector space T_pM. Using the canonical isomorphism

$$\iota \colon T_pM \times T_pM \to TT_pM$$

described in Example C.2.2, we can write $\dot{\alpha} = \iota(\alpha, \alpha')$ (see the proof of Lemma C.3.4). Since

$$\dot{\alpha}(0) = ((\exp_p \upharpoonright \widetilde{\mathcal{U}})^{-1} \circ \gamma)^{\cdot}(0) \overset{(3.1.11)}{=} (((\exp_p)_*)_{0_p})^{-1}(\dot{\gamma}(0))$$

$$\overset{\text{Lemma } 5.1.45}{=} \iota_{0_p}(\dot{\gamma}(0)) = \iota(0_p, v),$$

it follows that $\alpha'(0) = v$. Thus

$$F(v) = F(\alpha'(0)) = F\left(\lim_{t \to 0^+} \frac{1}{t}\alpha(t)\right) \overset{(F_1),(F_2)}{=} \lim_{t \to 0^+} \frac{1}{t}F(\alpha(t))$$

$$\overset{(9.5.5)}{=} \lim_{t \to 0^+} \frac{1}{t}\varrho(p, \exp_p(\alpha(t))) = \lim_{t \to 0^+} \frac{1}{t}\varrho(p, \gamma(t)),$$

as was to be shown. $\qquad\square$

9.5.3 The Myers – Steenrod Theorem

In this subsection we assume all Finsler manifolds to be connected.

Definition 9.5.18. Let (M, F) and $(N, \bar{F})$ be Finsler manifolds with Finslerian distances ϱ and $\bar{\varrho}$, respectively. A mapping $\varphi \colon M \to N$ is called

distance decreasing if $\bar{\varrho}(\varphi(p), \varphi(q)) \le \varrho(p, q)$,
distance preserving if $\bar{\varrho}(\varphi(p), \varphi(q)) = \varrho(p, q)$,

for all $p, q \in M$.

Lemma 9.5.19. *Local Finslerian isometries are distance decreasing, Finslerian isometries are distance preserving mappings.*

Proof. We use the same notation as in the preceding definition. Let $\varphi \colon M \to N$ be a local Finslerian isometry, and let $\gamma \colon I \to M$ be a piecewise smooth curve. Then $\varphi \circ \gamma$ has the same Finslerian length as γ because

$$\bar{F}((\varphi \circ \gamma)^{\cdot}(t)) = \bar{F}\big((\varphi_*)_{\gamma(t)}(\dot{\gamma}(t))\big) = F(\dot{\gamma}(t)),$$

whenever γ is differentiable at $t \in I$. From this it follows that

$$\inf\{\ell_{\bar{F}}(\widetilde{\gamma}) \in \mathbb{R} \mid \widetilde{\gamma} \in \Omega(\varphi(p), \varphi(q))\} \le \inf\{\ell_F(\gamma) \in \mathbb{R} \mid \gamma \in \Omega(p, q)\},$$

as asserted.

If φ is a Finslerian isometry, then both φ and φ^{-1} are distance decreasing, so φ must preserve the Finslerian distance.   □

Lemma and Definition 9.5.20. *Let (M, F) be an n-dimensional Finsler manifold. Given a point p in M, there is a family $(p_i)_{i=1}^n$ of points of M such that the mapping $u_\varrho := (\varrho_{p_1}, \ldots, \varrho_{p_n})$, where ϱ_{p_i} is the distance function from p_i ($i \in J_n$) is a diffeomorphism from an open neighbourhood $\mathcal{U}$ of p onto an open subset of $\mathbb{R}^n$. We say that $(\mathcal{U}, u_\varrho)$ is a* distance chart *and $(\varrho_{p_1}, \ldots, \varrho_{p_n})$ is a* distance coordinate system *for (M, F) at p. Then the points $p_1, \ldots, p_n$ are called the* base points *of the chart (or of the coordinate system).*

Proof. We construct a family $(p_i)_{i=1}^n$ of base points in M and a basis $(v_i)_{i=1}^n$ of $T_p M$ such that

$$\begin{cases} v_j(\varrho_{p_j}) > 0 \text{ for all } j \in J_n, \\ v_j(\varrho_{p_i}) = 0 \text{ if } i < j; \; i, j \in J_n. \end{cases} \tag{$*$}$$

First we show that $v(\varrho_a) > 0$ always holds if a is an emanating point for $v \in T_p M$ (see Remark 5.1.57). Indeed, by Proposition 5.1.56,

$$v = \lambda \dot{\gamma}_w(\delta) \text{ for some } w \in T_a M \setminus \{0\}, \; \lambda, \delta \in \mathbb{R}_+^*,$$

and hence

$$v(\varrho_a) = \lambda \dot{\gamma}_w(\delta)(\varrho_a) \overset{(3.1.9)}{=} \lambda(\varrho_a \circ \gamma_w)'(\delta).$$

Here, for any $s \in I_w$,

$$\varrho_a \circ \gamma_w(s) = \varrho(a, \gamma_w(s)) \overset{(5.1.21)}{=} \varrho(a, \exp_a(sw)) \overset{(9.5.5)}{=} F(sw) = sF(w),$$

therefore $(\varrho_a \circ \gamma_w)'(\delta) = F(w) > 0$.

Now choose any nonzero vector $v_1 \in T_p M$ and let p_1 be an emanating point for v_1. Since $(d\varrho_{p_1})_p(v_1) = v_1(\varrho_{p_1}) > 0$, we have $\dim(\operatorname{Im}(d\varrho_{p_1})_p) = 1$ and hence $\dim(\operatorname{Ker}(d\varrho_{p_1})_p) = n - 1$.

In the next step we select a vector $v_2 \in \operatorname{Ker}(d\varrho_{p_1})_p \setminus \{0\}$ together with an emanating point p_2 for v_2. Then

$$v_2(\varrho_{p_2}) > 0 \text{ and } v_2(\varrho_{p_1}) = (d\varrho_{p_1})_p(v_2) = 0.$$

Furthermore, we have

$$\dim(\operatorname{Ker}(d\varrho_{p_i})_p) = n - 1, \quad i \in \{1, 2\},$$

and the Sylvester inequality (Remark 1.1.14(iii)) gives

$$\dim\left(\operatorname{Ker}(d\varrho_{p_1})_p \cap \operatorname{Ker}(d\varrho_{p_2})_p\right) \geq (n-1) + (n-1) - n = n - 2.$$

We proceed by induction. Let $k \in J_n$, and suppose that we have a sequence $(v_i)_{i=1}^k$ of nonzero tangent vectors in T_pM together with a sequence $(p_i)_{i=1}^k$ of corresponding emanating points such that $v_j(\varrho_{p_i}) = 0$ if $i < j$, $i, j \in J_k$. As above, by the Sylvester inequality,

$$\dim\left(\bigcap_{i=1}^k \operatorname{Ker}(d\varrho_{p_k})_p\right) \geq n - k,$$

so, as long as $k < n$, we can choose a vector

$$v_{k+1} \in \bigcap_{i=1}^k \operatorname{Ker}(d\varrho_{p_k})_p \setminus \{0\},$$

together with an emanating point p_{k+1} for v_{k+1}. Then $v_{k+1}(\varrho_{p_{k+1}}) > 0$ and $v_{k+1}(\varrho_{p_i}) = 0$ for all $i < k + 1$. This proves that there exist families of vectors $(v_i)_{i=1}^n$ in T_pM and points $(p_i)_{i=1}^n$ in M satisfying $(*)$.

Now consider the mapping $u_\varrho := (\varrho_{p_1}, \ldots, \varrho_{p_n})$. Then, by Proposition 5.1.56 and Remark 9.5.16(b), u_ϱ is smooth in an open neighbourhood of p. We claim that $\left(((u_\varrho)_*)_p(v_i)\right)_{i=1}^n$ is a basis of $T_{u_\varrho(p)}\mathbb{R}^n$, and hence $(v_i)_{i=1}^n$ is a basis of T_pM. Let $(e^i)_{i=1}^n$ be the canonical coordinate system on $\mathbb{R}^n$. Since

$$((u_\varrho)_*)_p(v_j)(e^i) = v_j(e^i \circ u_\varrho) = v_j(\varrho_{p_i})$$

for all $i, j, \in J_n$, the basis theorem gives

$$((u_\varrho)_*)_p(v_j) = ((u_\varrho)_*)_p(v_j)(e^i)\left(\frac{\partial}{\partial e^i}\right)_{u_\varrho(p)} = v_j(\varrho_{p_i})\left(\frac{\partial}{\partial e^i}\right)_{u_\varrho(p)}.$$

By $(*)$, the matrix $(v_j(\varrho_{p_i}))$ is invertible, whence our claim. Then the inverse mapping theorem guarantees that there exists an open neighbourhood $\mathcal{U}$ of p such that $u_\varrho \restriction \mathcal{U}$ is a diffeomorphism onto its image. This concludes the proof. $\qquad\square$

Theorem 9.5.21 (Finslerian Myers–Steenrod). *A surjective distance preserving mapping between Finsler manifolds is a Finslerian isometry.*

Proof. Let (M, F) and $(N, \bar{F})$ be Finsler manifolds with Finslerian distances ϱ and $\bar{\varrho}$, respectively. Suppose that $\varphi \colon M \to N$ is a surjective distance preserving mapping. Then, obviously, φ is injective, so $\varphi \in \operatorname{Bij}(M, N)$, and φ^{-1} is also distance preserving.

(a) We show that φ is smooth. Then φ^{-1} is also smooth, so it follows that $\varphi \in \mathrm{Diff}(M, N)$. Choose and fix a point $p \in M$, and let $q := \varphi(p)$. Forward balls (see Definition and Lemma B.2.1) constitute a basis for the topology of M, so we can find a positive number r such that $B_r^+(p)$ is contained in a totally normal neighbourhood of p. Since φ is a distance preserving bijection, we have $\bar{B}_r^+(q) = \varphi(B_r^+(p))$. Let

$$(\bar{\mathcal{U}}, u_{\bar{\varrho}}) = (\bar{\mathcal{U}}, (\bar{\varrho}_{q_1}, \ldots, \bar{\varrho}_{q_n}))$$

be a distance chart for N at q such that the base points $q_1, \ldots, q_n$ are elements of $\bar{B}_r^+(q)$. This can be achieved because for any nonzero tangent vector in $T_q N$ we can find an emanating point in N which is arbitrarily close to q. Let

$$p_i := \varphi^{-1}(q_i), \quad i \in J_n.$$

Then $\{p_1, \ldots, p_n\} \subset B_r^+(p)$, and for every point $a \in \varphi^{-1}(\bar{\mathcal{U}}) \subset M$ we have

$$\varrho_{p_i}(a) = \varrho(p_i, a) = \bar{\varrho}(\varphi(p_i), \varphi(a)) = \bar{\varrho}(q_i, \varphi(a)) = \bar{\varrho}_{q_i} \circ \varphi(a).$$

Therefore,

$$u_{\bar{\varrho}} \circ \varphi = (\bar{\varrho}_{q_1} \circ \varphi, \ldots, \bar{\varrho}_{q_n} \circ \varphi) = (\varrho_{p_1}, \ldots, \varrho_{p_n}).$$

Since $B_r^+(p)$ is contained in a totally normal neighbourhood of p, the function ϱ_{p_i} is smooth on $B_r^+(p) \setminus \{p_i\}$ for each $i \in J_n$. Furthermore, $u_{\bar{\varrho}}$ is a diffeomorphism on a neighbourhood of q, hence φ is smooth at p.

(b) Now we turn to show that $\bar{F} \circ \varphi_* = F$. Let v be any tangent vector to M. Choose a smooth curve $\alpha: I \to M$ such that $\dot{\alpha}(0) = v$. Then $\varphi \circ \alpha$ is a smooth curve in N and $(\varphi \circ \alpha)'(0) = \varphi_*(v)$. So, applying Proposition 9.5.17 twice, we obtain

$$F(v) = \lim_{t \to 0^+} \frac{1}{t} \varrho(\alpha(0), \alpha(t)) = \lim_{t \to 0^+} \frac{1}{t} \bar{\varrho}(\varphi(\alpha(0)), \varphi(\alpha(t))) = \bar{F}(\varphi_*(v)).$$

This concludes the proof. $\qquad\square$

Remark 9.5.22. In the Riemannian case, Theorem 9.5.21 is due to Myers and Steenrod [77]. The original proof was simplified by several authors; for an elegant recent treatment see Petersen's book [81], Ch. 5.10. The Finslerian version of the theorem is due to Brickell [19], see also [40]. Our proof is based on the ideas of Petersen's book.

9.6 Projective Relatedness Again

Theorem 9.6.1 (A. Rapcsák). *Suppose that* $(M, \bar{F})$ *is a Finsler manifold with canonical spray* $\bar{S}$. *Let* S *be a further spray over* M. *Then the following assertions are equivalent:*

(i) *The sprays* S *and* $\bar{S}$ *have the same pregeodesics.*

(ii) *The sprays* S *and* $\bar{S}$ *are projectively related, namely*

$$\bar{S} = S - \frac{S\bar{F}}{F}C. \tag{9.6.1}$$

(iii) *The Finsler function* $\bar{F}$ *satisfies one, and hence all, of the Rapcsák equations* (R_1)–(R_8) *with respect to* S.

Proof. Assertions (i) and (ii) are equivalent by Proposition 8.4.6. Note that if S and $\bar{S}$ are projectively related, say $\bar{S} = S - 2PC$ $(P \in C^\infty(\mathring{T}M))$ then

$$0 \stackrel{\text{Remark } 9.2.19}{=} \bar{S}\bar{F} = S\bar{F} - 2PC(\bar{F}) = S\bar{F} - 2P\bar{F},$$

whence $P = \frac{1}{2}\frac{\bar{S}\bar{F}}{F}$, so we have (9.6.1).

Now we recall that, in view of Proposition 9.2.21, the nullspace of $\omega_{\bar{F}} = d\mathcal{L}_J\bar{F}$ is span$(C, \bar{S})$. Thus assertion (ii) holds if, and only if, $i_S d\mathcal{L}_J\bar{F} = 0$, i.e., if $\bar{F}$ satisfies the Rapcsák equation (R_1) with respect to S. This concludes the proof. $\qquad\square$

Lemma and Definition 9.6.2. *Let* (M, F) *be a Finsler manifold,* β *a one-form on* M, *and consider the function* $\bar{\beta}$ *determined by* β *according to* (7.4.6). *Suppose that*

$$\{|\bar{\beta}(v)| \in \mathbb{R} \mid v \in TM, \ F(v) = 1\} \subset [0, 1[. \tag{9.6.2}$$

Then $\bar{F} := F + \bar{\beta}$ *is again a Finsler function, and the change from* F *to* $\bar{F}$ *is called a* Randers change. *If, in particular,* F *is a Riemannian Finsler function, then we say that* $\bar{F}$ *is a* Randers function *and* $(M, \bar{F})$ *is a* Randers manifold.

Proof. We show that $\bar{F}$ satisfies conditions (F_1)–(F_4). The first two of them hold automatically: the function $\bar{\beta}$ is smooth, and, as we have already seen, 1^+-homogeneous (Lemma and Definition 7.4.2).

To verify (F_3), let $v \in \mathring{T}M$. Then $F(v) > 0$, $\frac{1}{F(v)}v \in F^{-1}(1)$, so

$$\bar{F}\left(\frac{1}{F(v)}v\right) = 1 + \bar{\beta}\left(\frac{1}{F(v)}v\right) \stackrel{(9.6.2)}{>} 0.$$

Multiplying both sides by $F(v)$, and using the 1^+-homogeneity of $\bar{F}$, we get that $\bar{F}(v) > 0$.

We showed in the proof of Proposition 8.4.23 that $\nabla^\mathsf{v}\nabla^\mathsf{v}\bar{\beta} = 0$. Thus $\nabla^\mathsf{v}\nabla^\mathsf{v}\bar{F} = \nabla^\mathsf{v}\nabla^\mathsf{v}F$, which implies by Corollary 9.2.16(iv) that condition (F_4) is also satisfied by $\bar{F}$. $\qquad\square$

Corollary 9.6.3. *Let η be the angular tensor field of the Finsler manifold (M, F) (see (9.2.20)). Then the tensor field $F^{-1}\eta$ is invariant under any Randers change.*

Proof. Indeed, $F^{-1}\eta \overset{(9.2.21)}{=} \nabla^\mathsf{v}\nabla^\mathsf{v}F$, and we have just seen that $\nabla^\mathsf{v}\nabla^\mathsf{v}F$ remains invariant under Randers changes. $\qquad\square$

Proposition 9.6.4 (M. Hashiguchi and Y. Ichijyō [57]). *Let $(M, \bar{F})$ be a Finsler manifold, obtained from a Finsler manifold (M, F) by a Randers change*

$$\bar{F} = F + \bar{\beta}; \quad \bar{\beta} := i_S\beta^\mathsf{v}, \quad \beta \in \mathfrak{X}^*(M).$$

Then $(M, \bar{F})$ and (M, F) have the same pregeodesics if, and only if, the one-form β is closed, i.e., $d\beta = 0$.

Proof. Let S be the canonical spray of (M, F). By Rapcsák's theorem 9.6.1, (M, F) and $(M, \bar{F})$ have common pregeodesics if, and only if, $\bar{F}$ satisfies a Rapcsák equation with respect to S; e.g., equation (R_1). Since in our case

$$i_S d\mathcal{L}_\mathbf{J}\bar{F} = i_S d\mathcal{L}_\mathbf{J}F + i_S d\mathcal{L}_\mathbf{J}\bar{\beta} \overset{\text{Proposition } 9.2.21}{=} i_S d\mathcal{L}_\mathbf{J}\bar{\beta} \overset{(7.4.7)}{=} i_S d\beta^\mathsf{v} = i_S(d\beta)^\mathsf{v},$$

(R_1) takes the form $i_S(d\beta)^\mathsf{v} = 0$. If $d\beta = 0$, this evidently holds. Conversely, if $i_S(d\beta)^\mathsf{v} = 0$, then

$$(d\beta)^\mathsf{v} \overset{(7.4.10)}{=} \frac{1}{2}\mathcal{L}_\mathbf{J} i_S(d\beta)^\mathsf{v} = 0,$$

therefore β is closed. $\qquad\square$

Remark 9.6.5. It is reasonable to call a Randers change $\bar{F} = F + \bar{\beta}$ *projective* if (M, F) and $(M, \bar{F})$ have the same pregeodesics. It follows from the above proposition and from the Poincaré lemma (Lemma 3.3.38) that *a Randers change is projective if, and only if, locally it is of the form*

$$\bar{F} = F + \overline{df} \overset{(7.4.9)}{=} F + f^{\mathsf{c}},$$

where f is a smooth function on an open subset of M.

Definition 9.6.6. A Finsler function F on TM is called *projectively reversible* if for any geodesic $\gamma \colon I \to M$ of (M, F), the *reversed curve*

$$\gamma^{\downarrow} \colon -I \to M, \ t \mapsto \gamma^{\downarrow}(t) := \gamma(-t)$$

is a pregeodesic of (M, F).

Lemma 9.6.7. *Let (M, F) be a Finsler manifold with canonical spray S. Then the canonical spray of the Finsler manifold $(M, F^{\downarrow})$ is the reverse* $S^{\downarrow} \overset{(5.1.13)}{:=} -\varrho_{\#}S$ *of S.*

Proof. We use the same argument as in the proof of Proposition 9.5.5: we are going to show that

$$S^{\downarrow}F^{\downarrow} = 0 \text{ and, for every } X \in \mathfrak{X}(M), \ S^{\downarrow}(X^{\mathsf{v}}F^{\downarrow}) - X^{\mathsf{c}}F^{\downarrow} = 0;$$

then Lemma 9.2.20 and the equivalence $(\mathrm{R}_1) \Leftrightarrow (\mathrm{R}_2)$ imply our claim. Notice first that

$$\varrho_{\#}X^{\mathsf{c}} = X^{\mathsf{c}} \text{ and } \varrho_{\#}X^{\mathsf{v}} = -X^{\mathsf{v}}. \tag{$*$}$$

Indeed, for any $v \in TM$ we have

$$\varrho_{\#}X^{\mathsf{c}}(v) = \varrho_* \circ X^{\mathsf{c}}(-v) \overset{(4.1.24)}{=} -1 \boxdot X^{\mathsf{c}}(-v) \overset{(4.1.63)}{=} -1 \boxdot \kappa(X_*(-v))$$

$$\overset{(4.1.21)}{=} \kappa(-X_*(-v)) = \kappa(X_*(v)) =: X^{\mathsf{c}}(v),$$

which proves the first relation in $(*)$. We verify the second relation by a coordinate calculation. Given a chart $(\mathcal{U}, (u^i)_{i=1}^n)$ on M, let $X \underset{(\mathcal{U})}{=} X^i \frac{\partial}{\partial u^i}$.

Then $X^{\mathsf{v}} \overset{(4.1.48)}{\underset{(\mathcal{U})}{=}} (X^i \circ \tau)\frac{\partial}{\partial y^i}$, and we have

$$\varrho_{\#}X^{\mathsf{v}} = \varrho_* \circ X^{\mathsf{v}} \circ \varrho \underset{(\mathcal{U})}{=} \varrho_* \circ \left((X^i \circ \tau \circ \varrho)\left(\frac{\partial}{\partial y^i} \circ \varrho\right)\right)$$

$$= (X^i \circ \tau)\varrho_* \circ \frac{\partial}{\partial y^i} \circ \varrho \overset{(3.1.25)}{=} -(X^i \circ \tau)\frac{\partial}{\partial y^i} \underset{(\mathcal{U})}{=} -X^{\mathsf{v}},$$

as was to be shown.

Now we turn to the point. Taking into account Lemma 3.1.49, we find that

$$S^{\downarrow}F^{\downarrow} = S^{\downarrow}(F \circ \varrho) = -(\varrho_{\#}S)(F \circ \varrho) = -(SF) \circ \varrho,$$

$$X^{\mathsf{c}}F^{\downarrow} = X^{\mathsf{c}}(F \circ \varrho) \overset{(*)}{=} (\varrho_{\#}X^{\mathsf{c}})(F \circ \varrho) = (X^{\mathsf{c}}F) \circ \varrho,$$

$$S^{\downarrow}(X^{\mathsf{v}}F^{\downarrow}) \overset{(*)}{=} -S^{\downarrow}(\varrho_{\#}X^{\mathsf{v}}(F \circ \varrho)) = -S^{\downarrow}((X^{\mathsf{v}}F) \circ \varrho)$$

$$= (\varrho_{\#}S)((X^{\mathsf{v}}F) \circ \varrho) = S(X^{\mathsf{v}}F) \circ \varrho.$$

Thus $SF = 0$ and $S(X^{\mathsf{v}}F) - X^{\mathsf{c}}F = 0$ imply the desired equalities $S^{\downarrow}F^{\downarrow} = 0$ and $S^{\downarrow}(X^{\mathsf{v}}F^{\downarrow}) - X^{\mathsf{c}}F^{\downarrow} = 0$. $\square$

Corollary 9.6.8. *Let (M, F) be a Finsler manifold. The geodesics of $(M, F^\downarrow)$ are precisely the reverses of the geodesics of (M, F), so the following are equivalent:*

  (i) (M, F) *is projectively reversible.*
 (ii) (M, F) *and $(M, F^\downarrow)$ have the same pregeodesics.*
(iii) *The canonical spray of (M, F) is projectively reversible.*

Proof. The statement is an immediate consequence of the previous lemma and Lemma 5.1.36. □

Proposition 9.6.9. *Let (M, F) be a Randers manifold obtained from a Riemannian Finsler manifold (M, F_M) by a Randers change $F = F_M + \bar{\beta}$, $\beta \in \mathfrak{X}^*(M)$. Then the following are equivalent:*

  (i) F *is projectively reversible.*
 (ii) *The one-form β is closed.*
(iii) *The canonical spray of (M, F) is projectively related to an affine spray.*
(iv) *The Douglas tensor of (M, F) vanishes.*

Proof. We have already seen in subsection 8.4.6 that assertions (iii) and (iv) are equivalent. Our next argument will be divided into three steps.
Step 1. We show the equivalence of (i) and (ii). Since F_M is clearly reversible and the function $\bar{\beta}$ is fibrewise linear, the reverse of F is

$$F^\downarrow = F \circ \varrho = F_M \circ \varrho + \bar{\beta} \circ \varrho = F_M - \bar{\beta} = F - 2\bar{\beta},$$

so $F^\downarrow$ can be obtained from F by a Randers change. Then, by Proposition 9.6.4, F and $F^\downarrow$ have the same pregeodesics if, and only if, β is closed. Taking into account Corollary 9.6.8, we infer that (i) and (ii) are equivalent.
Step 2. We prove that (ii) implies (iii).

Let S be the canonical spray of (M, F). If the one-form β is closed, then F_M and F have the same pregeodesics (Proposition 9.6.4), therefore their canonical sprays are projectively related (Proposition 8.4.6). The canonical spray of F_M is an affine spray because it is the geodesic spray of the Levi-Civita derivative on a Riemannian manifold (see Proposition and Definition 7.5.13 and Corollary 9.3.7). Thus, we conclude that S is projectively related to an affine spray.
Step 3. To complete the proof, we show that (iii) implies (i). Note first that every affine spray is reversible. Indeed, the spray coefficients of an affine spray are completely homogeneous (see (5.1.11)), therefore, by Remark 5.1.22(a), our claim follows.

Now suppose that (iii) is satisfied. Let S_0 be an affine spray projectively related to S, and let γ be a pregeodesic of S. Then γ is a pregeodesic also of S_0. However, since S_0 is reversible, the reverse curve $\gamma^{\downarrow}$ is also a pregeodesic of S_0 by Lemma 5.1.36. Thus $\gamma^{\downarrow}$ is a pregeodesic of S by our condition, therefore F is projectively reversible. $\qquad\square$

Remark 9.6.10. *Every reversible Randers manifold is a Riemannian manifold.* Namely, if (M, F) is a Randers manifold, where $F = F_M + \bar{\beta}$, then $F = F_M$. Indeed, by the reversibility,

$$F = F \circ \varrho = F_M \circ \varrho + \bar{\beta} \circ \varrho = F_M - \bar{\beta} = F - 2\bar{\beta},$$

whence $\bar{\beta} = 0$.

9.7 Projective Metrizability

Who of us would not be glad to lift the veil behind which the future lies hidden; to cast a glance at the next advances of our science and at the secrets of its development during future centuries?

David Hilbert

Definition 9.7.1. A spray is called *metrizable in a broad sense* or *projectively Finsler* if there is a Finsler function whose canonical spray is projectively related to the given spray.

Having this metrizability concept, we can formulate the following metrizability problem:

Rapcsák's Problem. *Determine all Finsler functions whose canonical spray belongs to the projective class of a given spray.*

This is, in fact, an old problem, which goes back to the second half of the 19th century. For surfaces, it was solved by G. Darboux [37]. He proved that (locally) *every spray over a two-dimensional manifold is projectively Finsler*, and he gave an explicit construction of the solution functions. In 1995, M. Matsumoto [73] extended Darboux's result to the so-called parametric case.

By Theorem 9.6.1, the Rapcsák equations provide a system of partial differential equations for the Finsler function to be determined. Here we present a useful reformulation of the problem.

Lemma 9.7.2. *A spray S over a manifold M is projectively Finsler if, and only if, there is a Finsler one-form $\tilde{\lambda} \in \mathcal{F}_1^0(\mathring{T}M)$ such that*

(i) $\nabla_C \widetilde{\lambda} = 0$, *i.e.*, $\widetilde{\lambda}$ *is* 0^+*-homogeneous;*

(ii) $\nabla^v \widetilde{\lambda}(\widetilde{X}, \widetilde{Y}) = \nabla^v \widetilde{\lambda}(\widetilde{Y}, \widetilde{X})$ *for all* $\widetilde{X}, \widetilde{Y} \in \Gamma(\mathring{\pi})$;

(iii) $\nabla^h \widetilde{\lambda}(\widetilde{X}, \widetilde{Y}) = \nabla^h \widetilde{\lambda}(\widetilde{Y}, \widetilde{X})$ *for all* $\widetilde{X}, \widetilde{Y} \in \Gamma(\mathring{\pi})$, *where* ∇^h *is the h-Berwald derivative induced by* S;

(iv) $\widetilde{\lambda}(\widetilde{\delta})(v) > 0$ *if* $v \neq 0$;

(v) *the rank of* $\nabla^v \nabla^v (\widetilde{\lambda}(\widetilde{\delta}))$ *is* $n - 1$, *where* $n = \dim M$.

Proof. Suppose first that S is projectively Finsler, and let $\widetilde{F}$ be a Finsler function whose canonical spray is projectively related to S. Consider the Finsler one-form

$$\widetilde{\ell}_\flat \stackrel{(9.2.19)}{=} \nabla^v \widetilde{F}.$$

Then, by Lemma 6.2.22, $\widetilde{\ell}_\flat$ is 0^+-homogeneous, since $\widetilde{F}$ is 1^+-homogeneous. We know from Lemma 6.2.28(ii) that the vertical Hessian $\nabla^v \widetilde{\ell}_\flat = \nabla^v \nabla^v \widetilde{F}$ is symmetric, so (ii) is also satisfied. Relation (iii) is just the Rapcsák equation (R_7), so it holds by Theorem 9.6.1. Since

$$\widetilde{\ell}_\flat(\widetilde{\delta}) = \nabla^v \widetilde{F}(\widetilde{\delta}) = C\widetilde{F} = \widetilde{F},$$

the function $\widetilde{\ell}_\flat(\widetilde{\delta})$ is positive on $\mathring{T}M$ by the positive definiteness of $\widetilde{F}$. Finally, (v) is satisfied by Corollary 9.2.16(iv).

Conversely, suppose that there is a one-form $\widetilde{\lambda} \in \mathcal{F}_1^0(\mathring{T}M)$ with the properties (i)–(v). Define a function on $\mathring{T}M$ by $\widetilde{F}_0 := \widetilde{\lambda}(\widetilde{\delta})$. Then, obviously, $\widetilde{F}_0$ is smooth. Since

$$C\widetilde{F}_0 = C(\widetilde{\lambda}(\widetilde{\delta})) = (\nabla_C \widetilde{\lambda})(\widetilde{\delta}) + \widetilde{\lambda}(\nabla_C \widetilde{\delta}) \stackrel{(i),(6.2.33)}{=} \widetilde{\lambda}(\widetilde{\delta}) =: \widetilde{F}_0,$$

$\widetilde{F}_0$ is 1^+-homogeneous. Thus, by Lemma 4.2.11, $\widetilde{F}_0$ can be continuously extended to a function $\widetilde{F}$ on TM. Then $\widetilde{F}$ satisfies (F_1) and (F_2). It is positive definite by (iv), and satisfies the ellipticity condition (F_4) by (v) and by Corollary 9.2.16(iv). Therefore $\widetilde{F}$ is a Finsler function. Observe that

$$\nabla^v \widetilde{F} = \widetilde{\lambda}. \tag{$*$}$$

Indeed, for every vector field X on M,

$$\nabla^v \widetilde{F}(\widehat{X}) = X^v \widetilde{F} = X^v(\widetilde{\lambda}(\widetilde{\delta})) = (\nabla_{X^v} \widetilde{\lambda})(\widetilde{\delta}) + \widetilde{\lambda}(\nabla_{X^v} \widetilde{\delta})$$

$$= \nabla^v \widetilde{\lambda}(\widehat{X}, \widetilde{\delta}) + \widetilde{\lambda}(\nabla_{X^v} \widetilde{\delta}) \stackrel{(ii)}{=} \nabla^v \widetilde{\lambda}(\widetilde{\delta}, \widehat{X}) + \widetilde{\lambda}(\nabla_{X^v} \widetilde{\delta})$$

$$= \nabla_C \widetilde{\lambda}(\widehat{X}) + \widetilde{\lambda}(\nabla_{X^v} \widetilde{\delta}) \stackrel{(i)}{=} \widetilde{\lambda}(\nabla_{X^v} \widetilde{\delta}) \stackrel{(6.2.33)}{=} \widetilde{\lambda}(\widehat{X}),$$

whence $(*)$. We verify finally that $\widetilde{F}$ satisfies the Rapcsák equation (R_7) with respect to S. For any sections $\widetilde{X}, \widetilde{Y} \in \Gamma(\mathring{\pi})$,

$$\nabla^h \nabla^v \widetilde{F}(\widetilde{X}, \widetilde{Y}) \stackrel{(*)}{=} \nabla^h \widetilde{\lambda}(\widetilde{X}, \widetilde{Y}) \stackrel{(iii)}{=} \nabla^h \widetilde{\lambda}(\widetilde{Y}, \widetilde{X}) = \nabla^h \nabla^v \widetilde{F}(\widetilde{Y}, \widetilde{X}),$$

as was to be shown. $\qquad\qquad\qquad\qquad\qquad\qquad\qquad\qquad\qquad\qquad\qquad\square$

Remark 9.7.3. For the investigation of (formal) integrability of the system of partial differential equations (i)–(iii) (in a little different setting) we refer to [22].

Definition 9.7.4. A Finsler manifold is called *projectively flat* if its canonical spray has this property.

Theorem 9.7.5. *A Finsler manifold (M, F) is projectively flat if, and only if, at every point of M there is a chart $(\mathcal{U}, (u^i)_{i=1}^n)$ on M with induced chart $(\tau^{-1}(\mathcal{U}), ((x^i)_{i=1}^n, (y^i)_{i=1}^n))$ on TM such that in the induced coordinates F satisfies one (and hence both) of the following equivalent partial differential equations:*

$$\frac{\partial^2 F}{\partial x^i \partial y^j} = \frac{\partial^2 F}{\partial x^j \partial y^i}; \quad i, j \in J_n; \tag{H_1}$$

$$y^k \frac{\partial^2 F}{\partial x^k \partial y^i} = \frac{\partial F}{\partial x^i}; \quad i \in J_n. \tag{H_2}$$

Then $(\mathcal{U}, (u^i)_{i=1}^n)$ is a rectilinear chart for the canonical spray of (M, F).

Proof. Let S be the canonical spray of (M, F), and let S_0 be a flat spray over M. By Corollary 8.5.8, at every point of M there is a chart $(\mathcal{U}, (u^i)_{i=1}^n)$ with induced chart $(\tau^{-1}(\mathcal{U}), ((x^i)_{i=1}^n, (y^i)_{i=1}^n))$ such that $S_0 \underset{(\mathcal{U})}{=} y^i \frac{\partial}{\partial x^i}$. In view of Theorem 9.6.1, S_0 and S are projectively related if, and only if, F satisfies the Rapcsák equations (R_1)–(R_8) with respect to S_0. Now the local form (9.2.32) of (R_7) gives (H_1), and the local form (9.2.33) of (R_1) leads to (H_2). Then

$$S \underset{(\mathcal{U})}{=} y^i \frac{\partial}{\partial x^i} - 2Py^i \frac{\partial}{\partial y^i},$$

where $P \in C^\infty(\mathring{T}M)$ is 1^+-homogeneous, so the chart $(\mathcal{U}, (u^i)_{i=1}^n)$ is rectilinear. Thus the proof is complete. $\square$

Remark 9.7.6. As it is well-known, at the Second International Congress of Mathematicians in Paris in 1900, David Hilbert delivered a talk with the title 'Mathematical Problems' [58]. He listed and commented 23 problems, which became a challenge and inspiration for generations of mathematicians. The *fourth problem*,

'Problem of the straight line as the shortest distance between two points' was immediately related to Hilbert's actual research on the axiomatic foundations of geometry. In Busemann's words [26], the essence of the problem is this: 'Omit from his axioms for the foundations of geometry besides the parallel axioms all those which contain the concept of angle, and replace them by the triangle inequality, which follows from the congruence axiom for triangles (CT).

(1) Determine all geometries satisfying these conditions.
(2) Study the individual ones.

This is not quite Hilbert's formulation because with respect to angles he only omits the CT. The remaining angle axioms have no significant applications without CT.'

To put the problem into a more general context, let us call a distance function

$$d \colon \mathbb{R}^n \times \mathbb{R}^n \to \mathbb{R}, \ (a, b) \mapsto d(a, b)$$

projective if it is continuous with respect to the canonical topology of $\mathbb{R}^n$, and if d is 'additive along the straight lines', i.e., $d(a, b) + d(b, c) = d(a, c)$, where a, b, c are points on a straight line in the given order. Then the 'fourth problem requests, among many other things, a method of construction for the cone of such projective metrics, and this task lies at the heart of the problem' (R. Alexander [3]).

We see, as Álvarez Paiva explained, '...the problem is very vaguely written. At first I thought this was a weak point, but now I think that the strength of the problem is that it admits a number of different formulations' ([100], p. 142). From our viewpoint, an appropriate reformulation of the problem is the following:

Finslerian version of Hilbert's fourth problem. *Construct and study all Finsler functions over $\mathbb{R}^n$ whose geodesics, as points sets, are the straight lines of $\mathbb{R}^n$. More generally, given a manifold M, find and study the projectively flat Finsler functions over M.*

At first sight, this is a sweeping generalization of Hilbert's 'purely geometric' problem. However, the first thorough approach to the problem conceived in this spirit in G. Hamel's thesis [56]. Hamel's work was supervised by Hilbert, immediately after his Paris lecture. This indicates that the reformulation of the problem in differential geometric terms was found relevant by Hilbert himself. In our terminology and notation, Hamel proved (among others) the following result:

Theorem (G. Hamel [56]) *Suppose that F is a Finsler function on $\overset{\circ}{T}\mathbb{R}^2 \cong \mathbb{R}^2 \times (\mathbb{R}^2 \setminus \{0\})$. Let (e^1, e^2) be the canonical coordinate system on $\mathbb{R}^2$, and (x^1, x^2, y^1, y^2) the induced coordinate system on $T\mathbb{R}^2$. The parametrized straight lines of $\mathbb{R}^2$ are the pregeodesics of F if, and only if, F satisfies the partial differential equation*

$$\frac{\partial^2 F}{\partial x^1 \partial y^2} = \frac{\partial^2 F}{\partial x^2 \partial y^1}. \tag{H_1'}$$

We recognize at once that (H_1') is just equation (H_1) in this special situation. Therefore, we call equations (H_1), (H_2) the *Hamel equations*. We note that Hamel solved the differential equation (H_1') (this is not too difficult), so he provided a complete solution of the Finslerian version of Hilbert's problem 4 in two dimensions.

Unfortunately, we have to interrupt and finish this exciting story: it needs a whole book to explain. We mention only that some of the further, main contributors to the field are J. C. Álvarez Paiva [4], H. Busemann, A. V. Pogorelov [82] and Z. I. Szabó [91]. We find the recent papers of M. Crampin and his collaborators [33], [35], [36] also very important.

9.8 Berwald Manifolds

In 1926 Berwald [14] introduced an important special class of Finsler manifolds. In our terms, he called a Finsler manifold an *affinely connected Finsler space* if the Christoffel symbols of the Finslerian Berwald derivative 'depend only on the position'. To translate this definition into a more rigorous form, consider a Finsler manifold (M, F), and choose a chart $(\mathcal{U}, (u^i)_{i=1}^n)$ on M with induced chart $(\tau^{-1}(\mathcal{U}), ((x^i)_{i=1}^n, (y^i)_{i=1}^n))$ on TM. If the spray coefficients of the canonical spray of (M, F) are the functions G^i, then the Christoffel symbols of the Finslerian Berwald derivative are $G^i_{jk} = \frac{\partial^2 G^i}{\partial y^j \partial y^k}$. The condition

'the G^i_{jk}'s depend only on the position'

means that

'the Christoffel symbols G^i_{jk} are vertical lifts'

or, equivalently, $\frac{\partial G^i_{jk}}{\partial y^l} = 0$ $(i, j, k, l \in J_n)$. Since the functions $B^i_{jkl} \overset{(7.13.4)}{=} \frac{\partial G^i_{jk}}{\partial y^l}$ are just the components of the Berwald curvature of (M, F), we can also say that the Berwald curvature of (M, F) vanishes.

Although the name 'affinely connected Finsler space' is felicitous and expressive, we shall instead use the term 'Berwald manifold', which is commonly accepted today. So our official definition is the following:

> A Finsler manifold is called a Berwald manifold *if its Berwald curvature vanishes.*

In the previous chapters we obtained several properties equivalent to the vanishing of the Berwald curvature of an Ehresmann connection or of a spray. These also yield different characterizations of Berwald manifolds. However, we begin our overview with a result which has no counterpart for spray manifolds.

Proposition 9.8.1. *A Finsler manifold is a Berwald manifold if, and only if, the h-Berwald differential of its Cartan tensor vanishes.*

Proof. Let (M, F) be a Finsler manifold with canonical spray S, Cartan tensor $\mathcal{C}_\flat := \nabla^v g$ and Landsberg tensor $\mathbf{L} := \nabla^h g$. Using the mixed Ricci formula (7.13.9), for any vector fields X, Y, Z, U on M we have

$$\nabla^h \mathcal{C}_\flat(\widehat{Y}, \widehat{X}, \widehat{Z}, \widehat{U}) = \nabla^h \nabla^v g(\widehat{Y}, \widehat{X}, \widehat{Z}, \widehat{U})$$
$$= \nabla^v \nabla^h g(\widehat{X}, \widehat{Y}, \widehat{Z}, \widehat{U}) + g(\mathbf{B}(\widehat{X}, \widehat{Y})\widehat{Z}, \widehat{U}) + g(\mathbf{B}(\widehat{X}, \widehat{Y})\widehat{U}, \widehat{Z}). \quad (*)$$

If (M, F) is a Berwald manifold, then $(*)$ reduces to

$$\nabla^h \mathcal{C}_\flat(\widehat{Y}, \widehat{X}, \widehat{Z}, \widehat{U}) = \nabla^v \nabla^h g(\widehat{X}, \widehat{Y}, \widehat{Z}, \widehat{U}) = \nabla^v \mathbf{L}(\widehat{X}, \widehat{Y}, \widehat{Z}, \widehat{U})$$
$$\overset{(9.3.4)}{=} \nabla^v(\nabla^v E \circ \mathbf{B})(\widehat{X}, \widehat{Y}, \widehat{Z}, \widehat{U}) = 0,$$

whence $\nabla^h \mathcal{C}_\flat = 0$.

Conversely, suppose that $\nabla^h \mathcal{C}_\flat = 0$. Then

$$0 = \nabla^h \mathcal{C}_\flat(\widetilde{\delta}) = \nabla_{\mathcal{H}\widetilde{\delta}} \mathcal{C}_\flat = \nabla_S \mathcal{C}_\flat \overset{(9.3.6)}{=} -\mathbf{L} = -\nabla^h g,$$

so from $(*)$ we obtain that

$$g(\mathbf{B}(\widehat{X}, \widehat{Y})\widehat{Z}, \widehat{U}) + g(\mathbf{B}(\widehat{X}, \widehat{Y})\widehat{U}, \widehat{Z}) = 0.$$

Similarly,

$$g(\mathbf{B}(\widehat{X}, \widehat{Z})\widehat{Y}, \widehat{U}) + g(\mathbf{B}(\widehat{X}, \widehat{Z})\widehat{U}, \widehat{Y}) = 0,$$
$$-g(\mathbf{B}(\widehat{X}, \widehat{U})\widehat{Z}, \widehat{Y}) - g(\mathbf{B}(\widehat{X}, \widehat{U})\widehat{Y}, \widehat{Z}) = 0.$$

Adding the three equalities and using the total symmetry of the Berwald curvature, we find that $2g(\mathbf{B}(\widehat{X}, \widehat{Y})\widehat{Z}, \widehat{U}) = 0$. By the non-degeneracy of g this implies that $\mathbf{B} = 0$.

Thus we have proved our statement for the type $(0,3)$ Cartan tensor $\mathcal{C}_\flat$. From this it follows easily that the assertions $\mathbf{B} = 0$ and $\nabla^h \mathcal{C} = 0$ are also equivalent. We leave this part to the reader. $\qquad\square$

Now we turn to the results which are strongly related to our characterizations of linear Ehresmann connections and affine sprays.

Proposition 9.8.2. *Let (M, F) be a Finsler manifold. The following are equivalent:*

($\mathbf{B}_1$) *The Berwald curvature of (M, F) vanishes, i.e., (M, F) is a Berwald manifold.*

($\mathbf{B}_2$) *The canonical spray of (M, F) is an affine spray.*

($\mathbf{B}_3$) *The canonical connection of (M, F) is a linear connection.*

($\mathbf{B}_4$) *There exists a torsion-free covariant derivative on M which has the same geodesics as (M, F).*

($\mathbf{B}_5$) *There exists a torsion-free covariant derivative D on M such that the D-parallel translations preserve the Finsler norms of tangent vectors to M, i.e., for every smooth curve $\gamma\colon I \to M$ and any scalars $s, t \in I$,*

$$F_{\gamma(s)} \circ P_t^s(\gamma) = F_{\gamma(t)},$$

where $P_t^s(\gamma)\colon T_{\gamma(t)}M \to T_{\gamma(s)}M$ is the parallel translation along γ with respect to D.

Proof. In the usual manner, let S, $\mathcal{H}$ and ∇ denote the canonical spray, the canonical connection and the Finslerian Berwald derivative on (M, F), respectively.

(a) Assertions ($\mathbf{B}_1$) and ($\mathbf{B}_2$) are equivalent by Lemma 8.4.27, while the equivalence of ($\mathbf{B}_1$) and ($\mathbf{B}_3$) is an immediate consequence of Proposition 7.15.1.

(b) By Lemma 9.3.4 and Corollary 7.6.6, a covariant derivative D on M satisfies the compatibility condition formulated in ($\mathbf{B}_5$) if, and only if, the Ehresmann connection induced by D is compatible with F. Since every linear, torsion-free Ehresmann connection is induced by a torsion-free covariant derivative (Proposition 7.15.4), it follows that ($\mathbf{B}_3$) and ($\mathbf{B}_5$) are equivalent.

(c) We show that ($\mathbf{B}_2$) implies ($\mathbf{B}_4$). Referring to Proposition 7.15.4 again, let D be the torsion-free covariant derivative which induces $\mathcal{H}$. Consider a smooth curve $\gamma\colon I \to M$. Applying Lemma 7.7.3 and Corollary 7.6.6, we obtain that the following assertions are equivalent:

(i) γ is a geodesic of S.

(ii) $\dot{\gamma}$ is $\mathcal{H}$-parallel along γ.

(iii) $\dot{\gamma}$ is parallel with respect to D.

(iv) γ is a geodesic of D.

Therefore, the covariant derivative D has the same geodesics as (M, F).

(d) To complete the proof, we show that (B$_4$) implies (B$_2$). Let D be a covariant derivative on M which has common geodesics with (M, F), and let $\mathcal{H}^D$ be the Ehresmann connection induced by D. Then, by Proposition and Definition 7.5.13, the associated semispray $\bar{S} := \mathcal{H}^D(\tilde{\delta})$ of $\mathcal{H}^D$ is an affine spray, which has the same geodesics as D. Thus S and $\bar{S}$ have common geodesics; hence $S = \bar{S}$, so S is an affine spray.   □

Remark 9.8.3. It turns out from the above proof that the covariant derivative D appearing in (B$_5$) and (B$_4$) is unique: it is just the torsion-free covariant derivative which induces the canonical connection of (M, F). We say that D is the *base covariant derivative* of the Berwald manifold (M, F). We emphasize that *Berwald manifolds are the Finsler manifolds whose base manifold admits a torsion-free covariant derivative compatible with the Finsler function.* Propositions 7.15.1, 7.15.4 and 5.1.44 provide further conditions for characterizing Berwald manifolds among Finsler manifolds.

Corollary 9.8.4. *Every Riemannian Finsler manifold is a Berwald manifold. The base covariant derivative is the Levi-Civita derivative of the Riemannian metric from which the Finsler function arises.*

Proof.   This is an immediate consequence of Corollary 9.3.7 and the proposition above.   □

Lemma 9.8.5. *Let (M, D) be an affinely connected manifold, and consider the geodesic spray of D. Given a point p in M, let $\mathcal{U}$ be a normal neighbourhood of p. Then for any $b \in T_pM$, there exist a positive real number t_0 and an open neighbourhood $\mathcal{U}_0 \subset \mathcal{U}$ of p such that the mapping*

$$\begin{cases} X : \mathcal{U}_0 \to TM, \quad q \mapsto X(q) := P_0^{t_0}(\gamma_q)(b), \\ \gamma_q \text{ is the unique geodesic segment in } \mathcal{U}_0 \text{ from } p \text{ to } \gamma_q(t_0) = q \end{cases} \tag{9.8.1}$$

is a (smooth) vector field on $\mathcal{U}_0$.

Proof.   To prove this subtle technical result, we are going to apply the smooth dependence theorem 3.2.28. Let $\mathcal{H}^D$ be the linear Ehresmann connection induced by D, and let S^D be the geodesic spray of D. Consider the exponential mapping $\exp \colon \widetilde{TM} \to M$ determined by S^D and its restriction

$\exp_p$ to $\widetilde{TM} \cap T_pM$. Let $\mathcal{V} := (\exp_p)^{-1}(\mathcal{U})$; then $\mathcal{V}$ is an open subset of T_pM.

Step 1. Define a mapping

$$Z: \mathcal{V} \times \mathcal{U} \subset T_pM \times M \to \tau^{-1}(\mathcal{U}) \subset TM$$

by

$$Z(v,q) := (\exp_p)_* \big(\iota((\exp_p)^{-1}(q), v) \big); \quad v \in \mathcal{V}, q \in \mathcal{U},$$

where $\iota: \mathcal{V} \times T_pM \to TV$ is the canonical isomorphism described in Example C.2.2. We show that for any fixed $v \in \mathcal{V}$, *the mapping*

$$Z(v): \mathcal{U} \to T\mathcal{U}, \quad q \mapsto Z(v)(q) = Z(v,q)$$

is a vector field on $\mathcal{U}$ whose integral curve starting at p is just the geodesic γ_v of D with initial velocity v.

Indeed, the smoothness of $Z(v)$ is guaranteed by the construction, and it is easy to check that $\tau \circ Z(v) = 1_{\mathcal{U}}$, so $Z(v) \in \mathfrak{X}(\mathcal{U})$. Further, let t be an element of the domain of γ_v, and define $\alpha: s \in \mathbb{R} \mapsto sv \in T_pM$. Then we have $\gamma_v = \exp_p \circ \alpha$ by (5.1.21), and

$$Z(v)(\gamma_v(t)) = (\exp_p)_* \big(\iota((\exp_p)^{-1}(\gamma_v(t)), v) \big) = (\exp_p)_* (\iota(\alpha(t), v))$$

$$= (\exp_p)_*(\dot{\alpha}(t)) \stackrel{(3.1.11)}{=} (\exp_p \circ \alpha)^{\cdot}(t) = \dot{\gamma}_v(t),$$

as desired.

Step 2. With the help of the mapping Z, define a further mapping

$$\widetilde{Z}: \mathcal{V} \times \tau^{-1}(\mathcal{U}) \subset T_pM \times TM \to T(\tau^{-1}(\mathcal{U})), \quad (v,w) \mapsto \widetilde{Z}(v,w)$$

by

$$\widetilde{Z}(v,w) := (Z(v))^{\mathsf{h}}(w) \stackrel{(7.2.17)}{=} \mathcal{H}^D\big(w, Z(v)(\tau(w))\big).$$

Then $\widetilde{Z}$ satisfies the conditions of Theorem 3.2.28, and so there exist

an open neighbourhood $\mathcal{U}_1$ of b in TM,
an open interval I containing 0,
an open neighbourhood W of 0_p in $\mathcal{V}$,
a smooth mapping $\varphi: I \times \mathcal{U}_1 \times W \to \tau^{-1}(\mathcal{U})$

such that for each $(w,v) \in \mathcal{U}_1 \times W$ the curve

$$\varphi_w^v: I \to T\mathcal{U}, \quad t \mapsto \varphi_w^v(t) := \varphi(t, w, v)$$

is an integral curve of $(Z(v))^{\mathsf{h}}$ starting at w, i.e.,

$$(Z(v))^{\mathsf{h}} \circ \varphi_w^v = \dot{\varphi}_w^v, \quad \varphi_w^v(0) = w. \tag{$*$}$$

Step 3. Consider, in particular, the curve φ_b^v, where b is the given vector in T_pM and $v \in W$. We show that

$$\tau \circ \varphi_b^v = \gamma_v. \qquad (**)$$

To see this, we calculate the velocity vector of $\tau \circ \varphi_b^v$ at a point $t \in I$. Taking into account that $(Z(v))^{\mathsf{h}} \underset{\tau}{\sim} Z(v)$, we find that

$$(\tau \circ \varphi_b^v)\,\dot{}\,(t) \overset{(3.1.11)}{=} \tau_*(\dot{\varphi}_b^v(t)) \overset{(*)}{=} \tau_* \circ (Z(v))^{\mathsf{h}} \circ \varphi_b^v(t) = Z(v) \circ \tau \circ \varphi_b^v(t),$$

therefore $\tau \circ \varphi_b^v$ is the integral curve of $Z(v)$ starting at p. Thus, by Step 1, we obtain $(**)$. From this it follows that φ_b^v is a vector field along γ_v.

Step 4. We show that φ_b^v *is an* $\mathcal{H}^D$-*parallel vector field along* γ_v. Indeed, for every $t \in I$ we have

$$\dot{\varphi}_b^v(t) \overset{(*)}{=} (Z(v))^{\mathsf{h}}(\varphi_b^v(t)) = \mathcal{H}^D(\varphi_b^v(t), Z(v) \circ \tau \circ \varphi_b^v(t))$$

$$\overset{(**)}{=} \mathcal{H}^D(\varphi_b^v(t), Z(v) \circ \gamma_v(t)) \overset{\text{Step 1}}{=} \mathcal{H}^D(\varphi_b^v(t), \dot{\gamma}_v(t)).$$

On the other hand, by Corollary 7.6.6, $\mathcal{H}^D$ and D have the same parallel vector fields along a smooth curve, so for any fixed $t_0 \in I$ we have

$$P_0^{t_0}(\gamma_v)(b) = \varphi_b^v(t_0).$$

Now let $\mathcal{U}_0 := \exp_p(\mathcal{W})$. Then for every $q = \exp_p(v) \in \mathcal{U}_0$,

$$X(q) := P_0^{t_0}(\gamma_q)(b) = P_0^{t_0}(\gamma_v)(b) = \varphi_b^v(t_0)$$
$$:= \varphi(t_0, b, v) = \varphi(t_0, b, (\exp_p)^{-1}(q)),$$

which proves the smoothness of X. $\qquad\qquad\qquad\qquad\qquad\qquad\Box$

Theorem 9.8.6 (Z. I. Szabó). *Let* (M, F) *be a Berwald manifold. Then there exists a Riemannian metric* g *on* M *such that the Levi-Civita derivative defined by* g *is the base covariant derivative of* (M, F).

Proof. At each point p in M, let g_p be the averaged scalar product on the Finsler vector space (T_pM, F_p) constructed in Proposition and Definition 9.1.51. Then

$$g \colon M \to T_2^0(TM), \quad p \mapsto g_p \in T_2^0(T_pM)$$

is a section of $T_2^0(TM)$; we show that this section is smooth.

Choose a fixed point p_0 in M, and let $(b_i)_{i=1}^n$ be a basis of $T_{p_0}M$. Let D be the base covariant derivative on M. For every $i \in J_n$, let X_i be the local vector field constructed from b_i by (9.8.1). Then $(X_i)_{i=1}^n$ is a frame field of TM over an open neighbourhood $\mathcal{U}$ of p_0 because the parallel translations are linear isomorphisms. Let $(\theta^i)_{i=1}^n$ be the dual coframe of $(X_i)_{i=1}^n$, i.e., the family of one-forms θ^i on $\mathcal{U}$ such that $\theta^i(X_j) = \delta_j^i$, $i, j \in J_n$. Then for each $p \in \mathcal{U}$, there exist unique scalars $(g_p)_{ij} \in \mathbb{R}$ such that

$$g_p = (g_p)_{ij} \theta_p^i \otimes \theta_p^j.$$

We show that

$$(g_p)_{ij} = (g_{p_0})_{ij}; \quad i, j \in J_n.$$

Indeed, since D is compatible with F by (B$_5$), for any smooth curve segment $\gamma \colon [0, t] \to \mathcal{U}$ from p_0 to p, the parallel translation $P_0^t(\gamma)$ is an F-norm preserving linear mapping from $T_{p_0} M$ onto $T_p M$. Therefore, by Proposition 9.1.53, $P_0^t(\gamma)$ is an orthogonal mapping with respect to the scalar products g_{p_0} and g_p. So we obtain

$$(g_{p_0})_{ij} = g_{p_0}(X_i(p_0), X_j(p_0)) = g_p(X_i(p), X_j(p)) = (g_p)_{ij},$$

as we claimed. Thus $g \underset{(\mathcal{U})}{=} (g_{p_0})_{ij} \theta^i \otimes \theta^j$, which proves the smoothness of g.

We know from the construction that the scalar products g_p are positive definite, so g is a Riemannian metric on M. By our above argument, the parallel translations with respect to D are orthogonal mappings. Thus from Proposition 6.1.62 we infer that D is compatible with g. Since D is torsion-free, it follows that D is the Levi-Civita derivative on (M, g). $\square$

Remark 9.8.7. Let F be a Riemannian Finsler function arising from a Riemannian metric g_M on M. Then it follows from Example 9.1.52 that the Riemannian metric constructed from F by the method of the previous proof is just the starting Riemannian metric g_M.

Proposition 9.8.8 (M. Crampin's Elegant Theorem [31]). *A Randers manifold is a Berwald manifold if, and only if, its canonical spray is reversible.*

Proof. The condition is necessary because the canonical spray of a Berwald manifold is affine, and affine sprays are reversible (see, e.g., the proof of Proposition 9.6.9).

Now suppose that (M, F) is a Randers manifold whose canonical spray S is reversible. Let F_M be the Riemannian Finsler function and $\beta \in \mathcal{A}_1(M)$ the one-form for which $F = F_M + \bar{\beta}$. Then we have

$$F^{\downarrow} = F_M - \bar{\beta} = F - 2\bar{\beta}. \tag{$*$}$$

By the reversibility of S, we infer from Lemma 9.6.7 that S is the canonical spray of $(M, F^\downarrow)$ as well. Therefore, applying Lemma 9.2.20, we obtain that

$$0 = i_S \omega_{F^\downarrow} \overset{(*)}{=} i_S d\mathcal{L}_\mathbf{J}(F - 2\bar\beta) = i_S d\mathcal{L}_\mathbf{J} F - 2i_S d\mathcal{L}_\mathbf{J}\bar\beta = -2i_S d\mathcal{L}_\mathbf{J}\bar\beta,$$

$$0 = SF^\downarrow \overset{(*)}{=} SF - 2S\bar\beta = -2S\bar\beta.$$

Thus

$$i_S d\mathcal{L}_\mathbf{J}\bar\beta = 0 \quad \text{and} \quad S\bar\beta = 0,$$

and hence

$$SF_M \overset{(*)}{=} SF - S\bar\beta = 0, \quad i_S d\mathcal{L}_\mathbf{J} F_M \overset{(*)}{=} i_S d\mathcal{L}_\mathbf{J} F - i_S d\mathcal{L}_\mathbf{J}\bar\beta = 0.$$

Referring again to Lemma 9.2.20, it follows that S is the canonical spray of the Riemannian Finsler manifold (M, F_M). However, by Corollary 9.8.4, (M, F_M) is a Berwald manifold, so its canonical spray is an affine spray by (B$_2$). Therefore, from the same reason, (M, F) itself is a Berwald manifold. $\qquad\square$

Crampin's original proof of the proposition is similarly simple; he derives the result as a consequence of Proposition 9.6.9.

9.9   Oriented Finsler Surfaces

9.9.1   *Berwald Frames*

Definition 9.9.1. Let (M, F) be a two-dimensional Finsler manifold. A *Berwald frame* of the slit Finsler bundle $\mathring{\pi}\colon \mathring{T}M \times_M TM \to \mathring{T}M$ or of the Finsler manifold (M, F) is a pair (ℓ, m), where $\ell := \frac{1}{F}\tilde\delta$ is the canonical unit section of $\mathring{\pi}$, and $m \in \Gamma(\mathring{\pi})$ is a Finsler vector field such that

$$g(\ell, m) = 0, \quad g(m, m) = 1, \tag{9.9.1}$$

i.e., for every $u \in \mathring{T}M$, $(\ell(u), m(u))$ is a g_u-orthonormal basis of the vector space $\{u\} \times T_{\mathring{\tau}(u)}M \cong T_{\mathring{\tau}(u)}M$.

Lemma 9.9.2. *If (ℓ, m) is a Berwald frame of the two-dimensional Finsler manifold (M, F), then the Finsler vector field m is 0^+-homogeneous:*

$$\nabla^{\mathsf{v}}_{\tilde\delta} m = \nabla_C m = 0. \tag{9.9.2}$$

Proof. Since (ℓ, m) restricts to an orthonormal basis on each fibre, Fourier expansion is available. So we get

$$\nabla_C m = g(\nabla_C m, \ell)\ell + g(\nabla_C m, m)m.$$

From (9.9.1),

$$0 = C(g(m, \ell)) = (\nabla_C g)(m, \ell) + g(\nabla_C m, \ell) + g(m, \nabla_C \ell)$$

$$\overset{(9.2.8),(9.2.18)}{=} g(\nabla_C m, \ell).$$

Similarly,

$$0 = C(g(m, m)) = (\nabla_C g)(m, m) + 2g(\nabla_C m, m) \overset{(9.2.8)}{=} 2g(\nabla_C m, m),$$

therefore $\nabla_C m = 0$. $\qquad\square$

Proposition 9.9.3. *Let (M, F) be a two-dimensional Finsler manifold. Then the slit Finsler bundle $\mathring{\pi}\colon \mathring{T}M \times_M TM \to \mathring{T}M$ admits a Berwald frame if, and only if, the manifold M is orientable.*

Proof. Suppose first that M is orientable, and let ω be a volume form on M. This induces a volume form $\widehat{\omega}$ on $\mathring{\pi}$ such that

$$\widehat{\omega}(\widehat{X}, \widehat{Y}) := (\omega(X, Y))^{\mathrm{v}} \restriction \mathring{T}M; \quad X, Y \in \mathfrak{X}(M).$$

Let $\widehat{\omega}^\sharp$ be the type $(1, 1)$ tensor metrically equivalent to $\widehat{\omega}$, that is, the endomorphism of $\Gamma(\mathring{\pi})$ given by

$$g(\widehat{\omega}^\sharp(\widetilde{X}), \widetilde{Y}) = \widehat{\omega}(\widetilde{X}, \widetilde{Y}); \quad \widetilde{X}, \widetilde{Y} \in \Gamma(\mathring{\pi}); \qquad (*)$$

cf. Remark 2.2.55 or (6.2.59). Then, by the non-degeneracy of $\widehat{\omega}$, the endomorphism $\widehat{\omega}^\sharp \in \mathrm{End}(\Gamma(\mathring{\pi}))$ is in fact an isomorphism. We show that

$$\psi := \frac{1}{\sqrt{\det \widehat{\omega}^\sharp}}\widehat{\omega}^\sharp$$

preserves the scalar product g, i.e.,

$$g(\psi(\widetilde{X}), \psi(\widetilde{Y})) = g(\widetilde{X}, \widetilde{Y}) \text{ for all } \widetilde{X}, \widetilde{Y} \in \Gamma(\mathring{\pi}). \qquad (**)$$

Indeed,

$$g(\psi(\widetilde{X}), \psi(\widetilde{Y})) = \frac{1}{\det \widehat{\omega}^\sharp} g(\widehat{\omega}^\sharp(\widetilde{X}), \widehat{\omega}^\sharp(\widetilde{Y})) \overset{(*)}{=} \frac{1}{\det \widehat{\omega}^\sharp} \widehat{\omega}(\widetilde{X}, \widehat{\omega}^\sharp(\widetilde{Y}))$$

$$\overset{(1.3.24)}{=} \frac{\det \widehat{\omega}^\sharp}{\det \widehat{\omega}^\sharp} \widehat{\omega}((\widehat{\omega}^\sharp)^{-1}(\widetilde{X}), \widetilde{Y}) \overset{(*)}{=} g(\widetilde{X}, \widetilde{Y}).$$

Observe that for every $\widetilde{X} \in \Gamma(\mathring{\pi})$, $\psi(\widetilde{X})$ is g-orthogonal to $\widetilde{X}$:

$$g(\psi(\widetilde{X}), \widetilde{X}) = \frac{1}{\sqrt{\det \widehat{\omega}^\sharp}} g(\widehat{\omega}^\sharp(\widetilde{X}), \widetilde{X}) = \frac{1}{\sqrt{\det \widehat{\omega}^\sharp}} \widehat{\omega}(\widetilde{X}, \widetilde{X}) = 0.$$

Now let $m := \psi(\ell)$. Then $g(\ell, m) = 0$ and

$$g(m,m) = g(\psi(\ell), \psi(\ell)) \overset{(**)}{=} g(\ell, \ell) = 1,$$

therefore $(\ell, m) := (\ell, \psi(\ell))$ is a Berwald frame for $\mathring{\pi}$.

Conversely, suppose that $\mathring{\pi}$ admits a Berwald frame (ℓ, m). Let (μ_1, μ_2) be its dual coframe, and define a two-form $\omega := \mu_1 \wedge \mu_2$. Let $\underline{\ell}$, $\underline{m}$ and $\underline{\omega}$ denote the principal parts of ℓ, m and ω, respectively (see Remark 4.1.2(a),(f)).

Given any point $p \in M$, we show that the orientation represented by $\underline{\omega}(u) \in A_2(T_pM)$ is independent of the choice of $u \in \mathring{T}_pM$. Indeed, we have $\underline{\omega}(u)(\underline{\ell}(u), \underline{m}(u)) = 1$ for all $u \in \mathring{T}_pM$. Since $\underline{\ell} \restriction \mathring{T}_pM$ and $\underline{m} \restriction \mathring{T}_pM$ are continuous, and

$$\underline{\omega}(u)(\underline{\ell}(v), \underline{m}(v)) \neq 0 \text{ for all } u, v \in \mathring{T}_pM,$$

it follows that $\underline{\omega}(u)(\underline{\ell}(v), \underline{m}(v)) > 0$ for all $u, v \in \mathring{T}_pM$, whence our claim.

Now let $(\mathcal{U}_\alpha, u_\alpha)_{\alpha \in \mathcal{A}}$ be an atlas of M, and choose a nowhere vanishing vector field Z_α on each set $\mathcal{U}_\alpha$. Then $\omega_\alpha := \underline{\omega} \circ Z_\alpha$ is a volume form on $\mathcal{U}_\alpha$ for all $\alpha \in \mathcal{A}$. If, for some indices $\alpha, \beta \in \mathcal{A}$, $\mathcal{U}_\alpha$ and $\mathcal{U}_\beta$ are overlapping, then for any point $p \in \mathcal{U}_\alpha \cap \mathcal{U}_\beta$, $\underline{\omega}_\alpha(p)$ and $\underline{\omega}_\beta(p)$ represent the same orientation of T_pM, so one of them is a positive multiple of the other. Therefore, if $(f_\alpha)_{\alpha \in \mathcal{A}}$ is a partition of unity subordinate to the covering $(\mathcal{U}_\alpha)_{\alpha \in \mathcal{A}}$, then $\omega_M := \sum_{\alpha \in \mathcal{A}} f_\alpha \omega_\alpha$ is a volume form on M. $\square$

9.9.2   The Fundamental Equations of Finsler Surfaces

Throughout this subsection we assume that (M, F) is a two-dimensional Finsler manifold endowed with a Berwald frame (ℓ, m) of the slit Finsler bundle. Briefly, (M, F) is a Finsler surface with Berwald frame (ℓ, m). Thus we implicitly require that M is orientable. However, the topological consequences of this condition will nowhere be used. Alternatively (and essentially equivalently) we may work locally.

Definition and Proposition 9.9.4. Let (M, F) be a Finsler surface with Berwald frame (ℓ, m).

(i) The *Gaussian curvature* of (M, F) is the function

$$\kappa := g(\mathcal{R}(\ell, m), m), \tag{9.9.3}$$

where g is the fundamental tensor and $\mathcal{R}$ is the fundamental affine curvature of (M, F). The only nonzero components of $\mathcal{R}$ with respect to (ℓ, m) are κ and $-\kappa$:

$$\mathcal{R}(\ell, m) = \kappa m, \quad \mathcal{R}(m, \ell) = -\kappa m. \tag{9.9.4}$$

(ii) The Cartan tensor and the vector-valued Cartan tensor of (M, F) can be represented in the form

$$\mathcal{C}_\flat = \mathfrak{I}\nabla^{\mathrm{v}}_m \theta_g \otimes \nabla^{\mathrm{v}}_m \theta_g \otimes \nabla^{\mathrm{v}}_m \theta_g \tag{9.9.5}$$

and

$$\mathcal{C} = \mathfrak{I}\nabla^{\mathrm{v}}_m \theta_g \otimes \nabla^{\mathrm{v}}_m \theta_g \otimes m, \tag{9.9.6}$$

respectively, where θ_g is the Finsler one-form defined by (9.2.12), and

$$\mathfrak{I} := g(\mathcal{C}(m, m), m) = \mathcal{C}_\flat(m, m, m). \tag{9.9.7}$$

The function $\mathfrak{I}$ is called the *main scalar* of the Finsler surface (M, F).

Proof. (i) We apply Fourier expansion with respect to (ℓ, m). Then

$$\mathcal{R}(\ell, m) = g(\mathcal{R}(\ell, m), \ell)\ell + g(\mathcal{R}(\ell, m), m)m.$$

To prove (9.9.4) is is sufficient to show that $g(\mathcal{R}(\ell, m), \widetilde{\delta}) = 0$. However, this can be seen by an immediate calculation:

$$g(\mathcal{R}(\ell, m), \widetilde{\delta}) \overset{(9.2.12)}{=} \theta_g(\mathcal{R}(\ell, m)) \overset{(9.2.13)}{=} F\nabla^{\mathrm{v}}F(\mathcal{R}(\ell, m)) = Fi\mathcal{R}(\ell, m)(F)$$

$$\overset{(7.8.1)}{=} F\mathbf{v}[\mathcal{H}\ell, \mathcal{H}m](F) = F([\mathcal{H}\ell, \mathcal{H}m](F) - \mathbf{h}[\mathcal{H}\ell, \mathcal{H}m](F)) = 0,$$

since the horizontal vector fields kill the Finsler function by (CC_3).

(ii) To prove (9.9.5) and (9.9.6), note first that by (9.2.40), the only nonzero component of $\mathcal{C}_\flat$ is $\mathfrak{I} := \mathcal{C}_\flat(m, m, m)$. Thus, for any Finsler vector fields $\widetilde{X}, \widetilde{Y}$ on $\overset{\circ}{T}M$,

$$\mathcal{C}(\widetilde{X}, \widetilde{Y}) = \mathcal{C}\big(g(\widetilde{X}, \ell)\ell + g(\widetilde{X}, m)m, g(\widetilde{Y}, \ell)\ell + g(\widetilde{Y}, m)m\big)$$

$$= g(\widetilde{X}, m)g(\widetilde{Y}, m)\mathcal{C}(m, m).$$

Here

$$\mathcal{C}(m, m) = g(\mathcal{C}(m, m), \ell)\ell + g(\mathcal{C}(m, m), m)m$$

$$= \mathcal{C}_\flat(m, m, \ell)\ell + \mathcal{C}_\flat(m, m, m)m = \mathfrak{I}m,$$

$$g(\widetilde{X}, m) = g(m, \widetilde{X}) := \nabla^{\mathrm{v}}\nabla^{\mathrm{v}}E(m, \widetilde{X}) = (\nabla^{\mathrm{v}}_m \nabla^{\mathrm{v}}E)(\widetilde{X}) \overset{(9.2.13)}{=} (\nabla^{\mathrm{v}}_m \theta_g)(\widetilde{X}),$$

$$g(\widetilde{Y}, m) = (\nabla^{\mathrm{v}}_m \theta_g)(\widetilde{Y}),$$

therefore

$$\mathcal{C}(\widetilde{X}, \widetilde{Y}) = (\mathfrak{J}\nabla^v_m \theta_g \otimes \nabla^v_m \theta_g \otimes m)(\widetilde{X}, \widetilde{Y}), \qquad (*)$$

which proves (9.9.6). This implies (9.9.5) because

$$\mathcal{C}_b(\widetilde{X}, \widetilde{Y}, \widetilde{Z}) = g(\mathcal{C}(\widetilde{X}, \widetilde{Y}), \widetilde{Z}) \overset{(*)}{=} (\mathfrak{J}\nabla^v_m \theta_g \otimes \nabla^v_m \theta_g)(\widetilde{X}, \widetilde{Y})g(m, \widetilde{Z})$$
$$= (\mathfrak{J}\nabla^v_m \theta_g \otimes \nabla^v_m \theta_g \otimes \nabla^v_m \theta_g)(\widetilde{X}, \widetilde{Y}, \widetilde{Z}).$$

Thus the proof is complete.   □

Lemma 9.9.5. *The Gaussian curvature of a Finsler surface is 1^+-homogeneous, and the main scalar is $(-1)^+$-homogeneous.*

Proof. The notation is as above. We see by direct computation that

$$C\kappa = C(g(\mathcal{R}(\ell, m), m)) = (\nabla_C g)(\mathcal{R}(\ell, m), m)$$
$$\qquad + g(\nabla_C(\mathcal{R}(\ell, m)), m) + g(\mathcal{R}(\ell, m), \nabla_C m)$$
$$\overset{(9.2.8),(9.9.2)}{=} g(\nabla_C(\mathcal{R}(\ell, m)), m) \overset{(9.2.18),(9.9.2)}{=} g((\nabla_C \mathcal{R})(\ell, m), m)$$
$$\overset{\text{Lemma 7.8.4}}{=} g(\mathcal{R}(\ell, m), m) =: \kappa,$$

so κ is 1^+-homogeneous.

Since the Cartan tensor $\mathcal{C}_b$ is $(-1)^+$-homogeneous (see Definition and Lemma 9.2.25(i)) and m is 0^+-homogeneous, it follows immediately that the main scalar $\mathfrak{J} = \mathcal{C}_b(m, m, m)$ is $(-1)^+$-homogeneous.   □

Corollary 9.9.6. *A Finsler surface with Berwald frame is Riemannian if, and only if, its main scalar vanishes.*

Proof. In the proof of Definition and Proposition 9.9.4 we saw that $(\nabla^v_m \theta_g)(\widetilde{X}) = g(\widetilde{X}, m)$ for any $\widetilde{X} \in \Gamma(\mathring{\pi})$. In particular, $(\nabla^v_m \theta_g)(m) = g(m, m) = 1$, thus $\nabla^v_m \theta_g$ does not vanish on $\mathring{T}M$. Now the assertion follows from (9.9.5) and Proposition 9.2.26(iv).   □

Lemma 9.9.7. *Keeping the notation above, let S be the canonical spray, $\mathcal{H}$ the canonical connection of (M, F), and let ∇ be the Finslerian Berwald* *derivative determined by* F. *Then we have the following formulae:*

$$\mathbf{i}\ell(F) = 1, \quad \mathbf{i}m(F) = 0; \tag{9.9.8a,b}$$

$$\nabla_S \ell = 0, \quad \nabla_S m = 0; \tag{9.9.9a,b}$$

$$\nabla_{\mathbf{i}\ell}\ell = 0, \quad \nabla_{\mathcal{H}\ell}\ell = 0; \tag{9.9.10a,b}$$

$$\nabla_{\mathbf{i}m}\ell = \frac{1}{F}m, \quad \nabla_{\mathcal{H}m}\ell = 0; \tag{9.9.11a,b}$$

$$\nabla_{\mathbf{i}\ell}m = 0, \quad \nabla_{\mathcal{H}\ell}m = 0; \tag{9.9.12a,b}$$

$$\nabla_{\mathbf{i}m}m = -\frac{1}{F}\ell - \frac{1}{2}\Im m; \tag{9.9.13}$$

$$\nabla_{\mathcal{H}m}m = \frac{1}{2}(S\Im)m. \tag{9.9.14}$$

Proof.   By immediate calculation,

$$(\mathbf{i}\ell)F = \frac{1}{F}CF = 1;$$

$$(\mathbf{i}m)F = \nabla^{\mathrm{v}}F(m) = \frac{1}{F}\nabla^{\mathrm{v}}E(m) = \frac{1}{F}\theta_g(m) = g(m, \ell) = 0;$$

$$\nabla_S \ell = \nabla_S\left(\frac{1}{F}\widetilde{\delta}\right) = S\left(\frac{1}{F}\right)\widetilde{\delta} + \frac{1}{F}\nabla_S\widetilde{\delta} = \mathrm{h}S\left(\frac{1}{F}\right)\widetilde{\delta} + \frac{1}{F}\nabla_{\mathrm{h}S}\widetilde{\delta}$$

$$\overset{(8.1.4), (CC_3)}{=} 0,$$

which prove (9.9.8a,b) and (9.9.9a). To show (9.9.9b), we use Fourier expansion:

$$\nabla_S m = g(\nabla_S m, \ell)\ell + g(\nabla_S m, m)m. \tag{$*$}$$

Since

$$0 = Sg(\ell, m) = (\nabla_S g)(\ell, m) + g(\nabla_S \ell, m) + g(\ell, \nabla_S m)$$

$$\overset{\text{Cor. 9.3.16}, (9.9.9a)}{=} g(\ell, \nabla_S m),$$

the first term is zero at the right-hand side of ($*$). Similarly,

$$0 = Sg(m, m) = (\nabla_S g)(m, m) + 2g(\nabla_S m, m) = 2g(\nabla_S m, m),$$

so we indeed have $\nabla_S m = 0$.

Now we can proceed with some easy steps:

$$\nabla_{\mathbf{i}\ell}\ell = \frac{1}{F}\nabla_C \ell \overset{(9.2.18)}{=} 0; \qquad \nabla_{\mathcal{H}\ell}\ell = \frac{1}{F}\nabla_S \ell \overset{(9.9.9a)}{=} 0;$$

$$\nabla_{\mathbf{i}m}\ell \overset{(B\nabla_1)}{=} \mathbf{j}[\mathbf{i}m, \mathcal{H}\ell] = \mathbf{j}\left[\mathbf{i}m, \frac{1}{F}S\right] = \frac{1}{F}\mathbf{j}[\mathbf{i}m, S] + \mathbf{i}m\left(\frac{1}{F}\right)\widetilde{\delta}$$

$$\overset{(5.1.4),(9.9.8b)}{=} \frac{1}{F}m;$$

$$\nabla_{\mathcal{H}m}\ell = \nabla_{\mathcal{H}m}\left(\frac{1}{F}\widetilde{\delta}\right) = \mathcal{H}m\left(\frac{1}{F}\right)\widetilde{\delta} + \frac{1}{F}\nabla_{\mathcal{H}m}\widetilde{\delta} \overset{(CC_3),(8.1.4)}{=} 0;$$

$$\nabla_{i\ell}m = \frac{1}{F}\nabla_{\mathcal{C}}m \overset{(9.9.2)}{=} 0; \quad \nabla_{\mathcal{H}\ell}m = \frac{1}{F}\nabla_{\mathcal{S}}m \overset{(9.9.9b)}{=} 0.$$

Thus it remains to show (9.9.13) and (9.9.14). By Fourier expansion again,

$$\nabla_{im}m = g(\nabla_{im}m, \ell)\ell + g(\nabla_{im}m, m)m.$$

Using the same trick as above,

$$0 = im(g(m, \ell)) = (\nabla^{\mathsf{v}}_m g)(m, \ell) + g(\nabla_{im}m, \ell) + g(m, \nabla_{im}\ell)$$

$$\overset{(9.2.39),(9.9.11a)}{=} \mathcal{C}_\flat(m, m, \ell) + g(\nabla_{im}m, \ell) + \frac{1}{F} \overset{(9.2.40)}{=} g(\nabla_{im}m, \ell) + \frac{1}{F},$$

whence $g(\nabla_{im}m, \ell) = -\frac{1}{F}$. Similarly,

$$0 = im(g(m, m)) = (\nabla^{\mathsf{v}}_m g)(m, m) + 2g(\nabla_{im}m, m)$$

$$= \mathcal{C}_\flat(m, m, m) + 2g(\nabla_{im}m, m),$$

so we get $g(\nabla_{im}m, m) = -\frac{1}{2}\mathcal{C}_\flat(m, m, m) =: -\frac{1}{2}\mathcal{J}$. Thus

$$\nabla_{im}m = -\frac{1}{F}\ell - \frac{1}{2}\mathcal{J}m,$$

as we claimed. Finally,

$$\nabla_{\mathcal{H}m}m = g(\nabla_{\mathcal{H}m}m, \ell)\ell + g(\nabla_{\mathcal{H}m}m, m)m. \tag{$**$}$$

Since

$$0 = \mathcal{H}m(g(m, \ell)) = (\nabla^{\mathsf{h}}_m g)(m, \ell) + g(\nabla_{\mathcal{H}m}m, \ell) + g(m, \nabla_{\mathcal{H}m}\ell)$$

$$\overset{(9.3.2),(9.9.11b)}{=} \mathbf{L}(m, m, \ell) + g(\nabla_{\mathcal{H}m}m, \ell) \overset{Cor. 9.3.15(ii)}{=} g(\nabla_{\mathcal{H}m}m, \ell),$$

the first term is zero on the right-hand side of ($**$). The second Fourier coefficient can be calculated as follows:

$$\mathbf{L}(m, m, m) := (\nabla^{\mathsf{h}}g)(m, m, m) = (\nabla_{\mathcal{H}m}g)(m, m)$$

$$= \mathcal{H}m(g(m, m)) - 2g(\nabla_{\mathcal{H}m}m, m) = -2g(\nabla_{\mathcal{H}m}m, m),$$

whence $g(\nabla_{\mathcal{H}m}m, m) = -\frac{1}{2}\mathbf{L}(m, m, m)$. Thus

$$\nabla_{\mathcal{H}m}m = -\frac{1}{2}\mathbf{L}(m, m, m)m \overset{(9.3.6)}{=} \frac{1}{2}(\nabla_{\mathcal{S}}\mathcal{C}_\flat)(m, m, m)m$$

$$= \frac{1}{2}\big(S(\mathcal{C}_\flat(m, m, m)) - 3\mathcal{C}_\flat(\nabla_{\mathcal{S}}m, m, m)\big)m$$

$$\overset{(9.9.9b)}{=} \frac{1}{2}S(\mathcal{C}_\flat(m, m, m))m = \frac{1}{2}(S\mathcal{J})m,$$

as was to be shown.

This concludes the proof. $\square$

Theorem 9.9.8. *Let (M, F) be a Finsler surface with Berwald frame* (ℓ, m). *Then we have the following commutator formulae:*

$$[im, S] = \mathcal{H}m, \tag{9.9.15}$$

$$[\mathcal{H}\ell, \mathcal{H}m] = \kappa(im), \tag{9.9.16}$$

$$[\mathcal{H}m, im] = \frac{1}{F}\mathcal{H}\ell + \frac{1}{2}\mathfrak{I}(\mathcal{H}m) + \frac{1}{2}(S\mathfrak{I})im, \tag{9.9.17}$$

$$[im, \mathcal{H}\ell] = \frac{1}{F}\mathcal{H}m. \tag{9.9.18}$$

Proof. First we go back to Corollary 7.10.4:

$$[im, S] = [im, \mathcal{H}\widetilde{\delta}] \stackrel{(7.10.16)}{=} \mathcal{H}\nabla_{im}\widetilde{\delta} - i\nabla_S m \stackrel{(6.2.33),(9.9.9b)}{=} \mathcal{H}m.$$

Next we use the horizontal Ricci formula (**R1**): for every smooth function $f \in C^\infty(\mathring{T}M)$,

$$\nabla^h \nabla^h f(\ell, m) - \nabla^h \nabla^h f(m, \ell) = i\mathcal{R}(\ell, m)(f).$$

Here

$$\nabla^h \nabla^h f(\ell, m) = (\nabla_{\mathcal{H}\ell}\nabla^h f)(m) = \mathcal{H}\ell(\mathcal{H}m(f)) - \nabla^h f(\nabla_{\mathcal{H}\ell}m)$$

$$\stackrel{(9.9.12b)}{=} \mathcal{H}\ell(\mathcal{H}m(f)),$$

and, similarly,

$$\nabla^h \nabla^h f(m, \ell) = \mathcal{H}m(\mathcal{H}\ell(f)) - \nabla^h f(\nabla_{\mathcal{H}m}\ell) \stackrel{(9.9.11b)}{=} \mathcal{H}m(\mathcal{H}\ell(f)),$$

therefore

$$[\mathcal{H}\ell, \mathcal{H}m] = i\mathcal{R}(\ell, m) \stackrel{(9.9.4)}{=} \kappa(im),$$

as desired.

To prove (9.9.17), we use again Corollary 7.10.4:

$$[\mathcal{H}m, im] \stackrel{(7.10.16)}{=} i\nabla_{\mathcal{H}m}m - \mathcal{H}\nabla_{im}m$$

$$\stackrel{(9.9.13),(9.9.14)}{=} \frac{1}{F}\mathcal{H}\ell + \frac{1}{2}\mathfrak{I}(\mathcal{H}m) + \frac{1}{2}(S\mathfrak{I})im.$$

Finally,

$$[im, \mathcal{H}\ell] = \left[im, \mathcal{H}\left(\frac{1}{F}\widetilde{\delta}\right)\right] = \left[im, \frac{1}{F}S\right]$$

$$= \frac{1}{F}[im, S] + im\left(\frac{1}{F}\right)S \stackrel{(9.9.8b),(9.9.15)}{=} \frac{1}{F}\mathcal{H}m,$$

as was to be shown. $\square$

Remark 9.9.9. Identities (9.9.15)–(9.9.17) correspond to the box formula (7.6) in Berwald's paper [15]. Berwald calls these relations *Cartan's permutation formulas.*

Corollary 9.9.10. *Let* (M, F) *be a Finsler surface with Berwald frame* (ℓ, m). *Then the Gaussian curvature and the main scalar of* (M, F) *satisfy the* Berwald identity

$$\frac{1}{2}\Im\kappa + \mathrm{i}m(\kappa) + \frac{1}{2F}S(S\Im) = 0. \tag{9.9.19}$$

Proof. We apply the Jacobi identity to the vector fields $\mathcal{H}\ell$, $\mathcal{H}m$ and $\mathrm{i}m$. Taking into account (9.9.15)–(9.9.18) and (CC$_3$), we find that

$$
\begin{aligned}
0 &= [\mathcal{H}\ell, [\mathcal{H}m, \mathrm{i}m]] + [\mathcal{H}m, [\mathrm{i}m, \mathcal{H}\ell]] + [\mathrm{i}m, [\mathcal{H}\ell, \mathcal{H}m]] \\
&= \left[\mathcal{H}\ell, \frac{1}{F}\mathcal{H}\ell + \frac{1}{2}\Im(\mathcal{H}m) + \frac{1}{2}(S\Im)\mathrm{i}m\right] + \left[\mathcal{H}m, \frac{1}{F}\mathcal{H}m\right] + [\mathrm{i}m, \kappa(\mathrm{i}m)] \\
&= \frac{1}{2}\big(\mathcal{H}\ell(\Im)\mathcal{H}m + \Im[\mathcal{H}\ell, \mathcal{H}m] + \mathcal{H}\ell(S\Im)\mathrm{i}m + S\Im[\mathcal{H}\ell, \mathrm{i}m]\big) + \mathrm{i}m(\kappa)\mathrm{i}m \\
&= \frac{S\Im}{2F}\mathcal{H}m + \frac{1}{2}(\Im\kappa)\mathrm{i}m + \frac{1}{2F}S(S\Im)\mathrm{i}m - \frac{S\Im}{2F}\mathcal{H}m + \mathrm{i}m(\kappa)\mathrm{i}m \\
&= \left(\frac{1}{2}\Im\kappa + \frac{1}{2F}S(S\Im) + \mathrm{i}m(\kappa)\right)\mathrm{i}m,
\end{aligned}
$$

whence our assertion. □

9.9.3 *Surviving Curvature Components*

We continue to assume that (M, F) *is a Finsler surface with Berwald frame* (ℓ, m).

Proposition 9.9.11. *The only surviving component of the Berwald curvature of* (M, F) *is*

$$\mathbf{B}(m, m)m = -\frac{S\Im}{F}\ell + \frac{1}{2}(\mathrm{i}m(S\Im) + \mathcal{H}m(\Im))m. \tag{9.9.20}$$

Proof. By Corollary 7.13.7, $\mathbf{B}$ vanishes if one of its arguments is $\ell = \frac{1}{F}\widetilde{\delta}$. Using (9.9.11b), (9.9.12b), (9.9.13), (9.9.14) and (9.9.17), we calculate the only surviving component:

$$
\begin{aligned}
\mathbf{B}(m, m)m &\overset{(7.13.1)}{:=} \nabla_{\mathrm{i}m}\nabla_{\mathcal{H}m}m - \nabla_{\mathcal{H}m}\nabla_{\mathrm{i}m}m - \nabla_{[\mathrm{i}m, \mathcal{H}m]}m \\
&= \frac{1}{2}\nabla_{\mathrm{i}m}((S\Im)m) + \nabla_{\mathcal{H}m}\left(\frac{1}{F}\ell + \frac{1}{2}\Im m\right) + \nabla_{\frac{1}{F}\mathcal{H}\ell + \frac{1}{2}\Im(\mathcal{H}m) + \frac{1}{2}(S\Im)\mathrm{i}m}m
\end{aligned}
$$

$$= \frac{1}{2}\mathrm{im}(S\mathfrak{I})m - \frac{1}{2F}(S\mathfrak{I})\ell - \frac{1}{4}(S\mathfrak{I})\mathfrak{I}m + \frac{1}{2}\mathcal{H}m(\mathfrak{I})m + \frac{1}{4}\mathfrak{I}(S\mathfrak{I})m$$

$$+ \frac{1}{4}\mathfrak{I}(S\mathfrak{I})m - \frac{1}{2F}(S\mathfrak{I})\ell - \frac{1}{4}(S\mathfrak{I})\mathfrak{I}m = -\frac{S\mathfrak{I}}{F}\ell + \frac{1}{2}(\mathrm{im}(S\mathfrak{I}) + \mathcal{H}m(\mathfrak{I}))m.$$

This is what was to be shown. $\qquad\square$

Corollary and Definition 9.9.12. *Let (M, F) be a Finsler surface with Berwald frame (ℓ, m). Then the following are equivalent:*

(i) *(M, F) is a Berwald manifold.*
(ii) *The Berwald curvature of (M, F) vanishes.*
(iii) *$S\mathfrak{I} = 0$ and $\mathrm{im}(S\mathfrak{I}) + \mathcal{H}m(\mathfrak{I}) = 0$.*
(iv) *$(\mathcal{H}\ell)\mathfrak{I} = \nabla^h\mathfrak{I}(\ell) = 0$ and $(\mathcal{H}m)\mathfrak{I} = \nabla^h\mathfrak{I}(m) = 0$.*
(v) *$\nabla^h\mathfrak{I} = 0$.*

If one (and hence all) of these conditions is satisfied, then we say that (M, F) is a Berwald surface. $\qquad\square$

Corollary 9.9.13. *The only nonzero component of the trace of the Berwald curvature of (M, F) is the function*

$$(\mathrm{tr}\,\mathbf{B})(m, m) = \frac{1}{2}(\mathrm{im}(S\mathfrak{I}) + \mathcal{H}m(\mathfrak{I})). \tag{9.9.21}$$

Proof. We calculate $\mathrm{tr}\,\mathbf{B}$ with the help of the g-orthonormal frame (ℓ, m) as follows:

$$(\mathrm{tr}\,\mathbf{B})(m, m) \overset{(1.2.21)}{=} \mathrm{tr}(\widetilde{X} \mapsto \mathbf{B}(\widetilde{X}, m)m)$$

$$= g(\mathbf{B}(\ell, m)m, \ell) + g(\mathbf{B}(m, m)m, m) = g(\mathbf{B}(m, m)m, m)$$

$$\overset{(9.9.20)}{=} \frac{1}{2}g\big((\mathrm{im}(S\mathfrak{I}) + \mathcal{H}m(\mathfrak{I}))m, m\big) = \frac{1}{2}(\mathrm{im}(S\mathfrak{I}) + \mathcal{H}m(\mathfrak{I})). \quad\square$$

Proposition 9.9.14. *The only surviving component of the Douglas tensor of (M, F) is*

$$\mathbf{D}(m, m)m$$

$$= -\frac{1}{6}\left(\frac{3S\mathfrak{I}}{E} + \mathrm{im}(\mathrm{im}(S\mathfrak{I})) + \mathrm{im}(\mathcal{H}m(\mathfrak{I})) + \mathfrak{I}(\mathrm{im}(S\mathfrak{I}) + \mathcal{H}m(\mathfrak{I}))\right)\widetilde{\delta}.$$

Proof. By Lemma 8.4.26, $\mathbf{D}$ vanishes if some of its arguments is $\ell = \frac{1}{F}\widetilde{\delta}$. We calculate the surviving component $\mathbf{D}(m, m)m$. From (8.4.23),

$$\mathbf{D}(m, m)m = \mathbf{B}(m, m)m - (\mathrm{tr}\,\mathbf{B})(m, m)m - \frac{1}{3}(\nabla^v\,\mathrm{tr}\,\mathbf{B})(m, m, m)\widetilde{\delta}.$$

Owing to Proposition 9.9.11 and Corollary 9.9.13, our only task is to calculate the last term of the right-hand side:

$$(\nabla^{\mathrm{v}} \operatorname{tr} \mathbf{B})(m, m, m) = (\nabla_{\mathbf{i}m} \operatorname{tr} \mathbf{B})(m, m) = \mathbf{i}m(\operatorname{tr} \mathbf{B}(m, m))$$

$$- 2 \operatorname{tr} \mathbf{B}(\nabla_{\mathbf{i}m} m, m) \stackrel{(9.9.13),(9.9.21)}{=} \frac{1}{2} \mathbf{i}m(\mathbf{i}m(S\mathfrak{I}))$$

$$+ \frac{1}{2} \mathbf{i}m(\mathcal{H}m(\mathfrak{I})) + \frac{\mathfrak{I}}{2}(\mathbf{i}m(S\mathfrak{I}) + \mathcal{H}m(\mathfrak{I})).$$

Since

$$\mathbf{B}(m, m)m - (\operatorname{tr} \mathbf{B})(m, m)m = -\frac{S\mathfrak{I}}{F}\ell = -\frac{S\mathfrak{I}}{2E}\widetilde{\delta},$$

we obtain the desired relation.   □

Proposition 9.9.15. *The only surviving component of the Landsberg tensor of (M, F) is*

$$\mathbf{L}(m, m, m) = -S\mathfrak{I}, \tag{9.9.22}$$

therefore the Landsberg tensor of a Finsler surface with Berwald frame vanishes if, and only if, the main scalar is a first integral of the canonical spray.

Proof. By Corollary 9.3.15(ii), $\mathbf{L}(\widetilde{X}, \widetilde{Y}, \widetilde{Z}) = 0$ if $\ell = \frac{1}{F}\widetilde{\delta} \in \{\widetilde{X}, \widetilde{Y}, \widetilde{Z}\}$. In the last step of the proof of Lemma 9.9.7 we saw that

$$\nabla_{\mathcal{H}m} m = -\frac{1}{2}\mathbf{L}(m, m, m)m = \frac{1}{2}(S\mathfrak{I})m,$$

whence (9.9.22).   □

Corollary 9.9.16. *A Finsler surface with Berwald frame (ℓ, m) is a Berwald surface if, and only if, its Berwald curvature is trace-free, and its Landsberg tensor vanishes.*

Proof. We have already known that all Berwald manifolds have the mentioned properties, so the condition is necessary.

Conversely, suppose that (M, F) is such that $\operatorname{tr} \mathbf{B} = 0$ and $\mathbf{L} = 0$. Then from (9.9.22) and (9.9.21) it follows that

$$S\mathfrak{I} = 0 \quad \text{and} \quad \mathbf{i}m(S\mathfrak{I}) + \mathcal{H}m(\mathfrak{I}) = 0,$$

so (M, F) is a Berwald surface by Corollary and Definition 9.9.12.   □

9.9.4 Szabó's 'Rigidity Theorem'

In this concluding subsection we show that if a connected Finsler surface is a Berwald surface, then it is either flat or Riemannian. This surprising result is due to Z. I. Szabó [90]. As a matter of fact, Szabó proved a general classification theorem for connected Berwald manifolds, which gives the two-dimensional result as a special case. Following the ideas of Bao, Chern and Shen ([8], 10.6 A), we present here a simple direct proof. To do this, we need some preparations.

Lemma 9.9.17. *Let* (M, F) *be a Finsler surface with Berwald frame* (ℓ, m). *Choose a unit vector* $v \in F^{-1}(1)$, *and let* $\gamma_v \colon I_v \to TM$ *be the maximal integral curve of* $\mathbf{i}m \in \mathfrak{X}^v(\mathring{T}M)$ *starting at* v. *Then* $I_v = \mathbb{R}$, *and* γ_v *is a simple periodic curve from* $\mathbb{R}$ *onto the unit circle*

$$\mathbb{S}_{\tau(v)} := \{u \in T_{\tau(v)}M \mid F(u) = 1\}$$

of $T_{\tau(v)}M$.

Proof. **Step 1.** We show that $F \circ \gamma_v$ and $\tau \circ \gamma_v$ are both constant mappings. Indeed, for every $t \in I_v$,

$$(F \circ \gamma_v)'(t) \overset{(3.1.9)}{=} \dot\gamma_v(t)(F) = (\mathbf{i}m \circ \gamma_v)(t)(F) = ((\mathbf{i}m)F)(\gamma_v(t)) \overset{(9.9.8\mathrm{b})}{=} 0,$$

$$(\tau \circ \gamma_v)\dot{\,}(t) \overset{(3.1.8)}{=} ((\tau \circ \gamma_v)_*)_t \left(\frac{d}{dr}\right)_t \overset{(3.1.6)}{=} (\tau_*)_{\gamma_v(t)} \circ ((\gamma_v)_*)_t \left(\frac{d}{dr}\right)_t$$

$$= (\tau_*)_{\gamma_v(t)} \dot\gamma_v(t) = (\tau_*)_{\gamma_v(t)} \big(\mathbf{i}m(\gamma_v(t))\big) = 0,$$

since $\mathbf{i}m$ is vertical. Thus γ_v takes its values in

$$\tau^{-1}(\tau(v)) \cap F^{-1}(1) = \mathbb{S}_{\tau(v)}.$$

Step 2. We show that γ_v *maps the whole real line onto* $\mathbb{S}_{\tau(v)}$.

Since γ_v is also an integral curve of the vector field $\mathbf{i}m \restriction \mathbb{S}_{\tau(v)}$, and $\mathbb{S}_{\tau(v)}$ is a (one-dimensional) compact manifold, it follows by Corollary 3.2.20 that $I_v = \mathbb{R}$. Clearly, the vector field $\mathbf{i}m$ does not vanish anywhere. Thus, the integral curves of $\mathbf{i}m$ are immersions, and hence, by Proposition 3.1.23(ii), they are local diffeomorphisms. This means that every point of $\mathbb{S}_{\tau(v)}$ has an open neighbourhood (in $\mathbb{S}_{\tau(v)}$) which is the image of an integral curve of $\mathbf{i}m$. So we may conclude that $\mathrm{Im}(\gamma_v) = \gamma_v(\mathbb{R})$ is an open subset in $\mathbb{S}_{\tau(v)}$. However, it is also a closed subset of the unit circle. Indeed, suppose that $u \in \mathbb{S}_{\tau(v)}$ is a boundary point of $\gamma_v(\mathbb{R})$. Then there is an integral curve of $\mathbf{i}m$ through u such that its image intersects $\gamma_v(\mathbb{R})$. Thus $\gamma_v(\mathbb{R})$ must contain u, therefore it is a closed subset of $S_{\tau(v)}$. Now, by the connectedness of $\mathbb{S}_{\tau(v)}$, we conclude that $\gamma_v(\mathbb{R}) = \mathbb{S}_{\tau(v)}$. By Corollary 3.2.18, the integral curve $\gamma_v \colon \mathbb{R} \to \mathbb{S}_{\tau(v)}$ is either injective, simple periodic, or constant. The last possibility can clearly be excluded because $\mathbf{i}m$ is nowhere vanishing. The curve γ_v is not injective either, otherwise it would be a bijective local diffeomorphism, i.e., a diffeomorphism of $\mathbb{R}$ onto $S_{\tau(v)}$, which is impossible.

$\square$

Lemma 9.9.18. *Let (M, F) be a Berwald surface with Berwald frame (ℓ, m), Gaussian curvature κ and main scalar $\mathfrak{I}$. Given a unit vector v in TM, let $\gamma_v \colon \mathbb{R} \to \mathbb{S}_{\tau(v)}$ be the parametrization of the unit circle of $T_{\tau(v)}M$ described by the preceding lemma. Then we have*

$$\kappa \circ \gamma_v(t) = \kappa(\gamma(0)) \exp\left(-\frac{1}{2} \int_0^t (\mathfrak{I} \circ \gamma_v) \right), \quad t \in \mathbb{R} \qquad (9.9.23)$$

and

$$\int_0^\lambda (\mathfrak{I} \circ \gamma_v) = 0, \qquad (9.9.24)$$

where λ is the period of γ_v.

Proof. By Corollary and Definition 9.9.12, $S\mathfrak{I} = 0$, so the Berwald identity (9.9.19) reduces to

$$\mathfrak{I}\kappa + 2(\mathbf{i}m)\kappa = 0. \qquad (*)$$

Calculating as above in Step 1, we find that

$$(\kappa \circ \gamma_v)'(t) = ((\mathbf{i}m)\kappa)(\gamma_v(t)), \quad t \in \mathbb{R}.$$

Thus, composing both sides of $(*)$ on the right with γ_v, we obtain that

$$(\mathfrak{I} \circ \gamma_v)(\kappa \circ \gamma_v) + 2(\kappa \circ \gamma_v)' = 0.$$

Therefore, the function $\kappa \circ \gamma_v \colon \mathbb{R} \to \mathbb{R}$ solves the differential equation

$$(\mathfrak{I} \circ \gamma_v)x + 2x' = 0.$$

This implies (9.9.23). Now (9.9.24) is an immediate consequence of the periodicity of γ_v. $\square$

Theorem 9.9.19 (Z. I. Szabó). *Let (M, F) be a connected Berwald surface. Then*

(i) *(M, F) is flat if its Gaussian curvature vanishes identically.*

(ii) *(M, F) is a Riemannian Finsler surface if its Gaussian curvature is not identically zero.*

Proof. Let (ℓ, m) be a Berwald frame of (M, F).

(i) If $\kappa = 0$, then $0 = \kappa m \overset{(9.9.4)}{=} \mathcal{R}(\ell, m)$, therefore, by Definition and Proposition 9.9.4, $\mathcal{R} = 0$, and hence $\mathbf{H} = 0$ (Corollary and Definition 8.2.7). Since (M, F) is a Berwald surface, we also have $\mathbf{B} = 0$; thus the Finsler surface (M, F) is flat.

(ii) Now we turn to the case when κ is not identically zero. Let $v \in \mathring{T}M$ be such that $\kappa(v) \neq 0$. By the (1^+-)homogeneity of κ, we may suppose that $v \in \mathbb{S}_{\tau(v)}$. Then, in view of (9.9.23), κ cannot vanish at any point of the circle $\mathbb{S}_{\tau(v)}$. Observe that

$$\kappa(\mathbf{i}m)\mathfrak{I} = 0. \tag{$*$}$$

Indeed, since $\nabla^{\mathbf{h}}\mathfrak{I} = 0$ by Corollary and Definition 9.9.12, we have

$$\kappa(\mathbf{i}m)\mathfrak{I} = \mathbf{i}\mathcal{R}(\ell, m)\mathfrak{I} = \mathbf{v}[\mathcal{H}\ell, \mathcal{H}m]\mathfrak{I} = [\mathcal{H}\ell, \mathcal{H}m]\mathfrak{I} - \mathbf{h}[\mathcal{H}\ell, \mathcal{H}m]\mathfrak{I} = 0.$$

Let $\gamma_v \colon \mathbb{R} \to \mathbb{S}_{\tau(v)}$ be the maximal integral curve of $\mathbf{i}m$ starting at v. Then for all $t \in \mathbb{R}$ we have

$$(\mathbf{i}m\mathfrak{I}) \circ \gamma_v(t) = \mathbf{i}m(\gamma_v(t))\mathfrak{I} = \dot{\gamma}_v(t)\mathfrak{I} \overset{(3.1.9)}{=} (\mathfrak{I} \circ \gamma_v)'(t),$$

and composing both sides of equality $(*)$ on the right with γ_v, we find that

$$0 = (\kappa \circ \gamma_v)((\mathbf{i}m\mathfrak{I}) \circ \gamma_v) = (\kappa \circ \gamma_v)(\mathfrak{I} \circ \gamma_v)'.$$

Since $\kappa \circ \gamma_v \restriction \mathbb{S}_{\tau(v)}$ is nowhere zero, from this it follows that $(\mathfrak{I} \circ \gamma_v)'$ vanishes identically. Thus the function $\mathfrak{I} \circ \gamma_v$ is constant, and its constant value is zero by (9.9.24). Therefore, the main scalar $\mathfrak{I}$ vanishes on $\mathbb{S}_{\tau(v)}$. Owing to its $(-1)^+$-homogeneity, $\mathfrak{I}$ is zero on the whole $\mathring{T}_{\tau(v)}M$, hence $F_{\tau(v)}$ is a Euclidean norm. The tangent spaces of a connected Berwald manifold are linearly isometric by Proposition 9.8.2(B_5'), so we conclude that (M, F) is Riemannian. $\qquad\square$

Definition 9.9.20. A Finsler manifold (M, F) is called a *generalized Berwald manifold* if there exists a covariant derivative D on the base manifold such that D-parallel translations preserve the Finsler norms of tangent vectors to M. More precisely, for any smooth curve $\gamma \colon I \to M$ and for any $s, t \in I$,

$$F_{\gamma(s)} \circ P_t^s(\gamma) = F_{\gamma(t)},$$

where $P_t^s(\gamma) \colon T_{\gamma(t)}M \to T_{\gamma(s)}M$ is the parallel translation along γ with respect to D.

Notice that the characterizing property of a generalized Berwald manifold is just condition (B$_5$) from Proposition 9.8.2 without assuming that the covariant derivative is torsion-free. This justifies the term 'generalized'.

In the remainder of this subsection, we show that *Szabó's 'Rigidity Theorem' does not hold for generalized Berwald manifolds*, namely we construct a connected orientable generalized Berwald manifold of dimension two which is neither flat nor Riemannian.

We begin with a general remark. When we want to prove, for example, that a given Finsler manifold is not flat, a straightforward method is to calculate the curvature of its canonical connection, and to show that it does not vanish. However, such calculations are tedious, even in simple cases. So, instead of this direct method, we shall use a more conceptual approach, which is based on the following observation.

If a Finsler manifold is flat, then its affine and Berwald curvature vanish. Thus, a flat Finsler manifold is necessarily a Berwald manifold. In the proof of Proposition 9.8.2 we saw that the covariant derivative in (B$_5$) induces the canonical connection. Therefore, the curvature of this covariant derivative also vanishes if the canonical connection is flat (Lemma 7.15.3). As a result, *if a Finsler manifold is flat, then there exists a flat torsion-free covariant derivative on the base manifold satisfying the compatibility condition* (B$_5$).

After these preparatory remarks, we give an example of a non-Riemannian Finsler function on $\mathbb{R}^2$ for which there exists a unique covariant derivative with non-zero torsion satisfying condition (B$_5$).

Step 1. *Construction of the Finsler function.* Let us consider the two-dimensional manifold $\mathbb{R}^2$ together with its standard global chart $(\mathbb{R}^2, (u^1, u^2))$ (Example 2.1.9). Define a frame field (E_1, E_2) of $T\mathbb{R}^2$ by

$$E_1 := u^1 \frac{\partial}{\partial u^1} + \frac{\partial}{\partial u^2}, \quad E_2 := -\frac{\partial}{\partial u^1},$$

and let (E^1, E^2) be its dual coframe. Then

$$E^1 = du^2, \quad E^2 = -du^1 + u^1 \, du^2.$$

Consider the Finsler norm $f := \sqrt{4(u^1)^2 + 12(u^2)^2} - u^1$ on $\mathbb{R}^2$, and define the desired Finsler function F by

$$F(v) := f(E^1_{\tau(v)}(v), E^2_{\tau(v)}(v)), \quad v \in T\mathbb{R}^2.$$

Then for any $p \in \mathbb{R}^2$ and $(a^1, a^2) \in \mathbb{R}^2$ we have

$$F(a^1 E_1(p) + a^2 E_2(p)) = f(a^1, a^2). \tag{1}$$

Explicitly,

$$F := \sqrt{4(y^2)^2 + 12(-y^1 + x^1 y^2)^2} - y^2,$$

where (x, y) is the induced chart on $T\mathbb{R}^2$. We see that F is actually a Randers function on $\mathbb{R}^2$, and it is not Riemannian.

Step 2. *Construction of a suitable covariant derivative.* Any vector field Y on $\mathbb{R}^2$ can be written in the form $Y = Y^i E_i$, where Y^1, Y^2 are smooth functions on $\mathbb{R}^2$. Let

$$D_X Y = X(Y^i) E_i, \quad X \in \mathfrak{X}(\mathbb{R}^2). \tag{2}$$

Then D is a covariant derivative on $\mathbb{R}^2$ (see, e.g., Example 6.1.5). For any point $p \in \mathbb{R}^2$ and vector $v \in T_p \mathbb{R}^2$, the mapping

$$X_v \colon q \in \mathbb{R}^2 \mapsto X_v(q) := E_p^1(v) E_1(q) + E_p^2(v) E_2(q) \in T_q \mathbb{R}^2$$

is a vector field on $\mathbb{R}^2$ satisfying $DX_v = 0$. Given a smooth curve $\gamma \colon I \to \mathbb{R}^2$ such that $\gamma(0) = p$, we have

$$D^\gamma(X_v \circ \gamma)(t) \overset{(6.1.62)}{=} D_{\gamma(t)} X_v = 0.$$

Therefore, $X_v \circ \gamma$ is the parallel vector field along γ satisfying $X_v(\gamma(0)) = v$. Thus

$$P_0^t(\gamma)(v) = X_v(\gamma(t)) = E_p^1(v) E_1(\gamma(t)) + E_p^2(v) E_2(\gamma(t))$$

holds for every $t \in I$, and it follows from (1) that $F \circ P_0^t(\gamma)(v) = F(v)$. Since the vector v and the curve γ were both arbitrary, we find that D satisfies condition (B$_5$) with respect to F.

Finally, the torsion of D does not vanish. For example,

$$T(E_1, E_2) \overset{(6.1.46)}{=} D_{E_1} E_2 - D_{E_2} E_1 - [E_1, E_2] \overset{(2)}{=} -[E_1, E_2]$$

$$= \left[u^1 \frac{\partial}{\partial u^1} + \frac{\partial}{\partial u^2}, -\frac{\partial}{\partial u^1} \right] \overset{(3.1.39)}{=} \frac{\partial}{\partial u^1}.$$

Step 3. *There is no other covariant derivative on $\mathbb{R}^2$ satisfying* (B$_5$) *with respect to F.* Notice first that there are only two linear isomorphisms $A \in \mathrm{End}(\mathbb{R}^2)$ satisfying $f \circ A = f$. These are represented by the matrices

$$\begin{pmatrix} 1 & 0 \\ 0 & 1 \end{pmatrix} \quad \text{and} \quad \begin{pmatrix} 1 & 0 \\ 0 & -1 \end{pmatrix}. \tag{3}$$

Indeed, the four conditions that A preserves the Finsler norms of the vectors $(1, 0)$, $(-1, 0)$, $(0, 1)$ and $(0, -1)$ imply that A is either the identity or the reflection on the axis $\mathrm{span}(e_1)$.

Now suppose that a covariant derivative $\bar{D}$ also satisfies (B$_5$) with respect to F, and let $\gamma \colon I \to \mathbb{R}^2$ be a smooth curve. Then the linear isomorphism

$$(\bar{P}_t^s(\gamma))^{-1} \circ P_t^s(\gamma) \in \mathrm{End}(T_{\gamma(t)} \mathbb{R}^2) \tag{4}$$

is norm preserving, therefore, taking into account (1), in the basis $(E_1(\gamma(t)), E_2(\gamma(t)))$ it is represented by one of the two matrices in (3). The parallel translation is smooth, hence the linear isomorphism (4) depends continuously on s. Since $(\bar{P}_t^t(\gamma))^{-1} \circ P_t^t(\gamma) = 1_{T_{\gamma(t)}\mathbb{R}^2}$, it follows that $\bar{P}_t^s(\gamma) = P_t^s(\gamma)$ for all $s \in I$. This implies that that D and $\bar{D}$ are identical, because a covariant derivative is determined by its induced parallel translations (by Proposition 6.1.59).

Finally, we note that our construction above was inspired by L. Tamássy's paper [92].

Appendix A

Sets, Mappings and Operations

A.1 Set Notations and Concepts

I have the ability to think *things*, and to designate them by simple signs $(a, b, \ldots, X, Y, \ldots)$ in such a completely characteristic way that I can always recognize them again without doubt. My thinking operates with these designated things in certain ways, according to certain laws, and I am able to recognize these laws through self-observation, and to describe them perfectly.

David Hilbert

A *set* consists of *elements* which are capable of possessing certain *properties* and having certain *relations* between themselves or with elements of other sets.

Nicolas Bourbaki

As in any presentation of mathematics in our days, we use the language of naive set theory, avoiding, however, the symbols of formal logic; with one natural exception: the equality sign cannot be omitted. As abbreviations, we also use the symbols $\Rightarrow$ ('implies') and $\Leftrightarrow$ ('if, and only if'). The reason is obvious: a *theorem* typically consists of an implication $P \Rightarrow Q$, where P is the *hypothesis* of the theorem and Q is its *conclusion*. We indicate the end of a proof by the symbol $\square$, and we write $\triangle$ if the proof is omitted.

For the convenience of the reader, we collect here certain notations, terminology and conventions used throughout the text.

A.1.1. As we have just mentioned, *equality*, denoted by $=$, is regarded as a logical concept. Heuristically, the relation $a = b$ means that the objects denoted by the symbols a and b are the same; its negation is written $a \neq b$.

More generally, we often denote the negation of a relation $x \, R \, y$ by $x \, \not{R} \, y$. If an object a equals b by definition, we write $a := b$ or $b =: a$.

The meaning of the symbols

$$\in, \, \cup, \, \cap, \, \backslash, \, \emptyset$$

is common, so they do not need any comment. We denote by $\mathbb{N}$ the set $\{0, 1, 2, \dots\}$ of natural numbers; $\mathbb{N}^* := \mathbb{N} \backslash \{0\}$ is the set of positive integers. For each $n \in \mathbb{N}^*$,

$$J_n := \{1, \dots, n\}.$$

For sets A and B, $A \subset B$ means that every element of A is an element of B, that is, $a \in A$ implies $a \in B$. We then say that A is *contained* in B, or that A is a *subset* of B. If $A \subset B$ and $B \subset A$, then $A = B$.

A.1.2. Let S be a set. Given a property P, there exists a unique subset of S whose elements are all elements $x \in S$ for which $P(x)$ is true. This subset of S is denoted by $\{x \in S \mid P(x)\}$. We have a unique set $\mathcal{P}(S)$ whose elements are the subsets of S, that is,

$$A \in \mathcal{P}(S) \text{ if, and only if, } A \subset S.$$

This set is called the *power set* of S.

A.1.3. Let a be an object. There exists a unique set $\{a\}$ with the following property:

$$x \in \{a\} \text{ if, and only if, } x = a.$$

A set of this type is called a *set with one element* or a *singleton*.

For any two objects a, b there exists a unique set $\{a, b\}$ such that

$$x \in \{a, b\} \text{ if, and only if, } x = a \text{ or } x = b.$$

A set of this type is called a *set with two elements* if $a \neq b$. Notice that $\{a, a\} = \{a\}$.

Sets of three, four, ... elements may be defined similarly. The sets that can be obtained in this way and the empty set are called *finite sets*, and all other sets are said to be *infinite sets*.

A.1.4. To any objects a and b corresponds an object (a, b), called an *ordered pair*, defined by the following requirement:

$$(a, b) = (u, v) \text{ if, and only if, } a = u \text{ and } b = v.$$

Warning: The ordered pair (a, b) and the set $\{a, b\}$ are sharply different objects.

Given two sets A and B (different or not), there exists a unique set $A \times B$ characterized by the following property:

$c \in A \times B$ if, and only if, there exist $a \in A$ and $b \in B$ such that $c = (a, b)$.

The set $A \times B$ is called the *Cartesian product*, or simply the *product* of A and B. Thus, by a slight abuse of notation,

$$A \times B = \{(a, b) \mid a \in A \text{ and } b \in B\}.$$

It follows easily from the definition that

$$A \times B = \emptyset \text{ if, and only if, } A = \emptyset \text{ or } B = \emptyset.$$

A.2 Mappings

Let E and F be two sets, which may or may not be distinct. A relation between a variable element x of E and a variable element y of F is called a *functional relation in y* if, *for all $x \in E$, there exists a unique $y \in F$ which is in the given relation with x.*

Nicolas Bourbaki

A.2.1. Let S and T be sets. A *mapping* or a *map* from S to T is a triple $f = (S, T, \Gamma)$ satisfying the following conditions:

(i) Γ is a subset of $S \times T$;

(ii) for each $s \in S$ there exists a unique $t \in T$ such that $(s, t) \in \Gamma$.

Then we use the following terminology and notation:

1. The set S is the *domain* (or *initial set*) of f; the set T is the *range* (or *final set*) of f; $\Gamma \subset S \times T$ is the *graph* of f.
2. If $(s, t) \in \Gamma$, then $f(s) := t$ is the *value* of f at s. We also say that f *sends* s to $f(s) := t$ or f *assigns* $f(s) := t$ to s. With this notation, the graph of f is $\Gamma = \{(s, f(s)) \in S \times T \mid s \in S\}$.
3. A mapping $f = (S, T, \Gamma)$ will usually be written in the more expressive form

$$f \colon S \to T, \ s \mapsto f(s) := t,$$

meaning that the element t is assigned to s by the 'rule' represented by the set Γ. Occasionally, the notation $S \xrightarrow{f} T$ will also be used instead of $f \colon S \to T$.

4. The set of mappings from S to T is denoted by $\mathrm{Map}(S,T)$. The elements of $\mathrm{Map}(S,S)$ are called the *transformations* of S.

If one of the sets S and T is empty, then $S \times T = \emptyset$ by A.1.4, so there is at most one mapping from S to T:

if $S = \emptyset$, the only mapping from S to T is $f = (\emptyset, T, \emptyset)$;
if $S \neq \emptyset$ and $T = \emptyset$, we have no mapping from S to T.

A.2.2. Let $f \colon S \to T$ be a mapping. If A is a subset of S, the mapping

$$f \upharpoonright A \colon A \to T, \ a \mapsto (f \upharpoonright A)(a) := f(a)$$

is called the *restriction* of f to A. If a mapping $g \colon S \to T$ has the same restriction to A as f, then f and g are said to *coincide on* A. We say, finally, that f is an *extension* of a mapping $h \colon A \to T$ *to* S if $f \upharpoonright A = h$.

The *identity transformation* of a set S is the mapping

$$1_S \colon S \to S, \ s \mapsto 1_S(s) := s.$$

More pedantically, $1_S := (S, S, \Gamma)$, where $\Gamma := \{(s,s) \in S \times S \mid s \in S\}$ is the *diagonal* in the product set $S \times S$. When it is clear from the context which set we mean, we write simply 1 or **1**. If $A \subset S$, the mapping $j_A := 1_S \upharpoonright A$ is called the *insertion mapping* (or the *canonical inclusion*) of A into S.

For any sets S, T, the mappings

$$\mathrm{pr}_1 \colon S \times T \to S, \ (s,t) \mapsto \mathrm{pr}_1(s,t) := s$$

and

$$\mathrm{pr}_2 \colon S \times T \to T, \ (s,t) \mapsto \mathrm{pr}_2(s,t) := t$$

are called the *canonical projections* of $S \times T$ onto S and T, respectively. We also say that pr_1 is the *first projection* and pr_2 is the *second projection* in $S \times T$.

If A, A_1, A_2 are further sets and

$$\varphi \colon A \to S, \ \psi \colon A \to T, \ \varphi_1 \colon A_1 \to S, \ \varphi_2 \colon A_2 \to T$$

are mappings, then we define two further mappings

$$(\varphi, \psi) \colon A \to S \times T \text{ and } \varphi_1 \times \varphi_2 \colon A_1 \times A_2 \to S \times T$$

by the following rules:

$$(\varphi, \psi)(a) := (\varphi(a), \psi(a)), \quad a \in A; \tag{A.2.1}$$

$$(\varphi_1 \times \varphi_2)(a_1, a_2) := (\varphi_1(a_1), \varphi_2(a_2)), \quad (a_1, a_2) \in A_1 \times A_2. \tag{A.2.2}$$

A.2.3. A mapping $f\colon S \to T$ induces two mappings

$$f_*\colon \mathcal{P}(S) \to \mathcal{P}(T) \text{ and } f^*\colon \mathcal{P}(T) \to \mathcal{P}(S)$$

as follows:

$$f_*(A) := \{b \in T \mid \text{there exists an } a \in S \text{ such that } f(a) = b\}, \ A \in \mathcal{P}(S);$$
$$f^*(B) := \{a \in S \mid f(a) \in B\}, \ B \in \mathcal{P}(T).$$

The set $f_*(A)$ is called the *image* of A by f, and the set $f^*(B)$ is called the *inverse image* of B by f. In particular,

$$\mathrm{Im}(f) := f_*(S) \text{ is the } image \text{ of } f,$$
$$f^*(b) := f^*(\{b\}) \text{ is the } fibre \text{ of } f \text{ over } b \ (b \in T).$$

In the following, as it is a common practice in mathematics, we denote f_* also by f and f^* also by f^{-1}. Strictly speaking, these notations are incorrect, since they can, and do, lead to confusion (see A.2.6 below). However, we want to preserve the 'star' notation for other purposes.

A.2.4. A mapping $f\colon S \to T$ is said to be *injective* if $s_1 \neq s_2$ implies $f(s_1) \neq f(s_2)$ (or, in contrapositive form, $f(s_1) = f(s_2)$ implies $s_1 = s_2$); *surjective* if $\mathrm{Im}(f) = T$; *bijective* if it is both injective and surjective.

We denote by $\mathrm{Bij}(S,T)$ the subset of $\mathrm{Map}(S,T)$ formed by the bijective mappings from S onto T. In particular, we write

$$\mathrm{Bij}(S) := \mathrm{Bij}(S,S), \quad S_n := \mathrm{Bij}(J_n) \quad (n \in \mathbb{N}^*).$$

The elements of $\mathrm{Bij}(S)$ are also mentioned as the *permutations* of S. This terminology for the elements of S_n is almost exclusive.

A.2.5. Let $f\colon S \to T$ and $g\colon T \to U$ be mappings. The *composition* of f and g is the mapping

$$g \circ f\colon S \to U, \ s \mapsto g \circ f(s) := g(f(s)).$$

Sometimes, if we want to emphasize not only the domain and range of the composition $g \circ f$, but the 'intermediate' set as well, we write

$$g \circ f\colon S \to T \to U.$$

Diagrams of mappings

$$
\begin{array}{ccc}
S & \xrightarrow{f} & T \\
 & \searrow{\scriptstyle h} & \downarrow{\scriptstyle g} \\
 & & U
\end{array}
\quad \text{and} \quad
\begin{array}{ccc}
S & \xrightarrow{f} & T \\
{\scriptstyle h}\downarrow & & \downarrow{\scriptstyle g} \\
U & \xrightarrow{k} & V
\end{array}
$$

are *commutative* if $g \circ f = h$ and $g \circ f = k \circ h$, respectively.

If the mappings $f \colon S \to T$ and $\sigma \colon T \to S$ are such that $f \circ \sigma = 1_T$, then f is surjective and σ is injective. The injective mapping σ is said to be a *section* of the surjective mapping f. It 'selects' for each $t \in T$ an element of the fibre of f over t.

Sections appear everywhere in differential geometry, so they will be re-defined in the context of fibre bundles (Definition 2.2.7).

We assume that any surjective mapping has a section. This assumption is just a version of the

Axiom of choice. *Given a mapping* $F \colon S \to \mathcal{P}(T)$ *such that* $F(s) \neq \emptyset$ *for every* $s \in S$, *there exists a mapping* $f \colon S \to T$ *such that*

$$f(s) \in F(s) \text{ for all } s \in S.$$

We accept this axiom.

A.2.6. If $f \colon S \to T$ is a bijective mapping, then there exists a unique mapping $g \colon T \to S$ such that

$$f \circ g = 1_T \text{ and } g \circ f = 1_S.$$

The mapping g is bijective; it is called the *inverse* of f and denoted (also) by f^{-1}.

Conversely, if $f \in \mathrm{Map}(S, T)$, $g \in \mathrm{Map}(T, S)$ such that $f \circ g = 1_T$ and $g \circ f = 1_S$, then f and g are bijective mappings inverse to each other.

A.2.7. A set S is said to be *countable* if either $S = \emptyset$ or there exists a surjective mapping $\mathbb{N}^* \to S$. The latter condition is equivalent to the existence of an injective mapping $S \to \mathbb{N}^*$. If S is not countable, it is called *uncountable*. *A nonempty set* S *is countable if, and only if, either*

there is a bijective mapping $S \to \mathbb{N}^*$

or

there is a bijective mapping $S \to J_n$ *for some positive integer* n.

In the first case S is said to be *countably infinite* or *denumerable*.

A.2.8. Let I and A be sets. A mapping

$$f \colon I \to A, \ i \mapsto f(i) =: a_i$$

is sometimes also called a *family of elements* in A, and it is denoted by $(a_i)_{i \in I}$. Then the domain I of f is called the *index set*. Instead of $(a_i)_{i \in I}$ we often simply write (a_i) if the intended index set is clear from the context.

If J is a subset of I, the restriction of f to J is called a *subfamily* of $(a_i)_{i \in I}$ having J as the index set, and written as $(a_i)_{i \in J}$.

A family of elements of A is said to be a *sequence* (of elements of A) if the index set I of the family is a subset of $\mathbb{N}$. A sequence is *finite* or *infinite* according to whether I is finite or infinite. If $I := J_n = \{1, \ldots, n\}$ ($n \in \mathbb{N}^*$), we denote the sequence $(a_i)_{i \in J_n}$ also by $(a_i)_{i=1}^n$ or $(a_1, \ldots, a_n)$. Then we say that $(a_i)_{i=1}^n$ is an *(ordered) n-tuple* of elements of A whose ith *coordinate* is the term a_i. The set of all n-tuples of elements of A is the *Cartesian product* $A^n := \underbrace{A \times \cdots \times A}_{n \text{ times}}$.

A.2.9. Let S be a nonempty set, and let m and n be positive integers. A *matrix with m rows and n columns*, briefly an $m \times n$ *matrix*, over S is a mapping

$$A \colon J_m \times J_n \to S, \ (i,j) \mapsto A(i,j) =: a^i_j,$$

i.e., a 'doubly indexed family' $(a^i_j)_{(i,j) \in J_m \times J_n}$ of elements of S. When the index sets J_m, J_n are clear from the context, we write more concisely $A = (a^i_j)$. The elements a^i_j are also called the *entries* of A. The subfamily

$$\left(a^i_j\right)_{(i,j) \in \{i\} \times J_n} =: \left(a^i_1 \ \ldots \ a^i_n \right) \text{ is the ith } row,$$

the subfamily

$$\left(a^i_j\right)_{(i,j) \in J_m \times \{j\}} =: \begin{pmatrix} a^1_j \\ \vdots \\ a^m_j \end{pmatrix} \text{ is the jth } column$$

of the matrix. Accordingly, an $m \times n$ matrix $A = \left(a^i_j\right)$ is also written as a 'rectangular array'

$$A = \begin{pmatrix} a^1_1 & a^1_2 & \ldots & a^1_n \\ a^2_1 & a^2_2 & \ldots & a^2_n \\ \vdots & \vdots & & \vdots \\ a^m_1 & a^m_2 & \ldots & a^m_n \end{pmatrix}.$$

Note that we use the following convention:

superscripts count the rows, subscripts count the columns.

We also write an $m \times n$ matrix in the form $A = (a_{ij})$; then the index i indicates the row and the index j indicates the column in which the entry a_{ij} occurs.

The set of all $m \times n$ matrices over S will be denoted by $M_{m \times n}(S)$; when $m = n$, we abbreviate $M_{n \times n}(S)$ to $M_n(S)$. The elements of $M_n(S)$ are called *square matrices* or *quadratic matrices of order n*.

The *transpose* of an $m \times n$ matrix $A = (a_{ij})$ is the $n \times m$ matrix ${}^t A = (b_{ij})$, where $b_{ij} = a_{ji}$ for all $(i, j) \in J_n \times J_m$. A (necessarily square) matrix is *symmetric* if it equals its own transpose.

A.2.10. If S is a set and $(A_i)_{i \in I}$ is a family of elements of the power set $\mathcal{P}(S)$, we often say that $(A_i)_{i \in I}$ is a *family of subsets of S* (or simply a *family of sets*). We define the *union* $\bigcup_{i \in I} A_i$ and the *intersection* $\bigcap_{i \in I} A_i$ of a family $(A_i)_{i \in I}$ of subsets of S as follows:

$a \in \bigcup_{i \in I} A_i$ if, and only if, there exists an $i \in I$ such that $a \in A_i$;

$a \in \bigcap_{i \in I} A_i$ if, and only if, $a \in A_i$ for all $i \in I$.

If $I = \emptyset$, then $\bigcup_{i \in I} A_i = \emptyset$ and $\bigcap_{i \in I} A_i = S$. When $I = \{1, 2\}$, the union and intersection are the familiar $A_1 \cup A_2$ and $A_1 \cap A_2$. The *Cartesian product* of the family $(A_i)_{i \in I}$, denoted by $\prod_{i \in I} A_i$, is the set of all families $(a_i)_{i \in I}$ such that $a_i \in A_i$ for all $i \in I$.

A.2.11. Let S be a set. If B is a subset of S, a *covering* of B is a family $(A_i)_{i \in I}$ of subsets of S such that $B \subset \bigcup_{i \in I} A_i$. A *partition* of S is a covering $(A_i)_{i \in I}$ of S such that $A_i \neq \emptyset$ for each $i \in I$ and $A_i \cap A_j = \emptyset$ for $i \neq j$.

Let $\mathcal{A} = (A_i)_{i \in I}$ and $\mathcal{B} = (B_j)_{j \in J}$ be two families of subsets of S. We say that $\mathcal{B}$ is *finer* than $\mathcal{A}$ (or a *refinement* of $\mathcal{A}$) if for every $j \in J$ there is an $i \in I$ (depending on j) such that $B_j \subset A_i$.

A.3   Groups and Group Actions

There are few notions in Mathematics more primitive than that of a law of composition: it seems inseparable from the first rudiments of calculations on natural numbers and measurable quantities. ... what is essential in a group, that is its law of composition and not the nature of the objects which constitute the group...

Nicolas Bourbaki

A.3.1. A *group* is a set G together with a mapping

$$*\colon G \times G \to G, \quad (a, b) \mapsto a * b,$$

called a *binary operation* (or *law of composition*) satisfying the following axioms:

(G1) *(associativity)* for all $a, b, c \in G$, $(a * b) * c = a * (b * c)$;

(G2) *(existence of a neutral element)* there exists an element $e \in G$ such that $e * a = a = a * e$ for all $a \in G$;

(G3) *(existence of inverses)* for each $a \in G$ there exists an element $b \in G$ such that $a * b = b * a = e$.

A group $(G, *)$ is called a *commutative group* if

(G4) $a * b = b * a$ for all $a, b \in G$.

A *homomorphism* from a group $(G_1, *)$ to a group $(G_2, \diamond)$ is a mapping $\varphi \colon G_1 \to G_2$ such that

$$\varphi(a * b) = \varphi(a) \diamond \varphi(b) \quad \text{for all } a, b \in G_1$$

('φ is compatible with the group operations'). An *isomorphism* is a bijective homomorphism (i.e., a bijective mapping $\varphi \colon G_1 \to G_2$ which 'preserves the group operations'.) Two groups $(G_1, *)$ and $(G_2, \diamond)$ are *isomorphic* if there exists an isomorphism between them; then we write $G_1 \cong G_2$. An isomorphism from a group G onto itself is called an *automorphism* of G. The set of all automorphisms of G is denoted by $\text{Aut}(G)$.

Notation. We usually abbreviate $(G, *)$ to G. Using *multiplicative notation*, we write $a \cdot b$ or ab for $a * b$ and 1 for e. Then the group is called *multiplicative*, and we say that ab is the *product* of a and b, and 1 is an *identity element* of G. Alternatively, in *additive notation* we write $a + b$ for $a * b$ and 0 for e. Then the group itself is also called *additive*, $a + b$ is mentioned as the *sum* of a and b, and 0 as a *zero element* of G. By convention, *additive notation is used only for commutative groups*. For a general ('abstract') group which may not be commutative we shall use the multiplicative notation and usually write e for its identity element.

If $(G, *)$ is a group, then it has a *unique* neutral element (if e_1 and e_2 are neutral elements, then $e_1 = e_1 * e_2 = e_2$ by assumption). Also, the element b in axiom (G3) is uniquely determined by a. We call it the *inverse* of a, and we denote it by a^{-1}. In additive notation we write $-a$, and we say that $-a$ is the *negative* of a. A sum of the form $a + (-b)$ is abbreviated to $a - b$, and called the *difference* between a and b.

These observations enable us to speak about *the* neutral element of a group and *the* inverse of an element of a group.

A.3.2. Let H be a nonempty subset of a (multiplicative) group G. If

$$a, b \in H \text{ implies } ab \in H \text{ and } a \in H \text{ implies } a^{-1} \in H, \qquad (*)$$

then H is a group with respect to the operation defined on G. The group so obtained is said to be a *subgroup* of G. Every group G has at least two (but not necessarily distinct) subgroups: G itself and the *trivial subgroup* $\{e\}$ consisting only of the identity element. A subgroup of G different to G and $\{e\}$ is called a *proper subgroup*.

For any subset S of a group G there is a smallest subgroup of G containing S. It consists of all finite products of elements of S and their inverses.

Indeed, let H be the intersection of all subgroups of G containing S. Then H is again a subgroup containing S and, obviously, the smallest such group. Clearly, H also contains all finite products of elements of S and their inverses. The set of such products satisfies $(*)$, hence it is a subgroup containing S, and therefore equals H.

The subgroup of H obtained in this way is called the *subgroup generated by S* and is denoted by $\langle S \rangle$. In particular, $\langle \emptyset \rangle = \{e\}$. If $\langle S \rangle = G$, we say that S *generates the group* G.

A.3.3. Let G and H be groups. We denote the identity element of both groups by e. If $\varphi \colon G \to H$ is a homomorphism, then the *kernel*

$$\mathrm{Ker}(\varphi) := \{g \in G \mid \varphi(g) = e\} = \varphi^{-1}(e)$$

of φ is a subgroup of G, and the image of φ (see A.2.3) is a subgroup of H.

Suppose that φ is surjective. Then the family

$$(G_h)_{h \in H} := \left(\varphi^{-1}(h)\right)_{h \in H}$$

of the fibres of φ is a partition of G. Denote the set of these fibres by $G/\mathrm{Ker}(\varphi)$, i.e., let

$$G/\mathrm{Ker}(\varphi) := \{A \in \mathcal{P}(G) \mid A \text{ is a fibre of } \varphi\}.$$

Then $G/\mathrm{Ker}(\varphi)$ with the binary operation defined by

$$G_{h_1} \cdot G_{h_2} := G_{h_1 h_2}; \quad h_1, h_2 \in H$$

is a group. The identity element of this group is $G_e = \mathrm{Ker}(\varphi)$, and the inverse of G_h is $G_{h^{-1}}$. The group $G/\mathrm{Ker}(\varphi)$ is said to be the *quotient* of G by $\mathrm{Ker}(\varphi)$.

A.3.4. Now we present some basic examples of groups which we need, explicitly or implicitly, throughout the text.

(a) **The group** $\mathrm{Map}(S, G)$. Let S be a nonempty set, G a group, and consider the set $\mathrm{Map}(S, G)$. Define the product fg of two mappings f, g in $\mathrm{Map}(S, G)$ by the rule

$$(fg)(s) := f(s)g(s) \quad \text{for all } s \in S.$$

It may be checked immediately that $\mathrm{Map}(S,G)$ is a group under this law of composition. The identity element is the constant mapping from S to G with value e, and the inverse of $f \in \mathrm{Map}(S,G)$ is the mapping f^{-1} given by

$$f^{-1}(s) := (f(s))^{-1} \quad \text{for all } s \in S.$$

If G is commutative, so is $\mathrm{Map}(S,G)$, and we use the additive notation for the group operation in $\mathrm{Map}(S,G)$ if G is written additively.

(b) **Symmetric groups.** Let S be a nonempty set. *The set* $\mathrm{Bij}(S)$ *of permutations of S is a group under the operation* $\circ$ *defined in* A.2.5. Since composition of bijective mappings is a bijective mapping, $\circ$ is indeed a binary operation on $\mathrm{Bij}(S)$. Associativity holds automatically (the composition of mappings is associative in general). The identity transformation 1_S of S is the identity element of the group, and the inverse mapping $\alpha^{-1} \in \mathrm{Bij}(S)$ is the inverse of $\alpha \in \mathrm{Bij}(S)$ with respect to the binary operation $\circ$ (see A.2.6). The group $\mathrm{Bij}(S)$ is said to be the *symmetric group on the set S.*

In the special case when $S = J_n$, the symmetric group on S is denoted by S_n and called the *symmetric group of degree n.* It is a finite group of *order $n!$,* i.e., the number of elements of S_n is $n!$. A permutation $\tau \in S_n$ is called a *transposition* if there exist $i, j \in J_n$ such that $i \neq j$ and

$$\tau(i) = j, \ \tau(j) = i; \ \tau(k) = k \text{ if } k \in J_n \setminus \{i,j\},$$

i.e., τ interchanges i and j and leaves the other elements of J_n fixed. It may be shown by induction on n that *every element of S_n can be expressed as a product of transpositions.* It will be useful to define for every $k \in \{0, \ldots, n-1\}$ the permutation $\sigma_k \in S_n$ as follows:

$$\begin{cases} \sigma_0 := 1_{J_n}; \quad \sigma_1(j) := \begin{cases} j+1 & \text{if } j < n, \\ 1 & \text{if } j = n \end{cases}; \\ \sigma_{i+1} := \sigma_i \circ \sigma_1, \quad i \in \{2, \ldots, n-2\}. \end{cases} \quad \text{(A.3.1)}$$

(c) **The alternating group.** As usual, we denote the set of integers by $\mathbb{Z}$. Informally, $\mathbb{Z} = \{0, \pm 1, \pm 2, \ldots\}$. Under the ordinary addition $\mathbb{Z}$ is a commutative group. The multiplication in $\mathbb{Z}$ makes the subset $\{-1, 1\}$ of $\mathbb{Z}$ into a commutative group.

Let n be a positive integer, and consider the symmetric group S_n. *There exists a unique homomorphism* $\varepsilon \colon S_n \to \{-1, 1\}$ *such that for every transposition* $\tau \in S_n$ *we have* $\varepsilon(\tau) = -1$. We say that $\varepsilon(\sigma)$ is the *sign* of the permutation $\sigma \in S_n$. A permutation is called *even* if its sign is 1, and it is called *odd* if its sign is -1. The *alternating group of degree n*, denoted by A_n, is the kernel of the homomorphism ε (i.e., the set of even permutations in S_n).

(d) **The group of automorphisms.** Let G be a group. The set $\text{Aut}(G)$ of automorphisms of G is a subgroup of the symmetric group $\text{Bij}(G)$ on G. Thus, in its own right, $\text{Aut}(G)$ is a group, called the *automorphism group* of G.

Each element a in G determines three transformations λ_a, ϱ_a and $\mathbf{c}_a$ of G given by

$$\lambda_a(g) := ag, \quad \varrho_a(g) := ga \quad (g \in G) \quad \text{and} \quad \mathbf{c}_a := \lambda_a \circ \varrho_a^{-1}.$$

As immediate consequences of the group axioms, we obtain

$$\lambda_a \circ \lambda_b = \lambda_{ab}, \quad \varrho_a \circ \varrho_b = \varrho_{ba} \quad \text{and} \quad \lambda_a \circ \varrho_b = \varrho_b \circ \lambda_a$$

for all $a, b \in G$. Obviously, $\lambda_e = \varrho_e = 1_G$. In particular, λ_a and ϱ_a have inverses $\lambda_{a^{-1}}$ and $\varrho_{a^{-1}}$, therefore $\lambda_a, \varrho_a \in \text{Bij}(G)$. Notice, however, that left and right translations are not homomorphisms, they are only permutations of G. Conjugations behave nicely: for each $a \in G$ we have $\mathbf{c}_a \in \text{Aut}(G)$. Indeed,

$$\mathbf{c}_a(g_1 g_2) = \lambda_a \circ \varrho_a^{-1}(g_1 g_2) = a(g_1 g_2)a^{-1} = ag_1\left(a^{-1}a\right)g_2 a^{-1}$$
$$= \left(ag_1 a^{-1}\right)\left(ag_2 a^{-1}\right) = \mathbf{c}_a(g_1)\mathbf{c}_a(g_2)$$

for all $g_1, g_2 \in G$. Automorphisms of G of the form $\mathbf{c}_a$ ($a \in G$) are called *inner*. The mapping

$$\mathbf{c}\colon G \to \text{Aut}(G), \ a \mapsto \mathbf{c}(a) := \mathbf{c}_a$$

is a homomorphism: applying the above rules of calculation, we find that

$$\mathbf{c}(ab) := \mathbf{c}_{ab} := \lambda_{ab} \circ \varrho_{(ab)^{-1}} = \lambda_a \circ \lambda_b \circ \varrho_{a^{-1}} \circ \varrho_{b^{-1}}$$
$$= \lambda_a \circ \varrho_{a^{-1}} \circ \lambda_b \circ \varrho_{b^{-1}} = \mathbf{c}_a \circ \mathbf{c}_b,$$

for all $a, b \in G$. Thus $\text{Im}(\mathbf{c})$ is a subgroup of $\text{Aut}(G)$. We denote this subgroup by $\text{Inn}(G)$, and we call it the *group of inner automorphisms* of G. Since

$$\text{Ker}(\mathbf{c}) := \{a \in G \mid \mathbf{c}_a = 1_G\} = \left\{a \in G \mid aga^{-1} = g \text{ for all } g \in G\right\}$$
$$= \{a \in G \mid ag = ga \text{ for all } g \in G\},$$

the kernel of $\mathbf{c}$ consists of all elements of G which commute with every element of G. This subgroup of G is called the *centre* of G and denoted by $Z(G)$.

Two subgroups H and K of G are said to be *conjugate* to each other if there exists an element a of G such that $K = \mathbf{c}_a(H)$; then $H = \mathbf{c}_{a^{-1}}(K)$. A subgroup H of G is *normal* (or *invariant*) if it is conjugate to itself for each $a \in G$, i.e.,

$$\mathbf{c}_a(H) = \{aha^{-1} \in G \mid h \in H\} =: aHa^{-1} = H \quad \text{for all } a \in H.$$

It is easy to show that the kernel of a homomorphism is a normal subgroup. Moreover, by a basic result of group theory, *every normal subgroup occurs as the kernel of a homomorphism.* Thus 'normal subgroups and (surjective) homomorphisms are two sides of one coin' ([50], p. 138).

To prove that a subgroup H of G is normal, it suffices to check that $aHa^{-1} \subset H$ for all $a \in G$. Indeed, rewriting this with a^{-1} for a, we get $a^{-1}Ha \subset H$. Multiplying this relation on the left with a and on the right with a^{-1}, we find that $H \subset aHa^{-1}$, whence $aHa^{-1} = H$.

Applying this observation, we show that $\text{Inn}(G)$ *is a normal subgroup of* $\text{Aut}(G)$. Let $\varphi \in \text{Aut}(G)$ and $\mathbf{c}_a \in \text{Inn}(G)$. Then for every $g \in G$ we have

$$\varphi \circ \mathbf{c}_a \circ \varphi^{-1}(g) = \varphi\left(a\varphi^{-1}(g)a^{-1}\right) = \varphi(a)g\varphi\left(a^{-1}\right)$$
$$= \varphi(a)g(\varphi(a))^{-1} = \mathbf{c}_{\varphi(a)}(g),$$

therefore $\varphi \circ \mathbf{c}_a \circ \varphi^{-1} = \mathbf{c}_{\varphi(a)} \in \text{Inn}(G)$, as desired.

A.3.5. Let S be a nonempty set and let G be a group. A *left action* of G on S is a mapping

$$G \times S \to S, \quad (g, p) \mapsto g \cdot p$$

such that

(i) $g_1 \cdot (g_2 \cdot p) = (g_1 g_2) \cdot p$ for all $g_1, g_2 \in G$, $p \in S$;
(ii) $e \cdot p = p$ for all $p \in S$.

Then we also say that G *acts* (on the left) on S, and we simply write gp instead of $g \cdot p$ when there is no danger of confusion. A set together with a (left) action of a group G is called a (left) *G-set.*

Let G act on S. For any fixed $g \in G$, we define the mapping $\vartheta_g : S \to S$ by $\vartheta_g(p) = g \cdot p$, $p \in S$. Then we have:

(i') $\vartheta_g \in \text{Bij}(S)$;
(ii') the mapping $\vartheta : G \to \text{Bij}(S)$, $g \mapsto \vartheta_g$ is a homomorphism.

Indeed, for each $p \in S$,

$$\vartheta_{g^{-1}} \circ \vartheta_g(p) := \vartheta_{g^{-1}}(g \cdot p) := g^{-1} \cdot (g \cdot p) \overset{(i)}{=} (g^{-1}g) \cdot p = e \cdot p \overset{(ii)}{=} p,$$

hence $\vartheta_{g^{-1}} \circ \vartheta_g = 1_S$. Interchanging the role of g and g^{-1} we find that $\vartheta_g \circ \vartheta_{g^{-1}} = 1_S$. Thus ϑ_g has $\vartheta_{g^{-1}}$ as its inverse, and therefore $\vartheta_g \in \text{Bij}(S)$. Relation (ii') is a similarly easy consequence of (i). The homomorphism from G to $\text{Bij}(S)$ so defined is said to be the *permutation representation* associated to the given action. Conversely, every homomorphism

$$\vartheta \colon G \to \text{Bij}(S), \ g \mapsto \vartheta_g$$

yields an action of G on S given by

$$(g, p) \in G \times S \mapsto g \cdot p := \vartheta_g(p) \in S.$$

Thus *a left action of a group G on a set S and a homomorphism $G \to \text{Bij}(S)$ are the same notion*, phrased in different terminology. The action is said to be *faithful* (or *effective*) if its permutation representation is injective.

A *right action* of G on S is defined analogously: it is a mapping

$$S \times G \to S, \ (p, g) \mapsto p \cdot g$$

that satisfies the axioms

(i*) $(p \cdot g_1) \cdot g_2 = p \cdot (g_1 g_2)$ for all $p \in S$ and $g_1, g_2 \in G$;
(ii*) $p \cdot e = p$ for all $p \in S$.

If we are given a left action of G on S, then the mapping

$$S \times G \to S, \ (p, g) \mapsto p \cdot g := g^{-1} \cdot p$$

is a right group action. Conversely, given a right action of G on S, we can form a left action by $g \cdot p := p \cdot g^{-1}$. Thus everything that we can say about left actions may be translated to right actions – and vice versa.

A.3.6. Let a left action of a group G on a set S be given, and consider the associated permutation representation $\vartheta \colon G \to \text{Bij}(S)$, $g \mapsto \vartheta_g$.

(a) By the *kernel of the action* we mean the kernel of the associated permutation representation, i.e., the normal subgroup

$$\text{Ker}(\vartheta) = \{g \in G \mid \vartheta_g = 1_S\} = \{g \in G \mid g \cdot p = p \text{ for all } p \in S\}$$

of G.

(b) The *stabilizer* (or *isotropy group*) of an element $p \in S$ is

$$\text{Stab}(p) := \{g \in G \mid g \cdot p = p\}.$$

It is a subgroup of G, but it need not be a normal subgroup. Clearly

$$\bigcap_{p \in S} \operatorname{Stab}(p) = \operatorname{Ker}(\vartheta),$$

which is a normal subgroup of G. Thus the action is effective if, and only if, $\bigcap_{p \in S} \operatorname{Stab}(p) = \{e\}$. The action is said to be *free* if

$$\operatorname{Stab}(p) = \{e\} \text{ for all } p \in S.$$

For any $g \in G$ and $p \in S$ we have

$$\operatorname{Stab}(g \cdot p) = g \operatorname{Stab}(p) g^{-1} = \mathbf{c}_g(\operatorname{Stab}(p)).$$

(c) For each element $p \in S$ the set $G \cdot p := \{g \cdot p \in S \mid g \in G\}$ is called the *orbit of G through p*. A subset A of G is a *G-orbit* (or simply an *orbit*) if $A = G \cdot p$ for some $p \in S$. The G-orbits form a partition of S. The action of G is said to be *transitive* if there is only one orbit, i.e., for any two points $p, q \in S$ there exists a $g \in G$ such that $g \cdot p = q$. Then we also say that G acts transitively on the set S, and S itself is called a *homogeneous G-set*. Relation $\operatorname{Stab}(g \cdot p) = \mathbf{c}_g(\operatorname{Stab}(p))$ implies that if the action is transitive, then any two stabilizers are conjugate to each other.

A.3.7. Let U and V be nonempty sets and k a positive integer. Consider the set U^k of all k-tuples of elements of U and the symmetric group S_k of degree k. Then S_k acts on the set $\operatorname{Map}(U^k, V)$ by the rule

$$(\sigma, f) \in S_k \times \operatorname{Map}(U^k, V) \mapsto \sigma f \in \operatorname{Map}(U^k, V),$$
$$(\sigma f)(u_1, \dots, u_k) := f(u_{\sigma(1)}, \dots, u_{\sigma(k)}) \text{ for all } (u_1, \dots, u_k) \in U^k. \tag{A.3.2}$$

Indeed, if $\iota \in S_k$ is the identity element, then $\iota f = f$ $(f \in \operatorname{Map}(U^k, V))$ holds obviously. We show that

$$\varrho(\sigma f) = (\varrho \circ \sigma) f \text{ for all } \sigma, \varrho \in S_k, \ f \in \operatorname{Map}(U^k, V). \tag{A.3.3}$$

Let $(a_1, \dots, a_k) \in U^k$, and define the elements $b_i := a_{\varrho(i)} \in U$, $i \in J_k$. If $\sigma(i) = j$, then we find $b_{\sigma(i)} = b_j := a_{\varrho(j)} = a_{\varrho \circ \sigma(i)}$, $i \in J_k$, so it follows that

$$(\varrho(\sigma f))(a_1, \dots, a_k) := (\sigma f)(a_{\varrho(1)}, \dots, a_{\varrho(k)}) = (\sigma f)(b_1, \dots, b_k)$$
$$= f(b_{\sigma(1)}, \dots, b_{\sigma(k)}) = f(a_{\varrho \circ \sigma(1)}, \dots, a_{\varrho \circ \sigma(k)}) =: ((\varrho \circ \sigma) f)(a_1, \dots, a_k),$$

as required.

A.3.8. Let S be a nonempty set, k an integer, G a commutative group (written additively), and let $A \in \mathrm{Map}(S^k, G)$. Then we sometimes write

$$\sum_{\mathrm{cyc}} A(s_1, \ldots, s_k) := \sum_{i=0}^{k-1} (\sigma_i A)(s_1, \ldots, s_k)$$

where the σ_i's are the cyclic permutations defined by (A.3.1), while $\sigma_i A$ is given by (A.3.2). For example, if $k = 3$, we have

$$\sum_{\mathrm{cyc}} A(s_1, s_2, s_3) := A(s_1, s_2, s_3) + A(s_2, s_3, s_1) + A(s_3, s_1, s_2).$$

Writing (X, Y, Z) instead of (s_1, s_2, s_3),

$$\sum_{\mathrm{cyc}} A(X, Y, Z) := A(X, Y, Z) + A(Y, Z, X) + A(Z, X, Y).$$

Observe that $\sum_{\mathrm{cyc}} A(X, Y, Z)$ is invariant under cyclic permutations:

$$\sum_{\mathrm{cyc}} A(X, Y, Z) = \sum_{\mathrm{cyc}} A(Y, Z, X) = \sum_{\mathrm{cyc}} A(Z, X, Y).$$

In general, $\sum_{\mathrm{cyc}} A(s_1, \ldots, s_k)$ has the same invariance property. We note, finally, that for cyclic sums of the form

$$A(X, Y, Z, U) + A(Y, Z, X, U) + A(Z, X, Y, U)$$

we use the abbreviation

$$\sum_{\mathrm{cyc}(X,Y,Z)} A(X, Y, Z, U).$$

A.4 Rings

A.4.1. A *ring* is a set R together with two binary operations $+$ and $\cdot$ (called addition and multiplication) satisfying the following axioms:

(RI 1) $(\mathsf{R}, +)$ is a commutative group, mentioned as the additive group of the ring;

(RI 2) the multiplication is associative and has an identity element, i.e., there exists an element $1 \in \mathsf{R}$ such that $1 \cdot a = a \cdot 1$ for all $a \in \mathsf{R}$;

(RI 3) the distributive laws

$$a(b + c) = ab + ac \text{ and } (a + b)c = ac + bc$$

hold for all $a, b, c \in \mathsf{R}$.

If, in addition,

(RI 4) $ab = ba$ for all $a, b \in \mathsf{R}$,

then R is said to be a *commutative ring*.

We follow Bourbaki in requiring that rings have an identity element 1. If $1 = 0$, the resulting ring $\mathsf{R} = \{0\}$ is called the *zero ring*.

An element a of a ring R is *invertible* if there exist $b, c \in \mathsf{R}$ such that

$$ab = 1 \text{ and } ca = 1.$$

Then $cab = b$ (from the second relation), whence $c = b$ (by the first relation); therefore $a^{-1} := b = c$ is the inverse of a with respect to the multiplication. We denote by $\mathsf{R}^\times$ the set of invertible elements of R. Obviously, $\mathsf{R}^\times$ satisfies all the axioms of a (multiplicative) group. We say that $\mathsf{R}^\times$ is the *group of invertible elements* or the *group of units* of R.

A nonzero ring in which every nonzero element is a unit, i.e., invertible, is called a *division ring* (or *skew field*). A *field* is a commutative division ring.

A.4.2. Here are a few examples of rings.

(a) The most important example is the commutative ring $\mathbb{Z}$ of integers with the usual addition and multiplication. The group of units of $\mathbb{Z}$ is $\mathbb{Z}^\times = \{-1, 1\}$, cf. A.3.4(c).

The other familiar number systems $\mathbb{Q}$, $\mathbb{R}$, $\mathbb{C}$, the sets of rational, real and complex numbers, together with the usual addition and multiplication, are fields. If $\mathsf{K} \in \{\mathbb{Q}, \mathbb{R}, \mathbb{C}\}$, we write $\mathsf{K}^* := \mathsf{K} \setminus \{0\}$. Then $\mathsf{K}^\times = \mathsf{K}^*$. For the sake of uniformity, we also write $\mathbb{Z}^* := \mathbb{Z} \setminus \{0\}$. Of course, $\mathbb{Z}^* \neq \mathbb{Z}^\times$.

The field $\mathbb{R}$ of real numbers is an *ordered field* (see, e.g., [42], p. 17). We write $\mathbb{R}_+$ for the set of nonnegative real numbers. Then

$$\mathbb{R}_+^* := \mathbb{R}_+ \setminus \{0\} = \{a \in \mathbb{R} \mid a > 0\}$$

is the set of *positive real numbers* (or the *positive numbers*).

(b) Let S be a nonempty set and R a ring. Then the set $\mathrm{Map}(S, \mathsf{R})$ of mappings from S into R is a ring under the usual pointwise addition and multiplication of mappings:

$$(f + g)(p) := f(p) + g(p), \ (fg)(p) := f(p)g(p); \quad f, g \in \mathrm{Map}(S, \mathsf{R}), \ p \in S.$$

The zero element and the identity element in $\mathrm{Map}(S, \mathsf{R})$ are the constant mappings with values $0 \in \mathsf{R}$ and $1 \in \mathsf{R}$, respectively. The ring $\mathrm{Map}(S, \mathsf{R})$ is commutative if, and only if, R is commutative. In particular, the elements of the ring $\mathrm{Map}(S, \mathbb{R})$, i.e., the real-valued mappings on a set S, will usually be mentioned as *functions on S*.

If the set S and the ring R have more structure, we may form other rings of R-valued mappings which respect the additional structures. A typical example is the ring of C^r functions on a manifold, see Remark 2.1.15(c) in the text.

(c) Fix a ring R. Let m and n be positive integers, and consider first the set $M_{m \times n}(\mathsf{R})$ of all $m \times n$ matrices with entries from R. We define addition in $M_{m \times n}(\mathsf{R})$ componentwise:
$$A + B = \left(a_j^i\right) + \left(b_j^i\right) := \left(a_j^i + b_j^i\right) \in M_{m \times n}(\mathsf{R}).$$
Now let k also be a positive integer. The *product* of an $m \times k$ matrix $A = \left(a_j^i\right) \in M_{m \times k}(\mathsf{R})$ and a $k \times n$ matrix $B = \left(b_j^i\right) \in M_{k \times n}(\mathsf{R})$ is the matrix
$$AB := \left(\sum_{r=1}^k a_r^i b_j^r\right) \in M_{m \times n}(\mathsf{R}).$$
These operations make the set $M_n(\mathsf{R})$ of square matrices into a ring. The identity element of this ring is the $n \times n$ *identity matrix* $1_n = \left(\delta_j^i\right)$, where δ_j^i is the *Kronecker delta* defined by
$$\delta_i^i := 1; \quad \delta_j^i := 0 \text{ if } i \neq j \quad (i, j \in J_n). \tag{A.4.1}$$
Note that if R is a nonzero ring and $n \geq 2$, then $M_n(\mathsf{R})$ is not commutative, even if R is commutative.

The group of invertible elements of $M_n(\mathsf{R})$ is called the *general linear group* $\mathrm{GL}_n(\mathsf{R})$:
$$\mathrm{GL}_n(\mathsf{R}) := (M_n(\mathsf{R}))^\times.$$
Now suppose that *the ring* R *is commutative*. Then we may define the *determinant* of a square matrix $A = \left(a_j^i\right) \in M_n(\mathsf{R})$ by the formula
$$\det(A) := \sum_{\sigma \in S_n} \varepsilon(\sigma) a_1^{\sigma(1)} a_2^{\sigma(2)} \ldots a_n^{\sigma(n)}. \tag{A.4.2}$$
It is easy to see that the elementary facts about determinants which can be found in any text on classical linear algebra remain valid in this generality.

In terms of determinants, we have an important criterion for a square matrix to be invertible:

If R *is a commutative ring, a matrix* $A \in M_n(\mathsf{R})$ *is invertible if, and only if, its determinant is invertible in* R, *i.e.,*
$$\mathrm{GL}_n(\mathsf{R}) = \left\{A \in M_n(\mathsf{R}) \mid \det(A) \in \mathsf{R}^\times\right\} = \det^{-1}\left(\mathsf{R}^\times\right).$$
In particular, *if* F *is a field, then* $A \in M_n(\mathsf{F})$ *is invertible if, and only if,* $\det(A) \neq 0$. So in this case
$$\mathrm{GL}_n(\mathsf{F}) = \{A \in M_n(\mathsf{F}) \mid \det(A) \neq 0\}.$$

Appendix B

Topological Concepts

Only the countable exists at infinity.

Jean Dieudonné

In this appendix we briefly summarize the topological concepts and facts which we use throughout the book. To keep the appendix reasonably short, we state most of these facts without proofs. A few of these proofs are easy exercises, which the reader may try to reconstruct alone, but some of them are rather difficult, so we refer the interested reader to references [2], [42], [63], [68]. However, one has to be careful with some terms. For example, many authors (including us) suppose compact topological spaces and topological vector spaces to be of Hausdorff type, while others do not. Many authors define a neighbourhood to be an open set, while others (including us) do not.

B.1 Basic Definitions and Constructions

Definition B.1.1. (a) A *topological space* is a set S together with a set $\mathcal{T}$ of subsets of S, or, in other words, a subset $\mathcal{T}$ of the power set $\mathcal{P}(S)$, which satisfies the following axioms:

(T1) $\emptyset$ and S belong to $\mathcal{T}$;
(T2) the union of any family of sets in $\mathcal{T}$ also belongs to $\mathcal{T}$;
(T3) the intersection of a *finite* family of sets in $\mathcal{T}$ also belongs to $\mathcal{T}$.

The elements of $\mathcal{T}$ are called *open* sets, and $\mathcal{T}$ itself is called a *topology* on the set S. A subset of S is said to be *closed* if its complement is open.

643

A *neighbourhood* of a point $p \in S$ is a set $V \subset S$ which contains an open set $\mathcal{U}$ containing p, i.e., $\mathcal{U} \in \mathcal{T}$ and $p \in \mathcal{U} \subset V$. Similarly, a neighbourhood of a set $A \subset S$ is a set $V \subset S$ which contains an open set $\mathcal{U}$ containing A.

The *interior* of a set $A \subset S$, denoted by int(A), is the union of all open sets contained in A. The *closure* of a set $A \subset S$, denoted by cl(A), is the intersection of all closed sets containing A. The *boundary* of $A \subset S$ is bd(A) := cl(A) \ int(A).

We say that a sequence $(p_n)_{n \in \mathbb{N}}$ in S *converges to* $p \in S$ if, for any neighbourhood $\mathcal{U}$ of p, there is a number $N \in \mathbb{N}$ such that $p_n \in \mathcal{U}$ whenever $n \geq N$. (Note that p is not uniquely determined in general.)

(b) A subset $\mathcal{B} \subset \mathcal{T}$ is said to be a *basis* for the topology $\mathcal{T}$ if every element of $\mathcal{T}$ can be obtained as a union of sets in $\mathcal{B}$. Then we also say that the topology $\mathcal{T}$ is *generated* by $\mathcal{B}$.

(c) Let S and T be topological spaces. A mapping $f \colon S \to T$ is said to be *continuous* if, for every open set $\mathcal{U} \subset T$, the inverse image $f^{-1}(\mathcal{U})$ is open in S. A bijective mapping $f \colon S \to T$ is called a *homeomorphism* if both f and f^{-1} are continuous. We say that the topological spaces S and T are *homeomorphic* if there exists a homeomorphism of S onto T.

Example B.1.2. (a) If S is an arbitrary set, then $\mathcal{T} := \mathcal{P}(S)$ is a topology on S, called the *discrete topology*.

(b) If $(S, \mathcal{T})$ is a topological space, and $\mathcal{U} \subset S$, then we define the *subspace topology* $\mathcal{T}_{\mathcal{U}}$ on $\mathcal{U}$ by declaring a set $V \subset \mathcal{U}$ to be open if there is an open set $W \subset S$ such that $V = \mathcal{U} \cap W$. It is easy to see that we obtain a topological space $(\mathcal{U}, \mathcal{T}_{\mathcal{U}})$ in this way, which is called a *topological subspace* of $(S, \mathcal{T})$. Unless otherwise stated, we always assume that every subset of a topological space is endowed with the subspace topology.

(c) Let $S_1, \ldots, S_n$ be topological spaces, and consider the Cartesian product $S := S_1 \times \cdots \times S_n$ (see A.1.4). There is a unique topology on S for which the sets of the form $\mathcal{U}_1 \times \cdots \times \mathcal{U}_n$, where each $\mathcal{U}_i$ is open in S_i for $i \in J_n$, form a basis. We call this the *product topology* on S. The Cartesian product S, endowed with the product topology, is called the *product space* of $S_1, \ldots, S_n$.

Remark B.1.3. (a) By (T2), int(A) is an open set, namely the largest open set contained in A.

(b) It can easily be seen that the intersection of any family of closed sets is closed. Thus, in particular, cl(A) is a closed set, namely the smallest closed set containing A.

(c) We have $\mathrm{bd}(A) = \mathrm{cl}(A) \cap \mathrm{cl}(S \setminus A)$, so by (T3), $\mathrm{bd}(A)$ is a closed set.

B.2 Metric Topologies and the Contraction Principle

Definition and Lemma B.2.1. Let S be a set and $\varrho \colon S \times S \to \mathbb{R}$ a function such that

(M1) $\varrho(p, q) \geq 0$ for all $p, q \in S$, and $\varrho(p, q) = 0$ if, and only if, $p = q$;
(M2) $\varrho(p, s) \leq \varrho(p, q) + \varrho(q, s)$ for all $p, q, s \in S$ *(triangle inequality)*.

If $a \in S$ and r is a positive real number, then we say that the sets

$$B_r^+(a) := \{p \in S \mid \varrho(a, p) < r\} \quad \text{and} \quad B_r^-(a) := \{p \in S \mid \varrho(p, a) < r\}$$

are the *forward* and *backward balls*, respectively, with centre a and radius r. There exists a unique topology $\mathcal{T}^+$ on S for which the set of all forward balls constitutes a basis, and another (maybe different) unique topology $\mathcal{T}^-$ for which the set of all backward balls constitutes a basis. We say that ϱ is a *quasi-distance* on S if, besides (M1) and (M2), it also satisfies

(M3) the topologies $\mathcal{T}^+$ and $\mathcal{T}^-$ generated by the forward and backward balls coincide.

Then S is also said to be a *quasi-metric space*. Thus, every quasi-metric space is a topological space in a natural way. The topology so obtained is also called a *metric topology*.

If, in addition to (M1)–(M3), we also have

(M4) $\varrho(p, q) = \varrho(q, p)$ for all $p, q \in S$,

then ϱ is said to be a *distance function* or a *metric*, and S, together with the function ϱ, a *metric space*. In this case, we simply write $B_r(a) := B_r^+(a) = B_r^-(a)$, and we call it the *(metric) ball* with centre a and radius r. $\triangle$

Remark B.2.2. (a) If ϱ is a metric, then, of course, (M3) is redundant, since it is a consequence of (M4).

(b) If (S, ϱ) is a quasi-metric space, and $\mathcal{U} \subset S$ is an arbitrary subset, then the restriction $\varrho_{\mathcal{U}} := \varrho \restriction (\mathcal{U} \times \mathcal{U})$ makes $\mathcal{U}$ into a quasi-metric space. The subspace topology on $\mathcal{U}$ coincides with the topology induced by $\varrho_{\mathcal{U}}$. If there is no risk of confusion, $\varrho_{\mathcal{U}}$ can be simply abbreviated to ϱ. If S is a metric space, then $\mathcal{U}$ is a metric space as well.

Definition B.2.3. Let (S, ϱ) be a quasi-metric space. We say that a sequence $(p_n)_{n \in \mathbb{N}}$ in S is a *forward Cauchy sequence* if for any positive ε there is a number $N \in \mathbb{N}$ such that

$$\varrho(p_m, p_n) < \varepsilon \quad \text{whenever } N \leq m \leq n,$$

and we say that $(p_n)_{n \in \mathbb{N}}$ is a *backward Cauchy sequence* if for any positive ε there is a number $N \in \mathbb{N}$ such that

$$\varrho(p_n, p_m) < \varepsilon \quad \text{whenever } N \leq m \leq n.$$

The quasi-metric space (S, ϱ) is *forward (backward) complete* if every forward (backward) Cauchy sequence converges, respectively, and it is *complete* if it is both forward and backward complete.

In a metric space the two types of Cauchy sequences coincide, thus we simply speak of *Cauchy sequences* in that case, and, of course, the two types of completeness coincide as well.

Lemma B.2.4 (Contraction Principle). *Let (S, ϱ) be a complete metric space, and suppose that $\varphi \colon S \to S$ is a contraction mapping, i.e., there is a real number c with $0 \leq c < 1$ such that*

$$\varrho(\varphi(p), \varphi(q)) \leq c \, \varrho(p, q) \tag{$*$}$$

whenever $p, q \in S$. Then φ has a unique fixed point in S.

Proof. Let $p_0 \in S$ be an arbitrary point. We define a sequence $(p_n)_{n \in \mathbb{N}}$ in S recursively by letting $p_{n+1} := \varphi(p_n)$, $n \in \mathbb{N}$. If $m, n \in \mathbb{N}$, $m \leq n$, then we have

$$\varrho(p_m, p_n) = \varrho(\varphi^m(p_0), \varphi^n(p_0)) \overset{(*)}{\leq} c^m \varrho(p_0, \varphi^{n-m}(p_0))$$

$$\overset{\text{(M2)}}{\leq} c^m \left(\varrho(p_0, \varphi(p_0)) + \varrho(\varphi(p_0), \varphi^2(p_0)) + \cdots + \varrho(\varphi^{n-m-1}(p_0), \varphi^{n-m}(p_0)) \right)$$

$$\overset{(*)}{\leq} c^m \left(\sum_{i=0}^{n-m-1} c^i \right) \varrho(p_0, \varphi(p_0)) \leq c^m \left(\sum_{i=0}^{\infty} c^i \right) \varrho(p_0, \varphi(p_0))$$

$$= \frac{c^m}{1-c} \varrho(p_0, \varphi(p_0)),$$

which implies that $(p_n)_{n \in \mathbb{N}}$ is a Cauchy sequence, and it converges to a point $p := \lim_{n \to \infty} p_n$ by the completeness of S. (See also Remark B.3.2 below.) Since φ is a contraction mapping, it is continuous, and

$$\varphi(p) = \varphi(\lim_{n \to \infty} p_n) = \lim_{n \to \infty} \varphi(p_n) = \lim_{n \to \infty} p_{n+1} = p,$$

thus p is indeed a fixed point of φ. Finally, if p and q are both fixed points of φ, then

$$0 \le \varrho(p,q) = \varrho(\varphi(p), \varphi(q)) \le c\, \varrho(p,q).$$

Since $c < 1$, it follows that $\varrho(p,q) = 0$, therefore $p = q$, and so the fixed point is unique. □

Example B.2.5. The most important example for a metric space is the Euclidean n-space $\mathbb{R}^n$ (see Example 1.1.13) for some fixed $n \in \mathbb{N}^*$, with the so-called *Euclidean distance function*. The Euclidean distance of two points $p = (p^1, \ldots, p^n)$, $q = (q^1, \ldots, q^n) \in \mathbb{R}^n$ is defined by

$$\varrho(p,q) := \sqrt{(p^1 - q^1)^2 + \cdots + (p^n - q^n)^2}.$$

In this way, as noted above, every subset of $\mathbb{R}^n$ becomes a metric space. Unless otherwise stated, we regard each subset of $\mathbb{R}^n$ as a metric space with the restriction of the Euclidean distance and a topological space with the induced topology.

B.3 More Topological Concepts

Definition B.3.1. A topological space S is said to be a *Hausdorff space* if any two distinct points in S have disjoint neighbourhoods.

Remark B.3.2. In a Hausdorff space each convergent sequence $(p_n)_{n \in \mathbb{N}}$ converges to a unique point, denoted by $\lim_{n \to \infty} p_n$. Every quasi-metric space is a Hausdorff space. Any subspace of a Hausdorff space is a Hausdorff space, and the product space of a finite number of Hausdorff spaces is a Hausdorff space as well.

Definition B.3.3. We say that a topological space satisfies the *second axiom of countability*, or it is *second countable* for short, if there is a countable basis for its topology.

Remark B.3.4. (a) There also exists a first axiom of countability, but we do not need that in this book.

(b) The metric topology of $\mathbb{R}^n$ is second countable, since the set of all open balls with rational centre points and rational radii is a topological basis.

(c) Any subspace of a second countable space is second countable, and the product space of a finite number of second countable spaces is second countable.

Definition B.3.5. A topological space S is said to be *connected* if there are no disjoint, nonempty, open subsets $\mathcal{U}, \mathcal{V} \subset S$ such that $\mathcal{U} \cup \mathcal{V} = S$. A subset of S is said to be connected if it is a connected space with the subspace topology.

Remark B.3.6. (a) Any finite product of connected topological spaces is connected.

(b) The connected subsets of the real line $\mathbb{R}$ are precisely the intervals.

(c) If S and T are topological spaces, $f \colon S \to T$ is a continuous mapping, and S is connected, then $f(S)$ is connected.

From (b) and (c), the following *intermediate value theorem* can be easily deduced: if S is a connected topological space, $f \colon S \to \mathbb{R}$ is a continuous function, $p, q \in S$, and $t \in \mathbb{R}$ such that $f(p) < t < f(q)$, then there is a point $r \in S$ such that $f(r) = t$.

Definition B.3.7. If S is a topological space, and $p, q \in S$, then a *path* in S from p to q is a continuous mapping $\gamma \colon [0,1] \to S$ such that $\gamma(0) = p$ and $\gamma(1) = q$. A topological space is said to be *path connected* if for every $p \in S$ and $q \in S$ there is a path in S from p to q.

Remark B.3.8. Every path connected topological space is connected. The converse is not true. Now we discuss a sufficient condition for the equivalence of connectedness and path connectedness of a topological space.

Definition B.3.9. A topological space is said to be *locally path connected* if it admits a basis of path connected open sets, or, equivalently, if any open neighbourhood of a given point contains a path connected open neighbourhood of that point.

Proposition B.3.10. *Let S be a locally path connected topological space. Then S is connected if, and only if, it is path connected.*   △

Remark B.3.11. Every topological manifold (Definition 2.1.1) is locally path connected.

Definition B.3.12. (a) An *open covering* of a topological space is a covering (see A.2.11) which consists of open sets.

(b) A Hausdorff topological space S is said to be *compact* if every open covering of S has a finite subcovering. A subset of S is said to be compact if it is a compact topological space with the subspace topology. A subset $\mathcal{U} \subset S$ is said to be *relatively compact* (in S) if its closure $\mathrm{cl}(\mathcal{U})$ is compact.

Remark B.3.13. We list below some basic results concerning compactness.

(i) *Every closed subset of a compact topological space is compact.*

(ii) *Every compact subset of a Hausdorff space is closed.*

(iii) *Continuous images of compact spaces are compact.* In other words, if f is a continuous mapping from a compact space S into a topological space T, then $f(S) \subset T$ is compact.

(iv) *If S and T are compact topological spaces, then their product space $S \times T$ is also compact.* From this it follows by induction that *the product of finitely many compact topological spaces is always compact.* Surprisingly, a much stronger result is also true: *the product of any family of compact spaces is compact.* This is the *Tychonoff product theorem*, 'probably the most important single theorem of general topology' (J. L. Kelley). For the definition of the product of an arbitrary family of topological spaces and for a proof of the theorem see, e.g., [63], Ch. II. §3.

To formulate the next results, we need the concept of boundedness in a metric space. To define it, let A be a nonempty subset of a metric space (S, ϱ). The *diameter* of A is

$$\mathrm{diam}(A) := \sup\{\varrho(p,q) \in \mathbb{R} \mid p, q \in A\}.$$

A *bounded set* in S is a nonempty subset of S whose diameter is finite. Then, in particular, the bounded sets of the real line $\mathbb{R}$ are also defined: these are the bounded sets of the metric space $(\mathbb{R}, \varrho)$, where $\varrho(s,t) := |s-t|$ for all $s, t \in \mathbb{R}$ (cf. Example B.2.5).

(v) (Heine–Borel–Lebesgue theorem). *A subset of the real line $\mathbb{R}$ is compact if, and only if, it is closed and bounded.*

(vi) (Extreme value theorem). *Let S be a nonempty compact metric space, and let f be a continuous real-valued function on S. Then $f(S)$ is bounded, and f attains both a minimum and a maximum value on S.* Indeed, $f(S) \subset \mathbb{R}$ is compact by (iii), hence it is bounded and closed by (v). Thus

$$m_1 := \inf\{f(s) \in \mathbb{R} \mid s \in S\} \quad \text{and} \quad m_2 := \sup\{f(s) \in \mathbb{R} \mid s \in S\}$$

exist and belong to $f(S)$. Let s_1 and s_2 be any points in S such that $f(s_1) = m_1$, $f(s_2) = m_2$. Then $f(s_1) \leq f(s) \leq f(s_2)$ for all $s \in S$.

(vii) (Generalized Heine–Borel–Lebesgue theorem). Let the Euclidean n-space $\mathbb{R}^n$ be equipped with its Euclidean distance function. Then *a subset of $\mathbb{R}^n$ is compact if, and only if, it is closed and bounded.*

We note that any compact set in any metric space is both closed and bounded. However, the converse of this statement does not hold in general.

B.4   Topological Vector Spaces

Definition B.4.1. A *(real) topological vector space* is a real vector space V equipped with a Hausdorff topology such that both the addition

$$V \times V \to V, \ (v, w) \mapsto v + w$$

and the scalar multiplication

$$\mathbb{R} \times V \to V, \ (\alpha, v) \mapsto \alpha v$$

are continuous mappings. Here the products $V \times V$ and $\mathbb{R} \times V$ are endowed with the product topology (Example B.1.2(c)). In this case we also say that the given topology is a *vector topology* for the real vector space V.

Definition B.4.2. A *norm* on a real vector space V is a function

$$\| \ \| : V \to \mathbb{R}, \quad v \in V \mapsto \|v\| \in \mathbb{R}$$

which satisfies the following axioms:

(N1) $\|v\| \geq 0$ for any $v \in V$, and $\|v\| = 0$ if, and only if, $v = 0$;
(N2) $\|\lambda v\| = |\lambda| \|v\|$ for any $\lambda \in \mathbb{R}$ and $v \in V$;
(N3) $\|v + w\| \leq \|v\| + \|w\|$ for any $v, w \in V$.

A *(real) normed space* is a real vector space endowed with a norm.

Remark B.4.3. In a normed space we define the distance of two points $p, q \in V$ by $\varrho(p, q) := \|p - q\|$. From (N1)–(N3) it follows easily that V becomes a metric space with ϱ. A normed space will always be considered as a metric space with respect to the distance function ϱ.

In particular, the Euclidean norm of $\mathbb{R}^n$ (see Lemma and Definition 1.4.2) induces in this way the Euclidean distance function defined in Example B.2.5.

Definition B.4.4. A normed space is called a *Banach space* if it is a complete metric space.

Proposition B.4.5. *Each normed space is a topological vector space with the topology induced by the distance function.* $\triangle$

Definition B.4.6. Two norms on the same vector space are said to be *equivalent* if they generate the same topology.

Proposition B.4.7. *Two norms* $\| \ \|_1$ *and* $\| \ \|_2$ *on a vector space* V *are equivalent if, and only if, there are positive constants* c *and* C *such that*

$$c\|v\|_2 \leq \|v\|_1 \leq C\|v\|_2 \text{ for each } v \in V.$$

For a proof, see [2], 2.1.9 Proposition.

Proposition and Definition B.4.8. *Let* V *and* W *be two normed spaces. Then the set* $B(V, W)$ *of all* continuous *linear mappings from* V *to* W *is a real vector space, and the function*

$$\| \ \| : B(V, W) \to \mathbb{R}, \ \varphi \mapsto \|\varphi\| := \sup \left\{ \frac{\|\varphi(v)\|}{\|v\|} \in \mathbb{R} \ \middle| \ v \in V \setminus \{0\} \right\} \quad (B.4.1)$$

is a norm on $B(V, W)$, *called the* operator norm. *(For simplicity, all norms are denoted by the same symbol.) The operator norm satisfies the inequality*

$$\|\varphi(v)\| \leq \|\varphi\|\|v\|, \quad (B.4.2)$$

for every $\varphi \in B(V, W)$ *and* $v \in V$.

For a proof, we refer to [2], §2.2 or [42], Section 5.7. We note that the operator norm defined by (B.4.1) depends on the norms on V and on W. However, if we replace these norms by equivalent norms, the new norm on $B(V, W)$ will be equivalent to the old one. If V is finite-dimensional, then every linear mapping of V into W is continuous, thus $B(V, W) = L(V, W)$ in this case.

Remark B.4.9. It can be easily shown that every finite-dimensional normed space is a Banach space and that any two norms on a finite-dimensional vector space are equivalent, thus *every finite-dimensional vector space carries a canonical vector topology.* Moreover, we have the following fundamental result.

Theorem B.4.10 (A. Tychonoff). *On every finite-dimensional real vector space* V *there is a unique vector topology, which is generated by an arbitrarily given norm on* V.

For a nice proof, see [93], section 1.2.

Appendix C

Calculus in Vector Spaces

C.1 Differentiation in Vector Spaces

> This 'intrinsic' formulation of calculus, due to its greater 'abstraction', and in particular to the fact that again and again, one has to leave the initial spaces and to climb higher and higher to new 'function spaces' (especially when dealing with the theory of higher derivatives), certainly requires some mental effort, contrasting with the comfortable routine of the classical formulas. But we believe that the result is well worth the labor ...
>
> Jean Dieudonné ([42], p. 147)

In this section we shall work on finite-dimensional real vector spaces. As we have already mentioned, they can always be considered as topological vector spaces with their canonical vector topology (Theorem B.4.10). We have pointed out that this topology can be derived from an arbitrary norm, thus, whenever convenient, we shall assume that a norm is also given in our vector space. The notions obtained in this way will not depend on the choice of the norm.

Definition C.1.1. Let V and W be finite-dimensional real vector spaces, $\mathcal{U} \subset V$ an open set, and $f : \mathcal{U} \to W$ a mapping.

(i) We say that f is *differentiable* at a point $p \in \mathcal{U}$ if there exists a (necessarily unique) linear mapping $f'(p) \in L(V, W) = \mathrm{Hom}(V, W)$ such that

$$\lim_{v \to 0} \frac{f(p + v) - f(p) - f'(p)(v)}{\|v\|} = 0, \qquad (\text{C.1.1})$$

where $\| \ \|$ is an arbitrary norm on V. The mapping f is *differentiable over* $\mathcal{U}$ if it is differentiable at each point of $\mathcal{U}$. In this case, the

653

derivative of f is the mapping

$$f' : \mathcal{U} \to L(V, W), \quad p \in \mathcal{U} \mapsto f'(p).$$

If, in addition, f' is continuous, then f is said to be *of class* C^1.

(ii) If f' is differentiable at $p \in \mathcal{U}$, then we say that f is *twice differentiable* at p. Its derivative is denoted by $f''(p)$ and called the *second derivative* of f at p. Then $f''(p)$ is an element of the vector space

$$L(V, L(V, W)) \cong L^2(V, W).$$

We say that f is *twice differentiable over* $\mathcal{U}$ if it is twice differentiable at each point of $\mathcal{U}$, and in this case the second derivative of f is

$$f'' : \mathcal{U} \to L(V, L(V, W)) \cong L^2(V, W), \quad p \in \mathcal{U} \mapsto f''(p).$$

If f'' is continuous, then f is *of class* C^2.

(iii) Now we proceed inductively. Suppose that f is $r-1$ times differentiable in an open neighbourhood of $p \in \mathcal{U}$, and its derivative of order $r-1$ is denoted by $f^{(r-1)}$. If $f^{(r-1)}$ is differentiable at $p \in \mathcal{U}$, then we say that f is r *times differentiable* at p, and its r*th derivative* at p is $f^{(r)}(p) := \left(f^{(r-1)}\right)'(p)$. It is an element of the vector space

$$L\left(V, L^{r-1}(V, W)\right) \cong L^r(V, W).$$

If f is r times differentiable at each point of $\mathcal{U}$, then f is r *times differentiable over* $\mathcal{U}$, and its rth derivative is

$$f^{(r)} : \mathcal{U} \to L\left(V, L^{r-1}(V, W)\right) \cong L^r(V, W), \quad p \in \mathcal{U} \mapsto f^{(r)}(p).$$

If $f^{(r)}$ is continuous, then f is *of class* C^r. Finally, f is *of class* C^∞ if it is of class C^r for each $r \in \mathbb{N}^*$. Mappings of class C^∞ are also mentioned as *smooth mappings*.

(iv) Let $\mathcal{V} \subset W$ be also an open set and $r \in \mathbb{N}^* \cup \{\infty\}$. We say that a bijective mapping $f : \mathcal{U} \to \mathcal{V}$ is a *diffeomorphism of class* C^r if both f and f^{-1} are of class C^r. Smooth diffeomorphisms, i.e., diffeomorphisms of class C^∞ are simply mentioned as diffeomorphisms.

Remark C.1.2. If $V = \mathbb{R}$, then $f'(p) \in L(\mathbb{R}, W)$, which may be identified with the vector $f'(p)(1) \in W$. In this case, the open set $\mathcal{U} \subset \mathbb{R}$ is typically an open interval, which we denote by I rather than by $\mathcal{U}$. Thus, if f is differentiable on I, we can write $f' : I \to W$.

Remark C.1.3. The isomorphism $L(V, L(V, W)) \cong L^2(V, W)$ may be established in two different canonical ways. One possibility is

$$f''(p)(v, w) := (f''(p)(v))(w),$$

and the other one would be obtained with the role of v and w interchanged. Later we shall see that it is of no importance which one we choose.

Example C.1.4. If $\varphi \in L(V, W)$, then φ is differentiable, and $\varphi'(p) = \varphi$ for each $p \in V$. If $r \geq 2$, then $\varphi^{(r)}(p) = 0$ at every point $p \in V$. Thus φ is of class C^{∞}.

Lemma C.1.5. *Let V_1, V_2 and W be finite-dimensional real vector spaces. If $b: V_1 \times V_2 \to W$ is a bilinear mapping, then b is differentiable at every point $(p_1, p_2) \in V_1 \times V_2$, and its derivative is given by*

$$b'(p_1, p_2)(v_1, v_2) = b(p_1, v_2) + b(v_1, p_2); \quad (v_1, v_2) \in V_1 \times V_2. \quad \text{(C.1.2)}$$

Proof. Suppose that $V_1 \times V_2$ and W are equipped with norms, both denoted by the same symbol $\| \; \|$. First we show that

$$\lim_{(v_1, v_2) \to 0} \frac{\|b(v_1, v_2)\|}{\|(v_1, v_2)\|} = 0. \quad (*)$$

Since b is continuous at 0, for each positive number ε we can find a positive number δ such that $\|b(v_1, v_2)\| \leq \varepsilon$ whenever $\|(v_1, v_2)\| \leq \delta$. Equivalently,

$$\left\| b\left(\frac{\delta(v_1, v_2)}{\|(v_1, v_2)\|} \right) \right\| \leq \varepsilon \quad \text{for every } (v_1, v_2) \in V_1 \times V_2 \setminus \{0\}. \quad (**)$$

Now let $(v_1, v_2) \in V_1 \times V_2$ such that $\|(v_1, v_2)\| < \delta^2$. Then

$$\begin{aligned}
\frac{\|b(v_1, v_2)\|}{\|(v_1, v_2)\|} &= \|(v_1, v_2)\| \frac{\|b(v_1, v_2)\|}{\|(v_1, v_2)\|^2} \\
&= \frac{\|(v_1, v_2)\|}{\delta^2} \left\| b\left(\frac{\delta v_1}{\|(v_1, v_2)\|}, \frac{\delta v_2}{\|(v_1, v_2)\|} \right) \right\| \\
&= \frac{\|(v_1, v_2)\|}{\delta^2} \left\| b\left(\frac{\delta(v_1, v_2)}{\|(v_1, v_2)\|} \right) \right\| \overset{(**)}{\leq} \frac{\|(v_1, v_2)\|}{\delta^2} \varepsilon < \varepsilon,
\end{aligned}$$

and this proves $(*)$. Thus, we find that

$$\begin{aligned}
\lim_{(v_1, v_2) \to 0} & \frac{b((p_1, p_2) + (v_1, v_2)) - b(p_1, p_2) - b(p_1, v_2) - b(v_1, p_2)}{\|(v_1, v_2)\|} \\
&= \lim_{(v_1, v_2) \to 0} \frac{b(p_1 + v_1, p_2 + v_2) - b(p_1, p_2) - b(p_1, v_2) - b(v_1, p_2)}{\|(v_1, v_2)\|} \\
&= \lim_{(v_1, v_2) \to 0} \frac{b(v_1, v_2)}{\|(v_1, v_2)\|} \overset{(*)}{=} 0,
\end{aligned}$$

as was to be shown. $\square$

Proposition C.1.6. *If f is differentiable at $p \in \mathcal{U}$, then for any $v \in V$ we have*

$$f'(p)(v) = \lim_{t \to 0} \frac{f(p + tv) - f(p)}{t}. \quad \text{(C.1.3)}$$

Proof.   We start from the right-hand side as follows:

$$\lim_{t \to 0^+} \frac{f(p + tv) - f(p)}{t} = \lim_{t \to 0^+} \left(\frac{f(p + tv) - f(p) - f'(p)(tv)}{t} + f'(p)(v) \right)$$

$$= \|v\| \lim_{t \to 0^+} \frac{f(p + tv) - f(p) - f'(p)(tv)}{\|tv\|} + f'(p)(v) = f'(p)(v)$$

by the definition of $f'(p)$. A similar calculation shows that

$$\lim_{t \to 0^-} \frac{f(p + tv) - f(p)}{t} = f'(p)(v)$$

also holds, from which the assertion follows.   □

Lemma C.1.7. *The mapping $f : \mathcal{U} \subset V \to W$ is differentiable at $p \in \mathcal{U}$ if, and only if, there is a mapping $\varphi : \mathcal{U} \to L(V, W)$ such that it is continuous at p and*

$$f(q) - f(p) = \varphi(q)(q - p)$$

for any $q \in \mathcal{U}$.

Proof.   First suppose that there exists such a mapping φ. We show that f satisfies the definition of differentiability with $f'(p) = \varphi(p)$. Indeed,

$$\lim_{v \to 0} \frac{f(p + v) - f(p) - \varphi(p)(v)}{\|v\|} = \lim_{q \to p} \frac{f(q) - f(p) - \varphi(p)(q - p)}{\|q - p\|}$$

and

$$\left\| \frac{f(q) - f(p) - \varphi(p)(q - p)}{\|q - p\|} \right\| = \left\| \frac{(\varphi(q) - \varphi(p))(q - p)}{\|q - p\|} \right\| \leq \|\varphi(q) - \varphi(p)\|,$$

which tends to 0 if $q \to p$ by the continuity of φ in p. (We used the operator norm on the finite-dimensional real vector space $L(V, W)$, see (B.4.1).) From this it follows that f is differentiable at p with $f'(p) = \varphi(p)$.

   Conversely, let f be differentiable at p. We may consider an auxiliary positive definite scalar product $\langle \, , \, \rangle$ on the vector space V and define a mapping $\varphi : \mathcal{U} \to L(V, W)$ by

$$\varphi(q)(v) := \begin{cases} f'(p)(v) + \dfrac{\langle q - p, v \rangle}{\|q - p\|^2}(f(q) - f(p) - f'(p)(q - p)) & \text{if } q \neq p \\ f'(p)(v) & \text{if } q = p, \end{cases}$$

where v is an arbitrary vector in V. An easy calculation shows that the required equality holds indeed. Thus it remains to show that φ is continuous at p. We have

$$\|\varphi(q)(v) - \varphi(p)(v)\| = \left\| \frac{\langle q - p, v \rangle}{\|q - p\|^2}(f(q) - f(p) - f'(p)(q - p)) \right\|$$

$$\leq \frac{\|f(q) - f(p) - f'(p)(q - p)\|}{\|q - p\|}\|v\|,$$

therefore

$$\|\varphi(q) - \varphi(p)\| \leq \frac{\|f(q) - f(p) - f'(p)(q-p)\|}{\|q-p\|}.$$

Here the right-hand side tends to 0 as $q \to p$, from which the continuity of φ at p follows. $\qquad\square$

Proposition C.1.8 (Chain Rule). *Let U, V, W be finite-dimensional real vector spaces, $\mathcal{U} \subset U$, $\mathcal{V} \subset V$ open sets, $f\colon \mathcal{U} \to \mathcal{V}$, $g\colon \mathcal{V} \to W$ mappings, and $p \in \mathcal{U}$. If f is differentiable at p, and g is differentiable at $f(p)$, then $g \circ f$ is also differentiable at p, and*

$$(g \circ f)'(p) = g'(f(p)) \circ f'(p). \tag{C.1.4}$$

Proof. By the previous lemma, there is a mapping $\varphi\colon \mathcal{U} \to L(U,V)$, continuous at p, such that

$$f(q) - f(p) = \varphi(q)(q-p) \quad (q \in \mathcal{U}),$$

and there is a mapping $\psi\colon \mathcal{V} \to L(V,W)$, continuous at $f(p)$, such that

$$g(r) - g(f(p)) = \psi(r)(r - f(p)) \quad (r \in \mathcal{V}).$$

Thus, for any $q \in \mathcal{U}$, we have

$$g(f(q)) - g(f(p)) = \psi(f(q))(f(q) - f(p)) = \psi(f(q))(\varphi(q)(q-p)).$$

Now define the mapping

$$\chi\colon \mathcal{U} \to L(U,W), \ \chi(q)(u) := \psi(f(q))(\varphi(q)(u)); \quad q \in \mathcal{U}, u \in U.$$

Then

$$g(f(q)) - g(f(p)) = \chi(q)(q-p) \quad (q \in \mathcal{U}),$$

and since the composition of mappings

$$\varepsilon\colon L(V,W) \times L(U,V) \to L(U,W), \ (\alpha, \beta) \mapsto \alpha \circ \beta$$

is bilinear and hence continuous, $\chi = \varepsilon \circ (\psi \circ f, \varphi)$ is also continuous at p, thus, again by the previous lemma, $g \circ f$ is differentiable at p with derivative

$$(g \circ f)'(p) = \chi(p) = \psi(f(p)) \circ \varphi(p) = g'(f(p)) \circ f'(p). \qquad\square$$

Corollary C.1.9. *If f and g are as in the previous proposition, and both are of class C^1, then $g \circ f$ is also of class C^1.*

Proof. Under these conditions, $g \circ f$ is differentiable at each point of $\mathcal{U}$. In addition, if ε is the composition of linear mappings as in the previous proof, then

$$(g \circ f)' = \varepsilon \circ (g' \circ f, f'),$$

thus $(g \circ f)' \colon \mathcal{U} \to L(U, W)$ is also continuous, and $g \circ f$ is of class C^1. $\square$

Example C.1.10. Let V be a finite-dimensional real vector space and $b \colon V \times V \to \mathbb{R}$ a symmetric bilinear form. If Q is the quadratic form associated with b (see 1.4.2), then Q is differentiable at every point $p \in V$, and

$$Q'(p)(v) = 2b(p, v), \quad v \in V. \tag{C.1.5}$$

Indeed, Q can be written in the form $Q = b \circ (1_V \times 1_V)$, where the product $\times$ is defined by (A.2.2). Since $1'_V(p) = 1_V$ for every $p \in V$ (Example C.1.4), using the chain rule we find that

$$Q'(p)(v) = (b \circ (1_V \times 1_V))'(p)(v) = b'(p, p)(v, v)$$
$$\stackrel{\text{(C.1.2)}}{=} b(p, v) + b(v, p) = 2b(p, v),$$

as we claimed.

Example C.1.11. Let $\mathcal{U}$ be an open subset of a (finite-dimensional, real) vector space. Let $f \colon \mathcal{U} \to \mathbb{R}$ be a C^2 function and $c \colon I \to \mathbb{R}$ a C^2 curve. Then for every $t \in I$,

$$(f \circ c)''(t) = f'(c(t))(c''(t)) + f''(c(t))(c'(t), c'(t)). \tag{C.1.6}$$

To see this, we use the chain rule twice. First,

$$(f \circ c)'(t) = f'(c(t))(c'(t)) = \delta \circ (f' \circ c, c')(t),$$

where $\delta \colon V^* \times V \to \mathbb{R}$ is the Kronecker tensor. Secondly,

$$(f \circ c)''(t) = (\delta \circ (f' \circ c, c'))'(t)$$
$$= \delta'\big(f'(c(t)), c'(t)\big)\big(f''(c(t))(c'(t)), c''(t)\big)$$
$$\stackrel{\text{(C.1.2)}}{=} \delta(f'(c(t)), c''(t)) + \delta(f''(c(t))(c'(t)), c'(t))$$
$$\stackrel{\text{Example 1.2.3}}{=} f'(c(t))(c''(t)) + f''(c(t))(c'(t), c'(t)),$$

as we claimed.

Now we turn to the study of differentiable mappings between open subsets of the Euclidean n- and m-spaces $\mathbb{R}^n$ and $\mathbb{R}^m$.

Proposition C.1.12. *Let $\mathcal{U} \subset \mathbb{R}^n$ be an open set. A mapping $f \colon \mathcal{U} \to \mathbb{R}^m$ is differentiable at a point $p \in \mathcal{U}$ if, and only if, each of its coordinate functions $f^i := e^i \circ f$ $(i \in J_m)$ is differentiable at p, where $\left(e^i\right)_{i=1}^m$ is the dual of the canonical basis of $\mathbb{R}^m$ (see Example 1.1.35). In this case, its derivative acts as*

$$f'(p)(v) = \begin{pmatrix} (f^1)'(p)(v) \\ \vdots \\ (f^m)'(p)(v) \end{pmatrix} \quad (v \in \mathbb{R}^n).$$

The mapping f is of class C^1 if, and only if, each of its coordinate functions is of class C^1.

Proof. By Example C.1.4, each member e^i of the canonical dual basis is differentiable, and its derivative at every point of $\mathbb{R}^m$ is itself. Thus, if f is differentiable at p, then, by the chain rule,

$$(f^i)'(p) = (e^i \circ f)'(p) = (e^i)'(f(p))(f'(p)) = e^i(f'(p)).$$

Conversely, suppose that each component function f^i is differentiable at p, and let

$$c_i \colon \mathbb{R} \to \mathbb{R}^m, \ t \in \mathbb{R} \mapsto te_i \in \mathbb{R}^m$$

be the ith canonical embedding of $\mathbb{R}$ into $\mathbb{R}^m$ for each $i \in J_m$. Then

$$f = \sum_{i=1}^m c_i \circ f^i.$$

Each mapping c_i is linear, thus its derivative at any point is itself, which may be identified with the vector e_i by Remark C.1.3. Thus we have

$$f'(p) = \sum_{i=1}^m c_i{}' \left(f^i(p)\right)\left((f^i)'(p)\right) = \sum_{i=1}^m (f^i)'(p)e_i,$$

from which it follows that f is differentiable at p, and the formula for its derivative can also be easily seen. From this formula the necessary and sufficient condition for f to be of class C^1 also follows. $\square$

Definition C.1.13. Let $\mathcal{U} \subset \mathbb{R}^n$ be an open set, $f \colon \mathcal{U} \to \mathbb{R}^m$ a mapping, and $p \in \mathcal{U}$. If the limit

$$D_i f(p) := \lim_{t \to 0} \frac{f(p + te_i) - f(p)}{t}$$

exists for some $i \in J_n$, then it is said to be the *ith partial derivative of f* at p.

Remark C.1.14. If f is differentiable at p, then, by Proposition C.1.6, each of its partial derivatives exists at p, and $D_j f(p) = f'(p)(e_j)$ for $j \in J_n$. Moreover, if $f^i := e^i \circ f$ $(i \in J_m)$, then each partial derivative of each function f^i exists at p, and $D_j f(p) = \big(D_j f^1(p), \ldots, D_j f^m(p)\big)$.

Definition C.1.15. Let $\mathcal{E}_n := (e_1, \ldots, e_n)$ be the canonical basis of $\mathbb{R}^n$ and $\mathcal{E}_m := (\bar{e}_1, \ldots, \bar{e}_m)$ the canonical basis of $\mathbb{R}^m$. If a mapping $f \colon \mathcal{U} \subset \mathbb{R}^n \to \mathbb{R}^m$ is differentiable at $p \in \mathcal{U}$, then the matrix of the linear mapping $f'(p)$ with respect to the bases $\mathcal{E}_n$ and $\mathcal{E}_m$ (Example 1.1.22), i.e., the matrix

$$J_f(p) := [f'(p)]^{\mathcal{E}_n}_{\mathcal{E}_m}$$

is called the *Jacobian matrix* of f at p.

Proposition C.1.16. If $f \colon \mathcal{U} \subset \mathbb{R}^n \to \mathbb{R}^m$ is differentiable at $p \in \mathcal{U}$, and $f^i := e^i \circ f$ $(i \in J_m)$, then the Jacobian matrix of f at p can be calculated as

$$J_f(p) = \begin{pmatrix} D_1 f^1(p) & \cdots & D_n f^1(p) \\ \vdots & & \vdots \\ D_1 f^m(p) & \cdots & D_n f^m(p) \end{pmatrix},$$

or, in a more succinct notation, $(J_f(p))^i_j = D_j f^i(p)$ $(i \in J_m,\ j \in J_n)$.

Proof. As we saw in Example 1.1.22, the jth column of $J_f(p)$ contains the components of the vector $f'(p)(e_j) = D_j f(p)$ with respect to the basis $\mathcal{E}_m$ for each $j \in J_n$, and, by the previous remark, these components are just $D_j f^1(p), \ldots, D_j f^m(p)$. $\qquad\square$

Proposition C.1.17. A mapping $f \colon \mathcal{U} \subset \mathbb{R}^n \to \mathbb{R}^m$ is of class C^1 if, and only if, each of its partial derivatives $D_j f$ exists and each one is continuous on $\mathcal{U}$ for $j \in J_n$.

For a proof, see [48], Theorem 4.2. or [42], sections 8.9 and 8.10.

Proposition C.1.18. Let U, V, W be finite-dimensional real vector spaces, $\mathcal{U} \subset U$, $\mathcal{V} \subset V$ open sets, $f \colon \mathcal{U} \to \mathcal{V}$, $g \colon \mathcal{V} \to W$ mappings, and $r \in \mathbb{N}^*$.

(i) If $p \in \mathcal{U}$, f is r times differentiable at p, and g is r times differentiable at $f(p)$, then $g \circ f$ is also r times differentiable at p.

(ii) If f and g are of class C^r, then $g \circ f$ is also of class C^r.

(iii) If f and g are of class C^∞, then $g \circ f$ is also of class C^∞.

Proof. (i) We prove the assertion by induction on r. For $r = 1$, this is just the chain rule. Now let $r \geq 2$, and suppose that the assertion is true for $r-1$. Then f is differentiable in a neighbourhood of p, g is differentiable in a neighbourhood of $f(p)$, and

$$(g \circ f)' = \varepsilon \circ (g' \circ f, f') \qquad (*)$$

in a neighbourhood of p, where

$$\varepsilon \colon L(V, W) \times L(U, V) \to L(U, W), \ (\alpha, \beta) \mapsto \alpha \circ \beta$$

is the composition of linear mappings, which is bilinear, hence of class C^∞. Thus, the right-hand side of $(*)$ is $r - 1$ times differentiable at p, and $g \circ f$ is r times differentiable at p.

(ii) We proceed again by induction on r. If f and g are of class C^r, then $(*)$ holds on the whole of $\mathcal{U}$, and it is easy to see that the right-hand side is of class C^{r-1} by the induction hypothesis. Thus $(g \circ f)'$ is of class C^{r-1}, and $g \circ f$ is of class C^r.

(iii) Since f and g are of class C^r for each $r \in \mathbb{N}^*$, $g \circ f$ is also of class C^r for each $r \in \mathbb{N}^*$ by (ii), so $g \circ f$ is of class C^∞. $\qquad\square$

Theorem C.1.19 (Schwarz). *Let V be a finite-dimensional real vector space, $\mathcal{U} \subset V$ an open set, and suppose that the function $f \colon \mathcal{U} \to \mathbb{R}$ is twice differentiable at a point $p \in \mathcal{U}$. Then the second derivative of f at p is a symmetric bilinear function, i.e.,*

$$f''(p)(u, v) = f''(p)(v, u)$$

for any vectors $u, v \in V$.

Proof. By the condition, f is differentiable in an open neighbourhood of p. Without loss of generality, we may assume that f is differentiable on $\mathcal{U}$.

Since $\mathcal{U}$ is open, there is a positive number a such that $p + su + tv \in \mathcal{U}$ whenever $s, t \in [0, a]$. Define a function $\Delta \colon [0, a] \to \mathbb{R}$ by

$$\Delta(t) := f(p + tu + tv) - f(p + tu) - f(p + tv) + f(p), \quad t \in [0, a].$$

We shall show that

$$\lim_{t \to 0} \frac{\Delta(t)}{t^2} = f''(p)(u, v) = f''(p)(v, u).$$

First, observe that

$$\Delta(t) := f(p + tu + tv) - f(p + tu) - (f(p + tv) - f(p)),$$

thus, applying the mean value theorem ([11], p. 130) to the function

$$s \in [0, t] \mapsto f(p + su + tv) - f(p + su),$$

we obtain that there is a number $\tau \in \,]0, t[$ (depending, of course, on the choice of t) such that

$$\Delta(t) = t(f'(p + \tau u + tv)(u) - f'(p + \tau u)(u)).$$

Since f is twice differentiable at p, this implies, in particular, that the function $q \mapsto f'(q)(u)$ is differentiable at p, i.e.,

$$\lim_{w \to 0} \frac{f'(p + w)(u) - f'(p)(u) - f''(p)(w, u)}{\|w\|} = 0.$$

Let

$$\eta(w) := \begin{cases} \dfrac{f'(p + w)(u) - f'(p)(u) - f''(p)(w, u)}{\|w\|} & \text{if } w \in (\mathcal{U} - p) \setminus \{0\}, \\ 0 & \text{if } w = 0. \end{cases}$$

Then η is continuous at 0, and

$$f'(p + w)(u) = f'(p)(u) + f''(p)(w, u) + \|w\|\eta(w)$$

whenever $w \in \mathcal{U} - p$. Now we substitute this into the expression of Δ above with the choices $w := \tau u + tv$ and $w := \tau u$ to obtain

$$\begin{aligned} \Delta(t) &= t(f'(p)(u) + f''(p)(\tau u + tv, u) + \|\tau u + tv\|\eta(\tau u + tv) \\ &\quad - f'(p)(u) - f''(p)(\tau u, u) - \|\tau u\|\eta(\tau u)) \\ &= t(f''(p)(tv, u) + \|\tau u + tv\|\eta(\tau u + tv) - \|\tau u\|\eta(\tau u)) \\ &= t^2 f''(p)(v, u) + t(\|\tau u + tv\|\eta(\tau u + tv) - \|\tau u\|\eta(\tau u)). \end{aligned}$$

Thus

$$\lim_{t \to 0} \frac{\Delta(t)}{t^2} = f''(p)(v, u) + \lim_{t \to 0} \frac{\|\tau u + tv\|}{t}\eta(\tau u + tv)$$

$$- \lim_{t \to 0} \frac{\|\tau u\|}{t}\eta(\tau u) = f''(p)(v, u),$$

since $0 < \tau < t$ (thus both fractions are bounded from above) and η is continuous at 0. Interchanging the role of u and v in the deduction above, we obtain the same result with u and v interchanged, which implies that the assertion is true. $\qquad\qquad\qquad\qquad\qquad\qquad\qquad\qquad\qquad\quad \square$

Corollary C.1.20. *If V and W are finite-dimensional real vector spaces, $\mathcal{U} \subset V$ is an open set, $r \in \mathbb{N}^*$, and the mapping $f \colon \mathcal{U} \to W$ is r times differentiable at a point $p \in \mathcal{U}$, then its rth derivative $f^{(r)}(p)$ is a symmetric r-linear mapping.*

Proof. The statement in this corollary exceeds that of the previous theorem in two aspects: first, the range of f is not necessarily the real line; second, we state here that the derivative of any order is symmetric (if it exists). The first obstacle is surmounted more easily by taking a dual basis on W and applying the theorem of Schwarz to each component function of f. As to the second, one has to proceed by induction on r; we leave the details to the reader. $\qquad\square$

Lemma C.1.21 (Mean Value Inequality). *Let V and W be finite-dimensional real vector spaces. Let f be a continuous mapping of a neighbourhood of a segment of $\overline{p(p+v)} := \{p + tv \in V \mid t \in [0,1]\}$ into W. If f is differentiable at every point of $\overline{p(p+v)}$, then*

$$\|f(p+v) - f(p)\| \le \|v\| \sup_{t \in [0,1]} \|f'(p+tv)\|.$$

For a proof (in the context of Banach spaces) we refer to [42], Section 8.5. See also [2], 2.4.8 and [63], XVII, §4.

Theorem C.1.22 (Taylor's Formula). *Let V be a finite-dimensional real vector space, $\mathcal{U} \subset V$ an open set. Let $p \in \mathcal{U}$ and $v \in V$ such that the line segment $\overline{p(p+v)}$ is contained in $\mathcal{U}$. If the function $f : \mathcal{U} \to \mathbb{R}$ is $r + 1$ times differentiable at each point of $\overline{p(p+v)}$ for some $r \in \mathbb{N}$, then there exists a number $\vartheta \in {]0,1[}$ such that*

$$f(p+v) = f(p) + \sum_{k=1}^{r} \frac{f^{(k)}(p)(v,\ldots,v)}{k!} + \frac{f^{(r+1)}(p+\vartheta v)(v,\ldots,v)}{(r+1)!}.$$

Proof. Define a function $F : [0,1] \to \mathbb{R}$ by

$$t \in [0,1] \mapsto F(t) := f(p+tv).$$

Using the chain rule and the interpretation in Remark C.1.2 for the inner mapping $t \mapsto p + tv$, we obtain by induction that

$$F^{(k)}(t) = f^{(k)}(p+tv)(v,\ldots,v) \quad (t \in [0,1], k \in J_{r+1}). \qquad (*)$$

Now let us define a function $g : [0,1] \to \mathbb{R}$ by

$$g(t) = F(1) - \sum_{k=0}^{r} \frac{F^{(k)}(t)}{k!}(1-t)^k - \left(F(1) - \sum_{k=0}^{r} \frac{F^{(k)}(0)}{k!} \right)(1-t)^{(r+1)}.$$

An easy calculation shows that $g(0) = g(1) = 0$, thus, by Rolle's classical mean value theorem (see, e.g., [11], p. 129), there is a point $\vartheta \in {]0,1[}$ such that $g'(\vartheta) = 0$. Another easy calculation shows that

$$g'(\vartheta) = \left((r+1) \left(F(1) - \sum_{k=0}^{r} \frac{F^{(k)}(0)}{k!} \right) - \frac{F^{(r+1)}(\vartheta)}{r!} \right) (1 - \vartheta)^r.$$

Here the second factor cannot be zero, thus the first factor must be zero, which is equivalent to

$$F(1) = \sum_{k=0}^{r} \frac{F^{(k)}(0)}{k!} + \frac{F^{(r+1)}(\vartheta)}{(r+1)!}.$$

Substituting the expression $(*)$ into the last equality for the values $t = 0$ and $t = \vartheta$, we obtain the desired formula. $\qquad\square$

Lemma C.1.23. *Let $\mathcal{U} \subset \mathbb{R}^n$ be an open set and let $[a,b] \subset \mathbb{R}$ be a closed interval. Let $f \colon \mathcal{U} \times [a,b] \to \mathbb{R}$ be a continuous function such that the partial derivatives $D_i f \colon \mathcal{U} \times [a,b] \to \mathbb{R}$ exist and are continuous for each $i \in J_n$. For any fixed $p \in \mathcal{U}$, let $j_p \colon [a,b] \to \mathcal{U} \times [a,b]$ be the inclusion map opposite p given by $j_p(t) := (p,t)$, $t \in [a,b]$. Define a function $F \colon \mathcal{U} \to \mathbb{R}$ by*

$$F(p) := \int_a^b f \circ j_p, \quad p \in \mathcal{U}.$$

Then F is of class C^1, and

$$D_i F(p) = \int_a^b D_i f \circ j_p.$$

For a proof, see [64], Ch. XVII, §8.

The following theorem is of great importance. However, its proof is rather laborious, thus we refer the interested reader to [64] or [48]; see also [1] and [42], where the theorem is stated and proved in the more general setting of Banach spaces.

Theorem C.1.24 (Inverse Mapping Theorem). *Let V and W be finite-dimensional real vector spaces, $\mathcal{U} \subset V$ an open set, $r \in \mathbb{N}^* \cup \{\infty\}$, $f \colon \mathcal{U} \to W$ a mapping of class C^r, and $p \in \mathcal{U}$. If the derivative $f'(p) \in L(V,W)$ is a linear isomorphism, then p has an open neighbourhood $\mathcal{V} \subset \mathcal{U}$ and $f(p)$ has an open neighbourhood $\mathcal{W} \subset W$ such that the restriction $f \restriction \mathcal{V}$ is a diffeomorphism of class C^r between $\mathcal{V}$ and $\mathcal{W}$.*

Of course, the condition of the theorem can hold only if $\dim V = \dim W$. Moreover, from the chain rule it can be also seen that there can be a diffeomorphism of class C^r between two open subsets of two vector spaces only if they have the same dimension.

C.2 Canonical Constructions

Throughout this section, by a vector space we mean a (non-trivial) finite-dimensional real vector space equipped with its canonical smooth structure described in Example 2.1.9. In the following we present some constructions of sections 3.1 and 4.1 in the more specific context of vector spaces.

C.2.1 Tangent Bundle and Derivative

Lemma C.2.1. *Let $\mathcal{U}$ be an open subset (and hence an open submanifold) of a vector space V, and let $f \in C^\infty(\mathcal{U})$. Given any point p in $\mathcal{U}$, for every $v \in V$ we have*

$$\iota_p(v)(f) = f'(p)(v), \qquad (C.2.1)$$

where ι_p is the canonical isomorphism of V onto $T_pV \cong T_p\mathcal{U}$ given by Lemma 3.1.19.

Proof. Consider, as in the proof of Lemma 3.1.19, the curve

$$\alpha \colon \mathbb{R} \to V, \ t \mapsto \alpha(t) := p + tv.$$

Then

$$\iota_p(v)(f) = \dot\alpha(0)(f) \overset{(3.1.9)}{=} (f \circ \alpha)'(0) = \lim_{t \to 0} \frac{f(\alpha(t)) - f(\alpha(0))}{t}$$

$$= \lim_{t \to 0} \frac{f(p + tv) - f(p)}{t} \overset{(C.1.3)}{=} f'(p)(v)$$

as we claimed. □

Example C.2.2. Let, as above, $\mathcal{U}$ be (a nonempty) open subset of the vector space V. *The tangent bundle*

$$\tau_\mathcal{U} \colon T\mathcal{U} \to \mathcal{U} \qquad (*)$$

of $\mathcal{U}$ and the trivial V-bundle

$$\mathrm{pr}_1 \colon \mathcal{U} \times V \to \mathcal{U} \qquad (**)$$

are canonically isomorphic: the mapping

$$\iota \colon \mathcal{U} \times V \to T\mathcal{U}, \ (p, v) \mapsto \iota(p, v) := \iota_p(v) \qquad (C.2.2)$$

is a strong bundle isomorphism. Its inverse is the strong bundle map

$$T\mathcal{U} \to \mathcal{U} \times V, \ v \mapsto (p, (\iota_p)^{-1}(v)) \quad \text{if } v \in T_p\mathcal{U}. \qquad (C.2.3)$$

In particular, we have the natural identifications

$$T_p\mathcal{U} \cong \{p\} \times V, \ v \mapsto (p, (\iota_p)^{-1}(v)), \quad p \in \mathcal{U}. \qquad (C.2.4)$$

We freely identify the isomorphic vector bundles $(*)$ and $(**)$, and omit the identifying map ι whenever there is no danger of confusion.

Using this identification, any vector field X on $\mathcal{U}$ can be regarded as a mapping of the form

$$X = (1_{\mathcal{U}}, \underline{X}), \quad \underline{X} \in C^\infty(\mathcal{U}, V);$$

see Example 2.2.27. We recall that $\underline{X}$ is said to be the *principal part* of X. Then

$$X(p) = (1_{\mathcal{U}}, \underline{X})(p) \overset{(A.2.1)}{=} (p, \underline{X}(p)), \quad p \in \mathcal{U}. \qquad (C.2.5)$$

If, in particular, $X(p) = (p, v)$, $p \in \mathcal{U}$ for some fixed $v \in V$, then X is called the *constant vector field corresponding to v*.

Remark C.2.3. Let V and W be vector spaces, and let $\mathcal{U} \subset V$ be an open subset. Consider a differentiable mapping $\varphi \colon \mathcal{U} \to W$. Sometimes it proves convenient to write

$$\varphi'(p, v) := \varphi'(p)(v), \quad (p, v) \in \mathcal{U} \times V,$$

and, by an abuse of notation, to define the mapping

$$\varphi' \colon \mathcal{U} \times V \to W, \ (p, v) \mapsto \varphi'(p, v) = \varphi'(p)(v). \qquad (C.2.6)$$

We present some illustrations of this formalism.

(a) Let $X = (1_{\mathcal{U}}, \underline{X})$ be a vector field on $\mathcal{U}$. Then

$$\begin{cases} \varphi' \circ X \colon \mathcal{U} \to \mathcal{U} \times V \to W, \\ \varphi' \circ X(p) = \varphi'(p, \underline{X}(p)) \overset{(C.2.6)}{=} \varphi'(p)(\underline{X}(p)). \end{cases} \qquad (C.2.7)$$

By a slight abuse of notation, sometimes we write $\varphi'(X)$ instead of $\varphi' \circ X$. Then, in particular, we have

$$f'(X) = Xf, \quad f \in C^\infty(\mathcal{U}). \qquad (C.2.8)$$

Indeed, from (C.2.7), $f'(X)(p) = f'(p)(\underline{X}(p))$, for all $p \in \mathcal{U}$. On the other hand

$$Xf(p) \overset{(3.1.15)}{=} X(p)(f) = (p, \underline{X}(p))(f) \overset{(C.2.2)}{=} \iota_p(\underline{X}(p))(f) \overset{(C.2.1)}{=} f'(p)(\underline{X}(p)),$$

which justifies our claim.

If X is the constant vector field corresponding to $v \in V$, then we obtain that

$$Xf(p) = f'(p)(v) = f'(p, v); \quad f \in C^\infty(\mathcal{U}), \ p \in \mathcal{U}. \qquad (C.2.9)$$

(b) Suppose that φ is a linear mapping of V into W. Then

$$\varphi' = \varphi \circ \mathrm{pr}_2 \colon V \times V \to V \to W. \tag{C.2.10}$$

Indeed, for every $(p, v) \in V \times V$,

$$\varphi'(p, v) := \varphi'(p)(v) \stackrel{\text{Example C.1.4}}{=} \varphi(v) = \varphi \circ \mathrm{pr}_2(p, v).$$

Lemma C.2.4. *Let V and W be vector spaces, and let φ be a smooth mapping from an open subset $\mathcal{U}$ of V to an open subset $\mathcal{V}$ of W.*

(i) *The derivative*

$$\varphi_* \colon T\mathcal{U} \cong \mathcal{U} \times V \to T\mathcal{V} \cong \mathcal{V} \times W$$

of φ defined by Lemma 3.1.33 and (3.1.4) can be represented in the form

$$\varphi_* = (\varphi \circ \tau_{\mathcal{U}}, \varphi'), \tag{C.2.11}$$

where φ' is given by (C.2.6).

(ii) *Let $X \in \mathfrak{X}(\mathcal{U})$ and $Y \in \mathfrak{X}(\mathcal{V})$ be vector fields with principal parts $\underline{X}$ and $\underline{Y}$, respectively. Then X and Y are φ-related if, and only if,*

$$\varphi' \circ X = \underline{Y} \circ \varphi, \tag{C.2.12}$$

i.e., the following diagram commutes:

$$
\begin{array}{ccc}
\mathcal{U} & \stackrel{X}{\longrightarrow} & \mathcal{U} \times V \\
\varphi \downarrow & & \downarrow \varphi' \\
\mathcal{V} & \underset{\underline{Y}}{\longrightarrow} & W
\end{array}
.
$$

Equivalently,

$$\varphi'(p)(\underline{X}(p)) = \underline{Y}(\varphi(p)), \quad p \in \mathcal{U}. \tag{C.2.13}$$

Proof. (i) Let $(p, v) \in \mathcal{U} \times V \cong T\mathcal{U}$. Then, for every $f \in C^\infty(\mathcal{U})$,

$$(\varphi_*)_p(p, v)(f) \stackrel{(3.1.4)}{:=} (p, v)(f \circ \varphi) \stackrel{(C.2.2)}{=} \iota_p(v)(f \circ \varphi) \stackrel{(C.2.1)}{=} (f \circ \varphi)'(p)(v)$$

$$\stackrel{(C.1.4)}{=} f'(\varphi(p))(\varphi'(p)(v)) \stackrel{(C.2.1)}{=} \iota_{\varphi(p)}(\varphi'(p)(v))(f)$$

$$\stackrel{(C.2.2)}{=} (\varphi(p), \varphi'(p)(v))(f) = (\varphi \circ \tau_{\mathcal{U}}, \varphi')(p, v)(f),$$

whence (C.2.11).

(ii) On the one hand,

$$\varphi_* \circ X \stackrel{(C.2.11)}{=} (\varphi \circ \tau_{\mathcal{U}}, \varphi') \circ X = (\varphi, \varphi' \circ X);$$

on the other hand,

$$Y \circ \varphi = (1_{\mathcal{V}}, \underline{Y}) \circ \varphi = (\varphi, \underline{Y} \circ \varphi).$$

Thus $X \underset{\varphi}{\sim} Y$ if, and only if, (C.2.12) is satisfied. The equivalence of (C.2.12) and (C.2.13) can also be seen immediately. $\qquad\square$

Lemma C.2.5. *Let V be a vector space and $\mathcal{U}$ an open subset of V. Given two vectors u, v in V, let X and Y be the constant vector fields on $\mathcal{U}$ corresponding to u and v, respectively. Then $[X, Y] = 0$.*

Proof. Choose a point $p \in \mathcal{U}$ and a function $f \in C^\infty(\mathcal{U})$. Define the function

$$f'_u : \mathcal{U} \to \mathbb{R}, \quad p \mapsto f'_u(p) := f'(p, u) := f'(p)(u).$$

Then, by (C.2.9), $Xf = f'_u$. Therefore,

$$(Xf)'(p)(v) = (f'_u)'(p)(v) \overset{\text{(C.1.3)}}{=} \lim_{t \to 0} \frac{f'_u(p + tv) - f'_u(p)}{t}$$

$$= \left(\lim_{t \to 0} \frac{f'(p + tv) - f'(p)}{t} \right)(u) \overset{\text{(C.1.3)}}{=} f''(p)(v)(u) \overset{\text{Remark C.1.3}}{=} f''(p)(v, u).$$

Thus we obtain that

$$Y(Xf)(p) \overset{\text{(C.2.9)}}{=} (Xf)'(p, v) = f''(p)(v, u) \overset{\text{Theorem C.1.19}}{=} f''(p)(u, v)$$

$$= (Yf)'(p, u) = X(Yf)(p),$$

whence our claim. $\square$

Proposition C.2.6. *Let $\mathcal{U}$ be a nonempty open subset of a vector space V. If $X = (1_\mathcal{U}, \underline{X})$ and $Y = (1_\mathcal{U}, \underline{Y})$ are vector fields on $\mathcal{U}$, then*

$$[X, Y](p) = \big(p, \underline{Y}'(p)(\underline{X}(p)) - \underline{X}'(p)(\underline{Y}(p))\big), \quad p \in \mathcal{U}. \tag{C.2.14}$$

Equivalently,

$$[X, Y] = (1_\mathcal{U}, \underline{Y}' \circ X - \underline{X}' \circ Y) =: (1_\mathcal{U}, \underline{Y}'(X) - \underline{X}'(Y)). \tag{C.2.15}$$

Proof. **Step 1.** Choose a basis $(u_i)_{i=1}^n$ of V. For every $i \in J_n$, let U_i be the constant vector field corresponding to u_i. Then $(U_i)_{i=1}^n$ is a frame field of $T\mathcal{U} \cong \mathcal{U} \times V$, so X and Y can be represented in the form $X = X^i U_i$ and $Y = Y^i U_i$, where $X^i, Y^i \in C^\infty(\mathcal{U})$. For every $p \in \mathcal{U}$,

$$(p, \underline{X}(p)) = X(p) = (X^i U_i)(p) = (p, X^i(p)u_i),$$

therefore $\underline{X}(p) = X^i(p)u_i$. In the same way, $\underline{Y}(p) = Y^i(p)u_i$. Now we calculate the derivative of the mapping $\underline{X} : \mathcal{U} \to V$ at a point $p \in \mathcal{U}$.

For every $v \in V$,

$$\underline{X}'(p)(v) \overset{\text{(C.1.3)}}{=} \lim_{t \to 0} \frac{\underline{X}(p + tv) - \underline{X}(p)}{t}$$

$$= \left(\lim_{t \to 0} \frac{X^i(p + tv) - X^i(p)}{t} \right) u_i \overset{\text{(C.1.3)}}{=} (X^i)'(p)(v)u_i.$$

An analogous formula is valid for the derivative $\underline{Y}'(p)$. So we have

$$\underline{X}'(p)(v) = (X^i)'(p)(v)u_i, \quad \underline{Y}'(p)(v) = (Y^i)'(p)(v)u_i. \tag{C.2.16}$$

Step 2. Using the preceding lemma and (3.1.39), we find that

$$[X, Y] = [X^j U_j, Y^i U_i] = X^j(U_j Y^i)U_i - Y^i(U_i X^j)U_j$$
$$= (X^j(U_j Y^i) - Y^j(U_j X^i))U_i.$$

Evaluating both sides at a point $p \in \mathcal{U}$ gives

$$[X, Y](p) = (p, X^j(p)(U_j Y^i)(p)u_i - Y^j(p)(U_j X^i)(p)u_i)$$
$$\overset{(\text{C.2.9})}{=} (p, X^j(p)(Y^i)'(p)(u_j)u_i - Y^j(p)(X^i)'(p)(u_j)u_i)$$
$$= (p, (Y^i)'(p)(X^j(p)u_j)u_i - (X^i)'(p)(Y^j(p)u_j)u_i)$$
$$\overset{\text{Step 1}}{=} (p, (Y^i)'(p)(\underline{X}(p))u_i - (X^i)'(p)(\underline{Y}(p))u_i)$$
$$\overset{(\text{C.2.16})}{=} (p, \underline{Y}'(p)(\underline{X}(p)) - \underline{X}'(p)(\underline{Y}(p))),$$

which proves (C.2.14). The equivalence of (C.2.14) and (C.2.15) is clear. $\square$

C.2.2 *Lifts of Functions*

We recall that the product $V \times W$ of two vector spaces has a natural linear structure defined by the formulae

$$(v, w) + (v_1, w_1) := (v + v_1, w + w_1), \qquad \lambda(v, w) := (\lambda v, \lambda w)$$

for any $(v, w), (v_1, w_1) \in V \times W$ and any scalar $\lambda \in \mathbb{R}$ (cf. Example 1.1.18). We note that this is the only linear structure for $V \times W$ which makes the canonical projections

$$\mathrm{pr}_1 \colon V \times W \to V \quad \text{and} \quad \mathrm{pr}_2 \colon V \times W \to W$$

defined in A.2.2 linear.

In the following we assume that $V \times W$ is equipped with the product topology (Example B.1.2(c)) and the product manifold structure (Example 2.1.8). These constructions are extended to the product of any positive number of vector spaces. We denote the first and the second projection of the product of several vector spaces also by pr_1 and pr_2. This is an abuse of notation again, but it seems tolerable.

Let $\mathcal{U}$ be an open subset of V. Then, as we have seen above, the tangent bundle of $\mathcal{U}$ is $T\mathcal{U} \cong \mathcal{U} \times V$ with the projection $\tau_{\mathcal{U}} = \mathrm{pr}_1$. Here $\mathcal{U} \times V$ is an open subset of $V \times V$, so the tangent bundle of $T\mathcal{U}$, i.e., the double tangent bundle of $\mathcal{U}$ is

$$TT\mathcal{U} \cong T(\mathcal{U} \times V) \cong \mathcal{U} \times V \times V \times V \tag{C.2.17}$$

with the projection

$$\tau_{TU} = (\mathrm{pr}_1, \mathrm{pr}_2)\colon (p, v, w_1, w_2) \mapsto (p, v). \tag{C.2.18}$$

Lemma C.2.7. *Suppose that $\mathcal{U}$ is an open subset of a vector space V. Let f be a smooth function on $\mathcal{U}$. Let*

$$f^{\mathsf{v}}\colon \mathcal{U} \times V \to \mathbb{R} \quad and \quad f^{\mathsf{c}}\colon \mathcal{U} \times V \to \mathbb{R}$$

be the vertical and the complete lifts of f as introduced in Definition 4.1.3. Then, for any $p \in \mathcal{U}$ and any $v, w_1, w_2 \in V$,

$$f^{\mathsf{v}}(p, v) = f(p), \quad f^{\mathsf{c}}(p, v) = f'(p)(v), \tag{C.2.19}$$

$$(f^{\mathsf{v}})'(p, v)(w_1, w_2) = f'(p)(w_1), \tag{C.2.20}$$

$$(f^{\mathsf{c}})'(p, v)(w_1, w_2) = f'(p)(w_2) + f''(p)(w_1, v). \tag{C.2.21}$$

Proof. The first relation in (C.2.19) is evident from the definition of f^{v}. Since

$$f^{\mathsf{c}}(p, v) := \iota_p(v)(f) \stackrel{\text{(C.2.1)}}{=} f'(p)(v),$$

the second relation is also true. Obviously,

$$f^{\mathsf{v}} = f \circ \mathrm{pr}_1\colon \mathcal{U} \times V \to \mathcal{U} \to \mathbb{R}.$$

Since the mapping pr_1 is linear on $V \times V$, it is everywhere differentiable, and we have $\mathrm{pr}'_1(p) = \mathrm{pr}_1$ for all $p \in \mathcal{U}$ (Example C.1.4). Thus

$$(f^{\mathsf{v}})'(p, v)(w_1, w_2) = (f \circ \mathrm{pr}_1)'(p, v)(w_1, w_2)$$

$$\stackrel{\text{(C.1.4)}}{=} f'(p)(\mathrm{pr}_1(w_1, w_2)) = f'(p)(w_1),$$

as desired.

To verify (C.2.21), we use (C.1.3):

$$(f^{\mathsf{c}})'(p, v)(w_1, w_2) = \lim_{t \to 0} \frac{f^{\mathsf{c}}((p, v) + t(w_1, w_2)) - f^{\mathsf{c}}(p, v)}{t}$$

$$= \lim_{t \to 0} \frac{f^{\mathsf{c}}(p + tw_1, v + tw_2) - f^{\mathsf{c}}(p, v)}{t}$$

$$\stackrel{\text{(C.2.19)}}{=} \left(\lim_{t \to 0} \frac{f'(p + tw_1) - f'(p)}{t} \right)(v) + \left(\lim_{t \to 0} \frac{tf'(p + tw_1)}{t} \right)(w_2)$$

$$= f''(p)(w_1)(v) + f'(p)(w_2) = f''(p)(w_1, v) + f'(p)(w_2).$$

This concludes the proof. $\square$

C.2.3   The Vector Bundle $\tau_* \colon TTU \to TU$

Lemma C.2.8. *Let U be an open subset of a vector space V and consider its tangent bundle $\tau \colon TU \cong U \times V \to U$. Then the derivative*

$$\tau_* \colon TTU \overset{(C.2.17)}{=} U \times V \times V \times V \to TU \cong U \times V$$

is given by

$$\tau_*(p, v, w_1, w_2) = (p, w_1), \tag{C.2.22}$$

briefly,

$$\tau_* = (\mathrm{pr}_1, \mathrm{pr}_3). \tag{C.2.23}$$

Proof.   As we have already discussed, the bundle projection τ can be identified with the first projection of $U \times V$. Thus, applying (C.2.11) to $\varphi := \tau = \mathrm{pr}_1$, we find that

$$\tau_* = (\mathrm{pr}_1 \circ \tau_{TU}, \mathrm{pr}_1').$$

Here $\mathrm{pr}_1 \circ \tau_{TU}$ is equal to the first projection of $U \times V \times V \times V$ (cf. (C.2.18)). From (C.2.10) it follows that

$$\mathrm{pr}_1' = \mathrm{pr}_1 \circ (\mathrm{pr}_3, \mathrm{pr}_4),$$

therefore

$$(\mathrm{pr}_1)'(p, v)(w_1, w_2) = \mathrm{pr}_1(w_1, w_2) = w_1 = \mathrm{pr}_3(p, v, w_1, w_2),$$

thus proving (C.2.22) and (C.2.23). $\square$

Lemma C.2.9. *With the notation as above, let $\kappa \colon TTU \to TTU$ be the canonical involution defined by Proposition 4.1.8. Then for every*

$$(p, v, w_1, w_2) \in TTU \cong U \times V \times V \times V,$$

we have

$$\kappa(p, v, w_1, w_2) = (p, w_1, v, w_2), \tag{C.2.24}$$

therefore

$$\kappa = (\mathrm{pr}_1, \mathrm{pr}_3, \mathrm{pr}_2, \mathrm{pr}_4). \tag{C.2.25}$$

Proof.   Write $\kappa = (K_{\mathcal{U}}, K_V, K_1, K_2)$ with

$$K_{\mathcal{U}} := \mathrm{pr}_1 \circ \kappa, \quad K_V := \mathrm{pr}_2 \circ \kappa; \quad K_i := \mathrm{pr}_{i+2} \circ \kappa, \quad i \in \{1, 2\}.$$

By conditions (i) and (ii) in Proposition 4.1.8,

$$(\mathrm{pr}_1, \mathrm{pr}_2) \overset{(\mathrm{C}.2.18)}{=} \tau_{T\mathcal{U}} = \tau_* \circ \kappa \overset{(\mathrm{C}.2.23)}{=} (\mathrm{pr}_1, \mathrm{pr}_3) \circ (K_{\mathcal{U}}, K_V, K_1, K_2)$$
$$= (K_{\mathcal{U}}, K_1),$$

$$(\mathrm{pr}_1, \mathrm{pr}_3) \overset{(\mathrm{C}.2.23)}{=} \tau_* = \tau_{T\mathcal{U}} \circ \kappa \overset{(\mathrm{C}.2.18)}{=} (\mathrm{pr}_1, \mathrm{pr}_2) \circ (K_{\mathcal{U}}, K_V, K_1, K_2)$$
$$= (K_{\mathcal{U}}, K_V).$$

Thus $\kappa = (\mathrm{pr}_1, \mathrm{pr}_3, \mathrm{pr}_2, K_2)$, therefore

$$\kappa(p, v, w_1, w_2) = (p, w_1, v, K_2(p, v, w_1, w_2)). \tag{$*$}$$

It remains to determine the mapping K_2. For this we use condition (iii) in Proposition 4.1.8. It requires that

$$\kappa(p, v, w_1, w_2) f^{\mathrm{c}} = (p, v, w_1, w_2) f^{\mathrm{c}}, \tag{$**$}$$

for any $f \in C^\infty(\mathcal{U})$ and any $(p, u, w_1, w_2) \in TT\mathcal{U}$. Using (C.2.1) and $(*)$, $(**)$ can be written in the form

$$(f^{\mathrm{c}})'(p, w_1)(v, K_2(p, v, w_1, w_2)) = (f^{\mathrm{c}})'(p, v)(w_1, w_2).$$

In view of (C.2.21), this leads to

$$f'(p)(K_2(p, v, w_1, w_2)) + f''(p)(v, w_1) = f'(p)(w_2) + f''(p)(w_1, v),$$

whence, by Theorem C.1.19,

$$f'(p)(K_2(p, v, w_1, w_2)) = f'(p)(w_2).$$

Since V^* has a basis of the form $(f_i'(p))_{i=1}^n$, $f_i \in C^\infty(\mathcal{U})$, from this it follows that $K_2 = \mathrm{pr}_4$, which concludes the proof. $\square$

Lemma C.2.10. *Every smooth second-order vector field*

$$S : T\mathcal{U} \cong \mathcal{U} \times V \to TT\mathcal{U} \cong \mathcal{U} \times V \times V \times V$$

can be represented in the form

$$S = (1_{T\mathcal{U}}, \mathrm{pr}_2, S_2) = (\mathrm{pr}_1, \mathrm{pr}_2, \mathrm{pr}_2, S_2),$$

where $S_2 : \mathcal{U} \times V \to V$ *is a smooth mapping. Thus*

$$S(p, v) = (p, v, v, S_2(p, v)) \quad \text{for all} \quad (p, v) \in \mathcal{U} \times V.$$

Proof.   Let $S = (f_1, f_2, S_1, S_2)$, where

$$f_i := \mathrm{pr}_i \circ S, \quad S_i := \mathrm{pr}_{i+2} \circ S; \quad i \in \{1, 2\}.$$

Then, on the one hand,

$$(\mathrm{pr}_1, \mathrm{pr}_2) = 1_{T\mathcal{U}} = \tau_{T\mathcal{U}} \circ S \overset{(\mathrm{C}.2.18)}{=} (\mathrm{pr}_1, \mathrm{pr}_2) \circ S = (f_1, f_2),$$

because S is a vector field on $T\mathcal{U}$. On the other hand, since S is a second-order vector field, by Lemma 5.1.3 we have

$$(\mathrm{pr}_1, \mathrm{pr}_2) = 1_{T\mathcal{U}} = \tau_* \circ S \overset{(\mathrm{C}.2.23)}{=} (\mathrm{pr}_1, \mathrm{pr}_3) \circ S = (f_1, S_1).$$

Therefore, $S = (\mathrm{pr}_1, \mathrm{pr}_2, \mathrm{pr}_2, S_2)$, as was to be shown. $\square$

C.2.4 Lifts of Vector Fields

Lemma C.2.11. *Let $\mathcal{U}$ be an open subset of a vector space V, and let X be a vector field on $\mathcal{U}$ with principal part $\underline{X}$. Then the vertical and the complete lift of X can be represented in the form*
$$X^{\mathsf{v}} = (1_{T\mathcal{U}}, 0, \underline{X} \circ \mathrm{pr}_1) = (\mathrm{pr}_1, \mathrm{pr}_2, 0, \underline{X} \circ \mathrm{pr}_1)$$
and
$$X^{\mathsf{c}} = (1_{T\mathcal{U}}, \underline{X} \circ \mathrm{pr}_1, \underline{X}') = (\mathrm{pr}_1, \mathrm{pr}_2, \underline{X} \circ \mathrm{pr}_1, \underline{X}'),$$
respectively. Therefore, for every $(p, v) \in \mathcal{U} \times V$,
$$X^{\mathsf{v}}(p, v) = (p, v, 0, \underline{X}(p)), \quad X^{\mathsf{c}}(p, v) = (p, v, \underline{X}(p), \underline{X}'(p)(v)).$$

Proof. Since X^{v} and X^{c} are vector fields on $T\mathcal{U} \cong \mathcal{U} \times V$, by our preceding argument they are of the form
$$X^{\mathsf{v}} = (\mathrm{pr}_1, \mathrm{pr}_2, X_1^{\mathsf{v}}, X_2^{\mathsf{v}}) \quad \text{and} \quad X^{\mathsf{c}} = (\mathrm{pr}_1, \mathrm{pr}_2, X_1^{\mathsf{c}}, X_2^{\mathsf{c}}),$$
where
$$X_i^{\mathsf{v}} \colon \mathcal{U} \times V \to V, \quad X_i^{\mathsf{c}} \colon \mathcal{U} \times V \to V, \quad i \in \{1, 2\}$$
are smooth mappings. Our task is to express them in terms of the principal part of X.

(a) By Lemma 4.1.28, $X^{\mathsf{v}} \underset{\tau}{\sim} o$. Thus
$$o \circ \tau = \tau_* \circ X^{\mathsf{v}} \overset{(\mathrm{C.2.23})}{=} (\mathrm{pr}_1, \mathrm{pr}_3) \circ X^{\mathsf{v}} = (\mathrm{pr}_1, X_1^{\mathsf{v}}),$$
whence $X_1^{\mathsf{v}} = 0$.

To determine X_2^{v}, we use (4.1.51). For any smooth function $f \in C^{\infty}(\mathcal{U})$ and any $(p, v) \in \mathcal{U} \times V$,
$$X^{\mathsf{v}} f^{\mathsf{c}}(p, v) = X^{\mathsf{v}}(p, v)(f^{\mathsf{c}}) = (p, v, 0, X_2^{\mathsf{v}}(p, v))(f^{\mathsf{c}}) = \iota_{(p,v)}(0, X_2^{\mathsf{v}}(p, v))(f^{\mathsf{c}})$$
$$\overset{(\mathrm{C.2.1})}{=} (f^{\mathsf{c}})'(p, v)(0, X_2^{\mathsf{v}}(p, v)) \overset{(\mathrm{C.2.21})}{=} f'(p)(X_2^{\mathsf{v}}(p, v));$$
$$(Xf)^{\mathsf{v}}(p, v) \overset{(\mathrm{C.2.19})}{=} (Xf)(p) = (p, \underline{X}(p))(f) \overset{(\mathrm{C.2.1})}{=} f'(p)(\underline{X}(p)),$$
so (4.1.51) yields
$$f'(p)(X_2^{\mathsf{v}}(p, v)) = f'(p)(\underline{X}(p)).$$
As in the proof of Lemma C.2.9, from this we infer that $X_2^{\mathsf{v}} = \underline{X} \circ \tau = \underline{X} \circ \mathrm{pr}_1$, as desired.

(b) We derive the representation of X^{c}. Starting with the definition, we get
$$X^{\mathsf{c}} = \kappa \circ X_* = \kappa \circ (1_{\mathcal{U}}, \underline{X})_* = \kappa \circ ((1_{\mathcal{U}}, \underline{X}) \circ \mathrm{pr}_1, (1_{\mathcal{U}}, \underline{X})')$$
$$= \kappa \circ (\mathrm{pr}_1, \underline{X} \circ \mathrm{pr}_1, 1_{\mathcal{U}}', \underline{X}') \overset{(\mathrm{C.2.25})}{=} (\mathrm{pr}_1, 1_{\mathcal{U}}', \underline{X} \circ \mathrm{pr}_1, \underline{X}').$$
Here, at every $(p, v) \in \mathcal{U} \times V$,
$$1_{\mathcal{U}}'(p)(v) \overset{(\mathrm{C.1.3})}{=} \lim_{t \to 0} \frac{p + tv - p}{t} = v = \mathrm{pr}_2(p, v),$$
whence the desired result. $\qquad\square$

C.2.5 $\widetilde{\delta}$, i, j *and* J

Finsler bundles were introduced in Remark 4.1.1(a), in the setting of abstract manifolds. Now we consider the Finsler bundle

$$\pi \colon T\mathcal{U} \times_\mathcal{U} T\mathcal{U} \to T\mathcal{U} \tag{C.2.26}$$

over $T\mathcal{U} \cong \mathcal{U} \times V$, where $\mathcal{U}$ is an open subset of the vector space V. We recall that the fibre of π over $(p,v) \in \mathcal{U} \times V$ is

$$\pi^{-1}((p,v)) = \{(p,v)\} \times T_p\mathcal{U} \cong \{(p,v)\} \times \{p\} \times V \cong \{(p,v)\} \times V.$$

so the elements of $T\mathcal{U} \times_\mathcal{U} T\mathcal{U}$ can be regarded as triples of the form

$$(p,v,w) \in \mathcal{U} \times V \times V.$$

More precisely, the mapping

$$T\mathcal{U} \times_\mathcal{U} T\mathcal{U} \to \mathcal{U} \times V \times V, \quad ((p,v),(p,w)) \mapsto (p,v,w) \tag{C.2.27}$$

defines a canonical strong bundle isomorphism between π and the vector bundle

$$(\mathrm{pr}_1, \mathrm{pr}_2) \colon \mathcal{U} \times V \times V \to \mathcal{U} \times V.$$

In the following we freely identify these isomorphic bundles.

A section $\widetilde{X}$ of the bundle (C.2.27) (and hence of π) can be represented in the form

$$\widetilde{X} = (1_{T\mathcal{U}}, \underline{\widetilde{X}}) = (\mathrm{pr}_1, \mathrm{pr}_2, \underline{\widetilde{X}}),$$

where $\underline{\widetilde{X}}$, the principal part of $\widetilde{X}$, is a smooth mapping from $\mathcal{U} \times V$ to V. The *basic section* determined by a vector field $X = (1_\mathcal{U}, \underline{X}) \in \mathfrak{X}(\mathcal{U})$ is

$$\widehat{X} = (1_{T\mathcal{U}}, \underline{X} \circ \mathrm{pr}_1) = (\mathrm{pr}_1, \mathrm{pr}_2, \underline{X} \circ \mathrm{pr}_1).$$

According to (4.1.3), the *canonical section* $\widetilde{\delta}$ of π is the section whose principal part is also the identity transformation $1_{T\mathcal{U}} = (\mathrm{pr}_1, \mathrm{pr}_2)$. Thus

$$\widetilde{\delta} = (1_{T\mathcal{U}}, 1_{T\mathcal{U}}) = (\mathrm{pr}_1, \mathrm{pr}_2, \mathrm{pr}_1, \mathrm{pr}_2),$$

i.e., for every $(p,v) \in \mathcal{U} \times V$,

$$\widetilde{\delta}(p,v) = ((p,v),(p,v)) = (p,v,p,v).$$

Equivalently, using the interpretation (C.2.27),

$$\widetilde{\delta} = (1_{T\mathcal{U}}, \mathrm{pr}_2), \quad \widetilde{\delta}(p,v) = (p,v,v).$$

Lemma C.2.12. *Let $\mathcal{U}$ be an open subset of a vector space V. Given a point p in $\mathcal{U}$, consider the canonical inclusion*

$$i_p \colon T_p\mathcal{U} \cong \{p\} \times V \to T\mathcal{U} \cong \mathcal{U} \times V.$$

Then $i_p = 1_{\mathcal{U} \times V} \restriction \{p\} \times V$, and for every $(p,v), (p,w) \in \{p\} \times V$ we have

$$(i_p)'(p,v)(p,w) = (0,w). \tag{C.2.28}$$

Proof. Note first that here $\{p\} \times V$ is regarded as a vector space with the linear structure which makes it isomorphic to V. On the open subset $\mathcal{U} \times V$ of $V \times V$ we use the linear structure inherited from $V \times V$. Keeping these in mind, we calculate:

$$(i_p)'(p,v)(p,w) \overset{(C.1.3)}{=} \lim_{t \to 0} \frac{i_p((p,v) + t(p,w)) - i_p(p,v)}{t}$$

$$= \lim_{t \to 0} \frac{i_p(p, v + tw) - i_p(p,v)}{t} = \lim_{t \to 0} \frac{(p, v + tw) - (p,v)}{t} = (0, w). \quad \square$$

Corollary C.2.13. *We preserve the notation of the preceding lemma.*

(i) *For any $p \in \mathcal{U}$ and any $v, w \in V$,*

$$(i_p)_*(p, v, p, w) = (p, v, 0, w), \qquad (C.2.29)$$

therefore

$$(i_p)_* = (\mathrm{pr}_1, \mathrm{pr}_2, 0, \mathrm{pr}_4). \qquad (C.2.30)$$

(ii) *The vertical lift of $(p, w) \in T_p\mathcal{U} \cong \{p\} \times V$ to $(p, v) \in T_p\mathcal{U}$ is*

$$(p, w)^\uparrow(p, v) = (p, v, 0, w) = \iota_{(p,v)}(0, w) \in V_{(p,v)}T\mathcal{U}. \qquad (C.2.31)$$

Proof. (i) In view of (C.2.28), by a slight abuse of notation, $i'_p = (0, \mathrm{pr}_4)$. Using this, we find that

$$(i_p)_* \overset{(C.2.11)}{=} (i_p \circ \tau_{\mathcal{U} \times V}, (i_p)') = (i_p \circ (\mathrm{pr}_1, \mathrm{pr}_2), 0, \mathrm{pr}_4) = (\mathrm{pr}_1, \mathrm{pr}_2, 0, \mathrm{pr}_4).$$

So we have (C.2.30), which implies (C.2.29).

(ii) By definition,

$$(p, w)^\uparrow(p, v) := \ell^\uparrow_{(p,v)}(p, w) \overset{(4.1.30)}{=} (i_p)_* \circ \iota_{(p,v)}(p, w)$$

$$\overset{(C.2.2)}{=} (i_p)_*(p, v, p, w) \overset{(C.2.29)}{=} (p, v, 0, w),$$

as desired. $\square$

Corollary C.2.14. *We continue to assume that $\mathcal{U}$ is an open subset of a vector space V.*

(i) *The canonical surjection*

$$\mathbf{j} := (\tau_{T\mathcal{U}}, \tau_*) \colon TT\mathcal{U} \cong \mathcal{U} \times V \times V \times V \to T\mathcal{U} \times_{\mathcal{U}} T\mathcal{U} \cong \mathcal{U} \times V \times V$$

can be represented as

$$\mathbf{j} = (\tau_{T\mathcal{U}}, \mathrm{pr}_3). \qquad (C.2.32)$$

(ii) *The* canonical isomorphism

$$\mathbf{i} \colon TU \times_U TU \cong U \times V \times V \to VTU \cong U \times V \times \{0\} \times V$$

is of the form

$$\mathbf{i} = (\pi, 0, \mathrm{pr}_3). \tag{C.2.33}$$

(iii) *For the* vertical endomorphism

$$\mathbf{J} := \mathbf{i} \circ \mathbf{j} \colon U \times V \times V \times V \to U \times V \times V \times V$$

we have the representation

$$\mathbf{J} = (\tau_{TU}, 0, \mathrm{pr}_3). \tag{C.2.34}$$

Equivalently, for any $p \in U$ and any $v, w, w_1, w_2 \in V$,

$$\mathbf{j}(p, v, w_1, w_2) = (p, v, w_1), \quad \mathbf{i}(p, v, w) = (p, v, 0, w),$$
$$\mathbf{J}(p, v, w_1, w_2) = (p, v, 0, w_1).$$

Proof.   (i) By a direct calculation,

$$\mathbf{j}(p, v, w_1, w_2) = (\tau_{TU}, \tau_*)(p, v, w_1, w_2)$$
$$\overset{\text{(A.2.1),(C.2.22)}}{=} (p, v, p, w_1) \cong (p, v, w_1),$$

which shows (C.2.32).

(ii) First we note that, by (C.2.22),

$$VTU = \mathrm{Ker}(\tau_*) = \{(p, v, 0, w) \in TTU \mid p \in U; \ v, w \in V\},$$

so we indeed have $VTU \cong U \times V \times \{0\} \times V$. Thus

$$\mathbf{i}(p, v, w) \overset{(4.1.38)}{:=} \ell_{(p,v)}^{\uparrow}(p, w) \overset{(C.2.31)}{=} (p, v, 0, w).$$

(iii) As a consequence of the preceding observations,

$$\mathbf{J}(p, v, w_1, w_2) = \mathbf{i} \circ \mathbf{j}(p, v, w_1, w_2) = \mathbf{i}(p, v, w_1) = (p, v, 0, w_1). \qquad \square$$

C.3   The Standard Covariant Derivative

Throughout this section, V is a (non-trivial) real vector space of dimension n and U is a nonempty open subset of V. We denote by D the standard covariant derivative on U, more precisely, on the tangent bundle $TU \cong U \times V$ of U (see Example 6.1.5).

Lemma C.3.1. *For any vector fields* $X = (1_\mathcal{U}, \underline{X})$, $Y = (1_\mathcal{U}, \underline{Y})$ *on* $\mathcal{U}$,

$$D_X Y(p) = \left(p, \underline{Y}'(p)(\underline{X}(p))\right), \quad p \in \mathcal{U}. \tag{C.3.1}$$

Both the torsion and the curvature of D are zero.

Proof. Let $(U_i)_{i=1}^n$ be the frame field on $T\mathcal{U}$ determined by a basis $(u_i)_{i=1}^n$ of V, as in the proof of Proposition C.2.6. Then the vector field Y can be written in the form $Y = Y^i U_i$ with some smooth functions $Y^i \in C^\infty(\mathcal{U})$, and we obtain

$$D_X Y(p) \overset{(6.1.10)}{=} X(Y^i) U_i(p) = X(Y^i)(p)(p, u_i)$$

$$\overset{(C.2.8)}{=} (Y^i)'(p)(\underline{X}(p))(p, u_i) = (p, (Y^i)'(p)(\underline{X}(p))u_i)$$

$$\overset{(C.2.16)}{=} (p, \underline{Y}'(p)(\underline{X}(p))),$$

as desired.

By Example 6.1.40, the curvature of D is zero. It also has vanishing torsion because for all $p \in \mathcal{U}$,

$$T(X,Y)(p) \overset{(6.1.46)}{=} \left(D_X Y - D_Y X - [X,Y]\right)(p)$$

$$\overset{(C.3.1)}{=} \left(p, \underline{Y}'(p)(\underline{X}(p)) - \underline{X}'(p)(\underline{Y}(p))\right) - [X,Y](p) \overset{(C.2.14)}{=} 0.$$

This completes the proof. □

Lemma C.3.2. *Let f be a smooth function on $\mathcal{U}$. If X and Y are constant vector fields on $\mathcal{U}$ corresponding to $v \in V$ and $w \in V$, respectively, then*

$$\mathrm{Hess}(f)(X,Y)(p) \overset{(6.1.24)}{:=} D(Df)(X,Y)(p) = f''(p)(v,w), \quad p \in \mathcal{U}. \tag{C.3.2}$$

Proof. From (6.1.25),

$$\mathrm{Hess}(f)(X,Y)(p) = (X(Yf))(p) - (D_{(p,v)}Y)(f)$$

$$\overset{(C.3.1)}{=} (X(Yf))(p) - (p, \underline{Y}'(p)(v)) \overset{Y \text{ is constant}}{=} (X(Yf))(p) = f''(p)(v,w)$$

(for the last step, see proof of Lemma C.2.5). □

Proposition C.3.3. *Assume that V is a semi-Euclidean vector space with scalar product b. Then the mapping*

$$\begin{cases} g: p \in \mathcal{U} \mapsto g_p \in T_2^0(T_p\mathcal{U}) \cong T_2^0(\{p\} \times V), \\ g_p((p,v),(p,w)) := b(v,w) \end{cases}$$

is a semi-Riemannian metric on $\mathcal{U}$. The Levi-Civita derivative on $(\mathcal{U}, g)$ is the standard covariant derivative on $\mathcal{U}$.

Proof. We have only to check that D is metric; then the Ricci lemma (Proposition 6.1.43) and the vanishing of the torsion of D imply our claim.

Let X, Y, Z be vector fields on $\mathcal{U}$ with principal parts $\underline{X}, \underline{Y}, \underline{Z}$, respectively. Let p be a point of $\mathcal{U}$. Note first that

$$g(Y, Z)(p) = g_p(Y(p), Z(p)) = g_p\big((p, \underline{Y}(p)), (p, \underline{Z}(p))\big)$$

$$:= b(\underline{Y}(p), \underline{Z}(p)) \overset{(A.2.1)}{=} b \circ (\underline{Y}, \underline{Z})(p),$$

i.e.,

$$g(Y, Z) = b \circ (\underline{Y}, \underline{Z}). \tag{$*$}$$

Let $v := \underline{X}(p)$. Then

$$X(g(Y, Z))(p) \overset{(*)}{=} X(p)(b \circ (\underline{Y}, \underline{Z})) = \iota_p(v)(b \circ (\underline{Y}, \underline{Z}))$$

$$\overset{(C.2.1)}{=} (b \circ (\underline{Y}, \underline{Z}))'(p)(v) \overset{(C.1.4)}{=} b'(\underline{Y}(p), \underline{Z}(p)) \circ (\underline{Y}, \underline{Z})'(p)(v)$$

$$= b'(\underline{Y}(p), \underline{Z}(p))(\underline{Y}'(p)(v), \underline{Z}'(p)(v))$$

$$\overset{(C.1.2)}{=} b(\underline{Y}(p), \underline{Z}'(p)(v)) + b(\underline{Y}'(p)(v), \underline{Z}(p))$$

$$=: g_p\big((p, \underline{Y}(p)), (p, \underline{Z}'(p)(\underline{X}(p)))\big) + g_p\big((p, \underline{Y}'(p)(\underline{X}(p))), (p, \underline{Z}(p))\big)$$

$$= g(Y, D_X Z)(p) + g(D_X Y, Z)(p),$$

as was to be shown. $\qquad\qquad\square$

Lemma C.3.4. *Let* $\gamma \colon I \to \mathcal{U}$ *be a smooth curve, and let*

$$Z \colon I \to T\mathcal{U} \cong \mathcal{U} \times V, \ t \mapsto (\gamma(t), \underline{Z}(t)), \quad Z \in C^\infty(I, V)$$

be a vector field along γ. *Then*

$$D^\gamma Z = (\gamma, \underline{Z}'), \tag{C.3.3}$$

i.e., the principal part of $D^\gamma Z$ *is just the derivative of the principal part of* Z.

Proof. By Proposition 6.1.49, we may suppose that Z is of the form $Z = X \circ \gamma$, $X \in \mathfrak{X}(\mathcal{U})$. Then

$$D^\gamma Z \overset{(6.1.62)}{=} (DX) \circ \dot{\gamma}.$$

First we check that the seemingly obvious relation

$$\dot{\gamma} = (\gamma, \gamma') \tag{C.3.4}$$

is indeed true. Let $t \in I$. Observe that for every $h \in C^\infty(I)$,

$$\iota_t(1)(h) \overset{(C.2.1)}{=} h'(t)(1) \overset{\text{Remark C.1.2}}{=} h'(t) \overset{(3.1.7)}{=} \left(\frac{d}{dr}\right)_t (h),$$

whence

$$\left(\frac{d}{dr}\right)_t = \iota_t(1) = (t, 1). \tag{$*$}$$

Thus

$$\dot{\gamma}(t) \overset{(3.1.8)}{=} (\gamma_*)_t \left(\frac{d}{dr}\right)_t \overset{(*)}{=} (\gamma_*)_t(t, 1) \overset{(C.2.11)}{=} (\gamma(t), \gamma'(t)(1))$$
$$\overset{\text{Remark } C.1.2}{=} (\gamma(t), \gamma'(t)),$$

as claimed.

After these preparations, we find that

$$(DX) \circ \dot{\gamma}(t) = D_{\dot{\gamma}(t)}X \overset{(C.3.1),(C.3.4)}{=} \left(\gamma(t), \underline{X}'(\gamma(t))(\gamma'(t))\right)$$
$$\overset{(C.1.4)}{=} (\gamma(t), (\underline{X} \circ \gamma)'(t)) = (\gamma, (\underline{X} \circ \gamma)')(t).$$

This ends the proof of (C.3.3). $\qquad\qquad\square$

Proposition C.3.5. *Let* $\gamma \colon I \to \mathcal{U}$ *be a smooth curve. The parallel vector fields along* γ *are the constant vector fields along* γ, *i.e., they are of the form*

$$t \in I \mapsto (\gamma(t), v) \in \mathcal{U} \times V,$$

where $v \in V$ *is any fixed vector.*

Proof. Suppose that $Z \in \mathfrak{X}_\gamma(\mathcal{U})$ is a parallel vector field along γ such that $Z(0) = (\gamma(0), v) \in \{\gamma(0)\} \times V \cong T_{\gamma(0)}\mathcal{U}$. Let $0_\gamma \in \mathfrak{X}_\gamma(\mathcal{U})$ be the zero vector field along γ. Then, since the vector field Z is parallel,

$$0_\gamma = D^\gamma Z \overset{(C.3.3)}{=} (\gamma, \underline{Z}'),$$

so $\underline{Z}'(t) = 0$ for all $t \in I$. This implies immediately that Z is the constant vector field along γ corresponding to v. $\qquad\qquad\square$

Corollary C.3.6. *Preserving the notation above, let* $t_0, t_1 \in I$. *The parallel translation along* γ *from* $\gamma(t_0)$ *to* $\gamma(t_1)$ *is just the canonical isomorphism displayed by the diagram*

Proof.   For every $v \in V$, $P_{t_0}^{t_1}(\gamma)(\gamma(t_0), v) := Z(t_1) = (\gamma(t_1), \underline{Z}(t_1))$, where $Z \in \mathfrak{X}_\gamma(\mathcal{U})$ is the (unique) parallel vector field such that $Z(t_0) = (\gamma(t_0), v)$. Then, by the preceding lemma, $\underline{Z}(t) = v$ for all $t \in I$. Thus

$$P_{t_0}^{t_1}(\gamma)(\gamma(t_0), v) = (\gamma(t_1), v),$$

whence our claim.   □

Corollary C.3.7. *The geodesics of the standard covariant derivative on V are the affinely parametrized straight lines, i.e., the curves of the form*

$$t \in \mathbb{R} \mapsto p + tv \in V,$$

where $(p, v) \in V \times V$ is arbitrarily fixed.

Proof.   By Definition 6.1.63(a), a smooth curve $\gamma \colon \mathbb{R} \to V$ is a geodesic of D if, and only if,

$$0_\gamma = D^\gamma \dot\gamma \overset{(C.3.4)}{=} D^\gamma(\gamma, \gamma') \overset{(C.3.3)}{=} (\gamma, \gamma''),$$

i.e., if, and only if, $\gamma'' = 0$. Thus our claim follows from elementary calculus.   □

Bibliography

[1] Abraham, R. and Marsden, J. E. (1978). *Foundations of Mechanics (Second Edition)* (Benjamin/Cummings, Reading, MA).

[2] Abraham, R., Marsden, J. E. and Ratiu, T. (1988). *Manifolds, Tensor Analysis, and Applications (2nd edition)* (Springer-Verlag, New York).

[3] Alexander, R. (1988). Zonoid theory and Hilbert's fourth problem, *Geom. Dedicata* **28**, 2, pp. 199–211.

[4] Álvarez Paiva, J. C. (2005). Symplectic Geometry and Hilbert's fourth problem, *J. Diff. Geom.* **69**, 2, pp. 353–378.

[5] Amir, D. (1986). *Characterizations of Inner Product Spaces* (Birkhäuser Verlag).

[6] Anastasiei, M. (1996). A historical remark on the connections of Chern and Rund, in D. Bao, S.-S. Chern and Z. Shen (eds.), *Finsler Geometry*, Contemporary Mathematics (American Mathematical Society, Providence), pp. 171–176.

[7] Bácsó, S. and Szilasi, Z. (2010). On the projective theory of sprays, *Acta Math. Acad. Paed. Nyházi* **26**, pp. 171–207.

[8] Bao, D., Chern, S.-S. and Shen, Z. (2000). *An Introduction to Riemann–Finsler geometry* (Springer).

[9] Barden, B. and Thomas, Ch. (2003). *An Introduction to Differential Manifolds* (Imperial College Press, London).

[10] Belgun, F. (2012). Projective and conformal flatness, http://www.math.uni-hamburg.de/home/belgun/.

[11] Berberian, S. K. (1994). *A First Course in Real Analysis* (Springer-Verlag, New York).

[12] Berger, M. (2003). *A Panoramic View of Riemannian Geometry* (Springer).

[13] Berger, M. and Gostiaux, B. (1988). *Differential Geometry: Manifolds, Curves, and Surfaces* (Springer-Verlag, New York).

[14] Berwald, L. (1926). Untersuchung der Krümmung allgemeiner metrischer Räume auf Grund des in ihnen herrschenden Parallelismus, *Math. Z.* **25**, 1, pp. 40–73.

[15] Berwald, L. (1941). On Finsler and Cartan Geometries. III: Two-dimensional Finsler spaces with rectilinear extremals, *Ann. of Math.* **42**,

pp. 84–112.

[16] Berwald, L. (1947). Ueber Finslersche und Cartansche Geometrie IV: Projektivkrümmung allgemeiner affiner Räume und Finslersche Räume skalarer Krümmung, *Ann. of Math.* **48**, 3, pp. 755–781.

[17] Besse, A. L. (1978). *Manifolds all of whose Geodesics are Closed* (Springer-Verlag, Berlin).

[18] Bielecki, A. (1956). Une remarque sur la méthode de Banach – Cacciopoli – Tikhonov dans la théorie des équations différentielles ordinaires, *Bull. Acad. Polon. Sci. Cl. III.* **4**, pp. 261–264.

[19] Brickell, F. (1965). On the differentiability of affine and projective transformations, *Proc. Amer. Math. Soc.* **16**, pp. 567–574.

[20] Brickell, F. and Clark, R. S. (1970). *Differentiable Manifolds* (Van Nostrand, London).

[21] Bröcker, Th. and Jänich, K. (1982). *Introduction to differential topology* (Cambridge University Press).

[22] Bucataru, I. and Muzsnay, Z. (2011). Projective Metrizability and Formal Integrability, *SIGMA* **7**, 114.

[23] Burstall, F. and Rawnsley, J. (2007). Affine connections with W=0, arXiv: math/0702032v2 [math.DG] 26 Feb 2007.

[24] Busemann, H. (1950). The geometry of Finsler spaces, *Bull. Amer. Math. Soc.* **56**, pp. 5–16.

[25] Busemann, H. (1955). *The geometry of geodesics* (Academic Press, Inc, New York).

[26] Busemann, H. (1981). Review: A. V. Pogorelov, Hilbert's fourth problem, *Bull. Amer. Math. Soc.* **4**, 1, pp. 87–90.

[27] Busemann, H. and Mayer, W. (1941). On the foundations of calculus of variations, *Trans. Amer. Math. Soc.* **49**, 2, pp. 173–198.

[28] Cartan, É. (1934). *Les espaces de Finsler* (Hermann, Paris).

[29] Cartan, É. (1988). *Leçons sur la géométrie des espaces de Riemann* (Éditions Jacques Gabay, Sceaux), (reprint of the 2nd (1946) edition).

[30] Crampin, M. (1983). Generalized Bianchi identities for horizontal distributions, *Math. Proc. Camb. Phil. Soc.* **94**, pp. 125–132.

[31] Crampin, M. (2005). Randers spaces with reversible geodesics, *Publ. Math. Debrecen* **67**, 3-4, pp. 401–409.

[32] Crampin, M. (2007). Isotropic and R-flat sprays, *Houston J. Math.* **33**, 2, pp. 451–459.

[33] Crampin, M. (2011). Some remarks on the Finslerian version of Hilbert's fourth problem, *Houston J. Math.* **37**, 2, pp. 369–391.

[34] Crampin, M. (2012). On the construction of Riemannian metrics from Berwald spaces by averaging, to appear in Houston J. Math.

[35] Crampin, M., Mestdag, T. and Saunders, D. J. (2012). The multiplier approach to the projective Finsler metrizability problem, *Diff. Geom. Appl.* **30**, pp. 604–621.

[36] Crampin, M., Mestdag, T. and Saunders, D. J. (2013). Hilbert forms for a Finsler metrizable projective class of sprays, *Diff. Geom. Appl.* **31**, pp. 63–79.

[37] Darboux, G. (1894). *Leçons sur la théorie générale des surfaces III.* (Gauthier-Villars, Paris).

[38] Dazord, P. (1969). *Propriétés globales des géodésiques des Espaces de Finsler*, Thèse (575), Publ. Dep. Math. Lyon.

[39] del Castillo, L. (1976). Tenseurs de Weyl d'une gerbe de directions, *C.R. Acad. Sc. Paris Série A* **282**, pp. 595–598.

[40] Deng, S. and Hou, Z. (2002). The group of isometries of a Finsler space, *Pac. J. Math.* **207**, 1, pp. 149–155.

[41] Diaz, J.-G. (1972). *Etudes des tenseurs de courbure en géométrie finslérienne*, Thèse de 3ème cycle, n°118, Publ. Dep. Math. Lyon.

[42] Dieudonné, J. (1969). *Treatise on Analysis*, Vol. I (Academic Press, New York).

[43] Dieudonné, J. (1972). *Treatise on Analysis*, Vol. III (Academic Press, New York).

[44] Dieudonné, J. (1974). *Treatise on Analysis*, Vol. IV (Academic Press, New York).

[45] Dieudonné, J. (1982). *A Panorama of Pure Mathematics* (Academic Press, New York).

[46] Douglas, J. (1927-28). The general geometry of paths, *Ann. of Math.* **29**, 1/4, pp. 143–168.

[47] Eisenhart, L. P. (2005). *Non-Riemannian Geometry* (Dover edition, New York).

[48] Fleming, W. (1977). *Functions of Several Variables (2nd ed.)* (Springer-Verlag, New York).

[49] Flett, T. M. (1966). *Mathematical Analysis* (McGraw-Hill, New York).

[50] Goodman, F. M. (2012). *Algebra: abstract and concrete, ed. 2.5* (Semi Simple Press, Iowa City, IA).

[51] Greub, W. (1975). *Linear Algebra (Fourth Edition)* (Springer-Verlag, New York).

[52] Greub, W., Halperin, S. and Vanstone, R. (1972). *Connections, Curvature and Cohomology*, Vol. 1 (Academic Press, New York and London).

[53] Greub, W., Halperin, S. and Vanstone, R. (1973). *Connections, Curvature and Cohomology*, Vol. 2 (Academic Press, New York and London).

[54] Grifone, J. (1972). Structure presque-tangente et connexions, I, *Ann. Inst. Fourier Grenoble* **22**, 1, pp. 287–334.

[55] Gruber, P. M. (2007). *Convex and Discrete Geometry* (Springer).

[56] Hamel, G. (1903). Über die Geometrien, in denen die Geraden die Kürzesten sind, Dissertation, Göttingen, 1901, und *Math. Ann.* **57**, 2, pp. 231–264.

[57] Hashiguchi, M. and Ichijyō, Y. (1980). Randers spaces with rectilinear geodesics, *Rep. Fac. Sci. Kagoshima* **13**, pp. 33–40.

[58] Hilbert, D. (2000). Mathematical Problems, *Bull. Amer. Math. Soc. (N.S.)* **37**, 4, pp. 407–436.

[59] Hilgert, J. and Neeb, K.-H. (2012). *Structure and Geometry of Lie Groups* (Springer).

[60] Hunter, J. K. and Nachtergaele, B. (2001). *Applied Analysis* (World Scien-

tific).

[61] Jänich, K. (2001). *Vector Analysis* (Springer).

[62] Klein, J. (1982). Geometry of Sprays. Lagrangian case. Principle of Least Curvature, in *Proc. of the IUTAM-ISIMM Symp. of Torino*, pp. 177–196.

[63] Lang, S. (1993). *Real and Functional Analysis (3rd edition)* (Springer-Verlag, New York and Berlin).

[64] Lang, S. (1997). *Undergraduate Analysis (Second Edition)* (Springer).

[65] Lao Tsu (2011). *Tao Te Ching* (Vintage Books, New York), translated by Gia-Fu Feng and Jane English with Toinette Lippe.

[66] Lee, Jeffrey M. (2009). *Manifolds and Differential Geometry, Graduate Studies in Mathematics*, Vol. 107 (AMS).

[67] Lee, John M. (1997). *Riemannian Manifolds: An Introduction to Curvature* (Springer).

[68] Lee, John M. (2000). *Introduction to Topological Manifolds* (Springer).

[69] Lee, John M. (2003). *Introduction to Smooth Manifolds* (Springer).

[70] Madsen, I. H. and Tornehave, J. (1997). *From Calculus to Cohomology* (Cambridge University Press).

[71] Martínez, E., Cariñena, J. F. and Sarlet, W. (1992). Derivations of differential forms along the tangent bundle projection, *Diff. Geom. Appl.* **2**, 1, pp. 17–43.

[72] Matsumoto, M. (1986). *Foundations of Finsler Geometry and Special Finsler spaces* (Kaiseisha Press, Otsu).

[73] Matsumoto, M. (1995). Every Path Space of dimension Two is Projectively Related to a Finsler Space, *Open Systems & Information Dynamics* **3**, 3, pp. 291–303.

[74] Matsumoto, M. (2003). Finsler Geometry in the 20th Century, in P. L. Antonelli (ed.), *Handbook of Finsler geometry, Vol. 1–2* (Kluwer Academic Publishers, Dordrecht), pp. 557–966.

[75] Matveev, V. S., Rademacher, H.-B., Troyanov, M. and Zeghib, A. (2009). Finsler conformal Lichnerowicz–Obata conjecture, *Ann. Inst. Fourier (Grenoble)* **59**, 3, pp. 937–949.

[76] Michor, P. W. (2008). *Topics in Differential Geometry, Graduate Studies in Mathematics*, Vol. 93 (AMS).

[77] Myers, S. B. and Steenrod, N. (1939). The group of isometries of a Riemannian manifold, *Ann. of Math.* **40**, 2, pp. 400–416.

[78] Okada, T. (1982). Minkowskian product of Finsler spaces and Berwald connection, *J. Math. Kyoto Univ.* **22**, 2, pp. 323–332.

[79] O'Neill, B. (1983). *Semi-Riemannian Geometry* (Academic Press, New York).

[80] Papadopoulos, A. (2005). *Metric Spaces, Convexity and Nonpositive Curvature* (European Mathematical Society).

[81] Petersen, P. (2006). *Riemannian Geometry, 2nd ed.* (Springer).

[82] Pogorelov, A. V. (1979). *Hilbert's fourth problem* (Wiley, New York).

[83] Porteous, I. R. (1969). *Topological Geometry* (Van Nostrand-Reinhold, London).

[84] Rademacher, H.-B. (2004). Nonreversible Finsler Metrics of Positive Flag

Curvature, in *Riemann–Finsler Geometry* (MSRI Publications 50, Cambridge), pp. 261–302.

[85] Rapcsák, A. (1961). Über die bahntreuen Abbildungen metrischer Räume, *Publ. Math. Debrecen* **8**, pp. 285–290.

[86] Roberts, A. W. and Varberg, D. E. (1973). *Convex Functions* (Academic Press, New York).

[87] Roman, S. (2008). *Advanced Linear Algebra* (Springer).

[88] Shen, Z. (2001). *Differential Geometry of Spray and Finsler Spaces* (Kluwer Academic Publishers, Dordrecht).

[89] Spivak, M. (1965). *Calculus on Manifolds* (Addison-Wesley, Reading, Massachusetts).

[90] Szabó, Z. I. (1981). Positive definite Berwald spaces (Structure Theorems on Berwald spaces), *Tensor N.S.* **35**, pp. 25–39.

[91] Szabó, Z. I. (1986). Hilbert's fourth problem, I, *Adv. in Math.* **59**, 3, pp. 185–301.

[92] Tamássy, L. (2000). Point Finsler spaces with metrical linear connections, *Publ. Math. Debrecen* **56**, pp. 643–655.

[93] Thompson, A. C. (1996). *Minkowski Geometry* (Cambridge University Press).

[94] Thorpe, J. A. (1979). *Elementary Topics in Differential Geometry* (Springer-Verlag, New York, Heidelberg, Berlin).

[95] Traber, R. E. (1937). A fundamental lemma on normal coordinates and its applications, *Quart. J. Math. Oxford* **8**, 2, pp. 142–147.

[96] Tu, L. W. (2011). *An Introduction to Manifolds (Second edition)* (Springer).

[97] Warner, F. W. (1983). *Foundations of Differentiable Manifolds and Lie Groups* (Springer-Verlag, New York).

[98] Weyl, H. (1921). Zur Infinitesimalgeometrie: Einordnung der projektiven und der konformen Auffassung, *Nachrichten von der Gesellschaft der Wissenschaften zu Göttingen*, pp. 99–112.

[99] Whitehead, J. H. C. (1933). Convex regions in the geometry of paths – addendum, *Quart. J. Math. Oxford* **4**, pp. 226–227.

[100] Yandell, B. H. (2002). *The Honors Class: Hilbert's Problems and Their Solvers* (A K Peters, Natick, Massachusetts).

General Conventions

Notation Index

689

Index